CW00925647

Springer Series in Statistics

Advisors:
P. Diggle, S. Fienberg, K. Krickeberg,
I. Olkin, N. Wermuth

Springer Series in Statistics

Andersen/Borgan/Gill/Keiding: Statistical Models Based on Counting Processes.
Anderson: Continuous-Time Markov Chains: An Applications-Oriented Approach.
Andrews/Herzberg: Data: A Collection of Problems from Many Fields for the Student and Research Worker.
Anscombe: Computing in Statistical Science through APL.
Berger: Statistical Decision Theory and Bayesian Analysis, 2nd edition.
Bolfarine/Zacks: Prediction Theory for Finite Populations.
Brémaud: Point Processes and Queues: Martingale Dynamics.
Brockwell/Davis: Time Series: Theory and Methods, 2nd edition.
Choi: ARMA Model Identification.
Daley/Vere-Jones: An Introduction to the Theory of Point Processes.
Dzhaparidze: Parameter Estimation and Hypothesis Testing in Spectral Analysis of Stationary Time Series.
Fahrmeir/Tutz: Multivariate Statistical Modelling Based on Generalized Linear Models.
Farrell: Multivariate Calculation.
Federer: Statistical Design and Analysis for Intercropping Experiments.
Fienberg/Hoaglin/Kruskal/Tanur (Eds.): A Statistical Model: Frederick Mosteller's Contributions to Statistics, Science and Public Policy.
Fisher/Sen: The Collected Works of Wassily Hoeffding.
Good: Permutation Tests: A Practical Guide to Resampling Methods for Testing Hypotheses.
Goodman/Kruskal: Measures of Association for Cross Classifications.
Grandell: Aspects of Risk Theory.
Hall: The Bootstrap and Edgeworth Expansion.
Härdle: Smoothing Techniques: With Implementation in S.
Hartigan: Bayes Theory.
Heyer: Theory of Statistical Experiments.
Jolliffe: Principal Component Analysis.
Kotz/Johnson (Eds.): Breakthroughs in Statistics Volume I.
Kotz/Johnson (Eds.): Breakthroughs in Statistics Volume II.
Kres: Statistical Tables for Multivariate Analysis.
Leadbetter/Lindgren/Rootzén: Extremes and Related Properties of Random Sequences and Processes.
Le Cam: Asymptotic Methods in Statistical Decision Theory.
Le Cam/Yang: Asymptotics in Statistics: Some Basic Concepts.
Manoukian: Modern Concepts and Theorems of Mathematical Statistics.
Manton/Singer/Suzman (Eds.): Forecasting the Health of Elderly Populations.
Miller, Jr.: Simultaneous Statistical Inference, 2nd edition.
Mosteller/Wallace: Applied Bayesian and Classical Inference: The Case of *The Federalist Papers.*
Pollard: Convergence of Stochastic Processes.
Pratt/Gibbons: Concepts of Nonparametric Theory.

(continued after index)

Mark J. Schervish

Theory of Statistics

With 26 Illustrations

Springer-Verlag
New York Berlin Heidelberg London Paris
Tokyo Hong Kong Barcelona Budapest

Mark J. Schervish
Department of Statistics
Carnegie Mellon University
Pittsburgh, PA 15213
USA

Library of Congress Cataloging-in-Publication Data
Schervish, Mark J.
    Theory of Statistics / Mark J. Schervish
      p. cm. – (Springer series in statistics)
    Includes bibliographical references (p. -) and index.
    ISBN 0-387-94546-6 (alk. paper)
    1. Mathematical statistics. I. Title. II. Series.
QA276.S346   1995
519.5—dc20   95-11235

Printed on acid-free paper.

© 1995 Springer-Verlag New York, Inc.
All rights reserved. This work may not be translated or copied in whole or in part without the written permission of the publisher (Springer-Verlag New York, Inc., 175 Fifth Avenue, New York, NY 10010, USA), except for brief excerpts in connection with reviews or scholarly analysis. Use in connection with any form of information storage and retrieval, electronic adaptation, computer software, or by similar or dissimilar methodology now known or hereafter developed is forbidden. The use of general descriptive names, trade names, trademarks, etc., in this publication, even if the former are not especially identified, is not to be taken as a sign that such names, as understood by the Trade Marks and Merchandise Marks Act, may accordingly be used freely by anyone.

Production managed by Laura Carlson; manufacturing supervised by Joe Quatela.
Photocomposed pages prepared from the author's LaTeX files.
Printed and bound by Edwards Brothers, Inc., Ann Arbor, MI.
Printed in the United States of America.

9 8 7 6 5 4 3 2 1

ISBN 0-387-94546-6 Springer-Verlag New York Berlin Heidelberg

To Nancy, Margaret, and Meredith

Preface

This text has grown out of notes used for lectures in a course entitled *Advanced Statistical Theory* at Carnegie Mellon University over several years. The course (when taught by the author) has attempted to cover, in one academic year, those topics in estimation, testing, and large sample theory that are commonly taught to second year graduate students in a mathematically rigorous fashion. Most texts at this level fall into one of two categories. They either ignore the Bayesian point of view altogether or they cover Bayesian topics almost exclusively. This book covers topics in both classical[1] and Bayesian inference in a great deal of generality. My own point of view is Bayesian, but I believe that students need to learn both types of theory in order to achieve a fuller appreciation of the subject matter. Although many comparisons are made between classical and Bayesian methods, it is not a goal of the text to present a formal comparison of the two approaches as was done by Barnett (1982). Rather, the goal has been to prepare Ph.D. students to be able to understand and contribute to the literature of theoretical statistics with a broader perspective than would be achieved from a purely Bayesian or a purely classical course.

After a brief review of elementary statistical theory, the coverage of the subject matter begins with a detailed treatment of parametric statistical models as motivated DeFinetti's representation theorem for exchangeable random variables (Chapter 1). In addition, Dirichlet processes and other tailfree processes are presented as examples of infinite-dimensional parameters. Chapter 2 introduces sufficient statistics from both Bayesian and non-Bayesian viewpoints. Exponential families are discussed here because of the important role sufficiency plays in these models. Also, the concept of information is introduced together with its relationship to sufficiency. A representation theorem is given for general distributions based on sufficient statistics. Decision theory is the subject of Chapter 3, which includes discussions of admissibility and minimaxity. Section 3.3 presents an axiomatic derivation of Bayesian decision theory, including the use of conditional probability. Chapter 4 covers hypothesis testing, including unbiased tests, *P*-values, and Bayes factors. We highlight the contrasts between the traditional "uniformly most powerful" (UMP) approach to testing and decision theoretic approaches (both Bayesian and classical). In particular, we

---

[1] What I call classical inference is called frequentist inference by some other authors.

see how the asymmetric treatment of hypotheses and alternatives in the UMP approach accounts for much of the difference. Point and set estimation are the topics of Chapter 5. This includes unbiased and maximum likelihood estimation as well as confidence, prediction, and tolerance sets. We also introduce robust estimation and the bootstrap. Equivariant decision rules are covered in Chapter 6. In Section 6.2.2, we debunk the common misconception of equivariant rules as means for preserving decisions under changes of measurement scale. Large sample theory is the subject of Chapter 7. This includes asymptotic properties of sample quantiles, maximum likelihood estimators, robust estimators, and posterior distributions. The last two chapters cover situations in which the random variables are not modeled as being exchangeable. Hierarchical models (Chapter 8) are useful for data arrays. Here, the parameters of the model can be modeled as exchangeable while the observables are only partially exchangeable. We introduce the popular computational tool known as Markov chain Monte Carlo, Gibbs sampling, or successive substitution sampling, which is very useful for fitting hierarchical models. Some topics in sequential analysis are presented in Chapter 9. These include classical tests, Bayesian decisions, confidence sets, and the issue of sampling to a foregone conclusion.

The presentation of material is intended to be very general and very precise. One of the goals of this book was to be the place where the proofs could be found for many of those theorems whose proofs were "beyond the scope of the course" in elementary or intermediate courses. For this reason, it is useful to rely on measure theoretic probability. Since many students have not studied measure theory and probability recently or at all, I have included appendices on measure theory (Appendix A) and probability theory (Appendix B).[2] Even those who have measure theory in their background can benefit from seeing these topics discussed briefly and working through some problems. At the beginnings of these two appendices, I have given overviews of the important definitions and results. These should serve as reminders for those who already know the material and as groundbreaking for those who do not. There are, however, some topics covered in Appendix B that are not part of traditional probability courses. In particular, there is the material in Section B.3.3 on conditional densities with respect to nonproduct measures. Also, there is Section B.6, which attempts to use the ideas of gambling to motivate the mathematical definition of probability. Since conditional independence and the law of total probability are so central to Bayesian predictive inference, readers may want to study the material in Sections B.3.4 and B.3.5 also.

Appendix C lists purely mathematical theorems that are used in the text

---

[2]These two appendices contain sufficient detail to serve as the basis for a full-semester (or more) course in measure and probability. They are included in this book to make it more self-contained for students who do not have a background in measure theory.

without proof, and Appendix D gives a brief summary of the distributions that are used throughout the text. An index is provided for notation and abbreviations that are used at a considerable distance from where they are defined. Throughout the book, I have added footnotes to those results that are of interest mainly through their value in proving other results. These footnotes indicate where the results are used explicitly elsewhere in the book. This is intended as an aid to instructors who wish to select which results to prove in detail and which to mention only in passing. A single numbering system is used within each chapter and includes theorems, lemmas, definitions, corollaries, propositions, assumptions, examples, tables, figures, and equations in order to make them easier to locate when needed.

I was reluctant to mark sections to indicate which ones could be skipped without interrupting the flow of the text because I was afraid that readers would interpret such markings as signs that the material was not important. However, because there may be too much material to cover, especially if the measure theory and probability appendices are covered, I have decided to mark two different kinds of sections whose material is used at most sparingly in other parts of the text. Those sections marked with a plus sign (+) make use of the theory of martingales. A lot of the material in some of these sections is used in other such sections, but the remainder of the text is relatively free of martingales. Martingales are particularly useful in proving limit theorems for conditional probabilities. The remaining sections that can be skipped or covered out of order without seriously interrupting the flow of material are marked with an asterisk (*). No such system is foolproof, however. For example, even though essentially all of the material dealing with equivariance is isolated in Chapter 6, there is one example in Chapter 7 and one exercise that make reference to the material. Similarly, the material from other sections marked with the asterisk may occasionally appear in examples later in the text. But these occurrences should be inconsequential. Of course, any instructor who feels that equivariance is an important topic should not be put off by the asterisk. In that same vein, students really ought to be made aware of what the main theorems in Section 3.3 say (Theorems 3.108 and 3.110), even though the section could be skipped without interrupting the flow of the material.

I would like to thank many people who helped me to write this book or who read early drafts. Many people have provided corrections and guidance for clarifying some of the discussions (not to mention corrections to some proofs). In particular, thanks are due to Chris Andrews, Bogdan Doytchinov, Petros Hadjicostas, Tao Jiang, Rob Kass, Agostino Nobile, Shingo Oue, and Thomas Short. Morris DeGroot helped me to understand what is really going on with equivariance. Teddy Seidenfeld introduced me to the axiomatic foundations of decision theory. Mel Novick introduced me to the writings of DeFinetti. Persi Diaconis and Bill Strawderman made valuable suggestions after reading drafts of the book, and those suggestions are incorporated here. Special thanks go to Larry Wasserman, who taught from two early drafts of the text and provided invaluable feedback on the (lack of) clarity in various sections.

As a student at the University of Illinois at Urbana–Champaign, I learned statistical theory from Stephen Portnoy, Robert Wijsman, and Robert Bohrer (although some of these people may deny that fact after reading this book). Many of the proofs and results in this text bear startling resemblance to my notes taken as a student. Many, in turn, undoubtedly resemble works recorded in other places. Whenever I have essentially lifted, or cosmetically modified, or even only been deeply inspired by a published source, I have cited that source in the text. If results copied from my notes as a student or produced independently also resemble published results, I can only apologize for not having taken enough time to seek out the earliest published reference for every result and proof in the text. Similarly, the problems at the ends of each chapter have come from many sources. One source used often was the file of old qualifying exams from the Department of Statistics at Carnegie Mellon University. These problems, in turn, came from various sources unknown to me (even the ones I wrote). If I have used a problem without giving proper credit, please take it as a compliment. Some of the more challenging problems have been identified with an asterisk (*) after the problem number. Many of the plots in the text were produced using The New S Language and S-Plus [see Becker, Chambers, and Wilks (1988) and StatSci (1992)]. The original text processing was done using LaTeX, which was written by Lamport (1986) and was based on TeX by Knuth (1984).

Pittsburgh, Pennsylvania   MARK J. SCHERVISH
May 31, 1995

Contents

Preface   vii

Chapter 1: Probability Models   1
  1.1 Background 1
      1.1.1 General Concepts 1
      1.1.2 Classical Statistics 2
      1.1.3 Bayesian Statistics 4
  1.2 Exchangeability 5
      1.2.1 Distributional Symmetry 5
      1.2.2 Frequency and Exchangeability 10
  1.3 Parametric Models 12
      1.3.1 Prior, Posterior, and Predictive Distributions 13
      1.3.2 Improper Prior Distributions 19
      1.3.3 Choosing Probability Distributions 21
  1.4 DeFinetti's Representation Theorem 24
      1.4.1 Understanding the Theorems 24
      1.4.2 The Mathematical Statements 26
      1.4.3 Some Examples 28
  1.5 Proofs of DeFinetti's Theorem and Related Results* 33
      1.5.1 Strong Law of Large Numbers 33
      1.5.2 The Bernoulli Case 36
      1.5.3 The General Finite Case* 38
      1.5.4 The General Infinite Case 45
      1.5.5 Formal Introduction to Parametric Models* 49
  1.6 Infinite-Dimensional Parameters* 52
      1.6.1 Dirichlet Processes 52
      1.6.2 Tailfree Processes[+] 60
  1.7 Problems 73

Chapter 2: Sufficient Statistics   82
  2.1 Definitions 82
      2.1.1 Notational Overview 82
      2.1.2 Sufficiency 83
      2.1.3 Minimal and Complete Sufficiency 92
      2.1.4 Ancillarity 95
  2.2 Exponential Families of Distributions 102

---

*Sections and chapters marked with an asterisk may be skipped or covered out of order without interrupting the flow of ideas.

[+]Sections marked with a plus sign include results which rely on the theory of martingales. They may be skipped without interrupting the flow of ideas.

		2.2.1	Basic Properties 102
		2.2.2	Smoothness Properties 105
		2.2.3	A Characterization Theorem* 109
	2.3	Information 110	
		2.3.1	Fisher Information 111
		2.3.2	Kullback–Leibler Information 115
		2.3.3	Conditional Information* 118
		2.3.4	Jeffreys' Prior* 121
	2.4	Extremal Families* 123	
		2.4.1	The Main Results 124
		2.4.2	Examples 127
		2.4.3	Proofs$^+$ 129
	2.5	Problems 138	

Chapter 3: Decision Theory   144

	3.1	Decision Problems 144	
		3.1.1	Framework 144
		3.1.2	Elements of Bayesian Decision Theory 146
		3.1.3	Elements of Classical Decision Theory 149
		3.1.4	Summary 150
	3.2	Classical Decision Theory 150	
		3.2.1	The Role of Sufficient Statistics 150
		3.2.2	Admissibility 153
		3.2.3	James–Stein Estimators 163
		3.2.4	Minimax Rules 167
		3.2.5	Complete Classes 174
	3.3	Axiomatic Derivation of Decision Theory* 181	
		3.3.1	Definitions and Axioms 181
		3.3.2	Examples 186
		3.3.3	The Main Theorems 188
		3.3.4	Relation to Decision Theory 189
		3.3.5	Proofs of the Main Theorems* 190
		3.3.6	State-Dependent Utility* 205
	3.4	Problems 208	

Chapter 4: Hypothesis Testing   214

	4.1	Introduction 214	
		4.1.1	A Special Kind of Decision Problem 214
		4.1.2	Pure Significance Tests 216
	4.2	Bayesian Solutions 218	
		4.2.1	Testing in General 218
		4.2.2	Bayes Factors 220
	4.3	Most Powerful Tests 230	
		4.3.1	Simple Hypotheses and Alternatives 233
		4.3.2	Simple Hypotheses, Composite Alternatives 238
		4.3.3	One-Sided Tests 239
		4.3.4	Two-Sided Hypotheses 246
	4.4	Unbiased Tests 253	
		4.4.1	General Results 253

	4.4.2 Interval Hypotheses	255
	4.4.3 Point Hypotheses	257
4.5	Nuisance Parameters	265
	4.5.1 Neyman Structure	265
	4.5.2 Tests about Natural Parameters	268
	4.5.3 Linear Combinations of Natural Parameters	272
	4.5.4 Other Two-Sided Cases*	272
	4.5.5 Likelihood Ratio Tests	274
	4.5.6 The Standard F-Test as a Bayes Rule*	276
4.6	P-Values	279
	4.6.1 Definitions and Examples	279
	4.6.2 P-Values and Bayes Factors	283
4.7	Problems	285

Chapter 5: Estimation   296

5.1	Point Estimation	296
	5.1.1 Minimum Variance Unbiased Estimation	297
	5.1.2 Lower Bounds on the Variance of Unbiased Estimators	301
	5.1.3 Maximum Likelihood Estimation	307
	5.1.4 Bayesian Estimation	309
	5.1.5 Robust Estimation*	310
5.2	Set Estimation	315
	5.2.1 Confidence Sets	315
	5.2.2 Prediction Sets*	324
	5.2.3 Tolerance Sets*	325
	5.2.4 Bayesian Set Estimation	327
	5.2.5 Decision Theoretic Set Estimation*	328
5.3	The Bootstrap*	329
	5.3.1 The General Concept	329
	5.3.2 Standard Deviations and Bias	335
	5.3.3 Bootstrap Confidence Intervals	336
5.4	Problems	339

Chapter 6: Equivariance*   344

6.1	Common Examples	344
	6.1.1 Location Problems	344
	6.1.2 Scale Problems*	350
6.2	Equivariant Decision Theory	353
	6.2.1 Groups of Transformations	353
	6.2.2 Equivariance and Changes of Units	359
	6.2.3 Minimum Risk Equivariant Decisions	363
6.3	Testing and Confidence Intervals*	375
	6.3.1 P-Values in Invariant Problems	375
	6.3.2 Equivariant Confidence Sets	379
	6.3.3 Invariant Tests*	380
6.4	Problems	388

Chapter 7: Large Sample Theory — 394

- 7.1 Convergence Concepts 394
  - 7.1.1 Deterministic Convergence 394
  - 7.1.2 Stochastic Convergence 395
  - 7.1.3 The Delta Method 401
- 7.2 Sample Quantiles 404
  - 7.2.1 A Single Quantile 404
  - 7.2.2 Several Quantiles 408
  - 7.2.3 Linear Combinations of Quantiles* 410
- 7.3 Large Sample Estimation 412
  - 7.3.1 Some Principles of Large Sample Estimation 412
  - 7.3.2 Maximum Likelihood Estimators 415
  - 7.3.3 MLEs in Exponential Families 418
  - 7.3.4 Examples of Inconsistent MLEs 420
  - 7.3.5 Asymptotic Normality of MLEs 421
  - 7.3.6 Asymptotic Properties of M-Estimators* 424
- 7.4 Large Sample Properties of Posterior Distributions 428
  - 7.4.1 Consistency of Posterior Distributions+ 429
  - 7.4.2 Asymptotic Normality of Posterior Distributions 435
  - 7.4.3 Laplace Approximations to Posterior Distributions* . 446
  - 7.4.4 Asymptotic Agreement of Predictive Distributions+ .. 455
- 7.5 Large Sample Tests 458
  - 7.5.1 Likelihood Ratio Tests 458
  - 7.5.2 Chi-Squared Goodness of Fit Tests 461
- 7.6 Problems 467

Chapter 8: Hierarchical Models — 476

- 8.1 Introduction 476
  - 8.1.1 General Hierarchical Models 476
  - 8.1.2 Partial Exchangeability* 479
  - 8.1.3 Examples of the Representation Theorem* 480
- 8.2 Normal Linear Models 483
  - 8.2.1 One-Way ANOVA 483
  - 8.2.2 Two-Way Mixed Model ANOVA* 488
  - 8.2.3 Hypothesis Testing 491
- 8.3 Nonnormal Models* 495
  - 8.3.1 Poisson Process Data 495
  - 8.3.2 Bernoulli Process Data 497
- 8.4 Empirical Bayes Analysis* 500
  - 8.4.1 Naïve Empirical Bayes 500
  - 8.4.2 Adjusted Empirical Bayes 503
  - 8.4.3 Unequal Variance Case 504
- 8.5 Successive Substitution Sampling 505
  - 8.5.1 The General Algorithm 505
  - 8.5.2 Normal Hierarchical Models 512
  - 8.5.3 Nonnormal Models 517
- 8.6 Mixtures of Models 519
  - 8.6.1 General Mixture Models 519
  - 8.6.2 Outliers 521

		8.6.3 Bayesian Robustness 524
8.7	Problems 532	

Chapter 9: Sequential Analysis   536
9.1 Sequential Decision Problems 536
9.2 The Sequential Probability Ratio Test 548
9.3 Interval Estimation* 558
9.4 The Relevance of Stopping Rules 562
9.5 Problems 567

Appendix A: Measure and Integration Theory   570
A.1 Overview 570
    A.1.1 Definitions 570
    A.1.2 Measurable Functions 572
    A.1.3 Integration 573
    A.1.4 Absolute Continuity 574
A.2 Measures 575
A.3 Measurable Functions 582
A.4 Integration 587
A.5 Product Spaces 593
A.6 Absolute Continuity 597
A.7 Problems 602

Appendix B: Probability Theory   606
B.1 Overview 606
    B.1.1 Mathematical Probability 606
    B.1.2 Conditioning 607
    B.1.3 Limit Theorems 611
B.2 Mathematical Probability 612
    B.2.1 Random Quantities and Distributions 612
    B.2.2 Some Useful Inequalities 613
B.3 Conditioning 615
    B.3.1 Conditional Expectations 615
    B.3.2 Borel Spaces* 619
    B.3.3 Conditional Densities 623
    B.3.4 Conditional Independence 628
    B.3.5 The Law of Total Probability 632
B.4 Limit Theorems 634
    B.4.1 Convergence in Distribution and in Probability 634
    B.4.2 Characteristic Functions 639
B.5 Stochastic Processes 645
    B.5.1 Introduction 645
    B.5.2 Martingales$^+$ 645
    B.5.3 Markov Chains* 650
    B.5.4 General Stochastic Processes 651
B.6 Subjective Probability 654
B.7 Simulation* 659
B.8 Problems 661

Appendix C: Mathematical Theorems Not Proven Here — 665
- C.1 Real Analysis 665
- C.2 Complex Analysis 666
- C.3 Functional Analysis 667

Appendix D: Summary of Distributions — 668
- D.1 Univariate Continuous Distributions 668
- D.2 Univariate Discrete Distributions 672
- D.3 Multivariate Distributions 674

References — 675

Notation and Abbreviation Index — 689

Name Index — 691

Subject Index — 694

CHAPTER 1
Probability Models

1.1 Background

The purpose of this book is to cover important topics in the theory of statistics in a very thorough and general fashion. In this section, we will briefly review some of the basic theory of statistics with which many students are familiar. All that we do here will be repeated in a more precise manner at the appropriate place in the text.

1.1.1 General Concepts

Most paradigms for statistical inference make at least some use of the following structure. We suppose that some random variables $X_1, \ldots, X_n$ all have the same distribution, but we may be unwilling to say what that distribution is. Instead, we create a collection of distributions called a *parametric family* and denoted $\mathcal{P}_0$. For example, $\mathcal{P}_0$ might consist of all normal distributions, or just those normal distributions with variance 1, or all binomial distributions, or all Poisson distributions, and so forth. Each of these cases has the property that the collection of distributions can be indexed by a finite-dimensional real quantity, which is commonly called a *parameter*. For example, if the parametric family is all normal distributions, then the parameter can be denoted $\Theta = (M, \Sigma)$, where M stands for the mean and Σ stands for the standard deviation. The set of all possible values of the parameter is called the *parameter space* and is often denoted by Ω. When $\Theta = \theta$, the distribution of the observations is denoted by P_θ. Expected values are denoted as $\mathrm{E}_\theta(\cdot)$.

We will denote observed data X. It might be that X is a vector of observations that are mutually independent and identically distributed (IID), or X might be some general quantity. The set of possible values for X is the *sample space* and is often denoted as $\mathcal{X}$. The members P_θ of the parametric family will be distributions over this space $\mathcal{X}$. If X is continuous or discrete, then densities or probability mass functions[1] exist. We will denote the density or mass function for P_θ by $f_{X|\Theta}(\cdot|\theta)$. For example, if X is a single random variable with continuous distribution, then

$$P_\theta(a < X \le b) = \int_a^b f_{X|\Theta}(x|\theta)dx.$$

If $X = (X_1, \ldots, X_n)$, where the X_i are IID each with density (or mass function) $f_{X_1|\Theta}(\cdot|\theta)$ when $\Theta = \theta$, then

$$f_{X|\Theta}(x|\theta) = \prod_{i=1}^n f_{X_1|\Theta}(x_i|\theta), \tag{1.1}$$

where $x = (x_1, \ldots, x_n)$. After observing the data $X_1 = x_1, \ldots, X_n = x_n$, the function in (1.1), as a function of θ for fixed x, is called the *likelihood function*, denoted by $L(\theta)$. Section 1.3 is devoted to a motivation of the above structure based on the concept of *exchangeability* and DeFinetti's representation theorem 1.49. Exchangeability is discussed in detail in Section 1.2, and DeFinetti's theorem is the subject of Section 1.4.

1.1.2 Classical Statistics

Classical inferential techniques include tests of hypotheses, unbiased estimates, maximum likelihood estimates, confidence intervals and many other things. These will be covered in great detail in the text, but we remind the reader of a few of them here. Suppose that we are interested in whether or not the parameter lies in one portion Ω_H of the parameter space. We could then set up a *hypothesis* $H : \Theta \in \Omega_H$ with the corresponding *alternative* $A : \Theta \notin \Omega_H$. The simplest sort of *test* of this hypothesis would be to choose a subset $R \subseteq \mathcal{X}$, and then reject H if $x \in R$ is observed. The set R would be called the *rejection region* for the test. If $x \notin R$, we would say that we do not reject H. Tests are compared based on their power functions. The *power function* of a test with rejection region R is $\beta(\theta) = P_\theta(X \in R)$. The *size* of a test is $\sup_{\theta \in \Omega_H} \beta(\theta)$. Chapter 4 covers hypothesis testing in depth.

Example 1.2. Suppose that $X = (X_1, \ldots, X_n)$ and the X_i are IID with $N(\theta, 1)$ distribution under P_θ. The usual size α test of $H : \Theta = \theta_0$ versus $A : \Theta \ne \theta_0$ is

---

[1]Using the theory of measures (see Appendix A) we will be able to dispense with the distinction between densities and probability mass functions. They will both be special cases of a more general type of "density."

to reject H if $\overline{X} \in R$, where $\overline{X}$ is the sample average,

$$R = \left(-\infty, \theta_0 + \frac{1}{\sqrt{n}} \Phi^{-1}\left(\frac{\alpha}{2}\right)\right] \cup \left[\theta_0 + \frac{1}{\sqrt{n}} \Phi^{-1}\left(1 - \frac{\alpha}{2}\right), \infty\right),$$

and Φ is the standard normal cumulative distribution function (CDF).

The notation and terminology in Chapter 4 are different from the above because we consider a more general class of tests called *randomized tests*. These are special cases of randomized decision rules, which are introduced in Chapter 3. The following example illustrates the reason that randomized decisions are introduced.

Example 1.3. Let $X \sim \text{Bin}(5, \theta)$ given $\Theta = \theta$. Suppose that we wish to test $H : \Theta \leq 1/2$ versus $A : \Theta > 1/2$. It might seem that the best test would be to reject H if $X > c$, where c is chosen to make the test have the desired level. Unfortunately, only six different levels are available for tests of this form. For example, if $c \in [4, 5)$, the test has level $1/32$. If $c \in [3, 4)$, the test has level $3/16$, and so on. If you desire a level such as 0.05, you must use a more complicated test.

A function of the data which takes values in the parameter space is called a *(point) estimator* of Θ. Section 5.1 considers point estimation in depth.

Example 1.4. Suppose that $X = (X_1, \ldots, X_n)$ and the X_i are IID with $N(\theta, 1)$ distribution under P_θ, then $\phi(x) = \sum_{i=1}^n x_i/n = \overline{x}$ takes values in the parameter space and can be considered an estimator of Θ.

Sometimes we wish to estimate a function g of Θ. An estimator ϕ of $g(\Theta)$ is *unbiased* if $\mathrm{E}_\theta[\phi(X)] = g(\theta)$ for all $\theta \in \Omega$. An estimator ϕ of Θ is a *maximum likelihood estimator (MLE)* if

$$\sup_{\theta \in \Omega} L(\theta) = L(\phi(x)),$$

for all $x \in \mathcal{X}$. An estimator ψ of $g(\Theta)$ is an MLE if $\psi(X) = g(\phi(X))$, where ϕ is an MLE of Θ. The reader should verify that the estimator ϕ in Example 1.4 is both an unbiased estimator and an MLE of Θ.

If the parameter Θ is real-valued, it is common to provide interval estimates of Θ. If (A, B) is a pair of random variables with $A \leq B$, and if

$$P_\theta(A \leq \theta \leq B) \geq \gamma,$$

for all $\theta \in \Omega$, then $[A, B]$ is called a *coefficient γ confidence interval* for Θ. Section 5.2 covers the theory of set estimation, which includes confidence intervals, prediction intervals, and tolerance intervals as special cases.

Example 1.5 (Continuation of Example 1.4). Suppose that $X = (X_1, \ldots, X_n)$ and the X_i are IID with $N(\theta, 1)$ distribution under P_θ, and let

$$A = \overline{X} - \frac{c}{\sqrt{n}}, \quad B = \overline{X} + \frac{c}{\sqrt{n}},$$

where $c > 0$. Then $[A, B]$ is a coefficient $2\Phi(-c)$ confidence interval for Θ, where Φ is the standard normal CDF.

1.1.3 Bayesian Statistics

In the Bayesian paradigm, one treats all unknown quantities as random variables and constructs a joint probability distribution for all of them. Using the same setup as in Section 1.1.1, this would require that one construct a distribution for the parameter Θ in addition to the conditional distribution of X given $\Theta = \theta$, which was denoted by P_θ. The distribution of Θ is called the *prior distribution*. Together, the prior distribution and $\{P_\theta : \theta \in \Omega\}$ determine a joint distribution on the space $\mathcal{X} \times \Omega$. For example, suppose that the prior distribution has a density f_Θ, suppose that X is continuous, and let $B \subseteq \mathcal{X} \times \Omega$. Then

$$\Pr((X, \Theta) \in B) = \int \int I_B(x, \theta) f_{X|\Theta}(x|\theta) f_\Theta(\theta) dx d\theta,$$

where I_B is the indicator function of the set B. It will often be possible (although not necessary) to think of the space $\mathcal{X} \times \Omega$ as if it were the underlying probability space S which is introduced in Appendix B. In this way, X and Θ are both easily recognized as functions from S to their respective ranges. That is, if $s = (x, \theta)$, then $X(s) = x$ and $\Theta(s) = \theta$.

After observing the data $X = x$, one constructs the conditional distribution of Θ given $X = x$, which is called the *posterior distribution*, using Bayes' theorem:

$$f_{\Theta|X}(\theta|x) = \frac{f_{X|\Theta}(x|\theta) f_\Theta(\theta)}{\int_\Omega f_{X|\Theta}(x|t) f_\Theta(t) dt}. \tag{1.6}$$

A popular method of finding the posterior distribution is to note that the denominator of (1.6) is not a function of θ. (In fact, the denominator in (1.6) is called the *prior predictive density* of the data X, $f_X(x)$.) This means that we can find $f_{\Theta|X}(\theta|x)$ by calculating the numerator of (1.6) and then dividing it by whatever constant is required to make it a density as a function of θ.

Example 1.7 (Continuation of Example 1.4; see page 3). Suppose that $X = (X_1, \ldots, X_n)$ and the X_i are conditionally IID with $N(\theta, 1)$ given $\Theta = \theta$. Suppose that the prior distribution of Θ is $N(\theta_0, 1/\lambda)$, where θ_0 and λ are known constants. The likelihood function is

$$f_{X|\Theta}(x|\theta) = (2\pi)^{-\frac{n}{2}} \exp\left(-\frac{n}{2}[\theta - \bar{x}]^2 - \frac{1}{2}\sum_{i=1}^n [x_i - \bar{x}]^2\right),$$

and the prior density is $f_\Theta(\theta) = \sqrt{\lambda}(2\pi)^{-1/2} \exp(-\lambda[\theta - \theta_0]^2/2)$. Multiplying these together and simplifying yield the following expression for the numerator of (1.6):

$$k(x) \exp\left(-\frac{n+\lambda}{2}[\theta - \theta_1]^2\right), \tag{1.8}$$

where $\theta_1 = (\lambda \theta_0 + n\bar{x})/(\lambda + n)$, and $k(x)$ does not depend on θ. The expression in (1.8) is easily recognized as being proportional to the $N(\theta_1, 1/[\lambda+n])$ density as a function of θ. So, the posterior distribution of Θ given $X = x$ is $N(\theta_1, 1/[\lambda+n])$.

Inferences about Θ, in the Bayesian paradigm, are based on the posterior distribution. For example, one might use the posterior mean or median of Θ as an estimate of Θ. In Example 1.7 on page 4, the posterior mean and median are both θ_1. The Bayesian paradigm also accommodates inference about future observables. If Y denotes some future observations that are conditionally independent of X given Θ, such as $Y = (X_{n+1}, \ldots, X_{n+m})$, then the *posterior predictive density* of Y is

$$f_{Y|X}(y|x) = \int_\Omega f_{Y|\Theta}(y|\theta) f_{\Theta|X}(\theta|x) d\theta.$$

Example 1.9 (Continuation of Example 1.7; see page 4). Let $Y = X_{n+1}$, the next observation. The posterior predictive density of Y is

$$\begin{aligned}
f_{Y|X}(y|x) &= \int \frac{1}{\sqrt{2\pi}} \exp\left(-\frac{[y-\theta]^2}{2}\right) \frac{\sqrt{n+\lambda}}{\sqrt{2\pi}} \exp\left(-\frac{n+\lambda}{2}[\theta - \theta_1]^2\right) d\theta \\
&= \frac{\sqrt{n+\lambda}}{\sqrt{2\pi(n+\lambda+1)}} \exp\left(-\frac{n+\lambda}{2(n+\lambda+1)}[y-\theta_1]^2\right),
\end{aligned}$$

which is the density of the $N(\theta_1, 1 + 1/[n+\lambda])$ distribution.

The theory of prior, posterior, and predictive distributions is introduced in Section 1.3.1. Many Bayesian inferential techniques tend to be decision theoretic, so the theory of decisions is introduced in Chapter 3. In the text, Bayesian techniques are usually introduced at locations nearby those at which corresponding classical techniques are introduced.

1.2 Exchangeability

1.2.1 Distributional Symmetry

When one performs a statistical analysis, there are usually several quantities about which one is uncertain. For example, when conducting a political poll, one never knows in advance which of several answers each respondent will provide. In addition, even after the responses are in, one does not know the answers that would have been supplied by all of the people who were not polled. If one is interested in the proportions of the population who would provide each of the available responses, then all of the would-be responses of all members of the population are potentially of interest. The most complete specification of a probability distribution would give the joint distribution of all of these responses. From this joint distribution, the distributions of the various proportions of interest could also be calculated.

The quantities of interest can be more complicated than counts and proportions without changing the basic considerations. For example, a company may keep track of the total amount of a sample of its sale to a sample of its customers at a sample of its stores on a sample of days. It may be interested in various average sales amounts across different stores in a single department or across different departments in a single store, or across different days, and so on. Once again, the joint distribution of all vectors of total sale, register, store, and day would facilitate answering the questions of interest.

How does (or should) one construct the probability distributions needed in such examples, and how does one draw inferences from the various types of data? Some of the more common ways to draw inferences were described briefly in Section 1.1. In order better to understand probability and inference, let us take a very simplistic example, which should not be too encumbered by considerations of available scientific knowledge. Consider an old-fashioned thumbtack [2] (one of the metal ones with a round, curved head, not the colored plastic ones). We will toss this thumbtack onto a soft surface[3] and keep track of whether it comes to stop with the point up or with the point down. In the absence of any information to distinguish the tosses or to suggest that tosses occurring close together in time are any more or less likely to be similar to or different from each other than those that are far apart in time, it seems reasonable to treat the different tosses symmetrically. We might also believe that although we might only toss the thumbtack a few times, if we were to toss it many more times, the same judgment of symmetry would continue to apply to the future tosses. Under these conditions, it is traditional to model the outcomes of the tosses as independent and identically distributed (IID) random variables with $X_i = 1$ meaning that toss i is point up and $X_i = 0$ meaning that toss i is point down. In the classical framework, one invents a parameter, say θ, which is assumed to be a fixed value not yet known to us.[4] Then one says that the X_i are IID with $\Pr(X_i = 1) = \theta$. Within a Bayesian framework, one might

---

[2] This example is described in detail by Lindley and Phillips (1976). Other interesting examples of how exchangeability aids in the understanding of inference problems were given by Lindley and Novick (1981). This example is used, in preference to tossing of coins, because most readers will not have particularly strong prior opinions about how a thumbtack will land. On the other hand, most people believe that the typical coin selected from one's pocket or purse has probability pretty near 1/2 of landing head up.

[3] This is done to avoid damaging the thumbtack. This is the last scientific consideration we will make.

[4] A great deal of controversy in statistics arises out of the question of the meaning of such quantities. DeFinetti (1974) argues persuasively that one need not assume the existence of such things. Sometimes they are just assumed to be undefined properties of the experimental setup which magically make the outcomes behave according to our probability models. Sometimes they are defined in terms of the sequence of observations themselves (such as limits of relative frequencies). This last is particularly troublesome because the sequence of observations does not yet exist and hence the limit of relative frequency cannot be a fixed value yet.

construct a probability distribution μ for this unknown θ and say that

$$\Pr(X_1 = x_1, \ldots, X_n = x_n) = \int \theta^{x_1 + \cdots + x_n}(1-\theta)^{n - x_1 - \cdots - x_n} d\mu(\theta). \quad (1.10)$$

It seems unfortunate that so much machinery as assumptions of mutual independence and the existence of a mysterious fixed but unknown θ must be introduced to describe what seems, on the surface, to be a relatively simple situation. One purpose of this chapter is to show how to replace the heavy probabilistic assumptions of IID and "fixed but unknown θ" with a minimal assumption that reflects nothing more than the symmetry expressed in the problem. At the same time, we will be able to understand when models like that of (1.10) are appropriate and why relative frequency is such a popular device for thinking about probabilities. For example, when considering the tosses of the thumbtack, we said that we would treat the information to be obtained from any one toss in exactly the same way as we would treat the information from any other toss. Similarly, we would treat the information to be obtained from any two tosses in exactly the same way as we would treat the information from any other two tosses regardless of where they appear in the sequence of tosses, and so on for three or more tosses. This may seem like a heavy probabilistic assumption in itself. But it really is nothing more than an explicit expression of the symmetry amongst the tosses. Anything less would imply asymmetric treatment of the observations. Note that assuming the tosses to be IID assumes this symmetry and more. The symmetry is quite explicit in formula (1.10). Every permutation of the numbers $x_1, \ldots, x_n$ leads to the same value of the right-hand side of (1.10). If we assume nothing more than this permutation symmetry for a potentially infinite sequence of possible tosses of the thumbtack, then Theorem 1.49[5] will imply that there exists μ such that (1.10) holds. In a sense, the quantity θ is given an implicit meaning as a random variable Θ, rather than a fixed value, without having to explicitly give it meaning in advance. (See Example 1.45 on page 25.) Furthermore, the observations are not necessarily mutually independent, but they will be conditionally independent given Θ.

The minimal assumption of symmetry is known as *exchangeability*, and it is no more complicated than the permutation symmetry noticed in (1.10).

Definition 1.11. A finite set $X_1, \ldots, X_n$ of random quantities is said to be *exchangeable* if every permutation of $(X_1, \ldots, X_n)$ has the same joint distribution as every other permutation. An infinite collection is exchangeable if every finite subcollection is exchangeable.

For example, suppose that $X_1, \ldots, X_{100}$ are exchangeable. It follows easily from the definition that they all have the same marginal distribution.

---

[5]Theorem 1.47 is a simpler version that applies only to Bernoulli random variables.

Also, (X_1, X_2) has the same joint distribution as (X_{99}, X_1), (X_5, X_2, X_{48}) has the same joint distribution as (X_{13}, X_{100}, X_3), and so on. The following fact is easy to prove.

Proposition 1.12. *A collection C of random quantities is exchangeable if and only if, for every finite n less than or equal to the size of the collection C, every n-tuple of distinct elements of C has the same joint distribution as every other such n-tuple.*

As an example, we stated earlier that the assumption off IID random variables entailed symmetry and more.

Example 1.13. Consider a collection $X_1, X_2, \ldots$ (finite or infinite) of IID random variables. Clearly, $(X_{i_1}, \ldots, X_{i_n})$ has the same distribution as $(X_{j_1}, \ldots, X_{j_n})$ so long as $i_1, \ldots, i_n$ are all distinct and $j_1, \ldots, j_n$ are all distinct. Hence, every collection of IID random variables is exchangeable.

The motivation for the definition of exchangeability is to express symmetry of beliefs about the random quantities in the weakest possible way. The definition, as stated, does not require any judgment of independence or that any limit of relative frequencies will exist. It merely says that the labeling of the random quantities is immaterial. There are many situations in which this assumption is deemed reasonable, and many where it is not. For example, consider the company that sampled sales on various days at various stores. It might seem reasonable to declare that the sales at a particular store on a particular day are exchangeable. But the collection of all sales on all days at all stores might be modeled less symmetrically. In Chapter 8, we will discuss in more detail cases with less symmetry.

Back in the old days, before probability theory was overrun by σ-fields and the like, the concept of symmetry was central to most calculations of probabilities. Consider, for example, the first paragraph of the book by DeMoivre (1756):

> The Probability of an Event is greater or less according to the number of Chances by which it may happen, compared with the whole number of Chances by which it may either happen or fail.

DeMoivre was describing a judgment of symmetry amongst the possible outcomes of some experiment. But other authors, such as Venn (1876), rely on symmetry amongst a collection of random quantities to define probabilities as frequencies.[6] Although we now realize that symmetry is not essential to the definition of probability, it nevertheless is a widely used assumption that can help facilitate the construction of distributions. In addition, Theorem 1.49 helps to explain why frequencies are relevant to the calculation of

---

[6]The reader interested in an in-depth study of the early days of statistics and statistical reasoning should read Stigler (1986).

probabilities even though probabilities are not defined as frequencies. (See the discussion in Section 1.2.2.)

In Example 1.13 on page 8, we saw that IID random variables are exchangeable. Exchangeability is more general than IID, however. A very common case of exchangeable random quantities is the following. Suppose that $X_1, X_2, \ldots$ are conditionally IID given Y. Then the X_i are exchangeable. (See Problem 4 on page 73.)

Example 1.14. Suppose that $\{X_n\}_{n=1}^{\infty}$ are conditionally independent with density $f(x|y)$ given $Y = y$ and that Y has density $g(y)$. Then the joint density of any ordered n-tuple $(X_{i_1}, \ldots, X_{i_n})$ is

$$f_{X_{i_1},\ldots,X_{i_n}}(x_1,\ldots,x_n) = \int \prod_{j=1}^{n} f(x_j|y)g(y)dy.$$

Note that the right-hand side does not depend on $i_1, \ldots, i_n$.

The case of conditionally IID random quantities will turn out to be one of only two general forms of exchangeability. Theorem 1.49 will say that infinitely many random quantities are exchangeable if and only if they are conditionally IID given something.

Although an infinite sequence of exchangeable random variables is conditionally IID, sometimes the description of their joint distribution does not make this fact transparent. Example 1.15 is the famous *Polya urn scheme*. It is not obvious from the example that the random variables constructed are conditionally IID. Theorem 1.49, however, says that they are conditionally IID because they are exchangeable.[7]

Example 1.15. Let $\mathcal{X} = \{1, \ldots, k\}$, and let $u_1, \ldots, u_k$ be nonnegative integers such that $u = \sum_{i=1}^{k} u_i > 0$. Suppose that an urn contains u_i balls labeled i for $i = 1, \ldots, k$. We draw a ball at random[8] and record X_1 equal to the label. We then replace the ball and toss in one more ball with the same label. We then draw a ball at random again to get X_2 and repeat the process indefinitely. To prove that the sequence $\{X_i\}_{i=1}^{\infty}$ is exchangeable, let $n > 0$ be an integer and let $j_1, \ldots, j_n$ be elements of $\mathcal{X}$. For $i = 1, \ldots, k$, let $c_i(j_1, \ldots, j_n)$ be the number of times that i appears among $j_1, \ldots, j_n$. That is,[9] $c_i(j_1, \ldots, j_n) = \sum_{t=1}^{n} I_{\{i\}}(j_t)$. Define the notation

$$(a)_b = a(a-1)\cdots(a-b+1),$$

---

[7] Hill, Lane, and Sudderth (1987) prove that for $k = 2$, the Polya urn process is the only exchangeable urn process aside from IID processes and deterministic ones. (An urn process is deterministic if all balls drawn are the same. The common label for all balls can still be random.)

[8] What we mean by this is that every ball in the urn has the same probability of being drawn.

[9] We will often use the symbol $I_A(x)$ to stand for the indicator function of the set A. That is, $I_A(x) = 1$ if $x \in A$ and $I_A(x) = 0$ if $x \notin A$.

10   Chapter 1. Probability Models

where $(a)_0 = 1$ by convention. Then, we claim that

$$\Pr(X_1 = j_1, \ldots, X_n = j_n) = \frac{\prod_{i=1}^{k}(u_i + c_i(j_1, \ldots, j_n) - 1)_{c_i(j_1, \ldots, j_n)}}{(u + n - 1)_n}. \quad (1.16)$$

For $n = 1$, this reduces to $\Pr(X_1 = j_1) = u_{j_1}/u$, which is true. If we suppose that (1.16) is true for $n = 1, \ldots, m$, then $\Pr(X_1 = j_1, \ldots, X_{m+1} = j_{m+1})$ equals

$$\Pr(X_1 = j_1, \ldots, X_m = j_m) \Pr(X_{m+1} = j_{m+1} | X_1 = j_1, \ldots, X_m = j_m)$$

$$= \Pr(X_1 = j_1, \ldots, X_m = j_m) \frac{u_{j_{m+1}} + c_{j_{m+1}}(j_1, \ldots, j_{m+1}) - 1}{u + m}. \quad (1.17)$$

In replacing $\Pr(X_1 = j_1, \ldots, X_m = j_m)$ by (1.16) in (1.17), we note that

$$c_i(j_1, \ldots, j_{m+1}) = c_i(j_1, \ldots, j_m) \quad \text{if } i \neq j_{m+1},$$
$$c_{j_{m+1}}(j_1, \ldots, j_m) = c_{j_{m+1}}(j_1, \ldots, j_{m+1}) - 1.$$

The result now follows immediately.

The only other form of exchangeability, besides conditionally IID random quantities, is illustrated in a problem as simple as drawing balls without replacement from an urn.

Example 1.18. Suppose that an urn has 20 balls, 14 of which are red and 6 of which are blue. Suppose that we draw balls without replacement. Let X_i be 1 if the ith ball is red and 0 if it is blue. If we assume that all 20! possible ordered draws of the balls are equally likely, then it is not difficult to see that the X_i are exchangeable. To see that the draws are not conditionally IID, suppose that there were a random quantity Y such that the X_i were conditionally IID given Y. Since $0 < \Pr(X_1 = 0) = E(\Pr(X_1 = 0|Y))$ (by the law of total probability B.70), it follows that $\Pr(X_1 = 0|Y) = 0$ a.s. is impossible. Hence $\Pr(\Pr(X_1 = 0|Y) > 0) > 0$, from which it follows that

$$\Pr(\Pr(X_1 = 0, X_2 = 0, \ldots, X_7 = 0|Y) > 0) = \Pr\bigl((\Pr(X_1 = 0|Y))^7 > 0\bigr)$$
$$= \Pr(\Pr(X_1 = 0|Y) > 0) > 0.$$

Hence, $\Pr(X_1 = 0, \ldots, X_7 = 0) = E(\Pr(X_1 = 0, \ldots, X_7 = 0|Y)) > 0$. But this is absurd, since there are only 6 blue balls. It must be the case that the X_i, although exchangeable, are not conditionally IID.

Theorem 1.48 will say that a finite collection of random quantities is exchangeable if and only if they are like draws from an urn without replacement.

1.2.2   Frequency and Exchangeability

There was a time when people thought that probabilities had to be frequencies, and as such, we could not know what they were before collecting an infinite amount of data. [See Von Mises (1957) for an example.] Although it is still true that we cannot know frequencies (such as the limit of the proportion of successes in a sequence of exchangeable Bernoulli random variables) without collecting an infinite amount of data, DeFinetti's representation theorem for Bernoulli random variables 1.47 tells us that such a limit of frequencies Θ is only a conditional probability given information that we do not yet have. The probabilities themselves are calculated based on subjective judgments. The possibly surprising fact is that even though different people might calculate different probabilities for the same sequence of Bernoulli random variables, if they all believe the sequence to be exchangeable, then they all believe that there exists Θ such that conditional on $\Theta = \theta$, the random variables are IID $Ber(\theta)$. That is, the subjective judgment of exchangeability for a sequence of random variables entails certain consequences that are common to every specific instance of the judgment, even when the specific instances differ in other ways.

Example 1.19. Let $\{X_n\}_{n=1}^{\infty}$ be Bernoulli random variables. Suppose that two different people give them the following joint distributions. Let $i_1, i_2, \ldots$ stand for numbers in $\{0, 1\}$. One person believes

$$\Pr(X_1 = i_1, \ldots, X_n = i_n) = \frac{12}{x+2} \frac{1}{\binom{n+4}{x+2}},$$

where x stands for $\sum_{j=1}^{n} i_j$, and the other believes this probability to be $([n+1]\binom{n}{x})^{-1}$. The first person believes that $\Pr(X_1 = 1) = 0.4$, while the second believes $\Pr(X_1 = 1) = 0.5$. On the other hand, both of these distributions are exchangeable, and so Theorem 1.47 says that both persons believe that $\Theta = \lim_{N\to\infty} \sum_{n=1}^{N} X_i/N$ exists with probability 1, and that $\Pr(X_1 = 1|\Theta = \theta) = \theta$. They must disagree on the distribution of Θ. For example, the law of total probability B.70 says $\Pr(X_1 = 1) = E(\Theta)$, hence they must have different values of $E(\Theta)$.

If probabilities are not frequencies, then why are frequencies thought to be so important in calculating probabilities? The answer lies in careful examination of the implications of DeFinetti's representation theorem.

Example 1.20 (Continuation of Example 1.19). Suppose that the two people in this example both observe $X = (X_1, \ldots, X_{20}) = y$, and suppose that y consists of 14 1s and 6 0s. It is not difficult to calculate the conditional distribution of X_{21} given this data. For example, to get $\Pr(X_{21} = 1|X = y)$, we just divide the joint probability of $(X, X_{21}) = (y, 1)$ by the probability of $X = y$. The first person believes

$$\Pr(X_{21} = 1|X = y) = \frac{\frac{12}{17}\frac{1}{\binom{25}{17}}}{\frac{12}{16}\frac{1}{\binom{24}{16}}} = 0.64,$$

while the second person believes $\Pr(X_{21} = 1|X = y) = 17/22 = 0.68$. Notice how much closer these probabilities are to each other than were the prior probabilities of 0.4 and 0.5. Also, notice how close each of them is to the proportion of successes, 0.7.

In Example 1.57 on page 31, we will see the general method for finding the conditional distribution of Θ after observing some Bernoulli trials. But

Example 1.20 gives us some hint of what happens. In Example 1.20, after observing 20 Bernoulli trials, the mean of Θ changed to a number closer to the proportion of successes, regardless of what the prior mean of Θ was. The conditional mean of Θ given the observed data is the probability of a success on a future trial given the data. If we believe a sequence of Bernoulli random variables to be exchangeable, and we are not already certain about the limit of the proportion of successes, then after we observe some data, we will modify our opinion about future observations so that the probability of success is now closer to the observed proportion of successes. This phenomenon has nothing to do with frequencies being probabilities. It is merely a consequence of exchangeability.

1.3 Parametric Models

DeFinetti's representation theorem 1.49 says that infinitely many random quantities $\{X_n\}_{n=1}^\infty$ are exchangeable if and only if they are conditionally IID given the limit of their empirical probability measures. The *empirical probability measure* (or *empirical distribution*) of $X_1, \ldots, X_n$ is the random probability measure

$$\mathbf{P}_n(B) = \frac{1}{n} \sum_{i=1}^n I_B(X_i), \quad \text{for every } B \in \mathcal{B}. \tag{1.21}$$

For the case of random variables, the empirical distribution is equivalent to the empirical distribution function, $F_n(t) = \sum_{i=1}^n I_{(-\infty,t]}(X_i)/n$, the function which is 0 at $-\infty$ and has jumps of size $1/n$ at each observation.

If we are considering a sequence of exchangeable random quantities, let Θ be some one-to-one function of the limit of the empirical distributions, and let Ω be the set of possible values for Θ. Let P_θ denote the conditional distribution of X_n given $\Theta = \theta$. Then $\mathcal{P}_0 = \{P_\theta : \theta \in \Omega\}$ looks like a typical parametric family with which we are already familiar. Also, Θ is a measurable function of the entire sequence $\{X_n\}_{n=1}^\infty$, hence its distribution is induced (see Theorem A.81) from the distribution of the sequence. For this reason, it is natural to think of Θ as a random quantity in this situation.

Although DeFinetti's representation theorem 1.49 is central to motivating parametric models, it is not actually used in their implementation. Furthermore, the concept of parametric models extends to more general situations, albeit without the same justification. For this reason, we will postpone formal treatment of DeFinetti's theorem until Section 1.4. In Section 1.3.1, we introduce the framework for the use of parametric families in general situations. Most familiar examples will be of exchangeable random variables, but other examples will be given as well. In all cases, however, we will treat the parameter as a random quantity, just as we would if the data were exchangeable.

1.3.1 Prior, Posterior, and Predictive Distributions

We begin by making explicit the general concept of parameter and parametric family.

Definition 1.22. Let $(S, \mathcal{A}, \mu)$ be a probability space, and let $(\mathcal{X}, \mathcal{B})$ and (Ω, τ) be Borel spaces. Let $X : S \to \mathcal{X}$ and $\Theta : S \to \Omega$ be measurable. Then Θ is called a *parameter* and Ω is called a *parameter space*. The conditional distribution for X given Θ is called a *parametric family of distributions of* X. The parametric family is denoted by

$$\mathcal{P}_0 = \{P_\theta : \forall A \in \mathcal{B}, P_\theta(A) = \Pr(X \in A | \Theta = \theta), \text{ for } \theta \in \Omega\}.$$

We also use the symbol $P'_\theta(X \in A)$ to stand for $P_\theta(A)$.[10] The *prior distribution of* Θ is the probability measure μ_Θ over (Ω, τ) induced by Θ from μ.

Suppose that each P_θ, when considered as a measure on $(\mathcal{X}, \mathcal{B})$, is absolutely continuous with respect to a measure ν on $(\mathcal{X}, \mathcal{B})$. Let

$$f_{X|\Theta}(x|\theta) = \frac{dP_\theta}{d\nu}(x).$$

(It will be common in this text to denote the conditional density function of one random quantity X given another Y by $f_{X|Y}$.) We can assume that $f_{X|\Theta}$ is measurable with respect to the product σ-field $\mathcal{B} \otimes \tau$.[11] This will allow us to integrate this function with respect to measures on both $\mathcal{X}$ and Ω. The function $f_{X|\Theta}(x|\theta)$, considered as a function of θ after $X = x$ is observed, is often called the *likelihood function* $L(\theta)$.

For each $\theta \in \Omega$, the function $f_{X|\Theta}(\cdot|\theta)$ is the conditional density with respect to ν of X given $\Theta = \theta$. That is, for each $A \in \mathcal{B}$,

$$\Pr(X \in A | \Theta = \theta) = \int_A f_{X|\Theta}(x|\theta) d\nu(x).$$

We let μ_X denote the marginal distribution of X ($\mu_X(A) = \Pr(X \in A)$). Using Tonelli's theorem A.69, we can write

$$\mu_X(A) = \int_\Omega \int_A f_{X|\Theta}(x|\theta) d\nu(x) d\mu_\Theta(\theta) = \int_A \int_\Omega f_{X|\Theta}(x|\theta) d\mu_\Theta(\theta) d\nu(x).$$

It follows that μ_X is absolutely continuous with respect to ν with density

$$f_X(x) = \int_\Omega f_{X|\Theta}(x|\theta) d\mu_\Theta(\theta). \tag{1.23}$$

---

[10] In this manner, P'_θ is a probability measure on the space $(S, \mathcal{A})$ and P_θ is a probability measure on the space $(\mathcal{X}, \mathcal{B})$. This fine mathematical point could usually be ignored without causing much confusion, but we will try to be as precise as possible for the sake of those few cases where it matters.

[11] See Problem 9 on page 74 for a way to prove this.

This density is often called the *(prior) predictive density of X* or the *marginal density of X*.

For example, suppose that $X = (X_1, \ldots, X_n)$, where the X_i are exchangeable and conditionally independent given Θ, each with conditional density $f_{X_1|\Theta}(\cdot|\theta)$ with respect to a measure ν^1. Then the conditional joint density of $X_1, \ldots, X_n$ given $\Theta = \theta$ (the likelihood in this case) with respect to the n-fold product measure ν^n can be written as

$$f_{X_1,\ldots,X_n|\Theta}(x_1,\ldots,x_n|\theta) = \prod_{i=1}^{n} f_{X_1|\Theta}(x_i|\theta).$$

The unconditional joint (prior predictive) density of $X_1, \ldots, X_n$ is

$$f_{X_1,\ldots,X_n}(x_1,\ldots,x_n) = \int_\Omega \prod_{i=1}^{n} f_{X_1|\Theta}(x_i|\theta) d\mu_\Theta(\theta).$$

Example 1.24. Let $X = (X_1, \ldots, X_n)$, where the X_i are conditionally IID with $N(\mu, \sigma^2)$ distribution given $(M, \Sigma) = (\mu, \sigma)$. (Here the parameter is $\Theta = (M, \Sigma)$.) Let the prior distribution be that Σ^2 has inverse gamma distribution $\Gamma^{-1}(a_0/2, b_0/2)$ and M given $\Sigma = \sigma$ has $N(\mu_0, \sigma^2/\lambda_0)$ distribution with a_0, b_0, μ_0, and λ_0 constants. The likelihood function in this case can be written as

$$f_{X|\Theta}(x|\mu,\sigma) = (2\pi\sigma^2)^{-\frac{n}{2}} \exp\left(-\frac{1}{2\sigma^2}\left[n(\bar{x}-\mu)^2 + w\right]\right),$$

where $\bar{x} = \sum_{i=1}^{n} x_i/n$ and $w = \sum_{i=1}^{n}(x_i - \bar{x})^2$. The prior density with respect to Lebesgue measure is

$$f_\Theta(\mu,\sigma) = \frac{2\left(\frac{b_0}{2}\right)^{\frac{a_0}{2}}\sqrt{\lambda_0}}{\sqrt{2\pi}\Gamma\left(\frac{a_0}{2}\right)} \sigma^{-(a_0+2)} \exp\left(-\frac{1}{2\sigma^2}\left[\lambda_0(\mu-\mu_0)^2 + b_0\right]\right), \text{ for } \sigma > 0. \tag{1.25}$$

The prior predictive distribution of the observations can be calculated by multiplying together the two functions above and integrating out the parameter. After completing the square in the exponent, the product can be written as

$$\frac{2\left(\frac{b_0}{2}\right)^{\frac{a_0}{2}}\sqrt{\lambda_0}}{(2\pi)^{\frac{n+1}{2}}\Gamma\left(\frac{a_0}{2}\right)} \sigma^{-(a_1+2)} \exp\left(-\frac{1}{2\sigma^2}\left[\lambda_1(\mu-\mu_1)^2 + b_1\right]\right), \tag{1.26}$$

where

$$a_1 = a_0 + n, \qquad \lambda_1 = \lambda_0 + n,$$
$$b_1 = b_0 + w + \frac{n\lambda_0(\bar{x}-\mu_0)^2}{\lambda_0 + n}, \qquad \mu_1 = \frac{\lambda_0\mu_0 + n\bar{x}}{\lambda_0 + n}.$$

Note that, as a function of (μ, σ), this is in the same form as the prior density (1.25) with the four numbers $a_0, b_0, \mu_0, \lambda_0$ replaced by $a_1, b_1, \mu_1, \lambda_1$. Hence, the integral over (μ, σ) is just the constant factor that appears in (1.26) divided by

1.3. Parametric Models   15

the result of changing $a_0, b_0, \mu_0, \lambda_0$ to $a_1, b_1, \mu_1, \lambda_1$, respectively, in the constant factor in (1.25). That is,

$$f_X(x) = \frac{2\left(\frac{b_0}{2}\right)^{\frac{a_0}{2}}\sqrt{\lambda_0}}{(2\pi)^{\frac{n+1}{2}}\Gamma\left(\frac{a_0}{2}\right)} \frac{\sqrt{2\pi}\Gamma\left(\frac{a_1}{2}\right)}{2\left(\frac{b_1}{2}\right)^{\frac{a_1}{2}}\sqrt{\lambda_1}} = \frac{b_0^{\frac{a_0}{2}}\sqrt{\lambda_0}\Gamma\left(\frac{a_1}{2}\right)}{\pi^{\frac{n}{2}} b_1^{\frac{a_1}{2}}\sqrt{\lambda_1}\Gamma\left(\frac{a_0}{2}\right)}. \quad (1.27)$$

A specialized calculation of the preceding sort is often of interest in this example. Let Y_n be the average of the n observations. The conditional distribution of Y_n given $\Theta = (\mu, \sigma)$ is $N(\mu, \sigma^2/n)$. The prior predictive density of Y_n given $X = x$ can be calculated by integrating the $N(\mu, \sigma^2/n)$ density times (1.25) with respect to μ and σ. Alternatively, one can argue as follows. Using well-known features of the normal distribution, we can conclude that the conditional distribution of Y_n given $\Sigma = \sigma$ is $N(\mu_0, \sigma^2[1/n + 1/\lambda_0])$. If we multiply the corresponding normal density times the marginal density of Σ and integrate over σ, we get

$$\int_0^\infty \frac{\sqrt{2}\left(\frac{b_0}{2}\right)^{\frac{a_0}{2}}}{\Gamma\left(\frac{a_0}{2}\right)\sqrt{\pi\left(\frac{1}{n}+\frac{1}{\lambda_0}\right)}} \sigma^{-a_0-3}\exp\left(-\frac{1}{2\sigma^2}[b_0+(y-\mu_0)^2]\right)d\sigma$$

$$= \frac{\Gamma\left(\frac{a_0+1}{2}\right)}{\Gamma\left(\frac{a_0}{2}\right)\sqrt{b_0\pi\left(\frac{1}{n}+\frac{1}{\lambda_0}\right)}}\left(1+\frac{(y-\mu_0)^2}{b_0}\right)^{-\frac{a_0+1}{2}},$$

which is the density of the $t_{a_0}(\mu_0, \sqrt{(1/n+1/\lambda_0)b_0/a_0})$ distribution.

As we mentioned earlier, the use of parametric families does not require that the data be a collection of exchangeable random quantities. Here are some examples of nonexchangeable random quantities whose distributions could usefully be modeled using finite-dimensional parametric families.

Example 1.28. Let $\{X_n\}_{n=1}^\infty$ be a sequence of Bernoulli random variables that are not exchangeable. Instead, let $\mathcal{P}_0$ be the set of joint distributions for infinitely many Bernoulli random variables that form a Markov chain. For $P \in \mathcal{P}_0$, define

$$\Theta(P) = (\Pr(X_1=1|P), \Pr(X_{i+1}=1|X_i=1,P), \Pr(X_{i+1}=1|X_i=0,P))$$
$$= (P_1, P_{11}, P_{01}).$$

Let λ be any probability over $[0,1]^3$, and set

$$\Pr(X_1=i_1,\ldots,X_n=i_n) \quad (1.29)$$
$$= \int\int\int p_1^{i_1}(1-p_1)^{1-i_1} p_{11}^{k_{11}}(1-p_{11})^{k_{01}} p_{01}^{k_{10}}(1-p_{01})^{k_{00}} d\lambda(p_1,p_{11},p_{01}),$$

where $k_{s,t}$ is the number of times that s follows t in the sequence $i_1,\ldots,i_n$. Diaconis and Freedman (1980c) prove that, aside from pathological cases, if all finite sequences of 0s and 1s that have the same first element and the same values of $k_{s,t}$ have the same probability, then (1.29) must hold.

Example 1.30. This example is the simple linear regression problem. Suppose that $x_1, x_2, \ldots$ are fixed known numbers and $E_1, E_2, \ldots$ are exchangeable random variables that are conditionally independent given $\Sigma = \sigma$ with density $f_{E|\Sigma}(e|\sigma)$.

(Think of the E_i as the error or noise term in a regression model.) Define $Y_i = E_i + Bx_i$, where B is a random variable such that B and Σ have joint distribution $\mu_{B,\Sigma}$. The parameter now consists of $\Theta = (B, \Sigma)$. The random variables $Y_1, Y_2, \ldots$ are not exchangeable even though $E_1, E_2, \ldots$ are exchangeable. The reader should see Zellner (1971) for an in-depth discussion of Bayesian analysis of regression models.

The conditional distribution of Θ given $X = x$ is called the *posterior distribution of* Θ. The next theorem shows us how to calculate the posterior distribution of a parameter in the case in which there is a measure ν such that each $P_\theta \ll \nu$.

Theorem 1.31 (Bayes' theorem).[12] *Suppose that X has a parametric family $\mathcal{P}_0$ of distributions with parameter space Ω. Suppose that $P_\theta \ll \nu$ for all $\theta \in \Omega$, and let $f_{X|\Theta}(x|\theta)$ be the conditional density (with respect to ν) of X given $\Theta = \theta$. Let μ_Θ be the prior distribution of Θ. Let $\mu_{\Theta|X}(\cdot|x)$ denote the conditional distribution of Θ given $X = x$. Then $\mu_{\Theta|X} \ll \mu_\Theta$, a.s. with respect to the marginal of X, and the Radon–Nikodym derivative is*

$$\frac{d\mu_{\Theta|X}}{d\mu_\Theta}(\theta|x) = \frac{f_{X|\Theta}(x|\theta)}{\int_\Omega f_{X|\Theta}(x|t)d\mu_\Theta(t)}$$

for those x such that the denominator is neither 0 nor infinite. The prior predictive probability of the set of x values such that the denominator is 0 or infinite is 0, hence the posterior can be defined arbitrarily for such x values.

PROOF. First, we prove the claims about the denominator. Let

$$C_0 = \left\{ x : \int_\Omega f_{X|\Theta}(x|t)d\mu_\Theta(t) = 0 \right\},$$

$$C_\infty = \left\{ x : \int_\Omega f_{X|\Theta}(x|t)d\mu_\Theta(t) = \infty \right\}.$$

Let μ_X be the marginal distribution of X,

$$\mu_X(A) = \int_A \int_\Omega f_{X|\Theta}(x|\theta)d\mu_\Theta(\theta)d\nu(x).$$

It follows that

$$\mu_X(C_0) = \int_{C_0} \int_\Omega f_{X|\Theta}(x|\theta)d\mu_\Theta(\theta)d\nu(x) = 0,$$

$$\mu_X(C_\infty) = \int_{C_\infty} \int_\Omega f_{X|\Theta}(x|\theta)d\mu_\Theta(\theta)d\nu(x) = \int_{C_\infty} \infty d\nu(x).$$

---

[12]Theorem 1.31 applies equally well to infinite-dimensional parameters as to finite-dimensional parameters. In infinite-dimensional cases, however, the condition $P_\theta \ll \nu$ for all θ often fails. In fact, the proof applies even if (Ω, τ) is not a Borel space. In this last case, a regular conditional distribution is explicitly constructed without knowing in advance that one will exist.

1.3. Parametric Models   17

This last integral will equal ∞ if $\nu(C_\infty) > 0$. Since this is impossible, it must be that $\nu(C_\infty) = 0$, hence $\mu_X(C_\infty) = 0$.

The posterior distribution $\mu_{\Theta|X}$ must satisfy the following. For all sets $A \in \mathcal{B}$ and all $B \in \tau$,

$$\Pr(\Theta \in B, X \in A) = \int_A \mu_{\Theta|X}(B|x) d\mu_X(x). \quad (1.32)$$

Using Tonelli's theorem A.69 we can write

$$\begin{aligned}\Pr(\Theta \in B, X \in A) &= \int_B \int_A f_{X|\Theta}(x|\theta) d\nu(x) d\mu_\Theta(\theta) \\ &= \int_A \int_B f_{X|\Theta}(x|\theta) d\mu_\Theta(\theta) d\nu(x). \quad (1.33)\end{aligned}$$

Next, write

$$\int_A \mu_{\Theta|X}(B|x) d\mu_X(x) = \int_A \left[\mu_{\Theta|X}(B|x) \int_\Omega f_{X|\Theta}(x|\theta) d\mu_\Theta(\theta) \right] d\nu(x).$$

Combining this with (1.33) shows that (1.32) is satisfied for all A and B if and only if

$$\mu_{\Theta|X}(B|x) = \frac{\int_B f_{X|\Theta}(x|\theta) d\mu_\Theta(\theta)}{\int_\Omega f_{X|\Theta}(x|\theta) d\mu_\Theta(\theta)},$$

a.s. $[\mu_X]$. It follows that $\mu_{\Theta|X} \ll \mu_\Theta$ and that $d\mu_{\Theta|X}/d\mu_\Theta(\cdot|x)$ is as specified. $\square$

Example 1.34. Suppose that X has $Bin(n, \theta)$ given $\Theta = \theta$ and that the prior distribution of Θ is $Beta(a_0, b_0)$. The marginal density of X with respect to counting measure on the integers is

$$f_X(x) = \binom{n}{x} \frac{\Gamma(a_0 + b_0)\Gamma(a_0 + x)\Gamma(b_0 + n - x)}{\Gamma(a_0)\Gamma(b_0)\Gamma(a_0 + b_0 + n)}, \text{ for } x = 0, \ldots, n.$$

The posterior density of Θ with respect to the prior distribution of Θ is the ratio of $\binom{n}{x} \theta^x (1-\theta)^{n-x}$ to $f_X(x)$. The posterior density of Θ with respect to Lebesgue measure is

$$f_{\Theta|X}(\theta|x) = \frac{\Gamma(a_0 + b_0 + n)}{\Gamma(a_0 + x)\Gamma(b_0 + n - x)} \theta^{a_0 + x - 1}(1-\theta)^{b_0 + n - x}, \text{ for } 0 < \theta < 1,$$

which is easily seen to be the $Beta(a_0 + x, b_0 + n - x)$ density.

Example 1.35 (Continuation of Example 1.24; see page 14). For the case of conditionally IID $N(\mu, \sigma^2)$ random variables, the posterior density with respect to Lebesgue measure can be calculated by dividing the product of prior and likelihood (1.26) by the prior predictive density (1.27). The result is easily seen to be in the same form as the prior density (1.25) with the four constants $a_0, b_0, \mu_0, \lambda_0$ replaced by $a_1, b_1, \mu_1, \lambda_1$. In other words, the posterior distribution of Σ^2 is $\Gamma^{-1}(a_1/2, b_1/2)$, and the conditional posterior of M given $\Sigma = \sigma$ is $N(\mu_1, \sigma^2/\lambda_1)$.

Example 1.36. As an example in which Bayes' theorem does not apply, consider the case in which the conditional distribution of X given $\Theta = \theta$ is discrete with $P_\theta(\{\theta - 1\}) = P_\theta(\{\theta + 1\}) = 1/2$. Suppose that Θ has a density f_Θ with respect to Lebesgue measure. The P_θ distributions are not all absolutely continuous with respect to a single σ-finite measure. It is still possible to verify that the posterior distribution of Θ given $X = x$ is the discrete distribution with

$$\Pr(\Theta = x - 1 | X = x) = \frac{f_\Theta(x - 1)}{f_\Theta(x - 1) + f_\Theta(x + 1)},$$

and $\Pr(\Theta = x + 1 | X = x) = 1 - \Pr(\Theta = x - 1 | X = x)$. Note that the posterior is not absolutely continuous with respect to the prior.[13]

The *(posterior) predictive distribution of future data* is defined in the same way as the prior predictive distribution except that the posterior distribution of Θ is used instead of the prior distribution of Θ. For the case of conditionally IID random variables with conditional density $f_{X_1|\Theta}$ given Θ, we have

$$f_{X_{n+1},\ldots,X_{n+k}|X_1,\ldots,X_n}(x_{n+1},\ldots,x_{n+k}|x_1,\ldots,x_n) \quad (1.37)$$
$$= \int_\Omega \prod_{i=1}^k f_{X_1|\Theta}(x_{n+i}|\theta) d\mu_\Theta(\theta|x_1,\ldots,x_n).$$

Example 1.38 (Continuation of Example 1.35; see page 17). The posterior predictive distribution of future observations can be calculated after observing a sample of conditionally IID normal random variables. Let Y_m be the average of m future observations. Since the posterior distribution of Θ is in the same form as the prior (1.25) with $a_0, b_0, \mu_0, \lambda_0$ replaced by $a_1, b_1, \mu_1, \lambda_1$, it follows that the posterior predictive distribution of Y_m is of the same form as the prior predictive distribution. Using the result from the end of Example 1.24 on page 14, we get that the posterior predictive distribution of Y_m is $t_{a_1}(\mu_1, \sqrt{[1/m + 1/\lambda_1]b_1/a_1}\,)$.

To see how Bayes' theorem 1.31 applies to arbitrary random quantities whose distributions are modeled using parametric families, consider the following example.

Example 1.39. Consider two sequences $\{X_n\}_{n=1}^\infty$ and $\{Y_n\}_{n=1}^\infty$ of random variables that are each separately exchangeable. We can model them so that the parameters are related. For example, suppose that the X_i are IID $Exp(\theta)$ given $\Theta = \theta$ and the Y_j are IID $U(0, \theta)$ given $\Theta = \theta$, and we model the X_i and Y_j as conditionally independent given Θ. We may learn $X_1 = x_1, \ldots, X_n = x_n$ and then wish to make inference about Y_is. Let the prior for Θ be μ_Θ. The posterior is

$$\mu_\Theta(B|x_1,\ldots,x_n) = \frac{\int_B \theta^n \exp(-\bar{x}\theta) d\mu_\Theta(\theta)}{\int \psi^n \exp(-\bar{x}\psi) d\mu_\Theta(\psi)},$$

---

[13] Another example of this situation occurs in Problem 47 on page 80. In that example, Θ is an infinite-dimensional parameter, however.

where $x = \sum_{i=1}^{n} x_i$. The predictive density of $(Y_1, \ldots, Y_m)$ is

$$f_{Y_1,\ldots,Y_m|X_1,\ldots,X_n}(y_1,\ldots,y_m|x_1,\ldots,x_n) = \int_{\max y_i}^{\infty} \frac{1}{\theta^m} d\mu_\Theta(\theta|x_1,\ldots,x_n).$$

Since P_θ is a conditional distribution given another random variable $\Theta = \theta$, there exist conditional expectations given $\Theta = \theta$. Let E_θ stand for the expectation operator under P'_θ. That is, if Z is a random variable with finite absolute expectation, then $E_\theta(Z)$ means $E(Z|\Theta)(s)$ for all s such that $\Theta(s) = \theta$. By Theorem B.12, if $f : \mathcal{X} \to \mathbb{R}$ and $Z = f(X)$,

$$E_\theta(Z) = \int Z(x) dP_\theta(x).$$

Similarly, let $\mathrm{Var}_\theta(X)$ and $\mathrm{Cov}_\theta(X,Y)$ stand for the conditional variance of X given $\Theta = \theta$ and the conditional covariance between X and Y given $\Theta = \theta$, respectively.

There will be times when we wish to condition on other random variables in addition to Θ. Recall that for two random quantities $Z : S \to \mathbb{R}$ and $Y : S \to T$, the conditional expectation of Z given Y was defined to be an $\mathcal{A}_Y$ measurable function $E(Z|Y)(s)$ satisfying

$$E(ZI_B) = \int_B E(Z|Y)(s) d\mu(s),$$

for all $B \in \mathcal{A}_Y$. The conditional expectation of Z given Y and Θ will be an $\mathcal{A}_{(Y,\Theta)}$ measurable function $E(Z|Y,\Theta)$ satisfying

$$E(ZI_B) = \int_B E(Z|Y,\Theta)(s) d\mu(s),$$

for all $B \in \mathcal{A}_{(Y,\Theta)}$. It follows from Theorem B.75 that $E(Z|Y = y, \Theta = \theta) = E_\theta(Z|Y = y)$, where $E_\theta(\cdot|Y = y)$ is conditional expectation calculated from P'_θ. It follows from the law of total probability B.70 that

$$E(Z|Y = y) = \int_\Omega E(Z|Y = y, \Theta = \theta) d\mu_{\Theta|Y}(\theta|y),$$

where $\mu_{\Theta|Y}(\cdot|y)$ specifies the conditional distribution of Θ given $Y = y$.

1.3.2 Improper Prior Distributions

Two components are required to specify the distribution of a random quantity X by means of a parametric family. One is the choice of parametric family, and the other is the prior distribution over the parameter space. Both of these must be specified if one is to have a marginal distribution for X. Some people seem to think that choosing a prior distribution introduces subjectivity into the analysis of data but choosing a parametric family does not. These people are mistaken. Each choice one makes introduces subjectivity.

Philosophy aside, suppose that one finds it difficult to specify a prior distribution beacause one does not have much idea where the parameter is likely to be located. In such cases, one may wish to do calculations based on a prior distribution that spreads the probability very thinly over the parameter space. A problem that often arises is that, if we take the limit as the probability is spread more thinly, the prior distribution ceases to satisfy the axioms of probability theory.

Example 1.40. Suppose that we choose the parametric family of normal distributions with variance 1 and parameter Θ equal to the mean. The parameter space is the real line $\mathbb{R}$. Suppose that we want a normal prior distribution for Θ, but one with very high variance to indicate that we are not willing to say where we think Θ is with much certainty. The distribution $N(a, n)$ for large n has this property. But how can we choose n? If we let $n \to \infty$, there is no countably additive limit to the sequence of probability distributions. There is no normal distribution with infinite variance.

What has become common in problems like Example 1.40 is to choose a measure λ on (Ω, τ) which may not be a probability but still pretend that it is the prior distribution of Θ. That is, use λ in place of μ_Θ in Bayes' theorem 1.31. The "posterior" after observing $X = x$, if it exists, will have density with respect to λ,

$$\frac{f_{X|\Theta}(x|\theta)}{\int_\Omega f_{X|\Theta}(x|t) d\lambda(t)}. \tag{1.41}$$

The key is whether or not the denominator in (1.41) is finite and nonzero. If so, we can pretend that (1.41) is the posterior density of Θ given $X = x$ and then proceed with whatever analysis we want to perform. In this case, we call λ an *improper prior distribution*. If the denominator in (1.41) is 0 or infinite, one may need to choose another prior distribution.

Example 1.42. Suppose that $X \sim N(\theta, 1)$ given $\Theta = \theta$. We can use λ equal to Lebesgue measure as an improper prior. Suppose that we observe only one observation X. Since $f_{X|\Theta}(x|\theta) = (\sqrt{2\pi})^{-1} \exp(-[x - \theta]^2/2)$, it follows that the posterior density with respect to Lebesgue measure derived from Bayes' theorem 1.31 is equal to $f_{X|\Theta}(x|\theta)$ as a function of Θ. In other words, given $X = x$, Θ has $N(x, 1)$ distribution.

The above discussion of improper priors is not particularly precise mathematically. There are two traditional ways to make the concept of improper prior mathematically precise. Each of them opens its own particular can of worms, so we will only describe each very briefly and point the reader to relevant literature. First, one may remove the restriction that the probability of a set must be at most 1. Hartigan (1983) takes this approach and allows sets to have infinite probability. This makes improper priors "proper," but now many traditional theorems of probability theory which make implicit use of the upper bound on probabilities either must be reproved or fail to apply to infinite probabilities. The second approach is that of DeFinetti (1974), in which the requirement that probabilities be countably additive is relaxed. That is, probability is only required to be finitely additive.[14] Needless to say, most of the traditional results of probability theory need to be reproved or scrapped in this theory also.[15] The improper prior in Example 1.42, when thought of as a finitely additive prior, gives 0 probability to every compact set and still gives probability 1 to the whole real line. Hartigan (1983, Theorem 3.5) gives a version of Bayes' theorem for possibly infinite probabilities. Berti, Regazzini, and Rigo (1991) prove a Bayes' theorem for finitely additive probabilities, as do Heath and Sudderth (1989). An alternative to using improper priors is to do a robust Bayesian analysis, as described in Section 8.6.3.

1.3.3 Choosing Probability Distributions

We have assumed that probability distributions represent our (or someone else's) opinion about unknown quantities. At least a little thought should be given to how those probability distributions are chosen. The most common method for choosing a probability distribution might be called "availability." Most people who study statistics formally for no more than one academic year will only be able to describe one parametric family of distributions suitable for use with continuous data. The family of normal distributions is both computationally tractable and remarkably versatile as a model for many natural phenomena. Its versatility is due in part to the fact that many other distributions can easily be transformed to normal distributions so that the computational tractability of the normal distribution can be widely extended. The family of transformations introduced by Box and Cox (1964) is a classic example. Other methods for choosing probability distributions are based on data analytic techniques. Either the very data on which inference will be based or other seemingly relevant data are analyzed by various graphical techniques, hypothesis tests, or other procedures in order to try to select an appropriate probability model to use as a description of the uncertainty surrounding the data. The most direct methods for selecting distributions are those based on elicitation. In such

---

[14]A finitely additive probability is a function μ from a field $\mathcal{F}$ of subsets of a set S to $[0, 1]$ which satisfies $\mu(\emptyset) = 0$ and $\mu(A \cup B) = \mu(A) + \mu(B)$ if $A \cap B = \emptyset$. Kadane, Schervish, and Seidenfeld (1985) explore some of the implications of finitely additive probability for statistical inference. One well-known consequence of using improper priors is the famous *marginalization paradox* reported by Stone and Dawid (1972) and Dawid, Stone, and Zidek (1973).

[15]Schervish, Seidenfeld, and Kadane (1984) show how the law of total probability B.70 fails in the finitely additive theory. Stone (1976) gives an interesting example of this failure.

methods, an expert (a term to be left undefined) is questioned about his or her beliefs concerning relevant random quantities, and a probability model for those beliefs is inferred from the responses.

Each of the three types of methods for choosing probability distributions has its advantages and disadvantages. The availability method may seem silly as described above, but one usually does have a limited number of families of distributions that one is willing to consider. The methods described in Section 8.6 on mixtures of models can be useful in sorting out uncertainties amongst alternative models for a given data set. In particular, robust methods (Sections 5.1.5 and 8.6.3) are designed to assess or even limit the sensitivity of inferences to specifications of distributions.[16] In short, one may not be forced actually to choose a single probability distribution to represent his or her uncertainty. Comparing the effects of various possible choices may be sufficient for assessing the information content of the data. When one is determined to use a particular parametric family of distributions, and only the prior distribution for the parameter needs to be chosen, it may be the case that various alternatives make little difference and a choice by convenience (like an improper prior) will be sufficient. Whether or not this is true, considerations of Bayesian robustness will clearly be in order when such a choice is made.

Data-based techniques are particularly appealing when one is forced to analyze someone else's data without access to subject matter expertise. Also, if one must use one of the popular computer packages, which tend to be built exclusively around only one distribution for each type of data, it pays to be able to transform the data into something better suited to be modeled by that one distribution. Quantile plots and various graphical techniques [see, for example, Gnanadesikan (1977)] are very useful for helping to select such a transformation. Likelihood-based methods for choosing a distribution can be described as follows. Suppose that there is an index set $\aleph$, and for each $\alpha \in \aleph$, there is a possible distribution for the available data X. Let $f_{X,\alpha}$ denote the predictive density of the data X as calculated in (1.23) under the assumption that the distribution being used is the one corresponding to index α. One could then base a choice amongst the different values of α on the values of $g(\alpha) = f_{X,\alpha}(x)$ once $X = x$ is observed. (Typically, one chooses α to maximize $g(\alpha)$.) This is similar to what is done in empirical Bayes analysis (see Section 8.4 for more details). An obvious drawback to all such data-based methods is that they tend to understate the amount of uncertainty that remains about interesting unknown quantities. The reason is that one pretends to be sure of something (e.g., which parametric family or which value of α) of which one really is not sure.

Example 1.43. Suppose that X is a vector of 20 random variables and that we cannot decide whether to model them as IID $Lap(\mu, \sigma/\sqrt{2})$ or IID $N(\mu, \sigma^2)$ given

---

[16]Berger (1994) reviews the literature on robust Bayesian methods up to 1993.

$(M, \Sigma) = (\mu, \sigma)$. (In this way, μ and σ are the mean and standard deviation in both cases.) We could let $\aleph$ be the set of all triples (i, μ, σ), where $i = 1$ means Laplace and $i = 2$ means normal. Consider the following data values:

$$-0.0820, 1.3312, -1.3518, -1.4930, 0.0850, 0.7022, 1.735,$$
$$-0.3164, 2.1948, -0.0371, 0.3377, -0.3124, 0.6087, 0.7339,$$
$$-0.4632, 0.3398, -0.0352, 0.1597, -0.6344, -0.4435.$$

The value of α that leads to the largest $f_{X,\alpha}(x)$ is $(1, 0.0249, 0.9473)$. The largest value is 5.943×10^{-12}, which is only slightly larger than the value achieved at $\alpha = (2, 0.1530, 0.8909)$, namely 4.772×10^{-12}. If we decide to use the Laplace distribution model, we will be pretending that we were sure from the start that the data would be $Lap(\mu, \sigma/\sqrt{2})$, rather than taking into account the sizable amount of uncertainty that still remains about the underlying distribution.

In a classical setting, one might look at quantile plots to see whether the data looked more normal or more like a Laplace distribution. Figure 1.44 shows quantile plots for both Laplace and normal distributions. The two plots are about equally straight, although the Laplace plot is a little bit straighter. Choosing either distribution would surely be acting as if we knew something that was quite uncertain.

Elicitation techniques tend to lie on the interface between statistical theory and psychology. A series of questions must be designed for interrogation of the expert. The responses to these questions must then be reconciled with the axioms of probability theory, keeping in mind the expert's limited motivation and/or ability to answer accurately. Much has been written in the psychological literature about the ability of people to assess probabilities subjectively. Kahneman, Slovic, and Tversky (1982) have compiled an interesting collection of articles that, among other things, illustrate and

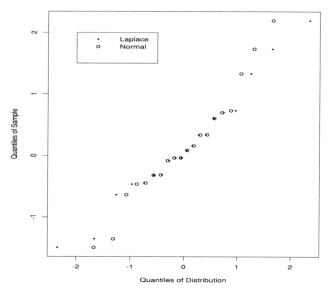

FIGURE 1.44. Quantile Plots for Laplace and Normal Distributions

describe various problems people have in quantifying uncertainty. Hogarth (1975) surveys the early literature on the assessment of subjective distributions. He concludes that humans are "ill-equipped for assessing subjective probability distributions." This conclusion might help to explain the number of tools that have emerged since that time to better equip statisticians and experts to choose distributions. These tools are directed primarily toward choosing a prior distribution for the parameters of a prespecified parametric family. For example, Kadane et al. (1980) and Garthwaite and Dickey (1988) give algorithms for specifying conjugate prior distributions for the parameters of normal linear models. Garthwaite and Dickey (1992) extend their method to deal with the selection of variables in multiple regression. Freedman and Spiegelhalter (1983) and Chaloner et al. (1993) describe methods for eliciting prior information for use in clinical trials. A common feature of most prior elicitation schemes is their reliance on the predictive distribution (1.23) to infer the prior. The reason for this is that experts are more likely to be comfortable thinking about the actual observables of their study rather than parameters of statistical models. Problems 18 and 19 at the end of this chapter give some simple examples of how this might be done. In order to take account of the fact that experts may not accurately respond to the elicitation inquiries, Dickey (1980) and later Gavasakar (1984) described probability models for the elicitation process itself. In these models, the responses U which the expert will give to elicitation questions are modeled as data with a distribution that depends on the subjective distribution P being elicited. One then tries to infer P from U using Bayesian or related methods. One must be careful not only to consider the possible errors in U as answers to the elicitation questions, but also to consider how sensitive the inference from U to P is. For example, does a small change in U produce a small or a large change in P? (See the last parts of Problem 18 on page 76.)

In the remainder of this text, as in most texts on the theory of statistics, little attention will be given to how the probability distributions are chosen. When a prior distribution is used for a specified parametric family, one can assume either that the prior was elicited by some method or other, or that the prior was chosen by convenience (a popular device), or that the prior is just one of many that will later be compared in a robustness study.

1.4 DeFinetti's Representation Theorem

1.4.1 Understanding the Theorems

In this section we will state some representation theorems for exchangeable random quantities and give a number of examples. The proofs of these theorems (and some related results of interest) are deferred to Section 1.5.[17] These theorems characterize all of the possible joint distributions for exchangeable random quantities which take values in a Borel space (think finite-dimensional Euclidean space), and they are essentially due to DeFinetti (1937). They can be summarized here as follows. If there is an infinite sequence of exchangeable random quantities $\{X_n\}_{n=1}^{\infty}$, then there must be some random quantity $\mathbf{P}$ such that the X_i are conditionally IID given $\mathbf{P}$. If the random quantities have Bernoulli distribution, then $\mathbf{P}$ can be taken to be the limit of the proportion of successes in the first n observations. In general, $\mathbf{P}$ will turn out to be the limit of the empirical distributions $\mathbf{P}_n$ of $X_1, \ldots, X_n$, which were defined in (1.21). This explains how DeFinetti's theorem helps to motivate models like (1.10).

Example 1.45. Consider the case of Bernoulli random variables $\{X_i\}_{i=1}^{\infty}$. Here, $\mathcal{X} = \{0, 1\}$ and a random probability measure $\mathbf{P}$ on $\mathcal{X}$ is equivalent to a random variable $\Theta \in [0, 1]$, where $\Theta = \mathbf{P}(\{1\})$. The empirical distribution $\mathbf{P}_n$ is equivalent to $\overline{X}_n$, the average of the first n observations, since $\mathbf{P}_n(\{1\}) = \overline{X}_n$. Theorem 1.47 (a special case of Theorem 1.49) will say that $\overline{X}_n$ converges to Θ a.s., and that, conditional on $\Theta = \theta$, the X_i are IID $Ber(\theta)$ random variables. This is what we meant on page 7 when we said that "θ is given an implicit meaning as a random variable Θ, rather than a fixed value." Also, the random variable Θ will have a distribution, which is the measure μ in (1.10).

The heavy mathematics in the proof of Theorem 1.49 is required to make precise what it means to have a random probability measure $\mathbf{P}$ and what it means to condition on such a thing. For random quantities that assume only finitely many different values, random probability measures are equivalent to finite-dimensional random vectors. For more general random quantities, random probability measures can be more complicated. For this reason, we prove Theorem 1.47 first, even though it is a special case of Theorem 1.49. The proof of Theorem 1.47 contains the essential ideas of the more complicated proof without being encumbered by so much mathematics.

If there are only finitely many exchangeable quantities $X_1, \ldots, X_N$, then all that we can prove is the following. Conditional on the empirical distribution $\mathbf{P}_N$ of $X_1, \ldots, X_N$, every ordered n-tuple (for $n \leq N$) of the X_i has the distribution of n draws *without replacement* from a finite population with distribution $\mathbf{P}_N$. It is the "without replacement" qualifier that prevents us from proving that $X_1, \ldots, X_N$ are conditionally independent. (See Example 1.18 on page 10.) It is possible for a finite collection of exchangeable random variables to be conditionally independent; however, it is not necessary. Looking at the Bernoulli case first might aid in understanding

---

[17]In most of this text, proofs are given immediately or almost immediately after the statements of results. Because DeFinetti's representation theorem 1.49 is so important for motivating statistical modeling, and because its proof involves some rather heavy mathematics, many readers may wish to forego reading the proofs on a first pass through this material. However, every reader should at least try to understand what Theorem 1.49 says.

26   Chapter 1. Probability Models

the finite case theorem.

Example 1.46. Let $X_1,\ldots,X_N$ be exchangeable Bernoulli random variables. Let the word "success" stand for $X_i = 1$, and let the word "failure" stand for $X_i = 0$. Let the word "trial" stand for one of the X_i. Since there are only finitely many (2^N) possible values for the vector $(X_1,\ldots,X_N)$, the entire joint distribution can be specified by giving probabilities to all of those 2^N vectors of 0s and 1s. If $X_1,\ldots,X_N$ are exchangeable, however, many of the vectors will have the same probability. For example, the N vectors with exactly one success and $N-1$ failures all have the same probability. Similarly, the $\binom{N}{2}$ vectors with exactly two successes and $N-2$ failures all have the same probability. In fact, for each $m = 0,\ldots,N$, all $\binom{N}{m}$ vectors with exactly m successes and $N-m$ failures have the same probability. Since the total number of successes in all N trials plays such an important role in the distribution, we give it a name, M. Let $p_M = \Pr(M = m)$ for $m = 0,\ldots,N$. Then, the probability of each vector with exactly m successes and $N-m$ failures is $p_M/\binom{N}{m}$. All probabilities associated with the joint distribution of $X_1,\ldots,X_N$ can be calculated from these values. For example, suppose that we let K equal the number of successes in a particular collection of n trials (for example, the first n, or the last n, or every other one from the first $2n$, etc.). Then $\Pr(K = k)$ can be calculated by adding up all the probabilities of the vectors for which the particular n trials include exactly k successes. This is nothing more than a straightforward counting argument, if we first partition the vectors according to the value of M. For each $m = k,\ldots,N-n+k$, there are $\binom{n}{k}\binom{N-n}{m-k}$ vectors with $M = m$ and with exactly k successes on the particular n trials of interest. It follows that

$$\Pr(K = k) = \sum_{m=k}^{N-n+k} \frac{\binom{n}{k}\binom{N-n}{m-k}}{\binom{N}{m}} p_m.$$

This last expression is easily recognized as a mixture of hypergeometric probabilities $Hyp(N,n,m)$ with mixing weights p_m. That is, it appears as if K has $Hyp(N,n,m)$ distribution conditional on $M = m$ and M has distribution $(p_0,\ldots,p_N)$. Note that the $Hyp(N,n,m)$ distribution is the distribution of the number of successes in n draws without replacement from an urn containing m successes and $N-m$ failures. Also, the random variable M is equivalent to the empirical distribution $\mathbf{P}_N$.

Example 1.46 above is the proof of the finite version of DeFinetti's theorem for Bernoulli random variables. It is also illustrative of the most general form of the finite version. The total number of successes must be replaced by the empirical distribution $\mathbf{P}_N$, and then we have that $X_1,\ldots,X_N$ are exchangeable if and only if the conditional distribution, given $\mathbf{P}_N$, of every finite subcollection $X_{i_1},\ldots,X_{i_n}$ is the distribution of n random draws *without replacement* from a population with distribution $\mathbf{P}_N$.

1.4.2   The Mathematical Statements

The Bernoulli case is simple enough to state without introduction.

Theorem 1.47 (DeFinetti's representation theorem for Bernoulli random variables). *An infinite sequence* $\{X_n\}_{n=1}^{\infty}$ *of Bernoulli random* variables is exchangeable if and only if there is a random variable Θ taking values in $[0, 1]$ such that, conditional on $\Theta = \theta$, $\{X_n\}_{n=1}^{\infty}$ are IID $Ber(\theta)$. Furthermore, if the sequence is exchangeable, then the distribution of Θ is unique and $\sum_{i=1}^{n} X_i/n$ converges to Θ almost surely.

For the more general cases, we need some more notation. Let $(\mathcal{X}, \mathcal{B})$ be a Borel space, and let $\mathcal{P}$ be the set of all probability measures on $(\mathcal{X}, \mathcal{B})$. The theorems stated below will give the conditional distributions of certain random quantities taking values in $\mathcal{X}$ given certain probability measures. To be mathematically precise, these probability measures must themselves be random quantities. That is, we will need a σ-field $\mathcal{C}_{\mathcal{P}}$ of subsets of $\mathcal{P}$ such that the appropriate probability measures can be thought of as measurable functions from some probability space $(S, \mathcal{A}, \mu)$ to $(\mathcal{P}, \mathcal{C}_{\mathcal{P}})$. Let $\mathcal{C}_{\mathcal{P}}$ be the smallest σ-field of subsets of $\mathcal{P}$ containing all sets of the form $A_{B,t} = \{P \in \mathcal{P} : P(B) \leq t\}$, for $B \in \mathcal{B}$ and $t \in [0, 1]$. This is the smallest σ-field for which the evaluation functions $g_B : \mathcal{P} \to \mathbb{R}$ are measurable,[18] where $g_B(P) = P(B)$. It is easy to show that $\mathbf{P}_n$, defined in (1.21), is a measurable function from the n-fold product space $(\mathcal{X}^n, \mathcal{B}^n)$ to $(\mathcal{P}, \mathcal{C}_{\mathcal{P}})$ (see Problem 24 on page 77). If $(S, \mathcal{A}, \mu)$ is a probability space, a measurable function $\mathbf{P} : S \to \mathcal{P}$ is called a *random probability measure*. In this way, $\mathbf{P}_n$ is a random probability measure for every n.

1.4.2.1 The Finite Version

In order to state the finite version of DeFinetti's theorem, we will find it convenient to refer to random samples from the empirical distribution of a collection of random variables $X_1, \ldots, X_N$. What we mean by this is the following. Suppose that $X_i = x_i$ for $i = 1, \ldots, N$. Create an urn with N balls labeled $x_1, \ldots, x_N$. A simple random sample of size n with/without replacement from the empirical distribution $\mathbf{P}_N$ of $X_1, \ldots, X_N$ is n draws with/without replacement from this urn such that, on each draw, every ball in the urn has equal probability of being drawn.

Theorem 1.48. *Suppose that $X_1, \ldots, X_N$ are random quantities taking values in a Borel space $(\mathcal{X}, \mathcal{B})$. Let $X = (X_1, \ldots, X_N)$, and for each $B \in \mathcal{B}$, let $\mathbf{P}_N(B) = \sum_{i=1}^{N} I_B(X_i)/N$ be the empirical distribution of X. The random quantities are exchangeable if and only if, for every ordered n-tuple*

---

[18]Those familiar with topological concepts will recognize the sets $A_{B,t}$ as a subbase for the topology of pointwise convergence of functions from $\mathcal{B}$ to $\mathbb{R}$, which is also the product topology when that set of functions is considered as the product space $\mathbb{R}^{\mathcal{B}}$. As such, $(\mathcal{P}, \mathcal{C}_{\mathcal{P}})$ is not a Borel space. This inconvenient circumstance will not cause problems for us, however. One of the steps in the proof of Theorem 1.49 is to show that the subset of $\mathcal{P}$ in which $\mathbf{P}$ lies is the image of a Borel space $(\mathcal{X}^{\infty}, \mathcal{B}^{\infty})$ under a measurable function. Hence regular conditional distributions are induced on $(\mathcal{P}, \mathcal{C}_{\mathcal{P}})$ by the corresponding distributions in $(\mathcal{X}^{\infty}, \mathcal{B}^{\infty})$.

$(i_1, \ldots, i_n)$ of distinct elements of $\{1, \ldots, N\}$, the joint distribution of $(X_{i_1}, \ldots, X_{i_n})$, conditional on $\mathbf{P}_N = P$ is that of a simple random sample without replacement from the distribution P.

1.4.2.2 The Infinite Version

The infinite version of DeFinetti's theorem is the following.

Theorem 1.49 (DeFinetti's representation theorem). *Let $(S, \mathcal{A}, \mu)$ be a probability space, and let $(\mathcal{X}, \mathcal{B})$ be a Borel space. For each n, let $X_n : S \to \mathcal{X}$ be measurable. The sequence $\{X_n\}_{n=1}^\infty$ is exchangeable if and only if there is a random probability measure $\mathbf{P}$ on $(\mathcal{X}, \mathcal{B})$ such that, conditional on $\mathbf{P} = P$, $\{X_n\}_{n=1}^\infty$ are IID with distribution P. Furthermore, if the sequence is exchangeable, then the distribution of $\mathbf{P}$ is unique, and $\mathbf{P}_n(B)$ converges to $\mathbf{P}(B)$ almost surely for each $B \in \mathcal{B}$.*

In Section 2.4, we present a more general theorem of Diaconis and Freedman (1984) and Lauritzen (1984, 1988), which applies to sequences of random quantities that are not exchangeable. This Theorem 2.111 will actually imply Theorem 1.49 as a special case, but its proof is far more complicated than the proof of Theorem 1.49, which we give in Section 1.5.

1.4.3 Some Examples

Here, we present more examples of exchangeable sequences and the implications of DeFinetti's theorem.

Example 1.50. Suppose that $X_1, \ldots, X_N$ are IID $Ber(r)$ random variables. If $M = \sum_{i=1}^N X_i$, then $\Pr(M = m) = \binom{N}{m} r^m (1-r)^{N-m}$ for $m = 0, \ldots, N$. That is, M has $Bin(N, r)$ distribution. Theorem 1.48 says that

$$\begin{aligned}
\Pr(X_{i_1} = j_1, \ldots, X_{i_n} = j_n) &= \sum_{m=k}^{N-n+k} \binom{N-n}{m-k} r^m (1-r)^{N-m} \\
&= \sum_{\ell=0}^{N-n} \binom{N-n}{\ell} r^{\ell+k} (1-r)^{N-\ell-k} \\
&= r^k (1-r)^{n-k} \sum_{\ell=0}^{N-n} \binom{N-n}{\ell} r^\ell (1-r)^{N-n-\ell} \\
&= r^k (1-r)^{n-k},
\end{aligned}$$

which corresponds to the X_i being IID $Ber(r)$. Hence, the probability of observing k ones in n trials is $\binom{n}{k} r^k (1-r)^{n-k}$, the binomial probability.

The following example helps to explain why Theorem 1.48 is not used very often with random variables having continuous distribution.

Example 1.51. Suppose that $X_1, \ldots, X_N$ are exchangeable with continuous joint CDF $F_{X_1,\ldots,X_N}$. The conditional distribution of $X_1, \ldots, X_n$ given $\mathbf{P}_N$ is that of a simple random sample without replacement from $\mathbf{P}_N$, but the distribution of $\mathbf{P}_N$ is not simple. If we let $B_1, \ldots, B_k$ be a partition of $\mathcal{X}$, the joint distribution of $V = (\mathbf{P}_N(B_1), \ldots, \mathbf{P}_N(B_k))$ can be expressed formally as follows. For each vector $(i_1, \ldots, i_k)$ such that $\sum_{j=1}^{k} i_j = N$,

$$\Pr\left(V = \left(\frac{i_1}{N}, \ldots, \frac{i_k}{N}\right)\right) = \int_A dF_{X_1,\ldots,X_N}(x_1, \ldots, x_N),$$

where A is the union of the $\binom{N}{i_1,\ldots,i_k}$ product sets of the form $B_{s_1} \times \cdots \times B_{s_N}$, where the subscripts $s_1, \ldots, s_N$ are integers from 1 to k with j appearing i_j times for each j. Needless to say, this formulation will not get us very far in general.

Example 1.52. This example is due to Bayes (1764). Suppose that $\{X_n\}_{n=1}^{\infty}$ are exchangeable Bernoulli random variables, and we set

$$\Pr(k \text{ successes in } n \text{ trials}) = \frac{1}{n+1}, \text{ for } k = 0, \ldots, n \text{ and } n = 1, 2, \ldots.$$

To check that this gives a consistent set of probabilities, we must show that, for every n and every n-tuple $(x_1, \ldots, x_n)$ of elements of $\{0, 1\}$,

$$\begin{aligned}\Pr(X_1 = x_1, \ldots, X_n = x_n) &= \Pr(X_1 = x_1, \ldots, X_n = x_n, X_{n+1} = 0) \\ &+ \Pr(X_1 = x_1, \ldots, X_n = x_n, X_{n+1} = 1).\end{aligned}$$

To show this, let $k = \sum_{i=1}^{n} x_i$. Then, the left-hand side equals $1/\left[(n+1)\binom{n}{k}\right]$. The right-hand side equals

$$\frac{1}{(n+2)\binom{n+1}{k}} + \frac{1}{(n+2)\binom{n+1}{k+1}} = \frac{1}{(n+1)\binom{n}{k}}.$$

To figure out what $\mathbf{P}$ is, recall that $\mathbf{P}_n(\{1\}) = \overline{X}_n$, the proportion of successes in the first n trials, and $\lim_{n \to \infty} \mathbf{P}_n(\{1\}) = \mathbf{P}(\{1\})$. Since $\overline{X}_n$ converges a.s. to Θ, it converges in distribution to Θ by Theorem B.90. Let $F_n(t) = \Pr(\overline{X}_n \leq t)$ be the CDF of $\overline{X}_n$. Write

$$F_n(t) = \Pr(\text{at most } nt \text{ successes in } n \text{ trials}) = \frac{\lfloor nt \rfloor + 1}{n+1},$$

where $\lfloor x \rfloor$ denotes the greatest integer less than or equal to x. It is trivial to see that $\lim_{n \to \infty} (\lfloor nt \rfloor + 1)/(n+1) = t$, for all $0 \leq t \leq 1$. Hence, $F(t) = t$ is the CDF of $\Theta = \lim_{n \to \infty} \overline{X}_n$. That is, the X_i are conditionally IID $Ber(\theta)$ given $\Theta = \theta$, and Θ has $U(0,1)$ distribution.

Example 1.53. Suppose that, for each n, the joint density of $X_1, \ldots, X_n$ is

$$f_{X_1,\ldots,X_n}(x_1, \ldots, x_n) = \frac{\Gamma(a+n)b^a}{\Gamma(a)(b + \sum_{i=1}^{n} x_i)^{a+n}}, \text{ for all } x_i > 0.$$

Clearly such random variables are exchangeable for each n. It can be seen that these densities are consistent also. (See Problem 20 on page 76.) Let us try to find the distribution of the limit of the empirical probability measures, $\mathbf{P}_n$.

$$\begin{aligned}
\Pr(\mathbf{P}_n((-\infty, c]) \leq t) &= \Pr(\text{at most } \lfloor tn \rfloor \ X_i \text{ are } \leq c) \\
&= \sum_{k=0}^{\ell} \Pr(\text{exactly } k \ X_i \text{ are } \leq c) \\
&= \sum_{k=0}^{\ell} \binom{n}{k} \Pr(X_1, \ldots, X_k \leq c, X_{k+1}, \ldots, X_n > c),
\end{aligned}$$

where $\ell = \lfloor tn \rfloor$. The probability that the first k X_i are at most c while the rest are greater is

$$\begin{aligned}
\int_0^c \cdots \int_0^c \int_c^\infty \cdots \int_c^\infty &\frac{\Gamma(a+n) b^a}{\Gamma(a)(b + \sum_{i=1}^n x_i)^{a+n}} dx_n \cdots dx_1 \\
&= \sum_{j=0}^k (-1)^j \binom{k}{j} \left(\frac{b}{b + (n-k+j)c} \right)^a \\
&= \sum_{j=0}^k (-1)^j \binom{k}{j} \int_0^\infty \frac{b^a}{\Gamma(a)} z^{a-1} \exp(-z[b + (n-k+j)c]) dz \\
&= \int_0^\infty \frac{b^a}{\Gamma(a)} z^{a-1} \exp(-z[b + cn])[\exp(cz) - 1]^k dz.
\end{aligned}$$

Multiplying this last expression by $\binom{n}{k}$ and summing over $k = 0, \ldots \ell$ give

$$\int_0^\infty \frac{b^a}{\Gamma(a)} z^{a-1} \exp(-bz) \Pr(Y_{n,z} \leq \ell) dz,$$

where $Y_{n,z}$ is a random variable with $Bin(n, 1 - \exp[-cz])$ distribution. For each z,

$$\lim_{n \to \infty} \Pr(Y_{n,z} \leq \ell) = \begin{cases} 1 & \text{if } 1 - \exp(-cz) < t \\ 0 & \text{if } 1 - \exp(-cz) > t \end{cases} = I_{(0, -\log(1-t)/c]}(z).$$

So, the CDF of the limit $\mathbf{P}((-\infty, c])$ of $\mathbf{P}_n((-\infty, c])$ is

$$\int_0^{-\frac{\log(1-t)}{c}} \frac{b^a}{\Gamma(a)} z^{a-1} \exp(-bz) dz,$$

which is the CDF of $1 - \exp(-c\Theta)$, where $\Theta \sim \Gamma(a, b)$. That is, it is as if there were a random variable $\Theta \sim \Gamma(a, b)$ and $\mathbf{P}((-\infty, c]) = 1 - \exp(-c\Theta)$. Put another way, it is as if the X_i were conditionally IID with $Exp(\theta)$ distribution given $\Theta = \theta$ and $\Theta \sim \Gamma(a, b)$. That this is indeed the case can be proven. (See Problem 22 on page 77.)

On page 18, we showed how to calculate conditional distributions for future observations given the ones that have already been observed using parametric families. For exchangeable random variables, we can often find these posterior predictive distributions without even introducing the parameter.

1.4. DeFinetti's Representation Theorem

Example 1.54 (Continuation of Example 1.52; see page 29). In the case of Bernoulli random variables, it is not difficult to calculate conditional probabilities. Suppose that we observe k^* successes in the first n^* trials, and we are interested in the probability of k successes in the next n trials. It is straightforward to calculate the probability of k successes in next n trials given k^* successes in the first n^* trials as

$$\frac{\binom{n^*}{k^*}\binom{n}{k}}{\binom{n^*+n}{k^*+k}} \frac{n^*+1}{n^*+n+1}.$$

For example, we get that the probability of k successes in the next n trials given two successes in the first five trials is

$$\frac{60\binom{n}{k}}{\binom{n+5}{k+2}} \frac{1}{n+6}. \tag{1.55}$$

It is easy to see that the future trials are still exchangeable given the past, and one could use the distribution in (1.55) to find the distribution of Θ given the observed trials, just as we found the original distribution of Θ to be $U(0,1)$. Alternatively, we have a theorem that applies to all exchangeable Bernoulli sequences.

Theorem 1.56. *Suppose that $\{X_i\}_{i=1}^{\infty}$ is an infinite sequence of exchangeable Bernoulli random variables. Let $\Theta = \lim_{n\to\infty} \sum_{i=1}^{n} X_i/n$, and let μ_Θ be the distribution of Θ. Conditional on seeing k^* successes in n^* trials, the distribution of Θ has CDF*

$$F^*(t) = \frac{\int_{[0,t]} \theta^{k^*}(1-\theta)^{n^*-k^*} d\mu_\Theta(\theta)}{\int \psi^{k^*}(1-\psi)^{n^*-k^*} d\mu_\Theta(\psi)}.$$

PROOF. We already know that the X_i are conditionally IID with $Ber(\theta)$ distribution given $\Theta = \theta$. It follows from the definition of conditional probability that, for every Borel subset B of $[0,1]$ and every n^* and $0 \leq k^* \leq n^*$,

$$\Pr(k^* \text{ successes in } n^* \text{ trials and } \Theta \in B) = \int_B \binom{n^*}{k^*} \theta^{k^*}(1-\theta)^{n^*-k^*} d\mu_\Theta(\theta).$$

Dividing this by

$$\Pr(k^* \text{ successes in } n^* \text{ trials}) = \int \psi^{k^*}(1-\psi)^{n^*-k^*} d\mu_\Theta(\psi)$$

completes the proof. □

Example 1.57 (Continuation of Example 1.54). After observing k^* successes in n^* trials, we can find the conditional distribution of Θ from Theorem 1.56. For example, if $n^* = 5$ and $k^* = 2$, Θ has conditional density

$$f^*(\theta) = 60\theta^2(1-\theta)^3$$

with respect to Lebesgue measure. The probability of a success on the sixth trial given that we observed two successes in the first five trials is then

$$\int_0^1 \theta 60\theta^2(1-\theta)^3 d\theta = \frac{3}{7},$$

which agrees with (1.55) with $n = 1$ and $k = 1$. After observing k^* successes in n^* trials, the distribution of Θ has density

$$\frac{(n^* + 1)!}{k^*!(n^* - k^*)!} \theta^{k^*} (1 - \theta)^{n^* - k^*},$$

which is the density of a $Beta(k^* + 1, n^* - k^* + 1)$ random variable. Given the data, the probability of success on the next trial is the conditional mean of Θ, namely $(k^* + 1)/(n^* + 2)$, which is approximately k^*/n^* if n^* is large.

This example helps to illustrate why observed frequencies are relevant for calculating probabilities, if the observations are thought to be exchangeable. Suppose that the distribution μ_Θ has a density f with respect to some measure on $[0, 1]$. Then the conditional density of Θ given k^* successes observed in n^* trials will be a constant times $\theta^{k^*}(1 - \theta)^{n^* - k^*} f(\theta)$. This density is higher for θ values near the observed k^*/n^* than is f. As n^* gets larger, this becomes more pronounced to the point of the density resembling a huge spike near k^*/n^*, if f is strictly positive in the vicinity of this value. This argument is the heuristic justification for the common practice of estimating Θ by k^*/n^*. The justification only applies when we believe the trials to be exchangeable. We do not have to believe that there exists a "fixed value" θ such that the trials are IID $Ber(\theta)$. We are just trying to estimate (or predict) what the limit of the relative frequencies will be.

Example 1.58 (Continuation of Example 1.53; see page 29). In the case of Bernoulli random variables, we saw that learning more data helped us to learn the value of Θ more precisely. If learning more data is supposed to help us to learn **P** in the present example, then the conditional distribution of X_{n+1} given the first n observations should approach **P**. The conditional density of X_{n+1} given the first n observations is

$$f_{X_{n+1}|X_1,\ldots,X_n}(x|x_1,\ldots,x_n) = \frac{\Gamma(a + n + 1)(b + \sum_{i=1}^n x_i)^{a+n}}{\Gamma(a + n)(b + x + \sum_{i=1}^n x_i)^{a+n+1}}$$

$$= \frac{a^*}{b^*}\left(1 + \frac{x}{b^*}\right)^{-(a^*+1)},$$

where $a^* = a + n$ and $b^* = b + \sum_{i=1}^n x_i$. Suppose that a^*/b^* converges to θ as $n \to \infty$. Then, the conditional density above converges to $\theta \exp(-x\theta)$, an $Exp(\theta)$ density. Once again, it is as if **P** is the CDF of the $Exp(\Theta)$ distribution for some random variable Θ. This is like saying that the distribution of **P** is concentrated on the set

$$E = \{P : P((-\infty, x]) = 1 - \exp(-x\theta), \text{ for some } \theta > 0\}.$$

In Problem 22 on page 77, you can prove that the probability is distributed over E as follows. Consider the mapping $V : \mathbb{R}^+ \to E$ defined by $V(\theta) = F_\theta$, where $F_\theta(x) = 1 - \exp(-x\theta)$ for all $x > 0$. To determine the appropriate probability measure on $\mathbb{R}^+$, we first note that the conditional distribution of the future given the past depends on the past only through $\sum_{i=1}^n X_i$. We suspect that some function of this converges in distribution to Θ. Solve Problem 21 on page 76 to determine the appropriate function and the limiting distribution. The measure induced on E by V is the distribution of **P**, $\mu_\mathbf{P}$. Integration in the space $\mathcal{P}$ is performed by integrating over $\mathbb{R}$:

$$\int_B g(P) d\mu_\mathbf{P}(P) = \int_{V^{-1}(B)} g(V(\theta)) d\theta.$$

The function V^{-1} gives us a way of dealing with P as if it were the real number $\theta = V^{-1}(P)$. This is a special case of a *parametric index*, to be defined in Definition 1.85.

1.5 Proofs of DeFinetti's Theorem and Related Results*

1.5.1 Strong Law of Large Numbers

Because the infinite forms of DeFinetti's theorem state that certain proportions converge almost surely, we will need to prove a strong law of large numbers for exchangeable random variables.[19] The strong law of large numbers for IID random variables 1.63 says that, for a sequence of IID random variables with finite mean, the sequence of averages of the first n of them converges almost surely to the mean. As stated, this result is clearly false for exchangeable random variables. For example, let X have finite mean but nondegenerate distribution, and suppose that $X_i = X$ for all i. Then $\{X_n\}_{n=1}^{\infty}$ is clearly an exchangeable sequence, and the average of the first n is equal to X for every n. Hence, the averages converge almost surely to X, not the mean. For exchangeable random variables, we can only prove that the averages converge to some random variable which might not be constant.

We will prove two versions of the strong law for two different sets of readers. The first version, Theorem 1.59, is based solely on elementary probability theory, but it concludes only that a subsequence of the sequence of sample averages converges almost surely,[20] and then only under the assumption of finite variance. But the restricted result of Theorem 1.59 is all that is needed in the proof of Theorem 1.49. The second version, Theorem 1.62, is based on the theory of reversed martingales and is a more complete statement of the strong law of large numbers than Theorem 1.59.

---

*This section may be skipped without interrupting the flow of ideas.

[19]One consequence of DeFinetti's representation theorem 1.49 will be that many theorems that apply to IID random quantities can be adapted to apply to exchangeable random quantities. Taylor, Daffer, and Patterson (1985) prove some examples of such theorems. See also Problem 31 on page 78 and Problem 39 on page 79. One such result would be the strong law of large numbers 1.62. Unfortunately, one of the steps in the proofs of Theorems 1.47 and 1.49 makes use of a strong law of large numbers for exchangeable random variables.

[20]Hartigan (1983, Theorem 4.6) simplifies his proof of DeFinetti's theorem using this fact. The statement of Theorem 1.59 can be extended to show that the entire sequence of sample averages converges almost surely as in Problem 38 on page 78.

1.5.1.1 An Elementary Version

We state and prove here an elementary version of the strong law of large numbers for exchangeable random variables, which is only general enough to allow us to prove Theorem 1.49. Those who desire a more complete statement of the strong law of large numbers and who have some familiarity with martingales may safely skip this theorem and proceed to the martingale version in Section 1.5.1.2.[21]

Theorem 1.59.[22] *Let $(S, \mathcal{A}, \mu)$ be a probability space and, for each n, let $X_n : S \to \mathbb{R}$ be exchangeable random variables. Assume that $-\infty < E(X_i X_j) = m_2 < \infty$ for all $i \neq j$ and $E(X_i^2) = \mu_2 < \infty$ for all i. Let $Y_n = \sum_{i=1}^n X_i/n$. Then the subsequence $\{Y_{8^k}\}_{k=1}^\infty$ converges almost surely.*

PROOF. We will prove that the subsequence converges almost surely by proving that it is a Cauchy sequence[23] almost surely. Use Tchebychev's inequality B.16 to write, for $m > n$,

$$\begin{aligned}
\Pr(|Y_m - Y_n| \geq c) &\leq \frac{E(Y_m - Y_n)^2}{c^2} \\
&= (E(Y_n^2) + E(Y_m^2) - 2E(Y_n Y_m)) \frac{1}{c^2} \\
&= \left(\frac{1}{n^2}[n\mu_2 + n^2(n^2 - 1)m_2] \right. \\
&\quad + \frac{1}{m^2}[m\mu_2 + m(m-1)m_2] \\
&\quad \left. - \frac{2}{mn}[n\mu_2 + n(n-1)m_2 + n(m-n)m_2] \right) \frac{1}{c^2} \\
&= \left(\frac{1}{n} - \frac{1}{m} \right) \frac{\mu_2 - m_2}{c^2} < \frac{\mu_2 - m_2}{nc^2}. \qquad (1.60)
\end{aligned}$$

Now, let $Z_k = Y_{8^k}$ and $A_k = \{s : |Z_{k+1} - Z_k| \geq 2^{-k}\}$. It follows easily from (1.60) (with $c = 2^{-k}$) that, for every k, $\Pr(A_k) < (\mu_2 - m_2)2^{-k}$. Now, let $A = \bigcap_{n=1}^\infty \bigcup_{k=n}^\infty A_k$. It follows from the first Borel–Cantelli lemma A.20 that $\Pr(A) = 0$. We finish the proof by showing that, for every $s \in A^C$, and every $\epsilon > 0$, there exists $N_{s,\epsilon}$ such that $n, m \geq N_{s,\epsilon}$ implies $|Z_n(s) - Z_m(s)| < \epsilon$.

---

[21] Two theorems (7.49 and 7.80) make use of the strong law of large numbers for IID random variables 1.63. This result is available as a consequence of Theorem 1.62, but not from Theorem 1.59.

[22] This theorem is used in the proof of Theorem 1.49. Its proof resembles the proof of a similar claim in Loève (1977, Section 6.3).

[23] A sequence of real numbers $\{x_n\}_{n=1}^\infty$ is *Cauchy* if, for every $\epsilon > 0$, there exists N such that $n, m \geq N$ implies $|x_n - x_m| < \epsilon$. Since $\mathbb{R}$ is a complete metric space, every Cauchy sequence converges.

Write $A^C = \cup_{n=1}^{\infty} \cap_{k=n}^{\infty} A_k^C$. For every $s \in A^C$, there exists c_s such that $s \in \cap_{k=c_s}^{\infty} A_k^C$. If $m > n \geq c_s$, it follows that

$$|Z_m(s) - Z_n(s)| \leq \sum_{i=n}^{m-1} |Z_{i+1}(s) - Z_i(s)| < 2^{-n+1} \leq 2^{-c_s+1}.$$

So, let $N_{s,\epsilon} > 1 + \max\{c_s, -\log_2 \epsilon\}$ to finish the proof. □

There are strong laws of large numbers for IID random variables also. Here is one which will help in the proof of Theorem 1.49.

Lemma 1.61 (Strong law of large numbers: bounded conditionally IID case). *Let $\{X_n\}_{n=1}^{\infty}$ be a sequence of bounded random variables, and let Θ be an arbitrary random quantity such that, conditional on Θ, the X_n are IID with mean $c(\Theta)$. Then $\sum_{i=1}^{n} X_i/n = Y_n$ converges almost surely to $c(\Theta)$.*

PROOF. Since Y_n converges almost surely to $c(\Theta)$ if and only if $Y_n - c(\Theta)$ converges almost surely to 0, and since $Y_n - c(\Theta)$ is the sample average of the first n of $X_i - c(\Theta)$, assume that $c(\Theta) = 0$ without loss of generality. Now, write

$$Y_n^4 = \frac{1}{n^4} \left(\sum_{i_1=1}^{n} \sum_{i_2=1}^{n} \sum_{i_3=1}^{n} \sum_{i=4=1}^{n} X_{i_1} X_{i_2} X_{i_3} X_{i_4} \right).$$

Each of the terms above, for which at least one of i_1, i_2, i_3, i_4 is not repeated, has mean 0 because the random variables are conditionally independent with mean 0. Let M be a bound for $|X_n|$. It follows that

$$E(Y_n^4 | \Theta) = \frac{1}{n^3} E(X_1^4 | \Theta) + 3 \frac{n-1}{n^3} (E(X_1^2 | \Theta))^2 \leq \frac{4M^4}{n^2}.$$

So, $E(Y_n^4) \leq 4M^4/n^2$ by the law of total probability B.70. It follows from the Markov inequality B.15 that, for each $\epsilon > 0$,

$$\Pr(|Y_n| > \epsilon) = \Pr(Y_n^4 > \epsilon^4) \leq \frac{4M^4}{\epsilon^4 n^2}.$$

So, $\sum_{n=1}^{\infty} \Pr(|Y_n| > \epsilon) < \infty$. The first Borel–Cantelli lemma A.20 implies that $\Pr(|Y_n| > \epsilon$ infinitely often$) = 0$. Since the event that Y_n converges to 0 is $\cap_{k=1}^{\infty} \{|Y_n| > 1/k$ infinitely often$\}^C$, it follows that Y_n converges to 0 almost surely. □

1.5.1.2 A Martingale Version[+]

A more complete proof of the strong law of large numbers for exchangeable random variables is borrowed from Kingman (1978).[24]

Theorem 1.62 (Strong law of large numbers).[25] *Let $(S, \mathcal{A}, \mu)$ be a probability space, and let $X_i : S \to \mathbb{R}$ be measurable for all i such that the X_i are exchangeable with $\mathrm{E}(|X_i|) < \infty$ for all i. Then there exists a σ-field $\mathcal{A}_\infty$ such that $Y_n = \sum_{i=1}^n X_i/n$ converges almost surely to $\mathrm{E}(X_1|\mathcal{A}_\infty)$.*

PROOF. Define $X = (X_1, X_2, \ldots)$. For $n > 0$, let $\mathcal{C}_n$ be the collection of all Borel subsets A of $\mathbb{R}^\infty$ which satisfy $x \in A$ if and only if $y \in A$ for all y that agree with x after coordinate n and such that the first n coordinates of y are a permutation of the first n coordinates of x. It is not difficult to show that $\mathcal{C}_n$ is a σ-field, and it is trivial to see that $f(x) = \sum_{i=1}^n x_i$ is measurable with respect to $\mathcal{C}_n$ (Problem 36 on page 78). Let $\mathcal{A}_n = X^{-1}(\mathcal{C}_n)$ and $Z_n = \mathrm{E}(X_1|\mathcal{A}_n)$. Since $\mathrm{E}(|X_1|) < \infty$ and $\{\mathcal{A}_n\}_{n=1}^\infty$ is a decreasing sequence of σ-fields, it follows from Part II of Lévy's theorem B.124 that $\lim_{n \to \infty} Z_n = \mathrm{E}(X_1|\mathcal{F}_\infty)$ and is finite a.s. We now show that $Y_n = Z_n$. Since $f(x) = \sum_{i=1}^n x_i$ is measurable with respect to $\mathcal{C}_n$, we need only prove that, for $A \in \mathcal{A}_n$, $\mathrm{E}(I_A Y_n) = \mathrm{E}(I_A X_1)$. But, $I_A X_i$ has the same distribution as $I_A X_j$ for all $i, j = 1, \ldots, n$ by the assumption of exchangeability and the permutation symmetry of the set A. Hence $\mathrm{E}(I_A X_1) = \sum_{i=1}^n \mathrm{E}(I_A X_i)/n = \mathrm{E}(I_A Y_n)$. □

As a corollary, we also mention the usual strong law of large numbers for IID random variables.

Corollary 1.63 (Strong law of large numbers: IID case).[26] *Suppose that $\{X_n\}_{n=1}^\infty$ is a sequence of IID random variables with $\mathrm{E}(X_i) = \mu$. Then $\sum_{i=1}^n X_i/n = Y_n$ converges almost surely to μ.*

1.5.2 The Bernoulli Case

The proof of DeFinetti's representation theorem for finitely many Bernoulli random variables $X_1, \ldots, X_N$ was given in Example 1.46. There, we saw that the conditional distribution of the number K of successes in n trials given $M = m$ was hypergeometric $Hyp(N, n, m)$, where $M = \sum_{i=1}^N X_i$. So,

---

[+]This section contains results that rely on the theory of martingales. It may be skipped without interrupting the flow of ideas.

[24]This proof is also similar to one given for the case of IID random variables by Doob (1953, Section VII, 6). Those who are unfamiliar with martingale theory may safely skip this section and study the elementary version given earlier. But these readers should be aware that two theorems (7.49 and 7.80) do make use of Corollary 1.63.

[25]This theorem can be used in the proof of Theorem 1.49.

[26]This theorem is used in the proofs of Theorems 7.49 and 7.80.

for example,
$$\Pr(K = k | M = m) = \frac{\binom{n}{k}\binom{N-n}{m-k}}{\binom{N}{m}}. \qquad (1.64)$$

Suppose that $N \to \infty$ in such a way that $M/N \to \Theta$. For fixed n and k, we can take limits in (1.64) as $N \to \infty$ and $m/n \to \theta$. Formally, we would get
$$\Pr(K = k | \Theta = \theta) = \binom{n}{k} \theta^k (1-\theta)^{n-k},$$
which is the model for $K \sim \text{Bin}(n, \theta)$. In fact, this is what Theorem 1.47 says is the case. The precise proof is a bit more complicated than the heuristic argument above, but the idea is the same.

PROOF OF THEOREM 1.47. The "if" direction is simple and is left to the reader. For the "only if" direction, assume that $\{X_n\}_{n=1}^{\infty}$ is an exchangeable Bernoulli sequence. Let $Y_\ell = \sum_{i=1}^{\ell} X_i / \ell$ for $\ell = 1, 2, \ldots$. By the strong law of large numbers 1.62 or 1.59, we know that Y_{8^n} converges almost surely. Let Θ denote the limit when the limit exists, and let $\Theta = 1/2$ when the limit does not exist. Let μ_Θ denote the distribution of Θ (a probability measure on $[0, 1]$.)

The main step in the proof is to show that for every integer k and every $j_1, \ldots, j_k \in \{0, 1\}$ and every Borel subset C of $[0, 1]$,
$$\Pr(X_1 = j_1, \ldots, X_k = j_k, \Theta \in C) = \int_C \theta^y (1-\theta)^{k-y} d\mu_\Theta(\theta), \qquad (1.65)$$
where $y = j_1 + \cdots + j_k$. To show this, let $Z_n = I_C(\Theta) Y_{8^n}^y (1 - Y_{8^n})^{k-y}$ and let $Z = I_C(\Theta) \Theta^y (1-\Theta)^{k-y}$. It is easy to see that $Z_n \to Z$ a.s., hence $Z_n \xrightarrow{D} Z$ by Theorem B.90. Since Z_n is uniformly bounded, $E(Z_n) \to E(Z)$. The right-hand side of (1.65) is just $E(Z)$. So, we need only show that $E(Z_n)$ converges to the left-hand side of (1.65). Let $m = 8^n$, and define $W_{\ell,t} = I_{\{j_t\}}(X_\ell)$, for each integer $t = 1, \ldots, k$. Then
$$\frac{1}{m} \sum_{\ell=1}^{m} W_{\ell,t} = \begin{cases} Y_m & \text{if } j_t = 1, \\ 1 - Y_m & \text{if } j_t = 0. \end{cases}$$

With this notation, we can write
$$Z_n = \frac{1}{m^k} I_C(\Theta) \sum_{i_1=1}^{m} \cdots \sum_{i_k=1}^{m} \prod_{t=1}^{k} W_{i_t,t}$$
$$= \frac{I_C(\Theta)}{m^k} \sum_{\text{all } i_t \text{ distinct}} \prod_{t=1}^{k} W_{i_t,t} + \frac{I_C(\Theta)}{m^k} \sum_{\text{at least two } i_t \text{ equal}} \prod_{t=1}^{k} W_{i_t,t}. \qquad (1.66)$$

The first sum on the right-hand side of (1.66) has $m!/(m-k)!$ terms. Since $E[I_C(\Theta) \prod_{t=1}^{k} W_{i_t,t}]$ equals the left-hand side of (1.65) when all i_t

are distinct, and since $m!/[(m-k)!m^k]$ converges to 1, the mean of the first term on the right of (1.66) converges to the left-hand side of (1.65). The second sum has $m^k - m!/(m-k)!$ terms, each of which is bounded between 0 and 1. Since $1 - m!/[(m-k)!m^k]$ converges to 0, so does the mean of the last expression in (1.66). This completes the proof of (1.65).

Equation (1.65) is exactly what it means to say that $X_1, \ldots, X_k$ are conditionally IID $Ber(\theta)$ given $\Theta = \theta$. To see that the distribution μ_Θ is unique, let $C = [0, 1]$ in (1.65) and note that this equation determines the means of all polynomial functions of Θ. Since polynomials are dense in the set of all bounded continuous functions on $[0, 1]$ by the Stone–Weierstrass theorem C.3, it follows that (1.65) determines the means of all bounded continuous functions of Θ, and Corollary B.107 says that the means of all bounded continuous functions determine the distribution. To finish the proof, we note that since $\{X_n\}_{n=1}^\infty$ are bounded and conditionally IID, Lemma 1.61 says that $\{Y_\ell\}_{\ell=1}^\infty$ converges a.s. Obviously, the limit must be Θ, a.s. □

1.5.3 The General Finite Case*

1.5.3.1 Proof of Theorem 1.48

Define a function $h : \mathcal{X}^N \to \mathcal{P}$ by $h(x) = P_x$, where

$$P_x(B) = \frac{1}{N} \sum_{i=1}^N I_B(x_i)$$

if $x = (x_1, \ldots, x_N)$ and $B \in \mathcal{B}$. We refer to P_x as the *empirical distribution* of x. It is easy to check that the function h is measurable (see Problem 24 on page 77.) Simple random samples with/without replacement from P_x can be defined in exactly the same way as they were for samples from $\mathbf{P}_N$ in Section 1.4.2.1. In fact, if $X = (X_1, \ldots, X_N)$, then $P_X = \mathbf{P}_N$.

Even though the space $(\mathcal{P}, \mathcal{C}_\mathcal{P})$ is quite complicated, the subset of $\mathcal{P}$ in which $\mathbf{P}_N$ lies is relatively simple. $\mathbf{P}_N$ concentrates all of its mass on at most N different points in $\mathcal{X}$. Hence, it is not nearly as complicated an object as it may appear. In fact, it is really nothing more than two vectors of equal length (at most N), where the coordinates of one of the vectors are elements of $\mathcal{X}$ and the coordinates of the other are nonnegative multiples of $1/N$ adding to 1.

The following lemma will be used to help prove Theorem 1.48 and an approximation in Theorem 1.70.

Lemma 1.67. *Suppose that $X = (X_1, \ldots, X_N)$ are exchangeable random quantities. Let Q_X be the probability measure giving their joint distribution on $(\mathcal{X}^N, \mathcal{B}^N)$. For $n \leq N$, $x = (x_1, \ldots, x_N)$, and $B \in \mathcal{B}^n$, let $H_n(B|x)$*

---

*This section may be skipped without interrupting the flow of ideas.

stand for the probability that n draws without replacement from an urn containing balls labeled $x_1, \ldots, x_N$ form a point in B. Also, let $M_n(B|x)$ stand for the probability that n draws with replacement from an urn containing balls labeled $x_1, \ldots, x_N$ form a point in B. Let $Y_1, \ldots, Y_N$ be conditionally IID given $X = x$ with distribution $M_N(\cdot|x)$. Then

$$\Pr((X_{i_1}, \ldots, X_{i_n}) \in B, X \in C) = \int_C H_n(B|x) dQ_X(x),$$

$$\Pr((Y_{i_1}, \ldots, Y_{i_n}) \in B, X \in C) = \int_C M_n(B|x) dQ_X(x),$$

for each $B \in \mathcal{B}^n$ and each $C \in \mathcal{B}^N$.

PROOF. The second equation above is immediate from the definition of conditional distribution. Next, notice that $H_n(B|x)$ can be written

$$H_n(B|x) = \frac{1}{\binom{N}{n}} \sum_{\text{All distinct } (j_1, \ldots, j_n)} I_B(x_{j_1}, \ldots, x_{j_n}).$$

By the exchangeability of $X_1, \ldots, X_N$, we have

$$\Pr((X_{i_1}, \ldots, X_{i_n}) \in B, X \in C)$$

$$= \frac{1}{\binom{N}{n}} \sum_{\substack{\text{distinct} \\ (j_1, \ldots, j_n)}} \Pr((X_{j_1}, \ldots, X_{j_n}) \in B, X \in C)$$

$$= \frac{1}{\binom{N}{n}} \sum_{\substack{\text{distinct} \\ (j_1, \ldots, j_n)}} \int_C I_B(x_{j_1}, \ldots, x_{j_n}) dQ_X(x)$$

$$= \int_C H_n(B|x) dQ_X(x). \qquad \square$$

We are now in position to prove Theorem 1.48.

PROOF OF THEOREM 1.48. The "if" part is fairly straightforward and left to the reader. Only the $n = N$ case is needed, since the others follow from it by taking marginal distributions.

For the "only if" part, assume that $X_1, \ldots, X_N$ are exchangeable, and let $\mathbf{P}_N$ be as defined. For each probability P with support on at most N points and probabilities of the form k/N, let $H'_n(B|P)$ be the probability that n draws without replacement from P forms a point in $B \in \mathcal{B}^n$. What we need to prove is that for all n and all distinct $i_1, \ldots, i_n$ and all $B \in \mathcal{B}^n$ and $C \in \mathcal{C}_\mathcal{P}$,

$$\Pr((X_{i_1}, \ldots, X_{i_n}) \in B, \mathbf{P}_N \in C) = \int_C H'_n(B|P) d\mu_N(P), \qquad (1.68)$$

where μ_N is the probability measure giving the distribution of $\mathbf{P}_N$. Let $h : \mathcal{X}^N \to \mathcal{P}$ be the function that maps a point $x = (x_1, \ldots, x_N)$ to the empirical distribution of x. That is, $h(x) = P_x$ and $h(X) = \mathbf{P}_N$. Let Q_X be the distribution of X. It follows that μ_N is the measure induced on $(\mathcal{P}, \mathcal{B}_\mathcal{P})$ from Q_X by h. Also, it follows that $H'_n(B|P) = H_n(B|x)$ for all x such that $h(x) = P$. This means that $H_n(B|x)$ is a function of $h(x)$, namely $H'_n(B|P_x)$. Now, write the left-hand side of (1.68) as

$$\Pr((X_{i_1}, \ldots, X_{i_n}) \in B, X \in h^{-1}(C)) = \int_{h^{-1}(C)} H_n(B|x) dQ_X(x)$$
$$= \int_C H'_n(B|P) d\mu_N(P),$$

where the first equality follows from Lemma 1.67, and the second follows from Theorem A.81. This proves (1.68). □

1.5.3.2 An Approximation Theorem

Some questions naturally arise when comparing the cases of finitely many and infinitely many exchangeable random variables. First, what happens when N becomes infinite? Second, is there any sense in which the finite or infinite N cases approximate each other? The following lemma bounds the difference between probabilities calculated under sampling with and without replacement from a finite set and is useful in addressing these approximation questions.

Lemma 1.69. *Suppose that we have an urn with N balls labeled $y_1, \ldots, y_N$. Let $\mathcal{Y}$ be the set of distinct items in the set of labels. Let $X = (X_1, \ldots, X_n)$ be values of the labels for n draws without replacement from the urn. Let $Y = (Y_1, \ldots, Y_n)$ be values of the labels for n draws with replacement from the urn. Let P be the distribution of Y and let Q be the distribution of X. Then $\sup_{A \subseteq \mathcal{Y}^n} |P(A) - Q(A)| \leq n(n-1)/(2N)$.*

PROOF.[27] First, suppose that there are no duplicate labels. Then both P and Q are constant on the set $A = \{x : Q(\{x\}) > 0\}$ and on the complement of this set. It follows that this set and its complement are where the supremum of the difference occurs. The difference between $Q(A)$ and $P(A)$ is easily seen to be $1 - P(A) = 1 - n!\binom{N}{n}/N^n$. If $x_i \in (0, 1)$ for all $i = 1, \ldots, n$, then it is easy to show (say, by induction) that $1 - \sum_{i=1}^n x_i \leq \prod_{i=1}^n (1 - x_i)$. It is clear that $n!\binom{N}{n}/N^n = \prod_{i=1}^n (1 - i/N)$, hence

$$1 - \sum_{i=1}^n \frac{i}{N} = 1 - \frac{n(n-1)}{2N} \leq P(A).$$

The result now follows by subtracting both sides from 1.

---

[27] Freedman (1977) gives the first part of this proof.

For the general case, suppose that for each i, ball i has two labels (i, y_i), and let $\mathcal{Y}_0$ denote the set of these labels. Let X' and Y' record both labels, so that the first part of the proof applies to their distributions. Call these distributions Q' and P'. Assume that X and Y still only record the second parts of the labels. For each $A \subseteq \mathcal{Y}^n$ there exists a set $A' \subseteq \mathcal{Y}_0^n$ such that $X \in A$ if and only if $X' \in A'$, and $Y \in A$ if and only if $Y' \in A'$. In fact,

$$A' = \bigcup_{(x_1,\ldots,x_n) \in A} \prod_{j=1}^n \{(i, y_i) : y_i = x_j\}.$$

It now follows that

$$\sup_{A \subseteq \mathcal{Y}^n} |P(A) - Q(A)| \leq \sup_{A' \subseteq \mathcal{Y}_0^n} |P'(A') - Q'(A')| \leq \frac{n(n-1)}{2N}. \quad \square$$

Lemma 1.69 allows us to prove some approximation theorems for finite exchangeable sequences. The next theorem is borrowed from Diaconis and Freedman (1980a). It says that the joint distribution of finitely many exchangeable random quantities is uniformly approximated by the joint distribution of conditionally IID random quantities.

Theorem 1.70. *Suppose that $X_1, \ldots, X_N$ are exchangeable random quantities taking values in a Borel space $(\mathcal{X}, \mathcal{B})$. For each $P \in \mathcal{P}$ and each n and each $B \in \mathcal{B}^n$, let $P^n(B)$ stand for the probability that a vector of n IID random variables with distribution P lies in B. Let μ_N be the distribution of $\mathbf{P}_N$. Then, for all n and all distinct $i_1, \ldots, i_n$ in $\{1, \ldots, N\}$ and all $B \in \mathcal{B}^n$,*

$$\left| \Pr((X_{i_1}, \ldots, X_{i_n}) \in B) - \int_{\mathcal{P}} P^n(B) d\mu_N(P) \right| \leq \frac{n(n-1)}{2N}. \quad (1.71)$$

PROOF. Let Q_X stand for the joint distribution of $X_1, \ldots, X_N$. For $x = (x_1, \ldots, x_N)$ and $B \in \mathcal{B}^n$, let $H_n(B|x)$ and $M_n(B|x)$ be as in Lemma 1.67. (If P_x is the empirical distribution of x, then $M_n(B|x) = P_x^n(B)$.) By Lemma 1.69, we have $|H_n(B|x) - M_n(B|x)| \leq n(n-1)/2N$ for all x. From Lemma 1.67, we have

$$\left| \Pr((X_{i_1}, \ldots, X_{i_n}) \in B) - \int M_n(B|x) dQ_X(x) \right|$$

$$= \left| \int H_n(B|x) dQ_X(x) - \int M_n(B|x) dQ_X(x) \right| \leq \frac{n(n-1)}{2N}.$$

All that remains is to show that the distribution μ_N of $\mathbf{P}_N$ satisfies

$$\int P^n(B) d\mu_N(P) = \int M_n(B|x) dQ_X(x). \quad (1.72)$$

Consider, once again, the function $h : \mathcal{X}^N \to \mathcal{P}$ which maps a point $x \in \mathcal{X}^N$ to P_x. Note that $h(x) = M_1(\cdot|x)$ also. Since $h(X) = \mathbf{P}_N$, Theorem A.81 says that

$$\int P^n(B) d\mu_N(P) = \int P_x^n(B) dQ_X(x). \tag{1.73}$$

Since $P_x^n(B) = M_n(B|x)$, it follows that (1.73) is (1.72). □

Theorem 1.49 says that infinitely many exchangeable random quantities are conditionally IID. Theorem 1.70 says that there is continuity in passing from the finite to the infinite case.

1.5.3.3 Conditional Distributions*

There is a general form for the conditional distributions of finitely many exchangeable random quantities. If $X_1, \ldots, X_N$ are exchangeable, it is easy to prove that $X_{k+1}, \ldots, X_N$ are exchangeable conditional on $X_1, \ldots, X_k$.

Proposition 1.74. *If $X_1, \ldots, X_N$ are exchangeable, then $X_{k+1}, \ldots, X_N$ are exchangeable conditional on $X_1, \ldots, X_k$.*

If one uses the conditional distribution of $X_{k+1}, \ldots, X_N$ given $X_1, \ldots, X_k$ in place of the distribution of X in Theorem 1.48, one obtains the conditional distribution of each subset of $X_{k+1}, \ldots, X_N$ given $X_1, \ldots, X_k$.

Proposition 1.75. *Suppose that $X = (X_1, \ldots, X_N)$ are exchangeable, that $k+n \leq N$, and that $j_1, \ldots, j_n \in \{k+1, \ldots, N\}$ are distinct. The conditional joint distribution of $X_{k+j_1}, \ldots, X_{k+j_n}$ given $(X_1, \ldots, X_k) = (x_1, \ldots, x_k)$ and $\mathbf{P}_N = P$ is that of a simple random sample of size n without replacement from the distribution P with one ball for each of $x_1, \ldots, x_k$ removed first.*

Example 1.76. Suppose that $X_1, \ldots, X_N$ are exchangeable Bernoulli random variables. For simplicity, suppose that we are interested in the first $k+n$ X_is. If we will observe $X_1, \ldots, X_k$, then it suffices to be able to calculate $\Pr(Y = \ell | Y^* = \ell^*)$ for each ℓ, ℓ^* where $Y = \sum_{i=k+1}^{k+n} X_i$ and $Y^* = \sum_{i=1}^{k} X_i$. It follows from the exchangeability that

$$\frac{\Pr(Y^* = \ell^*, Y = \ell)}{\Pr(Y^* = \ell^*)}$$

$$= \binom{k}{\ell^*}\binom{n}{\ell} \frac{\Pr(X_1 = 1, \ldots, X_{\ell^*+\ell} = 1, X_{\ell^*+\ell+1} = 0, \ldots, X_{k+n} = 0)}{\Pr(Y^* = \ell^*)}$$

$$= \frac{1}{\Pr(Y^* = \ell^*)} \sum_{m=\ell^*+\ell}^{N-k-n+\ell^*+\ell} \frac{\binom{k}{\ell^*}\binom{N-k}{m-\ell^*}}{\binom{N}{m}} \frac{\binom{n}{\ell}\binom{N-k-n}{m-\ell^*-\ell}}{\binom{N-k}{m-\ell^*}} p_m$$

$$= \sum_{m^*=\ell}^{N^*-n+\ell} \frac{\binom{n}{\ell}\binom{N^*-n}{m^*-\ell}}{\binom{N^*}{m^*}} p^*_{m^*},$$

---

*This section may be skipped without interrupting the flow of ideas.

1.5. Proofs of DeFinetti's Theorem

where we have made the substitutions

$$N^* = N - k, \quad m^* = m - \ell^*, \quad p^*_{m^*} = \frac{\binom{k}{\ell^*}\binom{N-k}{m-\ell^*}}{\binom{N}{m}} \frac{p_m}{\Pr(Y^* = \ell^*)}.$$

The conditional probability that $Y = \ell$ given $Y^* = \ell^*$ is in precisely the same form as the marginal probability of $Y = \ell$, except that the distribution of M has been replaced by the conditional distribution of $M^* = M - \ell^*$ given $Y^* = \ell^*$.[28]

For example, suppose that $p_m = 1/(N+1)$ for $m = 0, \ldots, N$. Then, the probability of one success in one trial is

$$\sum_{m=1}^{N} \frac{\binom{1}{1}\binom{N-1}{m-1}}{\binom{N}{m}} \frac{1}{N+1} = \frac{1}{N+1} \sum_{m=1}^{N} \frac{m}{N} = \frac{N(N+1)}{2(N+1)N} = \frac{1}{2}.$$

After seeing one observation $X_1 = 1$, we calculate the conditional distribution of the number of remaining successes

$$p^*_{m^*} = \frac{\binom{N-1}{m^*}\binom{1}{1}\frac{1}{N+1}}{\binom{N}{m^*+1}\frac{1}{2}} = \frac{2(m^*+1)}{N(N+1)} \begin{cases} < \frac{1}{N} & \text{if } m^* < \frac{N-1}{2}, \\ > \frac{1}{N} & \text{if } m^* > \frac{N-1}{2}. \end{cases}$$

The probability has been shifted to higher values of m^* after seeing one success. Suppose now that we see two observations $X_1 = 1$ and $X_2 = 0$. The probability of this is

$$\sum_{m=1}^{N-1} \frac{\binom{2}{1}\binom{N-2}{m-1}}{\binom{N}{m}} \frac{1}{N+1} = \frac{2}{N+1} \sum_{m=1}^{N-1} \frac{m(N-m)}{N(N-1)}$$

$$= \frac{2}{(N-1)N(N+1)} \left[\frac{N^2(N-1)}{2} - \frac{N(N-1)(2N-1)}{6} \right] = \frac{1}{3},$$

and the conditional distribution of the future is

$$p^*_{m^*} = \frac{\binom{N-2}{m^*}\binom{2}{1}\frac{1}{N+1}}{\binom{N}{m^*+1}\frac{1}{3}} = \frac{6(m^*+1)(N-m^*-1)}{N(N-1)(N+1)}.$$

The maximum of this probability occurs at the value of m^* closest to $(N-2)/2$, and the probability drops off as m^* moves away from $(N-2)/2$.

We can also prove an approximation theorem that applies to the calculation of conditional probabilities. The theorem says that if we pretend that X_i are conditionally IID given $\mathbf{P}_N$, and the probability of repeats is 0, then the conditional probabilities we calculate for future observations are uniformly close to the correct conditional probabilities.

Theorem 1.77. *Suppose that* $X = (X_1, \ldots, X_N)$ *are exchangeable and that* $\Pr(X_i = X_j) = 0$ *for* $i \neq j$. *Let* $\mathbf{P}_N$ *be the empirical distribution of*

---

[28]The readers should convince themselves that $p^*_{m^*}$ is indeed equal to $\Pr(M^* = m^* | Y^* = \ell^*)$.

X. Let $Y_1, \ldots, Y_N$ be conditionally IID with distribution P given $\mathbf{P}_N = P$. Then, for $n < N - k$, and $B \in \mathcal{B}^n$,

$$|\Pr((X_{k+1}, \ldots, X_{k+n}) \in B | X_1 = x_1, \ldots, X_k = x_k)$$
$$- \Pr((Y_{k+1}, \ldots, Y_{k+n}) \in B | Y_1 = x_1, \ldots, Y_k = x_k)|$$
$$\leq \frac{n(n-1)}{2(N-k)} + 1 - \left(1 - \frac{k}{N}\right)^n, \quad \text{a.s.}$$

with respect to the distribution of X.

PROOF. First, we prove that the conditional distribution of X given $Y_1, \ldots, Y_k$ is the same as that of X given $X_1, \ldots, X_k$. Call the latter $Q_{X|X_1,\ldots,X_k}$. Let M_n and H_n be as in Lemma 1.67. Let $B \in \mathcal{B}^k$ be such that all points in B have k distinct coordinates. Then

$$M_k(B|x) = k! \binom{N}{k} H_k(B|x) / N^k$$

for all x with distinct coordinates. For each such B, Lemma 1.67 says that

$$\Pr((Y_1, \ldots, Y_k) \in B, X \in C) = \frac{k!\binom{N}{k}}{N^K} \Pr((X_1, \ldots, X_k) \in B, X \in C).$$

In particular, if C is the set of all x with distinct coordinates, then $\Pr(X \in C) = 1$ and

$$\Pr((Y_1, \ldots, Y_k) \in B) = \frac{k!\binom{N}{k}}{N^K} \Pr((X_1, \ldots, X_k) \in B).$$

Let $Q_{X_1,\ldots,X_k}$ and $Q_{Y_1,\ldots,Y_k}$ stand for the joint distributions of $(X_1, \ldots, X_k)$ and $(Y_1, \ldots, Y_k)$ respectively. It follows that for all integrable functions f and all $B \in \mathcal{B}^k$ such that all points have distinct coordinates:

$$\int_B f(x_1, \ldots, x_k) dQ_{Y_1,\ldots,Y_k}(x_1, \ldots, x_k) \tag{1.78}$$

$$= \frac{k!\binom{N}{k}}{N^K} \int_B f(x_1, \ldots, x_k) dQ_{Y_1,\ldots,Y_k}(x_1, \ldots, x_k). \tag{1.79}$$

From the definition of conditional distribution, we have

$$\Pr((X_1, \ldots, X_k) \in B, X \in C)$$
$$= \int_B Q_{X|X_1,\ldots,X_k}(C|x_1, \ldots, x_k) dQ_{X_1,\ldots,X_k}(x_1, \ldots, x_k).$$

If we set $f(x_1, \ldots, x_k) = Q_{X|X_1,\ldots,X_k}(C|x_1, \ldots, x_k)$ in (1.78), we get

$$\Pr((Y_1, \ldots, Y_k) \in B, X \in C)$$
$$= \int_B Q_{X|X_1,\ldots,X_k}(C|x_1, \ldots, x_k) dQ_{Y_1,\ldots,Y_k}(x_1, \ldots, x_k),$$

and the two conditional distributions are indeed the same for $(x_1,\ldots,x_k)$ vectors with distinct coordinates. Since such vectors have probability 1 under the distribution of X, the two conditional distributions are the same a.s.

Now, we apply Lemma 1.67 to both the conditional distributions given $X_1,\ldots,X_k$ and given $Y_1,\ldots,Y_k$. Let $B \in \mathcal{B}^n$ have all distinct coordinates, and let C be the set of all $x \in \mathcal{X}^N$ with distinct coordinates. Then, we get

$$\Pr((X_{k+1},\ldots,X_{k+n}) \in B | X_1 = x_1,\ldots,X_k = x_k)$$
$$= \int_C H_n(B|x_{k+1},\ldots,x_N) dQ_{X|X_1,\ldots,X_k}(x|x_1,\ldots,x_k),$$
$$\Pr((Y_{k+1},\ldots,Y_{k+n}) \in B | Y_1 = x_1,\ldots,Y_k = x_k)$$
$$= \int_C M_n(B|x) dQ_{X|X_1,\ldots,X_k}(x|x_1,\ldots,x_k).$$

Lemma 1.69 says that

$$|H_n(B|x_{k+1},\ldots,x_N) - M_n(B|x_{k+1},\ldots,x_N)| \le \frac{n(n-1)}{2(N-k)}.$$

We must now bound the difference $|M_n(B|x_{k+1},\ldots,x_N) - M_n(B|x)|$. As in the proof of Lemma 1.69, one of these probabilities is constant on $\{x_1,\ldots,x_N\}^n$ and the other is constant on a subset and is 0 elsewhere. The sets B with the largest difference will be the set A where the second probability is positive and its complement. Since $M_n(A|x_{k+1},\ldots,x_N) = 1$ and $M_n(A|x) = (1 - k/N)^n$, we get

$$|M_n(B|x) - M_n(B|x_{k+1},\ldots,x_N)| < 1 - \left(1 - \frac{k}{N}\right)^n.$$

The conclusion to the theorem now follows. □

Example 1.80. If $N = 1{,}000{,}000$, $n = 100$, and $k = 100$, then the bound in Theorem 1.77 is 0.0199, or about 2%. On the other hand, if $N = 1{,}000{,}000$, $n = 1000$, and $k = 1000$, then the bound is a useless 1.632.

1.5.4 The General Infinite Case

1.5.4.1 Approximation by the Finite Case*

Theorem 1.70 says that the probabilities concerning n random quantities calculated under the finite exchangeable distribution of $X_1,\ldots,X_N$ are uniformly approximated by those calculated under a conditionally IID distribution. As $N \to \infty$, one would expect that the joint distribution of

---

*This section may be skipped without interrupting the flow of ideas.

$X_1, \ldots, X_N$ would actually become that of conditionally IID random quantities. In examining the statement of Theorem 1.70, we note that the first term inside the absolute value in (1.71) does not depend on N, but μ_N clearly does depend on N. If we could show that there exists $\mu_{\mathbf{P}}$ and a subsequence $\{N_\ell\}_{\ell=1}^\infty$ such that, for all n and all $B \in \mathcal{B}^n$,

$$\lim_{\ell \to \infty} \int P^n(B) d\mu_{N_\ell}(P) = \int P^n(B) d\mu_{\mathbf{P}}(P), \qquad (1.81)$$

then we would have a representation theorem for infinite exchangeable sequences.[29] We will prove that this indeed is true in this section. In fact, (1.81) would follow from the continuous mapping theorem B.88 if we could prove that $P^n(B)$ is a bounded continuous function of P and that $\mathbf{P}_{N_\ell}$ converges in distribution to a random quantity with distribution $\mu_{\mathbf{P}}$.[30] In the case of an infinite sequence of exchangeable Bernoulli random variables, we could easily prove these facts.[31]

Example 1.82. Let $\{X_n\}_{n=1}^\infty$ be a sequence of exchangeable Bernoulli random variables. Then $\mathbf{P}_N$ is nothing more than the proportion of successes, $\overline{X}_N$, in the first N trials. (That is, $\mathbf{P}_N(\{1\}) = \overline{X}_N$ and $\mathbf{P}_N(\{0\}) = 1 - \overline{X}_N$.) The strong law of large numbers 1.59 or 1.62 will say that a subsequence $\overline{X}_{n_k}$ converges a.s. (hence in distribution by Theorem B.90) to something, call it Θ. Then $\mathbf{P}(\{1\}) = \Theta$, and $\mathbf{P}^n(B)$ is a bounded continuous function of Θ ($\mathbf{P}^n(B) = \sum_{y \in B} \Theta^{t(y)}(1-\Theta)^{n-t(y)}$, where $t(y) = \sum_{i=1}^n y_i$). It follows that $\{X_n\}_{n=1}^\infty$ is a sequence of exchangeable Bernoulli random variables if and only if there exists a distribution μ_Θ such that for all n, all distinct $i_1, \ldots, i_n$, and all $x_1, \ldots, x_n \in \{0, 1\}$,

$$\Pr(X_{i_1} = x_1, \ldots, X_{i_n} = x_n) = \int \theta^k (1-\theta)^{n-k} d\mu_\Theta(\theta),$$

where $k = \sum_{i=1}^n x_i$. This is the representation portion of Theorem 1.47.

In general, if we wish to show that the X_i are actually conditionally IID given some random probability measure $\mathbf{P}$, we will need to prove more than

---

[29]Since the first term on the left-hand side of (1.71) does not depend on N, the limit in (1.81) must be the same for all convergent subsequences.

[30]Diaconis and Freedman (1980a) offer a sketch of an abstract proof showing directly that $\mathbf{P}_N$ converges in distribution for general random quantities. We will not actually prove that $\mathbf{P}_N$ converges in distribution to $\mathbf{P}$. Rather, we prove that the finite-dimensional joint distributions of $\mathbf{P}_N$ converge to those of $\mathbf{P}$. Billingsley (1968), which contains an in-depth discussion of convergence in distribution, shows that convergence in distribution requires a condition called *tightness* in addition to convergence of finite dimensional joint distributions. The work of the tightness condition is done in that part of the proof of Theorem 1.49 in which we prove equation (1.84). An alternative proof is given by Aldous (1985, Section 7). A very general theorem is proven by Hewitt and Savage (1955).

[31]Heath and Sudderth (1976) give an alternative proof in the Bernoulli case which relies on a different subsequence argument.

just (1.81). We will need to show that for every measurable subset C of $\mathcal{P}$,

$$\Pr((X_{i_1},\ldots,X_{i_n}) \in B, \mathbf{P} \in C) = \int_C P^n(B) d\mu_{\mathbf{P}}(P). \qquad (1.83)$$

In effect, we need to prove that $\mathbf{P}$ exists and that

$$\mathbf{P}_N^n(B) I_C(\mathbf{P}) \xrightarrow{\mathcal{D}} \mathbf{P}^n(B) I_C(\mathbf{P})$$

for all B and C.

The distribution μ_N in Theorem 1.70 was seen to be the distribution of the empirical probability measure $\mathbf{P}_N$ of $X = (X_1,\ldots,X_N)$. If (1.83) is going to hold, it would stand to reason that $\mu_{\mathbf{P}}$ would equal the distribution of the limit of $\mathbf{P}_N$ as $N \to \infty$, if the limit exists. That this limit exists will follow from the strong law of large numbers 1.59 or 1.62. We will use this fact to prove that (1.83) holds.

1.5.4.2 Proof of Theorem 1.49

The proof of Theorem 1.49 is a bit complicated, so a brief outline will be given first. We use the strong law of large numbers to conclude that (at least a subsequence of) the empirical probability measures at each set B, $\{\mathbf{P}_n(B)\}_{n=1}^\infty$ converges to something, which we call $\mathbf{P}(B)$. To show that $\mathbf{P}(\cdot)$ is a random probability measure, we show that $\mathbf{P}(B) = \Pr(X_1 \in B|\mathcal{A}_\infty)$ for some σ-field $\mathcal{A}_\infty$. Since $\mathcal{X}$ is a Borel space, there is a regular conditional distribution of which $\mathbf{P}(B)$ is a version for all B. It is easy to show that $\mathbf{P}: S \to \mathcal{P}$ is a measurable function. The same calculation that lets us prove that $\mathbf{P}(B) = \Pr(X_1 \in B|\mathcal{A}_\infty)$, namely equation (1.84) in the proof, also leads to the conclusion that the X_i are conditionally IID.

PROOF OF THEOREM 1.49. The "if" direction is straightforward, and its proof is left to the reader. (See Problem 4 on page 73.) For the "only if" direction, assume that $\{X_n\}_{n=1}^\infty$ are exchangeable. Let $\mathbf{P}_n$ be the empirical distribution of $X_1,\ldots,X_n$. For each $B \in \mathcal{B}$, $\lim_{n\to\infty} \mathbf{P}_{8^n}(B) = \mathbf{P}(B)$ exists a.s., as either Theorem 1.59 or 1.62 says.

For each $B \in \mathcal{B}$ and each integer i, $\mathbf{P}(B) = \lim_{n\to\infty} \sum_{m=i}^{8^n} I_B(X_m)/8^n$, a.s. It follows that for each i, $\mathbf{P}(B)$ is measurable with respect to the σ-field generated by $\{X_n\}_{n=i}^\infty$. Hence, it is measurable with respect to the intersection (over i) of all of these σ-fields, which is the tail σ-field of $\{X_n\}_{n=1}^\infty$. Call the tail σ-field $\mathcal{A}_\infty$.

We next prove that for every k, all distinct $i_1,\ldots,i_k$, all $B_1,\ldots,B_k \in \mathcal{B}$, and every $C \in \mathcal{A}_\infty$,

$$\Pr(\{X_{i_1} \in B_1,\ldots,X_{i_k} \in B_k\} \cap C) = \int_C \prod_{j=1}^k \mathbf{P}(B_j)(s) d\mu(s). \qquad (1.84)$$

To do this, let $Z_n = I_C \prod_{j=1}^k \mathbf{P}_{8^n}(B_j)$ and $Z = I_C \prod_{j=1}^k \mathbf{P}(B_j)$. Note that $Z_n \to Z$, a.s. as $n \to \infty$, hence $Z_n \xrightarrow{\mathcal{D}} Z$ by Theorem B.90. Since Z_n

is uniformly bounded, $E(Z_n) \to E(Z)$. Since $\mathbf{P}(B_i)$ is measurable with respect to $\mathcal{A}_\infty$, the integral on the right-hand side of (1.84) is $E(Z)$. All that remains to the proof of (1.84) is to show that $E(Z_n)$ converges to the left-hand side of (1.84). To do this, let $m = 8^n$, and write

$$I_C \prod_{j=1}^{k} \mathbf{P}_m(B_j) = I_C \frac{1}{m^k} \sum_{i_1=1}^{m} \cdots \sum_{i_k=1}^{m} \prod_{j=1}^{k} I_{B_j}(X_{i_j})$$

$$= I_C \frac{1}{m^k} \left[\sum_{\text{all } i_j \text{ distinct}} \prod_{j=1}^{k} I_{B_j}(X_{i_j}) + \sum_{\text{at least two } i_j \text{ equal}} \prod_{j=1}^{k} I_{B_j}(X_{i_j}) \right].$$

In the last expression above, the first sum has $m!/(m-k)!$ terms, each of which has mean equal to the left-hand side of (1.84). Since $m!/[(m-k)!m^k]$ converges to 1, the mean of $1/m^k$ times this first sum converges to the left-hand side of (1.84). The second sum has $m^k - m!/(m-k)!$ terms, each of which is bounded between 0 and 1. Since $1 - m!/[(m-k)!m^k]$ converges to 0, so does the mean of the second sum. This completes the proof of (1.84).

Apply (1.84) with $k = 1$, $i_1 = 1$, and $B_1 = B$ to get that $E(I_B(X_1)I_C) = E(\mathbf{P}(B)I_C)$. This means that $\mathbf{P}(B)$ is a version of $\Pr(X_1 \in B|\mathcal{A}_\infty)$ for every B. Since $(\mathcal{X}, \mathcal{B})$ is a Borel space, this can be assumed to be part of a regular conditional distribution, and we can assume that $\mathbf{P}(B) = \Pr(X_1 \in B|\mathcal{A}_\infty)$. In this way $\mathbf{P}$ becomes a random probability measure so long as we can prove that it is a measurable function from $(S, \mathcal{A})$ to $(\mathcal{P}, \mathcal{C}_\mathcal{P})$. The σ-field $\mathcal{C}_\mathcal{P}$ was set up so that $\mathbf{P}$ is measurable if and only if $\mathbf{P}(B)$ is measurable for all B. Since $\mathcal{A}_\infty \subseteq \mathcal{A}$, $\mathbf{P}(B)$ is measurable for all B. Also, since $\Pr(X_1 \in B|\mathcal{A}_\infty) = \mathbf{P}(B)$ is a function of $\mathbf{P}$ for each B and since $\mathbf{P}$ is $\mathcal{A}_\infty$ measurable, it follows from Theorem B.73 that $\Pr(X_1 \in B|\mathbf{P}) = \mathbf{P}(B)$.

Now, let $\mu_\mathbf{P}$ denote the distribution of $\mathbf{P}$. To prove that the X_i are conditionally independent given $\mathbf{P} = P$ with distribution P, apply (1.84) with $C = \{\mathbf{P} \in A\}$ for arbitrary $A \in \mathcal{C}_\mathcal{P}$. The result is

$$\Pr(X_{i_1} \in B_1, \ldots, X_{i_k} \in B_k, \mathbf{P} \in A) = \int_{\{\mathbf{P} \in A\}} \prod_{j=1}^{k} \mathbf{P}(B_j)(s) d\mu(s)$$

$$= \int_A \prod_{j=1}^{k} P(B_j) d\mu_\mathbf{P}(P) = \int_A \prod_{j=1}^{k} \Pr(X_{i_j} \in B_j|\mathbf{P} = P) d\mu_\mathbf{P}(P),$$

where the first equation is immediate from (1.84), the second follows from Theorem A.81, and the third follows from the fact that $\Pr(X_1 \in B|\mathbf{P} = P) = P(B)$. This completes the proof of conditional independence given $\mathbf{P}$.

For the uniqueness, suppose that μ_1 and μ_2 are possible distributions for a random probability measure $\mathbf{P}$ such that the X_i are conditionally IID with distribution P given $\mathbf{P} = P$. We will prove that the finite-dimensional distributions of μ_1 and μ_2 agree, and then Theorem B.131 says that $\mu_1 = \mu_2$. Let $B_1, \ldots, B_n \in \mathcal{B}$, and let $k_1, \ldots, k_n$ be positive integers. We have already proven that

$$\Pr(X_1 \in B_1, \ldots, X_{k_1} \in B_1, X_{k_1+1} \in B_2, \ldots, X_{k_1+k_2} \in B_2, \ldots,$$

$$X_{k_1+\cdots+k_n} \in B_n) = \int \prod_{i=1}^{n} P(B_i)^{k_i} d\mu_1(P) = \int \prod_{i=1}^{n} P(B_i)^{k_i} d\mu_2(P).$$

Hence, the means of all polynomial functions of $(\mathbf{P}(B_1), \ldots, \mathbf{P}(B_n))$ are the same according to μ_1 and μ_2. By the Stone–Weierstrass theorem C.3 the means of all bounded continuous functions of finitely many $\mathbf{P}(B)$ values are determined by the means of polynomial functions. Hence, the means of all bounded continuous functions of $(\mathbf{P}(B_1), \ldots, \mathbf{P}(B_n))$ are the same according to μ_1 and μ_2. Corollary B.107 says that μ_1 and μ_2 give the same joint distribution to $(\mathbf{P}(B_1), \ldots, \mathbf{P}(B_n))$, and the proof of uniqueness is complete.

The convergence claim follows from Theorem 1.62 or from Lemma 1.61 and the fact that the bounded random variables $I_B(X_i)$ are conditionally IID. □

1.5.5 Formal Introduction to Parametric Models*

The infinite version of DeFinetti's representation theorem 1.49 says that if an infinite sequence of random quantities is exchangeable, then specifying the joint distribution of all of them can be done by specifying a distribution for the limit of the empirical probability measures. Every probability measure is a limit of empirical probability measures, and so the space $\mathcal{P}$, on which the distribution must be placed, is quite large. There are (at least) two problems involved in specifying a distribution over $\mathcal{P}$:

1. How do you perform the general integration $\int h(P) d\mu_{\mathbf{P}}(P)$?

2. How do you get the conditional distribution of $\mathbf{P}$ given data?[32]

These two problems are related, and the usual method of solving them is to say that $\mu_{\mathbf{P}}$ assigns probability 1 to a relatively small subset of $\mathcal{P}$. In Example 1.53 on page 29, we saw a case in which $\mu_{\mathbf{P}}$ assigned all of its probability to the set of exponential distributions. An alternative is to assign all probability to normal distributions or t distributions, and so forth.

---

*This section may be skipped without interrupting the flow of ideas.

[32] We have not tried to prove that $(\mathcal{P}, \mathcal{C}_\mathcal{P})$ is a Borel space. However, since $(\mathcal{X}^\infty, \mathcal{B}^\infty)$ is a Borel space and $\mathbf{P}: \mathcal{X}^\infty \to \mathcal{P}$ is measurable, regular conditional distributions exist on $\mathcal{X}^\infty$ and they induce regular conditional distributions on $\mathcal{P}$.

These cases all have something in common, namely that the set of distributions is finitely parameterized. That is, there exists a one-to-one mapping between the set of distributions and a subset of a finite-dimensional Euclidean space. In the normal example, the mapping associates the $N(m, s^2)$ distribution with $(m, s) \in \mathbb{R} \times \mathbb{R}^+$. With such a *parameter mapping*, we can switch the problem of integration over subsets of $\mathcal{P}$ to integration over Euclidean space. The problem of finding conditional distributions is resolved the same way. The conditional distribution in Euclidean space induces the appropriate conditional distribution in $\mathcal{P}$. (See Theorem B.28 on page 617.) There are cases (see Sections 1.6.1 and 1.6.2) in which we want the range of the parameter mapping to be an infinite-dimensional space. In such cases, we will need to develop special methods for calculating integrals.

Now, let $\mathcal{P}_0$ be a subset of $\mathcal{P}$ and let $\Theta' : \mathcal{P}_0 \to \Omega$ be a bimeasurable function, where Ω is a set with σ-field τ. The σ-field of subsets of $\mathcal{P}$ which we need to consider is $\mathcal{C}_0 = \{A \cap \mathcal{P}_0 : A \in \mathcal{C}_\mathcal{P}\}$. Let $\mu_\mathbf{P}$ be a probability measure on $(\mathcal{P}_0, \mathcal{C}_0)$ and let μ_Θ be the probability on (Ω, τ) induced by Θ' from $\mu_\mathbf{P}$. That is, for each $A \in \mathcal{C}_0$, $\mu_\mathbf{P}(A) = \mu_\Theta(\Theta'(A))$, and for each $B \in \tau$, $\mu_\Theta(B) = \mu_\mathbf{P}(\Theta'^{-1}(B))$. To integrate a measurable function $h : \mathcal{P}_0 \to \mathbb{R}$, we note that

$$\int h(P) d\mu_\mathbf{P}(P) = \int h(\Theta'^{-1}(\theta)) d\mu_\Theta(\theta),$$

where θ is used to stand for an arbitrary element of Ω and $P = \Theta'^{-1}(\theta) \in \mathcal{P}$. For example, if $\mathcal{X} = \mathbb{R}$ and P is the $N(m, s^2)$ distribution, and $\Theta'(P) = (m, s)$, then for $\theta = (m, s)$, $\Theta'^{-1}(\theta) = P$. As another example, let $\mathcal{X} = \{0, 1, 2\}$. In this case $\mathcal{P}$ is already a finite-dimensional set. We can let

$$\Omega = \{(p_0, p_1, p_2) : p_i \geq 0, p_0 + p_1 + p_2 = 1\}.$$

If P is the distribution that says $P(\{i\}) = p_i$ for $i = 0, 1, 2$ with $p_0 + p_1 + p_2 = 1$ and $p_i \geq 0$ for all i, then we can let $\Theta'(P) = (p_0, p_1, p_2)$. In this case $\mathcal{P}_0 = \mathcal{P}$.

In general, we can make the above discussion precise as follows. Let $(S, \mathcal{A}, \mu)$ be a probability space and let $(\mathcal{X}^1, \mathcal{B}^1)$ be a Borel space. Suppose that $\{X_n\}_{n=1}^\infty$ is an infinite sequence of exchangeable random quantities, $X_n : S \to \mathcal{X}^1$. Let $\mathcal{X}^n$ and $\mathcal{X}^\infty$ be finite and infinite products of $\mathcal{X}^1$ with σ-fields $\mathcal{B}^n$ and $\mathcal{B}^\infty$, respectively. Let X^∞ denote the function mapping S to $\mathcal{X}^\infty$ by $X^\infty(s) = (X_1(s), X_2(s), \ldots)$. Similarly, let $X^n : S \to \mathcal{X}^n$ be $X^n(s) = (X_1(s), \ldots, X_n(s))$. Let $\mathbf{P} : \mathcal{X}^\infty \to \mathcal{P}$ denote the "limit of empirical probabilities" function. Next we introduce a general definition.

Definition 1.85. A bimeasurable mapping Θ' from a subset $\mathcal{P}_0$ of $\mathcal{P}$ to a subset Ω of some Borel space with σ-field τ is called a *parametric index*. The parametric index is denoted by $\Theta' : \mathcal{P}_0 \to \Omega$. The set Ω is called the *parameter space*, and the set $\mathcal{P}_0$ is called the *parametric family*.

1.5. Proofs of DeFinetti's Theorem   51

Let $\Theta' : \mathcal{P} \to \Omega$ denote a parametric index. We have constructed the following sequence of functions:

$$S \xrightarrow{X^\infty} \mathcal{X}^\infty \xrightarrow{\mathbf{P}} \mathcal{P} \xrightarrow{\Theta'} \Omega.$$

Let the function $\Theta : S \to \Omega$ be defined by $\Theta(s) = \Theta'(\mathbf{P}(X^\infty(s)))$. (Note that the value of Θ is the same as the value of Θ', hence we will often find it convenient to use the symbol Θ to refer to both Θ and Θ'.) We call Θ the *parameter*. Let μ_Θ be the probability induced on (Ω, τ) by Θ from μ. Let $\mathcal{A}_X$ be the sub-σ-field of $\mathcal{A}$ generated by X^∞. Since $(\mathcal{X}^\infty, \mathcal{B}^\infty)$ is also a Borel space, regular conditional distributions given Θ exist. For each $A \in \mathcal{A}_X$, let $P'_\theta(A) = \Pr(A|\Theta)(s)$ for all s such that $\Theta(s) = \theta$. For each $B \in \mathcal{B}^\infty$, let $P_\theta(B) = P'_\theta(X^{\infty-1}(B))$. In words, $\{P_\theta : \theta \in \Omega\}$ specifies the conditional distribution of X^∞ given Θ.

Example 1.86. Let $\mathcal{X}^1 = \mathbb{R}$ and let $\mathcal{P}_0$ be the set of all normal distributions. Assume that $\mu_{\mathbf{P}}$ assigns probability 1 to the set $\mathcal{P}_0$. We can let $\Theta'(P)$ be the vector consisting of the mean and standard deviation of the normal distribution P. Then $\Theta(s)$ is the vector consisting of the mean and standard deviation of the limit of the empirical distribution of a sequence $\{X_n\}_{n=1}^\infty$ of exchangeable random variables. By the strong law of large numbers 1.63 and the fact that the X_n are conditionally IID given $\mathbf{P}$, $\Theta(s)$ is also the limit (a.s.) of the sample average $\overline{X}_n = \sum_{i=1}^n X_i/n$ and the sample standard deviation $\sqrt{\sum_{i=1}^n (X_i - \overline{X}_n)^2/(n-1)}$ of the data sequence. If $\theta = (\mu, \sigma)$, then P_θ is the distribution that says that $\{X_n\}_{n=1}^\infty$ are IID $N(\mu, \sigma^2)$ random variables. The notation P'_θ stands for the probability measure on $(S, \mathcal{A}_X)$ defined by $P'_\theta(X^{\infty-1}(B)) = P_\theta(B)$ for $B \in \mathcal{B}^\infty$.

The probability measures P_θ for $\theta \in \Omega$ are on the space $(\mathcal{X}^\infty, \mathcal{B}^\infty)$. They induce probabilities on all of the spaces $(\mathcal{X}^n, \mathcal{B}^n)$, for $n = 1, 2, \ldots$ via the obvious projections. It will prove convenient to refer to all of these induced probabilities by the same name, P_θ. That is, if $A \in \mathcal{B}^n$, let $P_\theta(A)$ denote $P_\theta(A \times \mathcal{X}^1 \times \mathcal{X}^1 \times \cdots)$. This will be very convenient without causing any confusion. If it becomes important to know over which space, $(\mathcal{X}^n, \mathcal{B}^n)$ or $(\mathcal{X}^\infty, \mathcal{B}^\infty)$, P_θ is defined, we will be explicit.

Sometimes the parameter Θ can be expressed as a meaningful function of the distribution P, say $H(P)$ which is also defined for distributions outside of the parametric family. For example, $H(P) = \int x dP(x)$, the mean of the distribution, is defined for every distribution with finite mean whether or not that distribution is a member of a parametric family of interest. When this occurs, it may be that H is continuous in the sense that $H(\mathbf{P}_n) \xrightarrow{\mathcal{D}} H(\mathbf{P})$ if $\lim_{n \to \infty} \mathbf{P}_n = \mathbf{P}$. The distribution of Θ can then be considered as an approximation to the distribution of $H(\mathbf{P}_n)$, where $\mathbf{P}_n$ is the empirical probability measure of the first n observations.

Example 1.87 (Continuation of Example 1.53; see page 29). In the $Exp(\theta)$ distribution, θ is one over the mean of the distribution. So, $H(P) = \left(\int x dP(x)\right)^{-1}$ and $H(\mathbf{P}_n) = \left(\sum_{i=1}^n X_i/n\right)^{-1} = 1/\overline{X}_n$. Indeed, $1/\overline{X}_n \xrightarrow{\mathcal{D}} \Theta$, so that we can take the distribution of Θ to be an approximation to the distribution of $1/\overline{X}_n$. (See Problem 21 on page 76.)

Example 1.88 (Continuation of Example 1.38; see page 18). The marginal posterior for M can be calculated by integrating σ out of (1.25) (after changing the 0 subscripts to 1), or by using the fact that M is the limit as m goes to ∞ of Y_m, so the distribution of M is the limit of the distributions of the Y_m. Since the $t_{a_1}(\mu_1, \sqrt{[1/m + 1/\lambda_1]b_1/a_1})$ densities converge to the $t_{a_1}(\mu_1, \sqrt{b_1/[a_1\lambda_1]})$ density as $m \to \infty$, Scheffé's theorem B.79 says that this limit is the distribution of M.

1.6 Infinite-Dimensional Parameters*

An alternative to the use of finite-dimensional parameter spaces is to attempt to place a probability distribution on an infinite-dimensional space $\mathcal{P}$. It is common to call such models *nonparametric*. We will consider two types of probability measures on infinite-dimensional parameter spaces, Dirichlet processes and tailfree processes.

1.6.1 Dirichlet Processes

Ferguson (1973) gives a probability measure on an infinite-dimensional space $\mathcal{P}$ for which certain calculations are simple. We can think of **P** as a stochastic process as in Section B.5 with index set $\mathcal{B}$, a σ-field of subsets of $\mathcal{X}$. To specify a distribution for **P**, we need to specify the joint distribution of $(\mathbf{P}(B_1), \ldots, \mathbf{P}(B_n))$ for all n and all $B_1, \ldots, B_n \in \mathcal{B}$ in such a way that the distributions are consistent according to Definition B.132. One way to do this is as follows. Let α be a finite measure on $(\mathcal{X}, \mathcal{B})$. For each integer $n > 0$ and partition $B_1, \ldots, B_n$ of $\mathcal{X}$, define the joint distribution of $(\mathbf{P}(B_1), \ldots, \mathbf{P}(B_n))$ to be the Dirichlet distribution $Dir_n(\alpha_1, \ldots, \alpha_n)$, where $\alpha_i = \alpha(B_i)$ for $i = 1, \ldots, n$. This is a distribution for a vector $(Y_1, \ldots, Y_n)$ such that $\sum_{i=1}^{n} Y_i = 1$ and such that $(Y_1, \ldots, Y_n)$ have joint density

$$\frac{\Gamma(\alpha_1 + \cdots + \alpha_n)}{\Gamma(\alpha_1) \cdots \Gamma(\alpha_n)} y_1^{\alpha_1 - 1} \cdots y_{n-1}^{\alpha_{n-1} - 1} (1 - y_1 - \cdots - y_{n-1})^{\alpha_n - 1}.$$

The Dirichlet distribution is a multivariate generalization of the Beta distribution. To avoid having to deal separately with the cases in which some of the sets B_i have $\alpha(B_i) = 0$, we will extend the definition of the Dirichlet distribution to allow some of the α_i to be 0. In this case, those coordinates corresponding to $\alpha_i = 0$ are equal to 0 with probability 1 and the rest of the coordinates have the usual Dirichlet distribution.

We prove next that this specification of the distribution of **P** is consistent.

---

*This section may be skipped without interrupting the flow of ideas.

1.6. Infinite-Dimensional Parameters

Theorem 1.89. *Let* $\mathbf{P}$ *be a random probability measure on a Borel space* $(\mathcal{X}, \mathcal{B})$, *and let* $B_1, \ldots, B_n \in \mathcal{B}$ *partition* $\mathcal{X}$. *Let* α *be a finite measure on* $(\mathcal{X}, \mathcal{B})$ *with* $\alpha(\mathcal{X}) > 0$, *and let* $\alpha_i = \alpha(B_i)$ *for all* i. *To say that*

$$(\mathbf{P}(B_1), \ldots, \mathbf{P}(B_n)) \sim Dir_n(\alpha_1, \ldots, \alpha_n)$$

specifies a consistent set of distributions in the sense of Definition B.132.

PROOF. Let $A_1, \ldots, A_p$ be elements of $\mathcal{B}$. Set $B_1, \ldots, B_n$ equal to the partition consisting of the constituents of $A_1, \ldots, A_p$. That is, each B_i equals one of the 2^p sets $C_1 \cap \cdots \cap C_p$, where each C_i is either A_i or A_i^C for $i = 1, \ldots, p$ (e.g., $A_1 \cap A_2^C \cap A_3 \cdots \cap A_p^C$). Let $G_i = \mathbf{P}(A_i)$ for $i = 1, \ldots, p$. We need to show that for each $i = 1, \ldots, p$ and each set of numbers $t_1, \ldots, t_{i-1}, t_{i+1}, \ldots, t_p$,

$$\lim_{t_i \to \infty} F_{G_1, \ldots, G_p}(t_1, \ldots, t_p) = F_{G_1, \ldots, G_{i-1}, G_i, \ldots, G_p}(t_1, \ldots, t_{i-1}, t_{i+1}, \ldots, t_p). \tag{1.90}$$

Let $Y_j = \mathbf{P}(B_j)$ for $j = 1, \ldots, n$. Let $c_i = \{j : B_j \subseteq A_i\}$. Then, the expression whose limit is being taken on the left-hand side of (1.90) is

$$\Pr\left(\sum_{j \in c_i} Y_j \leq t_i, \text{ for } i = 1, \ldots, p\right) = \Pr((Y_1, \ldots, Y_n) \in C(t)), \tag{1.91}$$

where

$$C(t) = \left\{(y_1, \ldots, y_n) : \sum_{j \in c_i} y_j \leq t_i, \text{ for } i = 1, \ldots, p\right\}.$$

Now, fix an i, and let $B_1^i, \ldots, B_m^i$ be the constituents of $\{A_j : j \neq i\}$. Each B_s^i is the union of two of the B_j. (For example, if $i = 1$ and $p = 3$, then $(A_1 \cap A_2^C \cap A_3) \cup (A_1^C \cap A_2^C \cap A_3)$ is one of the B_j^1.) Let $Z_s = \mathbf{P}(B_s^i)$ for $s = 1, \ldots, m$. The proposed distribution implies that $(Z_1, \ldots, Z_m)$ has $Dir_m(\beta_1, \ldots, \beta_m)$ distribution, where $\beta_s = \alpha_{j_1} + \alpha_{j_2}$ when $B_s^i = B_{j_1} \cup B_{j_2}$. For $j \neq i$, let $d_j = \{s : B_s^i \subseteq A_j\}$. The limit as $t_i \to \infty$ of the expression in (1.91) is

$$\lim_{t_i \to \infty} \Pr((Y_1, \ldots, Y_n) \in C(t)) = \Pr\left(Y_1, \ldots, Y_n) \in \bigcup_{t_i \in \mathbb{R}} C(t)\right)$$
$$= \Pr((Z_1, \ldots, Z_m) \in D_i(t)), \tag{1.92}$$

where

$$D_i(t) = \left\{(z_1, \ldots, z_m) : \sum_{s \in d_j} \beta_s \leq t_j, \text{ for } j \neq i\right\}.$$

It is easy to see that (1.92) is the same as the right-hand side of (1.90). □

Since the distributions specified are consistent, we can use them for the distribution of $\mathbf{P}$.

Definition 1.93. If α is a finite (not identically 0) measure on $(\mathcal{X}, \mathcal{B})$ and $\mathbf{P}$ is a random distribution such that, for each n, and each partition $\{B_1, \ldots, B_n\}$ of $\mathcal{X}$,

$$(\mathbf{P}(B_1), \ldots, \mathbf{P}(B_n)) \sim Dir_n(\alpha_1, \ldots, \alpha_n),$$

where $\alpha_i = \alpha(B_i)$ for $i = 1, \ldots, n$, then we say that $\mathbf{P}$ has *Dirichlet process distribution with base measure* α, denoted by $Dir(\alpha)$.

The Dirichlet process is useful only if we can do the necessary calculations for making inference. The most crucial is updating in the light of data.

Theorem 1.94. *Suppose that $\{X_n\}_{n=1}^{\infty}$ is a sequence of exchangeable random quantities, that they are conditionally independent with distribution P given $\mathbf{P} = P$, and that $\mathbf{P}$ has $Dir(\alpha)$ distribution. Then the marginal distribution of each X_i is the probability measure $\alpha/\alpha(\mathcal{X})$ and the conditional distribution of $\mathbf{P}$ given $X_1 = x_1, \ldots, X_k = x_k$ is $Dir(\beta)$, where β is the measure defined by $\beta(C) = \alpha(C) + \sum_{i=1}^{k} I_C(x_i)$ for each C.*

PROOF. First, we prove the claim about the marginal distribution of X_i. For $B \in \mathcal{B}$,

$$\Pr(X_i \in B) = \mathrm{E}(\Pr(X_i \in B | \mathbf{P})) = \mathrm{E}(\mathbf{P}(B)) = \frac{\alpha(B)}{\alpha(\mathcal{X})},$$

where the first equality follows from the law of total probability B.70, and the last follows from the fact that each coordinate of a Dirichlet distribution has Beta distribution.

By the form of the purported posterior, it is clear that if we can prove the result for $k = 1$, we can extend it to arbitrary k by induction. Let $\mu_\mathbf{P}$ denote the $Dir(\alpha)$ measure on the space of probability measures $\mathcal{P}$. For arbitrary n and partition $B_1, \ldots, B_n$, and arbitrary $B \in \mathcal{B}$ and $t_1, \ldots, t_n$, let $A = \{P : P(B_i) \leq t_i,$ for $i = 1, \ldots, n\}$. Assume that $\alpha(B) < \alpha(\mathcal{X})$.[33] Define, for $i = 1, \ldots, n$, $B_i^0 = B_i \cap B$ and $B_i^1 = B_i \cap B^C$. Then $B_1^0, \ldots, B_n^0, B_1^1, \ldots, B_n^1$ form a partition of at most $2n$ nonempty sets.[34] In particular, we can write $\mathbf{P}(B) = \sum_{i=1}^{n} \mathbf{P}(B_i^0)$. Define

$$A_B = \{(z_1, \ldots, z_{2n}) : z_i + z_{n+i} \leq t_i, \text{ for } i = 1, \ldots, n\},$$

and note that $P \in A$ if and only if $(P(B_1^0), \ldots, P(B_n^1)) \in A_B$. Let $\beta_i = \alpha(B_i^1)$ for $i = 1, \ldots, n$, and let $\beta_i = \alpha(B_i^0)$ for $i = n+1, \ldots, 2n$. Let $c = \{i :$

---

[33]If $\alpha(B) = \alpha(\mathcal{X})$, it is trivial to prove that $\int_B \mu_{P|X_1}(A|x) d\mu_{X_1}(x) = \Pr(\mathbf{P} \in A, X_1 \in B)$, where $\mu_{P|X_1}$ is the conditional distribution of $\mathbf{P}$ given X_1 to be defined later in this proof, and μ_{X_1} is the marginal distribution of X_1 already determined.

[34]Recall the extended definition of the Dirichlet distribution in which $\alpha_i = 0$ means that the ith coordinate is 0 with probability 1.

1.6. Infinite-Dimensional Parameters

$\beta_i \neq 0\}$, and let k be the highest number in c. Let $c' = \{i : \beta_i \neq 0\} \setminus \{k\}$. If $\beta_j = 0$, let $z_j = 0$ in the following equations. Then we can write

$$\begin{aligned}
&\Pr(\mathbf{P} \in A, X_1 \in B) \\
&= \Pr(\mathbf{P}(B_1) \leq t_1, \ldots, \mathbf{P}(B_n) \leq t_n, X_1 \in B) \\
&= \int_A P(B) d\mu_{\mathbf{P}}(P) \quad (1.95) \\
&= \int_{A_B} \sum_{j=1}^n z_j \frac{\Gamma(\alpha(\mathcal{X}))}{\prod_{i \in c} \Gamma(\beta_i)} \prod_{i \in c'} z_i^{\beta_i - 1} \left(1 - \sum_{i \in c'} z_i\right)^{\beta_k - 1} \prod_{i \in c'} dz_i \\
&= \sum_{j=1}^n \frac{\beta_i}{\alpha(\mathcal{X})} \int_{A_B} \frac{\Gamma(\alpha(\mathcal{X}) + 1)}{\prod_{i \in c} \Gamma(\beta_i^j)} \prod_{i \in c'} z_i^{\beta_i^j - 1} \left(1 - \sum_{i \in c'} z_i\right)^{\beta_k^j - 1} \prod_{i \in c'} dz_i,
\end{aligned}$$

where $\beta_j^j = \beta_j + 1$ and $\beta_i^j = \beta_i$ for $i \neq j$.

For each $x \in \mathcal{X}$, let α_x denote the measure defined by $\alpha_x(C) = \alpha(C) + I_C(x)$ for each $C \in \mathcal{B}$. Let $\mu_{P|X_1}(\cdot|x)$ denote the $Dir(\alpha_x)$ measure on $\mathcal{P}$. It is easy to see that for $x \in B$, $\mu_{P|X_1}(\cdot|x)$ says that the joint distribution of $\{\mathbf{P}(B_i^j)\}$, for $i = 1, \ldots, n$ and $j = 0, 1\}$ is $Dir_{2n}(\beta_1^j, \ldots, \beta_{2n}^j)$, where j is such that $x \in B_j^1$. Hence, $\mu_{P|X_1}(A|x)$ equals

$$\sum_{j=1}^n I_{B_j^1}(x) \int_{A_B} \frac{\Gamma(\alpha(\mathcal{X}) + 1)}{\prod_{i \in c} \Gamma(\beta_i^j)} \prod_{i \in c'} z_i^{\beta_i^j - 1} \left(1 - \sum_{i \in c'} z_i\right)^{\beta_k^j - 1} \prod_{i \in c'} dz_i.$$

It follows that

$$\begin{aligned}
&\int_B \mu_{P|X_1}(A|x) d\mu_{X_1}(x) \\
&= \frac{1}{\alpha(\mathcal{X})} \int_B \mu_{P|X_1}(A|x) d\alpha(x) \\
&= \sum_{j=1}^n \frac{\beta_i}{\alpha(\mathcal{X})} \int_{A_B} \frac{\Gamma(\alpha(\mathcal{X}) + 1)}{\prod_{i \in c} \Gamma(\beta_i^j)} \prod_{i \in c'} z_i^{\beta_i^j - 1} \left(1 - \sum_{i \in c'} z_i\right)^{\beta_k^j - 1} \prod_{i \in c'} dz_i,
\end{aligned}$$

which is the same as the last expression in (1.95). That is

$$\int_B \mu_{P|X_1}(A|x) d\mu_{X_1}(x) = \Pr(\mathbf{P} \in A, X_1 \in B),$$

which is what it means to say that the conditional distribution of $\mathbf{P}$ given $X_1 = x$ is $Dir(\alpha_x)$. □

By combining Theorems 1.94 and 1.89, we see that the posterior distribution found in Theorem 1.94 is a regular conditional distribution.

Example 1.96. If α is a continuous measure (i.e., every singleton has 0 measure), then the posterior measure β is a mixture of discrete and continuous parts. There is mass 1 at every observed data value, but no other values have positive measure.

Ferguson (1973) and Blackwell (1973) prove that there is a set of discrete distributions $\mathcal{P}_0 \subseteq \mathcal{P}$ such that the $Dir(\alpha)$ distribution assigns probability 1 to $\mathcal{P}_0$. Sethuraman (1994) proves an alternative theorem, which not only shows that the Dirichlet process is a probability on discrete distributions, but also gives an algorithm for approximately simulating a CDF with $Dir(\alpha)$ distribution. The result of Sethuraman (1994) is that the set of points on which the $Dir(\alpha)$ distribution concentrates its mass is an infinite IID sample $Y_1, Y_2, \ldots$ from the probability $\alpha/\alpha(\mathcal{X})$, and the probability assigned to Y_n is P_n, where $P_1 = Q_1$, and for $n > 1$, $P_n = Q_n \prod_{i=1}^{n-1}(1 - Q_i)$, where the Q_i are IID with $Beta(1, \alpha(\mathcal{X}))$ distribution. What we prove here is a very simple theorem of Krasker and Pratt (1986) which implies that the $Dir(\alpha)$ distribution assigns probability 1 to a set of discrete distributions.

Theorem 1.97. *Let $\{X_n\}_{n=1}^{\infty}$ be conditionally IID with distribution P given $\mathbf{P} = P$. For $n > 1$, define*

$$a_n = \Pr(X_n \text{ is distinct from } X_1, \ldots, X_{n-1}).$$

If $\lim_{n \to \infty} a_n = 0$, then $\mathbf{P}$ is a discrete distribution with probability 1.

PROOF. Define

$$B_\epsilon = \{P : \exists A \in \mathcal{B} \text{ such that } \mathbf{P}(A) > \epsilon \text{ and } \Pr(\{x\}) = 0 \text{ for all } x \in A\}.$$

It suffices to prove that $\Pr(B_\epsilon) = 0$ for all $\epsilon > 0$. The conditional probability, given $\mathbf{P} = P$ and $X_1, \ldots, X_{n-1}$, that X_n is distinct from $X_1, \ldots, X_{n-1}$ is at least ϵ for all $P \in B_\epsilon$. It follows that

$$\begin{aligned} a_n &= \text{E}\Pr(X_n \text{ is distinct from } X_1, \ldots, X_{n-1}|X_1, \ldots, X_{n-1}, \mathbf{P}) \\ &\geq \text{E}\left[\Pr(X_n \text{ is distinct from } X_1, \ldots, X_{n-1}|X_1, \ldots, X_{n-1}, \mathbf{P})I_{B_\epsilon}(\mathbf{P})\right] \\ &\geq \epsilon \Pr(B_\epsilon). \end{aligned}$$

Since $\lim_{n \to \infty} a_n = 0$, $\Pr(B_\epsilon) = 0$ for all $\epsilon > 0$ is necessary. □

For the $Dir(\alpha)$ distribution, it is easy to calculate $a_n = \alpha(\mathcal{X})/[\alpha(\mathcal{X}) + n - 1]$.

The posterior predictive distribution of a future observation is a weighted average of the prior measure $\alpha/\alpha(\mathcal{X})$ and the empirical probability measure.

Proposition 1.98. *Assume that $\{X_n\}_{n=1}^{\infty}$ are conditionally IID with distribution P given $\mathbf{P} = P$ and that $\mathbf{P}$ has $Dir(\alpha)$ distribution. The posterior predictive distribution of a future X_i given $X_1 = x_1, \ldots, X_n = x_n$ is $\beta/[\alpha(\mathcal{X}) + n]$, where $\beta(C) = \alpha(C) + \sum_{i=1}^{n} I_C(x_i)$.*

The predictive joint distribution of several future observations can be obtained by applying Proposition 1.98 several times, each time after conditioning on one more random variable. This gives a straightforward way to generate a sample whose conditional distribution is $\mathbf{P}$, which itself has a Dirichlet process distribution. The joint distribution can also be described as follows.

1.6. Infinite-Dimensional Parameters

Lemma 1.99.[35] *Assume that $\{X_n\}_{n=1}^{\infty}$ are conditionally IID with distribution P given $\mathbf{P} = P$ and that $\mathbf{P}$ has $Dir(\alpha)$ distribution. Let $n > 0$. If p is a partition of $\{1,\ldots,n\}$, let $g(p)$ be the number of nonempty sets in p, and let $k_1(p),\ldots,k_{g(p)}(p)$ be the numbers of elements of the $g(p)$ sets. (Note that $\sum_{i=1}^{g(p)} k_i(p) = n$ for all p.) For each $x \in \mathcal{X}^n$, let $R(x)$ be the partition of $\{1,\ldots,n\}$ which matches x. (That is, x has $g(R(x))$ distinct coordinates, and for each set A in the partition $R(x)$, those coordinates of x whose subscripts are in A are all equal to each other.) For each $x \in \mathcal{X}^n$, define $Z(x) \in \mathcal{X}^{g(R(x))}$ to be the vector of distinct coordinates such that $Z(x)_i$ is repeated $k_i(R(x))$ times in x. For each p and each subset B of $\mathcal{X}^n$, define B_p to be that subset of $\mathcal{X}^{g(p)}$ which consists of the set of distinct coordinates of points in $B \cap R^{-1}(p)$. (That is, $B_p = Z(B \cap R^{-1}(p))$.) Define the measure ν on $\mathcal{X}^n$ by*

$$\nu(B) = \sum_{\text{All } p} \alpha^{g(p)}(B_p).$$

The joint distribution of $X_1,\ldots,X_n$ has the following density with respect to the measure ν:

$$f_X(x) = \prod_{i=1}^{n}(\alpha(\mathcal{X})+i-1)^{-1} \sum_{\text{All } p} I_{R^{-1}(p)}(x) \prod_{i=1}^{g(p)} \prod_{j=2}^{k_i(p)} (\alpha(\{Z(x)_i\})+j-1),$$

where an empty product is taken to be 1.

PROOF. Let $X = (X_1,\ldots,X_n)$. We need to show that $\Pr(X \in B) = \int_B f_X(x)d\nu(x)$ for all $B \subseteq \mathcal{X}^n$. Let $B \subseteq \mathcal{X}^n$. We will show that

$$\Pr(X \in B \cap R^{-1}(p)) = \int_{B \cap R^{-1}(p)} f_X(x)d\nu(x)$$

for every partition p, and the result will then follow by adding up finitely many terms. It is easy to see that $\nu(C) = \alpha^{g(p)}(C_p)$ for each p and each subset C of $R^{-1}(p)$, and that f_X is a function of Z. It follows that

$$\int_{B \cap R^{-1}(p)} f_X(x)d\nu(x)$$
$$= \int_{B_p} f_X(Z^{-1}(z))d\alpha^{g(p)}(z) \qquad (1.100)$$
$$= \prod_{i=1}^{n}(\alpha(\mathcal{X})+i-1)^{-1} \int_{B_p} \prod_{i=1}^{g(p)} \prod_{j=2}^{k_i(p)} (\alpha(\{z_i\})+j-1)d\alpha^{g(p)}(z).$$

---

[35]This lemma is used in Examples 1.102 and 1.103.

Fix p and write $B_p = B_1 \cup B_2$, where every coordinate of every point in B_1 has 0 α measure. The points in B_2 have at least one coordinate with positive α measure. There are at most countably many values of y such that $\alpha(\{y\}) > 0$, say they are $y_1, y_2, \ldots$. For $k = 1, \ldots, g(p)$, $i_1, \ldots, i_k$ distinct elements of $\{1, \ldots, g(p)\}$, and $\ell_1, \ldots, \ell_k$ distinct integers, let $B_{2;i_1,\ldots,i_k;\ell_1,\ldots,\ell_k}$ be the subset of B_2 in which $z_{i_t} = y_{\ell_t}$ for $t = 1, \ldots, k$ and all other coordinates of z are distinct points with 0 α measure. These sets are disjoint, and their union is B_2. On each of these sets, and on B_1, the integrand in the far right-hand side of (1.100) is constant. Hence, the far right-hand side of (1.100) can be written as

$$\prod_{i=1}^{n}(\alpha(\mathcal{X})+i-1)^{-1}\left\{\alpha^{g(p)}(B_1)\prod_{i=1}^{g(p)}\prod_{j=2}^{k_i(p)}(j-1)\right.\tag{1.101}$$

$$+\sum \alpha^{g(p)}(B_{2;i_1,\ldots,i_k;\ell_1,\ldots,\ell_k})\left(\prod_{t=1}^{k}\prod_{j=2}^{k_{i_t}(p)}(\alpha(\{y_{\ell_t}\})j-1)\right)$$

$$\left.\times\left(\prod_{i\notin\{i_1,\ldots,i_k\}}\prod_{j=2}^{k_i(p)}(j-1)\right)\right\}.$$

Now, we will show that (1.101) is the probability that $X \in B \cap R^{-1}(p)$. Let $j_1, \ldots, j_{g(p)}$ be $g(p)$ coordinates that are distinct for all x in $R^{-1}(p)$. Let

$$W = Z(X) = (X_{j_1}, \ldots, X_{j_{g(p)}}) \in B_p.$$

The first term in (1.101) is the probability that $W \subset B_1$ and that the other coordinates of X all match the coordinates they need to match in order for $R(X) = p$. Also, each of the summands in the second term in (1.101) is the probability that $W \in B_{2;i_1,\ldots,i_k;\ell_1,\ldots,\ell_k}$ and that the the other coordinates of X all match the coordinates they need to match. The sum is then the probability that $X \in B \cap R^{-1}(p)$. □

Example 1.102. As a simple example of Lemma 1.99, suppose that $\mathcal{X} = \mathbb{R}$ and α is some finite continuous (no point masses) measure. The measure ν is then the sum of the various k-dimensional product measures of α for $k = 1, \ldots, n$ over the sets where there are exactly k distinct coordinates. For example, if $n = 3$, then the partitions are

$$p_1 = \{\{1\},\{2\},\{3\}\}, \quad p_2 = \{\{1,2\},\{3\}\}, \quad p_3 = \{\{1\},\{2,3\}\},$$
$$p_4 = \{\{1,3\},\{2\}\}, \quad\quad\quad\quad p_5 = \{\{1,2,3\}\}.$$

So, $g(p_1) = 3$, and $k_i(p_1) = 1$, while $g(p_2) = g(p_3) = g(p_4) = 2$, and so on. Also,

$$\begin{aligned}R^{-1}(p_1) &= \{(x_1,x_2,x_3): x_1 \neq x_2, x_1 \neq x_3, x_2 \neq x_3\},\\ R^{-1}(p_2) &= \{(x_1,x_2,x_3): x_1 = x_2, x_1 \neq x_3\},\\ R^{-1}(p_3) &= \{(x_1,x_2,x_3): x_2 = x_3, x_1 \neq x_3\},\\ R^{-1}(p_4) &= \{(x_1,x_2,x_3): x_1 = x_3, x_1 \neq x_2\},\\ R^{-1}(p_5) &= \{(x_1,x_2,x_3): x_1 = x_2 = x_3\}.\end{aligned}$$

1.6. Infinite-Dimensional Parameters

The measure ν is α^3 on $R^{-1}(p_1)$ plus α^2 on $R^{-1}(p_2) \cup R^{-1}(p_3) \cup R^{-1}(p_4)$ plus α on $R^{-1}(p_5)$. Also,

$$f_X(x) = \frac{1}{\alpha(\mathcal{X})[\alpha(\mathcal{X})+1][\alpha(\mathcal{X})+2]} \begin{cases} 2 & \text{if } x \in R^{-1}(p_5), \\ 1 & \text{otherwise.} \end{cases}$$

To calculate the probability that X is in the unit cube B, say, we must add up five integrals, one for each partition.

$$\Pr(0 \leq X_i \leq 1, \text{ for } i = 1, 2, 3)$$
$$= \frac{1}{\alpha(\mathcal{X})[\alpha(\mathcal{X})+1][\alpha(\mathcal{X})+2]} \bigg(\alpha^3(B_{p_1}) + \alpha^2(B_{p_2}) + \alpha^2(B_{p_3}) + \alpha^2(B_{p_4})$$
$$+ 2\alpha(B_{p_5}) \bigg).$$

For concreteness, suppose that $\mathcal{X}$ is $[-1, 1]$ and α is Lebesgue measure. Then $\alpha(\mathcal{X}) = 2$ and $\alpha^{g(p)}(B_p) = 1$ for all p. So, $\Pr(X \in B) = 0.25$, substantially above the product probability $\alpha^3(B)/\alpha(\mathcal{X})^3 = 0.125$. The negative unit cube (all X_i between -1 and 0) also has probability 0.25, while the six other subcubes each has probability $1/12$.

Straightforward applications of Dirichlet process priors to one-sample problems are singularly uninteresting, except in cases in which one might use the *bootstrap* technique (see Section 5.3). There are, however, ways to make use of Dirichlet process priors in less straightforward fashion.

Example 1.103. Suppose that $\{X_n\}_{n=1}^{\infty}$ are conditionally IID with distribution P given $\mathbf{P} = P$, and we are sure that there exists a finite number θ such that $P((-\infty, \theta]) = 1$. Unless θ is known, it is not possible that $\mathbf{P}$ has Dirichlet process distribution. If we let Θ be the unknown least upper bound on the support of the X_is, we can suppose that $\mathbf{P}$ given $\Theta = \theta$ has $Dir(\alpha_\theta)$ distribution, where α_θ is a finite measure on $(-\infty, \theta]$. Let Θ have prior density f_Θ. Let $c_\theta = \alpha_\theta((-\infty, \theta])$. Suppose also that α_θ is absolutely continuous with respect to Lebesgue measure with Radon–Nikodym derivative a_θ. Using Lemma 1.99, the likelihood function for Θ after observing $X_1, \ldots, X_n$ to obtain g distinct values $y_1, \ldots, y_g$ with k_i repetitions of y_i is

$$L(\theta) = \frac{\prod_{i=1}^{g} \left(a_\theta(y_i) \prod_{j=2}^{k_i} (j-1) \right)}{\prod_{i=1}^{n} (c_\theta + i - 1)} I_{[\max\{y_1, \ldots, y_g\}, \infty)}(\theta).$$

Hence, the posterior distribution for Θ can be found. Conditional on $\Theta = \theta$, the posterior for $\mathbf{P}$ is a Dirichlet process with measure β_θ equal to α_θ plus point masses at the observed values. The marginal posterior of $\mathbf{F}$ is a mixture of Dirichlet processes. Antoniak (1974) studied mixtures of Dirichlet processes and describes many of their properties.

Example 1.103 can be somewhat deceiving if one is really trying to model data from a continuous distribution. If $g = n$ in that example, then all of the $k_i = 1$. If c_θ is the same for all θ, then the likelihood function is the same as one would obtain by modeling the data as conditionally IID given

$\Theta = \theta$ with density $a_\theta(\cdot)/c_\theta$. This is probably not the effect one thought one was achieving by using a Dirichlet process. That is, there is nothing the least bit nonparametric about the analysis one ends up performing in this situation. In fact, this phenomenon is quite general.

Lemma 1.104.[36] *Suppose that person 1 believes that $\{X_n\}_{n=1}^\infty$ are IID with a continuous distribution. For each $\theta \in \Omega$, let α_θ be a continuous finite measure with $\alpha_\theta(\mathcal{X}) = c$ for all θ. Suppose that person 2 models the data as conditionally IID given $\mathbf{P} = P$ and $\Theta = \theta$ with distribution P and that $\mathbf{P}$ given $\Theta = \theta$ has $Dir(\alpha_\theta)$ distribution. Suppose that person 3 models the data as conditionally IID given $\Theta = \theta$ with distribution α_θ/c. Assume that $\alpha_\theta \ll \eta$ for all θ. Suppose that person 2 and person 3 use the same prior distribution for Θ. Then person 1 believes that, with probability 1, for every n, person 2 and person 3 will calculate exactly the same posterior distributions for Θ given $X_1, \ldots, X_n$.*

PROOF. First, note that the density f_X in Lemma 1.99 is constant in θ for every data set that has no observed values at points where α_θ puts positive mass. Such a data set will occur with probability 1 according to person 1. Let $a_\theta = d\alpha_\theta/d\eta$. With probability 1 (according to person 1) person 2 will then have likelihood function proportional to $\prod_{i=1}^n a_\theta(x_i)$. This is the same as the likelihood function that person 3 will have. Hence, persons 2 and 3 will calculate the same posterior. □

Example 1.105. Since the Dirichlet process assigns probability 1 to discrete CDFs, it may not be considered suitable for cases in which one really wants a continuous CDF. One possibility is to model the observable data $\{X_n\}_{n=1}^\infty$ as $X_i = Y_i + Z_i$ where $\{Y_n\}_{n=1}^\infty$ are conditionally IID with CDF G given $\mathbf{G} = G$, where $\mathbf{G}$ has $Dir(\alpha)$ distribution, and $\{Z_n\}_{n=1}^\infty$ are independent of $\{Y_n\}_{n=1}^\infty$ and of $\mathbf{G}$ and of each other with a distribution having density f. The posterior distribution of $\mathbf{G}$ is not easy to obtain in this case, but a method that can be used to approximate it will be given in Section 8.5. Escobar (1988) gives an algorithm for implementing this method.

1.6.2 Tailfree Processes[+]

In this section, we introduce a second class of distributions over an infinite-dimensional space of probabilities. This time it will be possible for the random probability measure $\mathbf{P}$ not to be discrete.

Definition 1.106. Let $(\mathcal{X}, \mathcal{B})$ be a Borel space. For each integer $n > 0$, let π_n be a countable partition of $\mathcal{X}$ whose elements are in $\mathcal{B}$. Suppose that

---

[36] This lemma is used to show why it may not be sensible to use a Dirichlet process for the prior if there will also be an additional finite-dimensional parameter of interest.

[+] This section contains results that rely on the theory of martingales. It may be skipped without interrupting the flow of ideas.

π_{n+1} is a refinement of π_n for each n. Let the trivial partition be $\pi_0 = \{\mathcal{X}\}$. Let $\mathcal{C} = \cup_{n=1}^{\infty} \pi_n$. Suppose that $\mathcal{B}$ is the smallest σ-field containing $\mathcal{C}$. For each n, let $\{V_{n;B} : B \in \pi_n\}$ be a collection of nonnegative random variables such that the collections are mutually independent. For each $n \geq 1$ and each $B_1 \supseteq \cdots \supseteq B_n$ with $B_i \in \pi_i$, define

$$\mathbf{P}(B_n) = \prod_{i=1}^{n} V_{i;B_i}.$$

Then we say that the stochastic process $\mathbf{P} = \{\mathbf{P}(B) : B \in \mathcal{C}\}$. is *tailfree with respect to* $(\{\pi_n\}_{n=1}^{\infty}, \{V_{n;B} : n \geq 1, B \in \pi_n\})$. For each $n \geq 1$ and $B \in \pi_n$, define $ps(B) = C$, where $C \in \pi_{n-1}$ and $B \subset C$. Call this the *most recent superset of B*. For each $x \in \mathcal{X}$ and each n, define

$$\begin{aligned} C_n(x) &= \text{that } B \in \pi_n \text{ such that } x \in B, \\ V_n(x) &= V_{n;B_n(x)}. \end{aligned}$$

Note that the random variables in $\{V_{n;B} : B \in \pi_n\}$ do not have to be independent of each other, but they must be independent of those in $\{V_{m;B} : B \in \pi_m\}$ for $m \neq n$. The class of tailfree processes was introduced by Freedman (1963) and Fabius (1964). Also, see Ferguson (1974).

A necessary condition for $\mathbf{P}$ to be a random probability measure is that

$$\sum_{\text{All } C \text{ such that } ps(C) = B} V_{n;C} = 1. \qquad (1.107)$$

Another necessary condition is that if B_n is a union of elements of π_n for each n, $B_1 \supseteq B_2 \supseteq \cdots$, and $\cap_{n=1}^{\infty} B_n = \emptyset$, then

$$\prod_{n=1}^{\infty} \left(\sum_{\{B : B \in \pi_n, B \subseteq B_n\}} V_{n;B} \right) = 0, \text{ a.s.} \qquad (1.108)$$

These two conditions are also sufficient. (See Problem 51 on page 81.) For the remainder of this book, when we refer to a tailfree process, we will assume that it is a random probability measure.

As an example we can show that Dirichlet processes are tailfree with respect to every sequence of partitions.

Example 1.109. Let $\mathbf{P}$ have $Dir(\alpha)$ distribution. Let $\{\pi_n\}_{n=1}^{\infty}$ be a sequence of countable partitions such that π_{n+1} is a refinement of π_n for all n. We can prove that $\mathbf{P}$ is tailfree with respect to $\{\pi_n\}_{n=1}^{\infty}$. For each n and each $B \in \pi_n$, set $V_{n;B} = \mathbf{P}(B)/\mathbf{P}(ps(B))$. The fact that the collections $\{V_{n;B} : B \in \pi_n\}$ are independent follows from a well-known fact about Dirichlet distributions. (See Problem 52 on page 81.)

As another example, we can place a tailfree process distribution on the class of distributions symmetric around a point.

Example 1.110. Let $\mathcal{X} = \mathbb{R}$, and let $\pi_1 = \{(-\infty, 0), \{0\}, (0, \infty)\}$. Let $\{\pi_n^-\}_{n=2}^{\infty}$ be a sequence of nested partitions of $(-\infty, 0)$. For each $n > 1$, let π_n^+ be the partition of $(0, \infty)$ formed by the negatives of the sets in π_n^-. Let $\pi_n = \pi_n^- \cup \pi_n^+ \cup \{0\}$. So long as $V_{n;B} = V_{n;C}$ whenever $B = -C$, **P** will be symmetric around 0.

When $X \sim \mathbf{P}$ given **P** and **P** is a tailfree process, the predictive distribution of X can be computed.

Proposition 1.111. *Let* **P** *be a random probability measure that is tailfree with respect to* $(\{\pi_n\}_{n=1}^{\infty}, \{V_{n;B} : n \geq 1, B \in \pi_n\})$ *and* $X \sim \mathbf{P}$ *given* **P**. *Let* $A \in \pi_m$, *and let* $A = \cap_{i=1}^{m} B_i$, *where* $B_i \in \pi_i$ *for each* $i \leq m$. *Then the predictive probability that* $X \in A$ *is*

$$\mu_X(A) = \mathrm{E}\mathbf{P}(A) = \prod_{i=1}^{m} \mathrm{E}(V_{i;B_i}). \tag{1.112}$$

It is sometimes possible to find a density for the predictive distribution of X with respect to some measure ν of interest (like Lebesgue measure).

Lemma 1.113.[37] *Let* $(\mathcal{X}, \mathcal{B}, \nu)$ *be a σ-finite measure space. Assume that* **P** *is tailfree with respect to* $(\{\pi_n\}_{n=1}^{\infty}, \{V_{n;B} : n \geq 1, B \in \pi_n\})$. *Assume that each element of each π_n has positive ν measure. For each $x \in \mathcal{X}$, let*

$$f_n(x) = \frac{1}{\nu(C_n(x))} \prod_{i=1}^{n} \mathrm{E}(V_i(x)).$$

If $\lim_{n \to \infty} f_n(x) = f(x)$, *a.e.* $[\nu]$, *and* $\int f(x)d\nu(x) = 1$, *then* $f = d\mu_X/d\nu$.

PROOF. We need to prove that for each $B \in \mathcal{C}$, $\mu_X(B) = \int_B f(x)d\nu(x)$. The extensions to the smallest field containing $\mathcal{C}$ and the smallest σ-field containing $\mathcal{C}$ are straightforward. Let $B \in \pi_n$, and let $B \subseteq B_i \in \pi_i$ for $i = 1, \ldots, n$. Then $V_i(x) = V_{i,B_i}$ for all $x \in B$ and $i = 1, \ldots, n$. By (1.112), we have, for each $x_0 \in B$,

$$\mu_X(B) = \prod_{i=1}^{n} \mathrm{E}(V_i(x_0)) = \nu(C_n(x_0))f_n(x_0) = \int_B f_n(x)d\nu(x),$$

since $f_n(x) = f_n(x_0)$ for all $x \in B$ and $B = C_n(x_0)$. For $k > n$, write $B = \cup_{\alpha \in A} D_\alpha$ as the partition of B by elements of π_k. Since f_k is constant on each D_α (call the value $f_k(x_\alpha)$ for $x_\alpha \in D_\alpha$), we can write

$$\int_B f_k(x)d\nu(x) = \sum_{\alpha \in A} \int_{D_\alpha} f_k(x)d\nu(x) = \sum_{\alpha \in A} \prod_{i=1}^{k} \mathrm{E}(V_i(x_\alpha))$$
$$= \sum_{\alpha \in A} \mu_X(D_\alpha) = \mu_X(B).$$

---

[37]This lemma is used in Example 1.114.

Hence, we have $\int_B f_k(x)d\nu(x) = \mu_X(B)$ for all $k \geq n$. So,

$$\lim_{k \to \infty} \int_B f_k(x)d\nu(x) = \mu_X(B).$$

It follows from Scheffé's theorem B.79 that

$$\lim_{k \to \infty} \int_B f_k(x)d\nu(x) = \int_B f(x)d\nu(x). \qquad \square$$

Example 1.114. Suppose that ν is a finite measure and, for each n and B, $E(V_{n;B}) = \nu(B)/\nu(ps(B))$. In the notation of Lemma 1.113, $f_n(x) = 1/\nu(\mathcal{X})$ for all n and x. In this case, $\mu_X = \nu/\nu(\mathcal{X})$, and the density is constant. In fact, this gives a convenient way to force a tailfree process to have a desired predictive distribution for X.

Tailfree processes are conjugate in the sense that the posterior is tailfree if the prior is tailfree.

Theorem 1.115. *Let $\mathbf{P}$ be a random probability measure that is tailfree with respect to $(\{\pi_n\}_{n=1}^\infty, \{V_{n;B} : n \geq 1, B \in \pi_n\})$ and $X \sim \mathbf{P}$ given $\mathbf{P}$, then $\mathbf{P}$ given X is tailfree with respect to $(\{\pi_n\}_{n=1}^\infty, \{V_{n;B} : n \geq 1, B \in \pi_n\})$.*

PROOF. Fix k and $n_1 < \ldots < n_k$. Let V^i, for $i = 1, \ldots, k$, be a finite (say, with size s_i) collection of the elements of π_{n_i}, and let F_{V^i} denote their joint CDF. Let $F_{V^i|X}(\cdot|x)$ denote the conditional CDF of V^i given $X = x$. Let $A \in \mathcal{B}$, and let $C_i \subseteq \mathbb{R}^{s_i}$, for $i = 1, \ldots, k$. We must show that

$$\Pr(X \in A, V^i \in C_i, \text{ for } i = 1, \ldots, k) = \int_A \prod_{i=1}^k F_{V^i|X}(C_i|x)d\mu_X(x), \tag{1.116}$$

where μ_X is given by (1.112). If we can prove (1.116) for all $A \in \mathcal{C}$, it will be true for all $A \in \mathcal{B}$ by Theorem A.26.

First, we find $F_{V^i|X}$. By definition, for $C \in \mathbb{R}^{s_i}$, $F_{V^i|X}(C|\cdot)$ is any measurable function h such that

$$\int_A h(x)d\mu_X(x) = \Pr(X \in A, V^i \in C),$$

for all $A \in \mathcal{B}$. Once again, the equation need only hold for all $A \in \mathcal{C}$. We propose the following function:

$$h(x) = \frac{E(I_C(V^i)V_{n_i}(x))}{E(V_{n_i}(x))}, \tag{1.117}$$

where $V_m(x)$ is defined in Definition 1.106 to be the random variable corresponding to that element of partition π_m which contains x. Note that h is constant on each element of π_{n_i}. We find it convenient to let $h(B)$ stand for that constant value if $x \in B \in \pi_{n_i}$. Let $A = B_n \in \pi_n$, let $m = \max\{n_i, n\}$, and define, for $j = 1, \ldots, m$,

$$U_j = \begin{cases} V_{j;B_j} & \text{if } j \le n, j \ne n_i, A \subseteq B_j \in \pi_j, \\ 1 & \text{if } j > n, j \ne n_i, \\ (V_{n_i;B_{n_i}}, V^i) & \text{if } j = n_i \le n, A \subseteq B_{n_i} \in \pi_{n_i}, \\ (B^*, V^i) & \text{if } j = n_i > n, \end{cases}$$

$$D_j = \begin{cases} \mathbb{R} & \text{if } j \le n, j \ne n_i, A \subseteq B_j \in \pi_j, \\ \{1\} & \text{if } j > n, j \ne n_i, \\ \mathbb{R} \times C & \text{if } j = n_i, \end{cases}$$

where B^* is the union of all elements of π_{n_i} which are subsets of A. If $j = n_i \le n$, it is possible that the first coordinate of U_j is repeated later in the vector. With these definitions, it is clear that

$$\Pr(X \in A, V^i \in C)$$
$$= \Pr(X \in A, U_j \in D_j, j = 1, \ldots, m)$$
$$= \int_{D_1} \cdots \int_{D_m} u_{n_i,1} \prod_{j \ne n_i}^{m} u_j dF_{U_m}(u_m) \cdots dF_{U_1}(u_1)$$
$$= \mathrm{E}(U_{n_i,1} I_C(V^i)) \prod_{j \ne i} \mathrm{E}(U_j)$$
$$= \begin{cases} \mathrm{E}(V_{n_i, B_{n_i}} I_C(V^i)) \prod_{j \ne n_i} \mathrm{E}(V_{j;B_j}) & \text{if } n_i \le n, \\ \sum_{\substack{B \text{ such} \\ \text{that } B \subseteq A}} \mathrm{E}(I_C(V^i) V_{n_i;B}) \prod_{j=1}^{n_i-1} \mathrm{E}(V_{j;B_j(B)}) & \text{if } n_i > n. \end{cases}$$

If we let $h(x)$ be as in (1.117), then

$$\int_A h(x) d\mu_X(x) = \sum_{B \in \pi_{n_i}} \int_{A \cap B} h(x) d\mu_X(x). \qquad (1.118)$$

If $n_i \le n$, there is only one term in the sum in (1.118). It is the term with $B = B_{n_i}$ such that $A \subseteq B_{n_i}$, and the integral equals

$$h(B_{n_i}) \mu_X(A) = \frac{\mathrm{E}(I_C(V^i) V_{n_i;B_{n_i}})}{\mathrm{E}(V_{n_i;B_{n_i}})} \prod_{j=1}^{n} \mathrm{E}(V_{j;B_j})$$
$$= \mathrm{E}(I_C(V^i) V_{n_i;B_{n_i}}) \prod_{j \ne n_i} \mathrm{E}(V_{j;B_j}),$$

which is what we needed to show. Similarly, if $n_i > n$, the only terms in the sum in (1.118) which appear are those for which $B \subseteq A$, and A is the union of these sets. The sum becomes

$$\sum_{B \text{ such that } B \subseteq A} h(B) \mu_X(B) \qquad (1.119)$$

$$= \sum_{B \text{ such that } B \subseteq A} \frac{\mathrm{E}(I_C(V^i)V_{n_i;B})}{\mathrm{E}(V_{n_i;B})} \prod_{j=1}^{n_i} \mathrm{E}(V_{j;B_j(B)}),$$

where $B_j(B) \in \pi_j$ and $B \subseteq B_j(B)$ for $j = 1, \ldots, n_i$. The right-hand side of (1.119) can be written as

$$\sum_{B \text{ such that } B \subseteq A} \mathrm{E}(I_C(V^i)V_{n_i;B}) \prod_{j=1}^{n_i-1} \mathrm{E}(V_{j;B_j(B)}),$$

and it follows that h is a version of $F_{V^i|X}(C|\cdot)$.

To prove (1.116), let $A \in \pi_n$. If $n < n_k$, break A into its intersections with all elements of π_{n_k} and then add up the sides of (1.116) over the disjoint intersections to see that (1.116) holds. Hence we need only prove (1.116) if $n \geq n_k$. As before, let B_j be the element of partition π_j such that $A \subseteq B_j$ for $j = 1, \ldots, n$. Since $n \geq n_k$, A is a subset of one and only one of the partition elements for each π_{n_i} partition. Define, for $j = 1, \ldots, n$,

$$W_j = \begin{cases} B_j & \text{if } j \notin \{n_1, \ldots, n_k\}, \\ (V_{j;B_j}, V^i) & \text{if } j = n_i, \end{cases}$$

$$D_j = \begin{cases} [0,1] & \text{if } j \notin \{n_1, \ldots, n_k\}, \\ \mathbb{R} \times C_i & \text{if } j = n_i. \end{cases}$$

In what follows, $W_{j,1}$ will be the first coordinate of W_j. It is easy to see that

$$\Pr(X \in A, V^i \in C_i, \text{ for } i = 1, \ldots, k)$$
$$= \Pr(X \in A, W_j \in D_j, \text{ for } j = 1, \ldots, n)$$
$$= \int_{D_1} \cdots \int_{D_n} \prod_{j=1}^{n} w_{j,1} dF_{W_n}(w_n) \cdots dF_{W_1}(w_1)$$
$$= \prod_{j=1}^{n} \mathrm{E}(W_{j,1} I_{D_j}(W_j))$$
$$= \prod_{j \notin \{n_1, \ldots, n_k\}} \mathrm{E}(V_{j;B_j}) \prod_{i=1}^{k} \mathrm{E}(V_{n_i;B_{n_i}} I_{C_i}(V^i))$$
$$= \prod_{j=1}^{n} \mathrm{E}(V_{j;B_j}) \prod_{i=1}^{k} \frac{\mathrm{E}(V_{n_i;B_{n_i}} I_{C_i}(V^i))}{\mathrm{E}(V_{n_i;B_{n_i}})}$$
$$= \int_A \prod_{i=1}^{k} F_{V^i|X}(C_i|x) d\mu_X(x). \qquad \square$$

It is not difficult to check that the posterior distribution found in Theorem 1.115 still satisfies the two conditions (1.107) and (1.108). Hence, it is a regular conditional distribution.

Example 1.120. Suppose that $\mathcal{X} = \mathbb{R}$ and each set B is an interval. In the proof of Theorem 1.115, suppose that one of the V^i is just the single coordinate $V_{n_i}(x)$. Also, suppose that this $V_{n_i}(x)$ has a prior density $f(v)$ with respect to some measure (like Lebesgue measure). Then (1.117) says that the posterior density of $V_{n_i}(x)$ (aside from a normalizing constant) is $vf(v)$. If V' is another random variable from partition n_i, and if $(V', V_{n_i}(x))$ have joint prior density $g(v', v)$, then their joint posterior is proportional to $vg(v', v)$.

Tailfree processes are more general than Dirichlet processes. In fact, they can be continuous or even absolutely continuous with respect to nondiscrete measures with probability one. Mauldin, Sudderth, and Williams (1992) prove a result giving conditions under which **P** is continuous with probability 1. (See Problem 57 on page 81.) Kraft (1964) and Métivier (1971) proved theorems that said that if $\mathcal{X} = (0, 1]$ and π_n has 2^n sets, and if a few other conditions hold, then **P** has a density with respect to Lebesgue measure with probability 1. We generalize these latter theorems to arbitrary tailfree processes.

Theorem 1.121. *Let $(\mathcal{X}, \mathcal{B}, \nu)$ be a probability space. Assume that **P** is tailfree with respect to $(\{\pi_n\}_{n=1}^{\infty}, \{V_{n;B} : n \geq 1, B \in \pi_n\})$. Assume that each element of each π_n has positive ν measure and that, for all n and $B \in \pi_n$, $\mathrm{E}(V_{n;B}) = \nu(B)/\nu(ps(B))$. For each n and each $x \in \mathcal{X}$, define*

$$\mathbf{f}_n(x) = \frac{1}{\nu(C_n(x))} \prod_{i=1}^{n} V_i(x).$$

If $\sup_n \int_{\mathcal{X}} \mathrm{E}[\mathbf{f}_n^2(x)] d\nu(x) < \infty$, then, with probability 1,

1. $\lim_{n \to \infty} \mathbf{f}_n(x) = \mathbf{f}(x)$ *exists and is finite a.e.* $[\nu]$, *and*

2. $\mathbf{P}(A) = \int_A \mathbf{f}(x) d\nu(x)$, *for all* $A \in \mathcal{B}$.

Before proving Theorem 1.121, we should say a little about its statement. The condition that $\mathrm{E}(V_{n;B}) = \nu(B)/\nu(ps(B))$ is equivalent to saying that $\nu = \mu_X$ in (1.112). (See Problem 50 on page 81.) The formula for $\mathbf{f}_n(x)$ is nothing more than the formula for $\mathbf{P}(C_n(x))/\nu(C_n(x))$. Hence, $\mathbf{f}_n$ is the density (with respect to ν) corresponding to an approximation to **P** which ignores the fine detail on sets in partitions past n. In other words, $\mathbf{f}_n$ is constant on all sets in π_n and is a density for **P** restricted to the smallest σ-field containing all sets in partitions up to n. Since $\mathrm{E}(\mathbf{f}(x)) = 1$ for all n and x, the theorem says that if there is not too much variation in the approximate densities, then the approximate densities converge to a density for **P**.

PROOF OF THEOREM 1.121.[38] Consider the probability space $(S \times \mathcal{X}, \mathcal{A} \otimes \mathcal{B}, \mu \times \nu)$. For part (1), let $\mathcal{F}_n$ be the product σ-field of the σ-field generated

---

[38] The proof makes use of martingale theory.

1.6. Infinite-Dimensional Parameters 67

by $\{V_{n;B} : B \in \pi_i, i \leq n\}$ with the σ-field $\mathcal{B}$. Then, as a function from $S \times \mathcal{X}$ to $\mathbb{R}$, $\mathbf{f}_n$ is measurable with respect to $\mathcal{F}_n$. Also, since the $V_i(x)$ are independent for fixed x,

$$E(\mathbf{f}_n(x)|\mathcal{F}_{n-1}) = \frac{1}{\nu(C_{n-1}(x))} \prod_{i=1}^{n-1} V_i(x) \left(EV_n(x) \frac{\nu(C_{n-1}(x))}{\nu(C_n(x))} \right) = \mathbf{f}_{n-1}(x).$$

So, the stochastic process $\{(\mathbf{f}_n(x), \mathcal{F}_n)\}_{n=1}^{\infty}$ is a martingale. Since

$$\int \mathbf{f}_n(x) d\mu \times \nu(s, x) = 1 \text{ for all } n,$$

the Martingale convergence theorem B.117 implies that $\lim_{n \to \infty} \mathbf{f}_n = \mathbf{f}$ exists and is finite a.s. $[\mu \times \nu]$, which means (in terms of the original probability space) that with probability 1, $\lim_{n \to \infty} \mathbf{f}_n(x) = \mathbf{f}(x)$ exists and is finite a.e. $[\nu]$.

For part (2), we first show that the sequence $\{\mathbf{f}_n\}_{n=1}^{\infty}$ is uniformly integrable with respect to $\mu \times \nu$:

$$\int_{\{(s,x):\mathbf{f}_n(x)>m\}} \mathbf{f}_n(x) d\mu \times \nu(s, x)$$
$$= \int_{\mathcal{X}} \int_S \mathbf{f}_n(x) I_{(m,\infty)}(\mathbf{f}_n(x)) d\mu(s) d\nu(x)$$
$$< \int_{\mathcal{X}} \int_S \frac{1}{m} \mathbf{f}_n^2(x) d\mu(s) d\nu(x)$$
$$= \frac{1}{m} \int_{\mathcal{X}} E[\mathbf{f}_n^2(x)] d\nu(x), \tag{1.122}$$

where the first equation follows from Tonelli's theorem A.69, and the inequality follows since $I_{(m,\infty)}(\mathbf{f}_n(x)) < \mathbf{f}_n(x)/m$. By assumption, the supremum over n of the last expression in (1.122) is a finite number divided by m, which goes to 0 with m, so the sequence is uniformly integrable.

Next, we prove that $\mathbf{f}$ is a density with respect to ν with probability 1. By Theorem A.60, we have

$$\lim_{n \to \infty} \int \mathbf{f}_n(x) d\mu \times \nu(x) = \int \mathbf{f}(x) d\mu \times \nu(x). \tag{1.123}$$

Since the left-hand side is 1, we have that $\mathbf{f}$ is integrable. It follows that if $A \in \mathcal{A} \otimes \mathcal{B}$, then $I_A(s, x)|\mathbf{f}_n(x) - \mathbf{f}(x)|$ is uniformly integrable. Let $A = B \times \mathcal{X}$, where

$$B = \left\{ s : \int_{\mathcal{X}} \mathbf{f}(x) d\nu(x) > 1 \right\}.$$

Then $\lim_{n \to \infty} \int_A \mathbf{f}_n(x) d\mu \times \nu(s, x) = \int_A \mathbf{f}(x) d\mu \times \nu(s, x)$. The left-hand side is just $\mu(B) = \Pr(B)$ since $\mathbf{f}_n$ is a density. But the right-hand side is greater than $\mu(B)$ if $\mu(B) > 0$, so the integral of $\mathbf{f}$ is at most 1 with probability 1. But $\int \mathbf{f}(x) d\mu \times \nu(s,x) = 1$ from (1.123), so $\int_{\mathcal{X}} \mathbf{f}(x) d\nu(x) = 1$ with probability 1. It follows from Scheffé's theorem B.79 that part (2) is true. □

There is a convenient way to check the condition $\sup_n \int_{\mathcal{X}} \mathrm{E}[\mathbf{f}_n^2(x)] d\nu(x) < \infty$ in Theorem 1.121.

Lemma 1.124. If $\sum_{n=1}^{\infty} \sup_{B \in \pi_n} \mathrm{Var}(V_{n;B})/(\mathrm{E}V_{n;B})^2 < \infty$, then

$$\sup_n \int_{\mathcal{X}} \mathrm{E}[\mathbf{f}_n^2(x)] d\nu(x) < \infty.$$

PROOF. Since the set in π_n to which x belongs is not random and $\mathrm{E}(V_n(x)) = \mathrm{E}(V_{n;B})$ for all $x \in B$, we have

$$\begin{aligned}
0 &\leq \sum_{n=1}^{\infty} \sup_x \left\{ \frac{\mathrm{E}[V_n^2(x)]}{(\mathrm{E}V_n(x))^2} - 1 \right\} \\
&= \sum_{n=1}^{\infty} \sup_x \frac{\mathrm{Var} V_n(x)}{(\mathrm{E}V_n(x))^2} \\
&= \sum_{n=1}^{\infty} \sup_{B \in \pi_n} \frac{\mathrm{Var}(V_{n;B})}{(\mathrm{E}V_{n;B})^2} < \infty.
\end{aligned}$$

Note that $\log(y) < y - 1$ for all $y > 0$. With $y = \sup_x \mathrm{E}[V_n^2(x)]/(\mathrm{E}V_n(x))^2$, we get, for each n,

$$\sup_x \log \frac{\mathrm{E}[V_n^2(x)]}{(\mathrm{E}V_n(x))^2} < \sup_x \frac{\mathrm{E}[V_n^2(x)]}{(\mathrm{E}V_n(x))^2} - 1,$$

hence

$$\begin{aligned}
\log \left(\prod_{n=1}^{\infty} \sup_x \frac{\mathrm{E}[V_n^2(x)]}{(\mathrm{E}V_n(x))^2} \right) &= \sum_{n=1}^{\infty} \sup_x \log \frac{\mathrm{E}[V_n^2(x)]}{(\mathrm{E}V_n(x))^2} \\
&\leq \sum_{n=1}^{\infty} \sup_x \left\{ \frac{\mathrm{E}[V_n^2(x)]}{(\mathrm{E}V_n(x))^2} - 1 \right\} < \infty.
\end{aligned}$$

Since the $Z_n(B)$ are independent, we get that $\prod_{i=1}^{n} \mathrm{E}V_n(x)^2/(\mathrm{E}V_n(x))^2$ equals $\mathrm{E}\mathbf{f}_n^2(x)$. Hence, $\sup_x \mathrm{E}\mathbf{f}_n^2(x)$ is integrable. □

The following simple corollaries follow from Tonelli's theorem A.69.

Corollary 1.125. Let $X \sim \mathbf{P}$ given $\mathbf{P}$. If $\mathbf{P} \ll \nu$ with probability 1, then the predictive distribution of X has density with respect to ν equal to the mean of $d\mathbf{P}/d\nu$.

Corollary 1.126. If $\{X_n\}_{n=1}^{\infty}$ are conditionally IID with distribution $\mathbf{P}$ given $\mathbf{P}$ and $\mathbf{P} \ll \nu$ with probability 1, then conditional on $X_1, \ldots, X_n$ $\mathbf{P} \ll \nu$ with probability 1, a.s. with respect to the joint distribution of $X_1, \ldots, X_n$.

At this point, we will introduce one special class of tailfree processes which includes Dirichlet processes as special cases but also includes cases that satisfy the conditions of Theorem 1.121. The class is called *Polya tree distributions*. A good introduction to these processes is contained in the papers by Mauldin, Sudderth, and Williams (1992) and Lavine (1992). [See also Mauldin and Williams (1990)].

Definition 1.127. Let **P** be tailfree with respect to $(\{\pi_n\}_{n=1}^{\infty}, \{V_{n;B} : n \geq 1, B \in \pi_n\})$ and suppose that

- for each $n \geq 0$ and each $B \in \pi_n$, there are exactly k sets in π_{n+1}, $B_1(B), \ldots, B_k(B)$ such that $B = ps(B_i(B))$,

- for each $n \geq 0$ and each $B \in \pi_n$, the joint distribution of $\{V_{n+1;B_i(B)} : i = 1, \ldots, k\}$ is Dirichlet, and they are independent for different $B \in \pi_n$,

then **P** has *Polya tree distribution*.

Note that each partition π_n has k^n elements. It is possible to allow some of the partition elements to be $\emptyset$ so that there are fewer than k^n nonempty elements of π_n, but then the Dirichlet distributions would have to be partially degenerate in the sense that some coordinates would have to be 0 with probability 1.

The posterior distribution of a Polya tree process **P** given an observation X can be determined by examining the step in the proof of Theorem 1.115 in which the posterior is given, namely (1.117). For each n and each $x \in \mathcal{X}$, let

$$W_n(x) = (V_{n;B_1(C_{n-1}(x))}, \ldots, V_{n;B_k(C_{n-1}(x))}), \quad (1.128)$$

in the notation of Definitions 1.127 and 1.106. This is the vector of random variables for partition π_n corresponding to the subsets of $C_{n-1}(x)$. According to (1.117), the posterior distribution of $W_n(x)$ given $X = x$ has a density with respect to the prior distribution equal to $v_x/\mathrm{E}(V_n(x))$, where v_x is the dummy variable for the coordinate corresponding to $V_n(x)$, which is a coordinate of $W_n(x)$. If $Dir_k(a_{n,1}(x), \ldots, a_{n,k}(x))$ is the prior distribution of $W_n(x)$, then the posterior is Dirichlet with the ith parameter equal to

$$b_{n,i}(x) = a_{n,i}(x) + I_{B_i(C_{n-1}(x))}(x). \quad (1.129)$$

For all of the other V random variables corresponding to partition π_n, the posterior distributions are the same as the prior distributions. In summary, the posterior distributions of the V random variables are the same as the priors for all Vs corresponding to sets such that x is not in the most recent superset. For those sets such that x is in the most recent superset, the distribution of the vector of V random variables is Dirichlet with the same parameters as in the prior except for the set in which x lies, whose parameter is one higher than in the prior. Note that this is the same thing that happens in the Dirichlet process. The difference is that, for a Dirichlet process, the above argument applies to every set, and every set is in the first partition π_1 for the Dirichlet process. Given this description of the posterior, we can use Corollary 1.125 to construct the predictive density of a future observation X_{n+1} given the observed values of $X_1, \ldots, X_n$.

Example 1.130. Let **P** have a Polya tree distribution. Suppose that the conditions of Theorem 1.121 hold and that $\mathbf{P} \ll \nu$ with probability 1. Let $x_1, \ldots, x_m$ be the observed values of the first m random quantities. For each $x \in \mathcal{X}$ and each n such that x is not in the same element of π_{n-1} as one of the x_i,

$$E[V_n(x)|X_1 = x_1, \ldots, X_m = x_m] = E[V_n(x)] = \frac{\nu(C_n(x))}{\nu(ps(C_n(x)))}.$$

It follows that for all such n and x,

$$E\mathbf{f}_n(x) = \frac{1}{\nu(C_r(x))} \prod_{i=1}^{r} g_i(x), \qquad (1.131)$$

where r is the first integer such that x is not in the same element of π_{r-1} with any of the x_i, and $g_i(x)$ is the posterior mean of $V_i(x)$ given the observed data. It follows from Tonelli's theorem A.69 that $E\mathbf{f}(x)$ equals (1.131).

We can actually find an explicit formula for $g_i(x)$. Using the same notation as above, suppose that $V_n(x)$ is coordinate $i_n(x)$ of $W_n(x)$ and that $W_n(x)$ has $Dir_k(a_{n,1}(x), \ldots, a_{n,k}(x))$ prior distribution. Then, the posterior distribution of $V_n(x)$ is $Beta(a, b)$ with

$$a = a_{n,i_n(x)} + \sum_{j=1}^{m} I_{C_n(x)}(x_j),$$

$$b = \sum_{\ell \neq i_n(x)} a_{n,\ell}(x) + \sum_{j=1}^{m} I_{C_{n-1}(x) \setminus C_n(x)}(x_j).$$

That is, the first parameter of the posterior beta distribution equals the prior parameter plus the number of observations that are in the same partition set as x. The second parameter of the posterior beta distribution equals the prior parameter plus the number of observations that were in the same partition set as x in the most recent partition but now are not in the same partition set as x. It follows that $g_n(x) = a/(a+b)$.

For the special case of $\mathcal{X} = [0, 1]$, $k = 2$ and $a_{n,1}(x) = c_1$, $a_{n,2}(x) = c_2$ for all n and x, Dubins and Freedman (1963) show that **P**, although continuous, has no density. That is, the distribution is not absolutely continuous with respect to Lebesgue measure. For Polya trees with $a_{n,i}(x) = c_n$ for all n, i, and x, $\text{Var}(V_n(x)) = (k-1)/[k^2(c_n + 1)]$. Lemma 1.124 says that if $\sum_{n=1}^{\infty} 1/c_n$ converges, then **P** is absolutely continuous with respect to ν with probability 1. If ν is absolutely continuous with respect to Lebesgue measure, this gives us an easy way to construct Polya tree processes that have densities with respect to Lebesgue measure.

Example 1.132. Let $\{X_n\}_{n=1}^{\infty}$ be an exchangeable sequence such that the prior marginal distribution of each X_i is $N(0, 100)$. Let $Y_i = \Phi(X_i/10)$, where Φ is the $N(0,1)$ CDF. The prior marginal distribution of the Y_i is $U(0,1)$. Suppose that we model the Y_i as conditionally IID given $\mathbf{P}$ and that $\mathbf{P}$ has a Polya tree process distribution on $[0,1]$ with $k = 2$ and $a_{n,i}(x) = n^2/2$ for all n, i, and x. This is a special case of Example 1.114 on page 63, hence each Y_i has marginal distribution $U(0,1)$. Fifty observations $X_1, \ldots, X_{50}$ were simulated from a Laplace distribution $Lap(1,1)$, which does not look much like the prior marginal distribution $N(0,100)$. The posterior mean of the density of $\mathbf{P}(10\Phi^{-1})$ was computed and is plotted in Figure 1.133 together with the prior marginal mean of the density and a histogram of the data values. The posterior mean of the density of $\mathbf{P}(10\Phi^{-1})$ is high where the data values are close together, as is to be expected. The posterior mean smooths out some of the ups and downs in the histogram, especially those in the tails. The reason it smooths the tails a bit more than the center of the distribution is that the partition sets in the tail which do and do not contain observations only belong to partitions π_n for relatively large n. A few observations do not have much impact on the posterior distribution of $V_{n;B}$ for large n because the prior is $Beta(n^2/2, n^2/2)$. In the center of the distribution, however, the different partition sets belong to the same π_n for smaller values of n.

It is interesting to note that there is a similarity between the posterior distributions from Polya tree processes and Dirichlet processes. Let $\mathbf{P}_1$ have $Dir(\alpha)$ distribution, and let $\mathbf{P}_2$ have Polya tree process with $a_{1,i} = \alpha(B_{1,i})$ for $i = 1, \ldots, k$. Then, it is easy to check that for every data set, the posterior distributions of $\mathbf{P}_j(B_{1,1}), \ldots, \mathbf{P}_j(B_{1,k})$ are the same for $j = 1, 2$. That is, for sets in the first partition π_1, the Polya tree process looks just like a Dirichlet process. For Dirichlet processes, two disjoint sets

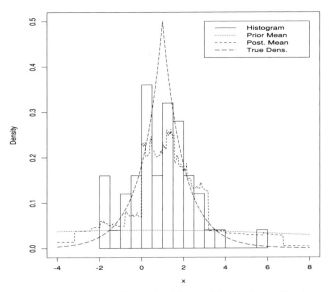

FIGURE 1.133. Posterior Mean of Polya Tree Density

have posterior distributions that depend in no way on where the two sets are located relative to each other. The same is true for elements of the first partition in a Polya tree process. For Polya tree processes, two sets in the nth partition will have their probabilities more closely related when they share more superset partition sets. For example, two subsets of $B_{1,1}$ will be more closely related than a subset of $B_{1,1}$ and a subset of $B_{1,2}$. Two subsets of $B_{2,1} \subseteq B_{1,1}$ will be more closely related than a subset of $B_{2,1}$ and a subset of $B_{2,2} \subseteq B_{1,1}$, even though both are subsets of $B_{1,1}$.

One potential problem with tailfree process priors is the dependence on the sequence of partitions. One consequence of this dependence is easily seen in Figure 1.133. The tall vertical lines in the posterior mean plot occur at boundaries of sets in early partitions. The following example explores this in more detail.

Example 1.134. Suppose that $\mathcal{X} = [0,1]$ and we use a Polya tree prior with $k = 2$ and $a_{n,i}(x) = n^2/2$ for all n, i, and x. Suppose that $X_1 = 0.49$. The predictive density of X_2 at the value $x = 0.51$ is calculated as in Example 1.130 on page 70. It is

$$f_{X_2|X_1}(0.51|0.49) = \frac{0.5}{2}\frac{1}{0.5} = 0.5.$$

On the other hand, the predictive density of X_2 at $x = 0.47$ is

$$f_{X_2|X_1}(0.47|0.49) = \left(\prod_{n=1}^{5} \frac{\frac{n^2}{2}+1}{n^2+1}\right) \frac{\frac{6^2}{2}}{6^2+1} \frac{1}{\frac{1}{2^6}} = 2.1183.$$

Note that in each of these cases, the proposed value of X_2 differs from the observed X_1 by 0.02 and they are all in the vicinity of 0.5, and yet the first predictive density is so much smaller than the second. The reason is the following. In the first case, the two data values share no partition sets in common, not even in π_1. In the second case, the two data values share the same partition set for the first five partitions. In symbols, $C_1(0.49) \neq C_1(0.51)$ while $C_n(0.47) = C_n(0.49)$ for $n = 1, \ldots, 5$. Sharing partition sets is what makes predictive densities large.

One way to reduce the effect of the problem illustrated in Example 1.134 is to use a mixture of tailfree priors with partitions that have no common boundaries.

Example 1.135 (Continuation of Example 1.134). Suppose that we use a half-and-half mixture of two Polya tree priors with $k = 2, 3$ and $a_{n,i}(x) = n^2/k$. After some tedious algebra, one calculates the two predictive densities as

$$f_{X_2|X_1}(0.51|0.49) = 1.8312,$$
$$f_{X_2|X_1}(0.47|0.49) = 2.3191.$$

The reason that the first density is now almost as high as the second is that 0.49 and 0.51 appear together in one more partition set in the $k = 3$ prior than do 0.49 and 0.47. The densities are higher than with $k = 2$ alone because a prior with larger k tends to let the density track the data more. With values so close together as the ones in this example, the prior with $k = 3$ has very high posterior mean of $\mathbf{f}(x)$ for x near 0.49 and very low mean for x not near 0.49. (Using the $k = 3$ prior alone, the two predictive density values would have been 3.1624 and 2.5200, respectively.)

1.7 Problems

Throughout this text problems are given and the following type of expression is often used: "Suppose that (some random quantities) are conditionally independent given $\Theta = \theta$." This will mean that, for all θ in some parameter space (implicit or explicit), the random quantities are conditionally independent given $\Theta = \theta$ with some distributions to be specified later in the problem. Some of the more challenging problems throughout the text have been identified with an asterisk (*) after the problem number.

Section 1.2:

1. Let X_1, X_2, X_3 be random variables whose joint distribution is given by
$$\Pr(X_1 = 1, X_2 = 1, X_3 = 0) = \Pr(X_1 = 1, X_2 = 0, X_3 = 1)$$
$$= \Pr(X_1 = 0, X_2 = 1, X_3 = 1) = \frac{1}{3}.$$

   (a) Prove that X_1, X_2, X_3 are exchangeable.
   (b) Prove that if $X_4 \in \{0, 1\}$ is another random variable, then it cannot be that X_1, X_2, X_3, X_4 are exchangeable.

2. For each positive integer n, let F_n be the joint CDF of n random variables. Suppose that the following two conditions hold:
   - The sequence of n-dimensional CDFs $\{F_n\}_{n=1}^{\infty}$ is consistent. (See Definition B.132 on page 652.)
   - For each n, each n-tuple $(x_1, \ldots, x_n)$, and each permutation $(y_1, \ldots, y_n)$ of $(x_1, \ldots, x_n)$, $F_n(x_1, \ldots, x_n) = F_n(y_1, \ldots, y_n)$.

   Prove that there is a sequence of random variables $\{X_n\}_{n=1}^{\infty}$ that are exchangeable and such that F_n is the joint CDF of $X_1, \ldots, X_n$ for every n.

3. Suppose that $\{X_n\}_{n=1}^{\infty}$ are exchangeable. Let $Y_i = X_{n+i}$ for $i = 1, 2, \ldots$. Show that $\{Y_n\}_{n=1}^{\infty}$ are exchangeable conditional on $X_1, \ldots, X_n$.

4. Suppose that $\{X_n\}_{n=1}^{\infty}$ are conditionally IID given Y. Prove that they are exchangeable.

5. Suppose that $\{X_n\}_{n=1}^{\infty}$ are IID $N(\mu, 1)$ conditional on $M = \mu$, and $M \sim N(0, 1)$. Find the joint distribution of every subset of size k of the X_i and show that the X_i are exchangeable. Also, find the conditional distribution of $\{X_{n+k}\}_{k=1}^{\infty}$ given $X_1 = x_1, \ldots, X_n = x_n$.

6. In Example 1.15 on page 9, prove that the limit (as $n \to \infty$) of the empirical probability measures of $X_1, \ldots, X_n$ is the Dirichlet distribution $Dir_k(u_1, \ldots, u_k)$.

7. Let $\{Y_n\}_{n=1}^{\infty}$ be IID random variables with CDF F. Let Z be independent of $\{Y_n\}_{n=1}^{\infty}$ with CDF G, and let $X_n = Y_n + Z$ for every n.
   (a) Prove that $\{X_n\}_{n=1}^{\infty}$ are exchangeable.
   (b) Write the joint CDF of $(X_1, \ldots, X_n)$ in terms of F and G.

Section 1.3:

8. Let $X_1, \ldots, X_m$ be numerical characteristics of m individuals in a finite population. Suppose that we are interested in $S = \sum_{i=1}^{m} X_i$, the population total. We model the X_i as exchangeable as follows. Let Θ be a parameter such that conditional on $\Theta = \theta$, the X_i are IID with $Exp(\theta)$ distribution and Θ has $\Gamma(a, b)$ distribution. Suppose that we observe $X_1 = x_1, \ldots, X_n = x_n$ for some $n < m$. Find the predictive distribution of S given this data. Also, find the mean of this predictive distribution.

9. Let Θ be a parameter with parameter space (Ω, τ), and let $f_{X|\Theta}(x|\theta) = dP_\theta/d\nu$ for every θ. Let μ_Θ be a prior on (Ω, τ). Let Q be the joint distribution of (X, Θ). Show that $f_{X|\Theta} = dQ/d\nu \times \mu_\Theta$, a.s. $[\nu \times \mu_\Theta]$. This says that, for every prior distribution, we can find a version of $f_{X|\Theta}$ which is jointly measurable in (x, θ).

10. Prove that the formula on the right-hand side of (1.37) on page 18 is the same as
$$\frac{f_{X_1, \ldots, X_{n+k}}(x_1, \ldots, x_{n+k})}{f_{X_1, \ldots, X_n}(x_1, \ldots, x_n)}.$$

11. Suppose that for every $m = 1, 2, \ldots$, $f_{X_1, \ldots, X_m}(x_1, \ldots, x_m)$ equals
$$\begin{cases} \frac{1}{10^m} \sum_{i=0}^{10} a_i i^x (10-i)^{m-x} & \text{if all } x_i \in \{0, 1\}, \\ 0 & \text{otherwise,} \end{cases}$$
where $x = \sum_{j=1}^{m} x_j$, and the numbers a_i are nonnegative and add to 1. Let $\Theta = \lim_{n \to \infty} \sum_{i=1}^{n} X_i/n$. Prove that the prior distribution of Θ is $\Pr(\Theta = i/10) = a_i$ for $i = 0, \ldots, 10$.

12. Suppose that for every $m = 1, 2, \ldots$,
$$f_{X_1, \ldots, X_m}(x_1, \ldots, x_m) = \frac{2}{(m+1)c_m(x_1, \ldots, x_m)^{m+1}}, \quad \text{if all } x_i \geq 0,$$
where $c_m(x_1, \ldots, x_m) = \max\{2, x_1, \ldots, x_m\}$.

   (a) Prove that the X_i are exchangeable and that these distributions are consistent.

   (b) Let $Y_n = c_n(X_1, \ldots, X_n)$. Find the distribution of Y_n and the limit of this distribution as $n \to \infty$.

   (c) Find the conditional density of X_{n+1} given $X_1 = x_1, \ldots, X_n = x_n$, and assume that $\lim_{n \to \infty} c_n(x_1, \ldots, x_n) = \theta$. Find the limit of the conditional density as $n \to \infty$.

   (d) Use DeFinetti's representation theorem to show that the prior (the answer to part (b)) and likelihood (the answer to part (c)) combine to give the original joint distribution.

13. Let Θ be a random variable. Suppose that $\{X_n\}_{n=1}^{\infty}$ are IID $N(0, 1)$. Let $T(0) = 0$. For each $i > 0$, let $T(i)$ be the first $j > T(i-1)$ such that $X_j \geq \Theta$. Let $Y_i = X_{T(i)}$ for $i = 1, 2, \ldots$. If we use Lebesgue measure as an improper prior distribution for Θ, how many observations must we observe before the posterior becomes proper?

1.7. Problems   75

14. An observation X is to be made in hopes of learning something about a parameter Θ. The prior distribution of Θ has some density f_Θ, but we are not certain what is the appropriate distribution to use for X given Θ. Suppose that we have k choices with densities $f_1(\cdot|\theta), \ldots, f_k(\cdot|\theta)$. Let $\pi_1, \ldots, \pi_k$ be nonnegative numbers adding to 1, and set

$$f_{X|\Theta}(x|\theta) = \sum_{i=1}^{k} \pi_i f_i(x|\theta).$$

Show that there are numbers $\pi_1^*(x), \ldots, \pi_k^*(x)$ adding to 1 such that

$$f_{\Theta|X}(\theta|x) = \sum_{i=1}^{k} \pi_i^*(x) p_i(\theta|x),$$

where $p_i(\cdot|x)$ is the posterior density of Θ that we would have calculated if we had used $f_i(\cdot|\theta)$ as the conditional density for X given $\Theta = \theta$.

15. Let $\{X_n\}_{n=1}^{\infty}$ be IID $Ber(\theta)$ random variables given $\Theta = \theta$. Define a prior μ_Θ for Θ by $\mu_\Theta(B) = [\delta(B) + \mu(B)]/2$ where μ is Lebesgue measure and $\delta(B) = I_B(1/2)$.

    (a) Find the marginal distribution for $X_{i_1}, \ldots, X_{i_n}$ for distinct integers $i_1, \ldots, i_n$.

    (b) Find the posterior distribution for Θ given $X_1 = x_1, \ldots, X_n = x_n$, that is, find $\Pr(\Theta \leq \theta | X_1 = x_1, \ldots, X_n = x_n)$.

16. Suppose that $\{X_n\}_{n=1}^{\infty}$ are conditionally independent random variables with $X_i \sim N(\mu_i, 1)$ given $(M, M_1, \ldots, M_n, \ldots) = (\mu, \mu_1, \ldots, \mu_n, \ldots)$. Suppose also that given $M = \mu$,

$$\begin{pmatrix} M_1 \\ \vdots \\ M_n \end{pmatrix} \sim N_n \left(\begin{bmatrix} \mu \\ \vdots \\ \mu \end{bmatrix}, \begin{bmatrix} 2 & 1 & \cdots & 1 \\ 1 & 2 & \cdots & 1 \\ \vdots & & \ddots & \\ 1 & \cdots & 1 & 2 \end{bmatrix} \right)$$

for each n, and $M \sim N(\mu_0, 1)$.

    (a) Prove that $\{X_n\}_{n=1}^{\infty}$ are exchangeable.
    (b) Find a one-dimensional random variable Θ such that the X_i are conditionally IID given $\Theta = \theta$, and find the distribution of Θ.
    (c) Show that $\overline{X}_n = \sum_{i=1}^n X_i/n$ does not converge in probability to M.

17. Let c be a constant, and let X, Y be conditionally independent given $\Theta = \theta$ with $X \sim Poi(\theta)$, $Y \sim Poi(c\theta)$. Let $\Theta \sim \Gamma(\alpha_0, \beta_0)$.

    (a) Find the posterior distribution of Θ given $X = x$.
    (b) Find the posterior predictive distribution of Y given $X = x$, and show that it is a member of the negative binomial family.

18. Suppose that an expert believes that $\{X_n\}_{n=1}^\infty$ are exchangeable Bernoulli random variables. Let $\Theta = \lim_{n\to\infty} \sum_{i=1}^n X_i/n$, and assume that a statistician wishes to model $\Theta \sim Beta(a,b)$. The statistician tries to elicit the values of a and b from the expert by asking questions like, "What is the probability that $X_1 = 1$?" and "How many $X_i = 1$ in a row would you have to observe before you would raise the probability that the next $X_j = 1$ up to q?" Suppose that the answer to the first question is p, and suppose that q in the second question is chosen to be $(1+p)/2$. Let the answer to the second question be m.

   (a) Find values for a and b which are consistent with the model.

   (b) Find the partial derivatives of a and b with respect to p, and find the effects of a change of ± 1 in m on both a and b.

   (c) Suppose that the second question above was changed to "If you were to observe $X_1 = \cdots = X_{10} = 1$, what would you give as the $\Pr(X_{11} = 1)$?" Let the answer to this question be r. Find values of a and b consistent with the values of p and r, and find the partial derivatives of a and b with respect to p and r.

19. Suppose that an expert believes that $\{X_n\}_{n=1}^\infty$ are conditionally IID with $N(\mu,\sigma^2)$ distribution given $(M, \Sigma) = (\mu,\sigma)$. A statistician wishes to model $M \sim N(\mu_0, \sigma^2/\lambda_0)$ given $\Sigma = \sigma$ and $\Sigma^2 \sim \Gamma^{-1}(a_0/2, b_0/2)$. The statistician tries to elicit the values of μ_0, λ_0, a_0, and b_0 from the expert by asking a sequence of questions such as:

   - What is the median of the distribution of X_1? (Suppose that the answer is u_1.)
   - Given that $X_1 \geq u_1$, what is the conditional median of the distribution of X_1? (Suppose that the answer is u_2.)
   - Given that $X_1 \geq u_2$, what is the conditional median of the distribution of X_1? (Suppose that the answer is u_3.)
   - If $X_1 = u_2$ is observed, what would be the conditional median of the distribution of X_2? (Suppose that the answer is u_4.)

   (a) Prove that the following constraints on u_1, u_2, u_3, and u_4 are sufficient for there to exist a prior distribution of the desired form consistent with the responses: $u_1 < u_4 < u_2$ and $(u_3 - u_1)/(u_2 - u_1) > 1.705511$. (The constraints are actually necessary as well.)

   (b) Suppose that the following answers are given: $u_1 = 14.56$, $u_2 = 21.34$, $u_3 = 29.47$, and $u_4 = 19.25$. Find values of μ_0, λ_0, a_0, and b_0 which are consistent with these answers.

Section 1.4:

20. For the joint density in Example 1.53 on page 29, prove that the distributions are consistent (as n changes).

21. Consider the joint density in Example 1.53 on page 29, and define $Y_n = n/\sum_{i=1}^n X_i$. Find the distribution of Y_n. Also, let $n \to \infty$ and prove that the limit of the distribution of Y_n is $\Gamma(a,b)$.

22. For the joint density in Example 1.53, prove that the conditional distribution of the X_i given $\Theta = \theta$, namely $Exp(\theta)$, and the marginal distribution of Θ, found in Problem 21, namely $\Gamma(a,b)$, do indeed induce the joint distribution for $X_1, \ldots, X_n$ for all n. How do we know that no other combination of distributions will induce the joint distribution of the X_i?

23. Let $X_1, \ldots, X_{14}$ be exchangeable Bernoulli random variables, and let $M = \sum_{i=1}^{14} X_i$. Let the distribution of M be given by the mass function (density with respect to counting measure)

$$f_M(m) = \begin{cases} 0.3 & \text{if } m = 2, \\ 0.2 & \text{if } m = 8, \\ 0.5 & \text{if } m = 13. \end{cases}$$

   (a) Find the probability that in four specific trials, we observe three successes and one failure (without regard to which of the trials is a failure and which are successes).

   (b) Suppose that we observe three successes in the first four trials. Find all the probabilities of k successes in n future trials for $n = 1, \ldots, 10$ and $k = 0, \ldots, n$. (Give a formula.)

24. Suppose that $X_1, \ldots, X_n$ are exchangeable and take values in the Borel space $(\mathcal{X}, \mathcal{B})$. Prove that the empirical probability measure $\mathbf{P}_n$ is a measurable function from the n-fold product space $(\mathcal{X}^n, \mathcal{B}^n)$ to $(\mathcal{P}, \mathcal{C}_\mathcal{P})$.

25.*Refer to Problem 29 on page 664.

   (a) Find the distribution of $\Theta = \lim_{n \to \infty} \sum_{i=1}^n X_i/n$.

   (b) Assume that we observe $X_1 = 1$ and $X_2 = 0$. Find the conditional distribution of $X_3, \ldots, X_n$ given this data for all $n = 3, 4, \ldots$.

   (c) Using the same data, find the conditional distribution of Θ.

26. Suppose that $\{X_n\}_{n=1}^\infty$ is an infinite sequence of exchangeable random variables with finite variance.

   (a) Prove that the covariance of X_i with X_j is nonnegative for $i \neq j$.

   (b) Give an example of such a sequence in which $\text{Cov}(X_i, X_j) = 0$ but the random variables are not mutually independent.

27.*Let X_1, X_2, X_3 be IID $U(0,1)$ random variables. After observing $X_1 = x_1, X_2 = x_2, X_3 = x_3$, define Y_1, Y_2 to be the results of drawing two numbers at random without replacement from the set $\{x_1, x_2, x_3\}$. Prove that Y_1 and Y_2 are IID $U(0,1)$.

28. Let $X_1, \ldots, X_n$ be IID with some distribution P on a Borel space $(\mathcal{X}, \mathcal{B})$. Let the conditional distribution of $Y_1, \ldots, Y_k$ (for $k < n$) given $X_1 = x_1, \ldots, X_n = x_n$ be that of k draws without replacement from the set $\{x_1, \ldots, x_n\}$. Prove that the joint distribution of $Y_1, \ldots, Y_k$ is that of IID random quantities with distribution P.

29. Let $(\mathcal{X}, \mathcal{B})$ be a Borel space, and let X_i take values in $\mathcal{X}$ for $i = 1, \ldots, n$. Suppose that $X_1, \ldots, X_n$ are exchangeable. Let the conditional distribution of $Y_1, \ldots, Y_k$ (for $k < n$) given $X_1 = x_1, \ldots, X_n = x_n$ be that of k draws without replacement from the set $\{x_1, \ldots, x_n\}$. Prove that the joint distribution of $Y_1, \ldots, Y_k$ is the same as the joint distribution of $X_1, \ldots, X_k$.

30. In the setup of Problem 3 on page 73, let $\mathbf{P}$ be the limit of the empirical probability measures of $X_1, \ldots, X_n$ as $n \to \infty$. Show that $\mathbf{P}$ is also the limit of the empirical probability measures of the Y_i.

31. State and prove a central limit theorem for exchangeable random variables. You may use Theorems B.97 and 1.49.

Section 1.5:

32. Prove Corollary 1.63 on page 36 using Theorem 1.62. (*Hint:* Prove that $\lim Y_n$ is measurable with respect to the tail σ-field of $\{X_n\}_{n=1}^{\infty}$. Then apply the Kolmogorov zero–one law B.68.)

33. Refer to Example 1.76 on page 42. Let $M^* = M - X_1$. Take the conditional distribution of M^* given $X_1 = x_1$ as a prior distribution for M^* after learning that $X_1 = x_1$.

    (a) Find the probability that $X_2 = 0$ (conditional on $X_1 = 1$) using this new prior distribution.

    (b) Find the posterior distribution for M^* given $X_2 = 0$ (and $X_1 = 1$.)

34. Let $\{X_n\}_{n=1}^{\infty}$ be exchangeable Bernoulli random variables, and let

$$Y = \min\left\{n : \sum_{i=1}^{n} X_i \geq 2\right\},$$

that is, Y is the time until the second success (e.g., if $X_1 = 1, X_2 = 0, X_3 = 1$, then $Y = 3$).

    (a) Find the distribution of Y using the form of DeFinetti's representation theorem in Example 1.82 on page 46.

    (b) Find the conditional distribution of $\{X_{n+k}\}_{k=1}^{\infty}$ given $Y = n$.

    (c) Show that the distribution in part (refp202) is the same as the conditional distribution of $\{X_{n+1}\}_{k=1}^{\infty}$ given $\sum_{i=1}^{n} X_i = 2$.

35. Suppose that $\{X_n\}_{n=1}^{\infty}$ are bounded, exchangeable random variables. Let $\Theta = \lim_{n \to \infty} \sum_{i=1}^{n} X_i/n$, a.s. Prove that $\mathrm{Var}(\Theta) = \mathrm{Cov}(X_1, X_2)$.

36. Prove that the collection $\mathcal{C}_n$ in the proof of Theorem 1.62 is a σ-field. Also prove that $f: \mathrm{I\!R}^{\infty} \to \mathrm{I\!R}$ is measurable with respect to $\mathcal{C}_n$ if and only if $f(y) = f(x)$ for all y that agree with x after coordinate n and such that the first n coordinates of y are a permutation of the first n coordinates of x.

37. Let $\{X_n\}_{n=1}^{\infty}$ be IID nonnegative random variables with $\mathrm{E}(X_i) = \infty$. Show that $\sum_{i=1}^{N} X_i/N = Y_N$ diverges to ∞ almost surely.

38.*Under the conditions of Theorem 1.59 it is possible to prove that Y_n converges almost surely, rather than just the subsequence $\{Y_{n_k}\}_{k=1}^{\infty}$.

    (a) Let $v = \sum_{i=1}^{\infty} i^{-3/2}$ and $c_{i,k} = 1/(kvi^{3/2})$ for all i and k. Define $V_{i,k} = \{s : |Y_{(i+1)^4}(s) - Y_{i^4}(s)| < c_{i,k}\}$. Use the second to last equation in (1.60) to prove that $\sum_{i=1}^{\infty} \Pr(V_{i,k}^C) < \infty$.

(b) Let $A_{k,n} = \bigcap_{i=n}^{\infty} V_{i,k}$. Show that for each $\epsilon > 0$ and k, there exists n_k such that $\Pr(A_{k,n_k}^C) < \epsilon/2^k$.

(c) For each i, j, k with $i^4 \leq k < (i+1)^4$, define $G_{i,j,k} = \{s : |Y_j(s) - Y_{i^4}(s)| > 1/k\}$. Use the second to last equation in (1.60) to prove that $\Pr(G_{i,j,k})$ is at most a fixed multiple of k^2/i^5.

(d) Define $H_{i,k} = \bigcup_{j=i^4}^{(i+1)^4 - 1} G_{i,j,k}$. Prove that $\Pr(H_{i,k})$ is at most a fixed multiple of k^2/i^2.

(e) Define $J_{k,n} = \bigcup_{i=n}^{\infty} H_{i,k}$. Prove that for each k and $\epsilon > 0$, there exists m_k such that $\Pr(J_{k,m_k}) < \epsilon/2^k$.

(f) Prove that for every pair of sequences $\{n_k\}_{k=1}^{\infty}$ and $\{m_k\}_{k=1}^{\infty}$,

$$\left(\bigcap_{k=1}^{\infty} J_{k,m_k}^C\right) \cap \left(\bigcap_{k=1}^{\infty} A_{k,n_k}\right) \subseteq \{s : \{Y_n(s)\}_{n=1}^{\infty} \text{ is a Cauchy sequence}\}.$$

(g) Prove that for every $0 < c < 1$, the probability that $\{Y_n\}_{n=1}^{\infty}$ is a Cauchy sequence is at least c, hence it must be 1.

39. State and prove a weak law of large numbers for exchangeable random variables. You may use Theorems B.95 and 1.49. (Don't use Theorem 1.62.)

40.*Suppose that $\{X_n\}_{n=1}^{\infty}$ is a sequence of exchangeable random variables with finite mean. Let $\{n_k\}_{k=1}^{\infty}$ be a subsequence of $\{1, 2, \ldots\}$. Prove that

$$\lim_{n \to \infty} \frac{1}{n} \sum_{i=1}^{n} X_i = \lim_{k \to \infty} \frac{1}{k} \sum_{i=1}^{k} X_{n_i}, \text{ a.s.}$$

(*Hint:* Use DeFinetti's representation theorem and the strong law of large numbers 1.62 to express the two limits as the same function of **P**.)

41.*In this problem, you will prove the following generalization of a theorem of Aldous (1981): An infinite sequence of random quantities $\{X_n\}_{n=1}^{\infty}$ taking values in a Borel space $\mathcal{X}$ is exchangeable if and only if there exists a measurable function $f : [0,1]^2 \to \mathcal{X}$ such that $X = (X_1, X_2, \ldots)$ has the same distribution as $(f(Z_0, Z_1), f(Z_0, Z_2), \ldots)$, where $Z_0, Z_1, \ldots$ are IID $U(0,1)$.

(a) Let **P** be the limit of the empirical distributions of $\{X_n\}_{n=1}^{\infty}$, and let $\mathbf{P}^*$ be the limit of the empirical distributions of $\{X_{2n}\}_{n=1}^{\infty}$. Use Problem 40 on this page and Problem 31 on page 664 to show that $\mathbf{P}^* = \mathbf{P}$, a.s.

(b) Note that **P** is a measurable function on $\mathcal{X}^{\infty}$. Use Proposition B.145 and Lemma B.41 to show that there exists a measurable function $g : [0,1] \to \mathcal{P}$ such that $g(Z_0)$ has the same distribution as **P**.

(c) Map $\mathcal{X}$ to $\mathbb{R}$ and for each z map $g(z)$ to a CDF F_z. Then define

$$F_z^{-1}(q) = \begin{cases} \inf\{x : g(z)(x) \geq q\} & \text{if } q > 0, \\ \sup\{x : g(z)(x) > 0\} & \text{if } q = 0. \end{cases}$$

Find a function $\psi : \mathbb{R} \to \mathcal{X}$ such that $\{\psi(F_{Z_0}^{-1}(Z_i))\}_{i=1}^{\infty}$ has the same joint distribution as $\{X_{2i+1}\}_{i=1}^{\infty}$.

80   Chapter 1. Probability Models

(d) Show that $f(z,w) = \psi(F_z^{-1}(w))$ is the desired function.

Section 1.6.1:

42. Prove Proposition 1.98 on page 56.

43. Suppose that $\{X_n\}_{n=1}^{\infty}$ are conditionally independent with distribution P given $\mathbf{P} = P$, and $\mathbf{P}$ has $Dir(\alpha)$ distribution where α is a finite measure on $(\mathcal{X}, \mathcal{B})$. Let K_n be the number of distinct values amongst $X_1, \ldots, X_n$. Prove that $\lim_{n \to \infty} \mathrm{E}(K_n)/\log(n) = \alpha(\mathcal{X})$.

44. Consider the situation described in Example 1.103 on page 59. Consider two different choices for α_θ. The first is α_θ equal to the $U(0, \theta)$ density. The second is Lebesgue measure on $[0, \theta]$.

   (a) If no repeats occur in the data, show that the likelihood function for Θ is the same for both choices of α_θ.

   (b) Explain the differences between the two likelihood functions in the case in which repeats do occur in the data.

45. Suppose that $\mathbf{P}$ has $Dir(\alpha)$ distribution and that X_1 and X_2 are random variables that are conditionally IID with distribution P given $\mathbf{P} = P$. Suppose that α is absolutely continuous with respect to Lebesgue measure with Radon–Nikodym derivative $a(\cdot)$. Find the joint density of (X_1, X_2) with respect to the measure, which is two-dimensional Lebesgue measure on $A = \{(x_1, x_2) : x_1 \neq x_2\}$ plus one-dimensional Lebesgue measure on $B = \{(x_1, x_2) : x_1 = x_2\}$. (For definiteness, calculate one-dimensional Lebesgue measure of a subset $C \subseteq B$ by calculating the Lebesgue measure of the set $C_1 = \{x : (x, x) \in C\}$. The most natural alternative would be to multiply this measure by $\sqrt{2}$ so that it equaled length for line segments.)

46. Suppose that $\mathbf{P}$ has $Dir(\alpha)$ distribution, where α is a finite measure on $(\mathbb{R}, \mathcal{B})$. Let Z be the median of $\mathbf{P}$, namely $Z = \inf\{x : \mathbf{P}((-\infty, x]) \geq 1/2\}$. Show that the median of Z is the median of the distribution $\alpha/\alpha(\mathbb{R})$. (*Hint:* Use the result of Problem 30 on page 664.)

47. Let $\mathbf{P} \sim Dir(\alpha)$, where α is continuous (no point masses). Prove that the posterior distribution of $\mathbf{P}$ is not absolutely continuous with respect to the prior. In fact they are mutually singular (that is, the posterior assigns probability 1 to a set to which the prior assigns probability 0).

Section 1.6.2:

48. Prove Proposition 1.111 on page 62.

49. Let $\mathbf{P}$ be a random probability measure on $(\mathcal{X}, \mathcal{B})$. Let $X_i : S \to \mathcal{X}$ be random quantities that are conditionally IID given $\mathbf{P}$ with distribution $\mathbf{P}$. Let μ_X be the marginal distribution of X_i. We will say that a probability measure P is *continuous* if $P(X_1 = X_2) = 0$. Prove that $\mathbf{P}$ is continuous with probability 1 if and only if $\Pr(X_2 = x | X_1 = x) = 0$, a.s. $[\mu_X]$.

50. Let $(\mathcal{X}, \mathcal{B}, \nu)$ be a probability space. Assume that $\mathbf{P}$ is tailfree with respect to $(\{\pi_n\}_{n=1}^{\infty}, \{V_{n;B} : n \geq 1, B \in \pi_n\})$. Assume that each element of each π_n has positive ν measure. Show that $\mathrm{E}(V_{n;B}) = \nu(B)/\nu(ps(B))$ for all n and B if and only if $\mu_X(B) = \nu(B)$ for all B, where μ_X is defined in (1.112).

51.*Prove that (1.107) and (1.108) are necessary and sufficient in order that a tailfree process be a probability measure with probability 1. (*Hint:* That the two conditions are necessary is straightforward once one realizes that countable additivity of a measure μ is equivalent to $\lim_{n\to\infty} \mu(D_n) = 0$ when $D_1 \supseteq D_2 \supseteq \cdots$ and $\cap_{n=1}^{\infty} D_n = \emptyset$. Next prove that (1.107) is sufficient for the process to be countably additive on the smallest σ-field containing all sets in π_n for each n. Then show that the union of all these σ-fields is a field and that (1.108) is sufficient for the process to be countably additive on this field.)

52. Let $(X_1, \ldots, X_{n+1}) \sim Dir_{n+1}(a_1, \ldots, a_{n+1})$. Define $Y_i = \sum_{j=1}^{i} X_i$ for $i = 1, \ldots, n+1$, and set $Z_i = Y_i/Y_{i+1}$ for $i = 1, \ldots, n$. Prove that the Z_i are mutually independent with $Z_i \sim Beta(a_1 + \cdots + a_i, a_{i+1})$.

53. Prove Corollary 1.125 on page 68.

54. Prove Corollary 1.126 on page 68.

55.*Assume that the conditions of Theorem 1.121 hold. Prove that the posterior distribution of $\mathbf{P}$ as found in Theorem 1.115 is the same thing that Bayes' theorem 1.31 gives, where we let the Θ in Bayes' theorem 1.31 be $\mathbf{P}$.

56. Let $X \sim \mathbf{P}$ given $\mathbf{P}$. Suppose that $\mathrm{E}|g(X)| < \infty$, where $g : \mathcal{X} \to \mathbb{R}$. Prove that $\mathrm{E}(|g(X)| \| \mathbf{P}) < \infty$, a.s.

57.*Let $\mathbf{P}$ have a Polya tree process distribution with each set partitioned into k sets at the next level. For each x and n, let $i_n(x)$ be such that $C_n(x) = B_{i_n(x)}(B_{n-1}(x))$. Let $Dir_k(a_{n,1}(x), \ldots, a_{n,k}(x))$ be the prior distribution of $W_n(x)$ in (1.128), and define $s_n(x) = \sum_{i=1}^{k} a_{n,i}(x)$.

   (a) Prove that $\Pr(X = x) = \prod_{n=1}^{\infty} (a_{n,i_n(x)}/s_n(x))$.

   (b) If X_1, X_2 are conditionally IID with distribution $\mathbf{P}$ given $\mathbf{P}$, prove that
   $$\Pr(X_2 = x | X_1 = x) = \prod_{n=1}^{\infty} \frac{b_{n,i_n(x)}(x)}{s_n(x) + 1},$$
   where $b_{n,i}$ is defined in (1.129).

   (c) If $\inf_{x,n} s_n(x) > 0$ and $\sup_{x,n,i} a_{n,i}(x)/s_n(x) < 1$, then $\mathbf{P}$ is continuous with probability 1.

   (*Hint:* Use Problem 49 on page 80.)

58. Consider a Dirichlet process $Dir(\alpha)$ as a Polya tree with $k = 2$ and $a_{n,i}(x) = c_n$ for all n, i, and x.

   (a) If $a = \alpha(\mathcal{X})$, show that $c_n = a/2^n$ for all n. (*Hint:* Use the result of Problem 52 above, which implies that the product of a $Beta(b, c)$ times an independent $Beta(b + c, d)$ is $Beta(b, c + d)$.)

   (b) Show that a condition for $\mathbf{P}$ to be continuous in Problem 57 above is violated.

CHAPTER 2
Sufficient Statistics

We now turn our attention to the broad area of *statistics*. This will concern the manner in which one learns from data. In this chapter, we will study some of the basic properties of probability models and how data can be used to help us learn about the parameters of those models.

2.1 Definitions

2.1.1 Notational Overview

We assume that there is a probability space $(S, \mathcal{A}, \mu)$ underlying all probability calculations. It will be common to refer to probabilities calculated under μ using the symbol $\Pr(\cdot)$. Conditional probabilities will be denoted $\Pr(\cdot|\cdot)$. We also assume that there is a random quantity $X: S \to \mathcal{X}$, where $\mathcal{X}$ is called the *sample space* (with σ-field $\mathcal{B}$), which will usually be some subset of a Euclidean space, but will be a Borel space in any event. We will often refer to X as the *data*. Let $\mathcal{A}_X$ stand for the sub-σ-field of $\mathcal{A}$ generated by X (that is, $\mathcal{A}_X = X^{-1}(\mathcal{B})$). Since $(\mathcal{X}, \mathcal{B})$ is a Borel space, $\mathcal{B}$ contains all singletons. This will allow us to claim that random quantities are functions of X if and only if they are measurable with respect to $\mathcal{A}_X$ by Theorem A.42. Generic elements of $\mathcal{X}$ will usually be denoted by x, y, or z, or $x_1, x_2, \ldots$, depending on how many we need at once.

Assume that there is a parametric family $\mathcal{P}_0$ of distributions for X, and let the parameter be $\Theta: S \to \Omega$, where Ω is a parameter space with σ-field τ. Usually, Ω will be a subset of some finite-dimensional Euclidean space, but not always. X will usually be a vector of exchangeable coordinates, but this is not required. When the coordinates of X are exchangeable, then the

2.1. Definitions

elements of $\mathcal{P}_0$ will usually be distributions that say that the coordinates of X are IID each with distribution P_θ for some $\theta \in \Omega$. As mentioned in Section 1.5.5, we will use the symbol P_θ to stand for the conditional distribution of X given $\Theta = \theta$ (a distribution on $(\mathcal{X}, \mathcal{B})$) as well as the conditional distribution of each coordinate of X in the case in which X has exchangeable coordinates. We use the symbol $P'_\theta(\cdot)$ to stand for $\Pr(\cdot|\Theta = \theta)$, a probability on $(S, \mathcal{A}_X)$. That is, for $A \in \mathcal{A}_X$ with $A = X^{-1}(B)$ for some $B \in \mathcal{B}$,

$$P'_\theta(A) = P'_\theta(X \in B) = \Pr(X \in B|\Theta = \theta) = P_\theta(B).$$

Example 2.1. Let $\{X_n\}_{n=1}^\infty$ be conditionally IID random variables with $N(\theta, 1)$ distribution given $\Theta = \theta$. Let $X = (X_1, \ldots, X_n)$. If $B \in \mathcal{B}^1$, the one-dimensional Borel σ-field, then

$$P_\theta(B) = \int_B \frac{1}{\sqrt{2\pi}} \exp\left(-\frac{(x-\theta)^2}{2}\right) dx = P'_\theta(X_i \in B) = \Pr(X_i \in B|\Theta = \theta).$$

Similarly, if $C \in \mathcal{B}^n$, the n-dimensional Borel σ-field, then

$$\begin{aligned} P_\theta(C) &= \int_C (2\pi)^{-\frac{n}{2}} \exp\left(-\frac{1}{2}\sum_{i=1}^n (x_i - \theta)^2\right) dx_1 \cdots dx_n \\ &= P'_\theta(X \in C) = \Pr(X \in C|\Theta = \theta). \end{aligned}$$

Let μ_Θ be the distribution of Θ, and let $D \in \tau$ and $B \in \mathcal{B}$. Let $A = X^{-1}(B)$ and $E = \Theta^{-1}(D)$. Then we have

$$\Pr(X \in B, \Theta \in D) = \mu(A \cap E) = \int_E \Pr(A|\Theta)(s) d\mu(s) = \int_D P_\theta(B) d\mu_\Theta(\theta),$$

where the last equality follows from Theorem A.81.

Example 2.2 (Continuation of Example 2.1). Suppose that $\Theta \sim N(0, 1)$. Then, for each $D \in \tau$ and $B \in \mathcal{B}$, $\Pr(\Theta \in D, X \in B)$ equals

$$\int_D \int_B (2\pi)^{-\frac{n+1}{2}} \exp\left(-\frac{1}{2}\left[\sum_{i=1}^n (x_i - \theta)^2 + \theta^2\right]\right) dx_1 \cdots dx_n d\theta.$$

2.1.2 Sufficiency

A *statistic* is virtually any measurable function of the data, X.

Definition 2.3. Let $(\mathcal{T}, \mathcal{C})$ be a measurable space such that $\mathcal{C}$ contains all singletons. If $T : \mathcal{X} \to \mathcal{T}$ is measurable, then T is called a *statistic*.

It appears that almost anything can be a statistic. The only requirement is that it be a measurable function of X to a space in which singletons are measurable sets, such as a Borel space. It will prove convenient, when $T : \mathcal{X} \to \mathcal{T}$, to refer to T as a random quantity. When we do this, we will mean the random quantity $T(X) : S \to \mathcal{T}$. When we need to refer to the specific value that T assumes when $X = x$, we write $T(x)$. We will let $P_{\theta,T}$ stand for the probability measure induced on the space $(\mathcal{T}, \mathcal{C})$ from P_θ by the function T. In this way $P'_\theta(T(X) \in C) = P_{\theta,T}(C)$. We will often write $P'_\theta(T \in C)$ to stand for this quantity as well.

There is a special class of statistics that are very useful in statistical inference. These are statistics that provide a summary of the data sufficient for performing all inferences of interest.

Definition 2.4. Let $\mathcal{P}_0$ be a parametric family of distributions on $(\mathcal{X}, \mathcal{B})$. Let (Ω, τ) be a parameter space and $\Theta : \mathcal{P}_0 \to \Omega$ be the parameter. Let $T : \mathcal{X} \to \mathcal{T}$ be a statistic. We say that T is a *sufficient statistic for* Θ *(in the Bayesian sense)* if, for every prior μ_Θ, there exist versions of the posteriors $\mu_{\Theta|X}$ and $\mu_{\Theta|T}$ such that, for every $B \in \tau$ $\mu_{\Theta|X}(B|x) = \mu_{\Theta|T}(B|T(x))$, a.s. $[\mu_X]$, where μ_X is the marginal distribution of X.

It appears that once one has settled on a parametric family of distributions for the data X, one need only calculate a sufficient statistic, because the posterior distribution of Θ given the sufficient statistic is the same as given X, no matter which prior distribution one uses. So long as one sticks with the chosen parametric family, the sufficient statistic is *sufficient* for making inference about Θ, and thereby about future observations (conditionally independent of X given Θ) through (1.37) on page 18.[1]

Example 2.5. Let $\{X_n\}_{n=1}^\infty$ be exchangeable Bernoulli random variables, and let $\mathcal{P}_0$ be the set of all IID distributions (the largest parametric family available). Let $X = (X_1, \ldots, X_n)$, and let P_θ be the distribution that says the coordinates of X are IID $Ber(\theta)$ random variables. We have already seen (Theorem 1.56) that if the prior is μ_Θ, then the posterior for Θ has Radon–Nikodym derivative

$$\frac{d\mu_{\Theta|X}}{d\mu_\Theta}(\theta|x) = \frac{\theta^{\sum_{i=1}^n x_i}(1-\theta)^{n-\sum_{i=1}^n x_i}}{\int \psi^{\sum_{i=1}^n x_i}(1-\psi)^{n-\sum_{i=1}^n x_i}d\mu_\Theta(\psi)}.$$

Next, treat $T(X) = \sum_{i=1}^n X_i$ as the data. The density of T given $\Theta = \theta$ (with respect to counting measure on the nonnegative integers) is $f_{T|\Theta}(t|\theta) = \binom{n}{t}\theta^t(1-\theta)^{n-t}$ for $t = 0, \ldots, n$. It follows from Bayes' theorem 1.31 that the posterior given $T = t = \sum_{i=1}^n x_i$ has derivative

$$\frac{d\mu_{\Theta|T}}{d\mu_\Theta}(\theta|t) = \frac{\binom{n}{t}\theta^t(1-\theta)^{n-t}}{\int \binom{n}{t}\psi^t(1-\psi)^{n-t}d\mu_\Theta(\psi)}.$$

This is the same as the other posterior, hence T is sufficient according to Definition 2.4.

---

[1] See Problem 24 on page 141 for an example of observations not conditionally independent given Θ for which a sufficient statistic is not sufficient for making predicitive inference.

In Example 2.5, for every prior the posterior distribution of Θ given $X = x$ was a function of $T(x)$. The following lemma says that this fact is enough to conclude that T is sufficient.

Lemma 2.6.[2] *Let T be a statistic and let $\mathcal{B}_T$ be the sub-σ-field of $\mathcal{B}$ generated by T. Then T is sufficient in the Bayesian sense if and only if, for every prior distribution μ_Θ, there exists a version of the posterior distribution given X, $\mu_{\Theta|X}$, such that for all $B \in \tau$, $\mu_{\Theta|X}(B|\cdot)$ is measurable with respect to $\mathcal{B}_T$.*

PROOF. This result is an immediate consequence of Theorem B.73 if we make the following correspondences:

Theorem B.73	$\mathcal{B}$	$\mathcal{C}$	$\mathcal{Z}$
Lemma 2.6	$X^{-1}(\mathcal{B})$	$X^{-1}(\mathcal{B}_T)$	$I_B(\Theta)$

□

Example 2.7. Let $\mathcal{P}_0$ be the set of all IID exponential distributions, and let P_θ say that $\{X_n\}_{n=1}^\infty$ are IID $Exp(\theta)$ random variables. If μ_Θ is the prior distribution and $X = (X_1, \ldots, X_n)$, then the posterior has density

$$\frac{d\mu_{\Theta|X}}{d\mu_\Theta}(\theta|x) = \frac{\theta^n \exp\left(-\theta \sum_{i=1}^n x_i\right)}{\int \psi^n \exp\left(-\psi \sum_{i=1}^n x_i\right) d\mu_\Theta(\psi)}$$

with respect to the prior. Notice that this is a function of $T(x) = \sum_{i=1}^n x_i$, hence Lemma 2.6 says that T is sufficient.

There is a more commonly used definition of sufficient statistic, which does not refer to prior distributions. Loosely speaking, this definition says that T is sufficient if the conditional distribution of X given $\Theta = \theta$ and T does not depend on θ.

Definition 2.8. Let $\mathcal{P}_0$ be a parametric family of distributions on $(\mathcal{X}, \mathcal{B})$. Let (Ω, τ) be a parameter space and $\Theta : \mathcal{P}_0 \to \Omega$ be the parameter. Let $T : \mathcal{X} \to \mathcal{T}$ be a statistic. Suppose that there exist versions of $P_\theta(\cdot|T)$ and a function $r : \mathcal{B} \times \mathcal{T} \to [0, 1]$ such that $r(\cdot, t)$ is a probability on $(\mathcal{X}, \mathcal{B})$ for every $t \in \mathcal{T}$, $r(A, \cdot)$ is measurable for every $A \in \mathcal{B}$, and for every $\theta \in \Omega$ and every $B \in \mathcal{B}$

$$P_\theta(B|T = t) = r(B, t), \quad \text{a.e. } [P_{\theta,T}].$$

Then we say that T is *a sufficient statistic for* Θ *(in the classical sense)*.

This definition says that a statistic T is sufficient if and only if, after one observes the value $T(X) = t$, one can generate data X' with conditional distribution $r(\cdot|t)$, and then the conditional distribution of X' given Θ is the same as the conditional distribution of X given Θ. It will be common to use the symbol $\mathrm{E}(\cdot|T)$ in place of $\mathrm{E}_\theta(\cdot|T)$ when T is sufficient. In such a case, if $\mathrm{E}_\theta|g(X)| < \infty$ for all θ, then $\mathrm{E}(g(X)|T = t) = \int g(x) dr(x, t)$.

---

[2] This lemma is used in the proofs of Lemma 2.15 and Theorem 2.29.

Example 2.9 (Continuation of Example 2.5; see page 84). The X_i are IID $Ber(\theta)$ given $\Theta = \theta$, and $X = (X_1, \ldots, X_n)$. Let $T(x) = \sum_{i=1}^{n} x_i$. We need to compute $P'_\theta(X = x | T(X) = t)$ for all θ and all x such that $t = T(x)$. Since both X and T are discrete random variables,

$$P'_\theta(X = x | T(X) = t) = \frac{\theta^t (1-\theta)^{n-t}}{\binom{n}{t} \theta^t (1-\theta)^{n-t}} = \binom{n}{t}^{-1}.$$

Set $r(\cdot | t)$ to be the distribution that is uniform on the set of all x such that $\sum_{i=1}^{n} x_i = t$ (probability $\binom{n}{t}^{-1}$ for each such x.) We now see that T is sufficient according to Definition 2.8.

Example 2.10 (Continuation of Example 2.7; see page 85). If $X = (X_1, \ldots, X_n)$ with the X_i having $Exp(\theta)$ distribution given $\Theta = \theta$, let $T(x) = \sum_{i=1}^{n} x_i$. We need to find the conditional distribution of X given $T = t$. By Corollary B.55, the conditional distribution of X given $T = t$ and $\Theta = \theta$ has density

$$\frac{f_{X|\Theta}(x_1, \ldots, x_n | \theta)}{f_{T|\Theta}(t|\theta)} = \frac{\theta^n \exp(-\theta t)}{\frac{1}{(n-1)!} \theta^n t^{n-1} \exp(-t\theta)} = \frac{(n-1)!}{t^{n-1}}$$

with respect to a measure $\nu_{X|T}(\cdot | t)$, which does not depend on θ. Since this distribution is the same for all θ, T is sufficient in the classical sense.

In general, sufficient statistics need not be much simpler than the entire data set.

Definition 2.11. Let $X_1, \ldots, X_n$ be random variables. Define $X_{(1)}$ to be $\min\{X_1, \ldots, X_n\}$, and for $k > 1$, define

$$X_{(k)} = \min\left(\{X_1, \ldots, X_n\} \setminus \{X_{(1)}, \ldots, X_{(k-1)}\}\right).$$

The vector $(X_{(1)}, \ldots, X_{(n)})$ is called the *order statistics* of $X_1, \ldots, X_n$.

Proposition 2.12. *Let $X = (X_1, \ldots, X_n)$, and suppose that $X_1, \ldots, X_n$ are exchangeable random variables. Then the order statistics are sufficient.*

Example 2.13. Let $X_1, \ldots, X_n$ be conditionally IID with Cauchy distribution $Cau(\theta, 1)$ given $\Theta = \theta$. In this case, one can show that all sufficient statistics are at least as complicated as the order statistics. To do this, however, some theorems from Section 2.1.3 will prove useful.

In all cases of interest to us, the two definitions of sufficient statistic are equivalent.

Theorem 2.14.[3] *Let $(\mathcal{T}, \mathcal{C})$ be a Borel space, and let $T : \mathcal{X} \to \mathcal{T}$ be a statistic. The following are both true:*

---

[3] The proof of part 1 is reminiscent of the proof of Theorem 1 of Halmos and Savage (1949). The proof of part 2 is due to Blackwell and Ramamoorthi (1982).

1. If there is a σ-finite measure ν such that for all θ, $P_\theta \ll \nu$ and T is sufficient in the Bayesian sense, then T is sufficient in the classical sense.

2. If T is sufficient in the classical sense, then T is sufficient in the Bayesian sense.

The proof of part 1 of Theorem 2.14 requires a lemma that will also be used in the proof of Theorem 2.21.

Lemma 2.15. *Let ν be a σ-finite measure such that $P_\theta \ll \nu$ for all θ. If T is sufficient in the Bayesian sense, then there exists a probability measure ν^* such that $P_\theta \ll \nu^*$ for all θ and $dP_\theta/d\nu^*(x)$ is a function $h(\theta, T(x))$. Also, $\nu^* \ll \nu$.*

PROOF. Let $dP_\theta/d\nu(x) = f_{X|\Theta}(x|\theta)$. Since each $P_\theta \ll \nu$, Theorem A.78 says that there exist countable sequences $\{\theta_i\}_{i=1}^\infty$ and $\{c_i\}_{i=1}^\infty$ such that $c_i \geq 0$, $\sum_{i=1}^\infty c_i = 1$, and $P_\theta \ll \nu^*$, for every $\theta \in \Omega$, where $\nu^* = \sum_{i=1}^\infty c_i P_{\theta_i}$. Note that $\nu^* \ll \nu$. For $\theta \in \Omega$ such that θ is not one of the θ_i, specify the following prior distribution over Ω: $\Pr(\Theta = \theta) = 1/2$, and $\Pr(\Theta = \theta_i) = c_i/2$, for $i = 1, 2, \ldots$. The posterior probability of $\Theta = \theta$ given $X = x$ is

$$\Pr(\Theta = \theta | X = x) = \frac{f_{X|\Theta}(x|\theta)}{f_{X|\Theta}(x|\theta) + \sum_{i=1}^\infty c_i f_{X|\Theta}(x|\theta_i)}$$

$$= \left(1 + \frac{\sum_{i=1}^\infty c_i f_{X|\Theta}(x|\theta_i)}{f_{X|\Theta}(x|\theta)}\right)^{-1}.$$

According to Theorem 2.6, for each θ, this is a function of $T(x)$. That is, there exists a function h such that, for each θ,

$$\frac{f_{X|\Theta}(x|\theta)}{\sum_{i=1}^\infty c_i f_{X|\Theta}(x|\theta_i)} = h(\theta, T(x)). \tag{2.16}$$

By the chain rule A.79, it can be seen that the left-hand side of (2.16) is equal to $dP_\theta/d\nu^*(x)$.

For $\theta \in \{\theta_i\}_{i=1}^\infty$, replace the prior above by one that has $\Pr(\Theta = \theta_i) = c_i$ for all i. We still have that (2.16) is $dP_\theta/d\nu^*(x)$. Also, Theorem 2.6 says that $\Pr(\Theta = \theta_j | X = x)$ is still a function of $T(x)$ for all j. But $\Pr(\Theta = \theta_j | X = x)$ is just c_j times the left-hand side of (2.16). □

PROOF OF THEOREM 2.14. For part 1, define the function r to be the conditional probability function on $\mathcal{X}$ given $T = t$ calculated from the probability ν^* in Lemma 2.15. That is, for every $C \in \mathcal{C}$ and every $B \in \mathcal{B}$,

$$\nu^*(T^{-1}(C) \cap B) = \int_C r(B, t) d\nu_T^*(t),$$

where ν_T^* is the probability on $(\mathcal{T}, \mathcal{C})$ induced by T from ν^*. It is easy to see that this implies that for every integrable $g : \mathcal{T} \to \mathbb{R}$ and $B \in \mathcal{B}$,

$$\int g(T(x))I_B(x)d\nu^*(x) = \int g(t)r(B,t)d\nu_T^*(t). \tag{2.17}$$

We now wish to show that this function r can serve as the conditional distribution of X given $T = t$ and $\Theta = \theta$ for all θ. To see that this is true, note that for all $B \in \mathcal{B}$, $P_\theta(B|T=t)$ is any function $m : \mathcal{T} \to [0,1]$ satisfying

$$P'_\theta(X \in B, T(X) \in C) = \int_C m(t)dP_{\theta,T}(t), \quad \text{for all } C \in \mathcal{C}, \tag{2.18}$$

where $P_{\theta,T}$ is the probability on $(\mathcal{T}, \mathcal{C})$ induced by T from P_θ. According to Lemma 2.15, we have that

$$\frac{dP_{\theta,T}}{d\nu_T^*}(t) = h(\theta, t). \tag{2.19}$$

The left-hand side of (2.18) can be written as

$$\int I_B(x)I_C(T(x))h(\theta, T(x))d\nu^*(x) = \int I_C(t)r(B,t)h(\theta,t)d\nu_T^*(t)$$
$$= \int_C r(B,t)dP_{\theta,T}(t),$$

where the first equality follows from (2.17) and the second follows from (2.19). It follows that $r(B,t)$ can play the role of $m(t)$ in (2.18), and the proof of part 1 is complete.

To prove part 2, let r be as in Definition 2.8, and let μ_Θ be a prior for Θ. By the law of total probability B.70 (conditional on T), the conditional distribution of X given $T = t$, $\mu_{X|T}(\cdot|t)$, is given for every $B \in \mathcal{B}$ by

$$\mu_{X|T}(B|T=t) = \int_\Omega P_\theta(B|T=t)d\mu_{\Theta|T}(\theta|t)$$
$$= \int_\Omega r(B,t)d\mu_{\Theta|T}(\theta|t) = r(B,t),$$

where $\mu_{\Theta|T}$ is the posterior distribution of Θ given T. Hence, we have that the conditional distribution of X given T and Θ is the conditional distribution of X given T. According to Theorem B.64, this means that X and Θ are conditionally independent given T. According to Theorem B.61, we have that the posterior given T and X is the same as the posterior given T. Corollary B.74 says that the posterior given T and X is the same as the posterior given X. □

Blackwell and Ramamoorthi (1982) give an example in which the extra condition in part 1 of Theorem 2.14 fails and there exists a statistic sufficient in the Bayesian sense but not sufficient in the classical sense. There is, however, the following result.

Theorem 2.20. *If T is sufficient in the Bayesian sense, then for every prior distribution μ_Θ, there exists a version of the conditional distribution of X given T, $\mu_{X|T}$ such that for every $B \in \mathcal{B}$,*

$$\mu_\Theta(\{\theta : P_\theta(B|T=t) = \mu_{X|T}(B|t), \text{ a.e. } [\mu_T]\}) = 1,$$

where μ_T is the marginal distribution of T.

PROOF. Let μ_Θ be a prior distribution for Θ. Let $\mu_{X|T}(B|T)$ be the conditional probability of $\{X \in B\}$ given $T = t$ calculated from the marginal distribution of X and T (not conditional on Θ.) Since T is sufficient in the Bayesian sense, the conditional distribution of Θ given T is the same as the conditional distribution of Θ given X (and T). This means that Θ is independent of X given T according to Theorem B.64. Theorem B.61 then says that the conditional distribution of X given Θ and T is the same as the conditional distribution given T, which means that for all $B \in \mathcal{B}$, $P_\theta(B|T=t) = \mu_{X|T}(B|t)$, a.s. with respect to the joint distribution of Θ and T. The result now follows. □

Theorem 2.20 says that every prior distribution assigns probability 1 to a subset of the parameter space which, if it were the entire parameter space, would allow us to conclude that sufficiency in the Bayesian sense implied sufficiency in the classical sense without the added condition of absolute continuity.

There is an easier way to characterize sufficiency in the case in which all conditional distributions given Θ are absolutely continuous with respect to a single σ-finite measure.

Theorem 2.21 (Fisher–Neyman factorization theorem).[4] *Assume that $\{P_\theta : \theta \in \Omega\}$ is a parametric family such that $P_\theta \ll \nu$ (σ-finite) for all θ and $dP_\theta/d\nu(x) = f_{X|\Theta}(x|\theta)$. Then $T(X)$ is sufficient for Θ if and only if there are functions m_1 and m_2 such that*

$$f_{X|\Theta}(x|\theta) = m_1(x)m_2(T(x), \theta), \text{ for all } \theta.$$

PROOF. First, we do the "if" part. Let $f_{X|\Theta}(x|\theta) = m_1(x)m_2(T(x),\theta)$ for all θ, and let μ_Θ be an arbitrary prior for Θ. Bayes' theorem 1.31 says that the posterior distribution of Θ given $X = x$ is absolutely continuous with respect to the prior, and the Radon–Nikodym derivative is

$$\frac{d\mu_{\Theta|X}}{d\mu_\Theta}(\theta|x) = \frac{m_1(x)m_2(T(x),\theta)}{\int_\Omega m_1(x)m_2(T(x),\psi)d\mu_\Theta(\psi)} = \frac{m_2(T(x),\theta)}{\int_\Omega m_2(T(x),\psi)d\mu_\Theta(\psi)},$$

which is a function of $T(x)$. It follows from Lemma 2.6 that $T(X)$ is sufficient.

---

[4] Versions of this theorem originated with Fisher (1922, 1925) and Neyman (1935).

For the "only if" part, assume that $T(X)$ is sufficient. According to Lemma 2.15, there is a measure ν^* such that $P_\theta \ll \nu^*$ for all θ, $dP_\theta/d\nu^*(x)$ is a function h of θ and $T(x)$, and $\nu^* \ll \nu$. It follows that

$$f_{X|\Theta}(x|\theta) = h(\theta, T(x))\frac{d\nu^*}{d\nu}(x).$$

If we set $m_1(x)$ equal to the second factor on the right and $m_2(T(x), \theta)$ equal to the first factor on the right, we are done. □

Example 2.22. Let P_θ say that $\{X_n\}_{n=1}^\infty$ are IID $U(0,\theta)$, $\Omega = (0,\infty)$, and let $X = (X_1, \ldots, X_n)$. Then

$$f_{X|\Theta}(x|\theta) = \frac{1}{\theta^n} I_{[0,\infty)}(\min_i x_i) I_{[0,\theta]}(\max_i x_i).$$

By Theorem 2.21, $T(X) = \max_i X_i$ is sufficient. If μ_Θ is a prior, then the posterior has derivative $(d\mu_{\Theta|T}/d\mu_\Theta)(\theta|t) = I_{[t,\infty)}(\theta)/\theta^n$.

The case not covered by Theorem 2.21, in which not all P_θ are absolutely continuous with respect to the same σ-finite measure, is more complicated. One case in which the conclusion to Theorem 2.21 still applies is that of discrete random variables.

Proposition 2.23. *If $\{P_\theta : \theta \in \Omega\}$ is a parametric family such that each P_θ is a discrete distribution, then $T(X)$ is sufficient in the classical sense for Θ if and only if there are functions m_1 and m_2 such that*

$$\Pr(X = x|\Theta = \theta) = m_1(x)m_2(T(x), \theta), \text{ for all } \theta.$$

This proposition is needed only to handle cases in which $C = \cup_{\theta \in \Omega}\{x : P_\theta(x) > 0\}$ is an uncountable set; otherwise all P_θ are absolutely continuous with respect to counting measure on C.

The following lemma tells us that when a statistic T is sufficient and the distributions of X are all dominated by a common σ-finite measure, then we can replace X by T and the distributions of T are still dominated by a σ-finite measure. In fact, we can give a formula for the density of T.

Lemma 2.24.[5] *Assume the conditions of Theorem 2.21, and assume that $T : \mathcal{X} \to \mathcal{T}$ is sufficient. Then there exists a measure ν_T on $(\mathcal{T}, \mathcal{C})$ such that $P_{\theta,T} \ll \nu_T$ and $dP_{\theta,T}/d\nu_T(t) = m_2(t, \theta)$.*

PROOF. Apply Theorem A.78 to find a probability $\nu^* = \sum_{i=1}^\infty c_i P_{\theta_i}$ such that $P_\theta \ll \nu^*$ for all θ. Then

$$\frac{dP_\theta}{d\nu^*}(x) = \frac{f_{X|\Theta}(x|\theta)}{\sum_{i=1}^\infty c_i f_{X|\Theta}(x|\theta_i)} = \frac{m_2(T(x), \theta)}{\sum_{i=1}^\infty c_i m_2(T(x), \theta_i)}.$$

---

[5]This lemma is used in the proof of Lemma 2.58.

Since this density is a function of $T(x)$, we can write

$$P_{\theta,T}(B) = \int_{T^{-1}(B)} \frac{dP_\theta}{d\nu^*}(x) d\nu^*(x)$$

$$= \int_{T^{-1}(B)} \frac{m_2(T(x),\theta)}{\sum_{i=1}^\infty c_i m_2(T(x),\theta_i)} d\nu^*(x) = \int_B \frac{m_2(t,\theta)}{\sum_{i=1}^\infty c_i m_2(t,\theta_i)} d\nu_T^*(t),$$

where ν_T^* is the measure on $(\mathcal{T},\mathcal{C})$ induced by T from ν^*. Define ν_T by $d\nu_T/d\nu_T^*(t) = \sum_{i=1}^\infty c_i m_2(t,\theta_i)$ to complete the proof. $\square$

In many of the examples that we have considered and will consider, there exists a sufficient statistic whose dimension is the same for all sample sizes. In such cases, there might exist a particularly convenient family of prior distributions available for the parameter.

Theorem 2.25. *Suppose that there exists a sufficient statistic of fixed dimension k for all sample sizes. That is, suppose that there exist functions T_n (with image $\mathcal{T} \subseteq \mathbb{R}^k$ for all n), $m_{1,n}$, and $m_{2,n}$ such that*

$$f_{X_1,\ldots,X_n|\Theta}(x_1,\ldots,x_n|\theta) = m_{1,n}(x_1,\ldots,x_n)m_{2,n}(T_n(x_1,\ldots,x_n),\theta).$$

Suppose also, that for all n and all $t \in \mathcal{T}$,

$$0 < c(t,n) = \int_\Omega m_{2,n}(t,\theta) d\lambda(\theta) < \infty,$$

for some measure λ. Then the family of densities with respect to λ

$$\wp = \left\{ \frac{m_{2,n}(t,\cdot)}{c(t,n)} : t \in \mathcal{T}, n = 1, 2, \ldots \right\}$$

forms a conjugate family in the sense that the posterior density with respect to λ is a member of this class if the prior is a member of the class.

PROOF.[6] Let $f_\Theta(\theta) = m_{2,\ell}(t,\theta)/c(t,\ell)$ for some $t \in \mathcal{T}$ and some ℓ. Let $(y_1,\ldots,y_\ell)$ be such that $T_m(y_1,\ldots,y_\ell) = t$. Suppose that the data are $X_1 = x_1,\ldots,X_n = x_n$ for some n. We note that

$$\begin{aligned} & f_{X_1,\ldots,X_{\ell+n}|\Theta}(x_1,\ldots,x_n,y_1,\ldots,y_\ell|\theta) \\ &= m_{1,n}(x_1,\ldots,x_n)m_{2,n}(T_n(x_1,\ldots,x_n),\theta) \\ &\quad \times m_{1,\ell}(y_1,\ldots,y_\ell)m_{2,\ell}(t,\theta) \qquad (2.26) \\ &= m_{1,n+\ell}(x_1,\ldots,x_n,y_1,\ldots,y_\ell)m_{2,n+\ell}(t',\theta), \end{aligned}$$

where $t' = T_{n+\ell}(x_1,\ldots,x_n,y_1,\ldots,y_\ell)$. The posterior density of Θ with respect to the measure λ would be

$$f_{\Theta|X_1,\ldots,X_n}(\theta|x_1,\ldots,x_n)$$

---

[6]This proof follows the presentation of Section 9.3 of DeGroot (1970).

$$= \frac{m_{2,\ell}(t,\theta)m_{1,n}(x_1,\ldots,x_n)m_{2,n}(T_n(x_1,\ldots,x_n),\theta)}{\int_\Omega m_{2,\ell}(t,\psi)m_{1,n}(x_1,\ldots,x_n)m_{2,n}(T_n(x_1,\ldots,x_n),\psi)d\lambda(\psi)}$$

$$= \frac{m_{2,n+\ell}(t',\theta)}{c(t',n+\ell)},$$

by (2.26). □

The family of prior densities $\wp$ and their corresponding distributions is called a *natural conjugate family of priors*.

Example 2.27. Let $\{X_n\}_{n=1}^\infty$ be a sequence of conditionally IID $Ber(\theta)$ random variables given $\Theta = \theta$. Let $T_n = \sum_{i=1}^n X_i$. Then $m_{2,n}(t,\theta) = \theta^t(1-\theta)^{n-t}$ and $c(t,n) = t!(n-t)!/(n+1)!$. The family of natural conjugate priors is a subset of the family of *Beta* distributions. In particular, $m_{2,n}(t,\theta)/c(t,n)$ is the $Beta(t+1, n-t+1)$ density as a function of θ. Actually, the entire collection of Beta distributions has the property that if the prior is in the Beta family, then the posterior is as well. Theorem 2.25 only tells us that Beta distributions with integer parameters are natural conjugate.

2.1.3 Minimal and Complete Sufficiency

The entire data set is always sufficient, so there is not always a savings in using a sufficient statistic. However, there are often times when a simpler statistic than the entire data set is also sufficient. There is a sense in which a sufficient statistic can be as simple as possible.

Definition 2.28. A sufficient statistic $T : \mathcal{X} \to \mathcal{T}$ is called *minimal sufficient* if for every sufficient statistic $U : \mathcal{X} \to \mathcal{U}$, there is a measurable function $g : \mathcal{U} \to \mathcal{T}$ such that $T = g(U)$, a.s. $[P_\theta]$ for all θ.

Clearly, a bimeasurable function of a minimal sufficient statistic is also minimal sufficient. The following theorem says that the mapping from data values to the likelihood function is minimal sufficient.

Theorem 2.29. *Suppose that there exist versions of $f_{X|\Theta}(\cdot|\theta)$ for every θ and a measurable function[7] $T : \mathcal{X} \to \mathcal{T}$ such that $T(y) = T(x)$ if and only if $y \in \mathcal{D}(x)$, where $\mathcal{D}(x)$ is the set*

$$\{y \in \mathcal{X} : f_{X|\Theta}(y|\theta) = f_{X|\Theta}(x|\theta)h(x,y), \forall \theta \text{ and some } h(x,y) > 0\},$$

then $T(X)$ is a minimal sufficient statistic.

PROOF. First, we show that the distinct sets $\mathcal{D}(x)$ form a partition. If $y \in \mathcal{D}(x)$, then we can set $h(y,x) = 1/h(x,y)$ and we get that $f_{X|\Theta}(x|\theta) = f_{X|\Theta}(y|\theta)h(y,x)$ for all θ. So $x \in \mathcal{D}(y)$. With $h(x,x) = 1$ for all x, we see that $x \in \mathcal{D}(x)$, so the distinct $\mathcal{D}(x)$ form a partition.

---

[7]In most examples it is relatively easy to construct the function T, but you can actually prove that such a function exists in general. See Problem 15 on page 139.

Next, we show that $T(X)$ is sufficient. If μ_Θ is an arbitrary prior, the posterior after learning $X = x$ is absolutely continuous with respect to the prior with Radon–Nikodym derivative

$$\frac{d\mu_{\Theta|X}}{d\mu_\Theta}(\theta|x) = \frac{f_{X|\Theta}(x|\theta)}{\int f_{X|\Theta}(x|\psi)d\mu_\Theta(\psi)},$$

according to Bayes' theorem 1.31. If $y \in \mathcal{D}(x)$, the posterior after learning $X = y$ has Radon–Nikodym derivative

$$\frac{d\mu_{\Theta|X}}{d\mu_\Theta}(\theta|y) = \frac{h(x,y)f_{X|\Theta}(x|\theta)}{\int h(x,y)f_{X|\Theta}(x|\psi)d\mu_\Theta(\psi)},$$

which is the same as the posterior after learning $X = x$. Hence, the posterior is a function of $T(x)$. Lemma 2.6 says that $T(X)$ is sufficient.

Finally, we prove that $T(X)$ is minimal. Let $U(X)$ be another sufficient statistic. Use the Fisher–Neyman factorization theorem 2.21 to write

$$f_{X|\Theta}(x|\theta) = m_1(x)m_2(U(x),\theta). \tag{2.30}$$

Since $P_\theta(\{x : m_1(x) = 0\}) = 0$ for all θ, we can safely assume that $m_1(x) > 0$ for all x. We need to show that if $U(x) = U(y)$ for some $x, y \in \mathcal{X}$, then $y \subset \mathcal{D}(x)$. It would then follow that $T(y) = T(x)$, and this would make T a function of U. Suppose that $U(x) = U(y)$. Use (2.30) to write

$$f_{X|\Theta}(y|\theta) = \frac{m_1(y)}{m_1(x)}f_{X|\Theta}(x|\theta),$$

for all θ. With $h(x,y) = m_1(y)/m_1(x)$, we see that $y \in \mathcal{D}(x)$. $\square$

Example 2.31. Suppose that P_θ says that $\{X_n\}_{n=1}^\infty$ are IID $Ber(\theta)$ random variables. Let $X = (X_1, \ldots, X_n)$. Then

$$f_{X|\Theta}(x|\theta) = \theta^{\sum_{i=1}^n x_i}(1-\theta)^{n-\sum_{i=1}^n x_i},$$

for all $x_i \in \{0,1\}$. So the ratio

$$\frac{f_{X|\Theta}(y|\theta)}{f_{X|\Theta}(x|\theta)} = \left(\frac{\theta}{1-\theta}\right)^{\sum_{i=1}^n y_i - \sum_{i=1}^n x_i}$$

is the same for all θ if and only if $\sum_{i=1}^n x_i = \sum_{i=1}^n y_i$, in which case $h(x,y) = 1$ and $\mathcal{D}(x) = \{y : \sum_{i=1}^n y_i = \sum_{i=1}^n x_i\}$. Then $T(X) = \sum_{i=1}^n X_i$ is the minimal sufficient statistic.

Example 2.32. Suppose that P_θ says that $\{X_n\}_{n=1}^\infty$ are IID $U(0,\theta)$ random variables. Let $X = (X_1, \ldots, X_n)$ and suppose that the sample space is $\mathbb{R}^{+n}$. Then $f_{X|\Theta}(x|\theta) = \theta^{-n}I_{[0,\theta]}(\max x_i)$. Now suppose that

$$\theta^{-n}I_{[0,\theta]}(\max x_i) = h(x,y)\theta^{-n}I_{[0,\theta]}(\max y_i),$$

for all θ. This is true if and only if

$$I_{[0,\theta]}(\max x_i) = h(x,y)I_{[0,\theta]}(\max y_i),$$

for all θ, which in turn is true if and only if $\max x_i = \max y_i$, in which case $h(x,y) = 1$ and $\mathcal{D}(x) = \{y : \max y_i = \max x_i\}$. Then $T(X) = \max X_i$ is the minimal sufficient statistic.

Example 2.33. Suppose that P_θ says that $\{X_n\}_{n=1}^\infty$ are IID with density

$$f_{X_1|\Theta}(y|\theta) = \frac{\theta^y}{y! \sum_{t=0}^\theta \frac{\theta^t}{t!}}, \text{ for } y = 0, \ldots, \theta,$$

with respect to counting measure on the positive integers. Here Ω is the set of positive integers. If $X = (X_1, \ldots, X_n)$, then

$$f_{X|\Theta}(x|\theta) = \frac{\theta^x}{\prod_{i=1}^n (x_i!) \left(\sum_{t=0}^\theta \frac{\theta^t}{t!}\right)^n}, \text{ if } \max x_i \leq \theta,$$

where $x = \sum_{i=1}^n x_i$. Set $h(x,y) = \prod_{i=1}^n (x_i!)/\prod_{i=1}^n (y_i!)$ for all x and y. It follows that $f_{X|\Theta}(x|\theta) = h(x,y)f_{X|\Theta}(y|\theta)$ for all θ if and only if $\sum_{i=1}^n x_i = \sum_{i=1}^n y_i$ and $\max x_i = \max y_i$. Set $T(x) = \left(\sum_{i=1}^n x_i, \max x_i\right)$ and note that $x \in \mathcal{D}(y)$ if and only if $T(x) = T(y)$. So $T(X)$ is minimal sufficient.

There are some cases in which we need a sufficient statistic to satisfy an additional property. These situations are difficult to describe at the present time, but they arise in several places later in this text (in Chapters 3, 4 and 5, in particular).

Definition 2.34. A statistic T is *complete* if for every measurable, real-valued function g, $\mathrm{E}_\theta(g(T)) = 0$ for all $\theta \in \Omega$ implies $g(T) = 0$, a.s. $[P_\theta]$ for all θ.

A statistic T is *boundedly complete* if for every bounded, measurable, real-valued function g, $\mathrm{E}_\theta(g(T)) = 0$ for all $\theta \in \Omega$ implies $g(T) = 0$, a.s. $[P_\theta]$ for all θ.

Example 2.35. Suppose that P_θ says that $T \sim Poi(\theta)$. Suppose also that

$$\mathrm{E}_\theta(g(T)) = \sum_{t=0}^\infty g(t) \frac{\theta^t \exp(-\theta)}{t!} = 0$$

for all θ. Then $\sum_{t=0}^\infty g(t)\theta^t/t! = 0$ for all θ. This expression is a power series representation of the analytic function $h(\theta) = 0$. Since power series for analytic functions are unique, it must be that $g(t)/t! = 0$ for all nonnegative integers t. There are many functions g with this property, such as $g(t) = \sin(2\pi t)$. All such g satisfy $P_\theta(g(T) = 0) = 1$ for all θ. So T is complete.

Theorem 2.36 (Bahadur's theorem).[8] *If U is a boundedly complete sufficient statistic and finite-dimensional, then it is minimal sufficient.*

---

[8] See Bahadur (1957).

PROOF. Let T be another sufficient statistic. We need to show that U is a function of T. Express $U = (U_1(X), \ldots, U_k(X))$, where k is the dimension of U. Let $V_i(U) = (1 + \exp(U_i))^{-1}$, so that $V = (V_1, \ldots, V_k)$ is a one-to-one measurable function of U and each V_i is bounded. Define

$$H_i(t) = E_\theta(V_i(U)|T = t),$$
$$L_i(u) = E_\theta(H_i(T)|U = u).$$

Since U and T are sufficient, these conditional means given $\Theta = \theta$ do not depend on θ. Since the V_i are bounded, so are the H_i and L_i. Note that

$$E_\theta(V_i(U)) = E_\theta(E_\theta(V_i(U)|T)) = E_\theta(H_i(T))$$
$$= E_\theta(E_\theta(H_i(T)|U)) = E_\theta(L_i(U)).$$

It follows that $E_\theta(V_i(U) - L_i(U)) = 0$ for all θ. Since U is boundedly complete, it follows that $P_\theta(V_i = L_i) = 1$ for all θ. So, $E_\theta(L_i(U)|T) = H_i(T)$. Since L_i and H_i are bounded, they have finite variance, and Proposition B.78 says that

$$\text{Var}_\theta(L_i(U)) = E_\theta \text{Var}_\theta(L_i(U)|T) + \text{Var}_\theta(H_i(T)),$$
$$\text{Var}_\theta(H_i(T)) = E_\theta \text{Var}_\theta(H_i(T)|U) + \text{Var}_\theta(L_i(U)).$$

It follows easily from these equations that $\text{Var}_\theta(L_i(U)|T) = 0$, a.s. $[P_\theta]$, hence $\text{Var}_\theta(V_i(U)|T) = 0$, a.s. $[P_\theta]$. So $V_i(U) = E(V_i(U)|T) = H_i(T)$, a.s. $[P_\theta]$. Since V_i is one-to-one, we get $U_i = V_i^{-1}(H_i(T))$ for each i, and U is a function of T, as needed. □

2.1.4 Ancillarity

At the other extreme from sufficiency lie statistics that are independent of the parameter.

Definition 2.37. A statistic U is called *ancillary* if the conditional distribution of U given $\Theta = \theta$ is the same for all θ.

Example 2.38. Let X_1, X_2 be conditionally independent given $\Theta = \theta$, each with conditional distribution $N(\theta, 1)$. Let $U = X_2 - X_1$. The conditional density of U given $\Theta = \theta$ is $N(0, 2)$. Since this distribution is the same for all θ, U is ancillary.

Sometimes the two extremes meet and a minimal sufficient statistic contains a coordinate that is ancillary.

Definition 2.39. If a minimal sufficient statistic is $T = (T_1, T_2)$ and T_2 is ancillary, then T_1 is called *conditionally sufficient given T_2*.

Example 2.40. Suppose that $X_1, \ldots, X_n$ are conditionally IID given $\Theta = \theta$ with $U(\theta - 1/2, \theta + 1/2)$ distribution. Let $X = (X_1, \ldots, X_n)$. Then

$$f_{X|\Theta}(x|\theta) = I_{[\theta - \frac{1}{2}, \infty)}(\min x_i) I_{(-\infty, \theta + \frac{1}{2}]}(\max x_i).$$

Let $T_1 = \max X_i$ and $T_2 = \max X_i - \min X_i$. Then $T = (T_1, T_2)$ is minimal sufficient and T_2 is ancillary (see Problem 25 on page 141). So T_1 is conditionally sufficient given T_2. In particular, if $n = 2$, the density of T_2 with respect to Lebesgue measure is $f_{T_2}(t_2) = 2(1 - t_2) I_{[0,1]}(t_2)$.

When a statistic is ancillary, it does not mean that you should ignore it. It only means that if you learned nothing but the ancillary, you would not change your mind about Θ. You might, however, change your mind about everything else, including the conditional distribution of other data given Θ.

Example 2.41 (Continuation of Example 2.40; see page 95). The joint density of T given $\Theta = \theta$ is

$$f_{T_1,T_2|\Theta}(t_1,t_2|\theta) = n(n-1) t_2^{n-2} I_{[0,1]}(t_2) I_{[\theta-\frac{1}{2}+t_2, \theta+\frac{1}{2}]}(t_1).$$

Since T_1 is the maximum of n IID uniform random variables given Θ, it follows that the distribution of T_1 given $\Theta = \theta$ is $Beta(n,1)$ shifted by $\theta - 1/2$. It is not hard to show that the conditional distribution of T_1 given $(\Theta, T_2) = (\theta, t_2)$ is $U(\theta - 1/2 + t_2, \theta + 1/2)$. The bigger t_2 is, the more concentrated the distribution of T_1 given Θ is. Even though T_2 is ancillary and (by itself) tells us nothing about Θ, it tells us something about the conditional distribution of T_1 given Θ.

A common (but not universal) suggestion, in classical inference, is to perform inference conditional on ancillaries. The reason for this is that when one performs classical inference conditional on a statistic, the statistic does not count as data in the inference; it merely counts as background information that we supposedly knew before we collected the data. Since the ancillary does not contain (in itself) any information about Θ, no information is being lost by not treating it as part of the data. This allows the classical statistician a convenient way to condition on at least some of the data. In the Bayesian framework, one could construct a joint distribution for Θ and the data, then condition on all of the data, and whether a statistic is ancillary becomes irrelevant. In fact, whether a statistic is sufficient becomes irrelevant. Fisher (1934) proposed inference conditional on ancillaries because he claimed that it made better use of the information available in the actual sample obtained. Example 2.52 on page 100 illustrates Fisher's point, as does the first half of the following example.

Example 2.42. Suppose that X_1, X_2 are conditionally IID with $U(\theta - 1/2, \theta + 1/2)$ distribution given $\Theta = \theta$. Let $T_1 = \max\{X_1, X_2\}$ and $T_2 = |X_1 - X_2|$. It is traditional, in the classical literature, to interpret the statement

$$P_\theta(T_1 - T_2 \leq \Theta \leq T_1) = \frac{1}{2} \text{ for all } \theta$$

as meaning that one is 50 percent confident that the random interval

$$[T_1 - T_2, T_1] = [\min X_i, \max X_i] \tag{2.43}$$

will contain Θ.[9] However, we already saw that the conditional distribution of T_1 given $T_2 = t_2$ and $\Theta = \theta$ is $U(\theta - 1/2 + t_2, \theta + 1/2)$. It follows that

$$P_\theta(T_1 - T_2 \leq \Theta \leq T_1 | T_2 = t) = \begin{cases} \frac{t}{1-t} & \text{if } t < \frac{1}{2}, \\ 1 & \text{if } t \geq \frac{1}{2}. \end{cases}$$

If, for example, $T_2 \geq 1/2$ is observed, then we know that Θ is in the interval between $\min X_i$ and $\max X_i$. It would seem that knowledge of the ancillary gives us a better idea of how much "confidence" we should have that Θ is in the interval.

Alternatively, we could choose our interval using the conditional distribution given the ancillary T_2. For example, we can easily show that

$$P_\theta\left(T_1 - \frac{1}{4}(1+t_2) \leq \Theta \leq T_1 + \frac{1}{4} - \frac{3}{4}t_2 \,\middle|\, T_2 = t_2\right) = \frac{1}{2}, \text{ for all } \theta. \quad (2.44)$$

In the classical theory, one would be 50 percent confident that the random interval $[T_1 - (1+T_2)/4, T_1 + 1/4 - 3T_2/4]$ covers Θ conditional on T_2. In fact, since the probability in (2.44) is the same for all T_2 values, one would be 50 percent confident that the random interval covers Θ marginally. If one desires an interval in which one can place 50 percent confidence after seeing the data, then the interval in (2.44) makes far more sense than the one in (2.43). If T_2 is observed to be small, then we have not learned much about Θ, and the conditional interval is wide to reflect the uncertainty. The unconditional interval in (2.43) is very short, however, which is counterintuitive. Similarly, when T_2 is observed to be large, we have learned a lot about Θ and the second interval is short, while the first one is wide.

Suppose that we have a prior distribution for Θ with density $f_\Theta(\theta)$. Then the posterior density of Θ is a constant times $f_\Theta(\theta)I_{[t_1-1/2+t_2,t_1+1/2]}(\theta)$. If $f_\Theta(\theta)$ is almost constant over the interval $[t_1 - 1/2, t_1 - t_2 + 1/2]$, then the posterior is approximately

$$f_{\Theta|X}(\theta|x) \approx \frac{1}{1-t_2}I_{[t_1-\frac{1}{2}+t_2,t_1+\frac{1}{2}]}(\theta).$$

The posterior probability that Θ is in the interval in (2.44) is nearly $1/2$. If one uses the improper prior with constant density, then the posterior for Θ is $U(t_1 - 1/2 + t_2, t_1 + 1/2)$, and the fact that the posterior probability is $1/2$ that Θ is in the interval (2.44) will turn out to be a special case of Theorem 6.78.

Sometimes there is more than one ancillary statistic available. Some principle is needed to choose between them.

Definition 2.45. An ancillary U is *maximal* if every other ancillary is a function of U.

Example 2.46. Let P_θ say that $\{Y_n\}_{n=1}^\infty$ are IID with density (with respect to counting measure on the set $\{(0,0),(0,1),(1,0),(1,1)\}$)

$$f_{Y_1|\Theta}(y|\theta) = \begin{cases} \frac{1}{6}(1-\theta) & \text{if } y = (0,0), \\ \frac{1}{6}(1+\theta) & \text{if } y = (0,1), \\ \frac{1}{6}(2+\theta) & \text{if } y = (1,0) \\ \frac{1}{6}(2-\theta) & \text{if } y = (1,1). \end{cases}$$

---

[9] This is an example of a *confidence interval* statement. We will discuss confidence intervals in more depth in Section 5.2.1.

Here, $\Omega = [0,1]$. Now, let $X = (Y_1, \ldots, Y_N)$. Let the observable counts be N_{ij} equal to the number of Ys with the first coordinate i and the second coordinate j. Let M_i be the number of vectors with the first coordinate i, and let N_j be the number with the second coordinate j.

		First Coordinate		
		0	1	
Second	0	N_{00}	N_{10}	N_0
Coordinate	1	N_{01}	N_{11}	N_1
		M_0	M_1	

Then $N = N_{00} + N_{10} + N_{01} + N_{11}$ and

$$f_{X|\Theta}(x|\theta) = \left(\frac{1}{6}\right)^N (1-\theta)^{N_{00}} (1+\theta)^{N_{01}} (2+\theta)^{N_{10}} (2-\theta)^{N_{11}}.$$

Any three of the N_{ij} is minimal sufficient. We also see that

$$M_0 \sim \text{Bin}\left(N, \frac{1}{3}\right), \text{ given } \Theta = \theta,$$

$$N_0 \sim \text{Bin}\left(N, \frac{1}{2}\right), \text{ given } \Theta = \theta.$$

Both M_0 and N_0 are ancillary, but neither is maximal. The conditional inference will depend on which ancillary one chooses.

For example, $E_\theta(1 - 3N_{00}/N_0|N_0) = \theta$, and $E_\theta(1 - 2N_{00}/M_0|M_0) = \theta$. If one wanted to estimate Θ in the classical framework, there would seem to be two natural estimators available depending on which ancillary one chooses. (See Problems 31 and 32 on page 142.)

Sometimes we need to condition on a statistic even if it is not ancillary. The following example was given by Morris DeGroot (personal communication). A similar example can be found in Pratt (1962).

Example 2.47. Consider a meter that is trying to measure a quantity Θ. Suppose that the meter gives a reading Z, which has $N(\theta, 1)$ distribution given $\Theta = \theta$ if $Z < 2$, but if $Z \geq 2$, the reading is always 2. Let $X = \min\{Z, 2\}$ be the reading. Then $P_\theta(X = 2) = 1 - \Phi(2 - \theta)$, where Φ is the standard normal distribution function. For $x < 2$, $f_{X|\Theta}(x|\theta)$ is the $N(\theta, 1)$ density. The event $\{X = 2\}$ is not ancillary but is obviously important for what inference to perform. For example, trying to construct an unbiased estimator is difficult since

$$E_\theta(X) = \Phi(2-\theta)\theta + 2[1 - \Phi(2-\theta)] - \frac{1}{\sqrt{2\pi}} \exp\left\{-\frac{(2-\theta)^2}{2}\right\}.$$

On the other hand, if $X < 2$ is observed, the inference should be the same as if we had merely observed Z, since we actually did observe Z and the fact that $Z \geq 2$ could have occurred but didn't is irrelevant. If $X = 2$ is observed, the inference should be based on the fact that all we know is $Z \geq 2$, since the fact that $X < 2$ had been possible is now irrelevant.

A possible Bayesian solution to this problem would be to let Θ have a conjugate prior distribution, say $\Theta \sim N(\theta_0, \sigma_0^2)$ for known values of θ_0 and σ_0^2. The conditional distribution of Θ given $X = x$ is $N(\theta_1, \sigma_1^2)$ if $x < 2$, where

$$\mu_1 = \frac{\theta_0 + \sigma_0^2 x}{1 + \sigma_0^2}, \quad \sigma_1^2 = \frac{\sigma_0^2}{1 + \sigma_0^2}.$$

Inference would then proceed as if no truncation had been possible. On the other hand, if $X = 2$ is observed, the conditional distribution of Θ given $X = 2$ has density

$$f_{\Theta|X}(\theta|2) = \frac{\exp\left(-\frac{1}{2\sigma_0^2}(\theta - \theta_0)^2\right)[1 - \Phi(2 - \theta)]}{\sigma_0\sqrt{2\pi}\left[1 - \Phi\left(\frac{2-\theta_0}{\sqrt{1+\sigma_0^2}}\right)\right]}.$$

If we want the posterior mean of Θ, we can integrate to get

$$E(\Theta|X = 2) = \theta_0 + \frac{\exp\left(-\frac{1}{2(1+\sigma_0^2)}(\theta_0 - 2)^2\right)}{\sqrt{2\pi}\sqrt{1+\sigma_0^2}\left[1 - \Phi\left(\frac{2-\theta_0}{\sqrt{1+\sigma_0^2}}\right)\right]}.$$

Brown (1967) and Buehler and Fedderson (1963) prove that there are other statistics that are not ancillary but on which it might pay to condition when making inferences. In particular, they consider the case in which $X_1, \ldots, X_n$ are conditionally IID with $N(\mu, \sigma^2)$ distribution, conditional on $\Theta = \theta = (\mu, \sigma)$. Let $\overline{X} = \sum_{i=1}^n X_i/n$ and $S^2 = \sum_{i=1}^n (X_i - \overline{X})^2/(n-1)$. It is well known that $P_\theta(|\overline{X} - \mu|/S > k)$ depends only on k and n, call it $\alpha(k, n)$. What these authors show is that there is a set C and a number $a < \alpha(k, n)$ such that $P_\theta(|\overline{X} - \mu|/S > k|(\overline{X}, S) \in C) \leq a$ for all θ. Pierce (1973), Wallace (1959), and Buehler (1959) give conditions under which such examples can and cannot arise.

Ancillaries are only useful if there is no boundedly complete sufficient statistic.

Theorem 2.48 (Basu's theorem).[10] *If T is a boundedly complete sufficient statistic and U is ancillary, then U and T are independent given $\Theta = \theta$, and they are marginally independent no matter what prior one uses.*

PROOF. Let A be some measurable set of possible values of U. Since U is ancillary, $P'_\theta(U \in A) = \Pr(U \in A)$ for all θ. But, $P'_\theta(U \in A) = \int P'_\theta(U \in A|T = t)dP_{\theta,T}(t)$. So

$$\int [\Pr(U \in A) - \Pr(U \in A|T = t)]\, dP_{\theta,T}(t) = 0, \qquad (2.49)$$

for all θ, since T is sufficient. Let $g(t) = \Pr(U \in A) - \Pr(U \in A|T = t)$, which is a bounded measurable function. Equation (2.49) says that $E_\theta(g(T)) = 0$ for all θ. Since T is boundedly complete, we have $P'_\theta(g(T) = 0) = 1$ for all θ. This means that $P'_\theta(U \in A) = P'_\theta(U \in A|T = t)$, a.s. $[P_{\theta,T}]$ for all θ, which implies that U and T are conditionally independent given $\Theta = \theta$, for all θ.

---

[10] See Basu (1955, 1958).

Let μ_Θ be an arbitrary prior, and let B be a measurable set of possible values of T.

$$\Pr(U \in A, T \in B) = \int_\Omega \int_B \Pr(U \in A | T = t) dP_{\theta,T}(t) d\mu_\Theta(\theta)$$

$$= \int_\Omega \Pr(U \in A) P_{\theta,T}(B) d\mu_\Theta(\theta) = \Pr(U \in A) \Pr(T \in B),$$

which says that U and T are marginally independent. □

Basu's theorem 2.48 says that if T is a boundedly complete sufficient statistic, conditioning on an ancillary is not going to change the joint distribution of T and Θ. Both Bayesian and classical statisticians would ignore the ancillary in such a case.

Example 2.50. Suppose that P_θ says that $\{X_n\}_{n=1}^\infty$ are IID $N(\theta, 1)$. Let $X = (X_1, \ldots, X_n)$, $T = \overline{X}$, and $U = \sum_{i=1}^n (X_i - T)^2/(n-1)$. Then T is a complete sufficient statistic and U is ancillary. They are independent given $\Theta = \theta$ and are marginally independent no matter what prior we use.

Example 2.51. Suppose that P_θ says that $\{X_n\}_{n=1}^\infty$ are IID $N(\mu, \sigma^2)$, where $\theta = (\mu, \sigma)$. Let $X = (X_1, \ldots, X_n)$, $T_1 = \overline{X}$, and $T_2 = \sqrt{\sum_{i=1}^n (X_i - T_1)^2/(n-1)}$. The fact that $T = (T_1, T_2)$ is a complete sufficient statistic will follow most easily from Theorem 2.74, to be proven later. Let

$$U = \left(\frac{X_1 - T_1}{T_2}, \ldots, \frac{X_n - T_1}{T_2} \right).$$

Then, U is ancillary and independent of T given $\Theta = \theta$. Also T and U are marginally independent no matter what prior we use. The distribution of U is uniform on a sphere of radius 1 in an $(n-1)$-dimensional hyperplane. (See Problem 28 on page 141.)

One reason that some people give for conditioning on an ancillary is that they get a better measure of the precision of the inference. Here is an example due to Basu.

Example 2.52. Let $\Theta = (\Theta_1, \ldots, \Theta_N)$, where N is the (known) size of a population and Θ_i is some characteristic of unit i in the population. Select a set of labels $i_1, \ldots, i_n$ from $\{1, \ldots, N\}$ with replacement, with $n \leq N$. Let $X_j = \Theta_{i_j}$ be observed for $j = 1, \ldots, n$. Let $X = (X_1, \ldots, X_n)$. If the selection is random, then $f_{X|\Theta}(x|\theta) = 1/N^n$ for all x compatible with θ. (Notice that the distribution of X is dependent on Θ even though the distribution of the labels is not.) Let M be the number of distinct labels drawn. Then $P_\theta(M = m)$ is the same for all θ, so M is ancillary. Let $X_1^*, \ldots, X_M^*$ be the distinct observed values. One possible estimate of the population average is $\overline{X}^* = \sum_{i=1}^M X_i^*/M$. The conditional variance of $\overline{X}^*$ given $\Theta = \theta$ and $M = m$ is

$$\mathrm{Var}(\overline{X}^* | \Theta = \theta, M = m) = \frac{N-m}{N-1} \frac{\sigma^2}{m},$$

where $\sigma^2 = \sum_{i=1}^N (\theta_i - \overline{\theta})^2/N$.

To see that this is a better measure of the variance of $\overline{X}^*$ than is the marginal variance, consider the simple case $n = 3$. The distribution of M is

$$f_M(m) = \begin{cases} \frac{1}{N^2} & \text{if } m = 1, \\ \frac{2(N-1)}{N^2} & \text{if } m = 2, \\ \frac{(N-1)(N-2)}{N^2} & \text{if } m = 3, \\ 0 & \text{otherwise.} \end{cases}$$

Since $\mathrm{E}(\overline{X}^* | \Theta = \theta, M = m) = \overline{\theta}$ for all m, it follows that

$$\mathrm{Var}(\overline{X}^* | \Theta = \theta) = \mathrm{E}\left(\frac{N-M}{N-1}\frac{\sigma^2}{M}\bigg|\Theta = \theta\right)$$

$$= \frac{\sigma^2}{N^2}\left(1 + (N-2) + \frac{(N-2)(N-3)}{3}\right) = \frac{\sigma^2}{3}\frac{N^2 - 2N + 3}{N^2}.$$

If $M = 3$, the marginal variance is larger than the conditional variance, while if $M = 1$, the marginal variance is too small.

To execute a Bayesian solution, we need a distribution for Θ. Suppose that we model the Θ_i as exchangeable random variables with Θ_i conditionally independent $N(\psi, \sigma^2)$ given $(\Psi, \Sigma) = (\psi, \sigma)$. The distribution of Ψ given $\Sigma = \sigma$ is $N(\psi_0, \sigma^2/\lambda_0)$, while the distribution of Σ^2 is $\Gamma^{-1}(a_0/2, b_0/2)$.[11] The data consist of observing $M = m$ and $\Theta_{i_j} = x_j^*$, for $j = 1, \ldots, m$. The unobserved Θ_i are still exchangeable, and their conditional distribution given $(\Psi, \Sigma) = (\psi, \sigma)$ is as in the prior. The distribution of Ψ given $\Sigma = \sigma$ and $X - x$ is $\Psi \sim N(\psi_1, \sigma^2/\lambda_1)$, and the distribution of Σ^2 given $X = x$ is $\Gamma^{-1}(a_1/2, b_1/2)$, where

$$\lambda_1 = \lambda_0 + m, \qquad \psi_1 = \frac{\lambda_0 \psi_0 + m\overline{x}^*}{\lambda_1},$$

$$a_1 = a_0 + m, \qquad b_1 = b_0 + \sum_{i=1}^m (x_i^* - \overline{x}^*)^2 + \frac{m\lambda_0}{m + \lambda_0}(\psi_0 - \overline{x}^*)^2.$$

The posterior distribution of the population average $\overline{\Theta}$ is obtained in stages. First, conditional on $(\Psi, \Sigma) = (\psi, \sigma)$,

$$\overline{\Theta} \sim N\left(\frac{m\overline{x}^* + (N-m)\psi}{N}, \sigma^2 \frac{N-m}{N^2}\right).$$

Integrating ψ out of this, we get that conditional on $\Sigma = \sigma$,

$$\overline{\Theta} \sim N\left(\frac{m\overline{x}^* + (N-m)\psi_1}{N}, \frac{\sigma^2}{N^2}\left[N - m + \frac{(N-m)^2}{\lambda_1}\right]\right).$$

Finally, integrating σ out, we get that $\overline{\Theta}$ has distribution

$$t_{a_1}\left(\frac{m\overline{x}^* + (N-m)\psi_1}{N}, \frac{b_1}{N^2 a_1}\left[N - m + \frac{(N-m)^2}{\lambda_1}\right]\right).$$

---

[11] This is an example of a *hierarchical model*, which will be discussed in more detail in Chapter 8.

If we use an improper prior ($\lambda_0 = 0$, $b_0 = 0$, $a_0 = -1$), then the location becomes $\bar{x}^*$ and the squared scale factor becomes

$$\frac{N-m}{Nm} \frac{1}{m-1} \sum_{i=1}^{m} (x_i^* - \bar{x}^*)^2.$$

The latter is very close to the traditional finite population sampling theory variance estimate.

We conclude this section with two examples that are similar on their surface, but in one example the ancillary is part of the sufficient statistic and in the other it is not.

Example 2.53. Let $Z \sim Ber(1/2)$ (independent of Θ), and let Y and W be conditionally independent given $\Theta = \theta$ (and independent of Z) with $Y \sim N(\theta, 1)$ and $W \sim N(\theta, 2)$. If $Z = 0$, we will observe $X = (Y, Z)$. If $Z = 1$, we will observe $X = (W, Z)$. Let X_1 stand for the first coordinate of X. The likelihood function is

$$f_{X|\Theta}(x_1, z|\theta) = \frac{1}{2\sqrt{2\pi}} \left[\exp\left(-\frac{1}{2}(x_1 - \theta)^2\right) \right]^{1-z} \left[\frac{1}{\sqrt{2}} \exp\left(-\frac{1}{4}(x_1 - \theta)^2\right) \right]^{z}.$$

It is possible to show that X_1 is not a sufficient statistic. (See Problem 8 on page 138.) In this case, it makes perfect sense to perform inference conditional on the ancillary Z.

Example 2.54. Let $Z \sim Ber(1/2)$ (independent of Θ), and let $\{Y_n\}_{n=1}^{\infty}$ be conditionally IID with $Ber(\theta)$ random variables given $\Theta = \theta$ (and independent of Z.) If $Z = 1$, we will observe $Y_1, \ldots, Y_n$ for fixed n. If $Z = 0$, we will observe the Y_i until we see k successes with $k < n$. In Problem 9 on page 138, we will see that a sufficient statistic is (N, M), where N is the number of observed Y_i and M is the number of successes among the observed Y_i. Clearly, Z is ancillary, but it is difficult to justify conditioning on Z since it is not part of the sufficient statistic. That is, if we observe n of the Y_i and there are k successes among them, then it does not matter to us whether $Z = 0$ or 1. (Of course, in all other cases we can figure out what Z was from the rest of the data.)

2.2 Exponential Families of Distributions

2.2.1 Basic Properties

There is a special class of distributions for which complete sufficient statistics with fixed dimension always exist. This class includes some, but not all, of the commonly used distributions.

Definition 2.55. A parametric family with parameter space Ω and density $f_{X|\Theta}(x|\theta)$ with respect to a measure ν on $(\mathcal{X}, \mathcal{B})$ is called an *exponential family* if

$$f_{X|\Theta}(x|\theta) = c(\theta) h(x) \exp\left\{ \sum_{i=1}^{k} \pi_i(\theta) t_i(x) \right\},$$

2.2. Exponential Families of Distributions

for some measurable functions $\pi_1, \ldots, \pi_k, t_1, \ldots, t_k$ and some integer k.

Example 2.56. Suppose that P_θ says that $\{X_n\}_{n=1}^\infty$ are IID $N(\mu, \sigma^2)$, where $\theta = (\mu, \sigma)$. Let $X = (X_1, \ldots, X_n)$.

$$
\begin{aligned}
f_{X|\Theta}(x|\theta) &= \frac{1}{(2\pi)^{n/2}\sigma^n} \exp\left\{-\frac{1}{2\sigma^2}\sum_{i=1}^n (x_i - \mu)^2\right\} \\
&= \frac{1}{(2\pi)^{n/2}}\sigma^{-n} \exp\left\{-\frac{n\mu^2}{2\sigma^2}\right\} \exp\left\{-\frac{1}{2\sigma^2}\sum_{i=1}^n x_i^2 + \frac{\mu}{\sigma^2}n\bar{x}\right\}.
\end{aligned}
$$

In this form, we see that $k = 2$ and

$$h(x) = \frac{1}{(2\pi)^{n/2}}, \quad t_1(x) = n\bar{x}, \quad t_2(x) = \sum_{i=1}^n x_i^2,$$

$$c(\theta) = \sigma^{-n} \exp\left\{-\frac{n\mu^2}{2\sigma^2}\right\}, \quad \pi_1(\theta) = \frac{\mu}{\sigma^2}, \quad \pi_2(\theta) = -\frac{1}{2\sigma^2}.$$

The function $c(\theta)$ in Definition 2.55 can be written as

$$c(\theta) = \left(\int_{\mathcal{X}} h(x) \exp\left\{\sum_{i=1}^k \pi_i(\theta) t_i(x)\right\} d\nu(x)\right)^{-1},$$

so that the dependence on θ is through the vector $\pi = (\pi_1(\theta), \ldots, \pi_k(\theta)) \in \mathbb{R}^k$. We might as well let π be the parameter.

Definition 2.57. In an exponential family, the *natural parameter* is the vector $\Pi = (\pi_1(\Theta), \ldots, \pi_k(\Theta))$, and

$$\Gamma = \left\{\pi \in \mathbb{R}^k : \int_{\mathcal{X}} h(x) \exp\left(\sum_{i=1}^k \pi_i t_i(x)\right) d\nu(x) < \infty\right\}$$

is called the *natural parameter space*.

The mapping $\Pi : \Omega \to \Gamma$ need not be one-to-one, nor need it be onto. It is common, however, to use the symbol Θ for the natural parameter and assume that $\Omega = \Gamma$. It is obvious from the form of the exponential family density and the Fisher–Neyman factorization theorem 2.21 that $T(X) = (t_1(X), \ldots, t_k(X))$ is a sufficient statistic. This statistic is sometimes called the *natural sufficient statistic*.

The sufficient statistic from an exponential family sample also has an exponential family distribution.

Lemma 2.58.[12] *If X has an exponential family distribution, then so does the natural sufficient statistic $T(X)$, and the natural parameter for T is the*

---

[12] This lemma is used in the proof of Theorem 2.62.

same as for X. In particular, there is a measure ν_T such that $P_{\theta,T} \ll \nu_T$ for all θ and $dP_{\theta,T}/d\nu_T(t) = c(\theta)\exp(\theta^\top t)$.

PROOF. Apply Lemma 2.24 with $m_2(T(x),\theta) = c(\theta)\exp(\sum_{i=1}^k \theta_i t_i(x))$ and $m_1(x) = h(x)$. □

Example 2.59 (Continuation of Example 2.56; see page 103). In the case of n conditionally IID $N(\mu,\sigma^2)$ random variables, the natural sufficient statistics are $T_1 = \sum_{i=1}^n X_i$ and $T_2 = \sum_{i=1}^n X_i^2$. It is well known that T_1 and $W = T_2 - T_1^2/n$ are independent, with T_1 having $N(n\mu, n\sigma^2)$ distribution and W having $\Gamma([n-1]/2, 1/[2\sigma^2])$ distribution. It follows that the joint density of (T_1, T_2) is

$$\left[\sqrt{\pi}\sigma^n \Gamma\left(\frac{n-1}{2}\right) 2^n\right]^{-1} \left(t_2 - \frac{t_1^2}{n}\right)^{\frac{n-1}{2}} \exp\left(-\frac{1}{2\sigma^2}\left[\frac{(t_1 - n\mu)^2}{n} + t_2 - \frac{t_1^2}{n}\right]\right).$$

This can be simplified to $c(\theta)h(t_1,t_2)\exp(\theta_1 t_1 + \theta_2 t_2)$, with $c(\theta)$ the same as in Example 2.56 and $h(t_1,t_2)$ a constant times $(t_2 - t_1^2/n)^{[n-1]/2}$.

There are degenerate exponential families. That is, it is possible for some linear function of X to be constant (the same constant for all θ) with probability 1 given $\Theta = \theta$ for all θ. For example, let Y_1 be k_1-dimensional and have conditional density given $\Theta = \theta$ (with respect to a measure ν_1) $c(\theta)\exp(y_1^\top \theta)$. Define $\Gamma^\top = (\Theta^\top, \Psi^\top)$, where Ψ is k_2-dimensional. Let ν_2 be the measure that puts a mass of 1 on the point $r \in \mathbb{R}^{k_2}$ and puts 0 mass on the rest of $\mathbb{R}^{k_2}$. Define $\nu = \nu_1 \times \nu_2$, and let $Y^\top = (Y_1^\top, r^\top)$. Then Y has conditional density given $\Gamma = (\theta, \psi) = \gamma$ with respect to ν,

$$c^*(\gamma)\exp(y^\top \gamma) = c(\theta)\exp(y_1^\top \theta),$$

where $c^*(\gamma) = c(\theta)\exp(-r^\top \psi)$. The natural parameter space of Γ values is $\Omega \times \mathbb{R}^{k_2}$, where Ω is the natural parameter space of Θ values. For this reason, we introduce a definition.

Definition 2.60. An exponential family of distributions for X is *degenerate* if there exists at least one vector α and a scalar r such that $P_\theta(\alpha^\top X = r) = 1$ for all θ. If the exponential family is not degenerate, it is called *nondegenerate*.

Example 2.61. Let $X \sim \text{Mult}_k(n; p_1, \ldots, p_k)$ given $P = (p_1, \ldots, p_k)$. The natural parameter is $\Theta = (\log P_1, \ldots, \log P_k)$. We know that $P_\theta(\mathbf{1}^\top X = n) = 1$ for all θ, where $\mathbf{1}$ is a vector of k 1s.

When the exponential family is degenerate, the natural parameter space is a subset of a $(k-1)$-dimensional linear manifold in $\mathbb{R}^k$, hence it has empty interior. Some theorems in Section 2.2.2 will require that the natural parameter space have a nonempty interior. However, degenerate families of distributions are easily converted into nondegenerate families by means of linear transformations. For example, with the multinomial distribution, we could just delete the last coordinates of both X and the natural parameter. For this nondegenerate family, the natural parameter space does contain an open subset of $\mathbb{R}^{k-1}$.

2.2.2 Smoothness Properties

The means of functions of exponential family random variables tend to be smooth functions of the natural parameter. In fact, the natural parameter space is itself a nice subset of Euclidean space.

Theorem 2.62. *The natural parameter space Ω of an exponential family is convex and $1/c(\theta)$ is a convex function.*

PROOF. We will work in the sufficient statistic space. Write $1/c(\theta) = \int \exp\{t^\top \theta\} d\nu_T(t)$, where ν_T is the measure from Lemma 2.58. Since $\exp(\cdot)$ is a convex function, we get, for $\theta_1, \theta_2 \in \Omega$ and $0 < \alpha < 1$,

$$\frac{1}{c(\alpha\theta_1 + [1-\alpha]\theta_2)} = \int \exp\{t^\top[\alpha\theta_1 + (1-\alpha)\theta_2]\} d\nu_T(t)$$

$$\leq \int \left(\alpha \exp\{t^\top \theta_1\} + (1-\alpha) \exp\{t^\top \theta_2\}\right) d\nu_T(t)$$

$$= \alpha \int \exp\{t^\top \theta_1\} d\nu_T(t) + (1-\alpha) \int \exp\{t^\top \theta_2\} d\nu_T(t)$$

$$= \alpha \frac{1}{c(\theta_1)} + (1-\alpha)\frac{1}{c(\theta_2)} < \infty.$$

This proves that $\alpha\theta_1 + (1-\alpha)\theta_2 \in \Omega$, so Ω is convex. It also proves that $1/c$ is convex. □

Example 2.63. The family of exponential distributions, $Exp(\psi)$, with densities $f_{X|\Psi}(x|\psi) = \psi \exp(-\psi x)$ for $x > 0$ has $h(x) = I_{(0,\infty)}(x)$ and natural parameter $\theta = -\psi$. So $c(\theta) = -\theta$ and $1/c(\theta) = -1/\theta$ is convex. The natural parameter space is $(-\infty, 0)$, a convex set.

The following theorem is used in several places in the remainder of the text to establish smoothness properties of various conditional means, given the natural parameters, of functions of random variables with exponential family distributions.

Theorem 2.64. *Let the density of $T(X)$ with respect to a measure ν_T be $c(\theta) \exp\{t^\top \theta\}$. If $\phi : \mathcal{T} \to \mathbb{R}$ is measurable and $\int |\phi(t)| \exp\{t^\top \theta\} d\nu_T(t) < \infty$, for θ in the interior of the natural parameter space, then*

$$f(z) = \int \phi(t) \exp\{t^\top z\} d\nu_T(t)$$

is an analytic function[13] of z in the region where the real part of z is interior to the natural parameter space, and

$$\frac{\partial}{\partial z_i} f(z) = \int t_i \phi(t) \exp\{t^\top z\} d\nu_T(t).$$

---

[13] By *analytic function*, we mean a complex-valued function of a complex (vector) argument that is differentiable with respect to that complex argument.

PROOF. We will do the $k = 1$ case, as the others follow by induction. Let $z_0 = a + ib$ and $\delta = \delta_1 + i\delta_2$ for some a in the interior of Ω. Then

$$\frac{f(z_0 + \delta) - f(z_0)}{\delta} = \int \phi(t) \exp(tz_0) \frac{\exp\{t\delta\} - 1}{\delta} d\nu_T(t). \quad (2.65)$$

The maximum modulus theorem C.8 says that an analytic function on a closed bounded set achieves its maximum on the boundary of the set. For $0 < \gamma < \epsilon$, consider the set $C(\gamma, \epsilon) = \{z : \gamma \leq |z| \leq \epsilon\}$. For fixed ϵ and every $0 < \gamma < \epsilon$,

$$\max_{z \in C(\gamma,\epsilon)} \left| \frac{\exp(tz) - 1}{z} \right| \leq \max\left\{ \frac{\exp(|t|\epsilon) - 1}{\epsilon}, \frac{\exp(|t|\gamma) - 1}{\gamma} \right\}.$$

Since the limit as $\gamma \to 0$ of the last term above is $|t|$ and $\exp(|t|\epsilon) - 1 > |t|\epsilon$, it follows that $|(\exp(tz) - 1)/z| \leq \exp(|t|\epsilon)/\epsilon$ for all $|z| \leq \epsilon$. Thus, we have that if $|\delta| \leq \epsilon$, the absolute value of the integrand in (2.65) is no more than $|\phi(t)| \exp(at) \exp(|t|\epsilon)/\epsilon$. Thus, the integral of the absolute value is at most $\int |\phi(t)| (\exp\{t(a+\epsilon)\} + \exp\{t(a-\epsilon)\}) d\nu_T(t)/\epsilon$. Choose ϵ small enough so that $a \pm \epsilon$ are in the interior of Ω. By the dominated convergence theorem,

$$\lim_{\delta \to 0} \frac{f(z_0 + \delta) - f(z_0)}{\delta} = \int \phi(t) t \exp(tz_0) d\nu_T(t). \qquad \square$$

Theorem 2.64 allows us to calculate moments of the sufficient statistics in exponential families by taking derivatives of the function $\log c(\theta)$.

Example 2.66. Let $\phi(t) = 1$. Then we can calculate

$$\begin{aligned}
E_\theta(T_i) &= \int c(\theta) t_i \exp(t^\top \theta) d\nu_T(t) = c(\theta) \frac{\partial}{\partial \theta_i} \int \exp(t^\top \theta) d\nu_T(t) \\
&= c(\theta) \frac{\partial}{\partial \theta_i} \frac{1}{c(\theta)} = -\frac{1}{c(\theta)} \frac{\partial c(\theta)}{\partial \theta_i} = -\frac{\partial}{\partial \theta_i} \log c(\theta).
\end{aligned}$$

Example 2.67 (Continuation of Example 2.63; see page 105). Consider the $Exp(\psi)$ distribution with $\theta = -\psi$. Here $\log c(\theta) = \log(-\theta)$. So the partial with respect to θ is $1/\theta = -E_\theta(T)$.

Example 2.68 (Continuation of Example 2.56; see page 103). Consider the case of $N(\mu, \sigma^2)$ distributions. Here, the natural parameter is $(\theta_1, \theta_2) = (\mu/\sigma^2, -1/[2\sigma^2])$, and the natural sufficient statistic is $(T_1, T_2) = (n\overline{X}, \sum_{i=1}^n X_i^2)$. So,

$$\log c(\theta) = \frac{n}{2} \log(-2\theta_2) + \frac{n}{4} \frac{\theta_1^2}{\theta_2}. \quad (2.69)$$

The partial derivative with respect to θ_1 is

$$\frac{n}{2} \frac{\theta_1}{\theta_2} = -n\mu = -E_\theta(T_1).$$

The partial derivative with respect to θ_2 is

$$\frac{n}{2\theta_2} - \frac{n\theta_1^2}{4\theta_2^2} = -n(\sigma^2 + \mu^2) = -E_\theta(T_2).$$

The method illustrated in Example 2.66 is actually quite general.

Proposition 2.70.[14] *Let $T = (T_1, \ldots, T_k)$. Suppose that the conditional density of T given $\Theta = \theta$ is $f_{T|\Theta}(t|\theta) = c(\theta) \exp(\theta^\top t)$. Let $\ell_1, \ldots, \ell_k \geq 0$ be such that $\ell = \ell_1 + \cdots + \ell_k$. Then*

$$E_\theta \left(\prod_{i=1}^k T_i^{\ell_i} \right) = c(\theta) \frac{\partial^\ell}{\partial \theta_1^{\ell_1} \cdots \partial \theta_k^{\ell_k}} \frac{1}{c(\theta)}.$$

In particular, $E_\theta(T_i) = -\partial/\partial \theta_i \log c(\theta)$, and

$$\mathrm{Cov}_\theta(T_i, T_j) = -\frac{\partial^2}{\partial \theta_i \partial \theta_j} \log c(\theta).$$

Example 2.71 (Continuation of Example 2.68; see page 106). For the $N(\mu, \sigma^2)$ case, $\log c(\theta)$ is given in (2.69). The covariance of $\overline{X}$ and $\sum_{i=1}^n X_i^2$ is

$$-\frac{\partial^2}{\partial \theta_1 \partial \theta_2} \left[\frac{n}{2} \log(-2\theta_2) + \frac{n}{4} \frac{\theta_1^2}{\theta_2} \right] = \frac{n\theta_1}{2\theta_2^2} = 2n\mu\sigma^2,$$

as can be verified directly.

A similar result holds for the posterior means of polynomial functions of Θ, if we use a conjugate prior.

Proposition 2.72. *Let $X = (X_1, \ldots, X_n)$ where the X_i are conditionally IID, given $\Theta = \theta$, with density equal to $c(\theta) \exp(\theta^\top T(x))$, where Θ is a k-dimensional parameter. Suppose that the prior for Θ is proportional to $c(\theta)^a \exp(\theta^\top b)$, where $a > 0$ and b is a k-dimensional vector (a natural conjugate prior). Suppose that $\ell_1, \ldots, \ell_k \geq 0$ and $\ell = \ell_1 + \cdots + \ell_k$. Write the predictive density of X as $f_X(x) = g(t_1, \ldots, t_k)$, where $t_i = \sum_{j=1}^n T_i(x_j)$. Then*

$$E\left(\prod_{i=1}^k \Theta_i^{\ell_i} \middle| X = x \right) = \frac{1}{f_X(x)} \frac{\partial^\ell}{\partial t_1^{\ell_1} \cdots \partial t_k^{\ell_k}} g(t_1, \ldots, t_k).$$

Example 2.73. Suppose that $X_1, \ldots, X_n$ are conditionally IID with $N(\mu, \sigma^2)$ distribution given $M = \mu$ and $\Sigma = \sigma$. Let the prior be natural conjugate as in Example 1.24 on page 14. The marginal density of the data is given in (1.27) in that example. Rewriting (1.27) in terms of the natural sufficient statistics $T_1 = n\overline{X}$ and $T_2 = W + n\overline{X}^2$, we get

$$g(t_1, t_2) = \text{constant} \times \left(b_0 + t_2 + \frac{t_1^2}{n} + \frac{n\lambda_0}{\lambda_0 + n} \left[\frac{t_1}{n} - \mu_0 \right]^2 \right)^{-\frac{a_1}{2}}.$$

---

[14]This proposition is used in the proofs of Theorems 3.44 and 7.57.

The partial derivative of this with respect to t_1 divided by $g(t_1, t_2)$ equals

$$-\frac{a_1}{2}\left(b_0 + t_2 + \frac{t_1^2}{n} + \frac{n\lambda_0}{\lambda_0 + n}\left[\frac{t_1}{n} - \mu_0\right]^2\right)^{-1}\left(\frac{2\lambda_0}{\lambda_0 + n}\left[\frac{t_1}{n} - \mu_0\right] - \frac{2t_1}{n}\right),$$

which simplifies to $\mu_1 a_1/b_1$, the posterior mean of $\Theta_1 = M/\Sigma^2$.

Diaconis and Ylvisaker (1979) prove other interesting results about posterior means of parameters when conjugate priors are used for exponential families.

The following theorem is used to show that certain estimators and hypothesis tests have classical optimality properties when the data come from an exponential family.

Theorem 2.74. *If the natural parameter space Ω of an exponential family contains an open set in $\mathbb{R}^k$, then $T(X)$ is a complete sufficient statistic.*

PROOF. We will prove the $k = 1$ case, and the others follow by induction. Let $T(X)$ have density $c(\theta)\exp\{t\theta\}$ with respect to a measure ν_T. Let g be a function such that $E_\theta(g(T)) = 0$ for all θ. Then

$$\int g(t)c(\theta)\exp\{t\theta\}d\nu_T(t) = 0$$

for all θ. This says that

$$\int g^+(t)\exp\{t\theta\}d\nu_T(t) = \int g^-(t)\exp\{t\theta\}d\nu_T(t), \qquad (2.75)$$

where g^+ and g^- are respectively the positive and negative parts of g. Since $E_\theta(g(T))$ exists for all θ, both sides of (2.75) are finite for all θ. Let θ_0 be interior to Ω, and let the common value of both sides of (2.75) be r when $\theta = \theta_0$. Define two probability measures:

$$P(A) = \frac{1}{r}\int_A g^+(t)\exp\{t\theta_0\}d\nu_T(t),$$

$$Q(A) = \frac{1}{r}\int_A g^-(t)\exp\{t\theta_0\}d\nu_T(t).$$

The two sides of (2.75) are

$$\int \exp(t[\theta - \theta_0])dP(t) = \int \exp(t[\theta - \theta_0])dQ(t).$$

By Theorem 2.64, these are analytic functions of $\psi = \theta - \theta_0$. According to Theorem C.7, these functions equal their power series expansions in a neighborhood, say $(-\psi_0, \psi_0)$, of $\psi = 0$. The fact that they agree for all real values near 0 implies that they have the same derivatives at 0, hence they have the same power series expansion around 0, hence they are also equal

2.2. Exponential Families of Distributions

at imaginary values of ψ near 0, hence they are also equal at all imaginary values because they are analytic in the region where the real part of ψ is in $(-\psi_0, \psi_0)$. For $\psi = iu$, we get that the characteristic function of P equals the characteristic function of Q in a neighborhood of 0. By Corollary B.106, it follows that $P = Q$, hence $g^+(t) = g^-(t)$ a.e. $[\nu_T]$. This ensures that $P_\theta(g(T) = 0) = 1$ for all θ. □

As examples, the sufficient statistics from normal, exponential, Poisson, and Bernoulli distributions are complete.

2.2.3 A Characterization Theorem*

The following theorem characterizes one-parameter exponential families essentially as those families of distributions with smooth densities on a common set with one-dimensional sufficient statistics for all sample sizes.

Theorem 2.76. *Suppose that $X_1, \ldots, X_n$ are conditionally IID given $\Theta = \theta$ each with density $f_{X_1|\Theta}(\cdot|\theta)$. Let T be a one-dimensional sufficient statistic. Write*

$$\prod_{i=1}^{n} f_{X_1|\Theta}(x_i|\theta) = m_1(x) m_2(t, \theta).$$

Define

$$K_\theta(t) = \frac{\partial}{\partial \theta} \log m_2(t, \theta).$$

Assume the following conditions:

1. *The set of y such that $f_{X_1|\Theta}(y|\theta) > 0$ is the same for all θ.*
2. *$f_{X_1|\Theta}(y|\theta)$ is differentiable with respect to θ for each y.*
3. *$f_{X_1|\Theta}(y|\theta)$ is differentiable with respect to y for each θ.*
4. *There exists θ_0 such that $K_{\theta_0}(t)$ has an inverse.*

Then, X has an exponential family distribution with a one-dimensional natural parameter.

PROOF. Write

$$\sum_{i=1}^{n} \log f_{X_1|\Theta}(x_i|\theta) = \log m_2(t, \theta) + \log m_1(x),$$

and define

$$v(x_i) = \left. \frac{\partial}{\partial \theta} \log f_{X_1|\Theta}(x_i|\theta) \right|_{\theta=\theta_0},$$

$$q_\theta(r) = K_\theta \left[K_{\theta_0}^{-1}(r) \right], \qquad r(x) = \sum_{i=1}^{n} v(x_i).$$

---

*This section may be skipped without interrupting the flow of ideas.

110   Chapter 2. Sufficient Statistics

Since $K_\theta(t) = q_\theta(K_{\theta_0}(t)) = q_\theta(r(x))$, it follows that

$$\frac{\partial^2}{\partial \theta \partial x_i} \log f_{X_1|\Theta}(x_i|\theta) = \frac{\partial}{\partial r} q_\theta(r) \frac{\partial}{\partial x_i} r(x).$$

Thus, we get $\partial r(x)/\partial x_i = v'(x_i)$, and

$$\frac{1}{v'(x_i)} \frac{\partial^2}{\partial \theta \partial x_i} \log f_{X|\Theta}(x|\theta) = \frac{\partial}{\partial r} q_\theta(r). \tag{2.77}$$

Since r is invariant under permutations of the coordinates of x and the left-hand side of (2.77) depends on x through x_i alone, both sides must depend only on θ. So, we get

$$\frac{\partial}{\partial r} q_\theta(r) = c_1(\theta), \qquad q_\theta(r) = r c_1(\theta) + c_2(\theta),$$

$$K_\theta(t) = K_{\theta_0}(t) c_1(\theta) + c_2(\theta) = \frac{\partial}{\partial \theta} \log m_2(t, \theta),$$

$$\log m_2(t, \theta) = K_{\theta_0}(t) \phi_1(\theta) + \phi_2(\theta) + s(t),$$

where $\phi_i(\theta) = \int_{\theta_0}^{\theta} c_i(u) du$, for $i = 1, 2$, and $s(t)$ is determined by boundary conditions. It follows that

$$m_2(t, \theta) = \exp\{s(t)\} \exp\{\phi_2(\theta)\} \exp\{K_{\theta_0}(t) \phi_1(\theta)\}.$$

Thus, we see that the density is in the form of an exponential family with $k = 1$ and

$$h(x) = m_1(x) \exp\{s(t)\}, \qquad c(\theta) = \exp\{\phi_2(\theta)\},$$
$$t_1(x) = K_{\theta_0}(t), \qquad \pi_1(\theta) = \phi_1(\theta). \qquad \square$$

There are similar theorems in multiparameter cases, but they have even more conditions. We give a different type of theorem characterizing exponential families by their sufficient statistics in Theorem 2.114. The importance of the existence of a fixed-dimensional sufficient statistic is twofold. First, it means that there is a fixed amount of information that must be stored for making inference about Θ regardless of the sample size. Second, there is the possibility of using natural conjugate prior distributions as in Theorem 2.25.

2.3 Information

It seems intuitively sensible to expect more data to provide more information about a parameter or a distribution. Similarly, if a statistic is sufficient, it should contain all of the information about the parameter, and vice versa. To make these ideas precise, we need to define information. There are two popular definitions of information: Fisher information and Kullback–Leibler information.

2.3.1 Fisher Information

Fisher information is designed to provide a measure of how much information a data set provides about a parameter in a parametric family with some smoothness properties.

Definition 2.78. Suppose that Θ is k-dimensional and that $f_{X|\Theta}(x|\theta)$ is the density of X with respect to ν. The following conditions will be known as the *FI regularity conditions*:

1. There exists B with $\nu(B) = 0$ such that for all θ, $\partial f_{X|\Theta}(x|\theta)/\partial \theta_i$ exists for $x \notin B$ and each i.

2. $\int f_{X|\Theta}(x|\theta) d\nu(x)$ can be differentiated under the integral sign with respect to each coordinate of θ.

3. The set $C = \{x : f_{X|\Theta}(x|\theta) > 0\}$ is the same for all θ.

Definition 2.79. Assume that the three FI regularity conditions above hold. Then the matrix $\mathcal{I}_X(\theta) = ((I_{X,i,j}(\theta)))$ with elements

$$I_{X,i,j}(\theta) = \mathrm{Cov}_\theta \left(\frac{\partial}{\partial \theta_i} \log f_{X|\Theta}(X|\theta), \frac{\partial}{\partial \theta_j} \log f_{X|\Theta}(X|\theta) \right)$$

is called *the Fisher information matrix about Θ based on X*. The random vector with coordinates $\partial \log f_{X|\Theta}(X|\theta)/\partial \theta_i$ is called the *score function*. If T is a statistic, the *conditional score function* is the vector whose ith coordinate is $\partial \log f_{X|T,\Theta}(X|t,\theta)/\partial \theta_i$. The *conditional Fisher information given $T = t$*, denoted by $\mathcal{I}_{X|T}(\theta|t)$, is the conditional covariance matrix of the conditional score function.

Here are some examples.

Example 2.80. Let b be known, and suppose that $X \sim N(\theta, b)$ given $\Theta = \theta$. Then, the FI regularity conditions are satisfied and

$$f_{X|\Theta}(x|\theta) = \frac{1}{\sqrt{2\pi b}} \exp\left\{-\frac{1}{2b}(x-\theta)^2\right\},$$

$$\frac{\partial}{\partial \theta} \log f_{X|\Theta}(x|\theta) = \frac{x-\theta}{b},$$

and $\mathcal{I}_X(\theta) = 1/b$. Here we see that the smaller the known variance is, the more information there is in the data about Θ. This is intuitively sensible.

Example 2.81. Suppose that $X \sim U(0,\theta)$ given $\Theta = \theta$. That is, $f_{X|\Theta}(x|\theta) = \theta^{-1} I_{(0,\theta)}(x)$. In this case FI regularity conditions 1 and 3 fail, but we can still calculate $\partial \log f_{X|\Theta}(x|\theta)/\partial \theta = -1/\theta$. We could then try to define the Fisher information to be $\mathcal{I}_X(\theta) = 1/\theta^2$. But this function will not have the properties that Fisher information has under all three FI regularity conditions.

Example 2.82. Suppose that $X \sim \text{Bin}(n,p)$ given $P = p$. Then

$$f_{X|P}(x|p) = \binom{n}{x} p^x (1-p)^{n-x}, \text{ for } x = 0, \ldots, n,$$

$$\log f_{X|P}(x|p) = \log \binom{n}{x} + x \log(p) + (n-x) \log(1-p),$$

$$\frac{\partial}{\partial p} \log f_{X|P}(x|p) = \frac{x}{p} - \frac{n-x}{1-p} = \frac{x}{p(1-p)} - \frac{n}{1-p},$$

$$\mathcal{I}_X(p) = \frac{n}{p(1-p)}.$$

The more extreme P is, the more information an observation has about P.

Example 2.83. Suppose that P_θ says $X \sim N(\mu, \sigma^2)$, where $\theta = (\mu, \sigma)$. Then

$$f_{X|\Theta}(x|\theta) = \frac{1}{\sigma\sqrt{2\pi}} \exp\left\{-\frac{1}{2\sigma^2}(x-\mu)^2\right\},$$

$$\frac{\partial}{\partial \mu} \log f_{X|\Theta}(x|\theta) = \frac{x-\mu}{\sigma^2},$$

$$\frac{\partial}{\partial \sigma} \log f_{X|\Theta}(x|\theta) = -\frac{1}{\sigma} + \frac{(x-\mu)^2}{\sigma^3},$$

$$\text{Var}_\theta\left(\frac{X-\mu}{\sigma^2}\right) = \frac{1}{\sigma^2},$$

$$\text{Var}_\theta\left(-\frac{1}{\sigma} + \frac{(X-\mu)^2}{\sigma^3}\right) = \frac{2}{\sigma^2},$$

$$\text{Cov}_\theta\left(-\frac{1}{\sigma} + \frac{(X-\mu)^2}{\sigma^3}, \frac{X-\mu}{\sigma^2}\right) = 0,$$

$$\mathcal{I}_X(\theta) = \begin{pmatrix} \frac{1}{\sigma^2} & 0 \\ 0 & \frac{2}{\sigma^2} \end{pmatrix}.$$

A useful result about the score function is the following.

Proposition 2.84. *When the FI regularity conditions hold, the mean of the score function is 0. If, in addition, T is a statistic, then the conditional mean given T of the conditional score function is 0, a.s.* $[P_{\theta,T}]$.

If we can differentiate twice under the integral signs (as in exponential families), we obtain

$$0 = \int \frac{\partial^2}{\partial \theta_i \partial \theta_j} f_{X|\Theta}(x|\theta) d\nu(x) = \mathbb{E}_\theta \left[\frac{\frac{\partial^2}{\partial \theta_i \partial \theta_j} f_{X|\Theta}(X|\theta)}{f_{X|\Theta}(X|\theta)}\right].$$

Now, use the fact that

$$\frac{\partial^2}{\partial \theta_i \partial \theta_j} \log f_{X|\Theta}(X|\theta)$$

$$= \frac{\left(\frac{\partial^2}{\partial \theta_i \partial \theta_j} f_{X|\Theta}(X|\theta)\right) f_{X|\Theta}(X|\theta) - \left(\frac{\partial}{\partial \theta_i} f_{X|\Theta}(X|\theta)\right)\left(\frac{\partial}{\partial \theta_j} f_{X|\Theta}(X|\theta)\right)}{f^2_{X|\Theta}(X|\theta)}$$

in order to conclude

$$E_\theta\left[\frac{\partial^2}{\partial\theta_i\partial\theta_j}\log f_{X|\Theta}(X|\theta)\right]$$
$$= 0 - \text{Cov}_\theta\left(\frac{\partial}{\partial\theta_i}\log f_{X|\Theta}(X|\theta), \frac{\partial}{\partial\theta_j}\log f_{X|\Theta}(X|\theta)\right) = -\mathcal{I}_{X,i,j}(\theta).$$

This gives an alternative method for calculating $\mathcal{I}_X(\theta)$ when we can differentiate twice under the integral sign. In exponential families with the natural parameterization, the situation is even simpler, since the second derivative of the logarithm of the density does not depend on the data, hence no expectation need be calculated. In this case,

$$\mathcal{I}_X(\theta) = -\left(\left(\frac{\partial^2}{\partial\theta_i\partial\theta_j}\log c(\theta)\right)\right).$$

Example 2.85 (Continuation of Example 2.82; see page 112). The derivative of the score function is

$$\frac{\partial^2}{\partial p^2}\log f_{X|P}(x|p) = -\frac{x}{p^2(1-p)} + \frac{x-np}{p(1-p)^2}.$$

The mean of this is $-n/[p(1-p)^2] = -\mathcal{I}_X(p)$.

Suppose that $X_1,\ldots,X_n$ are conditionally IID given $\Theta = \theta$ with density $f_{X_1|\Theta}(x|\theta)$. Let $X = (X_1,\ldots,X_n)$. In this case, $\log f_{X|\Theta}(X|\theta) = \sum_{i=1}^n \log f_{X_1|\Theta}(X_i|\theta)$, a sum of IID random variables conditional on $\Theta = \theta$. It follows that the covariance matrix for the sum, namely $\mathcal{I}_X(\theta)$, is n times the covariance matrix of one of them, namely $\mathcal{I}_{X_1}(\theta)$. That is, $\mathcal{I}_X(\theta) = n\mathcal{I}_{X_1}(\theta)$, so Fisher information adds up over IID observations. In fact it is additive over any finite collection of conditionally independent data sets (see Problem 39 on page 143). In this sense it measures how much information we have in a data set. Also, the more information the data provide, the better we should be able to estimate functions of Θ. Two such results, which will be proven later, are Theorems 5.13 and 7.57.

There is another sense in which Fisher information measures the information in a data set. Let $Y = g(X)$ be an arbitrary statistic. We will see that $\mathcal{I}_X(\theta)$ is at least as large as $\mathcal{I}_Y(\theta)$.

Theorem 2.86. *Let $Y = g(X)$. Suppose that Θ is k-dimensional and $P_\theta \ll \nu_X$ for all θ. Then $\mathcal{I}_X(\theta) - \mathcal{I}_Y(\theta)$ is positive semidefinite. The matrix is all 0s if and only if Y is sufficient.*

PROOF. Define $Q_\theta(C) = P'_\theta[(X,Y) \in C]$. By Corollary B.55, $Q_\theta \ll \nu$, where $\nu(C) = \nu_X(\{x : (x,g(x)) \in C\})$, with Radon–Nikodym derivative

$$f_{X,Y|\Theta}(x,y|\theta) = f_{X|\Theta}(x|\theta) = f_{Y|\Theta}(y|\theta)f_{X|Y,\Theta}(x|y,\theta).$$

It follows that

$$\frac{\partial}{\partial \theta_i} \log f_{X|\Theta}(x|\theta) = \frac{\partial}{\partial \theta_i} \log f_{Y|\Theta}(y|\theta) + \frac{\partial}{\partial \theta_i} \log f_{X|Y,\Theta}(x|y,\theta), \text{ a.s. } [Q_\theta],$$
(2.87)

for all θ. We will prove that the two terms on the right-hand side of (2.87) are uncorrelated and that the last term is 0 a.s. if and only if Y is sufficient.

Proposition 2.84 says that the first two expressions in (2.87) have mean 0 and that the last one has 0 conditional mean given Y, a.s. $[P_{\theta,Y}]$. It follows from the law of total probability B.70 that for all i and j,

$$\text{Cov}_\theta \left(\frac{\partial}{\partial \theta_i} \log f_{Y|\Theta}(Y|\theta), \frac{\partial}{\partial \theta_j} \log f_{X|Y,\Theta}(X|Y,\theta) \right)$$

$$= \text{E}_\theta \left\{ \frac{\partial}{\partial \theta_i} \log f_{Y|\Theta}(Y|\theta) \frac{\partial}{\partial \theta_i} \log f_{X|Y,\Theta}(X|Y,\theta) \right\}$$

$$= \text{E}_\theta \left\{ \text{E}_\theta \left[\frac{\partial}{\partial \theta_i} \log f_{Y|\Theta}(Y|\theta) \frac{\partial}{\partial \theta_i} \log f_{X|Y,\Theta}(X|Y,\theta) \middle| Y \right] \right\}$$

$$= \text{E}_\theta \left\{ \frac{\partial}{\partial \theta_i} \log f_{Y|\Theta}(Y|\theta) \text{E}_\theta \left(\frac{\partial}{\partial \theta_i} \log f_{X|Y,\Theta}(X|Y,\theta) \middle| Y \right) \right\} = 0,$$

Hence, the two terms on the right-hand side of (2.87) are uncorrelated. Since the conditional mean of the conditional score function is 0, a.s., Proposition B.78 says that

$$\mathcal{I}_X(\theta) = \mathcal{I}_Y(\theta) + \text{E}_\theta \mathcal{I}_{X|Y}(\theta|Y).$$

It follows that $\mathcal{I}_X(\theta) - \mathcal{I}_Y(\theta)$ is positive semidefinite. The difference is all 0s if and only if, for all i, the conditional score function equals 0 a.s. $[Q_\theta]$. This happens if and only if $f_{X|Y,\Theta}(x|y,\theta)$ is constant in θ, which means if and only if Y is sufficient. $\square$

One feature of Fisher information, which is worth noting, is that it depends on which of several equivalent parameterizations one chooses.

Example 2.88. Suppose that P_θ says $X \sim N(\mu, \sigma^2)$, where $\theta = (\mu, \sigma)$. The Fisher information matrix was seen in Example 2.83 on page 112 to be

$$\mathcal{I}_X(\theta) = \begin{pmatrix} \frac{1}{\sigma^2} & 0 \\ 0 & \frac{2}{\sigma^2} \end{pmatrix}.$$

Now suppose that we chose the natural parameterization of the exponential family, namely

$$\eta_1 = \frac{\mu}{\sigma^2} = \frac{\theta_1}{\theta_2^2}, \quad \eta_2 = -\frac{1}{2\sigma^2} = -\frac{1}{2\theta_2^2}.$$

The function c is

$$c(\eta) = \frac{1}{\sqrt{2\pi}(-2\eta_2)^{-\frac{1}{2}}} \exp\left(\frac{\eta_1^2}{4\eta_2}\right).$$

Taking the negative of the matrix of second partial derivatives, we get

$$I^*(\eta) = \begin{pmatrix} -\frac{1}{2\eta_2} & \frac{\eta_1}{2\eta_2^2} \\ \frac{\eta_1}{2\eta_2^2} & \frac{1}{2\eta_2^2} - \frac{\eta_1^2}{2\eta_2^3} \end{pmatrix}.$$

This is clearly not the same as $\mathcal{I}_X(g^{-1}(\eta))$.

In general, when changing parameters to $H = g(\Theta)$, we can use the chain rule as follows. If $f(\theta)$ is a function of k variables and g is one-to-one, then

$$\frac{\partial}{\partial \eta_i} f(g^{-1}(\eta)) = \sum_{j=1}^{k} \frac{\partial}{\partial \theta_j} f(\theta) \frac{\partial}{\partial \eta_i} g_j^{-1}(\eta),$$

where $\theta_j = g_j^{-1}(\eta)$ is the jth coordinate of g^{-1}. In our case, we need to consider $f(\theta) = \log f_{X|\Theta}(X|\theta)$. It follows that the Fisher information about H is the matrix

$$\mathcal{I}_X^*(\eta) = \Delta(\eta) \mathcal{I}_X(g^{-1}(\eta)) \Delta^\top(\eta),$$

where $\Delta(\eta)$ is a matrix whose (i,j) entry is $\partial g_j^{-1}(\eta)/\partial \eta_i$. The reader can verify that this method also works in Example 2.88 above.

2.3.2 Kullback–Leibler Information

There is another measure of information that has similar properties to Fisher information. This measure of information is designed to measure how far apart two distributions are in the sense of likelihood. That is, if an observation were to come from one of the distributions, how likely is it that you could tell that the observation did not come from the other distribution?

Definition 2.89. Let P and Q be probability measures on the same space. Let p and q be their densities with respect to a common measure ν on that space, for example, $P+Q$. The *Kullback–Leibler information* in X is defined as

$$\mathcal{I}_X(P;Q) = \int \log \frac{p(x)}{q(x)} p(x) d\nu(x).$$

In the case of parametric families, let θ and ψ be two elements of Ω. The Kullback–Leibler information is then

$$\mathcal{I}_X(\theta;\psi) = \mathrm{E}_\theta \left\{ \log \frac{f_{X|\Theta}(X|\theta)}{f_{X|\Theta}(X|\psi)} \right\}.$$

If T is a statistic, let p_t and q_t denote conditional densities for P and Q given $T = t$ with respect to a measure ν_t. Then the *conditional Kullback–Leibler information* is

$$\mathcal{I}_{X|T}(P;Q|t) = \int \log \frac{p_t(x)}{q_t(x)} p_t(x) d\nu_t(x).$$

In general, $\mathcal{I}_X(P;Q) \neq \mathcal{I}_X(Q;P)$, so Kullback–Leibler information is not a metric. The sum $\mathcal{I}_X(P;Q) + \mathcal{I}_X(Q;P)$ is sometimes called the *Kullback–Leibler divergence* [see Kullback (1959)]. Even divergence fails the triangle inequality in general, so it is not a metric.

Example 2.90. Suppose that $X \sim N(\theta, 1)$ given $\Theta = \theta$. Then

$$\log \frac{f_{X|\Theta}(x|\theta)}{f_{X|\Theta}(x|\psi)} = \frac{1}{2}\left[(x-\psi)^2 - (x-\theta)^2\right].$$

It follows that $\mathcal{I}_X(\theta;\psi) = (\psi - \theta)^2/2$. This time $\mathcal{I}_X(\theta;\psi) = \mathcal{I}_X(\psi;\theta)$.

Example 2.91. Suppose that $X \sim Ber(\theta)$ given $\Theta = \theta$. Then

$$\log \frac{f_{X|\Theta}(x|\theta)}{f_{X|\Theta}(x|\psi)} = x \log \frac{\theta}{\psi} + (1-x)\log \frac{1-\theta}{1-\psi}.$$

It follows that

$$\mathcal{I}_X(\theta;\psi) = \theta \log \frac{\theta}{\psi} + (1-\theta)\log \frac{1-\theta}{1-\psi}.$$

This time $\mathcal{I}_X(\theta;\psi) \neq \mathcal{I}_X(\psi;\theta)$.

Kullback–Leibler information measures the information in a data set in some of the same ways that Fisher information does.

Proposition 2.92. *The Kullback–Leibler information $\mathcal{I}_X(P;Q) \geq 0$, and it equals 0 if and only if $P = Q$. The conditional Kullback–Leibler information $\mathcal{I}_{X|T}(P;Q|t) \geq 0$, a.s. $[P_T]$, and it equals 0 a.s. $[P_T]$ if and only if $p_t(x) = q_t(x)$, a.s. $[P]$. (See Definition 2.89.) Also, if X and Y are conditionally independent given Θ and $\theta, \psi \in \Omega$, then*

$$\mathcal{I}_{X,Y}(\theta;\psi) = \mathcal{I}_X(\theta;\psi) + \mathcal{I}_Y(\theta;\psi).$$

Theorem 2.93. *If $Y = g(X)$, then $\mathcal{I}_X(\theta;\psi) \geq \mathcal{I}_Y(\theta;\psi)$ with equality for all θ and ψ if and only if Y is sufficient.*

PROOF. Use the same setup as in Theorem 2.86.

$$\begin{aligned}\mathcal{I}_X(\theta;\psi) &= E_\theta \log \frac{f_{X|\Theta}(X|\theta)}{f_{X|\Theta}(X|\psi)} \\ &= E_\theta \log \frac{f_{Y|\Theta}(Y|\theta)}{f_{Y|\Theta}(Y|\psi)} + E_\theta \log \frac{f_{X|Y,\Theta}(X|Y,\theta)}{f_{X|Y,\Theta}(X|Y,\psi)} \\ &= \mathcal{I}_Y(\theta;\psi) + E_\theta\left[\mathcal{I}_{X|Y}(\theta;\psi|Y)\right] \geq \mathcal{I}_Y(\theta;\psi),\end{aligned}$$

where the last line follows from Proposition 2.92. To make the inequality into equality, Proposition 2.92 says that we must have

$$f_{X|Y,\Theta}(X|Y,\theta) = f_{X|Y,\Theta}(X|Y,\psi), \text{ a.s. } [P_\theta].$$

But this is true for all θ and ψ if and only if Y is sufficient. □

The Kullback–Leibler information tells us how far one distribution is from another in terms of likelihood.

Example 2.94. Let $\Omega = (0,1)$ and suppose that P_θ says that $\{X_n\}_{n=1}^\infty$ are IID $Ber(\theta)$. Let $X = (X_1, \ldots, X_n)$. Let $\psi > \theta$, and let Θ be discrete with

$$f_\Theta(y) = \begin{cases} \pi_0 & \text{if } y = \theta, \\ 1 - \pi_0 & \text{if } y = \psi. \end{cases}$$

Then

$$\Pr(\Theta = \theta | X = x) = \frac{\pi_0 \theta^x (1-\theta)^{n-x}}{\pi_0 \theta^x (1-\theta)^{n-x} + (1-\pi_0) \psi^x (1-\psi)^{n-x}}$$

$$= \left(1 + \frac{1-\pi_0}{\pi_0} \left(\frac{\psi}{\theta}\right)^x \left(\frac{1-\psi}{1-\theta}\right)^{n-x}\right)^{-1},$$

where $x = \sum_{i=1}^n x_i$. Let $p_n = x/n$. Then

$$\Pr(\Theta = \theta | X = x) = \left(1 + \frac{1-\pi_0}{\pi_0} \left[\left(\frac{\psi}{\theta}\right)^{p_n} \left(\frac{1-\psi}{1-\theta}\right)^{1-p_n}\right]^n\right)^{-1}$$

$$= \left(1 + \frac{1-\pi_0}{\pi_0} \exp\{(\mathcal{I}_X(p_n; \theta) - \mathcal{I}_X(p_n; \psi)) n\}\right)^{-1}.$$

So, the probability of either θ or ψ increases with more data, depending on to which one p_n is closer in Kullback–Leibler information.

One advantage Kullback–Leibler information has over Fisher information is that it is not affected by changes in parameterization. Another advantage is that Kullback–Leibler information can be used even if the distributions under consideration are not all members of a parametric family.

Example 2.95. Suppose that P is the standard normal $N(0,1)$ distribution and Q is the Laplace distribution $Lap(0,1)$. Then

$$p(x) = \frac{1}{\sqrt{2\pi}} \exp\left(-\frac{x^2}{2}\right), \quad E_P(X^2) = 1, \quad E_P(|X|) = \sqrt{\frac{2}{\pi}},$$

$$q(x) = \frac{1}{2} \exp(-|x|), \quad E_Q(X^2) = 2, \quad E_Q(|X|) = 1.$$

It follows that

$$\log \frac{p(x)}{q(x)} = \frac{1}{2} \log \frac{2}{\pi} - \frac{1}{2} x^2 + |x|,$$

$$\mathcal{I}_X(P; Q) = \frac{1}{2} \log \frac{2}{\pi} - \frac{1}{2} + \sqrt{\frac{2}{\pi}} = 0.07209,$$

$$\mathcal{I}_X(Q; P) = -\frac{1}{2} \log \frac{2}{\pi} + \frac{2}{2} - 1 = 0.22579.$$

If data come from a Laplace distribution, it is easier to tell that they don't come from a normal distribution than vice versa.

Another advantage to Kullback–Leibler information is that no smoothness conditions on the densities (like the FI regularity conditions) are needed.

Example 2.96. Suppose that P_θ says that X has a uniform $U(0,\theta)$ distribution. For $\delta > 0$,

$$\mathcal{I}_X(\theta; \theta + \delta) = \int_0^\theta \log\left(\frac{\theta + \delta}{\theta}\right) \frac{1}{\theta} dx = \log\left(1 + \frac{\delta}{\theta}\right),$$

$$\mathcal{I}_X(\theta + \delta; \theta) = \int_0^\theta \log\left(\frac{\theta}{\theta + \delta}\right) \frac{1}{\theta + \delta} dx + \int_\theta^{\theta+\delta} \infty \frac{1}{\theta + \delta} dx = \infty.$$

If an observation has a $U(0,\theta)$ distribution, there is some information to distinguish the observation from one with a $U(0,\theta+\delta)$ distribution. On the other hand, if an observation has a $U(0,\theta+\delta)$ distribution, then there is infinite information to distinguish the observation from one with a $U(0,\theta)$ distribution. The reason for this is that there is positive probability that the $U(0,\theta)$ distribution can be ruled out entirely. In a sense, this is the most powerful kind of information possible for distinguishing distributions.

There is at least one connection between Kullback–Leibler information and Fisher information when they both exist and when two derivatives can be passed under the integral sign. In this case,

$$\frac{\partial^2}{\partial \theta_i \partial \theta_j} \mathcal{I}_X(\theta_0; \theta)\bigg|_{\theta=\theta_0} = \frac{\partial^2}{\partial \theta_i \partial \theta_j} \int \log \frac{f_{X|\Theta}(x|\theta_0)}{f_{X|\Theta}(x|\theta)} f_{X|\Theta}(x|\theta_0) d\nu(x)\bigg|_{\theta=\theta_0}$$

$$= \int -\frac{\partial^2}{\partial \theta_i \partial \theta_j} \log f_{X|\Theta}(x|\theta)\bigg|_{\theta=\theta_0} f_{X|\Theta}(x|\theta_0) d\nu(x)$$

$$= -\mathrm{E}_{\theta_0}\left(\frac{\partial}{\partial \theta_i} \log f_{X|\Theta}(X|\theta_0) \frac{\partial}{\partial \theta_j} \log f_{X|\Theta}(X|\theta_0)\right) = I_{X,i,j}(\theta_0),$$

the (i,j) element of the Fisher information matrix.

Example 2.97 (Continuation of Example 2.91; see page 116). The second partial derivative of the Kullback–Leibler information is

$$\frac{\partial^2}{\partial \psi^2} \mathcal{I}_X(\theta; \psi) = \frac{\theta}{\psi^2} + \frac{1-\theta}{(1-\psi)^2}.$$

If one plugs in $\psi = \theta$, one gets $1/[\theta(1-\theta)] = \mathcal{I}_X(\theta)$, the Fisher information.

2.3.3 Conditional Information*

We defined conditional Fisher information in Definition 2.79 as the conditional covariance matrix of the conditional score function. We also defined conditional Kullback–Leibler information in Definition 2.89 as the conditional mean of the logarithm of the ratio of the conditional densities. We used these conditional information measures in the proofs of Theorems 2.86 and 2.93 to show that sufficient statistics contain all of the information in a

---

*This section may be skipped without interrupting the flow of ideas.

sample. However, in Section 2.1.4, it was suggested that performing inference conditional on ancillary statistics makes better use of the information available in the actual sample obtained. We can make this idea more precise by considering conditional Fisher and Kullback–Leibler information given an ancillary.

Theorem 2.98. *Let U be an ancillary statistic. Both Fisher and Kullback–Leibler information have the property that the information is the mean of the conditional information given U.*

PROOF. Suppose that X has a density $f_{X|\Theta}$ given Θ with respect to a measure ν. If $u = U(x)$, then we can write $f_{X|\Theta}(x|\theta) = f_U(u) f_{X|U,\Theta}(x|u,\theta)$, since U is independent of Θ. If the FI regularity conditions hold, then

$$\frac{\partial}{\partial \theta_i} \log f_{X|\Theta}(x|\theta) = \frac{\partial}{\partial \theta_i} \log f_{X|U,\Theta}(x|u,\theta).$$

Since the mean of the conditional score function is 0, a.s., the mean of the conditional covariance matrix equals the marginal covariance matrix by Proposition B.78. In symbols, $\mathcal{I}_X(\theta) = \mathrm{E}_\theta \mathcal{I}_{X|U}(\theta|U)$. Similarly, for Kullback–Leibler information,

$$\frac{f_{X|\Theta}(x|\theta)}{f_{X|\Theta}(x|\psi)} = \frac{f_{X|,U\Theta}(x|u,\theta)}{f_{X|U,\Theta}(x|u,\psi)},$$

so that $\mathcal{I}_X(\theta;\psi) = \mathrm{E}\mathcal{I}_{X|U}(\theta;\psi|U)$. □

Some data sets have more information and some have less depending on the value of the ancillary U. Theorem 2.98 says that the amount of information averages out to the marginal information over the distribution of U, but we can make use of the observed value of U to tell us whether we have one of the data sets with more or less information.

Example 2.99 (Continuation of Example 2.38; see page 95). In this example, $X = (X_1, X_2)$ with the X_i being IID $N(\theta, 1)$ given $\Theta = \theta$. We had $U = X_2 - X_1$. The conditional distribution of X given U can be obtained from the conditional distribution of X_1 given U, which is $N(\theta + u/2, 1/2)$. The conditional score function is $2(X_1 - \theta - u/2)$, which has conditional variance equal to 2 for all u. Similarly, the conditional Kullback–Leibler information is $\mathcal{I}_{X|U}(\theta;\psi|u) = (\theta - \psi)^2$ for all u. Hence, this ancillary does not help distinguish data sets from each other.

Example 2.100 (Continuation of Example 2.52; see page 100). In this problem, there were two ancillaries, M_0 and N_0. We can write the second derivative of the logarithm of the density as

$$\frac{\partial^2}{\partial \theta^2} \log f_{X|\Theta}(X|\theta) = -\frac{N_{00}}{(1-\theta)^2} - \frac{N_{01}}{(1+\theta)^2} - \frac{N_{10}}{(2+\theta)^2} - \frac{N_{11}}{(2-\theta)^2}. \quad (2.101)$$

The Fisher information is $\mathcal{I}_X(\theta) = N(2-\theta^2)/[(1-\theta^2)(4-\theta^2)]$. According to Problem 43 on page 143, we can find the conditional Fisher information by calculating minus the conditional mean of (2.101). Conditional on $M_0 = m_0$, we get

$$\mathcal{I}_{X|M_0}(\theta|m_0) = \frac{3m_0 + N(1 - \theta^2)}{(1 - \theta^2)(4 - \theta^2)}.$$

It is clear that this is an increasing function of m_0, so the more observations we get with first coordinate equal to 0, the more conditional information we have given M_0. Conditional on $N_0 = n_0$, we get

$$\mathcal{I}_{X|N_0}(\theta|n_0) = \frac{2n_0\theta + 2N(1 - \theta) - N\theta^2}{(1 - \theta^2)(4 - \theta^2)}.$$

This is an increasing function of n_0. It is easy to verify that the means of these two are both equal to the marginal information, since $E(M_0) = 1/3$ and $E(N_0) = 1/2$.

In Example 2.100, one might ask which of the ancillaries does a better job of distinguishing data sets from each other in terms of information. This might be answered by looking at how spread out is the distribution of the conditional information.

Example 2.102 (Continuation of Example 2.100; see page 119). We can compute $\text{Var}(M_0) = 2N/9$ and $\text{Var}(N_0) = N/4$, so that the variance of $\mathcal{I}_{X|M_0}(\theta|M_0)$ is $2/\theta^2 (> 1)$ times as large as the variance of $\mathcal{I}_{X|N_0}(\theta|N_0)$. Suppose that we are interested in the statistic N_{00}. We can calculate any aspect we wish of the conditional distribution of N_{00} given either M_0 or N_0 (and Θ). To see how much more M_0 distinguishes data sets than does N_0, Figure 2.103 shows the distribution of the conditional mean of N_{00} given M_0 and N_0 for $\theta = 0.1$ and $N = 50$. It is easy to see how much more the distribution is spread out conditional on M_0 than on N_0. Since the variance of the conditional mean of N_{00} is greater given M_0 than given N_0 (2.25 versus 1.125), it follows from Proposition B.78 that the mean of the conditional variance given M_0 must be smaller (by the same amount) given

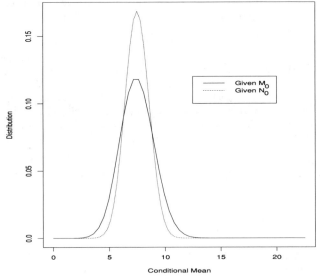

FIGURE 2.103. Distribution of Conditional Means of N_{00} given M_0 and N_0

M_0 than given N_0. In fact, the values are 4.125 and 5.25, respectively. Because this example is sufficiently simple, one can even calculate the probability that the conditional variance of N_{00} given M_0 will be smaller than the conditional variance given N_0. The probability is 0.8346.

2.3.4 Jeffreys' Prior*

Fisher information turns out to have a role to play in one popular method for choosing prior distributions. Suppose that one desires a method for choosing a prior density with the following property. If the parameter Θ were to be transformed by a one-to-one differentiable function g with differentiable inverse, then the prior for $\Psi = g(\Theta)$ obtained by the method would be the same as the usual transformation of the prior obtained for Θ by the method. For example, suppose that Θ is a positive parameter and that the method produces a prior with density $f_\Theta(\cdot)$ with respect to Lebesgue measure. Let $\Psi = \Theta^2$. The usual method of transformations would make the prior for Ψ equal to

$$f_\Psi(\psi) = f_\Theta\left(\sqrt{\psi}\right) \frac{1}{2\sqrt{\psi}}.$$

We want the method, when applied directly to Ψ to produce this same prior.

A class of methods that have this property is the following. Let $h : \mathcal{X} \times \Omega \to \mathbb{R}$ be a function, and define

$$f_\Theta(\theta) = \sqrt{\operatorname{Var}_\theta \frac{\partial}{\partial \theta} h(X, \theta)}. \tag{2.104}$$

To see that this works, let $\Psi = g(\Theta)$ and let the new parameter space be Ω'. Note that h must be modified to $h' : \mathcal{X} \times \Omega' \to \mathbb{R}$ by

$$h'(x, \psi) = h(x, g^{-1}(\psi)),$$

or else the expression on the right of (2.104) makes no sense. Now

$$\frac{\partial}{\partial \psi} h'(x, \psi) = \frac{\partial}{\partial \psi} h(x, g^{-1}(\psi)) = \left.\frac{\partial}{\partial \theta} h(x, \theta)\right|_{\theta=g^{-1}(\psi)} \frac{\partial}{\partial \psi} g^{-1}(\psi),$$

$$\operatorname{Var}_\theta \frac{\partial}{\partial \psi} h'(X, \psi) = \operatorname{Var}_\theta \left(\left.\frac{\partial}{\partial \theta} h(X, \theta)\right|_{\theta=g^{-1}(\psi)}\right) \left(\frac{\partial}{\partial \psi} g^{-1}(\psi)\right)^2$$

$$= (f_\Theta(g^{-1}(\psi)))^2 \left(\frac{\partial}{\partial \psi} g^{-1}(\psi)\right)^2,$$

$$f_\Psi(\psi) = f_\Theta(g^{-1}(\psi)) \left|\frac{\partial}{\partial \psi} g^{-1}(\psi)\right|,$$

---

*This section may be skipped without interrupting the flow of ideas.

which is just what transformation of variables would give.

The most popular function h to use for such a method is the logarithm of the conditional density $h(x, \theta) = \log f_{X|\Theta}(x|\theta)$. In this case,

$$\frac{\partial}{\partial \theta} h(x, \theta) = \frac{\partial}{\partial \theta} \log f_{X|\Theta}(x|\theta),$$

which is the score function. We have already seen that under the FI regularity conditions, $\mathrm{Var}_\theta \partial h(X, \theta)/\partial \theta = \mathcal{I}_X(\theta)$. So, the method says to use $f_\Theta(\theta) = c\sqrt{\mathcal{I}_X(\theta)}$ as the prior density, where c is chosen to make the integral of $f_\Theta(\theta)$ equal to 1, if possible. If no such c exists, then $f_\Theta(\theta) = \sqrt{\mathcal{I}_X(\theta)}$ is often used as an improper prior. This type of prior is called *Jeffreys' prior* after Harold Jeffreys, who proposed it in Jeffreys (1961, p. 181).

Example 2.105. Suppose that $X \sim Bin(n, p)$ given $P = p$. Then we saw in Example 2.82 on page 112 that the Fisher information is $\mathcal{I}_X(p) = n(p[1-p])^{-1}$. This makes the Jeffreys' prior proportional to $p^{-1/2}(1-p)^{-1/2}$. This is the $Beta(1/2, 1/2)$ distribution, which is a proper prior.

Example 2.106. Suppose that $X \sim Negbin(a, p)$ given $P = p$. It is not difficult to show that the Fisher information is $\mathcal{I}_X(p) = a(p^2[1-p])^{-1}$. This makes the Jeffreys' prior proportional to $p^{-1}(1-p)^{-1/2}$. This is not a proper prior.

Interestingly also, Jeffreys' prior in this example is not the same as the prior for the case of binomial sampling (Example 2.105). This means that choosing a prior by Jeffreys' method has the unfortunate characteristic that it depends on something that would normally not be taken into account in a Bayesian analysis. For example, suppose that one were to be exposed to a sequence of exchangeable Bernoulli random variables one at a time. If one were asked to calculate the predictive distribution of each observation before it is observed, one would have to ask whether sampling was to continue to a fixed size or to a fixed number of successes (or failures) before one could even choose a prior distribution. This stopping criterion should be irrelevant before the first observation arrives,[15] but the method of Jeffreys' prior must take it into account.

Example 2.107. Suppose that the density with respect to Lebesgue measure of X given $\Theta = \theta$ is $f(x - \theta)$, where f is differentiable. Then

$$\frac{\partial}{\partial \theta} \log f(X - \theta) = \frac{f'(X - \theta)}{f(X - \theta)}.$$

---

[15] One could imagine situations in which the stopping criterion is chosen by someone who has information not available to us. In such a situation, it is possible that when we learn what this other person has chosen for the stopping criterion, we believe that the choice tells us some additional information that we would like to incorporate into our model. For example, suppose that this other person decides to stop as soon as five successes are observed. We might then say, "Aha! We will use the prior $p^{-1}(1-p)^{-1/2}$ to reflect this information." But then, we discover that we only have time to collect four observations. Jeffreys' rule says that we have to change back to the prior $p^{-1/2}(1-p)^{-1/2}$ because we will have a fixed sample size, even if we believe that the reason the sample size is being fixed has nothing to do with P.

The distribution of this quantity given $\Theta = \theta$ is the same as the distribution of $f'(X)/f(X)$ given $\Theta = 0$, hence the variance is constant as a function of θ. This means that Jeffreys' prior would be constant. If Ω is an unbounded set, Jeffreys' prior is improper.

For multiparameter problems, a similar derivation is possible. Let $f_\Theta(\theta)$ be proportional to the square root of the determinant of the covariance matrix of the gradient vector of $h(X,\theta)$ with respect to θ. It is easy to check that the gradient of $h(x, g^{-1}(\psi))$ with respect to ψ is equal to the matrix whose determinant is the Jacobian times the gradient of $h(x,\theta)$ with respect to θ evaluated at $\theta = g^{-1}(\psi)$. In the special case of Jeffreys' prior, h is the log of the density and $f_\Theta(\theta)$ becomes the square root of the determinant of the Fisher information matrix, $\mathcal{I}_X(\theta)$.

Example 2.108. In Example 2.83 on page 112, we found that if $X \sim N(\mu,\sigma^2)$ given $\Theta = (\mu,\sigma)$, the Fisher information matrix was diagonal with entries $1/\sigma^2$ and $2/\sigma^2$. The determinant is $2/\sigma^4$ and Jeffreys' prior is a constant over σ^2, an improper prior. The usual improper prior in this problem is a constant over σ.

One interesting feature of Jeffreys' prior is that its definition did not depend on the parameter space (except that we were able to take derivatives). That is, if the parameter space is actually an open subset of the set $\{\theta : f_{X|\Theta}(x|\theta)$ is a density$\}$, then Jeffreys' prior has the same form. Obviously, a different normalizing constant will be required if the prior is proper.

Example 2.109 (Continuation of Example 2.107; see page 122). Suppose that the parameter space is actually only the open interval (a,b), but the conditional density of X given $\Theta = \theta$ is still $f(x - \theta)$. Then Jeffreys' prior is the $U(a,b)$ distribution, which is proper.

2.4 Extremal Families*

In this chapter we have shown how one can determine a sufficient statistic once one has chosen a family of models indexed by a parameter. Lauritzen (1984, 1988) has developed a theory in which a family of probability models is determined once one chooses a sequence of sufficient statistics. Lauritzen's theory is general enough to apply to collections of random quantities that are not exchangeable and to collections more general than sequences. We will consider only the case of sequences here. Diaconis and Freedman (1984) also prove results of a similar nature, and Theorem 2.111 below is based on their work.

---

*This section may be skipped without interrupting the flow of ideas.

124   Chapter 2. Sufficient Statistics

2.4.1 The Main Results

Obviously, it takes more than just a sufficient statistic to identify an interesting class of probability models. For example, $T_n = \sum_{i=1}^n X_i$ is sufficient whether the X_i are conditionally IID $N(\theta,1)$ or $Ber(\theta)$ given $\Theta = \theta$. But these two models would not both be considered appropriate for the same data. The conditional distribution of $X_1, \ldots, X_n$ given T_n would also be useful in identifying a class of models. This conditional distribution can be described by a *transition kernel*.

Definition 2.110. Let $\mathcal{X}$ and $\mathcal{T}$ be topological spaces and let $\mathcal{B}$ and $\mathcal{C}$ be the Borel σ-fields. A function $r : \mathcal{B} \times \mathcal{T} \to [0,1]$ is called a *transition kernel* if, $r(\cdot, t)$ is a probability on $(\mathcal{X}, \mathcal{B})$ for every $t \in \mathcal{T}$, and $r(A, \cdot)$ is measurable for every $A \in \mathcal{B}$.

A transition kernel r is like a regular conditional distribution, except that it need not satisfy an equation like $\int_C r(B,t)d\mu_T(t) = \Pr(X \in B, T \in C)$, because there is no mention of a marginal distribution for T. Our goal, in this section, is to prove a representation theorem for the joint distribution of $\{X_n\}_{n=1}^\infty$ under the assumptions that a particular sequence of statistics $\{T_n\}_{n=1}^\infty$ are sufficient and that the conditional distributions given the sufficient statistics are a particular collection of transition kernels.

The basic structure we will consider here is the following. Let $(S, \mathcal{A})$ be a measurable space. For each n, let $(\mathcal{X}_n, \mathcal{B}_n)$ and $(\mathcal{T}_n, \mathcal{C}_n)$ be Borel spaces. The set $\mathcal{X}_n$ is the space in which all data available at time n lie, and $\mathcal{T}_n$ is the space in which the sufficient statistic at time n lies. Let $T_n : \mathcal{X}_n \to \mathcal{T}_n$ be measurable, and let $p_{n-1,n} : \mathcal{X}_n \to \mathcal{X}_{n-1}$ be onto and measurable. Then T_n is the sufficient statistic at time n, and $p_{n-1,n}$ is the function that extracts the data available at time $n-1$ from the data available at time n. Let $\mathcal{X}$ be the following subset of $\prod_{n=1}^\infty \mathcal{X}_n$:

$$\mathcal{X} = \{x = (x_1, x_2, \ldots) \in \prod_{n=1}^\infty \mathcal{X}_n : p_{n-1,n}(x_n) = x_{n-1}, \text{ for all } n > 1\}.$$

(It is easy to see that $\mathcal{X}$ is in the product σ-field.) The set $\mathcal{X}$ is the set of sequences of possible data which are consistent in the sense that the data at time n are an extension of the data at time $n-1$ for all $n > 1$. Define, for $k < n$,

$$p_{k,n} = p_{k,k+1}(p_{k+1,k+2}(\cdots(p_{n-1,n}(\cdot))))$$.

Let $\mathcal{B}$ be the Borel σ-field of $\mathcal{X}$. Let $p_n : \mathcal{X} \to \mathcal{X}_n$ be the nth coordinate projection function $p_n(x_1, x_2, \ldots) = x_n$. Let $X : S \to \mathcal{X}$ be measurable, and define $X_n = p_n(X)$. The definition of $\mathcal{X}$ makes it clear that, for all $x \in \mathcal{X}$, all n, and all $k < n$, $p_{k,n}(p_n(x)) = x_k$. That is, $X_k = p_{k,n}(X_n)$ for all n and all $k < n$. Let Σ_n be the sub-σ-field of $\mathcal{B}$ generated by $\{T_\ell(p_\ell)\}_{\ell=n}^\infty$, and set

$$\Sigma = \bigcap_{n=1}^\infty \Sigma_n.$$

2.4. Extremal Families

For brevity, we will use the symbol T_n to stand for $T_n(p_n(X))$ or for $T_n(p_n)$. This makes Σ the tail σ-field of the sequence of statistics $\{T_n\}_{n=1}^\infty$.

Theorem 2.111. *For each n, let $r_n : \mathcal{B}_n \times \mathcal{T}_n \to [0,1]$ be a transition kernel such that*

$$r_n(T_n^{-1}(\{t\}), t) = 1, \text{ for all } t \in \mathcal{T}_n. \tag{2.112}$$

Suppose that the following is true for each n and $t \in \mathcal{T}_{n+1}$:[16]

Condition S: *Assume that the distribution of X_{n+1} is $r_{n+1}(\cdot, t)$. Then $r_n(\cdot, s)$ is a regular conditional distribution for X_n given $T_n = s$ for all $s \in T_n(p_{n,n+1}(T_{n+1}^{-1}(\{t\})))$.*

Let $\mathcal{M}$ be the set of all distributions on $(\mathcal{X}, \mathcal{B})$ such that r_n is a version of the conditional distribution of X_n given T_n for all n. Then $\mathcal{M}$ is a convex set. Let $\mathcal{E}$ be the extreme points of $\mathcal{M}$. Suppose that $\mathcal{M}$ is nonempty. Then, there exists a set $E \in \Sigma$ and a transition kernel $Q : \mathcal{B} \times E \to [0,1]$ such that

1. $P(E) = 1$ for all $P \in \mathcal{M}$,

2. for each $x \in E$, $r_n(\cdot, T_n(x_n)) \xrightarrow{W} Q(\cdot, x)$,

3. for each $P \in \mathcal{M}$, Q is a regular conditional distribution for X given Σ,

4. for each $x \in E$ and $A \in \Sigma$, $Q(A, x) \in \{0, 1\}$,

5. for each $P \in \mathcal{M}$, there is a unique probability R on $(\mathcal{X}, \Sigma)$ such that

$$P = \int_E Q(\cdot, x) dR(x),$$

6. the R in part 5 is the restriction of P to Σ,

7. for each $P \in \mathcal{M}$, $P \in \mathcal{E}$ if and only if $P(\{x : P = Q(\cdot, x)\}) = 1$.

If the distribution of X is in the class $\mathcal{M}$, we say that $\{X_n\}_{n=1}^\infty$ is *partially exchangeable*[17] relative to the sequences $\{T_n\}_{n=1}^\infty$ and $\{r_n\}_{n=1}^\infty$. The family of distributions in $\mathcal{E}$ is called the *extremal family*. It is helpful to comment on the conditions in Theorem 2.111. Equation (2.112) says that $r_n(\cdot, t)$ puts

---

[16] Condition S is a way of saying that X_n is conditionally independent of T_{n+1} given T_n. The problem with saying it this way is that such a statement requires an explicit distribution for X_{n+1}, and such a distribution has not yet been defined.

[17] Dawid (1982) refers to the type of models considered here as *intersubjective models*. The reason is that all of the distributions in $\mathcal{M}$ have a common conditional distribution for the data X_n given T_n, and they only disagree on the marginal distribution of T_n.

all of its mass on the set of points where $T_n = t$, so that it really looks like a conditional distribution given $T_n = t$. Condition S is a way of expressing the idea that T_n is sufficient without introducing parameters. It says that conditioning X_{n+1} on T_{n+1} and then looking at the conditional distribution of X_n given T_n is the same as conditioning X_n on T_n from the start.

The most common situation in which the above conditions hold is that in which $\mathcal{X}_n = \mathcal{Y}^n$ and $\mathcal{B}_n = \mathcal{D}^n$ for some Borel space $(\mathcal{Y}, \mathcal{D})$. In this case $p_{n,n+1}(y_1, \ldots, y_{n+1}) = (y_1, \ldots, y_n)$, and there is a bimeasurable function $w : \mathcal{X} \to \mathcal{Y}^\infty$ given by $w((y_1, (y_1, y_2), \ldots) = (y_1, y_2, \ldots)$. If, in addition, the Y_i are exchangeable, we have a special version of Theorem 2.111.

Theorem 2.113. *Let $(\mathcal{Y}, \mathcal{D})$ be a Borel space and let $\mathcal{X}_n = \mathcal{Y}^n$, $\mathcal{B}_n = \mathcal{D}^n$. Let $Y_k : S \to \mathcal{Y}$ be measurable for all k. Define $\mathcal{X}$ as above and define $X : S \to \mathcal{X}$ by*[18]

$$X = (Y_1, (Y_1, Y_2), (Y_1, Y_2, Y_3), \ldots).$$

Let $w : \mathcal{X} \to \mathcal{Y}^\infty$ be defined by

$$w((y_1, (y_1, y_2), (y_1, y_2, y_3), \ldots)) = (y_1, y_2, y_3, \ldots).$$

Assume the conditions of Theorem 2.111, but replace condition S by the stronger

> Condition T: *Assume that the distribution of X_{n+1} is $r_{n+1}(\cdot, t)$. Then $r_n(\cdot, s)$ is a regular conditional distribution for X_n given Y_{n+1} and $T_n = s$ for all $s \in T_n(p_{n,n+1}(T_{n+1}^{-1}(\{t\})))$.*

Suppose that for every n and t, $r_n(\cdot, t)$ is the distribution of n exchangeable coordinates, and that T_n is a symmetric (with respect to permutations of the arguments) function of $Y_1, \ldots, Y_n$. Let $\mathcal{M}^*$ and $\mathcal{E}^*$ be the sets of distributions on $(\mathcal{Y}^\infty, \mathcal{D}^\infty)$ induced by w from those in $\mathcal{M}$ and $\mathcal{E}$, respectively. Then all elements of $\mathcal{M}^*$ are distributions of exchangeable random quantities. Also, $\mathcal{E}^*$ is the set of all IID distributions in $\mathcal{M}^*$ for which the coordinates have distribution equal to limits (as $n \to \infty$) of $r_n(p_{1,n}(\cdot), T_n(x_n))$ for $(x_1, x_2, \ldots) \in E$.

Exponential families are a special case which can be characterized by their transition kernels. A generalization of the following theorem was proven by Diaconis and Freedman (1990). What this theorem says is that if an extremal family has the same sufficient statistics and conditional distributions as an exponential family, then those members of the extremal family with densities are the exponential family distributions. There may also be degenerate distributions in the extremal family which would not be part of the exponential family. (See Example 2.117 on page 128.)

---

[18] In this way $X_n = (Y_1, \ldots, Y_n)$.

Theorem 2.114. *Let $h : \mathbb{R}^k \to [0, \infty)$ be strictly positive on a set of positive Lebesgue measure. Suppose that there exists θ such that*

$$c(\theta) = \int h(x) \exp(\theta^\top x) dx < \infty. \tag{2.115}$$

Let $\mathcal{Y} = \mathbb{R}^k = \mathcal{T}_n$ for all n. Suppose that $T_n(y_1, \ldots, y_n) = \sum_{i=1}^n y_i$. Let $h^{(1)} = h$ and

$$h^{(n)}(t) = \int h^{(n-1)}(t-y)h(y)dy,$$

for $n > 1$. For $B \subseteq \mathcal{X}_n$ and $t \in \mathbb{R}^k$, let

$$B(t) = \left\{ (y_1, \ldots, y_{n-1}) : \left(y_1, \ldots, y_{n-1}, t - \sum_{i=1}^{n-1} y_i \right) \in B \right\}.$$

Let

$$r_n(B,t) = \frac{1}{h^{(n)}(t)} \int_{B(t)} h\left(t - \sum_{i=1}^{n-1} y_i\right) \prod_{i=1}^{n-1} h(y_i) dy_1 \ldots dy_{n-1}.$$

Then the conditions of Theorem 2.113 are satisfied. Also, the members of the extremal family with distributions absolutely continuous with respect to Lebesgue measure are the members of the exponential family with the Y_i being IID with density $h(y) \exp(\theta^\top y)/c(\theta)$, for some θ satisfying (2.115).

2.4.2 Examples

In this section, we present examples of the above theorems for exchangeable sequences. For more general distributions, some examples are given in Section 8.1.3. The theorems can be summarized as follows. Suppose that we specify a sequence of sufficient statistics and conditional distributions for $\{Y_n\}_{n=1}^\infty$ given the sufficient statistics in such a way that the sequence is exchangeable. Then the X_n are conditionally IID with distribution being one of the limits of $r_n(p_{1,n}(\cdot), T_n(x_n))$. Examination of these limits should reveal the collection of extremal distributions.

The most straightforward example of Theorem 2.113 is to show that it implies DeFinetti's representation theorem 1.49 for random variables.

Example 2.116. Let $\mathcal{X}_n = \mathbb{R}^n$. Let $\mathcal{T}_n$ be the subset of $\mathbb{R}^n$ with the coordinates in nondecreasing order. Let $r_n(A, t)$ equal $1/n!$ times the number of permutations of the coordinates of t which are elements of A. Let $p_{n-1,n}(x_1, \ldots, x_n) = (x_1, \ldots, x_{n-1})$. Clearly, every IID distribution is in $\mathcal{M}$, so $\mathcal{M}$ is nonempty. Now, suppose that X_{n+1} has distribution $p_{n+1}(\cdot, t)$. The set $Y(t) = p_{n,n+1}(T_{n+1}^{-1}(\{t\}))$ for $t \in \mathcal{T}_{n+1}$ is just the set of vectors of length n whose coordinates are n draws without replacement from the coordinates of t, or equivalently, the set of vectors consisting of the first n coordinates of the permutations of the coordinates of t. The distribution of X_n is uniform over these $(n+1)!$ points (with repeats counted as more than one point), and the distribution of T_n is uniform on $T_n(Y(t))$, which consists of the $n+1$ vectors obtained by removing one coordinate from t. The conditional distribution of X_n given $T_n = s$ is clearly $r_n(\cdot, s)$ for each $s \in T_n(Y(t))$. Hence the conditions of Theorem 2.111 are satisfied.

A combinatorial argument like the one used in the proof of Theorem 1.49 shows that each limit point of a sequence $\{r_n(p_{k,n}(\cdot), T_n(x_n))\}_{n=1}^\infty$ of probabilities (for fixed k and x) has IID coordinates. Hence, $Q(\cdot, x)$ is a distribution of IID random variables for each x. We can determine which IID distribution by looking at the first coordinate. Since the $r_n(p_{1,n}(\cdot), T_n(x_n))$ distributions are just the empirical probability measures of the first n coordinates of x, $Q(\cdot, x)$ is the limit of the empirical probability measures for those x such that the limit exits. Since all CDFs on $\mathbb{R}$ are such limits, we get DeFinetti's representation theorem 1.49 out of Theorem 2.111.

When $(\mathcal{T}_n, \mathcal{C}_n)$ is the same space $(\mathcal{T}, \mathcal{C})$ for all n, it may be possible to identify the extremal distributions with elements of $\mathcal{T}$.

Example 2.117. Suppose that $\mathcal{X} = \mathbb{R}$ and $\mathcal{T}_n = \mathbb{R} \times \mathbb{R}^{+0}$ with $T_n(x_1, \ldots, x_n) = \left(\sum_{i=1}^n x_i, \sum_{i=1}^n x_i^2\right)$, and $r_n(\cdot, (t_1, t_2))$ the uniform distribution on the surface of the sphere of radius $\sqrt{t_2 - t_1^2/n}$ around $(t_1, \ldots, t_1)/n$. If an n-dimensional vector Y is uniformly distributed on the sphere of radius 1 around 0, then r_n is the distribution of $t_1/n + \sqrt{t_2 - t_1^2/n} Y$. So, we will find the distribution of Y_1. The conditional distribution of $(Y_2, \ldots, Y_n)$ given $Y_1 = y_1$ is uniform on the sphere in the $(n-1)$-dimensional space in which the first coordinate is y_1 with radius $\sqrt{1 - y_1^2}$ around the point $(y_1, 0, \ldots, 0)$. The marginal density of Y_1 is then the ratio of the surface areas of these two spheres. The surface area of sphere of radius r in $n > 1$ dimensions is $2\pi^{n/2} r^{n-1}/\Gamma(n/2)$. So

$$f_{Y_1}(y_1) = \frac{\Gamma\left(\frac{n}{2}\right)}{\Gamma\left(\frac{n-1}{2}\right)\sqrt{\pi}} (1 - y_1^2)^{\frac{n}{2}-1}.$$

Let $\sigma_n^2 = (t_2 - t_1^2/n)/n$. Then, the density of X_1 given $T_n = (t_1, t_2)$ is

$$f_{X_1|T_n}(x|t_1, t_2) = \frac{\Gamma\left(\frac{n}{2}\right)}{\Gamma\left(\frac{n-1}{2}\right)\sqrt{n\pi}\sigma_n} \left(1 - \frac{\left(x - \frac{t_1}{n}\right)^2}{n\sigma_n^2}\right)^{\frac{n}{2}-1}.$$

Since

$$\lim_{n\to\infty} \frac{\Gamma\left(\frac{n}{2}\right)}{\Gamma\left(\frac{n-1}{2}\right)\sqrt{n}} = \frac{1}{\sqrt{2}},$$

we have that

$$f_{X_1|T_n}(x|t_1, t_2) \sim (2\pi)^{-\frac{1}{2}} \sigma_n^{-1} \left(1 - \frac{\left(x - \frac{t_1}{n}\right)^2}{n\sigma_n^2}\right)^{\frac{n}{2}-1}.$$

If σ_n converges to σ and t_1/n converges to μ, this function converges uniformly on compact sets to the $N(\mu, \sigma^2)$ density. If σ_n goes to ∞, the limit is 0 and does not correspond to a probability distribution. If σ_n converges to 0 and t_1/n converges to μ, the density goes to 0 uniformly outside of every open interval around μ; hence the limit distribution is concentrated at μ. If σ_n converges to 0 but t_1/n

goes to ±∞, the limit is not a probability distribution. Hence the set $\mathcal{E}$ consists of all IID distributions in which the coordinates are either normally distributed or constant.

Theorem 2.111 is actually so general that it applies to all joint distributions.

Example 2.118. Let $\{\mathcal{Y}_n\}_{n=1}^{\infty}$ be a sequence of arbitrary Borel spaces. Let $\mathcal{X}_n = \mathcal{T}_n = \prod_{i=1}^{n} \mathcal{Y}_i$, and let T_n be the identity transformation on $\mathcal{X}_n$. Let $r_n(A, t) = I_A(t)$. Let $p_{n-1,n}(y_1, \ldots, y_n) = (y_1, \ldots, y_{n-1})$. Then the conditions of Theorem 2.111 are satisfied, and the extreme points of $\mathcal{M}$ are the point mass distributions $Q(A, x) = I_A(x)$ for all $A \in \mathcal{B}$. The tail σ-field is the whole σ-field $\mathcal{B}$, and the representation probability for P is P itself. Needless to say, this is not an interesting example of the representation, but it is an example.

2.4.3 Proofs[+]

The proof of Theorem 2.111 will proceed by means of a sequence of lemmas. The following simple proposition implies that $\mathcal{M}$ is a convex set.

Proposition 2.119. *Let P and Q be probability measures on a measurable space $(\mathcal{Y}, \mathcal{C})$, and let $R = \lambda P + (1 - \lambda)Q$ with $0 < \lambda < 1$. Let $\mathcal{D}$ be a sub-σ-field of $\mathcal{C}$ such that $P(\cdot|\mathcal{D}) = Q(\cdot|\mathcal{D})$. Then $R(\cdot|\mathcal{D}) = P(\cdot|\mathcal{D})$.*

Next, we prove that the conditional distribution of X_n given $\{T_\ell\}_{\ell=n}^{\infty}$ is the same as the conditional distribution given T_n alone.[19]

Lemma 2.120. *For each n and each $P \in \mathcal{M}$, X_n is conditionally independent of $\{T_{n+i}\}_{i=1}^{\infty}$ given T_n.*

PROOF. We will prove this by showing that the conditional distribution of X_n given $T_n, \ldots, T_m$ is the same as the conditional distribution of X_n given T_n for all m. It will follow from the result in Problem 13 on page 663 that this is also the conditional distribution of X_n given $\{T_\ell\}_{\ell=n}^{\infty}$.
For each $P \in \mathcal{M}$, $r_{n+1}(\cdot, t)$ is the conditional distribution of X_{n+1} given $T_{n+1} = t$. Condition S says that $r_n(\cdot, s)$ is the conditional distribution of X_n given $T_n = s$ and $T_{n+1} = t$. But r_n is the conditional distribution of X_n given T_n, so the result is true if $m = n+1$. We finish the proof by induction on m. Suppose that the conditional distribution of X_n given $T_n, \ldots, T_m$ is r_n. Now, find the conditional distribution of X_n given $T_n, \ldots, T_{m+1}$. The conditional distribution of X_{m+1} given T_{m+1} is r_{m+1}, so condition S says that the conditional distribution of X_m given (T_m, T_{m+1}) is r_m,

---

[+]This section contains results that rely on the theory of martingales. It may be skipped without interrupting the flow of ideas.
[19]This is a stronger statement of the fact that T_n is sufficient. If we think of $\{T_\ell\}_{\ell=n}^{\infty}$ as the parameter, then Lemma 2.120 is the usual classical concept of sufficiency.

the conditional distribution of X_m given T_m.[20] Since $(X_n, T_n, \ldots, T_m)$ is a function of X_m, its conditional distribution given (T_m, T_{m+1}) is the same as its conditional distribution given T_m. According to Theorem B.75, we can use this last conditional distribution to find the conditional distribution of X_n given $T_n, \ldots, T_{m+1}$ by conditioning X_n on $T_n, \ldots, T_m$. By the induction hypothesis, this just produces r_n, and the proof is complete. □

It follows from Lemma 2.120 that for every n, $r_n(\cdot, t)$ can be used as a version of the conditional distribution of X_n given $T_n = t, T_{n+1} = u, \ldots$.

Next, we find the conditional distributions of X_k given Σ for each k. These distributions are limits of the conditional distributions given T_n as $n \to \infty$.

Lemma 2.121. *For each $x \in \mathcal{X}$, and each n and $k < n$, let $R_{k,n,x}$ be the probability on $(\mathcal{X}_k, \mathcal{B}_k)$ induced by $p_{n,k}$ from $r_n(\cdot, T_n(p_n(x)))$.[21] Define L to be the set of all $x \in \mathcal{X}$ such that $R_{k,n,x}$ converges in distribution (as $n \to \infty$) for all k (denote the limit by $R_{k,x}$). Then $L \in \Sigma$, and $P(L) = 1$ for all $P \in \mathcal{M}$. Also, the function $f(x) = R_{k,x}(A)$ is measurable for all $A \in \mathcal{B}_k$ and is a version of $P(p_k^{-1}(A)|\Sigma)$ for all $P \in \mathcal{M}$.*

PROOF. For each k, let $\phi_k : \mathcal{X}_k \to [0,1]$ be a bimeasurable function. (See Definition B.31.) Let $Y_k = \phi_k(X_k)$. Let $Q_{k,n,x}$ be the probability induced on $[0,1]$ from $R_{k,n,x}$ by ϕ_k. Let $f : [0,1] \to \mathbb{R}$ be a bounded continuous function. By Lemma 2.120,

$$\begin{aligned} \mathrm{E}(f(Y_k)|T_n, T_{n+1}, \ldots) &= \mathrm{E}(f(\phi_k(p_{n,k}(X_n)))|T_n, T_{n+1}, \ldots) \\ &= \mathrm{E}(f(Y_k)|T_n). \end{aligned}$$

Now, define $g_{k,n}(x; f) = \int f(y) dQ_{k,n,x}(y)$. It follows that $\mathrm{E}(f(Y_k)|T_n = t) = g_{k,n}(x; f)$ if $t = T_n(p_n(x))$. According to part II of Lévy's theorem B.124, $\mathrm{E}(f(Y_k)|T_n, T_{n+1}, \ldots)$ converges almost surely to $\mathrm{E}(f(Y_k)|\Sigma)$. In terms of the points $x \in \mathcal{X}$, we say this as follows. First, we note that, as in Lemma 2.120, a version of the conditional distribution of X_n given $\{T_\ell\}_{\ell=n}^\infty$ is r_n for all $P \in \mathcal{M}$. Hence, versions of the conditional distribution can be chosen so that the set of x for which $g_{k,n}(x, f)$ converges does not depend on P. Let $G_{k,f}$ be the set of x for which $g_{k,n}(x; f)$ converges, and call the limit $\lambda_k(x; f)$. (Hence, λ_k does not depend on P either.) Then $G_{k,f} \in \Sigma$ and $P(G_{k,f}) = 1$ for all $P \in \mathcal{M}$. Also, $\lambda_k(\cdot; f)$ is measurable (with respect to Σ), and it is a version of $\mathrm{E}(f(Y_k)|\Sigma)$, that is, for all $A \in \Sigma$,

$$\mathrm{E}(f(Y_k)I_A) = \int_A \lambda_k(x; f) dP(x),$$

for all $P \in \mathcal{M}$.

---

[20] Note that we now have that X_m is conditionally independent of T_{m+1} given T_m because a marginal distribution of X_{m+1} has been identified.

[21] In symbols, $R_{k,n,x}(B) = r_n(p_{n,k}(B), T_n(p_n(x))) = r_n(p_{n,k}(B), T_n(x_n))$.

Let C_0 be a countable dense subset of the set of bounded continuous functions from $[0,1]$ to $\mathbb{R}$ (see Lemma B.42) using the uniform metric on functions. Let

$$G_k = \bigcap_{f \in C_0} G_{k,f},$$

so that $G_k \in \Sigma$ and $P(G_k) = 1$ for all $P \in \mathcal{M}$. Since the elements of C_0 are dense in the uniform metric, we have that for every bounded continuous $f : [0,1] \to \mathbb{R}$ and each $x \in G_k$, $\lim_{n \to \infty} g_{k,n}(x; f) = \lambda_k(x; f)$. This can also be written as

$$\lim_{n \to \infty} \int f(y) dQ_{k,n,x}(y) = \int f(y) dQ_{k,x}(y),$$

where $Q_{k,x}$ is the conditional distribution of Y_k given Σ calculated from the probability space $(\mathcal{X}, \mathcal{B}, P)$, which is the same for every $P \in \mathcal{M}$. In short, $Q_{k,n,x} \xrightarrow{W} Q_{k,x}$. Also, the function $g_k(x; f) = \int f(y) dQ_{k,x}(y)$ is measurable with respect to Σ and is a version of $E(f(Y_k)|\Sigma)$. Because $Q_{k,x}$ is a probability on $[0,1]$ rather than on $\mathcal{X}_k$, we need to show that $Q_{k,x}(\phi_k(\mathcal{X}_k)) = 1$, a.s. Since $g_k(x; f)$ is measurable even if f is only bounded and measurable (i.e., not necessarily continuous), it follows that $g_k(x; f)$ is also the conditional mean of $f(Y_k)$ given Σ for all bounded measurable f. Set $f = I_{\phi_k(\mathcal{X}_k)}$ to get

$$\int_{G_k} g_k(x; I_{\phi_k(\mathcal{X}_k)}) dP(x) = \int_{G_k} I_{\mathcal{X}_k}(p_k(x)) dP(x) = P(G_k \cap p_k^{-1}(\mathcal{X}_k)) = 1,$$

from which it follows that $Q_{k,x}(\phi_k(\mathcal{X}_k)) = 1$, a.s. To complete the proof, set

$$H_k = \{x \in G_k : Q_{k,x}(\phi_k(\mathcal{X}_k)) = 1\}$$

and $L = \cap_{k=1}^{\infty} H_k$, and let $R_{k,x}$ be the probability induced on $(\mathcal{X}_k, \mathcal{B}_k)$ by ϕ_k^{-1} from $Q_{k,x}$. □

The next lemma says that we can arrange for the $Q_{k,x}$ probabilities to be a consistent set of distributions as k varies.

Lemma 2.122. *In the notation of Lemma 2.121, let*

$$C = \{x \in L : R_{k,x}(p_{k-1,k}^{-1}(\cdot)) = R_{k-1,x}(\cdot), \text{ for all } k\}.$$

Then $C \in \Sigma$ and $P(C) = 1$ for all $P \in \mathcal{M}$.

PROOF. As in the proof of Lemma 2.121, let C_0 be a countable dense set of bounded continuous functions from $[0,1]$ to $\mathbb{R}$, and let

$$g_k(x; f) = \int f(y) dQ_{k,x}(y)$$

for each k and each bounded measurable f. Then both $g_{k-1}(x;f)$ and $g_k(x;f(p_{k-1,k}))$ are versions of $\mathrm{E}(Y_{k-1}|\Sigma)$. Let $H_{k,f}$ be the set of $x \in L$ such that the two versions are equal, and let

$$C = \bigcap_{k=1}^{\infty} \bigcap_{f \in C_0} H_{k,f}.$$

Each $H_{k,f} \in \Sigma$ and $P(H_{k,f}) = 1$ for all $P \in \mathcal{M}$. Since C_0 is a dense set, $x \in C$ implies $g_{k-1}(x;f) = g_k(x;f(p_{k-1,k}))$ for all bounded continuous f. □

Next, we combine the consistent conditional distributions on the $\mathcal{X}_k$ spaces into a conditional distribution on the space $\mathcal{X}$ given Σ.

Lemma 2.123. *There exists a transition kernel $Q : \mathcal{B} \times C \to [0,1]$ such that $Q(A,x)$ is a version of $P(A|\Sigma)$ for all $A \in \mathcal{B}$ and all $P \in \mathcal{M}$.*

PROOF. Lemma 2.122 says that the finite-dimensional distributions $R_{k,x}$ are consistent for each fixed $x \in C$. Theorem B.133 says that for each $x \in C$, there is a unique probability $Q(\cdot,x)$ on $(\mathcal{X}, \mathcal{B})$ with $R_{k,x}$ as the k-dimensional marginal for every k. But surely, for each $P \in \mathcal{M}$, $P(\cdot|\Sigma)$ is such a probability, so $Q(\cdot,x)$ is a version of $P(\cdot|\Sigma)$ for every $P \in \mathcal{M}$. □

If $E \subseteq C$, then we have established parts 2 and 3 of Theorem 2.111 (see Lemma 2.128 below.) Next we show that the probabilities $Q(\cdot,x)$ are mostly in $\mathcal{M}$.

Lemma 2.124. *Let $V = \{x \in C : Q(\cdot,x) \in \mathcal{M}\}$. Then $V \in \Sigma$ and $P(V) = 1$ for all $P \in \mathcal{M}$.*

PROOF. A point $x \in V$ if and only if, for all n, all $A \in \mathcal{B}_n$, and all $B \in C_n$,

$$\int_{p_n^{-1}(T_n^{-1}(B))} r_n(A, T_n(p_n(y))) dQ(y,x) = Q(\{T_n(X_n) \in B, X_n \in A\}, x). \tag{2.125}$$

Both sides of (2.125) are Σ measurable functions of x, so the set of x for which (2.125) holds for fixed n, A, and B is in Σ. Since $\mathcal{X}_n$ and $\mathcal{T}_n$ are Borel spaces, there exist countable fields of sets which generate their respective Borel σ-fields (Proposition B.43). The set of all x such that (2.125) holds for all A and B in those countable collections and all n simultaneously is therefore in Σ. But it is easy to see that if (2.125) holds for all A in a field that generates $\mathcal{B}_n$ (for fixed B and fixed n), then it holds for all $A \in \mathcal{B}_n$, and similarly for all B. Hence $V \in \Sigma$. To show that $P(V) = 1$ for all $P \in \mathcal{M}$, let $P \in \mathcal{M}$ and let $G \in \Sigma$. Since $Q(\cdot,x)$ is a version of $P(\cdot|\Sigma)$, it follows that the integral of the right-hand side of (2.125) over G is

$$\int_G Q(\{T_n(X_n) \in B, X_n \in A\}, x) dP(x) = P(G, T_n(X_n) \in B, X_n \in A).$$

Similarly, the integral of the left-hand side of (2.125) over G is

$$\int_G \int_{p_n^{-1}(T_n^{-1}(B))} r_n(A, T_n(p_n(y))) dQ(y,x) dP(x)$$
$$= \mathrm{E}(I_G I_B(T_n(X_n)) r_n(A, T_n(X_n))) = P(G, T_n(X_n) \in B, X_n \in A),$$

where the last equality follows from the fact that G and $\{T_n(X_n) \in B\}$ are both in Σ_n and r_n is a conditional distribution for X_n given Σ_n. Now, let G be the set of x such that the left-hand side of (2.125) is strictly greater than the right-hand side. If $P(G) > 0$, we have a contradiction, and similarly if G is the set of x such that the left-hand side is strictly less than the right-hand side. □

There may be many x for which $Q(\cdot, x)$ are the same, and it would be useful not to distinguish these if we want the representation to be unique.

Lemma 2.126. *Let Σ' be the smallest σ-field of subsets of V such that all of the functions $f_A(x) = Q(A,x)$ (as functions of x) are measurable. The σ-field Σ' is countably generated.*

PROOF. The σ-field Σ' is generated by all sets of the form $\{x \in V : Q(A,x) > q\}$, where q is a rational and A is an element of a countable field that generates $\mathcal{A}$ (Proposition B.43). □

Next, we show that for each $P \in \mathcal{M}$, Σ and Σ' differ only by probability zero sets.

Lemma 2.127. *For each $A \in \Sigma$, let $A' = \{x \in V : Q(A,x) = 1\}$. Then $A' \in \Sigma'$ and $P(A \Delta A') = 0$ for all $P \in \mathcal{M}$.*

PROOF. Since $Q(A, \cdot)$ is measurable with respect to Σ', $A' \in \Sigma'$. Now $A \Delta A' = (A \setminus A') \cup (A' \setminus A)$. Since $P(Q(A, X) = I_A(X)) = 1$ for all $A \in \Sigma$, $Q(A,x) = 0$, a.s. for $x \in A'$, and we get

$$P(A \setminus A') = \int_{A'^c} Q(A,x) dP(x) = 0.$$

Since $Q(A,x) = 1$ for all $x \in A'$,

$$P(A' \setminus A) = \int_{A'} [1 - Q(A,x)] dP(x) = 0. \qquad \square$$

Next, we identify the set E.

Lemma 2.128. *For each $x \in V$, let $S(x)$ be the atom in Σ' containing x, that is,*

$$S(x) = \{y \in V : Q(A,y) = Q(A,x), \text{ for all } A \in \Sigma'\}.$$

Let $E = \{x \in V : Q(S(x), x) = 1\}$. Then $E \in \Sigma'$ and $P(E) = 1$ for all $P \in \mathcal{M}$. Also,

$$E = \{x \in V : Q(x, A) = I_A(x), \text{ for all } A \in \Sigma'\}. \tag{2.129}$$

PROOF. First, we prove (2.129). Suppose that $x \in V$ and $Q(A, x) = I_A(x)$ for all $A \in \Sigma'$. Since $S(x) \in \Sigma'$, we have $Q(S(x), x) = I_{S(x)}(x) = 1$, since $x \in S(x)$. So $x \in E$, and the right-hand side of (2.129) is contained in E. If $x \in E$ and $A \in \Sigma'$, then $S(x) \subseteq A$ if and only if $x \in A$. It follows that $Q(A, x) = I_A(x)$, and E is a subset of the right-hand side of (2.129).

Next, we prove that $E \in \Sigma'$. Note that $Q(A, x) = I_A(x)$ for all $A \in \Sigma'$ if and only if $Q(A, x) = I_A(x)$ for all $A \in \mathcal{D}$, where $\mathcal{D}$ is a countable field generating Σ'. So,

$$E = \bigcap_{A \in \mathcal{D}} \{x \in V : Q(A, x) = I_A(x)\}, \qquad (2.130)$$

which is in Σ' because each of the sets in the intersection is in Σ'.

Finally, we prove $P(E) = 1$ for all $P \in \mathcal{M}$. Since $Q(A, x)$ is a version of $P(A|\Sigma)$ for all $x \in C$ by Lemma 2.123, we have that for all $A \in \Sigma$,

$$P(Q(A, X) = I_A(X)) = 1.$$

Now use (2.130) again, to conclude $P(E) = 1$. □

Since $E \subseteq C$, we have now established parts 1, 2, and 3 of Theorem 2.111. Next, we establish part 4.

Lemma 2.131. *If $x \in E$ and $A \in \Sigma$, then $Q(A, x) \in \{0, 1\}$.*

PROOF. Let $x \in E$ and $A \in \Sigma$. Then $Q(\cdot, x) \in \mathcal{M}$ by Lemma 2.124. By Lemma 2.127, there is $A' \in \Sigma'$ such that $Q(A, x) = Q(A', x)$. But $Q(A', x) = I_{A'}(x)$ by Lemma 2.128. □

Now, we are ready to prove parts 5 and 6 of Theorem 2.111.

Lemma 2.132. *Let Σ^* be the σ-field of subsets of E defined by*

$$\Sigma^* = \{A \cap E : A \in \Sigma'\}.$$

For each $P \in \mathcal{M}$, there is a unique probability R on (E, Σ^*) such that $P = \int_E Q(\cdot, x) dR(x)$ and R is the restriction of P to Σ^* as well as the restriction of P to Σ.

PROOF. Since $Q(A, x)$ is a version of $P(A|\Sigma)$, it follows that R equal to the restriction of P to Σ^* satisfies the representation. To show uniqueness, let R be a probability on (E, Σ^*) which satisfies the representation and let $A \in \Sigma^*$. Then

$$R(A) = \int_E I_A(x) dR(x) = \int_E Q(A, x) dR(x) = P(A),$$

where the second equality follows from Lemma 2.128, and the third follows from the representation. Since $\Sigma^* \subseteq \Sigma$, the restriction of P to Σ agrees with the restriction of P to Σ^* on Σ^*. □

The following result follows easily from Lemma 2.132.

Corollary 2.133. *If P and P' are in $\mathcal{M}$ and they agree on Σ^*, then they are the same.*

Finally, we can prove part 7 of Theorem 2.111.

Lemma 2.134. *A probability $P \in \mathcal{M}$ is extreme if and only if it is a zero–one measure on Σ. Also, $P \in \mathcal{M}$ is extreme if and only if*

$$P(\{x \in L : Q(\cdot, x) = P\}) = 1. \tag{2.135}$$

PROOF. According to Lemma 2.127, $P \in \mathcal{M}$ is a zero–one measure on Σ if and only if its restriction to Σ^* is a zero–one measure. P is a zero–one measure on Σ^* if and only if it is concentrated on one of the atoms, which are sets of the form $\{x \in E : Q(\cdot, x) = R\}$, for some R. But the representation in Lemma 2.132 implies that $R = P$. So, P is a zero–one measure on Σ^* if and only if (2.135) holds.

Next, we prove that if $P \in \mathcal{E}$, then P is a zero–one measure on Σ^*. Suppose, to the contrary, that there is $A \in \Sigma^*$ such that $0 < P(A) = \alpha < 1$. Let

$$P_1 = \frac{1}{\alpha} \int_A Q(\cdot, x) dR(x), \quad P_2 = \frac{1}{1-\alpha} \int_{E \setminus A} Q(\cdot, x) dR(x),$$

where R is the restriction of P to Σ^*. Clearly, $P_1(A) = 1$ and $P_2(A) = 0$, so $P_1 \neq P_2$. But $P = \alpha P_1 + (1-\alpha) P_2$, so P is not extreme.

Finally, we prove that if P is a zero–one measure on Σ^*, then P is extreme. Suppose, to the contrary, that $P = \alpha P_1 + (1-\alpha) P_2$ for some $0 < \alpha < 1$ and $P_1 \neq P_2$ in $\mathcal{M}$. Since $P_i \ll P$ on Σ^* for $i = 1, 2$, it follows that P, P_1, and P_2 all concentrate on the same atom in Σ^*, hence they agree on Σ^*. By Corollary 2.133, $P = P_1 = P_2$, a contradiction. □

In particular, Lemma 2.134 says that we can locate the extreme points by finding all of the $Q(\cdot, x)$ measures for $x \in E$.

Lemma 2.136. *A probability $P \in \mathcal{M}$ is in $\mathcal{E}$ if and only if it is a $Q(\cdot, x)$ for some $x \in E$.*

PROOF. Lemma 2.134 says that $x \in E$ implies $Q(\cdot, x)$ is extreme. (Check the definition of E in Lemma 2.128.) Conversely, if P is extreme, then P is a zero–one measure on Σ^* and it equals $Q(\cdot, x)$ for x in the atom on which P concentrates, according to (2.135). □

The proofs of Theorems 2.113 and 2.114 require a lemma first.

Lemma 2.137. *Let $(\mathcal{Y}, \mathcal{D})$ be a Borel space, and assume the conditions of Theorem 2.111 hold with $\mathcal{X}_n = \mathcal{Y}^n$. Let X_n be $(Y_1, \ldots, Y_n)$. Then all IID distributions in $\mathcal{M}^*$ are in $\mathcal{E}^*$.*

PROOF. Let $\mathcal{A}_n$ be the σ-field generated by all functions from $\mathcal{X}_n$ to $\mathcal{T}_n$ which are symmetric with respect to permutations of the coordinates. Let $\mathcal{A}_\infty = \cap_{n=1}^\infty \mathcal{A}_n$. Then $\Sigma^* \subseteq \mathcal{A}_\infty$, where Σ^* is the image of Σ under the bimeasurable mapping w. We will prove that the IID distributions are zero–one on $\mathcal{A}_\infty$. We do this by proving that IID distributions are conditional distributions given $\mathcal{A}_\infty$. Let $\mathbf{P}$ stand for the distribution of the data which says that the Y_i are IID with distribution equal to the limit of the empirical CDFs. Since the empirical CDF based on $Y_1, \ldots, Y_n$ is $\mathcal{A}_n$ measurable, and since $\mathcal{A}_n \subseteq \mathcal{A}_{n+1}$ for all n, it follows that $\mathbf{P}$ is $\mathcal{A}_n$ measurable for every n, hence it is $\mathcal{A}_\infty$ measurable. To see that $\mathbf{P}(B) = \Pr((Y_{i_1}, \ldots, Y_{i_k}) \in B | \mathcal{A}_\infty)$, we need to prove that, for every $C \in \mathcal{A}_\infty$,

$$\Pr(\{X_{i_1}, \ldots, X_{i_k}\} \in B\} \cap C) = \int_C \mathbf{P}(B)(s) d\mu(s).$$

The proof of this is virtually identical to the proof of (1.84) on page 47 and will not be repeated here. This means that the IID distributions are conditional distributions given $\mathcal{A}_\infty$, hence they are 0-1 on Σ^* and in the extremal family. □

PROOF OF THEOREM 2.113. To see that each distribution in $\mathcal{M}^*$ is the distribution of exchangeable random quantities, let $i_1, \ldots, i_k$ be distinct, let $n = \max\{i_1, \ldots, i_k\}$, let $B \in \mathcal{B}^k$, and let $\eta \in \mathcal{M}^*$. Let

$$\begin{aligned} A &= \{(x_1, \ldots, x_n) : (x_{i_1}, \ldots, x_{i_k}) \in B\}, \\ A' &= \{(x_1, \ldots, x_n) : (x_1, \ldots, x_k) \in B\}. \end{aligned}$$

Since $r_n(\cdot, t)$ is the distribution of exchangeable random quantities for all n and t, we have $r_n(A, t) = r_n(A', t)$ for all n and t. Let η_{T_n} be the distribution of T_n. Then

$$\begin{aligned} \eta((X_{i_1}, \ldots, X_{i_k}) \in B) &= \int r_n(A, t) d\eta_{T_n}(t) \\ &= \int r_n(A', t) d\eta_{T_n}(t) = \eta((X_1, \ldots, X_k) \in B). \end{aligned}$$

Next, we want to prove that the IID distributions are the extremal distributions. Lemma 2.137 says that the IID distributions in $\mathcal{M}^*$ are contained in $\mathcal{E}^*$. We now prove that all extremal distributions are IID It follows from DeFinetti's representation theorem 1.49 that every distribution η in $\mathcal{M}^*$ is a mixture of IID distributions. We show next that if $\eta \in \mathcal{E}^*$, then the mixture must be trivial. Let $\eta \in \mathcal{E}^*$, and represent

$$\eta_n(\cdot) = \int_{\mathcal{P}} P_n(\cdot) d\mu_{\mathbf{P}}(P), \qquad (2.138)$$

as in DeFinetti's representation theorem 1.49, where P_n is the distribution that says that $X_n = (Y_1, \ldots, Y_n)$ are IID with distribution P. Let $q(P)$ be the joint distribution on $(\mathcal{Y}^\infty, \mathcal{D}^\infty)$ which says that $\{Y_n\}_{n=1}^\infty$ is IID with distribution P. Since condition T says that X_n is conditionally independent of $\{Y_{n+i}\}_{i=1}^\infty$ given T_n, and since $\mathbf{P}$ is a function of $\{Y_{n+i}\}_{i=1}^\infty$, it follows that

2.4. Extremal Families   137

X_n is conditionally independent of **P** given T_n. This means that the conditional distribution of X_n given $T_n = t$ and **P** is $r_n(\cdot, t)$. Hence, $q(\mathbf{P}) \in \mathcal{M}^*$ with probability 1. This makes (2.138) a representation of η as a mixture of elements of $\mathcal{M}^*$. Since $\eta \in \mathcal{E}^*$, the mixture must be trivial, that is, $q(\mathbf{P}) = \eta$ with probability one. So, we have that all distributions in $\mathcal{E}^*$ are IID distributions.

The last claim in the theorem follows from the fact that each element of $\mathcal{E}$ is the limit of r_n probabilities. □

PROOF OF THEOREM 2.114. Since $r_n(\mathbb{R}^k, t) = 1$ for all t, it is clear that r_n is a transition kernel. Since $B(t) = B \cap T_n^{-1}(\{t\})$, r_n satisfies condition 2.112. Since every member of the exponential family in question has the conditional distribution of $(Y_1, \ldots, Y_n)$ given T_n equal to r_n, it follows that condition S holds and that every member of the exponential family is also in $\mathcal{M}$. Since $Y_{n+1} = T_{n+1} - T_n$, Proposition B.28 says that the conditional distribution of X_n given $(T_n, Y_{n+1}) = (t, y)$ is the same as that given $(T_n, T_{n+1}) = (t, t + y)$, so condition T holds. So the conditions of Theorem 2.113 hold. Lemma 2.137 says that every member of the exponential family is in the extremal family. We now prove that all extremal distributions are in the exponential family. We may assume, without loss of generality, that $h(0) > 0$. (If not, find c such that $h(c) > 0$ and subtract c from all X_i. Then replace $h(y)$ by $v(y) = h(y + c)$ and note that r_n has the same form in terms of v as it does in terms of h.) Let f be the density of a distribution in the extremal family, and let

$$f^{(2)}(t) = \int f(t-y)f(y)dy.$$

Then $f^{(2)}$ is the density of $Y_1 + Y_2$, since Y_1 and Y_2 are IID in the extremal family. Since f leads to the same conditional distributions as h, we get

$$f(y) = \int \frac{h(y)h(t-y)}{h^{(2)}(t)} f^{(2)}(t) dt,$$

hence $f(0) > 0$, since $h(0) > 0$. It also follows that

$$\frac{f(t-y)f(y)}{f^{(2)}(t)} = \frac{h(t-y)h(y)}{h^{(2)}(t)}, \quad \text{a.e.,} \qquad (2.139)$$

since both sides give the conditional distribution of Y_1 given $Y_1 + Y_2 = t$. Define

$$\lambda(y) = \log \frac{f(y)}{h(y)} - \log \frac{f(0)}{h(0)},$$

$$\psi(t) = \log \frac{f^{(2)}(t)}{h^{(2)}(t)} - 2\log \frac{f(0)}{h(0)}.$$

By taking the log of both sides of (2.139), we get $\lambda(t - y) + \lambda(y) = \psi(t)$. Now, set $y = t$ and note that $\lambda(0) = 0$, so that $\lambda(t) = \psi(t)$. It follows that

$\lambda(t-y) + \lambda(y) = \lambda(t)$. According to Theorem C.9, $\lambda(y) = a + b^\top y$ for some scalar a and vector b. Hence, $f(y)$ is a constant times $h(y) \exp(b^\top y)$ for all y such that $f(y) > 0$. To see that $f(y) > 0$ whenever $h(y) > 0$, note that (2.139) implies that $f(y) = 0$ and $h(y) > 0$ means that $h(t-y)/h^{(2)}(t) = 0$ for all t. But this would contradict

$$h(y) = \int \frac{h(y)h(t-y)}{h^{(2)}(t)} h^{(2)}(t) dt.$$
□

2.5 Problems

Section 2.1.2:

1. Suppose that P_θ says that $X_1, \ldots, X_n$ are IID $N(\theta, 1)$, for $\theta \in \mathbb{R}$. Let $X = (X_1, \ldots, X_n)$ and find a one-dimensional sufficient statistic T. Also, find the conditional distribution of X given $T = t$.

2. Refer to the definition of $P_{\theta,T}$ on page 84. If $\{P_\theta : \theta \in \Omega\}$ is a regular conditional distribution on $(\mathcal{X}, \mathcal{B})$ given Θ, then prove that $\{P_{\theta,T} : \theta \in \Omega\}$ is a regular conditional distribution on $(T, \mathcal{C})$ given Θ.

3. Suppose that $X_1, \ldots, X_n$ are conditionally IID with $N(\mu, \sigma^2)$ distribution given $\Theta = (\mu, \sigma)$. Find a two-dimensional sufficient statistic.

4. Let $X_1, \ldots, X_n$ be conditionally independent given $P = p$ with X_i having conditional density (with respect to counting measure on the nonnegative integers)

$$f_{X_i|P}(x|p) = \frac{\Gamma(\alpha_i + x)}{\Gamma(\alpha_i) x!} p^x (1-p)^{\alpha_i},$$

where $\alpha_1, \ldots, \alpha_n$ are known strictly positive numbers. (These are generalized negative binomial random variables.) Define $T = \sum_{i=1}^n X_i$. Find the conditional distribution of $(X_1, \ldots, X_n)$ given $T = t$ and $P = p$.

5. Let P_θ say that $X_1, \ldots, X_n$ are IID $Poi(\theta)$. Show that $T = \sum_{i=1}^n X_i$ is sufficient by both Definitions 2.4 and 2.8.

6. Prove Proposition 2.12 on page 86 and find the conditional distribution of $(X_1, \ldots, X_n)$ given the order statistics.

7. Prove Proposition 2.23 on page 90.

8. For the experiment in Example 2.53 on page 102, find the conditional distribution of X given X_1 and Θ and show that X_1 is not sufficient. Show that X is minimal sufficient.

9.*Consider the experiment described in Example 2.54 on page 102. Let N be the number of observed Y_i, $\mathcal{X} = \cup_{m=2}^\infty \{0,1\}^m$, and $X = (Z, Y_1, \ldots, Y_N)$.

   (a) Find the density X given $\Theta = \theta$ with respect to counting measure on $(\mathcal{X}, 2^\mathcal{X})$.

(b) Let M be the number of observed successes among the Y_i, that is, $M = \sum_{i=1}^{N} Y_i$. Show that (N, M) is sufficient. In particular, we do not need to keep track of Z.

(c) Find the conditional distribution of Z given (N, M, Θ), and show that it does not depend on Θ.

10. (Nonexchangeable example) Let P_θ say that $\{X_n\}_{n=1}^{\infty}$ are Bernoulli random variables with the following joint distribution:

$$P_\theta(X_1 = 1) = \theta_1$$

$$P_\theta(X_i = 1 | X_1, \ldots, X_{i-1}) = \begin{cases} \theta_{11} & \text{if } X_{i-1} = 1, \\ \theta_{10} & \text{if } X_{i-1} = 0, \end{cases}$$

where $\theta = (\theta_1, \theta_{11}, \theta_{10})$.

(a) Let $X = (X_1, \ldots, X_n)$. Find a four-dimensional sufficient statistic.

(b) Suppose that $X = (X_1, \ldots, X_N)$, where N is the number of observations until k successes (1s) have been observed, where k is known. Find a three-dimensional sufficient statistic.

Section 2.1.3:

11. Prove that the sufficient statistic T found in Problem 1 on page 138 is minimal sufficient.

12. Show that T is a complete sufficient statistic in Problem 4 on page 138.

13. (Logistic regression) Let $\{Y_i\}_{i=1}^{\infty}$ be Bernoulli random variables, but assume that each Y_i comes with a known vector x_i of k covariates. Conditional on $\Theta = \theta$ (a vector of length k), the Y_i are independent with

$$\log\left\{\frac{P_\theta(Y_i = 1)}{P_\theta(Y_i = 0)}\right\} = \theta^\top x.$$

Let $X = (Y_1, \ldots, Y_n)$. Find a minimal sufficient statistic (vector).

14. *Let $(X_1, Y_1), \ldots, (X_n, Y_n)$ be conditionally IID with uniform distribution on the disk of radius r centered at (θ_1, θ_2) in $\mathbb{R}^2$ given $(\Theta_1, \Theta_2, R) = (\theta_1, \theta_2, r)$.

(a) If (Θ_1, Θ_2) is known, find a minimal sufficient statistic for R.

(b) If all parameters are unknown, show that the convex hull of the sample points is a sufficient statistic.

15. *Here, we will construct a function T as needed in Theorem 2.29 for the general case. The function will turn out to be essentially the likelihood function.

(a) Let $\mathcal{F}$ be the space of functions $f : \Omega \to [0, \infty)$ with the product σ-field $\mathcal{C}$. Prove that the function $T_1 : \mathcal{X} \to \mathcal{F}$ defined by $T_1(x) = f_{X|\Theta}(x|\cdot)$ is measurable.

(b) Consider the relation $\sim$ on $\mathcal{F}$ defined by $f \sim g$ if there exists $c > 0$ such that $g(\theta) = cf(\theta)$ for all θ. Prove that $\sim$ is an equivalence relation. (That is, prove that (i) $f \sim f$, (ii) $f \sim g$ implies $g \sim f$, and (iii) ($f \sim g$ and $g \sim h$) implies $f \sim h$.)

(c) Let $\mathcal{T}$ be the set of all equivalence classes $[f] = \{g : g \sim f\}$. Let the σ-field of subsets of $\mathcal{T}$ be the smallest σ-field containing sets of the form $A_{\theta,\psi,c} = \{[f] : f(\theta) \leq cf(\psi)\}$. Prove that $[f] \in A_{\theta,\psi,c}$ if and only if $[g] \in A_{\theta,\psi,c}$ for all $g \in [f]$.

(d) Let $T_2 : \mathcal{F} \to \mathcal{T}$ be defined by $T_2(f) = [f]$. Prove that T_2 is measurable.

(e) Prove that $T = T_2(T_1)$ satisfies $T(x) = T(y)$ if and only if $y \in \mathcal{D}(x)$ in the notation of Theorem 2.29.

16. Suppose that $\{(X_n, Y_n)\}_{n=1}^{\infty}$ are conditionally IID given $\Theta = \theta$ with distribution uniform on the disk of radius θ centered at $(0,0)$, that is,

$$f_{X_i, Y_i | \Theta}(x_i, y_i | \theta) = \frac{1}{2\pi\theta^2} I_{[0,\theta]}\left(\sqrt{x_i^2 + y_i^2}\right).$$

Let $X = [(X_1, Y_1), \ldots, (X_n, Y_n)]$. Find a complete sufficient statistic and its distribution.

17. Suppose that $\Omega = \{(\theta_1, \theta_2) : \theta_2 > \theta_1\}$ and P_{θ_1, θ_2} says that $X_1, \ldots, X_n$ are IID with $U(\theta_1, \theta_2)$ distribution. Find minimal sufficient statistics.

18.*Let X be a discrete random variable, and let

$$f_{X|\Theta}(x|\theta) = \begin{cases} \theta & \text{if } x = 0, \\ (1-\theta)^2 \theta^{x-1} & \text{if } x = 1, 2, \ldots, \\ 0 & \text{otherwise} \end{cases}$$

be the density of X (conditional on $\Theta = \theta$) with respect to counting measure on the integers. Let $\Omega = (0,1)$. Prove that X is boundedly complete, but not complete.

19. Suppose that P_θ says that $\{X_n\}_{n=1}^{\infty}$ are IID $Ber(\theta)$. Let $X = (X_1, \ldots, X_n)$ and $T = \sum_{i=1}^{n} X_i$. Prove that T is a complete sufficient statistic without using Theorem 2.74.

20. Suppose that P_θ says that $\{X_n\}_{n=1}^{\infty}$ are IID $U(0, \theta)$. Let $X = (X_1, \ldots, X_n)$ and $T = \max_{i=1,\ldots,n} X_i$. Prove that T is a complete sufficient statistic.

21.*Suppose that $X_1, \ldots, X_n$ are IID given $\Theta = \theta$ with conditional distribution uniform on the set $[0, \theta] \cup [2\theta, 3\theta]$. That is,

$$f_{X_i|\Theta}(x|\theta) = \frac{1}{2\theta}\left(I_{[0,\theta]}(x) + I_{[2\theta, 3\theta]}(x)\right).$$

Find a minimal sufficient statistic (dimension at most 3).

22. Let $Z = (X_1, \ldots, X_n, Y_1, \ldots, Y_n)$ where the X_i and Y_i are all conditionally independent with $X_i \sim N(\mu, \sigma_X^2)$ and $Y_i \sim N(\mu, \sigma_Y^2)$ given $\Theta = (\mu, \sigma_X, \sigma_Y)$. Let $T(Z) = (\overline{X}, \overline{Y}, S_X^2, S_Y^2)$, the usual sample means and variances. Show that T is minimal sufficient but that T is not boundedly complete.

23.*Let $\Omega = \{1, 2, 3\}$, and let $X = (X_1, \ldots, X_n)$ be conditionally IID given $\Theta = \theta$, with density $f_\theta(\cdot)$. Suppose that P_1, P_2, and P_3 have density functions (with respect to Lebesgue measure): $f_1(x) = I_{(-1,0)}(x)$, $f_2(x) = I_{(0,1)}(x)$, and $f_3(x) = 2xI_{(0,1)}(x)$. Thus, the model has only three members. Let $S(x) = \{(j, k); \prod_i f_j(x_i) + \prod_i f_k(x_i) > 0\}$. Let

$$T(x) = \left\{ \frac{\prod_i f_k(x_i)}{\prod_i f_j(x_i)}; j < k, (j, k) \in S(x) \right\}.$$

Show that T is minimal sufficient.

24. (Nonexchangeable example) Suppose that $\{X_n\}_{n=1}^\infty$ is a sequence of random variables, that $\Theta \in (-1, 1)$ is a parameter such that the conditional distribution of X_1 given $\Theta = \theta$ is $N(0, [1-\theta^2])$, and that, for $i > 1$, the conditional distribution of X_i given $\Theta = \theta$ and $(X_1, \ldots, X_{i-1}) = (x_1, \ldots, x_{i-1})$ is $N(\theta x_{i-1}, 1)$.[22]

   (a) If $X = (X_1, \ldots, X_n)$, find a three-dimensional minimal sufficient statistic.

   (b) Find the conditional distribution of X_{n+1} given $X = x$, and show that it depends on more of the data X than the minimal sufficient statistic.

Section 2.1.4:

25. Suppose that P_θ says that $\{X_n\}_{n=1}^\infty$ are IID $U(\theta - 1/2, \theta + 1/2)$. Let $X = (X_1, \ldots, X_n)$. Find minimal sufficient statistics and find a nonconstant function of the sufficient statistic which is ancillary.

26. Prove that if S is ancillary, then S and Θ are independent no matter what prior one uses for Θ.

27. (A vector example) Suppose that P_θ says

$$\begin{pmatrix} X_1 \\ Y_1 \end{pmatrix}, \ldots, \begin{pmatrix} X_n \\ Y_n \end{pmatrix} \stackrel{\text{IID}}{\sim} N_2\left(\begin{bmatrix} 0 \\ 0 \end{bmatrix}, \begin{bmatrix} 1 & \theta \\ \theta & 1 \end{bmatrix}\right)$$

with $\Omega = (-1, 1)$.

   (a) Find a two-dimensional minimal sufficient statistic.

   (b) Prove that the minimal sufficient statistic found above is not complete.

   (c) Prove that $Z_1 = \sum_{i=1}^n X_i^2$ and $Z_2 = \sum_{i=1}^n Y_i^2$ are both ancillary but that (Z_1, Z_2) is not ancillary.

28.*Consider the situation in Example 2.51 on page 100. Prove that the distribution of U is uniform on the sphere centered at the vector $\mathbf{0}$ of n 0s in the hyperplane defined by $\mathbf{1}^\top u = 0$, where $\mathbf{1}$ is the vector of n 1s. (Hint: Let A be an orthogonal $n \times n$ matrix that maps the hyperplane to itself. Prove that AU has the same distribution as U.)

---

[22] Such a sequence is often called an *autoregression of order 1*.

29. Suppose that $\Pr(X > 0) = 1$ and that the conditional distribution of Y given $X = x$ is $U(0, x)$. Let $Z = X - Y$ and suppose that Y and Z are independent. Let $f_X(x)$, $f_Y(y)$, and $f_Z(z)$ be differentiable.

   (a) Prove that $\Pr(X > c) > 0$ for all $c > 0$.

   (b) Prove $f_X(x) = a^2 x \exp(-ax)$, for $x > 0$.

30. Let $X_1, \ldots, X_n$ be conditionally IID given $\Theta = \theta$ each with density $g(x - \theta)$ for some function g. Prove that $\max\{X_1, \ldots, X_n\} - \min\{X_1, \ldots, X_n\}$ is ancillary, but not maximal ancillary if $n > 2$.

31. Consider the situation in Example 2.46 on page 97. Suppose that we wish to condition on N_0 if
$$\text{Var}_\theta\left(1 - 3\frac{N_{00}}{N_0}\bigg| N_0\right) \leq \text{Var}_\theta\left(1 - 2\frac{N_{00}}{M_0}\bigg| M_0\right),$$
and we wish to condition on M_0 otherwise. For which data sets would we choose N_0, and for which would we choose M_0?

32. Consider the situation in Example 2.46 on page 97. Suppose that we need to choose upon which ancillary to condition before we see the data. Suppose that we decide to condition on N_0 if
$$\text{Var}_\theta\left(1 - 3\frac{N_{00}}{N_0}\right) \leq \text{Var}_\theta\left(1 - 2\frac{N_{00}}{M_0}\right),$$
and we will condition on M_0 otherwise. On which ancillary will we condition?

33. Call a statistic U *ignorable* if there exists a sufficient statistic T such that T and U are conditionally independent given Θ. Prove that an ignorable statistic is ancillary.

Section 2.2:

34. Express the family of Poisson distributions in exponential family form. Find the natural parameter, natural parameter space, and sufficient statistic. Use Theorem 2.64 to find the mean and the variance of the sufficient statistic.

35. Express the family of Beta distributions in exponential family form. Find the natural parameter, natural parameter space, and sufficient statistic(s). Use Theorem 2.64 to find the mean and the variance of the sufficient statistics. (*Hint:* The derivative of the log of the gamma function is called the *digamma* function ψ. The second derivative is called the trigamma function ψ'.)

36. In Problem 9 on page 138, show that the family of distributions for observed data is an exponential family but that the sufficient statistic is not complete. How do you reconcile this with Theorem 2.74?

37. Prove Proposition 2.70 on page 107.

38. Prove Proposition 2.72 on page 107.

2.5. Problems

Section 2.3:

39. Suppose that X and Y are conditionally independent given Θ with conditional densities $f_{X|\Theta}(x|\theta)$ and $f_{Y|\Theta}(y|\theta)$, respectively. Suppose that Θ is k-dimensional. Prove that $\mathcal{I}_{X,Y}(\theta) = \mathcal{I}_X(\theta) + \mathcal{I}_Y(\theta)$.
40. Prove Proposition 2.84 on page 112.
41. Prove Proposition 2.92 on page 116.
42. Suppose that P_θ says that $X \sim Poi(\theta)$.
    (a) Find the Fisher information $\mathcal{I}_X(\theta)$.
    (b) Find Jeffreys' prior.
43. Suppose that the FI regularity conditions hold and that two derivatives can be passed under integral signs. Let T be an ancillary statistic. Prove that $\mathcal{I}_{X|T}(\theta|t)$ has (i,j) entry equal to $-E_\theta \left(\partial^2 \log f_{X|T,\Theta}(X|t,\theta)/\partial\theta_i\partial\theta_j \right)$.
44. Let $\Omega = \{(p_1, p_2, p_3) : p_i \geq 0, \sum_{i=1}^{3} p_i = 1\}$ and
$$f_{X|\Theta}(x|p_1, p_2, p_3) = \begin{cases} p_1 & \text{if } x = 1, \\ p_2 & \text{if } x = 2, \\ p_3 & \text{if } x = 3, \\ 0 & \text{otherwise.} \end{cases}$$
Let $\theta_0 = (1/3, 1/3, 1/3)$, and find the value of θ such that $E_\theta(X) = 2.5$ and $\mathcal{I}_X(\theta_0; \theta)$ is minimized.
45. Suppose that $X \sim U(0, \theta)$ given $\Theta = \theta$. Find the Kullback–Leibler information $\mathcal{I}_X(\theta_1; \theta_2)$ for all pairs (θ_1, θ_2).
46. Suppose that person 1 believes $\Pr(\Theta = 1/3) = \pi_0$ and $\Pr(\Theta = 1/2) = 1 - \pi_0$ and person 2 believes $\Pr(\Theta = q) = 1$. Both persons believe that $\{X_n\}_{n=1}^{\infty}$ are IID $Ber(\theta)$ given $\Theta = \theta$. Let $Y_n = \sum_{i=1}^{n} X_i/n$.
    (a) Find $\Pr(\Theta = 1/3|Y_n = q)$ for person 1.
    (b) For each possible value of q, describe person 2's beliefs about how the value of $\Pr(\Theta = 1/3|Y_n)$ (calculated by person 1) will behave as $n \to \infty$.
47. Let Y be the number of patients (out of n) who survive for one year after an operation. Let Z be the number of patients who survive for five years. Let $\Theta = (P, Q)$, and suppose that we model $Y \sim Bin(n, p)$ given $\Theta = (p, q)$ and $Z \sim Bin(y, q)$ given $Y = y$ and $\Theta = (p, q)$. Let $X = (Y, Z)$. Find $\mathcal{I}_X(\theta)$ and Jeffreys' prior.

Section 2.4:

48. *Let $T_n(x_1, \ldots, x_n) = \sum_{i=1}^{n} x_i$, and suppose that $(X_1, \ldots, X_n)$ given $T_n = t$ is distributed uniformly on the portion of the hyperplane $\sum_{i=1}^{n} x_i = t$ with all coordinates nonnegative. Find the extremal family of distributions.
49. *Let $\mathcal{X}^1 = \{0, 1\}$. Let $T_n(x_1, \ldots, x_n) = \sum_{i=1}^{n} x_i$, and suppose that the distribution of $X_1, \ldots, X_n$ given $T_n = t$ is that of draws without replacement from an urn containing t 1s and $n - t$ 0s. Find the extremal family of distributions.

CHAPTER 3

Decision Theory

A major use of statistical inference is its application to decision making under uncertainty. When the costs and/or benefits of our actions depend on quantities we will not know until after we make our decisions, we need to be able to weigh the costs against the uncertainties intelligently.

3.1 Decision Problems

3.1.1 Framework

Suppose that one can determine ahead of time a set of actions from which one will have to choose. We name this set $\aleph$ and call it the *action space*. This set will contain all of the actions under consideration. We will occasionally need to introduce a measure over this set, so let α be a σ-field of subsets of $\aleph$. In the most general type of decision problem we will consider, we suppose that there is a not yet observed quantity V (taking values in a set $\mathcal{V}$) on which depends the amount we lose as a function of our action.

Example 3.1. Suppose that we are trying to decide whether to keep a store open for an extra hour during a busy shopping season. We might be able to determine the extra costs of overhead and payroll associated with staying open, but the amount of additional net sales V is as yet unknown. The final profit or loss associated with the decision depends on V.

Definition 3.2. Let $(S, \mathcal{A}, \mu)$ be a probability space, $\aleph$ an action space, and $V : S \to \mathcal{V}$ a function. A *loss function* is a function $L : \mathcal{V} \times \aleph \to \mathbb{R}$. $L(v, a)$ measures "how much" we lose by choosing action a when $V = v$.

Consider the following simple example.

3.1. Decision Problems

Example 3.3. Let $V \sim N(1,1)$, and suppose that $\aleph = \mathbb{R}$ and $L(v,a) = (v-a)^2$. In this case, the amount I lose when I choose action a is the squared distance between a and the unknown V. Alternatively, we might have $L(v,a) = 3|v-a|$. In this case, I lose three times the distance between V and a.

The conditional distribution of V given $\Theta = \theta$ will be denoted by $P_{\theta,V}$. For convenience we will assume that V and X are conditionally independent given Θ. Most often, in discussions of statistical theory, the function V is Θ, so that $P_{\theta,V}(B) = I_B(\theta)$, and $L : \Omega \times \aleph \to \mathbb{R}$. But this is not actually necessary. The goal of decision theory is to make $L(V,a)$ as small as possible by choice of $a \in \aleph$. Unfortunately, this would normally require that we know the value V. For example, in Example 3.3, both losses are smallest if $a = V$. If V is a function that depends on the future (coordinates later than those observed or the parameter), then we will not know V at the time a decision will need to be made. When $V = \Theta$, we will not even know V after the decision is made.

The tools we use to make decisions are called *decision rules*.

Definition 3.4. A *randomized decision rule* δ is a mapping from $\mathcal{X}$ to probability measures on $(\aleph, \alpha)$ such that for every $A \in \alpha$, $\delta(\cdot)(A)$ is measurable. A *nonrandomized decision rule* δ is a randomized decision rule that for each x assigns probability 1 to a single action, denoted by $\delta(x)$. That is, a randomized decision rule δ is a nonrandomized rule if, for each $x \in \mathcal{X}$, there exists $a_x \in \aleph$ such that $\delta(x)(A) = I_A(a_x)$, and in such a case a_x is denoted by $\delta(x)$.

Note that the definition of a randomized decision rule makes it a regular conditional distribution over $(\aleph, \alpha)$ given X. Of course, if one were actually to use a randomized decision rule, one would need to choose an action in $\aleph$, not just a probability measure over $(\aleph, \alpha)$. To do this, one takes the observed x and simulates (see Section B.7) a pseudorandom element of $\aleph$ according to the probability measure $\delta(x)(\cdot)$. Hence, an alternative method for specifying a randomized rule is to specify, for each possible x, the way in which one will simulate the action from $\aleph$.

Example 3.5. Suppose that m and n are even integers. Suppose that P_θ says that $X_1, \ldots, X_{n+m}$ are IID $Ber(\theta)$ random variables. Let $X = (X_1, \ldots, X_n)$ and $V = \sum_{i=1}^{m} X_{n+i}$. Let the action space be $\aleph = \{a_0, a_1\}$, and suppose that the loss function is

$$L(v,a) = \begin{cases} 0 & \text{if } \left(v < \frac{m}{2} \text{ and } a = a_0\right) \text{ or if } \left(v > \frac{m}{2} \text{ and } a = a_1\right), \\ \frac{1}{2} & \text{if } v = \frac{m}{2}, \\ 1 & \text{otherwise.} \end{cases}$$

Let $Y = \sum_{i=1}^{n} X_i$ and $y = \sum_{i=1}^{n} x_i$. Here is a plausible randomized decision rule:

$$\delta(x) = \begin{cases} \text{probability 1 on } a_0 & \text{if } y < \frac{n}{2}, \\ \text{probability 1 on } a_1 & \text{if } y > \frac{n}{2}, \\ \text{probability } \frac{1}{2} \text{ on each} & \text{if } y = \frac{n}{2}. \end{cases}$$

If $Y = n/2$ is observed, one could flip a fair coin to decide between the two actions.

If δ is a randomized rule, set

$$L(v, \delta(x)) = \int_{\aleph} L(v, a) d\delta(x)(a).$$

This then allows us to talk about the loss incurred by either a randomized or a nonrandomized rule without regard to the result of the auxiliary randomization in the randomized rule.

Example 3.6 (Continuation of Example 3.5; see page 145). If $y = n/2$, then one can easily show that $L(v, \delta(x)) = 1/2$ for all v.

3.1.2 Elements of Bayesian Decision Theory

In the Bayesian paradigm, one calculates the *posterior risk*

$$r(\delta|x) = \int_{\mathcal{V}} L(v, \delta(x)) d\mu_{V|X}(v|x)$$

for each decision rule and chooses the one with the smallest posterior risk. Here, $d\mu_{V|X}$ denotes the conditional distribution of V given X. If we do this for every x and the posterior risk is never $+\infty$, the resulting rule is called a *formal Bayes rule*.

Definition 3.7. If δ_0 is such that $r(\delta_0|x) < \infty$ for all x and $r(\delta_0|x) \leq r(\delta|x)$ for all x and all decision rules δ, then δ_0 is called a *formal Bayes rule*.

The use of formal Bayes rules is based on the following principle.

> **The Expected Loss Principle:** When one compares two rules after observing data, the better rule is the one with the smaller posterior risk.

A justification for this principle will be given in Section 3.3. One feature of that justification, which we do not use here, however, is that the loss function needs to be bounded.

Example 3.8. Let $\mathcal{V} \subseteq \mathbb{R}$, $\aleph \subseteq \mathbb{R}$, and $L(v, a) = (v - a)^2$. Then

$$r(\delta|x) = \int_{\mathcal{V}} (v - a)^2 d\mu_{V|X}(v|x) = \mathrm{E}(V^2|X = x) - 2\delta(x)\mathrm{E}(V|X = x) + \delta(x)^2.$$

Assuming that $\mathrm{E}(V^2|X = x) < \infty$, we can easily minimize the posterior risk by setting $\delta(x) = \mathrm{E}(V|X = x)$. This result is very general. So long as the posterior variance of V is finite, a formal Bayes rule with squared-error loss is the posterior mean of V.

It is possible that there exist x values such that the posterior risk given $X = x$ is $+\infty$ for every decision rule. Also, it is possible that although there exist rules with posterior risk $< \infty$ given $X = x$, there is no rule that achieves the minimum of the posterior risk. In these cases, there is no formal Bayes rule as we have defined it, although there may exist x values such that, conditional on $X = x$, the posterior risk can be minimized at a value $< \infty$. In this latter case, we call a rule that minimizes the posterior risk at all values of x for which a minimum $< \infty$ can be achieved a *partial Bayes rule*.

Example 3.9. Suppose that $\{Y_n\}_{n=1}^{\infty}$ are conditionally IID with Cauchy distribution $Cau(\theta, 1)$ given $\Theta = \theta$, where $\Omega = \mathbb{R} = \aleph$, $V = \Theta$, and $L(\theta, a) = (a-\theta)^2$. Let the prior distribution of Θ be $Cau(0,1)$. Let $t > 0$ and let $X_i = \min\{t, Y_i\}$. Define $X = (X_1, X_2, X_3)$. If at least one of the X_i is strictly less than t, then the posterior risk will be finite for some decision rule. But if all three $X_i = t$, the posterior risk is infinite for all decision rules. In this example, a partial Bayes rule is any rule that chooses the action minimizing the posterior risk for those data in which all at least one $X_i < t$. As we saw in Example 3.8, the action to choose in those cases is the posterior mean of Θ. For those data x such that the posterior risk is infinite (all $X_i = t$), it might still make sense to choose $\delta(x)$ in such a way that the posterior distribution of $L(V, \delta(x))$ is stochastically small. That is, if we define $Z_\delta = L(V, \delta(x))$, we should prefer δ_1 to δ_2 if the CDF of Z_{δ_1} is everywhere larger than the CDF of Z_{δ_2}.

Example 3.10. Let $\aleph = (0,1) = \Omega$, and let $X = (X_1, \ldots, X_{10})$ where the X_i are IID with $Ber(\theta)$ distribution given $\Theta = \theta$. Let $V = \Theta$ and $L(\theta, a) = (\theta - 0.1 - a)^2$. Let the prior distribution of Θ be $Beta(1,1)$ so that the posterior given $X = x$ is $Beta(x+1, 11-x)$. If $X = x > 0$ is observed, then the posterior risk is minimized at $\delta_0(x) = (x+1)/11 - 0.1$. However, if $X = 0$ is observed, the posterior risk is an increasing function of a for $a \in \aleph$, so we would like to choose $\delta_0(0)$ as small as possible. But the action space is not closed, so there is no smallest possible value. Any decision rule δ such that $\delta(x) = \delta_0(x)$ for $x > 0$ is a partial Bayes rule.

If the posterior risk of a randomized rule is finite, or the loss function is nonnegative, we can write the posterior risk as

$$r(\delta|x) = \int_\aleph \int_\mathcal{V} L(v, a) dF_{V|X}(v|x) d\delta(x)(a). \tag{3.11}$$

In this case, if the inner integral, $h_x(a) = \int_\mathcal{V} L(v, a) dF_{V|X}(v|x)$, considered as a function of a for fixed x, does not achieve a minimum at some value of a, then it is easy to see that (3.11) does not achieve its minimum at any probability $\delta(x)$. This leads us to state the following result.

Theorem 3.12. *If a formal Bayes rule exists with finite posterior risk, then there is a nonrandomized formal Bayes rule. If, at a particular value of x, the posterior risks of nonrandomized rules are unbounded below, then there exists a randomized rule with posterior risk $-\infty$ at that x (whether or not there is such a nonrandomized rule).*

PROOF. Before proving the first part, let $h_x(a) = \int_V L(v,a)dF_{V|X}(v|x)$, the posterior risk of a nonrandomized rule with $\delta(x) = a$. Then, if a formal Bayes rule exists with finite posterior risk, $h_x(a)$ (as a function of a) must be bounded below. Furthermore, if for some x, $h_x(a) = \infty$ for all a, then every decision rule has infinite posterior risk at x. Hence, we can assume that $c(x) = \inf_{a \in \aleph} h_x(a) < \infty$ for all x.

We will prove the contrapositive of the first part of the theorem, namely that if, at some value of x, there is no nonrandomized formal Bayes rule, then the formal Bayes rule does not exist. Suppose that there is no nonrandomized formal Bayes rule at a value x. Then there exists no $b \in \aleph$ such that $h_x(b) = c(x)$. Suppose that δ is a randomized rule with finite posterior risk (so $c(x) > -\infty$ also.) Let $A_n = \{a : h_x(a) \geq c(x) + 1/n\}$. Since $\{a : h_x(a) = c(x)\} = \emptyset$, we can choose n large enough so that $\delta(x)(A_n) > 0$. It follows that

$$\int h_x(a)d\delta(x)(a) \geq c(x)\delta(x)(A_n^C) + \int_{A_n} h_x(a)d\delta(x)(a)$$
$$\geq c(x)\delta(x)(A_n^C) + \delta(x)(A_n)\left(c(x) + \frac{1}{n}\right)$$
$$= c(x) + \frac{\delta(x)(A_n)}{n} > c(x).$$

Since $c(x) = \inf_{a \in \aleph} h_x(a)$, there exists a such that

$$c(x) < h_x(a) < \int h_x(a)d\delta(x)(a).$$

It follows that δ is not a formal Bayes rule.

For the second part, suppose that $c(x) = \inf_{a \in \aleph} h_x(a) = -\infty$. For each $k = 1, 2, \ldots$, let a_k be such that $h_x(a_k) \leq -2^k$ and for $k > 1$, $h_x(a_k) < h_x(a_{k-1})$. Let $\delta(x)$ assign probability 2^{-k} to a_k for each k. Then, it is easy to see that δ has posterior risk $-\infty$ even if $h_x(a) > -\infty$ for all a. □

Although Theorem 3.12 says that there are cases in which one need only consider nonrandomized rules in order to find formal Bayes rules, it may be that there are still some randomized formal Bayes rules as well.

Example 3.13. Suppose that P_θ says that $\{X_n\}_{n=1}^\infty$ are IID $Ber(\theta)$. Let $\aleph = \{a_0, a_1\}$ and

$$L(\theta, a) = \begin{cases} 0 & \text{if } (\theta \leq \frac{1}{2} \text{ and } a = a_0) \text{ or if } (\theta > \frac{1}{2} \text{ and } a = a_1), \\ 1 & \text{otherwise.} \end{cases}$$

Let $X = (X_1, \ldots, X_n)$ and suppose that n is even. Let $Y = \sum_{i=1}^n X_i$.

Suppose that the prior is η equal to Lebesgue measure. If $Y = y$ successes are observed in n trials, the posterior is $Beta(y+1, n-y+1)$. The posterior risk for choosing $a = a_0$ is $r(y) = \Pr(\Theta > 1/2|Y = y)$, and the posterior risk for choosing $a = a_1$ is $\Pr(\Theta \leq 1/2|Y = y) = 1 - r(y)$. The formal Bayes rule will be to choose a_0 if $r(y) < 1 - r(y)$ (that is, if $r(y) < 1/2$) and to choose a_1 if

$r(y) > 1/2$. If, however, $y = n/2$, the posterior will be $Beta(n/2+1, n/2+1)$, which is symmetric about $1/2$, so $r(y) = 1/2$. Randomized rules of the following form are formal Bayes rules:

$$\delta(X) = \begin{cases} \text{probability 1 on } a_0 & \text{if } Y < \frac{n}{2}, \\ \text{probability 1 on } a_1 & \text{if } Y > \frac{n}{2}, \\ \text{probability } \frac{1}{2} \text{ on each} & \text{if } Y = \frac{n}{2}. \end{cases}$$

Next, we illustrate the case in which losses are unbounded below.

Example 3.14. Suppose that $\aleph = \Omega = \mathbb{R}$ and that $L(\theta, a) = -(\theta-a)^2$. In words, we are trying to choose a as far away from Θ as possible. Suppose that the posterior distribution of Θ has finite variance σ^2 and mean μ. Then the posterior risk for a nonrandomized rule δ is $-(\mu - \delta(x))^2 - \sigma^2$. So, every nonrandomized rule has finite posterior risk. If, for each x, $\delta(x)(\cdot)$ is a randomized rule such that the distribution over $\aleph$ has infinite variance, then δ will have posterior risk $-\infty$.

3.1.3 Elements of Classical Decision Theory

In the classical paradigm, one conditions on $\Theta = \theta$ (assuming that there was a parametric family specified) and calculates the *risk function*

$$R(\theta, \delta) = \int_\mathcal{X} \int_\mathcal{V} L(v, \delta(x)) dP_{\theta, V}(v) dP_\theta(x).$$

For the case in which V is not Θ, we can define[1]

$$L(\theta, a) = \int_\mathcal{V} L(v, a) dP_{\theta, V}(v). \tag{3.15}$$

In either case ($V = \Theta$ or not), the risk function becomes

$$R(\theta, \delta) = \int_\mathcal{X} L(\theta, \delta(x)) dP_\theta(x).$$

There is usually no way to choose δ to make $R(\theta, \delta)$ as small as possible for all θ simultaneously. One possibility is to choose a probability measure η over Ω and try to minimize

$$r(\eta, \delta) = \int_\Omega R(\theta, \delta) d\eta(\theta),$$

which is called the *Bayes risk*. Let μ_X denote the marginal probability measure over $\mathcal{X}$, namely $\mu_X(A) = \int_\Omega P_\theta(A) d\eta(\theta)$. Suppose that $P_\theta \ll \nu$ for every θ. If $L \geq 0$, we can use Tonelli's theorem A.69, or if $L(\theta, \delta(x)) f_{X|\Theta}(x|\theta)$

---

[1] Notice that a predictive decision problem has been replaced, in the classical setting, by a parametric decision problem with loss $L(\theta, a)$, which does not depend on the future observable V.

is integrable with respect to $\nu \times \eta$, we can use Fubini's theorem A.70 to conclude that

$$r(\eta, \delta) = \int_{\mathcal{X}} r(\delta|x) d\mu_X(x).$$

Each δ that minimizes $r(\eta, \delta)$ is called a *Bayes rule*, assuming that $r(\eta, \delta)$ is finite. Otherwise no Bayes rule exists. So, we can prove the following.

Proposition 3.16. *If a Bayes rule δ exists, then there is a partial Bayes rule that equals δ a.s. $[\mu_X]$.*

3.1.4 Summary

We now summarize the last several definitions in the case where $V = \Theta$.

Definition 3.17. Suppose that we have a decision problem with action space $\aleph$, parameter space Ω, sample space $\mathcal{X}$, and loss function $L : \Omega \times \aleph \to \mathbb{R}$. Let δ be a randomized rule. Then $L(\theta, \delta(x)) = \int_\aleph L(\theta, a) d\delta(x)(a)$. The *posterior risk of δ* is

$$r(\delta|x) = \int_\Omega L(\theta, \delta(x)) d\mu_{\Theta|X}(\theta|x).$$

Let A be the set of all x for which there exists a_x that achieves a finite minimum posterior risk. Then a decision rule such that $\delta_0(x) = a_x$ for all $x \in A$ is called a *partial Bayes rule*. If $A = \mathcal{X}$, then a partial Bayes rule is called a *formal Bayes rule*. The *risk function of a rule δ* is $R(\theta, \delta) = \int_\mathcal{X} L(\theta, \delta(x)) dP_\theta(x)$. If η is a prior distribution for Θ, the *Bayes risk of δ with respect to η* is $r(\eta, \delta) = \int_\Omega R(\theta, \delta) d\eta(\theta)$. If there is a δ that minimizes this quantity at a finite value, then that rule is called the *Bayes rule with respect to η*.

3.2 Classical Decision Theory

3.2.1 The Role of Sufficient Statistics

As defined, decision rules can be arbitrary functions of the data. We learned in Chapter 2 that all we needed from the data were sufficient statistics, so it should be the case that decision rules should only be functions of sufficient statistics. Of course, formal Bayes rules will only be functions of the sufficient statistic, since the posterior distribution is a function of the sufficient statistic. The next theorem says that if a choice of decision rules will be based solely on the risk functions, then even in classical statistics, decision rules need only be functions of sufficient statistics.

Theorem 3.18.[2] *Let δ_0 be a randomized rule and T be a sufficient statistic. Then there exists a rule δ_1 that is a function of the sufficient statistic and has the same risk function.*

PROOF. For $A \in \alpha$, define

$$\delta_1(t)(A) = \mathrm{E}(\delta_0(X)(A)|T=t).$$

(Since T is sufficient, this expectation does not depend on θ.) It follows easily that for any $\delta_0(x)$ integrable function $h: \aleph \to \mathbb{R}$,

$$\mathrm{E}\left(\int h(a)d\delta_0(X)(a)\bigg| T=t\right) = \int h(a)d\delta_1(t)(a). \quad (3.19)$$

(Just check the equation for indicators, simple functions, nonnegative functions, then integrable functions.) Then,

$$\begin{aligned}
R(\theta, \delta_1) &= \int_{\mathcal{X}} L(\theta, \delta_1(T(x)))dP_\theta(x) \\
&= \int_{\mathcal{X}}\int_{\aleph} L(\theta, a)d\delta_1(T(x))(a)dP_\theta(x), \\
R(\theta, \delta_0) &= \int_{\mathcal{X}}\int_{\aleph} L(\theta, a)d\delta_0(x)(a)dP_\theta(x).
\end{aligned}$$

It follows from (3.19) that

$$\int_{\aleph} L(\theta,a)d\delta_1(T(x))(a) = \mathrm{E}\left\{\int_{\aleph} L(\theta,a)d\delta_0(X)(a)\bigg| T=T(x)\right\}.$$

Now, use the Law of total probability B.70 to write $R(\theta, \delta_1)$ as

$$\begin{aligned}
&\int_{\mathcal{X}}\int_{\aleph} L(\theta,a)d\delta_1(T(x))(a)dP_\theta(x) \\
&= \mathrm{E}_\theta\left(\mathrm{E}\left\{\int_{\aleph} L(\theta,a)d\delta_0(X)(a)\bigg| T\right\}\right) = \mathrm{E}_\theta\left(\int_{\aleph} L(\theta,a)d\delta_0(X)(a)\right) \\
&= \int_{\mathcal{X}}\int_{\aleph} L(\theta,a)d\delta_0(x)(a)dP_\theta(x) = R(\theta, \delta_0). \quad \square
\end{aligned}$$

Note that if δ_0 in Theorem 3.18 is nonrandomized, then δ_1 will still be randomized if T is not one-to-one. (See Problem 8 on page 209.)

There are cases in which nonrandomized rules are all we need.

Theorem 3.20.[3] *Suppose that $\aleph$ is a convex subset of $\mathbb{R}^m$ and that for all $\theta \in \Omega$, $L(\theta, a)$ is a convex function of a. Let δ be a randomized rule and let*

---

[2] This theorem is used in the proof of the Rao–Blackwell theorem 3.22.
[3] This theorem is used in the proof of Theorem 3.22.

$B \subseteq \mathcal{X}$ be the set of all x such that $\int_{\aleph} |a| d\delta(x)(a) < \infty$. Let the mean of the distribution δ, considered as a nonrandomized rule, be

$$\delta_0(x) = \int_{\aleph} a d\delta(x)(a), \quad \text{for } x \in B.$$

Then $L(\theta, \delta_0(x)) \leq L(\theta, \delta(x))$ for all $x \in B$ and for all θ.

PROOF. Since $\aleph$ is convex, Theorem B.17 says that $\delta_0(x) \in \aleph$ for all $x \in B$. It follows that

$$L(\theta, \delta_0(x)) = L\left(\theta, \int_{\aleph} a d\delta(x)(a)\right) \leq \int_{\aleph} L(\theta, a) d\delta(x)(a) = L(\theta, \delta(x)),$$

for all $x \in B$. The inequality follows from Jensen's inequality B.17. □

If $B = \mathcal{X}$ in Theorem 3.20, then the posterior risk for the nonrandomized rule δ_0 will be no larger than that for the randomized rule δ. If $P_\theta(B) = 1$ for all θ, then the risk function of the nonrandomized rule will be no larger than that of the randomized rule.

Example 3.21. Suppose that P_θ says that $X_1, \ldots, X_n$ are IID $Ber(\theta)$. Let $X = (X_1, \ldots, X_n)$, $\aleph = \Omega = [0, 1]$, and $L(\theta, a) = (\theta - a)^2$. Let $y = \sum_{i=1}^n x_i$, and set

$$\delta(y) = \begin{cases} \frac{y}{n} & \text{with probability } \frac{1}{2}, \\ \frac{y+1}{n+2} & \text{with probability } \frac{1}{2}. \end{cases}$$

This is like flipping a coin between the proportion of successes and the posterior mean from a $U(0, 1)$ prior. Then

$$\delta_0(y) = \frac{y}{2n} + \frac{y+1}{2n+4},$$

$$L(\theta, \delta(y)) = \frac{1}{2}\left(\theta - \frac{y}{n}\right)^2 + \frac{1}{2}\left(\theta - \frac{y+1}{n+2}\right)^2$$

$$= \theta^2 - \theta\left(\frac{y}{n} + \frac{y+1}{n+2}\right) + \frac{1}{2}\left(\frac{y^2}{n^2} + \frac{(y+1)^2}{(n+2)^2}\right),$$

$$L(\theta, \delta_0(y)) = \left(\theta - \frac{y}{2n} - \frac{y+1}{2n+4}\right)^2$$

$$= \theta^2 - \theta\left(\frac{y}{n} + \frac{y+1}{n+2}\right) + \frac{1}{4}\left(\frac{y}{n} + \frac{y+1}{n+2}\right)^2.$$

Since $(x + z)^2/2 < x^2 + z^2$ for $x \neq z$, it follows that $L(\theta, \delta_0(y)) < L(\theta, \delta(y))$.

The theory of hypothesis testing (see Chapter 4) is one in which $\aleph$ is not a convex set, and indeed, randomized rules figure prominently in the classical theory of hypothesis testing.

Theorem 3.22 (Rao–Blackwell theorem).[4] *Suppose that $\aleph$ is a convex subset of $\mathbb{R}^m$ and that for all $\theta \in \Omega$, $L(\theta, a)$ is a convex function*

---

[4] This theorem originated with Rao (1945) and Blackwell (1947).

of a. Suppose also that T is sufficient and δ_0 is nonrandomized such that $E_\theta(\|\delta_0(X)\|) < \infty$. Define

$$\delta_1(t) = E(\delta_0(X)|T=t).$$

Then $R(\theta, \delta_1) \leq R(\theta, \delta_0)$ for all θ.

PROOF. Consider δ_0 as the randomized rule $\delta_3(x)(A) = I_A(\delta_0(x))$, for $A \in \alpha$. For $A \in \alpha$, let

$$\delta_4(t)(A) = E[\delta_3(X)(A)|T=t], \quad \delta_2(t) = \int_\aleph a d\delta_4(t)(a).$$

By Theorems 3.20 and 3.18,

$$R(\theta, \delta_2) \leq R(\theta, \delta_4) = R(\theta, \delta_3) = R(\theta, \delta_0).$$

All that remains is to show that $\delta_2 = \delta_1$. Using the law of total probability B.70, we can write

$$\delta_2(t) = \int_\aleph a d\delta_4(t)(a) = \int_\mathcal{X} \int_\aleph a d\delta_3(x)(a) dF_{X|T}(x|t),$$

where $F_{X|T}$ is the conditional distribution function of X given T. Since $\delta_3(x)(\cdot)$ is a point mass at $\delta_0(x)$, we get

$$\delta_2(t) = \int_\mathcal{X} \delta_0(x) dF_{X|T}(x|t) = E(\delta_0(X)|T=t) = \delta_1(t). \quad \square$$

Example 3.23. Suppose that P_θ says that $X_1, \ldots, X_n$ are IID $N(\theta, 1)$. Let $X = (X_1, \ldots, X_n)$. Let $\aleph = [0, 1]$ and

$$L(\theta, a) = (a - \Phi(c - \theta))^2,$$

for some fixed $c \in \mathbb{R}$. A naïve decision rule is $\delta_0(x) = \sum_{i=1}^n I_{(-\infty, c]}(x_i)/n$. But $T = \overline{X}$ is sufficient and δ_0 is not a function of T. Since $\aleph$ is convex and the loss function is a convex function of a, we should calculate

$$E(\delta_0(X)|T=t) = \frac{1}{n} \sum_{i=1}^n E(I_{(-\infty, c]}(X_i)|T=t) = \Pr(X_1 \leq c|T=t) = \Phi\left(\frac{c-t}{\sqrt{\frac{n-1}{n}}}\right),$$

since the distribution of X_1 given $T = t$ is $N(t, [n-1]/n)$.

3.2.2 Admissibility

The Rao–Blackwell theorem 3.22 tells us that under some conditions, the risk function of one decision rule is no larger than that of another. Similarly, Theorem 3.20 tells us that under some conditions, the loss incurred from one decision rule is no larger than that from another. These theorems have a common theme. That is, sometimes we can tell that one decision rule is better than another no matter what Θ equals.

Definition 3.24. A decision rule δ is *inadmissible* if there is another decision rule δ_1 such that $R(\theta, \delta_1) \le R(\theta, \delta)$ for all θ with strict inequality for some θ. If there is such a δ_1, we say that δ_1 *dominates* δ. If there is no such δ_1, then we say δ is *admissible*.

Example 3.25. Suppose that P_θ says that $X_1, \ldots, X_n$ are IID $N(\mu, \sigma^2)$, where $\theta = (\mu, \sigma)$. Let $X = (X_1, \ldots, X_n)$, $\aleph = [0, \infty)$, and $L(\theta, a) = (a - \sigma^2)^2$. Define

$$\delta(x) = \frac{1}{n-1} \sum_{i=1}^n (x_i - \bar{x})^2, \quad \delta_1(x) = \frac{1}{n+1} \sum_{i=1}^n (x_i - \bar{x})^2.$$

Then it can be shown that δ_1 dominates δ (see Problem 11 on page 210).

The criterion of admissibility may seem too severe if some values of θ are deemed to be virtually impossible.

Definition 3.26. Let λ be a measure on (Ω, τ) and let δ be a decision rule. For every decision rule δ_1, let $A_{\delta_1} = \{\theta : R(\theta, \delta_1) < R(\theta, \delta)\}$. Suppose that for every decision rule δ_1, if $R(\theta, \delta_1) \le R(\theta, \delta)$ a.e. $[\lambda]$ then $\lambda(A_{\delta_1}) = 0$. Then δ is λ-*admissible*.

A Bayes rule with respect to a probability measure λ is λ-admissible.

Theorem 3.27.[5] *Suppose that λ is a probability and δ is a Bayes rule with respect to λ. Then δ is λ-admissible.*

PROOF. Let δ_1 be a decision rule. If $R(\theta, \delta_1) \le R(\theta, \delta)$ a.s. $[\lambda]$ with strict inequality for all $\theta \in A$ with $\lambda(A) > 0$, then

$$\int_\Omega R(\theta, \delta_1) d\lambda(\theta) < \int_\Omega R(\theta, \delta) d\lambda(\theta),$$

which contradicts δ being a Bayes rule with respect to λ. □

A λ-admissible rule will be admissible if λ is a probability that is spread out appropriately. Theorems 3.28, 3.29, 3.31, and 3.32 say this in different ways.

Theorem 3.28. *If Ω is discrete, λ is a probability that gives positive probability to each element of Ω, and δ is Bayes with respect to λ, then δ is admissible.*

PROOF. Suppose that δ_1 dominates δ. Then $R(\theta, \delta_1) \le R(\theta, \delta)$ for all θ, and for some θ_0 we have $R(\theta_0, \delta_1) < R(\theta_0, \delta)$. It follows that

$$r(\lambda, \delta_1) = \sum_{\text{All } \theta} \lambda(\{\theta\}) R(\theta, \delta_1) < \sum_{\text{All } \theta} \lambda(\{\theta\}) R(\theta, \delta) = r(\lambda, \delta),$$

since $\lambda(\{\theta_0\}) > 0$. This contradicts that δ is Bayes. □

---

[5] This theorem is used in the proof of Theorem 3.31.

3.2. Classical Decision Theory

Theorem 3.29. *If every Bayes rule with respect to a prior λ has the same risk function, then they are all admissible.*

PROOF. Let δ be a Bayes rule with respect to the prior λ, and let $g(\theta)$ be the risk function of every such Bayes rule. Suppose that δ_0 dominates δ. Then $R(\theta, \delta_0) \leq g(\theta)$ for all θ with strict inequality for some θ. But then $\int_\Omega R(\theta, \delta_0) d\lambda(\theta) \leq \int_\Omega g(\theta) d\lambda(\theta)$. Since δ is a Bayes rule, the inequality must be an equality. This means that δ_0 is also a Bayes rule, hence it has risk function $g(\theta)$, which is a contradiction. □

Here is an example in which the condition of Theorem 3.29 does not hold.

Example 3.30. Let $\Omega = (0, \infty)$ and $\aleph = [0, \infty)$. Let $L(\theta, a) = (\theta - a)^2$. Let $X \sim U(0, \theta)$ given $\Theta = \theta$, and let λ be the $U(0, c)$ distribution for $c > 0$. Then $\Theta | X = x$ has density $\log(c/x) I_{(x,c)}(\theta)$. The formal Bayes rules are of the form

$$\delta(x) = \begin{cases} (c-x) \log\left(\frac{c}{x}\right) & \text{if } x < c, \\ \text{arbitrary} & \text{if } x \geq c. \end{cases}$$

Clearly, $R(\theta, \delta)$ for $\theta > c$ will depend on the arbitrary part of the definition of δ. For example, $\delta_0(x) = c$ for $x \geq c$ will have a larger risk function than $\delta_1(x) = x$ for $x \geq c$, even if $\delta_1(x) = \delta_0(x)$ for $x < c$.

The following theorem may apply when the parameter space is an open subset of $\mathbb{R}^k$ or when it is the closure of an open set.

Theorem 3.31. *Let Ω be a subset of $\mathbb{R}^k$ such that every neighborhood of every point in Ω intersects the interior of Ω. Let λ be a measure on (Ω, τ) such that Lebesgue measure on Ω is absolutely continuous with respect to λ. Suppose that δ_0 is λ-admissible and that it has finite risk function. Suppose that $R(\theta, \delta)$ is continuous in θ for all δ with finite risk function. Then δ_0 is admissible.*

PROOF. If δ_0 were inadmissible, then there would be δ_1 such that $R(\theta, \delta_1) \leq R(\theta, \delta_0)$ for all θ with strict inequality for some θ_0. By continuity of risk functions, $R(\theta, \delta_1) < R(\theta, \delta_0)$ for all θ in some neighborhood N of θ_0, which intersects the interior of Ω. Since Lebesgue measure is absolutely continuous with respect to λ, $\lambda(N) > 0$. Hence δ_0 is not λ-admissible. This contradiction proves the result. □

Consider an exponential family with natural parameter space Ω containing an open set. Let the loss be $L(\theta, a) = (a - g(\theta))^2$ for g a continuous function. The risk function of each decision rule with finite variance for all θ will be continuous in θ according to Theorem 2.64. If δ is a Bayes rule with respect to a prior λ that has a strictly positive density with respect to Lebesgue measure, then δ is λ-admissible by Theorem 3.27. Since the natural parameter space is convex, it satisfies the conditions of Theorem 3.31 and δ is admissible.

The following theorem says that with a strictly convex loss function, every Bayes rule is admissible.

156   Chapter 3. Decision Theory

Theorem 3.32. *Suppose that $\aleph$ is a convex subset of $\mathbb{R}^m$ and that all P_θ are absolutely continuous with respect to each other. If $L(\theta, \cdot)$ is strictly convex for all θ and δ_0 is λ-admissible for some λ, then δ_0 is admissible.*

PROOF. If δ_0 were inadmissible, then there would be δ_1 such that $R(\theta, \delta_1) \leq R(\theta, \delta_0)$ for all θ with strict inequality for some θ_0. Define $\delta_2(x) = [\delta_0(x) + \delta_1(x)]/2$. Then, for every θ,

$$\begin{aligned}
R(\theta, \delta_2) &= \int_{\mathcal{X}} L\left(\theta, \frac{\delta_0(x) + \delta_1(x)}{2}\right) dP_\theta(x) \\
&\leq \frac{1}{2} \int_{\mathcal{X}} [L(\theta, \delta_0(x)) + L(\theta, \delta_1(x))] dP_\theta(x) \\
&= \frac{1}{2} R(\theta, \delta_0) + \frac{1}{2} R(\theta, \delta_1) \leq R(\theta, \delta_0).
\end{aligned}$$

The first inequality above will be strict unless $P'_\theta(\delta_1(X) = \delta_0(X)) = 1$. Since all P_θ are absolutely continuous with respect to each other, it follows that $P'_\theta(\delta_1(X) = \delta_0(X)) = 1$ for one θ if and only if $P'_\theta(\delta_1(X) = \delta_0(X)) = 1$ for all θ. Hence the first inequality will be strict unless the distribution of $\delta_1(X)$ is the same as the distribution of $\delta_0(X)$ given $\Theta = \theta$ for all θ. In this case, δ_1 could not dominate δ_0, hence the inequality must be strict for all θ. This would imply that δ_0 is not λ-admissible, no matter what λ is. □

Example 3.33. Suppose that P_θ says that $X_1, \ldots, X_n$ are IID $Ber(\theta)$, and let $X = (X_1, \ldots, X_n)$. Suppose that $\aleph = [0, 1]$ and that the loss is $L(\theta, a) = (\theta - a)^2/[\theta(1-\theta)]$. Define $Y = \sum_{i=1}^n X_i$, and let the prior be Lebesgue measure on $[0, 1]$. The posterior given $X = x$ would be $Beta(y+1, n-y+1)$, where $y = \sum_{i=1}^n x_i$. Then

$$E(L(\Theta, a)|X = x) = \frac{\Gamma(n+2)}{\Gamma(y+1)\Gamma(n-y+1)} \int_0^1 (\theta - a)^2 \theta^{y-1}(1-\theta)^{n-y-1} d\theta$$

is minimized at $a = y/n$ for all x and all $n > 0$. So, $\delta_0(x) = y/n$ is a Bayes rule with respect to λ and it is admissible by Theorem 3.32.

Theorem 3.32 even applies to λ, which put 0 mass on large portions of the parameter space. See Problem 17 on page 210 for examples.

The concept of λ-admissibility did not require that λ be a probability measure. It is common to try to find "Bayes" rules with respect to non-probability measures.

Definition 3.34. Let $dP_\theta/d\nu(x) = f_{X|\Theta}(x|\theta)$. Suppose that λ is a measure on (Ω, τ) and that for every x there exists $\delta(x)$ such that

$$\int_\Omega L(\theta, \delta(x)) f_{X|\Theta}(x|\theta) d\lambda(\theta) = \min_{a \in \aleph} \int_\Omega L(\theta, a) f_{X|\Theta}(x|\theta) d\lambda(\theta). \quad (3.35)$$

The rule δ is called a generalized Bayes rule with respect to λ.

If
$$0 < c = \int_\Omega f_{X|\Theta}(x|\theta)d\lambda(\theta) < \infty, \tag{3.36}$$
then, after observing x, one can pretend that the "prior" distribution of θ has density $f_{X|\Theta}(x|\theta)/c$ with respect to λ and that there are no data. If a formal Bayes rule exists in this problem, it is a generalized Bayes rule. For this reason, generalized Bayes rules with respect to λ are also called formal Bayes rule with respect to λ.

Example 3.37. Suppose that P_θ says that $X_1, \ldots, X_n$ are IID $U(0,\theta)$. Let $X = (X_1, \ldots, X_n)$, $\Omega = (0, \infty)$, and $\aleph = [0, \infty)$. Let λ be Lebesgue measure on $(0, \infty)$ and $L(\theta, a) = (\theta - a)^2$. Then
$$f_{X|\Theta}(x|\theta) = \theta^{-n} I_{(0,\theta)}(\max x_i) = \theta^{-n} I_{(\max x_i, \infty)}(\theta).$$
We get c in (3.36) equal to $[(n-1)(\max x_i)^{n-1}]^{-1}$. So, we could invent the "prior" density
$$\frac{(n-1)(\max x_i)^{n-1}}{\theta^n} I_{(\max x_i, \infty)}(\theta).$$
The Bayes rule with respect to this prior is the mean, which is $\delta(x) = (n-1)\max x_i/(n-2)$. This is a generalized Bayes rule.

It sometimes happens that the integral with respect to λ of the risk function of a generalized Bayes rule with respect to λ is finite. In such cases, there is an analog to Theorem 3.31.

Theorem 3.38. *If Ω is a subset of $\mathbb{R}^k$ such that every neighborhood of every point in Ω intersects the interior of Ω, $R(\theta, \delta)$ is continuous in θ for all δ, Lebesgue measure on Ω is absolutely continuous with respect to λ, δ_0 is a generalized Bayes rule with respect to λ, and $L(\theta, \delta_0(x))f_{X|\Theta}(x|\theta)$ is $\nu \times \lambda$ integrable, then δ_0 is λ-admissible and admissible.*

PROOF. All we need to show is that δ_0 is λ-admissible and then apply Theorem 3.31. For each decision rule δ, $R(\theta, \delta) = \int_{\mathcal{X}} L(\theta, \delta(x))f_{X|\Theta}(x|\theta)d\nu(x)$. If $L(\theta, \delta(x))f_{X|\Theta}(x|\theta)$ is $\nu \times \lambda$ integrable, then
$$\int_\Omega R(\theta, \delta)d\lambda(\theta) = \int_\Omega \int_{\mathcal{X}} L(\theta, \delta(x))f_{X|\Theta}(x|\theta)d\nu(x)d\lambda(\theta)$$
$$= \int_{\mathcal{X}} \int_\Omega L(\theta, \delta(x))f_{X|\Theta}(x|\theta)d\lambda(\theta)d\nu(x),$$
where the last equality follows from Fubini's theorem A.70. If δ_1 is any other rule, then
$$\int_\Omega L(\theta, \delta_0(x))f_{X|\Theta}(x|\theta)d\lambda(\theta) \leq \int_\Omega L(\theta, \delta_1(x))f_{X|\Theta}(x|\theta)d\lambda(\theta),$$
for all x, since δ_0 is a generalized Bayes rule with respect to λ. Hence,
$$\int_\Omega R(\theta, \delta_0)d\lambda(\theta) \leq \int_\Omega R(\theta, \delta_1)d\lambda(\theta).$$

158   Chapter 3. Decision Theory

So, it cannot be the case that $R(\theta, \delta_1) \leq R(\theta, \delta_0)$ for all θ with strict inequality for $\theta \in A$ with $\lambda(A) > 0$.   □

Example 3.39. Suppose that $X_1, \ldots, X_n$ are IID $Ber(\theta)$ given $\Theta = \theta$, and let $X = (X_1, \ldots, X_n)$. Let $\aleph = [0, 1]$ and let the loss be $L(\theta, a) = (\theta - a)^2$. Define $Y = \sum_{i=1}^{n} X_i$, and let λ have Radon–Nikodym derivative $1/[\theta(1-\theta)]$ with respect to Lebesgue measure on $(0, 1)$. The posterior given $X = x$ would be $Beta(y, n-y)$, where $y = \sum_{i=1}^{n} x_i$ unless $y = 0$ or $y = n$. For $1 \leq y \leq n-1$, the generalized Bayes rule is $\delta(x) = y/n$. For $y = 0$ or $y = n$, the only values of $\delta(x)$ which make (3.35) finite are $\delta(x) = y/n$. So δ is a generalized Bayes rule with respect to λ. Now $L(\theta, \delta(x)) f_{X|\Theta}(x|\theta) = (\theta - y/n)^2 \theta^y (1-\theta)^{n-y}$, which has integral $1/n$ with respect to counting measure times λ. Hence, δ is λ-admissible and admissible.

Sometimes the integral of the risk function with respect to an infinite measure is not finite. For this reason, Blyth (1951) proved the following theorem, which makes use of a sequence of generalized Bayes rules to conclude that a rule is admissible.

Theorem 3.40.[6] *Let δ be a decision rule. Let $\{\lambda_n\}_{n=1}^{\infty}$ be a sequence of measures on (Ω, τ) such that a generalized Bayes rule δ_n with respect to λ_n exists for every n with*

$$r(\lambda_n, \delta_n) = \int R(\theta, \delta_n) d\lambda_n(\theta),$$

$$\lim_{n \to \infty} r(\lambda_n, \delta) - r(\lambda_n, \delta_n) = 0. \qquad (3.41)$$

Suppose that either of the following conditions holds:

- *All P_θ are absolutely continuous with respect to each other; $\aleph$ is a convex set; $L(\theta, a)$ is strictly convex in a for all θ; and there exist c, a set C, and a measure λ such that $\lambda_n \ll \lambda$ and $d\lambda_n/d\lambda(\theta) \geq c$ for $\theta \in C$ with $\lambda(C) > 0$.*

- *Every neighborhood of every point in Ω intersects the interior of Ω, for every open subset $C \subseteq \Omega$ there exists a number c such that $\lambda_n(C) \geq c$ for all n, and the risk function of every decision rule is continuous in θ.*

Then δ is admissible.

PROOF. Suppose that δ is inadmissible. Then there is δ' such that $R(\theta, \delta') \leq R(\theta, \delta)$ for all θ and $R(\theta_0, \delta') < R(\theta_0, \delta)$.

If the first condition holds, set $\delta'' = (\delta + \delta')/2$ and get that $L(\theta, \delta''(x)) < [L(\theta, \delta(x)) + L(\theta, \delta'(x))]/2$ for all θ and all x for which $\delta(x) \neq \delta'(x)$. Since $P'_{\theta_0}(\delta(X) = \delta'(X)) < 1$ and all P_θ are absolutely continuous with respect to

---

[6]This theorem is used in Example 3.43 and in the proof of Theorem 3.44.

each other, we have $P'_\theta(\delta(X) = \delta'(X)) < 1$ for all θ and $R(\theta, \delta'') < R(\theta, \delta)$ for all θ. So, for each n,

$$r(\lambda_n, \delta) - r(\lambda_n, \delta_n) \geq r(\lambda_n, \delta) - r(\lambda_n, \delta'') \geq \int_C [R(\theta, \delta) - R(\theta, \delta'')] d\lambda_n(\theta)$$
$$\geq c \int_C [R(\theta, \delta) - R(\theta, \delta'')] d\lambda(\theta) > 0.$$

This contradicts (3.41).

If the second condition holds, there exists $\epsilon > 0$ and open $C \subseteq \Omega$ such that $R(\theta, \delta') < R(\theta, \delta) - \epsilon$ for all $\theta \in C$. Now note that for each n,

$$r(\lambda_n, \delta) - r(\lambda_n, \delta_n) \geq r(\lambda_n, \delta) - r(\lambda_n, \delta')$$
$$\geq \int_C [R(\theta, \delta) - R(\theta, \delta')] d\lambda_n(\theta) \geq \epsilon \lambda_n(C) \geq \epsilon c,$$

where c is guaranteed in the second condition. This contradicts (3.41). □

We can use Theorem 3.40 together with the following lemma to prove that some common estimators are admissible.

Lemma 3.42.[7] *Suppose that $\Theta = (\Theta_1, \Theta_2)$. Also, suppose that, for each possible value $\theta_{2,0}$ of Θ_2, δ is admissible when the parameter space is $\Omega_0 = \{\theta = (\theta_1, \theta_{2,0}) \in \Omega\}$. Then δ is admissible.*

PROOF. Suppose that δ were inadmissible. Then there exists δ^* such that $R(\theta, \delta^*) \leq R(\theta, \delta)$ for all θ and $R(\theta_0, \delta^*) < R(\theta_0, \delta)$ for some $\theta_0 \in \Omega$. Let $\theta_0 = (\theta_{1,0}, \theta_{2,0})$, and let $\Omega_0 = \{\theta = (\theta_1, \theta_{2,0}) \in \Omega\}$. We now have a contradiction to δ's being admissible when the parameter space is Ω_0. □

Example 3.43. Suppose that P_θ says that X has $N(\mu, \sigma^2)$ distribution, where $\theta = (\mu, \sigma)$. Let the loss be $L(\theta, a) = (\mu - a)^2$. We now prove that $\delta(x) = x$ is admissible.

Denote $\Theta = (M, \Sigma)$. It is easy to calculate $R(\theta, \delta) = \sigma^2$. For each value σ_0 of Σ, we will show that δ is admissible for the parameter space $\Omega_0 = \{(\mu, \sigma_0) : \mu \in \mathbb{R}\}$. Let λ_n be the measure on Ω_0 having density $\sqrt{n}$ times the $N(0, \sigma_0^2 n)$. The generalized Bayes rule with respect to λ_n is $\delta_n(x) = nx/(n+1)$. The integral of the risk function of δ_n with respect to λ_n is $r_n = n^{3/2} \sigma_0^2/(n+1)$. The integral of the risk function of δ with respect to λ_n is $\sqrt{n} \sigma_0^2$. Note that

$$\sqrt{n}\sigma_0^2 - \frac{n^{\frac{3}{2}}\sigma_0^2}{n+1} = \frac{\sqrt{n}\sigma_0^2}{n+1},$$

which goes to 0 as $n \to \infty$. If C is an open subset of Ω_0, then $\lambda_n(C) \geq \lambda_1(C)$ for all n. Since L is strictly convex in a for all θ, the conditions of Theorem 3.40 apply in the parameter space Ω_0, so δ is admissible with this parameter space. By Lemma 3.42, it is admissible for the entire parameter space.

---

[7]This lemma is used in Example 3.43.

The following theorem is a simplified version of a theorem of Brown and Hwang (1982). It will allow us to extend Example 3.43 to two dimensions.

Theorem 3.44. *Suppose that X has a k-dimensional exponential family distribution given $\Theta = \theta$ with density $f_{X|\Theta}(x|\theta) = c(\theta)\exp(\theta^\top x)$ with respect to a measure ν. Let the natural parameter space Ω be a rectangular region $I_1 \times \cdots \times I_k$. Let $I_i = (a_{1,i}, a_{2,i})$ with $a_{j,i}$ possibly infinite. Suppose that $\lim_{\theta_i \to a_{j,i}} f_{X|\Theta}(x|\theta) = 0$ for all i, j, x. Suppose that there exist a set $S \subseteq \Omega$ with positive Lebesgue measure and a sequence of almost everywhere differentiable functions $\{h_n\}_{n=1}^\infty$ such that*

- $h_n : \Omega \to [0, 1]$,
- $h_n(\theta) = 1$ *if* $\theta \in S$,
- $\lim_{n\to\infty} \|\nabla\sqrt{h_n(\theta)}\| = 0$ *for all θ (where ∇ denotes the gradient),*
- $\int_\Omega \sup_n \|\nabla\sqrt{h_n(\theta)}\|^2 d\theta < \infty.$

Then $\delta(x) = x$ is admissible as an estimator of $g(\theta) = \mathrm{E}_\theta(X)$ with loss $L(\theta, a) = \sum_{i=1}^k (g_i(\theta) - a_i)^2$.

PROOF. Let λ_n be the measure on (Ω, τ) with Radon–Nikodym derivative $h_n(\theta)$ with respect to Lebesgue measure (λ). Then $d\lambda_n/d\lambda(\theta) = 1$ for all $\theta \in S$ with $\lambda(S) > 0$ and the loss is strictly convex in a, so the first of the two alternative conditions of Theorem 3.40 is met. We need to find the generalized Bayes rule δ_n with respect to λ_n and show that $\lim_{n\to\infty} r(\lambda_n, \delta) - r(\lambda_n, \delta_n) = 0$. Since $g(\theta) = -\nabla \log c(\theta)$ by Proposition 2.70, the generalized Bayes rule with respect to λ_n will be

$$\begin{aligned}
\delta_n(x) &= \frac{\int_\Omega (-\nabla \log c(\theta)) c(\theta) \exp(\theta^\top x) h_n(\theta) d\theta}{\int_\Omega c(\theta) \exp(\theta^\top x) h_n(\theta) d\theta} \\
&= \frac{\int_\Omega (-\nabla c(\theta)) \exp(\theta^\top x) h_n(\theta) d\theta}{\int_\Omega c(\theta) \exp(\theta^\top x) h_n(\theta) d\theta} \\
&= \frac{\int_\Omega c(\theta) \exp(\theta^\top x) [x h_n(\theta) + \nabla h_n(\theta)] d\theta}{\int_\Omega c(\theta) \exp(\theta^\top x) h_n(\theta) d\theta} \\
&= x + \frac{\int_\Omega (\nabla h_n(\theta)) f_{X|\Theta}(x|\theta) d\theta}{\int_\Omega f_{X|\Theta}(x|\theta) h_n(\theta) d\theta},
\end{aligned} \tag{3.45}$$

where the third equality follows by doing integration by parts with respect to θ_i for the ith coordinate of the integral (with $u = \exp(\theta^\top x) h_n(\theta)$ and $dv = -\partial c(\theta)/\partial \theta_i$) and using $\lim_{\theta_i \to a_{j,i}} f_{X|\Theta}(x|\theta) = 0$ to drop the integrated term $uv = f_{X|\Theta}(x|\theta)$). Now write

$$r(\lambda_n, \delta) - r(\lambda_n, \delta_n)$$
$$= \int_\Omega \int_\mathcal{X} \left(\|\delta(x) - g(\theta)\|^2 - \|\delta_n(x) - g(\theta)\|^2 \right) f_{X|\Theta}(x|\theta) h_n(\theta) d\nu(x) d\lambda(\theta)$$

$$= \int_{\mathcal{X}} \int_{\Omega} (x - \delta_n(x))^\top (x + \delta_n(x) - 2g(\theta)) f_{X|\Theta}(x|\theta) h_n(\theta) d\lambda(\theta) d\nu(x).$$

According to (3.45), we have that

$$\int_{\Omega} g(\theta) f_{X|\Theta}(x|\theta) h_n(\theta) d\lambda(\theta) = \delta_n(x) \int_{\Omega} f_{X|\Theta}(x|\theta) h_n(\theta) d\theta.$$

For convenience, define

$$H_n(x) = \int_{\Omega} f_{X|\Theta}(x|\theta) h_n(\theta) d\theta, \text{ and } J_n(x) = \int_{\Omega} (\nabla h_n(\theta)) f_{X|\Theta}(x|\theta) d\theta.$$

Then $x - \delta_n(x) = -J_n(x)/H_n(x)$ and

$$r(\lambda_n, \delta) - r(\lambda_n, \delta_n) = \int_{\mathcal{X}} \frac{\|J_n(x)\|^2}{H_n(x)} d\nu(x).$$

Use the fact that $\nabla h_n(\theta) = 2\sqrt{h_n(\theta)} \nabla \sqrt{h_n(\theta)}$ and the Cauchy–Schwarz inequality B.19 to conclude that

$$\|J_n(x)\|^2 \leq 4 H_n(x) \int_{\Omega} \|\nabla \sqrt{h_n(\theta)}\|^2 f_{X|\Theta}(x|\theta) d\theta.$$

This means that

$$\begin{aligned} r(\lambda_n, \delta) - r(\lambda_n, \delta_n) &\leq 4 \int_{\mathcal{X}} \int_{\Omega} \|\nabla \sqrt{h_n(\theta)}\|^2 f_{X|\Theta}(x|\theta) d\theta d\nu(x) \\ &= 4 \int_{\Omega} \|\nabla \sqrt{h_n(\theta)}\|^2 d\theta, \end{aligned}$$

which goes to 0 as $n \to \infty$ because of the last two conditions in the theorem and the dominated convergence theorem A.57. □

Example 3.46. Suppose that $X \sim N_2(\mu, \sigma)$ given $\Theta = (\mu, \sigma)$ where σ is a 2×2 positive definite diagonal matrix and μ is two-dimensional.[8] Let $\aleph = \mathbb{R}^2$, let $L(\theta, a) = (a_1 - \mu_1)^2 + (a_2 - \mu_2)^2$, and let $\delta(x) = x$. For each value

$$\sigma_0 = \begin{pmatrix} \sigma_{0,1}^2 & 0 \\ 0 & \sigma_{0,2}^2 \end{pmatrix},$$

let $\Omega_0 = \{\theta = (\mu, \sigma_0) : \mu \in \mathbb{R}^2\}$ be a subparameter space. The natural parameter of this exponential family is $\psi = (\mu_1/\sigma_{0,1}^2, \mu_2/\sigma_{0,2}^2)$, where $\sigma_{0,1}^2$ and $\sigma_{0,2}^2$ are the diagonal elements of σ_0. Consider the following sequence of functions:

$$h_n(\psi) = \begin{cases} 1 & \text{if } \|\psi\|^2 \leq 1, \\ \left(1 - \frac{\log \|\psi\|}{\log(n)}\right)^2 & \text{if } 1 \leq \|\psi\| \leq n, \\ 0 & \text{if } \|\psi\| \geq n. \end{cases}$$

---

[8] This example was compiled from material in Brown and Hwang (1982) and Section 8.9 of Berger (1985).

Here $S = \{\theta : \|\psi\| \leq 1\}$ in Theorem 3.44. It is easy to see that

$$\|\sqrt{h_n(\psi)}\|^2 = (\|\psi\| \log(n))^{-2} I_{[1,n]}(\|\psi\|)$$
$$\leq (\|\psi\| \log(\max\{\|\psi\|, 2\}))^{-2} I_{[1,\infty)}(\|\psi\|).$$

It follows that $\lim_{n\to\infty} \|\nabla \sqrt{h_n(\psi)}\| = 0$ for all ψ. To verify the last condition of Theorem 3.44, we need only show that

$$\int_{\{\psi:\|\psi\|\geq 2\}} (\|\psi\| \log \|\psi\|)^{-2} d\psi < \infty.$$

By transforming to polar coordinates, we see that this integral is a finite constant plus a constant times

$$\int_2^\infty \frac{r^{-1}}{\log^2(r)} dr = \int_{\ln(2)}^\infty \frac{dz}{z^2} < \infty.$$

The following result allows us to translate admissibility in one decision problem to admissibility in a different decision problem if the loss functions are related.

Proposition 3.47.[9]

- Let Ω be an open subset of $\mathbb{R}^k$. If $c(\theta) > 0$ for all $\theta \in \Omega$, then δ is admissible with loss $L(\theta, a)$ if and only if δ is admissible with loss $c(\theta)L(\theta, a)$.

- Let $\Omega \subseteq \mathbb{R}$ be an interval. Suppose that $c(\theta) \geq 0$ for all θ and is strictly positive except for $\theta \in A$, where A consists solely of points in Ω which are isolated from each other. Suppose also that, with loss function $L(\theta, a)$, the risk function of every decision rule is continuous from the left at every $\theta \in A$. (Alternatively, suppose that all risk functions are continuous from the right at every $\theta \in A$.) Then δ is admissible with loss $L(\theta, a)$ if and only if δ is admissible with loss $c(\theta)L(\theta, a)$.

- Let $d(\theta)$ be a real-valued function of θ. Then δ is admissible with loss $L(\theta, a)$ if and only if it is admissible with loss $L(\theta, a) + d(\theta)$.

The proof of this proposition is simple and is left for the reader.

Example 3.48 (Continuation of Example 3.33; see page 156). Let $c(\theta) = \theta(1-\theta)$ and $\Omega = (0, 1)$. By Proposition 3.47, $\delta_0(x) = y/n$ is admissible with loss $L(\theta, a) = (\theta - a)^2$. Since c and all risk functions are continuous, even at the endpoints, it is easy to show that δ_0 is also admissible if $\Omega = [0, 1]$.

Proposition 3.49. Suppose that δ_0 is λ-admissible with loss $L(\theta, a)$. Then δ_0 is λ-admissible with loss $c(\theta)L(\theta, a)$ if $c(\theta) > 0$ a.e. $[\lambda]$.

---

[9]This proposition is used in Examples 3.48 and 3.59. It is also used to simplify the class of loss functions in hypothesis testing in Chapter 4.

3.2.3 James–Stein Estimators

We have seen some examples of simple decision rules that are admissible, but there is a notorious example of a simple decision rule that is inadmissible. This example has spawned a great amount of study. It begins with Stein (1956) and James and Stein (1960), who showed the following.

Theorem 3.50. *Suppose that the conditional distribution of $X_1, \ldots, X_n$ given $\Theta = (\mu_1, \ldots, \mu_n)$ is that they are independent with $X_i \sim N(\mu_i, 1)$. Let $X = (X_1, \ldots, X_n)$, $\aleph = \Omega = \mathbb{R}^n$, and let the loss be $L(\theta, a) = \sum_{i=1}^{n} (\mu_i - a_i)^2$. Then, if $n > 2$, $\delta(x) = x$ is inadmissible. In fact, a rule that dominates δ is*

$$\delta_1(x) = \delta(x) \left[1 - \frac{n-2}{\sum_{i=1}^{n} x_i^2}\right].$$

The proof we give here requires a few lemmas. The first is due to Stein (1981).

Lemma 3.51. *Let $g : \mathbb{R} \to \mathbb{R}$ be a differentiable function with derivative g'. Suppose that X has $N(\mu, 1)$ distribution and $\mathrm{E}(|g'(X)|) < \infty$. Then $\mathrm{E}(g'(X)) = \mathrm{Cov}(X, g(X))$.*

PROOF. Let $\phi(t) = \exp(-t^2/2)/\sqrt{2\pi}$ be the standard normal density function. Use integration by parts to show that

$$\phi(x - \mu) = \int_x^\infty (z - \mu)\phi(z - \mu)dz = -\int_{-\infty}^x (z - \mu)\phi(z - \mu)dz.$$

We will use these facts in what follows.

$$\begin{aligned}
\mathrm{E}(g'(X)) &= \int_{-\infty}^\infty g'(x)\phi(x - \mu)dx \\
&= \int_0^\infty g'(x)\phi(x - \mu)dx + \int_{-\infty}^0 g'(x)\phi(x - \mu)dx \\
&= \int_0^\infty g'(x) \int_x^\infty (z - \mu)\phi(z - \mu)dzdx - \int_{-\infty}^0 g'(x) \int_{-\infty}^x (z - \mu)\phi(z - \mu)dzdx \\
&= \int_0^\infty (z - \mu)\phi(z - \mu) \int_0^z g'(x)dxdz - \int_{-\infty}^0 (z - \mu)\phi(z - \mu) \int_z^0 g'(x)dxdz \\
&= \int_0^\infty (z - \mu)\phi(z - \mu)[g(z) - g(0)]dz - \int_{-\infty}^0 (z - \mu)\phi(z - \mu)[g(0) - g(z)]dz \\
&= \int_{-\infty}^\infty [g(z) - g(0)](z - \mu)\phi(z - \mu)dz \\
&= \int_{-\infty}^\infty g(z)(z - \mu)\phi(z - \mu)dz = \mathrm{Cov}(X, g(X)). \quad \square
\end{aligned}$$

Lemma 3.52. *Let $g : \mathbb{R}^n \to \mathbb{R}^n$ be a vector of differentiable functions, $(g_1, \ldots, g_n)$. Let X have $N_n(\theta, I)$ distribution. For each i, define $h_i(y)$ to be the expected value of $g_i(X_1, \ldots, X_{i-1}, y, X_{i+1}, \ldots, X_n)$ and*

$$h_i'(y) = \mathrm{E}\frac{d}{dy}g_i(X_1, \ldots, X_{i-1}, y, X_{i+1}, \ldots, X_n). \tag{3.53}$$

Suppose that, for all i, $\mathrm{E}(|h_i'(X_i)|) < \infty$. Then

$$\mathrm{E}\|X + g(X) - \theta\|^2 = n + \mathrm{E}\left(\|g(X)\|^2 + 2\sum_{i=1}^{n}\frac{\partial}{\partial x_i}g_i(x)\bigg|_{x=X}\right).$$

PROOF. Write

$$\begin{aligned}\mathrm{E}\|X + g(X) - \theta\|^2 &= \mathrm{E}\|X - \theta\|^2 + \mathrm{E}\|g(X)\|^2 + 2\mathrm{E}[(X-\theta)^\top g(X)]\\ &= n + \mathrm{E}\|g(X)\|^2 + 2\mathrm{E}\sum_{i=1}^{n}[(X_i - \theta_i)g_i(X)].\end{aligned}$$

All we need to prove is that for each i,

$$\mathrm{E}[(X_i - \theta_i)g_i(X)] = \mathrm{E}\left(\frac{\partial}{\partial x_i}g_i(x)\bigg|_{x=X}\right).$$

By integrating out the X_j for $j \neq i$, the left-hand side can be written as

$$\mathrm{E}[(X_i - \theta_i)g_i(X)] = \mathrm{E}[(X_i - \theta_i)h_i(X_i)]$$
$$= \mathrm{Cov}(X_i, h_i(X_i)) = \mathrm{E}(h_i'(X_i)) = \mathrm{E}\left(\frac{\partial}{\partial x_i}g_i(x)\bigg|_{x=X}\right),$$

where the first equality follows from the definition of h_i, the third follows from Lemma 3.51, and the fourth follows from (3.53) and then integrating out X_i. $\square$

PROOF OF THEOREM 3.50. Now, let $g(x) = -x(n-2)/\sum_{j=1}^{n} x_j^2$, and use the notation of Lemma 3.52. This makes $g_i(x) = -(n-2)x_i/\sum_{j=1}^{n} x_j^2$. For each $x \neq 0$, the second partial derivative of g_i with respect to the ith coordinate of x is uniformly bounded in a neighborhood of x_i for each set of values of x_j for $j \neq i$. A simple application of Taylor's theorem C.1 with remainder shows that h_i can be differentiated under the expectation. We can write

$$\mathrm{E}_\theta(|h_i'(X_i)|) \leq (n-2)\int\cdots\int \frac{\left|\sum_{j=1}^{n} x_j^2 - 2x_i^2\right|}{\left[\sum_{j=1}^{n} x_j^2\right]^2} f_{X|\Theta}(x|\theta)dx_1\cdots dx_n.$$

This can be bounded by three times the expected value of one over a χ_n^2 random variable. The expected value of one over a χ_n^2 random variable is

$n/(n-2)$ if $n > 2$, hence $E_\theta(|h_i'(x)|) < \infty$. Lemma 3.52 says that the risk function for δ_1 is

$$n + E_\theta\left[\|g(X)\|^2 + 2\sum_{i=1}^n \frac{\partial}{\partial x_i}g_i(x)\Big|_{x=X}\right].$$

We can write

$$\|g(x)\|^2 = (n-2)^2 \frac{\sum_{i=1}^n x_i^2}{\left[\sum_{j=1}^n x_j^2\right]^2} = \frac{(n-2)^2}{\sum_{i=1}^n x_i^2},$$

$$\sum_{i=1}^n \frac{\partial}{\partial x_i}g_i(x) = -(n-2)^2 \frac{\sum_{i=1}^n x_i^2}{\left[\sum_{j=1}^n x_j^2\right]^2} = \frac{-(n-2)^2}{\sum_{i=1}^n x_i^2},$$

$$\|g(x)\|^2 + 2\sum_{i=1}^n \frac{\partial}{\partial x_i}g_i(x) = \frac{-(n-2)^2}{\sum_{j=1}^n x_j^2} < 0, \tag{3.54}$$

for all x. It follows that the risk function is less than n for all θ. □

From (3.54), in order to calculate the risk for δ_1, we need the mean of $1/\sum_{j=1}^n X_j^2$. Note that $Z = \sum_{j=1}^n X_j^2$ has noncentral χ^2 distribution, $NC\chi_n^2(\lambda)$ with $\lambda = \sum_{j=1}^n \mu_j^2$. From the form of the $NC\chi^2$ density, it is clear that Z has the same distribution as Y, where $Y \sim \chi_{n+2k}^2$ given $K = k$ and $K \sim Poi(\lambda)$. The mean of $1/Z$ is

$$E\left(\frac{1}{Z}\right) = E\left(\frac{1}{Y}\right) = EE\left(\frac{1}{Y}\Big|K\right) = E\left(\frac{1}{n+2K-2}\right).$$

Notice that when $\lambda = 0$, $K = 0$, a.s. and $R(\theta, \delta_1) = 2$. This is where the risk function is smallest. A plot of $R(\theta, \delta_1)$ as a function of λ for $n = 6$ is given in Figure 3.56. There is no reason why the smallest value of the risk function must occur when $\theta = 0$. We could subtract a vector θ_0 from X and then add θ_0 back on to δ_1 to get an estimator that has the minimum of its risk function at θ_0. This would give the decision rule

$$\delta_2(X) = \theta_0 + \delta_1(X - \theta_0). \tag{3.55}$$

It may be that we cannot decide which vector θ_0 to subtract. It is possible to choose based on the data. If $n \geq 4$, then we could use the decision rule

$$\delta_3(X) = (X - \overline{X}\mathbf{1})\left(1 - \frac{n-3}{\sum_{i=1}^n(X_i - \overline{X})^2}\right) + \overline{X}\mathbf{1},$$

where $\mathbf{1}$ denotes a vector whose coordinates are all 1. (See Problem 20 on page 211.)

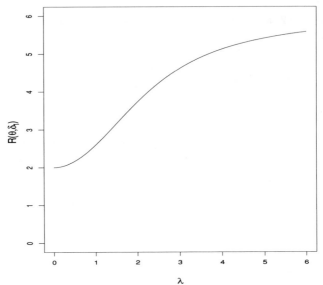

FIGURE 3.56. Risk Function of James–Stein Estimator for $n = 6$

There is a way to derive the James–Stein estimator from an empirical Bayes argument.[10] This was done by Efron and Morris (1975). Suppose that $\Theta \sim N_n(\theta_0, \tau^2 I)$. The Bayes estimate for Θ is

$$\theta_0 + (X - \theta_0)\frac{\tau^2}{\tau^2 + 1}.$$

The empirical Bayes approach tries to estimate τ from the marginal distribution of X. The marginal distribution of X is $N_n(\theta_0, (1 + \tau^2)I)$. So, we could estimate $1 + \tau^2$ by $\sum_{i=1}^{n}(X_i - \theta_{0i})^2/c$ for some c. An estimate of $\tau^2/(\tau^2 + 1)$ is

$$1 - \frac{c}{\sum_{i=1}^{n}(X_i - \theta_{0i})^2}.$$

The empirical Bayes estimator is then $\delta_2(X)$ if $c = n - 2$. If we take the empirical Bayes approach one step further and also try to estimate θ_0, we could use $\overline{X}\mathbf{1}$ as an estimate, and the estimate of $1 + \tau^2$ would be $\sum_{i=1}^{n}(X_i - \overline{X})^2/c$. With $c = n - 3$, we get $\delta_3(X)$.

Another way to arrive at estimators like these is through hierarchical models (to be discussed in more detail in Chapter 8). For example, $\Theta_1, \ldots, \Theta_n$ could be modeled as conditionally IID $N(\mu, \tau^2)$ given $M = \mu$ and $T = \tau$. Then M and T could have some distribution, rather than merely being estimated as in the empirical Bayes approach. Strawderman (1971) finds a class of Bayes rules that dominate δ_0 when $p \geq 5$ and are admissible by Theorems 3.27 and 3.32.

---

[10]See Section 8.4 for more detail on empirical Bayes analysis.

The estimator $\delta_1(X)$ is actually inadmissible as can be shown in Problem 22 on page 211. Brown (1971) considers the problem of finding necessary and sufficient conditions for an estimator to be admissible in this setting.

3.2.4 Minimax Rules

There are usually lots of admissible rules. Unless one is willing to choose one by choosing a Bayes rule with respect to some prior distribution, then one needs some other criterion by which to choose a rule. One such criterion is the *minimax principle*.

> **The Minimax Principle:** In comparing rules, the rule δ with the smallest value of $\sup_\theta R(\theta, \delta)$ is best.

The minimax principle says to prepare for the worst possible value θ of Θ. When playing a game against an opponent who is trying to make things bad for you, there may be good reason to prepare for the worst. When it makes sense to consider how likely are various alternative value of Θ, the worst value may turn out not to be of such a concern.

Definition 3.57. A rule δ_0 is called *minimax* if, for all δ, $\sup_{\theta \in \Omega} R(\theta, \delta_0) \leq \sup_{\theta \in \Omega} R(\theta, \delta)$, alternatively, $\sup_{\theta \in \Omega} R(\theta, \delta_0) = \inf_\delta \sup_{\theta \in \Omega} R(\theta, \delta)$.

Proposition 3.58. *If δ has constant risk and it is admissible, then it is minimax.*

Example 3.59. We saw earlier that when $X \sim N(\mu, \sigma^2)$ given $\Theta = (\mu, \sigma)$, $\delta(x) = \bar{x}$ is admissible when the loss function was $L(\theta, a) = (\mu - a)^2$. By Proposition 3.47, it is also admissible with loss $L'(\theta, a) = (\mu - a)^2/\sigma^2$. The risk function for this new loss is constant $R(\theta, \delta) = 1/n$. Hence δ is minimax with loss L'. Every other decision rule will have to have a risk function that approaches or surpasses $1/n$ for some θ values.

Theorem 3.60. *Let $\{\lambda_n\}_{n=1}^\infty$ be a sequence of probability measures on the parameter space (Ω, τ) with δ_n being the Bayes rule with respect to λ_n. Suppose that $\lim_{n \to \infty} r(\lambda_n, \delta_n) = c < \infty$. If there is δ_0 such that $R(\theta, \delta_0) \leq c$ for all θ, then δ_0 is minimax.*

PROOF. Assume δ_0 is not minimax. Then there is δ' and $\epsilon > 0$ such that

$$R(\theta, \delta') \leq \sup_{\phi \in \Omega} R(\phi, \delta_0) - \epsilon \leq c - \epsilon,$$

for all θ. Choose n_0 so that $r(\lambda_n, \delta_n) \geq c - \epsilon/2$, for all $n \geq n_0$. Then, for $n \geq n_0$,

$$\begin{aligned} r(\lambda_n, \delta') &= \int R(\theta, \delta') d\lambda_n(\theta) \leq (c - \epsilon) \int d\lambda_n(\theta) \\ &= c - \epsilon < c - \frac{\epsilon}{2} \leq r(\lambda_n, \delta_n), \end{aligned}$$

which contradicts that δ_n is the Bayes rule with respect to λ_n. □

Example 3.61. Suppose that P_θ says that $X_1, \ldots, X_m$ are independent with $X_i \sim N(\theta_i, 1)$, where $\theta = (\theta_1, \ldots, \theta_m)$. Let $\delta_0(X) = X$, $\aleph = \mathbb{R}^m$, and $L(\theta, a) = \sum_{i=1}^m (\theta_i - a_i)^2$. Let λ_n be the probability measure that says Θ has distribution $N_m(0, nI)$. The Bayes rules are $\delta_n(X) = nX/(n+1)$, and the Bayes risks are $r(\lambda_n, \delta_n) = mn/(n+1)$, which go to m as $n \to \infty$. Also, $R(\theta, \delta_0) = m$, so δ_0 is minimax. We see that minimax rules need not be admissible.

Example 3.62. Suppose that P_θ says that $X \sim Bin(n, \theta)$ and that $L(\theta, a) = (\theta - a)^2/[\theta(1-\theta)]$, where $\Omega = (0,1)$ and $\aleph = [0,1]$. A rule with constant risk is $\delta_0(X) = X/n$. The risk is $R(\theta, \delta_0) = 1/n$. We saw earlier that δ_0 is admissible, so it is minimax.

Now, suppose that we use the loss $L'(\theta, a) = (\theta - a)^2$. We will see that no analog to Proposition 3.47 operates here. If the prior for Θ is $Beta(\alpha, \beta)$, then the Bayes rule is $\delta(x) = (\alpha + x)/(\alpha + \beta + n)$. The risk function for this rule is

$$R(\theta, \delta) = \frac{1}{\alpha + \beta + n} \left\{ \theta^2 \left[(\alpha + \beta)^2 - n \right] + \theta \left[n - 2\alpha(\alpha + \beta) \right] + \alpha^2 \right\},$$

which is constant if $\alpha = \beta = \sqrt{n}/2$. The constant risk is $1/(4 + 8\sqrt{n} + 4n)$. So δ is minimax. The rule can be expressed as $\delta(x) = (x + \sqrt{n}/2)/(n + \sqrt{n})$. Notice that this is like changing your prior distribution as the sample size changes.

Since minimax rules are designed to prepare for the worst, we should see if there are prior distributions that make the worst θ just likely enough so that the corresponding Bayes rules are minimax.[11]

Definition 3.63. A prior distribution λ_0 for Θ is *least favorable* if

$$\inf_\delta r(\lambda_0, \delta) = \sup_\lambda \inf_\delta r(\lambda, \delta).$$

Such a λ_0 is sometimes called a *maximin strategy* for nature.

Let λ_0 and δ_0 be a fixed probability and decision rule, respectively. It is true that $\inf_\delta r(\lambda_0, \delta) \leq \sup_\lambda r(\lambda, \delta_0)$. So, it follows that

$$\underline{V} \equiv \sup_\lambda \inf_\delta r(\lambda, \delta) \leq \inf_\delta \sup_\lambda r(\lambda, \delta) = \inf_\delta \sup_\theta R(\theta, \delta) \equiv \overline{V}.$$

Definition 3.64. The number $\underline{V}$ above is called the *maximin value* of the decision problem, and the number $\overline{V}$ is called the *minimax value*.

Theorem 3.65. *If δ_0 is Bayes with respect to λ_0 and $R(\theta, \delta_0) \leq r(\lambda_0, \delta_0)$ for all θ, then δ_0 is minimax and λ_0 is least favorable.*

---

[11] There is a delicate balance between how likely the bad θ values are and how likely the rest of the parameter space is. In the second part of Example 3.62, the worst θ values, in some sense, are those near $1/2$ because the data are most variable given $\Theta = 1/2$. The prior $Beta(\sqrt{n}/2, \sqrt{n}/2)$ puts enough mass near $1/2$ to force us to take seriously the possibility that the data will be highly variable. However, it still spreads enough mass around the remainder of the parameter space so that we cannot ignore other θ values. If the prior put probability 1 on $\Theta = 1/2$, for example, then the Bayes rule would be $\delta(x) = 1/2$ for all x.

PROOF. Since $\overline{V} \leq \sup_\theta R(\theta, \delta_0) \leq r(\lambda_0, \delta_0) = \inf_\delta (\lambda_0, \delta) \leq \underline{V} \leq \overline{V}$, it follows that all of the inequalities are equalities. □

Example 3.66 (Continuation of Example 3.62; see page 168). We saw that the minimax rule with loss $(\theta - a)^2$ was Bayes with respect to the $Beta(\sqrt{n}/2, \sqrt{n}/2)$ prior, λ_0. Since the risk function is constant, $R(\theta, \delta) = r(\lambda_0, \delta)$ for all θ. It follows that λ_0 is least favorable. The reason it is least favorable is that it puts a lot of mass on the θ values (near $1/2$) that have high variance for X. On the other hand, it does not put so much mass there that the estimator is drawn too close to $1/2$.

Definition 3.67. A rule δ_0 is *extended Bayes* if for every $\epsilon > 0$ there exists a prior λ_ϵ such that $r(\lambda_\epsilon, \delta_0) \leq \epsilon + \inf_\delta r(\lambda_\epsilon, \delta)$.

Example 3.68. Suppose that P_θ says that $X \sim N(\theta, 1)$. Let $L(\theta, a) = (\theta - a)^2$, and let λ_ϵ be the $N(0, (1-\epsilon)/\epsilon)$ prior for Θ. Let $\delta_0(x) = x$. The Bayes rule with respect to λ_ϵ is

$$\delta_\epsilon(x) = \frac{x}{1 + \frac{\epsilon}{1-\epsilon}} = (1-\epsilon)x.$$

The Bayes risk is $r(\lambda_\epsilon, \delta_\epsilon) = 1 - \epsilon$, while the Bayes risk of δ_0 is $r(\lambda_\epsilon, \delta_0) = 1$, which is no greater than $\epsilon + 1 - \epsilon$. So δ_0 is extended Bayes.

Theorem 3.69. *A constant-risk extended Bayes rule is minimax.*

PROOF. Let δ_0 be a constant-risk extended Bayes rule. Let $R(\theta, \delta_0) = c$ for all θ. Suppose that δ_0 is not minimax, but rather that there is a rule δ_1 such that $\sup_\theta R(\theta, \delta_1) = c - \epsilon$, for some $\epsilon > 0$. Let $\lambda_{\epsilon/2}$ be as in Definition 3.67. That is,

$$r(\lambda_{\frac{\epsilon}{2}}, \delta_0) \leq \frac{\epsilon}{2} + \inf_\delta r\left(\lambda_{\frac{\epsilon}{2}}, \delta\right).$$

Since δ_0 has constant risk c, its Bayes risk is also c. So

$$c \leq \frac{\epsilon}{2} + \inf_\delta r\left(\lambda_{\frac{\epsilon}{2}}, \delta\right).$$

The Bayes risk of δ_1 can be no greater than $c - \epsilon$, so $\inf_\delta r(\lambda_{\epsilon/2}, \delta) \leq c - \epsilon$. It follows that $c \leq \epsilon/2 + c - \epsilon = c - \epsilon/2$, which is a contradiction. □

Example 3.70. Suppose that P_θ says that $X \sim Poi(\theta)$, where $\Omega = (0, \infty)$ and $\aleph = [0, \infty)$. Let the loss function be $L(\theta, a) = (\theta - a)^2/\theta$, and let $\delta_0 = x$. The risk function is $R(\theta, \delta_0) = 1$, constant. Let λ_ϵ be the prior that says Θ has $\Gamma(1, \epsilon/[1-\epsilon])$ distribution. The posterior distribution is

$$\Theta | X = x \sim \Gamma\left(x + 1, \frac{1}{1-\epsilon}\right).$$

The Bayes rule is $\delta_\epsilon(x) = x(1-\epsilon)$, with Bayes risk $1 - \epsilon$. Since the Bayes risk of δ_0 is 1, δ_0 is extended Bayes, hence minimax.

There are certain situations in which minimax rules are known to exist. These involve finite parameter spaces. When Ω is finite, the risk function is just a vector in some Euclidean space. The set of all risk functions of all decision rules is just a convex set of vectors.[12]

Definition 3.71. Suppose that $\Omega = \{\theta_1, \ldots, \theta_k\}$. Let

$$R = \{z \in \mathbb{R}^k : z_i = R(\theta_i, \delta), \text{ for some decision rule } \delta \text{ and } i = 1, \ldots, k\}.$$

We call R, the *risk set*. The *lower boundary of set* $C \subseteq \mathbb{R}^k$ is the set

$$\{z \in \overline{C} : x_i \leq z_i \text{ for all } i \text{ and } x_i < z_i \text{ for some } i \text{ implies } x \notin C\}.$$

The lower boundary of the risk set is denoted ∂_L. The risk set is *closed from below* if $\partial_L \in R$.

Example 3.72. Consider a situation with $\Omega = \{0, 1\}$ and $\aleph = \{1, 2, 3\}$. Let the loss function be

$L(\theta, a)$

θ	1	2	3
0	0	1	0.5
1	1	0	0.2

Supposing that no data are available, the class of randomized rules consists of the set of all probability distributions over the action space $\aleph$. The risk function for a randomized rule with probabilities (p_1, p_2, p_3) for the three actions $(1, 2, 3)$ is just the point $(p_2 + 0.5p_3, p_1 + 0.2p_3)$. The set of all such points is the shaded region in Figure 3.73.

We can locate the minimax rule in Figure 3.73 by looking at all orthants of the form $O_s = \{(x_0, x_1) : x_0 \leq s, x_1 \leq s\}$ and finding the one with the smallest s that intersects the risk set. These orthants are shown in Figure 3.73. For all $s < 0.3846$, the orthant O_s fails to intersect the risk set. But $O_{0.3846}$ does intersect the risk set at the point $(0.3846, 0.3846)$. This point corresponds to the randomized rule with probabilities $(0.2308, 0, 0.7692)$ on the three available actions. It is interesting to note that one is required to randomize in order to achieve the minimax rule. This is somewhat disconcerting for the following reason (among others). After performing the randomization, one will then either choose action $a = 1$ or action $a = 3$. In either case, one is no longer using the minimax rule, and the risk point for the chosen decision is either $(0, 1)$ or $(0.5, 0.2)$, but not $(0.3846, 0.3846)$ as hoped.

Two lines are added to Figure 3.73 to show the Bayes rules with respect to two different priors. The line $0.5x_0 + 0.5x_1 = 0.35$ passes through the point $(0.5, 0.2)$ to indicate that the action $a = 3$ is Bayes with respect to the prior that puts equal probability on each parameter value. (The action $a = 3$ is also Bayes with respect to many other priors.) The prior with $\Pr(\Theta = 0) = .6154$ is least favorable, and the line $0.6154x_0 + 0.3845x_1 = 0.3846$ passes through all of the points corresponding to Bayes rules with respect to the least favorable distribution. (See Problem 25 on page 211 to see how the minimax principle is actually in conflict with the expected loss principle in this example.)

---

[12] The remainder of this section is devoted to proving the minimax theorem 3.77. The discussion of risk sets is used in the proofs of the minimax theorem 3.77 and the complete class theorem 3.95. It is also used briefly in the discussion of simple hypotheses and simple alternatives in Chapter 4. All of this material can be skipped without disrupting the flow of the remaining material.

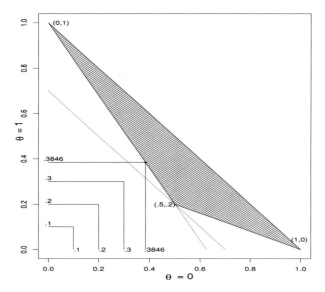

FIGURE 3.73. Risk Set for Example 3.72

Notice that the risk set in Figure 3.73 is convex. This is true in general.

Lemma 3.74. *The risk set is convex.*

PROOF. Let z_i be a point in the risk set that corresponds to a decision rule δ_i for $i = 1, 2$, and let $0 \le \alpha \le 1$. Then $\alpha z_1 + (1-\alpha) z_2$ corresponds to the randomized decision rule $\alpha \delta_1 + (1-\alpha)\delta_2$. □

There is a common misconception that a minimax rule can be located by finding that point in the risk set with all coordinates equal which lies closest to the origin. Here is an example in which the unique minimax rule corresponds to a point with distinct coordinates.

Example 3.75. Consider a situation with $\Omega = \{0, 1\}$ and $\aleph = \{1, 2, 3\}$. Let the loss function be

$$L(\theta, a)$$

θ	1	2	3
0	0	0.25	1
1	1	0.5	0.75

The class of randomized rules consists of the set of all probability distributions over the action space $\aleph$. The risk function for a randomized rule with probabilities (p_1, p_2, p_3) for the three actions $(1, 2, 3)$ is just the point $(0.25 p_2 + p_3, p_1 + 0.5 p_2 + 0.75 p_3)$. The risk set is illustrated in Figure 3.76 together with the point corresponding to the unique minimax rule, choose action 2. The point $(0.625, 0.625)$ is the closest point to the origin which has equal coordinates.

The following theorem gives conditions under which minimax rules and least favorable distributions exist.

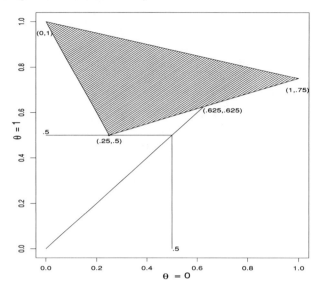

FIGURE 3.76. Minimax Rule with Unequal Risks

Theorem 3.77 (Minimax theorem). *Suppose that the loss function is bounded below and Ω is finite. Then $\sup_\lambda \inf_\delta r(\lambda, \delta) = \inf_\delta \sup_\theta R(\theta, \delta)$, and there exists a least favorable distribution λ_0. If R is closed from below, then there is a minimax rule that is a Bayes rule with respect to λ_0.*

The proof requires a few lemmas.

Lemma 3.78. *Suppose that Ω is finite. The loss function is bounded below if and only if the risk set is bounded below.*

PROOF. Suppose that the loss function is bounded below. That is, there exists c such that $L(\theta, a) \geq c$ for all θ and a. Then $R(\theta, \delta) \geq c$ for all θ and δ, since the risk function is just the integral of the loss function with respect to a probability measure.

Suppose that the loss function is unbounded below, that is, there exists a sequence $\{(\theta_n, a_n)\}_{n=1}^\infty$ such that $L(\theta_n, a_n) < -n$ for each n. Since Ω is finite, there exists a single θ and a sequence $\{b_n\}_{n=1}^\infty$ such that $L(\theta, b_n) < -n$ for all n. With $\delta_n(x) = b_n$ for all x, we have $R(\theta, \delta_n) < -n$, hence the risk set is not bounded below. □

Lemma 3.79. *If a set $C \subseteq \mathbb{R}^k$ is bounded below, then its lower boundary is nonempty.*

PROOF. First, note that the lower boundary of C is the same as the lower boundary of $\overline{C}$, the closure of C. Next, let $c_1 = \inf\{z_1 : z \in \overline{C}\}$, and for $i = 2, \ldots, k$, let

$$C_i = \{z \in \overline{C} : z_j = c_j, \text{ for } j = 1, \ldots, i-1\},$$

$$c_i = \inf\{z_i : z \in C_i\}.$$

Since $\overline{C}$ is closed, the point $c = (c_1,\ldots,c_k) \in \overline{C}$. Suppose that the lower boundary is empty. Then, there is a point x such that $x_i \leq c_i$ for all i with at least one inequality strict. This is clearly a contradiction to the way that c was constructed. □

Lemma 3.80. *Suppose that the loss function is bounded below. If there is a minimax rule, then there is a point on ∂_L whose maximum coordinate is the same as the minimax risk.*

PROOF. Let z be the risk function for a minimax rule, and let s be the minimax risk $\max\{z_1,\ldots,z_k\}$. Let $C = R \cap \{x \in \mathbb{R}^k : x_i \leq s, \text{ for all } i\}$. Since the loss function is bounded below, Lemma 3.78 says the risk set is bounded below, so C is bounded below. Lemma 3.79 says that the lower boundary of C is nonempty. Clearly, the lower boundary of C is a subset of the lower boundary of R. Since each point in C is the risk function of a minimax rule, the result is proven. □

PROOF OF MINIMAX THEOREM 3.77.[13] Let $\overline{R}$ denote the closure of R. For each real s, let

$$A_s = \{z \in \mathbb{R}^k : z_i \leq s \text{ for all } i\}.$$

Then A_s is closed and convex for each s. Let $s_0 = \inf\{s : A_s \cap R \neq \emptyset\}$.

We will prove first that there is a least favorable distribution. It is easy to see that

$$s_0 = \inf_\delta \sup_\theta R(\theta, \delta). \tag{3.81}$$

Note that the interior of A_{s_0} is convex and does not intersect R. By the separating hyperplane theorem C.5, there is a vector v and number c such that $v^\top z \geq c$ for all $z \in R$ and $v^\top x \leq c$ for all x in the interior of A_{s_0}. It is clear that each coordinate of v is nonnegative since, if $v_i < 0$, a sequence of points $\{x^n\}_{n=1}^\infty$ in the interior of A_{s_0} exists with $\lim_{n\to\infty} x_i^n = -\infty$ and all other $x_i^n = s_0 - \epsilon$ and $\lim_{n\to\infty} v^\top x^n = \infty > c$. So, let λ_0 be the probability that puts mass $\lambda_{0,i} = v_i / \sum_{j=1}^k v_j$ on θ_i for $i = 1,\ldots,k$. Since $(s_0,\ldots,s_0)$ is in the closure of the interior of A_{s_0}, it follows that $c \geq s_0 \sum_{j=1}^k v_j$. We can now calculate

$$\inf_\delta r(\lambda_0, \delta) = \inf_{z \in R} \lambda_0^\top z \geq \frac{c}{\sum_{j=1}^k v_j} \geq s_0 = \inf_\delta \sup_\theta R(\theta, \delta). \tag{3.82}$$

It follows that λ_0 is a least favorable distribution.

Next, we prove that if R is closed from below, there is a minimax rule. Let $\{s_n\}_{n=1}^\infty$ be a decreasing sequence converging to s_0. Note the following facts:

---

[13] This proof is similar to proofs of Ferguson (1967) and Berger (1985).

- For each n, $\overline{R} \cap A_{s_n} \neq \emptyset$ is closed and bounded.
- $A_{s_0} = \cap_{n=1}^{\infty} A_{s_n} \neq \emptyset$.

It follows from the Bolzano–Weierstrass theorem C.6 that $\overline{R} \cap A_{s_0}$ is closed and nonempty. It follows from (3.81) that every point in $R \cap A_{s_0}$ is the risk function of a minimax rule. Now, apply Lemma 3.79 with $C = \overline{R} \cap A_{s_0}$ to see that there is a point of ∂_L contained in $\overline{R} \cap A_{s_0}$. Since $\partial_L \subseteq R$, we have a point in R that is the risk function of a minimax rule.

Finally, we prove that a minimax rule δ whose risk function is on ∂_L is a Bayes rule with respect to λ_0. Since $R(\theta, \delta) \leq s_0$ for all θ, it follows that $r(\lambda_0, \delta) \leq s_0$. Combining this with (3.82) completes the proof. $\square$

3.2.5 Complete Classes

Sometimes minimax rules are too hard to find, or they are not good rules. It might be worthwhile just to find all admissible rules. Or, one could find a set of rules such that every other rule is dominated by one in the set.

Definition 3.83. A class of decision rules $\mathcal{C}$ is *complete* if, for every $\delta \notin \mathcal{C}$, there is $\delta_0 \in \mathcal{C}$ such that δ_0 dominates δ. A class $\mathcal{C}$ is *essentially complete* if for every $\delta \notin \mathcal{C}$, there is $\delta_0 \in \mathcal{C}$ such that $R(\theta, \delta_0) \leq R(\theta, \delta)$ for all θ. A class $\mathcal{C}$ is *minimal (essentially) complete* if no proper subclass is also (essentially) complete.

Lemma 3.84.[14] *If $\mathcal{C}$ is a complete class and A is the set of all admissible rules, then $A \subseteq \mathcal{C}$.*

PROOF. If $\delta_0 \notin \mathcal{C}$, then there is $\delta \in \mathcal{C}$ such that δ dominates δ_0, hence δ_0 is inadmissible, hence $\delta_0 \notin A$. $\square$

Proposition 3.85. *If $\mathcal{C}$ is an essentially complete class and there is an admissible $\delta \notin \mathcal{C}$, then there is $\delta_0 \in \mathcal{C}$ such that $R(\theta, \delta_0) = R(\theta, \delta)$ for all θ.*

Lemma 3.86. *If a minimal complete class exists, it consists of exactly the admissible rules.*

PROOF. Let $\mathcal{C}$ be a minimal complete class and be A the set of admissible rules. By Lemma 3.84, $A \subseteq \mathcal{C}$. We need to prove $\mathcal{C} \subseteq A$. Assume the contrary. That is, assume that there is $\delta_0 \in \mathcal{C}$ but $\delta_0 \notin A$. Then there is δ_1 that dominates δ_0. Either $\delta_1 \in \mathcal{C}$ or not. If $\delta_1 \in \mathcal{C}$, call $\delta_2 = \delta_1$. If $\delta_1 \notin \mathcal{C}$, there exists $\delta_2 \in \mathcal{C}$ such that δ_2 dominates δ_1. In either case, $\delta_2 \in \mathcal{C}$ dominates δ_0. If δ_0 dominates some other rule δ, then δ_2 also dominates δ, so $\mathcal{C}\setminus\{\delta_0\}$ is a complete class. But this contradicts the fact that $\mathcal{C}$ was minimal complete. $\square$

There is one famous case in which we can find a minimal complete class.[15]

---

[14] This lemma is used in the proof of Lemma 3.86.
[15] This theorem originated with Neyman and Pearson (1933).

3.2. Classical Decision Theory

Theorem 3.87 (Neyman–Pearson fundamental lemma). *Let Ω and $\aleph$ both be $\{0,1\}$, $L(0,0) = L(1,1) = 0$, and*

$$L(1,0) = k_1 > 0, \quad L(0,1) = k_0 > 0.$$

Let $f_i(x) = dP_i/d\nu(x)$ for $i = 0,1$, where $\nu = P_0 + P_1$. Let δ be a decision rule. Define $\phi(x) = \delta(x)(\{1\})$. (This function ϕ is called the test function *corresponding to δ.)*

Let $\mathcal{C}$ denote the class of all rules with the test functions of the following forms:

For each $k \in (0, \infty)$ and each function $\gamma: \mathcal{X} \to [0,1]$,

$$\phi_{k,\gamma}(x) = \begin{cases} 1 & \text{if } f_1(x) > kf_0(x), \\ \gamma(x) & \text{if } f_1(x) = kf_0(x), \\ 0 & \text{if } f_1(x) < kf_0(x). \end{cases} \quad (3.88)$$

For $k = 0$,

$$\phi_0(x) = \begin{cases} 1 & \text{if } f_1(x) > 0, \\ 0 & \text{if } f_1(x) = 0. \end{cases}$$

For $k = \infty$,

$$\phi_\infty(x) = \begin{cases} 1 & \text{if } f_0(x) = 0, \\ 0 & \text{if } f_0(x) > 0. \end{cases}$$

Then $\mathcal{C}$ is a minimal complete class.

Before giving the proof of this theorem, we will give an outline of the proof because there are so many steps. We need to prove that if δ is a rule not in $\mathcal{C}$, then there is a rule in $\mathcal{C}$ that dominates δ. For a rule δ not in $\mathcal{C}$, we find a rule $\delta^* \in \mathcal{C}$ that has the same value of the risk function at $\theta = 0$. Half of the proof is devoted to this step. We show that the risk functions of rules in $\mathcal{C}$ at $\theta = 0$ are decreasing in k, but they may not be continuous. However, by defining $\gamma(x)$ appropriately, we can find a rule $\delta^* \in \mathcal{C}$ such that $R(0, \delta^*) = R(0, \delta)$. We then show that $R(1, \delta^*) < R(1, \delta)$.

PROOF OF THEOREM 3.87. Let $\mathcal{C}'$ be $\mathcal{C}$ together with all rules whose test functions are of the form $\phi_{0,\gamma}$ in (3.88). Let $\delta \in \mathcal{C}' \setminus \mathcal{C}$. Then the test function for δ is $\phi_{0,\gamma}$ for some γ such that $P_0(\gamma(X) > 0) > 0$. Let δ_0 be the rule whose test function is ϕ_0. Since $f_1(x) = 0$ for all $x \in A = \{x : \phi_{0,\gamma}(x) \neq \phi_0(x)\}$, it follows that $R(1, \delta) = R(1, \delta_0)$. But

$$\begin{aligned} R(0, \delta) &= k_0 \left[E_0(\gamma(X) I_A(X)) + P_0(f_1(X) > 0) \right] \\ &= k_0 E_0(\gamma(X) I_A(X)) + R(0, \delta_0) > R(0, \delta_0). \end{aligned}$$

Hence δ is inadmissible and is dominated by δ_0. We will now proceed to prove that $\mathcal{C}'$ is a complete class. It will then follow from what we just proved that $\mathcal{C}$ is a complete class.

Next, let ϕ be a test function corresponding to a rule δ not in $\mathcal{C}'$. Let

$$\alpha = R(0, \delta) = \int k_0 \phi(x) f_0(x) d\nu(x).$$

Note that $\alpha \leq k_0$. We will now try to find a rule $\delta^* \in C'$ such that $R(0, \delta^*) = \alpha$ and $R(1, \delta^*) < R(1, \delta)$. To that end, we define the function

$$g(k) = \int_{\{f_1(x) \geq k f_0(x)\}} k_0 f_0(x) d\nu(x).$$

Note that if $\gamma(x) = 1$ for all x and δ^* has test function $\phi_{k,\gamma}$, then $g(k) = R(0, \delta^*)$. Since $\{x : f_1(x) \geq k f_0(x)\}$ becomes smaller as k increases, and $f_1(x) < \infty$ a.e. $[\nu]$, it is easy to see that $g(k)$ decreases to 0 as $k \to \infty$. Also, it is easy to see that $g(0) = k_0 \geq \alpha$. We now prove that g is continuous from the left and we find the limit from the right. First, note that

$$\bigcap_{k<m} \{x : f_1(x) \geq k f_0(x)\} = \{x : f_1(x) \geq m f_0(x)\}, \tag{3.89}$$

$$\bigcup_{k>m} \{x : f_1(x) \geq k f_0(x)\} = \{x : f_1(x) > m f_0(x)\} \cup \{x : f_0(x) = 0\}.$$

Because g is bounded, the monotone convergence theorem A.52 gives that

$$\lim_{k \uparrow m} g(k) = g(m)$$

$$\lim_{k \downarrow m} g(k) = \int_{\{x : f_1(x) > m f_0(x)\}} k_0 f_0(x) d\mu(x), \tag{3.90}$$

hence g is continuous from the left. Note that if $\gamma(x) = 0$ for all x and δ has test function $\phi_{m,\gamma}$, then $R(0, \delta^*) = \lim_{k \downarrow m} g(k)$. Since g is continuous from the left, either there is a largest k such that $g(k) > \alpha$ or there is a smallest k such that $g(k) = \alpha$. The first case occurs if g has a jump discontinuity and it jumps from a value greater than α down to a value at most α. The second case occurs if g drops continuously to α. In any case, let the guaranteed value of k be denoted k^*. If $\alpha = 0$, it may be that $k^* = \infty$. If $\alpha > 0$, we must have $k^* < \infty$ because g decreases to 0.

We will construct a decision rule called δ^* whose test function ϕ has the form of $\phi_{k^*,\gamma}$. We consider the three possible cases:

1. $\alpha = 0$ and $k^* < \infty$,
2. $\alpha = 0$ and $k^* = \infty$,
3. $\alpha > 0$ and $k^* < \infty$.

We begin by proving that by appropriate choice of the function γ, we can make $R(0, \delta^*) = R(0, \delta) = \alpha$.

In the first case, we can use (3.89) and (3.90) with $m = k^*$ to show that $\gamma(x) = 0$ makes $R(0, \delta^*) = 0 = \alpha$. In the second case,

$$R(0, \delta^*) = \int k_0 \phi_\infty(x) f_0(x) d\nu(x) = 0 = \alpha.$$

3.2. Classical Decision Theory

In the third case, if $g(k^*) = \alpha$, set $\gamma(x) = 1$ to make $R(0, \delta^*) = g(k^*) = \alpha$. Otherwise, $g(k^*) > \alpha$. Set the right-hand side of (3.90) with $m = k^*$ equal to $v \leq \alpha$. Because g has a jump discontinuity at k^*, it must be that

$$k_0 P_0(f_1(x) = k^* f_0(x)) = g(k^*) - v > \alpha - v \geq 0.$$

For those x such that $f_1(x) = k^* f_0(x)$, define

$$0 \leq \gamma(x) = \frac{\alpha - v}{g(k^*) - v} < 1.$$

It follows that

$$\begin{aligned}
R(0, \delta^*) &= \int k_0 \phi_{k^*}(x) f_0(x) d\nu(x) \\
&= v + \int_{\{f_1(x) = k^* f_0(x)\}} k_0 \frac{\alpha - v}{g(k^*) - v} f_0(x) d\nu(x) \\
&= v + \frac{\alpha - v}{g(k^*) - v} k_0 P_0(f_1(x) = k^* f_0(x)) = \alpha.
\end{aligned}$$

If $k^* < \infty$, define

$$h(x) = [\phi_{k^*, \gamma}(x) - \phi(x)][f_1(x) - k^* f_0(x)].$$

We know that $\phi_{k^*, \gamma}(x) = 1 \geq \phi(x)$ for all x such that $f_1(x) - k^* f_0(x) > 0$ and $\phi_{k^*, \gamma}(x) = 0 \leq \phi(x)$ for all x such that $f_1(x) - k^* f_0(x) < 0$. Since ϕ is not of the form of some $\phi_{k, \gamma}$, then there must be a set B such that $\nu(B) > 0$ and $h(x) > 0$ for all $x \in B$. Since $\nu = P_0 + P_1$, we get that $f_0(x) + f_1(x) = 1$ a.e. $[\nu]$. So,

$$\begin{aligned}
0 &< \int_B h(x) d\nu(x) \leq \int h(x) d\nu(x) \\
&= \int [\phi_{k^*}(x) - \phi(x)] f_1(x) d\nu(x) - k^* \int [\phi_{k^*}(x) - \phi(x)] f_0(x) d\nu(x) \\
&= \int [\phi_{k^*}(x) - \phi(x)] f_1(x) d\nu(x) + \frac{k^*}{k_0}(\alpha - \alpha) \\
&= \frac{1}{k_1}[R(1, \delta) - R(1, \delta^*)].
\end{aligned}$$

Hence $R(1, \delta) > R(1, \delta^*)$.

If $k^* = \infty$, then $R(0, \delta) = 0$, and hence $\phi(x) = 0$ for almost all x such that $f_0(x) > 0$. So

$$\begin{aligned}
R(1, \delta) &= k_1 P_0(f_0(X) > 0) + k_1 \int_{\{x : f_0(x) = 0\}} [1 - \phi(x)] f_1(x) d\nu(x) \\
&> k_1 P_0(f_0(X) > 0) = R(1, \delta^*),
\end{aligned}$$

where the inequality follows from the fact that

$$\int_{\{x:f_0(x)=0\}} [1-\phi(x)]f_1(x)d\nu(x) = 0$$

implies that $\phi(x) = \phi_\infty(x)$, a.e. $[\nu]$. Since δ was assumed not to be in $\mathcal{C}'$, this cannot happen.

What we have shown is that for every $\delta \notin \mathcal{C}'$, there is $\delta^* \in \mathcal{C}'$ such that δ^* dominates δ. Hence $\mathcal{C}'$ is complete. As claimed earlier, it now follows that $\mathcal{C}$ is complete.

It is easy to check (see Problem 29 on page 212) that no element of $\mathcal{C}$ dominates any other element of $\mathcal{C}$, so nothing can be removed without destroying the completeness of $\mathcal{C}$. Hence, $\mathcal{C}$ is minimal complete. □

Notice that $\mathcal{C}$ consists of all Bayes rules with respect to those priors with positive mass on each θ (see Theorem 3.91), plus only one Bayes rule with respect to each of the priors that put mass 0 on one of the θ values.

Proposition 3.91. *In the decision problem described in Theorem 3.87, each rule $\phi_{k,\gamma}$ is a Bayes rule with respect to a prior that assigns positive probability to each parameter value. The only admissible Bayes rule with respect to the prior that says $\Pr(\Theta = 0) = 1$ is ϕ_∞, and the only admissible Bayes rule with respect to the prior that says $\Pr(\Theta = 1) = 1$ is ϕ_0.*

Example 3.92. Let $\theta_1 > \theta_0$, and let f_0 and f_1 be the $N(\theta_0, 1)$ and $N(\theta_1, 1)$ densities, respectively. Then, for any k, $f_1(x)/f_0(x) > k$ if and only if

$$x > \frac{\theta_1 + \theta_0}{2} + \frac{\log k}{\theta_1 - \theta_0}.$$

There is no need to introduce $\gamma_k(x)$, since equality has zero probability.

Example 3.93. Let

$$f_0(x) = \binom{n}{x}p_0^x(1-p_0)^{n-x}, \quad f_1(x) = \binom{n}{x}p_1^x(1-p_1)^{n-x},$$

for some $1 > p_1 > p_0 > 0$. Then, for any k, $f_1(x)/f_0(x) > k$ if and only if

$$x > \frac{n\log\left(\frac{1-p_0}{1-p_1}\right) + \log k}{\log\left(\frac{p_1(1-p_0)}{p_0(1-p_1)}\right)}.$$

For example, if $p_1 = 0.9$, $p_0 = 0.5$, and $n = 10$, we get

$$x > \frac{16.09 + \log(k)}{2.197}.$$

If $k = 4.408$, for example, $x = 8$ is the cutoff, and $\gamma(8)$ must be chosen.

Example 3.94. Let

$$f_0(x) = \binom{12}{x}\left(\frac{1}{2}\right)^{12}, \quad f_1(x) = \frac{1}{12}I_{[0,12]}(x).$$

These are $Bin(12, 1/2)$ and $U(0, 12)$ distributions. The measure μ is Lebesgue measure plus counting measure on the integers. To make both distributions absolutely continuous, we must change $f_1(x)$ to

$$f_1(x) = \sum_{i=0}^{11} \frac{1}{12} I_{(i,i+1)}(x).$$

Then,

$$\frac{f_1(x)}{f_0(x)} = \begin{cases} \infty & \text{if } 0 < x < 12, x \text{ not an integer,} \\ 0 & \text{if } x = 0, 1, \ldots, 12, \\ \text{undefined} & \text{otherwise.} \end{cases}$$

There is only one admissible rule, namely, do what is obvious. If the observed x is an integer, choose the binomial distribution; otherwise, choose the uniform distribution.

There is an analog to the minimax theorem for complete classes. The proof is similar also and is adapted from Ferguson (1967). This theorem can also be thought of as a generalization of the Neyman–Pearson fundamental lemma to larger, but still finite, parameter spaces and more general action spaces. However, explicit forms for the decision rules are not given because the action space is unspecified.

Theorem 3.95 (Complete class theorem). *Suppose that Ω has only k points, the loss function is bounded below, and the risk set is closed from below. Then the set of all Bayes rules is a complete class, and the set of admissible Bayes rules is a minimal complete class. These are also the rules whose risk functions are on the lower boundary of the risk set.*

First, we need a lemma.

Lemma 3.96.[16] *Suppose that Ω has only k points and the loss function is bounded below. Let the risk set be R. Then*

- *every admissible rule is a Bayes rule, and*

- *for every point $z \in \partial_L$, there exists a prior $\lambda^\top = (\lambda_1, \ldots, \lambda_k)$ such that $\lambda^\top z = \inf_{y \in R} \lambda^\top y.$*

---

[16]This lemma is used in the proofs of Theorem 3.95 and Lemma 4.43. It is also used in the discussion of testing simple hypotheses against simple alternatives in Chapter 4.

PROOF. If a rule is admissible, it is clear that its risk function is a point in ∂_L. So, let $z \in \partial_L$, and define

$$A = \{x : x_i < z_i, \text{ for all } i\}.$$

Then A and R are disjoint convex sets, so the separating hyperplane theorem C.5 says that there exists a constant c and a vector v such that for all $x \in A$ and all $y \in R$, $v^\top y \geq c \geq v^\top x$. If a coordinate $v_i < 0$, we can find a point x such that $x_j = z_j - \epsilon$ for $j \neq i$ and x_i is sufficiently negative so that $v^\top x > c$, a contradiction. So we know that all coordinates of v are nonnegative. Set

$$\lambda_i = \frac{v_i}{\sum_{j=1}^k v_j}.$$

Since z is a limit point of A, there exists a point $x \in A$ such that $v^\top x$ is arbitrarily close to $v^\top z$. Hence $c = v^\top z$, and $\lambda^\top y \geq \lambda^\top z$ for all $y \in R$. So, if $z \in R$, then z is the risk function of a Bayes rule with respect to λ. If $z \notin R$, then it is still true that $\lambda^\top z = \inf_{y \in R} \lambda^\top y$. □

PROOF OF THEOREM 3.95. From the definition of the lower boundary of the risk set, it is clear that every point on ∂_L corresponds to an admissible rule. It is also clear that to every point in R not on ∂_L there corresponds a point in R that dominates it. Hence the lower boundary contains the risk functions of all and only admissible rules.

Next, we show that the rules whose risk functions are on ∂_L form a minimal complete class. For each $z \in R$, define

$$A_z = \{x : x_i \leq z_i, \text{ for all } i\}.$$

Let z not be on ∂_L. Then there exists $z' \in R$ such that z' dominates z and $A_{z'} \subset A_z$. If $z' \in \partial_L$, we are done; if not, apply Lemma 3.79 with $C = A_{z'} \cap R$ to conclude that there exists a point in ∂_L that is at least as good as z' and hence dominates z. This makes the admissible rules a complete class. Since no admissible rule can be dominated, it is also a minimal complete class.

Lemma 3.96 shows that ∂_L consists of the risk functions of all admissible Bayes rules, so these rules form a minimal complete class. Since the set of Bayes rules contains the set of admissible Bayes rules, the set of Bayes rules is a complete class also. □

Notice that the complete class theorem 3.95 says that the rules $\phi_{k,\gamma}$, ϕ_0, and ϕ_∞ in the Neyman–Pearson fundamental lemma 3.87 are the rules whose risk functions are on ∂_L in the risk set.

3.3 Axiomatic Derivation of Decision Theory*

In Section 3.1, we claimed that a Bayesian would choose that decision rule which minimized the expected loss with respect to the posterior distribution of Θ. (Recall the expected loss principle.) This may seem reasonable or ad hoc on the surface, but there is some justification for such a principle. Von Neumann and Morgenstern (1947) and Anscombe and Aumann (1963) set up a system of axioms for preferences among decisions, which lead to the conclusion that one should choose the decision that minimizes the expected loss. In this section, we present those axioms together with a proof of the important conclusion. For alternative derivations, see DeGroot (1970), Savage (1954), Fishburn (1970), or Ferguson (1967). In this section, we prove Theorem 3.108, which says that so long as preferences satisfy some axioms, there is a probability distribution and a utility function (like the negative of a loss function) such that, in any comparison of decisions, the one with higher expected utility is the more preferred decision. We also prove Theorem 3.110, which says that if data are to be observed before making decisions, then the comparison should be based on conditional expected utility given the data.

We begin with background information in the form of definitions and axioms. Then we present some examples of the axioms, the statements of the main results, and the proofs.

3.3.1 Definitions and Axioms

The setup we will consider is one in which there is uncertainty about which of several possibilities will occur. Each possibility will be called a *state (of Nature)*. Let R be the set of states of Nature. Let $\mathcal{A}_1$ be a σ-field of subsets of R. Assume that $(R, \mathcal{A}_1)$ is a Borel space. We will also assume that the final outcomes of our choices will be that we will receive some consequence or *prize*. In general, let the set of prizes be an arbitrary set P with σ-field of subsets $\mathcal{A}_2$ which contains all singletons. We assume that all prizes are available in all states.

In addition to states of Nature, we will assume that there is also uncertainty about the outcome of another experiment about which we are willing to specify probabilities. This experiment is assumed to be capable of producing events with arbitrary probabilities, and we are completely indifferent between the possible outcomes prior to choosing between the options available to us. We also assume that this experiment is independent of the state of Nature. One can think of this experiment as a spinner with arbitrarily fine precision which we believe to be fair. The purpose of this experiment is to allow us to refer to probability distributions over the

---

*This section may be skipped without interrupting the flow of ideas.

set of prizes.

Definition 3.97. A *Von Neumann–Morgenstern lottery (NM-lottery)* is a probability on the space $(P, \mathcal{A}_2)$ that is concentrated on finitely many prizes.[17] Call the set of all NM-lotteries $\mathcal{L}$.

For convenience, if $L \in \mathcal{L}$ gives probability 1 to a prize p, then we will often denote L by p. If the NM-lottery L awards prize p_i with probability α_i for $i = 1, \ldots, k$, we will denote it by $L = \alpha_1 p_1 + \cdots + \alpha_k p_k$, where $\alpha_i \geq 0$ for all i, $\sum_{i=1}^{k} \alpha_i = 1$. This NM-lottery is to be interpreted as meaning that the results of the experiment T are to be partitioned into k events $E_1, \ldots, E_k$ with probabilities $\alpha_1, \ldots, \alpha_k$, respectively, and prize p_i is awarded if event E_i occurs. We assume that the details of the partitioning of the results of T are irrelevant to us. We only care about the probabilities of the various prizes.

The choices we need to make will be among NM-lotteries and some more general gambles. In order to make choices among gambles, we need to be able to say which ones we like better than others.[18] We will use the symbol $\preceq$ to indicate weak preference, that is, $L \preceq L'$ means that we like L' at least as much as L. Assuming that a weak preference $\preceq$ is defined on $\mathcal{L}$, we can now define the more general class of gambles to which we will extend $\preceq$. These gambles will be defined as functions from R to $\mathcal{L}$.

Definition 3.98. A function $H : R \to \mathcal{L}$ is called a *horse lottery* if, for each NM-lottery L, $\{r : H(r) \preceq L\} \in \mathcal{A}_1$. Let $\mathcal{H}$ denote the set of all horse lotteries.

If $H(r) = L$ for all r, we will often denote H by L. A horse lottery is a gamble that pays off the result of a state-specific NM-lottery depending on which state occurs.

---

[17]The reason that we choose this restricted set of probability distributions rather than the set of all probability distributions on $(P, \mathcal{A}_2)$ is threefold. First, since our axioms will require that various relations hold for all NM-lotteries, the smaller the set of NM-lotteries is, the less restrictive are our axioms. Second, there is one useful result (Corollary 3.143) that fails without further assumptions if we allow all probabilities to be NM-lotteries. Third, since we will only consider measures on $(P, \mathcal{A}_2)$ which concentrate on finitely many points, the σ-field $\mathcal{A}_2$ can be quite arbitrary.

[18]There is one assumption implicit in all of this discussion which is not of a mathematical nature. The assumption is that our choices do not affect our opinions. As an example that violates this assumption, suppose that I am trying to decide whether or not to offer accidental death insurance to an individual. If I sell the insurance, the individual may be more inclined to act in a risky manner (such as mountain climbing or bungee jumping). If the individual does not have insurance, he or she may be less inclined toward the risky behavior. The theory described in this section assumes that these considerations are absent from our choices.

It is useful to have notation for strict preference and indifference also. If $H_1 \preceq H_2$ but it is not the case that $H_2 \preceq H_1$, then we say $H_1 \prec H_2$ and that H_2 is *strictly preferred to* H_1. If both $H_1 \preceq H_2$ and $H_2 \preceq H_1$, we say $H_1 \sim H_2$ and we are *indifferent between* H_1 and H_2, or H_1 and H_2 are *indifferent*.

The first axiom says that the relation $\preceq$ is a *weak order*, which we define now.

Definition 3.99. A binary relation $\preceq$ on a set A is a *weak order* if it satisfies the following conditions:

- For every $a \in A$, $a \preceq a$.
- For every $a, b \in A$, either $a \preceq b$ or $b \preceq a$, or both.
- If $a \preceq b$ and $b \preceq c$, then $a \preceq c$.

We say that $\preceq$ is *degenerate* if $a \preceq b$ for all $a, b \in A$. If $\preceq$ is not degenerate, we say it is *nondegenerate*.

Note that a preference is degenerate if and only if all horse lotteries are indifferent.

Axiom 1 (Weak order). *The relation of weak preference $\preceq$ is a weak order on the set $\mathcal{H}$ of horse lotteries.*

The first axiom does not put very many constraints on the possible preferences we could express. There is the constraint that preferences be transitive, but this is not a very controversial requirement. There is also the constraint that every pair of horse lotteries be affirmatively compared. That is, for every H_1 and H_2, either $H_1 \preceq H_2$ or $H_2 \preceq H_1$ (or both). It is not allowed that I refuse (or am unable) to compare H_1 with H_2. Seidenfeld, Schervish, and Kadane (1992) study, in depth, the problems that arise if one relaxes Axiom 1 to allow that certain horse lotteries are not compared.

Since the set $\mathcal{L}$ is convex, the set of all functions from R to $\mathcal{L}$ is convex with $H = \alpha H_1 + (1-\alpha) H_2$ defined by $H(r) = \alpha H_1(r) + (1-\alpha) H_2(r)$ for all r. Unfortunately, the requirement that $\{r : H(r) \preceq L\} \in \mathcal{A}_1$ for all $L \in \mathcal{L}$ prevents us from concluding immediately that $\alpha H_1 + (1-\alpha) H_2 \in \mathcal{H}$, even if both $H_1, H_2 \in \mathcal{H}$. However, all NM-lotteries satisfy this condition. That is, if $H_1(r) = L_1$ for all r and $H_2(r) = L_2$ for all r, then $\alpha H_1 + (1-\alpha) H_2 = \alpha L_1 + (1-\alpha) L_2 \in \mathcal{H}$. Eventually, we will be able to prove that $\mathcal{H}$ is convex. (See Lemma 3.131.) Until we do so, however, many results will have to be stated in such a way that they still apply without $\mathcal{H}$ being convex. This will be done by adding a condition that the result hold not necessarily for all horse lotteries, but only for every convex set of horse lotteries. The set of constant horse lotteries (identified with $\mathcal{L}$ as above) is convex, so the condition is not vacuous.

Another requirement of the same sort as transitivity is one that says that if I prefer one prize to another, then I should prefer a gamble that gives me that prize with some probability to a gamble that gives me the other prize with the same probability, all other things being equal. We make this precise with another axiom.

Axiom 2 (Sure-thing principle). *For each convex set $\mathcal{H}_0$ of horse lotteries; for every three horse lotteries $H_1, H_2, H \in \mathcal{H}_0$; and for every $0 < \alpha \leq 1$, $H_1 \preceq H_2$ if and only if $\alpha H_1 + (1-\alpha)H \preceq \alpha H_2 + (1-\alpha)H$.*

Two other axioms are often assumed for mathematical reasons. The first assures that no horse lottery is worth infinitely more than another. A definition is required first.

Definition 3.100. Let L be an NM-lottery and let $\{L_k\}_{k=1}^{\infty}$ be a sequence of NM-lotteries. We say that L_k *converges pointwise to L (denoted $L_k \to L$)* if and only if, for every $A \in \mathcal{A}_2$, $\lim_{k\to\infty} L_k(A) = L(A)$.

Since each NM-lottery is a probability distribution over $(P, \mathcal{A}_2)$, it is a function from $\mathcal{A}_2$ to $[0,1]$. The above definition of pointwise convergence agrees with the usual concept of pointwise convergence of functions.

Axiom 3 (Continuity). *Let H be a horse lottery, and let $\{H_k\}_{k=1}^{\infty}$ be a sequence of horse lotteries such that $H_k(r)$ converges pointwise to $H(r)$ for all r. Let H' be another horse lottery. If $H_k \preceq H'$ for all k, then $H \preceq H'$. If $H' \preceq H_k$ for all k, then $H' \preceq H$.*

The next axiom assures that the relative values of prizes do not vary from state to state. It is not obvious that such an axiom should be adopted. In fact, this axiom appears to be nothing more than a mathematical tool for ensuring that probability and utility can be separated. In Section 3.3.6, we will consider what happens if we do not assume Axiom 4. A definition is required before we can state Axiom 4.

Definition 3.101. If $E \in \mathcal{A}_1$ and $H_1, H_2 \in \mathcal{H}$, the preference between H_1 and H_2 is said to be *called-off when event E occurs* if $H_1(r) = H_2(r)$ for all $r \in E$. A set $B \in \mathcal{A}_1$ is called *null* if whenever the preference between H_1 and H_2 is called-off when B^C occurs, we have $H_1 \sim H_2$. A subset is called *nonnull* if it is not null. Similarly, a state s is *nonnull* if the singleton set $\{s\}$ is nonnull.

Axiom 4 (State independence). *Suppose that there is a nonnull set B such that the preference between H_1 and H_2 is called-off if B^C occurs. Suppose also that $H_1(r) = L_1$ and $H_2(r) = L_2$ for $r \in B$. Then $H_1 \preceq H_2$ if and only if $L_1 \preceq L_2$.*

An interesting discussion of state independence is given by Schervish, Seidenfeld, and Kadane (1990).

Next, we introduce an axiom that is only needed when there are infinitely many states of Nature. When there are infinitely many possible prizes and states, Seidenfeld and Schervish (1983) show that it is possible for H_1 to be preferred to H_2 merely because H_1 offers more possible prizes than H_2, even though H_2 offers more valuable prizes than H_1. To avoid this problem, we introduce an axiom.

Axiom 5 (Dominance). *If $H_1(r) \preceq H_2(r)$ for all r, then $H_1 \preceq H_2$.*

The dominance axiom ties the values of horse lotteries to the values of the NM-lotteries which they assume. It can be shown (see Problem 38 on page 213) that Axioms 1–4 imply dominance when R is finite. An example in which Axioms 1–4 hold but Axiom 5 does not is in Example 3.107.

Next, we define conditional preference and introduce an axiom to link preferences with conditional preferences.

Definition 3.102. Let $(\mathcal{X}, \mathcal{B})$ be a measurable space, and let $X : R \to \mathcal{X}$ be a random quantity. We define *a conditional preference relation given X* to be a set of binary relations on $\mathcal{H}$, $\{\preceq_x : x \in \mathcal{X}\}$, where $H_1 \preceq_x H_2$ is read as "we would like H_2 at least as much as H_1 if we were to learn that $X = x$ and nothing else of relevance," and

- for every $x \in \mathcal{X}, H \in \mathcal{H}, L \in \mathcal{L}$, $\{r : H(r) \preceq_x L\} \in \mathcal{A}_1$;

- $H_1 \preceq_x H_2$ if and only if, for every pair (H_1', H_2') of horse lotteries satisfying $H_i'(r) = H_i(r)$ for $i = 1, 2$ and all $r \in X^{-1}(\{x\})$, $H_1' \preceq_x H_2'$; and

- for all $H_1, H_2 \in \mathcal{H}$, $\{x : H_1 \preceq_x H_2\} \in \mathcal{B}$.

The first condition in the list is to ensure that the same horse lotteries are compared conditionally as unconditionally. The second condition guarantees that conditional on $X = x$, it does not matter what would have happened if $X = y \neq x$ had occurred. This makes conditional preference truly conditional. The measurability condition is added for mathematical convenience. None of the axioms stated for unconditional preference says anything about conditional preference. We now suppose that the following axiom holds.

Axiom 6 (Conditional preference). *If $(\mathcal{X}, \mathcal{B})$ is a measurable space, $X : R \to \mathcal{X}$ is a random quantity, and $\{\preceq_x : x \in \mathcal{X}\}$ is a conditional preference relation given X, then*

- *for all x, $\preceq_x$ satisfies Axioms 1–5;*

- *$H_1 \preceq_x H_2$ for all x implies $H_1 \preceq H_2$;*

- *if $H_1 \preceq_x H_2$ for all x and $H_1 \prec_x H_2$ for all $x \in B$, where $X^{-1}(B)$ is nonnull, then $H_1 \prec H_2$.*

This axiom says that if we know that we will prefer H_2 to H_1 after we observe X no matter what value we observe for X, then we should prefer H_2 to H_1 now. In the case in which there are only finitely many states of Nature, there is a means of deriving conditional preferences from unconditional preferences, but one still needs an axiom to say that the derived preferences should be used conditionally. The method is to make use of called-off preferences. If $X: R \to \mathcal{X}$ is a random quantity and $X^{-1}(\{x\}) = E$, then one can require that conditional preference given E agree with preference that are called-off when E^C occurs.

If we must condition on more than one quantity, there is an issue of consistency. For example, if Y is a function of X and we first condition on Y and then on X, do we get the same conditional preferences as if we had conditioned on X alone?

Definition 3.103. Let $X: R \to \mathcal{X}$ and $Y: R \to \mathcal{Y}$ be random quantities such that Y is a function of X, $Y = h(X)$. We say that conditional preference relations $\{\preceq_y: y \in \mathcal{Y}\}$ and $\{\preceq_x: x \in \mathcal{X}\}$ given Y and X are *consistent* if there exists a set $B \subseteq \mathcal{Y}$ such that $Y^{-1}(B)$ is null and for every $y \notin B$, $\{\preceq_x: x \in h^{-1}(y)\}$ is a conditional preference relation relative to $\preceq_y$ which satisfies Axiom 6.

3.3.2 Examples

Here are a few examples to illustrate how the axioms can be satisfied or violated.

Example 3.104. Suppose that there are only two states of Nature r_1 and r_2. Suppose also that there are only two prizes, p_1 and p_2. Then horse lotteries are characterized by the pair of numbers (q_1, q_2), where q_i is the probability of p_2 in state r_i for $i = 1, 2$. We give two examples of weak preference, one which satisfies the axioms and one which does not.

First, suppose that we claim that $(q_1, q_2) \preceq (q_1', q_2')$ if and only if $q_1 + q_2 \leq q_1' + q_2'$. It is straightforward to check that this satisfies all of the axioms. The representation of preference according to Theorem 3.108 below will be $\Pr(r_1) = \Pr(r_2) = 1/2$ and $U(p_2) > U(p_1)$. Consider two horse lotteries $H_1 = (q_1, q_2)$ and $H_2 = (q_1, q_3)$. The preference between H_1 and H_2 is called-off if $\{r_1\}$ occurs. It is easy to see in this example, and it can be shown in general that if r_2 is nonnull (see Lemma 3.148), then $H_1 \preceq H_2$ if and only if $H_3 \preceq H_4$ for all H_3, H_4 of the form $H_3 = (q_4, q_2)$, $H_4 = (q_4, q_3)$. That is, a pair of horse lotteries that differ only on a nonnull state are ranked the same as every other pair that differ the same way on the same state.

Second, suppose that we claim that $(q_1, q_2) \preceq (q_1', q_2')$ if and only if $q_1 < q_1'$ or ($q_1 = q_1'$ and $q_2 \leq q_2'$). This fails Axiom 3, and there is no expected utility representation for the preferences.

Example 3.105. Let $P = \{p_1, \ldots, p_m\}$ and $R = \{r_1, \ldots, r_n\}$. Let $q_1, \ldots, q_n$ be nonnegative numbers that add to 1. Let $u_1, \ldots, u_m$ be real numbers. For each NM-lottery $L = \alpha_1 p_1 + \cdots \alpha_m p_m$, define $U(L) = \sum_{i=1}^{m} \alpha_i u_i$. For each horse lottery $H = (L_1, \ldots, L_n)$, define $U(H) = \sum_{j=1}^{n} q_j U(L_j)$. Say that $H_1 \preceq H_2$ if and only if $U(H_1) \leq U(H_2)$. It is easy to see that this is a weak order. Since $U(\alpha L_1 + (1-\alpha)L_2) = \alpha_1 U(L_1) + (1-\alpha)U(L_2)$, it is easy to verify that Axiom 2 holds. Continuity follows since $L_k \to L$ implies $U(L_k) \to U(L)$. State independence follows easily from the definition of $U(H)$. Theorem 3.108 says that all examples in the finite case will be like this.

Let $X : R \to \mathcal{X}$ be a random quantity. Clearly, X can take on only finitely many values. For each value x of X, let $X^{-1}(\{x\}) = \{p_{k_1(x)}, \ldots, p_{k_{s_x}(x)}\}$. If $v_x = \sum_{j=1}^{s_x} q_{k_j(x)} = 0$, then $X^{-1}(\{x\})$ is null. Otherwise, define $U_x(H) = \sum_{j=1}^{s_x} q_{k_j(x)} U(L_{k_j(x)})$, and say that $H_1 \preceq_x H_2$ if and only if $U_x(H_1) \leq U_x(H_2)$. It is easy to verify that this is a conditional preference and that it satisfies Axiom 6. Theorem 3.110 says that all conditional preferences must be of this form in the finite case.

Example 3.106. Let $(R, \mathcal{A}_1)$ be an arbitrary Borel space with a probability Q, and let P be an arbitrary set. Let $U : P \to \mathbb{R}$ be a bounded function. For each NM-lottery $L = \alpha_1 p_1 + \cdots \alpha_k p_k$, define $U(L) = \sum_{i=1}^{k} \alpha_i U(p_i)$. Let $\mathcal{H}$ be the set of all functions $H : R \to \mathcal{L}$ such that $U(H(r))$ is a measurable function of r. For $H \in \mathcal{H}$, define $U(H) = \int U(H(r))dQ(r)$. Say that $H_1 \preceq H_2$ if and only if $U(H_1) \leq U(H_2)$. For each $H \in \mathcal{H}$ and each $L \in \mathcal{L}$, $\{r : H(r) \preceq L\} = \{r : U(H(r)) \leq U(L)\} \in \mathcal{A}_1$ because we assumed that $U(H(\cdot))$ is measurable. Axiom 1 clearly holds. To see that Axiom 2 holds, note that $U(\alpha H_1 + [1-\alpha]H_2) = \alpha U(H_1) + (1-\alpha)U(H_2)$. If $H_n(r) \to H(r)$ for all r, then $\lim_{n \to \infty} U(H_n(r)) = U(H(r))$ for all r. Since U is bounded, the dominated convergence theorem A.57 says that $\lim_{n \to \infty} U(H_n) = U(H)$. This implies that Axiom 3 holds. Note that $B \in \mathcal{A}_1$ is null if and only if $Q(B) = 0$. To see that Axiom 4 is satisfied, let the preference between H_1 and H_2 be called-off when B^C occurs, B is nonnull, and $H_1(r) = L_1$ and $H_2(r) = L_2$ for all $r \in B$. Then $U(H_1) - U(H_2) = Q(B)[U(L_1) - U(L_2)]$, so $H_1 \preceq H_2$ if and only if $L_1 \preceq L_2$. Axiom 5 follows easily from part 3 of Proposition A.49. Theorem 3.108 says that when R and/or P is infinite, this example describes all preference relations that satisfy the axioms.

Let $X : R \to \mathcal{X}$ be a random quantity. Let $\{Q(\cdot|x) : x \in \mathcal{X}\}$ be a regular conditional distribution given X. (Use Corollary B.55 to choose a version of $Q(\cdot|x)$ that gives probability 1 to $X^{-1}(\{x\})$.) For each $x \in \mathcal{X}$ and $H \in \mathcal{H}$, define $U_x(H) = \int U(H(r))dQ(r|x)$. If $H_1, H_2 \in \mathcal{H}$, Theorem B.46 can be used to show that $U_x(H_i)$ is a measurable function of x for each i, hence $\{x : H_1 \preceq_x H_2\} \in \mathcal{B}$, and $\{\preceq_x : x \in \mathcal{X}\}$ is a conditional preference. Axiom 6 follows from the law of total probability B.70. Theorem 3.110 says that except for differences on null sets, this example describes all conditional preferences that satisfy Axiom 6.

Example 3.107. This example is based on one of Seidenfeld and Schervish (1983), and it is designed to show why Axiom 5 is needed in the infinite case. Let $R = [0, 1]$ and let $\mathcal{A}_1$ be the Borel σ-field. Let Q be Lebesgue measure, and let $P = [0, 1]$. Let $V : P \to \mathbb{R}$ be defined by $V(p) = p$. For each NM-lottery $L = \alpha_1 p_1 + \cdots \alpha_k p_k$, define $V(L) = \sum_{i=1}^{k} \alpha_i V(p_i)$. For each function $H : R \to \mathcal{L}$, define $w_H(p, r) = H(r)(\{p\})$, that is, the probability that $H(r)$ assigns to the prize p. Let $\mathcal{H}$ be the set of all H such that $w_H(p, r)$ is a measurable function of r for every p and $V(H(r))$ is a measurable function of r. For $H \in \mathcal{H}$, define $V(H) = \int V(H(r))dQ(r)$. Note that V is the same as the U in Example 3.106. Define $w_H(p) = \int w_H(p, r)dQ(r)$. Since $\sum_{\text{All } p} w_H(p, r) = 1$ for all r, there can be at most countably many p such that $w_H(p) > 0$. Define $W(H) = 1 - \sum_{\text{All } p} w_H(p)$. The value $W(H)$ measures the extent to which more than countably many different prizes are assigned by H. For example, if the set of all prizes assigned by H is countable, then $W(H) = 0$. In particular, it is easy to see that $W(L) = 0$ for all $L \in \mathcal{L}$. Define $U(H) = V(H) + W(H)$ and say that $H_1 \preceq H_2$ if and only if $U(H_1) \leq U(H_2)$. Axiom 1 is clearly satisfied. To see that Axiom 2 is satisfied, note that $w_{\alpha H_1 + [1-\alpha]H_2}(p) = \alpha w_{H_1}(p) + (1-\alpha)w_{H_2}(p)$ for all p, so $W(\alpha H_1 + [1-\alpha]H_2) = \alpha W(H_1) + (1-\alpha)W(H_2)$. Now, use the fact that $V(\alpha H_1 + [1-\alpha]H_2) = \alpha V(H_1) + (1-\alpha)V(H_2)$ as shown in Example 3.106 to see that $U(\alpha H_1 + [1-\alpha]H_2) = \alpha U(H_1) + (1-\alpha)U(H_2)$. If $H_n(r) \to H(r)$ for all r, then $\lim_{n\to\infty} w_{H_n}(p,r) = w_H(p,r)$ for all p, r. Let $\{p_i\}_{i=1}^{\infty}$ be the prizes such that either $w_{H_n}(p_i) > 0$ for some n or $w_H(p_i) > 0$. Define $f(r) = \sum_{i=1}^{\infty} w_H(p_i, r)$ and $f_n(r) = \sum_{i=1}^{\infty} w_{H_n}(p_i, r)$. Then $W(H) = \int f(r)dQ(r)$ and $W(H_n) = \int f_n(r)dQ(r)$. Since $0 \leq f_n \leq 1$ and $\lim_{n\to\infty} f_n(r) = f(r)$ for all r, it follows from the dominated convergence theorem A.57 that $\lim_{n\to\infty} W(H_n) = W(H)$. As shown in Example 3.106, $\lim_{n\to\infty} V(H_n) = V(H)$, so $\lim_{n\to\infty} U(H_n) = U(H)$ and Axiom 3 holds. To see that Axiom 4 is satisfied, let the preference between H_1 and H_2 be called-off when B^C occurs, B is non-null, and $H_1(r) = L_1$ and $H_2(r) = L_2$ for all $r \in B$. Then $W(H_1) = W(H_2) = \sum_{\text{All } p} \int_{B^C} w_{H_i}(p,r)dQ(r)$, so $U(H_1) - U(H_2) = Q(B)[V(L_1) - V(L_2)]$. We see that $H_1 \preceq H_2$ if and only if $L_1 \preceq L_2$. To see that Axiom 5 is violated, let $H_1(r) = 1r$, that is, the NM-lottery that gives prize $p = r$ with probability 1. Let $H_2 = 1$, that is, the constant horse lottery that gives the prize 1 with probability 1 for all r. It is easy to calculate $V(H_1) = 1/2$, $W(H_1) = 1$, $V(H_2) = 1$, and $W(H_2) = 0$. So $U(H_1) = 3/2 > 1 = U(H_2)$. But $U(H_1(r)) = V(H_1(r)) = r$ for all r and $U(H_2(r)) = V(1) = 1$ for all r. So $H_1(r) \preceq H_2(r)$ for all r, but $H_2 \prec H_1$. Note that the ranking of horse lotteries by U is not an expected utility representation because of the added function W.

3.3.3 The Main Theorems

Since the proofs of the major theorems are very long and not particularly straightforward, we state here, for the interested reader, the main results. The proofs will be given in Section 3.3.5.

Theorem 3.108. *Assume Axioms 1–5, and assume that preference is non-degenerate. Then, there exists a bounded function $U : \mathcal{H} \to \mathbb{R}$ such that $U(H(r))$ is a measurable function of r for all $H \in \mathcal{H}$ and that satisfies*

$$U(\alpha H_1 + [1-\alpha]H_2) = \alpha U(H_1) + (1-\alpha)U(H_2), \tag{3.109}$$

for all $\alpha \in [0,1]$ and all H_1, H_2. Also, there exists a probability Q on $(R, \mathcal{A}_1)$ such that for every $H_1, H_2 \in \mathcal{H}$, $H_1 \preceq H_2$ if and only if $\int U(H_1(r))dQ(r) \leq \int U(H_2(r))dQ(r)$. The probability Q is unique, and U is unique up to positive affine transformation.

The function U in Theorem 3.108 is called a *utility function*.

We also prove a theorem linking preference and conditional preference.

Theorem 3.110. *Assume the conditions of Theorem 3.108. Let $(\mathcal{X}, \mathcal{B})$ be a Borel space, and let $X : R \to \mathcal{X}$ be a random quantity. Let Q be the probability from Theorem 3.108, and let $\{Q(\cdot|x) : x \in \mathcal{X}\}$ be a regular conditional distribution given X. Let $\{\preceq_x: x \in \mathcal{X}\}$ be a conditional preference relation given X which satisfies Axiom 6. Then there exists a set*

B such that $X^{-1}(B)$ is null and for all $x \notin B$, $H_1 \preceq_x H_2$ if and only if $\int U(H_1(r))dQ(r|x) \leq \int U(H_2(r))dQ(r|x)$.

In Theorem 3.110, if $Y: R \to \mathcal{Y}$ is a function of X, and $\{\preceq_y: y \in \mathcal{Y}\}$ is a conditional preference relation given Y, then $\{\preceq_y: y \in \mathcal{Y}\}$ is consistent with $\{\preceq_x: x \in \mathcal{X}\}$ because Theorem B.75 and Corollary B.74 say that conditioning on Y and then X is the same as conditioning on X alone.

3.3.4 Relation to Decision Theory

Earlier in this chapter we set up decision theory using action spaces and loss functions. There is a natural connection between these concepts and the concepts introduced in this section. Let the states of Nature be possible values of the parameter ($R = \Omega$), or of some future observable ($R = \mathcal{V}$), or possibly data and parameter together ($R = \mathcal{X} \times \Omega$). Let the actions (elements of $\aleph$) index functions from states to prizes (or, more generally, from states to NM-lotteries). That is, for each $a \in \aleph$, there exists a horse lottery $H_a : R \to \mathcal{L}$ such that $H_a(r)$ is the prize (NM-lottery) we get in state of Nature r. For example, if $R = \Omega$, then we can consider $L(\theta, a) = c - U(H_a(\theta))$ for arbitrary c. In this way bounded loss functions are like the negatives of utility functions. Unbounded loss functions, however, do not correspond to utilities that satisfy the axioms stated in this chapter.

Example 3.111. Suppose that $R = \Omega$ is the interval $[c_0, c_1]$. Let P, the set of prizes, be a bounded interval of monetary units containing my current fortune y and having half-width at least $(c_1 - c_0)^2$. Let $U(p) = p$ for each $p \in P$. If $\aleph = \Omega$, then for each action $a \in \aleph$ we construct the horse lottery $H_a(\theta) = y - (a - \theta)^2$, a function from Ω to P. Then $y - U(H_a(\theta)) = (a - \theta)^2$ is squared-error loss. Note that we used the bounded intervals to ensure that utility is bounded.

Axiomatic developments like the one given in this section have two main consequences. The obvious one is that, as stated in the main theorems, if one satisfies the axioms, then the preferences have an expected utility representation. The contrapositive is also true. If preferences do not have an expected utility representation, then at least one axiom must be violated.

Example 3.112 (Continuation of Example 3.72; see page 170). The minimax principle is often in conflict with Axiom 2. In this example, the minimax rule corresponds to the convex combination

$$(0.3846, 0.3846) = 0.7692(0.5, 0.2) + 0.2308(0, 1)$$

in Figure 3.73. According to the minimax principle, $(0, 1) \prec (0.5, 0.2)$ because $0.5 < 1$. If Axiom 2 were satisfied, then

$$\begin{aligned}(0.3846, 0.3846) &= 0.7692(0.5, 0.2) + 0.2308(0, 1)\\ &\prec 0.7692(0.5, 0.2) + 0.2308(0.5, 0.2)\\ &= (0.5, 0.2).\end{aligned}$$

But $(0.5, 0.2) \prec (0.3846, 0.3846)$ according to the minimax principle, because $0.3846 < 0.5$. Hence, the minimax principle violates Axiom 2 in this example.[19]

3.3.5 Proofs of the Main Theorems*

The theorems we prove pertain to the general case in which $(R, \mathcal{A}_1)$ is an arbitrary Borel space and P is an arbitrary set. Some readers may wish to focus on the *finite case* in which $R = \{r_1, \ldots, r_n\}$, $\mathcal{A}_1 = 2^R$, $P = \{p_1, \ldots, p_m\}$, and $\mathcal{A}_2 = 2^P$. In each of the results, we will point out how the proofs can be simplified (usually by skipping major portions) in the finite case. In the finite case, if $H(r_i) = L_i$ for $i = 1, \ldots, n$, then we will denote $H = (L_1, \ldots, L_n)$. Since $\mathcal{A}_1 = 2^R$, it follows that $\mathcal{H}$ is convex in the finite case. In the finite case Axiom 5 follows from Axioms 1–4. (See Problem 38 on page 213.)

Until we prove that $\mathcal{H}$ is convex in general (Lemma 3.131), all of the lemmas we prove will need to contain conditions like the one that appeared in Axiom 2 concerning an arbitrary convex set $\mathcal{H}_0$ of horse lotteries. Once we prove Lemma 3.131, then $\mathcal{H}_0$ can be taken equal to $\mathcal{H}$ in all of these results. For this reason, in this section we will assume that $\mathcal{H}_0$ is a convex set of horse lotteries. The theorems in this section will apply to every such set $\mathcal{H}_0$.[20] Because some of the lemmas in this section are also useful in Section 3.3.6, the hypotheses often include which axioms are assumed explicitly.

Axiom 2 has an "if and only if" clause preceded by a quantification. The implication in one direction is straightforward, namely that if $H_1 \preceq H_2$, then $\alpha H_1 + (1-\alpha)H \preceq \alpha H_2 + (1-\alpha)H$ for all H and all $\alpha \in (0, 1]$ (assuming that the mixtures are horse lotteries.) The other direction of implication has more striking consequences. In words, if a horse lottery appears in two mixtures on both sides of a preference, then the smaller amount can be "removed" from each mixture without changing the preference.

Lemma 3.113. *Assume Axioms 1 and 2. Let $H_1, H_2, H \in \mathcal{H}_0$, and let $\alpha, \beta \in (0, 1)$.*

- *Suppose that $\alpha H + (1 - \alpha)H_1 \preceq \beta H + (1 - \beta)H_2$. If $\alpha > \beta$, then*

$$\frac{\alpha - \beta}{1 - \beta} H + \frac{1 - \alpha}{1 - \beta} H_1 \preceq H_2.$$

If $\alpha < \beta$, then

$$H_1 \preceq \frac{\beta - \alpha}{1 - \alpha} H + \frac{1 - \beta}{1 - \alpha} H_2.$$

---

[19]See also Problem 25 on page 211.

*This section may be skipped without interrupting the flow of ideas.

[20]In the finite case, $\mathcal{H}_0$ can be taken equal to $\mathcal{H}$, since $\mathcal{H}$ is known to be convex in that case.

- Suppose that $\alpha H + (1-\alpha)H_1 \prec \beta H + (1-\beta)H_2$. If $\alpha > \beta$, then
$$\frac{\alpha - \beta}{1 - \beta} H + \frac{1 - \alpha}{1 - \beta} H_1 \prec H_2.$$

If $\alpha < \beta$, then
$$H_1 \prec \frac{\beta - \alpha}{1 - \alpha} H + \frac{1 - \beta}{1 - \alpha} H_2.$$

PROOF. The first two statements are proved in almost identical fashion. We will prove only the first one. Axiom 2 says that for arbitrary $0 \leq \eta < 1$,
$$\eta H + (1-\eta)[\gamma H + (1-\gamma)H_1] \preceq \eta H + (1-\eta)H_2$$
implies that $\gamma H + (1-\gamma)H_1 \preceq H_2$. Let $\eta = \beta$ and $\gamma = (\alpha - \beta)/(1-\beta)$.

The last two statements are proved in almost identical fashion. We will only prove the third one. Axiom 2 and the definition of $\prec$ say that for arbitrary $0 \leq \eta < 1$,
$$\eta H + (1-\eta)[\gamma H + (1-\gamma)H_1] \prec \eta H + (1-\eta)H_2$$
implies that $\gamma H + (1-\gamma)H_1 \preceq H_2$. In fact, we can conclude $\gamma H + (1-\gamma)H_1 \prec H_2$ because, if not, then $H_2 \preceq \gamma H + (1-\gamma)H_1$ and Axiom 2 implies
$$\eta H + (1-\eta)H_2 \preceq \eta H + (1-\eta)[\gamma H + (1-\gamma)H_1],$$
a contradiction. Now, let $\eta = \beta$ and $\gamma = (\alpha - \beta)/(1-\beta)$. $\square$

The next lemma says that a less preferred gamble can be substituted for a more preferred one on the left side of any $\preceq$ relation. Similarly, a more preferred gamble can be substituted for a less preferred one on the right side of any $\preceq$ relation.

Lemma 3.114. *Assume Axioms 1 and 2.*

- *Assume that $H_1, H_2 \in \mathcal{H}_0$ and that $H_1 \preceq H_2$. Then for every $0 < \alpha \leq 1$, every $H_3 \in \mathcal{H}_0$, and every H_4, if $\alpha H_2 + (1-\alpha)H_3 \preceq H_4$, then $\alpha H_1 + (1-\alpha)H_3 \preceq H_4$, and if $H_4 \preceq \alpha H_1 + (1-\alpha)H_3$, then $H_4 \preceq \alpha H_2 + (1-\alpha)H_3$.*

- *Assume that $H_1, H_2 \in \mathcal{H}_0$ and that $H_1 \prec H_2$. Then for every $0 < \alpha \leq 1$, every H_3, and every H_4, if $\alpha H_2 + (1-\alpha)H_3 \preceq H_4$, then $\alpha H_1 + (1-\alpha)H_3 \prec H_4$, and if $H_4 \preceq \alpha H_1 + (1-\alpha)H_3$, then $H_4 \prec \alpha H_2 + (1-\alpha)H_3$.*

PROOF. Suppose that $\alpha H_2 + (1-\alpha)H_3 \preceq H_4$ and $H_1 \preceq H_2$. Axiom 2 says that
$$\alpha H_1 + (1-\alpha)H_3 \preceq \alpha H_2 + (1-\alpha)H_3.$$
It follows from the transitivity of $\preceq$ that $\alpha H_1 + (1-\alpha)H_3 \preceq H_4$. The remaining cases are all similar. $\square$

It follows easily that two indifferent horse lotteries can be substituted for each other in all comparisons.

Corollary 3.115. *Assume Axioms 1 and 2. Let $H_1, H_2 \in \mathcal{H}_0$, and assume $H_1 \sim H_2$. Then for every $0 < \alpha \leq 1$, every $H_3 \in \mathcal{H}_0$, and every H_4, $\alpha H_1 + (1-\alpha)H_3 \preceq H_4$ if and only if $\alpha H_2 + (1-\alpha)H_3 \preceq H_4$, and $H_4 \preceq \alpha H_1 + (1-\alpha)H_3$ if and only if $H_4 \preceq \alpha H_2 + (1-\alpha)H_3$.*

The next lemma says that two different mixtures of the same two horse lotteries are ranked according to how much probability they give to the better of the two horse lotteries.

Lemma 3.116. *Assume Axioms 1 and 2. Let $H_1 \prec H_2 \in \mathcal{H}_0$. Suppose that*
$$H_3 \sim \alpha H_2 + (1-\alpha)H_1, \quad H_4 \sim \beta H_2 + (1-\beta)H_1.$$
Then $\alpha \leq \beta$ if and only if $H_3 \preceq H_4$.

PROOF. Suppose that $H_3 \preceq H_4$ but $\beta < \alpha$. Then
$$\alpha H_2 + (1-\alpha)H_1 \preceq \beta H_2 + (1-\beta)H_1,$$
by Corollary 3.115. Since $\alpha > \beta$, use Lemma 3.113 to conclude that $H_2 \preceq (\beta/\alpha)H_2 + ([\alpha-\beta]/\alpha)H_1$. Use Lemma 3.113 once again to conclude that $H_2 \preceq H_1$, which is a contradiction.

Next, suppose that $\alpha \leq \beta$ but $H_4 \prec H_3$. Then $\beta H_2 + (1-\beta)H_1 \prec \alpha H_2 + (1-\alpha)H_1$, by Corollary 3.115. A contradiction follows just as before. □

A useful consequence of the first three axioms is what is often called an *Archemedian condition*.[21]

Lemma 3.117 (Archemedian condition). *Assume Axioms 1–3, and assume that $H_1, H_3 \in \mathcal{H}_0$. If $H_1 \prec H_2 \prec H_3$, then there exists a unique $0 < \alpha < 1$ so that*
$$\alpha H_3 + (1-\alpha)H_1 \sim H_2.$$

PROOF. Suppose that $H_1 \prec H_2 \prec H_3$. Let $\aleph_0 = \{\alpha : \alpha H_3 + (1-\alpha)H_1 \preceq H_2\}$, and define $\beta_0 = \sup\{\alpha : \alpha \in \aleph_0\}$. Since $\aleph_0$ contains 0, it is nonempty and β_0 is well defined. Define $H = \beta_0 H_3 + (1-\beta_0)H_1$. For $k = 1, 2, \ldots$, let $\alpha_k \in \aleph_0$ be such that $\lim_{k \to \infty} \alpha_k = \beta_0$ and define $G_k = \alpha_k H_3 + (1-\alpha_k)H_1$. We have, for each r, $G_k(r) \to H(r)$ and, for each k, $G_k \preceq H_2$. By Axiom 3, $H \preceq H_2$. Next, let $\aleph_1 = \{\alpha : H_2 \preceq \alpha H_3 + (1-\alpha)H_1\}$, and define $\beta_1 = \inf\{\alpha : \alpha \in \aleph_1\}$. Since $\aleph_1$ contains 1, it is nonempty and β_1 is well defined. Define $G = \beta_1 H_3 + (1-\beta_1)H_1$. For $k = 1, 2, \ldots$, let $\gamma_k \in \aleph_1$ be such that $\lim_{k \to \infty} \gamma_k = \beta_1$ and define $R_k = \gamma_k H_3 + (1-\gamma_n)H_1$. We have, for each s, $R_k(r) \to G(r)$ and, for each k, $H_2 \preceq R_k$. By Axiom 3, $H_2 \preceq G$. By Lemma 3.116, $\beta_0 \leq \beta_1$. If $\beta_0 < \beta < \beta_1$, then neither $H_2 \preceq \beta H_3 + (1-\beta)H_1$

---

[21] In the proofs of results in the finite case, we do not explicitly use Axiom 3, but rather we only use the Archemedian condition. This fact would allow us to prove a converse to Lemma 3.117 in the finite case.

nor $\beta H_3 + (1-\beta)H_1 \preceq H_2$, which contradicts Axiom 1. It follows that $\beta_0 = \beta_1$ and α is the common value. Clearly, any value other than α is either in $\aleph_0$ or $\aleph_1$ but not in both, so α is unique. □

As an aside, some people prefer to take the Archemedian condition in Lemma 3.117 as an axiom instead of Axiom 3. In the finite case, they are equivalent.

Proposition 3.118. *Assume Axioms 1 and 2 and assume that P and R are finite. If the Archemedian condition from Lemma 3.117 holds, then continuity (Axiom 3) holds.*

Lemma 3.119. *Assume Axioms 1 and 2 and the Archemedian condition of Lemma 3.117. There exists a function $U: \mathcal{H}_0 \to \mathbb{R}$ such that*

$$U(H_1) \leq U(H_2) \text{ if and only if } H_1 \preceq H_2. \tag{3.120}$$

PROOF. If $H_1 \sim H_2$ for all $H_1, H_2 \in \mathcal{H}_0$, then just set $U(H) = 0$ for all H. For the rest of the proof, assume that there exist $H_*, H^* \in \mathcal{H}_0$ such that $H_* \prec H^*$. We will use the Archemedian condition in Lemma 3.117 to help define U. For each $H \in \mathcal{H}$ such that $H_* \preceq H \preceq H^*$, define $U(H)$ equal to the value of α such that $\alpha H^* + (1-\alpha)H_* \sim H$. Note that $U(H_*) = 0$ and $U(H^*) = 1$.[22] For each H such that $H \preceq H_*$, define $U(H)$ equal to $-\alpha/(1-\alpha)$ for that α such that $\alpha H^* + (1-\alpha)H \sim H_*$. For each H such that $H^* \preceq H$, define $U(H)$ equal to $1/\alpha$ for that value of α such that $\alpha H + (1-\alpha)H_* \sim H^*$. Next, we prove that (3.120) is true.

There are six possible arrangements of H_1 and H_2 relative to H_* and H^* (ignoring the permutations of H_1 and H_2 themselves). Lemma 3.116 shows that (3.120) is true if both H_1 and H_2 are between H_* and H^*.[23] It is easy to see that if only one of H_1 and H_2 is between H_* and H^* that (3.120) is true, since one value of U is between 0 and 1 and the other is not. Also, (3.120) is true if $H_i \preceq H_* \prec H^* \preceq H_{3-i}$, for $i = 1$ or 2. The only cases that remain are (i) that in which both H_1 and H_2 are preferred to H^* and (ii) that in which H_* is preferred to both H_1 and H_2. For case (i), we have

$$H^* \sim \frac{1}{U(H_1)} H_1 + \frac{U(H_1) - 1}{U(H_1)} H_*, \tag{3.121}$$

$$H^* \sim \frac{1}{U(H_2)} H_2 + \frac{U(H_2) - 1}{U(H_2)} H_*. \tag{3.122}$$

Lemma 3.116 and (3.122) say that $U(H_1) \leq U(H_2)$ if and only if

$$H^* \preceq \frac{1}{U(H_1)} H_2 + \frac{U(H_1) - 1}{U(H_1)} H_*.$$

---

[22] In the finite case, one can prove that there exist two NM-lotteries H_* and H^* such that $H_* \preceq H \preceq H^*$ for all $H \in \mathcal{H}$. (See Problem 33 on page 212.) For this reason, one can skip to the next paragraph in the finite case.

[23] The proof ends here in the finite case because we can choose H_* and H^* so that $H_* \preceq H \preceq H^*$ for all $H \in \mathcal{H}$, by Problem 33 on page 212.

194   Chapter 3. Decision Theory

This and (3.121) are true if and only if $H_1 \preceq H_2$ by Lemma 3.113. Case (ii) is similar.   □

Lemma 3.123. *Assume Axioms 1 and 2 and the Archemedian condition of Lemma 3.117. The function U constructed in the proof of Lemma 3.119 satisfies (3.109) for all $H_1, H_2 \in \mathcal{H}_0$.*

PROOF. If $H_1 \sim H_2$ for all $H \in \mathcal{H}_0$, the result is trivial. So, assume that there exist $H_* \prec H^* \in \mathcal{H}_0$. There are 10 cases to handle, depending both on how H_1 and H_2 compare to H_* and H^* and on the value of $c = \alpha U(H_1) + (1-\alpha)U(H_2)$. Without loss of generality, assume $H_1 \preceq H_2$, since they are arbitrary.[24]

Case 1. $H_* \preceq H_1, H_2 \preceq H^*$. Since

$$H_1 \sim U(H_1)H_* + [1 - U(H_1)]H^*,$$
$$H_2 \sim U(H_2)H_* + [1 - U(H_2)]H^*,$$

we can use Corollary 3.115 to conclude that

$$\alpha H_1 + (1-\alpha)H_2$$
$$\sim (\alpha U(H_1) + [1-\alpha]U(H_2))H^* + (1 - \alpha U(H_1) - [1-\alpha]U(H_2))H_*,$$

so that (3.109) holds.

Case 2. $H_1 \prec H_* \preceq H_2 \preceq H^*$ and $c \geq 0$. In this case, $c \leq 1$ is clear, and

$$H_2 \sim U(H_2)H^* + [1 - U(H_2)]H_*, \qquad (3.124)$$

$$H_* \sim -\frac{U(H_1)}{1-U(H_1)}H^* + \frac{1}{1-U(H_1)}H_1. \qquad (3.125)$$

Mix H_1 with weight α with both sides of (3.124) to obtain

$$\alpha H_1 + (1-\alpha)H_2$$
$$\sim \alpha H_1 + (1-\alpha)U(H_2)H^* + (1-\alpha)[1 - U(H_2)]H_*$$
$$= \beta\left[\frac{-U(H_1)}{1-U(H_1)}H^* + \frac{1}{1-U(H_1)}H_1\right]$$
$$+ (1-\beta)\left[\frac{\alpha U(H_1) + (1-\alpha)U(H_2)}{1-\alpha+\alpha U(H_1)}H^* + \frac{(1-\alpha)[1-U(H_2)]}{1-\alpha+\alpha U(H_1)}H_*\right],$$

where $\beta = \alpha[1 - U(H_1)]$, which is less than 1 because $c > 0$. Use (3.125) and Lemma 3.113 to see that the last expression is $\sim cH^* + (1-c)H_1$. This implies that (3.109) is true.

Case 3. $H_1 \prec H_* \preceq H_2 \preceq H^*$ and $c < 0$. In this case, (3.124) and (3.125) are still true. This time let $\beta = (\alpha - \alpha U(H_1))/(1 - \alpha U(H_1))$; mix the left-hand side of (3.124) with weight $1 - \beta$ with the right-hand side of (3.125)

---

[24]Only case 1 is needed in the finite case.

3.3. Axiomatic Derivation of Decision Theory

with weight β, and mix the other sides also. The result is

$$\frac{1}{1-\alpha U(H_1)}[\alpha H_1 + (1-\alpha)H_2] + \frac{-\alpha U(H_1)}{1-\alpha U(H_1)}H^*$$
$$\sim \frac{(1-\alpha)U(H_2)}{1-\alpha U(H_1)}H^* + \frac{1-c}{1-\alpha U(H_1)}H_*.$$

Now use Lemma 3.113 to remove the common H^* from both sides (there is more on the left than on the right) to get

$$\frac{1}{1-c}[\alpha H_1 + (1-\alpha)H_2] + \frac{-c}{1-c}H^* \sim H_*.$$

It follows that (3.109) is true.

Case 4. $H_1 \prec H_* \prec H^* \prec H_2$ and $c \in [0,1]$. In this case,

$$H_* \sim \frac{-U(H_1)}{1-U(H_1)}H^* + \frac{1}{1-U(H_1)}H_1, \qquad (3.126)$$

$$H^* \sim \frac{1}{U(H_2)}H_2 + \frac{U(H_2)-1}{U(H_2)}H_*. \qquad (3.127)$$

Let $\beta = \alpha(1-U(H_1))/[\alpha(1-U(H_1))+(1-\alpha)U(H_2)]$, and take the mixture of (3.126) with weight β and (3.127) with weight $1-\beta$. The result is

$$\frac{\alpha(1-U(H_1))}{\alpha(1-U(H_1))+(1-\alpha)U(H_2)}H_* + \frac{(1-\alpha)U(H_2)}{\alpha(1-U(H_1))+(1-\alpha)U(H_2)}H^*$$

$$\sim \frac{-\alpha U(H_1)}{\alpha(1-U(H_1))+(1-\alpha)U(H_2)}H^* + \frac{\alpha}{\alpha(1-U(H_1))+(1-\alpha)U(H_2)}H_1$$

$$+ \frac{(1-\alpha)}{\alpha(1-U(H_1))+(1-\alpha)U(H_2)}H_2 + \frac{(1-\alpha)(U(H_2)-1)}{\alpha(1-U(H_1))+(1-\alpha)U(H_2)}H_*.$$

One can now use Lemma 3.113 to remove a common component consisting of the two terms involving H^* and H_* on the right-hand side with weight

$$\frac{-\alpha U(H_1) + (1-\alpha)[U(H_2)-1]}{\alpha(1-U(H_1))+(1-\alpha)U(H_2)}.$$

The result says that (3.109) is true.

Case 5. $H_1 \prec H_* \prec H^* \prec H_2$ and $c > 1$. In this case, (3.126) and (3.127) are still true. This time, let $\beta = \alpha(1-U(H_1))/[\alpha(1-U(H_1))+(1-\alpha)U(H_2)]$, and mix (3.126) with weight β with (3.127) with weight $1-\beta$ to get

$$\gamma H^* + (1-\gamma)L \sim \gamma\left(\frac{1}{c}[\alpha H_1 + (1-\alpha)H_2] + \frac{c-1}{c}H_*\right) + (1-\gamma)L,$$

where
$$\gamma = \frac{c}{\alpha(1 - U(H_1)) + (1 - \alpha)U(H_2)},$$
$$L = \frac{-\alpha U(H_1)}{\alpha(1 - 2U(H_1))}H^* + \frac{-\alpha[U(H_1) - 1]}{\alpha(1 - 2U(H_1))}H_*.$$

Use Lemma 3.113 to remove the common component of L from both sides and the result is (3.109).

Case 6. $H_1, H_2 \prec H_*$. In this case, $c < 0$ is clear and we have (3.126) together with
$$H_* \sim \frac{-U(H_2)}{1 - U(H_2)}H^* + \frac{1}{1 - U(H_2)}H_2. \qquad (3.128)$$

Let $\beta = \alpha(1 - U(H_1))/[\alpha(1 - U(H_1)) + (1 - \alpha)(1 - U(H_2))]$, and mix (3.126) with weight β with (3.128) with weight $1 - \beta$ to get
$$H_* \sim \frac{-c}{1-c}H^* + \frac{1}{1-c}[\alpha H_1 + (1 - \alpha)H_2].$$

This implies that (3.109) is true.

Case 7. $H^* \prec H_1, H_2$. This is analogous to case 6.

Case 8. $H_1 \prec H_* \prec H^* \prec H_2$ and $c < 0$. This is analogous to case 5.

Case 9. $H_* \preceq H_1 \preceq H^* \prec H_2$ and $c > 1$. This is analogous to case 3.

Case 10. $H_* \preceq H_1 \preceq H^* \prec H_2$ and $c \leq 1$. This is analogous to case 2. □

Now, we can prove that U is bounded on $\mathcal{H}_0$.[25]

Lemma 3.129. *Assume Axioms 1–3. Then U is bounded on $\mathcal{H}_0$.*

PROOF. If $H_1 \sim H_2$ for all $H_1, H_2 \in \mathcal{H}_0$, the result is trivial, so assume that there exist $H_* \prec H^* \in \mathcal{H}_0$. Without loss of generality we can assume that $U(H_*) = 0$ and $U(H^*) = 1$ (otherwise, just replace U by $[U - U(H_*)]/[U(H^*) - U(H_*)]$ and the preferences and boundedness are not changed). Suppose, to the contrary, that $U(\cdot)$ is unbounded above. (A similar construction works if U is unbounded below.) Let $\{H_n\}_{n=1}^{\infty}$ be such that $U(H_n) > n$. Let $H'_n = (1 - 1/n)H_* + (1/n)H_n$ for each n. Then $H^* \prec H'_n$ for all n because $U(H^*) = 1$ and $U(H'_n) > 1$ for all n, but $H_n \to H_*$ and $H_* \prec H^*$. This contradicts Axiom 3. □

If $\mathcal{H}_0 \subseteq \mathcal{H}$ contains H_* and H^* with $H_* \prec H^*$, define [26]
$$\beta_1(\mathcal{H}_0) = \sup_{H \in \mathcal{H}_0} U(H), \qquad \beta_0(\mathcal{H}_0) = \inf_{H \in \mathcal{H}_0} U(H).$$

The following lemma is useful in allowing us to find NM-lotteries with arbitrary utilities.[27]

---

[25] The conclusions of Lemma 3.129 are obvious in the finite case.

[26] In the finite case, we can arrange for $\beta_1(\mathcal{H}) = 1$ and $\beta_0(\mathcal{H}) = 0$.

[27] In the finite case, Lemma 3.130 follows trivially from the fact that there exist NM-lotteries L^* and L_* that achieve the maximum (1) and minimum (0) values of U and the fact that $U(\alpha L^* + (1 - \alpha)L_*) = \alpha$ for every $\alpha \in [0, 1]$.

3.3. Axiomatic Derivation of Decision Theory   197

Lemma 3.130. *Assume Axioms 1–3. For each $\beta \in (\beta_0(\mathcal{H}_0), \beta_1(\mathcal{H}_0))$, there exists an NM-lottery L with $U(L) = \beta$.*

PROOF. If $H_1 \sim H_2$ for all $H_1, H_2 \in \mathcal{H}_0$, then $\beta_0(\mathcal{H}_0) = \beta_1(\mathcal{H}_0)$, and the result is vacuous, so assume that there exist $H_* \prec H^* \in \mathcal{H}_0$ with $U(H_*) = 0$ and $U(H^*) = 1$. Assume, to the contrary, that $a_0 = \inf_{L \in \mathcal{L}} U(L) > \beta_0(\mathcal{H}_0)$.[28] We know that $a_0 \leq 0$, since $U(H_*) = 0$. Let H be a horse lottery such that $U(H) < a_0$, which must exist since $\beta_0(\mathcal{H}_0)$ is the infimum of all utilities of horse lotteries. Let $\epsilon = a_0 - U(H)$, so that $U(H) = a_0 - \epsilon < 0$. Let $\alpha = a_0/[2(a_0 - \epsilon)]$, which is easily seen to be between 0 and 1/2. Let L be an NM-lottery such that $U(L) = a_0(1/2 + \alpha)/2$, which is in the open interval $(a_0/2, \alpha a_0)$. Define $H' = \alpha H + (1 - \alpha) H_*$. This means that $U(H') = a_0/2 < U(L)$, hence $H' \prec L$. But $H'(r) = \alpha H(r) + (1 - \alpha) H_*$. We have assumed that $U(H(r)) \geq a_0$ for all r, since $H(r) \in \mathcal{L}$. So $U(H'(r)) \geq \alpha a_0 > U(L)$. This implies that $L \preceq H'(r)$ for all r. Axiom 5 implies $L \preceq H'$, a contradiction. A similar contradiction holds if we assume that $\sup_L U(L) < \beta_1(\mathcal{H}_0)$. □

We are now in position to prove that $\mathcal{H}$ is itself convex.

Lemma 3.131.[29] *Assume Axioms 1–3. Let $\mathcal{H}_0$ be the set of all constant horse lotteries.*

- *For each horse lottery $H \in \mathcal{H}$, the function $g : R \to \mathbb{R}$ defined by $g(r) = U(H(r))$ is measurable.*

- *If $H_1, H_2 \in \mathcal{H}$ and $0 \leq \alpha \leq 1$, then $\alpha H_1 + (1 - \alpha) H_2 \in \mathcal{H}$.*

PROOF. For the first part, let $H \in \mathcal{H}$ and let $g(r) = U(H(r))$. We know that $\beta_0(\mathcal{H}_0) \leq g(r) \leq \beta_1(\mathcal{H}_0)$ for all r. To prove that g is measurable, we need to show that for every $c \in (\beta_0(\mathcal{H}_0), \beta_1(\mathcal{H}_0))$, $\{r : g(r) \leq c\} \in \mathcal{A}_1$. For each such c, let L_c be an NM-lottery with $U(L_c) = c$, as guaranteed by Lemma 3.130. Then

$$\{r : g(r) \leq c\} = \{r : U(H(r)) \leq U(L_c)\} = \{r : H(r) \preceq L_c\} \in \mathcal{A}_1,$$

where the second equality follows from Lemma 3.119, and the inclusion follows from the definition of a horse lottery.

For the second part, let $H_1, H_2 \in \mathcal{H}$ and $0 \leq \alpha \leq 1$. We need to prove that for all $L \in \mathcal{L}$, $\{r : \alpha H_1(r) + (1 - \alpha) H_2(r) \preceq L\} \in \mathcal{A}_1$. Let $L \in \mathcal{L}$. Lemma 3.123 says that

$$\{r : \alpha H_1(r) + (1 - \alpha) H_2(r) \preceq L\} \quad (3.132)$$
$$= \{r : \alpha U(H_1(r)) + (1 - \alpha) U(H_2(r)) \leq U(L)\}.$$

---

[28]This can only happen if $\beta_0(\mathcal{H}_0) < 0$. If $\beta_0(\mathcal{H}_0) = 0$, then $a_0 = \beta_0(\mathcal{H}_0)$ must occur.

[29]The conclusions of Lemma 3.131 are already known in the finite case.

198   Chapter 3. Decision Theory

But the first part of this lemma shows that both $U(H_1(\cdot))$ and $U(H_2(\cdot))$ are measurable functions. Hence the convex combination is measurable. It follows that the set on the right-hand side of (3.132) is in $\mathcal{A}_1$. □

From now on, so long as we assume Axioms 1–3, we can assume that $\mathcal{H}$ is closed under convex combination.

Lemma 3.133. *Assume Axiom 5 and that preference is nondegenerate. Then there exist two NM-lotteries L_* and L^* such that $L_* \prec L^*$.*

PROOF. Since the preference is nondegenerate, there exist horse lotteries $H_* \prec H^*$. If, to the contrary, $H_*(r) \sim H^*(r)$ for all r, then Axiom 5 says $H^* \preceq H_*$, a contradiction. □

Lemma 3.134. *Assume Axioms 1–3. Let H_* and H^* be horse lotteries such that $U(H_*) = 0$ and $U(H^*) = 1$. For each $B \in \mathcal{A}_1$, define H_B by*

$$H_B(r) = \begin{cases} H^*(r) & \text{if } r \in B, \\ H_*(r) & \text{if not.} \end{cases} \quad (3.135)$$

Let $Q(B) = U(H_B)$. Suppose that $H_* \preceq H_B$ for all B. Then Q is a probability.

PROOF. It is easy to see that H_B is a horse lottery. It follows from $H_* \preceq H_B$ that $Q(B) \geq 0$. It is easy to see that $Q(\emptyset) = 0$ and $Q(R) = 1$. If C and D are disjoint, define $H = \frac{1}{2}H_C + \frac{1}{2}H_D$, which equals $\frac{1}{2}H_{C \cup D} + \frac{1}{2}H_*$. According to (3.109),

$$\begin{aligned}
\frac{1}{2}[Q(C) + Q(D)] &= \frac{1}{2}[U(H_C) + U(H_D)] = U\left(\frac{1}{2}H_C + \frac{1}{2}H_D\right) \\
&= U\left(\frac{1}{2}H_{C \cup D} + \frac{1}{2}H_*\right) = \frac{1}{2}[U(H_{C \cup D}) + U(H_*)] \\
&= \frac{1}{2}Q(C \cup D),
\end{aligned}$$

from which it follows that $Q(C \cup D) = Q(C) + Q(D)$.[30] Next, let $\{A_n\}_{n=1}^{\infty}$ be mutually disjoint subsets of R, and let

$$A = \bigcup_{i=1}^{\infty} A_i, \quad B_n = \bigcup_{i=1}^{n} A_i.$$

For every n, we have

$$\frac{1}{2}H_{B_n} + \frac{1}{2}H_* \preceq \frac{1}{2}H_A + \frac{1}{2}H_* \preceq H^*,$$

---

[30]The proof ends here in the finite case, since there do not exist infinitely many disjoint subsets of R. Note that in the finite case Axiom 3 is not used, only the Archemedian condition of Lemma 3.117 is used.

hence, we can choose $\alpha_n \in [0,1]$ so that

$$H_n = (1-\alpha_n)\left[\frac{1}{2}H_{B_n} + \frac{1}{2}H_*\right] + \alpha_n H^* \sim \frac{1}{2}H_* + \frac{1}{2}H_A.$$

Since we just showed that Q is finitely additive, we have

$$\begin{aligned}
\frac{1}{2}Q(A) &= \frac{1}{2}U(H_A) = \frac{1}{2}U(H_*) + \frac{1}{2}U(H_A) = U(H_n) \\
&= \frac{1}{2}(1-\alpha_n)U(H_{B_n}) + \frac{1}{2}(1-\alpha_n)U(H_*) + \alpha_n U(H^*) \\
&= \frac{1}{2}(1-\alpha_n)\sum_{i=1}^{n} Q(B_n) + \alpha_n.
\end{aligned}$$

It follows that for all n, $Q(A) = (1-\alpha_n)\sum_{i=1}^{n} Q(B_n) + 2\alpha_n$. If we can show that $\lim_{n\to\infty} \alpha_n = 0$, then we have that Q is countably additive. Let $\{\alpha_{n_k}\}_{k=1}^{\infty}$ be a convergent subsequence of $\{\alpha_n\}_{n=1}^{\infty}$ with limit α. Then

$$H_{n_k}(r) \to (1-\alpha)\left[\frac{1}{2}H_A(r) + \frac{1}{2}H_*(r)\right] + \alpha H^*(r) = H(r).$$

It follows from Axiom 3 (with $H' = H'' = \frac{1}{2}H_* + \frac{1}{2}H_A$) that $H \sim \frac{1}{2}H_* + \frac{1}{2}H_A$. But $H = (1-\alpha)[\frac{1}{2}H_* + \frac{1}{2}H_A] + \alpha H^*$. It follows from Axiom 2 that either $\alpha = 0$ or $H^* \sim \frac{1}{2}H_* + \frac{1}{2}H_A$. Since this latter is clearly false, it must be that $\alpha = 0$ and $\alpha_n \to 0$. □

Lemma 3.136. *Assume the conditions of Theorem 3.108. In Lemma 3.134, let $H_* = L_*$ and $H^* = L^*$ from Lemma 3.133. Then $H_* \preceq H_B$ for all $B \in \mathcal{A}_1$. For all $B \in \mathcal{A}_1$, $Q(B) = 0$ if and only if B is null.*

PROOF. The fact that $Q(B) = 0$ if and only if B is null follows easily from Axiom 4 and is left to the reader (as Problem 37 on page 213). By Axiom 5, $L_* \preceq H_B \preceq L^*$.[31] □

Lemma 3.137. *Assume the conditions of Theorem 3.108. Let H be a horse lottery that takes on only finitely many different NM-lotteries. Then*

$$U(H) = \int U(H(r))dQ(r).$$

PROOF. Let $L'_1, \ldots, L'_n$ be the different NM-lotteries that H takes on. Let

$$\begin{aligned}
b_1 &= \max\{1, U(L'_1), \ldots, U(L'_n)\} \\
b_0 &= \min\{0, U(L'_1), \ldots, U(L'_n)\}.
\end{aligned}$$

---

[31] In the finite case, the fact that $L_* \preceq H_B \preceq L^*$ was already known without appeal to Axiom 5.

Define $c_1 = [b_1(1-b_0)]^{-1}$ and $c_2 = -b_0/(1-b_0)$. Clearly, $c_1 > 0, c_2 \geq 0$, and $c_1 + c_2 \leq 1$. Let $H'' = c_1 H + c_2 L^* + (1 - c_1 - c_2) L_*$.[32] Then

$$U(H'') = c_1 U(H) + c_2,$$
$$U(H''(r)) = c_1 U(H(r)) + c_2, \text{ for all } r. \quad (3.138)$$

Also, $0 \leq U(H''(r)) \leq 1$ follows from (3.109) and simple algebra. Since $c_1 \neq 0$ and

$$\int U(H''(r)) dQ(r) = c_1 \int U(H(r)) dQ(r) + c_2,$$

it is sufficient to prove the result for H''. Since $H''(r)$ is the same mixture of $H(r)$ and L_* and L^* for all r, H'' takes on only finitely many different NM-lotteries also. Let $H''(r) = L_i$ for $r \in B_i$ for $i = 1, \ldots, n$, where the $B_i \in \mathcal{A}_1$ form a finite partition of R. For each i, define H_i by

$$H_i(r) = \begin{cases} L_i & \text{if } r \in B_i, \\ L_* & \text{if not.} \end{cases}$$

It is easy to see that H_i is a horse lottery and that

$$\frac{1}{n} H'' + \frac{n-1}{n} L_* = \frac{1}{n} H_1 + \cdots + \frac{1}{n} H_n.$$

Hence, $U(H'') = \sum_{i=1}^n U(H_i)$. Since

$$\int U(H''(r)) dQ(r) = \sum_{i=1}^n U(L_i) Q(B_i),$$

we complete the proof by showing that $U(H_i) = U(L_i) Q(B_i)$ for each i.

Since $0 \leq U(L_i) \leq 1$, we know that $L_i \sim U(L_i) L^* + [1 - U(L_i)] L_*$. By Axiom 4, we can substitute the right-hand side of this expression for L_i in the definition of H_i and conclude that $H_i \sim H_i'$, where

$$H_i'(r) = \begin{cases} U(L_i) L^* + [1 - U(L_i)] L_* & \text{if } r \in B_i, \\ L_* & \text{if not.} \end{cases}$$

Hence, $U(H_i) = U(H_i')$. For each i, define the horse lottery H_{B_i} as in Lemma 3.134. So, $H_i' = U(L_i) H_{B_i} + [1 - U(L_i)] L_*$. It follows that $U(H_i') = U(L_i) Q(B_i)$, as desired. □

Lemma 3.139.[33] *Assume the conditions of Theorem 3.108. Let H be an arbitrary horse lottery. Then $U(H) = \int U(H(r)) dQ(r)$.*

---

[32] In the finite case, $c_1 = 1$, $c_2 = 0$, and $H'' = H$.
[33] This lemma is not needed in the finite case.

3.3. Axiomatic Derivation of Decision Theory

PROOF. First, suppose that $U(H(r)) \geq 0$ for all r. Let $H'' = \frac{1}{2}H + \frac{1}{2}L^*$. Since $U(H'') = \frac{1}{2}U(H) + \frac{1}{2}$ and $\int U(H''(r))dQ(r) = \frac{1}{2}\int U(H(r))dQ(r) + \frac{1}{2}$, it suffices to prove the result for H''. Let $b_1 = \sup_r U(H''(r))$. It follows from Lemma 3.130 that for all $x \leq b_1$ there exists an NM-lottery L with $U(L) = x$.

For each n and $k = 0, 1, \ldots, n2^n$, define

$$B_{n,k} = \left\{ r : \frac{(k-1)}{2^n} < U(H''(r)) \leq \frac{k}{2^n} \right\}.$$

Define the horse lotteries H_n for each n by

$$H_n(r) = L_{n,k} \text{ for all } r \in B_{n,k}, \ k = 0, 1, \ldots, n2^n,$$

where $L_{n,k}$ are chosen (see Lemma 3.130) so that $U(L_{n,k}) = \min\{b_1, (k-1)/2^n\}$ for $k \geq 1$ and $L_{n,0} = L_*$. It follows from Axiom 5 that $L_* \preceq L_{n,k} \preceq H''(r)$ for all n, k and all $r \in B_{n,k}$, hence $0 \leq U(H_n(r)) \leq U(H''(r))$ for all r, n. Since $U(H_n(r))$ converges to $U(H''(r))$ for all r, the monotone convergence theorem A.52 implies

$$\lim_{n \to \infty} \int U(H_n(r))dQ(r) = \int U(H''(r))dQ(r). \tag{3.140}$$

Lemma 3.137 says that the integrals on the left-hand side of (3.140) are $U(H_n)$. Since Axiom 5 says that $U(H_n) \leq U(H'')$ for all n,

$$\int U(H''(r))dQ(r) \leq U(H'').$$

Since U is bounded above, we can choose $M_{n,k}$ to be NM-lotteries such that $U(M_{n,k}) = \min\{b_1, k/2^n\}$. Just as above, let $H^n(r) = M_{n,k}$ for $r \in B_{n,k}$, so that $U(H''(r)) \leq U(H^n(r)) \leq b_1$ for all r, n. The dominated convergence theorem A.57 says that

$$U(H'') \leq \lim_{n \to \infty} \int U(H^n(r))dQ(r) = \int U(H''(r))dQ(r).$$

It follows that $\int U(H''(r))dQ(r) \geq U(H'')$, and the result is proven when $U(H(r)) \geq 0$ for all r.

A similar argument works if $U(H(r)) \leq 0$ for all r. For arbitrary H, let $H^+(r) = H(r)$ if $U(H(r)) \geq 0$ and $H^+(r) = L_*$ otherwise. Let $H^-(r) = H(r)$ if $U(H(r)) < 0$ and $H^-(r) = L_*$ otherwise. Then $\frac{1}{2}H^+ + \frac{1}{2}H^- = \frac{1}{2}H + \frac{1}{2}L_*$, and $U(H) = U(H^+) + U(H^-)$. The result now follows. □

The last two lemmas prove the essential uniqueness of U and the uniqueness of Q.

Lemma 3.141. *Assume the conditions of Theorem 3.108. The utility U from Lemma 3.119 is unique up to positive affine transformation.*

PROOF. Let U_1' and U_2' be two utilities. If preference is degenerate, then both U_1' and U_2' are constant and the result is trivial. So, suppose that there exist H_* and H^* with $H_* \prec H^*$. For $i = 1, 2$, define $U_i(H) = [U_i'(H) - U_i'(H_*)]/[U_i'(H^*) - U_i'(H_*)]$. This makes $U_i(H_*) = 0$ and $U_i(H^*) = 1$ for $i = 1, 2$ without affecting the other properties of each U_i. Now, suppose that there exists H such that $U_1(H) \neq U_2(H)$. Without loss of generality, assume $U_1(H) < U_2(H)$. There are five cases to consider.[34]

Case 1. $0 \leq U_1(H) < U_2(H) \leq 1$. Let $U_1(H) < \alpha < U_2(H)$ and let $H' = \alpha H^* + (1-\alpha)H_*$. Then $U_i(H') = \alpha$ for $i = 1, 2$. Now $U_1(H) > U_1(H')$, meaning $H \prec H'$, and $U_2(H') < U_2(H)$, meaning $H' \prec H$, a contradiction.

Case 2. $U_1(H) < U_2(H) < 0$. Let $U_1(H) < c < U_2(H)$, and define $H' = c/(c-1)H^* + (-1)/(c-1)H$. Then $U_1(H') < 0$, so $H' \prec H_*$, and $U_2(H') > 0$, so $H_* \prec H'$, a contradiction.

Case 3. $1 < U_1(H) < U_2(H)$. Let $U_1(H) < c < U_2(H)$, and define $H' = (1/c)H + (c-1)/cH_*$. Then $U_1(H') < 1$, so $H' \prec H^*$, and $U_2(H') > 1$, so $H^* \prec H'$, a contradiction.

Case 4. $U_1(H) < 0 < U_2(H)$. Then $H \prec H_* \prec H$, a contradiction.

Case 5. $U_1(H) < 1 < U_2(H)$. Then $H \prec H^* \prec H$, a contradiction.

Finally, note that if $U_1 = U_2$, then U_1' and U_2' are positive affine transformations of each other. □

Lemma 3.142. *Under the conditions of Theorem 3.108, the probability Q is unique.*

PROOF. Lemma 3.141 shows that the utility is unique up to positive affine transformation, so suppose that there are two different probabilities Q_1 and Q_2 such that for both $i = 1$ and $i = 2$, $H_1 \preceq H_2$ if and only if $\int U(H_1(r))dQ_i(r) \leq \int U(H_2(r))dQ_i(r)$. Pick two NM-Lotteries L_* and L^* such that $L_* \prec L^*$. Let B be an arbitrary subset of R and define H_B as in (3.135). It follows that $U(H_B) = Q_i(B)$ for $i = 1, 2$, so that $Q_1(B) = Q_2(B)$. □

Since NM-lotteries are concentrated on only finitely many prizes, the following is a simple consequence of (3.109).

Corollary 3.143. *Under the conditions of Theorem 3.108, if $L = \alpha_1 p_1 + \cdots + \alpha_k p_k$, then $U(L) = \sum_{i=1}^{k} \alpha_i U(p_i)$.*

The proof of Theorem 3.110 requires a lemma first.

Lemma 3.144. *Under the conditions of Theorem 3.110, if L_1, L_2 are NM-lotteries such that $L_1 \preceq (\prec) L_2$, then there exists a set B such that $X^{-1}(B)$ is null and, for all $x \notin B$, $L_1 \preceq_x (\prec_x) L_2$.*

PROOF. Let $L_1 \preceq L_2$, and let $B = \{x : L_2 \prec_x L_1\}$. Define $H_1(r) = L_2$ for all $r \in X^{-1}(B)$ and $H_1(r) = L_1$ for all $r \notin X^{-1}(B)$. Then Axiom 6 says

---

[34] Only case 1 is needed in the finite case.

3.3. Axiomatic Derivation of Decision Theory

that $H_1 \preceq L_1$, but Axiom 4 says that $L_1 \preceq H_1$ if $X^{-1}(B)$ is nonnull. It follows that $X^{-1}(B)$ must be null. A similar proof works if $L_1 \prec L_2$. □

Before giving the proof of Theorem 3.110, we give a brief outline. We use Theorem 3.108 to represent conditional preference by expected utility separately for each value of x. We then use Lemma 3.144 to show that the utility function in the conditional preference representation must equal the utility function for unconditional preference except on a null set. We prove that the probability measure for the conditional preference representation must equal conditional probability calculated from the unconditional preference by showing that if it were not, we could construct a pair of horse lotteries that are conditionally ordered one way for all x, but that are marginally ordered the opposite way, contradicting Axiom 6.

PROOF OF THEOREM 3.110. Let L_* and L^* be as in Lemma 3.133. According to Theorem 3.108, since $\preceq_x$ satisfies Axioms 1–5 for each x, there is, for each x, a probability P_x on $(R, \mathcal{A}_1)$ and a utility U_x such that $H_1 \preceq_x H_2$ if and only if $\int U_x(H_1(r)) dP_x(r) \leq \int U_x(H_2(r)) dP_x(r)$. Let Q_X denote the distribution of X induced from Q. Lemma 3.144 says that each pair of NM-lotteries is ranked the same by $\preceq_x$ except possibly for x in a set with null inverse image. Since Lemma 3.136 says that a set C is null if and only if $Q(C) = 0$, we can assume that there is B_0 such that $Q(X^{-1}(B_0)) = 0$ and $U_x(L_*) < U_x(L^*)$, for all $x \notin B_0$. We can certainly assume that $U_x(L_*) = 0 = U(L_*)$ and $U_x(L^*) = 1 = U(L^*)$ for all $x \notin B_0$. Let B_- be the set of all x such that there exists L_x with $U_x(L_x) < U(L_x)$, and let B_+ be the set of all x such that there exists L_x with $U_x(L_x) > U(L_x)$. We will show next that for each $x \in B_+ \cup B_-$, we could choose L_x so that $U(L_x) = \frac{1}{2}$. For each $x \in B_+ \cup B_-$, let

$$L'_x = \frac{1}{b_1(1-b_0)} L_x + \frac{b_1 - 1}{b_1(1-b_0)} L_* + \frac{-b_0}{1-b_0} L^*,$$

where $b_0 = \min\{0, U(L_x)\}$ and $b_1 = \max\{1, U(L_x)\}$. Then $U_x(L'_x) \neq U(L'_x)$, but now $0 \leq U(L'_x) \leq 1$. By mixing L'_x with either L_* or L^* to create L''_x, we can have $U(L''_x) = \frac{1}{2}$ and either $U_x(L''_x) > \frac{1}{2}$ or $U(L''_x) < \frac{1}{2}$. That is, we can assume that $U(L_x) = \frac{1}{2}$. Let $L^{1/2} = \frac{1}{2} L_* + \frac{1}{2} L^*$. Define horse lotteries H_+ and H_- by

$$H_+(r) = \begin{cases} L_x & \text{if } X(r) = x \text{ and } x \in B_+, \\ L^{1/2} & \text{otherwise}, \end{cases}$$

$$H_-(r) = \begin{cases} L_x & \text{if } X(r) = x \text{ and } x \in B_-, \\ L^{1/2} & \text{otherwise}. \end{cases}$$

Since $\{r : H_+(r) \preceq L\}$ is either R or $\emptyset$ depending on whether or not $U(L) = \frac{1}{2}$, H_+ is a horse lottery, and similarly for H_-. By construction $H_- \preceq_x L^{1/2}$ for all x and $H_- \prec_x L^{1/2}$ for all $x \in B_-$. Also, by the measurability condition in the definition of conditional preference, $B_- = \{x : L^{1/2} \preceq_x H_-\}^C$, so $B_- \in \mathcal{B}$. It follows from Axiom 6 that $X^{-1}(B_-)$

is null. By a similar argument, we can show that $X^{-1}(B_+)$ is null. Let B' equal $B_0 \cup B_- \cup B_+$. Then $X^{-1}(B')$ is null and, for all $x \notin B'$, $U_x(L) = U(L)$ for all $L \in \mathcal{L}$.

Next,[35] we prove that $P_x(A)$ is a measurable function of x for all $A \in \mathcal{A}_1$. For each $A \in \mathcal{A}_1$, let $H_A(r) = L^*$ if $r \in A$ and $H_A(r) = L_*$ if $r \notin A$. For each $c \in [0, 1]$,

$$\{x : P_x(A) \leq c\} = \{x : H_A \preceq_x cL^* + (1-c)L_*\} \in \mathcal{B}$$

follows from the measurability assumption on conditional preference.

Finally, we prove that $Q(\cdot|x)$ and $P_x(\cdot)$ agree almost surely. Let $D = \{x : P_x(\cdot) \neq Q(\cdot|x)\}$ and $B = B' \cup D$. If we can prove that $D \in \mathcal{B}$ and $Q_X(D) = 0$, the proof is complete.[36] Since $(R, \mathcal{A}_1)$ is a Borel space, $\mathcal{A}_1$ is countably generated (Proposition B.43). Let $\{A_n\}_{n=1}^{\infty}$ generate $\mathcal{A}_1$. Then

$$D = \bigcup_{n=1}^{\infty} \left(\{x : P_x(A_n) < Q(A_n|x)\} \cup \{x : P_x(A_n) > Q(A_n|x)\} \right). \quad (3.145)$$

Since both $P_x(A_n)$ and $Q(A_n|x)$ are measurable functions, each of the sets in the union is in $\mathcal{B}$, so $D \in \mathcal{B}$.

If $Q_X(D) > 0$, then one of the sets in the union (3.145) must have strictly positive Q_X measure. Let $D' = \{x : P_x(A_1) < Q(A_1|x)\}$, and suppose that $Q_X(D') > 0$. For each rational $q \in [0, 1]$, let $D_q = \{x : P_x(A_1) < q < Q(A_1|x)\}$. Then D' is the union of all the D_q. Since this union is countable, there exists q such that $Q_X(D_q) > 0$. Define $H_1(r) = L^*$ for $r \in A_1 \cap X^{-1}(D_q)$ and $H_1(r) = L_*$ otherwise. Also, define $H_2(r) = L^*$ for $r \in A_1$ and $H_2(r) = L_*$ otherwise. Then $U_x(H_1) = U_x(H_2)$ according to the definition of conditional preference because $H_1(r) = H_2(r)$ for all $r \in X^{-1}(D_q)$. But $P_x(A_1) = U_x(H_1)$ by the uniqueness of probability and Lemma 3.134. Define $H_3(r) = qL^* + (1-q)L_*$ for all $r \in X^{-1}(D_q)$ and $H_3(r) = L_*$ otherwise. Then the definition of conditional preference implies that $U_x(H_3) = U_x(qL^* + (1-q)L_*) = q$. Since $U_x(H_3) = q > U_x(H_1)$ for all $x \in D_q$, we have $H_1 \prec_x H_3$ for all $x \in D_q$. But $H_1 \preceq_y H_3$ for all $y \notin D_q$, since $H_1(r) = H_3(r)$ for all $r \notin X^{-1}(D_q)$. It follows from Axiom 6 that $H_1 \prec H_3$. Now, note the following contradiction:

$$U(H_3) = qQ_X(D_q) < \int_{D_q} Q(A_1|x) dQ_X(x) = Q(\{X \in D_q\} \cap A_1) = U(H_1),$$

where the first and last equalities follow from Lemma 3.137, the inequality follows from the definition of D_q, and the other equality follows from the definition of conditional probability. □

---

[35]This paragraph is not needed in the finite case.
[36]Since $D \in \mathcal{B}$ is obvious in the finite case, the rest of this paragraph is not needed in the finite case.

3.3.6 State-Dependent Utility*

We mentioned earlier that Axiom 4 may not be reasonable to assume. It may be the case that when the state of Nature changes, the relative values of various prizes also change. For example, if the states of Nature involve different exchange rates between two currencies, then the relative values of fixed amounts of the two currencies will change according to the state of Nature. For this reason, we prove a theorem that does not assume Axiom 4. If Axiom 4 fails, then Axiom 5 may not even be desirable, as the next example shows.

Example 3.146. In this example, we will have the relative values of the prizes change drastically from one state to the next. Let $R = \{r_1, r_2\}$ and $P = \{p_1, p_2\}$. Let $U_i(p_i) = 1$ and $U_i(p_{3-i}) = 0$ for $i = 1, 2$. For NM-lottery $L = \alpha p_1 + (1-\alpha)p_2$, define $U_i(L) = \alpha U_i(p_1) + (1-\alpha)U_i(p_2)$. For horse lottery $H = (L_1, L_2)$, define $U(H) = 0.4 U_1(L_1) + 0.6 U_2(L_2)$. Consider the following two horse lotteries $H_1 = (L_{1,1}, L_{1,2})$ and $H_2 = (L_{2,1}, L_{2,2})$, where

$$L_{1,1} = p_2, \qquad L_{1,2} = p_2,$$
$$L_{2,1} = p_1, \qquad L_{2,2} = \frac{1}{2}p_1 + \frac{1}{2}p_2.$$

One can easily calculate $U(L_{1,1}) = U(L_{1,2}) = 0.6$, while $U(L_{2,1}) = 0.4$ and $U(L_{2,2}) = \frac{1}{2}$. So $H_2(r_i) \prec H_1(r_i)$ for $i = 1, 2$, but $U(H_1) = 0.6$ and $U(H_2) = 0.7$, thus $H_1 \prec H_2$. Even though each of the NM-lotteries awarded by H_1 is marginally preferred to the corresponding NM-lottery awarded by H_2, $U_1(H_2(r_1))$ is sufficiently higher than $U_1(H_1(r_1))$ to make up for the fact that $U_2(H_2(r_2))$ is a little lower than $U_2(H_1(r_2))$.

The functions U_i in Example 3.146 are called a *state-dependent utility*. Since we will have to abandon Axiom 5 (at least in its current form) in order to abandon Axiom 4, and since some version of dominance is essential for the infinite case, we will only deal with the case in which R and P are finite in this section.[37]

Theorem 3.147. *Assume Axioms 1 and 2 and the Archemedian condition of Lemma 3.117. Assume that preference is nondegenerate. Then, there exist a probability $Q = (q_1, \ldots, q_n)$ over $R = \{r_1, \ldots, r_n\}$ and a state-dependent utility function $(U_1, \ldots, U_n)$ such that for every $H_1 = (L_{1,1}, \ldots, L_{1,n})$ and $H_2 = (L_{2,1}, \ldots, L_{2,n})$, $H_1 \preceq H_2$ if and only if*

$$\sum_{i=1}^{n} U_i(L_{1,i})q_i \leq \sum_{i=1}^{n} U_i(L_{2,i})q_i.$$

---

*This section may be skipped without interrupting the flow of ideas.

[37] One possible approach to dealing with the infinite case would be to assume the existence of a conditional preference relation $\{\preceq_r : r \in R\}$ that satisfied Axiom 6. The type of dominance that we need in the state-dependent case is built into Axiom 6.

The U_i functions are unique up to positive affine transformation (one for each i). The only property of Q determined by the preferences is that non-null states must have positive probability.

The reader will note that Theorem 3.147 makes no claim of uniqueness for Q. It is easy to see why not. Suppose that $(q_1, \ldots, q_n)$ is a probability over the states and $(U_1, \ldots, U_n)$ is a state-dependent utility. Let $(t_1, \ldots, t_n)$ be another probability such that $t_i = 0$ if and only if $q_i = 0$. For each i such that $t_i > 0$, define $V_i = q_i U_i / t_i$. If $t_i = 0$, set $V_i = U_i$. Then

$$\sum_{i=1}^{n} q_i U_i(L_i) = \sum_{i=1}^{n} r_i V_i(L_i)$$

for all $(L_1, \ldots, L_n)$. If $(q_1, \ldots, q_n)$ and U are as guaranteed by Theorem 3.108 and $(t_1, \ldots, t_n)$ is as above, then $V_i = q_i U / t_i$ will satisfy

$$\sum_{i=1}^{n} q_i U(L_i) = \sum_{i=1}^{n} t_i V_i(L_i).$$

This same construction can be applied whether or not Axiom 4 holds. What Axiom 4 achieves is the ability to identify a unique probability and state-independent utility. It does not preclude the existence of alternative state-dependent representations of preference.

The proof of Theorem 3.147 resembles those parts of the proof of Theorem 3.108 that were relevant for the finite case. The first thing we do is define the state-dependent utility by means of called-off comparisons. Then we define a particular Q that makes $U(H) = \sum_{i=1}^{n} q_i U_i(H(r_i))$. We need a lemma first.

Lemma 3.148. *Assume Axioms 1 and 2. For each state r_j, each pair (L_1, L_2) of NM-lotteries, and each four horse lotteries H_1, H_2, H_3, H_4 satisfying the following conditions:*

- *the preference between H_1 and H_2 is called-off when $\{r_j\}^C$ occurs,*

- *the preference between H_3 and H_4 is called-off when $\{r_j\}^C$ occurs, and*

- $H_1(r_j) = H_3(r_j) = L_1$, $H_2(r_j) = H_4(r_j) = L_2$,

we have $H_1 \preceq H_2$ if and only if $H_3 \preceq H_4$.

PROOF. First, note that $\frac{1}{2} H_1 + \frac{1}{2} H_4 = \frac{1}{2} H_2 + \frac{1}{2} H_3$. Use Lemma 3.114 to see that $H_1 \preceq H_2$ implies $\frac{1}{2} H_1 + \frac{1}{2} H_3 \preceq \frac{1}{2} H_1 + \frac{1}{2} H_4$, which implies $H_3 \preceq H_4$ by Axiom 2. Similarly, $H_3 \preceq H_4$ implies $H_1 \preceq H_2$. □

PROOF OF THEOREM 3.147. For each state r_j and each $(n-1)$-tuple of NM-lotteries $(L_1, \ldots, L_{j-1}, L_{j+1}, \ldots, L_n)$, consider the set of horse lotteries of the form

$$(L_1, \ldots, L_{j-1}, L, L_{j+1}, \ldots, L_n),$$

where L is an arbitrary element of $\mathcal{L}$. According to Lemma 3.148, the ranking of these horse lotteries will be the same no matter what one chooses for the L_is. Hence, we can treat the set of these horse lotteries as the entire set of interest and apply Lemma 3.119 to obtain a utility function $U_j : L \to [0, 1]$ satisfying

$$(L_1, \ldots, L_{j-1}, L_1', L_{j+1}, \ldots, L_n) \preceq (L_1, \ldots, L_{j-1}, L_2', L_{j+1}, \ldots, L_n)$$

if and only if $U_j(L_1') \leq U_j(L_2')$, no matter what one chooses for the L_is. For each j such that r_j is nonnull, there are prizes p_j^* and p_{*j} such that $U_j(p_j^*) = 1$ and $U_j(p_{*j}) = 0$. (If r_j is null, $U_j(p_i)$ can be arbitrary, since there are no preferences among the horse lotteries.) It is easy to see that the best and worst horse lotteries are respectively

$$H^* = (p_1^*, \ldots, p_n^*), \quad H_* = (p_{*1}, \ldots, p_{*n}).$$

Now, set up the following horse lotteries:

$$H_i^*(r_j) = \begin{cases} p_j^* & \text{if } j = i, \\ p_{*j} & \text{if } j \neq i. \end{cases}$$

Define $q_i = U(H_i^*)$, where U is constructed by Lemma 3.119 based on all of $\mathcal{H}$ with $U(H^*) = 1$ and $U(H_*) = 0$. Clearly, $q_i \geq 0$ for all i. To see that $\sum_{i=1}^n q_i = 1$, note that the equal mixture of all the H_i^*s is

$$\frac{1}{n}H_1^* + \cdots + \frac{1}{n}H_n^* = \frac{1}{n}H^* + \frac{n-1}{n}H_*.$$

Evaluating U at both sides of this expression gives $1/n \sum_{i=1}^n q_i = 1/n$.

Finally, we prove that if $H = (L_1, \ldots, L_n)$, then $U(H) = \sum_{i=1}^n q_i U_i(L_i)$. This will complete the proof. Construct n horse lotteries

$$H_i(r_j) = \begin{cases} L_i & \text{if } i = j, \\ p_{*j} & \text{if } i \neq j. \end{cases}$$

By taking an equal mixture of all n of these, we get

$$\frac{1}{n}H + \frac{n-1}{n}H_* = \frac{1}{n}H_1 + \cdots + \frac{1}{n}H_n.$$

Evaluating U at both sides of this gives $U(H)/n = (1/n)\sum_{i=1}^n U(H_i)$. So, we need only prove that $U(H_i) = q_i U_i(L_i)$ for each i. From the definition of U_i, we see that $H_i \sim H^i$, where

$$H^i(r_j) = \begin{cases} U_i(L_i)p_i^* + (1 - U_i(L_i))p_{*i} & \text{if } i = j, \\ p_{*j} & \text{if } i \neq j. \end{cases}$$

Since $H^i = U_i(L_i)H_i^* + (1 - U_i(L_i))H_*$,

$$U(H_i) = U(H^i) = (1 - U_i(L_i))U(H_*) + U_i(L_i)U(H_i^*) = q_i U_i(L_i). \quad \square$$

3.4 Problems

Section 3.1:

1. *Consider the rule δ in Example 3.13 on page 148.
   (a) Find a formula for the risk function.
   (b) Find a formula for the Bayes risk with respect to Lebesgue measure.
   (c) Prove that the Bayes risk is strictly less than $1/2$ for all even n.
   (d) Find the exact value of the Bayes risk if $n = 2$.

2. Prove Proposition 3.16 on page 150.

3. Two firms are planning to make competing secret bids on the price at which they will supply a computer system to a government agency. The firm with the lower bid will get the job. (No cost overruns will be allowed.) One firm believes that its actual cost of supplying the system is sure to be c and it has a prior distribution on the bid Θ of the other firm. Let h be the cost of preparing and submitting the bid. For each situation below, find the bid this firm should make to maximize expected profit:
   (a) $h = 0$ and $f_\Theta(\theta) = \exp(-\theta/\mu)/\mu$, for $\theta > 0$ and μ known.
   (b) $h = 0$ and f_Θ is arbitrary.
   (c) $h > 0$ and $f_\Theta(\theta) = \exp(-\theta/\mu)/\mu$, for $\theta > 0$ and μ known.
   (d) $h > 0$ and f_Θ is arbitrary.

4. An actuary wants to estimate the mean number of claims for industrial injuries in a newly opened factory, in order to determine the premium for insurance. The actuary believes that, to a good approximation, the number of claims in any year by any one person is Poisson with mean θ conditional on a parameter $\Theta = \theta$. Different persons are assumed to be independent given Θ. Past experience with similar factories gives a prior density for Θ:

$$f_\Theta(\theta) = \frac{m(m\theta)^{r-1} e^{-m\theta}}{\Gamma(r)}$$

for fixed known values of m and $r > 1$. After n person-years have elapsed in this factory, s injuries are observed.

   (a) Using $f_\Theta(\theta)$ above, show that with a loss function

$$L(\theta, d) = \frac{(d-\theta)^2}{\theta},$$

the best choice of the premium d is

$$d^*(s) = \frac{r+s-1}{m+n}.$$

   (b) Now, assume that n (the number of person-years) is fixed and treat S (the number of injuries) as random (not yet observed). Find the risk function for d^*. Also find the Bayes risk for d^* with respect to the prior f_Θ and the posterior risk for $d^*(s)$ given $S = s$.

5. Let $X_1, \ldots, X_n$ be IID $Ber(\theta)$ random variables conditional on $\Theta = \theta$. Suppose that we have a loss function $L(\theta, a) = (\theta - a)^2 / [\theta^2 (1 - \theta)^5]$, where the action space is $\aleph = [0, 1]$. The prior distribution of Θ is $Beta(\alpha, \beta)$. Find conditions on α and β such that both of the following are true:

   - The formal Bayes rule exists and has finite posterior risk for all possible samples.
   - The Bayes rule exists and has finite risk.

6. Suppose that the conditional density of X given $\Theta = \theta$ is $\exp(-|x - \theta|)/2$ and that Θ has prior density $\exp(-|\theta - \eta|)/2$ for some number η. Let $\aleph = \mathbb{R}$ and $L(\theta, a) = (\theta - a)^2$. Find the formal Bayes rule.

Section 3.2.1:

7. Suppose that P_θ says that $X_1, \ldots, X_n$ are IID $N(\theta, 1)$. Let $\delta_0(X)$ be the median of the sample, and let $T = \delta_1(X)$ be the sample average. Find a randomized rule based on the mean, $\delta(T)$, which has the same risk function as δ_0 no matter what the loss function is. (You may wish to solve Problem 12 on page 663 or Problem 1 on page 138 first. You probably cannot write a closed-form solution for the randomized rule. You may either describe the probability distribution in words sufficiently precise to define it or give an algorithm for actually performing a randomization that will have the appropriate distribution.)

8. Let $\delta : \mathcal{X} \to \aleph$ be a nonrandomized rule, and let $T : \mathcal{X} \to \mathcal{T}$ be a sufficient statistic. Let δ_1 be the rule constructed in Theorem 3.18 on page 151. Show that for each t, the distribution $\delta_1(t)(\cdot)$ on $\aleph$ is the probability measure induced by δ from the conditional distribution of X given $T = t$.

9. *Let $\Omega = (0, \infty) \times (0, \infty)$, $\mathcal{X} = \mathbb{R}^3$, and $\aleph = \mathbb{R}^+$. Suppose that P_θ says X_1, X_2, X_3 are IID $U(\alpha, \beta)$, where $\theta = (\alpha, \beta)$. Let

$$L(\theta, a) = \left(\frac{\alpha + \beta}{2} - a \right)^2.$$

Let $\delta_0(X) = \overline{X}$.

   (a) Find a two-dimensional sufficient statistic, T.

   (b) Use the Rao–Blackwell theorem 3.22 to find a rule $\delta_1(T)$ whose risk function is at least as good as that of $\delta_0(X)$.

   (c) Find the risk functions $R(\theta, \delta_0)$ and $R(\theta, \delta_1)$, and show that there is at least one θ such that $R(\theta, \delta_1) < R(\theta, \delta_0)$.

10. *Let $\{X_n\}_{n=1}^\infty$ be conditionally IID $Ber(\theta)$ given $\Theta = \theta$, and let $X = (X_1, \ldots, X_n)$. Let $\aleph = \Omega = (0, 1)$ and $L(\theta, a) = (\theta - a)^2$. Let the prior distribution λ of Θ be $U(0, 1)$. Let $\delta_0(x)$ be the sample median, that is

$$\delta_0(x) = \begin{cases} 0 & \text{if more than half of the observations are 0,} \\ 1 & \text{if more than half of the observations are 1,} \\ \frac{1}{2} & \text{if exactly half of the observations are 0.} \end{cases}$$

(a) Find $R(\theta, \delta_0)$ and $r(\lambda, \delta_0)$.

(b) Let $T = \sum_{i=1}^{n} X_i$. Find the rule δ_1 guaranteed by the Rao–Blackwell theorem 3.22.

Section 3.2.2:

11. In Example 3.25 on page 154, find the risk function for both δ and δ_1 and show that δ_1 dominates δ.

12. Suppose that P_θ says that $X \sim Bin(n, \theta)$. Let $\Omega = (0, 1)$ and $\aleph = [0, 1]$. Let $L(\theta, a) = (\theta - a)^2$ and

$$\delta(x) = \begin{cases} \frac{x}{n} & \text{with probability } \frac{1}{2}, \\ \frac{1}{2} & \text{with probability } \frac{1}{2}. \end{cases}$$

Find a nonrandomized rule that dominates δ.

13. Find an example of a decision problem with a decision rule δ_0 and a probability λ on the parameter space such that δ_0 is λ-admissible but δ_0 is not a Bayes rule with respect to λ.

14. Suppose that $X \sim Exp(1/\theta)$ given $\Theta = \theta$. Let the action space be $[0, \infty)$, and let the loss function be $L(\theta, a) = (\theta - a)^2$.

    (a) Prove that $\delta(x) = x$ is inadmissible.

    (b) Find a nonconstant admissible rule.

15. Let $X = (X_1, \ldots, X_n)$, where the X_i are conditionally IID $N(\mu, \sigma^2)$ given $\Theta = (\mu, \sigma)$. Let $c_1 > 0$ and c_2 be constants. Let $\aleph = \mathbb{R}$ and $L(\theta, a) = (\mu - a)^2$. Define $\delta(x) = (n\bar{x} + c_1 c_2)/(n + c_1)$. Show that δ is admissible.

16. *Prove Proposition 3.47 on page 162.

17. Assume that $L(\theta, a) = (\theta - a)^2$ in each of the following questions.

    (a) Suppose that $X \sim N(\theta, 1)$ given $\Theta = \theta$. Show that for each constant c, $\delta(x) \equiv c$ is admissible.

    (b) Suppose that $X \sim U(0, \theta)$ given $\Theta = \theta$. Show that for each constant c, $\delta(x) \equiv c$ is inadmissible.

18. Let $\Omega = (0, 1)$, $\aleph = [0, 1]$, and $L(\theta, a) = (\theta - a)^2$. Suppose that P_θ says that $X \sim Geo(\theta)$, that is,

$$f_{X|\Theta}(x|\theta) = \begin{cases} (1 - \theta)\theta^x & \text{for } x = 0, 1, \ldots, \\ 0 & \text{otherwise} \end{cases}$$

is the density of X with respect to counting measure given $\Theta = \theta$. Show that $\delta(x) = x/(x + 1)$ is admissible.

19. Suppose that $X \sim N(\theta, 1)$ given $\Theta = \theta$, and let Θ have an $N(0, 1)$ prior. Suppose that the parameter space and the action space are both $(-\infty, \infty)$. Let $L(\theta, a) = 0$ if $a \geq \theta$ and $L(\theta, a) = 1$ if $a < \theta$.

    (a) Show that there is no Bayes rule.

    (b) Show that every decision rule is inadmissible.

(c) Show that if the action space is $[-\infty, \infty]$, then there is a Bayes rule and that it is the only admissible rule.

Section 3.2.3:

20. *Prove that the modified James–Stein estimator $\delta_3(X)$ has smaller risk function than $\delta(X)$ if $n \geq 4$. (*Hint:* Let Γ be an orthogonal transformation with first row proportional to $\mathbf{1}^\top$, and let Z be the last $n-1$ coordinates of ΓX. What does Theorem 3.50 say about estimating $\Gamma\Theta$ by $\Gamma\delta_3(X)$?)

21. *Say that a function $g : \mathbb{R} \to \mathbb{R}$ is *absolutely continuous* if there exists a function g' such that for all $x_1 < x_2$, $g(x_2) = g(x_1) + \int_{x_1}^{x_2} g'(y) dy$.[38]

    (a) Prove that the conclusion to Lemma 3.51 continues to hold if the assumption that g is differentiable is replaced by the assumption that g is absolutely continuous.

    (b) Prove that the conclusion to Lemma 3.52 continues to hold if the assumption that the coordinates of g are differentiable is replaced by the assumption that h_i is absolutely continuous for every i.

22. Let $g(x) = -x \min\{c, (n-2)/\sum_{i=1}^n x_i^2\}$ be a function from $\mathbb{R}^n$ to $\mathbb{R}^n$. Let $\delta^*(x) = x + g(x)$.

    (a) Using Problem 21 above, find all values of $c > 0$ such that $\delta^*(x)$ has smaller risk than $\delta(x) = x$ in the setting of Theorem 3.50.

    (b) Prove that for $c > (n-2)/(n+2)$, $\delta^*(x)$ has smaller risk than $\delta_1(x)$ in the setting of Theorem 3.50.

Section 3.2.4:

23. Prove Proposition 3.58 on page 167.

24. Let $X \sim Geo(\theta)$ given $\Theta = \theta$. Let $L(\theta, a) = (\theta - a)^2/[\theta(1-\theta)]$. Prove that $\delta(x) = I_{\{0\}}(x)$ is minimax.

25. In Example 3.72 (see page 170), let $p_i = \Pr(\Theta = i)$ for $i = 0, 1$ be a prior distribution. Prove that it is impossible for the Bayes risk of the minimax rule to be simultaneously stictly less than the Bayes risks of both action 3 and action 1. This example shows how the minimax principle can be in very serious conflict with the expected loss principle.

Section 3.2.5:

26. Prove Proposition 3.85 on page 174.

---

[38]Such functions are called absolutely continuous because they have a property similar to measures that are absolutely continuous with respect to Lebesgue measure. In particular, if g is nondecreasing, then $\eta((a, b]) = g(b) - g(a)$ defines a measure that is absolutely continuous with respect to Lebesgue measure.

27. Suppose that P_0 says $X \sim U(0,1)$ and P_1 says $X \sim U(0,7)$, and that the loss function is as in Theorem 3.87. Find all of the admissible rules under the conditions of that theorem. Express each rule by saying which intervals of X values lead to making each decision.

28. Suppose that an observation X is to be made and it is believed that X has one of two densities:
$$f_0(x) = \frac{1}{2}\exp(-|x|), \quad f_1(x) = \frac{1}{\sqrt{2\pi}}\exp\left(-\frac{1}{2}x^2\right).$$
Find all of the admissible procedures according to the Neyman–Pearson fundamental lemma 3.87 (using the loss function stated there). Express the rules in terms of intervals in which each decision is taken.

29. Prove the claim at the end of the proof of the Neyman–Pearson fundamental lemma 3.87 that no element of $\mathcal{C}$ dominates any other element of $\mathcal{C}$.

30. Prove Proposition 3.91 on page 178.

Section 3.3:

31. Suppose that there are $k \geq 2$ horses in a race and that a gambler believes that p_i is the probability that horse i will win ($\sum_{i=1}^{k} p_i = 1$). Suppose that the gambler has decided to wager an amount x to be divided among the k horses. If he or she wagers x_i on horse i and that horse wins, the utility of the gambler is $\log(c_i x_i)$, where $c_1, \ldots, c_k$ are known positive numbers. Find values $x_1, \ldots, x_k$ to maximize expected utility.

32. Suppose that two agents have a common strictly increasing utility function U for their fortunes in dollar amounts and that their current fortunes are the same, x_0. (So, for example, the utility of receiving an additional x dollars would be $U(x_0 + x)$.)

    (a) Let R be a random dollar amount that is strictly greater than $-x_0$. If one of our agents contemplates selling R, what would be the lowest price at which the agent would be willing to sell it? What would be the highest price that an agent who did not own R would be willing to buy it?

    (b) Suppose that one agent receives a gift consisting of a lottery ticket that will pay $r > 0$ dollars with probability $1/2$ and pays nothing with probability $1/2$ and that both agents agree on these probabilities. Construct a utility function U having the property that, as soon as an agent receives this gift, he or she is willing to sell it at some price less than $r/2$ and the other agent is willing to buy it at that same price.

33. Assume Axioms 1 and 2 and the Archemedian condition of Lemma 3.117. Let $R = \{r_1, \ldots, r_n\}$ and $P = \{p_1, \ldots, p_m\}$. Consider the set $\mathcal{H}'$ of all horse lotteries of the form $(p_{i_1}, \ldots, p_{i_n})$. (These are all horse lotteries whose NM-lotteries assign probability 1 to a single prize.) Let $H_*, H^* \in \mathcal{H}'$ be such that $H_* \preceq H \preceq H^*$ for all $H \in \mathcal{H}'$. Prove that $H_* \preceq H \preceq H^*$ for all $H \in \mathcal{H}$.

34. Prove Proposition 3.118 on page 193. (*Hint:* You can use Theorem 3.147 if you wish.)

35. Let $\gamma_1 < \gamma_2 < 1$, and suppose that H_1 and H_2 are horse lotteries such that
$$\gamma_1 H_1 + (1 - \gamma_1)H_2 \sim \gamma_2 H_1 + (1 - \gamma_2)H_2.$$
Assume Axioms 1–3 and prove that $H_1 \sim H_2$.

36. Assume all of the axioms, including Axiom 6. Show that conditional preference given R is the same as unconditional preference.

37. Prove the part of Lemma 3.134 that says that $Q(B) = 0$ if and only if B is null.

38.*Assume Axioms 1–4. Let R be finite. Prove that $H_1(r) \preceq H_2(r)$ for all $r \in R$ implies $H_1 \preceq H_2$. (*Hint:* Create a comparison between H_1 and H' that is called-off when $\{r_1\}^C$ occurs. Use induction on the number of states.)

CHAPTER 4
Hypothesis Testing

4.1 Introduction

4.1.1 A Special Kind of Decision Problem

Recall the setup used at the beginning of Chapter 3. We had a probability space $(S, \mathcal{A}, \mu)$ and a function $V : S \to \mathcal{V}$. One example of V is the parameter Θ. Other examples are measurable functions of Θ. Other V functions, which are not functions of Θ, are possible but are rarely seen in classical statistics. This is true to a greater extent in hypothesis testing for reasons that will become more apparent once we study the criteria used for selecting tests in classical statistics.

Definition 4.1. Suppose that we can partition $\mathcal{V}$ into $\mathcal{V} = \mathcal{V}_H \cup \mathcal{V}_A$, where $\mathcal{V}_H \cap \mathcal{V}_A = \emptyset$. The statement that $V \in \mathcal{V}_H$ is a *hypothesis* and is labeled H. The corresponding *alternative* is labeled A and is the statement that $V \in \mathcal{V}_A$. If $V = \Theta$, we have $\Omega = \Omega_H \cup \Omega_A$ with $\Omega_H \cap \Omega_A = \emptyset$ and $V \in \mathcal{V}_H$ if and only if $\Theta \in \Omega_H$. In this case, we write $H : \Theta \in \Omega_H$ and $A : \Theta \in \Omega_A$. A decision problem is called *hypothesis testing* if $\aleph = \{0, 1\}$ and $L(v, a)$ satisfies $L(v, 1) > L(v, 0)$ for $v \in \mathcal{V}_H$ and $L(v, 1) < L(v, 0)$ for $v \in \mathcal{V}_A$. The action $a = 1$ is called *rejecting the hypothesis*, and the action $a = 0$ is called *accepting the hypothesis*.[1] If we reject H but H is true, we made a *type I error*. If we accept H and it is false, we made a *type II error*.

---

[1] Some authors prefer to call action $a = 0$ *not rejecting the hypothesis*.

4.1. Introduction

A simple type of hypothesis testing loss function is

$$L(v, a) = \begin{cases} c_a & \text{if } v \in V_H, \\ b_a & \text{if } v \in V_A, \end{cases} \quad (4.2)$$

where $c_1 > c_0$ and $b_0 > b_1$. It is easy to see (see Problem 1 on page 285) that (4.2) is equivalent to a loss function of the same form with $c_0 = b_1 = 0$, $b_0 = 1$, and $c_1 = c > 0$. Such a loss function is called a *0–1–c loss function*. If, in addition, $c = 1$, it is called a *0–1 loss function*. More general loss functions than the 0–1–c loss might often seem appropriate for the type of problems in which hypothesis testing is used. For example, if the parameter is real, the hypothesis is that $\Theta \leq \theta_0$, and $c > 0$, an appropriate loss might be

$$L(\theta, a) = \begin{cases} \theta - \theta_0 & \text{if } \theta > \theta_0, \, a = 0, \\ (\theta_0 - \theta)c & \text{if } \theta \leq \theta_0, \, a = 1. \end{cases} \quad (4.3)$$

This loss provides for penalties for choosing the wrong decision that are commensurate with the inaccuracy of the decision. But this loss can be written as $|\theta - \theta_0|$ times the 0–1–c loss. By Proposition 3.47, so long as the risk functions of all decision rules are continuous from the left (or all are continuous from the right) at $\theta = \theta_0$, rules admissible under the 0–1–c loss will be admissible under this loss. One could begin the study of hypothesis testing by concentrating solely on which decision rules are admissible. For this purpose, the 0–1 loss is sufficient. The focus of hypothesis testing, however, is on finding tests that meet certain ad hoc criteria to be defined later.

A randomized decision rule δ in a hypothesis testing problem can be described by its *test function*, which is the measurable function $\phi : \mathcal{X} \to [0, 1]$ given by

$$\phi(x) = \delta(x)(1) = \Pr(\text{choose } a = 1 | X = x).$$

One should think of a randomized test ϕ as follows. First, observe $X = x$, and then flip a coin with probability of heads equal to $\phi(x)$. If the coin comes up heads, reject the hypothesis. Because of this interpretation, randomized tests are seldom used in practice.

Definition 4.4. Suppose that $V = \Theta$. The *power function* of a test ϕ is $\beta_\phi(\theta) = E_\theta \phi(X)$. The *operating characteristic curve* is $\rho_\phi = 1 - \beta_\phi$. The *size* of ϕ is $\sup_{\theta \in \Omega_H} \beta_\phi(\theta)$. A test is called *level* α, for some number $0 \leq \alpha \leq 1$, if its size is at most α. A hypothesis is *simple* if Ω_H is a singleton. Similarly, the alternative is *simple* if Ω_A is a singleton. The hypothesis (alternative) is *composite* if it is not simple. For symmetry, we also define the *base of the test* to be $\inf_{\theta \in \Omega_A} \beta_\phi(\theta)$. A test is said to have *floor* γ if the base is at least γ.

The definitions of power function, size, level, and operating characteristic are all standard in classical theory, but the definitions of *base* and *floor* are not. Some elaboration is in order. There is a duality between hypotheses and alternatives which is not respected in most of the classical hypothesis-testing literature. The definitions of base and floor are introduced to complete the duality among the concepts usually defined. For example, suppose that we decide to switch the names of alternative and hypothesis, so that Ω_H becomes Ω_A, and vice versa. Then we can switch tests from ϕ to $\psi = 1 - \phi$ and the "actions" accept and reject become switched. The power function of ϕ is the operating characteristic of ψ, and vice versa. The size of ϕ is one minus the base of ψ, and vice versa. The test ϕ has level α if and only if ψ has floor $1 - \alpha$. The classical optimality criteria for tests do not respect this duality. That is, a test ϕ may satisfy the appropriate classical optimality criterion for a specified hypothesis–alternative pair, but when the names of hypothesis and alternative are switched and the same optimality criterion is appropriate, $1 - \phi$ does not satisfy the same optimality criterion. (See Problem 31 on page 289 for an example.) For this reason, when appropriate, we will introduce new optimality criteria that are dual to the existing ones.

It is easy to see that the risk function for a hypothesis-testing problem is closely related to the power function. If the loss function is 0–1–c, then the risk function is

$$R(\theta, \phi) = \begin{cases} c\beta_\phi(\theta) & \text{if } \theta \in \Omega_H, \\ 1 - \beta_\phi(\theta) & \text{if } \theta \in \Omega_A. \end{cases} \quad (4.5)$$

Now suppose that we let $\Omega'_H = \Omega_A$ and $\Omega'_A = \Omega_H$, so that hypothesis and alternative are switched. Also, switch the names of the actions, set $\psi = 1 - \phi$, and let the loss be c times the 0–1–$1/c$ loss function. Then the risk function of ψ in this new problem is

$$R'(\theta, \psi) = \begin{cases} \beta_\psi(\theta) & \text{if } \theta \in \Omega'_H, \\ c(1 - \beta_\psi(\theta)) & \text{if } \theta \in \Omega'_A, \end{cases}$$

which is easily seen to equal $R(\theta, \phi)$. So, the risk function respects the duality between hypotheses and alternatives, as will considerations of admissibility.

4.1.2 Pure Significance Tests

A simpler framework for hypothesis testing dates back at least to Pearson (1900). In this simpler framework, one need only explicitly state the hypothesis (call it H as before), which is either a single distribution for the data or a class of distributions. One then creates a weak order $\preceq$ on the sample space $\mathcal{X}$, where $x \preceq y$ is intended to mean that y is more at odds with H than x is.[2]

---

[2] See Definition 3.99 on page 183. Basically, the binary relation $\preceq$ must be reflexive and transitive, and all pairs of data values must be compared.

4.1. Introduction

Example 4.6. Let H state that $X \sim N(0,1)$. We can say that $x \preceq y$ if $|x| \leq |y|$.

We are quite free to define $\preceq$ however we wish, so long as it is a weak order.

Example 4.7. Let H state that $X \sim N(0,1)$. We could define $x \preceq y$ by $|x| \geq |y|$.

The most common way to define $\preceq$ is in terms of a statistic $T : \mathcal{X} \to \mathbb{R}$. We would say that $x \preceq y$ if and only if $T(x) \leq T(y)$. In Example 4.6, $T(x) = |x|$. A *pure significance test* is obtained by calculating the *significance probability* $p_H(x)$ (Definition 4.8) and rejecting H if $p_H(x)$ is too small.

Definition 4.8. Let the hypothesis H be a set of distributions on $(\mathcal{X}, \mathcal{B})$. Suppose that the quantity $p_Q(x) = Q(\{y : x \preceq y\})$ is the same (or approximately the same) for all Q in H. Then the common value $p_H(x)$ is called the *significance probability of the data x relative to the weak order* $\preceq$, and the test that rejects H when $p_H(x)$ is small is called a *pure significance test*.

Example 4.9 (Continuation of Example 4.6). It is easy to see that

$$p_H(x) = \int_{-\infty}^{-|x|} \phi(y)dy + \int_{|x|}^{\infty} \phi(y)dy = 2\Phi(-|x|),$$

where ϕ and Φ are the standard normal density and CDF, respectively. This pure significance test would be the same as the usual test of the hypothesis that the mean of a normal distribution with variance 1 is 0 versus the alternative that the mean is not 0.

For the case of Example 4.7, we have

$$p_H(x) = \int_{-|x|}^{|x|} \phi(y)dy = 2\Phi(|x|).$$

This test would lead to rejecting H if the data are too consistent with H. This is similar to what Fisher (1936) did when considering how closely the data of Mendel (1866) matched a theory that Fisher later showed to be inaccurate.

Example 4.10. Suppose that H is the set of distributions that say that $X = (X_1, \ldots, X_n)$ are conditionally IID with $N(0, \sigma^2)$ distribution given $\Sigma = \sigma$. Let $T(x)$ be the usual t statistic for testing the hypothesis that the mean of a normal sample is 0, namely $T(x) = \sqrt{n}|\bar{x}|/s$, where $\bar{x} = \sum_{i=1}^{n} x_i/n$ and $s^2 = \sum_{i=1}^{n}(x_i - \bar{x})^2/(n-1)$. Then $P_\sigma(\{y : T(x) \leq T(y)\})$ is the same for all σ. In fact $p_H(x) = 2T_{n-1}(-|T(x)|)$, where T_{n-1} is the CDF of the $t_{n-1}(0,1)$ distribution. The usual t-test is a pure significance test.

The advantages to pure significance tests over general hypothesis tests are that one need not explicitly state the alternatives and one is free to choose the weak order $\preceq$ however one sees fit. Of course, one would normally choose $\preceq$ with some alternative in mind, but one need not say what the alternative is, nor need one calculate any probabilities conditional on the alternative.

A serious disadvantage is that one never knows, until one considers explicit alternatives, whether one should continue calculating probabilities as if the hypothesis were true or not. Just because $p_H(x)$ is large does not mean that H is a better probability model for the data than some other plausible distribution not part of H. Similarly, if $p_H(x)$ is quite small, it may be the case that many other distributions not part of H also give very small probability to the set $\{y : x \preceq y\}$. Berkson (1942) forcefully argues this point, but not forcefully enough for Fisher (1943).

We will not discuss pure significance tests any further in this book except to mention a few points.[3] First, all of the hypothesis tests developed in this chapter can be interpreted as pure significance tests if one feels compelled to do so, although the hypotheses may need to be modified in order to satisfy the definition of pure significance test. Second, the goodness of fit tests described in Section 7.5.2 were originally intended to be interpreted as pure significance tests. Third, pure significance tests have no role to play in the Bayesian framework as described in various parts of this text. If the hypothesis H describes all of the probability distributions one is willing to entertain, then one cannot reject H without rejecting probability models altogether. If one is willing to entertain models not in H, then one needs to take them into account, as well as their merits relative to H, before deciding whether or not to reject H.

4.2 Bayesian Solutions

4.2.1 Testing in General

The Bayesian solution to a hypothesis-testing problem with 0–1–c loss is straightforward theoretically. The posterior risk from choosing action $a = 1$ is $c\Pr(V \in \mathcal{V}_H | X = x)$, and the posterior risk of choosing action $a = 0$ is $\Pr(V \in \mathcal{V}_A | X = x)$. The optimal decision is to choose $a = 1$ if

$$c\Pr(V \in \mathcal{V}_H | X = x) < \Pr(V \in \mathcal{V}_A | X = x),$$

which is equivalent to

$$\Pr(V \in \mathcal{V}_H | X = x) < \frac{1}{1+c}. \tag{4.11}$$

So, the Bayesian solution is to reject the hypothesis if its posterior probability is too small, that is, smaller than $1/(1+c)$. Theoretically, that is all there is to Bayesian hypothesis testing with 0–1–c loss. In practical problems, it may be computationally difficult to calculate the posterior probability that $V \in \mathcal{V}_H$, but this is a numerical analysis problem.

---

[3]Cox and Hinkley (1974, Chapters 3–5) discuss pure significance tests and related topics in great detail. A nice review is contained in Cox (1977).

4.2. Bayesian Solutions   219

Example 4.12. Suppose that P_θ says that $\{X_n\}_{n=1}^\infty$ are IID $N(\mu, \sigma^2)$, where $\theta = (\mu, \sigma)$ and $X = (X_1, \ldots, X_n)$. Let $V = \Theta$ and $\Omega_H = \{(\mu, \sigma) : \mu \geq \mu_0\}$, and let L be a 0–1–c loss function. If we use the measure with Radon–Nikodym derivative $1/\sigma$ with respect to Lebesgue measure as an improper prior, then the posterior distribution of M is $t_{n-1}(\overline{x}, s/\sqrt{n})$. The formal Bayes rule is

$$\phi(x) = \begin{cases} 1 & \text{if } t < T_{n-1}^{-1}\left(\frac{1}{1+c}\right), \\ 0 & \text{if } t > T_{n-1}^{-1}\left(\frac{1}{1+c}\right), \\ \text{arbitrary} & \text{if } t = T_{n-1}^{-1}\left(\frac{1}{1+c}\right), \end{cases}$$

where $t = \sqrt{n}(\overline{x} - \mu_0)/s$ is the usual t statistic and T_{n-1} is the CDF of the $t_{n-1}(0,1)$ distribution. Note that this is the usual size $1/(1+c)$ t-test of H from every elementary statistics course.

The Bayesian solution, as stated above, applies to predictive hypothesis testing as well as to parametric testing. For example, note that (4.11) is formulated in terms of predictive probabilities. Classical theory is not as well equipped as Bayesian to deal with predictive hypothesis tests.[4] The closest the classical theory comes to dealing with predictive testing is as a predictive decision problem. The type of hypothesis constructed in Example 4.13 is closely related to tolerance sets as described in Section 5.2.3.

Example 4.13. In the classical setting (see (3.15) on page 149) the predictive loss function is first converted to a parametric loss function and then the parametric decision problem is solved. For the hypothesis-testing case with 0–1–c loss,

$$\begin{aligned} L(v, \delta(x)) &= cI_{\mathcal{V}_H}(v)\phi(x) + I_{\mathcal{V}_A}(v)[1 - \phi(x)], \\ L(\theta, \delta(x)) &= cP_{\theta, V}(\mathcal{V}_H)\phi(x) + \{1 - P_{\theta, V}(\mathcal{V}_H)[1 - \phi(x)]\} \\ &= \phi(x)[(c+1)P_{\theta, V}(\mathcal{V}_H) - 1] + 1 - P_{\theta, V}(\mathcal{V}_H), \\ R(\theta, \phi) &= \beta_\phi(\theta)[(c+1)P_{\theta, V}(\mathcal{V}_H) - 1] + 1 - P_{\theta, V}(\mathcal{V}_H). \quad (4.14) \end{aligned}$$

Now, define

$$\begin{aligned} \Omega_H &= \left\{\theta : P_{\theta, V}(\mathcal{V}_H) \geq \frac{1}{1+c}\right\}, \\ \Omega_A &= \Omega_H^C, \\ e(\theta) &= \begin{cases} 1 - P_{\theta, V}(\mathcal{V}_H) & \text{if } \theta \in \Omega_H, \\ -P_{\theta, V}(\mathcal{V}_H) & \text{if } \theta \in \Omega_A, \end{cases} \\ d(\theta) &= |(c+1)P_{\theta, V}(\mathcal{V}_H) - 1|. \end{aligned}$$

Now note that $R(\theta, \phi)$ in (4.14) is exactly equal to $e(\theta)$ plus $d(\theta)$ times the risk function from a 0–1 loss as given in (4.5) for the hypothesis $H : \Theta \in \Omega_H$.

---

[4] The interested reader should try to extend the classical definitions of level, power, and so forth to the case of predictive hypothesis testing and see what happens. The problem arises because one usually assumes that the future data are independent of the past data conditional on the parameters, and all classical inferences are conditional on the parameters. Hence, the past tells us nothing about the future, and vice versa.

If power functions are continuous at that θ such that $P_{\theta,V}(\mathcal{V}_H) = 1/(c+1)$, then the predictive testing problem has been converted into a parametric testing problem. In words, we have replaced a test concerning the observable V with a test concerning the conditional distribution of V given Θ.

Another area in which Bayesian and classical hypothesis testing differ dramatically is their treatment of more general loss functions. When the focus of classical testing is on admissible tests, then it does not matter which of several equivalent loss functions one uses. A Bayesian solution to a testing problem will depend on which loss one uses because one is trying to minimize the posterior risk. For example, with the loss function in (4.3), the posterior risk for choosing $a = 0$ is $\int I_{(\theta_0,\infty)}(\theta)(\theta-\theta_0)dF_{\Theta|X}(\theta|x)$, while the posterior risk for choosing $a = 1$ is $c \int I_{(-\infty,\theta_0]}(\theta)(\theta_0 - \theta)dF_{\Theta|X}(\theta|x)$. A little algebra shows that the formal Bayes rule is to choose $a = 1$ if

$$E(\Theta|X = x) - \theta_0 > (c-1) \Pr(\Theta \le \theta_0|X = x) \{\theta_0 - E(\Theta|\Theta \le \theta_0, X = x)\}.$$

It may turn out that this is the same decision rule as is optimal with a 0–1–c' loss for some number c'.

Example 4.15. Suppose that $X \sim N(\theta, 1)$ given $\Theta = \theta$ and that the hypothesis is $H : \Theta \le \theta_0$ with loss (4.3). It is easy to see that the formal Bayes rule with respect to a prior μ_Θ is to choose action $a = 1$ if

$$\int_{(\theta_0,\infty)} (\theta - \theta_0) \exp\left(x[\theta - \theta_0] - \frac{\theta^2}{2}\right) d\mu_\Theta(\theta)$$
$$> c \int_{(-\infty,\theta_0]} (\theta_0 - \theta) \exp\left(x[\theta - \theta_0] - \frac{\theta^2}{2}\right) d\mu_\Theta(\theta).$$

The expression on the left is increasing in x and the expression on the right is decreasing in x, so the formal Bayes rule is to choose action $a = 1$ if $x > k$ for some number k. This rule has the same form as the formal Bayes rules with respect to 0–1–c' loss functions.

In classical hypothesis testing, it is not common to recommend different tests depending on whether the loss is 0–1–c or of the form of (4.3) or anything else. In fact, very little attention is paid to what the loss function might be in classical testing. Were the focus solely on finding all admissible rules, this might not be a problem. However, once we advance beyond the simplest types of testing situations, the classical theory will tend to abandon the goal of finding all admissible rules and concentrate instead on finding all tests that satisfy certain ad hoc criteria.

4.2.2 Bayes Factors

The most striking difference between classical and Bayesian hypothesis testing arises in the treatment of point hypotheses of the form $H : \Theta = \theta_0$

versus $A : \Theta \neq \theta_0$. When the parameter space is uncountable, prior distributions are typically continuous. This means that the prior (and posterior) probability of $\Theta = \theta_0$ is 0. In order to take seriously the problem of testing a point hypothesis, one must use a prior distribution in which $\Pr(\Theta = \theta_0) > 0$. Alternatively, one can replace the hypothesis with (what might be more reasonable) an interval hypothesis of the form $H' : \Theta \in [\theta_0 - \epsilon, \theta_0 + \delta]$. This latter case is no different from anything considered already. The case of a point hypothesis has some interesting features, which we will explore in the remainder of this section.

Jeffreys (1961) suggests the use of what are now called Bayes factors for comparing a point hypothesis to a continuous alternative. Let $P_\theta \ll \nu$ for all θ, and suppose that one assigns probability p_0 to $\{\Theta = \theta_0\}$ and uses a prior distribution λ on $\Omega \setminus \{\theta_0\}$ for the conditional prior given $\Theta \neq \theta_0$. Then the joint density of the data and Θ (with respect to ν times the sum of λ and a point mass at θ_0) is

$$f_{X,\Theta}(x,\theta) = \begin{cases} p_0 f_{X|\Theta}(x|\theta_0) & \text{if } \theta = \theta_0, \\ (1-p_0) f_{X|\Theta}(x|\theta) & \text{if } \theta \neq \theta_0. \end{cases}$$

The marginal density of the data is

$$f_X(x) = p_0 f_{X|\Theta}(x|\theta_0) + (1-p_0) \int f_{X|\Theta}(x|\theta) d\lambda(\theta).$$

The posterior distribution of Θ has density (with respect to the sum of λ and a point mass at θ_0)

$$f_{\Theta|X}(\theta|x) = \begin{cases} p_1 & \text{if } \theta = \theta_0, \\ (1-p_1)\frac{f_{X|\Theta}(x|\theta)}{f_X(x)} & \text{if } \theta \neq \theta_0, \end{cases}$$

where

$$p_1 = \frac{p_0 f_{X|\Theta}(x|\theta_0)}{f_X(x)}$$

is the posterior probability of $\Theta = \theta_0$. It is easy to see that

$$\frac{p_1}{1-p_1} = \frac{p_0}{1-p_0} \frac{f_{X|\Theta}(x|\theta_0)}{\int f_{X|\Theta}(x|\theta) d\lambda(\theta)}. \tag{4.16}$$

The second factor on the right-hand side of (4.16) is the *Bayes factor*. It would be the posterior odds in favor of $\Theta = \theta_0$ if $p_0 = .5$. For other values of p_0, one needs to multiply the prior odds times the Bayes factor to calculate the posterior odds. The advantage of calculating a Bayes factor over the posterior odds $(p_1/[1-p_1])$ is that one need not state a prior odds in favor of the hypothesis. This might be useful if one is reporting the results of an experiment rather than trying to make a decision. One must still, however, state a prior distribution over the alternative given that the hypothesis is false.

Example 4.17. Suppose that $X \sim N(\theta, 1)$ given $\Theta = \theta$ and $\Omega_H = \{\theta_0\}$, $\Omega_A = (-\infty, \theta_0) \cup (\theta_0, \infty)$. Let the prior probability of the hypothesis be $\Pr(\Theta = \theta_0) = p_0 > 0$. Suppose that the conditional prior distribution of Θ given $\Theta \neq \theta_0$ is a measure λ. It is not difficult to show that

$$\frac{p_1}{1-p_1} = \frac{p_0}{1-p_0} \exp\left(-\frac{\theta_0^2}{2}\right) \left[\int \exp\left(x[\theta - \theta_0] - \frac{\theta^2}{2}\right) d\lambda(\theta)\right]^{-1}.$$

If λ puts positive mass on both sides of θ_0, then it is easily verified that $\int \exp(x[\theta - \theta_0] - \theta^2/2) d\lambda(\theta)$ is convex as a function of x and goes to ∞ as $x \to \pm\infty$. So all formal Bayes rules will be of the form "reject H if x is outside of some bounded interval."

When testing hypotheses of the form $H : \Theta = \theta_0$, the formal Bayes rule can be written in the form "reject H if the Bayes factor is less than something." It is possible to bound the Bayes factor from below when the likelihood function is bounded above. That is, we might be able to find a distribution λ that would lead to the smallest possible Bayes factor.[5] This lower bound would give a bound on how strongly the data conflict with the hypothesis.

Example 4.18 (Continuation of Example 4.17). The Bayes factor in this example is

$$\exp\left(-\frac{\theta_0^2}{2}\right) \left[\int \exp\left(x[\theta - \theta_0] - \frac{\theta^2}{2}\right) d\lambda(\theta)\right]^{-1},$$

which is minimized (over λ) by the distribution that puts probability 1 on that value of θ which maximizes the integrand, which is the likelihood function. In this case, that would be $\theta = x$, and the lower bound on the Bayes factor is $\exp(-[x-\theta_0]^2/2)$. For example, if $x = \theta_0 + 1.96$, which is the critical value for the usual two-sided level 0.05 test of H, we get a lower bound of 0.1465. This says that a data value that would just barely lead to rejecting H at level 0.05 could not possibly change one's odds against the hypothesis by more than a factor of 7, and then only in the extremely unlikely case that one believed before seeing the data that Θ was sure to equal $\theta_0 + 1.96$ if it was not θ_0. Put another way, in order for the posterior probability of H to be as low as 0.05, the prior probability p_0 would have to be lower than 0.2643, and much lower if a more reasonable prior on the alternative were used.

A more realistic expression of prior opinion might be that the prior, given $\Theta \neq \theta_0$, is a normal distribution with mean θ_0 and some variance τ^2. In this case, the Bayes factor is

$$\sqrt{1+\tau^2} \exp\left(-\frac{(x-\theta_0)^2 \tau^2}{2(1+\tau^2)}\right). \tag{4.19}$$

The prior in this class that leads to the smallest Bayes factor can easily be shown (see Problem 7 on page 286) to be the one with

$$\tau^2 = \begin{cases} (x-\theta_0)^2 - 1 & \text{if } |x-\theta_0| > 1, \\ 0 & \text{otherwise}. \end{cases}$$

---

[5]For more discussion of this technique, see Edwards, Lindman, and Savage (1963).

The minimum Bayes factor is $|x - \theta_0| \exp(\{-[x - \theta_0]^2 + 1\}/2)$ if $|x - \theta_0| > 1$. The minimum is 1 if $|x - \theta_0| \le 1$. At $x = \theta_0 + 1.96$, the minimum Bayes factor is 0.4734. This time, the prior probability p_0 would have to be lower than 0.1 in order for the posterior probability to be as low as 0.05.

Intermediate to the two bounds above is the bound obtained by supposing that Θ has distribution symmetric around θ_0, but not necessarily normal. Since a prior that is symmetric around θ_0 is a mixture of priors that put probability $1/2$ on two points symmetrically located around θ_0, the smallest value of the Bayes factor among all symmetric priors can be obtained by maximizing over priors that put probability $1/2$ on $\theta_0 \pm c$ for $c \ge 0$. For such a prior, the density of the data given that the hypothesis is false equals

$$\left(2\sqrt{2\pi}\right)^{-1} \left[\exp\left(-\frac{[x - \theta_0 + c]^2}{2}\right) + \exp\left(-\frac{[x - \theta_0 - c]^2}{2}\right) \right].$$

Maximizing this as a function of c leads to $c = 0$ if $|x - \theta_0| \le 1$. If $|x - \theta_0| > 1$, the maximum occurs at the solution to the equation

$$\frac{x - \theta_0 + c}{x - \theta_0 - c} = \exp(2c[x - \theta_0]).$$

For $|x - \theta_0| > 1.5$, the solution c is very nearly equal to $|x - \theta_0|$ (although it is always strictly smaller than $|x - \theta_0|$). If $x = \theta_0 + 1.96$, for example, then $c = 1.958$. The value of $f_X(x)$, when $c = |x - \theta_0|$, is $[1 + \exp(-2|x - \theta_0|^2)]/[2\sqrt{2\pi}]$. If $x = \theta_0 + 1.96$, for example, then the lower bound on the Bayes factor is 0.2928, approximately twice the global lower bound. This is not surprising, since the two-point distribution puts half of its probability very nearly at the same point as does the one-point distribution that led to the global lower bound. The other half of the probability is on a point that contributes nearly nothing because it is so far from x.

The global lower bound on the Bayes factor, namely

$$\frac{f_{X|\Theta}(x|\theta_0)}{\sup_{\theta \ne \theta_0} f_{X|\Theta}(x|\theta)}, \tag{4.20}$$

is closely related to the likelihood ratio test statistic, which is discussed in Section 4.5.5.

Upper bounds on Bayes factors are usually harder to come by. This is due to the fact that there are often priors (even conjugate priors) that place such high probability on the data being very far from what was observed that the hypothesis will be highly favored if such a prior is used for the alternative. For example, in Example 4.17, if the alternative prior is $N(\theta_0, \tau^2)$, the Bayes factor goes to ∞ as τ^2 goes to ∞. In this regard, it is important to note that improper priors are particularly inappropriate for the conditional distribution of Θ given $\Theta \ne \theta_0$. The limit as τ^2 goes to ∞ in Example 4.17 leads to an improper prior. As we just noted, the Bayes factor goes to ∞ because the improper prior for the alternative says that Θ has probability 1 of being outside of every bounded interval. Since the data will surely be inside some bounded interval, it will appear to be much more consistent with the hypothesis than the alternative. There are ways, however, to use limits of proper priors in Bayes factors.

Example 4.21 (Continuation of Example 4.18; see page 222). Suppose that we wish to let τ^2 go to ∞ in the $N(\theta_0, \tau^2)$ prior for Θ given the alternative. In order to use an improper prior to approximate a proper prior in this problem, we would have to let the prior on the hypothesis be improper also. This could be done by letting p_0 go to zero in such a way that $p_0\tau \to k$. In this case, $p_0/[1-p_0]$ times the Bayes factor converges to $k\exp(-[x-\theta_0]^2/2)$. It this way, k acts like the prior odds ratio, and $\exp(-[x-\theta_0]^2/2)$ acts like the Bayes factor. In fact, k is the limit (as $\tau \to \infty$) of the ratio of p_0 to the prior probability that Θ is in the interval $[-\sqrt{\pi/2}, \sqrt{\pi/2}]$ given the alternative. (See Problem 8 on page 287.)

By restricting the class of prior distributions, one can obtain useful upper bounds on Bayes factors. For example, in Example 4.17, one could restrict attention to priors with $\tau^2 \leq c$. Since the Bayes factor is increasing as a function of τ^2, we get that the maximum occurs at $\tau^2 = c$. For large c, one can easily compute the upper bound to be approximately $\sqrt{c}$ times the global minimum Bayes factor.

Bayes factors can also be calculated in cases in which the hypothesis is of the form $H: g(\Theta) = g(\theta_0)$ versus $A: g(\Theta) \neq g(\theta_0)$ for some function g. For example, the hypothesis might concern only one of several coordinates of Θ. In this case, global lower bounds on the Bayes factor are not particularly useful.

Example 4.22. Let $\Theta = (M, \Sigma)$, and suppose that $X_1, \ldots, X_n$ are conditionally IID given $\Theta = (\mu, \sigma)$ with $N(\mu, \sigma^2)$ distribution. Suppose that $H: M = \mu_0$ is the hypothesis. Given $M \neq \mu_0$, we suppose that $\Sigma^2 \sim \Gamma^{-1}(a_0/2, b_0/2)$ and that M given $\Sigma = \sigma$ has $N(\mu^0, \sigma^2/\lambda_0)$ distribution. This is the usual conjugate prior distribution. Conditional on $M = \mu_0$, we still need a prior distribution of Σ^2. We will use the conditional distribution given $M = \mu_0$ obtained from the joint distribution given $M \neq \mu_0$. Conditional on $M = \mu_0$, Σ^2 has $\Gamma^{-1}(a_0^*/2, b_0^*/2)$ distribution, where $a_0^* = a_0 + 1$ and $b_0^* = b_0 + \lambda_0(\mu_0 - \mu^0)^2$. The conditional density of (X, Σ) given $M = \mu_0$, $f_{X,\Sigma|M}(x, \sigma|\mu_0)$, equals

$$\frac{2\left(\frac{b_0^*}{2}\right)^{\frac{a_0^*}{2}}}{(2\pi)^{\frac{n}{2}}\Gamma\left(\frac{a_0^*}{2}\right)} \sigma^{-(n+a_0^*+1)} \exp\left\{-\frac{1}{2\sigma^2}\left[b_0^* + w + n(\bar{x}_n - \mu_0)^2\right]\right\}.$$

Given $M \neq \mu_0$, the joint density of (X, M, Σ) is

$$\frac{2\left(\frac{b_0}{2}\right)^{\frac{a_0}{2}}\sqrt{\lambda_0}}{(2\pi)^{\frac{n+1}{2}}\Gamma\left(\frac{a_0}{2}\right)} \sigma^{-n-a_0-2} \exp\left\{-\frac{1}{2\sigma^2}\left[b_0 + w + \lambda_1(\mu - \mu^1)^2 + \frac{n\lambda_0}{\lambda_1}(\bar{x}_n - \mu_0)^2\right]\right\},$$

where

$$\bar{x}_n = \frac{1}{n}\sum_{i=1}^n x_i, \qquad w = \sum_{i=1}^n (x_i - \bar{x}_n)^2,$$

$$\lambda_1 = \lambda_0 + n, \qquad \mu^1 = \frac{n\bar{x}_n + \lambda_0\mu^0}{\lambda_1}.$$

If we integrate the parameters out of the two densities above and take the ratio, we get the Bayes factor:

$$\frac{\sqrt{\lambda_1}}{\sqrt{\lambda_0}} \frac{b_1^{\frac{a_1}{2}}}{(b_1^*)^{\frac{a_1^*}{2}}} \frac{b_0^{*-\frac{a_0^*}{2}} \Gamma\left(\frac{a_0}{2}\right) \Gamma\left(\frac{a_1^*}{2}\right)}{b_0^{-\frac{a_0}{2}} \Gamma\left(\frac{a_0^*}{2}\right) \Gamma\left(\frac{a_1}{2}\right)}, \quad (4.23)$$

where

$$a_1 = a_0 + n, \qquad b_1 = b_0 + w + \frac{n\lambda_0}{\lambda_1}(\bar{x}_n - \mu^0)^2,$$
$$a_1^* = a_0^* + n, \qquad b_1^* = b_0^* + w + n(\bar{x}_n - \mu_0)^2.$$

To put a lower bound on the Bayes factor, we first note that the conditional distribution of M given Σ and $M \neq \mu_0$ which will lead to the largest marginal density for X given $M \neq \mu_0$ is the one that says $M = \bar{x}$ with probability 1. We are then left with the problem of finding distributions for Σ given $M = \mu_0$ and given $M \neq \mu_0$. It is easy to see that if we let the distribution of Σ be concentrated at the same value c for both the hypothesis and the alternative, then the Bayes factor is $\exp(-n(\bar{x} - \mu_0)^2/[2c^2])$, which goes to 0 as c goes to 0, unless $\bar{x} = \mu_0$. If $\bar{x} = \mu_0$ (a probability 0 event given Θ), the lower bound on the Bayes factor is still 0, but one achieves this by letting the priors for Σ be different under the hypothesis and alternative.

If one wished to use improper priors, one would have to let λ_0 go to 0 while $p_0/\sqrt{\lambda_0}$ converges to some finite strictly positive number k.[6] In this case $a_0^* = a_0$ instead of $a_0 + 1$ because Σ and M are independent in the improper prior. To convert (4.23) to the case of the improper prior, we set $a_0 = -1$ and $b_0 = 0$. The product of the prior odds and the Bayes factor becomes

$$k\sqrt{n}\left(1 + \frac{t^2}{n-1}\right)^{-\frac{n-1}{2}},$$

where $t = \sqrt{n}(\bar{x} - \mu_0)/\sqrt{w/[n-1]}$ is the usual t statistic used to test $H : M = \mu_0$.

In general, minimizing a Bayes factor for a problem like the one in Example 4.22 would require choosing the prior for the alternative to maximize the predictive density and choosing the prior for the hypothesis to minimize the predictive density. But this latter problem was already seen to lead to the minimum being 0 in most cases. In short, the global lower bound on the Bayes factor, when the hypothesis concerns only a function of the parameter, will most likely be 0 and so is not useful. An alternative to the global lower bound is an approximate Bayes factor formed by maximizing the marginal density of X separately under the hypothesis and alternative

$$\frac{\sup_{\theta \in \Omega_H} f_{X|\Theta}(x|\theta)}{\sup_{\theta \in \Omega_A} f_{X|\Theta}(x|\theta)}. \quad (4.24)$$

---

[6]This approach was suggested in personal communication with Luke Tierney. It is also the approach taken by Robert (1993).

This approximate Bayes factor is also closely related to likelihood ratio tests (see Section 4.5.5).

Example 4.25 (Continuation of Example 4.22; see page 224). To maximize the marginal density of the data under the alternative, we choose the prior distribution to concentrate all of its probability on the values for σ and μ which provide a maximum for the likelihood function. These are clearly $\mu = \bar{x}$ and $\sigma = \sqrt{w/n}$. Under the hypothesis, we must choose σ to maximize the likelihood, and the appropriate value is $\sigma = \sqrt{w/n + (\bar{x} - \mu_0)^2}$. This approximation corresponds to letting $\lambda_0 = 0$, $b_0 = b_0^* = 0$, and $a_0 = a_0^* = 0$ in the analysis with the conjugate prior. The approximate Bayes factor would then equal

$$\left(\frac{w}{w + n(\bar{x} - \mu_0)^2}\right)^{\frac{n}{2}} = \left(1 + \frac{t^2}{n-1}\right)^{-\frac{n}{2}}, \qquad (4.26)$$

where $t = \sqrt{n}(\bar{x} - \mu_0)/\sqrt{w/[n-1]}$ is the usual t statistic used to test $H : M = \mu_0$.

An alternative approximation to the Bayes factor is available by approximating the marginal densities of the data under the alternative and hypothesis using the method of Laplace.[7] That is, approximate the product of likelihood times the prior by a multivariate normal density with the wrong normalizing constant. Then approximate the integral over the parameter space by the integral of a normal density. For example, suppose that under the hypothesis, the parameter is Ψ with prior density $f_\Psi(\psi)$ and that under the alternative, the parameter is Θ with prior density $f_\Theta(\theta)$. The likelihoods are $f_{X|\Psi}(x|\psi)$ and $f_{X|\Theta}(x|\theta)$, respectively. Assume that $\hat{\Psi}$ and $\hat{\Theta}$ provide the largest values of the two likelihood functions. If the maxima occur at points where the partial derivatives of the likelihoods are 0, and if the likelihoods have continuous second partial derivatives, then we can write

$$\log f_{X|\Psi}(x|\psi) \approx \log f_{X|\Psi}(x|\hat{\Psi}) + \frac{1}{2}[\psi - \hat{\Psi}]^\top A[\psi - \hat{\Psi}],$$

where A is the matrix of second partial derivatives of the logarithm of the likelihood evaluated at $\hat{\Psi}$.[8] This matrix will typically be negative definite. Let $\sigma_\psi = -A^{-1}$. A similar expression is obtained for Θ.

$$\log f_{X|\Theta}(x|\theta) \approx \log f_{X|\Theta}(x|\hat{\Theta}) + \frac{1}{2}[\theta - \hat{\Theta}]^\top B[\theta - \hat{\Theta}],$$

Let $\sigma_\theta = -B^{-1}$. If, in addition, the prior densities are relatively flat in the regions where the likelihoods attain their largest values, we can write

$$\int f_{X|\Psi}(x|\psi) f_\Psi(\psi) d\psi \approx f_\Psi(\hat{\Psi}) f_{X|\Psi}(x|\hat{\Psi})$$

---

[7] We will discuss the large sample properties of the method of Laplace in Section 7.4.3. (In particular, see Theorem 7.116 and the ensuing discussion.) Here, we give only a description of the method without any rigorous justification. The derivation presented here is based on Kass and Raftery (1995).

[8] The matrix $-A$ is sometimes called the *observed Fisher information*.

$$\times \int \exp\left(-\frac{1}{2}[\psi - \hat{\Psi}]^T \sigma_\psi^{-1}[\psi - \hat{\Psi}]\right) d\psi$$
$$= f_\Psi(\hat{\Psi}) f_{X|\Psi}(x|\hat{\Psi})(2\pi)^{\frac{k}{2}}|\sigma_\psi|^{\frac{1}{2}},$$

where k is the dimension of the vector Ψ. A similar expression is obtained for the integral over θ. The Bayes factor is

$$\frac{\int f_{X|\Psi}(x|\psi)f_\Psi(\psi)d\psi}{\int f_{X|\Theta}(x|\theta)f_\Theta(\theta)d\theta} \approx (2\pi)^{\frac{k-p}{2}} \frac{f_\Psi(\hat{\Psi})f_{X|\Psi}(x|\hat{\Psi})|\sigma_\psi|^{\frac{1}{2}}}{f_\Theta(\hat{\Theta})f_{X|\Theta}(x|\hat{\Theta})|\sigma_\theta|^{\frac{1}{2}}}. \tag{4.27}$$

The factor $f_\Psi(\hat{\Psi})/f_\Theta(\hat{\Theta})$ can be removed and multiplied times the prior odds $p_0/[1-p_0]$ to capture the prior input required. The rest of the approximate Bayes factor does not require the specification of any prior distributions. The removed factor, however, is not entirely prior-dependent. It also depends on the observed data.

Example 4.28 (Continuation of Example 4.22; see page 224). In the case of testing $H : M = \mu_0$, we have $k = 1$ and $p = 2$ in (4.27) because $\Theta = (M, \Sigma)$ and $\Psi = \Sigma$. The likelihood functions have their maxima at $\hat{\Psi} = \sqrt{[w + n(\bar{x} - \mu_0)^2]/n}$ and $\hat{\Theta} = (\bar{x}, \sqrt{w/n})$. The matrices σ_ψ and σ_θ are

$$\sigma_\psi = \frac{w + n(\bar{x} - \mu_0)^2}{2n^2}, \quad \sigma_\theta = \frac{w}{n^2}\begin{pmatrix} 1 & 0 \\ 0 & \frac{1}{2} \end{pmatrix}.$$

The approximate Bayes factor is $1/\sqrt{2\pi}$ times the factor $f_\Psi(\hat{\Psi})/f_\Theta(\hat{\Theta})$ times the expression in (4.26) times the ratio of the square roots of the determinants of the two matrices above. The result is

$$\frac{nf_\Psi(\hat{\Psi})}{f_\Theta(\hat{\Theta})\sqrt{2\pi w}}\left(1 + \frac{t^2}{n-1}\right)^{-\frac{n-1}{2}}, \tag{4.29}$$

where $t = \sqrt{n}(\bar{x} - \mu_0)/\sqrt{w/[n-1]}$ is the usual t statistic for testing H.

To see how the approximation compares with the actual Bayes factor, suppose that the prior distributions are conjugate, as on page 224, and we let f_Ψ be the conditional prior calculated from f_Θ given that H is true. Then

$$f_\Psi(\sigma) = \frac{2\left(\frac{b_0^*}{2}\right)^{\frac{a_0^*}{2}}}{\Gamma\left(\frac{a_0^*}{2}\right)}\sigma^{-(a_0^*+1)}\exp\left(-\frac{b_0^*}{2\sigma^2}\right),$$

$$f_\Theta(\mu, \sigma) = \frac{2\left(\frac{b_0}{2}\right)^{\frac{a_0}{2}}\sqrt{\lambda_0}}{\sqrt{2\pi}\Gamma\left(\frac{a_0}{2}\right)}\sigma^{-(a_0+2)}\exp\left(-\frac{b_0 + \lambda_0(\mu - \mu^0)^2}{2\sigma^2}\right).$$

Plugging $\sigma = \hat{\Psi}$ into the first of these and $(\mu, \sigma) = (\bar{x}, \sqrt{w/n})$ into the second and taking the ratio give

$$\frac{f_\Psi(\hat{\Psi})}{f_\Theta(\hat{\Theta})} = \frac{\sqrt{2\pi}\left(\frac{b_0^*}{2}\right)^{\frac{a_0^*}{2}}\Gamma\left(\frac{a_0}{2}\right)\left(\frac{w}{n}\right)^{\frac{a_0+2}{2}}}{\sqrt{\lambda_0}\left(\frac{b_0}{2}\right)^{\frac{a_0}{2}}\Gamma\left(\frac{a_0^*}{2}\right)\hat{\Psi}^{a_0^*+1}}\exp\left(\frac{b_0 + \lambda_0(\bar{x} - \mu^0)^2}{2\left(\frac{w}{n}\right)} - \frac{b_0^*}{2\hat{\Psi}^2}\right).$$

If n is large, the exponential term above can be approximated by

$$\left(1 + \frac{b_0 + \frac{n\lambda_0}{n+\lambda_0}(\bar{x} - \mu^0)^2}{w}\right)^{-\frac{a_0+n}{2}} \left(1 + \frac{b_0 + \lambda_0(\mu_0 - \mu^0)^2}{n\hat{\Psi}^2}\right)^{-\frac{a_0^*+n}{2}}.$$

If we substitute this into (4.29) and notice that $1 + t^2/[n-1] = \hat{\Psi}^2/(w/n)$, we get

$$\frac{n \left(\frac{b_0^*}{2}\right)^{\frac{a_0^*}{2}} \Gamma\left(\frac{a_0}{2}\right) \left(\frac{w}{n}\right)^{\frac{n-1}{2}} \left(\frac{w}{n}\right)^{\frac{a_0+2}{2}} \left(1 + \frac{b_0 + \frac{n\lambda_0}{n+\lambda_0}(\bar{x} - \mu^0)^2}{w}\right)^{\frac{a_0+n}{2}}}{\sqrt{w}\left(\frac{b_0}{2}\right)^{\frac{a_0}{2}} \sqrt{\lambda_0}\Gamma\left(\frac{a_0^*}{2}\right) \hat{\Psi}^{n-1} (\hat{\Psi}^2)^{\frac{a_0^*+1}{2}} \left(1 + \frac{b_0+\lambda_0(\mu_0-\mu^0)^2}{n\hat{\Psi}^2}\right)^{\frac{a_0^*+n}{2}}}$$

$$= \frac{n(b_0^*)^{\frac{a_0^*}{2}} \Gamma\left(\frac{a_0}{2}\right) b_1^{\frac{a_1}{2}}}{(b_0)^{\frac{a_0}{2}} \sqrt{2\lambda_0}\Gamma\left(\frac{a_0^*}{2}\right) (b_1^*)^{\frac{a_1^*}{2}}},$$

where b_1, b_1^*, a_1, and a_1^* are defined after (4.23). If λ_0 and a_0 are small relative to n, we can approximate $n/\sqrt{2}$ by $\sqrt{\lambda_0 + n}\Gamma(a_1^*/2)/\Gamma(a_1/2)$. With this approximation, the expression above becomes exactly (4.23). Although $f_\Psi(\hat{\Psi})/f_\Theta(\hat{\Theta})$ depends on the data, one could calculate values of the ratio for a range of plausible priors to see how much it could reasonably vary.

As an example, suppose that $\mu_0 = 1.5$ and that $n = 14$, $\bar{x} = 2.7$, and $w = 41$ are observed. Then $\hat{\Psi} = 2.0901$ and $\hat{\Theta} = (2.7, 1.7113)$. That portion of (4.29) that does not depend on the prior is

$$\frac{n}{\sqrt{2\pi w}}\left(1 + \frac{t^2}{n-1}\right)^{-\frac{n-1}{2}} = 0.0648.$$

Next, we let $\mu^0 = 1.5$ and let the other hyperparameters a_0, b_0, λ_0 be elements of the set $\{0.1, 1, 5, 10, 20\}$. Figure 4.30 shows the 125 different values of the logarithm of the ratio $f_\Psi(\hat{\Psi})/f_\Theta(\hat{\Theta})$ with λ_0 varying most rapidly and a_0 varying most slowly. Since $\log(0.0648) = -2.736$, those priors corresponding to values on the vertical axis greater than 2.736 (horizontal line) will lead to Bayes factors greater than 1, while the others lead to Bayes factors less than 1. Examining Figure 4.30, we see that many reasonable priors (those with small to moderate values of a_0 and λ_0 and values of b_0/a_0 in the vicinity of the observed sample variance $w/n = 2.93$) give values for the log of the ratio near 2.736. This suggests that the data will not dramatically alter anyone's opinion very much as to whether or not M = 1.5. The other approximate Bayes factor (4.26) is 0.0608, which suggests a significant reduction to the odds in favor of the hypothesis. The t statistic for the usual classical test would be 2.53, and the hypothesis would be rejected at level 0.05.

An interesting difference between Bayes factors and the results of classical hypothesis tests arises from the comparison of the various lower bounds on the Bayes factor to the significance probability (see Definition 4.8). We will generalize the concept of significance probability in Section 4.6 and

4.2. Bayesian Solutions   229

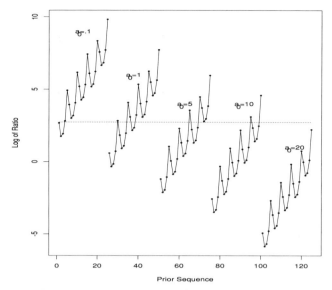

FIGURE 4.30. Logarithms of Ratios of Prior Densities

then make detailed comparisons with Bayes factors. At this point, we only mention that the results can often be in stark contrast. In particular, data with very small significance probability (supposedly suggesting that the data do not support the hypothesis) can have relatively large values for the lower bounds on the Bayes factor (suggesting that the data do not conflict with the hypothesis to a great extent). This conflict is sometimes called "Lindley's paradox" [see Lindley (1957) and Jeffreys (1961)].

If one believes that the probability is 0 that the parameter lies in a low-dimensional subset of the parameter space, then it is not appropriate to test the types of hypotheses we have considered in this section. As an alternative, one can calculate a measure of how far the parameter of interest is from the hypothesized low-dimensional subset. For example, if the parameter is $\Theta = (M, \Sigma)$ and the hypothesis is that M is near 0, then the posterior distribution of $|M|$ contains a great deal of information about how far M is from 0. Also, the posterior distribution of $|M|/\Sigma$ contains information of a similar sort.

Example 4.31. Suppose that $X \sim N(\theta, 1)$, given $\Theta = \theta$, and that our hypothesis is that Θ is near θ_0. If we consider all prior distributions of the form $\Theta \sim N(0, \tau^2)$, then the posterior distribution of Θ is $N([\theta_0 + \tau^2 x]/[1+\tau^2], \tau^2/[1+\tau^2])$. For each $\delta > 0$, we can calculate

$$\Pr(|\Theta - \theta_0| \leq \delta) = \Phi\left(\lambda[\theta_0 - x] + \frac{\delta}{\lambda}\right) - \Phi\left(\lambda[\theta_0 - x] - \frac{\delta}{\lambda}\right),$$

where $\lambda = \tau/\sqrt{1+\tau^2}$, and Φ is the standard normal CDF. There is no useful upper bound on this probability, but a lower bound is obtained by letting $\lambda \to$

1, which means $\tau^2 \to \infty$. This corresponds to the usual improper prior. After observing $X = x$, one could plot $\Pr(|\Theta - \theta_0| \leq \delta)$ (using an improper prior) as a function of δ to describe how far Θ is likely to be from θ_0. For example, if $x = \theta_0 + 1.96$, then $\Pr(|\Theta - \theta_0| \leq \delta) = 0.05$ for $\delta = 0.3983$.

For multiparameter problems, there may be many possible summaries of the parameters that measure the extent to which the parameter differs from the hypothesis. We will consider a very general class of such summaries in Section 8.2.3.

4.3 Most Powerful Tests

As we saw earlier, the power function of a test is closely associated with the risk function for a 0–1–c loss. It makes sense, then, that most attention in classical hypothesis testing focuses on the power function. The following definitions begin to introduce the criteria by which tests are evaluated in the classical framework.

Definition 4.32. Suppose that $\Omega = \Omega_H \cup \{\theta_1\}$, where $\theta_1 \notin \Omega_H$. A level α test ϕ of $H : \Theta \in \Omega_H$ versus $A : \Theta = \theta_1$ is called *most powerful (MP) level α* if, for every level α test ψ, $\beta_\psi(\theta_1) \leq \beta_\phi(\theta_1)$.

The corresponding dual criterion is the following.

Definition 4.33. Suppose that $\Omega = \Omega_A \cup \{\theta_0\}$, where $\theta_0 \notin \Omega_A$. A floor α test ϕ of $H : \Theta = \theta_0$ versus $A : \Theta \in \Omega_A$ is called *most cautious (MC) floor α* if, for every floor α test ψ, $\beta_\psi(\theta_0) \geq \beta_\phi(\theta_0)$.

For more general cases, we have the following definitions.

Definition 4.34. A level α test ϕ is *uniformly most powerful (UMP) level α* if, for every other level α test ψ, $\beta_\psi(\theta) \leq \beta_\phi(\theta)$ for all $\theta \in \Omega_A$. A floor α test ϕ is *uniformly most cautious (UMC) floor α* if, for every other floor α test ψ, $\beta_\psi(\theta) \geq \beta_\phi(\theta)$ for all $\theta \in \Omega_H$.

In some cases, both criteria (UMP and UMC) lead to the same optimal tests. In some cases, they do not. (See Problem 31 on page 289.) Either way, there is asymmetry in these definitions. A different criterion is used for protecting against one type of error than that used for protecting against the other. One argument given for the particular choice is that type I error is more costly than type II error, so we arrange for the maximum type I error probability to be small. However, what often happens is that the probability of type II error can become even smaller for most values of the parameter. Here is a simple example.

Example 4.35. Suppose that $X \sim Poi(\theta)$ given $\Theta = \theta$, and that $\Omega = \{1, 10\}$. We are interested in testing $H : \Theta = 1$ versus $A : \Theta = 10$. The MP level 0.05 test is (see Proposition 4.37 ahead)

$$\phi(x) = \begin{cases} 0 & \text{if } x \leq 2, \\ 0.5058 & \text{if } x = 3, \\ 1 & \text{if } x \geq 4. \end{cases}$$

The probability of type II error for this test is 0.0065, which is much smaller than the probability of type I error.

In Example 4.35, we protect ourselves more against the less costly error than against the more costly error. If type I error is more costly, it might make sense to minimize it using the UMC criterion and let the probability of type II error be a bit larger.

Example 4.36 (Continuation of Example 4.35; see page 230). The MC floor 0.05 test of $H : \Theta = 1$ versus $A : \Theta = 10$ is

$$\psi(x) = \begin{cases} 0 & \text{if } x \leq 4, \\ 0.4516 & \text{if } x = 5, \\ 1 & \text{if } x \geq 6. \end{cases}$$

The probability of type I error for this test is 0.00198. This test provides more protection against the more costly error, while keeping the probability of the other error at a low level.

Another alternative is to try to balance the costs of the two types of error in a deliberate fashion. For example, Lehmann (1958) offered the suggestion that one decrease the required level as the sample size increases so that the power would decrease also. Schervish (1983) suggested that the size of the test be matched to the power function at an alternative chosen based on substantive grounds.

Not many theorems can be proven about MP and UMP tests in general without some assumptions of additional structure. One general result is already familiar to us. The Neyman–Pearson lemma 3.87 provided a minimal complete class for this decision problem. For convenience, we restate that result here using the language of hypothesis testing.

Proposition 4.37 (**Neyman–Pearson fundamental lemma**). *Let $\Omega = \{\theta_0, \theta_1\}$ and let $P_\theta \ll \nu$, for some measure ν and both values of θ. Let $f_i(x) = dP_{\theta_i}/d\nu(x)$ for $i = 0, 1$. Let $H : \Theta = \theta_0$ and $A : \Theta = \theta_1$.*
For each $k \in (0, \infty)$ and each function $\gamma : \mathcal{X} \to [0, 1]$, define the test

$$\phi_{k,\gamma}(x) = \begin{cases} 1 & \text{if } f_1(x) > kf_0(x), \\ \gamma(x) & \text{if } f_1(x) = kf_0(x), \\ 0 & \text{if } f_1(x) < kf_0(x). \end{cases}$$

Also define the two tests

$$\phi_0(x) = \begin{cases} 1 & \text{if } f_1(x) > 0, \\ 0 & \text{if } f_1(x) = 0, \end{cases}$$

$$\phi_\infty(x) = \begin{cases} 1 & \text{if } f_0(x) = 0, \\ 0 & \text{if } f_0(x) > 0. \end{cases}$$

All of these tests are MP of their respective levels and MC of their respective floors.

Note that ϕ_∞ will have size 0 because it never rejects H when $f_0 > 0$. On the other hand, ϕ_0 will have the largest possible size for an admissible test, equal to 1 in many problems but not always.

The following result gives conditions under which MP tests are essentially unique.

Lemma 4.38. *Let $\Omega = \{\theta_0, \theta_1\}$ and let $P_\theta \ll \nu$, for some measure ν and both values of θ. Let $f_i(x) = dP_{\theta_i}/d\nu(x)$ for $i = 0, 1$, and let*

$$B_k = \{x : f_1(x) = kf_0(x)\}.$$

Suppose that for all $k \in [0, \infty]$, $P_{\theta_i}(B_k) = 0$ for $i = 0, 1$. Let ϕ be a test of $H : \Theta = \theta_0$ versus $A : \Theta = \theta_1$ of the form

$$\phi(x) = \begin{cases} 1 & \text{if } f_1(x) > kf_0(x), \\ 0 & \text{otherwise,} \end{cases}$$

and let ψ be another test such that $\beta_\phi(\theta_0) = \beta_\psi(\theta_0)$. Then either $\phi = \psi$, a.s. $[P_{\theta_i}]$ for $i = 0, 1$ or $\beta_\phi(\theta_1) > \beta_\psi(\theta_1)$.

PROOF. Let ϕ and ψ be as stated in the lemma. Define $A_> = \{x : \psi(x) > \phi(x)\}$ and $A_< = \{x : \psi(x) < \phi(x)\}$. Clearly, ϕ is in the form of a test from Proposition 4.37, and so it is MP of its size. Also, since ϕ only takes on the values 0 and 1, we have

$$A_> \subseteq \{x : \phi(x) = 0\} = \{x : f_1(x) \le kf_0(x)\},$$
$$A_< \subseteq \{x : \phi(x) = 1\} = \{x : f_1(x) > kf_0(x)\}.$$

It follows that

$$\int_{A_>} [\phi(x) - \psi(x)] f_1(x) d\nu(x) \ge k \int_{A_>} [\phi(x) - \psi(x)] f_0(x) d\nu(x),$$
$$\int_{A_<} [\phi(x) - \psi(x)] f_1(x) d\nu(x) \ge k \int_{A_<} [\phi(x) - \psi(x)] f_0(x) d\nu(x). \quad (4.39)$$

Because of the way $A_>$ and $A_<$ are defined, the sum of the two left-hand sides in (4.39) is $\beta_\phi(\theta_1) - \beta_\psi(\theta_1)$, and the sum of the two right-hand sides is $k[\beta_\phi(\theta_0) - \beta_\psi(\theta_0)] = 0$. Now, assume that $P_{\theta_1}(\phi(X) = \psi(X)) < 1$. It follows that $P_{\theta_1}(A_> \cup A_<) > 0$. Since $P_{\theta_1}(B_k) = 0$, we have that at least one of the inequalities in (4.39) is strict. In this case $\beta_\phi(\theta_1) > \beta_\psi(\theta_1)$. Finally, assume that $P_{\theta_0}(\phi(X) = \psi(X)) < 1$. It follows that either $P_{\theta_0}(A_>) > 0$ or $P_{\theta_0}(A_<) > 0$. In the latter case, $P_{\theta_1}(A_<) > 0$, and we have just proven that $\beta_\phi(\theta_1) > \beta_\psi(\theta_1)$. If $P_{\theta_0}(A_>) > 0$ and $P_{\theta_0}(A_<) = 0$, then $\psi \ge \phi$ with strict inequality on a set of positive P_{θ_0} probability, which would contradict $\beta_\psi(\theta_0) = \beta_\phi(\theta_0)$. □

Example 4.40. Let $X \sim N(\theta, 1)$ given $\Theta = \theta$, and let $\Omega = \{\theta_0, \theta_1\}$. Then B_k is a singleton set for every k and $P_{\theta_i}(B_k) = 0$ for every k and $i = 0, 1$.

Example 4.41. Let $X \sim U(0, \theta)$ given $\Theta = \theta$, and let $\Omega = \{\theta_0, \theta_1\}$. If $k = \theta_0/\theta_1$, then B_k is a set with positive probability under both P_{θ_0} and P_{θ_1}. So, the conditions of Lemma 4.38 are not met in this example.

4.3.1 Simple Hypotheses and Alternatives

In a simple–simple testing problem, the parameter space has only two points in it, and so the risk function has only two values. This makes it particularly easy to compare the risk functions of all tests at once. Each test corresponds to a point in two-dimensional space. Each coordinate is the risk function evaluated at one of the parameter values. For definiteness, let $\Omega = \{\theta_0, \theta_1\}$ and let $\Omega_H = \{\theta_0\}$, so that $\Omega_A = \{\theta_1\}$. Let α_0 stand for $\beta_\phi(\theta_0)$ and α_1 for $1 - \beta_\phi(\theta_1)$ for an arbitrary test ϕ. Then the risk function of ϕ is represented by the point $(\alpha_0, \alpha_1) \in [0, 1]^2$. The risk set, as defined in Definition 3.71, is the set of all possible (α_0, α_1) points.

Example 4.42. Suppose that $X \sim N(\theta, 1)$ given $\Theta = \theta$ and $\Omega = \{0, 1\}$. According to the Neyman–Pearson fundamental lemma 4.37, the MP tests of $H : \Theta = 0$ versus $A : \Theta = 1$ are those that reject H when $\exp(-[x-1]^2/2) > k \exp(-x^2/2)$ for various values of $k \in [0, \infty]$. This inequality simplifies to $x > c$ for arbitrary $c \in [-\infty, \infty]$. For each c, we get a point in the risk set with

$$\alpha_0 = P_0(X > c) = 1 - \Phi(c) = \Phi(-c),$$
$$\alpha_1 = P_1(X \leq c) = \Phi(c - 1).$$

A plot of these points is given in Figure 4.44.

Figure 4.44 has several features that are typical of all risk sets.

Lemma 4.43. *The risk set for a simple–simple hypothesis-testing problem is closed, convex, and symmetric about the point $(1/2, 1/2)$. It also contains that portion of the line $\alpha_1 = 1 - \alpha_0$ lying in the unit square.*

PROOF. The rule $\phi(x) \equiv \alpha_0$ corresponds to the point $(\alpha_0, 1-\alpha_0)$ for each α_0 between 0 and 1, so the risk set contains the portion of the line $\alpha_1 = 1 - \alpha_0$ which lies in the unit square.

Suppose that (a, b) is in the risk set. The symmetrically placed point about $(.5, .5)$ is $(1 - a, 1 - b)$. If ϕ produces the first point, then $1 - \phi$ produces the second point. Hence the risk set is symmetric about $(.5, .5)$. Lemma 3.74 shows that the risk set is convex.

To show that the risk set is closed, we need only show that it contains its lower boundary. The rest of the boundary is included by symmetry and convexity (and the fact that the line $\alpha_1 = 1 - \alpha_0$ is in). By Proposition 3.91, a Bayes rule exists for every prior. By Lemma 3.96, every point on ∂_L is the risk function for one of these Bayes rules; hence ∂_L is in the risk set. □

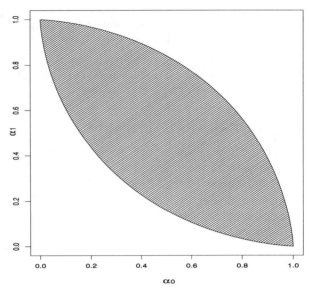

FIGURE 4.44. Risk Set for Testing $H : \Theta = 0$ versus $A : \Theta = 1$ with $X \sim N(\theta, 1)$

The definition of the lower boundary ∂_L of the risk set (see Definition 3.71) is designed so that exactly those tests which produce points on ∂_L are admissible. The lower boundary also consists solely of Bayes rules. (See Lemma 3.96.) Proposition 3.91 tells us that if a Bayes rule with respect to some prior is not in ∂_L, then one of the two prior probabilities is 0 and there is another Bayes rule with respect to that prior that is in ∂_L. These considerations lead to the following result.

Lemma 4.45.[9] *If a test ϕ is MP level α for testing $H : \Theta = \theta_0$ versus $A : \Theta = \theta_1$, then either $\beta_\phi(\theta_1) = 1$ or $\beta_\phi(\theta_0) = \alpha$.*

PROOF. We will prove the contrapositive. Suppose that both $\beta_\phi(\theta_1) < 1$ and $\beta_\phi(\theta_0) < \alpha$. Create another test ϕ' as follows. Let $A = \{x : \phi(x) < 1\}$. Note that $P_{\theta_1}(A) > 0$, since $\beta_\phi(\theta_1) < 1$. Let $g_c(x) = \min\{c, 1-\phi(x)\}$. Then, for $c \geq 0$, $h_i(c) = \mathrm{E}_{\theta_i} g_c(X) \geq 0$ and $h_i(c)$ is nondecreasing in c. Also, it is easy to see that $h_i(c)$ is continuous in c, since $|h_i(c) - h_i(d)| \leq |c - d|$. It follows that there exists c such that $h_0(c) = \alpha - \beta_\phi(\theta_0)$. Let $\phi' = \phi + g_c$. It follows that $\beta_{\phi'}(\theta_0) = \alpha$. Also, $\phi'(x) > \phi(x)$ for all $x \in A$. Since $P_{\theta_1}(A) > 0$, it follows that $h_1(c) > 0$ and $\beta_{\phi'}(\theta_1) > \beta_\phi(\theta_1)$. So ϕ is not MP level α. □

Lemma 4.45 says that a test that is MP level α must have size α unless all tests with size α are inadmissible. This result allows us to say when the two optimality criteria (MP and MC) are equivalent in the simple–simple testing situation.

---

[9]This lemma is used in the proofs of Lemmas 4.47 and 4.103.

Proposition 4.46. *Suppose that α_0 and α_1 are both strictly between 0 and 1. Suppose that a test ϕ of $H : \Theta = \theta_0$ versus $A : \Theta = \theta_1$ corresponds to the point (α_0, α_1) in the risk set. Then ϕ is MC floor $1 - \alpha_1$ if and only if it is MP level α_0.*

The reason for the restriction that α_0 and α_1 be strictly between 0 and 1 is that many tests with $\alpha_0 \in \{0, 1\}$ or $\alpha_1 \in \{0, 1\}$ are inadmissible even though they may satisfy one of the two optimality criteria.

The following lemma allows us to conclude that if ϕ is an MP level α test and we switch the names of hypothesis and alternative, then $1 - \phi$ becomes an MC floor $1 - \alpha$ test in the new problem (see Problem 30 on page 289 for a more general version.)

Lemma 4.47. *If ϕ is MP level α for testing $H : \Theta = \theta_0$ versus $A : \Theta = \theta_1$, then $1 - \phi$ has the smallest power at θ_1 among all tests with size at least $1 - \alpha$.*

PROOF. First, note that $1 - \phi$ has size at least $1 - \alpha$. Next, suppose that $\beta_\phi(\theta_1) = 1$. Then $1 - \phi$ has power 0 at θ_1 and is clearly the least powerful of any class to which it belongs. By Lemma 4.45, the only other case to consider is that in which $\beta_\phi(\theta_0) = \alpha$. In this case, $1 - \phi$ has level $1 - \alpha$. Suppose, to the contrary, that $\beta_\psi(\theta_0) \geq 1 - \alpha$ and $\beta_\psi(\theta_1) < \beta_{1-\phi}(\theta_1)$. Then $\beta_{1-\psi}(\theta_0) \leq \alpha$ and $\beta_{1-\psi}(\theta_1) > \beta_\phi(\theta_1)$, which contradicts the assumption that ϕ is MP level α. □

The following lemma says that in the comparison of two MP Neyman–Pearson tests, the one with the smaller level will also have the smaller power.

Lemma 4.48.[10] *Let $\{P_\theta : \theta \in \Omega\}$ be a parametric family. If ϕ_1 is a level α_1 test of the form of the Neyman–Pearson fundamental lemma 4.37 for testing $H : \Theta = \theta_0$ versus $A : \Theta = \theta_1$, and if ϕ_2 is a level α_2 test of that form with $\alpha_1 < \alpha_2$, then $\beta_{\phi_1}(\theta_1) < \beta_{\phi_2}(\theta_1)$.*

PROOF. By the Neyman–Pearson fundamental lemma 3.87, both ϕ_1 and ϕ_2 are admissible. If $\beta_{\phi_1}(\theta_1) \geq \beta_{\phi_2}(\theta_1)$, then ϕ_2 is inadmissible. □

The Neyman–Pearson fundamental lemma 4.37 tells us all of the admissible MP and MC tests. We also saw (Theorem 3.95 and Lemma 3.96) that these are the tests corresponding to points on ∂_L, the lower boundary of the risk set, and they are the Bayes rules with respect to positive priors and one Bayes rule for each of the priors that assign 0 probability to one of the parameter values. The usual classical approach to choosing one of the admissible tests is not to choose a prior distribution and then take the Bayes rule, but rather to choose a value of α and then choose the MP level α test. In cases with simple hypotheses and simple alternatives, the classical and Bayesian procedures will agree. That is, for every prior distribution, there

---

[10]This lemma is used in the proof of Theorem 4.56.

236   Chapter 4. Hypothesis Testing

is a formal Bayes rule and α such that the formal Bayes rule is MP level α. Similarly, for every α, there is a prior such that the MP level α test is a formal Bayes rule. Only in cases more complicated than those described so far can we distinguish these two approaches. Example 4.49 is one such case.

Example 4.49. Suppose that X_1 and X_2 are conditionally independent $U(0, \theta)$ random variables given $\Theta = \theta$ and $\Omega = \{1, 2\}$. Let $\Omega_H = \{1\}$ and $\Omega_A = \{2\}$. Suppose that Z is independent of X_1 and X_2 given Θ with distribution $Ber(1/2)$. Hence Z is ancillary. Suppose that we observe Z and $X = \max_{1 \leq i \leq n} X_i$ where $n = 1$ if $Z = 0$ and $n = 2$ if $Z = 1$. That is, the sample size is random $(n = Z + 1)$ but ancillary. The marginal densities of X (given $\Theta = 1, 2$) are

$$f_1(x) = \frac{1}{2} + x \text{ if } 0 < x < 1,$$

$$f_2(x) = \frac{1}{4} + \frac{x}{4} \text{ if } 0 < x < 2.$$

The MP level α test based solely on X is[11] $\phi(x) = 1$ if $f_2(x)/f_1(x) > c$ for some c. This becomes

$$\frac{\frac{1}{4} + \frac{x}{4}}{\frac{1}{2} + x} > c, \text{ or } x \geq 1.$$

For $1/3 \leq c \leq 1/2$, the first inequality is $x < (1/4 - c/2)(c - 1/4)$. For $c < 1/3$, the inequality is $x > 0$, and for $c > 1/2$, the inequality is $x \geq 1$. So, the MP level α test for $0 < \alpha < 1$ is $\phi(x) = 1$ if $x \geq 1$ or $x < (\sqrt{1 + 8\alpha} - 1)/2$. The power of this test is $\beta_\phi(2) = (9 + 4\alpha + \sqrt{1 + 8\alpha})/16$.

This test may seem odd because it picks $\theta = 2$ for small values of x. An alternative test is formed by conditioning on the ancillary Z. If $Z = 0$ $(n = 1)$, the MP level α conditional test is $\psi(x) = 1$ if $x > 1 - \alpha$. Actually, we could have chosen $x \geq 1$ together with any interval of length α, but the test is easier to write if we put the interval next to $x \geq 1$. Similarly, if $Z = 1$ $(n = 2)$, then the MP level α conditional test is $\psi(x) = 1$ if $x > \sqrt{1 - \alpha}$. (Once again, the ratio of densities is constant for all $x < 1$, so we could have chosen any set with probability α given $\Theta = 1$.) The power of this conditional test can be calculated to equal $(5 + 3\alpha)/8$, which is always smaller than the power of the test ϕ. So the test conditional on the ancillary is inadmissible.

The fact that the MP level α test conditional on the ancillary is inadmissible has led some [e.g., Bondar (1988)] to conclude that we should not condition on the ancillary in such problems. This mistaken conclusion is due to the fact that the conditional test being compared has conditional size α given all values of the ancillary. The lesson should be that we should not fix the size to be the same for all values of the ancillary, but we should continue to condition on the ancillary. If we have a prior (π_1, π_2) with $\pi_1 + \pi_2 = 1$, then the Bayes rule will be drastically different depending on whether $Z = 0$ or $Z = 1$ is observed.[12] A complete characterization of the Bayes rules ϕ_{π_1} for all values of π_1 is as follows.

---

[11] This test is not the MP level α test based on the joint distribution of the data (X, Z).

[12] These Bayes rules will be the MP level α tests based on the joint distribution of (X, Z) according to the Neyman–Pearson fundamental lemma 4.37.

4.3. Most Powerful Tests   237

We always have $\phi_{\pi_1}(x) = 1$ for $x \geq 1$. For $0 < x < 1$, the value of $\phi_{\pi_1}(x)$ is

π_1	$Z = 0$	$Z = 1$
$< \frac{1}{5}$	1	1
$\frac{1}{5}$	1	arbitrary
between $\frac{1}{5}$ and $\frac{1}{3}$	1	0
$\frac{1}{3}$	arbitrary	0
$> \frac{1}{3}$	0	0

Notice that the Bayes rule is never the same as either the unconditional level α test or the conditional level α test when $0 < \alpha < 1$. That is, there is no value of π_1 strictly between 0 and 1 such that the Bayes rule rejects H for some (but not all) $0 < x < 1$ for both $Z = 0$ and $Z = 1$. The Bayes rule either rejects H for all values of $0 < x < 1$ for at least one of the Z values or it rejects H for no values of $0 < x < 1$ for at least one of the Z values. The power functions of the Bayes rules are

π_1	$\beta(1)$	$\beta(2)$
$< \frac{1}{5}$	1	1
$\frac{1}{5}$	$\frac{1+a}{2}$	$\frac{7+a}{8}$
between $\frac{1}{5}$ and $\frac{1}{3}$	$\frac{1}{2}$	$\frac{7}{8}$
$\frac{1}{3}$	$\frac{a}{2}$	$\frac{5+2a}{8}$
$> \frac{1}{3}$	0	$\frac{5}{8}$

Here a is any number between 0 and 1 corresponding to the "arbitrary" parts of some of the Bayes rules. Note that for each α between 0 and 1, there is a Bayes rule with size α. (For example, to get $\alpha = 0.05$, let $a = 0.1$ in the fourth row of the table. One such test is the following. If $Z = 1$ is observed, $\phi_{1/3}(x) = 1$ for $x \geq 1$, and if $Z = 0$ is observed, $\phi_{1/3}(x) = 1$ for $x > 0.9$.) Since the Bayes rule with size α is the MP level α test, it has higher power than the unconditional size α test. (See Problem 15 on page 287.) On the other hand, it does not have conditional level α given the ancillary.

This example illustrates a conflict between two principles of classical statistics. The principle of conditioning on ancillaries together with the principle of choosing MP level α tests leads to the MP conditional size α test. This test is dominated by the MP unconditional size α test that ignores the ancillary. This test, in turn, is dominated by the unconditional size α test that makes use of the ancillary. The natural conclusion is to use this last test. But if we are to make use of the ancillary, aren't we supposed to condition on it? And if we condition on the ancillary, aren't we supposed to use the conditional size α test? The reason that it is difficult to justify (in the classical framework) using the size α test based on the whole data is that once the ancillary is observed, the conditional size of the test changes depending on the value of the ancillary. Why should an ancillary affect my choice of the size of the test? This begs the more important question, "How should the size of a test be chosen in a particular problem?" There are no general decision theoretic principles that lead one to be able to choose the size of a test based on a loss function or a prior distribution. There are cases in which one can find a simple correspondence between the size of a test and a loss function, but these seem to depend on additional structure not present in all problems, or they seem to be isolated instances not easily generalized. (See Theorem 6.74 on page 376 for a description of some additional structure and Example 4.61 on page 241 for an isolated example.)

4.3.2 Simple Hypotheses, Composite Alternatives

The next most complicated testing situation taken up by the classical theory is that of a simple hypothesis versus a composite alternative.[13] It is clear that, even from a decision theoretic perspective, a UMP level α test will have no larger Bayes risk than any other test whose size is α, no matter what prior distribution we use, so long as $\Omega_H = \{\theta_0\}$ and $\Omega = \{\theta_0\} \cup \Omega_A$. (See Problem 16 on page 287.)

Example 4.50. Suppose that $X \sim N(\theta, 1)$ given $\Theta = \theta$ and $\Omega = [\theta_0, \infty)$ with $\Omega_H = \{\theta_0\}$. For each $\theta_1 \in \Omega_A$, the MP level α test is $\phi(x) = 1$, if $f_{X|\Theta}(x|\theta_1)/f_{X|\Theta}(x|\theta_0) > k$. We can calculate the ratio

$$\frac{f_{X|\Theta}(x|\theta_1)}{f_{X|\Theta}(x|\theta_0)} = \exp\{x(\theta_1 - \theta_0)\} \exp\left\{-\frac{\theta_1^2 - \theta_0^2}{2}\right\}.$$

This is greater than k if and only if

$$x > t = \frac{\log k + \frac{1}{2}(\theta_1^2 - \theta_0^2)}{\theta_1 - \theta_0} = \frac{\theta_1 + \theta_0}{2} + \frac{\log k}{\theta_1 - \theta_0}.$$

(If we had a 0–1–c loss and the prior probability of θ_i were π_i, we would obtain this same test as long as $k = c\pi_0/\pi_1$.) The size of the test $\phi_t(x) = 1$ if $x > t$ is $\alpha_t = 1 - \Phi(t - \theta_0)$. So, $t = \theta_0 + \Phi^{-1}(1 - \alpha_t)$. For fixed α, the MP level α test of H versus $A_1 : \Theta = \theta_1$ is $\phi^\alpha(x) = 1$ if $x > \theta_0 + \Phi^{-1}(1 - \alpha)$. Notice that this is the same test for every θ_1. Hence this test is UMP level α for testing H versus A. Also notice that the conditions of Lemma 4.38 are met in this example, so that the UMP level α test is also the unique MP size α test for each $\theta_1 \in \Omega_A$ and hence the unique UMP level α test.

Now, consider a Bayesian approach in which the prior distribution satisfies $\Pr(\Theta = \theta_0) = p_0 > 0$. Suppose that the conditional prior distribution of Θ given $\Theta > \theta_0$ is a measure λ. It is not difficult to calculate the Bayes factor (see Section 4.2.2) in this case as

$$\frac{(1 - p_0) \Pr(\Theta = \theta_0 | X = x)}{p_0 \Pr(\Theta \neq \theta_0 | X = x)} = \exp\left(-\frac{\theta_0^2}{2}\right) \left[\int_{\theta_0}^\infty \exp\left(x[\theta - \theta_0] - \frac{\theta^2}{2}\right) d\lambda(\theta)\right]^{-1}.$$

Since $\theta > \theta_0$ in the exponent inside the integral, the Bayes factor is a decreasing function of x no matter what λ is. Hence, the formal Bayes rule (with 0–1–c

---

[13] By dealing with this case next, we postpone the issue that arises when the size of the test is the supremum of the power function over the hypothesis rather than just the value of the power function at the hypothesis. This subtle point is actually at the root of the asymmetry between hypotheses and alternatives.

loss) will be to reject H if $x > t$ is observed, where t is the largest x so that $\Pr(\Theta = \theta_0 | X = x) \geq 1/(1+c)$. In this case the prior distribution determines t, which in turn determines the size of the Bayes rule when thought of as a size α test. Each Bayes rule is a UMP level α test for some α, but the particular α (even for fixed c) depends on the prior. Alternatively, for a fixed prior, each value of c will lead to a different α, but the correspondence between α and c (although monotone for each prior) will be different for different priors.

The dual concept of UMC test would apply if the hypothesis were composite and the alternative were simple. It would then be the case that, if we switched the names of hypothesis and alternative, 1 minus the UMP level α test would become the UMC floor $1 - \alpha$ test. (See Problem 30 on page 289.) This case is interesting in that it is never discussed in the hypothesis-testing literature because the classical theory is not equipped to deal with it. If the power function is continuous (as it is in most of the examples that we can calculate) the power of an MP level α test of $H : \Theta < \theta_0$ versus $A : \Theta = \theta_0$ will be α. The optimality criterion gives us no way to choose between level α tests. They all have the same power on the alternative. Clearly, though, some are better than others. In fact, those that have smaller power functions for $\theta < \theta_0$ are better. But the MP criterion does not accommodate such comparisons.

4.3.3 One-Sided Tests

Next we consider the case in which both H and A are composite. For now, we will proceed along the classical UMP level α lines. We begin by introducing a concept that makes UMP and UMC tests come out the same.

Definition 4.51. If $\Omega \subseteq \mathbb{R}$, $\mathcal{X} \subseteq \mathbb{R}$, and $dP_\theta/d\nu(x) = f_{X|\Theta}(x|\theta)$ for some measure ν, then the parametric family is said to have *monotone likelihood ratio (MLR)* if, whenever $\theta_1 < \theta_2$, $f_{X|\Theta}(x|\theta_2)/f_{X|\Theta}(x|\theta_1)$ is a monotone function of x a.e. $[P_{\theta_1} + P_{\theta_2}]$ in the same direction for all pairs of θ_1 and θ_2. A parametric family has *increasing MLR* if the ratio is increasing for all $\theta_1 < \theta_2$, and it has *decreasing MLR* if the ratio is decreasing.

Example 4.52. Suppose that X has a one-parameter exponential family distribution with Θ being the natural parameter, $f_{X|\Theta}(x|\theta) = c(\theta)h(x)\exp(\theta x)$. Then

$$\frac{f_{X|\Theta}(x|\theta_2)}{f_{X|\Theta}(x|\theta_1)} = \frac{c(\theta_2)}{c(\theta_1)} \exp\{x(\theta_2 - \theta_1)\},$$

which is increasing in x for all $\theta_1 < \theta_2$.

Example 4.53. Suppose that $f_{X|\Theta}(x|\theta) = \left(\pi\left[1 + (x-\theta)^2\right]\right)^{-1}$, the Cauchy distribution with location shifts. Then

$$\frac{f_{X|\Theta}(x|\theta_2)}{f_{X|\Theta}(x|\theta_1)} = \frac{1 + (x-\theta_1)^2}{1 + (x-\theta_2)^2}.$$

This ratio goes to 1 as x approaches either ∞ or $-\infty$, but it is not constantly 1 and hence is not monotone. The same problem occurs with Cauchy distributions having only a scale parameter. However, the absolute value of a Cauchy random variable with a scale parameter does have MLR. (See Problem 17 on page 287.)

Example 4.54. Suppose that $f_{X|\Theta}(x|\theta) = 1/\theta$ for $0 < x < \theta$. For $\theta_2 > \theta_1$, write the likelihood ratio as

$$\frac{f_{X|\Theta}(x|\theta_2)}{f_{X|\Theta}(x|\theta_1)} = \begin{cases} \text{undefined} & \text{if } x \leq 0, \\ \frac{\theta_1}{\theta_2} & \text{if } 0 < x < \theta_1, \\ \infty & \text{if } \theta_1 \leq x < \theta_2, \\ \text{undefined} & \text{if } x \geq \theta_2. \end{cases}$$

The two "undefined" regions can be ignored because the ratio need only be monotone in x a.e. $[P_{\theta_1} + P_{\theta_2}]$. The "undefined" regions have 0 probability under both P_{θ_1} and P_{θ_2}. The likelihood ratio is then seen to be monotone increasing for every $\theta_1 < \theta_2$, although it is not strictly increasing. It only takes on two different values.

Proposition 4.55. *If g is measurable, monotonic, and one-to-one and $Y = g(X)$ and the family of distributions of X has MLR, then so does the family of distributions of Y.*

Theorem 4.56. *If $\{P_\theta : \theta \in \Omega\}$ is a parametric family with increasing MLR, then every test of the form*

$$\phi(x) = \begin{cases} 1 & \text{if } x > x_0, \\ \gamma & \text{if } x = x_0, \\ 0 & \text{if } x < x_0 \end{cases} \tag{4.57}$$

has nondecreasing (in θ) power function. Furthermore, each such test is UMP of its size for testing $H : \Theta \leq \theta_0$ versus $A : \Theta > \theta_0$, no matter what θ_0 is. Finally, for each $\alpha \in [0, 1]$ and each $\theta_0 \in \Omega$, there exists $x_0 \in \mathbb{R} \cup \{\pm\infty\}$ and $\gamma \in [0, 1]$ such that the test ϕ is UMP level α for testing H versus A.

PROOF. Let $\theta_1 < \theta_2 \in \Omega$. The Neyman–Pearson fundamental lemma 4.37 says that the MP test of $H_1 : \Theta = \theta_1$ versus $A_2 : \Theta = \theta_2$ is

$$\phi(x) = \begin{cases} 1 & \text{if } \frac{f_{X|\Theta}(x|\theta_2)}{f_{X|\Theta}(x|\theta_1)} > k, \\ \gamma(x) & \text{if } \frac{f_{X|\Theta}(x|\theta_2)}{f_{X|\Theta}(x|\theta_1)} = k, \\ 0 & \text{if } \frac{f_{X|\Theta}(x|\theta_2)}{f_{X|\Theta}(x|\theta_1)} < k. \end{cases}$$

Because the parametric family has MLR increasing, we can write ϕ as

$$\phi(x) = \begin{cases} 1 & \text{if } x > \bar{t}, \\ \gamma(x) & \text{if } \underline{t} \leq x \leq \bar{t}, \\ 0 & \text{if } x < \underline{t}. \end{cases} \tag{4.58}$$

(Note that we have put $x = \underline{t}$ and $x = \bar{t}$ in the $\phi(x) = \gamma(x)$ category. It may be required to set $\gamma(\underline{t}) = 0$ and/or $\gamma(\bar{t}) = 1$, depending on the

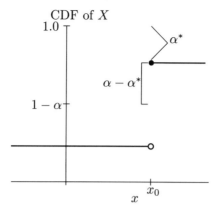

FIGURE 4.59. Step in Proof of Theorem 4.56

values of $f_{X|\Theta}(\underline{t}|\theta_2)/f_{X|\Theta}(\underline{t}|\theta_1)$ and $f_{X|\Theta}(\overline{t}|\theta_2)/f_{X|\Theta}(\overline{t}|\theta_1)$ if X does not have continuous distribution.) Let ϕ be of the form (4.58) and set $\alpha' = \beta_\phi(\theta_1)$. Let $\phi_{\alpha'}(x) \equiv \alpha'$, and since ϕ is MP, it follows that $\beta_\phi(\theta_2) \geq \alpha'$. Hence, we have shown that ϕ has nondecreasing power function.

Now, let $\alpha \in [0,1]$ be given, and let

$$x_0 = \begin{cases} \inf\{x : P_{\theta_0}(-\infty, x] \geq 1-\alpha\} & \text{if } \alpha < 1, \\ \inf\{x : P_{\theta_0}(-\infty, x] > 0\} & \text{if } \alpha = 1. \end{cases}$$

Then $\alpha^* = P_{\theta_0}(x_0, \infty) \leq \alpha$ and $P_{\theta_0}(\{x_0\}) \geq \alpha - \alpha^*$. (See Figure 4.59.) Let

$$\gamma^* = \begin{cases} 0 & \text{if } P_{\theta_0}(\{x_0\}) = 0, \\ \frac{\alpha - \alpha^*}{P_{\theta_0}(\{x_0\})} & \text{if } P_{\theta_0}(\{x_0\}) > 0. \end{cases}$$

(Note that $x_0 = -\infty$ is possible for $\alpha = 1$ and an unbounded distribution. In this case, $P_{\theta_0}(\{x_0\}) = 0$, $\alpha^* = 1$, and $\gamma^* = 0$.) Let ϕ be of the form (4.58) with $\underline{t} = x_0 = \overline{t}$ and $\gamma(x_0) = \gamma^*$. Then $\beta_\phi(\theta_0) = \alpha$ and ϕ is MP level α for testing $H_0 : \Theta = \theta_0$ versus $A_1 : \Theta = \theta_1$ for every $\theta_1 > \theta_0$, since it is the same test for all θ_1. Hence, ϕ is UMP level α for testing H_0 versus A.

Finally, let ψ be any test having level α for H. Then ψ has level at most α for H_0, and by Lemma 4.48, ϕ is at least as powerful as ψ at all $\theta \in A$. Since $\beta_\phi(\theta)$ is increasing, ϕ has level α for H, so it is UMP level α for testing H versus A. □

Definition 4.60. A test such as ϕ in Theorem 4.56 is called a *one-sided test*. A hypothesis of the form $H : \Theta \leq \theta_0$ or $H : \Theta \geq \theta_0$ is called a *one-sided hypothesis*.

Example 4.61. Suppose that $X \sim Poi(\theta)$ given $\Theta = \theta$. Then $f_{X|\Theta}(x|\theta) = \exp(-\theta)\theta^x/x!$ for $x = 0, 1, \ldots$ is the conditional density of X with respect to counting measure. For $\theta_1 < \theta_2$,

$$\frac{f_{X|\Theta}(x|\theta_2)}{f_{X|\Theta}(x|\theta_1)} = \exp(\theta_1 - \theta_2)\left(\frac{\theta_2}{\theta_1}\right)^x,$$

which increases in x. Hence, this family has MLR increasing. The UMP level 0.05 test of $H : \Theta \leq 1$ versus $A : \Theta > 1$ is

$$\phi(x) = \begin{cases} 1 & \text{if } x > v, \\ \gamma & \text{if } x = v, \\ 0 & \text{if } x < v, \end{cases}$$

where v and γ are chosen so that

$$\gamma \exp(-1)\frac{1^v}{v!} + \sum_{x=v+1}^{\infty} \exp(-1)\frac{1^x}{x!} = 0.05.$$

A little algebra shows us that $v = 3$ and $\gamma = 0.506$.

As a possible Bayesian solution to this problem, suppose that we have a 0–1–c loss and the prior for Θ is $\Gamma(a, b)$. Then the posterior given $X = x$ is $\Gamma(a+x, b+1)$. Since Gamma distributions are stochastically larger the larger the first parameter is (for fixed second parameter), we conclude that $\Pr(\Theta \leq 1|X = x) < 1/(1+c)$ if and only if $x \geq x_0$ for some x_0. For the improper prior with $a = b = 0$ (corresponding to density $d\theta/\theta$) and with $c = 19$ (so $1/(1+c) = 0.05$), the value of x_0 is 4. This is the same as the UMP level 0.05 test except for the randomization at $x = 3$.

There are also versions of Theorem 4.56 for decreasing MLR and for hypotheses of the form $H : \Theta \geq \theta_0$.

Proposition 4.62. *If $\{P_\theta : \theta \in \Omega\}$ is a parametric family with decreasing MLR, then any test of the form*

$$\phi(x) = \begin{cases} 1 & \text{if } x < x_0, \\ \gamma & \text{if } x = x_0, \\ 0 & \text{if } x > x_0 \end{cases} \tag{4.63}$$

has a nondecreasing (in θ) power function. Furthermore, it is UMP of its size for testing $H : \Theta \leq \theta_0$ versus $A : \Theta > \theta_0$, no matter what θ_0 is. Finally, for each $\alpha \in [0, 1]$ and each $\theta_0 \in \Omega$, there exists $x_0 \in \mathbb{R} \cup \{\pm\infty\}$ and $\gamma \in [0, 1]$ such that the test ϕ is UMP level α for testing H versus A.

Proposition 4.64. *If $\{P_\theta : \theta \in \Omega\}$ is a parametric family with increasing MLR, then any test of the form (4.63) has a nonincreasing (in θ) power function. Furthermore, it is UMP of its size for testing $H : \Theta \geq \theta_0$ versus $A : \Theta < \theta_0$, no matter what θ_0 is. Finally, for each $\alpha \in [0, 1]$ and each $\theta_0 \in \Omega$, there exists $x_0 \in \mathbb{R} \cup \{\pm\infty\}$ and $\gamma \in [0, 1]$ such that the test ϕ is UMP level α for testing H versus A.*

Proposition 4.65. *If $\{P_\theta : \theta \in \Omega\}$ is a parametric family with decreasing MLR, then any test of the form (4.57) has a nonincreasing (in θ) power function. Furthermore, it is UMP of its size for testing $H : \Theta \geq \theta_0$ versus $A : \Theta < \theta_0$, no matter what θ_0 is. Finally, for each $\alpha \in [0,1]$ and each $\theta_0 \in \Omega$, there exists $x_0 \in \mathbb{R} \cup \{\pm\infty\}$ and $\gamma \in [0,1]$ such that the test ϕ is UMP level α for testing H versus A.*

The proofs of these propositions are all very similar to the proof of Theorem 4.56. The tests in these propositions are also called *one-sided tests*.

The following simple result shows that one-sided tests also minimize the power function on the hypothesis among tests with the same size.

Corollary 4.66.[14] *Let the family of distributions have MLR, and suppose that the hypothesis is either $H : \Theta \leq \theta_0$ or $H : \Theta \geq \theta_0$. A one-sided UMP level α test ϕ satisfies $\beta_\phi(\theta) \leq \beta_\psi(\theta)$ for all $\theta \in \Omega_H$ and all tests ψ such that $\beta_\psi(\theta_0) \geq \alpha$.*

PROOF. Let ψ satisfy $\beta_\psi(\theta_0) \geq \alpha$, and let $\theta \in \Omega_H$. As in the proof of Theorem 4.56, we can show that a MP level $1 - \alpha$ test of $H' : \Theta = \theta_0$ versus $A' : \Theta = \theta$ is any one-sided test with size $1 - \alpha$. Since $1 - \phi$ is a one-sided test with size $1 - \alpha$ and ψ has level $1 - \alpha$, it follows that $\beta_{1-\phi}(\theta) \geq \beta_{1-\psi}(\theta)$, hence $\beta_\phi(\theta) \leq \beta_\psi(\theta)$. □

The following result is a simple consequence of Lemma 4.38, and it gives conditions under which UMP tests are essentially unique. It will also be useful in showing that there are situations in which there are no UMP tests of $H : \Theta = \theta_0$ versus $A : \Theta \neq \theta_0$.

Proposition 4.67. *Suppose that $\{P_\theta : \theta \in \Omega\}$ is a parametric family with $P_\theta \ll \nu$ for all θ and with increasing MLR. Let $f_{X|\Theta}(\cdot|\theta)$ denote $dP_\theta/d\nu$. Let $\theta_0 \in \Omega$, and define*

$$B_{\theta,k} = \{x : f_{X|\Theta}(x|\theta) = kf_{X|\Theta}(x|\theta_0)\}.$$

Suppose that for all $\theta > \theta_0$ and all $k \in [0, \infty]$, $P_\theta(B_{\theta,k}) = 0$. Let ϕ be a test of $H : \Theta = \theta_0$ versus $A : \Theta > \theta_0$ of the form (4.57), and let ψ be another test such that $\phi(\theta_0) = \beta_\psi(\theta_0)$. Then either $\phi = \psi$, a.s. $[P_\theta]$ for all $\theta \geq \theta_0$, or there exists $\theta > \theta_0$ such $\beta_\phi(\theta) > \beta_\psi(\theta)$.

There are versions of Proposition 4.67 for decreasing MLR and for alternatives of the form $A : \Theta < \theta_0$, but we will not state them here. An example of one of these propositions is the first part of Example 4.50 on page 238.

The following is a complete class theorem for the case of MLR.

---

[14]This corollary is used in the proof of Theorem 4.68.

Theorem 4.68. *Let the action space be $\aleph = \{0,1\}$. Suppose that a parametric family has MLR and that there exists $\theta_0 \in \Omega$ such that*

$$[L(\theta,1) - L(\theta,0)](\theta_0 - \theta) \qquad (4.69)$$

has the same sign for all $\theta \neq \theta_0$. Then the appropriate family of one-sided tests is an essentially complete class.

PROOF. Consider the case in which (4.69) is always positive and the family has increasing MLR. Let the hypotheses be $H : \Theta \leq \theta_0$. Let ϕ be any test. Then

$$\begin{aligned}R(\theta,\phi) &= \int \{\phi(x)L(\theta,1) + [1 - \phi(x)]L(\theta,0)\} f_{X|\Theta}(x|\theta)d\nu(x) \\ &= L(\theta,0) + [L(\theta,1) - L(\theta,0)] \int \phi(x) f_{X|\Theta}(x|\theta)d\nu(x).\end{aligned}$$

So, if ϕ and ψ are any two tests, then

$$\begin{aligned}R(\theta,\psi) - R(\theta,\phi) &= [L(\theta,1) - L(\theta,0)] \int [\psi(x) - \phi(x)] f_{X|\Theta}(x|\theta)d\nu(x) \\ &= [L(\theta,1) - L(\theta,0)][\beta_\psi(\theta) - \beta_\phi(\theta)].\end{aligned}$$

Now let ψ be an arbitrary test. We must show that there is a one-sided test ϕ which is at least as good as ψ. Let ϕ be a one-sided test with $\beta_\phi(\theta_0) = \beta_\psi(\theta_0)$. Then Theorem 4.56 and Corollary 4.66 give us that $\beta_\psi(\theta) - \beta_\phi(\theta)$ has the same sign as $\theta_0 - \theta$. This implies that $R(\theta,\psi) \geq R(\theta,\phi)$ for all θ.

For the other cases (decreasing MLR and/or (4.69) always negative), use one of Propositions 4.62–4.65 together with Corollary 4.66 to obtain a similar result. □

We can use Theorem 4.68 to help prove that in an MLR family with a one-sided hypothesis, one-sided tests are UMC as well as UMP. (See Problems 23 and 24 on page 288.)

Proposition 4.70.[15] *Let ϕ be a one-sided test as in Theorem 4.56 or in Propositions 4.62–4.65 for a one-sided hypothesis versus the corresponding one-sided alternative in an MLR family. Suppose that the base of ϕ is γ. Then ϕ is UMC floor γ.*

Suppose that we have a prior distribution μ_Θ on the parameter space. If there is a test with finite Bayes risk and the loss function is bounded below, then Theorem 4.68 allows us to conclude that one-sided tests are formal Bayes rules. (See Problem 33 on page 289.) For the case of 0–1–c

---

[15] This is not precisely the same as Corollary 4.66. Corollary 4.66 says that the power function is minimized on the hypothesis subject to the power function at θ_0 being at least α. If a power function is monotone but not continuous, the base of the test might be different from its size.

loss, the result follows in a simpler fashion from the fact that the posterior probability of a semi-infinite interval is a monotone function of the data in MLR families.

Theorem 4.71. *Suppose that $\{P_\theta : \theta \in \Omega\}$ is a parametric family of distributions for X with MLR and that μ_Θ is an arbitrary prior distribution on (Ω, τ). Then the posterior probability given $X = x$ that Θ is in a semi-infinite interval is a monotone function of x.*

PROOF. We will prove that if the family has increasing MLR then the posterior probability that $\Theta \geq \theta_0$ is a nondecreasing function of x. The other cases are all similar. Let $x_1 < x_2$. Then

$$\frac{\Pr(\Theta \geq \theta_0 | X = x_2)}{\Pr(\Theta < \theta_0 | X = x_2)} - \frac{\Pr(\Theta \geq \theta_0 | X = x_1)}{\Pr(\Theta < \theta_0 | X = x_1)}$$

$$= \frac{\int_{[\theta_0,\infty)} f_{X|\Theta}(x_2|\theta) d\mu_\Theta(\theta)}{\int_{(-\infty,\theta_0)} f_{X|\Theta}(x_2|\theta) d\mu_\Theta(\theta)} - \frac{\int_{[\theta_0,\infty)} f_{X|\Theta}(x_1|\theta) d\mu_\Theta(\theta)}{\int_{(-\infty,\theta_0)} f_{X|\Theta}(x_1|\theta) d\mu_\Theta(\theta)}$$

$$= \left(\int_{(-\infty,\theta_0)} f_{X|\Theta}(x_2|\theta) d\mu_\Theta(\theta) \int_{(-\infty,\theta_0)} f_{X|\Theta}(x_1|\theta) d\mu_\Theta(\theta) \right)^{-1}$$

$$\times \left[\int_{[\theta_0,\infty)} \int_{(-\infty,\theta_0)} \left[f_{X|\Theta}(x_2|\theta_2) f_{X|\Theta}(x_1|\theta_1) \right. \right.$$

$$\left. \left. - f_{X|\Theta}(x_2|\theta_1) f_{X|\Theta}(x_1|\theta_2) \right] d\mu_\Theta(\theta_1) d\mu_\Theta(\theta_2) \right] \quad (4.72)$$

Since the family of distributions has increasing MLR, it follows that

$$f_{X|\Theta}(x_2|\theta_2) f_{X|\Theta}(x_1|\theta_1) - f_{X|\Theta}(x_2|\theta_1) f_{X|\Theta}(x_1|\theta_2) \geq 0,$$

for all $x_1 < x_2$ and all $\theta_1 < \theta_2$. This makes the last expression in the last line of (4.72) nonnegative, and the result is proven. □

Corollary 4.73. *Suppose that $\{P_\theta : \theta \in \Omega\}$ is a parametric family of distributions for X with MLR and that μ_Θ is an arbitrary prior distribution on (Ω, τ). Suppose that we are testing a one-sided hypothesis against the corresponding one-sided alternative with a 0–1–c loss. Then one-sided tests are formal Bayes rules.*

When no UMP level α test exists, one option that has been suggested is to find a *locally most powerful* test.

Definition 4.74. *Let d be a strictly positive function on Ω_A. We say that a level α test ϕ is a locally most powerful level α relative to d (LMP) if, for every level α test ψ, there exists c such that $\beta_\phi(\theta) \geq \beta_\psi(\theta)$ for all θ such that $0 < d(\theta) < c$.*

For a one-dimensional parameter Θ and $H : \Theta \le \theta_0$, if power functions are continuously differentiable and ϕ is the unique level α test with maximum derivative for the power function at θ_0, then ϕ is LMP level α relative to $d(\theta) = \theta - \theta_0$ for $\theta > \theta_0$. (See Problem 34 on page 289.)

4.3.4 Two-Sided Hypotheses

Two-sided testing situations come in two forms.

Definition 4.75. If $H : \theta_1 \le \Theta \le \theta_2$ and $A : \Theta > \theta_2$ or $\Theta < \theta_1$, then the *alternative is two-sided*. If $H : \Theta \ge \theta_2$ or $\Theta \le \theta_1$ and $A : \theta_1 < \Theta < \theta_2$, then the *hypothesis is two-sided*.

The only difference, from a decision theoretic viewpoint, between two-sided alternatives and two-sided hypotheses is where the endpoints go. The tradition in classical hypothesis testing is to put the endpoints into the hypothesis. That is, the hypothesis is closed and the alternative is open. There is no need to require this, especially in cases in which the power functions are continuous. However, the treatments of these two cases are drastically different in the classical framework. The difference has nothing to do with where the endpoints go, but rather with the asymmetric treatment of hypotheses and alternatives in the optimality criteria.

In this section, we consider the case of two-sided hypotheses. We put off the case of two-sided alternatives until Section 4.4. Some mathematical lemmas are needed first.

Lemma 4.76 (Lagrange multipliers).[16] *Let T be any set, let $f : T \to \mathbb{R}$ and $g_i : T \to \mathbb{R}$, for $i = 1, \ldots, n$ be functions, and let $\lambda_1, \ldots, \lambda_n$ be real numbers. If t_0 minimizes $f(t) + \sum_{i=1}^{n} \lambda_i g_i(t)$ and satisfies $g_i(t_0) = c_i$ for $i = 1, \ldots, n$, then t_0 minimizes $f(t)$ subject to $g_i(t) \le c_i$ for each $\lambda_i > 0$ and $g_i(t) \ge c_i$ for each $\lambda_i < 0$.*

PROOF. Suppose, to the contrary, that there exists t_1 such that $f(t_1) < f(t_0)$, while $g_i(t_1) \le c_i$ for each $\lambda_i > 0$ and $g_i(t_1) \ge c_i$ for each $\lambda_i < 0$. Then

$$f(t_1) + \sum_{i=1}^{n} \lambda_i g_i(t_1) < f(t_0) + \sum_{i=1}^{n} \lambda_i c_i = f(t_0) + \sum_{i=1}^{n} \lambda_i g_i(t_0).$$

This contradicts the assumption that t_0 minimizes $f(t) + \sum_{i=1}^{n} \lambda_i g_i(t)$. □

Corollary 4.77.[17] *Let T be any set, let $f : T \to \mathbb{R}$ and $g_i : T \to \mathbb{R}$, for $i = 1, \ldots, n$ be functions, and let $\lambda_1, \ldots, \lambda_n$ be real numbers. If t_0 maximizes $f(t) + \sum_{i=1}^{n} \lambda_i g_i(t)$ and satisfies $g_i(t_0) = c_i$ for $i = 1, \ldots, n$,*

---

[16] This lemma is used in the proof of Lemma 4.78.
[17] This corollary is used in the proof of Lemma 4.78.

then t_0 maximizes $f(t)$ subject to $g_i(t) \geq c_i$ for each $\lambda_i > 0$ and $g_i(t) \leq c_i$ for each $\lambda_i < 0$.

The following lemma is sometimes called the generalized Neyman–Pearson lemma due to the resemblance it bears to Proposition 4.37.

Lemma 4.78.[18] *Let $p_0, p_1, \ldots, p_n$ be integrable (and not a.e. 0) functions (with respect to a measure ν), and let*

$$\phi_0(x) = \begin{cases} 1 & \text{if } p_0(x) > \sum_{i=1}^n k_i p_i(x), \\ \gamma(x) & \text{if } p_0(x) = \sum_{i=1}^n k_i p_i(x), \\ 0 & \text{if } p_0(x) < \sum_{i=1}^n k_i p_i(x), \end{cases}$$

where $0 \leq \gamma(x) \leq 1$ and the k_i are constants. Then ϕ_0 minimizes $\int [1 - \phi(x)] p_0(x) d\nu(x)$ subject to

$$\int \phi(x) p_j(x) d\nu(x) \leq \int \phi_0(x) p_j(x) d\nu(x), \text{ for those } j \text{ such that } k_j > 0,$$

$$\int \phi(x) p_j(x) d\nu(x) \geq \int \phi_0(x) p_j(x) d\nu(x), \text{ for those } j \text{ such that } k_j < 0.$$

PROOF. Let ϕ be an arbitrary measurable function taking values between 0 and 1 which satisfies the preceding inequality constraints. Since $\phi(x) \leq \phi_0(x)$ whenever $p_0(x) - \sum_{i=1}^n k_i p_i(x) > 0$ and $\phi(x) \geq \phi_0(x)$ whenever $p_0(x) - \sum_{i=1}^n k_i p_i(x) < 0$, it is clear that

$$\int [\phi(x) - \phi_0(x)] \left[p_0(x) - \sum_{i=1}^n k_i p_i(x) \right] d\nu(x) \leq 0.$$

It follows from this that

$$\int [1 - \phi_0(x)] p_0(x) d\nu(x) + \sum_{i=1}^n k_i \int \phi_0(x) p_i(x) d\nu(x) \quad (4.79)$$

$$\leq \int [1 - \phi(x)] p(x) d\nu(x) + \sum_{i=1}^n k_i \int \phi(x) p_i(x) d\nu(x).$$

Now, let T be the set of all measurable functions from $\mathcal{X}$ to $[0, 1]$, and let

$$f(t) = \int [1 - t(x)] p_0(x) d\nu(x),$$

$$g_i(t) = \int t(x) p_i(x) d\nu(x), \text{ for } i = 1, \ldots, n.$$

Equation (4.79) says that ϕ_0 minimizes $f(\phi) + \sum_{i=1}^n k_i g_i(\phi)$. Lemma 4.76 implies that ϕ_0 minimizes $f(\phi)$ subject to the constraints. □

There is also a version of this lemma corresponding to Corollary 4.77.

---

[18]This lemma is used in the proofs of Theorems 4.82 and 4.104.

Corollary 4.80.[19] *Let $p_0, p_1, \ldots, p_n$ be integrable (and not a.e. 0) functions (with respect to a measure ν), and let*

$$\phi_0(x) = \begin{cases} 0 & \text{if } p_0(x) > \sum_{i=1}^n k_i p_i(x), \\ \gamma(x) & \text{if } p_0(x) = \sum_{i=1}^n k_i p_i(x), \\ 1 & \text{if } p_0(x) < \sum_{i=1}^n k_i p_i(x), \end{cases}$$

where $0 \leq \gamma(x) \leq 1$ and the k_i are constants. Then ϕ_0 maximizes $\int [1 - \phi(x)] p_0(x) d\nu(x)$ subject to

$$\int \phi(x) p_j(x) d\nu(x) \geq \int \phi(x) p_j(x) d\nu(x), \text{ for those } j \text{ such that } k_j > 0,$$

$$\int \phi(x) p_j(x) d\nu(x) \leq \int \phi(x) p_j(x) d\nu(x), \text{ for those } j \text{ such that } k_j < 0.$$

The theorems we can prove assume that we are dealing with a one-parameter exponential family.

Lemma 4.81.[20] *Assume that the parametric family has monotone likelihood ratio. If ϕ is an arbitrary test and $\theta_1 < \theta_2$, define $\alpha_i = \beta_\phi(\theta_i)$ for $i = 1, 2$. Then there is a test of the form*

$$\psi(x) = \begin{cases} 1 & \text{if } c_1 < x < c_2, \\ \gamma_i & \text{if } x = c_i, \\ 0 & \text{if } c_1 > x \text{ or } c_2 < x, \end{cases}$$

with $c_1 \leq c_2$ such that $\beta_{\psi_0}(\theta_i) = \alpha_i$ for $i = 1, 2$.

PROOF. Define ϕ_w to be the UMP level w test of $H : \Theta \leq \theta_1$ versus $A : \Theta > \theta_1$, and for each $0 \leq u \leq 1 - \alpha_1$, set

$$\phi'_u(x) = \phi_{\alpha_1 + u}(x) - \phi_u(x).$$

First, note that since $\alpha_1 + u \geq u$ for all u, $0 \leq \phi'_u(x) \leq 1$ for all u and all x. This means that ϕ'_u is really a test. By design, $\beta_{\phi'_u}(\theta_1) = \alpha_1$. Also, ϕ'_u has the form of ψ for each u (with c_1 or c_2 possibly infinite). This is true since all $\phi_w(x)$ are 0 for small x and 1 for large x. By construction, $\phi'_0 = \phi_{\alpha_1}$ is the MP level α_1 test of $H' : \Theta = \theta_1$ versus $A' : \Theta = \theta_2$, and $\phi'_{1-\alpha_1} = 1 - \phi_{1-\alpha_1}$ is the least powerful such test. Since ϕ is also a level α_1 test of H' versus A', it follows that

$$\beta_{\phi'_{1-\alpha_1}}(\theta_2) \leq \alpha_2 \leq \beta_{\phi'_0}(\theta_2).$$

If we can show that $\beta_{\phi_w}(\theta_2)$ is continuous in w, we can conclude that there exists w such that $\beta_{\phi_w}(\theta_2) = \alpha_2$. The proof of this is left to the reader. (See Problem 40 on page 290.) It follows that ϕ'_w has the form of ψ and $\beta_{\phi'_w}(\theta_i) = \alpha_i$ for $i = 1, 2$. □

---

[19]This corollary is used in the proof of Theorem 4.82.
[20]This lemma is used in the proofs of Theorem 4.82 and Lemma 4.99. It resembles part of Theorem 1 on p. 217 of Ferguson (1967).

4.3. Most Powerful Tests

Theorem 4.82. *In a one-parameter exponential family with natural parameter, if $\Omega_H = (-\infty, \theta_1] \cup [\theta_2, \infty)$ and $\Omega_A = (\theta_1, \theta_2)$, with $\theta_1 < \theta_2$, a test of the form*

$$\phi_0(x) = \begin{cases} 1 & \text{if } c_1 < x < c_2, \\ \gamma_i & \text{if } x = c_i, \\ 0 & \text{if } c_1 > x \text{ or } c_2 < x, \end{cases}$$

with $c_1 \leq c_2$ minimizes $\beta_\phi(\theta)$ for all $\theta < \theta_1$ and for all $\theta > \theta_2$, and it maximizes $\beta_\phi(\theta)$ for all $\theta_1 < \theta < \theta_2$ subject to $\beta_\phi(\theta_i) = \alpha_i$ for $i = 1, 2$ where $\alpha_i = \beta_{\phi_0}(\theta_i)$ for $i = 1, 2$. If $c_1, c_2, \gamma_1, \gamma_2$ are chosen so that $\alpha_1 = \alpha_2 = \alpha$, then ϕ_0 is UMP level α.

PROOF. Suppose that ϕ_0 is of the form stated in the theorem, and let $f_{X|\Theta}(x|\theta) = c(\theta)\exp(\theta x)$, so that $h(x)$ is incorporated into the measure ν. Let θ_1 and θ_2 be as in the statement of the theorem, and let θ_0 be another element of Ω. Define $p_i(x) = c(\theta_i)\exp(\theta_i x)$, for $i = 0, 1, 2$.

First, suppose that $\theta_1 < \theta_0 < \theta_2$. Set $b_i = \theta_i - \theta_0$ for $i = 1, 2$. Next, note that the function $a_1\exp(b_1 x) + a_2\exp(b_2 x)$ is strictly monotone if a_1 and a_2 have opposite signs, and it is always negative if both a_1 and a_2 are negative. Suppose that we try to solve the following two equations for a_1, a_2:

$$\begin{align} 1 &= a_1\exp(b_1 c_1) + a_2\exp(b_2 c_1), \\ 1 &= a_1\exp(b_1 c_2) + a_2\exp(b_2 c_2), \end{align} \quad (4.83)$$

where c_1 and c_2 are from the definition of ϕ_0. The reader can easily verify that this system of linear equations has nonzero determinant and hence has a solution. The solution must have both $a_1, a_2 > 0$. Set $k_i = a_i c(\theta_0)/c(\theta_i)$. With these values of k_1, k_2, apply Lemma 4.78. Note that minimizing $\int [1 - \phi(x)]p_0(x)d\nu(x)$ is equivalent to maximizing $\beta_\phi(\theta_0)$. The test that maximizes $\beta_\phi(\theta_0)$ subject to $\beta_\phi(\theta_i) \leq \beta_{\phi_0}(\theta_i)$ for $i = 1, 2$ is

$$\phi(x) = 1, \text{ if } c(\theta_0)\exp(\theta_0 x) > k_1 c(\theta_1)\exp(\theta_1 x) + k_2 c(\theta_2)\exp(\theta_2 x) \quad (4.84)$$

if this test satisfies the constraints. The test in (4.84) can be rewritten as

$$\phi(x) = 1 \text{ if } 1 > a_1\exp(b_1 x) + a_2\exp(b_2 x),$$

where a_1 and a_2 are the solutions to the two linear equations above. Since $a_1\exp(b_1 x) + a_2\exp(b_2 x)$ goes to infinity as $x \to \pm\infty$, it follows that $\phi(x) = 1$ for $c_1 < x < c_2$, which leads to $\phi(x) = \phi_0(x)$. This same argument applies for all θ_0 between θ_1 and θ_2, hence the same ϕ_0 maximizes $\beta_\phi(\theta)$ for all $\theta_1 < \theta < \theta_2$.

Next, try to minimize $\beta_\phi(\theta_0)$ for $\theta_0 < \theta_1$. (An identical argument works for $\theta_0 > \theta_2$.) This time, we will use Corollary 4.80. Set $b_i = \theta_i - \theta_0$ for $i = 1, 2$. Next, note that the function $a_1\exp(b_1 x) + a_2\exp(b_2 x)$ is strictly monotone if a_1 and a_2 have the same signs. If $a_1 < 0 < a_2$, the derivative has only one zero and the function goes to 0 as $x \to -\infty$ and to ∞ as $x \to \infty$. This means that the function equals 1 for only one value of x. Hence, the solution to the equations in (4.83) must have $a_1 > 0 > a_2$. Solve the equations and set $k_i = a_i c(\theta_0)/c(\theta_i)$. Since maximizing $\int [1 - \phi(x)]p_0(x)d\nu(x)$ is the same as minimizing $\beta_\phi(\theta_0)$, the test that minimizes $\beta_\phi(\theta_0)$ subject to $\beta_\phi(\theta_1) \geq \alpha_1$ and $\beta_\phi(\theta_2) \leq \alpha_2$ is

$$\phi(x) = 1 \text{ if } c(\theta_0)\exp(\theta_0 x) < k_1 c(\theta_1)\exp(\theta_1 x) + k_2 c(\theta_2)\exp(\theta_2 x),$$

with $k_1 > 0$ and $k_2 < 0$. This can be rewritten as

$$\phi(x) = 1 \text{ if } 1 < a_1 \exp(b_1 x) + a_2 \exp(b_2 x),$$

with $a_1 > 0 > a_2$ and $b_2 > b_1 > 0$. Since $a_1 \exp(b_1 x) + a_2 \exp(b_2 x)$ goes to 0 as $x \to -\infty$ and goes to $-\infty$ as $x \to \infty$, it follows that $\phi(x) = 1$ for $c_1 < x < c_2$. Once again, we get the same test for all θ_0 and the same test as before.

Finally, consider the test $\phi^\alpha(x) \equiv \alpha$, and now suppose that $\alpha_1 = \alpha_2 = \alpha$. Lemma 4.81 guarantees that c_1, c_2, γ_1, and γ_2 can be chosen so that ϕ_0 has the stated form with $\alpha_1 = \alpha_2 = \alpha$. The power function of ϕ^α is the constant α. It must be that $\beta_{\phi_0}(\theta) \leq \alpha$ for every $\theta \in \Omega_H$. Hence ϕ_0 has level α. Since every level α test ψ must satisfy $\beta_\psi(\theta_i) \leq \alpha$ ($i = 1, 2$), and ϕ_0 maximizes the power on the alternative subject these constraints, it follows that ϕ_0 is UMP level α. □

Example 4.85. Suppose that $Y \sim Exp(\theta)$ given $\Theta = \theta$. Let $X = -Y$ so that Θ is the natural parameter. Let $\Omega_H = (0, 1] \cup [2, \infty)$, $\Omega_A = (1, 2)$, and $\alpha = 0.1$. We must solve the equations

$$\begin{aligned} \exp(c_2) - \exp(c_1) &= 0.1, \\ \exp(2c_2) - \exp(2c_1) &= 0.1. \end{aligned}$$

If we let $a = \exp(c_2)$ and $b = \exp(c_1)$, these equations simplify to $a - b = 0.1$ and $a^2 - b^2 = 0.1$, respectively. The solution is easily calculated to be $a = 0.55$ and $b = 0.45$. So the solution to the original equations is $c_1 = -0.7985$ and $c_2 = -0.5978$. Since the distribution is continuous, $\gamma_1 = \gamma_2 = 0$. We reject H if $0.7985 > Y > 0.5978$.

Example 4.86. Suppose that $X \sim Bin(n, p)$ given $P = p$. Let $\Theta = \log(P/(1 - P))$, the natural parameter. Then $f_{X|\Theta}(x|\theta) = c(\theta)\exp(\theta x)$, where $c(\theta) = (1 + \exp(\theta))^{-n}$, and ν is $\binom{n}{x}$ times counting measure on $\{0, \ldots, n\}$. The hypothesis $H : P \leq 1/4$ or $P \geq 3/4$ corresponds to $\Omega_H = (-\infty, -1.099] \cup [1.099, \infty)$ in Θ space. If $n = 10$ and $\alpha = 0.1$, we get the UMP level α test by choosing $c_1 = 4$ and $c_2 = 6$ with $\gamma_1 = \gamma_2 = 0.2565$.

Suppose that we have a prior distribution μ_Θ on the parameter space. If there is a test with finite Bayes risk and the loss function is bounded below, then Theorem 4.82 allows us to conclude that tests of the form given in that theorem are formal Bayes rules. (See Problem 38 on page 290.)

Of course, if we switch the names of hypothesis and alternative, then the formal Bayes rule will be 1 minus the test in Theorem 4.82. This will be even more apparent after we see Lemma 4.99. One interesting difference between Bayes rules and UMP level α tests is that not all Bayes rules for testing two-sided hypotheses need to have the same value for the power function at the two endpoints of the alternative.

Example 4.87 (Continuation of Example 4.85; see page 250). Once again, suppose that $Y \sim Exp(\theta)$ given $\Theta = \theta$. Suppose that the prior distribution for Θ is $Exp(1)$. The posterior distribution will be $\Gamma(2, 1 + y)$. Let the loss function be 0–1–c with $c = 0.5$. Solving numerically for the formal Bayes rule, we get

$$\phi(y) = \begin{cases} 1 & \text{if } 0.02133 < y < 0.83685, \\ 0 & \text{otherwise.} \end{cases}$$

The power function of this test is 0.454 at $\theta = 1$ and 0.229 at $\theta = 2$. The level of the test is 0.454, but it is not UMP level 0.454. The UMP level 0.454 test would have power function 0.454 at $\theta = 2$, but would not be a Bayes rule for the stated decision problem. The intuitive reason for the lopsided power function of the formal Bayes rule is that the prior puts so much more mass below 1 than above 2 (0.3679 versus 0.1353). It makes sense that the test should protect more against alternatives with small Θ than those with large Θ.

Curiously, however, if we use the improper prior with Radon–Nikodym derivative $1/\theta$ with respect to Lebesgue measure, the formal Bayes rule will be a UMP level α test for all 0–1–c losses. To see this, note that the posterior distribution of Θ is $Exp(y)$. The posterior probability that the hypothesis is true is now $1 - \exp(-y) + \exp(-2y)$. To find the formal Bayes rule with 0–1–c loss, we set this expression equal to $1/(1+c)$ and solve for y. There will be two solutions[21] $c_1 < c_2$ (the endpoints of the rejection region for the test), and they will satisfy

$$1 - \exp(-c_1) + \exp(-2c_1) = 1 - \exp(-c_2) + \exp(-2c_2) = \frac{1}{1+c}.$$

Rearranging terms in this expression leads to

$$\exp(-c_1) - \exp(-c_2) = \exp(-2c_1) - \exp(-2c_2).$$

The right-hand side of this last equation is the power function of the test at $\theta = 2$, and the left-hand side is the power function at $\theta = 1$. If α is the common value of the two sides, then the test is UMP level α.

Theorem 4.68 says that the class of UMP level α tests is essentially complete for decision problems that include hypothesis-testing loss functions for one-sided hypotheses. The first part of Example 4.87 shows that for two-sided hypotheses, the class of UMP level α tests is not essentially complete. The formal Bayes rule given there is admissible and the risk function is not the same as that of any UMP level α test. This, then, is the first point

---

[21] There may actually be only one solution or no solutions, because the posterior probability of the hypothesis is bounded below. In these cases, the formal Bayes rule always accepts H.

at which classical hypothesis-testing theory has departed from the decision theoretic approach to hypothesis testing. When we had simple hypotheses and alternatives or one-sided hypotheses and alternatives with MLR, the class of (U)MP level α tests included (essentially) all admissible procedures. When we move to two-sided hypotheses, we lose that property. One can prove, however (see Problem 37 on page 290), that the class of tests of the form given in Theorem 4.82 is essentially complete. The restriction to tests that are UMP for their level does not follow from considerations of admissibility. The argument to justify such a restriction might be, "In comparing two tests with the same level, I want to choose, if possible, the test with higher power function on the alternative." This would make perfect sense if the hypothesis consisted of a single point. (See Problem 16 on page 287.) However, the hypothesis is not a single point in general. The classical theory treats the hypothesis as if it were a single point and does not distinguish between tests based on their power functions on the hypothesis so long as they have the same level. To put the case more succinctly, the level of a test does not completely describe the power function on the hypothesis, but the classical theory pretends as if it did. That is why the formal Bayes rule in the first part of Example 4.87 on page 251 is lumped together with all level 0.454 tests even though it has advantages over some other level 0.454 tests. These advantages are simply ignored when the level of a test is taken as the entire summary of the power function on the hypothesis.

The restriction of attention to UMP level α tests, rather than all tests of the form of Theorem 4.82, has another consequence that is even more surprising, perhaps, than the fact that the tests do not form an essentially complete class. A simple example will illustrate the general situation.

Example 4.88. Let $X \sim N(\mu, 1)$ given $M = \mu$. Suppose that we are considering two different hypotheses about M with $\Omega_{H_1} = (-\infty, -0.5] \cup [0.5, \infty)$ and $\Omega_{H_2} = (-\infty, -0.7] \cup [0.51, \infty)$. The UMP level 0.05 test of H_1 versus $A_1 : M \in (-0.5, 0.5)$ is to reject H_1 if $X \in (-0.071, 0.071)$. The UMP level 0.05 test of H_2 versus $A_2 : M \in (-0.7, 0.51)$ is to reject H_2 if $X \in (-0.167, -0.017)$. Since $\Omega_{H_2} \subset \Omega_{H_1}$, it makes sense that if we reject H_1, then *a fortiori* we should be able to reject H_2. However, if $X \in [-0.017, 0.071)$, we would reject H_1 at level 0.05 but accept H_2 at the same level.

The type of contradictory conclusions we were able to obtain in Example 4.88 actually occurs quite generally in level α testing, once we leave the one-sided situation. (See also Problem 41 on page 290.) We will see them again in Section 4.4.2 and in Section 4.6. Gabriel (1969) introduced the concept of *coherent* tests of several hypotheses. A collection of tests of various hypotheses is coherent if rejecting one hypothesis H always leads to rejecting every hypothesis that implies H. Testing several hypotheses at the same level is not always coherent, as Example 4.88 shows. The problem lies in choosing tests based on their level rather than on decision theoretic criteria. For example, if one were to reject hypotheses whose posterior probabilities were less than some number γ, then rejecting H_1 would always lead to rejecting H_2 when H_2 implies H_1 (as in Example 4.88).

Example 4.89 (Continuation of Example 4.88; see page 252). Suppose that we use the usual improper prior (Lebesgue measure). Then the posterior distribution of M is $N(x, 1)$. The level 0.05 test of H_1 corresponds to rejecting H_1 if the posterior probability of H_1 is less than 0.618. The posterior probability of H_2 is less than 0.618 whenever $x \in (-0.72, 0.535)$. Notice that this last interval strictly contains the rejection region for H_1, $(-0.071, 0.071)$, so that rejection of H_1 will always imply rejection of H_2 when rejection means "posterior probability less than 0.618."

There is a natural sense in which incoherent tests are inadmissible. See Problem 42 on page 291.

4.4 Unbiased Tests

4.4.1 General Results

For the cases not previously considered, there generally do not exist UMP level α tests. This is not to say that there are no good tests in other situations, but rather that the criterion of UMP level α needs to be relaxed if we are going to find the good tests. Consider the following example, which is typical of what happens when the alternative is two-sided.

Example 4.90. Suppose that $X \sim N(\theta, 1)$ given $\Theta = \theta$ and that we wish to test $H : \Theta = \theta_0$ versus $A : \Theta \neq \theta_0$ at some level $\alpha \in (0, 1)$. The UMP level α test ϕ' of H versus $A' : \Theta < \theta_0$ and the UMP level α test ϕ'' of H versus $A'' : \Theta > \theta_0$ are both level α tests of H versus A. If there is a UMP level α test ϕ of H versus A, then its power function must be at least as large as that of ϕ' for $\theta < \theta_0$ and at least as large as that of ϕ'' for $\theta > \theta_0$. But such a ϕ would also be a level α test of H versus A', so Proposition 4.67 says that either $\phi = \phi'$ a.s.[22] or there is $\theta < \theta_0$ such that $\beta_\phi(\theta) < \beta_{\phi'}(\theta)$. Since we are assuming that the latter is false, we must have $\phi = \phi'$ a.s. The same argument applied to A'' and ϕ'' implies that $\phi = \phi''$ a.s. $[P_\theta]$ for all θ. But this is impossible since $\phi'(x) = I_{(-\infty, c']}(x)$, for some finite number c', and $\phi''(x) = I_{[c'', \infty)}(x)$, for some finite number c''. It follows that no test ϕ is UMP level α for testing H versus A.

The way to circumvent the lack of UMP level α tests in cases like Example 4.90 is to create a new criterion that one-sided tests fail to satisfy when the alternative is two-sided.[23] The rationale is that even though the power function of a one-sided test is high in one part of the alternative, it is very low in the other part. The new optimality criterion requires that the power function be higher on the alternative than on the hypothesis.

---

[22]Since all $N(\theta, 1)$ distributions are mutually absolutely continuous, a.s. with respect to one of them means a.s. with respect to all of them.

[23]When the conditions of Proposition 4.67 fail, there may be UMP level α tests for two-sided alternatives. See Problem 27 on page 288 for an example that even has MLR.

Definition 4.91. A test ϕ is *unbiased level α* if it has level α and if $\beta_\phi(\theta) \geq \alpha$ for all $\theta \in \Omega_A$. If $\Omega \subseteq \mathbb{R}^k$, a test ϕ is called *α-similar* if $\beta_\phi(\theta) = \alpha$ for each $\theta \in \overline{\Omega}_H \cap \overline{\Omega}_A$. More generally, ϕ is α-similar on $B \subseteq \Omega$ if $\beta_\phi(\theta) = \alpha$ for each $\theta \in B$. If ϕ is UMP among all unbiased level α tests, then ϕ is *uniformly most powerful unbiased (UMPU) level α*.

The concepts of unbiased level α and α-similar are closely related.

Proposition 4.92.[24] *If a test ϕ is unbiased level α and $\beta_\phi(\cdot)$ is continuous, then ϕ is α-similar.*

Proposition 4.93. *If ϕ is a UMP level α test, then ϕ is unbiased level α.*

Since ϕ being unbiased level α implies that ϕ has floor α, the dual concept to unbiased level α is simply *unbiased floor α*.

Definition 4.94. A test ϕ is *unbiased floor α* if it has floor α and if $\beta_\phi(\theta) \leq \alpha$ for all $\theta \in \Omega_H$. If ϕ is UMC among all unbiased floor α tests, then ϕ is *uniformly most cautious unbiased (UMCU) floor α*.

It is interesting to note that the collection of unbiased test may not be essentially complete. The test in the first part of Example 4.87 on page 251 is admissible with 0–1–0.5 loss, but it is not unbiased and it does not have the same risk function as an unbiased test. The restriction to unbiased tests, just like the restriction to UMP level α tests in the previous section, does not follow from considerations of admissibility. It is true that the restriction to unbiased tests rules out the use of one-sided tests in problems like Example 4.90 on page 253, but it also rules out many admissible tests.

Example 4.95. Suppose that $Y \sim Exp(\theta)$ given $\Theta = \theta$ with $\Omega_H = [1, 2]$ and $\Omega_A = (-\infty, 1) \cup (2, \infty)$. Suppose that the loss function is asymmetric in the following way.

$$L(\theta, a) = \begin{cases} 3 & \text{if } \theta \in \Omega_H \text{ and } a = 1, \\ \frac{1}{2} & \text{if } \theta > 2 \text{ and } a = 0, \\ 1 & \text{if } \theta < 1 \text{ and } a = 0, \\ 0 & \text{otherwise.} \end{cases}$$

We will use the usual improper prior with Radon–Nikodym derivative $1/\theta$ with respect to Lebesgue measure so that the posterior distribution is $Exp(y)$. The formal Bayes rule will minimize the posterior risk. The posterior risks for the two possible decisions are

$a = 0$	$a = 1$
$3(\exp(-y) - \exp(-2y))$	$\frac{1}{2}\exp(-2y) + 1 - \exp(-y)$

Solving to see when the risk for $a = 1$ is smaller, we see that this occurs when $y < 0.2569$ or $y > 0.9959$. The test that rejects when one of these conditions holds has power function 0.5959 at $\theta = 1$ and 0.5382 at $\theta = 2$. Since it is more

---

[24]This proposition is used in the proofs of Lemma 4.96 and of Theorems 4.123 and 4.124.

important to reject H when θ is small, the power is higher for small θ values. The test has level 0.5959, but it is biased.

One technique for finding UMPU tests will be to restrict attention first to α-similar tests. The following lemma shows why this will work in many cases.

Lemma 4.96.[25] *Suppose that $\beta_\phi(\cdot)$ is continuous in θ for every ϕ. If ϕ_0 is UMP among α-similar tests and has level α, then ϕ_0 is UMPU level α.*

PROOF. Since $\psi(x) \equiv \alpha$ is α-similar and ϕ_0 is UMP α-similar, it follows that $\beta_{\phi_0}(\theta) \geq \alpha$ for every $\theta \in \Omega_A$. Since ϕ_0 has level α, it is unbiased level α. Every unbiased level α test is α-similar by Proposition 4.92. So, the test that is UMP among α-similar tests, namely ϕ_0, has power function at least as high (on the alternative) as the test that is UMPU level α. But ϕ_0 is also unbiased level α. Hence ϕ_0 is UMPU level α. □

Proposition 4.97.[26] *Suppose that $\beta_\phi(\cdot)$ is continuous in θ for every ϕ. If ϕ_0 is UMC among α-similar tests and ϕ_0 has base α, then ϕ_0 is UMCU floor α.*

4.4.2 Interval Hypotheses

In this section, we will consider the case in which the alternative is two-sided and the hypothesis is a nondegenerate compact interval. That is, the case $H : \Theta \in [\theta_1, \theta_2]$ versus $A : \Theta \notin [\theta_1, \theta_2]$, with $\theta_1 < \theta_2$. It turns out that there is no UMP level α test of H versus A in one-parameter exponential families (for $\alpha > 0$). One would suspect that if ϕ were the optimal level $1 - \alpha$ test of A versus H as derived in Theorem 4.82, then $1 - \phi$ would be the appropriate level α test for H versus A.[27] This will, in fact, turn out to be the case, but the test is no longer UMP level α, but only UMPU level α.[28]

Example 4.98 (Continuation of Example 4.87; see page 251). In this example, $Y \sim Exp(\theta)$ given $\Theta = \theta$. For consistency with the classical approach, suppose that we use the improper prior with Radon–Nikodym derivative $1/\theta$ with respect to Lebesgue measure. The posterior distribution of Θ given $Y = y$ is $Exp(y)$. At the end of Example 4.87 on page 251, we saw that the formal Bayes rule for 0–1–c loss would be a UMP level α test. Suppose now that we switch the hypothesis and alternative, so that $\Omega_H = [1, 2]$ and $\Omega_A = (-\infty, 1) \cup (2, \infty)$. Suppose that we use the 0–1–c loss with $c = 3.04$. The posterior probability

---

[25] This lemma is used in the proof of Theorem 4.100.
[26] This proposition is used in the proof of Theorem 4.100.
[27] One can prove (see Problem 30 on page 289) that $1 - \phi$ is UMC floor α for testing H versus A. But this is not the most popular optimality criterion.
[28] One should also be aware that there is no UMC floor α test in the case of two-sided hypotheses in exponential families. The concepts of UMP and UMC really are dual to each other. Neither of them is the unique best optimality criterion.

of Ω_H is $\exp(-y) - \exp(-2y)$. We then reject H, that is, we choose $a = 1$, if $\exp(-y) - \exp(-2y) < 1/4.04$, which is true if $y > 0.7985$ or $y < 0.5978$. This is the same (a.s.) as 1 minus the UMP level 0.1 test in the earlier example. Note that the conditions of Proposition 4.67 are met in this example, and so no test will be UMP level 0.9. The test we have just constructed will be UMPU level 0.9, however.

We are now in position to begin to prove that the UMPU level α test for the case of two-sided alternatives in a one-parameter exponential family is just 1 minus the UMP level $1 - \alpha$ test for the two-sided hypothesis.

Lemma 4.99.[29] *In a one-parameter exponential family with natural parameter, if ϕ is any test of $H : \Theta \in [\theta_1, \theta_2]$ versus $A : \Theta \notin [\theta_1, \theta_2]$ with $\theta_1 < \theta_2$, then there is a test of the form*

$$\psi(x) = \begin{cases} 1 & \text{if } x < c_1 \text{ or } x > c_2, \\ \gamma_i & \text{if } x = c_i, \\ 0 & \text{if } c_1 < x < c_2, \end{cases}$$

with $\beta_\psi(\theta_i) = \beta_\phi(\theta_i)$ for $i = 1, 2$ and

$$\beta_\psi(\theta) \begin{cases} \leq \beta_\phi(\theta) & \text{if } \theta \in \Omega_H, \\ \geq \beta_\phi(\theta) & \text{if } \theta \in \Omega_A. \end{cases}$$

PROOF. Lemma 4.81 says that $1 - \psi$ can be chosen to have $\beta_{1-\psi}(\theta_i) = \beta_{1-\phi}(\theta_i)$ and thus that ψ is in the desired form. Theorem 4.82 then shows that $1 - \psi$ minimizes and maximizes power in just the opposite regions from where we want ψ to minimize and maximize power under the same conditions. □

The tests in Lemma 4.99 are called *two-sided tests*. It is easy to see that when the conditions of Lemma 4.99 hold, the class of two-sided tests is essentially complete for hypothesis-testing loss functions. (See Problem 48 on page 292.) The next theorem says that the UMPU level α tests are a subset of this essentially complete class. We could show, as in Example 4.87 on page 251, however, that this subset is not essentially complete.

Theorem 4.100. *Assume the same conditions as Lemma 4.99. Also suppose that $\beta_\psi(\theta_i) = \alpha$ for $i = 1, 2$. A test of the form ψ is UMPU level α and UMCU floor α.*

PROOF. By comparing ψ with $\phi^\alpha(x) \equiv \alpha$ and using Lemma 4.99, we see that ψ is unbiased level α and unbiased floor α. Also, Lemma 4.99 shows that ψ is UMP α-similar and UMC α-similar. Lemma 4.96 and Proposition 4.97 can be applied since the power functions are continuous in an exponential family. □

---

[29] This lemma is used in the proof of Theorem 4.100 and to show that the class of two-sided tests is essentially complete.

Example 4.101. Suppose that $X \sim N(\mu, 1)$ given $\Theta = \mu$. Let $\Omega_H = [-1, 1]$ and $\alpha = 0.1$. Set $c_2 = 2.286$ and $c_1 = -2.286$. $\theta_1 = -1$ and $\theta_2 = 1$. So

$$\psi(x) = \begin{cases} 1 & \text{if } x < -2.286 \text{ or } x > 2.286, \\ 0 & \text{otherwise}; \end{cases}$$

$$\begin{aligned}
\beta_\psi(\theta_2) &= \Pr(N(1,1) \notin [-2.286, 2.286]), \\
&= 1 - (\Phi(1.286) - \Phi(-3.286)) = 0.1, \\
&= 1 - (\Phi(3.286) - \Phi(-1.286)), \\
&= \Pr(N(-1,1) \notin [-2.286, 2.286]) = \beta_\psi(\theta_1).
\end{aligned}$$

Hence, ψ is UMPU level 0.1 and UMCU floor 0.1.

Lest the reader think that UMPU and UMCU tests are always the same, note that the UMP and UMC tests in Problem 31 on page 289 are both unbiased, but they are not the same. Those tests are one-sided however.

We should also note that the type of contradictory conclusions drawn in Example 4.88 on page 252 also arises for interval hypotheses. The following example is adapted from Schervish (1994b).

Example 4.102. Suppose that $X \sim N(\mu, 1)$ given $M = \mu$. We wish to consider two different hypotheses, $H_1 : M \in [-0.5, 0.5]$ versus $A_1 : M \notin [-0.5, 0.5]$ and $H_2 : M \in [-0.82, 0.52]$ versus $A_2 : M \notin [-0.82, 0.52]$. The UMPU level 0.05 test of H_1 is to reject H_1 if $X \notin [-2.185, 2.185]$. The UMPU level 0.05 test of H_2 is to reject H_2 if $X \notin [-2.475, 2.175]$. So, if $X \in (2.175, 2.185]$, we would reject H_2 and accept H_1, even though H_1 implies H_2.

If we had used Lebesgue measure for an improper prior, the posterior probability of H_1 given $X = x$ is less than 0.0424 if $x \notin [-2.185, 2.185]$, the rejection region for the UMPU level 0.05 test. The posterior probabilty of H_2 given $X = x$ is less than 0.0424 if $x \notin [-2.531, 2.231]$. So, if the decision rule is to reject the hypothesis if the posterior probability is less than 0.0424, then we would reject H_1 whenever we reject H_2.

4.4.3 Point Hypotheses

In this section, we will deal with the case in which $\Omega_H = \{\theta_0\}$ and $\Omega_A = \Omega \setminus \{\theta_0\}$. This is like the case of an interval hypothesis with a two-sided alternative, except that the interval is degenerate. The proofs of some of the results for two-sided alternatives relied on the fact that the two endpoints of the hypothesis were distinct. When the endpoints are the same, some changes are required.

Lemma 4.103.[30] *Suppose that the power functions of all tests are differentiable. If ϕ is the UMP level α test of $H : \Theta = \theta_0$ versus $A : \Theta < \theta_0$, then the derivative of the power function of ϕ at θ_0 is smallest among all tests with size α. Similarly, if ϕ is the UMP level α test of $H : \Theta = \theta_0$ versus $A : \Theta > \theta_0$, then the derivative of the power function of ϕ at θ_0 is largest among all tests with size α.*

---

[30]This lemma is used in the proof of Theorem 4.104.

PROOF. We prove only the first part, since the second is very similar. By Lemma 4.45, it follows that $\beta_\phi(\theta_0) = \alpha$ if ϕ is UMP level α. Let ψ be another size α test. Since ϕ is UMP level α, for every $\epsilon > 0$,

$$\beta_\phi(\theta_0 - \epsilon) \geq \beta_\psi(\theta_0 - \epsilon),$$
$$\alpha - \beta_\phi(\theta_0 - \epsilon) \leq \alpha - \beta_\psi(\theta_0 - \epsilon),$$
$$\frac{\beta_\phi(\theta_0) - \beta_\phi(\theta_0 + \epsilon)}{\epsilon} \leq \frac{\beta_\psi(\theta_0) - \beta_\psi(\theta_0 + \epsilon)}{\epsilon}.$$

Since the derivatives are the limits of the quantities in the last inequality as ϵ goes to 0, the result follows. □

Theorem 4.104.[31] *In a one-parameter exponential family with natural parameter, let $\Omega_H = \{\theta_0\}$, where θ_0 is in the interior of Ω. Let ϕ be any test of H versus $A : \Theta \neq \theta_0$. Then there is a test of the form of ψ in Lemma 4.99 such that*

$$\beta_\psi(\theta_0) = \beta_\phi(\theta_0),$$
$$\left.\frac{d}{d\theta}\beta_\psi(\theta)\right|_{\theta=\theta_0} = \left.\frac{d}{d\theta}\beta_\phi(\theta)\right|_{\theta=\theta_0}, \qquad (4.105)$$

and, for every $\theta \neq \theta_0$, $\beta_\psi(\theta)$ is maximized among all tests ψ satisfying the two equalities above.

PROOF. Let $\alpha = \beta_\phi(\theta_0)$ and $\gamma = d\beta_\phi(\theta)/d\theta|_{\theta=\theta_0}$. Let ϕ_w be the UMP level w test of $H : \Theta \geq \theta_0$ versus $A : \Theta < \theta_0$, and for each $0 \leq u \leq \alpha$, set

$$\phi'_u(x) = \phi_u(x) + 1 - \phi_{1-\alpha+u}(x).$$

By design, $\beta_{\phi'_u}(\theta_0) = \alpha$ for all u. Also, ϕ'_u has the form of ψ for every u (with c_1 or c_2 possibly infinite). By construction, $\phi'_0 = 1 - \phi_{1-\alpha}$ is the UMP level α test of $H' : \Theta = \theta_0$ versus $A' : \Theta > \theta_0$ and $\phi'_\alpha = \phi_\alpha$ is the least powerful such test. It follows from Lemma 4.103 that the derivatives of the power functions of ϕ'_0 and ϕ'_α at θ_0 are respectively the smallest possible and the largest possible among all tests with power function α at θ_0. Hence

$$\left.\frac{d}{d\theta}\beta_{\phi'_\alpha}(\theta)\right|_{\theta=\theta_0} \leq \gamma \leq \left.\frac{d}{d\theta}\beta_{\phi'_0}(\theta)\right|_{\theta=\theta_0}.$$

To prove that there is a ψ satisfying (4.105), we need only show that $d\beta_w(\theta)/d\theta|_{\theta=\theta_0}$ is continuous in w.[32] Recall that

$$\phi_w(x) = \begin{cases} 1 & \text{if } x < c_w, \\ \gamma_w & \text{if } x = c_w, \\ 0 & \text{if } x > c_w, \end{cases}$$

---

[31] This theorem is used in the proofs of Corollary 4.109 and Theorem 4.124. It is also used to show that two-sided tests form an essentially complete class.

[32] The proof follows part of the proof of Theorem 2 on pp. 220–221 of Ferguson (1967).

for some numbers c_w and γ_w such that

$$\beta_{\phi_w}(\theta_0) = P_{\theta_0}(X \leq c_w) + \gamma_w P_{\theta_0}(X = c_w) = w. \quad (4.106)$$

Define $h(x,g) = P_{\theta_0}(X \leq x) + g P_{\theta_0}(X = x)$, and define the random variable $V = h(X, G)$, where G has $U(0,1)$ distribution and is independent of X and Θ. For $0 < w \leq 1$, we note that

$$c_w = \inf\{u : F_{X|\Theta}(u|\theta) \geq w\}$$

and

$$\gamma_w P_{\theta_0}(X = c_w) = [w - P_{\theta_0}(X < c_w)].$$

For $w = 0$, $c_0 = \sup\{u : F_{X|\Theta}(u|\theta) = 0\}$ and $\gamma_0 = 0$. It follows that for all t, $h(x,g) \leq t$ if and only if either $x < c_t$ or $x = c_t$ and $g \leq \gamma_t$. For $t \geq 0$, we have

$$\begin{aligned}
F_{V|\Theta}(t|\theta_0) &= \int\int I_{[0,t]}(h(x,g)) f_{X|\Theta}(x|\theta_0) dg d\nu(x) \\
&= \int\int [I_{(-\infty,c_t)}(x) + I_{\{c_t\}}(x) I_{[0,\gamma_t]}(g)] dg f_{X|\Theta}(x|\theta_0) d\nu(x) \\
&= \int [I_{(-\infty,c_t)}(x) + \gamma_t I_{\{c_t\}}(x)] f_{X|\Theta}(x|\theta_0) d\nu(x) \quad (4.107) \\
&= \int \phi_t(x) f_{X|\Theta}(x|\theta_0) d\nu(x) = \beta_{\phi_t}(\theta_0) = t,
\end{aligned}$$

by (4.106). Hence V has $U(0,1)$ distribution given $\Theta = \theta_0$. From (4.107), we can write $\phi_w(x) = E\{I_{[0,w]}(V)|X = x\}$. It follows from Theorem 2.64 that for every test η,

$$\begin{aligned}
\frac{d}{d\theta}\beta_\eta(\theta)\bigg|_{\theta=\theta_0} &= \frac{d}{d\theta}\int \eta(x) c(\theta) \exp(\theta x) d\nu(x)\bigg|_{\theta=\theta_0} \\
&= \int \eta(x) \frac{d}{d\theta} c(\theta) \exp(\theta x)\bigg|_{\theta=\theta_0} \\
&= \int \eta(x)[xc(\theta) + c'(\theta)] \exp(\theta x) d\nu(x) \\
&= E_{\theta_0}[X\eta(X)] - \beta_\eta(\theta_0) E_{\theta_0}(X).
\end{aligned}$$

It follows that

$$\begin{aligned}
\frac{d}{d\theta}\beta_{\phi_w}(\theta)\bigg|_{\theta=\theta_0} &= E_{\theta_0}\{X\phi_w(X)\} - w E_{\theta_0}(X) \\
&= E_{\theta_0}\{XI_{[0,w]}(V)\} - w E_{\theta_0}(X).
\end{aligned}$$

Since w is continuous and V has continuous distribution, it follows that the above expression is continuous in w.

What remains to be proven is that ψ maximizes the power function among all tests with a fixed size and a fixed value of the derivative of the power function. As in the proof of Theorem 4.82, let $f_{X|\Theta}(x|\theta) = c(\theta)\exp(\theta x)$, so that $h(x)$ is incorporated into the measure ν. For a test η with $\beta_\eta(\theta_0) = \alpha$, the derivative of the power function at θ_0 will equal γ if and only if $E_{\theta_0}[X\eta(X)] = \gamma + \alpha E_{\theta_0}(X)$. Let $\theta_1 \neq \theta_0$. We now apply Lemma 4.78 with

$$\begin{aligned} p_0(x) &= c(\theta_1)\exp(\theta_1 x), \\ p_1(x) &= c(\theta_0)\exp(\theta_0 x), \\ p_2(x) &= x c(\theta_0)\exp(\theta_0 x). \end{aligned}$$

The test η with the largest power at θ_1 subject to

$$\begin{aligned} \beta_\eta(\theta_0) &\leq (\geq) \ \alpha, \\ \left.\frac{d}{d\theta}\beta_\eta(\theta)\right|_{\theta=\theta_0} &\leq (\geq) \ \gamma \end{aligned}$$

is $\eta(x) = 1$ if

$$\exp(\theta_1 x) > k_1 \exp(\theta_0 x) + k_2 x \exp(\theta_0 x), \qquad (4.108)$$

where the signs of k_1 and k_2 depend on which inequalities we use. The inequality (4.108) simplifies to $\exp([\theta_1 - \theta_0]x) > k_1 + k_2 x$. This inequality is satisfied for x outside of a bounded interval or for x in a semi-infinite interval. We already know (Theorem 4.56 and Propositions 4.62–4.64) that tests with $\psi_0(x) = 1$ for x in a semi-infinite interval are one-sided and they minimize the power function on one side of the hypothesis. Hence, we need $\psi_0(x) = 1$ for x outside of a bounded interval, and ψ_0 has the form of ψ. Furthermore, the same ψ_0 works whether $\theta_1 > \theta_0$ or $\theta_1 < \theta_0$, by choosing k_1 and k_2 correctly. □

When the conditions of Theorem 4.104 hold and the loss function is of the hypothesis-testing type, then it follows that the class of two-sided tests is essentially complete. Corollary 4.109 says that the class of UMPU level α tests is a subset of this essentially complete class. At the end of Example 4.111 on page 261, we will see that the class of UMPU tests is not essentially complete.

Corollary 4.109.[33] *In a one-parameter exponential family with natural parameter, let $\Omega_H = \{\theta_0\}$, where θ_0 is in the interior of Ω. If ψ is a size α test of the form of Lemma 4.99 with*

$$\left.\frac{d}{d\theta}\beta_\psi(\theta)\right|_{\theta=\theta_0} = 0,$$

then it is UMPU level α.

---

[33] This corollary is used in the proof of Theorem 4.124.

4.4. Unbiased Tests 261

PROOF. Since the test $\phi_\alpha(x) \equiv \alpha$ has size α and 0 derivative at θ_0, Theorem 4.104 tells us that ψ is unbiased level α. In light of Theorem 4.104, all we need to show is that all unbiased level α tests must have power functions with 0 derivative at θ_0. Any test ϕ with $\beta_\phi(\theta_0) = \alpha$ but with nonzero derivative will have power strictly less than α on one side or the other of θ_0 because the power function is differentiable. Such a test could not be unbiased level α. □

Example 4.110. Suppose that $X \sim N(\theta, 1)$ given $\Theta = \theta$ and that we wish to test $H : \Theta = \theta_0$ versus $A : \Theta \neq \theta_0$. To make the test ψ unbiased, we need

$$
\begin{aligned}
0 &= \left.\frac{d}{d\theta} \beta_\psi(\theta)\right|_{\theta=\theta_0} \\
&= \left.\frac{d}{d\theta}\left[1 - \int_{c_1}^{c_2} \frac{1}{\sqrt{2\pi}} \exp\left(-\frac{(x-\theta)^2}{2}\right) dx\right]\right|_{\theta=\theta_0} \\
&= -\int_{c_1}^{c_2} \frac{x-\theta_0}{\sqrt{2\pi}} \exp\left(-\frac{(x-\theta_0)^2}{2}\right) dx \\
&= -\int_{c_1-\theta_0}^{c_2-\theta_0} \frac{x}{\sqrt{2\pi}} \exp\left(-\frac{x^2}{2}\right) dx
\end{aligned}
$$

which is true if and only if $-(c_1 - \theta_0) = c_2 - \theta_0 = c$. In this case $\beta_\psi(\theta_0) = 2[1 - \Phi(c)] = \alpha$ if and only if $c = \Phi^{-1}(1 - \alpha/2)$. This gives the usual equal-tailed, two-sided test, which is UMPU level α.

In Example 4.110, if a Bayesian used a proper continuous prior distribution, then $\Pr(\Theta = \theta_0) = 0$ both before and after observing X. There are at least two ways to treat point hypotheses from a Bayesian perspective. One is to treat them as surrogates for interval hypotheses in which the length of the interval has not been stated. Another is to assign positive probability to the point hypothesis.

Example 4.111 (Continuation of Example 4.110). Suppose that I really want to test $H' : |\Theta - \theta_0| \leq \delta$ versus $A' : |\Theta - \theta_0| > \delta$. Suppose that the prior distribution of Θ is $N(\theta_0, \tau^2)$. Then the posterior distribution of Θ given $X = x$ is $N(\theta_1, \tau^2/(1+\tau^2))$, where

$$\theta_1 = \frac{\theta_0 + x\tau^2}{1+\tau^2}.$$

If we use a 0–1–c loss function, the Bayes rule is to reject H' if its posterior probability is less than $1/(1+c)$. The posterior probability of H' is

$$
\begin{aligned}
&\int_{\theta_0-\delta}^{\theta_0+\delta} \frac{\sqrt{1+\frac{1}{\tau^2}}}{\sqrt{2\pi}} \exp\left\{-\frac{1}{2}\left(1+\frac{1}{\tau^2}\right)(\theta-\theta_1)^2\right\} d\theta \\
&= \int_{\theta_0-\theta_1-\delta}^{\theta_0-\theta_1+\delta} \frac{\sqrt{1+\frac{1}{\tau^2}}}{\sqrt{2\pi}} \exp\left\{-\frac{1}{2}\left(1+\frac{1}{\tau^2}\right)\theta^2\right\} d\theta,
\end{aligned}
$$

which is clearly a decreasing function of

$$|\theta_0 - \theta_1| = \left|\theta_0 - \frac{\theta_0 + x\tau^2}{1+\tau^2}\right| = \frac{\tau^2}{1+\tau^2}|\theta_0 - x|.$$

So, the Bayes rule is to reject H' if $|x - \theta_0| > d$ for some d. This has the same form as the UMPU level α test.

Alternatively, suppose that $\Pr(\Theta = \theta_0) = p_0 > 0$. Conditional on $\Theta \neq \theta_0$, suppose that $\Theta \sim N(\theta_0, \tau^2)$. We computed the Bayes factor for this case in Example 4.18 on page 222. The Bayes factor was given in (4.19) as

$$\sqrt{1+\tau^2}\exp\left(-\frac{(x-\theta_0)^2\tau^2}{2(1+\tau^2)}\right).$$

(See Problem 6 on page 286 for the entire posterior distribution of Θ.) If we use a 0–1–c loss function, then the Bayes rule is to reject H if the probability that it is true is less than $1/(1+c)$. This corresponds to the Bayes factor being less than some number, which in turn is easily seen to correspond to $|x - \theta_0| > d$ for some d. This is in the same form as the UMPU level α test.

Finally, suppose that we continue to let $\Pr(\Theta = \theta_0) = p_0 > 0$, but that the conditional prior given $\Theta \neq \theta_0$ is $N(\theta', \tau^2)$ with $\theta' \neq \theta_0$. Then the same kind of calculation as above leads to the Bayes factor being small when x is far from $[(1+\tau^2)\theta_0 - \theta']/\tau^2$. Such a test is two-sided but is not UMPU of its level. In fact, the test is biased, even though it is admissible. We see once again that the class of UMPU tests is not essentially complete.

In Example 4.111, two different types of prior distributions both led to Bayes rules that were of the same form as the UMPU level α test. Unfortunately, there did not appear to be any transparent connection between the size α and the loss function or the prior distribution. The reason for this is related to the inadmissibility of incoherent tests as illustrated in Problem 42 on page 291. We will discuss this matter more in Section 4.6.

Example 4.112. Suppose that $X \sim Bin(n, p)$ given $\Theta = \log(p/(1-p))$. Then the density of X with respect to counting measure on $\{0, \ldots, n\}$ is

$$f_{X|\Theta}(x|\theta) = \binom{n}{x}[1+\exp(\theta)]^{-n}\exp(x\theta).$$

Let $\Omega_H = \{\theta_0\}$ and $\Omega_A = \mathbb{R}\setminus\{\theta_0\}$. The UMPU level α test is

$$\phi(x) = \begin{cases} 1 & \text{if } x < c_1 \text{ or } x > c_2, \\ \gamma_i & \text{if } x = c_i, \\ 0 & \text{otherwise,} \end{cases} \quad (4.113)$$

where $c_1 \leq c_2$. Supposing that $c_1 < c_2$, we have

$$\beta_\phi(\theta) = 1 - [1+\exp(\theta)]^{-n}\left[(1-\gamma_1)\binom{n}{c_1}\exp(c_1\theta)\right.$$
$$\left. + (1-\gamma_2)\binom{n}{c_2}\exp(c_2\theta) + \sum_{x=c_1+1}^{c_2-1}\binom{n}{x}\exp(x\theta)\right].$$

It follows that

$$\frac{d}{d\theta}\beta_\phi(\theta) = n\left[(1-\gamma_1)\binom{n}{c_1}\exp(c_1\theta) + (1-\gamma_2)\binom{n}{c_2}\exp(c_2\theta)\right.$$
$$\left. + \sum_{x=c_1+1}^{c_2-1}\binom{n}{x}\exp(x\theta)\right]\exp(\theta)[1+\exp(\theta)]^{-n-1}$$
$$- [1+\exp(\theta)]^{-n}\left[(1-\gamma_1)c_1\binom{n}{c_1}\exp(c_1\theta)\right.$$
$$\left. + (1-\gamma_2)c_2\binom{n}{c_2}\exp(c_2\theta) + \sum_{x=c_1+1}^{c_2-1}\binom{n}{x}x\exp(x\theta)\right].$$

Once c_1 and c_2 are determined, solving for γ_1 and γ_2 amounts to solving two linear equations. Now, suppose that $\theta_0 = \log(0.25/0.75)$. Then, with $\alpha = 0.05$ and $n = 10$, we get (after some numerical calculation)

$$c_1 = 0, \quad \gamma_1 = 0.52804,$$
$$c_2 = 5, \quad \gamma_2 = 0.00918.$$

Most people who want a level 0.05 test of this hypothesis would not bother to compute the UMPU level 0.05 test but rather would perform what is called an *equal-tailed test*. Since Theorem 4.104 says that the two-sided tests are admissible, we could try to find a two-sided test of the form (4.113) such that the probability of rejecting for small X equals the probability of rejecting for large X (both equal to 0.025). In this case, the test would have

$$c_1 = 0, \quad \gamma_1 = 0.44394,$$
$$c_2 = 5, \quad \gamma_2 = 0.09028.$$

This test is biased because the derivative of the power function is 0.0236 at θ_0. In other words, the probability of rejecting the hypothesis will be slightly less than 0.05 given $\Theta = \theta$ for a short interval of θ values below θ_0.

One possible Bayesian solution would be to set $\Pr(P = p_0) = q_0$ and let $P \sim \text{Beta}(\alpha_0, \beta_0)$ otherwise, where $P = \exp(\Theta)/(1+\exp(\Theta))$. Then, the Bayes factor will be

$$\frac{p_0^x(1-p_0)^{n-x}\prod_{i=0}^{n-1}(\alpha_0+\beta_0+i)}{\prod_{i=0}^{x-1}(\alpha_0+i)\prod_{j=0}^{n-x-1}(\beta_0+j)}. \quad (4.114)$$

In the special case with $\alpha_0 = \beta_0 = 1$, the Bayes factor is

$$(n+1)\binom{n}{x}p_0^x(1-p_0)^{n-x}. \quad (4.115)$$

These values have been calculated for $n = 10$ and $p_0 = 1/4$ in Table 4.116 together with the posterior probability when $q_0 = 1/2$. Note that if we used a 0–1–c loss function with $c = 19$ (so that $1/(1+c) = 0.05$), we would still accept H even when $X = 6$ was observed.

As we noted in Section 4.2.2, we would run into trouble if we naïvely tried to use an improper prior (with $\alpha_0 = \beta_0 = 0$) for the alternative. The Bayes factor

TABLE 4.116. Bayes Factor and Posterior Probability in Binomial Example

x	Bayes Factor	Posterior Prob.
0	0.619	0.3825
1	2.064	0.6737
2	3.097	0.7559
3	2.753	0.7335
4	1.606	0.6163
5	0.642	0.3911
6	0.178	0.1514
7	0.034	0.0329
8	0.004	0.0042
9	3×10^{-4}	0.0003
10	1×10^{-5}	1×10^{-5}

in (4.114) would become ∞ in this case. On the other hand, if we let q_0 go to zero at a rate such that

$$\lim \frac{q_0(\alpha_0 + \beta_0)}{\alpha_0 \beta_0} = k,$$

then the product of the prior odds ratio $q_0/(1 - q_0)$ and the Bayes factor would converge to

$$k(n-1)\binom{n-2}{x-1}p_0^x(1-p_0)^{n-x}$$

if both $x > 0$ and $n - x > 0$. This has a form similar to (4.115).

Another Bayesian solution would be to replace H and A by $H' : |\Theta - \theta_0| \leq \delta$ and $A' : |\Theta - \theta_0| > \delta$. Suppose that $P \sim Beta(\alpha_0, \beta_0)$. The posterior distribution of P given $X = x$ is $Beta(\alpha_0 + x, \beta_0 + n - x)$. It is an easy matter to calculate $\Pr(H' \text{ true} | X = x)$ for various values of δ and x. Figure 4.119 gives plots of $\Pr(|P - 1/4| \leq \delta | X = x)$ for $\alpha_0 = \beta_0 = 1$ for all values of $x = 0, \ldots, 10$ when $n = 10$. For example, suppose that $\delta = 0.1$. We see that for $x = 0, \ldots, 5$, the posterior probability of the hypothesis is greater than 0.05. So, if $c = 19$ and we use the 0–1–c loss function, we would accept H' if $X \leq 5$ and would reject otherwise.

Notice that the condition that the derivative of the power function be 0 in Example 4.112 was equivalent to

$$\alpha E_{\theta_0} X = E_{\theta_0}(X\phi(X)) \tag{4.117}$$

if the size is α. This is true in general in exponential families.

Proposition 4.118.[34] *If X has a one-parameter exponential family distribution with natural parameter Θ and ϕ is a test of $H : \Theta = \theta_0$ with size α, then β_ϕ has 0 derivative at θ_0 if and only if (4.117) holds.*

---

[34] This proposition is used in the proof of Theorem 4.124.

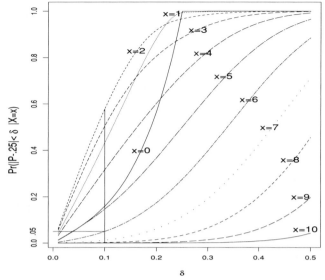

FIGURE 4.119. $\Pr\left(\left|P - \frac{1}{4}\right| \leq \delta \,\Big|\, X = x\right)$ for all $\delta \leq \frac{1}{2}$ and all x

When UMPU level α tests do not exist, one can try to find LMPU (locally most powerful unbiased) level α tests.[35] When power functions are continuously differentiable and ϕ is the unique unbiased level α test of $H : \Theta = \theta_0$ with maximum second derivative for the power function, then ϕ is LMPU level α relative to $d(\theta) = |\theta - \theta_0|$. (See Problem 50 on page 292.)

4.5 Nuisance Parameters

When the parameter is multidimensional and Ω_H is a smaller-dimensional space, the remaining dimensions are often called *nuisance parameters* for reasons that will become apparent shortly. In a Bayesian analysis, one must integrate nuisance parameters out of the posterior joint distribution of the parameters and base inference on the marginal distribution of the parameters of interest. This can be a nuisance also.

4.5.1 Neyman Structure

The approach that we will take to finding UMPU tests in the presence of nuisance parameters is to find a statistic T such that the conditional distribution of the data given T has a one-dimensional parameter. In many cases, it will then turn out that the UMPU test among all tests that have

---

[35] We leave it to the interested reader to write a formal definition of LMPU.

266   Chapter 4. Hypothesis Testing

level α conditional on T will also be UMPU unconditionally. If a test is α-similar conditional on T, we say that it has *Neyman structure*.

Definition 4.120. Let $G \subseteq \Omega$ be a subparameter space corresponding to a subfamily $\mathcal{Q}_0$ of $\mathcal{P}_0$, and let $\Psi : \mathcal{Q}_0 \to G$ be the subparameter. If T is a sufficient statistic for Ψ in the classical sense, then a test ϕ has *Neyman structure relative to G and T* if $E_\theta[\phi(X)|T = t]$ is constant in t a.s. $[P_\theta]$ for all $\theta \in G$.

It is easy to see that if $\mathcal{Q}_0 = \{P_\theta : \theta \in \overline{\Omega}_H \cap \overline{\Omega}_A\}$ and if ϕ has Neyman structure, then ϕ is α-similar. We will prove later (Lemma 4.122) that under certain conditions all α-similar tests have Neyman structure. In these cases, one can find UMP α-similar (hence UMPU level α by Proposition 4.92) tests by restricting attention to tests with Neyman structure. Consider the following example.

Example 4.121. Suppose that $X_1, \ldots, X_n$ are IID $N(\mu, \sigma^2)$ random variables conditional on $(M, \Sigma) = (\mu, \sigma)$. The usual two-sided t-test of $H : M = \mu_0$ versus $A : M \neq \mu_0$ is

$$\phi(x) = \begin{cases} 1 & \text{if } |\overline{x} - \mu_0| > \frac{s}{\sqrt{n}} T_{n-1}^{-1}\left(1 - \frac{\alpha}{2}\right), \\ 0 & \text{otherwise,} \end{cases}$$

where T_{n-1} is the CDF of the $t_{n-1}(0,1)$ distribution and $s^2 = \sum_{i=1}^n (x_i - \overline{x})^2/(n-1)$. Here, the intersection of $\overline{\Omega}_H$ with $\overline{\Omega}_A$ is $\Omega_H = \{(\mu, \sigma) : \mu = \mu_0\}$. It is easy to see that ϕ is α-similar, as follows. Given $(M, \Sigma) = (\mu_0, \sigma) \in \Omega_H$, the conditional distribution of $T = (\overline{X} - \mu_0)/(S/\sqrt{n})$ is $t_{n-1}(0,1)$ for all σ. Hence

$$\beta_\phi(\gamma) = P'_\gamma\left[\sqrt{n}\frac{|\overline{X} - \mu_0|}{S} > T_{n-1}^{-1}\left(1 - \frac{\alpha}{2}\right)\right] = \alpha.$$

Let the subparameter space be Ω_H itself. A sufficient statistic for the subparameter Σ is

$$U = \sum_{i=1}^n (X_i - \mu_0)^2 = (n-1)S^2 + n(\overline{X} - \mu_0)^2.$$

We can write

$$W = \frac{\overline{X} - \mu_0}{\sqrt{U}} = \frac{\text{sign}(T)}{\sqrt{n}\sqrt{\frac{n-1}{T^2} + 1}},$$

so that W is a one-to-one increasing function of T and $\phi(X)$ is a function of W. We need to show that the conditional distribution of W given $(M, \Sigma) = \gamma$ and $U = u$ is the same for all $\gamma \in \Omega_H$ and all u. If this is true, then W is independent of U given $(M, \Sigma) = \gamma \in \Omega_H$ and

$$E_\gamma[\phi(X)|U = u] = E_\gamma[\phi(X)] = \alpha,$$

for all $\gamma \in \Omega_H$ and ϕ would have Neyman structure relative to the Ω_H and U.

Since the distribution of $(X_1 - \mu_0, \ldots, X_n - \mu_0)$ is spherically symmetric, we showed in Examples B.56 and B.60 (see pages 627 and 628) that the conditional distribution of

$$\left(\frac{X_1 - \mu_0}{\sqrt{U}}, \ldots, \frac{X_n - \mu_0}{\sqrt{U}}\right)$$

given U is uniform on the sphere of radius 1 and is independent of U. Hence W is independent of U given $(M, \Sigma) = \gamma \in \Omega_H$.

Lemma 4.122.[36] *If T is a boundedly complete sufficient statistic for the subparameter space $G \subseteq \Omega$, then every α-similar test on G has Neyman structure relative to G and T.*

PROOF. By α-similarity, $E_\theta\{E_\theta[\phi(X)|T] - \alpha\} = 0$, for all $\theta \in G$. Since T is boundedly complete, it must be that $E_\theta[\phi(X)|T] = \alpha$, a.s. $[P_\theta]$ for all $\theta \in G$. □

We can now combine this result with Proposition 4.92 to conclude a useful result for identifying cases in which UMPU tests exist.

Theorem 4.123. *Let $G = \overline{\Omega_H} \cap \overline{\Omega_A}$. Let I be an index set, and suppose that $G = \cup_{i \in I} G_i$, where the subsets G_i form a partition of G. Suppose that there exists a statistic T that is a boundedly complete sufficient statistic for each subparameter space G_i, $i \in I$. Also, assume that the power function of every test is continuous. If there is a UMPU level α test ϕ among those which have Neyman structure relative to G_i and T for all $i \in I$, then ϕ will be UMPU level α.*

PROOF. Because the power functions are continuous, Proposition 4.92 says that all unbiased level α tests are α-similar. Lemma 4.122 says that because there is a boundedly complete sufficient statistic T for each subparameter space G_i, every α-similar test has Neyman structure relative to G_i and T. The result now follows. □

The way that Theorem 4.123 is generally used is the following. We suppose that power functions are continuous and that there exists a partition of $G = \overline{\Omega_H} \cap \overline{\Omega_A}$ into one or two sets ($G = G_0$ or $G = G_1 \cup G_2$) and a statistic T that is a boundedly complete sufficient for each G_i. We also suppose that the conditional distribution of X given T is a one-parameter family with parameter $g(\Theta)$ for some function g. We also suppose that Ω_H can be written as $\{\theta : g(\theta) \leq b_0\}$ or $\{\theta : g(\theta) \in [b_0, b_1]\}$ or one of the other forms with which we are already familiar. So, for example, if $\Theta = (M, \Sigma)$ and $\Omega_H = \{(\mu, \sigma) : b_0 \leq \mu \leq b_1\}$, then $g(\mu, \sigma) = \mu$, $G_1 = \{(\mu, \sigma) : \mu = b_0\}$, and $G_2 = \{(\mu, \sigma) : \mu = b_1\}$. For the one-sided cases, we assume that the family of distributions of X given T has MLR, but for the other cases, the conditional distribution of X given T should be a one-parameter exponential family with natural parameter $g(\Theta)$. We then find the UMP or UMPU level α test of H conditional on T. For all of the cases, except for the case in which the hypothesis is $H : g(\Theta) = b_0$, the UMP or UMPU level α test conditional on T will also be UMP or UMPU among tests with Neyman structure. The reason is that these tests are derived as UMP or UMPU among all tests that satisfy conditions that are equivalent to

---

[36]This lemma is used in the proofs of Theorems 4.123 and 4.124.

having Neyman structure. For example, when $H : g(\Theta) \in [b_0, b_1]$, the two-sided test is conditionally UMPU level α among all tests that have level α and have conditional power function equal to α at $g(\theta) = b_0$ and $g(\theta) = b_1$. This is exactly what it means to have Neyman structure relative to $G_0 = \{\theta : g(\theta) = b_0\}$ and relative to $G_0 = \{\theta : g(\theta) = b_1\}$. For the case of $H : g(\Theta) = b_0$, the two-sided test is conditionally UMPU level α among those tests with conditional power function α at $g(\theta) = b_0$ and with derivative of the conditional power function equal to 0 at $g(\theta) = b_0$. This last condition is not part of the definition of Neyman structure. Hence, in these cases, we need to prove that every Neyman structure test also satisfies this last condition. Problem 58 on page 293 is an example of this situation. In multiparameter exponential families, when $g(\Theta)$ is just one of the coordinates of Θ, every α-similar level α test will have zero derivative for the conditional power function, as we prove in Theorem 4.124.

4.5.2 Tests about Natural Parameters

The case in which we can prove the most complete result is that of an exponential family in which the hypothesis concerns one of the parameters.

Theorem 4.124. *Let $(X_1, \ldots, X_k)$ have a k-parameter exponential family distribution with natural parameter $\Theta = (\Theta_1, \ldots, \Theta_k)$, and let $U = (X_2, \ldots, X_k)$.*

1. *Suppose that the hypothesis is one-sided or two-sided concerning only Θ_1. Then there is a UMP level α test conditional on U, and it is UMPU level α.*

2. *If the hypothesis concerns only Θ_1 and the alternative is two-sided, then there is a UMPU level α test conditional on U, and it is UMPU level α.*

PROOF. Suppose that the density of X with respect to a measure ν is written

$$f_{X|\Theta}(x|\theta) = c(\theta)h(x)\exp\left(\sum_{i=1}^{k} x_i \theta_i\right).$$

Let $G = \overline{\Omega}_H \cap \overline{\Omega}_A$, the intersection of the closures of the hypothesis and alternative sets. The conditional density of X_1 given $(X_2, \ldots, X_k) = (x_2, \ldots, x_k)$ with respect to the measure $d\nu_{\mathcal{X}|\mathcal{U}}(x_1|u)$ (from Theorem B.46 with $\mathcal{X} = \mathbb{R}$ and $\mathcal{U} = \mathbb{R}^{k-1}$) is

$$f_{X_1|\Theta,U}(x_1|\theta,u) = \frac{\exp(\theta_1 x_1)h(x)}{\int h(x)\exp(\theta_1 x_1) d\nu_{\mathcal{X}|\mathcal{U}}(x_1|u)},$$

which can be seen to depend on θ only through θ_1. So, for each vector u, the conditional distribution of X_1 is a one-parameter exponential family with natural parameter Θ_1. For the hypotheses considered in this theorem, the subparameter space G is either the set $G_0 = \{\theta : \theta_1 = \theta_1^0\}$ for some θ_1^0 or the union of two such sets, $G_1 = \{\theta : \theta_1 = \theta_1^1\}$ and $G_2 = \{\theta : \theta_1 = \theta_1^2\}$. For each such subset of Ω, the subparameter $\Psi = (\Theta_2, \ldots, \Theta_k)$ has complete sufficient statistic $U = (X_2, \ldots, X_k)$. Let η be an unbiased level α test. It follows from Proposition 4.92 that η is α-similar on G_0, or on G_1 and on G_2, whichever is appropriate. By Lemma 4.122, η has Neyman structure. Also, for every test η, $\beta_\eta(\theta) = E_\theta(E_\theta[\eta(X)|U])$, so that a test that maximizes the conditional power function uniformly for $\theta \in \Omega_A$ subject to constraints also maximizes the marginal power function subject to the same constraints.

For part (1), in the conditional problem given $U = u$, there is a level α test ϕ that maximizes the conditional power function uniformly on Ω_A subject to having Neyman structure (see Theorem 4.56, Propositions 4.62–4.65, and Theorem 4.82). Since every unbiased level α test has Neyman structure, and the power function is the expectation of the conditional power function, ϕ is UMPU level α.

For part (2), we consider two cases. First, suppose that $\Omega_H = \{\theta : c_1 \leq \theta_1 \leq c_2\}$ with $c_2 > c_1$. Then, Lemma 4.99 shows that there is a test ϕ whose conditional power function is maximized uniformly on Ω_A subject to having Neyman structure. It follows as before that ϕ is UMPU level α. Finally, suppose that $\Omega_H = \{\theta : \theta_1 = \theta_1^0\}$. If η is unbiased level α, then β_η must have zero partial derivative with respect to θ_1 evaluated at every point in G. Using Theorem 2.64, just as in the proof of Theorem 4.104, we get, for every $\theta_* \in G$,

$$0 = \frac{\partial}{\partial \theta_1} \beta_\eta(\theta) \bigg|_{\theta=\theta_*} = E_{\theta_*}[X_1 \eta(X) - \alpha X_1].$$

By the law of total probability B.70, $E_{\theta_*}(E_{\theta_*}[X_1\eta(X) - \alpha X_1|U]) = 0$, for every $\theta_* \in G$. Since U is complete for $\Theta \in G$, it follows that

$$E_{\theta_*}[X_1 \eta(X) - \alpha X_1 | U] = 0, \quad \text{a.s. } [P_{\theta_*}]. \tag{4.125}$$

So every unbiased level α test η must satisfy (4.125). Proposition 4.117 says that (4.125) is equivalent to the derivative of the conditional power function with respect to θ_1 at $\theta_1 = \theta_1^0$ being zero. Subject to this condition and having Neyman structure, there is a test that maximizes the conditional power function at all $\theta \in \Omega_A$ according to Corollary 4.109. This test is then UMPU level α. □

Example 4.126. Suppose that (X_1, U) has a multiparameter exponential family and $\Omega_H = \{\theta : \theta_1 \leq \theta_1^0\}$ with $\Omega_A = \{\theta : \theta_1 > \theta_1^0\}$. The conditional UMP level α test is

$$\phi(x_1|u) = \begin{cases} 1 & \text{if } x_1 > d(u), \\ \gamma(u) & \text{if } x_1 = d(u), \\ 0 & \text{if } x_1 < d(u), \end{cases}$$

where d and γ are chosen so that the conditional size is α a.s. This test is UMPU level α by Theorem 4.124.

Example 4.127 (Continuation of Example 4.121; see page 266). The usual two-sided t-test is an example of Theorem 4.124. To see this, write the joint density of $(\overline{X}, S^2)$ as

$$\begin{aligned}
f_{\overline{X}, S^2 | M, \Sigma}(\overline{x}, s^2 | \mu, \sigma) &= \frac{\sqrt{n} \left(\frac{n-1}{2\sigma^2}\right)^{\frac{n-1}{2}}}{\sqrt{2\pi}\Gamma\left(\frac{n-1}{2}\right)\sigma} \\
&\quad \times \exp\left(-\frac{1}{2\sigma^2}\left[n(\overline{x} - \mu_0 - [\mu - \mu_0])^2 + (n-1)s^2\right]\right) \\
&= r(\mu, \sigma) h(\overline{x}, s^2) \exp\left(\theta_1 v + \theta_2 u\right),
\end{aligned}$$

where $r(\mu, \sigma)$ depends solely on the parameters and

$$\theta_1 = \frac{\mu - \mu_0}{\sigma^2}, \qquad \theta_2 = -\frac{1}{2\sigma^2},$$

$$v = n(\overline{x} - \mu_0), \qquad u = n(\overline{x} - \mu_0)^2 + (n-1)s^2.$$

(Note that u is the observed value of the statistic U from Example 4.121.) Theorem 4.124 says that the UMPU level α test of $H : \Theta_1 = 0$ versus $A : \Theta_1 \neq 0$ is the conditional UMPU level α test of H versus A given U. Note that $\Theta_1 = 0$ is equivalent to $M = \mu_0$. Since V has a one-parameter exponential family distribution with natural parameter Θ_1 given U, the test will be a two-sided test of the form

$$\phi(v|u) = \begin{cases} 1 & \text{if } v < d_1(u) \text{ or if } v > d_2(u), \\ 0 & \text{if } d_1(u) \leq v \leq d_2(u), \end{cases}$$

where $d_1(u)$ and $d_2(u)$ are chosen so that the conditional power function equals α at $\theta_1 = 0$ for all u and so that the derivative of the conditional power function equals 0 at $\theta_1 = 0$ for all u. As we saw in Example 4.121, the two-sided t-test has the above form with $d_1(u) = -c\sqrt{u}$ and $d_2 = c\sqrt{u}$, where $c > 0$ is a constant. We also saw that the t-test has conditional level α given U. The fact that the derivative of the conditional power function is 0 follows, as in the proof of Theorem 4.124, from the fact that the partial derivative of the marginal power function is 0. Hence, the usual two-sided t-test is UMPU level α.

One possible Bayesian approach is to put positive probability on Ω_H. This was done in Example 4.22 on page 224, where Bayes factors were computed. It is not the case that the usual two-sided t-test is equivalent to rejecting H when the Bayes factor is less than a constant. However, with a special type of improper prior, the two tests are equivalent. In Example 4.22, we showed that if $\lambda_0 \to 0$ and $p_0 \to 0$ in such a way that the ratio $p_0/\sqrt{\lambda_0} \to k$, a constant, then the posterior odds in favor of the hypothesis converge to

$$k\sqrt{n}\left(1 + \frac{t^2}{n-1}\right)^{-\frac{n-1}{2}},$$

where t is the usual t statistic.

Example 4.128. Suppose that X and Y are conditionally independent given $\Gamma = (\Lambda, M) = (\lambda, \mu)$ with $X \sim Poi(\lambda)$ and $Y \sim Poi(\mu)$. We wish to test $H : \Lambda = 2M$ versus $A : \Lambda \neq 2M$. Set $G = \{(\lambda, \mu) : \lambda = 2\mu\}$. Set $\Psi = \Lambda = 2M$ for the subparameter space G, and note that

$$f_{X,Y|\Psi}(x, y|\psi) = \exp(-1.5\psi)\frac{\psi^{x+y}2^{-y}}{x! y!},$$

4.5. Nuisance Parameters

for $x, y = 0, 1, \ldots$. It follows that $T = X + Y$ is a complete sufficient statistic for the subparameter space. The distribution of T is

$$f_{T|\Gamma}(t|\lambda, \mu) = \exp(-\lambda - \mu)\mu^t \sum_{a=0}^{t} \frac{\left(\frac{\lambda}{\mu}\right)^a}{a!(t-a)!},$$

for $t = 0, 1, \ldots$. The conditional distribution of (X, Y) given T and Γ is one-dimensional and can be represented by X alone. It is

$$f_{X|T,\Gamma}(x|t, \lambda, \mu) = \frac{\left(\frac{\lambda}{\mu}\right)^x}{x!(t-x)!} \left(\sum_{a=0}^{t} \frac{\left(\frac{\lambda}{\mu}\right)^a}{a!(t-a)!}\right)^{-1},$$

for $x = 0, \ldots, t$. This is easily seen to be a one-parameter exponential family with natural parameter $\Theta = g(\Gamma) = \log(\Lambda/M)$. The hypothesis can be written as $H : \Theta = \log(2)$ versus $A : \Theta \neq \log(2)$. All we need to show is that every α-similar test with level α has zero derivative for the conditional power function at $\Theta = \log(2)$. To do this, first we reparameterize to Θ and M. Then

$$f_{X,Y|\Theta,M}(x, y|\theta, \mu) = \exp(-\mu[\exp(\theta) + 1]) \frac{\exp(x\theta)\mu^{x+y}}{x!y!}.$$

Every unbiased α-similar level α test must have partial derivative of the power function with respect to θ equal to 0 at every point $(\log(2), \mu)$ in G, otherwise the power function would dip below α on the alternative. The partial derivative of the power function of a test ϕ with respect to θ is

$$\sum_{x=0}^{\infty} \sum_{y=0}^{\infty} \phi(x, y) \exp(-\mu[\exp(\theta) + 1]) \frac{\exp(x\theta)\mu^{x+y}}{x!y!} [x - \mu \exp(\theta)]$$

$$= E_{\theta,\mu}(X\phi(X, Y)) - \mu \exp(\theta)\beta_\phi(\theta, \mu).$$

Now, plug in $\theta = \log(2)$ and set this equal to 0. Note that 2μ is the mean of X and the power function at $(\log(2), \mu)$ is α for every μ for an α-similar test. Hence, every α-similar level α test ϕ must satisfy

$$0 = E_{\log(2),\mu}\left(X\phi(X, Y) - X\alpha\right) = E_{\log(2),\mu}\left(E\left(X\phi(X, Y) - X\alpha|T\right)\right),$$

for all μ. Let $h(t) = E(X\phi(X, Y) - X\alpha|T = t)$. Since T is complete for the subparameter space, $E_{\log(2),\mu}(h(T)) = 0$ for all μ implies $h(T) = 0$, a.s. $[P_{\log(2),\mu}]$ for all μ. By Proposition 4.118, this is equivalent to the derivative of the conditional power function being 0 at $\theta = \log(2)$.

None of the theorems of this section provides an essentially complete class of tests. To find an essentially complete class, we could piece together conditional tests given $U = (X_2, \ldots, X_k)$. For example, let ψ be a test of $H : \Theta_1 \leq \theta_1^0$ versus $A : \Theta_1 > \theta_1^0$. Let $\beta_\psi(\theta_1|u)$ be the conditional power function of ψ given $U = u$. For each u, there is a one-sided test of the form

$$\phi(x) = \begin{cases} 1 & \text{if } x_1 > c(u), \\ \gamma(u) & \text{if } x_1 = c(u), \\ 0 & \text{if } x_1 < c(u), \end{cases} \tag{4.129}$$

which has maximum conditional power function for $\theta_1 > \theta_1^0$ and minimum conditional power function for $\theta_1 < \theta_1^0$ subject to the power being $\beta_\psi(\theta_1^0)$ at $\theta_1 = \theta_1^0$. Since ϕ in (4.129) minimizes and maximizes the power in precisely the right places uniformly in u among all tests in a class that contains ψ, it follows that $R(\theta, \phi) \leq R(\theta\psi)$ for all θ if a hypothesis-testing loss function is being used. It follows that tests of the form (4.129) form an essentially complete class. Other hypotheses can be handled in a similar way.

4.5.3 Linear Combinations of Natural Parameters

Most of the popular tests in the theory of normal distributions and linear models involve linear combinations of the natural parameters of an exponential family. In a k-parameter exponential family with natural parameter Θ, let $\Psi_1 = \sum_{i=1}^k c_i \theta_i$ with $c_1 \neq 0$. Let $\Psi_i = \Theta_i$ for $i = 2, \ldots, k$, and set $Y_1 = X_1/c_1$ and $Y_i = X_i - c_i X_1/c_1$ for $i = 2, \ldots, k$. Then $Y = (Y_1, \ldots, Y_k)$ has exponential family distribution with natural parameter Ψ. If we want to test a hypothesis concerning Ψ_1, we can proceed as above.

Example 4.130 (Continuation of Example 4.127; see page 270). The natural parameters are $\Theta_1 = M/\Sigma^2$ and $\Theta_2 = -1/[2\Sigma^2]$. Testing $M = \mu_0$ is then equivalent to testing $\Theta_1 + 2\mu_0 \Theta_2 = 0$. So, we set $\Psi_1 = \Theta_1 + 2\mu_0 \Theta_2$ and the usual two-sided t-test will result exactly as in Example 4.127.

Example 4.131. Suppose that X and Y are conditionally independent given $(\Lambda, M) = (\lambda, \mu)$ with $X \sim Poi(\lambda)$ and $Y \sim Poi(\mu)$. Suppose that $H : \Lambda = M$. In natural parameter form, $\Theta_1 = \log \Lambda$ and $\Theta_2 = \log M$. So $H : \Theta_1 - \Theta_2 = 0$. Set $\Psi_1 = \Theta_1 - \Theta_2$ and $\Psi_2 = \Theta_2$. Let $Z_1 = X$ and $Z_2 = X + Y$. Then (Z_1, Z_2) has exponential family distribution with natural parameter Ψ. We need the conditional distribution of Z_1 given Z_2. For $z_1 = 0, \ldots, z_2$, and $z_2 = 0, 1 \ldots$, we have

$$f_{Z_1, Z_2 | \Psi}(z_1, z_2 | \psi) = c(\psi) \frac{1}{z_1!(z_2 - z_1)!} \exp(z_1 \psi_1 + z_2 \psi_2),$$

$$f_{Z_2 | \Psi}(z_2 | \psi) = d(\psi) \frac{1}{z_2!} \exp(\psi_2 z_2),$$

$$f_{Z_1 | Z_2, \Psi}(z_1 | z_2, \psi) = r(\psi) k(z_1, z_2) \exp(z_1 \psi_1)$$

$$= r(\psi) \binom{z_2}{z_1} \left(\frac{\lambda}{\mu}\right)^{z_1} = \binom{z_2}{z_1} \left(\frac{\lambda}{\lambda + \mu}\right)^{z_1} \left(\frac{\mu}{\lambda + \mu}\right)^{z_2 - z_1},$$

for $z_1 = 0, \ldots, z_2$. So Z_1 given $Z_2 = z_2$ and $(\Theta_1, \Theta_2) = (\log \lambda, \log \mu)$ has $Bin(z_2, \lambda/(\lambda + \mu))$ distribution. The UMPU level α test for the binomial parameter equal to $1/2$, conditional on Z_2 is the UMPU level α test of $\Lambda = M$.

4.5.4 Other Two-Sided Cases*

In Theorem 4.124, we saw how to deal with cases in which the hypothesis or alternative is two-sided in a natural parameter. But other two-sided cases

---

*This section may be skipped without interrupting the flow of ideas.

arise.

Example 4.132. Suppose that $X_1, \ldots, X_n$ are conditionally IID $N(\mu, \sigma^2)$ given $(M, \Sigma) = (\mu, \sigma)$. This is an exponential family with natural parameters $\Theta_1 = M/\Sigma^2$ and $\Theta_2 = -1/[2\Sigma^2]$. Suppose that we wish to test $H : a \leq M \leq b$ versus $A :$ not H. We can rewrite H in terms of the natural parameters as

$$H : \Theta_1 + 2b\Theta_2 \leq 0 \text{ and } \Theta_1 + 2a\Theta_2 \geq 0.$$

We can reparameterize to $\Psi_1 = \Theta_1 + 2b\Theta_2$ and $\Psi_2 = \Theta_1 + 2a\Theta_2$, and the hypothesis becomes $H : \Psi_1 \leq 0$ and $\Psi_2 \geq 0$. (If $a < b$, the parameter space is $\{(\psi_1, \psi_2) : \psi_1 < \psi_2\}$.) This is not of any of the forms we have studied so far.

Suppose that ϕ is an unbiased level α test of H versus A. Then ϕ must be α-similar. This means that $\beta_\phi(0, \psi_2) = \beta_\phi(\psi_1, 0) = \alpha$ for all $\psi_1 < \psi_2$. It is not easy to construct nontrivial tests with these properties. (Of course, the trivial test $\phi(x) = \alpha$ for all x has these properties and is unbiased.)

We can address this problem simply within the Bayesian framework. To stay as close to the classical solutions as possible, suppose that we use the usual improper prior, so that the posterior distribution of M is $M \sim t_{n-1}(\bar{x}_n, s_n^2/n)$, where $\bar{x}_n = \sum_{i=1}^n x_i/n$ and $s_n^2 = \sum_{i=1}^n (x_i - \bar{x}_n)^2/(n-1)$. The posterior probability that H is true is

$$p = T_{n-1}\left(\sqrt{n}\frac{b - \bar{x}_n}{s_n}\right) - T_{n-1}\left(\sqrt{n}\frac{a - \bar{x}_n}{s_n}\right).$$

The formal Bayes rule with a 0–1–c loss would be to reject H if $p < 1/(1+c)$. To see what this test looks like, note that for each value of s_n, p is a decreasing function of $|t|$, where

$$t = \sqrt{n}\frac{\bar{x}_n - \frac{a+b}{2}}{s_n}. \tag{4.133}$$

In fact,

$$p = T_{n-1}\left(\sqrt{n}\frac{b-a}{2s_n} - t\right) - T_{n-1}\left(-\sqrt{n}\frac{b-a}{2s_n} - t\right).$$

So the formal Bayes rule is to reject H if $|t| > d(s_n)$, where it is also easy to see that $d(s_n)$ is a decreasing function of s_n.

One possible classical solution is to abandon the UMPU criterion and just use the usual t-test for testing that $M = (a+b)/2$. That is, reject H if $|t| > d$, where t is defined in (4.133) and d is determined to make the level α. Unfortunately, the conditional distribution of $\sqrt{n}(\bar{X}_n - [a+b]/2)/S_n$ given $(M, \Sigma) = (\mu, \sigma)$ is noncentral t $NCt_{n-1}(\sqrt{n}[\mu - (a+b)/2]/\sigma)$. The power function of the usual t-test at (μ, σ) for $\mu \neq (a+b)/2$ goes to 1 as σ goes to 0 for fixed μ. Hence the usual t-test has level 1 as a test of H versus A. To get a test with level $\alpha < 1$, one could let d depend on S_n as in the formal Bayes rule (although one should note that the formal Bayes rule also has level 1—the problem occurs as $\sigma \to \infty$ for the formal Bayes rule). Calculating the power function of the resulting test would require a separate two-dimensional integration over the space of $(\bar{x}_n, s_n)$ values for each (μ, σ) pair.

Another classical solution would be to add together the UMPU level $\alpha/2$ test of $H' : \Theta \leq b$ versus $A' : \Theta > b$ and the UMPU level $\alpha/2$ test of $H'' : \Theta \geq a$ versus $A'' : \Theta < a$. It can be shown (see Problem 54 on page 292) that this test has size α.[37] The power function is easy to calculate. It is just the sum of

---

[37]This test is the *likelihood ratio test*, to be defined in Section 4.5.5.

the two power functions for the two one-sided tests. If $a < b$, this test is biased, since there exists $\theta \in \Omega_A$ such that the power function at θ is close to $\alpha/2$. Note, however, that if $a = b$, then this test is exactly the same as the UMPU level α test of $\Theta = a$ versus $\Theta \neq a$.

One could change the hypothesis in Example 4.132 to make it more like Example 4.127.

Example 4.134 (Continuation of Example 4.127; see page 270). Suppose that we wish to test $H : M \in [\mu_0 + a\Sigma, \mu_0 + b\Sigma]$ versus $A : M \notin [\mu_0 + a\Sigma, \mu_0 + b\Sigma]$. This is a test about a linear combination of natural parameters, namely $\Psi_1 = \Theta_1 + 2\mu_0\Theta_2$. The hypothesis is $H : \Psi_1 \in [a,b]$ versus $A : \Psi_1 \notin [a,b]$. We can let $\Psi_2 = \Theta_2 = -1/[2\Sigma^2]$. We would now need to work with the conditional distribution of $V = n(\overline{X} - \mu_0)$ given $U = n(\overline{X} - \mu_0)^2 + (n-1)S^2$. This conditional distribution will have a density equal to a constant (function of u and ψ_1) times $\exp(-v\psi_1)(u - v^2/n)^{[n-1]/2-1}$ for $0 < v < \sqrt{nu}$. For each value of u, one would have to find $d_1(u)$ and $d_2(u)$ so that

$$\alpha = \Pr\left(n[\overline{X} - \mu_0] \notin [d_1(u), d_2(u)] \,\middle|\, U = u, \Psi_1 = \psi_1\right),$$

for $\psi_1 = a$ and for $\psi_1 = b$. Of course, one could wait until the data were observed and then do it only for the observed value of U.

4.5.5 Likelihood Ratio Tests

A popular approach to forming tests, when no obvious UMP or UMPU test is available, is to use the *likelihood ratio criterion* (LR). The idea is to start with the Neyman–Pearson concept of likelihood ratios. In the Neyman–Pearson setup, the likelihood under H is just a single number, as it is under A. In general, however, the likelihoods are functions of θ. In the Bayesian approach, we integrate those functions with respect to a distribution. In LR tests, we maximize those functions over θ. The LR criterion is

$$LR = \frac{\sup_{\theta \in \Omega_H} f_{X|\Theta}(X|\theta)}{\sup_{\theta \in \Omega} f_{X|\Theta}(X|\theta)}.$$

To test a hypothesis H using the LR criterion, choose a number c and reject H if $LR < c$. The number c is usually chosen to make the level of the test equal to some prespecified value. Sometimes the distribution of LR is recognizable, and sometimes it is not. If the distribution is not recognizable, then an approximate distribution is provided in Section 7.5.1.

If $\sup_{\theta \in \Omega} f_{X|\Theta}(x|\theta) = \sup_{\theta \in \Omega_A} f_{X|\Theta}(x|\theta)$, then LR is the same as the approximate Bayes factor in (4.24) when $X = x$ is observed.[38] If, in addition, Ω_H is a single point, then LR is the same as the global lower bound on the Bayes factor (4.20).

---

[38] For example, if every point in Ω_H is the limit of a sequence of points in Ω_A and $f_{X|\Theta}(x|\theta)$ is continuous in θ, this condition will hold.

4.5. Nuisance Parameters   275

Example 4.135. Suppose that $X_1,\ldots,X_n$ are conditionally IID with $Ber(\theta)$ distribution given $\Theta = \theta$. Let the hypothesis be $H : \Theta = \theta_0$ versus $A : \Theta \neq \theta_0$. Let $Y = \sum_{i=1}^n X_i$. Then

$$LR = \begin{cases} \dfrac{\theta_0^Y(1-\theta_0)^{n-Y}}{(\frac{Y}{n})^Y(1-\frac{Y}{n})^{n-Y}} & \text{if } Y \notin \{0,n\}, \\ \theta_0^n & \text{if } Y = n, \\ (1-\theta_0)^n & \text{if } Y = 0, \end{cases}$$

since $\theta^Y(1-\theta)^{n-Y}$ is largest if $\theta = Y/n$. The LR test would be to reject H if LR is smaller than some specified value. As a function of Y, LR increases until Y/n reaches θ_0 and then decreases. For example, if $\theta_0 = 1/4$ and $n = 10$, we have a case similar to Example 4.112 on page 262. The UMPU level $\alpha = 0.05$ test of $H : \Theta = 1/4$ was found there to be a test with rejected H if $Y \geq 6$ and randomized if $Y \in \{0,5\}$. If $Y = 0$ is observed, then $LR = 0.0563$. If $Y = 6$ is observed, then $LR = 0.0647$. It follows that the level $\alpha = 0.05$ LR test is not the same as the UMPU level 0.05 test. The reason is that no LR test can reject for $Y = 6$ without rejecting also for $Y = 0$, since LR is smaller at $Y = 0$ than at $Y = 6$. The level 0.05 LR test will reject H for $Y \geq 7$ and will randomize if $Y = 0$ with probability 0.8259 of rejecting H. Note that the LR test is of the form of Theorem 4.104, so that it is admissible, but it is not UMPU.

Example 4.136. Suppose that $X_1,\ldots,X_n$ are conditionally IID given $\Theta = (M,\Sigma) = (\mu,\sigma)$ with $N(\mu,\sigma^2)$ distribution. Suppose that $H : M = \mu_0$ is the hypothesis. The formula in (4.26) gives the observed value of LR, namely $(1 + t^2/[n-1])^{-n/2}$, where t is the test statistic for the usual t-test. Since LR is a decreasing function of $|t|$, the level α LR test will be the same as the UMPU level α test for all α.

As mentioned earlier, the main reason for introducing LR tests is that they can be used in situations in which UMPU tests do not exist. In the following example, $\sup_{\theta \in \Omega} f_{X|\Theta}(x|\theta)$ is not equal to $\sup_{\theta \in \Omega_A} f_{X|\Theta}(x|\theta)$ for all possible data values x, so that there will exist data sets for which LR is not equal to the approximate Bayes factor (4.24).

Example 4.137. Suppose that $X_1,\ldots,X_n$ are conditionally IID $N(\mu,\sigma^2)$ given $(M,\Sigma) = (\mu,\sigma)$, and $H : a \leq M \leq b$ versus $A :$ not H. We can easily calculate the LR criterion:

$$LR = \begin{cases} 1 & \text{if } \overline{X} \in [a,b], \\ \left(1 + \dfrac{(\overline{X}-a)^2}{nS_n^2}\right)^{-n/2} & \text{if } \overline{X} < a, \\ \left(1 + \dfrac{(\overline{X}-b)^2}{nS_n^2}\right)^{-n/2} & \text{if } \overline{X} > b. \end{cases}$$

We can easily see that the level α LR test will be the sum of two one-sided tests. Since Problem 54 on page 292 shows that the level of the sum of two one-sided tests in this problem is the sum of the levels, and since LR decreases equally fast as $\overline{X}$ drops below a or rises above b, it follows that the two tests should each have level $\alpha/2$. The LR test becomes the test described at the end of Example 4.132.

In Section 7.5.1, we will prove some large sample properties of LR tests.

4.5.6 The Standard F-Test as a Bayes Rule*

In many of the examples in this chapter, we compared various classical tests to Bayesian procedures. In particular, we found that by using improper priors, we could make many classical tests into formal Bayes rules. For the case of normal distributions, it is actually possible to find a proper prior distribution that leads to a similar result. Consider one of the general linear models, such as analysis of variance or regression. In this section, we will find a proper prior distribution for the parameters such that the standard F-test emerges as a Bayes rule and is seen to be admissible with 0–1–c loss.

General linear models can be transformed into the following. The parameters are $\Theta = (M, \Psi, \Sigma)$, with M an r-dimensional vector and Ψ an s-dimensional vector. Let $k = s + r$. The sufficient statistics are (Y, W, U), which are conditionally independent given the parameters with distributions

$$Y \sim N_r(\mu, \sigma^2 I), \quad W \sim \sigma^2 \chi_d^2, \quad U \sim N_s(\psi, \sigma^2 I). \tag{4.138}$$

Example 4.139. Suppose that $X_{1,i}, i = 1, \ldots, n_1$ are IID $N(\mu_1, \sigma^2)$ independent of $X_{2,i}, i = 1, \ldots, n_2$ with $N(\mu_2, \sigma^2)$ distribution given $M_1 = \mu_1, M_2 = \mu_2, \Sigma = \sigma$. A popular hypothesis is $H : M_1 = M_2$. One can write this as $H : M_1 - M_2 = 0$. In the above notation, $r = 1$, Y is the difference between the two sample averages times $\sqrt{n_1 n_2}/(n_1 + n_2)$, and $M = M_1 - M_2$. Also, $s = 1$, U is the sum of all the observations divided by $\sqrt{n_1 + n_2}$, and $\Psi = (n_1 M_1 + n_2 M_2)/\sqrt{n_1 + n_2}$. Finally, $d = n_1 + n_2 - 2$ and W is the pooled sum of squared deviations.

Example 4.140. Suppose that $Y_1, \ldots, Y_n$ are conditionally independent given $B = \beta, \Sigma = \sigma$ with $Y_i \sim N(x_i^\top \beta, \sigma^2)$ distribution, where the x_i are known k-dimensional vectors. This is the standard linear regression model. A typical hypothesis is of the form $H : CB = c$ versus $A : CB \neq c$, where C is an $r \times k$ matrix of rank $r \leq k$ and c is an r-dimensional vector. Define the matrix $G = \sum_{i=1}^n x_i x_i^\top$, and assume that G is nonsingular. The usual least-squares estimator of B is $\hat{B} = G^{-1} \sum_{i=1}^n x_i Y_i$. Its conditional distribution given the parameters is $N_k(\beta, \sigma^2 G^{-1})$. The conditional distribution of $C\hat{B}$ given the parameters is $N_r(C\beta, \sigma^2 C G^{-1} C^\top)$. Let D be a $(k-r) \times k$ matrix whose $k - r$ rows are all orthogonal to the r rows of CG^{-1}.[39] The sufficient statistics can then be written as

$$Y = (CG^{-1}C^\top)^{-\frac{1}{2}} C\hat{B}, \quad W = \sum_{i=1}^n (Y_i - x_i^\top \hat{B})^2, \quad U = (DG^{-1}D^\top)^{-\frac{1}{2}} D\hat{B}.$$

With $\Psi = (DG^{-1}D^\top)^{-1/2} DB$, $M = (CG^{-1}C^\top)^{-1/2} CB$, $s = k-r$, and $d = n-k$, we see that (Y, W, U) have the distributions given in (4.138).

For the general situation, construct new random variables $B > 0$, $\Gamma \in \mathbb{R}^r$, and $H \in \{0, 1\}$, which are conditionally independent of (Y, W, U) given Θ.

---

*This section may be skipped without interrupting the flow of ideas.
[39] If $k = r$, we don't need the U vector.

4.5. Nuisance Parameters

The distribution of Θ and the new parameters is as follows. Given $\Psi = \psi$, $\Sigma = \sigma$, $B = \beta$, $\Gamma = \gamma$, and $H = h$; $M = h\gamma\beta/(1+\gamma^\top\gamma)$ with probability 1. Given $\Sigma = \sigma$, $B = \beta$, $\Gamma = \gamma$, and $H = h$,

$$\Psi \sim N_s\left(\mathbf{0}, \frac{\gamma^\top\gamma}{1+\gamma^\top\gamma}I\right).$$

Given $B = \beta$, $\Gamma = \gamma$, and $H = h$, $\Sigma = 1/\sqrt{1+\gamma^\top\gamma}$ with probability 1. This makes the conditional distribution of Ψ into $N_s(\mathbf{0}, (1-\sigma^2)I)$, which we shall call P_σ. Given $B = \beta$ and $H = h$, the density of Γ is

$$f_{\Gamma|B,H}(\gamma|\beta, h) = \pi_{h,\beta}(\gamma)$$
$$= \begin{cases} c_0(1+\gamma^\top\gamma)^{-\frac{r+d}{2}} & \text{if } h = 0, \\ c_1(\beta)(1+\gamma^\top\gamma)^{-\frac{r+d}{2}} \exp\left(\frac{1}{2}\frac{\gamma^\top\gamma\beta^2}{1+\gamma^\top\gamma}\right) & \text{if } h = 1. \end{cases}$$

Finally, B and H are independent, with B having some density $f(\beta)$ strictly positive on all of $(0, \infty)$, and $\Pr(H = 0) = p_0$.

The Bayes rule with respect to 0–1–c loss is to reject $H = 0$ if $\Pr(H = 1|Y = y, U = u, W = w)$ is large.

$$\frac{\Pr(H = 1|Y = y, U = u, W = w)}{\Pr(H = 0|Y = y, U = u, W = w)} = \qquad (4.141)$$

$$\frac{\int\int\int\int f_{Y,U,W|\Theta}(y,u,w|\mu,\psi,\sigma)dP_\sigma(\psi)dQ_{1,\gamma,\beta}(\sigma,\mu)\pi_{1,\beta}(\gamma)d\gamma f(\beta)d\beta}{\int\int\int\int f_{Y,U,W|\Theta}(y,u,w|\mu,\psi,\sigma)dP_\sigma(\psi)dQ_{0,\gamma}(\sigma,\mu)\pi_{0,\beta}(\gamma)d\gamma f(\beta)d\beta},$$

where $Q_{1,\gamma,\beta}$ is the distribution for (M, Σ) that puts all of its mass on $\mu = \gamma\beta/(1+\gamma^\top\gamma)$ and $\sigma = 1/\sqrt{1+\gamma^\top\gamma}$; $Q_{0,\gamma}$ is the distribution of (M, Σ) that puts all of its mass on $\mu = \mathbf{0}$ and $\sigma = 1/\sqrt{1+\gamma^\top\gamma}$; and $f_{Y,U,W|\Theta}(y,u,w|\mu,\psi,\sigma)$ is proportional to

$$\sigma^{-s-r-d}w^{\frac{d}{2}-1}\exp\left(-\frac{1}{2\sigma^2}\left[w + \sum_{i=1}^r(y_i - \mu_i)^2 + \sum_{j=1}^s(u_j - \psi_j)^2\right]\right).$$

To find the Bayes rule, we begin with the innermost integration (over ψ) in both numerator and denominator (since they are the same). The integral to be performed is

$$\int \sigma^{-s}(1-\sigma^2)^{-\frac{s}{2}}\exp\left(-\frac{A}{2}\right)d\psi,$$

where

$$A = \frac{\sum_{j=1}^s(u_j - \psi_j)^2}{\sigma^2} + \sum_{j=1}^s\psi_j^2 = \sum_{j=1}^s\left\{u_j^2 + \frac{[\psi_j - (1-\sigma^2)u_j]^2}{\sigma^2(1-\sigma^2)}\right\}.$$

278   Chapter 4. Hypothesis Testing

It follows that $\sigma(1-\sigma^2)^{1/2}$ is a scale factor for each coordinate of ψ. Hence the integral is proportional to $\exp(-u^\top u/2)$. Since this depends on the data alone, it cancels out of the numerator and the denominator along with $w^{d/2-1}$ and the constant in the data density. What remains of (4.141) is

$$\frac{\int\int\int \sigma^{-r-d}\exp\left(\frac{-1}{2\sigma^2}[w+\sum_{i=1}^r(y_i-\mu_i)^2]\right)dQ_{1,\gamma,\beta}(\sigma,\mu)\pi_{1,\beta}(\gamma)d\gamma f(\beta)d\beta}{\int\int\int \sigma^{-r-d}\exp\left(\frac{-1}{2\sigma^2}[w+\sum_{i=1}^r(y_i-\mu_i)^2]\right)dQ_{0,\gamma}(\sigma,\mu)\pi_{0,\beta}(\gamma)d\gamma f(\beta)d\beta}.$$

The next innermost integrals are with respect to point mass probabilities, so they merely evaluate the integrands at the points where they put their mass. The result is

$$\frac{\int\int (1+\gamma^\top\gamma)^{\frac{r+d}{2}}\exp\left\{-\frac{1+\gamma^\top\gamma}{2}\left[w+\left\|y-\frac{\gamma\beta}{1+\gamma^\top\gamma}\right\|^2\right]\right\}\pi_{1,\beta}(\gamma)d\gamma f(\beta)d\beta}{\int\int (1+\gamma^\top\gamma)^{\frac{r+d}{2}}\exp\left\{-\frac{1+\gamma^\top\gamma}{2}(w+y^\top y)\right\}\pi_0(\gamma)d\gamma f(\beta)d\beta}.$$

Next, we integrate over γ. In the denominator, we get

$$\int c_0 \exp\left\{-\frac{1+\gamma^\top\gamma}{2}(y^\top y + w)\right\}d\gamma$$
$$= c_0 \exp\left\{-\frac{1}{2}(y^\top y + w)\right\}(y^\top y + w)^{-\frac{r}{2}}(2\pi)^{\frac{r}{2}}.$$

Call this last expression K. In the numerator, we get

$$\int c_1(\beta)\exp\left\{-\frac{1}{2}\left[(1+\gamma^\top\gamma)(w+y^\top y)-2y^\top\gamma\beta\right]\right\}d\gamma$$
$$= c_1(\beta)\exp\left\{-\frac{y^\top y+w}{2}\right\}\exp\left\{\frac{1}{2}\frac{y^\top y\beta^2}{y^\top y+w}\right\}$$
$$\times \int \exp\left\{-\frac{y^\top y+w}{2}\left\|\gamma-\frac{y\beta}{w+y^\top y}\right\|^2\right\}d\gamma$$
$$= \text{constant} \times c_1(\beta) \times K \times \exp\left\{\frac{1}{2}\frac{y^\top y\beta^2}{y^\top y+w}\right\}.$$

So, the ratio is

$$\frac{\text{constant}\times \int c_1(\beta)f(\beta)\exp\left\{\frac{1}{2}\frac{y^\top y\beta^2}{y^\top y+w}\right\}d\beta}{\int f(\beta)d\beta}.$$

This is a one-to-one increasing function of

$$\frac{y^\top y}{y^\top y + w} = \frac{rF}{rF+d},$$

where F is the classical F statistic. Hence the Bayes rule is to reject H when $F > c$ for some number c, which is the classical F-test. Because of the way the prior distribution was constructed, we can show that the Bayes rule is admissible (see Problem 60 on page 294), so the F-test is admissible. Notice that the prior distribution depends on the sample size, so it could not be used as a real prior distribution unless we knew for sure what the sample size would be before observing the data.

4.6 P-Values

4.6.1 Definitions and Examples

A common criticism of hypothesis-testing methodology is that the decision to "reject" or "accept" a hypothesis is not informative enough. One should also provide a measure of the strength of evidence in favor of (or against) the hypothesis. The posterior probability of the hypothesis is an obvious candidate to provide the strength of evidence in favor of the hypothesis, but the posterior is not available in a classical analysis. In fact, there is no theory for strength of evidence or degree of support in the classical theory. Instead, some alternatives to testing hypotheses are available. The alternative considered here is to give the set of all levels for which a specific hypothesis would be rejected.[40] For most of the tests that we will consider in this book, the set of α values such that the level α test would reject H will be an interval starting at some lower endpoint and extending to 1. The lower endpoint will be called the *P-value* of the observed data relative to the collection of tests.

Definition 4.142. Let H be a hypothesis, and let Γ be a set indexing nonrandomized tests of H (i.e., $\{\phi_\gamma : \gamma \in \Gamma\}$ is a set of nonrandomized tests of H). For each $\gamma \in \Gamma$, let $\varphi(\gamma)$ be the size of the test ϕ_γ. Define

$$p_H(x) = \inf\{\varphi(\gamma) : \phi_\gamma(x) = 1\}.$$

We call $p_H(x)$ the *P-value* of the observed data x relative to the set of tests for the hypothesis H.

Usually, when the data have continuous distribution, we can arrange for $\Gamma = [0, 1]$ and $\varphi(\gamma) = \gamma$. That is, there is one and only one size γ test in the set for each $\gamma \in [0, 1]$. If the data have a discrete distribution, it may be impossible to achieve certain sizes with nonrandomized tests. Often, it is understood implicitly which set of tests is under consideration and what is

---

[40] Another alternative is to provide interval (or set) estimates for parameters (see Section 5.2). For example, a coefficient $1 - \alpha$ confidence set (Definition 5.47) is defined in such a way that it contains all of the values of θ such that the hypothesis $H : \Theta = \theta$ would be accepted at level α (see Proposition 5.48).

the hypothesis. In these cases, $p_H(x)$ is called the *P*-value without reference to the set of tests or the hypothesis.

Example 4.143. Suppose that $X \sim N(\theta, 1)$ given $\Theta = \theta$ and $H_1 : \Theta \in [-0.5, 0.5]$. The UMPU level α test of H_1 is $\phi_\alpha(x) = 1$ if $|x| > c_\alpha$, for some number c_α. If $X = 2.18$ is observed, ϕ_α will reject H_1 if and only if $2.18 > c_\alpha$. Since c_α increases as α decreases, the *P*-value is that α such that $c_\alpha = 2.18$. If $c_\alpha = 2.18$, then the test is $\phi_\alpha(x) = 1$ if $|x| > 2.18$, so $\alpha = \Phi(-2.68) + 1 - \Phi(1.68) = 0.0502$ is the *P*-value.

The reader will note that we used the same notation $p_H(x)$ to denote the *P*-value as we used for the significance probability in Definition 4.8. The reason is that they are almost always the same thing.

Proposition 4.144. *Let $\{\phi_\gamma : \gamma \in \Gamma\}$ be a collection of tests. Suppose that $\Gamma \subseteq [0,1]$ and $\gamma_1 > \gamma_2$ implies that for all x, $\phi_{\gamma_1}(x) \geq \phi_{\gamma_2}(x)$. Define the binary relation $\preceq$ on the sample space by $x \preceq y$ if and only if the P-value for x is at least as large as the P-value for y. Then $\preceq$ is a weak order and the P-value always equals the significance probability.*

The conditions of Proposition 4.144 say that for every possible observation x, if x leads to rejection of H at one level, then it leads to rejection at every higher level. This is just a precise way of saying what we said earlier about the set of levels at which a hypothesis would be rejected being an interval running from some lower bound up to 1. Although the conditions of Proposition 4.144 are met in most situations, the following example [from Lehmann (1986)[41]] is a case in which they are not.

Example 4.145. Suppose that $\Omega = \{1, 2, 3\}$ and $\mathcal{X} = \{1, 2, 3, 4\}$. Consider the following conditional distribution for X given Θ:

$f_{X\|\Theta}(x\|\theta)$		1	2	3	4
	1	$\frac{2}{13}$	$\frac{4}{13}$	$\frac{3}{13}$	$\frac{4}{13}$
θ	2	$\frac{4}{13}$	$\frac{2}{13}$	$\frac{1}{13}$	$\frac{6}{13}$
	3	$\frac{4}{13}$	$\frac{3}{13}$	$\frac{2}{13}$	$\frac{4}{13}$

Consider the hypothesis $H : \Theta \leq 2$ versus $A : \Theta = 3$. One can show that the MP level $5/13$ test of H is $\phi_{5/13}(x) = 1$ if $x \in \{1, 3\}$ and that the MP level $6/13$ test is $\phi_{6/13}(x) = 1$ if $x \in \{1, 2\}$. For $\alpha = 1$, $\phi_\alpha(x) = 1$ for $x \in \{1, 2, 3, 4\}$. So $X = 3$ leads us to reject H at some high values of α and at some low values of α, but not at certain values in between. The infimum of the set of all α such that $\phi_\alpha(3) = 1$ no longer tells us all of the levels at which we would reject H. In particular, one of the conditions of Proposition 4.144 is violated.

Because *P*-values are between 0 and 1 and because the smaller the *P*-value is the smaller α would have to be before one could accept the hypothesis, people like to think of the *P*-value as if it were the probability that the

---

[41] The example appears in Problem 34 on page 121 of that text and in Problem 29 on page 116 of the 1959 edition.

hypothesis is true. Those who are more careful with their terminology will still suggest that it is the degree to which the data support the hypothesis. Sometimes this is approximately true, as in the next two examples.[42]

Example 4.146. Suppose that $X \sim Bin(n, p)$ given $P = p$, and let $H : P \leq p_0$. The UMP level α test rejects H when $X > c_\alpha$, where c_α increases as α decreases. The P-value of an observed value x is the value of α such that $c_\alpha = x - 1$ unless $x = 0$, in which case the P-value is 1. The P-value can then be calculated as

$$p_H(x) = \sum_{i=x}^{n} \binom{n}{i} p_0^i (1-p_0)^{n-i}.$$

This formula is also correct when $x = 0$.

Next, suppose that we used an improper prior for P of the form $Beta(0, 1)$. The posterior distribution of P would be $Beta(x, n+1-x)$. If $x > 0$, the posterior probability that H is true is $\Pr(Y \leq p_0)$, where $Y \sim Beta(x, n+1-x)$. This is the probability that at least x out of n IID $U(0, 1)$ random variables are less than or equal to p_0, because Y has the distribution of the xth order statistic from a sample of n IID $U(0, 1)$ random variables. The probability that a single $U(0, 1)$ is less than or equal to p_0 is p_0, and the n of them are IID so the probability that at least x of them are less than or equal to p_0 is

$$\sum_{i=x}^{n} \binom{n}{i} p_0^i (1-p_0)^{n-i} = p_H(x).$$

So, the P-value is equal to the posterior probability that the hypothesis is true (using an improper prior), at least when the posterior is proper. If $x = 0$, then the posterior is still improper $Beta(0, n + 1)$.[43]

Consider next what happens if $H : P \geq p_0$. It turns out that the improper prior must change to $Beta(1, 0)$. (See Problem 64 on page 294.) Because two different priors are needed to obtain the "degree of support" for the two different hypotheses, we get the following anomaly. If we take the two hypotheses together, $\{P \leq p_0\} \cup \{P \geq p_0\}$, the total degree of support is

$$1 + \binom{n}{x} p_0^x (1-p_0)^{n-x}.$$

One can easily check that this is not due to the fact that $\{P = p_0\}$ is included in both hypotheses. One could leave it out of either one and the results would be the same.

A similar situation occurs with Poisson data.

Example 4.147 (Continuation of Example 4.61; see page 241). The P-value of an observed data value x is $\Pr(X \geq x|\Theta = 1)$. This can also be written as

$$\begin{aligned} p_H(x) &= \Pr(\text{at least } x \text{ events in one time unit of a rate 1 Poisson process}) \\ &= \Pr(\text{time until } x\text{th event is } \leq 1) = \Pr(Y \leq 1) = \Pr(\Theta \leq 1|X = x), \end{aligned}$$

---

[42] Berkson (1942) carefully examines the use of significance probabilities as evidence in favor of hypotheses.

[43] There is a sense in which $\Pr(P \leq p_0 | X = 0) = 1$ even in this case, but it requires the notion of finitely additive probability.

where $Y \sim \Gamma(x, 1)$ and assuming that the "prior" for Θ is the improper $d\theta/\theta$. So, the P-value is the posterior probability that H is true if the prior is improper. This is actually true in general for Poisson distributions and hypotheses of the form $H : \Theta \leq \theta_0$. The implications of this result include the following. If a Bayesian uses the improper prior $d\theta/\theta$ and has a 0–1–c loss function, then he or she will reject H if the P-value is less than $1/(1+c)$. This is the UMP level α test if $\alpha = 1/(1+c)$.

Next, suppose that $H : \Theta \geq 1$ and $A : \Theta < 1$. The UMP level α test is

$$\phi(x) = \begin{cases} 1 & \text{if } x < c, \\ \gamma & \text{if } x = c, \\ 0 & \text{if } x > c, \end{cases}$$

where c and γ are chosen so that ϕ has size α. The P-value of an observed data value x is $\Pr(X \leq x | \Theta = 1)$. As we did earlier, we can write this as

$$\begin{aligned} p_H(x) &= \Pr(\text{at most } x \text{ events in one time unit of a rate 1 Poisson process}) \\ &= \Pr(\text{time until event } x+1 \text{ is } > 1) = \Pr(Y > 1) = \Pr(\Theta > 1 | X = x), \end{aligned}$$

where $Y \sim \Gamma(x+1, 1)$ and assuming that the "prior" for Θ is the improper $d\theta$.

If we modify Example 4.143 slightly, we discover a case in which it is simply impossible to use P-values for measuring degree of support. [See also Schervish (1994b).]

Example 4.148 (Continuation of Example 4.143; see page 280). Let $H_2 : \Theta \in [-0.82, 0.52]$. The UMPU level α test is $\psi_\alpha(x) = 1$ if $|x + 0.15| > d_\alpha$. If $X = 2.18$ is observed, then $d_\alpha = 2.33$ and

$$\alpha = \Phi(-3) + 1 - \Phi(1.66) = 0.0498.$$

This is smaller than the "degree of support" for the smaller hypothesis H_1. It does not make any sense to have a concept of degree of support that gives more support to a smaller hypothesis than it gives to a larger one. In the one-sided testing case, this does not happen. (See Problem 62 on page 294.)

In Example 4.148, we saw that the P-value of a data value relative to the class of UMPU tests behaved strangely as the hypothesis varied. This example is closely related to the incoherent tests discovered in Example 4.88 on page 252 and Example 4.102 on page 257. Problems 61 and 62 on page 294 show that for one-sided hypotheses with known or unknown variance, the P-value always equals the posterior probability of the hypothesis calculated from the usual improper prior. In the case of interval hypotheses with unknown variance, the situation is somewhat different.

Example 4.149. Suppose that $X_1, \ldots, X_n$ are conditionally IID with $N(\mu, \sigma^2)$ distribution given $\Theta = (\mu, \sigma)$. For a hypothesis of the form $H : a \leq M \leq b$ with $a < b$, we have not found UMPU tests. We do, however, have the collection of likelihood ratio (LR) tests. (See Examples 4.132 and 4.137.) The UMPU tests of one-sided hypotheses (like $M \leq b$ or $M \geq a$) and point hypotheses (like $M = c$ versus $M \neq c$) are also LR tests. So, we might try to compare the P-values for various hypotheses relative to the families of LR tests. If $\overline{X} = \overline{x} > b$ and

$S^2 = \sum_{i=1}^{n}(X_i - \overline{X})^2/(n-1) = s^2$ are observed, the P-value for $H : a \leq \mathrm{M} \leq b$ will be the level of the LR test that rejects H when $\sqrt{n}(\overline{X} - b)/S > \sqrt{n}(\overline{x} - b)/s$ or when $\sqrt{n}(\overline{X} - a)/S < -\sqrt{n}(\overline{x} - b)/s$. Call this P-value $p_H(x)$. It is easy to see that $p_H(x)$ is precisely the same as the P-value for the hypothesis $H_b : \mathrm{M} = b$ versus $A_b : \mathrm{M} \neq b$ relative to the collection of UMPU (two-sided) tests. Also $p_H(x)$ is precisely twice the size of the P-value for the hypothesis $H'_b : \mathrm{M} \leq b$ versus $A'_b : \mathrm{M} > b$ relative to the collection of UMPU (one-sided) tests.

Since the one-sided P-values equal posterior probabilities of the hypotheses when using improper priors (see Problem 62 on page 294), we find that the one-sided P-value for the hypothesis $H'_c : \mathrm{M} \leq c$ is a continuous function of c. It follows that there exists $c > b$ such that the one-sided P-value for H'_c satisfies $p_H(x) > p_{H'_c}(x) > p_H(x)/2$. If we are to interpret the P-values relative to the collections of LR tests as degrees of support for the respective hypotheses, then if $x > b$, the degree of support for every hypothesis of the form $H_a : \mathrm{M} \in [a, b]$ (for varying a but fixed b) is the same number $p_H(x)$ (even if $a = b$ or if a is much less than b). But the degree of support for the hypothesis $H'_c : \mathrm{M} \in (-\infty, c]$ (where $c > b$ is chosen as above) is $p_{H'_c}(x) < p(x)$. In words, the data offer more support for every hypothesis of the form $\mathrm{M} \in [a, b]$ than they do for $\mathrm{M} \in (\infty, c]$ even though $[a, b] \subseteq (-\infty, c]$ for every a.

We see that there are cases (usually one-sided testing) in which P-values can correspond to a degree of support for the hypothesis, but there are other cases (e.g., two-sided alternatives) when they cannot. It is possible, for example, with normal data, to express certain P-values as weighted averages over the corresponding hypotheses of P-values for testing point hypotheses. (See Problem 66 on page 294.) To generalize this idea beyond normal distributions (or symmetric location families), one needs to consider tests that may not be UMPU. Spjøtvoll (1983) defines a measure of "acceptability" of point hypotheses, which has the property that for many distributions and certain hypotheses, the weighted average of the acceptability over the hypothesis equals something closely related to the P-value.

In Section 6.3.1, we give some general conditions under which P-values are equal to the posterior probabilities of hypotheses. Casella and Berger (1987) study the problem of testing a one-sided hypothesis–alternative pair and find that in many cases, the P-value is approximately a limit of posterior probabilities. Examples 4.143 and 4.148 point out that the P-value cannot be taken as a method for providing a "degree of support" for general hypotheses, however.

4.6.2 P-Values and Bayes Factors

In Section 4.2.2, we introduced Bayes factors as ways to quantify the degree of support for a hypothesis in a data set. In particular, there are lower bounds on Bayes factors which indicate the smallest amount of support one could coherently say that the data supply to the hypothesis. When the lower bound is not particularly small, one would be hard pressed to argue that the data are highly inconsistent with the hypothesis. Since P-values have also been suggested as measures of support for the hypothesis the data offer, it seems natural to compare the two. In one-sided cases, we found that posterior probabilities (when using improper priors) often corresponded to P-values. In this section we will only compare Bayes factors to P-values for testing hypotheses of the form $H : \Theta = \theta_0$ versus $A : \Theta \neq \theta_0$. Edwards, Lindman, and Savage (1963) and later Berger and Sellke (1987) made comparisons of P-values with lower bounds on Bayes factors, and the following two examples are inspired by the presentations in those sources.

Example 4.150. Suppose that $X_1, \ldots, X_n$ are conditionally IID with $N(\theta, 1)$ distribution given $\Theta = \theta$, and we are interested in testing $H : \Theta = \theta_0$. Let $p_0 = \Pr(\Theta = \theta_0) > 0$. If we let the prior distribution λ of Θ given that $\Theta \neq \theta_0$ be unrestricted, then the global lower bound on the Bayes factor is easily calculated to be $\exp(-n[\overline{x} - \theta_0]^2/2)$. The lower bound on the Bayes factor for λ being a normal distribution centered at θ_0 is 1 if $|\overline{x} - \theta_0| \leq 1/\sqrt{n}$, and it equals

$$\sqrt{n}|\overline{x} - \theta_0| \exp\left(-\frac{n}{2}(\overline{x} - \theta_0)^2 + \frac{1}{2}\right)$$

if $|\overline{x} - \theta_0| > 1/\sqrt{n}$. The UMPU level α test of H is to reject H if $\sqrt{n}|\overline{x} - \theta_0| > \Phi^{-1}(1 - \alpha/2)$, where Φ is the standard normal CDF, so the P-value for an observation $\overline{x}$ is $1 - 2\Phi(\sqrt{n}|\overline{x} - \theta_0|)$. All of these, the P-value and the two lower bounds, are monotone decreasing functions of $\sqrt{n}|\overline{x} - \theta_0|$. Table 4.152 compares the two lower bounds with the P-value and lists what the prior probability of the hypothesis would have to be in order for the posterior probability to be as low as the P-value. Notice how small the prior probability would have to be in order for there to exist even a prior distribution on the alternative which would allow the posterior probability to equal the P-value. For the normal distribution priors and small P-values, the required p_0 is quite small.

The discrepancy between the P-value and posterior probability of the hypothesis, as described in Example 4.150, is sometimes called "Lindley's paradox" [see Lindley (1957) and Jeffreys (1961)]. The contrast between P-values and posterior probabilities is even more striking when one considers more reasonable prior distributions on the alternative rather than the lower bounds.

Example 4.151. Suppose that $X_1, \ldots, X_n$ are conditionally IID with $N(\mu, \sigma^2)$ distribution given $(M, \Sigma) = (\mu, \sigma)$. Suppose that we use a prior distribution that

TABLE 4.152. Comparison of P-Values and Lower Bounds in Example 4.150

P-value	Global Bound	Prior[a]	Normal Bound	Prior[a]
0.1	0.2585	0.3006	0.7011	0.1368
0.05	0.1465	0.2643	0.4734	0.1001
0.01	0.0362	0.2179	0.1539	0.0616
0.001	0.0045	0.1835	0.0242	0.0398
0.0001	0.0005	0.1622	0.0033	0.0293

[a]This is the largest possible value of p_0 which is consistent with the posterior probability being equal to the P-value.

is conjugate as in Example 4.22 on page 224. The Bayes factor for the hypothesis $H : M = \mu_0$ was given in (4.23). Consider what happens to this expression as $n \to \infty$. First, suppose that the usual t statistic converges to a constant t_0. That is, assume that $\sqrt{n}(\overline{x}_n - \mu_0)/s_n$ converges to t_0, where $s_n^2 = w/(n-1)$. In this case, the formula in (4.23) behaves asymptotically (as $n \to \infty$) like $\sqrt{n}$ times a constant. This means that the Bayes factor goes to ∞ as n increases, hence the posterior probability of the hypothesis goes to 1. What happens to the P-value for this same sequence of data sets? Since the t statistic is converging to a constant t_0, the P-value is converging to $1 - 2\Phi(t_0)$. For example, if $t_0 = 1.96$, the P-value will go to 0.05, while the posterior probability of the hypothesis goes to 1. This is an extreme example of Lindley's paradox. Once again, the suggestion of Lehmann (1958) to let α decrease as n increases would seem appropriate.

The situation is much the same if one uses the approximate Bayes factor as calculated in (4.29). This formula does not require that the prior be natural conjugate, but merely smooth in some sense. Since both $\hat{\Psi}$ and $\hat{\Theta}$ converge to finite values almost surely given Θ (by the strong law of large numbers 1.62), the expression in (4.29) also behaves asymptotically like $\sqrt{n}$ times a constant if the t statistic converges to t_0.

Of course, the t statistic will converge to a finite value with positive probability if only if the hypothesis is true. So, it is comforting that virtually any smooth prior distribution will lead to eventually discovering that the hypothesis is true, if it is indeed true.[44] On the other hand, it is a bit disconcerting that the P-value will stay bounded away from 1 with positive probability no matter how much data we observe. (See Problem 65 on page 294 to see how to prove that the P-value has $U(0,1)$ distribution given that the hypothesis is true.) If the hypothesis is false, it is easy to check that the P-value will go to 0, as will the Bayes factor.

The irreconcilability of P-values and posterior probabilities as illustrated in Examples 4.150 and 4.151 is quite typical of cases in which Ω_H is a lower-dimensional set than Ω. [See Schervish (1994b) for some examples with distributions other than normal.] Together with the strange behavior of P-values in Examples 4.148 and 4.149, it becomes difficult to justify their use to measure strength of evidence in favor of the hypothesis for two-sided problems.

4.7 Problems

Section 4.1:

1. Prove that the loss function in (4.2) is equivalent to a 0–1–c loss function. (By "equivalent" we mean that both the posterior risk and the risk function will rank all decision rules the same way regardless of which loss is used.)

2. Prove that the general form of the hypothesis-testing loss function (in Definition 4.1) can be written as $d(\theta)$ times a 0–1 loss function for some function $d > 0$ if the loss is 0 whenever a correct decision is made.

---

[44] In Section 7.4, we will prove some results that make more precise this limiting ability of posterior distributions to identify the value of a parameter.

3. Let $X = (X_1, \ldots, X_n)$ be such that the X_i are conditionally IID with $N(0, \sigma^2)$ distribution given $\Sigma = \sigma$ under the hypothesis H. Let $T(x) = \sqrt{n}\bar{x}/s$, where $\bar{x} = \sum_{i=1}^n x_i/n$ and $s^2 = \sum_{i=1}^n (x_i - \bar{x})^2/(n-1)$. Define $x \preceq y$ if $T(x) \leq T(y)$. Find $p_H(x)$, showing that it is the same for all σ.

Section 4.2:

4. Suppose that $X \sim N(\theta, 1)$ given $\Theta = \theta$. Suppose that $L(\theta, 1) < L(\theta, 0)$ for all $\theta < \theta_0$ and that $L(\theta, 1) > L(\theta, 0)$ for all $\theta > \theta_0$. Prove that, for every prior there exists k such that the formal Bayes rule will be to choose action $a = 1$ if $X < k$.

5. Suppose that $X_1, \ldots, X_n$ are conditionally IID with $N(\mu, \sigma^2)$ distribution given $\Theta = (\mu, \sigma)$. Use the improper prior having Radon–Nikodym derivative $1/\sigma$ with respect to Lebesgue measure on $(0, \infty) \times \mathbb{R}$. Let μ_0 and d be known values, and suppose that the loss function is

$$L(\theta, a) = \begin{cases} c & \text{if } a = 1 \text{ and } |\mu - \mu_0| \leq d\sigma, \\ 1 & \text{if } a = 0 \text{ and } |\mu - \mu_0| > d\sigma, \\ 0 & \text{otherwise.} \end{cases}$$

Prove that the formal Bayes rule will be of the following form: Choose $a = 1$ if $|T| > k$ for some constant k, where $T = \sqrt{n}(\bar{X}_n - \mu_0)/S_n$ and

$$\bar{X}_n = \frac{1}{n} \sum_{i=1}^n X_i, \quad S_n^2 = \frac{1}{n-1} \sum_{i=1}^n (X_i - \bar{X}_n)^2.$$

6. Suppose that $X \sim N(\theta, 1)$ given $\Theta = \theta$ and Θ has $\Pr(\Theta = \theta_0) = p_0$ and given $\Theta \neq \theta_0$, $\Theta \sim N(\theta_0, \tau^2)$. Prove that the posterior density of Θ with respect to the measure $\nu(A) = I_A(\theta_0) + \lambda(A)$, where λ is Lebesgue measure, is given by

$$f_{\Theta|X}(\theta|x) = \begin{cases} p_1 & \text{if } \theta = \theta_0, \\ \frac{(1-p_1)\tau}{\sqrt{2\pi(1+\tau^2)}} \exp\left[-\frac{\tau^2}{2(1+\tau^2)}(\theta - \theta_1)^2\right] & \text{if } \theta \neq \theta_0, \end{cases}$$

where

$$\theta_1 = \frac{x\tau^2 + \theta_0}{1 + \tau^2},$$

$$\frac{p_1}{1-p_1} = \frac{p_0}{(1-p_0)}\sqrt{1+\tau^2} \exp\left\{-\frac{1}{2}\left[\frac{\tau^2}{1+\tau^2}\right](x - \theta_0)^2\right\}.$$

7. Suppose that $X \sim N(\theta, 1)$ given $\Theta = \theta$. Let $H : \Theta = \theta_0$ and $A : \Theta \neq \theta_0$. Let the conditional prior given $\Theta \neq \theta_0$ be $N(\theta_0, \tau^2)$.

   (a) Prove that the Bayes factor is minimized if

   $$\tau^2 = \begin{cases} (x - \theta_0)^2 - 1 & \text{if } |x - \theta_0| > 1, \\ 0 & \text{otherwise.} \end{cases}$$

(b) Show that the minimum Bayes factor is $|x-\theta_0|\exp(\{-[x-\theta_0]^2+1\}/2)$ if $|x-\theta_0| > 1$, and is 1 if $|x - \theta_0| \leq 1$.

8. In Example 4.21 on page 224, prove that

$$k = \lim_{\tau \to \infty} \frac{p_0}{\Phi\left(\frac{\sqrt{\pi/2}}{\tau}\right) - \Phi\left(\frac{-\sqrt{\pi/2}}{\tau}\right)}.$$

Section 4.3.1:

9. Let (α_0, α_1) be a point on the lower boundary of the risk set for a simple–simple hypothesis-testing problem. Prove that $\alpha_0 + \alpha_1 \leq 1$.

10. In a simple–simple hypothesis-testing problem, prove that the minimax rule for a 0–1 loss function is any test that corresponds to the point where ∂_L intersects the line $y = x$.

11. Let $\Omega = \{0, 1\}$. Suppose that P_0 says that $X \sim U(-\sqrt{3}, \sqrt{3})$ and P_1 says that $X \sim N(0, 1)$. Let $\Omega_H = \{0\}$. Draw the risk set for a 0–1 loss function, and find the minimax rule.

12. In a simple–simple hypothesis-testing problem with 0–1 loss, show that a MP level α test has size α unless all tests with size α are inadmissible.

13. Return to the situation in Problem 28 on page 212. Consider the hypothesis $H : X \sim f_0$ versus $A : X \sim f_1$. Find all α such that the MP level α test is of the form "Reject H if $-d \leq X \leq d$," and write d as a function of α.

14.*Prove Proposition 4.46 on page 235.

15. In Example 4.49 on page 236, prove that the Bayes rule with size α has higher power than the unconditional size α test.

Section 4.3.2:

16. Suppose that the loss function is 0–1–c, that $\Omega_H = \{\theta_0\}$, and that $\Omega = \{\theta_0\} \cup \Omega_A$. Prove that a UMP level α test has no larger Bayes risk than any other size α test, no matter what prior we use.

Section 4.3.3:

17. Let $Y = |X|$ where $f_{X|\Theta}(x|\theta) = 1/(\pi\theta[1 + (x/\theta)^2])$. Suppose that $\Theta > 0$ for sure. Prove that the family of distributions for Y has MLR.

18. Let X have Cauchy distribution $Cau(\theta, 1)$ given $\Theta = \theta$.

   (a) Prove that the MP level α test of $H : \Theta = \theta_0$ versus $A : \Theta = \theta_1$ for $\theta_1 > \theta_0$ is essentially unique. That is, if ϕ and ψ are both MP level α tests, then $P_\theta(\phi(X) = \psi(X)) = 1$ for all θ.

   (b) Prove that there is no UMP level α test of $H : \Theta = \theta_0$ versus $A : \Theta > \theta_0$ for $0 < \alpha < 1$.

19. Let the parameter space be the open interval $\Omega = (0, 100)$. Let X_1 and X_2 be conditionally independent given $\Theta = \theta$ with $X_1 \sim Poi(\theta)$ and $X_2 \sim Poi(100 - \theta)$. We are interested in the hypothesis $H : \Theta \leq c$ versus $A : \Theta > c$.

(a) Show that there is no UMP level α test of H versus A.

(b) Show that $T = X_1 + X_2$ is ancillary.

(c) Find the conditional UMP level α test given T.

(d) Find a prior distribution for Θ such that the conditional UMP level α test given T is to reject H if $\Pr(H$ is true$|X_1 = x_1, X_2 = x_2) < \alpha$.

20. Let $\{P_\theta : \theta \in \Omega\}$ be a parametric family, and let $p(x, \theta) = (dP_\theta/dx)(x)$ be the density of a member of the family with respect to Lebesgue measure. Assume that $\partial^2 \log p(x, \theta)/\partial x \partial \theta$ exists for all x and θ. Prove that the family has increasing MLR if and only if $\partial^2 \log p(x, \theta)/\partial x \partial \theta \geq 0$ for all x and θ.

21. Prove Proposition 4.55 on page 240.

22. Let $\Omega = \{\theta_1, \theta_2, \theta_3\}$ with $\theta_1 < \theta_2 < \theta_3$. Suppose that given $\Theta = \theta$, $X \sim N(\theta, 1)$. Let $H : \Theta \in \{\theta_1, \theta_2\}$ and $A : \Theta = \theta_3$. Show that each test ϕ satisfying $\beta_\phi(\theta_1) = \beta_\phi(\theta_2) = \alpha$ is inadmissible if $0 < \alpha < 1$ and the loss function is 0–1.

23. Suppose that the parametric family has MLR increasing and that $H : \Theta \leq \theta_0$ is the hypothesis of interest. Let the alternative be $A : \Theta > \theta_0$. Suppose that the power function of every test is continuous. Prove that the UMP level α test is the UMC floor α test.

24. Suppose that the parametric family has MLR increasing and that $H : \Theta \leq \theta_0$ is the hypothesis of interest. Let the alternative be $A : \Theta > \theta_0$. Suppose that the UMP level α test has base γ. Prove that the UMP level α test is the UMC floor γ test.

25. Show that the family of $U(0, \theta)$ distributions for $\theta > 0$ does not satisfy the conditions of Proposition 4.67. Find two UMP level α tests of $H : \Theta = 1$ versus $A : \Theta > 1$ which are not almost surely equal.

26. Show that the family of $Poi(\theta)$ distributions for $\theta > 0$ does not satisfy the conditions of Proposition 4.67. Nevertheless, prove that the collection of one-sided tests of the form (4.57) for $H : \Theta \leq \theta_0$ versus $A : \Theta > \theta_0$ forms a minimal complete class when the loss function is of hypothesis-testing type (see Definition 4.1.)

27. *Suppose that $\Omega = (0, \infty)$ and that $X_1, \ldots, X_n$ are IID $U(0, \theta)$ given $\Theta = \theta$. For $0 \leq \alpha < 1$, find a UMP level α test for each of the hypothesis–alternative pairs below:

    (a) $H : \Theta \leq \theta_0$ versus $A : \Theta > \theta_0$.

    (b) $H : \Theta \geq \theta_0$ versus $A : \Theta < \theta_0$.

    (c) $H : \Theta = \theta_0$ versus $A : \Theta \neq \theta_0$.

    (d) In part (a), find a second UMP level α test that differs from the one you found for part (a) with positive probability given $\Theta = \theta$ for θ in a set of positive Lebesgue measure.

28. (a) Suppose that $X \sim U(0, \theta)$ given $\Theta = \theta$. Find the conditional distribution of $Y = X^{-a}$ given $\Theta = \theta$.

    (b) Let $a > 0$ be fixed, and suppose that $X \sim Par(a, \theta)$ given $\Theta = \theta$. Find UMP level α tests for each of the three hypothesis–alternative pairs in Problem 27 above.

4.7. Problems

29. *The density of the $NCB(\alpha, \beta, \psi)$ distribution is given in Appendix D.

    (a) Prove that, for fixed α and β, this family of distributions has increasing MLR in x where ψ is the parameter. (Feel free to pass derivatives under summations.)

    (b) Use the result of the previous part to show that the noncentral F distribution has increasing MLR also.

    (c) Also show that the noncentral t distribution has increasing MLR.

30. Prove that if ϕ is UMP level α for testing $H : \Theta \in \Omega_H$ versus $A : \Theta \in \Omega_A$, then $1 - \phi$ is UMC floor $1 - \alpha$ for testing $H' : \Theta \in \Omega_A$ versus $A' : \Theta \in \Omega_H$.

31. *Let $X_1, \ldots, X_n$ be IID $U(\theta - 1/2, \theta + 1/2)$ given $\Theta = \theta$. Let $H : \Theta \geq \theta_0$ and $A : \Theta < \theta_0$.

    (a) Find the UMP level α test of H versus A.

    (b) Find the UMC floor α test of H versus A.

    (c) Suppose that we begin with a uniform improper prior for Θ. Let the loss be 0–1–c with $1/(1 + c) = \alpha$. Find the formal Bayes rule.

    (d) Calculate the power functions of the three tests above. Compare them and explain the differences intuitively.

    (e) Find the UMP level α test of H versus A conditional on the ancillary $U = \max X_i - \min X_i$.

    (f) Find the UMC floor α test of H versus A conditional on the ancillary $U = \max X_i - \min X_i$, and show that this is the same as the UMP level α test conditional on the ancillary.

32. Let $\Omega = (-\infty, \theta_0]$, and let $X_1, \ldots, X_n$ be IID $U(\theta - 1/2, \theta + 1/2)$ given $\Theta = \theta$. Let $H : \Theta = \theta_0$ and $A : \Theta < \theta_0$. Suppose that we have a prior distribution such that $\Pr(\Theta = \theta_0) = p_0 > 0$ and that the conditional distribution of Θ given $\Theta < \theta_0$ has strictly positive density $g(\theta)$ for $\theta < \theta_0$. Assume that the loss is 0–1–c. Find the formal Bayes rule for data values such that $\max\{x_1, \ldots, x_n\} < \theta_0 + 1/2$. (This condition assures that the data are consistent with the parameter space.) Does this test match any of the tests in Problem 31 above?

33. Suppose that μ_Θ is a probability measure on (Ω, τ). Suppose that the conditions of Theorem 4.68 are satisfied and that there is a test with finite Bayes risk with respect to μ_Θ. Suppose that the loss function is bounded below. Show that there is a one-sided test that is a formal Bayes rule.

34. Let $\Omega \subseteq \mathbb{R}$ and $H : \Theta \leq \theta_0$. If power functions are continuously differentiable and ϕ is the unique level α test with maximum derivative for the power function at θ_0, show that ϕ is LMP level α relative to $d(\theta) = \theta - \theta_0$ for $\theta > \theta_0$.

35. Let $X_1, \ldots, X_n$ be conditionally IID with $Cau(\theta, 1)$ distribution given $\Theta = \theta$.

    (a) Find the LMP level α test of $H : \Theta \leq \theta_0$ versus $A : \Theta > \theta_0$ for $0 < \alpha < 1$. (Hint: Pass derivatives under integral signs.)

(b) Prove that the power function of this test goes to 0 as $\theta \to \infty$.

36. Let $\Omega = (0, \infty)$, and let

$$f_{X|\Theta}(x|\theta) = \begin{cases} (2\theta)^{-1} & \text{if } x \leq \theta, \\ \theta(2x^2)^{-1} & \text{if } x > \theta. \end{cases}$$

(a) Show that this family of distributions has MLR.
(b) Show that the conditions of Proposition 4.67 are not met.
(c) Show that the UMP level α test of $H : \Theta = \theta_0$ versus $A : \Theta > \theta_0$ is unique if $\alpha < 1/2$.
(d) Show that the UMP level α test of $H : \Theta = \theta_0$ versus $A : \Theta < \theta_0$ is unique if $\alpha < 1/2$.
(e) If $\alpha > 1/2$, find two UMP level α tests of $H : \Theta = \theta_0$ versus $A : \Theta > \theta_0$ that are not almost surely equal.

Section 4.3.4:

37. Suppose that the conditions of Theorem 4.82 are met. Let the loss function be of the hypothesis-testing type for a two-sided hypothesis. Show that the class of tests of the form given in Theorem 4.82 is essentially complete.

38. Suppose that μ_Θ is a probability measure on (Ω, τ). Let the loss function be of the hypothesis-testing type for a two-sided hypothesis. Suppose that the conditions of Theorem 4.82 are satisfied and that there is a test with finite Bayes risk with respect to μ_Θ. Suppose that the loss function is bounded below. Show that there is a test of the form given in Theorem 4.82 which is a formal Bayes rule.

39. Suppose that $X \sim N(\theta, 1)$ given $\Theta = \theta$. Let Ω_H be the set of rational numbers, and let Ω_A be the set of irrational numbers. Prove that the UMP level α test of $H : \Theta \in \Omega_H$ versus $A : \Theta \in \Omega_A$ is the trivial test $\phi(x) \equiv \alpha$.

40.*Let X be a random variable, and define

$$\phi_w(x) = \begin{cases} 1 & \text{if } x < c_w, \\ \gamma_w & \text{if } x = c_w, \\ 0 & \text{if } x > c_w. \end{cases}$$

Let P_1 and P_2 be two different possible distributions of X with corresponding expectations E_1 and E_2 such that $P_1 \ll P_2$ and $P_2 \ll P_1$. Suppose that $E_1 \phi_w(X) = w$ for all w. Prove that $g(w) = E_2 \phi_w(X)$ is continuous. (*Hint:* To prove that g is continuous at w, consider three cases. First, suppose that $1 > \gamma_w > 0$, and prove that so long as z is close enough to w so that $1 > \gamma_z > 0$ and $c_z = c_w$, then $g(z)$ is close to $g(w)$. Second, suppose that $\gamma_w = 0$. When w increases, either γ_w or c_w must increase (or both). Either way, the increase can be made small enough so that $g(w)$ does not change much. This is similar if w decreases. The third case, $\gamma_w = 1$, is similar to the second case.)

41. Let $X \sim Exp(\theta)$ given $\Theta = \theta$. Consider the two hypotheses $H_1 : \Theta \leq 1$ versus $A_1 : \Theta > 1$ and $H_2 : \Theta \notin (1, 2)$ versus $A_2 : \Theta \in (1, 2)$.

(a) Find the UMP level 0.05 tests of the two hypotheses.

(b) Find the set of all x values such that the UMP level 0.05 test of H_2 rejects H_2 but the UMP level 0.05 test of H_1 accepts H_1.

42. Let $\Omega_1 \subset \Omega_2$ be strictly nested subsets of the parameter space. Let $H_i : \Theta \in \Omega_i$ and $A_i : \Theta \notin \Omega_i$ for $i = 1, 2$. Suppose that L_i is 0–1–c loss for testing H_i versus A_i for $i = 1, 2$ (with the same c for both cases). Consider the problem of simultaneously testing both hypotheses with action space $\{0, 1\}^2$, the first coordinate being the action for the first hypothesis, and the second coordinate being the action for the second hypothesis. (That is, for example, $a = (a_1, a_2) = (0, 1)$ means to reject H_2 but accept H_1.) Suppose that the loss function for the simultaneous tests is $L(\theta, a) = L_1(\theta, a_1) + L_2(\theta, a_2)$. A pair of tests (ϕ_1, ϕ_2) can be thought of as a randomized decision rule in this problem. In this case, $\phi_i(x) = \Pr(\text{reject } H_i | X = x)$. We can say that a pair of tests (ϕ_1, ϕ_2) is *incoherent* if there exists $\theta \in \Omega_2 \setminus \Omega_1$ such that
$$P_\theta(\{x : \phi_1(x) < \phi_2(x)\}) > 0.$$

(a) Prove that an incoherent pair of tests is inadmissible. (*Hint:* Switch the two tests for all x such that $\phi_1(x) < \phi_2(x)$.)

(b) Consider the special case in that $X \sim N(\theta, 1)$ given $\Theta = \theta$, $\Omega_1 = \{\theta_0\}$, and $\Omega_2 = (-\infty, \theta_0]$. Let ϕ_1 and ϕ_2 be the UMPU level α tests of their respective hypotheses. Find a pair of tests that dominates (ϕ_1, ϕ_2) in the decision problem defined above.

Section 4.4:

43. For each $k = 1, 2, \ldots$, let $\gamma_{k,\alpha}$ denote the $1 - \alpha$ quantile of the χ_k^2 distribution. Let Y_k have $NC\chi_k^2(c^2)$ distribution. Prove that $\Pr(Y_k > \gamma_\alpha) \leq \Pr(Y_1 > \gamma_\alpha)$ for all $k \geq 1$. (*Hint:* Let $X_1, \ldots, X_k$ be IID $N(\theta, 1)$ given $\Theta = \theta$, and consider two tests of $H : \Theta = 0$ based on $\sum_{i=1}^k X_i^2$ and $(\sum_{i=1}^k X_i)^2$.)

44. Prove Proposition 4.92 on page 254.

45. Prove Proposition 4.93 on page 254.

46. Let $X_1, \ldots, X_n$ be IID $U(\theta - 1/2, \theta + 1/2)$ given $\Theta = \theta$. Let $H : \Theta = \theta_0$ and $A : \Theta \neq \theta_0$.

(a) Let ϕ be a test with size α. Prove that ϕ is unbiased level α if $\phi(x) = 1$ for all x which satisfy $f_{X|\Theta}(x|\theta_0) = 0$. (*Hint:* Use the fact that if $f_{X|\Theta}(x|\theta_0)f_{X|\Theta}(x|\theta_1) > 0$ for all $x \in B$, then $P_{\theta_0}(B) = P_{\theta_1}(B)$.)

(b) Prove that there does not exist a UMPU level α test for $\alpha > 0$. (*Hint:* You can slightly modify the UMP and UMC tests for the one-sided case in Problem 31 on page 289 to produce unbiased level α tests with maximum power for $\theta < \theta_0$ and $\theta > \theta_0$, respectively. Then prove that it is impossible for a single test to achieve both maxima.)

47. Prove Proposition 4.97 on page 255.

48. Suppose that the conditions of Lemma 4.99 hold and that the loss function is of the hypothesis-testing type. Prove that the class of two-sided tests is essentially complete.

49. In a one-parameter exponential family with natural parameter Θ, prove that the condition
$$\frac{d}{d\theta}\beta_\phi(\theta)\bigg|_{\theta=\theta_0} = 0$$
can be written as (4.117) if $\beta_\phi(\theta_0) = \alpha$.

50. Let $\Omega \subseteq \mathbb{R}$ and $H : \Theta = \theta_0$ versus $A : \Theta \neq \theta_0$. If power functions are twice continuously differentiable and ϕ is the unique unbiased level α test with maximum second derivative for the power function at θ_0, show that ϕ is LMPU level α relative to $d(\theta) = |\theta - \theta_0|$ for $\theta \neq \theta_0$.

Section 4.5:

51.*Let the joint density of (X_1, X_2) given $(\Theta_1, \Theta_2) = (\theta_1, \theta_2)$ be
$$f_{X_1,X_2|\Theta_1,\Theta_2}(x_1, x_2|\theta_1, \theta_2) = \frac{\phi(x_2 - \theta_2)\phi(x_1 - \theta_1)}{\Phi(x_1 - \theta_2)} I_{(-\infty,x_1)}(x_2),$$
where ϕ and Φ are respectively the standard normal density and CDF. Find the UMPU level α test of $H : \Theta_2 \leq c$ versus $A : \Theta_2 > c$.

52. Let the parameter space be $\Omega = (0, 1) \times \mathbb{R}$. Conditional on $(P, \Theta) = (p, \theta)$, $N \sim \mathrm{Bin}(10, p)$, and conditional on $N = n$, $X_1, \ldots, X_{n+1}$ are IID $N(\theta, 1)$. You get to observe $N, X_1, \ldots, X_{N+1}$. We are interested in the hypothesis $H : \Theta \leq 0$ versus $A : \Theta > 0$.

   (a) Find the UMPU level α test of H versus A.

   (b) Suppose that $P = p$ is actually known. Show that there is no UMPU level α test of H versus A.

53. In the framework of Example 4.121 on page 266, let $H : \Gamma_1 \leq \mu_0$ and $A : \Gamma_1 > \mu_0$. Prove that the usual size α one-sided t-test is UMPU level α.

54.*Suppose that $X_1, \ldots, X_n$ are conditionally IID $N(\mu, \sigma^2)$ given $\Theta = (\mu, \sigma)$. Define
$$\overline{x}_n = \frac{1}{n}\sum_{i=1}^n x_i, \qquad s_n = \sqrt{\frac{1}{n-1}\sum_{i=1}^n (x_i - \overline{x}_n)^2},$$
$$t_1 = \sqrt{n}\frac{\overline{x}_n - a}{s_n}, \qquad t_2 = \sqrt{n}\frac{\overline{x}_n - b}{s_n}.$$
Let $\alpha_1, \alpha_2 \geq 0$ and $\alpha = \alpha_1 + \alpha_2 \leq 1$. Define
$$\phi(x) = \begin{cases} 1 & \text{if } t_1 < T_{n-1}^{-1}(\alpha_1) \text{ or if } t_2 > T_{n-1}^{-1}(1 - \alpha_2), \\ 0 & \text{otherwise.} \end{cases}$$

   (a) Prove that ϕ has size α as a test of $H : a \leq M \leq b$ versus $A :$ not H. (*Hint:* For each $\mu \in [a, b]$ find the distributions of t_1 and t_2 given $\Theta = (\mu, \sigma)$ and see what happens as $\sigma \to \infty$.)

(b) Prove that ϕ is not an unbiased level α test if both α_1 and α_2 are strictly positive. (*Hint:* Show that the limit of the power function at (a, σ) is α_1 as $\sigma \to 0$. Since the power function is continuous, what does this say about the power function at $(a+\epsilon, \sigma)$ for σ and ϵ small?)

55. Suppose that $X_1 \sim Bin(n_1, p_1)$ independent of $X_2 \sim Bin(n_2, p_2)$ conditional on $(P_1, P_2) = (p_1, p_2)$. Find the UMPU level α test of $H : P_1 = P_2$ versus $A : P_1 \neq P_2$.

56.*Let $X_1, \ldots, X_n$ be IID given $\Theta = (\theta_1, \theta_2)$ with density
$$f_{X|\Theta}(x|\theta_1, \theta_2) = \exp\{-\psi(\theta_1) + \theta_1 \cos(x - \theta_2)\} I_{[0, 2\pi]}(x),$$
where
$$\psi(y) = \log \int_0^{2\pi} \exp(y \cos(t)) dt.$$
Let $\Omega = \{(\theta_1, \theta_2) : \theta_1 \geq 0, 0 \leq \theta_2 < 2\pi\}$.

(a) Let $H_1 = \Theta_1 \cos \Theta_2$ and $H_2 = \Theta_1 \sin \Theta_2$. Let $H : \Theta_2 = 0$ versus $A : \Theta_2 \neq 0$ and let $H^* : H_2 = 0$ versus $A^* : H_2 \neq 0$. Prove that the UMPU level α test of H^* is also UMPU level α for H. (You need not actually find the test to prove this.)(*Hint:* Use the fact that if f and g are analytic functions of one variable and $f(x) = g(x)$ for all x on some smooth curve, then $f = g$.)

(b) Find the form of the UMPU level α test of H as closely as you can. (You will not be able to find the cutoffs in closed form.)

(c) Why is this a reasonable test of H^* but not of H?

57.*Let X and Y have joint density given $(M, \Lambda) = (\mu, \lambda)$,
$$f_{X,Y|M,\Lambda}(x, y|\mu, \lambda) = \mu\lambda \exp(-\mu x - \lambda y) I_{[0,\infty)}(x) I_{[0,\infty)}(y).$$
Find UMPU level $\alpha = 0.2$ tests for the following hypotheses:

(a) $H : \Lambda \leq M + 1$ versus $A : \Lambda > M + 1$.
(b) $H : \Lambda = M$ versus $A : \Lambda \neq M$.

58.*Consider a breeding experiment in which each observation is a classification into one of three groups. Suppose that the observations are conditionally independent given $(P_1, P_2, P_3) = (p_1, p_2, p_3)$ with conditional probability p_i of being classified as group i.

(a) Find the form of the UMPU level α test of $H : P_2 = 3P_1$ versus $A : P_2 \neq 3P_1$ based on n observations, and say how you would determine the exact rejection region.

(b) For the case $n = 2$, find the precise form of the UMPU level 0.1 test.

(c) Suppose that (P_1, P_2, P_3) has a prior distribution of the Dirichlet form with density
$$f_{P_1, P_2}(p_1, p_2) = \frac{\Gamma(\alpha_1 + \alpha_2 + \alpha_3)}{\Gamma(\alpha_1)\Gamma(\alpha_2)\Gamma(\alpha_3)} p_1^{\alpha_1 - 1} p_2^{\alpha_2 - 1} (1 - p_1 - p_2)^{\alpha_3 - 1},$$
for all $p_i \geq 0$, $p_1 + p_2 + p_3 = 1$, where all $\alpha_i > 1$. Find the posterior mean of P_2/P_1.

294   Chapter 4. Hypothesis Testing

59. We will observe $Y_1, \ldots, Y_k$ where the conditional distribution of Y_i given $(B_0, B_1) = (\beta_0, \beta_1)$ is $Bin(n_i, p_i)$, where $p_i = [1 + \exp(\beta_0 + \beta_1 x_i)]^{-1}$, the Y_i are conditionally independent, and the n_i and x_i are all known.

    (a) Find minimal sufficient statistics.
    (b) Find the form of the UMPU level α test of $H : B_1 \leq c$ versus $A : B_1 > c$.
    (c) For the special case $k = 2$, $n_1 = 3$, $n_2 = 2$, $x_1 = 1$, $x_2 = 2$, $c = 0$, $\alpha = 0.1$, find the exact test for all possible data values.

60. Consider the proof in Section 4.5.6 that the F-test is a proper Bayes rule.

    (a) Prove that the prior distribution given in Section 4.5.6 puts positive probability on every open subset of the original parameter space for Θ.
    (b) Prove that the Bayes rule is admissible.

Section 4.6:

61. Suppose that X_i are IID $N(\mu, 1)$ given $M = \mu$ for $i = 1, \ldots, n$ and that we use Lebesgue measure as an improper prior for M. If $H : M \leq c$, show that the posterior probability that H is true equals the P-value associated with the family of one-sided tests.

62. Suppose that X_i are IID $N(\mu, \sigma^2)$ given $(M, \Sigma) = (\mu, \sigma)$ for $i = 1, \ldots, n$ and that we use the usual improper prior with Radon–Nikodym derivative $1/\sigma$ as an improper prior. If $H : M \leq c$, show that the posterior probability that H is true equals the P-value associated with the family of one-sided t-tests.

63. Prove Proposition 4.144 on page 280.

64. In Example 4.146 on page 281, suppose that we change H to $H : P \geq p_0$. Prove that the P-value, associated with the family of UMP level α tests, of an observed $x < n$ is the posterior probability that the hypothesis is true based on an improper prior $Beta(1, 0)$.

65. Let $\{P_\theta : \theta \in \Omega\}$ be a parametric family. For each α, let $\phi_\alpha(x) = I_{S_\alpha}(x)$ be a size α test of $H : \Theta = \theta_0$ versus some alternative such that, for $0 < \alpha < 1$, $S_\alpha = \cap_{\beta > \alpha} S_\beta$. Suppose that $S_0 = \emptyset$ and $S_1 = \mathcal{X}$.

    (a) Prove that if $\alpha < \beta$, then $S_\alpha \subseteq S_\beta$.
    (b) Let $p_H(x)$ be the P-value of an observed x. Show that, given $\Theta = \theta_0$, $p_H(X) \sim U(0, 1)$.
    (c) Suppose that ϕ_α is unbiased for each α. Prove that
    $$P'_\theta(p_H(X) \leq \alpha) \geq P'_{\theta_0}(p_H(X) \leq \alpha).$$

66.*Let $X \sim N(\theta, 1)$ given $\Theta = \theta$. Let $g(\theta, x)$ be the P-value for testing $H_\theta : \Theta = \theta$ versus $A_\theta : \Theta \neq \theta$ for data $X = x$.

    (a) If the hypothesis is $H : \Theta \in [a, b]$ versus $A : \Theta \notin [a, b]$, and $X = x > (a + b)/2$ is observed, prove that the P-value is $\Phi(a - x) + \Phi(b - x)$, where Φ is the standard normal CDF.

(b) Assume that Lebesgue measure is used as an improper prior for Θ and that data $X = x > b$ are observed. Prove that for all hypotheses of the form $\Theta \leq b$, $\Theta = b$, and $\Theta \in [a,b]$, the P-value equals $E(g(\Theta, x)|H$ is true, $X = x)$.

67. Let $X \sim N(\theta, 1/n)$ given $\Theta = \theta$, where n is known. Let the prior for Θ be a mixture of a point mass at 0 and an $N(0,1)$ distribution. Consider the hypothesis $H : \Theta = 0$ versus $A : \Theta \neq 0$. Draw a graph of the Bayes factor as a function of x and a graph of the P-value as a function of x for $n = 1, 10, 100$.

68. Return to Problem 31 on page 289.

    (a) Find the three P-values relative to the three sets of tests in parts (a), (b), and (c).

    (b) Find the posterior probability that H is true in part (c). Does it equal any of the three P-values?

69. Let X have $N(\mu, 1)$ distribution given $M = \mu$, and let Lebesgue measure be an improper prior for M. For hypotheses of the form $M \leq c$, $M \geq c$, or $M = c$, we will show that the P-value for the usual family of tests is the posterior probability that M is farther from X (in the direction of the alternative) than X is from c.

    (a) Let $H : M \leq c$ versus $A : M > c$. Prove that the P-value equals the posterior probability that $M - X > X - c$. State and prove a similar result for $H : M \geq c$.

    (b) Let $H : M = c$ versus $A : M \neq c$. Prove that the P-value equals the posterior probability that $|M - X| > |X - c|$.

    (c) Can you extend this interpretation to the case of $H : a \leq M$ versus $A : M \notin [a,b]$?

Chapter 5
Estimation

In Chapter 3, we discussed methods for choosing decision rules in problems with specified loss functions. In Section 3.3, we gave an axiomatic derivation of some of those methods. This derivation led to the conclusion that there is a probability and a loss function, and one should minimize the expected loss. There are decision problems in which $\aleph$ and Ω are the same (or nearly the same) space and the loss function $L(\theta, a)$ is an increasing function of some measure of distance between θ and a. Such problems are often called *point estimation problems*. The classical framework makes no use of the probability over the parameter space provided by the axiomatic derivation. One can also try to ignore the loss function as well. To estimate Θ without a specific loss function, one can adopt ad hoc criteria to decide if an estimator is good. In this chapter, we will study some of these criteria as well as some criteria for the problem of *set estimation*. In set estimation, the action space is a collection of subsets of the parameter space (or the closure of the parameter space). The idea is to find a set that is likely to contain the parameter without being "too big" in some sense.

5.1 Point Estimation

A *point estimator* of a function g of a parameter Θ is a statistic that takes its values in the same set (or at least a similar set) as does $g(\Theta)$. One popular type of point estimator is an *unbiased estimator*.

Definition 5.1. Let Ω be the parameter space for a parametric family with P_θ and E_θ specifying the conditional distribution of X given $\Theta = \theta$. Let $g : \Omega \to G$ be some measurable function. Let $G' \supseteq G$. A measurable function $\phi : \mathcal{X} \to G'$ is called an *estimator of* $g(\Theta)$. An estimator ϕ of $g(\Theta)$ is called *unbiased* if $E_\theta(\phi(X)) = g(\theta)$, for all $\theta \in \Omega$. The *bias of* ϕ is defined as

$$b_\phi(\theta) = E_\theta(\phi(X)) - g(\theta).$$

The next example is one of several that led some early researchers to believe that unbiased estimators may not be bad to use.

Example 5.2. Suppose that P_θ says that $\{X_n\}_{n=1}^\infty$ are IID $N(\mu, \sigma^2)$, where $\theta = (\mu, \sigma)$. If $X = (X_1, \ldots, X_n)$, then we define $\overline{X} = \sum_{i=1}^n X_i/n$. It is easy to see that $E_\theta(\overline{X}) = \mu$. So, if $g(\theta) = \mu$, we see that $\phi(X) = \overline{X}$ is unbiased.

The following example shows that restricting attention to unbiased estimators may lead to an impasse.

Example 5.3. Suppose that P_θ says that $X \sim Exp(\theta)$. If $\phi(X)$ is to be an unbiased estimator of Θ, then

$$E_\theta \phi(X) = \int_0^\infty \phi(x) \theta \exp(-\theta x) dx = \theta,$$

for all θ. This happens if and only if $\int_0^\infty \phi(x) \exp(-\theta x) dx = 1$, for all θ. By Theorem 2.64, we can differentiate the left-hand side with respect to θ under the integral and get $\int_0^\infty x \phi(x) \exp(-\theta x) dx = 0$ for all θ. This means that $E_\theta(X\phi(X)) = 0$ for all θ. Since X is a complete sufficient statistic, $\phi(x) = 0$, a.s. $[P_\theta]$ for all θ. This contradicts $\phi(X)$ being unbiased. Hence, there are no unbiased estimators of Θ.

5.1.1 Minimum Variance Unbiased Estimation

It is natural to check how unbiased estimators fare under certain loss functions. The most common one to use is squared-error loss $L(\theta, a) = (g(\theta) - a)^2$. The risk function of an estimator ϕ is

$$R(\theta, \phi) = E_\theta \left\{ (g(\theta) - \phi(X))^2 \right\} = b_\phi^2(\theta) + \operatorname{Var}_\theta \phi(X).$$

If an estimator is unbiased, the risk function is just the variance. This suggests the following "optimality" criterion for unbiased estimators.

Definition 5.4. An unbiased estimator ϕ is a *uniformly minimum variance unbiased estimator (UMVUE)* if ϕ has finite variance and, for every unbiased estimator ψ, $\operatorname{Var}_\theta \phi(X) \leq \operatorname{Var}_\theta \psi(X)$ for all $\theta \in \Omega$.

UMVUEs are not necessarily good, as we will see later. The criterion of unbiasedness only means that the average of $\phi(X)$ with respect to P_θ is $g(\theta)$ for all θ. It does not mean that you expect $\phi(X)$ to be near $g(\theta)$ nor does it mean that you expect $g(\Theta)$ to be near $\phi(x)$ after you have seen $X = x$.

We mentioned earlier that the concept of complete sufficient statistic would play a role in unbiased estimation. The following theorem is due to Lehmann and Scheffé (1955).

298   Chapter 5. Estimation

Theorem 5.5 (Lehmann–Scheffé theorem). *If T is a complete statistic, then all unbiased estimators of $g(\Theta)$, that are functions of T alone, are equal, a.s. $[P_\theta]$ for all θ. If there exists an unbiased estimator that is a function of a complete sufficient statistic, then it is a UMVUE.*

PROOF. Suppose that $\phi_1(T)$ and $\phi_2(T)$ are unbiased estimators of $g(\Theta)$. Then $E_\theta[\phi_1(T) - \phi_2(T)] = 0$ for all θ. Since T is a complete statistic, it follows that $\phi_1(T) = \phi_2(T)$, a.s. $[P_\theta]$. Now, suppose that there is an unbiased estimator $\phi(X)$ with finite variance. Define $\phi_3(T) = E(\phi(X)|T)$. Then $\phi_3(T)$ is unbiased by the law of total probability B.70. Using squared-error loss, the Rao–Blackwell theorem 3.22 says $R(\theta, \phi_3) \leq R(\theta, \phi)$ for all θ. Since the risk function is the variance for unbiased estimators, this makes ϕ_3 a UMVUE. □

Example 5.6. Suppose that $\{X_n\}_{n=1}^\infty$ are IID $N(\mu, \sigma^2)$ given $\Theta = (M, \Sigma) = (\mu, \sigma)$. Let $X = (X_1, \ldots, X_n)$. Then

$$\overline{X} = \frac{1}{n}\sum_{i=1}^n X_i, \text{ and } S^2 = \frac{1}{n-1}\sum_{i=1}^n (X_i - \overline{X})^2$$

are complete sufficient statistics. Since they are unbiased, they are UMVUE of M and Σ^2, respectively. Notice that S^2 does not minimize mean squared error, even among estimators of the form $c\sum_{i=1}^n (X_i - \overline{X})^2$. (See Example 3.25 on page 154 and Problem 11 on page 210.)

Example 5.7. Suppose that P_θ says that X has $Poi(\theta)$ distribution. We know that X is a complete sufficient statistic. Let $g(\Theta) = \exp(-3\Theta)$. We will find the UMVUE of $g(\Theta)$. The required condition is

$$E_\theta \phi(X) = \sum_{x=0}^\infty \phi(x) \exp(-\theta) \frac{\theta^x}{x!} = \exp(-3\theta),$$

for all θ. It follows from the uniqueness of Taylor series expansions for analytic functions that ϕ is unbiased if and only if $\phi(x) = (-2)^x$ for $x = 0, 1, 2, \ldots$. This ϕ, although UMVUE, is an abominable estimator of $g(\Theta)$. We will see some better estimators later in this chapter. (See Examples 5.29 and 5.32.)

The following results are useful when there is no complete sufficient statistic.

Proposition 5.8. *Let δ_0 be an unbiased estimator of $g(\Theta)$, and let*

$$\mathcal{U} = \{U : E_\theta U(X) = 0, \text{ for all } \theta\}.$$

Then, the set of all unbiased estimators of $g(\Theta)$ is $\{\delta_0 + U : U \in \mathcal{U}\}$.

Theorem 5.9. *An estimator δ is UMVUE of $E_\theta \delta(X)$ if and only if, for every $U \in \mathcal{U}$, $\mathrm{Cov}_\theta(\delta(X), U(X)) = 0$.*

PROOF. For the "only if" part, suppose that δ is UMVUE. It is clear that if $\text{Var}_\theta U(X) = 0$ for all θ, then $\text{Cov}_\theta(\delta(X), U(X)) = 0$. So, let $U \in \mathcal{U}$ be such that $\text{Var}_\theta U(X) > 0$ for some θ. Let $\lambda \in \mathbb{R}$, and define $\delta_\lambda = \delta + \lambda U$. Then δ_λ is unbiased also, and for every λ,

$$\begin{aligned}\text{Var}_\theta \delta(X) &\leq \text{Var}_\theta \delta_\lambda(X) \\ &= \text{Var}_\theta \delta(X) + 2\lambda \text{Cov}_\theta(\delta(X), U(X)) + \lambda^2 \text{Var}_\theta U(X).\end{aligned}$$

This is true for all λ and all θ if and only if

$$\lambda^2 \text{Var}_\theta U(X) \geq -2\lambda \text{Cov}_\theta(\delta(X), U(X)),$$

which, in turn is true for all λ and θ if and only if $\text{Cov}_\theta(\delta(X), U(X)) = 0$ for all θ.

For the "if" part, assume that for all $U \in \mathcal{U}$, $\text{Cov}_\theta(\delta(X), U(X)) = 0$ for all θ. Now, let $\delta_1(X)$ be an unbiased estimator of $E_\theta \delta(X)$. Then there exists $U \in \mathcal{U}$ such that $\delta_1 = \delta + U$. It follows that

$$\begin{aligned}\text{Var}_\theta(\delta_1(X)) &= \text{Var}_\theta(\delta(X)) + 2\text{Cov}_\theta(\delta(X), U(X)) + \text{Var}_\theta(U(X)) \\ &= \text{Var}_\theta(\delta(X)) + \text{Var}_\theta(U(X)) \geq \text{Var}_\theta(\delta(X)),\end{aligned}$$

hence $\delta(X)$ is UMVUE. □

Sometimes unbiased estimators exist, but none is UMVUE.

Example 5.10. Suppose that P_θ says that $Y_1, Y_2, \ldots$ are IID $Ber(\theta)$. Set

$$X = \begin{cases} 1 & \text{if } Y_1 = 1, \\ \text{\# of trials until 2nd failure} & \text{otherwise}, \end{cases}$$

and suppose that we observe only X. Then

$$f_{X|\Theta}(x|\theta) = \begin{cases} \theta & \text{if } x = 1, \\ \theta^{x-2}(1-\theta)^2 & \text{if } x = 2, 3, \ldots. \end{cases}$$

Define the estimator δ_0 to be $\delta_0(x) = 1$ if $x = 1$ and $\delta_0(x) = 0$ if not. Then δ_0 is an unbiased estimator of Θ. We will now try to find a UMVUE. Assume $E_\theta U(X) = 0$, for all θ. Then

$$\begin{aligned}E_\theta U(X) &= U(1)\theta + \sum_{x=2}^{\infty} \theta^{x-2}(1-\theta)^2 U(x) \\ &= U(2) + \sum_{k=1}^{\infty} \theta^k \left[U(k) - 2U(k+1) + U(k+2)\right] = 0\end{aligned}$$

if and only if $U(2) = 0$ and $U(k) = -(k-2)U(1)$ for all $k \geq 3$. This characterizes all functions in $\mathcal{U}$ according to the value $t = U(1)$. That is,

$$\mathcal{U} = \{U_t : U_t(x) = (x-2)t, \text{ for all } x\}.$$

Every unbiased estimator of Θ is $\delta_t(x) = \delta_0(x) + (x-2)t$, for some t. In order for δ_t to be UMVUE, it must have 0 covariance with every $U_s \in \mathcal{U}$. That is, for all s and all θ,

$$0 = \sum_{x=1}^{\infty} f_{X|\Theta}(x|\theta) \delta_t(x) U_s(x) = \theta(-s)(1-t) + \sum_{x=2}^{\infty} (1-\theta)^2 \theta^{x-2} ts(x-2)^2.$$

Divide both sides by $(1-\theta)^2$ to get

$$s(1-t) \frac{\theta}{(1-\theta)^2} = \sum_{x=2}^{\infty} ts\theta^{x-2}(x-2)^2.$$

By rewriting $\theta/[1-\theta]^2$ as an infinite sum, we get

$$s(1-t) \sum_{k=1}^{\infty} k\theta^k = \sum_{k=1}^{\infty} tsk^2 \theta^k.$$

Since these two series in this last equation are analytic functions of θ, it must be that $s(1-t)k = tsk^2$ for all s, k. This is not possible, hence there is no t such that these equations hold. And there is no UMVUE.

Oddly enough, there is a *locally minimum variance unbiased estimator* in Example 5.10.

Definition 5.11. An unbiased estimator $\delta_0(X)$ is *locally minimum variance unbiased, LMVUE*, at θ_0 if for every other unbiased estimator $\delta(X)$, $\mathrm{Var}_{\theta_0} \delta_0(X) \leq \mathrm{Var}_{\theta_0} \delta(X)$.

Example 5.12 (Continuation of Example 5.10; see page 299). First, note that

$$\mathrm{Var}_\theta \delta_t(X) = E_\theta \delta_t^2(X) - \theta^2,$$

so a LMVUE at θ_0 can be found by minimizing

$$E_{\theta_0} \delta_t^2(X) = E_{\theta_0}\left([\delta_0(X) + U_t(X)]^2\right)$$
$$= \theta_0(1-t)^2 + \sum_{x=2}^{\infty} \theta_0^{x-2}(1-\theta_0)^2(x-2)^2 t^2.$$

This expression is quadratic in t and can be minimized by choosing

$$t = \left[1 + \sum_{k=1}^{\infty} \theta_0^{k-1} k^2 (1-\theta_0)^2\right]^{-1},$$

which is different for each θ_0.

A LMVUE at θ_0 is not the "best" estimator if $\Theta = \theta_0$; it is merely an unbiased estimator such that the conditional variance given $\Theta = \theta_0$ is smaller than that of any other unbiased estimator.

5.1.2 Lower Bounds on the Variance of Unbiased Estimators

Suppose that one is interested in unbiased estimators. It would be nice to know how low the variances of such estimators can be. Under some regularity conditions, there exist lower bounds for the variances of unbiased estimators. The Fisher information plays an important role in these lower bounds.[1]

Theorem 5.13 (Cramér–Rao lower bound). *Suppose that the three FI regularity conditions hold (see Definition 2.78 on page 111), and let $\mathcal{I}_X(\theta)$ be the Fisher information. Suppose that $\mathcal{I}_X(\theta) > 0$, for all θ. Let $\phi(X)$ be a one-dimensional statistic with $E_\theta|\phi(X)| < \infty$ for all θ. Suppose also that $\int \phi(x) f_{X|\Theta}(x|\theta) d\nu(x)$ can be differentiated under the integral sign with respect to θ. Then*

$$\mathrm{Var}_\theta \phi(X) \geq \frac{\left(\frac{d}{d\theta} E_\theta \phi(X)\right)^2}{\mathcal{I}_X(\theta)}.$$

Before proving this theorem, we should look at some examples.

Example 5.14. Suppose that $X \sim N(\theta, b)$ given $\Theta = \theta$ and $\phi(x) = x$. Then, the conditions of Theorem 5.13 are satisfied, and we calculated $\mathcal{I}_X(\theta) = 1/b$ in Example 2.80 on page 111. So

$$E_\theta \phi(X) = \theta, \quad \frac{d}{d\theta} E_\theta \phi(X) = 1, \quad \mathrm{Var}_\theta \phi(X) = b.$$

In this case, the Cramér–Rao lower bound is met exactly.

Example 5.15. Suppose that $X \sim U(0, \theta)$ given $\Theta = \theta$. In this case $f_{X|\Theta}(x|\theta) = \theta^{-1} I_{(0,\theta)}(x)$. We saw in Example 2.81 on page 111 that the conditions of Theorem 5.13 are not met. Nevertheless, we were able to calculate something that could have been called Fisher information, namely $\mathcal{I}_X(\theta) = 1/\theta^2$. Let $\phi(x) = x$. Then $E_\theta \phi(X) = \theta/2$ and $\mathrm{Var}_\theta \phi(X) = \theta^2/12$. We can calculate

$$\frac{d}{d\theta} E_\theta \phi(X) = \frac{1}{2}.$$

If the Cramér–Rao lower bound held here, it would say that $\mathrm{Var}_\theta \phi(X) \geq \theta^2/4$, which is clearly false.

PROOF OF THEOREM 5.13. Let B and C be the sets described in the FI regularity conditions. (See Definition 2.78 on page 111.) Let $D = C \cap B^C$, so that, for all θ, $P_\theta(D) = 1$ and $\int_D f_{X|\Theta}(x|\theta) d\nu(x) = 1$. Taking the derivative with respect to θ of this integral, we get

$$\begin{aligned}
0 &= \int_D \frac{\partial}{\partial \theta} f_{X|\Theta}(x|\theta) d\nu(x) = \int_D \frac{\frac{\partial}{\partial \theta} f_{X|\Theta}(x|\theta)}{f_{X|\Theta}(x|\theta)} f_{X|\Theta}(x|\theta) d\nu(x) \\
&= E_\theta \left[\frac{\partial}{\partial \theta} \log f_{X|\Theta}(X|\theta) \right].
\end{aligned}$$

---

[1] The first lower bound is due to Rao (1945) and Cramér (1945, Chapter 32; 1946).

Also, we can differentiate to obtain

$$\frac{d}{d\theta}E_\theta\phi(X) = \int \phi(x)\frac{\partial}{\partial\theta}f_{X|\Theta}(x|\theta)d\nu(x) = E_\theta\left[\phi(X)\frac{\partial}{\partial\theta}\log f_{X|\Theta}(X|\theta)\right]$$
$$= E_\theta\left\{[\phi(X) - E_\theta\phi(X)]\frac{\partial}{\partial\theta}\log f_{X|\Theta}(X|\theta)\right\},$$

since the term being added on has zero mean. Now take the absolute value and use the Cauchy–Schwarz inequality B.19:

$$\left|\frac{d}{d\theta}E_\theta\phi(X)\right| \leq \sqrt{E_\theta\left[\phi(X) - E_\theta\phi(X)\right]^2}\sqrt{E_\theta\left[\frac{\partial}{\partial\theta}\log f_{X|\Theta}(X|\theta)\right]^2}$$
$$= \sqrt{\mathrm{Var}_\theta\phi(X)}\sqrt{\mathcal{I}_X(\theta)}.$$

Now square the extreme ends of this string and divide by $\mathcal{I}_X(\theta)$. □

For an unbiased estimator $\phi(X)$ of Θ, the smallest possible variance is $1/\mathcal{I}_X(\theta)$, since $dE_\theta\phi(X)/d\theta = 1$. A necessary and sufficient condition for the lower bound to be achieved is that the $\leq$ become an $=$ in Theorem 5.13. The $\leq$ in Theorem 5.13 was introduced by the Cauchy–Schwarz inequality B.19, which provides for equality if and only if the two factors are linearly related. (That is, if $E(X) = 0$ and $E(Y) = 0$, then $|E(XY)| = \sqrt{E(X^2)}\sqrt{E(Y^2)}$ if and only if there exist a and b such that $aX + bY = 0$, a.s. and $ab \neq 0$.) So, the Cramér–Rao lower bound is an equality if and only if $\phi(X)$ and the score function $\partial \log f_{X|\Theta}(X|\theta)/\partial\theta$ are linearly related, that is,

$$\frac{\partial}{\partial\theta}\log f_{X|\Theta}(x|\theta) = a(\theta)\phi(x) + d(\theta), \text{ a.s. } [P_\theta],$$

for all θ. If we solve this differential equation, we get

$$f_{X|\Theta}(x|\theta) = c(\theta)h(x)\exp\{\pi(\theta)\phi(x)\}.$$

This means that the Cramér–Rao lower bound can be sharp only in a one-parameter exponential family with $\phi(X)$ being a sufficient statistic.

Example 5.16. Suppose that $X \sim Exp(\lambda)$ given $\Lambda = \lambda$. Set $\Theta = 1/\Lambda$. Then

$$f_{X|\Theta}(x|\theta) = \frac{1}{\theta}\exp\left\{-\frac{1}{\theta}x\right\}I_{(0,\infty)}(x),$$
$$\frac{\partial}{\partial\theta}\log f_{X|\Theta}(x|\theta) = -\frac{1}{\theta} + \frac{x}{\theta^2}.$$

Since the score function is a linear function of x, it follows that $\phi(x)$ must also be a linear function of x if $\phi(X)$ is to achieve the lower bound. If $\phi(x) = a + bx$, then $E_\theta\phi(X) = a + b\theta$, so $a = 0$ and $b = 1$ gives an unbiased estimator that achieves the Cramér–Rao lower bound. The reader should verify that this is indeed the case.

Example 5.17. Outside of exponential families, the Cramér–Rao lower bound cannot be achieved. For example, suppose that

$$f_{X|\Theta}(x|\theta) = \frac{\Gamma\left(\frac{a+1}{2}\right)}{\Gamma\left(\frac{a}{2}\right)\sqrt{a\pi}}\left(1 + \frac{1}{a}(x-\theta)^2\right)^{-\frac{a+1}{2}}.$$

This is the family of $t_a(\theta, 1)$ distributions. In order for the variance to exist, suppose that $a \geq 3$. Then

$$\frac{\partial^2}{\partial \theta^2} \log f_{X|\Theta}(x|\theta) = -\frac{a+1}{a} \frac{1 - \frac{1}{a}(x-\theta)^2}{\left[1 + \frac{1}{a}(x-\theta)^2\right]^2}.$$

Call this $g(x)$. Then $\mathcal{I}_X(\theta) = -E_\theta g(X)$.

$$\mathcal{I}_X(\theta) = \int \frac{\Gamma\left(\frac{a+1}{2}\right)}{\Gamma\left(\frac{a}{2}\right)\sqrt{a\pi}} \frac{a+1}{a} \frac{1 - \frac{1}{a}(x-\theta)^2}{\left(1 + \frac{1}{a}(x-\theta)^2\right)^{\frac{a+5}{2}}}.$$

Since the denominator of the integrand looks like part of the t_{a+4} density, we will perform the following transformation:

$$\frac{z-\theta}{\sqrt{a+4}} = \frac{x-\theta}{\sqrt{a}},$$

$$x = \theta + \sqrt{\frac{a}{a+4}}(z - \theta), \qquad dx = \sqrt{\frac{a}{a+4}} dz.$$

The integral becomes

$$\frac{a+1}{a}\sqrt{\frac{a}{a+4}}\frac{\Gamma\left(\frac{a+1}{2}\right)}{\Gamma\left(\frac{a}{2}\right)\sqrt{a\pi}} \int \frac{1 - \frac{1}{a+4}(z-\theta)^2}{\left(1 + \frac{1}{a+4}(z-\theta)^2\right)^{\frac{a+5}{2}}} dz.$$

Except for the constant, the integral is $E(1 - U^2/[a+4])$, where $U \sim t_{a+4}$. The correct constant to make the integral equal to this expected value is

$$\frac{\Gamma\left(\frac{a+5}{2}\right)}{\Gamma\left(\frac{a+4}{2}\right)\sqrt{(a+4)\pi}},$$

and the expected value is $(a+1)/(a+2)$, so the result is

$$\mathcal{I}_X(\theta) = \frac{a+1}{a+2}\frac{a+1}{a}\sqrt{\frac{a}{a+4}}\frac{\Gamma\left(\frac{a+1}{2}\right)}{\Gamma\left(\frac{a}{2}\right)\sqrt{a\pi}} \frac{\Gamma\left(\frac{a+4}{2}\right)\sqrt{(a+4)\pi}}{\Gamma\left(\frac{a+5}{2}\right)} = \frac{a+1}{a+3}.$$

This means the Cramér–Rao lower bound is $(a+3)/(a+1)$. We know that $\phi(X) = X$ is UMVUE because X is a complete sufficient statistic, and $\text{Var}_\theta(X) = a/(a-2)$, which is always larger than the Cramér–Rao lower bound.

There is another lower bound that applies in more general cases, such as when the set of possible values for X depends on Θ. This next lower bound is due to Chapman and Robbins (1951).

Chapter 5. Estimation

Theorem 5.18 (Chapman–Robbins lower bound). *Let*

$$m(\theta) = E_\theta \phi(X),$$
$$supp(\theta) = \text{closure of } \{x : f_{X|\Theta}(x|\theta) > 0\}.$$

Assume that for each $\theta \in \Omega$, there is $\theta' \neq \theta$ such that $supp(\theta') \subseteq supp(\theta)$. Then,

$$Var_\theta \phi(X) \geq \sup_{\{\theta' : supp(\theta') \subseteq supp(\theta)\}} \left\{ \frac{[m(\theta) - m(\theta')]^2}{E_\theta \left[\frac{f_{X|\Theta}(X|\theta')}{f_{X|\Theta}(X|\theta)} - 1 \right]^2} \right\}.$$

PROOF. Let θ' be such that $supp(\theta') \subseteq supp(\theta)$. Let

$$U(X) = \frac{f_{X|\Theta}(X|\theta')}{f_{X|\Theta}(X|\theta)} - 1.$$

Then $E_\theta U(X) = 1 - 1 = 0$, and

$$Cov_\theta(U(X), \phi(X)) = \int_{supp(\theta)} [\phi(x) f_{X|\Theta}(x|\theta') - \phi(x) f_{X|\Theta}(x|\theta)] \, d\nu(x)$$
$$= m(\theta') - m(\theta),$$

since $supp(\theta') \subseteq supp(\theta)$. By the Cauchy–Schwarz inequality B.19, the square of the covariance is at most the product of the variances, so

$$[m(\theta') - m(\theta)]^2 \leq Var_\theta \phi(X) Var_\theta U(X). \qquad \square$$

Example 5.19. This is a case in which the Cramér–Rao lower bound does not apply. Let P_θ say that $\{X_n\}_{n=1}^\infty$ are IID with density $f_{X|\Theta}(x|\theta) = \exp(\theta - x) I_{[\theta, \infty)}$. Let $X = (X_1, \ldots, X_n)$. Then $supp(\theta) = [\theta, \infty)$ and $supp(\theta') \subseteq supp(\theta)$ so long as $\theta' \geq \theta$. From the proof of Theorem 5.18,

$$U(X) = \exp\{n(\theta' - \theta)\} I_{[\theta', \infty)}(\min X_i) - 1.$$

If $\phi(X)$ is an unbiased estimator of Θ, then $[m(\theta) - m(\theta')]^2 = (\theta - \theta')^2$, and

$$E_\theta U(X)^2 = \exp\{2n(\theta' - \theta)\} P_\theta(\min X_i \geq \theta')$$
$$\quad - 2 \exp\{n(\theta' - \theta)\} P_\theta(\min X_i \geq \theta') + 1,$$
$$P_\theta(\min X_i \geq \theta') = (P_\theta(X_1 \geq \theta'))^n$$
$$= \left[\int_{\theta'}^\infty \exp(\theta - x) dx \right]^n = \exp\{-n(\theta' - \theta)\},$$
$$E_\theta U(X)^2 = \exp\{n(\theta' - \theta)\} - 1.$$

The Chapman–Robbins lower bound is

$$Var_\theta \phi(X) \geq \sup_{\theta' \geq \theta} \frac{(\theta - \theta')^2}{\exp\{n(\theta' - \theta)\} - 1} \approx \frac{0.1619}{n^2}.$$

A simple unbiased estimator is $\phi(X) = \min X_i - 1/n$, which has variance $1/n^2$.

Another way to improve the Cramér–Rao lower bound is to raise it if it is unattainable. We know that it is attained if $\phi(X)$ is perfectly correlated with the score function $\partial \log f_{X|\Theta}(X|\theta)/\partial\theta$, that is, if the regression of $\phi(X)$ on the score function has 0 residual. If this is not possible, the residual might be made smaller by regressing $\phi(X)$ on more than just the score function.

Lemma 5.20. *Let $\phi(X)$ be an unbiased estimator of $g(\Theta)$ and let $\psi_i(x,\theta)$, $i = 1, \ldots, k$, be functions that are not linearly related and such that*

$$\gamma^\top = (\gamma_1, \ldots, \gamma_k), \qquad \gamma_i = \mathrm{Cov}_\theta(\phi(X), \psi_i(X,\theta)),$$
$$C = ((c_{ij})), \qquad c_{ij} = \mathrm{Cov}_\theta(\psi_i(X,\theta), \psi_j(X,\theta)).$$

Then $\mathrm{Var}_\theta \phi(X) \geq \gamma^\top C^{-1} \gamma$.

PROOF. The covariance matrix of $(\phi(X), \psi_1(X,\theta), \ldots, \psi_k(X,\theta))^\top$ is

$$\begin{pmatrix} \mathrm{Var}_\theta \phi(X) & \gamma^\top \\ \gamma & C \end{pmatrix}.$$

The inequality follows from the fact that the covariance matrix is positive semidefinite and C is nonsingular. □

Lemma 5.20 has two corollaries, one of which is an improvement on the Cramér–Rao lower bound and the other of which is a multiparameter version of the Cramér–Rao lower bound.

The first corollary to Lemma 5.20 is one that attempts to improve the Cramér–Rao lower bound by making use of the fact that the inequality can fail to be an equality because the score function is not linearly related to the estimator. To put this another way, if the residual from the linear regression of the estimator on the score function has nonzero variance, then it might be possible to get the residual variance down by regression on more than just the score function. This is the approach taken in Corollary 5.21.

Corollary 5.21 (Bhattacharyya system of lower bounds). *Assume the conditions of Theorem 5.13, assume that k partial derivatives with respect to θ can be passed under the integral sign, and assume that $\mathcal{J}(\theta)$ (defined below) is nonsingular. Then $\mathrm{Var}_\theta \phi(X) \geq \gamma^\top(\theta) \mathcal{J}^{-1}(\theta) \gamma(\theta)$, where*

$$\gamma(\theta) = \begin{pmatrix} \gamma_1(\theta) \\ \vdots \\ \gamma_k(\theta) \end{pmatrix}, \qquad \gamma_i(\theta) = \frac{d^i}{d\theta^i} \mathrm{E}_\theta \phi(X),$$
$$\mathcal{J}(\theta) = ((J_{ij}(\theta))), \qquad J_{ij}(\theta) = \mathrm{Cov}(\psi_i(X,\theta), \psi_j(X,\theta)),$$
$$\psi_i(x,\theta) = \frac{1}{f_{X|\Theta}(x|\theta)} \frac{\partial^i}{\partial \theta^i} f_{X|\Theta}(x|\theta).$$

PROOF. All we need to do, in order to apply Lemma 5.20, is note that

$$\frac{d^i}{d\theta^i} \mathrm{E}_\theta \phi(X) = \mathrm{Cov}_\theta\left(\phi(X), \frac{1}{f_{X|\Theta}(X|\theta)} \frac{\partial^i}{\partial \theta^i} f_{X|\Theta}(X|\theta)\right),$$

which follows for higher derivatives in just the same way that it did for the first derivative in the proof of Theorem 5.13. □

The Cramér–Rao lower bound is the special case with $k = 1$.

Example 5.22. Suppose that $X \sim Exp(\lambda)$ given $\Lambda = \lambda$. Set $\Theta = 1/\Lambda$. Then

$$f_{X|\Theta}(x|\theta) = \frac{1}{\theta}\exp\left\{-\frac{1}{\theta}x\right\} I_{(0,\infty)}(x),$$

$$\frac{\partial}{\partial \theta} f_{X|\Theta}(x|\theta) = \left(-\frac{1}{\theta} + \frac{x}{\theta^2}\right) f_{X|\Theta}(x|\theta),$$

$$\frac{\partial^2}{\partial \theta^2} f_{X|\Theta}(x|\theta) = \left(\frac{2}{\theta^2} - \frac{4x}{\theta^3} + \frac{x^2}{\theta^5}\right) f_{X|\Theta}(x|\theta).$$

Let $\phi(x) = x^2$. We can easily calculate $E_\theta(\phi(X)) = 2\theta^2$ and $Var_\theta(\phi(X)) = 20\theta^4$. Since $\mathcal{I}_X(\theta) = 1/\theta^2$, and $dE_\theta(\phi(X))/d\theta = 4\theta$, the Cramér–Rao lower bound on the variance of $\phi(X)$ is $16\theta^4$. A little bit of calculation yields

$$\mathcal{J}(\theta) = \begin{pmatrix} \frac{1}{\theta^2} & 0 \\ 0 & \frac{4}{\theta^4} \end{pmatrix}, \quad \gamma(\theta) = \begin{pmatrix} 4\theta \\ 4 \end{pmatrix}.$$

So, $\gamma(\theta)^\top \mathcal{J}^{-1}(\theta)\gamma(\theta) = 20\theta^4$, and the Bhattacharyya lower bound is achieved.

Corollary 5.23 (Multiparameter Cramér–Rao lower bound). *Assume that the FI regularity conditions on page 111 hold. Let $\mathcal{I}_X(\theta) = ((I_{ij}(\theta)))$ be the Fisher information matrix, and suppose that it is positive definite. Suppose that $\int \phi(x) f_{X|\Theta}(x|\theta) d\nu(x)$ can be twice differentiated under the integral sign with respect to coordinates of θ and that*

$$E_\theta |\phi(X)| < \infty \text{ and } \gamma^\top(\theta) = \left(\cdots, \frac{\partial}{\partial \theta_i} E_\theta \phi(X), \cdots\right).$$

Then $Var_\theta \phi(X) \geq \gamma^\top(\theta) \mathcal{I}_X^{-1}(\theta) \gamma(\theta)$.

PROOF. All we need to do, in order to apply Lemma 5.20, is note that

$$\frac{\partial}{\partial \theta_i} E_\theta \phi(X) = Cov_\theta \left(\phi(X), \frac{\partial}{\partial \theta_i} \log f_{X|\Theta}(X|\theta)\right),$$

just as in the proof of Theorem 5.13. □

Example 5.24. Suppose that P_θ says $X \sim N(\mu, \sigma^2)$, where $\theta = (\mu, \sigma)$. This is the same as Example 2.83 on page 112. There we calculated

$$\mathcal{I}_X(\theta) = \begin{pmatrix} \frac{2}{\sigma^2} & 0 \\ 0 & \frac{1}{\sigma^2} \end{pmatrix}.$$

Now, set $\phi(X) = X^2$. Then $E_\theta \phi(X) = \mu^2 + \sigma^2$ and

$$\gamma_1(\theta) = 2\mu, \qquad Var_\theta \phi(X) = 2\sigma^4 + 4\mu^2\sigma^2,$$
$$\gamma_2(\theta) = 2\sigma, \qquad \gamma^\top(\theta)\mathcal{I}_X(\theta)^{-1}\gamma(\theta) = 4\mu^2\sigma^2 + 2\sigma^4.$$

The Cramér–Rao lower bound is met exactly.

5.1.3 Maximum Likelihood Estimation

If the posterior density of Θ is very high at some value, say θ_0, and relatively low everywhere else, it means that we are quite sure that Θ is near θ_0. If the prior density of Θ is fairly flat near θ_0 and is not orders of magnitude larger at other values of θ than it is at θ_0, then the posterior density will differ from the likelihood function only by a constant factor near θ_0.[2]

Definition 5.25. Let X be a random quantity with conditional density $f_{X|\Theta}(x|\theta)$ given $\Theta = \theta$. If $X = x$ is observed, then the function $L(\theta) = f_{X|\Theta}(x|\theta)$ considered as a function of θ for fixed x is called the *likelihood function*. Any random quantity $\hat{\Theta}$ such that

$$\max_{\theta \in \Omega} f_{X|\Theta}(X|\theta) = f_{X|\Theta}(X|\hat{\Theta})$$

is called a *maximum likelihood estimator (MLE) of* Θ. If $\Theta = (\Theta_1, \Theta_2)$ and $\hat{\Theta} = (\hat{\Theta}_1, \hat{\Theta}_2)$, then $\hat{\Theta}_1$ is called an *MLE of* Θ_1.

The idea of maximizing the likelihood function in order to estimate a parameter dates back to Fisher (1922).

Example 5.26. Suppose that $X_1, \ldots, X_n$ are conditionally independent given $\Theta = \theta$ with $U(0, \theta)$ distribution. Then,

$$f_{X|\Theta}(x|\theta) = \frac{1}{\theta^n} I_{[0,\theta]}(\max_i x_i) I_{[0, \max_i x_i]}(\min_i x_i).$$

As a function of θ, the maximum of this function is at $\theta = \max_i x_i$. Hence the MLE is $\hat{\Theta} = \max_i X_i$.

Suppose that we had defined the density of each X_i to be $I_{(0,\theta)}(x)$ instead of using the closed interval. In this case there is no value of θ at which the maximum is achieved. At first, it would seem that this could easily be fixed by replacing max by sup in the definition of MLE. This would also require some continuity condition on the likelihood function. It turns out to be very inconvenient to do this. Rather, we should use the closed interval for the density.

The following example shows that the MLE may exist but not be unique.

Example 5.27. Suppose that $X_1, \ldots, X_n$ are conditionally independent given $\Theta = \theta$ with $U(\theta - 1/2, \theta + 1/2)$ distribution. Then,

$$f_{X|\Theta}(x|\theta) = I_{[\theta - \frac{1}{2}, \theta + \frac{1}{2}]}(\min_i x_i) I_{[\min_i x_i, \theta + \frac{1}{2}]}(\max_i x_i).$$

As a function of θ, this is constant for $\max_i x_i - 1/2 \leq \theta \leq \min_i x_i + 1/2$. Any random variable $\hat{\Theta}$ between $\max_i X_i - 1/2$ and $\min_i X_i + 1/2$ is an MLE.

---

[2] This observation has led some people to try to base inference on the likelihood function alone, rather than the posterior distribution. See Barndorff-Nielsen (1988) for an in-depth study of likelihood.

Theorem 5.28. *Let g be a measurable function from Ω to some space G. Suppose that there exists another space U and a one-to-one measurable function $h : \Omega \to G \times U$ such that $h(\theta) = (g(\theta), g^*(\theta))$ for some function g^*. If $\hat{\Theta}$ is an MLE of Θ, then $g(\hat{\Theta})$ is an MLE of $g(\Theta)$.*

PROOF. Since h is one-to-one, the parameter might just as well be $\Psi = h(\Theta)$. The likelihood for Ψ is $f_{X|\Psi}(x|\psi) = f_{X|\Theta}(x|h^{-1}(\psi))$. For fixed x, if the maximum of $f_{X|\Theta}(x|\theta)$ occurs at $\theta = \hat{\theta}$, define $\hat{\psi} = h(\hat{\theta})$. Then the maximum of $f_{X|\Theta}(x|\theta)$ occurs at $\theta = h^{-1}(\hat{\psi})$. Now, suppose that the maximum of $f_{X|\Psi}(x|\psi)$ occurs at $\psi = \psi_0$. If $f_{X|\Psi}(x|\psi_0) > f_{X|\Psi}(x|\hat{\psi})$, then $\theta = h^{-1}(\psi_0)$ would provide a higher value for $f_{X|\Theta}(x|\theta)$ than $\hat{\theta}$. This would be a contradiction. It follows that $\psi = \hat{\psi}$ provides a maximum for $f_{X|\Psi}(x|\psi)$. It follows that the MLE of Ψ is $h(\hat{\Theta})$ and the MLE of the first coordinates of Ψ, namely $g(\Theta)$, is $g(\hat{\Theta})$, the first coordinates of $h(\hat{\Theta})$. □

Example 5.29 (Continuation of Example 5.7; see page 298). Suppose that X given $\Theta = \theta$ has $Poi(\theta)$ distribution and $g(\theta) = \exp(-3\theta)$. Since the MLE of Θ is X, and g is one-to-one, the MLE of $g(\Theta)$ is $\exp(-3X)$. This is far more reasonable than the UMVUE $(-2)^X$.

If the loss function is $L(\theta, a) = \left(\theta + \frac{1}{3}\log a\right)^2$ and the (improper) prior distribution has Radon–Nikodym derivative $1/\theta$ with respect to Lebesgue measure, then the formal Bayes rule is also $\exp(-3X)$.

Example 5.30. Let $X_1, \ldots, X_n$ be IID $N(\mu, \sigma^2)$ given $\Theta = (M, \Sigma) = (\mu, \sigma)$. It is not difficult to see that the MLE of Θ is $(\overline{X}, \sqrt{W/n})$, where $\overline{X} = \sum_{i=1}^{n} X_i/n$ and $W = \sum_{i=1}^{n}(X_i - \overline{X})^2$. Suppose that we want the MLE of M^2. The function $g(M, \Sigma) = M^2$ is not a one-to-one function of either coordinate, but $g^*(\mu, \sigma) = (\sigma, \text{sign}(\mu))$ will satisfy the conditions of Theorem 5.28. So $\overline{X}^2$ is the MLE of M^2. The UMVUE of M^2 is $\overline{X}^2 - W/n^2$, which is negative with positive probability.

In exponential families, there is a simple method for finding MLEs in most cases. The logarithm of the likelihood function will be $\log L(\theta) = \log c(\theta) + x^\top \theta$. If the MLE occurs in the interior of the parameter space, it occurs where the partial derivatives of $\log L(\theta)$ are 0. That is, $x_i = -\partial \log c(\theta)/\partial \theta_i$. By using the method of Example 2.66, we see that the MLE is that θ such that $x = E_\theta X$.

Example 5.31 (Continuation of Example 2.68; see page 106). Let $X_1, \ldots, X_n$ be IID $N(\mu, \sigma^2)$ given $\Psi = (\mu, \sigma)$. The natural parameter of this exponential family is $\Theta_1 = M/\Sigma^2$ and $\Theta_2 = -1/[2\Sigma^2]$. The natural sufficient statistic is $X = (n\overline{X}, \sum_{i=1}^{n} X_i^2)$. Now $\log c(\theta) = n\log(-2\theta_2)/2 + n\theta_1^2/[4\theta_2]$. The partial derivative with respect to θ_1 is $n\theta_1/[2\theta_2]$ and the partial with respect to θ_2 is $n/[2\theta_2] - n\theta_1^2/[4\theta_2^2]$. Setting these equal to the negatives of the two coordinates of X and solving for θ_1 and θ_2 give $\hat{\Theta}_1 = n\overline{X}/\sum_{i=1}^{n}(X_i - \overline{X})^2$ and $\hat{\Theta}_2 = -n/[2\sum_{i=1}^{n}(X_i - \overline{X})^2]$. In terms of the usual parameterization $M = -\Theta_1/[2\Theta_2]$ and $\Sigma^2 = -1/[2\Theta_2]$, we get $\hat{M} = \overline{X}$ and $\hat{\Sigma}^2 = \sum_{i=1}^{n}(X_i - \overline{X})^2/n$ by Theorem 5.28.

5.1.4 Bayesian Estimation

Bayesian estimation tends to be somewhat more decision theoretic than classical estimation. If g is a function on the parameter space, $\aleph$ is the closure of $g(\Omega)$, and the loss function $L(\theta, a)$ increases as a moves away from $g(\theta)$, then one could reasonably be said to be estimating $g(\Theta)$. In Example 3.8 on page 146, we saw that if Θ is one-dimensional and if $L(\theta, a) = (\theta - a)^2$, then the formal Bayes rule is to use the posterior mean of Θ (so long as the posterior variance is finite).

Example 5.32 (Continuation of Example 5.7; see page 298). Suppose that P_θ says that X has $Poi(\theta)$ distribution, and $g(\Theta) = \exp(-3\Theta)$. If the prior distribution for Θ is $\Gamma(a, b)$, then the posterior after learning $X = x$ is $\Gamma(a+x, b+1)$. The posterior mean of $\exp(-3\Theta)$ is

$$\int_0^\infty \exp(-3\theta) \frac{(b+1)^{a+x}}{\Gamma(a+x)} \theta^{a+x-1} \exp(-(b+1)\theta) d\theta = \left(\frac{b+1}{b+4}\right)^{a+x}.$$

Another popular loss function is $L(\theta, a) = |\theta - a|$. The formal Bayes rule in this case is a special case of the following result.

Theorem 5.33. *Suppose that Θ has finite posterior mean. For the loss*

$$L(\theta, a) = \begin{cases} c(a - \theta) & \text{if } a \geq \theta, \\ (1 - c)(\theta - a) & \text{if } a < \theta, \end{cases} \quad (5.34)$$

a formal Bayes rule is any $1 - c$ quantile of the posterior distribution of Θ.

PROOF. Suppose that a' is chosen to be a $1 - c$ quantile of the posterior distribution of Θ. Then

$$\Pr(\Theta \leq a' | X = x) \geq 1 - c, \quad \Pr(\Theta \geq a' | X = x) \geq c.$$

If $a > a'$, then

$$L(\theta, a) - L(\theta, a') = \begin{cases} c(a - a') & \text{if } a' \geq \theta, \\ c(a - a') - (\theta - a') & \text{if } a \geq \theta > a', \\ (1 - c)(a' - a) & \text{if } \theta > a \end{cases}$$

$$= c(a - a') + \begin{cases} 0 & \text{if } a' \geq \theta, \\ a' - \theta & \text{if } a \geq \theta > a', \\ (a' - a) & \text{if } \theta > a. \end{cases}$$

It follows that the difference in the posterior risks is

$$\begin{aligned} r(a|x) - r(a'|x) &= c(a - a') + \int_{(a', a]} (a' - \theta) f_{\Theta|X}(\theta|x) d\lambda(\theta) \\ &\quad + (a' - a) \Pr(\Theta > a | X = x) \\ &\geq c(a - a') + (a' - a) \Pr(\Theta > a' | X = x) \\ &= (a - a')[c - \Pr(\Theta > a' | X = x)]. \end{aligned}$$

Since $\Pr(\Theta > a'|X = x) \leq c$, it follows that $r(a|x) \geq r(a'|x)$. Similarly, if $a < a'$, then

$$L(\theta, a) - L(\theta, a') = c(a - a') + \begin{cases} 0 & \text{if } a \geq \theta, \\ \theta - a & \text{if } a' \geq \theta > a, \\ (a' - a) & \text{if } \theta > a'. \end{cases}$$

It follows that $r(a|x) - r(a'|x) \geq (a' - a)[\Pr(\Theta > a|X = x) - c]$. Since $\Pr(\Theta > a|X = x) \geq \Pr(\Theta \geq a'|X = x) \geq c$, it follows that $r(a|x) \geq r(a'|x)$, so a' provides the minimum posterior risk. □

Notice that Theorem 5.33 remains true if the loss in (5.34) is replaced by

$$L(\theta, a) = \begin{cases} c(a - \theta) & \text{if } a > \theta, \\ (1 - c)(\theta - a) & \text{if } a \leq \theta, \end{cases}$$

even if Θ has a discrete distribution. The reason is that the loss is 0 when $\theta = a$ for both loss functions. As a corollary, we can let $c = 1/2$ in (5.34), and we get that the median is the formal Bayes rule for absolute error loss.

5.1.5 Robust Estimation*

A pragmatic approach to statistical inference will often allow for the possibility that probability distributions used in modeling are not to be taken too seriously. For example, when we model data as conditionally IID with $N(\mu, \sigma^2)$ distribution given $\Theta = (\mu, \sigma)$, we might not be saying that this description is a precise specification of our beliefs, but rather it is an approximation that we hope will be sufficient for most purposes. Occasionally, the approximation is not sufficient. For example, if there is some small chance that one observation will be generated by a process much different from the others, we might wish to use a model that makes this belief explicit. Alternatively, we might want to use a procedure for estimating a parameter that is not sensitive to the occasional observation that comes from the different process. This latter is the approach that leads to *robust estimation*. Of course, robust estimation is not concerned solely with occasional aberrant observation, but it is also concerned with general misspecification of distributions. The approach to robust estimation outlined here originated with Huber (1964).

Suppose that we will be estimating some functional T of the distribution P of the data. For example, if P is a distribution with finite mean, then $T(P) = \int x dP(x)$ is the mean expressed as a function of P. Similarly, the median of a one-dimensional distribution P with continuous, strictly increasing CDF F is $T(P) = F^{-1}(1/2)$. The *influence function* of a functional T is a means of assessing the sensitivity of the functional to small changes in the distribution P.

---

*This section may be skipped without interrupting the flow of ideas.

5.1. Point Estimation 311

Definition 5.35. Let $\mathcal{P}_0$ be a collection of distributions on a Borel space $(\mathcal{X}, \mathcal{B})$ and let $T : \mathcal{P}_0 \to \mathbb{R}^k$ be a functional. For each $x \in \mathcal{X}$, $P \in \mathcal{P}_0$, $t \in [0, 1)$, and $B \in \mathcal{B}$, we define $P_{x,t}(B) = (1-t)P(B) + tI_B(x)$. The *influence function of T at P* is the following function of x:

$$IF(x; T, P) = \lim_{t \downarrow 0} \frac{T(P_{x,t}) - T(P)}{t},$$

for those x such that the limit exists.

In particular, the influence function of T at P gives, for each x, the rate of change in T when P is contaminated by an infinitesimal mass at x. In other words, it is the right-hand derivative of $T(P_{x,t})$ with respect to t at $t = 0$.

Example 5.36. If T is the mean functional, then $T(P_{x,t}) = [1-t]T(P) + tx$. It follows that $IF(x; T, P) = x - T(P)$ for all x and P with finite mean $T(P)$. Clearly, if P is contaminated by some mass at $x = T(P)$, the mean will not change. Otherwise, the mean changes proportionally to how far x is from $T(P)$. In a finite sample setting, suppose that we obtain n observations $x_1, \ldots, x_n$ and we contemplate one further observation x_{n+1}. We can think of the empirical CDF of the first n observations as a probability measure P and then with $t = 1/(n+1)$, $P_{x_{n+1},t}$ will be the empirical CDF of all $n+1$ observations. In this case, the difference between the sample averages $T(P_{x_{n+1},t}) - T(P)$ is exactly $[x_{n+1} - T(P)]/(n+1)$.

Example 5.37. For the median functional, where F is the CDF of P,

$$T(P_{x,t}) = \begin{cases} F^{-1}\left(\frac{\frac{1}{2}-t}{1-t}\right) & \text{if } x < F^{-1}\left(\frac{\frac{1}{2}-t}{1-t}\right), \\ x & \text{if } F^{-1}\left(\frac{\frac{1}{2}-t}{1-t}\right) \leq x \leq F^{-1}\left(\frac{1}{2[1-t]}\right), \\ F^{-1}\left(\frac{1}{2[1-t]}\right) & \text{if } x > F^{-1}\left(\frac{1}{2[1-t]}\right). \end{cases}$$

If we let $\mathcal{P}_0$ be the class of distributions that have CDF with derivative at the median, then the influence function exists at all of $\mathbb{R}$ except the median of P. For each P (with derivative of the CDF being f) and all x not equal to the median,

$$IF(x; t, P) = \frac{1}{2f(F^{-1}(\frac{1}{2}))} \times \begin{cases} 1 & \text{if } x > F^{-1}(\frac{1}{2}), \\ -1 & \text{if } x < F^{-1}(\frac{1}{2}). \end{cases}$$

For all x less than the median, the effect of a contamination at x is essentially the same. This is similar, for all x greater than the median. In a finite sample setting, suppose that we obtain n observations $x_1, \ldots, x_n$ and we contemplate one further observation x_{n+1}. We can think of the empirical CDF of the first n observations as a probability measure P and then with $t = 1/(n+1)$, $P_{x_{n+1},t}$ will be the empirical CDF of all $n+1$ observations. In this case, the difference between the sample medians $T(P_{x_{n+1},t}) - T(P)$ has slightly different forms depending on whether n is odd or even. For the case of n odd, $T(P) = x_{([n+1]/2)}$, the $[n+1]/2$ order statistic. If $x_{n+1} > x_{([n+1]/2+1)}$, then $T(P_{x_{n+1},t}) = [x_{([n+1]/2)} + x_{([n+1]/2+1)}]/2$, which is independent of the value of x_{n+1} (so long as it is larger than $x_{([n+1]/2+1)}$). A similar expression holds for $x_{n+1} < x_{([n+1]/2-1)}$. So $T(P_{x_{n+1},t}) - T(P)$ is approximately the difference between $x_{([n+1]/2)}$ and the next observation either above or below it. If the distribution is continuous, then this difference is approximately one over two times the density at the median times $1/n$, $1/[2nf(F^{-1}(1/2))]$, which is approximately $tIF(x_{n+1}; t, P)$.

One way to summarize the influence function is by the *gross error sensitivity*. This is defined as $\gamma^*(T, P) = \sup_x |IF(x; T, P)|$. If $\gamma^*(T, P)$ is infinite, then there is no bound on how much T can change when P is contaminated by even a small amount of mass at an arbitrary point.

Example 5.38. For the mean functional T, $\gamma^*(T, P)$ is the largest absolute deviation possible from the mean. For distributions with unbounded support, $\gamma^* = \infty$. For the median, on the other hand, $\gamma^*(T, P) = [2f(F^{-1}(1/2))]^{-1}$, which is finite. For this reason, we say that the median is more robust with respect to gross errors than the mean.

There is a way to derive estimators with specified influence functions. Let $\mathcal{P}_0$ be a class of distributions on $(\mathcal{X}, \mathcal{B})$, and let $T : \mathcal{P}_0 \to \mathbb{R}^k$ be a k-dimensional functional of interest. Let $\psi : \mathcal{X} \times \mathbb{R}^k \to \mathbb{R}^k$ be a vector-valued function. Assume that $\psi(x, \theta)$ is differentiable with respect to θ at $\theta = T(P)$ a.s. $[P]$ for each $P \in \mathcal{P}_0$. Suppose that the mean of $\psi(X, T(P))$ is $\mathbf{0}$ for all distributions $P \in \mathcal{P}_0$. That is,

$$\int \psi(y, T(P)) dP(y) = \mathbf{0}, \tag{5.39}$$

for all $P \in \mathcal{P}_0$. For each $x \in \mathcal{X}$, $P \in \mathcal{P}_0$, $t \in [0, 1)$, and $B \in \mathcal{B}$, we define $P_{x,t}(B) = (1-t)P(B) + tI_B(x)$, and we assume that the mean of $\psi(X, T(P_{x,t}))$ is $\mathbf{0}$ also. That is,

$$(1-t)\int \psi(y, T(P_{x,t})) dP(y) + t\psi(x, T(P_{x,t})) = \mathbf{0}. \tag{5.40}$$

Subtracting (5.40) from (5.39) gives

$$\int [\psi(y, T(P)) - \psi(y, T(P_{x,t}))] \, dP(y) \tag{5.41}$$
$$= t \left[\psi(x, T(P_{x,t})) - \int \psi(y, T(P_{x,t})) dP(y)\right].$$

Suppose that we can differentiate $\int \psi(y, T(P_{x,t})) dP(y)$ with respect to t by differentiating under the integral sign. Dividing both sides of (5.41) by t and taking the limit as $t \to 0$ gives the derivative at $t = 0$. Since ψ is continuous in its second argument, and since $T(P_{x,t})$ is continuous at $t = 0$ for all x at which the influence function exists,

$$\lim_{t \to 0} \int \psi(y, T(P_{x,t})) dP(y) = 0,$$
$$\lim_{t \to 0} \psi(x, T(P_{x,t})) = \psi(x, T(P))$$

if the influence function of T exists at x. So, the limit as $t \to 0$ of $1/t$ times the right-hand side of (5.41) is $\psi(x, T(P))$. Since T is k-dimensional, the influence function will be a vector with coordinates $IF(x; T, P)_j$, for $j = 1, \ldots, k$. The limit as $t \to 0$ of $1/t$ times the ith coordinate of the left-hand side of (5.41) is

$$-\int \sum_{j=1}^{k} \frac{\partial}{\partial \theta_j} \psi_i(y, \theta)\bigg|_{\theta=T(P)} IF(x; T, P)_j dP(y). \tag{5.42}$$

Define the matrix $M = ((m_{i,j}))$, where

$$m_{i,j} = \int \frac{\partial}{\partial \theta_j} \psi_i(y, \theta)\bigg|_{\theta=T(P)} dP(y).$$

If we assume that the matrix M is finite and nonsingular, we can set (5.42) equal to $\psi_i(x, T(P))$ for each i and collect the resulting equations into a vector equation $-M[IF(x; T, P)] = \psi(x, T(P))$, so that

$$IF(x; T, P) = -M^{-1}\psi(x, T(P)).$$

For an empirical distribution P_n, $T(P_n) = T_n$, where T_n solves the equation

$$\frac{1}{n} \sum_{i=1}^{n} \psi(X_i, T_n) = \mathbf{0}. \tag{5.43}$$

Estimators that solve equations like (5.43) are called *M-estimators* because they are generalizations of maximum likelihood estimators in the following sense. If ρ is a function such that $\partial \rho(x, \theta)/\partial \theta_i = \psi_i(x, \theta)$, then to maximize $\sum_{k=1}^{n} \rho(X_k, T_n)$ it is necessary (but not sufficient in general) that (5.43) hold (if the maximum does not occur at a boundary point). One can think of $\rho(x, \theta)$ as a replacement for $\log f_{X|\Theta}(x|\theta)$. In this way, an M-estimator is a generalization of an MLE.

Example 5.44. Let $X_1, \ldots, X_n$ be conditionally IID with density $f_{X_1|\Theta}(x|\theta)$ given $\Theta = \theta$. Let $\mathcal{P}_0 = \{P_\theta : \theta \in \Omega\}$ and let $T(P) = \theta$ if P is P_θ. Suppose that $\Omega \subseteq \mathbb{R}^k$. Let $\psi_i(x, \theta) = \partial \log f_{X_1|\Theta}(x|\theta)/\partial \theta_i$, the score function. If $f_{X_1|\Theta}$ is sufficiently smooth, we have that the M-estimator corresponding to ψ is the MLE. The matrix M is the Fisher information matrix, so the influence function for the MLE, in the smooth case, is the inverse of the information matrix times the score function.

As an example, if $\Theta = (M, \Sigma)$ and $f_{X_1|\Theta}(\cdot|\mu, \sigma)$ is the $N(\mu, \sigma^2)$ distribution, then the score function is

$$\psi(x, \theta) = \frac{1}{\sigma^2}\left(\begin{array}{c} x - \mu \\ \frac{[x-\mu]^2}{\sigma} - \sigma \end{array}\right).$$

(See Example 2.83 on page 112.) The Fisher information matrix is

$$\mathcal{I}_X(\theta) = \left(\begin{array}{cc} \frac{1}{\sigma^2} & 0 \\ 0 & \frac{2}{\sigma^2} \end{array}\right).$$

So, the influence function of the MLE T is

$$IF(x;T,P) = \begin{pmatrix} x-\mu \\ \frac{[x-\mu]^2}{2\sigma} - \frac{\sigma}{2} \end{pmatrix} \quad (5.45)$$

if P is the $N(\mu,\sigma^2)$ distribution. In fact, one can verify directly that if P is any distribution with finite variance and $T'(P)$ is the standard deviation, then $IF(x;T',P)$ is the second coordinate in (5.45). (See Problem 32 on page 342.)

Example 5.46. For an example that does not meet the smoothness criteria, consider $X_1,\ldots,X_n$ conditionally IID with $U(0,\theta)$ distribution given $\Theta = \theta$. Let $\mathcal{P}_0$ be the class of distributions on $(\mathbb{R},\mathcal{B}^1)$ with bounded support, and let $T(P)$ be the supremum of the support. The MLE is the maximum of the sample, which is the supremum of the support of the empirical CDF, so the MLE is T of the empirical CDF. The influence function for T is $IF(x;T,P) = 0$ if $x \leq T(P)$ and ∞ if $x > T(p)$. (See Problem 33 on page 342.)

One famous M-estimator of a one-dimensional location parameter is based on the function $\psi(x,\theta) = h(x-\theta)$, where

$$h(t) = \begin{cases} -b & \text{if } t < -b, \\ t & \text{if } -b \leq t \leq b, \\ b & \text{if } t > b. \end{cases}$$

If $\mathcal{P}_0$ is a class of continuous distributions, it is not difficult to see that $T(P)$ will satisfy $\int \psi(y,T(P))dP(y) = 0$ if

$$T(P) = \frac{\int_v^{v+2b} x\,dP(x)}{P([v,v+2b])},$$

where v is such that $P((-\infty,v)) = P((v+2b,\infty))$. This is a *trimmed mean*, where the trimming removes all probability that is outside of an interval of fixed width with equal probability in the tails. If P is a continuous distribution, then $\psi(y,\theta)$ is differentiable at $\theta = T(P)$ with probability 1. The influence function will be

$$IF(x;T,P) = \frac{\psi(x,T(P))}{P([T(P)-b,T(P)+b])}.$$

Notice that as $b \to 0$, the influence function approaches the influence function of the median. The finite sample version is the estimator T_n which solves $\sum_{i=1}^n \psi(X_i,T_n) = 0$. It is not difficult to see that T_n is the average of $X_{(k+1)},\ldots,X_{(n-k)}$, where $X_{(1)} \leq \cdots \leq X_{(n)}$ are the order statistics and $k = \max\{m : X_{(n-m+1)} - X_{(m)} > 2b\}$. If $k+1 > n-k$, then T_n can be any number between $X_{(k)} + b$ and $X_{(n-k+1)} - b$. If $X_{(n)} - X_{(1)} \leq 2b$, then T_n is the sample average.

There are other types of trimmed means, where the trimming is not to an interval of fixed width, but rather the probabilities in the tails are fixed. For example, the $100\alpha\%$ trimmed mean of a distribution is the conditional mean given that the observation falls between the α and $1-\alpha$ quantiles of the distribution. That is, if F is the CDF corresponding to P,

$$T(P) = \frac{\int_{F^{-1}(\alpha)}^{F^{-1}(1-\alpha)} x dP(x)}{1 - 2\alpha}$$

for $\alpha < 1/2$. For $\alpha = 1/2$, the tradition is to call the median the 50% trimmed mean. The influence function of a trimmed mean at a continuous distribution is bounded and it has a shape similar to that of the trimmed mean, which we considered earlier. (See Problem 34 on page 342.)

In Section 7.3.6, we will give some results concerning large sample properties of M-estimators. More detailed discussion of robust estimators can be found in the books by Huber (1977, 1981) and Hampel et al. (1986). In Section 8.6.3, we discuss robustness considerations that are peculiar to the Bayesian perspective.

5.2 Set Estimation

A set estimator of a function g of a parameter Θ is a function from the data space $\mathcal{X}$ to a collection of subsets of the space in which $g(\Theta)$ lies.

5.2.1 Confidence Sets

In Section 4.6, we introduced the P-value as an alternative to testing a hypothesis at a fixed level. In nice problems, the P-value gave us the set of all levels at which we could accept the hypothesis. Another alternative is to fix the level and ask for the set of all hypotheses that we could accept at that level. This leads to the concept of a *confidence set*.

Definition 5.47. Let $g : \Omega \to G$ be a function, let η be the collection of all subsets of G, and let $R : \mathcal{X} \to \eta$ be a function. The function R is a *coefficient γ confidence set for $g(\Theta)$* if for every $\theta \in \Omega$,

- $\{x : g(\theta) \in R(x)\}$ is measurable, and
- $P'_\theta(g(\theta) \in R(X)) \geq \gamma$.

The confidence set R is *exact* if $P'_\theta(g(\theta) \in R(X)) = \gamma$ for all $\theta \in \Omega$. If $\inf_{\theta \in \Omega} P'_\theta(g(\theta) \in R(X)) > \gamma$, the confidence set is called *conservative*.[3]

The following result shows how confidence sets relate to nonrandomized tests in general. Its proof is left to the reader.

---

[3]Some authors require that $P'_\theta(R(X) = \emptyset) = 0$ for all θ before calling R a confidence set. Some require that $\inf_{\theta \in \Omega} P'_\theta(g(\theta) \in R(X)) = \gamma$ before saying that the coefficient is γ.

Proposition 5.48. Let $g : \Omega \to G$ be a function.

- For each $y \in G$, let ϕ_y be a level α nonrandomized test of $H : g(\Theta) = y$. Let $R(x) = \{y : \phi_y(x) = 0\}$. Then R is a coefficient $1 - \alpha$ confidence set for $g(\Theta)$. The confidence set R is exact if and only if ϕ_y is α-similar for all y.

- Let R be a coefficient $1 - \alpha$ confidence set for $g(\Theta)$. For each $y \in G$, define
$$\phi_y(x) = \begin{cases} 0 & \text{if } y \in R(x), \\ 1 & \text{otherwise}. \end{cases}$$
Then, for each y, ϕ_y has level α as a test of $H : g(\Theta) = y$. The test ϕ_y is α-similar for all y if and only if R is exact.

Example 5.49. Let $X_1, \ldots, X_n$ be conditionally IID with $N(\mu, \sigma^2)$ distribution given $(M, \Sigma) = (\mu, \sigma)$. Let $X = (X_1, \ldots, X_n)$. The usual UMPU level α test of $H : M = y$ is $\phi_y(x) = 1$ if $\sqrt{n}(\overline{x} - y)/s > T_{n-1}^{-1}(1 - \alpha/2)$, where T_{n-1} is the CDF of the $t_{n-1}(0,1)$ distribution. This translates into the confidence interval $[\overline{x} - T_{n-1}^{-1}(1 - \alpha/2)s/\sqrt{n}, \overline{x} + T_{n-1}^{-1}(1 - \alpha/2)s/\sqrt{n}]$.

Example 5.49 is typical of the most popular way to form confidence sets, namely the use of *pivotal* quantities. A pivotal is a function $h : \mathcal{X} \times \Omega \to \mathbb{R}$ whose distribution does not depend on the parameter. That is, for all c, $P_\theta(h(X, \theta) \leq c)$ is constant as a function of θ. In Example 5.49, the pivotal is $\sqrt{n}(\overline{X} - M)/S$, which has $t_{n-1}(0, 1)$ distribution given Θ. The general method of using a pivotal $h(X, \Theta)$ to form a confidence set is to set $R(x) = \{\theta : h(x, \theta) \leq F_h^{-1}(\gamma)\}$, where F_h is the CDF of $h(X, \Theta)$.

We can define randomized confidence sets if we want a correspondence to randomized tests.

Definition 5.50. Let $g : \Omega \to G$ be a function, and let $R^* : \mathcal{X} \times G \to [0, 1]$ be a function such that

- $R^*(\cdot, y) : X \to [0, 1]$ is measurable for all $y \in G$, and
- $E_\theta[R^*(X, g(\theta))] \geq \gamma$, for all $\theta \in \Omega$.

Then R^* is called a *coefficient γ randomized confidence set* for $g(\Theta)$.

The number $R^*(x, y)$ is to be thought of as the probability that y is included in the confidence set given that $X = x$ is observed.

Example 5.51. Suppose that $X \sim Bin(2, \theta)$ given $\Theta = \theta$. Let $g(\theta) = \theta$ for all θ. Define
$$R^*(x, \theta) = \begin{cases} \max\left\{0, 1 - \frac{0.05}{(1-\theta)^2}\right\} & \text{if } x = 0, \\ 1 & \text{if } x = 1 \text{ and } \theta \leq 1 - \sqrt{0.05}, \\ \max\left\{0, 1 - \frac{0.05 - (1-\theta)^2}{2\theta(1-\theta)}\right\} & \text{if } x = 1 \text{ and } \theta > 1 - \sqrt{0.05}, \\ \min\left\{1, 1 - \frac{\theta^2 - 0.95}{\theta^2}\right\} & \text{if } x = 2. \end{cases}$$

It is easy to check that
$$E_\theta[R^*(X,\theta)] = 0.95,$$
for all θ. So, if $X = 1$ is observed, the confidence set consists of the interval $[0, 1 - \sqrt{0.05}]$ together with possibly some θ values between $1 - \sqrt{0.05}$ and $\sqrt{0.95}$ chosen with decreasing probability as θ increases. For convenience, we could select a single $U \sim U(0,1)$ independent of X and, if $X = 1$ is observed, include θ in the confidence set if
$$1 - \frac{0.05 - (1-\theta)^2}{2\theta(1-\theta)} > U.$$
For example, if $U = 0.5$ is observed, the confidence set becomes $[0, 0.95]$. This example is a special case of Proposition 5.52 below.

Randomized tests correspond to randomized confidence sets in a manner similar to Proposition 5.48. The randomized confidence set in Example 5.51 was constructed according to Proposition 5.52 using the UMP level 0.05 tests of $H : \Theta = \theta$ versus $A : \Theta < \theta$.

Proposition 5.52. *Let $g : \Omega \to G$ be a function.*

- *For each $y \in G$, let ϕ_y be a level α test of $H : g(\Theta) = y$. Let $R^*(x,y) = 1 - \phi_y(x)$. Then R^* is a coefficient $1 - \alpha$ randomized confidence set for $g(\Theta)$. The randomized confidence set R^* is exact if and only if ϕ_y is α-similar for all y.*

- *Let R^* be a coefficient $1 - \alpha$ randomized confidence set for $g(\Theta)$. For each $y \in G$, define $\phi_y(x) = 1 - R^*(x,y)$. Then, for each y, ϕ_y has level α as a test of $H : g(\Theta) = y$. The test ϕ_y is α-similar for all y if and only if R^* is exact.*

The concept of UMP test corresponds to the concept of *uniformly most accurate* confidence set.

Definition 5.53. *Let $g : \Omega \to G$ be a function, and let R be a coefficient γ confidence set for $g(\Theta)$. Let η be the collection of all subsets of G, and let $B : G \to \eta$ be a function such that $y \notin B(y)$. Then R is uniformly most accurate (UMA) coefficient γ against B if for each $\theta \in \Omega$ and each $y \in B(g(\theta))$ and each coefficient γ confidence set T for $g(\Theta)$, $P'_\theta(y \in R(X)) \leq P'_\theta(y \in T(X))$.*

If R^* is a coefficient γ randomized confidence set for $g(\Theta)$, then R^* is UMA coefficient γ randomized against B if for every coefficient γ randomized confidence set T^* and every $\theta \in \Omega$ and each $y \in B(g(\theta))$,
$$E_\theta[R^*(X,y)] \leq E_\theta[T^*(X,y)].$$

The accuracy of a confidence set against B is its probability of not covering parameter values in $B(g(\theta))$ given $\Theta = \theta$. The set $B(g(\theta))$ is the set of values you wish not to have in your confidence set if $\Theta = \theta$. It is not exactly analogous to the alternative in hypothesis testing, but it is related, as we will see in Theorem 5.54.

One can consider a confidence set R as a randomized confidence set by setting
$$R^*(x,y) = \begin{cases} 1 & \text{if } y \in R(x), \\ 0 & \text{if not.} \end{cases}$$
For this reason, we state the following result in terms of randomized confidence sets.

Theorem 5.54. *Let $g(\theta) = \theta$ for all θ and let $B : \Omega \to \eta$ be as in Definition 5.53. Define*
$$B^{-1}(\theta) = \{\theta' : \theta \in B(\theta')\}.$$
Suppose that $B^{-1}(\theta)$ is nonempty for every θ. For each $\theta \in \Omega$, let ϕ_θ be a test. Define $R^*(x,\theta) = 1 - \phi_\theta(x)$. Then ϕ_θ is UMP level α for testing $H : \Theta = \theta$ versus $A : \Theta \in B^{-1}(\theta)$ for all θ if and only if R is UMA coefficient $1 - \alpha$ randomized against B.

PROOF. For the "only if" part, suppose that for each θ, ϕ_θ is UMP level α for testing $H_\theta : \Theta = \theta$ versus $A_\theta : \Theta \in B^{-1}(\theta)$. Let T^* be another coefficient $1 - \alpha$ randomized confidence set. Let $\theta \in \Omega$ and $\theta' \in B(\theta)$. All that remains is to show that $E_\theta[R^*(X,\theta')] \le E_\theta[T^*(X,\theta')]$. First, note that $\theta \in B^{-1}(\theta')$. Now, define a test $\psi(x) = 1 - T^*(x,\theta')$. This test ψ has level α as a test of $H_{\theta'}$, according to Proposition 5.48. Since $\phi_{\theta'}$ is UMP as a test of $H_{\theta'}$ against the alternative $A_{\theta'} : \Theta \in B^{-1}(\theta')$, and $\theta \in B^{-1}(\theta')$, it follows that $\beta_\psi(\theta) \le \beta_{\phi_{\theta'}}(\theta)$. We can rewrite this as
$$\beta_\psi(\theta) = E_\theta[\psi(X)] = 1 - E_\theta[T^*(X,\theta')] \le \beta_{\phi_{\theta'}}(\theta) = E_\theta[\phi_{\theta'}(X)]$$
$$= 1 - E_\theta[R^*(X,\theta')],$$
which establishes the result.

For the "if" part, suppose that R^* is a UMA coefficient $1 - \alpha$ randomized confidence set against B. For each $\theta \in \Omega$, let ψ_θ be a level α test of $H_\theta : \Theta = \theta$ and define $T^*(x,\theta) = 1 - \psi_\theta(x)$. Then Proposition 5.52 shows that T^* is a coefficient $1 - \alpha$ randomized confidence set. Let
$$\Omega' = \{(\theta',\theta) : \theta' \in \Omega, \theta \in B(\theta')\}$$
$$= \{(\theta,\theta') : \theta \in \Omega, \theta' \in B^{-1}(\theta)\},$$
where the second equality follows since $B^{-1}(\theta)$ is nonempty for all $\theta \in \Omega$. For each $(\theta,\theta') \in \Omega'$, we know that $E_{\theta'}[R^*(X,\theta)] \le E_{\theta'}[T^*(X,\theta)]$. This is the same as $\beta_{\phi_\theta}(\theta') \ge \beta_{\psi_\theta}(\theta')$, for all $\theta \in \Omega$ and all $\theta' \in B^{-1}(\theta)$. This last claim means that ϕ_θ is UMP level α for testing H_θ versus $A_\theta : \Theta \in B^{-1}(\theta)$. □

Example 5.55. Suppose that $X_1, \ldots, X_n$ are IID $N(\mu, 1)$ given $M = \mu$. Due to the continuity of this distribution, we will not need to consider randomized tests and confidence sets. Let
$$R(x) = \left(-\infty, \bar{x} + \frac{1}{\sqrt{n}}\Phi^{-1}(1-\alpha)\right].$$

We note that

$$P'_\mu(\mu \in R(X)) = P'_\mu\left(\mu \leq \overline{X} + \frac{1}{\sqrt{n}}\Phi^{-1}(1-\alpha)\right) = 1-\alpha,$$

so that R is an exact coefficient $1-\alpha$ confidence set. Consider the test $\phi_\mu(x) = 1$ if $\overline{x} < \mu - \Phi^{-1}(1-\alpha)/\sqrt{n}$. Then $R(x) = \{\mu : \phi_\mu(x) = 0\}$, and ϕ_μ is the UMP level α test of $H : M = \mu$ versus $A : M < \mu$. So $B^{-1}(\mu) = (-\infty, \mu)$, $B(\mu) = (\mu, \infty)$, and R is UMA coefficient $1 - \alpha$ against B. That is, if $\mu < \mu'$, then R has a smaller chance of covering μ' than does any other coefficient $1 - \alpha$ confidence set, conditional on $M = \mu$.

The following proposition (whose proof is left to the reader) illustrates why we do not need to introduce a dual concept to UMA confidence sets corresponding to UMC tests.

Proposition 5.56. *For each $\theta \in \Omega$, let ϕ_θ be a floor γ test of $H : \Theta \in \Omega_\theta$ versus $A : \Theta = \theta$, where $\theta \notin \Omega_\theta$. Let $R^*(x, \theta) = 1 - \phi_\theta(x)$. If ϕ_θ is UMC floor γ for testing $H : \Theta \in B^{-1}(\theta)$ versus $A : \Theta = \theta$ for all $\theta \in \Omega$, then R^* is UMA coefficient γ randomized against B.*

In other words, UMA confidence sets correspond to both UMP and UMC tests, just in different ways.[4]

The following example is due to Pratt (1961) and it illustrates an inadequacy in the theory of confidence intervals as described above.

Example 5.57. Suppose that $X_1, \ldots, X_n$ are IID $U(\theta - 1/2, \theta + 1/2)$ given $\Theta = \theta$. Minimal sufficient statistics are $T_1 = \min X_i$ and $T_2 = \max X_i$.

$$f_{T_1,T_2|\Theta}(t_1, t_2|\theta) = n(n-1)(t_2-t_1)^{n-2}, \text{ for } \theta - \tfrac{1}{2} \leq t_1 \leq t_2 \leq \theta + \tfrac{1}{2}.$$

Suppose that $B(\theta) = (-\infty, \theta)$ and that we want the UMA coefficient $1 - \alpha$ confidence set against B. If, for each θ_0, we find the UMP level α test of $\Omega_H = (-\infty, \theta_0]$ versus $\Omega_A = (\theta_0, \infty) = B^{-1}(\theta_0)$, we can use these tests to construct the UMA coefficient $1 - \alpha$ confidence set. This is not an exponential family and it does not have MLR, since the sufficient statistic is two-dimensional. To find the UMP level α test (and hence the UMA coefficient $1 - \alpha$ confidence set), we use the Neyman–Pearson lemma. First, let $\theta_1 > \theta_0$. For $k < 1$ and $\theta_1 < \theta_0 + 1$,

$$f_{T_1,T_2|\Theta}(t_1, t_2|\theta_1) > k f_{T_1,T_2|\Theta}(t_1, t_2|\theta_0) \quad (5.58)$$

if $t_1 > \theta_1 - 1/2$ or if $t_2 > \theta_0 + 1/2$ (see Figure 5.59). If $k = 1$, then (5.58) changes to equality on the shaded set in Figure 5.59. If $\theta_1 \geq \theta_0 + 1$, then (5.58) holds for $t_1 \geq \theta_1 - 1/2$ for every $k \geq 0$. To make the test have size α and to make it be the same for all $\theta_1 > \theta_0$, we must set $\phi = 1$ in the upper corner (shaded region in Figure 5.59), filling in a large enough area to have probability α. This would be

$$\phi(t_1, t_2) = \begin{cases} 1 & \text{if } t_2 > \theta_0 + \tfrac{1}{2} \text{ or } t_1 > \theta_0 + \tfrac{1}{2} - \alpha^{\frac{1}{n}}, \\ 0 & \text{if } t_2 \leq \theta_0 + \tfrac{1}{2} \text{ and } t_1 \leq \theta_0 + \tfrac{1}{2} - \alpha^{\frac{1}{n}}. \end{cases}$$

---

[4] Note that the hypothesis and alternative need to be switched in order for the same confidence set to correspond to both a UMP and a UMC test.

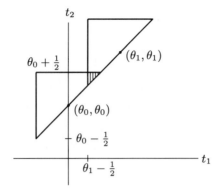

FIGURE 5.59. Construction of UMP Test in Uniform Example

(To see that this is MP for each $\theta_1 > \theta_0$, note that we choose $k = 1$ in (5.58) if $\theta_1 - 1/2 < \theta_0 + 1/2 - \alpha^{1/n}$, and we choose $k = 0$ if $\theta_1 - 1/2 > \theta_0 + 1/2 - \alpha^{1/n}$.)

The UMA coefficient $1 - \alpha$ confidence set against B is $[T_*, \infty)$, where $T_* = \max\{T_1 - 1/2 + \alpha^{1/n}, T_2 - 1/2\}$. This means that $P'_\theta(\theta' \geq T_*)$ is minimized for all $\theta' < \theta$ among all coefficient $1-\alpha$ confidence sets. Note, however, that $\Theta \geq T_2 - 1/2$ for sure, and $T_* \leq T_2 - 1/2$ whenever $T_1 - 1/2 + \alpha^{1/n} \leq T_2 - 1/2$, that is, when $T_2 - T_1 \geq \alpha^{1/n}$. So, we are 100% confident that $\Theta \geq T_*$ whenever $T_2 - T_1 \geq \alpha^{1/n}$, rather than $100(1 - \alpha)\%$ confident. (The probability that $T_2 - T_1 \geq \alpha^{1/n}$ is $1 - [n/\alpha^{1/n} - n + 1]$. If $\alpha = 0.05$ and $n = 10$, then $P'_\theta(T_2 - T_1 > \alpha^{1/n}) = 0.77$ for all θ. So the understatement of confidence will occur with probability 0.77 no matter what Θ is.)

But there is more. Suppose that we switch to $B'(\theta) = (\theta, \infty)$, so that the corresponding hypothesis and alternative are $H': \Theta \geq \theta_0$ versus $A': \Theta < \theta_0$. The UMA coefficient $1-\alpha$ confidence set against B' is $(-\infty, T^*]$, where $T^* = \min\{T_1 + 1/2, T_2 + 1/2 - \alpha^{1/n}\}$. (See Problem 31 on page 289.) If $T_2 - T_1 < 2\alpha^{1/n} - 1$, then the two intervals do not even overlap. For example, if $\alpha = 0.05$ and $n = 10$, the probability of this occurrence is 0.008. As an example, if $T_1 = 1$ and $T_2 = 1.3$, then $T_* = 1.24$ and $T^* = 1.06$. So we are 95% confident that $\Theta \leq 1.06$ and we are 95% confident that $\Theta \geq 1.24$. That makes us 190% confident, and we haven't even covered all the possible values of Θ. The most straightforward way out of these dilemmas in the classical framework is to condition on the ancillary $T_2 - T_1$. This will produce more sensible results, which will also be more closely in line with a Bayesian analysis. The confidence set produced will not be UMA, however. Another alternative is to use the UMA confidence interval but assert a confidence coefficient $\alpha(t_2 - t_1)$ that depends on the observed value of the ancillary.[5]

---

[5] A similar approach was suggested by Barnard (1976) in response to the claim by Welch (1939) that conditional confidence intervals are inefficient. Welch was pointing out (as we noted above) that conditional intervals, with the same confidence coefficient for every value of the ancillary, are less efficient (e.g., not UMA, or longer on average) than intervals based on the marginal distribution of the data. What Barnard showed was that one can fix the marginal confidence coefficient to be $1 - \alpha$ and then choose the conditional confidence coefficient in a way to optimize whatever criterion one desires.

Example 5.57 is a situation in which the distributions satisfy a condition called *invariance*, which will be discussed in Chapter 6. In Section 6.3.2, we will prove a theorem that says that in such cases, the posterior probability that a parameter lies in a confidence set is equal to the conditional confidence coefficient given the ancillary.

For two-sided or multiparameter confidence sets, we need to extend the concept of unbiasedness.

Definition 5.60. Let $g : \Omega \to G$ be a function, and let R be a coefficient γ confidence set for $g(\Theta)$. Let η be the collection of all subsets of G, and let $B : G \to \eta$ be a function such that $y \notin B(y)$. We say that R is *unbiased against B* if, for each $\theta \in \Omega$, $P'_\theta(y \in R(X)) \leq \gamma$ for all $y \in B(g(\theta))$. We say R is a *uniformly most accurate unbiased (UMAU) coefficient γ confidence set for $g(\Theta)$ against B* if it is UMA against B among unbiased coefficient γ confidence sets.

Proposition 5.61. *For each $\theta \in \Omega$, let $B(\theta)$ be a subset of Ω such that $\theta \notin B(\theta)$, and let ϕ_θ be a level α test of $H : \Theta = \theta$ versus $A : \Theta \in B^{-1}(\theta)$ such that ϕ_θ is nonrandomized. Let $R(x) = \{\theta : \phi_\theta(x) = 0\}$. Then R is a UMAU coefficient $1 - \alpha$ confidence set against B if ϕ_θ is UMPU level α for testing H versus A.*

The following example shows that the phenomenon of Example 5.57 can occur in exponential families with UMA unbiased confidence sets. This example is due to Fieller (1954).

Example 5.62. Let $\Theta = (B_0, B_1, \Sigma)$, and suppose that $Y_1, \ldots, Y_n$ are conditionally independent given $\Theta = (\beta_0, \beta_1, \sigma)$ with $Y_i \sim N(\beta_0 + \beta_1 x_i, \sigma^2)$, where $x_1, \ldots, x_n$ are known numbers. Sufficient statistics are

$$\overline{Y} = \frac{1}{n}\sum_{i=1}^n Y_i, \quad \hat{B}_1 = \frac{\sum_{i=1}^n (Y_i - \overline{Y})(x_i - \overline{x})}{\sum_{i=1}^n (x_i - \overline{x})^2},$$

$$W = \sum_{i=1}^n (Y_i - \overline{Y} - \hat{B}_1(x_i - \overline{x}))^2.$$

If we let $S_{xx} = \sum_{i=1}^n (x_i - \overline{x})^2$, then the joint density of the sufficient statistics is

$$f_{\overline{Y}, \hat{B}_1, W | \Theta}(\overline{y}, \hat{\beta}_1, w) = \left(\frac{1}{2\pi\sigma^2}\right)^{\frac{n}{2}} \frac{\sqrt{nS_{xx}}}{\Gamma(\frac{n}{2} - 1)} w^{\frac{n}{2} - 2}$$
$$\times \exp\left(-\frac{1}{2\sigma^2}\left\{n(\overline{y} - \beta_0 - \beta_1\overline{x})^2 + S_{xx}(\hat{\beta}_1 - \beta_1)^2 + w\right\}\right).$$

Suppose that we want a confidence interval for the value of x that makes $B_0 + B_1 x = 0$. Let $H = -B_0/B_1$ be that value. We will test $H : H = x_0$ by testing $(B_0 + B_1 x_0)/\Sigma^2 = 0$. The natural parameters and corresponding sufficient statistics are

322   Chapter 5. Estimation

Natural Parameter			Sufficient Statistic		
Θ_1	=	$\frac{B_0+B_1\bar{x}}{\Sigma^2}$	T_1	=	$n\bar{Y}$
Θ_2	=	$\frac{B_1}{\Sigma^2}$	T_2	=	$S_{xx}\hat{B}_1$
Θ_3	=	$-\frac{1}{2\Sigma^2}$	T_3	=	$W + S_{xx}\hat{B}_1^2 + n\bar{Y}^2$

Now, set $\Psi_2 = \Theta_2$, $\Psi_3 = \Theta_3$, and $\Psi_1 = (B_0 + B_1 x_0)/\Sigma^2 = \Theta_1 + \Theta_2(x_0 - \bar{x})$. The new sufficient statistics are

$$U_1 = n\bar{Y}, \quad U_2 = S_{xx}\hat{B}_1 - (x_0 - \bar{x})n\bar{Y}, \quad \text{and} \quad U_3 = T_3.$$

The inverse of the transformation from $(\bar{Y}, \hat{B}_1, W)$ to (U_1, U_2, U_3) is

$$w = u_3 - \frac{[u_2 + (x_0 - \bar{x})u_1]^2}{S_{xx}} - \frac{u_1^2}{n}, \quad \bar{y} = \frac{u_1}{n}, \quad \hat{\beta}_1 = \frac{u_2 + (x_0 - \bar{x})u_1}{S_{xx}},$$

and the Jacobian is $1/S_{xx}$. The joint density of the U_is given $\Psi = \psi$ is

$$\left(\frac{1}{2\pi\sigma^2}\right)^{\frac{n}{2}} \frac{1}{\sqrt{nS_{xx}}\Gamma\left(\frac{n}{2}-1\right)} \left[u_3 - \frac{u_1^2}{n} - \frac{[u_2 + (x_0 - \bar{x})u_1]^2}{S_{xx}}\right]^{\frac{n}{2}-2}$$

$$\times \exp(u_1\psi_1)g(u_2, u_3, \psi)I_{(0,\infty)}(w),$$

where g is some function that will not concern us. The conditional density of U_1 given $\Psi_1 = 0$ and $(U_2, U_3) = (u_2, u_3)$ is

$$\left[u_3 - \frac{u_1^2}{n} - \frac{[u_2 + (x_0 - \bar{x})u_1]^2}{S_{xx}}\right]^{\frac{n}{2}-2} I_{(0,\infty)}(w) h(u_2, u_3),$$

where h is a function that will not concern us. If we expand out the formula for w and complete the square in this function as a function of u_1, we get

$$r(u_2, u_3) \left[1 - \left[\frac{1}{n} + \frac{(x_0 - \bar{x})^2}{S_{xx}}\right] \left(\frac{u_1 + \frac{u_2 \frac{x_0 - \bar{x}}{S_{xx}}}{\frac{1}{n} + \frac{(x_0 - \bar{x})^2}{S_{xx}}}}{\left[u_3 - \frac{\frac{u_2^2}{nS_{xx}}}{\frac{1}{n} + \frac{(x_0 - \bar{x})^2}{S_{xx}}}\right]^{\frac{1}{2}}}\right)^2\right] I_{(0,\infty)}(w).$$

So, the conditional distribution of U_1 given U_2 and U_3 is a location and scale family and

$$\frac{U_1 + \frac{U_2 \frac{x_0 - \bar{x}}{S_{xx}}}{\frac{1}{n} + \frac{(x_0-\bar{x})^2}{S_{xx}}}}{\left[U_3 - \frac{\frac{U_2^2}{nS_{xx}}}{\frac{1}{n} + \frac{(x_0-\bar{x})^2}{S_{xx}}}\right]^{\frac{1}{2}}}$$

is independent of U_2 and U_3. The numerator of this expression is

$$\frac{\hat{B}_0 + \hat{B}_1 x_0}{\frac{1}{n} + \frac{(x_0 - \bar{x})^2}{S_{xx}}}, \qquad (5.63)$$

where $\hat{B}_0 = \overline{Y} - \hat{B}_1\overline{x}$. The denominator is

$$\left[W + \frac{\hat{B}_0 + \hat{B}_1 x_0}{\frac{1}{n} + \frac{(x_0-\overline{x})^2}{S_{xx}}}\right]^{\frac{1}{2}}. \tag{5.64}$$

The usual t statistic is

$$t = \frac{\hat{B}_0 + \hat{B}_1 x_0}{\sqrt{\frac{W}{n-2}}\sqrt{\frac{1}{n} + \frac{(x_0-\overline{x})^2}{S_{xx}}}}.$$

The ratio of (5.63) to (5.64) is $\left[t/\sqrt{n-2}\right]/\sqrt{1+t^2/(n-2)}$, whose absolute value is an increasing function of $|t|$. Hence, the usual t-test of $H: B_0 + B_1 x_0 = 0$ is the UMPU level α test of $\Psi_1 = 0$. Hence a UMAU coefficient $1 - \alpha$ confidence set for H is

$$\left\{z: \frac{(\hat{B}_0 + \hat{B}_1 z)^2}{\frac{W}{n-2}\left(\frac{1}{n} + \frac{(z-\overline{x})^2}{S_{xx}}\right)} \leq F^{-1}_{1,n-2}(1-\alpha)\right\},$$

where $F_{1,n-2}$ is the CDF of the F distribution with 1 and $n-2$ degrees of freedom. We will now find all z that satisfy this inequality. Define

$$v = \frac{\sqrt{n}(z-\overline{x})}{\sqrt{S_{xx}}}, \qquad v^* = \frac{-\overline{y}\sqrt{n}}{\hat{\beta}_1\sqrt{S_{xx}}},$$

$$F = \frac{(n-2)\hat{\beta}_1^2 S_{xx}}{W}, \qquad c = F^{-1}_{1,n-2}(1-\alpha).$$

Clearly v is a one-to-one function of z and v^* is its naïve estimator. The usual F statistic for testing $B_1 = 0$ is F. We can write

$$\hat{\beta}_0 + \hat{\beta}_1 z = \overline{y} + \hat{\beta}_1(z - \overline{x}) = \sqrt{\frac{S_{xx}}{n}}\hat{\beta}_1(v - v^*),$$

$$\frac{W}{n-2}\left(\frac{1}{n} + \frac{(z-\overline{x})^2}{S_{xx}}\right) = \frac{W}{n-2}\cdot\frac{1+v^2}{n}.$$

So, the confidence set is $\{z: F(v-v^*)^2/(1+v^2) \leq c\}$. Now, $F(v-v^*)^2/(1+v^2) \leq c$ if and only if

$$v^2(F-c) - 2vFv^* + (Fv^{*2} - c) \leq 0. \tag{5.65}$$

We have three cases depending on the sign of $F - c$.

1. $F = c$. Then $\hat{B}_1$ is just barely significant at level α. In this case (5.65) holds if $v \geq (v^{*2} - 1)/2v^*$ and $v^* > 0$ or if $v \leq (v^{*2} - 1)/2v^*$ and $v^* < 0$. The confidence set will be a semi-infinite interval.

2. $F < c$. Then $\hat{B}_1$ is insignificant at level α. In this case the quadratic has a negative coefficient for v^2. The maximum occurs at $v = Fv^*/(F-c)$, and the maximum value is $c[c - F(v^{*2} + 1)]/(F-c)$. If this maximum value is positive, then the confidence set is the exterior of a bounded open interval. If the maximum value is negative (equivalently if $F \leq c/(1+v^{*2})$, which can happen), the confidence set is $(-\infty, \infty)$, which is absurd.

324   Chapter 5. Estimation

3. $F > c$. Then $\hat{B}_1$ is significantly different from 0 at level α. The minimum of the quadratic is always negative and the coefficient of v^2 is positive, so the confidence set will be a closed and bounded interval.

In this example, even though the confidence set satisfies the conditions of being UMAU, it would not be sensible to use it after observing data.

5.2.2 Prediction Sets*

One attempt to do predictive inference in the classical setting is the construction of *prediction sets*.

Definition 5.66. Let $V : S \to \mathcal{V}$ be a random quantity. Let η be the collection of all subsets of $\mathcal{V}$, and let $R : \mathcal{X} \to \eta$ be a function. This function is a *coefficient γ prediction set for V* if

- $\{(x, v) : v \in R(x)\}$ is measurable, and
- $P'_\theta(V \in R(X)) \geq \gamma$, for every $\theta \in \Omega$.

The prediction set R is *exact* if $P'_\theta(V \in R(X)) = \gamma$ for all $\theta \in \Omega$. If $\inf_{\theta \in \Omega} P'_\theta(V \in R(X)) > \gamma$, the prediction set is called *conservative*.[6]

Example 5.67. Suppose that $\{X_n\}_{n=1}^\infty$ are IID $N(\mu, \sigma^2)$ given $(M, \Sigma) = (\mu, \sigma)$. Suppose that we will observe $X = (X_1, \ldots, X_n)$ and that we are interested in $V = \sum_{i=n+1}^{n+m} X_i/m$. Let $\overline{X}_n = \sum_{i=1}^n X_i/n$ and $S_n^2 = \sum_{i=1}^n (X_i - \overline{X}_n)^2/(n-1)$. Since

$$V - \overline{X}_n \sim N\left(0, \sigma^2 \left[\frac{1}{n} + \frac{1}{m}\right]\right)$$

and is independent of $S_n^2 \sim \Gamma([n-1]/2, [n-1]/[2\sigma^2])$, it follows that

$$\frac{V - \overline{X}_n}{S_n \sqrt{\frac{1}{n} + \frac{1}{m}}} \sim t_{n-1}(0, 1).$$

Define the set function $R(X)$ to be the interval

$$\left[\overline{X}_n - T_{n-1}^{-1}\left(1 - \frac{\alpha}{2}\right) S_n \sqrt{\frac{1}{n} + \frac{1}{m}}, \overline{X}_n + T_{n-1}^{-1}\left(1 - \frac{\alpha}{2}\right) S_n \sqrt{\frac{1}{n} + \frac{1}{m}}\right].$$

It is easy to see now that $P'_\theta(V \in R(X)) = 1 - \alpha$ for all $\theta \in \Omega$, so R is an exact coefficient $1 - \alpha$ prediction set for V.

There are also one-sided prediction sets. For example,

$$R'(X) = \left[\overline{X}_n - T_{n-1}^{-1}(1 - \alpha) S_n \sqrt{\frac{1}{n} + \frac{1}{m}}, \infty\right)$$

also satisfies $P'_\theta(V \in R'(X)) = 1 - \alpha$ for all $\theta \in \Omega$.

---

*This section may be skipped without interrupting the flow of ideas.
[6]Some authors require that $P'_\theta(R(X) = \emptyset) = 0$ for all θ before calling R a prediction set. Also, some authors require that $\inf_{\theta \in \Omega} P'_\theta(V \in R(X)) = \gamma$ before calling γ the coefficient.

Since tests and confidence sets correspond in a natural way, we might expect prediction sets to correspond to predictive tests in a similar way.

Example 5.68 (Continuation of Example 5.67; see page 324). Suppose that we set up the predictive hypothesis $H : V \leq v_0$ versus $A : V > v_0$. To parallel the relationship between parametric tests and confidence intervals, we should reject H if $v_0 \notin R'(X)$. What properties does this predictive test have? If we try to generalize the type I error probability, we might try to calculate

$$P_\theta(\text{Reject } H | V = v_0) = P_\theta(\text{Reject } H)$$
$$= P_\theta\left(\sqrt{n}\frac{\overline{X}_n - g}{S_n} > T_{n-1}^{-1}(1-\alpha)\sqrt{1+\frac{n}{m}}\right),$$

where the first equality follows from the fact that V and X are conditionally independent given Θ. This probability can be calculated using the noncentral t distribution. For $\mu = v_0$, the probability is $1 - T_{n-1}\left(T_{n-1}^{-1}(1-\alpha)\sqrt{1+n/m}\right) < \alpha$. The test is easily seen to be a UMPU test (level strictly less than α) for $M \leq v_0$, but it is not clear why it should be considered a test of $V \leq v_0$.

5.2.3 Tolerance Sets*

A classical alternative to a prediction set is a *tolerance set* as introduced by Wilks (1941).[7]

Definition 5.69. Let $V : S \to \mathcal{V}$ be a random quantity. Let η be the collection of all subsets of $\mathcal{V}$, and let $R : \mathcal{X} \to \eta$ be a function. This function is a δ tolerance set with confidence coefficient γ for V if

- $\{(x, v) : v \in R(x)\}$ is measurable, and
- $P'_\theta(P'_\theta[V \in R(X)|X] \geq \delta) \geq \gamma$, for every $\theta \in \Omega$.

The number δ is called the *tolerance coefficient*. The tolerance set R is *exact* if $P'_\theta(P'_\theta[V \in R(X)|X] \geq \delta) = \gamma$ for all $\theta \in \Omega$. One might wish to require that $P'_\theta(R(X) = \emptyset) = 0$ for all θ and/or $\inf_{\theta \in \Omega} P'_\theta(P'_\theta[V \in R(X)|X] \geq \delta) = \gamma$. If this last condition fails, the tolerance set would be called *conservative*.

Rather than making a single probability statement concerning the joint distribution of the data X and the future observable V (as is done in a prediction set), a tolerance set tries to separate the probability statements about X and V. A conditional probability statement about V is made given X (the tolerance coefficient) and then a statement is made about the distribution of this conditional probability (the confidence coefficient).

Example 5.70 (Continuation of Example 5.67; see page 324). Here the data are conditionally IID $N(\mu, \sigma^2)$ given $\Theta = (\mu, \sigma)$. Suppose that we want $R(x)$ to have

---

*This section may be skipped without interrupting the flow of ideas.
[7]For more detail on tolerance sets, see Aitchison and Dunsmore (1975).

the same form as the prediction set, namely $R_d(X) = [\overline{X}_n - dS_n, \overline{X}_n + dS_n]$, where d is yet to be determined. Now,

$$P'_\theta(V \in R_d(X)|X) = \Phi\left(\sqrt{m}\frac{\overline{X}_n + dS_n - \mu}{\sigma}\right) - \Phi\left(\sqrt{m}\frac{\overline{X}_n - dS_n - \mu}{\sigma}\right). \tag{5.71}$$

Call the right-hand side of (5.71) $H_d(X, \mu, \sigma)$. It is easy to see that the distribution of $H_d(X, \mu, \sigma)$ given $\Theta = (\mu, \sigma)$ is the same as the distribution of $H_d(X, 0, 1)$ given $\Theta = (0, 1)$. Let $y_{d,\gamma}$ be the $1 - \gamma$ quantile of $H_d(X, 0, 1)$. Since the distributions of $H_d(X, 0, 1)$ are stochastically increasing in d, it is clear that $y_{d,\gamma}$ is an increasing continuous function of d. Also, $y_{0,\gamma} = 0$ and $\lim_{d \to \infty} y_{d,\gamma} = 1$. Let d be such that $y_{d,\gamma} = \delta$. With this choice of d, we have $P'_\theta(H_d(X, \mu, \sigma) \geq \delta) = \gamma$. It follows that R_d is a δ tolerance set with confidence coefficient γ for V.[8]

There are also one-sided tolerance sets. For example,

$$R'_d(X) = [\overline{X}_n - dS_n, \infty) \tag{5.72}$$

satisfies $P'_\theta(P'_\theta[V \in R'_d(X)|X] \geq \delta) = \gamma$ for all $\theta \in \Omega$ if d is chosen so that δ is the $1 - \gamma$ quantile of $1 - \Phi(\sqrt{m}[\overline{X}_n - dS_n])$ given $\Theta = (0, 1)$.

One might think that tolerance sets could be used to construct predictive tests in the classical framework where prediction sets failed. In the sense of (3.15), they can.

Example 5.73 (Continuation of Example 5.67; see page 324). Consider the hypothesis $H : V \leq v_0$ and the tolerance set R'_d in (5.72). (Recall that V is the average of m future observations.) Here d is chosen so that δ is the $1 - \gamma$ quantile of the distribution of $1 - \Phi(\sqrt{m}[\overline{X}_n - dS_n])$ given $\Theta = (0, 1)$. We could reject H if $v_0 \notin R'_d(X)$. We now calculate

$$\begin{aligned}
P_\theta(\text{Reject H}) &= P'_\theta(v_0 < \overline{X}_n - dS_n) = P'_\theta\left(\frac{v_0 - \mu}{\sigma} < \frac{\overline{X}_n - \mu - dS_n}{\sigma}\right) \\
&= P'_{(0,1)}\left(\frac{v_0 - \mu}{\sigma} < \overline{X}_n - dS_n\right) \\
&= P'_{(0,1)}\left(1 - \Phi\left(\sqrt{m}\left[\overline{X}_n - dS_n\right]\right) < 1 - \Phi\left(\sqrt{m}\frac{v_0 - \mu}{\sigma}\right)\right).
\end{aligned}$$

This, in turn, is less than or equal to $1 - \gamma$ if and only if

$$1 - \Phi\left(\sqrt{m}\frac{v_0 - \mu}{\sigma}\right) < \delta. \tag{5.74}$$

Note that the left-hand side of (5.74) is $1 - P_\theta(V \leq v_0)$. So, we have replaced the hypothesis $H : V \leq v_0$ by the hypothesis $H' : P_\theta(V \leq v_0) > 1 - \delta$. (For $\delta = c/(1 + c)$, H' turns out to be the same as the hypothesis constructed in Example 4.13 on page 219.)

---

[8]Eberhardt, Mee, and Reeve (1989) give a program to compute the number d in examples like this.

5.2.4 Bayesian Set Estimation

In a Bayesian framework, set estimation is a type of inverse to the problem of computing posterior and/or predictive probabilities.[9] That is, rather than specifying a set and determining its posterior or predictive probability, you specify a probability and determine a set that has that probability. The problem is that there are usually many sets with the same probability. For example, suppose that the predictive distribution of V given $X = x$ is continuous with CDF $F_{V|X}(\cdot|x)$. Suppose that you want an interval T such that $\Pr(V \in T | X = x) = \gamma$. One such interval is $(-\infty, F_{V|X}^{-1}(\gamma|x)]$, and another is $[F_{V|X}^{-1}(1-\gamma|x), \infty)$, and there are many bounded intervals.

To choose between the many possible sets, it might make sense to have a loss function and then choose the set with the smallest posterior expected loss. This approach is discussed in Section 5.2.5. More commonly, one of the following approaches is taken:

- If V has a density $f_{V|X}(\cdot|x)$, determine a number t such that $T = \{v : f_{V|X}(v|x) \geq t\}$ satisfies $\Pr(V \in T | X = x) = \gamma$. This choice is called the *highest posterior density* region, or *HPD region*.

- For the case in which V is real-valued and one desires a bounded interval, choose the endpoints of the interval to be the $(1-\gamma)/2$ and $(1+\gamma)/2$ quantiles of the distribution of V.

HPD regions are sensitive to the dominating measure. That is, if the conditional distribution of V given $X = x$ is absolutely continuous with respect to two different measures, the HPD regions constructed from the two different densities might be different.

Example 5.75. Suppose that V given $X = x$ has $N(x, 1)$ distribution. This distribution is absolutely continuous with respect to Lebesgue measure with density $(2\pi)^{-1/2} \exp(-[v-x]^2/2)$. The corresponding HPD regions will be symmetric intervals around x because the density is a decreasing function of $|v - x|$. The $N(x, 1)$ distribution is also absolutely continuous with respect to the $N(0, 1)$ distribution. The density is $\exp(xv - x^2/2)$, which is an increasing function of v for $x > 0$ and is decreasing for $x < 0$. The HPD region would be a semi-infinite interval in either of these cases. If $x = 0$, then the dominating measure is the same as the distribution of V and the density is the constant 1. In this case, every set is an HPD region.

Even if there is no issue of which dominating measure to choose, HPD regions can be strange if the density of V is multimodal. In particular, they can be the union of several disconnected subsets. In such cases, one might prefer just to choose a reasonable shape for the region T and choose the particular region of that shape so that it is convenient to demonstrate that $\Pr(V \in T | X = x) = \gamma$.

---

[9] Sets with specified posterior probability are often called *credible sets* in the statistical literature.

5.2.5 Decision Theoretic Set Estimation*

Just as we could choose point estimates to minimize a loss function, we could also choose set estimates to minimize a loss function. The problem is that there are many possible loss functions and one rarely suffers a loss according to one of the tractable ones. Nevertheless, we will derive the optimal rules for some simple loss functions.

For a simple situation, suppose that we will form a semi-infinite interval of the form $(-\infty, a]$ for a one-dimensional parameter Θ with the loss function being

$$L(\theta, a) = \begin{cases} c(a - \theta) & \text{if } a \geq \theta, \\ (1-c)(\theta - a) & \text{if } a < \theta. \end{cases} \quad (5.76)$$

If $c < 1 - c$, this loss penalizes overly long intervals that contain the parameter less (per unit of length) than it penalizes short intervals that miss the parameter. If the posterior distribution of Θ has density $f_{\Theta|X}(\theta|x)$ with respect to a measure λ, and the posterior mean of Θ is finite, then Theorem 5.33 says that the optimal a is any $1 - c$ quantile of the posterior distribution of Θ.

For bounded intervals, the action space can be considered to be the set of ordered pairs (a_1, a_2) in which $a_1 \leq a_2$. Consider a loss like (5.76) that penalizes excessive length above, below, and around Θ differently:

$$L(\theta, [a_1, a_2]) = a_2 - a_1 + \begin{cases} c_1(a_1 - \theta) & \text{if } \theta < a_1, \\ 0 & \text{if } a_1 \leq \theta \leq a_2, \\ c_2(\theta - a_2) & \text{if } a_2 < \theta. \end{cases} \quad (5.77)$$

The optimal interval is the interval between two quantiles of the posterior distribution.

Theorem 5.78. *Suppose that the posterior mean of Θ is finite and the loss is as in (5.77) with $c_1, c_2 > 1$. The formal Bayes rule is the interval between the $1/c_1$ and $1 - 1/c_2$ quantiles of the posterior distribution of Θ.*

PROOF. We can rewrite the loss in (5.77) as $L_1(\theta, a_1) + L_2(\theta, a_2)$, where

$$L_1(\theta, a_1) = \begin{cases} (c_1 - 1)(a_1 - \theta) & \text{if } a_1 > \theta, \\ (\theta - a_1) & \text{if } a_1 \leq \theta, \end{cases}$$

$$L_2(\theta, a_2) = \begin{cases} (a_2 - \theta) & \text{if } a_2 \geq \theta, \\ (c_2 - 1)(\theta - a_2) & \text{if } a_2 < \theta. \end{cases}$$

Since each of these loss functions depends on only one action, the posterior means can be minimized separately. If we divide L_1 by c_1, then Theorem 5.33 says that the posterior mean of $L_1(\Theta, a_1)/c_1$ is minimized at a_1

---

*This section may be skipped without interrupting the flow of ideas.

equal to the $1/c_1$ quantile of the posterior. Similarly, if we divide L_2 by c_2, Theorem 5.33 says that the posterior mean of $L_2(\Theta, a_2)/c_2$ is minimized at the $(c_2 - 1)/c_2$ quantile of the posterior. □

In the special case in which $c_1 = c_2 = 2/\alpha > 1$, the optimal interval runs from the $\alpha/2$ quantile of the posterior to the $1 - \alpha/2$ quantile. This would be the usual equal-tailed, two-sided posterior probability interval for Θ.

There are other loss functions that do not penalize differently for how short the interval is when it misses the parameter. For example,

$$L_l(\theta, [a_1, a_2]) = a_2 - a_1 + c\left(1 - I_{[a_1,a_2]}(\theta)\right),$$
$$L_q(\theta, [a_1, a_2]) = (a_2 - a_1)^2 + c\left(1 - I_{[a_1,a_2]}(\theta)\right).$$

The existence of optimal rules in these cases actually requires weaker assumptions than those already considered, because the posterior mean of the loss is finite in all cases. That is, we need not assume that Θ has finite posterior mean to find the formal Bayes rules. If the posterior distribution of Θ has a continuous density with respect to Lebesgue measure, then calculus can be used to minimize the posterior risk. For example, the following result is easy to prove.

Proposition 5.79. *Suppose that $\Omega \subseteq \mathbb{R}$ and that the action space is the Borel σ-field τ of subsets of Ω. Suppose that the posterior distribution of Θ has a density $f_{\Theta|X}$ with respect to Lebesgue measure λ and that the loss is $L(\theta, B) = \lambda(B) + c(1 - I_B(\theta))$. Then, the formal Bayes rule is an HPD region of the form $B(x) = \{\theta : f_{\Theta|X}(\theta|x) \geq 1/c\}$.*

For the case in which the density $f_{\Theta|X}$ is *strongly unimodal* (that is, $\{\theta : f_{\Theta|X}(\theta|x) > a\}$ is an interval for all a), the formal Bayes rule for loss L_l is an HPD region. In the strongly unimodal case, one can also find the formal Bayes rule for loss L_q. (See Problem 40 on page 343.)

5.3 The Bootstrap*

5.3.1 The General Concept

There are many situations in which it is very difficult to work out analytically some feature of the distribution of some statistic in which we are interested. The idea of the *bootstrap*[10] is to suppose that a CDF $\hat{F}_n$ calculated from an observed sample $X_1, \ldots, X_n$ is sufficiently like the unknown CDF F so that one can use a calculation performed using $\hat{F}_n$ as an estimate of the calculation that we would like to perform using F. Two types of $\hat{F}_n$ are commonly used. For the *nonparametric bootstrap*, $\hat{F}_n$ is the empirical

---

*This section may be skipped without interrupting the flow of ideas.
[10] For a good overview of bootstrap methodology, see Young (1994).

CDF of the data. For the *parametric bootstrap*, one assumes a parametric model (with each X_i having CDF $F_{X_1|\Theta}(\cdot|\theta)$) and $\hat{F}_n$ is $F_{X_1|\Theta}(\cdot|\hat{\Theta}_n)$ for some estimate $\hat{\Theta}_n$ of Θ. To be more precise, we follow Efron (1979, 1982). Let $X = (X_1, \ldots, X_n)$ and let $\mathcal{F}$ be a space of CDFs in which we suppose that F lies. Let $R : \mathcal{X} \times \mathcal{F} \to \mathbb{R}$ be some function of interest. For example, R might be the difference between the sample median of $X_1, \ldots, X_n$ and the median of F:

$$R(X, F) = \frac{1}{2}\left[X_{(\lfloor \frac{n+1}{2} \rfloor)} + X_{(\lfloor \frac{n+2}{2} \rfloor)}\right] - F^{-1}\left(\frac{1}{2}\right),$$

where $F^{-1}(q)$ is understood to mean $\inf\{x : F(x) \geq q\}$, and $X_{(k)}$ is the kth order statistic. The bootstrap replaces $R(X, F)$ by $R(X^*, \hat{F}_n)$, where X^* is an IID sample of size n from $\hat{F}_n$. If we are interested in the conditional mean of $R(X, F)$ given $\mathbf{P} = P$ where P has CDF F, we try to calculate the mean of $R(X^*, \hat{F}_n)$. The success or failure of the bootstrap will depend upon the extent to which $\hat{F}_n$ is "like" F for the purposes of calculating the distribution of R.

Example 5.80. Suppose that $X_1, \ldots, X_n$ are conditionally IID $U(0, \theta)$ given $\Theta = \theta$. Here, F is the $U(0, \theta)$ CDF. First, take $R(X, F) = \sum_{i=1}^n X_i/n - \int x dF(x)$. The mean of this quantity is 0. In fact, the mean of $R(X, F)$ is 0 if F is any distribution with finite mean and X is an IID sample from F. In particular, the mean of $R(X^*, \hat{F}_n)$ given $\hat{F}_n$ is 0. The parametric and nonparametric bootstraps do just fine here.

Next, take $R(X, F) = n(F^{-1}(1) - X_{(n)})/F^{-1}(1)$, where $X_{(n)}$ is the largest coordinate of X. The distribution of $X_{(n)}/F^{-1}(1)$ is $Beta(n, 1)$ with CDF t^n for $0 \leq t \leq 1$. So, the CDF of $R(X, F)$ is $1 - (1 - t/n)^n$. For large n, this is approximately $1 - \exp(-t)$. For example, if $t = 0.1$, then $P_\theta(R(X, F) \geq 0.1) \approx \exp(-0.1) = 0.905$. On the other hand, for the nonparametric bootstrap, $R(X^*, \hat{F}_n) = n(X_{(n)} - X^*_{(n)})/X_{(n)}$ and

$$P_\theta(R(X^*, \hat{F}_n) = 0|\hat{F}_n) = 1 - \left(1 - \frac{1}{n}\right)^n,$$

which is approximately $1 - \exp(-1) = 0.632$ for large n. The nonparametric bootstrap will perform poorly here.[11]

Why is the bootstrap good for the first half of Example 5.80 but not for the second half? In the first half, we are only interested in the mean of $R(X, F)$, which is 0 no matter what F is. In the second half, virtually everything about the distribution of $R(X, F)$ depends very much on F. Even the mean of $R(X, F)$ is not the same for all F. For example, if $F(x)$ has a density that drops to 0 at $F^{-1}(1)$ like a power of $F^{-1}(1) - x$, then

---

[11] The second part of this example was given by Bickel and Freedman (1981). See Problem 41 on page 343 to see how the parametric bootstrap performs in this example.

$R(X, F)$ goes to ∞ with n. (See Theorem 7.32.) At the other extreme, if F has positive mass at $F^{-1}(1)$ (as does the empirical CDF), then $R(X, F)$ has positive probability of equaling 0.[12] The success or failure of the bootstrap in individual problems depends on the degree to which $\hat{F}_n$ approximates F for the specific purpose of calculating $R(X, F)$. For some Rs, $\hat{F}_n$ may be a wonderful approximation while for other Rs (even with the same data) $\hat{F}_n$ is a miserable approximation. One needs to be careful, when using the bootstrap, not to assume automatically that it will be suitable for one's specific purpose without doing some checking first.

In the nonparametric bootstrap, the distribution of $R(X^*, \hat{F}_n)$ will usually be a combinatorial nightmare, and Efron (1979) suggests using simulations to approximate features of its distribution. For example, if one is interested in the probability that $|R(X, F)| \leq 2$, one can generate many samples from $\hat{F}_n$, $X^{*,1}, \ldots, X^{*,m}$ and calculate the proportion of times that $|R(X^{*,j}, \hat{F}_n)| \leq 2$. Similarly, if one is interested in the mean of $R(X, F)$, one can generate many $X^{*,j}$ and calculate the average of $R(X^{*,j}, \hat{F}_n)$. For the parametric bootstrap, the distribution of $R(X^*, \hat{F}_n)$ is generally of the same form as the distribution of $R(X, F)$. Once again, simulation may be useful when this distribution is intractable. [13]

Example 5.81. Suppose that we are interested in the mean of $R(X, F) = n(Y_{1/2} - F^{-1}(1/2))^2$, where $Y_{1/2}$ is the median of a sample of size n from a distribution with CDF F. In Section 7.2 we will work out the theory for the asymptotic distributions of sample quantiles conditional on the CDF $\mathbf{P} = P$ when P has CDF F. But what if one does not know F? According to the assumptions of the bootstrap procedure, if $\hat{F}_n$ is sufficiently like F, we could sample n observations X^* from the distribution $\hat{F}_n$ and calculate the sample median $Y_{1/2}^*$. We could then subtract the original observed sample median (equal to $\hat{F}_n^{-1}(1/2)$) from $Y_{1/2}^*$, square the result, and multiply by n. We could then repeat this many times and calculate the average of the squared values. In the case of the nonparametric bootstrap, if n is only moderate in size (a few hundred or less), exact calculation of the bootstrap distribution of $R(X^*, \hat{F}_n)$ is possible using simple combinatorial arguments. In particular, if $x_{(k)}$ denotes the kth order statistic of the original sample, then for odd n,

$$\Pr(Y_{1/2}^* = x_{(k)}) = \frac{1}{n^n} \sum_{\ell=0}^{(n-1)/2} \sum_{q=(n+1)/2-\ell}^{n} \binom{n}{\ell, q, n-\ell-q} (k-1)^\ell (n-k)^{n-\ell-q}, \tag{5.82}$$

---

[12]Bickel and Freedman (1981) claim that even if one were to smooth the bootstrap by sampling from a continuous approximation to the empirical CDF, there would still be a problem in the second half of Example 5.80. Singh (1981) and Bickel and Freedman (1981) prove some large sample properties of the nonparametric bootstrap as it pertains to estimating central (not extreme) quantiles of a distribution.

[13]See Section B.7 for some ideas on how to simulate in general.

TABLE 5.83. Summary of Bootstrap Results for Median

Data Type	Sample Size	$ER(X,F)$	Bootstrap Average	RMS Error
Laplace	51	1.242	1.578	0.949
Laplace	101	1.162	1.423	0.637
Normal	51	1.551	1.898	1.045
Normal	101	1.561	1.702	0.924
Uniform	51	0.241	0.272	0.156
Uniform	101	0.245	0.272	0.115

for $k = 2, \ldots, n-1$. For $k = 1, n$, we have

$$\Pr(Y^*_{\frac{1}{2}} = x_{(k)}) = \frac{1}{n^n} \sum_{q=\frac{n+1}{2}}^{n} \binom{n}{q} (n-1)^{n-q}.$$

We simulated 100 data sets of size $n = 51$ and another 100 data sets of size $n = 101$ from the $Lap(0,1)$ distribution. Then we repeated the exercise with data from $N(0,1)$ and $U(0,1)$ distributions. We approximated the true mean of $R(X,F)$ for each of the cases using a simulation of 100,000 data sets (except for the uniform case in which the true value is just n times the variance of the $Beta([n+1]/2, [n+1]/2)$ distribution). The results are summarized in Table 5.83. The last column of Table 5.83 gives the square root of the average squared difference between the true value of $ER(X,F)$ (third column) and the 100 simulated averages of $R(X^*, \hat{F}_n)$. The fourth column gives the average of those 100 simulated averages. It appears that the bootstrap estimate (fourth column) is slightly high in all cases, but less so for larger sample sizes. The root mean squared error (last column) is very large compared to the true value, indicating that the bootstrap estimate of $ER(X,F)$ based on a single data set may not be particularly useful.

A Bayesian, faced with a difficult analytical problem, might also wish to resort to some form of computational procedure to replace the analysis. Rubin (1981) introduced a *Bayesian bootstrap*. This can be described as follows. First, simulate a CDF F with Dirichlet process distribution $Dir(\hat{F}_n)$ (see Section 1.6.1). Second, simulate an IID sample from the CDF F and compute the observed value of whatever function is of interest. Repeat this pair of simulations as many times as desired to obtain a sample of values for the function of interest. This procedure [as Rubin (1981) noted] suffers from a flaw it has in common with the nonparametric bootstrap. The flaw is that the only data values ever simulated are the same ones that were originally observed. One never simulates from a distribution with support larger than the observed sample. Unless the sample is incredibly large, this can make quite unrealistic the assumption that the CDF from which the bootstrap samples are drawn is like the one from which the original data were generated. Who would ever argue that the observed values were the only ones that could have been observed, unless the distribution has known finite support? It might make sense instead to use one of the tailfree process priors from Section 1.6.2 concentrated on continuous distributions.

5.3. The Bootstrap   333

Example 5.84 (Continuation of Example 5.81; see page 331).To be as non-parametric as possible, suppose that we model the data $\{X_n\}_{n=1}^{\infty}$ as IID with distribution P conditional on $\mathbf{P} = P$ and we give $\mathbf{P}$ a tailfree process prior (see Section 1.6.2). We now observe $X_1 = x_1, \ldots, X_n = x_n$. Suppose that we are interested in the mean of $R(X', \mathbf{F}) = (Y_{1/2} - \mathbf{F}^{-1}(1/2))^2$, where X' is a future sample of size n from $\mathbf{P}$, $\mathbf{F}$ is the CDF for $\mathbf{P}$, and $Y_{1/2}$ is the median of the observed sample. We could simulate a collection of distributions $P_1, \ldots, P_m$ from the posterior distribution of $\mathbf{P}$. For each P_j (having CDF F_j), we could simulate an IID sample X^{*j} of size n and find the sample median $Y_{1/2}^{*j}$. We also would need to find the median of P_j (call it $F_j^{-1}(1/2)$) and then average $(Y_{1/2}^{*j} - F_j^{-1}(1/2))^2$.

As an example, we might use the Bayesian bootstrap, which corresponds to a tailfree prior with improper prior distribution as well as to a Dirichlet process with improper prior. To simulate $\mathbf{F}$, we need only simulate a vector T with $Dir_n(1, \ldots, 1)$ distribution, and let $\mathbf{F}(x)$ be the sum of T_i for those i such that $x_{(i)} \leq x$. Then, the exact distribution of the median of a sample of size n drawn from the distribution $\mathbf{F}$ can be computed as in (5.82). For instance, for $k = 2, \ldots, n-1$ and odd n, we have

$$\Pr(Y^*_{\frac{1}{2}} = x_{(k)}) = \sum_{\ell=0}^{\frac{n-1}{2}} \sum_{q=\frac{n+1}{2}-\ell}^{n} \binom{n}{\ell, q, n-\ell-q} T^{\ell}_{<,k} T^q_k T^{n-\ell-q}_{>,k}, \qquad (5.85)$$

where $T_{<,k} = \sum_{i=1}^{k-1} T_i$ and $T_{>,k} = \sum_{i=k+1}^{n} T_i$. Using the same data sets as in Example 5.81 on page 331, we obtained the results summarized in Table 5.86. The Bayesian bootstrap seems to estimate $ER(X, F)$ to be even higher than does the bootstrap. This seems natural due to the additional variance introduced in the Bayesian bootstrap.

The particular choice of $R(X, F)$ in Example 5.84 was chosen to match the choice from Example 5.81. From a Bayesian viewpoint, however, a more interesting use of the bootstrap technology might be to try to predict the median of a future sample. In this case, we could use the bootstrap (or Bayesian bootstrap) distribution of the median of the sample X^* as a predictive distribution for the median of a future sample. In fact, the Bayesian bootstrap distribution of the median of the sample X^* is precisely the predictive distribution of the median of a future sample if $\mathbf{F}$ is modeled as a Dirichlet process with improper prior. Similarly, following the bootstrap logic, if $\hat{F}_n$ is sufficiently like $\mathbf{F}$, then the median of a sample drawn from

TABLE 5.86. Summary of Bayesian Bootstrap Results for Median

Data Type	Sample Size	$ER(X, F)$	Bootstrap Average	RMS Error
Laplace	51	1.242	1.988	1.151
Laplace	101	1.162	1.638	0.735
Normal	51	1.551	2.183	1.117
Normal	101	1.561	1.916	0.864
Uniform	51	0.241	0.308	0.153
Uniform	101	0.245	0.306	0.121

334   Chapter 5. Estimation

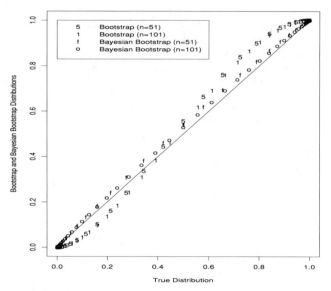

FIGURE 5.87. Bootstrap and Bayesian Bootstrap Distributions of Sample Median

$\hat{F}_n$ should have a distribution like that of the median of a sample drawn from **F**.

Example 5.88 (Continuation of Example 5.84; see page 333). We can use the bootstrap distribution of the median in (5.82) or the mean of the conditional Bayesian bootstrap distribution given T in (5.85) as predictive distributions for the medians of future samples. The mean of (5.85) is easy to compute in closed form since $(T_{<,k}, T_k, T_{>,k})$ has $Dir_3(k-1, 1, n-k)$ distribution. To see how closely these distributions approximate the distribution of the median of a future sample, we note that (for odd n) if data X comes from a distribution F, then the median has the same distribution as $F^{-1}(U)$, where $U \sim Beta([n+1]/2, [n+1]/2)$. Hence, we will only simulate data from the $U(0, 1)$ distribution in order to compare the two bootstrap distributions to the true distribution of the sample median. We will do this by calculating (for $k = 1, \ldots, n$) the probability that the median of a future sample lies below $x_{(k)}$ and comparing this to (5.82) and the mean of (5.85). Figure 5.87 shows plots of the bootstrap and Bayesian bootstrap distributions of a future median against the true distribution for sample sizes of 51 and 101. The bootstrap consistently assigns too little probability to both the upper and lower tails of the distribution. The Bayesian bootstrap, however, reproduces the true distribution (the diagonal line in Figure 5.87) remarkably well due to the additional variance it adds to the predictive distribution compared to the bootstrap.

One might be tempted to use tailfree priors or the Bayesian bootstrap in the second part of Example 5.80 to try to overcome the problems the nonparametric bootstrap had.[14] It is clear that the Bayesian bootstrap will fare no better than the nonparametric bootstrap for nearly the same

---

[14]This example is examined in detail by Schervish (1994a).

reason. Trying to use a tailfree process (like a Polya tree distribution) quickly leads to the realization that the problem as described cannot be solved satisfactorily without further modeling assumptions. For example, the typical **P** with Polya tree distribution on an interval $[0, \theta]$ will, with high posterior probability, have a density very close to zero on an interval $[a, \theta]$ if all of the observed data are less than or equal to a. The distribution of $n(\Theta - X^*_{(n)})/\Theta$ is likely to be concentrated on very large values because the probability that $X^*_{(n)} \leq a$ will be quite high. This suggests that one might wish to restrict attention to distributions whose densities stay above a certain level near θ. Alternatively, one might wish to replace $F^{-1}(1)$ with $F^{-1}(1 - \epsilon)$ for some small ϵ. There are any number of possible alternative formulations of the problem. One should give serious consideration to what one really wants to know before choosing a procedure that may not solve the problem of interest.

5.3.2 Standard Deviations and Bias

The bootstrap was originally [see Efron (1979)] designed as a tool for estimating the bias and standard error of a statistic.

Example 5.89. Suppose that conditional on $\mathbf{P} = P$, $X_1, \ldots, X_n$ are conditionally IID with CDF F, where F is the CDF of distribution P. We assume only that $\int x^2 dF(x) < \infty$. Let

$$R(X, F) = \left(\frac{1}{n}\sum_{i=1}^{n} X_i\right)^2 - \left(\int x\, dF(x)\right)^2,$$

and suppose that we are interested in the mean of R. This is the bias of the square of the sample average as an estimator of the square of the mean. If the observed sample average is $\bar{x}_n$ and we use $s_n^2 = \sum_{i=1}^{n}(x_i - \bar{x}_n)^2/n$ as an estimate of variance, then

$$R(X^*, \hat{F}_n) = \left(\frac{1}{n}\sum_{i=1}^{n} X_i^*\right)^2 - (\bar{x}_n)^2.$$

The mean of this, given the data, is s_n^2/n. The mean of $R(X, F)$ given $\mathbf{P} = P$ is σ^2/n, where $\sigma^2 = \int(x - \mu)^2 dF(x)$. Since s_n^2 is supposed to be close to σ^2 for large n, the bootstrap is thought to behave well in this case. If, instead, we had used

$$R'(X, F) = \frac{\left(\frac{1}{n}\sum_{i=1}^{n} X_i\right)^2 - \left(\int x\, dF(x)\right)^2}{\int(x - \mu)^2 dF(x)},$$

then the mean of $R'(X^*, \hat{F}_n)$ would have been $1/n$, which is exactly the mean of $R'(X, F)$.

How would one use the fact that $R'(X, F)$ has mean $1/n$ to "correct" $\overline{X}_n^2$ as an estimator of $\left(\int x\, dF(x)\right)^2$? Presumably, one would subtract s_n^2/n. Although this would usually do well, if s_n^2/n happens to be larger than $\overline{X}_n^2$, one would get a very silly result.

Similarly, if one were interested in the standard deviation of $\overline{X}^2$, one could assume that $\int x^4 dF(x) < \infty$, and apply the bootstrap to

$$R(X,F) = \left(\frac{1}{n}\sum_{i=1}^{n} X_i\right)^4 - \left(\int \left[\frac{1}{n}\sum_{i=1}^{n} x_i\right]^2 dF(x_1)\cdots dF(x_n)\right)^2. \quad (5.90)$$

The bootstrap estimate of standard deviation would be the square root of the average of the values of $R(X^*, \hat{F}_n)$. Unfortunately, the term after the minus sign in (5.90) is not easy to evaluate even with $F = \hat{F}_n$. An obvious alternative is to use the average of the values of $\left(\sum_{i=1}^{n} X_i^*/n\right)^2$.

The bootstrap can be applied to all types of statistics whose means and/or variances are difficult to calculate analytically. Efron and Tibshirani (1993) give many examples of such statistics, like correlations, regression coefficients, and nonlinear functions of such things, whose sampling distributions are nontrivial but whose bootstrap distributions are very straightforward. (See some of the problems at the end of this chapter.)

5.3.3 Bootstrap Confidence Intervals

Suppose that we desire a confidence interval for some function g of a parameter Θ. From the bootstrap perspective, it is preferable to write $g(\Theta)$ as $h(F)$. We might then desire a confidence interval of the form $(-\infty, h(\hat{F}_n) + Y]$ or $[h(\hat{F}_n) - Y_1, h(\hat{F}_n) + Y_2]$. The problem is to find Y, or Y_1 and Y_2. For the one-sided case, in order for the interval to be a coefficient γ confidence interval, it must be that $P_\theta(h(\hat{F}_n) + Y \geq h(F)) = \gamma$. Equivalently, we need $P_\theta(h(F) - h(\hat{F}_n) \leq Y) = \gamma$. There might be available a formula $\sigma^2(F)$ for the approximate variance of $h(\hat{F}_n)$. For example, if $h(F) = \int x dF(x)$, then $\sigma^2(F) = \int [x - h(F)]^2 dF(x)/\sqrt{n}$. In this case, one might replace Y by $\sigma(F)Y'$ or by $\sigma(\hat{F}_n)Y'$. Suppose that we do the latter.[15] Then we want Y' to satisfy

$$P_\theta\left(\frac{h(F) - h(\hat{F}_n)}{\sigma(\hat{F}_n)} \leq Y'\right) = \gamma. \quad (5.91)$$

This makes Y' equal to the γ quantile of $R(X, F)$, where R is the function on the left of the inequality in (5.91). What is commonly called the *percentile-t bootstrap confidence interval* for $h(F)$ would be

$$(-\infty, h(\hat{F}_n) + \sigma(\hat{F}_n)\hat{Y}'], \quad (5.92)$$

---

[15] The reason for switching from Y to $\sigma(\hat{F}_n)Y'$ is that one would suspect that Y' depends less on the underlying distribution than does Y.

where $\hat{Y}'$ is an estimate of Y' obtained by simulation. One could simulate $X_1^*, \ldots, X_b^*$ with distribution $\hat{F}_n$ and calculate, for $i = 1, \ldots, b$,

$$R(X_i^*, \hat{F}_n) = \frac{h(\hat{F}_n) - h(\tilde{F}_i)}{\sigma(\tilde{F}_i)},$$

where $\tilde{F}_i$ is the empirical CDF of the bootstrap sample X_i^*. The sample γ quantile of the $R(X_i^*, \hat{F}_n)$ values could then serve for $\hat{Y}'$ in (5.92). Hall (1992) examines bootstrap confidence intervals in detail and finds asymptotic expressions for their actual coverage probabilities.

Example 5.93. Consider the same data used in Example 1.132 on page 71. The data were a sample of size $n = 50$ from a Laplace distribution $Lap(1,1)$. We are interested in $h(F) = \int x dF(x)$, the mean of F. We simulated 10,000 bootstrap samples, $X_1^*, \ldots, X_{10000}^*$, and for each one, we calculated $R(X_i^*, \hat{F}_{50})$, where $\hat{F}_{50}$ is the empirical CDF of the 50 observations. The curve labeled "Bootstrap" in Figure 5.94 shows the 10,000 values of $\overline{X}_{50} + R(X_i^*, \hat{F}_{50})S_{50}$, where $S_{50} = \sigma(\hat{F}_{50})$, the MLE of the standard deviation of $\overline{X}_{50}$. As an example of confidence intervals, one-sided 95% lower and upper bound confidence intervals for $h(F)$ based on the bootstrap are $[0.5735, \infty)$ and $(-\infty, 1.3199]$, respectively.

Using the same prior distribution as in Example 1.132, we also used a tailfree process of Polya tree type to simulate 10,000 values of the mean of the distribution **P**. The empirical CDF of these values is plotted as the curve labeled "Tailfree" in Figure 5.94. Not surprisingly, we see that the extreme quantiles of the tailfree sample are farther from the sample mean than those of the bootstrap sample. This is due to the fact that the bootstrap procedure ignores the uncertainty from not knowing F when calculating R values. That is, we must pretend

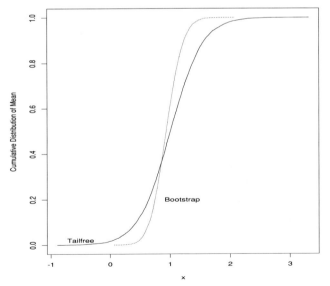

FIGURE 5.94. Distributions of Sampled Bootstrap and Tailfree Quantiles

that $F = \hat{F}_{50}$ when calculating the R values. The Bayesian solution takes into account the additional uncertainty from not knowing F. The 0.95 upper and lower bound posterior probability intervals corresponding to the confidence intervals calculated above are $[0.2574, \infty)$ and $(-\infty, 1.7628]$, respectively.

Hall (1992) shows that the percentile-t confidence interval has good frequency properties in terms of the conditional probability that the interval covers $h(F)$ given F. With a single sample, as in Example 5.93, frequency properties are neither apparent nor relevant. Suppose, however, that one were to use bootstrap confidence intervals in many applications (with many different Fs). From the classical perspective, one might actually be interested in the proportion of times that the interval covers $h(F)$ and how this compares to the nominal confidence coefficient.

Example 5.95. Suppose that we will sample data from several different $Lap(\mu, \sigma)$ distributions on several different occasions but we do not model the data this way. Rather, suppose that we use the bootstrap and a tailfree prior of Polya tree type as in Example 5.93 on page 337. As an example, 1000 data sets of size 50 each were simulated with many different Laplace distributions. The values of σ were generated as $\Gamma^{-1}(1,1)$ random variables and the locations were σ times $N(0,1)$ random variables. Location and scale changes do not affect the calculation of R in the bootstrap, but they do affect the variance of $\overline{X}$ and S. For each data set, 1000 bootstrap samples were formed and 1000 observations from the posterior mean of $\int x d\mathbf{P}(x)$ were simulated. We counted how many times μ was below each of the 1000 sample quantiles of the simulated values. Figure 5.96 shows these proportions for both the bootstrap and Polya tree samples. As expected, the bootstrap proportions match the nominal significance levels well.

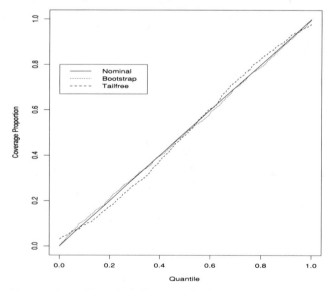

FIGURE 5.96. Empirical Coverage Probabilities for 1000 Samples

One final note is in order concerning the bootstrap. All you ever learn about by using the bootstrap, without further modeling assumptions, are properties of $\hat{F}_n$. Unless you have a way of saying how much and/or in what ways knowledge of $\hat{F}_n$ can be transformed into knowledge of **P**, the bootstrap can only tell you about $\hat{F}_n$, not about **P**.

5.4 Problems

Section 5.1.1:

1. Prove Proposition 5.8 on page 298.

2. Let X_1, X_2, X_3 be IID given $\Theta = \theta$ with $Exp(\theta)$ distribution (mean $= 1/\theta$). Find the UMVUE of $g(\Theta) = 1 - \exp(-x\Theta)$. (*Hint:* Use the Rao–Blackwell theorem 3.22.)

3. Suppose that $X_1, \ldots, X_n$ are IID given $\Theta = \theta$ with $Exp(1/\theta)$ distribution. Find the UMVUE of Θ and show that it is inadmissible with squared-error loss.

4. Suppose that $X \sim N(\theta, 1)$ given $\Theta = \theta$.

   (a) Find the UMVUE of Θ^2. What is wrong with this estimator?

   (b) Suppose that we have a decision problem with loss function $L(\theta, a) = (\theta^2 - a)^2$. Find the generalized Bayes rule with respect to Lebesgue measure. Show that this estimator is inadmissible.

5. Suppose that $X_1, \ldots, X_n$ are conditionally IID with $Ber(\theta)$ distribution given $\Theta = \theta$. Find the UMVUE of $\Theta(1 - \Theta)$.

6. Suppose that $X_1, \ldots, X_{20}$ are conditionally IID given $\Theta = \theta$ with $N(\theta, 1)$ distribution. We first collect $X_1, \ldots, X_{10}$ and compute $Y = \sum_{i=1}^{10} X_i/10$. If $Y < 5$, we set $Z = Y$, $N = 10$ and stop sampling. If $Y \geq 5$, we collect the other 10 observations and compute $Z = \sum_{i=1}^{20} X_i/20$. We then set $N = 20$. The data we report are (N, Y, Z).

   (a) Prove that Y is an unbiased estimator of Θ.

   (b) Prove that Z is biased and find its bias.

   (c) Show that N is not ancillary.

7. Let $V_i = (X_i, Y_i)$, $i = 1, \ldots, n$ be pairs of random variables, and let $\Theta = (\Lambda, M)$. Suppose that

   $$f_{V_i|\Theta}(x, y|\lambda, \mu) = \lambda\mu \exp(-\lambda x - \mu y), \text{ for } x > 0 \; y > 0.$$

   All we get to observe, for $i = 1, \ldots, n$, are

   $$Z_i = \min\{X_i, Y_i\}, \quad U_i = \begin{cases} 1 & \text{if } Z_i = X_i, \\ 0 & \text{if } Z_i = Y_i. \end{cases}$$

   (a) Prove that Z_i is conditionally independent of U_i given Θ.

(b) Find a complete sufficient statistic.

(c) Find a UMVUE of Λ.

8. Return to the situation of Problem 16 on page 140. Find the UMVUE of Θ.

9. Return to the situation of Problem 9 on page 138.

   (a) Find the conditional UMVUE of Θ given X_1.

   (b) Find an estimator with smaller unconditional variance than the conditional UMVUE.

   (c) Show that (N, M) is not a complete sufficient statistic.

10. Let $X_1, \ldots, X_n$ be conditionally IID with $N(\theta, 1)$ distribution given $\Theta = \theta$. Determine the UMVUE of $g(\Theta) = \Pr(|X_i| \leq c|\Theta)$, where $c > 0$ is fixed.

11. Suppose that $X_1, \ldots, X_n$ are conditionally IID given $\Theta = \theta$ with conditional density
$$f_{X|\Theta}(x|\theta) = \frac{\theta^\alpha \alpha}{x^{\alpha+1}} I_{[\theta, \infty)}(x),$$
where α is known and the parameter space is $\Omega = (0, \infty)$.

   (a) Find a complete sufficient statistic.

   (b) Find a UMVUE of Θ.

   (c) Prove that the UMVUE is inadmissible if the loss function is squared error.

12. Let Ω be the set of all integers, and suppose that X given $\Theta = \theta$ has discrete uniform distribution on the set $\{\theta - 1, \theta, \theta + 1\}$. Let $g : \Omega \to \mathbb{R}$ be a nonconstant function. Show that there is no UMVUE of $g(\Theta)$.

13. An alternative mode of estimation is the *method of moments*. Let $X = (X_1, \ldots, X_n)$ with the X_i being IID given Θ. Suppose that $\mu_k = E_\theta(X_i^k)$ is finite for $k = 1, \ldots, m$. Also suppose that there is a function h such that $g(\theta) = h(\mu_1, \ldots, \mu_m)$. Let $Y_k = \sum_{i=1}^n X_i^k / n$. Then $h(Y_1, \ldots, Y_m)$ is a method of moments estimator of $g(\Theta)$. Find method of moments estimators for each of the following situations:

   (a) $X_i \sim Exp(\theta)$ given $\Theta = \theta$, $g(\theta) = \theta$;

   (b) $X_i \sim N(\mu, \sigma^2)$ given $\Theta = (\mu, \sigma)$, $g(\theta) = \sigma^2$;

   (c) $X_i \sim Ber(\theta)$ given $\Theta = \theta$, $g(\theta) = \theta(1 - \theta)$.

Section 5.1.2:

14. Suppose that P_θ says that $X \sim Poi(\theta)$ and that we are trying to estimate $\exp(-3\Theta)$.

    (a) Find the Cramér–Rao lower bound for unbiased estimators.

    (b) If $\phi(X) = (-2)^X$, find $\text{Var}_\theta \phi(X)$.

    (c) Find both the Cramér–Rao lower bound and the Chapman–Robbins lower bound for the variance of unbiased estimators of Θ. Which is larger?

15. Suppose that $X_1, \ldots, X_n$ are conditionally IID $Poi(\theta)$ given $\Theta = \theta$.

    (a) Let r be a known integer. Find the UMVUE of $\exp(-\Theta)\Theta^r$.

    (b) Let $n = 1$ in part (a). Find the variance of the estimator and the Cramér–Rao lower bound.

16. Suppose that Y has $Exp(1)$ distribution and is independent of Θ. Suppose that $X_1, \ldots, X_n$ are IID conditional on $Y = y$ and $\Theta = \theta$ with $N(\theta, 1/y)$ distribution. We get to observe Y and $X_1, \ldots, X_n$.

    (a) Find the Cramér–Rao lower bound for the variance of unbiased estimators of Θ.

    (b) Show that no unbiased estimator achieves the Cramér–Rao lower bound.

    (c) Explain why the Cramér–Rao lower bound should not be taken seriously in the problem described here.

17. For the location family of t_a distributions in Example 5.17 on page 303, prove that the Bhattacharyya lower bound with $k = 2$ is the same as the Cramér–Rao lower bound.

18. Let $X \sim U(0, \theta)$ given $\Theta = \theta$.

    (a) Find the Chapman–Robbins lower bound on the variance of an unbiased estimator of Θ.

    (b) Find an unbiased estimator and find by how much its variance exceeds the lower bound.

19. In Example 5.19 on page 304, prove that $\min X_i - 1/n$ is the UMVUE.

20. Refer to the situation in Example 2.83 on page 112.

    (a) Prove that $\text{Cov}_\theta(\psi_1, \psi_2) = 0$.

    (b) Prove that $\text{Var}_\theta \phi(X) = 2\sigma^4 + 4\mu^2 \sigma^2$.

Section 5.1.3:

21. Consider the situation in Problem 6 on page 339. Find the MLE of Θ.

22. Find the maximum likelihood estimator of Θ if $X_1, \ldots, X_n$ are conditionally IID with distribution $Exp(\theta)$ given $\Theta = \theta$.

23. Suppose that $X \sim Cau(\theta, 1)$ given $\Theta = \theta$. Find the MLE of Θ.

24. Suppose that $X_1, \ldots, X_n$ are IID with Laplace distribution $Lap(\mu, \sigma)$ given $\Theta = (\mu, \sigma)$.

    (a) Prove that the value of μ that minimizes $\sum_{i=1}^n |x_i - \mu|$ is the median of the numbers $x_1, \ldots, x_n$.

    (b) Find the MLE of Θ.

25. Let X have a one-parameter exponential family distribution given Θ. Suppose that the MLE of Θ is interior to the parameter space. Prove that the MLE equals a method of moments estimator. (See Problem 13 on page 340.)

26. Consider the situation in Example 5.30 on page 308. Let squared error be the loss, that is, $L(\theta, a) = (\mu^2 - a)^2$. Show that the UMVUE dominates the MLE if $n \geq 2$. Find a formula for the difference in the risk functions. Also, find an estimator that dominates both the MLE and the UMVUE.

Section 5.1.4:

27. Consider the situation in Problem 6 on page 339. Find the likelihood function and show that a Bayesian would take the data at face value (that is, a Bayesian would calculate the same posterior as if the sample size had been fixed in advance at whatever value N turns out to be, no matter what the prior is).

28. Consider the situation in Problem 7 on page 339. If Λ and M are independent a priori with $\Lambda \sim \Gamma(a, b)$ and $M \sim \Gamma(c, d)$, find the posterior mean of Λ given the data.

29. In Example 5.10 on page 299, find the posterior mean of Θ given $X = x$ for all x assuming that the prior for Θ is $Beta(\alpha_0, \beta_0)$.

30. Return to the situation of Problem 16 on page 140. If Θ has a prior density $f_\Theta(\theta) = ac^a I_{[c,\infty)}(\theta)/\theta^{a+1}$, find the posterior mean of Θ.

31.*Suppose that, conditional on N, $\{X_i\}_{i=1}^\infty$ are independent with the first N of them having $Ber(1/3)$ distribution and the rest having $Ber(2/3)$ distribution. The prior for N is $f_N(n) = 2^{-n}$ for $n = 1, 2, \ldots$.

   (a) Find the posterior distribution of N given a finite sample $X_1, \ldots, X_n$, for known n.

   (b) If $X_1 = 0, \ldots, X_n = 0$ is the observed finite sample, find the posterior mean of N.

Section 5.1.5:

32. Let $\mathcal{P}_0$ be the set of distributions on $(\mathbb{R}, \mathcal{B}^1)$ with finite variance. Let $T(P)$ be the standard deviation of the distribution P. Show that $IF(x; T, P) = (x - \mu)^2/[2\sigma] - \sigma/2$, where μ is the mean of P.

33. Let $\mathcal{P}_0$ be the class of distributions on $(\mathbb{R}, \mathcal{B}^1)$ with bounded support, and let $T(P)$ be the supremum of the support. Prove that the influence function for T is $IF(x; T, P) = 0$ if $x \leq T(P)$ and ∞ if $x > T(p)$.

34. Find the influence function for the $100\alpha\%$ trimmed mean at a continuous distribution P.

Section 5.2:

35. Prove Proposition 5.48 on page 316.

36. Let the parameter space be Ω with σ-field of subsets τ. Let X be a random quantity taking values in a set $\mathcal{X}$, and let X have conditional density $f_{X|\Theta}$ given Θ. Let $v : \mathcal{X} \to [0, \infty)$ be a measurable function such that the set

$$L(x) = \{\theta \in \Omega : f_{X|\Theta}(x|\theta) \geq v(x)\}$$

is in τ for all x. Let Θ have a prior distribution μ_Θ, and let μ_X denote the prior predictive distribution of X. Let $C : \mathcal{X} \to \tau$ be another set function such that $\mu_\Theta(C(x)) \leq \mu_\Theta(L(x))$, a.s. $[\mu_X]$. Prove that $\Pr(\Theta \in C(x)|X = x) \leq \Pr(\Theta \in L(x)|X = x)$, a.s. $[\mu_X]$.

37. Prove Proposition 5.56 on page 319.

38. Prove Proposition 5.61 on page 321.

39. Prove Proposition 5.79 on page 329.

40. Suppose that $\Omega = \mathbb{R}$ and that the posterior density of Θ given X is strongly unimodal. Let the action space be the set of all closed and bounded intervals $[a_1, a_2]$ in $\mathbb{R}$.

    (a) Let the loss function be $L_l(\theta, [a_1, a_2]) = a_2 - a_1 + c\left(1 - I_{[a_1,a_2]}(\theta)\right)$. Prove that the formal Bayes rule is an HPD region.

    (b) Let the loss function be $L_q(\theta, [a_1, a_2]) = (a_2-a_1)^2 + c\left(1 - I_{[a_1,a_2]}(\theta)\right)$. Find the formal Bayes rule.

Section 5.3:

41. Suppose that one wished to construct a parametric bootstrap estimate in the second part of Example 5.80 on page 330.

    (a) Explain how to construct the parametric bootstrap estimate using the $U(0, \theta)$ parametric family.

    (b) Find the distribution of $R(X^*, \hat{F}_n)$ for the parametric bootstrap estimate.

    (c) Will the parametric bootstrap estimate have the same problem that the nonparametric bootstrap estimate has?

42. How would one use the nonparametric bootstrap to find the bias and standard deviation for the sample correlation coefficient from a sample of n pairs $(X_1, Y_1), \ldots, (X_n, Y_n)$?

43. Let $(x_1, Y_1), \ldots, (x_n, Y_n)$ be data pairs, and suppose that we entertain a regression model in which $E(Y_i|B_0 = \beta_0, B_1 = \beta_1) = \beta_0 + \beta_1 x_i$. The y-intercept of the regression line is $x = -\beta_0/\beta_1$. Let $(\hat{B}_0, \hat{B}_1)$ be the usual least-squares regression estimator.

    (a) How would one use the bootstrap to find the bias and standard deviation of the ratio $-\hat{B}_0/\hat{B}_1$?

    (b) Suppose that one used the following formula for the approximate variance of the ratio of two random variables Z_0/Z_1:

    $$\frac{\text{Var}(Z_0)}{Z_1^2} + \frac{Z_0^2 \text{Var}(Z_1)}{Z_1^4} - 2\frac{Z_0 \text{Cov}(Z_0, Z_1)}{Z_1^3}.$$

    Show how you would use this to find bootstrap confidence intervals for $-B_0/B_1$.

CHAPTER 6
Equivariance*

In Chapter 3, we introduced a few principles of classical decision theory (e.g., minimaxity, ancillarity) to help to choose among admissible rules. In Chapter 5, we introduced another principle called unbiasedness which could be used to select a subset of the class of all estimators. As we saw, sometimes none of the unbiased estimators was admissible. In this chapter, we introduce another ad hoc principle called *equivariance*,[1] which can also be used to select a subset of the class of all estimators. The principle of equivariance, in its most general form, relies on the algebraic theory of groups. However, the basic concept can be understood by means of a simple class of problems in which the principle can apply.

6.1 Common Examples

6.1.1 Location Problems

We will consider only parametric aspects of equivariance, since it is of interest primarily in the classical paradigm. Suppose that we have constructed a parametric family $\mathcal{P}_0$ with a parameter Θ.

Definition 6.1. First, let X and Θ be scalar random variables. If the conditional distribution of $X - \Theta$ given $\Theta = \theta$ is the same for all θ, then

---

*This chapter may be skipped without interrupting the flow of ideas.

[1]Some authors call the principle *invariance* rather than equivariance. We will see later why the term invariance is better used to mean something different, but related.

Θ is called a *location parameter* for X. If $\Theta > 0$, and the conditional distribution of X/Θ given $\Theta = \theta$ is the same for all θ, then Θ is called a *scale parameter* for X. If $\Theta = (\Theta_1, \Theta_2)$, where both Θ_i are scalar, $\Theta_2 > 0$, and the conditional distribution of $(X - \Theta_1)/\Theta_2$ given $\Theta = \theta$ is the same for all θ, then Θ is called a *location-scale parameter* for X.

Next, let X be a vector, and let Θ be a scalar. Let **1** denote the vector of the same length as X with every coordinate equal to 1. Then Θ is a *location parameter* for X if the conditional distribution of $X - \Theta \mathbf{1}$ given $\Theta = \theta$ is the same for all θ. If $\Theta > 0$ and the conditional distribution of X/Θ given $\Theta = \theta$ is the same for all θ, then Θ is a *scale parameter* for X. If $\Theta = (\Theta_1, \Theta_2)$, where both Θ_i are scalar, $\Theta_2 > 0$, and the conditional distribution of $(X - \Theta_1 \mathbf{1})/\Theta_2$ given $\Theta = \theta$ is the same for all θ, then Θ is called a *location-scale parameter* for X.

Next, let X be a vector, and let Θ be a vector of the same dimension. Then Θ is a *location parameter* for X if the conditional distribution of $X - \Theta$ given $\Theta = \theta$ is the same for all θ. If Θ is a nonsingular matrix parameter and the conditional distribution of $\Theta^{-1} X$ given $\Theta = \theta$ is the same for all θ, then Θ is a *scale parameter* for X. If $\Theta = (\Theta_1, \Theta_2)$, where Θ_1 is a vector of the same dimension as X, Θ_2 is a nonsingular matrix, and the conditional distribution of $\Theta_2^{-1}(X - \Theta_1)$ given $\Theta = \theta$ is the same for all θ, then Θ is called a *location-scale parameter*.

We will deal only with location parameters in this section. In fact, the only cases of location parameters we will consider are those in which X is a vector of exchangeable random variables that are conditionally IID given Θ, and Θ is scalar. That is, the conditional distribution of $X - \Theta \mathbf{1}$ given $\Theta = \theta$ is the same for all θ.

Theorem 6.2. *If Θ is a location parameter for X and $f_{X|\Theta}(x|\theta)$ is the Radon–Nikodym derivative of P_θ with respect to Lebesgue measure, then $f_{X|\Theta}(x|\theta) = g(x - \theta \mathbf{1})$ for some density function g.*

PROOF. The conditional joint CDF of $X - \Theta$ given $\Theta = \theta$ is

$$P'_\theta(X_i - \Theta \leq c_i, \text{ for } i = 1, \ldots, n)$$
$$= \int_{-\infty}^{c_1 + \theta} \cdots \int_{-\infty}^{c_n + \theta} f_{X|\Theta}(x|\theta) dx_n \cdots dx_1$$
$$= \int_{-\infty}^{c_1} \cdots \int_{-\infty}^{c_n} f_{X|\Theta}(y + \theta \mathbf{1}|\theta) dy_n \cdots dy_1,$$

which, for each $c_1, \ldots, c_n$, is the same for all θ if and only if $f_{X|\Theta}(y + \theta \mathbf{1}|\theta) = g(y)$ for some density g. This implies that $f_{X|\Theta}(x|\theta) = g(x - \theta \mathbf{1})$. □

Next, imagine two possible data vectors x and $y = x + c\mathbf{1}$. If $\theta + c \in \Omega$, then $f_{X|\Theta}(x|\theta) = f_{X|\Theta}(y|\theta + c)$. This says that if the data values were all shifted by the same amount, then the likelihood function would be translated by that same amount. The goal of equivariance is to take advantage of this "double shift." We make this idea more precise in a proposition.

Proposition 6.3. *If Θ is a location parameter for X, then the conditional distribution of $X+c\mathbf{1}$ given $\Theta = \theta$ is the same as the conditional distribution of X given $\Theta = \theta + c$.*

The word "equivariant" means that two (or more) things change in the same way. Proposition 6.3 says that two different changes to a problem produce the same change to a distribution. That is, the conditional distribution of X given Θ changes the same way whether we change X to $X + c\mathbf{1}$ or we change Θ to $\Theta + c$.

Definition 6.4. Consider a decision problem with parameter space $\mathbb{R}$ and action space $\mathbb{R}$. A loss function L is called *location invariant* if $L(\theta, a) = \rho(\theta - a)$, where ρ is some function. If ρ increases as its argument moves away from 0, such a decision problem is called *location estimation*.

A decision rule $\delta(x)$ is *location equivariant* if $\delta(x+c\mathbf{1}) = \delta(x)+c$, for all c and all x. A function $g(x)$ is *location invariant* if $g(x+c\mathbf{1}) = g(x)$, for all c and all x.

The word "invariant" means "does not change." Functions that satisfy $g(x+c\mathbf{1}) = g(x)$ have the property that their value does not change when their argument changes. Functions that satisfy $\delta(x+c\mathbf{1}) = \delta(x) + c$ have the property that their value changes the same way whether we change x to $x+c\mathbf{1}$ or we change $\delta(\cdot)$ to $\delta(\cdot)+c$.

Proposition 6.5. *If δ is location equivariant and Θ is a location parameter, then the conditional distribution of $\delta(X) - \Theta$ given $\Theta = \theta$ is the same for all θ.*

Note that Proposition 6.5 implies that the risk function of an equivariant estimator is constant if the loss is location invariant. (See Problem 3 on page 388.)

Lemma 6.6.[2] *Suppose that δ_0 is location equivariant. Then δ_1 is location equivariant if and only if there exists a location invariant function u such that $\delta_1 = \delta_0 + u$.*

PROOF. Clearly, if there is such a u, then $\delta_0 + u$ is equivariant. Equally clearly, if δ_1 is equivariant, then $u = \delta_1 - \delta_0$ is invariant. □

Lemma 6.7. *A function u is location invariant if and only if u is a function of x only through $(x_1 - x_n, \ldots, x_{n-1} - x_n)$.*

PROOF. The "if" part is trivial. For the "only if" part, let $c = -x_n$. Then

$$u(x + c\mathbf{1}) = u(x_1 - x_n, \ldots, x_{n-1} - x_n, 0) = u(x),$$

by invariance. Note that when $n = 1$, only constants are invariant. □

---

[2] This lemma is used in the proofs of Theorems 6.8 and 6.18.

Theorem 6.8. *Suppose that $Y = (X_1 - X_n, \ldots, X_{n-1} - X_n)$ and $L(\theta, a) = (\theta - a)^2$. Suppose that δ_0 is a location equivariant estimator with finite risk. Then, the equivariant estimator with smallest risk is $\delta_0(X) - E_0[\delta_0(X)|Y]$.*

PROOF. Let δ_0 be an arbitrary equivariant estimator with finite risk. By Lemma 6.6, all other equivariant estimators have the form $\delta_0(X) - v(Y)$. Since the risk function is constant for an equivariant δ,

$$\begin{aligned} R(\theta, \delta) &= R(0, \delta) = E_0[\delta_0(X) - v(Y)]^2 \\ &= E_0\left\{E_0[(\delta_0(X) - v(Y))^2|Y]\right\}, \end{aligned}$$

which is minimized by minimizing $E_0[(\delta_0(X) - v(Y))^2|Y = y]$ uniformly in y. This is accomplished by choosing $v(y) = E_0[\delta_0(X)|Y = y]$. □

The estimator in Theorem 6.8 is due to Pitman (1939). It is often called the *minimum risk equivariant (MRE) estimator* or *Pitman's estimator*. Throughout the rest of this section, we will use the symbol Y to stand for the vector defined in Theorem 6.8.

Example 6.9. Let $X_1, \ldots, X_n$ be IID $N(\theta, 1)$ given $\Theta = \theta$. Let

$$Y_*^\top = (Y^\top, X_n) = (Y_1, \ldots, Y_{n-1}, Y_n).$$

Let $\delta_0(X) = X_n = Y_n$. We can write

$$Y_* = \begin{pmatrix} 1 & 0 & \cdots & 0 & -1 \\ 0 & 1 & \cdots & 0 & -1 \\ & & \ddots & & \\ 0 & 0 & \cdots & 1 & -1 \\ 0 & 0 & \cdots & 0 & 1 \end{pmatrix} X.$$

Hence, given $\Theta = \theta$, $X \sim N_n(\theta \mathbf{1}, I_n)$ and

$$Y_* \sim N_n \left[\begin{pmatrix} 0 \\ \vdots \\ 0 \\ \theta \end{pmatrix}, \begin{pmatrix} 2 & 1 & 1 & -1 \\ 1 & \ddots & 1 & -1 \\ 1 & \cdots & 2 & -1 \\ -1 & \cdots & -1 & 1 \end{pmatrix} \right].$$

To get the conditional distribution of X_n given Y, we need the inverse of the upper left-hand corner of the covariance matrix of Y_*. This inverse is $I_n - J/n$, where J is a matrix of all 1s. So

$$X_n | Y = y, \Theta = 0 \sim N\left(\theta + [-1, \ldots, -1]\left(I_n - \frac{1}{n}J\right)y, c\right),$$

for some number c which we do not need, since the minimum risk equivariant estimator depends only on the mean of this distribution when $\theta = 0$. We can rewrite the mean as

$$-\sum_{i=1}^{n-1} y_i + \frac{1}{n}(n-1)\sum_{i=1}^{n-1} y_i = -\frac{1}{n}\sum_{i=1}^{n-1} y_i = -\overline{x} + x_n.$$

Pitman's estimator is then $X_n + \overline{X} - X_n = \overline{X}$, which is not surprising.

Pitman's estimator is often expressed in a different form.

Theorem 6.10. *In a location problem with one-dimensional parameter, Pitman's estimator can be written as* $\mathrm{E}(\Theta|X=x)$, *where we used a "uniform prior" for* Θ. *That is, the MRE estimator is the formal Bayes rule with respect to Lebesgue measure, if it has finite risk.*

PROOF. Suppose that $Y = (X_1 - X_n, \ldots, X_{n-1} - X_n)$ and $f_{X|\Theta}(x|\theta) = g(x - \theta\mathbf{1})$. Transform X to $(Y^\top, X_n)^\top$. The Jacobian of this transformation is 1, and we get

$$f_{Y,X_n|\Theta}(y, x_n|\theta) = g(y + (x_n - \theta)\mathbf{1}, x_n - \theta).$$

The marginal density of Y is

$$f_{Y|\Theta}(y|\theta) = \int g(y + (x_n - \theta)\mathbf{1}, x_n - \theta) dx_n = \int g(y + u\mathbf{1}, u) du.$$

This does not depend on θ because Y is ancillary (see Problem 4 on page 389). So, we can write the conditional density

$$f_{X_n|Y,\Theta}(x_n|y, 0) = \frac{g(y + x_n\mathbf{1}, x_n)}{\int g(y + u\mathbf{1}, u) du}.$$

Let $\delta_0(x) = x_n$ in Theorem 6.8. Then

$$v(y) = \mathrm{E}_0(X_n|Y = y) = \frac{\int u g(y + u\mathbf{1}, u) du}{\int g(y + u\mathbf{1}, u) du}.$$

Now change variables from u to $z = x_n - u$, so $u = x_n - z$. Then $y_i + u = x_i - z$, and

$$\delta(x) = x_n - v(y) = \frac{\int (x_n - u) g(y + u\mathbf{1}, u) du}{\int g(y + u\mathbf{1}, u) du}$$
$$= \frac{\int z g(x - z\mathbf{1}) dz}{\int g(x - z\mathbf{1}) dz} = \frac{\int \theta f_{X|\Theta}(x|\theta) d\theta}{\int f_{X|\Theta}(x|\theta) d\theta} = \mathrm{E}(\Theta|X = x),$$

if the prior for Θ is Lebesgue measure. □

The "uniform prior," or Lebesgue measure is closely related to location equivariance. The relation is that the Lebesgue measure of a set is invariant under location shifts. When we deal with more general types of equivariance, a generalization of Theorem 6.10 will emerge in that the MRE estimator will be the formal Bayes rule with respect to a "prior" distribution that is invariant in the appropriate sense.

Example 6.11 (Continuation of Example 6.9; see page 347). Let $X_1, \ldots, X_n$ be IID $N(\theta, 1)$ given $\Theta = \theta$. If $X = (X_1, \ldots, X_n)$ and $\delta_0(X) = X_n$, the posterior from a uniform prior is $\Theta \sim N(\bar{x}, 1/n)$, and the MRE is $\bar{X}$, as we saw earlier.

Example 6.12. Let X have Cauchy distribution $Cau(\theta, 1)$ given $\Theta = \theta$. The posterior from a uniform prior is $\Theta \sim Cau(x, 1)$, and there is no formal Bayes rule. In fact, it is easy to show that all equivariant estimators have infinite risk. Note that constant estimators (like $\delta(x) = c \in \Omega$ for all x) have finite risk but are not equivariant (and they have infinite posterior risk, but since the prior is improper, this is not surprising).

A maximin style theorem can be proven about equivariant estimation. Suppose that Nature knows that you will use an equivariant estimator and the loss function is squared error. Which location family should Nature choose to make your risk as large as possible? The answer is the family of normal distributions.

Theorem 6.13. *Suppose that* $L(\theta, a) = (\theta - a)^2$ *and*

$$\mathcal{F} = \left\{ f \geq 0 : \int f(x)dx = 1, \int xf(x)dx = 0, \int x^2 f(x)dx = 1 \right\}.$$

Suppose that $X_1, \ldots, X_n$ *are IID with conditional density* $f(x - \theta)$ *given* $\Theta = \theta$ *for some* $f \in \mathcal{F}$. *Define* $r_n(f)$ *to be the greatest lower bound of the risk over the set of all equivariant estimators. Then* $\sup_{f \in \mathcal{F}} r_n(f) = r_n(f_0)$, *where* f_0 *is the standard normal density.*

PROOF. If f_0 is the standard normal density, then $\overline{X}$ is MRE and $r_n(f_0) = 1/n$. Since $\overline{X}$ is always equivariant, it must be that $r_n(f) \leq 1/n$, since $1/n$ is the risk of $\overline{X}$ for all $f \in \mathcal{F}$.[3]   □

Example 6.14. Suppose that $X_1, \ldots, X_n$ are IID with conditional density $f(x - \theta)$ given $\Theta = \theta$ where $f(x) = \exp[-(x+1)]$ for $x \geq -1$. This is a family of shifted exponential distributions, rigged so that $\theta = 0$ has mean 0 also. The first-order statistic $X_{(1)} = \min X_i$ is equivariant and is a complete sufficient statistic. Since Y is ancillary, Theorem 2.48 says that $X_{(1)}$ and Y are independent given Θ. So

$$E_0(X_{(1)}|Y) = E_0(X_{(1)}) = \frac{1}{n} - 1,$$

and $X_{(1)} + (n-1)/n$ is MRE and its risk is $1/n^2$.

For more general loss functions, we have the following lemma.

Lemma 6.15. *If* ρ *is strictly convex and not monotone and* $L(\theta, a) = \rho(a - \theta)$, *then the MRE estimator of* Θ *exists if and only if there is some equivariant estimator with finite risk. The MRE is unique. The MRE is unbiased if* $\rho(t) = t^2$.

---

[3] There is a theorem of Kagan, Linnik, and Rao (1965) which shows that, for $n \geq 3$, $E_0(\overline{X}|Y) = 0$ if and only if $f = f_0$. This would imply that $r_n(f) < 1/n$ if $n \geq 3$ and $f \neq f_0$.

PROOF.[4] If no equivariant estimator has finite risk, then it makes no sense to talk about the MRE estimator. If ρ is strictly convex and δ_0 is equivariant with finite risk, let $Y = (X_1 - X_n, \ldots, X_{n-1} - X_n)$. Write

$$\phi(t; y) = E_0[\rho(\delta_0(X) - t)|Y = y].$$

We first show that ϕ is strictly convex as a function of t for fixed y:

$$\phi(\alpha t + (1-\alpha)u; y)$$
$$= E_0[\rho(\alpha \delta_0(X) - \alpha t + (1-\alpha)\delta_0(X) - (1-\alpha)u)|Y = y]$$
$$< E_0[\alpha \rho(\delta_0(X) - t) + (1-\alpha)\rho(\delta_0(X) - u)|Y = y]$$
$$= \alpha \phi(t; y) + (1-\alpha)\phi(u; y),$$

where the strict inequality holds because $\delta_0(X) - t$ cannot equal $\delta_0(X) - u$ a.s. $[P_0]$ if $t \neq u$. We can also show that ϕ is not monotone in t a.s. This follows from the fact that since ρ is strictly convex and not monotone, $\rho(x) \to \infty$ as $x \to \infty$, and as $x \to -\infty$ and $\delta_0(X) - t$ converges in probability to $-\infty$ or ∞ as $t \to \infty$ or $t \to -\infty$, respectively. Since convex functions are continuous on the interiors of their domains, $\phi(t; y)$ has a minimum at $t = v(y)$ for each y and the minimum is unique by strict convexity. It follows that $\delta(X) = \delta_0(X) - v(Y)$ is the MRE estimator, since it minimizes $E_0(\rho(\delta(X) - 0))$.

If $\rho(t) = t^2$, let $\delta(X)$ be any equivariant estimator:

$$E_\theta \delta(X) = E_0 \delta(X + \theta \mathbf{1}) = \theta + E_0 \delta(X) = \theta + c.$$

The risk of δ is its variance plus the bias squared, which is minimized by choosing $c = 0$. Hence the MRE estimator has $c = 0$ and is unbiased. □

6.1.2 Scale Problems*

It turns out that a scale problem with positive random variables and a positive parameter is identical to a location problem. All one needs to do is replace Θ by $\log \Theta$ and X_i by $\log X_i$. The general *scale problem* can be defined in a fashion similar to the general location problem.

Definition 6.16. Consider a decision problem with parameter space $\mathbb{R}^+$ and action space $\mathbb{R}^+ \cup \{0\}$. A loss function $L(\theta, a)$ is called *scale invariant* if $L(\theta, a) = \rho(a/\theta)$ for some function ρ. If ρ increases as its argument moves away from 1, such a decision problem is called *scale estimation*.

A decision rule $\delta(x)$ is *scale equivariant* if $\delta(cx) = c\delta(x)$, for all positive c and all x. A function $g(x)$ is *scale invariant* if $g(cx) = g(x)$, for all positive c and all x.

---

[4]This proof is based on the proof of Theorem 6.8 in Chapter 1 of Lehmann (1983).
*This section may be skipped without interrupting the flow of ideas.

A scale analog of Theorem 6.2 would say that $f_{X|\Theta}(x|\theta) = g(x/\theta)/\theta$. All of the other results concerning location problems have their counterparts in the case of scale problems with positive random variables. We will not restate them all here. It should be noted, however, that squared error must be changed to $(\log(a/\theta))^2$. Also, if one finds an estimator for $\log \Theta$, one should remember to exponentiate the result to produce an estimator of Θ.

Example 6.17. Suppose that $\{X_n\}_{n=1}^{\infty}$ are conditionally IID $U(0, \theta)$ given $\Theta = \theta$. Let $X = (X_1, \ldots, X_n)$. Then Θ is a scale parameter. Using the loss $L(\theta, a) = [\log(a/\theta)]^2$, we can find Pitman's estimator of $\log \Theta$ by finding the posterior mean of $\log \Theta$ based on a uniform prior for $\log \Theta$. This "prior" translates into the "prior" $1/\theta$ for Θ, since $\psi = \log \theta$ means $d\psi = d\theta/\theta$. The posterior density for Θ is then

$$f_{\Theta|X}(\theta|x) = \frac{nx_{(n)}^n}{\theta^{n+1}} I_{[x_{(n)}, \infty)}(\theta),$$

where $x_{(n)} = \max_i x_i$. (This is known as a *Pareto distribution*.) The mean of $\log \Theta$ is

$$\int_x^{\infty} \frac{nx^n \log \theta}{\theta^{n+1}} d\theta = \log x + \int_0^{\infty} nt \exp(-nt) dt = \log x + \frac{1}{n},$$

by making the transformation $t = \log(\theta/x)$. The MRE estimator of Θ becomes $\bar{X} \exp(1/n)$.

An alternative invariant loss function, which is more like squared-error loss, is $L(\theta, a) = (\theta - a)^2/\theta^2$. Here, we need not assume that the random variables are positive, because we will not have to take logarithms. An analog to Theorem 6.8 is proven next.

Theorem 6.18.[5] *Let Θ be a scale parameter and let $L(\theta, a) = (\theta - a)^2/\theta^2$ and $Y = (X_1/|X_n|, \ldots, X_n/|X_n|)$. Let δ_0 be an equivariant estimator with finite risk. Then the equivariant estimator with smallest risk is*

$$\delta_0(X) \frac{E_1[\delta_0(X)|Y]}{E_1[\delta_0^2(X)|Y]}.$$

PROOF. Let δ_0 be an arbitrary equivariant estimator with finite risk. By the scale analog to Lemma 6.6, all other equivariant estimators have the form $\delta_0(X)/v(Y)$, where v is scale invariant. Since the risk function is constant for an equivariant δ,

$$R(\theta, \delta) = R(1, \delta) = E_1[\delta_0(X)/v(Y) - 1]^2$$
$$= E_1\{E_1[(\delta_0(X)/v(Y) - 1)^2|Y]\},$$

which is minimized by minimizing $E_1[(\delta_0(X)/v(Y) - 1)^2|Y = y]$ uniformly in y. To do this, choose $v(y) = E_1[\delta_0^2(X)|Y = y]/E_1[\delta_0(X)|Y = y]$. □

It can be shown that the MRE estimator is also the formal Bayes rule with respect to the improper prior $1/\theta$.

---

[5] This theorem is used in Example 6.60.

Theorem 6.19. *In a scale problem with one-dimensional parameter, the MRE estimator can be written as the formal Bayes rule with respect to a prior having Radon–Nikodym derivative $1/\theta$ with respect to Lebesgue measure, if it has finite risk.*

PROOF. We begin with the equivariant estimator $|X_n|$. Let Y be as in Theorem 6.18. The transformation from x to $(y^\top, x_n)^\top$ has Jacobian $|x_n|^{n-1}$, so
$$f_{Y, X_n | \Theta}(y, x_n | 1) = f(|x_n|y) |x_n|^{n-1},$$
both for $y_n = +1$ and $y_n = -1$. So, the conditional density of X_n given $Y = y$ is
$$f_{X_n | Y, \Theta}(x_n | y, 1) = \frac{f(|x_n|y) |x_n|^{n-1}}{\int f(|u|y) |u|^{n-1} du}.$$

It follows that
$$v(y) = \frac{\int u^2 f(|u|y) |u|^{n-1} du}{\int |u| f(|u|y) |u|^{n-1} du} = \frac{\int_0^\infty f(uy) u^{n+1} du}{\int_0^\infty f(uy) u^n du},$$
because both integrands are symmetric around 0 as functions of u. Now make the change of variables $u = |x_n|/z$ with inverse $z = |x_n|/u$. Then $du = -|x_n| dz/z^2$. It follows that
$$v(y) = |x_n| \frac{\int f(|x_n| \frac{y}{z}) z^{-n-3} dz}{\int f(|x_n| \frac{y}{z}) z^{-n-2} dz} = |x_n| \frac{\int f(\frac{x}{z}) z^{-n-3} dz}{\int f(\frac{x}{z}) z^{-n-2} dz}.$$

Hence, the MRE estimator is $\delta(X)$, where
$$\delta(x) = \frac{|x_n|}{v(y)} = \frac{\int f(\frac{x}{z}) z^{-n-2} dz}{\int f(\frac{x}{z}) z^{-n-3} dz}.$$

To see that this is the formal Bayes rule with respect to the "prior" $1/\theta$, note that the posterior $f_{\Theta|X}(\theta|x)$ is proportional to $f(x/\theta)/\theta^{n+1}$. The expected loss is (aside from the proportionality constant)
$$\int_0^\infty (\theta - a)^2 \frac{1}{\theta^{n+3}} f\left(\frac{x}{\theta}\right) d\theta.$$

By expanding this as a function of a and taking the derivative, we find that the minimum occurs at $a = \delta(x)$. □

The reader should note that the measure $\lambda(A) = \int_A d\theta/\theta$ is invariant under scale changes. That is, the measure of a set A of positive numbers is the same as the measure of $|c|A$ for all real c.

6.2 Equivariant Decision Theory

6.2.1 Groups of Transformations

Equivariance occurs in more general situations than just location or scale problems. For example, it can occur in combined location-scale problems with a two-dimensional parameter. In fact, it can occur whenever there is a group of transformations that acts on the sample space, the parameter space, and the action space in the "same way." We will now make this notion more precise.

Definition 6.20. A *group* is a nonempty set G together with a binary operation $\circ$ called *composition* such that

- for each $g_1, g_2 \in G$, $g_1 \circ g_2 \in G$;
- there exists $e \in G$ such that $e \circ g = g$ for all $g \in G$;
- for each $g \in G$ there is $g^{-1} \in G$ such that $g^{-1} \circ g = e$;
- for each $g_1, g_2, g_3 \in G$, $g_1 \circ (g_2 \circ g_3) = (g_1 \circ g_2) \circ g_3$.

The element e is called the *identity* and for each g, g^{-1} is called the *inverse of* g. A group is *abelian* if $\circ$ is commutative, that is, if $g_1 \circ g_2 = g_2 \circ g_1$ for all g_1 and g_2.

There appears to be some asymmetry in the definition of the identity and inverses. This is illusory, however.

Lemma 6.21. *Let G be a group. For all $g \in G$, $g \circ e = g$ and $g \circ g^{-1} = e$. There is only one identity element, and for each $g \in G$, there is only one inverse of g. The inverse of g^{-1} is g.*

PROOF. Let $g \in G$. Let h be the inverse of g^{-1}. Since $g^{-1} \circ g = e$, $(g^{-1} \circ g) \circ g^{-1} = e \circ g^{-1} = g^{-1}$. It follows that

$$h \circ ((g^{-1} \circ g) \circ g^{-1}) = h \circ g^{-1} = e.$$

The left-hand side of this last equation can be rewritten using the associative property as

$$(h \circ g^{-1}) \circ (g \circ g^{-1}) = g \circ g^{-1}.$$

Hence $g \circ g^{-1} = e$. It follows that g satisfies the property required to be called the inverse of g^{-1}. Next, note that $e \circ e = e$ and $g^{-1} \circ g = e$, so

$$g \circ e = g \circ (e \circ e) = g \circ (e \circ (g^{-1} \circ g)) = g \circ ((e \circ g^{-1}) \circ g) = g \circ (g^{-1} \circ g) = g.$$

For the uniqueness claims, first, suppose that $h \circ g = e$. Then, using what we have just proved,

$$h = h \circ e = h \circ (g \circ g^{-1}) = (h \circ g) \circ g^{-1} = e \circ g^{-1} = g^{-1},$$

which means that the inverse of g is unique. It follows from what we proved above that g is the unique inverse of g^{-1}. Finally, let $h \circ g = g$ for all g. Then
$$h = h \circ (g \circ g^{-1}) = (h \circ g) \circ g^{-1} = g \circ g^{-1} = e.$$
Hence, the identity is unique. □

Sometimes two seemingly different groups are essentially the same.

Definition 6.22. Let G_1 and G_2 be groups with compositions $\circ_1$ and $\circ_2$, respectively. Let $\phi : G_1 \to G_2$ be a one-to-one onto function such that, for all $g, h \in G_1$, $\phi(g \circ_1 h) = \phi(g) \circ_2 \phi(h)$. Then ϕ is called a *group isomorphism*.

The following proposition is straightforward.

Proposition 6.23. *Let ϕ be a group isomorphism between G_1 and G_2. Then ϕ^{-1} is a group isomorphism between G_2 and G_1. Also, ϕ maps the identity in G_1 to the identity in G_2. Also, ϕ maps the inverse of each $g \in G_1$ to the inverse of $\phi(g) \in G_2$.*

The groups that will most interest us are groups of transformations. The set $\mathcal{U}$ in the following definition can be the sample space $\mathcal{X}$ or the parameter space Ω or the action space $\aleph$.

Definition 6.24. A measurable function $f : \mathcal{U} \to \mathcal{U}$ is called a *transformation of $\mathcal{U}$*. The function $e(u) = u$ for all $u \in \mathcal{U}$ is called the *identity transformation*.

Proposition 6.25. *Suppose that G is a set of transformations of a set $\mathcal{U}$ with $e \in G$ being $e(u) = u$ for all $u \in \mathcal{U}$. Let composition $\circ$ be the composition of functions. If G is a group with identity e, then every element of G is one-to-one.*

Example 6.26. Here are some examples of groups of transformations. When we refer to these examples in the future, we will call the ith one "Group i."

1. $\mathcal{U} = \mathbb{R}^n$, $g_c(x_1, \ldots, x_n) = (x_1 + c, \ldots, x_n + c)$ for each $c \in \mathbb{R}$. The identity is g_0, the inverse of g_c is g_{-c}, and the composition is $g_a \circ g_b = g_{a+b}$. This group is abelian.

2. $\mathcal{U} = \mathbb{R}^n$, $g_c(x_1, \ldots, x_n) = (cx_1, \ldots, cx_n)$ for each $c > 0$. The identity is g_1, the inverse of g_c is $g_{1/c}$, and the composition is $g_a \circ g_b = g_{ab}$. This group is abelian.

3. $\mathcal{U} = \mathbb{R}^n$, $g_{(a,b)}(x_1, \ldots, x_n) = (bx_1 + a, \ldots, bx_n + a)$ for each $b > 0$, $a \in \mathbb{R}$. The identity is $g_{(0,1)}$, the inverse of $g_{(a,b)}$ is $g_{(-a/b, 1/b)}$, and the composition is $g_{(a,b)} \circ g_{(c,d)} = g_{(bc+a, bd)}$. This group is not abelian.

4. $\mathcal{U} = \mathbb{R}^n$, $g_A(x) = Ax$, where $A \in GL(n)$, the set of nonsingular $n \times n$ matrices. The identity is g_I, the inverse of g_A is $g_{A^{-1}}$, and the composition is matrix multiplication $g_A \circ g_B = g_{AB}$. This group is not abelian. This is called the *general linear group of dimension n*.

5. $\mathcal{U} = \mathbb{R}^n$, $g_\rho(x_1, \ldots, x_n) = (x_{\rho(1)}, \ldots, x_{\rho(n)})$, where ρ is any permutation of $(1, \ldots, n)$. The identity is $g_{(1,\ldots,n)}$, the inverse of g_ρ is $g_{\rho^{-1}}$, and the composition is composition of permutations, $g_{\rho_1} \circ g_{\rho_2} = g_{\rho_1 \circ \rho_2}$. This group is not abelian.

6. $\mathcal{U}$ can be any measurable set and G can be the set of all one-to-one measurable functions whose inverses are also measurable.

If $A \subseteq \mathcal{U}$, we will use the shorthand notation gA to denote the set

$$gA = \{u \in \mathcal{U} : u = gy \text{ for some } y \in A\}.$$

Definition 6.27. Let $\mathcal{P}_0$ be a parametric family with parameter space Ω and sample space $(\mathcal{X}, \mathcal{B})$. Let G be a group of transformations of $\mathcal{X}$. We say that G *leaves $\mathcal{P}_0$ invariant* if for each $g \in G$ and each $\theta \in \Omega$ there exists $\theta^* \in \Omega$ such that $P_\theta(A) = P_{\theta^*}(gA)$ for every $A \in \mathcal{B}$.

It is easy to see that the θ^* in the Definition 6.27 is unique.

Lemma 6.28. *Suppose that G leaves $\mathcal{P}_0$ invariant. Then, for each $g \in G$ and $\theta \in \Omega$, θ^* in Definition 6.27 is unique.*

PROOF. Let $g \in G$ and $\theta \in \Omega$ be given. Suppose that both θ^* and θ' satisfy for every $A \in \mathcal{B}$,

$$P_\theta(A) = P_{\theta^*}(gA) = P_{\theta'}(gA).$$

It follows that for every $A \in \mathcal{B}$, $P_\theta(g^{-1}A) = P_{\theta^*}(A) = P_{\theta'}(A)$. Since distinct elements of the parameter space have to provide distinct probability measures, this last equation implies that $\theta^* = \theta'$. □

We will call the unique value θ^* by the name $\bar{g}\theta$ to indicate its connection to both g and θ. We can try to understand intuitively what it means to say that G leaves $\mathcal{P}_0$ invariant. Suppose that we believed that X had distribution P_θ given $\Theta = \theta$. We already know what the conditional distribution of gX is

$$P'_\theta(gX \in A) = P'_\theta(X \in g^{-1}A).$$

This has nothing to do with equivariance yet. It is a simple consequence of what we already know about the induced distribution of a function of a random quantity. What invariance of distributions means is that the second equation below holds (the others are all consequences of probability theory and group theory):

$$P'_\theta(X \in g^{-1}A) = P_\theta(g^{-1}A) = P_{\bar{g}\theta}(gg^{-1}A) = P_{\bar{g}\theta}(A) = P'_{\bar{g}\theta}(X \in A).$$

So, we see that the conditional distribution of gX given $\Theta = \theta$ is $P_{\bar{g}\theta}$, which is the conditional distribution of X given $\Theta = \bar{g}\theta$.

Proposition 6.29. *Suppose that G leaves $\mathcal{P}_0$ invariant, then, for each $g \in G$ the transformation $\bar{g} : \Omega \to \Omega$ is one-to-one and onto. Also $\overline{G} = \{\bar{g} : g \in G\}$ is a group, and $\overline{g^{-1}} = \bar{g}^{-1}$.*

356   Chapter 6. Equivariance

The proof of this proposition is straightforward and is good practice for those readers who have become rusty in group theory. See the proof of Lemma 6.31 below for some guidance.

Definition 6.30. If G leaves $\mathcal{P}_0$ invariant and $L(\theta, a)$ is a loss function on $\Omega \times \aleph$, we say that *the loss is invariant under G* if, for each $g \in G$ and each $a \in \aleph$, there exists a unique $a^* \in \aleph$ such that $L(\bar{g}\theta, a^*) = L(\theta, a)$ for all $\theta \in \Omega$. We will denote a^* by $\tilde{g}a$.

Lemma 6.31. *If the loss is invariant under G, then, for each $g \in G$, the transformation $\tilde{g} : \aleph \to \aleph$ is one-to-one and onto. Also $\tilde{G} = \{\tilde{g} : g \in G\}$ is a group, and $\widetilde{g^{-1}} = \tilde{g}^{-1}$.*

PROOF. First, we show that $\tilde{g}$ is onto for all $g \in G$. Let $a \in \aleph$ and $g \in G$. Recall (Proposition 6.29) that $\overline{g^{-1}} = \bar{g}^{-1}$. For all $\theta \in \Omega$, $L(\theta, a) = L(\bar{g}^{-1}\theta, \widetilde{g^{-1}}a)$. Since this is true for all θ and $\bar{g}^{-1}$ is one-to-one, it follows that $L(\psi, \widetilde{g^{-1}}a) = L(\bar{g}\psi, a)$, for all $\psi \in \Omega$ (just let $\psi = \bar{g}^{-1}\theta$). By the definition of $\tilde{g}$ it follows that $\tilde{g}\widetilde{g^{-1}}a = a$, hence $\tilde{g}$ is onto. Applying this same argument to g^{-1} gives that $\widetilde{g^{-1}}\tilde{g}a = a$. This also shows that $\tilde{g}$ is one-to-one and $\widetilde{g^{-1}} = \tilde{g}^{-1}$. Clearly, the identity transformation on $\aleph$ is $\tilde{e}$. The composition of $\tilde{g}$ and $\tilde{h}$ is clearly $\widetilde{gh}$, and the associative property follows directly from the associative property in $\bar{G}$. □

Example 6.32. Consider Group 3. Suppose that $X_1, \ldots, X_n$ are conditionally IID $N(\mu, \sigma^2)$ given $\Theta = (\mu, \sigma)$. Let

$$G = \{(a, b) : b > 0, a \in \mathbb{R}\}.$$

If $g = (a, b)$ and $\theta = (\mu, \sigma)$, then $\bar{g}\theta = (b\sigma, b\mu + a)$ and

$$P_{\bar{g}\theta}(gA) = \int_{bA+a} \frac{1}{(b\sigma\sqrt{2\pi})^n} \exp\left\{-\frac{1}{2b^2\sigma^2}\sum_{i=1}^n (x_i - a - b\mu)^2\right\} dx$$

$$= \int_{bA} \frac{1}{(b\sigma\sqrt{2\pi})^n} \exp\left\{-\frac{1}{2b^2\sigma^2}\sum_{i=1}^n (y_i - b\mu)^2\right\} dy$$

$$= \int_A \frac{1}{(\sigma\sqrt{2\pi})^n} \exp\left\{-\frac{1}{2\sigma^2}\sum_{i=1}^n (z_i - \mu)^2\right\} dz = P_\theta(A).$$

Suppose that the loss function is $L(\theta, d) = (d - \mu)^2/\sigma^2$. Then, if $\tilde{g}d = bd + a$, we get

$$L(\bar{g}\theta, \tilde{g}d) = \frac{(bd + a - b\mu - a)^2}{b^2\sigma^2} = L(\theta, d).$$

We might ask, "What is the set of all invariant loss functions?" That is, what is the set of all L such that, for every σ, μ, d, a, b,

$$L((\mu, \sigma), d) = L((b\mu + a, b\sigma), bd + a)?$$

Since this equation must hold for every σ, μ, d, a, b, it must hold for $b = 1/\sigma$ and $a = -\mu/\sigma$ no matter what μ and σ are. It then follows that

$$L((\mu, \sigma), d) = L\left((1,0), \frac{d - \mu}{\sigma}\right),$$

for all d, μ, and σ. That is, $L((\mu, \sigma), d) = \rho([d-\mu]/\sigma)$, for some arbitrary function ρ. It is clear that any L of this form is invariant, so we have found all invariant loss functions.

The method used at the end of Example 6.32 is actually a very general method for finding all invariant functions. The first step was to find a necessary condition for the function to be invariant. The second step is to check that the condition is also sufficient. A similar method works when trying to find all equivariant functions. (See Example 6.34 on page 357 for an illustration.)

Definition 6.33. A decision problem is *invariant under G* if $\mathcal{P}_0$ and the loss are invariant. In such a case, a nonrandomized decision rule $\delta(x)$ is *equivariant* if $\delta(gx) = \tilde{g}\delta(x)$ for all $g \in G$ and all $x \in \mathcal{X}$. A randomized rule $\delta^*(x)$ is *equivariant* if $\delta^*(gx)(\tilde{g}A) = \delta^*(x)(A)$ for all $A \in \alpha$, $x \in \mathcal{X}$, and $g \in G$. A function v is *invariant* if $v(gx) = v(x)$ for all $x \in \mathcal{X}$ and all $g \in G$.

We will rarely use randomized equivariant rules, except in discussions of invariant tests (c.f. Section 6.3.3).

Example 6.34. Consider Group 3. Let $\aleph = \mathbb{R}$ and suppose that we are only estimating the location parameter. For example, $L(\theta, d) = (\theta_1 - d)^2/\theta_2^2$. Here $\delta(gx) = \tilde{g}\delta(x)$ means that $\delta(bx_1 + a, \ldots, bx_n + a) = b\delta(x) + a$. Suppose that $a = -b\bar{x}$ and $b = 1/s$, where $s^2 = \sum_{i=1}^n (x_i - \bar{x})^2/(n-1)$. Then

$$\delta(x) = \bar{x} + s\delta\left(\frac{x_1 - \bar{x}}{s}, \ldots, \frac{x_n - \bar{x}}{s}\right).$$

Since every function of the above form is equivariant, we have found all equivariant rules.
Note that the function $v(x) = \delta([x_1 - \bar{x}]/s, \ldots, [x_n - \bar{x}]/s)$ is invariant and is an element of $\aleph$, while the function $h(x) = (\bar{x}, s)$ is equivariant and can be thought of as an element of $\tilde{G}$. With this notation, we have written $\delta(x) = h(x)v(x)$.

Example 6.34 suggests a generalization of Lemma 6.6.

Lemma 6.35. *Let $h : \mathcal{X} \to \tilde{G}$ be equivariant. Then $\delta : \mathcal{X} \to \aleph$ is equivariant if and only if $h^{-1}\delta$ is invariant. (Here h^{-1} means the element of $\tilde{G}$ that is the inverse of h, not the inverse of the function h, which might not even exist.)*

PROOF. For the "if" part, assume that $h^{-1}\delta = v$ is invariant. Then $v(x) \in \aleph$ and $\delta(x) = h(x)v(x)$. So

$$\delta(gx) = h(gx)v(gx) = \tilde{g}h(x)v(x) = \tilde{g}\delta(x),$$

and δ is equivariant.

For the "only if" part, assume that δ is equivariant. Let $v = h^{-1}\delta$. Then

$$v(gx) = h(gx)^{-1}\delta(gx) = (\tilde{g}h(x))^{-1}\tilde{g}\delta(x) = h(x)^{-1}\delta(x) = v(x),$$

so v is invariant. □

Example 6.36. Consider Group 1. Lemma 6.6 already says that δ is equivariant if and only if $-\delta_0 + \delta$ is invariant, where δ_0 is an arbitrary equivariant function.

Consider Group 3. If $\aleph = \mathbb{R}$, Example 6.34 showed how $\delta_0^{-1}\delta$ is invariant, where $\delta_0(x) = (\overline{x}, s)$. Now suppose that $\aleph = \Omega = G$ and

$$L((\mu, \sigma), d) = \frac{(d_2 - \mu)^2}{\sigma^2} + \left|\log \frac{d_1}{\sigma}\right|.$$

Then L is invariant and $\delta(x)$ is two-dimensional, say $(\delta_1(x), \delta_2(x))$. To say $\delta(gx) = \tilde{g}\delta(x)$ means that, for every a, b, x $\delta(bx_1 + a, \ldots, bx_n + a) = (b\delta_1(x) + a, b\delta_2(x))$. Now, we have already seen that $\delta_0(x) = (\overline{x}, s)$ is equivariant, so the most general equivariant estimator is $\delta(x) = \delta_0(x)v(x)$, where $v(x) \in \aleph$ is invariant. That is, let $v_2(x)$ be an arbitrary positive invariant function and let $v_1(x)$ be an arbitrary real-valued invariant function. Then $\delta(x) = (sv_1(x) + \overline{x}, sv_2(x))$ is the general form of an equivariant estimator.

Definition 6.37. An invariant function $v(x)$ is called *maximal invariant* if, for every invariant function $u(x)$, $v(x_1) = v(x_2)$ implies $u(x_1) = u(x_2)$ (i.e., u is a function of v). For each $x \in \mathcal{X}$, we call

$$O(x) = \{y : y = gx \text{ for some } g \in G\} \tag{6.38}$$

the *orbit of x*.

It is clear that an invariant function is always constant on orbits. In the statement of Theorem 6.8, Y is maximal invariant. Also, in the statement of Theorem 6.18, Y is maximal invariant. In invariant decision problems, the risk function of an equivariant decision rule is constant on orbits. This follows trivially from part 4 of the following lemma.

Lemma 6.39.[6] *In the notation of Definition 6.37,*

1. $y \in O(x)$ if and only if $x \in O(y)$;

2. orbits are equivalence classes;

3. a maximal invariant assumes distinct values on different orbits;

4. suppose that $m(x, \theta)$ is invariant under the group actions, that is, $m(gx, \overline{g}\theta) = m(x, \theta)$ for all x, g, θ. Then the distribution of $m(X, \theta)$ given $\Theta = \theta$ is a function of θ through the maximal invariant in Ω.

---

[6]This lemma is used in the proofs of Lemmas 6.65 and 6.66.

PROOF. Parts 1 and 2 are trivial.[7] For part 3, suppose that v is maximal invariant. Consider the function $O : \mathcal{X} \to 2^{\mathcal{X}}$ defined in (6.38). Clearly, O is invariant, hence it is a function of v. This means that if $O(x) \neq O(y)$, then $v(x) \neq v(y)$. That is, v assigns different values on different orbits. For part 4, let $r(\theta) = P'_\theta[m(X,\theta) \in B]$ for an arbitrary set B. Then,

$$P_{\bar{g}\theta}[m(X,\bar{g}\theta) \in B] = P_\theta[m(gX,\bar{g}\theta) \in B] = P_\theta[m(X,\theta) \in B],$$

by invariance of m. So, $r(\bar{g}\theta) = r(\theta)$, and r is invariant, hence it is a function of the maximal invariant in Ω. □

Corollary 6.40. *In an invariant decision problem, the risk function of an equivariant decision rule is constant on orbits in the parameter space.*

This corollary gives a plausible justification for restricting attention to equivariant decision rules. Since the risk function is constant on orbits in the parameter space when the loss is invariant, this makes it easier to compare equivariant rules by means of their risk functions. In particular, if the group acts transitively on the parameter space (i.e., there is only one orbit), then the problem of noncomparability of risk functions disappears altogether.

6.2.2 Equivariance and Changes of Units

One popular justification for the use of equivariant rules is that the result one obtains should not depend on the units in which the variables and parameters are measured. For example, suppose that we are estimating a length in feet and our measurements are in feet. If we were then to change our measurements (and the length) to inches, our estimate should be 12 times as large. This sounds like a requirement that the estimator be scale equivariant. Similarly, if we are estimating a temperature in °C and we convert all measurements to °K, then we should just add 273 to the °C estimate to get the °K estimate. This sounds like a requirement that the estimate be location equivariant. However, neither of these examples has anything to do with equivariance. First, we show that estimators need not be equivariant in order for the units of measurement to be converted correctly. Afterwards, we show why changes of measurement scale have nothing to do with equivariance.

Example 6.41. Suppose that Θ is tomorrow's temperature in °C, and suppose that I have a prior distribution for Θ which is $N(-12, 100)$.[8] I will observe $X \sim N(\theta, 25)$ (in °C) given $\Theta = \theta$, and the loss is $L(\theta, a) = (\theta - a)^2$. Ignoring absolute 0, this problem is a location problem. The only location equivariant rules are

---

[7] See Problem 15 on page 139 for the definition of equivalence class.
[8] This example was written during the winter.

$\delta_c(x) = x + c$. The Bayes rule (in °C) is

$$E(\Theta|X = x) = \frac{\frac{x}{25} - \frac{12}{100}}{\frac{1}{25} + \frac{1}{100}} = \frac{4x - 12}{5}.$$

This is clearly not equivariant. Does that mean that it will violate the rule of changing units? Of course not!

Suppose that we change all units to °K. The new parameter is Θ^* equal to tomorrow's temperature in °K and $\Theta^* = 273 + \Theta$, so the prior for Θ^* (in °K) is $N(261, 100)$. The datum we will observe is $X^* = X + 273 \sim N(\theta^*, 25)$ (in °K) given $\Theta^* = \theta^*$. The loss function is still assumed to be $L^*(\theta^*, a^*) = (\theta^* - a^*)^2$. Everything is now ready for finding the Bayes rule, which is the posterior mean of Θ^* (not the mean of Θ, which would clearly be to small by 273).

$$E(\Theta^*|X^* = x^*) = \frac{4x^* + 261}{5} = \frac{4x - 12}{5} + 273,$$

which is just what we would like.

Notice that, in Example 6.41, we have treated Θ, X, Θ^*, and X^* as pure numbers, but we were careful to say what the numbers stood for. A great deal of the confusion about changing units is caused by ignoring this simple, but vital, procedure. We now turn to a careful discussion of this point.

Changing units of measurement has nothing to do with equivariance. It is simply a reparameterization. When you reparameterize the problem, you must reparameterize the loss function, the prior, and the likelihood. Who would ever dream of using the same proper prior for °K as for °C? The same applies to any change of units. Surprisingly, the same applies even to more general transformations that do not correspond to changes of units.

Example 6.42. Let $X \sim U(0, \theta)$ given $\Theta = \theta$. Suppose that the loss function is $L(\theta, a) = (\theta - a)^2$ and the prior is $f_\Theta(\theta) = \theta^{-2} I_{[1,\infty)}(\theta)$. Then the posterior is

$$f_{\Theta|X}(\theta|x) = \frac{2c^2}{\theta^3} I_{[c,\infty)}(\theta),$$

where $c = \max\{1, x\}$. The Bayes rule is the posterior mean $E(\Theta|X = x) = 2c$.

Suppose that we reparameterize to $\Theta^* = \Theta^2$. If there is going to be a connection between the two decision problems, the loss function had better transform in such a way that we are essentially estimating the same thing. That is, the loss had better be $L^*(\theta^*, a^*) = \left(\sqrt{\theta^*} - \sqrt{a^*}\right)^2$. The prior for Θ^* is

$$f_{\Theta^*}(\theta^*) = \frac{1}{2\theta^{*\frac{3}{2}}} I_{[1,\infty)}(\theta^*),$$

and the conditional distribution of X given $\Theta^* = \theta^*$ is $U(0, \sqrt{\theta^*})$. As a red herring, we could also transform the data to $X^* = X^2$ and then

$$f_{X^*|\Theta^*}(x^*|\theta^*) = \frac{1}{2\sqrt{x^*\theta^*}} I_{[0,\theta^*]}(x^*).$$

(This makes the situation look more like the equivariance setup.) Now, we can find the posterior of Θ^*,

$$f_{\Theta^*|X^*}(\theta^*|x^*) = \frac{c^*}{\theta^{*2}} I_{[c^*,\infty)}(\theta^*),$$

where $c^* = \max\{x^*, 1\} = c^2$. The Bayes rule for the new loss function is *not* the posterior mean, rather it is the a^* that minimizes

$$\int_{c^*}^{\infty} (\sqrt{\theta^*} - \sqrt{a^*})^2 \frac{c^*}{\theta^{*2}} d\theta^*.$$

By expanding the square and differentiating with respect to a^*, we find that the posterior expected loss is minimized if $a^* = 4c^*$, which is precisely the square of $2c$.

In each of the above examples, the Bayes rule in the reparameterized problem is the reparameterization of the original Bayes rule. This is actually true in general.

Proposition 6.43. *Suppose that we reparameterize to $\Theta' = g(\Theta)$ where $g : \Omega \to \Omega'$ is bimeasurable and $\aleph' = \Omega'$. If the loss function changes from $L(\theta, a)$ to $L'(\theta', a') = L(g^{-1}(\theta'), g^{-1}(a'))$ and $\delta(x)$ is the formal Bayes rule for prior $f_\Theta(\theta)$ (based on data $X = x$ with conditional density $f_{X|\Theta}(x|\theta)$), then the formal Bayes rule in the reparameterized problem is $\delta'(x) = g(\delta(x))$.*

A note about transformations and loss functions is in order here. In Proposition 6.43, we transformed the loss function L to L'. In the usual equivariance setup, the loss function L is assumed to be invariant. That is, $L(g^{-1}(\theta'), g^{-1}(a')) = L(\theta', a')$, as was the case in Example 6.41. But this was a mere coincidence and had nothing to do with changing units. Even if the loss function had not been location invariant in Example 6.41, the Bayes rule would have respected the change of units, so long as the correct loss function were used after the change of units. Consider the following modification of Example 6.41.

Example 6.44 (Continuation of Example 6.41; see page 359). Suppose that the loss function for the °C problem is

$$L(\theta, a) = \begin{cases} (\theta - a)^2 & \text{if } \theta \geq 0, \\ 2(\theta - a)^2 & \text{if } \theta < 0. \end{cases}$$

This says that an error of a certain magnitude is twice as costly if the true temperature is below freezing than if it is freezing or above. If we try to use the same loss function in °K, then no such distinction is made in the costs of errors of the same magnitude. It is ludicrous to claim that these two decision problems are essentially the same problem with different units. Of course, the loss L here is not invariant. The transformed loss

$$L'(\theta', a') = \begin{cases} (\theta' - a')^2 & \text{if } \theta' \geq 273, \\ 2(\theta' - a')^2 & \text{if } \theta' < 273 \end{cases}$$

is the appropriate one to use in the °K scale.

Note that no transformation of the data is needed in Proposition 6.43. If one wishes to parallel the equivariance situation more completely, one can feel free to transform the data also, but it makes no difference to the conclusion of the proposition, so long as the transformation is bimeasurable, since the posterior distribution will be unchanged. The conclusion here is that there is no need for decision rules to be equivariant in order to obey conversion of units. Bayesian decision rules obey conversion of units without being equivariant.

Finally, we consider the root of the misconception that equivariant estimators are required in order to obey conversion of units.[9] Imagine that I will measure the length of a table with an inaccurate device. Let the measurement I hope to get in feet be denoted X, and let the sample space be $\mathcal{X} = \mathbb{R}^+$. That is, $\mathcal{X}$ is the set of possible measurements, in feet, which I might obtain. Let Θ denote the "true" length of the table (whatever that means) in feet, and suppose that the conditional distribution of X given $\Theta = \theta$ is denoted P_θ, where θ can be any positive number. That is, $\Omega = \mathbb{R}^+$ is the set of possible values of the "true" length of the table in feet. (Obviously some values of θ are far less likely than others as candidates for Θ, but we will ignore the Bayesian aspects of this problem for the time being.) If we convert our observed measurement to inches, we get $X' = 12X$. This situation resembles Group 2, scalar multiplication. Suppose that the parametric family $\mathcal{P}_0 = \{P_\theta : \theta \in \Omega\}$ is invariant under Group 2. We could mistakenly think of X' as $g_{12}X$ and then we could construct $\bar{g}_{12}\theta = 12\theta$. As we noted earlier, it is now perfectly correct to say that the distribution of $12X$ is $P_{12\theta}$. But here is where things get confused. The transformed θ, namely $\bar{g}_{12}\theta$, is supposed to be an element of Ω, which consists of the possible values of the "true" length of the table in feet, not inches! Although it is perfectly permissible to think of 12θ as the number of inches representing the true length of the table in inches, it is absolutely forbidden to think of $\bar{g}_{12}\theta$ as anything other than a possible length of the table in *feet*. In like manner, the sample space $\mathcal{X}$ is the set of possible measurements in feet, not inches. The transformed measurement $g_{12}X$ is 12 times as many feet as X, not the number of inches in X feet. Otherwise, how would we ever distinguish whether the number $12 \in \mathcal{X}$ stood for 12 feet or 1 foot converted to 12 inches? It cannot be both ways. We made it perfectly clear that $x \in \mathcal{X}$ stands for x feet and hence $12x \in \mathcal{X}$ stands for $12x$ feet, not x feet converted to inches. Hence $g_{12}X$ is *not* the converted measurement of the table to inches, but rather a measurement of the table in feet, which is 12 times as large as the measurement X. There is no other mathematically correct way to interpret these transformations. Hence, the invariance of the distributions has absolutely nothing to do with conversion of units from feet

---

[9]The author is indebted to Morris H. DeGroot for personal discussions about equivariance and invariance which helped immensely in clarifying these concepts.

to inches. That conversion is handled in a straightforward manner as a reparameterization, the way we did in Example 6.41 and Proposition 6.43.

6.2.3 Minimum Risk Equivariant Decisions

In location estimation with squared-error loss, Pitman's estimator was MRE and it was the generalized Bayes rule with respect to a uniform "prior" distribution. The feature of that prior distribution which made it the one to use in location estimation was the fact that it is invariant with respect to location shifts. Similarly, in scale estimation we discovered that the MRE, with a particular invariant loss, was also the Bayes rule with respect to an improper prior μ_Θ with Radon–Nikodym derivative $1/\theta$ with respect to Lebesgue measure. This measure is invariant with respect to scale changes. The pattern emerging here can be extended to more general groups, once we know how to find invariant measures.

Definition 6.45. Let G be a group with a σ-field of subsets Γ. Suppose that for each $g \in G$ and $A \in \Gamma$, $gA \in \Gamma$ and $A^{-1} \in \Gamma$. A measure λ on Γ is called *left Haar measure LHM (or left invariant measure)* if, for every $g \in G$ and every $A \in \Gamma$, $\lambda(gA) = \lambda(A)$. Similarly, ρ is called *right Haar measure RHM (or right invariant measure)* if $\rho(Ag) = \rho(A)$ for every $A \in \Gamma$ and every $g \in G$.

It should be noted that every positive multiple of an invariant measure is another invariant measure, so we may introduce an arbitrary constant multiple if we wish. It should also be easy to see that if G is abelian, then RHM=LHM.

Example 6.46. Group 1 is abelian, so LHM=RHM. Note that $\int_A dx = \int_{A+c} dx$, for all measurable A and all real c, so Lebesgue measure is invariant.

Group 2 is abelian, so LHM=RHM. Note that

$$\int_{cA} \frac{1}{x} dx = \int_A \frac{c\, dy}{y\, c} = \int_A \frac{1}{y} dy,$$

so the measure with Radon–Nikodym derivative $1/x$ with respect to Lebesgue measure is invariant.

Group 3 is not abelian, so we need to find LHM and RHM separately. The group action is $(a,b) \circ (c,d) = (bc + a, bd)$. Suppose that $h(x,y)$ is a Radon–Nikodym derivative for LHM with respect to Lebesgue measure. Then

$$\int_A h(x,y)dxdy = \int_{(a,b)A} h(x,y)dxdy.$$

Transform the left-hand side by $y = w/b$ and $x = (z-a)/b$. The Jacobian is $J = 1/b^2$, and we get

$$\int_A h(x,y)dxdy = \int_{(a,b)A} \frac{1}{b^2} h\left(\frac{z-a}{b}, \frac{w}{b}\right) dzdw = \int_{(a,b)A} h(z,w)dzdw.$$

Since this must hold for all a, b, A, we have

$$\frac{1}{b^2} h\left(\frac{z-a}{b}, \frac{w}{b}\right) = h(z, w),$$

for all a, b and a.e. $[dzdw]$. So, if $a = z$ and $b = w$, we must have $h(z, w) = h(0, 1)/w^2$. It follows that LHM has Radon–Nikodym derivative $1/y^2$ with respect to Lebesgue measure. Next, suppose that $h(x, y)$ is the Radon–Nikodym derivative for RHM. Then

$$\int_A h(x, y) dx dy = \int_{A(c,d)} h(x, y) dx dy.$$

Transform the left-hand side by $y = w/d$, and $x = z - wc/d$. Then the Jacobian is $J = 1/d$, and we get

$$\int_A h(x, y) dx dy = \int_{A(c,d)} \frac{1}{d} h\left(w - \frac{cw}{d}, \frac{w}{d}\right) dz dw = \int_{(a,b)A} h(z, w) dz dw.$$

Since this must hold for all c, d, A, we have

$$\frac{1}{d} h\left(z - \frac{cw}{d}, \frac{w}{d}\right) = h(z, w),$$

for all c, d and a.e. $[dzdw]$. So, if $c = z$ and $d = w$, we must have $h(z, w) = h(0, 1)/w$. It follows that RHM has Radon–Nikodym derivative $1/y$ with respect to Lebesgue measure. This is not the same as LHM.

Not only are measures of sets invariant under group operations when using LHM, but certain integrals are also invariant.

Lemma 6.47.[10] *If λ is LHM on G and f is integrable over G, then for all $g \in G$,*

$$\int_G f(g \circ h) d\lambda(h) = \int_G f(h) d\lambda(h).$$

PROOF. First let $f(h) = I_B(h)$ for some $B \in \Gamma$. Then

$$\int_G f(g \circ h) d\lambda(h) = \int I_B(g \circ h) d\lambda(h) = \int_{g^{-1}B} d\lambda(h)$$

$$= \lambda(g^{-1}B) = \lambda(B) = \int_G f(h) d\lambda(h).$$

By adding, we can extend to simple functions f. By the monotone convergence theorem A.52, we can extend to all nonnegative measurable functions, and by subtraction to all integrable functions. □

For a detailed discussion of Haar measure, see Nachbin (1965) or Halmos (1950, Chapter XI). For example, there are results giving conditions under which LHM exists. Since we will only use Haar measure explicitly when it does exist, we will not prove its existence. However, we will need to know that Haar measure is essentially unique.

---

[10]This lemma is used in the proofs of Lemmas 6.55 and 6.62.

Lemma 6.48.[11] *Let $(G, \circ)$ be a group, and let (G, Γ) be a topological space with the Borel σ-field. Suppose that λ is σ-finite and not identically 0 LHM on (G, Γ). Suppose that the function $f : G \times G \to G$ defined by $f(g, h) = g^{-1} \circ h$ is continuous. If λ' is also σ-finite and not identically 0 LHM on (G, Γ), then there exists a finite positive scalar c such that $\lambda' = c \cdot \lambda$.*

PROOF. The first step is to prove that $r(g) = I_{B^{-1}}(g)/\lambda(Ag)$ is a measurable function of g for each $A, B \in \Gamma$. Since $f(g, h)$ is continuous, it is continuous in g for fixed h, hence $f^*(g) = f(g, e) = g^{-1}$ is continuous, hence measurable. If $B \in \Gamma$, $f^{*-1}(B) = B^{-1}$, hence $B^{-1} \in \Gamma$. It follows that $I_{B^{-1}}(g)$ is measurable. The function $f'(g, h) = h \circ g = f(f^*(h), g)$ is also continuous, hence measurable. It follows that $v(g, h) = (g^{-1}, h \circ g)$ is continuous and measurable. It is easy to see that $v^{-1} = v$, so if $A \in \Gamma$, $G \times A \in \Gamma \otimes \Gamma$, hence $v(G \times A) \in \Gamma \otimes \Gamma$ and $m(g, h) = I_{v(G \times A)}(g^{-1}, h)$ is a measurable function. Define $\ell(g) = \int m(g, h) d\lambda(h)$. By Lemma A.67, this is a measurable function. Now notice that $I_{v(G \times A)}(g^{-1}, h) = I_{Ag}(h)$ and calculate

$$\ell(g) = \int I_{Ag}(h) d\lambda(h) = \lambda(Ag).$$

It follows that r is measurable.

Next, we prove that the following two one-to-one bicontinuous functions preserve measure in the product space $(G \times G, \Gamma \otimes \Gamma, \lambda' \times \lambda)$:

$$T_1(g, h) = (g, g \circ h),$$
$$T_2(g, h) = (h \circ g, h).$$

The proofs are similar, and we only prove that T_2 preserves measure. Note that $E \in \Gamma \otimes \Gamma$ implies that, for every $h \in G$,

$$\{g : (g, h) \in T_2(E)\} = h\{g : (g, h) \in E\} = hE^h,$$

where $E^h = \{g : (g, h) \in E\}$. It follows from Tonelli's theorem A.69 that

$$\lambda' \times \lambda(T_2(E)) = \int I_{T_2(E)}(g, h) d\lambda' \times \lambda(g, h) = \int \lambda'(hE^h) d\lambda(h)$$
$$= \int \lambda'(E^h) d\lambda(h) = \lambda' \times \lambda(E).$$

So T_2 preserves measure. Also, $T_1^{-1} T_2(g, h) = (h \circ g, g^{-1})$ preserves measure. Hence, for every nonnegative measurable function $v : G \times G \to \mathbb{R}$,

$$\int v(g, h) d\lambda' \times \lambda(g, h) = \int v(hg, g^{-1}) d\lambda' \times \lambda(g, h). \tag{6.49}$$

---

[11] This lemma is used to prove Corollary 6.52. The proof is adapted from Halmos (1950, Theorem 60.C).

This is proven by noting that it is true for indicators of events, hence for simple functions, hence for nonnegative measurable functions by the monotone convergence theorem A.52.

Let $A \in \Gamma$ have $0 < \lambda(A) < \infty$ and let $B \in \Gamma$. Define

$$r(g) = \frac{I_{B^{-1}}(g)}{\lambda(Ag)} = \frac{I_B(g^{-1})}{\lambda(Ag)},$$

which we have already shown to be measurable. We prove next that

$$\lambda'(B) = \lambda'(A) \int r(h) d\lambda(h). \tag{6.50}$$

Use Tonelli's theorem A.69 and (6.49) to write

$$\lambda'(A) \int r(h) d\lambda(h)$$
$$= \int I_A(g) r(h) d\lambda' \times \lambda(g, h) = \int I_A(h \circ g) r(g^{-1}) d\lambda' \times \lambda(g, h)$$
$$= \int \lambda(Ag^{-1}) r(g^{-1}) d\lambda'(g) = \int I_B(g) d\lambda'(g) = \lambda'(B),$$

where the second to last equation follows from $r(g^{-1})\lambda(Ag^{-1}) = I_B(g)$.

Next, apply (6.50) with $\lambda' = \lambda$ to get

$$\lambda(B) = \lambda(A) \int r(h) d\lambda(h).$$

Multiply both sides of this equation by $\lambda'(A)$ and apply (6.50) again to get $\lambda'(A)\lambda(B) = \lambda(A)\lambda'(B)$. Let $c = \lambda'(A)/\lambda(A)$. Since (6.50) is true for all B, it is true if $0 < \lambda'(B) < \infty$. It follows that $0 < \lambda'(A) < \infty$, hence $0 < c < \infty$, and the proof is complete. □

For the rest of this text, whenever LHM or RHM and groups are discussed, we will assume that the group satisfies the conditions of Lemma 6.48 and that the measures are σ-finite and not identically 0.

Lemma 6.51.[12] *If λ is LHM on a group G, then $\rho(A) = \lambda(A^{-1})$ is RHM, and we call ρ the RHM related to λ.*

PROOF. Note that $\rho(Ag) = \lambda(g^{-1}A^{-1}) = \lambda(A^{-1}) = \rho(A)$. □

The following corollary to Lemmas 6.48 and 6.51 now follows easily.

Corollary 6.52. *Assume the conditions of Lemma 6.48. If ρ and ρ' are both σ-finite and not identically 0 RHM on (G, Γ), then there exists a finite positive scalar c' such that $\rho' = c'\rho$.*

---

[12] This lemma is used to prove Corollary 6.52 and Lemma 6.54.

The following result, whose proof is based on the same concept as the proof of Lemma 6.47, is useful for converting between integrals with respect to LHM and RHM.

Proposition 6.53. [13] *If λ is LHM, ρ is the related RHM, and f is integrable with respect to ρ, then $\int f(g)d\rho(g) = \int f(g^{-1})d\lambda(g)$. If f is integrable with respect to λ, then $\int f(g)d\lambda(g) = \int f(g^{-1})d\rho(g)$.*

The following result gives a method for converting one LHM or RHM into many others.

Lemma 6.54. [14] *Let $\rho_g(B)$ be defined as $\rho(gB)$, and let $\lambda_g(B) = \lambda(Bg)$. Then ρ_g is RHM and λ_g is LHM for each $g \in G$.*

PROOF. Since $\lambda_g(hB) = \lambda(hBg) = \lambda(Bg) = \lambda_g(B)$, we have that λ_g is LHM, and a similar argument works for ρ_g. □

By Lemma 6.48, λ_g is a multiple of λ. Let the multiple be c_g. Similarly, by Corollary 6.52, ρ_g is a multiple of ρ, so define c'_g by $\rho_g(B) = c'_g \rho(B)$. In abelian groups, $c_g = c'_g = 1$ for all g, if λ and ρ are related. We introduce ρ_g because it will play an important role in the proof that Pitman's estimator is the formal Bayes rule with respect to an invariant measure.

We obtain interesting results if we replace λ by ρ in Lemma 6.47 and replace ρ by λ in Proposition 6.53.

Lemma 6.55. [15] *If ρ is RHM and $g \in G$ and f is integrable with respect to ρ, then*

$$\int f(gh)d\rho(h) = c'_{g^{-1}} \int f(h)d\rho(h).$$

If λ is LHM and $g \in G$ and f is integrable with respect to λ, then

$$\int f(hg)d\lambda(h) = c_{g^{-1}} \int f(h)d\lambda(h).$$

PROOF. In a manner similar to the proof of Lemma 6.47, we can prove that

$$\int f(h)d\rho_g(h) = \int f(g^{-1} \circ h)d\rho(h),$$

for all $g \in G$. It follows that

$$\int f(g \circ h)d\rho(h) = \int f(h)d\rho_{g^{-1}}(h) = c'_{g^{-1}} \int f(h)d\rho(h).$$

The proof of the other part is virtually identical, using Proposition 6.53. □
Actually, the numbers c_g and c'_g are related.

---

[13] This proposition is used in the proofs of Lemmas 6.55, 6.56, 6.65, 6.66, and Theorem 6.74.
[14] This lemma is used in the proof of Lemma 6.68.
[15] This lemma is used in the proofs of Lemmas 6.56, 6.65, and 6.66 and Theorem 6.74.

Lemma 6.56.[16] *If λ and ρ are related, then $c_g = 1/c'_g = c'_{g^{-1}}$ for all $g \in G$. Also $c'_g = c_{g^{-1}}$.*

PROOF. Let f be integrable with respect to ρ. Use Lemma 6.55 twice to write

$$\int f(h)d\rho(h) = \int f(g \circ g^{-1} \circ h)d\rho(h) = c'_g \int f(g \circ h)d\rho(h)$$
$$= c'_g c'_{g^{-1}} \int f(h)d\rho(h),$$

from which it follows that $c'_{g^{-1}} = 1/c'_g$. Next, use what we just proved and Proposition 6.53 and Lemma 6.55 to show that

$$\int f(h^{-1})d\lambda(h) = \int f(h)d\rho(h) = c'_g \int f(g \circ h)d\rho(h)$$
$$= c'_g \int f(g \circ h^{-1})d\lambda(h) = c'_g \int f([h \circ g^{-1}]^{-1})d\lambda(h)$$
$$= c'_g c_g \int f(h^{-1})d\lambda(h),$$

from which it follows that $c_g = 1/c'_g$. Then $c'_g = c_{g^{-1}}$ follows trivially. □

Example 6.57. Consider Group 3, so that $\rho(gB) = \int_{gB}(1/y)dxdy$. Let $g = (g_1, g_2)$ and transform by $y = g_2 w$, and $x = g_2 z - g_1$. Then $J = g_2^2$, and

$$\int_{gB} \frac{1}{y}dxdy = \int_B g_2^2 \frac{1}{g_2 w}dzdw = g_2 \int_B \frac{1}{w}dzdw = g_2\rho(B),$$

so $c'_g = g_2$.

There is a large class of examples in which the two groups of transformations G and $\overline{G}$ are isomorphic and are similar to both the parameter space and part of the sample space. We make this precise with the following condition.

Assumption 6.58. *Assume the following conditions:*

- *The distributions $\{P_\theta : \theta \in \Omega\}$ are invariant under the actions of groups G and $\overline{G}$.*

- *LHM λ and related RHM ρ on G exist.*

- *The conditions of Lemma 6.48 hold for G, so that LHM and RHM are essentially unique.*

- *The mapping $\phi : G \to \overline{G}$ defined by $\phi(g) = \overline{g}$ is a group isomorphism.*

---

[16]This lemma is used in the proof of Lemma 6.65.

6.2. Equivariant Decision Theory

- There is a bimeasurable mapping $\eta : \Omega \to \overline{G}$ which satisfies $\overline{g} \circ \eta(\theta) = \eta(\overline{g}\theta)$, for every $\overline{g} \in \overline{G}$ and $\theta \in \Omega$.

- There exists a bimeasurable function $t : \mathcal{X} \to G \times \mathcal{Y}$ for some space $\mathcal{Y}$ (where we write $t(X) = (H, Y)$), such that, for every $g \in G$ and $x \in \mathcal{X}$, $t(gx) = (g \circ h, y)$ if $t(x) = (h, y)$.

- For every θ, the distribution on $G \times \mathcal{Y}$ induced from P_θ by t has a density with respect to $\lambda \times \nu$, where ν is some measure on $\mathcal{Y}$.

Note that the Y part of $t(X) = (H, Y)$ is invariant when Assumption 6.58 holds. Also, since the function $t : \mathcal{X} \to G \times \mathcal{Y}$ is bimeasurable, it will be convenient to assume that $\mathcal{X} = G \times \mathcal{Y}$ and $X = t(X)$. Since t is one-to-one, the posterior of Θ given $t(X) = (h, y)$ will be the same as the posterior given $X = t^{-1}(h, y)$. Similarly, we will let P_θ stand for the induced distribution on $G \times \mathcal{Y}$, and we will let $f_{X|\Theta}(h, y|\theta)$ be the Radon–Nikodym derivative of P_θ with respect to $\lambda \times \nu$.

Theorem 6.59. *Assume Assumption 6.58. Let $\aleph$ be an action space and $L : \Omega \times \aleph \to \mathbb{R}$ be a loss function. Let $\tilde{G}$ be a group of transformations of $\aleph$ such that L is invariant. Then, if the formal Bayes rule with respect to ρ exists, it is the MRE rule, it is MRE conditional on Y, and Y is ancillary.*

Before proving Theorem 6.59, here are some examples.

Example 6.60. As we mentioned earlier, Pitman's estimator is MRE and it is the formal Bayes rule with respect to RHM on the location Group 1. Here $G = \mathbb{R}$ and we can map $\mathcal{X}$ to $G \times \mathcal{Y}$ if we let $\mathcal{Y} = \mathbb{R}^{n-1}$ and $t(x_1, \ldots, x_n) = (x_n, y)$ where $y = (x_1 - x_n, \ldots, x_{n-1} - x_n)$. Then $t(gx) = (gx_n, y)$. The loss $L(\theta, a) = (\theta - a)^2$ is invariant.

In the scale version, Group 2, $G = \mathbb{R}^+$ and we can let $\mathcal{Y} = \mathbb{R}^{n-1} \times \{-1, 1\}$ and $G = \{|x_n|\}$ so that $t(x) = (|x_n|, y)$, where $y = (x_1/|x_n|, \ldots, x_n/|x_n|)$. Then $t(gx) = (g|x_n|, y)$. RHM is dx/x. If we use the invariant loss $L(\theta, a) = (\theta - a)^2/\theta^2$, then Theorem 6.18 says that the MRE decision is indeed the formal Bayes rule with respect to RHM. Theorem 6.59 will also apply if the loss is $L(\theta, a) = [\log(\theta/a)]^2$.

With Group 3, we can write $t(x) = (x_{(n)}, x_{(n)} - x_{(1)}, y)$, where $x_{(i)}$ is the ith ordered element of x and

$$y = \left(\frac{x_{(2)} - x_{(1)}}{x_{(n)} - x_{(1)}}, \ldots, \frac{x_{(n-1)} - x_{(1)}}{x_{(n)} - x_{(1)}}, \pi \right),$$

where π is the permutation required to return the order statistic to the original data. Here $\mathcal{Y} = \mathbb{R}^{n-2} \times \Pi$, where Π is the set of permutations, and $G = \mathbb{R} \times \mathbb{R}^+$. Then $t(gx) = (g(x_{(n)}, x_{(n)} - x_{(1)}), y)$. There are several invariant losses. Here are three:

$$L_1(\theta, a) = \frac{(\theta_1 - a)^2}{\theta_2^2}, \quad \aleph = \mathbb{R};$$

$$L_2(\theta, a) = \frac{(\theta_2 - a)^2}{\theta_2^2}, \quad \aleph = \mathbb{R}^+;$$

$$L_3(\theta, a) = \frac{(\theta_1 - a_1)^2}{\theta_2^2} + \frac{(\theta_2 - a_2)^2}{\theta_2^2}, \quad \aleph = \mathbb{R} \times \mathbb{R}^+.$$

370   Chapter 6. Equivariance

The first is for location estimation, the second is for scale estimation, and the third is for simultaneous estimation of both. RHM has Radon–Nikodym derivative $1/\sigma$ with respect to Lebesgue measure. We can explicitly work through the normal distribution case. The likelihood function based on n observations is

$$\frac{1}{(2\pi)^n \sigma^n} \exp\left\{-\frac{1}{2\sigma^2}\sum_{i=1}^n (x_i - \mu)^2\right\} = \frac{1}{(2\pi)^n \sigma^n} \exp\left\{-\frac{1}{2\sigma^2}(w + n(\bar{x} - \mu)^2)\right\},$$

where $w = \sum_{i=1}^n (x_i - \bar{x})^2$. To find the posterior, we multiply by $1/\sigma$ and find the appropriate constant. If we let $\tau = \sigma^{-2}$, then $\sigma = \tau^{-1/2}$ and $d\sigma = -\tau^{-3/2} d\tau/2$. The posterior is, for some constant c,

$$c\tau^{(n-2)/2} \exp\left(-\frac{\tau}{2}[w + n(\mu - \bar{x})^2]\right).$$

This has the form of the product of an $N(\bar{x}, 1/(n\tau))$ density times a $\Gamma([n-1]/2, w/2)$ density. The posterior distribution of

$$\sqrt{n}\frac{\Theta_1 - \bar{X}}{\Theta_2} \tag{6.61}$$

is $N(0,1)$ given Θ_2. But since this distribution does not depend on Θ_2, it is also the marginal posterior. Also, the posterior distribution of W/Θ_2^2 is χ^2_{n-1}, where $W = \sum_{i=1}^n (X_i - \bar{X})^2$. These distributions parallel the prior conditional distributions of the sufficient statistics given Θ. That is, prior to seeing the data and conditional on Θ, (6.61) has $N(0,1)$ distribution and $W/\Theta_2^2 \sim \chi^2_{n-1}$. The posterior distributions were named *fiducial distributions* by Fisher (1935), because they seem to fall right out of the conditional distributions given Θ without any need for a prior on Θ. The quantity in (6.61) and W/Θ_2^2 are called *pivotal* quantities. These will be special cases of a more general result that will come later (Corollary 6.67).

The proof of Theorem 6.59 will proceed through a series of lemmas.

Lemma 6.62. *Assume Assumption 6.58. For every $\theta \in \Omega$ and every $g \in G$,*

$$f_{X|\Theta}(h, y|\theta) = f_{X|\Theta}(g \circ h, y|\bar{g}\theta), \quad a.e. \ [\lambda \times \nu]. \tag{6.63}$$

PROOF. Let $B \in \mathcal{A}$ be arbitrary. Since $P'_\theta(X \in B) = P'_{\bar{g}\theta}(X \in gB)$, we have, for every $g \in G$ and every $\theta \in \Omega$,

$$\int\int I_B(h,y) f_{X|\Theta}(h,y|\theta) d\lambda(h) d\nu(y)$$

$$= \int\int I_{gB}(h,y) f_{X|\Theta}(h,y|\bar{g}\theta) d\lambda(h) d\nu(y)$$

$$= \int\int I_B(g^{-1} \circ h, y) f_{X|\Theta}(h,y|\bar{g}\theta) d\lambda(h) d\nu(y)$$

$$= \int\int I_B(h,y) f_{X|\Theta}(g \circ h, y|\bar{g}\theta) d\lambda(h) d\nu(y),$$

where the last equality follows from Lemma 6.47. Since this is true for all $B \in \mathcal{A}$, the integrands of the first and last lines must be equal a.e. $[\lambda \times \nu]$. This immediately implies (6.63). □

A simple corollary to this result is obtained by letting $g = \phi^{-1}(\eta(\theta)^{-1})$, where ϕ and η are defined in Assumption 6.58.

Corollary 6.64. *Assume Assumption 6.58. There exists a function* $r : G \times \mathcal{Y} \to \mathbb{R}$ *such that, for every* $\theta \in \Omega$

$$f_{X|\Theta}(h, y|\theta) = r(\phi^{-1}(\eta(\theta)^{-1}) \circ h, y), \quad a.e. \ [\lambda \times \nu].$$

The formula given in the statement of Corollary 6.64 is particularly cumbersome due to the use of the notation $\phi^{-1}(\eta(\cdot))$. In fact, some of the proofs below would be almost unreadable if we continued to use this notation for the sake of mathematical precision. For this reason, we will take the following liberty with the notation for the remainder of the proof of Theorem 6.59. We will pretend that $\Omega = G = \overline{G}$ so that ϕ and η are just identity transformations, and we will not have to put the bar over elements of $\overline{G}$. This should not cause any confusion, since the sets really do behave virtually identically. For example, Corollary 6.64 now says

$$f_{X|\Theta}(h, y|\theta) = r(\theta^{-1} \circ h, y), \quad \text{a.e. } [\lambda \times \nu].$$

The following lemma will be useful both here and later.

Lemma 6.65.[17] *Under Assumption 6.58, Y is ancillary and the posterior density of Θ with respect to RHM is*

$$f_{\Theta|X}(\psi|h, y) = c_h f_{H|Y,\Theta}(h|y, \psi),$$

where the second factor on the right is the conditional density of H given Y and Θ.

PROOF. Since $\Omega = G$, there is only one orbit in the parameter space, hence the maximal invariant is constant. Since Y is invariant, Lemma 6.39, part 4 shows that Y is ancillary.

To calculate the posterior density of Θ given the data, we need the marginal "density" of the data:

$$\begin{aligned} f_X(h, y) &= \int f_{X|\Theta}(h, y|\psi) d\rho(\psi) = \int r(\psi^{-1} \circ h, y) d\rho(\psi) \\ &= \int r(\psi \circ h, y) d\lambda(\psi) = c_{h^{-1}} \int r(\psi, y) d\lambda(\psi) \\ &= c_{h^{-1}} \int f_{X|\Theta}(x, y|e) d\lambda(x) = f_Y(y) c'_h, \end{aligned}$$

---

[17] This lemma is used in the proofs of Lemma 6.66 and Theorem 6.74.

where the second and fifth equalities follow from Corollary 6.64, the third follows from Proposition 6.53, the fourth follows from Lemma 6.55, and the sixth follows from the fact that Y is ancillary and from Lemma 6.56. The posterior density of Θ given $X = (h, y)$ with respect to ρ is calculated via Bayes' theorem 1.31:

$$f_{\Theta|X}(\psi|h,y) = \frac{f_{X|\Theta}(h,y|\psi)}{f_X(h,y)} = \frac{f_{X|\Theta}(h,y|\psi)}{c'_h f_Y(y)} = c_h f_{H|Y,\Theta}(h|y,\psi). \quad \square$$

Lemma 6.66. *Assume the conditions of Theorem 6.59. If η is an equivariant rule, then the conditional risk function given $Y = y$ (constant as a function of θ) equals the posterior risk given $X = (h, y)$ (constant as a function of h).*

PROOF. Since Ω is isomorphic to $\overline{G}$, there is only one orbit and the risk function will be constant for every equivariant rule by Lemma 6.39, part 4. Also, the conditional risk function given Y will be constant in θ. The posterior risk given $X = (h', y)$ is

$$\int L(\theta, \eta(h', y)) f_{\Theta|X}(\theta|h', y) d\rho(\theta)$$

$$= c_{h'} \int L(\theta, \tilde{h}'\eta(e, y)) f_{H|\Theta,Y}(h'|\theta, y) d\rho(\theta)$$

$$= \frac{c_{h'}}{f_Y(y)} \int L(\overline{h'}^{-1}\theta, \eta(e, y)) f_{X|\Theta}(h', y|\theta) d\rho(\theta)$$

$$= \frac{c_{h'}}{f_Y(y)} \int L(\overline{h'}^{-1}\theta, \eta(e, y)) r(\theta^{-1} \circ h', y) d\rho(\theta)$$

$$= \frac{c_{h'}}{f_Y(y)} \int L(\overline{h'}^{-1}\theta, \eta(e, y)) r([(h')^{-1} \circ \theta]^{-1}, y) d\rho(\theta)$$

$$= \frac{1}{f_Y(y)} \int L(\theta, \eta(e, y)) r(\theta^{-1}, y) d\rho(\theta)$$

$$= \frac{1}{f_Y(y)} \int L(\theta^{-1}, \eta(e, y)) r(\theta, y) d\lambda(\theta)$$

$$= \frac{1}{f_Y(y)} \int L(h^{-1}, \eta(e, y)) f_{X|\Theta}(h, y|e) d\lambda(h)$$

$$= \int L(e, \eta(h, y)) f_{H|Y,\Theta}(h|y, e) d\lambda(h) = R(e, \eta|y),$$

where the first equality follows from Lemma 6.65 and equivariance of η, the second and eighth follow from invariance of L and the definition of conditional density, the third and seventh follow from Corollary 6.64, the fourth is elementary group theory, the fifth follows from Lemma 6.55 and Lemma 6.56, the sixth follows from Proposition 6.53, and the ninth follows from the definition of conditional risk function. $\square$

There is a useful corollary to Lemma 6.66.

Corollary 6.67. *Under Assumption 6.58, the conditional distribution of $\Theta^{-1}H$ given $Y = y$ is the same as the posterior distribution of $\Theta^{-1}H$.*

PROOF. Let $\aleph = G$, and for each $B \in \tau = \Gamma$, let $L(\theta, a) = I_B(\theta^{-1}a)$ in Lemma 6.66. The conclusion is that $P'_\theta(\Theta^{-1}H \in B|Y = y) = \Pr(\Theta^{-1}H \in B|(H, Y) = (h, y))$. □

The quantity $\Theta^{-1}H$ is called a *pivotal* quantity because we can switch back and forth between thinking of H or Θ as being the random variable and the other as fixed without changing the distribution. The common distribution is called the *fiducial distribution* by Fisher (1935).

Lemma 6.68. *Assume the conditions of Theorem 6.59. Assume that the formal Bayes rule with respect to ρ exists. Let $d(y)$ minimize the posterior risk if (e, y) is observed, where e is the identity in G. That is, $\min_a \int_\Omega L(\theta, a) f_{\Theta|X}(\theta|e, y) d\rho(\theta)$ occurs at $a = d(y)$. Define $\delta(h, y) = \tilde{h}d(y)$. Then δ is the formal Bayes rule, and it is equivariant.*

PROOF. First, note that δ is equivariant since, for $g \in G$,
$$\delta(g(h, y)) = \delta(g \circ h, y) = \widetilde{g \circ h} d(y) = \tilde{g}\tilde{h}d(y) = \tilde{g}\delta(h, y).$$

To see that δ is the formal Bayes rule, assume that (h, y) is observed. We must show that $\min_a \int_\Omega L(\theta, a) f_{\Theta|X}(\theta|h, y) d\rho(\theta)$ occurs at $a = \delta(h, y)$. We can write

$$\int_\Omega L(\theta, a) f_{\Theta|X}(\theta|h, y) d\rho(\theta)$$
$$= \frac{1}{c'_h f_Y(y)} \int_\Omega L(\theta, a) f_{X|\Theta}(h, y|\theta) d\rho(\theta)$$
$$= \frac{c_h}{f_Y(y)} \int_\Omega L(\overline{h}^{-1}\theta, \tilde{h}^{-1}a) r(\theta^{-1} \circ h, y) d\rho(\theta)$$
$$= \frac{c'_{h^{-1}}}{f_Y(y)} \int_\Omega L(\overline{h}^{-1}\theta, \tilde{h}^{-1}a) r([h^{-1} \circ \theta]^{-1}, y) d\rho(\theta)$$
$$= \frac{1}{f_Y(y)} \int_\Omega L(\theta, \tilde{h}^{-1}a) r(\theta^{-1}, y) d\rho(\theta)$$
$$= \frac{1}{f_Y(y)} \int_\Omega L(\theta, \tilde{h}^{-1}a) f_{X|\Theta}(e, y|\theta) d\rho(\theta)$$
$$= \int_\Omega L(\theta, \tilde{h}^{-1}a) f_{\Theta|X}(\theta|e, y) d\rho(\theta),$$

where the second equality uses Corollary 6.64 and the invariance of the loss function, the fourth equality follows from Lemma 6.54, the fifth follows from Corollary 6.64, and the sixth uses the fact that $c_e = 1$. Using the definition of d, the last integral above is minimized when $\tilde{h}^{-1}a = d(y)$, that is, when $a = \tilde{h}d(y) = \delta(h, y)$. □

Lemma 6.69. *Assume the conditions of Theorem 6.59. The δ defined in Lemma 6.68 is MRE and MRE conditional on Y.*

PROOF. From Lemma 6.68, we know that the posterior risk given $X = (e, y)$ is minimized for each y at the action $d(y)$. By equivariance, and the fact that the posterior risk given $X = (h, y)$ is constant in h (Lemma 6.66), it follows that the posterior risk is minimized at the action $\delta(x)$. Since Lemma 6.66 also shows that the risk function equals the posterior risk, the risk function is also minimized at δ, hence, δ is the MRE rule conditional on $Y = y$. The unconditional risk function of a rule η at $\theta = \bar{e}$ is

$$R(\bar{e}, \eta) = \int R(\bar{e}, \eta|y) f_Y(y) d\nu(y).$$

Since δ has minimum conditional risk function uniformly in y, the unconditional risk function of δ is clearly the minimum also. Hence δ is also the MRE rule. □

The conditions of Theorem 6.59 are often met when $\mathcal{Y}$ is the space of maximal invariants.

Example 6.70 (Continuation of Example 6.60; see page 369). Suppose that $X_1, \ldots, X_n$ are IID given $\Theta = \theta$ each with density $f([x - \theta_1]/\theta_2)/\theta_2$, for some density $f(\cdot)$. These distributions are invariant under Group 3. There are many possible invariant losses, as we saw earlier. The $\mathcal{Y}$ we calculated earlier was the space of maximal invariants. The MRE for loss L_1 is

$$\delta_1(x) = \frac{\int_\Omega \frac{\theta_1}{\theta_2^{n+3}} \prod_{i=1}^n f\left(\frac{x_i - \theta_1}{\theta_2}\right) d\theta_1 d\theta_2}{\int_\Omega \frac{1}{\theta_2^{n+3}} \prod_{i=1}^n f\left(\frac{x_i - \theta_1}{\theta_2}\right) d\theta_1 d\theta_2}.$$

The MRE for loss L_2 is

$$\delta_2(x) = \frac{\int_\Omega \frac{1}{\theta_2^{n+2}} \prod_{i=1}^n f\left(\frac{x_i - \theta_1}{\theta_2}\right) d\theta_1 d\theta_2}{\int_\Omega \frac{1}{\theta_2^{n+3}} \prod_{i=1}^n f\left(\frac{x_i - \theta_1}{\theta_2}\right) d\theta_1 d\theta_2}.$$

If f is the standard normal density, then $\delta_1(x) = \bar{x}$ and

$$\delta_2(x) = \frac{\Gamma\left(\frac{n}{2}\right)}{\Gamma\left(\frac{n+1}{2}\right)} \sqrt{\frac{w}{2}}.$$

It may be the case that all equivariant rules are inadmissible. For example, with Group 4 and one n-dimensional normal observation X, the MRE rule is to estimate Θ by X. But we saw in Section 3.2.3 that this is inadmissible if $n \geq 3$.

In later sections, we will see Theorems 6.74 and 6.78, which are like Theorem 6.59. The conclusions to those theorems say that certain formal Bayes inferences with respect to RHM priors agree with classical inferences conditional on the ancillary Y. This is why, in Theorem 6.59, we also showed that the MRE decision rule is MRE conditional on Y. Theorems 6.59, 6.74, and 6.78 parallel each other more this way.

Sometimes, the conclusions of Theorem 6.59 hold even when its conditions are not strictly met. For example, suppose that there is a nuisance parameter. It may be the case that for each fixed value of the nuisance parameter, the conditions of Theorem 6.59 apply to the problem with the appropriate subparameter space.

Example 6.71. Suppose that $X_1, \ldots, X_n$ are conditionally independent with $N(\mu, \sigma^2)$ distribution given $\Theta = (\mu, \sigma)$. Let $\aleph = \mathbb{R}$ and $L(\theta, a) = (\mu - a)^2$. This loss is not invariant under Group 3; however, it is invariant under Group 1. But the parameter space is not isomorphic to Group 1. For each value of σ, consider the subparameter space $\Omega_\sigma = \{(\mu, \sigma) : \mu \in \mathbb{R}\}$. The formal Bayes rule with respect to RHM on Ω_σ is $\delta(x) = \bar{x}$ for each σ. Since δ is the MRE rule for each σ, it is the MRE rule under Group 1 for the original problem.

It is not difficult to show that the situation of Example 6.71 generalizes to the following result.

Proposition 6.72. *Suppose that the parameter space is $\Omega = \Omega_1 \times \Omega_2$. Suppose that, for each $\theta_2 \in \Omega_2$, the conditions of Theorem 6.59 hold when $\Omega_1 \times \{\theta_2\}$ is taken as the parameter space. Then δ is the MRE rule if and only if it is MRE for each of the subproblems with fixed values of θ_2.*

There are situations in which there is no MRE.

Example 6.73 (Continuation of Example 6.71). This time, let $\aleph = [0, \infty)$ and $L(\theta, a) = (\sigma^2 - a)^2/\sigma^4$. Let the group be Group 2. For each value of μ, consider the subparameter space $\Omega_\mu = \{(\mu, \sigma) : \sigma > 0\}$. The formal Bayes rule with respect to RHM on Ω_μ is $\delta(x) = \sum_{i=1}^n (x_i - \mu)^2/(n+2)$. No single equivariant rule achieves the minimum risk for each μ.

6.3 Testing and Confidence Intervals*

6.3.1 *P*-Values in Invariant Problems

In Section 4.6, we introduced *P*-values as an alternative to testing hypotheses at preassigned levels. In Examples 4.146 (page 281) and 4.61 (page 241) we saw that sometimes the *P*-value relative to a collection of tests is the same as the posterior probability that the hypothesis is true based on an improper prior. A more general situation in which *P*-values correspond to posterior probabilities with improper priors arises when there is equivariance with respect to some group operating on the data and parameter spaces. The structure of the problem will need to be very much like that of Theorem 6.59. In addition, we will need to say something about the hypotheses of interest and how they interact with the group operation. We

---

*This section may be skipped without interrupting the flow of ideas.

also need to choose an appropriate set of tests with respect to which we calculate the P-value.

Theorem 6.74. *Assume Assumption 6.58 (see page 368). For each $\theta \in \Omega$, let Ω_θ be a subset of Ω such that the following conditions hold:*

1. *$\theta \in \Omega_\theta$;*

2. *for all $g \in G$ and all $\theta \in \Omega$, $g\Omega_\theta = \Omega_{\bar{g}\theta}$;*

3. *for all $\theta \in \Omega$ and all $\psi \in \Omega_\theta$, $\Omega_\psi \subseteq \Omega_\theta$;*

4. *for all $\theta \in \Omega$ and all $h \in \Omega_e$ (where e is the identity in G), $\Omega_\theta h \subseteq \Omega_\theta$.*

For each $\theta \in \Omega$, let G index a set of tests $\{\phi_{\theta,g} : g \in G\}$ of the hypothesis $H_\theta : \Theta \in \Omega_\theta$ versus $A_\theta : \Theta \notin \Omega_\theta$ defined by

$$\phi_{\theta,g}(h, y) = \begin{cases} 1 & \text{if } h \in \theta\Omega_{g^{-1}}^{-1}, \\ 0 & \text{if not.} \end{cases}$$

Suppose that we use ρ as a (possibly improper) prior for Θ. The posterior probability that H_θ is true given $t(X) = (h, y)$ is equal to the conditional P-value given $Y = y$ relative to the set of tests $\{\phi_{\theta,g} : g \in G\}$.

It should be noted, in the statement of Theorem 6.74, that the P-values must be calculated conditional on Y. But, Lemma 6.65 says that Y is ancillary. So, those who believe in conditioning on ancillaries would then want to calculate P-values conditional on Y anyway. Theorem 2.48 says that if there is a boundedly complete sufficient statistic, it will be independent of the ancillary. This leads to a simpler version of Theorem 6.74.

Corollary 6.75. *Under the conditions of Theorem 6.74, if H (the "group" part of $t(X)$) is a boundedly complete sufficient statistic, then the posterior probability that H_θ is true equals the P-value relative to the set of tests $\{\phi_{\theta,g} : g \in G\}$.*

Before proving Theorem 6.74, some explanation of the four conditions on Ω_θ is in order. The first condition is simply to connect θ with the corresponding hypothesis in a sensible way. The second condition ensures that the hypotheses are "equivariant" in some sense. The third condition is to ensure that the P-value is the size of the test $\phi_{\theta,g}$ when $H = g$. The fourth condition guarantees that the size of $\phi_{\theta,g}$ as a test of H_θ is equal to its power at θ. These last two conditions also capture the "one-sided" nature of the types of hypotheses to which this theorem applies. It will not apply to point hypotheses or to hypotheses such that Ω_θ has smaller dimension than Ω. The reason that the form of the test must be tied so closely to the form of the hypotheses is that there may be many classes of "equivariant" tests[18] and each class may lead to a different P-value. However, there is only one posterior probability that $\Theta \in \Omega_\theta$ with respect to RHM. Hence, we needed to identify exactly which class of tests has P-value equal to that posterior probability. This point will become clearer after Theorem 6.78.

As in the proof of Theorem 6.59, we will assume that $X = t(X)$ and that $G = \overline{G} = \Omega$, to make the notation simpler. The following lemma is also useful.

Lemma 6.76. *Under Assumption 6.58, the conditional distributions of the H part of X given Y are invariant.*

PROOF. Let B be a measurable subset of G, let $g \in G$, and let μ be the probability measure that gives the marginal distribution of Y. Define

$$v(\theta, g, y) = P'_\theta(gH \in B | Y = y).$$

We want to prove that $P'_{\overline{g}\theta}(H \in B | Y = y) = v(\theta, g, y)$, a.s. $[\mu]$ (for fixed θ and g). For every measurable subset A of $\mathcal{Y}$,

$$\int_A v(\theta, g, y) d\mu(y) = P'_\theta(Y \in A, gH \in B)$$
$$= P'_{\overline{g}\theta}(Y \in A, H \in B)$$
$$= \int_A P'_{\overline{g}\theta}(H \in B | Y = y) d\mu(y). \qquad \square$$

PROOF OF THEOREM 6.74. Let $\theta \in \Omega$ and let $\psi \in \Omega_\theta$. Then $\theta^{-1}\psi \in \Omega_e$ by condition 2. Also, we use conditions 2 and 4 to show that

$$P'_\psi(\phi_{\theta,g}(H, Y) = 1 | Y = y) = P'_\psi(H \in \theta\Omega_{g^{-1}}^{-1} | Y = y)$$
$$= P'_e(\psi H \in \theta\Omega_{g^{-1}}^{-1} | Y = y) = P'_e(H \in \psi^{-1}\theta\Omega_{g^{-1}}^{-1} | Y = y)$$
$$= P'_e(H^{-1} \in \Omega_{g^{-1}}\theta^{-1}\psi | Y = y) \le P'_e(H^{-1} \in \Omega_{g^{-1}} | Y = y)$$
$$= P'_e(H \in \Omega_{g^{-1}}^{-1} | Y = y) = P'_\theta(H \in \theta\Omega_{g^{-1}}^{-1} | Y = y)$$
$$= P'_\theta(\phi_{\theta,g}(H, Y) = 1 | Y = y).$$

This shows that the conditional size of the test $\phi_{\theta,g}$ (given Y) as a test of H_θ is equal to its conditional power function at θ. For each $g \in G$, define

$$Q(g, y) = P'_e(H \in \Omega_{g^{-1}}^{-1} | Y = y).$$

Then $P'_\theta(H \in \theta\Omega_{g^{-1}}^{-1} | Y = y) = Q(g, y)$, and it follows from what we just proved that the conditional size, given $Y = y$, of $\phi_{\theta,g}$ as a test of H_θ equals

---

[18] We put the word "equivariant" in quotes because it is not the test function itself that is equivariant, but rather the combination of the hypothesis and the test function. That is, if ψ_θ is a test of Ω_θ, then $\psi_\theta(h, y) = \psi_{\overline{g}\theta}(gh, y)$.

$Q(g, y)$. The conditional P-value given $Y = y$ can then be calculated as

$$p(h, y) = \inf_g \{Q(g, y) : \phi_{\theta, g}(h, y) = 1\}.$$

It is easy to see that $\phi_{\theta, \theta^{-1}h}(h, y) = 1$ by condition 1. It follows that $p(h, y) \leq Q(\theta^{-1}h, y)$. Next, suppose that $\phi_{\theta, g}(h, y) = 1$. It follows that $h \in \theta \Omega_{g^{-1}}^{-1}$, hence $h^{-1}\theta \in \Omega_{g^{-1}}$. Condition 3 implies that $\Omega_{h^{-1}\theta} \subseteq \Omega_{g^{-1}}$, from which it follows that $Q(\theta^{-1}h, y) \leq Q(g, y)$. It follows that $Q(\theta^{-1}h, y) \leq p(h, y)$, hence $Q(\theta^{-1}h, y) = p(h, y)$.

To complete the proof, we calculate the posterior probability that H_θ is true given $X = (h, y)$ and show that it equals $Q(\theta^{-1}h, y)$. Lemma 6.65 tells us that

$$f_{\Theta|X}(\psi|h, y) = c_h f_{H|Y,\Theta}(h|y, \psi).$$

In the following equalities, let (H', Y') have the same conditional distribution given Θ that X had before it was observed:

$$\Pr(\Theta \in \Omega_\theta | X = (h, y)) = c_h \int I_{\Omega_\theta}(\psi) f_{H|Y,\Theta}(h|y, \psi) d\rho(\psi)$$

$$= \int I_{\Omega_\theta}(\psi) r(\psi^{-1}h, y) d\rho(\psi) \frac{c_h}{f_Y(y)} = \int I_{\Omega_\theta}(\psi^{-1}) r(\psi \circ h, y) d\lambda(\psi) \frac{c_h}{f_Y(y)}$$

$$= \int I_{\Omega_\theta}(h\psi^{-1}) r(\psi, y) d\lambda(\psi) / f_Y(y) = \int I_{\Omega_{h^{-1}\theta}^{-1}}(\psi) r(\psi, y) d\lambda(\psi) / f_Y(y)$$

$$= \int I_{\Omega_{h^{-1}\theta}^{-1}}(g) f_{H|Y,\Theta}(g|y, e) d\lambda(g) = P'_e(H' \in \Omega_{h^{-1}\theta}^{-1} | Y' = y)$$

$$= Q(\theta^{-1}h, y),$$

where the first equality follows from Lemma 6.65, the second equality follows from Corollary 6.64, the third follows from Proposition 6.53, the fourth follows from Lemma 6.55, the fifth is just algebra, and the sixth follows from Corollary 6.64. □

Example 6.77. Let $X_1, \ldots, X_n$ be conditionally IID with $N(\mu, \sigma^2)$ distribution given $\Theta = (\mu, \sigma)$. The group is location and scale (Group 3). Consider the hypotheses $\Omega_\theta = \{(a, b) \in \Omega : a \leq \mu\}$ for $\theta = (\mu, \sigma)$. The corresponding tests are the usual one-sided t-tests. (The reader should check that the conditions of Theorem 6.74 are satisfied.) The associated P-values equal the posterior probabilities that the hypotheses are true if the prior is RHM, the measure with Radon–Nikodym derivative $1/\sigma$ with respect to Lebesgue measure.

As a less familiar example, let

$$\Omega_\theta = \{(a, b) \in \Omega : a \geq \mu, b \leq \sigma\}.$$

This is a simultaneous test of $H_{\mu, \sigma} : \mathrm{M} \geq \mu$ and $\Sigma \leq \sigma$. We will check condition 4 only. Since $e = (0, 1)$, $h = (m, s) \in \Omega_e$ satisfies $s \leq 1$ and $m \geq 0$. For such h,

$$\Omega_\theta h = \left\{(a, b) \in \Omega : b \leq \sigma s, a \geq \mu + \frac{bm}{s}\right\} \subseteq \Omega_\theta,$$

since $\sigma s \leq \sigma$ and $\mu + bm/s \geq \mu$. Suppose that data $(\overline{x}_n, s_n)$ are observed with $\overline{x}_n = \sum_{i=1}^{n} x_i/n$ and $s_n = \sqrt{\sum_{i=1}^{n}(x_i - \overline{x}_n)^2/(n-1)}$. The test $\phi_{\theta,g}$ rejects $H_{\mu,\sigma}$ if $\overline{x}_n \leq \mu + s_n g_1/g_2$ and $s_n \geq \sigma g_2$. The P-value is the size of the test $\phi_{\theta,([\overline{x}_n - \mu]/\sigma, s_n/\sigma)}$.

6.3.2 Equivariant Confidence Sets

In Section 5.2.1, we introduced confidence sets as an alternative to testing a single hypothesis about a parameter. In Example 5.57 on page 319, we saw that the confidence coefficient may not adequately express our degree of confidence that the parameter is in the set after seeing the data. That example is one in which the distributions are invariant under the action of the location group on the real numbers and the group is isomorphic to the parameter space. In addition, the sufficient statistic (T_1, T_2) can be transformed to $(T_1, T_2 - T_1)$ so that the group acts on T_1 and leaves $T_2 - T_1$ invariant. This is the same situation that arose in Theorems 6.59 and 6.74. In Theorem 6.74, we saw that posterior probabilities agreed with P-values conditional on the ancillary (invariant). A similar thing happens in Example 5.57, namely posterior probabilities (with respect to an improper prior) agree with conditional confidence coefficients. This is a special case of another theorem with conditions similar to the other two. This theorem is similar to one proved by Stein (1965). Chang and Villegas (1986) prove a similar theorem with slightly different conditions. Berger (1985, Section 6.6.3) also proves this theorem in a different way. Jaynes (1976, p. 181) gives a proof for the case of a location parameter.

Theorem 6.78. *Assume Assumption 6.58 (see page 368). For each $x \in \mathcal{X}$, let B_x be a measurable subset of Ω satisfying $B_{gx} = \overline{g} B_x$ for all $g \in G$. Let $C_\theta = \{x : \theta \in B_x\}$. Suppose that we use ρ as a (possibly improper) prior for Θ. Then, for all $x \in \mathcal{X}$ and all $\theta \in \Omega$,*

$$\Pr(\Theta \in B_x | X = x) = P'_\theta(X \in C_\theta | Y = y). \tag{6.79}$$

PROOF. As in the proofs of Theorems 6.59 and 6.74, we will assume that $X = t(X)$ and $G = \overline{G} = \Omega$ for ease of notation. Hence, we will write $x = (h, y)$. Now, write $B_{(h,y)} = h B_{(e,y)}$, and use Corollary 6.67 to say that

$$P'_\theta(\Theta^{-1} H \in B^{-1}_{(e,y)} | Y = y) = \Pr(\Theta^{-1} H \in B^{-1}_{(e,y)} | (H, Y) = (h, y)),$$

where $B^{-1}_{(e,y)} = \{g : g^{-1} \in B_{(e,y)}\}$. Since $\theta^{-1} h \in B^{-1}_{(e,y)}$ if and only if $\theta \in B_{(h,y)}$ if and only if $(h, y) \in C_\theta$, the result follows. □

If we think of $S(X) = B_X$ as a confidence set, then the left-hand side of (6.79) is the posterior probability that Θ is in the confidence set and the right-hand side is the conditional confidence coefficient given the ancillary.

At this point, we should examine the connection between Theorems 6.74 and 6.78. Since confidence sets and tests are equivalent, one would expect there to be some sort of equivalence between these two theorems. The problem is that Theorem 6.78 applies to all equivariant confidence sets. All such collections of confidence sets correspond to collections of test. All such tests satisfy the "equivariance" condition $\psi_\theta(h, y) = \psi_{g\theta}(gh, y)$ if ψ_θ means the corresponding test of Ω_θ. Furthermore, every such collection of tests leads to a P-value. Each such P-value will be the posterior probability of some set in the parameter space. That set may not equal Ω_θ, however. Here is how it works. For each α, suppose that we choose our $B_{x,\alpha}$ so that $C_{\theta,\alpha} = \{x : \theta \in B_{x,\alpha}\}$ satisfies $P'_\theta(X \in C_{\theta,\alpha} | Y = y) = 1 - \alpha$. This makes $B_{X,\alpha}$ a conditional coefficient $1 - \alpha$ confidence set given Y. Now define tests $\psi_{\theta,\alpha}$ to be 1 minus the indicator functions of the sets $C_{\theta,\alpha}$. Then the power function of $\psi_{\theta,\alpha}$ satisfies $\beta_{\psi_{\theta,\alpha}}(\theta) = \alpha$, and $\psi_{\theta,\alpha}$ is a level α test of the hypothesis

$$\Omega_\theta = \{\theta' \in \Omega : \beta_{\psi_{\theta,\alpha}}(\theta') \leq \alpha, \text{ for all } \alpha\}.$$

The conditional P-value relative to the set of tests $B_\theta = \{\psi_{\theta,\alpha} : \alpha \in [0,1]\}$ is

$$p(x) = \inf\{\alpha : x \in C^C_{\theta,\alpha}\}.$$

It follows that $P'_\theta(X \in C^C_{\theta,p(x)} | Y = y) = p(x)$. From (6.79), we conclude that

$$\Pr(\Theta \in B^C_{x,p(x)} | Y = y) = p(x).$$

In general, it might happen that $B^C_{x,p(x)} \neq \Omega_\theta$. Here is an example.

Example 6.80. Let $X_1, \ldots, X_n$ be conditionally IID with $N(\mu, \sigma^2)$ distribution given $\Theta = (\mu, \sigma)$. Here $(\overline{X}_n, S_n)$ is a complete sufficient statistic, where $\overline{X}_n = \sum_{i=1}^n X_i/n$, $S_n = \sqrt{\sum_{i=1}^n (X_i - \overline{X}_n)^2/(n-1)}$. So, we will ignore the Y part of the problem, since Y is independent of the complete sufficient statistic. Let

$$B_{X,\alpha} = \left[\overline{X}_n - \frac{1}{\sqrt{n}} S_n T^{-1}_{n-1}\left(\frac{\alpha}{2}\right), \overline{X}_n + \frac{1}{\sqrt{n}} S_n T^{-1}_{n-1}\left(\frac{\alpha}{2}\right)\right],$$

where T^{-1}_{n-1} is the inverse of the CDF of the $t_{n-1}(0,1)$ distribution. Then, $\psi_{\theta,\alpha}$ is the usual two-sided size α t-test of $H : \Theta = \theta$. The P-value is the α value $p_\theta(x)$ such that one of the endpoints of the interval $B_{X,\alpha}$ equals θ. This makes $B_{x,p_\theta(x)}$ equal to the interval centered at $\overline{X}_n$ and having half-width equal to $|\overline{X}_n - \theta|$. On the other hand, $\Omega_\theta = \{\theta\}$. The P-value is the posterior probability of some hypothesis, but not the hypothesis you thought you were testing.[19]

6.3.3 Invariant Tests*

In multiple parameter problems with hypotheses concerning several parameters at once, there may be many competing tests, none of which is

---

[19]See Problem 24 on page 392.
*This section may be skipped without interrupting the flow of ideas.

UMPU. Just as we used equivariance to reduce the collection of estimators to consider, we can try to reduce the number of tests to consider also. In hypothesis testing, the action space is $\aleph = \{0, 1\}$. There are only two groups that act on this set. One contains only an identity, while the other contains an identity and a "switch" operator, $g(i) = 1 - i$. If we were to construct groups such that $\tilde{G}$ were this second group, then there would have to be conditions under which we were willing to switch the hypothesis and the alternative. Due to the asymmetric treatment of hypotheses and alternatives in classical testing theory, this would not be advisable. Hence, we will only discuss cases in which $\tilde{G}$ consists of one element, namely an identity. Then a decision rule is equivariant if and only if it is invariant. That is, since tests are randomized rules,

$$\delta^*(gx)(\tilde{g}A) = \delta^*(gx)(A) = \delta^*(x)(A),$$

making $\delta^*(gx)(\cdot)$ the same probability as $\delta^*(x)(\cdot)$. So, each equivariant (invariant) test must be a function of the maximal invariant.

Example 6.81. Consider Group 3, namely one-dimensional location-scale. The maximal invariant is

$$\left(\frac{x_1 - \overline{x}}{\sqrt{w}}, \ldots, \frac{x_n - \overline{x}}{\sqrt{w}} \right),$$

where $w = \sum_{i=1}^n (r_i - \overline{r})^2$. Nobody would ever base a test on this alone, because it is ancillary in location-scale problems. In the normal distribution case, it is not even a function of the sufficient statistic.

If we first consider the sufficient statistic $(\overline{X}, W)$, we see that the maximal invariant is constant, hence only constant functions of the sufficient statistic are invariant.

This example raises the question of whether reduction of the set of tests by invariance is compatible with reduction by sufficiency. That is, suppose that we first reduce to the set of invariant tests and then find a sufficient statistic for the maximal invariant parameter and further reduce by considering only invariant tests that are a function of the sufficient function of the maximal invariant. Will we get the same tests as we would if we first reduced to only those tests that depend on the sufficient statistic and then reduced to only those that depend on the maximal invariant in the space of sufficient statistics? In Example 6.81 on page 381, the answer is yes, but only because both methods produce degenerate results. Hall, Wijsman, and Ghosh (1965) find conditions for this compatibility.

The following assumption is an obvious preliminary. It requires that the group operation is inherited by the sufficient statistic space.

Assumption 6.82. *If $T(X)$ is sufficient and, for each $g \in G$, we define $T_g(x) = T(gx)$, then $T_g(x)$ depends on x only through $T(x)$.*

Example 6.83. Suppose that $\mathcal{X} = \mathbb{R}^n$ and $T(x) = (\overline{x}, w)$, where $w = \sum_{i=1}^n (x_i - \overline{x})^2$. Then $g_{a,b}x = (\cdots, bx_i + c, \cdots)$ and $T(g_{a,b}x) = (b\overline{x} + a, b^2 w)$. This function satisfies Assumption 6.82, assuming it is sufficient. A function that does not satisfy the assumption is $H(x) = x_1 x_2$.

If Assumption 6.82 is satisfied, define g^* to be the transformation on $\mathcal{T}$ $g^*t = T_g(x)$ for any x such that $T(x) = t$. The set G^* of all such transformations is a group. Let $U : \mathcal{T} \to \mathcal{U}$ be the maximal invariant in the sufficient statistic space, and let $V : \mathcal{X} \to \mathcal{V}$ be the maximal invariant in the original data space. Then

$$U(T(gx)) = U(T_g(x)) = U(g^*t) = U(t) = U(T(x)).$$

So, $U(T(\cdot))$ is an invariant function in the original data space; hence it is a function of V. That is, there exists $H : \mathcal{V} \to \mathcal{U}$ such that $U(T(x)) = H(V(x))$ for all $x \in \mathcal{X}$.

Theorem 6.84 (Stein Theorem).[20] *Let T be sufficient and satisfy Assumption 6.82. Suppose that T has discrete distribution. Let U and V be maximal invariants in $\mathcal{T}$ and $\mathcal{X}$, respectively. Let $R(\Theta)$ be the maximal invariant in Θ. Then $U(T(X))$ is sufficient for $R(\Theta)$.*

PROOF. The proof proceeds through a series of claims.
 (a) $A = V^{-1}(B)$ for some $B \subseteq \mathcal{V}$ if and only if $gA = A$ for all g.
(Proof of a): Let $A = V^{-1}(B)$. Then

$$gA = \{gx : V(x) \in B\} = \{x : V(g^{-1}x) \in B\} = \{x : V(x) \in B\} = A,$$

since V is invariant. Now, let $gA = A$ for all g. Then

$$I_A(x) = I_{g^{-1}A}(x) = I_A(gx),$$

so $I_A(\cdot)$ is invariant and it must be a function of the maximal invariant, namely $I_A(x) = f(V(x))$. Let $B = f^{-1}(\{1\})$. Then $I_A(x) = I_B(V(x))$, so $A = \{x : V(x) \in B\}$. A set A that satisfies the conditions of this claim is called an *invariant set*.
 (b) $\Pr(X \in A | T = t)$ is an invariant function of t if A satisfies $gA = A$ for all $g \in G$.
(Proof of b): Choose any θ and t such that $\Pr(T(X) = t | \Theta = \theta) > 0$. Let $g \in G$.

$$\begin{aligned}
\Pr(X \in A | T(X) = t) &= \frac{\Pr(X \in A, T(X) = t | \Theta = \theta)}{\Pr(T(X) = t | \Theta = \theta)} \\
&= \frac{\Pr(X \in gA, T(X) = g^*t | \Theta = \overline{g}\theta)}{\Pr(T(X) = g^*t | \Theta = \overline{g}\theta)} \\
&= \Pr(X \in gA | T(X) = g^*t) \\
&= \Pr(X \in A | T(X) = g^*t),
\end{aligned}$$

---

[20]Hall, Wijsman, and Ghosh (1965) attribute this theorem to Stein.

where the second-to-last equality holds by sufficiency of T.

(c) If A is an invariant set, then $P_\theta(A|U(T(X)) = u)$ is constant in θ for each u.

(Proof of c): Write $P_\theta(A|U(T(X)) = u)$ as

$$\sum_{\{t \in U^{-1}(u)\}} \Pr(X \in A | T(X) = t) \Pr(T(X) = t | U(T(X)) = u, \Theta = \theta).$$

Since U is maximal invariant, $U^{-1}(u)$ is an orbit in $\mathcal{T}$. So, $U^{-1}(u) = \{t : t = g^* t_u\}$ for some $t_u \in \mathcal{T}$. It follows that $P_\theta(A|U = u)$ equals

$$\sum_{g^* \in G^*} \Pr(T(X) = g^* t_u | U = u, \Theta = \theta) \Pr(X \in A | T(X) = g^* t_u).$$

The last factor equals $\Pr(X \in A | T = t_u)$ by (b), so it factors out of the sum. Also,

$$\{x : U(T(x)) = u\} = \bigcup_{g^* \in G^*} \{x : T(x) = g^* t_u\},$$

so the remaining sum equals 1 and

$$P_\theta(A|U = u) = \Pr(X \in A | T(X) = t_u),$$

which is the same for all θ.

(d) Part (a) says that the invariant sets constitute the σ-field generated by V. Part (c) says that for each A in that σ-field, $P'_\theta(X \in A | U = u)$ is constant in θ. Hence U is sufficient. □

Hall, Wijsman, and Ghosh discuss conditions under which the Stein theorem 6.84 holds for continuous distributions.

Example 6.85. Consider Group 3 again. The maximal invariant is

$$V(x) = \left(\frac{x_1 - x_n}{x_{n-1} - x_n}, \ldots, \frac{x_{n-2} - x_n}{x_{n-1} - x_n}, \text{sign}(x_{n-1} - x_n) \right),$$

which is independent of Θ. So the sufficient statistic in the maximal invariant space is constant. If the sufficient statistic is $T(x) = (w, \bar{x})$, as with normal distributions, then the maximal invariant in the sufficient statistic space is also constant. Since there is only one orbit in the parameter space, the maximal invariant in the parameter space is also constant. In simple English, Group 3 equivariance is useless in hypothesis testing.

Definition 6.86. A function f on $\mathcal{X}$ is *almost invariant with respect to μ* if, for each $g \in G$, there exists $B_g \in \mathcal{B}$ such that $\mu(B_g) = 0$ and $f(x) = f(gx)$ for all $x \notin B_g$.

Proposition 6.87. *If $P_\theta \ll \mu$ for each θ, if $v(\theta)$ is maximal invariant in Ω, and if f is almost invariant with respect to μ, then the distribution of $f(X)$ given $\Theta = \theta$ depends on θ only through the $v(\theta)$.*

384 Chapter 6. Equivariance

The proof of this is very similar to the proof of part 4 of Lemma 6.39.

Definition 6.88. A test ϕ is *UMPU almost invariant (UMPUAI) level α* if it is UMPU among all almost invariant level α tests.

Theorem 6.89. *Suppose that $P_\theta \ll \mu$ for each θ and a hypothesis-testing problem is invariant under G and $\overline{G}$. Suppose that there exists ϕ^*, which is UMPU level α and ϕ^* is unique a.e. $[\mu]$. Suppose also that there exists ϕ_0, which is UMPUAI level α. Then ϕ_0 is also unique a.e. $[\mu]$ and $\phi_0 = \phi^*$ a.e. $[\mu]$.*

PROOF. Let U_α be the class of all unbiased level α tests. First, we show that $\phi \in U_\alpha$ if and only if $\phi_g \in U_\alpha$ for each g, where $\phi_g(x) = \phi(gx)$:

$$E_\theta \phi_g(X) = E_\theta \phi(gX) = E_{\overline{g}\theta} \phi(X),$$

which is greater than or equal to or less than or equal to α, respectively, according to $\overline{g}\theta \in \Omega_H$ or $\overline{g}\theta \in \Omega_A$, according to $\theta \in H$ or $\theta \in A$ by invariance. This makes ϕ_g unbiased level α.

Next, we show that ϕ_g^* is UMP in U_α. Since $\phi^* \in U_\alpha$, we have that $\phi_g^* \in U_\alpha$ by the first result. Let $\theta \in \Omega_A$. Then

$$\begin{aligned} E_\theta \phi_g^*(X) &= E_\theta \phi^*(gX) = E_{\overline{g}\theta} \phi^*(X) \\ &= \sup_{\phi \in U_\alpha} E_{\overline{g}\theta} \phi(X) = \sup_{\phi \in U_\alpha} E_\theta \phi(gX) \\ &= \sup_{\phi \in U_\alpha} E_\theta \phi_g(X) = \sup_{\phi \in U_\alpha} E_\theta \phi(X) = E_\theta \phi^*(X), \end{aligned}$$

since ϕ^* is UMP in U_α. So, $\phi_g^* = \phi^*$ a.e. $[\mu]$ for each g by the uniqueness of ϕ^*. This makes ϕ^* almost invariant. Since ϕ_0 is UMPUAI level α, $\beta_{\phi_0}(\theta) \geq \beta_{\phi^*}(\theta)$ for all $\theta \in \Omega_A$, so ϕ_0 is also UMPU level α and $\phi_0 = \phi^*$ a.e. $[\mu]$ also, and so it is also unique a.e. $[\mu]$. □

This theorem does not guarantee that the UMPUAI level α test is UMPU, but it provides insurance that if there is a unique UMPU level α test, we can find it by finding the UMPUAI level α test.

One-Way Analysis of Variance

Consider the one-way ANOVA (analysis of variance). That is, Y_{ij} are conditionally independent with $Y_{ij} \sim N(\mu_i, \sigma^2)$ given $M_i = \mu_i$ for $j = 1, \ldots, n_i$ and $i = 1, \ldots, k$ and $\Sigma = \sigma$. First, reduce by sufficiency to

$$\overline{Y}_i = \frac{1}{n_i} \sum_{j=1}^{n_i} Y_{ij}, \text{ for } i = 1, \ldots, k, \text{ and } W = \sum_{i=1}^{k} \sum_{j=1}^{n_i} (Y_{ij} - \overline{Y}_i)^2.$$

6.3. Testing and Confidence Intervals

Suppose that $\Omega_H = \{(\sigma, \mu) : BA\mu = \mathbf{0}\}$, where B is an $r \times k$ rank r matrix with $r \leq k$, A is the diagonal matrix

$$A = \begin{pmatrix} \sqrt{n_1} & \cdots & 0 \\ 0 & \ddots & 0 \\ 0 & \cdots & \sqrt{n_k} \end{pmatrix},$$

and $\mathbf{0}$ is the vector all of whose coordinates are 0. Without loss of generality, we can assume that B is the first r rows of an orthogonal matrix Γ. Let $\overline{Y}$ be the vector whose ith coordinate is $\overline{Y}_i$ for each i. Make the one-to-one transformation of the data to $X = \Gamma A \overline{Y}$, and W. Now, given the parameters, X is independent of W with

$$X \sim N_k(\gamma, \sigma^2 I), \quad W \sim \sigma^2 \chi_d^2,$$

where $d = n - k$ and $\gamma = \Gamma A \mu$. We can write

$$\Omega_H = \{(\gamma, \sigma) : \gamma_1 = \cdots = \gamma_r = 0\}.$$

Let the group G consist of triples (Λ, b, c), where Λ is $r \times r$ orthogonal, b is $(k - r)$-dimensional, and $c > 0$. Define

$$g_{\Lambda, b, c}(x, w) = \left[c \begin{pmatrix} \Lambda x_1 \\ x_2 + b \end{pmatrix}, c^2 w \right],$$

where we write $x^\top = (x_1^\top, x_2^\top)$ and x_1 is r-dimensional and x_2 is $(k - r)$-dimensional. In the parameter space

$$\overline{g}_{\Lambda, b, c}(\gamma, \sigma) = \left[c \begin{pmatrix} \Lambda \gamma_1 \\ \gamma_2 + b \end{pmatrix}, c\sigma \right],$$

with $\gamma^\top = (\gamma_1^\top, \gamma_2^\top)$ and γ_1 the first r coordinates, preserves the hypothesis. So the testing problem is invariant.

The maximal invariant in $\mathcal{X}$ is determined by $f(g(x, w)) = f(x, w)$ for all g, x, and w. So, for fixed x and w, let Λ have first row proportional to $x_1^\top$, $b = -x_2$, and let $c = 1/\sqrt{w}$. Then

$$f(g_{\Lambda, b, c}(x, w)) = f\left(\begin{bmatrix} \sqrt{\frac{x_1^\top x_1}{w}} \\ 0 \\ \vdots \\ 0 \end{bmatrix}, 1 \right) = f(x, w).$$

So $x_1^\top x_1 / w$ is maximal invariant, since it is clearly invariant. The usual F statistic for testing H is just d/r times the maximal invariant, and it has noncentral F distribution $NCF(r, d, \delta)$, where $\delta = \sum_{i=1}^r \gamma_i^2 / \sigma^2$ conditional on the parameters. The hypothesis H is equivalent to $\delta = 0$. Since the noncentral F distribution has MLR in the noncentrality parameter (see Problem 29 on page 289), the F-test is UMPUAI level α.

Multivariate Analysis of Variance

We now present an example of a case in which the number of tests available is so large that even a reduction by invariance still leaves too many tests to consider.[21] Imagine that the data consist of exchangeable p-dimensional observations $X_1, \ldots, X_n$. We will write the data matrix as $M = [M_1|M_2|M_3]$, where

$$M_1 = [X_1|\cdots|X_q], \quad M_2 = [X_{q+1}|\cdots|X_k], \quad M_3 = [X_{k+1}|\cdots|X_n],$$

where $n > k \geq q$. The parameter is $\Theta = (M_1, \ldots, M_k, \Sigma)$, where each M_i is a p-dimensional vector and Σ is a $p \times p$ positive definite matrix. The conditional distribution of the X_i given $\Theta = (\Sigma, \mu_1, \ldots, \mu_k)$ is that the X_i are independent with X_i having distribution $N_p(\mu_i, \Sigma)$ distribution for $i \leq k$ and X_i having $N_p(\mathbf{0}, \Sigma)$ distribution for $i > k$. The hypothesis of interest is $M_1 = \cdots = M_q = \mathbf{0}$.

The group we choose for this problem comes in four parts:

$$\begin{aligned}
G_1 &= \{g_A^1 : A \text{ is a } p \times k - q \text{ matrix}\}, \\
G_2 &= \{g_D^2 : D \text{ is } n - k \times n - k \text{ orthogonal}\}, \\
G_3 &= \{g_C^3 : C \text{ is } q \times q \text{ orthogonal}\}, \\
G_4 &= \{g_E^4 : E \text{ is } p \times p \text{ nonsingular}\}.
\end{aligned}$$

These groups are applied in sequence as follows:

$$g_{A,D,C,E}M = [EM_1 C | M_2 + A | EM_3 D].$$

The action on the parameter is $\bar{g}_{A,D,C,E}\Theta$ equal to

$$(E\Sigma E^\top, E[M_1|\cdots|M_q]C, [M_{q+1}|\cdots|M_k] + A, E[M_{k+1}|\cdots|M_n]D).$$

Note that the hypothesis is not altered by action of $\bar{g}$. That is,

$$M_1 = \cdots = M_q = \mathbf{0} \text{ if and only if } E[M_1|\cdots|M_q]C = [\mathbf{0}|\cdots|\mathbf{0}].$$

To find the maximal invariant, we set

$$f(M) = f(g_{A,D,C,E}M), \text{ for all } A, D, C, E, M.$$

In particular, suppose that $A = -M_2$, then $f(M) = f([EM_1 C | O | EM_3 D])$, where O is matrix of all zeros. Now, consider the following lemma.

Lemma 6.90. *Two $a \times b$ matrices R and T satisfy $RR^\top = TT^\top$ if and only if $T = RQ$ for some $b \times b$ orthogonal matrix Q.*

---

[21] For a good introduction to invariant tests in multivariate problems, see Anderson (1984, Chapter 8) or Kshirsagar (1972, Chapters 7–10).

6.3. Testing and Confidence Intervals

PROOF. First, suppose that $T = RQ$, then $TT^\top = (RQ)(RQ)^\top = RR^\top$. Next, suppose $RR^\top = TT^\top$. Write the singular-value decompositions of R and T as
$$R = \Gamma_R \Lambda_R \Omega_R^\top, \quad T = \Gamma_T \Lambda_T \Omega_T^\top,$$
where $\Gamma_R, \Gamma_T, \Omega_R,$ and Ω_T are orthogonal and Λ_R and Λ_T are "diagonal" matrices arranged so that the absolute values of the diagonal entries increase as you read down the diagonal. (The Λ matrices are not really diagonal because they are not square. Their only nonzero entries are (1,1), (2,2), etc., however.) Then
$$RR^\top = \Gamma_R \Lambda_R \Lambda_R^\top \Gamma_R = \Gamma_T \Lambda_T \Lambda_T^\top \Gamma_T = TT^\top.$$
Since these are two representations of the eigenvalue decomposition of the same matrix, it follows that $\Gamma_T = \Gamma_R$ and $\Lambda_T = \Lambda_R J$, where J is a diagonal matrix with only ± 1 in each diagonal entry. (If $RR^\top$ has eigenvalues with nonunit multiplicity, a permutation of the columns of Γ_T may be required to make it equal to Γ_R.) So, $T = R\Omega_R J \Omega_T^\top$. Since $\Omega_R J \Omega_T^\top$ is orthogonal, it follows that $T = RQ$, where Q is orthogonal. □

Now, let M_1^* be a $p \times q$ matrix such that $M_1 M_1^\top = M_1^* M_1^{*\top}$, and let C be orthogonal such that $M_1^* = M_1 C$. It follows that $f([M_1|M_2|M_3]) = f([M_1^*|M_2|M_3])$. Similarly, if M_3^* is such that $M_3 M_3^\top = M_3^* M_3^{*\top}$, then $f([M_1|M_2|M_3]) = f([M_1|M_2|M_3^*])$. It follows that f is a function of M_1 and M_3 through $M_1 M_1^\top$ and $M_3 M_3^\top$ only. Define $g(B, W) = f([M_1|O|M_3])$, where $B = M_1 M_1^\top$ and $W = M_3 M_3^\top$. It follows that $f(M) = g(B, W)$ if f is invariant. Also, $f(g_{A,C,D,E}M) = g(EBE^\top, EWE^\top)$. Finally, write the eigenvalue decomposition of

$$W^{-\frac{1}{2}} B W^{-\frac{1}{2}} = \Gamma \begin{pmatrix} \lambda_1 & 0 & \cdots & \cdots & \cdots & 0 \\ 0 & \ddots & 0 & \cdots & \cdots & 0 \\ 0 & \cdots & \lambda_s & 0 & \cdots & 0 \\ 0 & \cdots & \cdots & 0 & \cdots & 0 \\ 0 & \cdots & \cdots & \cdots & \ddots & 0 \\ 0 & \cdots & \cdots & \cdots & \cdots & 0 \end{pmatrix} \Gamma^\top = \Gamma \Lambda \Gamma^\top,$$

where s is the rank of B. Note that $s = \min\{p, q\}$ with probability 1. Then set $E = \Omega W^{-1/2}$. It follows that $f(M) = g(EBE^\top, EWE^\top) = g(\Lambda, I_p)$. Note that Λ is invariant; hence it is maximal invariant.

What we have just proven is that every invariant test in MANOVA must be a function of the nonzero eigenvalues of $W^{-1/2} B W^{-1/2}$, which are the same as the nonzero eigenvalues of $W^{-1} B$. A similar argument shows that the maximal invariant in the parameter space is the set of nonzero eigenvalues of $\Sigma^{-1} M$, where $M = [M_1|\cdots|M_q][M_1|\cdots|M_q]^\top$. The two special cases in which $s = 1$ are of interest. If $p = 1$, we have univariate ANOVA, and the only nonzero eigenvalue of $W^{-1} B$ is $q/(n-k)F$, where F is the

F statistic for testing H. If $q = 1$, then the only eigenvalue of $W^{-1}B$ is Hotelling's T^2. For cases in which $s > 1$, there is no UMPUAI test, but there are several well-known invariant tests based on the eigenvalues of $W^{-1}B$. One is based on the largest eigenvalue, another on the sum of the eigenvalues, and a third on the product of the nonzero eigenvalues.

A Test Based on Tolerance Sets

Let $\Theta = (M, \Sigma)$, and suppose that $\{X_n\}_{n=1}^{\infty}$ are conditionally IID with $N(\mu, \sigma^2)$ distribution given $\Theta = (\mu, \sigma)$. Let $X = (X_1, \ldots, X_n)$, and let $V = \sum_{i=n+1}^{n+m} X_i/m$. Suppose that we want to try to develop a test of the hypothesis $H : V \leq c$. First, we convert the hypothesis into a parametric hypothesis as in (3.15). For each $\delta \in (0, 1)$, let

$$\Omega_\delta = \{\theta = (\mu, \sigma) : P'_\theta(V \leq c) \geq \delta\}.$$

We might wish to choose values of δ and α and then require that for all $\theta \in \Omega_\delta$, $P_\theta(\text{reject } H) \leq \alpha$. This means that we are trying to test $H' : \Theta \in \Omega_\delta$ at level α. We will use a version of the group described in Problem 11 on page 389. An element g_a of the group acts on $\mathcal{X}$ by $g_a(x_1, \ldots, x_n) = (c + a(x_1 - c), \ldots, c + a(x_n - c))$. The maximal invariant in the sufficient statistic space is $T = \sqrt{n}(\overline{X} - c)/S$.[22] The maximal invariant in the parameter space is $B = (M - c)/\Sigma$. We know that $V \leq c$ if and only if $\sqrt{m}(V - M)/\Sigma \leq -\sqrt{m}B$, and so $P_\theta(V \leq c) \geq \delta$ if and only if $B \leq \Phi^{-1}(1-\delta)/\sqrt{m}$. So, $\Omega_\delta = \{\theta : \beta \leq \beta_0\}$, where $\beta = (\mu - c)/\sigma$ and $\beta_0 = \Phi^{-1}(1-\delta)/\sqrt{m}$. So the test we seek is equivalent to $H : B \leq \beta_0$. The conditional distribution of T given $B = \beta$ is noncentral t, $NCt_{n-1}(\sqrt{n}\beta)$. This distribution has increasing MLR in the noncentrality parameter β (see Problem 29 on page 289). The UMP invariant level α test is to reject H if T is greater than the $1 - \alpha$ quantile of the $NCt_{n-1}(\sqrt{n}\beta_0)$ distribution. Let this quantile be denoted d. Then $T > d$ is equivalent to $c \notin [\overline{X} - dS/\sqrt{n}, \infty)$, which in turn is equivalent to the test found in Example 5.73 on page 326.

6.4 Problems

Section 6.1.1:

1. Prove Proposition 6.3 on page 346.
2. Prove Proposition 6.5 on page 346.
3. Let Θ be a location parameter for X, let $\aleph = \Omega$, and suppose that $L(\theta, a)$ is a function of $\theta - a$. Prove that the risk function of a location equivariant rule δ is constant.

---

[22] The reader might wish to prove this in solving Problem 11 on page 389.

4. If Θ is a location parameter and $Y = g(X)$ is location invariant, then prove that Y is ancillary.

5. Suppose that $X_1, \ldots, X_n$ are IID given $\Theta = \theta$ each with density
$$f_{X_1|\Theta}(x|\theta) = \sqrt{\frac{2}{\pi}} \exp\left\{-\frac{1}{2}(x-\theta)^2\right\} I_{(\theta,\infty)}(x).$$
(This is called the *half-normal distribution*.) Let $L(\theta, a) = (\theta - a)^2$ and $\aleph = \Omega = \mathbb{R}$. Let G be the one-dimensional location group, $g_c(x_1, \ldots, x_n) = (x_1 + c, \ldots, x_n + c)$. Find the MRE estimator.

6. A function $g : \mathbb{R}^n \to \mathbb{R}$ is *even* if $g(-x_1, \ldots, -x_n) = g(x_1, \ldots, x_n)$. A function g is *odd* if $g(-x_1, \ldots, -x_n) = -g(x_1, \ldots, x_n)$. Suppose that S is odd and location equivariant and that T is even and location invariant. Suppose that $X_1, \ldots, X_n$ are IID with density f with respect to Lebesgue measure such that $f(c-x) = f(c+x)$ for some c and all x. (Such a density is called *symmetric* about c.) Suppose that the variances of $S(X_1, \ldots, X_n)$ and $T(X_1, \ldots, X_n)$ are both finite. Prove that the covariance between them is 0.

Section 6.1.2:

7. For each vector $x = (x_1, \ldots, x_n)$, let $k(x)$ denote the subscript of the last nonzero coordinate with $k(0, \ldots, 0) = 0$. Let $x_0 = 1$. Prove that a function u is scale invariant if and only if it is a function of x only through $y(x) = (x_1/|x_{k(x)}|, \ldots, x_n/|x_{k(x)}|)$.

8. Suppose that δ_0 is scale equivariant and not identically 0. Prove that δ_1 is scale equivariant if and only if $\delta_1 = u\delta_0$ for some scale invariant u.

Section 6.2.1:

9. Prove Proposition 6.25 on page 354.
10. Prove Proposition 6.29 on page 355.
11. *Let $X_1, \ldots, X_n$ be IID $N(\mu, \sigma^2)$ given $\Theta = (\mu, \sigma)$. Let $\aleph = \{0, 1\}$ and
$$L(\theta, a) = \begin{cases} R & \text{if } \mu \geq \mu_0 \text{ and } a = 1, \\ 1 & \text{if } \mu < \mu_0 \text{ and } a = 0, \\ 0 & \text{otherwise.} \end{cases}$$

   (a) Prove that the formal Bayes rule with respect to the improper prior with Radon–Nikodym derivative $1/\sigma$ with respect to Lebesgue measure is the usual level $1/(1+R)$ t-test.

   (b) Let G be a group that acts on $\mathcal{X}$ as follows:
   $$g_c(x_1, \ldots, x_n) = (c(x_1 - \mu_0) + \mu_0, \ldots, c(x_n - \mu_0) + \mu_0),$$
   for $c > 0$. Find $\overline{G}$ and $\tilde{G}$ so that this problem is invariant, and show that the t-test is equivariant.

Section 6.2.2:

12. Prove Proposition 6.43 on page 361. (*Hint:* The proof is very much like Example 6.42. There is no need to transform the data.)

Section 6.2.3:

13. *Let $\Omega = (0, \infty)$. Suppose that $X_1, \ldots, X_n$ are IID given $\Theta = \theta$ each with density
$$f_{X_1|\Theta}(x|\theta) = \frac{1}{\theta} f\left(\frac{x}{\theta}\right),$$
for some density function f. Let G be Group 2 and $\aleph = [0, \infty)$. Let $L(\theta, a) = (\theta^r - a)^2/\theta^{2r}$, for some $r > 0$.

    (a) Find $\tilde{G}$ so that this problem is invariant.
    (b) Characterize all equivariant rules.
    (c) Write a formula for the MRE rule.
    (d) If $f(x) = I_{[0,1]}(x)$, find the MRE rule.

14. Let $\Omega = (0, \infty)$. Suppose that $X_1, \ldots, X_n$ are IID given $\Theta = \theta$ each with density
$$f_{X_1|\Theta}(x|\theta) = \frac{1}{\theta} f\left(\frac{x}{\theta}\right),$$
for some density function f. Let G be Group 2 and $\aleph = [0, \infty)$. Let $L(\theta, a) = (k \log(\theta) - r \log(a))^2$, for some $k, r > 0$.

    (a) Find $\tilde{G}$ so that this problem is invariant.
    (b) Characterize all equivariant rules.
    (c) Write a formula for the MRE rule.

15. Let $X_1, \ldots, X_n$ be IID $U(0, \theta)$ random variables conditional on $\Theta = \theta$, and let the action space be $\aleph = [0, \infty)$. Let the loss function be $L(\theta, a) = (1 - a/\theta)^2$.

    (a) Show that this problem is invariant under the one-dimensional scale group, Group 2.
    (b) Find the MRE decision rule.

16. Prove Corollary 6.52 on page 366.

17. Prove Proposition 6.53 on page 367.

18. Suppose that $X_1, \ldots, X_n$ are IID $U(\theta_1, \theta_2 + \theta_1)$ given $\Theta = (\theta_1, \theta_2)$, where $\Omega = \mathbb{R} \times \mathbb{R}^+$. Let $\aleph = \Omega$ and
$$L(\theta, a) = \left(\frac{\theta_1 - a_1}{\theta_2}\right)^2 + \left(\frac{\theta_2 - a_2}{\theta_2}\right)^2.$$
Show that this problem is invariant under Group 3, and find the MRE decision rule.

19. Let $f : \mathbb{R} \to [0, \infty)$ be a function such that $\int |x| f(x) dx < \infty$. Suppose that $X_1, \ldots, X_n$ are conditionally IID given $\Theta = \theta$ each with density $f(x - \theta)$. Let the prior density of Θ be proportional to $f(c - \theta)$. Suppose that the loss function is $L(\theta, a) = \rho(\theta - a)$ for some function ρ. If the formal Bayes rule exists, show that it is the same as the MRE decision based on a sample containing one extra observation $X_{n+1} = c$.

20. Let $X_1, \ldots, X_n$ $(n \geq 2)$ be IID with $Exp(1/\theta)$ distribution given $\Theta = \theta$. Use the one-dimensional scale group, Group 2. Let the action space be the same as the parameter space, and let the loss be $L(\theta, a) = (\theta^2 + a^2)/(a\theta)$.

    (a) Find groups to act on the parameter and action spaces so that the decision problem is invariant.

    (b) Find the best equivariant rule.

21. Suppose that $X_1, \ldots, X_n$ are conditionally IID given $\Theta = \theta$ each with conditional density
$$f_{X_1|\Theta}(x|\theta) = \frac{\theta^\alpha \alpha}{x^{\alpha+1}} I_{[\theta, \infty)}(x),$$
where α is known and the parameter space is $\Omega = (0, \infty)$. Let Group 2 (the one-dimensional scale group) act on the data. Let the action space be the same as the parameter space.

    (a) Find groups acting on the parameter and action spaces so that the decision problem with loss $L(\theta, a) = (\theta - a)^2/\theta^2$ is invariant.

    (b) Find the MRE decision rule.

Section 6.3.1:

22. Show that Theorem 6.74 applies, and state the conclusions of the theorem in the situation described in Problem 31 on page 289.

Section 6.3.2:

23.*Each part of this question assumes the hypotheses of the preceding parts.

    (a) Let P and Q be probability measures on $(\mathbb{R}, \mathcal{B})$, where $\mathcal{B}$ is the Borel σ-field. Suppose that $X = (X_1, \ldots, X_n)$ is an IID sample from a distribution with probability measure P. Let Y be another real-valued random variable independent of X with distribution Q. Let $C = C(X)$ be a measurable subset of $\mathbb{R}$. Define the content of C by $Q(C)$. Prove that the expected value of the content equals the probability that $C(X)$ contains Y. You may assume all necessary measurability conditions.

    (b) Let Ω be a parameter space, and suppose now that P is only known to be an element of the parametric family $\{P_\theta | \theta \in \Omega\}$ and that Q is only known to be an element of the parametric family $\{Q_\theta | \theta \in \Omega\}$ (same parameter space). Let E_θ represent expectation with respect to the conditional distribution of X given $\Theta = \theta$. Suppose that we wish to choose C in order to maximize $E_\theta[Q_\theta(C)]$ uniformly in θ subject to $E_\theta[P_\theta(C)] \leq \beta$ for all θ. Prove that this is equivalent to finding a uniformly most powerful size β critical region for the hypothesis-testing problem :

H : $\quad X_1,\ldots,X_n,Y \quad$ are an IID sample from P_θ for some $\theta \in \Omega$,

A : $\quad X_1,\ldots,X_n \quad$ are an IID sample from P_θ independent of Y which has distribution Q_θ for some $\theta \in \Omega$.

(c) Suppose that $\theta = (\mu,\sigma) \in \mathbb{R} \times \mathbb{R}^+$, P_θ is the normal $N(\mu,\sigma^2)$ distribution, and Q_θ is the $N(\mu,\alpha\sigma^2)$ distribution for some known $\alpha \in (0,1)$. Show that the hypothesis-testing problem from (b) is invariant under the location-scale group.

(d) Let $S^2 = \sum_{i=1}^n (X_i - \overline{X})^2$, and show that $(Y, \overline{X}, S^2)$ is a sufficient statistic for this problem. Also, find a maximal invariant in the sufficient statistic space under the action of the location-scale group.

(e) Among all sets C as described in part (a) which are also equivariant under the action of the location-scale group on X, find the one that uniformly maximizes $E_\theta[Q_\theta(C)]$ subject to $E_\theta[P_\theta(C)] \leq \beta$ for all θ. (*Hint:* You may wish to use the form of the t density given on page 672.)

24. In Example 6.80 on page 380, prove that $p_\theta(x)$ equals the posterior probability that Θ is not in the interval $B_{x,p_\theta(x)}$.

25. Prove that Theorem 6.78 applies to the situation in Example 5.57 on page 319. For the case $\alpha = 0.05$ and $n = 10$, find the conditional confidence coefficients for the two intervals $(-\infty, T^*]$ and $[T_*, \infty)$ given the ancillary if the sufficient statistic is $(T_1, T_2) = (1, 1.3)$.

Section 6.3.3:

26. *Return to Problem 56 on page 293. Find a group of rotations and a loss function for estimating Θ_2 that make the decision problem invariant. Show that the hypothesis and alternative $H : \Theta_1 = 0$ and $A : \Theta_1 > 0$ are invariant, and find the form of the UMPUAI level α test as closely as you can. (I do not think you can find the cutoffs in closed form.)

27. *Suppose that X is distributed like $N_k(\mu, \Sigma)$ given $\Theta = (\mu, \Sigma)$. Let the group be Group 4 on page 354. Only one vector observation will be available.

   (a) Show that the family of distributions is invariant and show how a group element acts on the parameter space.

   (b) Suppose that we wish to test the hypothesis $H : \mathbf{M} = \mathbf{0}$ versus $A : \mathbf{M} \neq \mathbf{0}$. Show that the hypothesis-testing problem is invariant, and find the maximal invariant in the data space. Why are invariant tests useless in this case?

   (c) Suppose that we wish to estimate $\mathbf{M}$. Our action space is $\aleph = \mathbb{R}^k$, and our loss function is $L(\theta, a) = (\mu - a)^\top \Sigma^{-1} (\mu - a)$. Find a group $\tilde{G}$ operating on $\aleph$ so that the loss is invariant.

(d) For the estimation problem, show that all equivariant rules are of the form $\delta(x) = cx$ for some scalar c. (*Hint:* First, prove that for $i = 1, \ldots, k$, if x has 0 in coordinate i, then $\delta(x)$ has zero in coordinate i also. Finally, write $\delta(x) = \alpha(x)x + \beta(x)y(x)$, where $y(x)$ is orthogonal to x for all x and the representation is unique unless $\beta(x) = 0$. Then let A be an orthogonal matrix with first row proportional to $x^\top$ and second row proportional to $y(x)^\top$.)

28. Suppose that $X_1, \ldots, X_n$ are conditionally IID with $N(\mu, \sigma^2)$ distribution given $\Theta = (\mu, \sigma)$. Let G be the one-dimensional location group $g_c x = x + c\mathbf{1}$.

    (a) Show that Assumption 6.82 holds.

    (b) For what kinds of hypotheses can we find UMPUAI tests?

    (c) Will these tests be UMPU?

29. Suppose that $Y_{i,1}, \ldots, Y_{i,n_i}$ are conditionally distributed as $N_p(\mu_i, \sigma)$ given $M_i = \mu_i$ and $\Sigma = \sigma$ for $i = 1, \ldots, k$ and all $Y_{i,j}$ are conditionally independent. (Here Σ is a $p \times p$ positive definite matrix.) Suppose that the hypothesis to test is $H : MAC = O$, where M is the $p \times k$ matrix whose ith column is M_i, A is a $k \times k$ diagonal matrix with $\sqrt{n_i}$ in the ith diagonal element, C is a $k \times r$ matrix that equals the first r columns of an orthogonal matrix, and O is a $p \times r$ matrix of all zeros. (Compare to the one-way analysis of variance on page 384.) Transform the data in order to put this problem into the form of the multivariate analysis of variance, and find the matrices W and B in the discussion that begins on page 386.

Chapter 7

Large Sample Theory

7.1 Convergence Concepts

In calculus courses, the concept of convergence of sequences is introduced. In this section, we will generalize that concept to include different types of stochastic convergence.

7.1.1 Deterministic Convergence

We begin by defining types of deterministic convergence.

Definition 7.1. Let $\{x_n\}_{n=1}^\infty$ be a sequence in a normed linear space,[1] and let $\{r_n\}_{n=1}^\infty$ be a sequence of real numbers. We say that x_n *is small order of* r_n *(as* $n \to \infty$*)*, denoted $x_n = o(r_n)$, if for each $c > 0$ there exists N such that $\|x_n\| \leq c|r_n|$ for each $n \geq N$. We say that x_n *is large order of* r_n *(as* $n \to \infty$*)*, denoted $x_n = \mathcal{O}(r_n)$, if there exists $c > 0$ and N such that $\|x_n\| \leq c|r_n|$ for each $n \geq N$. If $\{y_n\}_{n=1}^\infty$ is a sequence of vectors and $x_n - y_n = o(r_n)$ (or $\mathcal{O}(r_n)$), then we write $x_n = y_n + o(r_n)$ (or $y_n + \mathcal{O}(r_n)$.)

What large order and small order allow us to do is to discuss limits of ratios without being explicit about the ratios as long as they stay bounded or go to zero. Large order means that the ratio of the quantities remains bounded. Small order means that the ratio goes to 0.

Example 7.2. Since $\lim_{n \to \infty} \log(n)/n = 0$, we have $\log(n) = o(n)$. Also, $n^r = o(n^p)$ if $p > r$. It is easy to prove that $\binom{n}{k} = \mathcal{O}(n^k)$ for fixed k.

---

[1] The norm of x is denoted by $\|x\|$. Note that a normed linear space is a metric space with metric $d(x, y) = \|x - y\|$.

Here are some simple consequences of the definitions:

- If $x_n = o(r_n)$, then $x_n = \mathcal{O}(r_n)$.
- If c is real and nonzero, then $x_n = \mathcal{O}(r_n)$ if and only if $x_n = \mathcal{O}(cr_n)$. Similarly, $x_n = o(r_n)$ if and only if $x = o(cr_n)$.
- Suppose that $y_n = o(r_n)$ and $x_n = \mathcal{O}(y_n)$. Then $x_n = o(r_n)$. If $x_n = o(y_n)$, then $x_n = o(r_n)$.
- If $x_n = o(r_n)$ and $y_n = o(s_n)$, then $x_n + y_n = o(|r_n| + |s_n|)$. Similarly, if $x_n = \mathcal{O}(r_n)$ and $y_n = \mathcal{O}(s_n)$, then $x_n + y_n = \mathcal{O}(|r_n| + |s_n|)$.
- If $x_n = o(r_n)$ and $y_n = \mathcal{O}(s_n)$, then $x_n + y_n = \mathcal{O}(|r_n| + |s_n|)$.
- If $x_n = o(r_n)$ and $y_n = o(s_n)$, then $x_n y_n = o(r_n s_n)$. Similarly, if $x_n = \mathcal{O}(r_n)$ and $y_n = \mathcal{O}(s_n)$, then $x_n y_n = \mathcal{O}(r_n s_n)$.
- If $x_n = o(r_n)$ and $y_n = \mathcal{O}(s_n)$, then $x_n y_n = o(r_n s_n)$.

There will be several situations in which we need to use the concepts of small order and large order. Let $\{r_n\}_{n=1}^\infty$ be a sequence of real numbers.

1. If $\limsup \|x_n/r_n\| < \infty$, then $x_n = \mathcal{O}(r_n)$.
2. If $\limsup \|x_n/r_n\| = 0$, then $x_n = o(r_n)$.
3. $x_n = o(1)$ if and only if $\lim_{n\to\infty} x_n = 0$.
4. If $r_n = o(1)$ and m is fixed, then $(1 + r_n)^m = 1 + o(1)$.
5. If $x_{n,k} = o(r_n)$ as $n \to \infty$ for each $k = 1, \ldots, m$, then $\sum_{k=1}^m x_{n,k} = o(r_n)$ if m is fixed.

This last example requires that m be fixed as $n \to \infty$. To see that it is false otherwise, consider $x_{n,k} = 2^k/n = o(1)$ as $n \to \infty$. But, $\sum_{k=1}^n x_{n,k} \to \infty$ as $n \to \infty$.

7.1.2 Stochastic Convergence

Next, we define stochastic versions of small order and large order. The setup requires a sequence of probability spaces $\{(\mathcal{X}_n, \mathcal{B}_n, P_n)\}_{n=1}^\infty$. Here, we assume that each space $\mathcal{X}_n$ is a normed linear space with norm $\|\cdot\|_n$ and that there are functions $X_n : S \to \mathcal{X}_n$ where $(S, \mathcal{A}, \mu)$ is an *underlying probability space*. (As before, $\mu(A)$ for $A \in \mathcal{A}$ will often be denoted $\Pr(A)$ and conditional probabilities derived from μ denoted $\Pr(\cdot|\cdot)$.) In this case, P_n is the probability induced on $(\mathcal{X}_n, \mathcal{B}_n)$ by X_n from μ. A common example is the one in which $S = \mathbb{R}^\infty$, $\mathcal{X}_n = \mathbb{R}^n$, and X_n is the first n coordinates. All of the results in this section and Section 7.2 apply equally well to cases in which the probabilities P_n are already conditional probabilities given some parameter Θ. Of course, in such cases, P_n would actually be $P_{\theta,n}$ and Pr would be denoted P'_θ. Problems 5 and 6 (see page 468) show how to convert certain limit theorems that are conditional on Θ into marginal limit theorems.

Definition 7.3. Let $\{X_n\}_{n=1}^\infty$ be a sequence of random quantities as above and let $\{r_n\}_{n=1}^\infty$ be a sequence of numbers. We say that X_n is *stochastically small order of* r_n *(as* $n \to \infty$*)*, denoted $X_n = o_P(r_n)$, if, for each $c > 0$ and each $\epsilon > 0$, there exists N such that $\Pr(\|X_n\|_n \le c|r_n|) \ge 1 - \epsilon$ for all $n \ge N$. We say X_n is *stochastically large order of* r_n *(as* $n \to \infty$*)*, denoted $X_n = \mathcal{O}_P(r_n)$, if, for each $\epsilon > 0$, there exists $c > 0$ and N such that $\Pr(\|X_n\|_n \le c|r_n|) \ge 1 - \epsilon$ for all $n \ge N$. If $\{Y_n\}_{n=1}^\infty$ is a sequence of random vectors and $X_n - Y_n = o_P(r_n)$ (or $\mathcal{O}_P(r_n)$), then we write $X_n = Y_n + o_P(r_n)$ (or $Y_n + \mathcal{O}_P(r_n)$.)

Proposition 7.4. $X_n = o_P(r_n)$ *if and only if, for each* $c > 0$,

$$\lim_{n \to \infty} \Pr(\|X_n\|_n \le c|r_n|) = 1.$$

Note that in the definition of $\mathcal{O}_p$, the c is allowed to vary with ϵ, so there is no obvious analog to Proposition 7.4 for $\mathcal{O}_P$. We will usually leave the subscript n off of the norm $\|\cdot\|_n$, since there is seldom any chance of confusing one norm with another.

Example 7.5. Let $\{Z_n\}_{n=1}^\infty$ be IID random variables with mean μ and variance σ^2. Let $X_n = \sqrt{n}(\overline{Z}_n - \mu)/\sigma$. So, $\mathcal{X}_n = \mathbb{R}$ for every n, and P_n, which is the distribution of X_n, is a probability measure on the Borel subsets of the real line. The central limit theorem B.97 (together with Problem 25 on page 664) says that $\lim_{n \to \infty} P_n((-\infty, t]) = \Phi(t)$ for all t, where Φ is the standard normal CDF. For each $\epsilon > 0$, there exists t such that $\Phi(t) - \Phi(-t) \ge 1 - \epsilon/2$. Choose N such that for each $n \ge N$,

$$P_n((-\infty, -t]) \le \Phi(-t) + \frac{\epsilon}{4}, \quad P_n((-\infty, t]) \ge \Phi(t) - \frac{\epsilon}{4}.$$

It follows that $\Pr(|X_n| \le t)$ equals

$$P_n((-\infty, t]) - P_n((-\infty, -t]) \ge \Phi(t) - \Phi(-t) - \frac{\epsilon}{2} \ge 1 - \epsilon.$$

Hence, $X_n = \mathcal{O}_P(1)$.[2] Also, $\overline{Z}_n - \mu = \mathcal{O}_P(1/\sqrt{n})$. If $0 \le \alpha < 1/2$, then $\overline{Z}_n - \mu = o_P(n^{-\alpha})$. In particular $(\alpha = 0)$, $\overline{Z}_n - \mu = o_P(1)$.

Stochastic convergence is closely related to the concept of convergence in probability. We restate Definition B.89 in the present context.

Definition 7.6. If $\{X_n\}_{n=1}^\infty$ and X are random quantities in a normed linear space, and if, for every $\epsilon > 0$, $\lim_{n \to \infty} \Pr(\|X_n - X\| > \epsilon) = 0$, then we say that X_n *converges in probability to* X, which is written $X_n \xrightarrow{P} X$.

---

[2]This phenomenon is quite general. See Problem 3 on page 467.

7.1. Convergence Concepts

Proposition 7.7. *Suppose that $Y_n = f_n(X_n)$ for each n where $f_n : \mathcal{X}_n \to R$, and R is a normed linear space with Borel σ-field. Assume that each f_n is measurable. Let $Y : S \to R$ be another random quantity. Then $\|Y_n - Y\| = o_P(1)$, if and only if Y_n converges in probability to Y.*

Example 7.8. Suppose that $\lim_{n\to\infty} E(Y_n - c)^2 = 0$. Then Tchebychev's inequality can be used to prove that $Y_n \xrightarrow{P} c$.

Definition 7.9. Let $\{P_\theta : \theta \in \Omega\}$ be a parametric family of distributions on a sequence space $\mathcal{X}^\infty$, and let $g : \Omega \to G$ be a measurable function to a metric space G with Borel σ-field. Let $\mathcal{X}_n = \mathcal{X}^n$, and let $Y_n : \mathcal{X}_n \to G$ be measurable. We say that Y_n is *consistent for* $g(\Theta)$ if $Y_n \xrightarrow{P} g(\theta)$ conditional on $\Theta = \theta$ for all $\theta \in \Omega$.

Example 7.10. Let $\{X_n\}_{n=1}^\infty$ be conditionally IID $N(\mu, \sigma^2)$ given $\Theta = (\sigma, \mu)$. Let $Y_n = \sum_{i=1}^n X_i/n$ and $g(\theta) = \mu$. Then Y_n is consistent for $g(\Theta)$ according to the weak law of large numbers B.95.

The following is a more general definition of "in probability."

Definition 7.11. Suppose that $\{(\mathcal{X}_n, \mathcal{B}_n, P_n)\}_{n=1}^\infty$ is a sequence of probability spaces. Define $\mathcal{Y} = \prod_{n=1}^\infty \mathcal{X}_n$. Let $T \subseteq \mathcal{Y}$. We say that T *occurs in probability*, denoted $\mathcal{P}(T)$, if, for each $\epsilon > 0$, there exists $T_n(\epsilon) \in \mathcal{B}_n$ for $n = 1, 2, \ldots$ such that $P_n(T_n(\epsilon)) \geq 1 - \epsilon$ for each n and $\prod_{n=1}^\infty T_n(\epsilon) \subseteq T$.

The following lemma essentially says that a sequence of random quantities $\{Y_n\}_{n=1}^\infty$ is $\mathcal{O}_P(r_n)$ or $o_P(r_n)$ if and only if the set of possible values for $(Y_1, Y_2, \ldots)$ which are $\mathcal{O}(r_n)$ or $o(r_n)$ occurs in probability.

Lemma 7.12. *Use the notation from Definition 7.11. Let $Y_n = f_n(X_n)$ and let*
$$T = \{(x_1, x_2, \ldots) \in \mathcal{Y} : f_n(x_n) = o(r_n)\}.$$
Then $\mathcal{P}(T)$ if and only if $Y_n = o_P(r_n)$. Similarly, if
$$T = \{(x_1, x_2, \ldots) \in \mathcal{Y} : f_n(x_n) = \mathcal{O}(r_n)\},$$
then $\mathcal{P}(T)$ if and only if $Y_n = \mathcal{O}_P(r_n)$.

PROOF. We will do only the o_P part since the $\mathcal{O}_P$ part is similar. First, for the "if" part, assume $Y_n = o_P(r_n)$ and let $\epsilon > 0$. Let $c_1 > c_2 > \cdots$ decrease to 0. For each $i > 1$, let $N(\epsilon, c_i) \geq N(\epsilon, c_{i-1})$ be such that for $n \geq N(\epsilon, c_i)$, $\Pr(\|Y_n\| \leq c_i|r_n|) \geq 1 - \epsilon$. Define $T_n(\epsilon) = \mathcal{X}_n$ for $n = 1, \ldots, N(\epsilon, c_1)$. For $N(\epsilon, c_{i-1}) < n \leq N(\epsilon, c_i)$, define
$$T_n(\epsilon) = \{x_n : \|f_n(x_n)\| \leq |r_n|c_{i-1}\}.$$

By construction, we have $P_n(T_n(\epsilon)) \geq 1 - \epsilon$ for every n. If $(x_1, x_2, \ldots) \in \prod_{n=1}^\infty T_n(\epsilon)$, then $\lim_{n\to\infty} \|f_n(x_n)\|/|r_n| = 0$ by construction. It follows that $(x_1, x_2, \ldots) \in T$ and we have proven $\mathcal{P}(T)$.

For the "only if" part, assume that $\mathcal{P}(T)$ and let $T_n(\epsilon)$ be as in Definition 7.11. Since $f_n(x_n) = o(r_n)$ for $(x_1, x_2, \ldots) \in T$, it follows that

$$z_n = \sup_{x_n \in T_n(\epsilon)} \frac{\|f_n(x_n)\|}{|r_n|} < \infty,$$

for all but finitely many n. Hence,

$$P_n\left(\left\{x_n : \frac{\|f_n(x_n)\|}{|r_n|} \leq z_n\right\}\right) \geq P_n(T_n(\epsilon)) \geq 1 - \epsilon.$$

Now, choose $x_n^* \in T_n(\epsilon)$ such that

$$z_n \leq \frac{\|f_n(x_n^*)\|}{|r_n|} + \frac{1}{n} \to 0, \text{ as } n \to \infty,$$

so $\lim_{n \to \infty} z_n = 0$. For each $\epsilon > 0$ and $c > 0$, choose N such that if $n \geq N$, then $z_n \leq c$. It follows that, if $n \geq N$, then $\Pr(\|Y_n\|/|r_n| \leq c) \geq 1 - \epsilon$. Hence, $Y_n = o_P(r_n)$. □

Example 7.13. Let $f_n : \mathcal{X}_n \to \mathbb{R}$ and $Y_n = f_n(X_n)$. Define

$$T = \{(x_1, x_2, \ldots) : \lim_{n \to \infty} f_n(x_n) = 0\}.$$

Then $Y_n = o_P(1)$ if and only if $\mathcal{P}(T)$, according to Lemma 7.12.

If countably many things occur in probability, then they simultaneously occur in probability.

Proposition 7.14.[3] *If $\mathcal{P}(S_i)$ for $i = 1, \ldots$, then $\mathcal{P}(\cap_{i=1}^\infty S_i)$. If $T \subseteq S$, then $\mathcal{P}(T)$ implies $\mathcal{P}(S)$.*

We are now in position to prove a theorem that says (in a more precise manner) that if you can prove a result involving o and $\mathcal{O}$, then you can replace o by o_P and $\mathcal{O}$ by $\mathcal{O}_P$ and prove a corresponding result.

Theorem 7.15.[4] *Let $\mathcal{Y}_0, \mathcal{Y}_{1,1}, \mathcal{Y}_{1,2}, \ldots, \mathcal{Y}_{2,1}, \mathcal{Y}_{2,2}, \ldots$ be metric spaces. Let $h_n : \mathcal{X}_n \to \mathcal{Y}_0$, $f_n^{(j)} : \mathcal{X}_n \to \mathcal{Y}_{1,j}$, for $j = 1, 2 \ldots$, and $g_n^{(k)} : \mathcal{X}_n \to \mathcal{Y}_{2,k}$ for $k = 1, 2, \ldots$. Suppose that $f_n^{(j)}(X_n) = \mathcal{O}_P(r_n^{(j)})$ and $g_n^{(k)}(X_n) = o_P(s_n^{(k)})$ for all j and k. Also, suppose that it is known that*

$$\left(f_n^{(j)}(x_n) = \mathcal{O}(r_n^{(j)}) \text{ and } g_n^{(k)}(x_n) = o(s_n^{(k)}) \text{ for all } j \text{ and } k\right) \text{ implies}$$

$$h_n(x_n) = \mathcal{O}(t_n) \ (or \ (h_n(x_n) = o(t_n)),$$

then $h_n(X_n) = \mathcal{O}_P(t_n)$ (or $h_n(X_n) = o_P(t_n)$).

---

[3]This proposition is used in the proof of Theorem 7.15.
[4]This theorem is used to help develop the delta method.

PROOF. We will only prove the $\mathcal{O}_P$ part. The o_P part is virtually identical. Let $S^{(2j-1)} = \{x : f_n^{(j)}(x_n) = \mathcal{O}(r_n^{(j)})\}$ for all j $S^{(2k)} = \{x : g_n^{(k)}(x_n) = o(s_n^{(k)})\}$ for all k. (If there are only finitely many $f_n^{(j)}$ or $g_n^{(k)}$, then just let $S^{(t)} = \prod_{n=1}^{\infty} \mathcal{X}_n$ after you run out of functions.) Define $T = \{x : h_n(x_n) = \mathcal{O}(t_n)\}$. The stated conditions imply that $\cap_{i=1}^{\infty} S^{(i)} \subseteq T$. Also, we have assumed that $\mathcal{P}(S^{(i)})$ for all i, so $\mathcal{P}(T)$ by Proposition 7.14. □

Example 7.16. Suppose that $w : \mathbb{R} \to \mathbb{R}$ has $k+1$ continuous derivatives at c. Define
$$T_k(x, c) = w(c) + (x - c)w'(c) + \cdots + \frac{1}{k!}(x - c)^k w^{(k)}(c),$$
where $g^{(k)}$ denotes the kth derivative of g. Taylor's theorem C.1 says (among other things) that
$$\lim_{x \to c} \frac{w(x) - T_k(x, c)}{(x - c)^k} = 0.$$
Suppose that $x_n - c = \mathcal{O}(r_n)$, where $r_n = o(1)$. Then $x_n - c = o(1)$, and we conclude that $w(x_n) - T_k(x_n, c) = o((x_n - c)^k)$, hence $w(x_n) = T_k(x_n, c) + o(r_n^k)$. Similarly, we can write $w(x_n) = T_{k-1}(x_n, c) + \mathcal{O}(r_n^k)$. Now, suppose that $X_n - c = \mathcal{O}_P(r_n)$. In the notation of Theorem 7.15, let $\mathcal{X}_n = \mathbb{R}$ for all n and let $\mathcal{Y}_0 = \mathcal{Y}_{1,1} = \mathbb{R}$. For each n, let $f_n^{(1)}(x) = x$ and $h_n(\cdot) = w(\cdot) - T_k(\cdot, c)$ or $w(\cdot) - T_{k-1}(\cdot, c)$. Suppose that there are no g functions. Then Theorem 7.15 says that
$$w(X_n) = T_k(X_n, c) + o_P(r_n^k) = T_{k-1}(X_n, c) + \mathcal{O}_P(r_n^k).$$
Furthermore, if w has $k+1$ continuous derivatives everywhere, then if $X_n - X_n^* = \mathcal{O}_P(r_n)$, then $w(X_n) = T_k(X_n, X_n^*) + o_P(r_n^k) = T_{k-1}(X_n, X_n^*) + \mathcal{O}_P(r_n^k)$.

Corollary 7.17.[5] *Let $\mathcal{Y}$ and $\mathcal{Z}$ be metric spaces. If $Y_n = f_n(X_n) \in \mathcal{Y}$ and $Y_n \xrightarrow{P} c \in \mathcal{Y}$ and $g : \mathcal{Y} \to \mathcal{Z}$ is continuous at c, then $g(Y_n) \xrightarrow{P} g(c)$.*

Another type of stochastic convergence is convergence in distribution. We restate Definition B.80 here.

Definition 7.18. Let $\{X_n\}_{n=1}^{\infty}$ be a sequence of random quantities and let X be another random quantity, all taking values in the same topological space $\mathcal{X}$. Suppose that
$$\lim_{n \to \infty} \mathrm{E}(f(X_n)) = \mathrm{E}(f(X))$$
for every bounded continuous function $f : \mathcal{X} \to \mathbb{R}$; then we say that X_n *converges in distribution to* X, which is written $X_n \xrightarrow{D} X$ or $\mathcal{L}(X_n) \to \mathcal{L}(X)$. If $X_n \xrightarrow{D} X$, we call the distribution of X the *asymptotic distribution* of X_n. If $X_n \xrightarrow{D} X$, and if R_n and R are the distributions of X_n and X, respectively, then we say that R_n *converges weakly to* R, denoted $R_n \xrightarrow{W} R$.

---

[5]This corollary is used to help prove that posterior distributions are asymptotically normal.

The portmanteau theorem B.83 gives several criteria that are equivalent to convergence in distribution. These can be used to derive a connection between convergence in distribution and o_P.

Lemma 7.19.[6] *Suppose that $\mathcal{X}$ is a metric space with metric d. If $X_n \xrightarrow{D} X$ and $d(X_n, Y_n) = o_P(1)$, then $Y_n \xrightarrow{D} X$.*

PROOF. Let R_n be the distribution of Y_n, and let P be the distribution of X. We must show that $R_n \xrightarrow{W} P$. (See Definition 7.18.) Let B be an arbitrary closed set. According to the portmanteau theorem B.83, it suffices to show that $\limsup R_n(B) \leq P(B)$. Define, for $C \in \mathcal{B}$,

$$d(x, C) = \inf_{y \in C} d(x, y).$$

Then
$$\{Y_n \in B\} \subseteq \{d(X_n, B) \leq \epsilon\} \cup \{d(X_n, Y_n) > \epsilon\}.$$

Define $C_\epsilon = \{x : d(x, B) \leq \epsilon\}$, which is a closed set. So,

$$\begin{aligned} R_n(B) &= \Pr(Y_n \in B) \\ &\leq \Pr(d(X_n, B) \leq \epsilon) + \Pr(d(X_n, Y_n) > \epsilon) \\ &= P_n(C_\epsilon) + \Pr(d(X_n, Y_n) > \epsilon). \end{aligned}$$

We have assumed that $\lim_{n \to \infty} \Pr(d(X_n, Y_n) > \epsilon) = 0$ and that $X_n \xrightarrow{D} X$, so we conclude $\limsup_{n \to \infty} R_n(B) \leq \limsup_{n \to \infty} P_n(C_\epsilon) \leq P(C_\epsilon)$. Since B is closed, $\lim_{\epsilon \to 0} P(C_\epsilon) = P(B)$. It follows then that

$$\limsup_{n \to \infty} R_n(B) \leq P(B),$$

hence $Y_n \xrightarrow{D} X$. □

Lemma 7.19 says that if $X_n \xrightarrow{D} X$, then so too does anything close to X_n, that is, anything that differs from X_n by $o_P(1)$.

Theorem 7.20. *If the σ-field on $\mathcal{X} \times \mathcal{Y}$ is the product σ-field, if $X_n \xrightarrow{D} X$ and $Y_n \xrightarrow{D} Y$, and if X_n is independent of Y_n for all n, then $(X_n, Y_n) \xrightarrow{D} (X, Y)$, where X and Y are independent.*

PROOF. Since X_n and Y_n are independent for each n, their joint characteristic function is

$$\phi(t, s) = \mathrm{E} \exp\left\{i \begin{pmatrix} t \\ s \end{pmatrix}^\top \begin{pmatrix} X_n \\ Y_n \end{pmatrix}\right\} = \mathrm{E} \exp\{it^\top X_n\} \mathrm{E} \exp\{is^\top Y_n\}.$$

---

[6]This lemma is used in the proofs of Theorems 7.22, 7.25, 7.35, and 7.63 and to help develop the delta method.

This product converges to $\mathrm{E}\exp\{it^\top X\}\, \mathrm{E}\exp\{is^\top Y\}$, which is the characteristic function of independent X and Y. Now apply Theorem B.93. □

Using the fact that a constant is independent of everything, we have the following simple corollary to Theorem 7.20.

Corollary 7.21.[7] *Suppose that $\{X_n\}_{n=1}^\infty$ take values in a metric space $\mathcal{X}$. If $X_n \xrightarrow{D} X$ and $b \in \mathcal{Y}$ is a constant, then $(X_n, b) \xrightarrow{D} (X, b)$.*

The conclusions of the following theorem are taken for granted in many calculations of asymptotic distributions.

Theorem 7.22. *1. Suppose that $\{X_n\}_{n=1}^\infty$ take values in a topological space $\mathcal{X}$ and that $\{Y_n\}_{n=1}^\infty$ take values in a topological space $\mathcal{Y}$. If $(X_n, Y_n) \xrightarrow{D} (X, Y)$, then $X_n \xrightarrow{D} X$.*

2. Suppose that $\{X_n\}_{n=1}^\infty$ take values in a metric space $\mathcal{X}$ and that $\{Y_n\}_{n=1}^\infty$ take values in a metric space $\mathcal{Y}$. Let $b \in \mathcal{Y}$. If $X_n \xrightarrow{D} X$ and $Y_n \xrightarrow{P} b$, then $(X_n, Y_n) \xrightarrow{D} (X, b)$.

PROOF. For part 1, let $g : \mathcal{X} \times \mathcal{Y} \to \mathcal{X}$ be defined by $g(x, y) = x$. Then g is continuous and the continuous mapping theorem B.88 says that $x_n = g(X_n, Y_n) \xrightarrow{D} g(X, Y) = X$.

For part 2, let d_1 be the metric in $\mathcal{X}$ and let d_2 be the metric in $\mathcal{Y}$. Then

$$d((x_1, y_1), (x_2, y_2)) = d_1(x_1, x_2) + d_2(y_1, y_2)$$

is a metric in $\mathcal{X} \times \mathcal{Y}$ and the product σ-field is the Borel σ-field. By Corollary 7.21, we have that $(X_n, b) \xrightarrow{D} (X, b)$. We have assumed that $d((X_n, Y_n), (X_n, b)) = d_2(Y_n, b) = o_P(1)$. So, by Lemma 7.19, $(X_n, Y_n) \xrightarrow{D} (X, b)$. □

7.1.3 The Delta Method

A method for finding the asymptotic distribution of a function of a random vector is based on Lemma 7.19 and is called *the delta method*. [See Rao (1973), Chapter 6.] As an example, let Y_n be the average of n IID random variables with mean μ and variance σ^2. The central limit theorem B.97 says that $\sqrt{n}(Y_n - \mu) \xrightarrow{D} N(0, \sigma^2)$.[8] Now, let g be a function with continuous derivative. We can write

$$g(t) = g(\mu) + (t - \mu)g'(\mu) + o(t - \mu).$$

---

[7]This corollary is used in the proof of Theorems 7.22.
[8]It is common to call an estimator Z_n, with the property that $\sqrt{n}(Z_n - \theta)$ converges in distribution to a nondegenerate distribution, $\sqrt{n}$-*consistent*.

If we are interested in $g(Y_n)$, we can write
$$g(Y_n) = g(\mu) + (Y_n - \mu)g'(\mu) + o(Y_n - \mu),$$
$$\sqrt{n}(g(Y_n) - g(\mu)) = \sqrt{n}(Y_n - \mu)g'(\mu) + \sqrt{n}o(Y_n - \mu).$$

Since $\sqrt{n}(Y_n - \mu)$ converges in distribution, $Y_n - \mu = \mathcal{O}_P(1/\sqrt{n})$. Hence $\sqrt{n}o(Y_n - \mu) = o_P(1)$ by Theorem 7.15. So,
$$\sqrt{n}(g(Y_n) - g(\mu)) = \sqrt{n}(Y_n - \mu)g'(\mu) + o_P(1).$$

By Lemma 7.19, we get a useful result when $g'(\mu) \neq 0$:
$$\sqrt{n}(g(Y_n) - g(\mu)) \xrightarrow{D} N(0, \sigma^2[g'(\mu)]^2).$$

The result in the example above suggests a valuable use for the delta method. If the variance of the asymptotic distribution of $\sqrt{n}(Y_n - \mu)$ is an undesirable quantity in the application for which it is intended, then a transformation of Y_n will have a different variance that may be more suitable. For example, suppose that nY_n has $Bin(n,p)$ distribution given $P = p$. The asymptotic distribution of $\sqrt{n}(Y_n - p)$ is $N(0, p(1-p))$ given $P = p$. For comparing several possible values of P, it might be nice if the only dependence of the random variable on P were through the mean. This can be arranged asymptotically by choosing a function g such that
$$g'(p) = \frac{1}{\sqrt{p(1-p)}}.$$

This is a simple differential equation to solve, and the solution is $g(t) = 2\arcsin(\sqrt{t})$. The asymptotic distribution of
$$\sqrt{n}\left[2\arcsin\left(\sqrt{Y_n}\right) - 2\arcsin\left(\sqrt{p}\right)\right]$$
is $N(0,1)$ given $P = p$. This is a special case of what is called a *variance stabilizing transformation*. The general method for constructing a variance stabilizing transformation is as follows. Suppose that $\sqrt{n}(Y_n - \mu)$ has asymptotic distribution $N(0, h(\mu))$. Then, choose a function $g(t)$ such that $g'(\mu) = 1/\sqrt{h(\mu)}$. That is,
$$g(t) = \int_c^t \frac{1}{\sqrt{h(\mu)}} d\mu,$$
where c is any constant such that the integral exists. The asymptotic distribution of $\sqrt{n}(g(Y_n) - g(\mu))$ will be $N(0,1)$. It is common, when $\sqrt{n}(Y_n - \mu) \xrightarrow{D} N(0, \sigma^2)$, to say that the asymptotic distribution of Y_n is $N(\mu, \sigma^2/n)$. In symbols, we may write $Y_n \sim AN(\mu, \sigma^2/n)$. In such cases we will call σ^2/n the *asymptotic variance* of Y_n.

There is also a multivariate delta method. If $g : \mathbb{R}^k \to \mathbb{R}$ has continuous first partial derivatives, let $\nabla g(\mu)$ be the gradient (vector of first partial derivatives) at μ. Then $g(t) = g(\mu) + (t-\mu)^\top \nabla g(\mu) + o(t-\mu)$. If $\sqrt{n}(Y_n - \mu) \xrightarrow{D} N_k(\mathbf{0}, \sigma)$, then

$$\sqrt{n}[g(Y_n) - g(\mu)] \xrightarrow{D} N(0, \nabla g(\mu)^\top \sigma \nabla g(\mu)).$$

Here are some multivariate applications of the delta method.

Example 7.23. Importance sampling (see Section B.7) is a means of approximating the ratio of integrals of the form $\int v(\theta)h(\theta)d\theta / \int h(\theta)d\theta$. Let $\{X_n\}_{n=1}^\infty$ be an IID sequence of pseudorandom numbers with density f, and let $W_i = h(X_i)/f(X_i)$ and $Z_i = v(X_i)W_i$. If these have finite variance, then the sample averages $(\overline{W}_n, \overline{Z}_n)$ will, by the multivariate central limit theorem B.99, be approximately bivariate normal with mean $(\omega, \xi) = (\int h(\theta)d\theta, \int v(\theta)h(\theta)d\theta)$ and covariance matrix equal to $1/n$ times the covariance matrix $\sigma = ((\sigma_{i,j}))$ of the (W_i, Z_i) pairs. Now, apply the delta method to find the asymptotic distribution of the ratio of the sample averages. The asymptotic mean is ξ/ω, the ratio we want to approximate, and the asymptotic variance is

$$\sigma_{1,1} \frac{\xi^2}{\omega^4} + \sigma_{2,2} \frac{1}{\omega^2} - 2\sigma_{1,2} \frac{\xi}{\omega^3},$$

In practice, it is common to approximate σ by the sample covariance matrix of the (W_i, Z_i) pairs.

The following example uses the reasoning behind the delta method without using the delta method itself.

Example 7.24. Suppose that we wish to find the asymptotic distribution of the roots of polynomials with random coefficients. Let $Y_n \sim AN_{k+1}(\mu, \Sigma/n)$, where $Y_n = (Y_{n0}, \ldots, Y_{nk})^\top$. Define the polynomial

$$p_n(u) = \sum_{j=0}^{k} Y_{nj} u^j.$$

Let U_n^* be the smallest root of $p_n(u)$. Define $p(u) = \sum_{j=0}^{k} \mu_j u^j$, and suppose that its smallest root is u_0 and this root has multiplicity one. That is, $p(u_0) = 0$ but $p'(u_0) \neq 0$. It is not difficult to show that the smallest root with odd multiplicity of a polynomial is a continuous function of the coefficients.[9] It follows

---

[9] The main reason is that a polynomial changes sign as the variable passes a root of odd multiplicity. There will be points arbitrarily close to the root at which the polynomial has opposite signs. If the coefficients don't change much, the signs will remain the same at these points, hence a root will be between them. If the root had even multiplicity, the polynomial would have constant sign in a neighborhood of the root, and small changes in the coefficients could remove all roots from the neighborhood.

from Theorem 7.15 that $U_n^* \xrightarrow{P} u_0$. To find the asymptotic distribution of U_n^*, write $p_n(U_n^*)$ as

$$0 = p_n(U_n^*) = p_n(u_0) + (U_n^* - u_0)p_n'(V_n^*),$$

where V_n^* is between u_0 and U_n^*. So, $S \subseteq \{V_n^* \to u_0\}$, and $V_n^* \xrightarrow{P} u_0$ also. Furthermore,

$$p_n'(V_n^*) = \sum_{j=1}^{k} j(V_n^*)^{j-1} Y_{jn} \xrightarrow{P} \sum_{j=1}^{k} j u_0^{j-1} \mu_j = p'(u_0) \neq 0.$$

So, $U_n^* - u_0 = [p_n(U_n^*) - p_n(u_0)]/W_n^*$, where $W_n^* = p_n'(V_n^*) \xrightarrow{P} p'(u_0)$. Now, let $u^\top = (1, u_0, \ldots, u_0^k)$ and write

$$\sqrt{n}(U_n^* - u_0) = -\sqrt{n} \frac{p_n(u_0)}{W_n^*} = -\sqrt{n} \frac{\sum_{j=0}^{k} Y_{nj} u_0^j}{W_n^*}$$

$$= -\sqrt{n} \frac{\sum_{j=0}^{k}(Y_{nj} - \mu_j) u_0^j}{W_n^*} = -\frac{\sqrt{n}}{W_n^*} u^\top (Y_n - \mu),$$

which converges in distribution to $N(0, u^\top \Sigma u / [p'(u_0)]^2)$ by Theorem 7.22.

7.2 Sample Quantiles

The reader interested in a thorough treatment of sample quantiles should read the book by David (1970). In this section, we present some of the more commonly used asymptotic results on the distribution of sample quantiles.

7.2.1 A Single Quantile

Suppose that $\{X_n\}_{n=1}^{\infty}$ are conditionally IID random variables with distribution P given $\mathbf{P} = P$ and suppose that P has a CDF F with derivative f (at least in a neighborhood of x_p where $F(x_p) = p$) and $\infty > f(x_p) > 0$. If the observed values of the first n X_i, when ordered from smallest to largest, are $x_{(1)}, \ldots, x_{(n)}$, define the *empirical CDF* by $F_n(x_{(i)}) = i/n$ and interpolate linearly in between. (Do something arbitrary, but continuous and strictly increasing below $x_{(1)}$.) Now, F_n is continuous and strictly increasing on $(-\infty, x_{(n)}]$.[10] Define the *sample p quantile* by $Y_p^{(n)} = F_n^{-1}(p)$, for $0 < p < 1$.

The goal of this section is to prove a theorem specifying the asymptotic distribution of a sample quantile.

---

[10]If $F(c) = 0$ and $F(x) > 0$ for $x > c$, then we only need F_n to be strictly increasing on $[c, x_{(n)}]$.

7.2. Sample Quantiles

Theorem 7.25. *Suppose that $\{X_n\}_{n=1}^{\infty}$ are conditionally IID with distribution P given $\mathbf{P} = P$ and suppose that P has CDF F with derivative f in a neighborhood of x_p, where $F(x_p) = p$, $0 < f(x_p) < \infty$, and $0 < p < 1$. Define $Y_p^{(n)} = F_n^{-1}(p)$, where F_n is the empirical CDF of $(X_1, \ldots, X_n)$. Then*

$$\sqrt{n}(Y_p^{(n)} - x_p) \xrightarrow{D} N\left(0, \frac{p(1-p)}{f^2(x_p)}\right).$$

The proof relies heavily on the following lemma.

Lemma 7.26. *For each $z \in \mathbb{R}$ there exists a sequence of random variables $\{A_n(z)\}_{n=1}^{\infty}$ such that $A_n(z) = o_P(1/\sqrt{n})$ and*

$$\sqrt{n}(Y_p^{(n)} - x_p) \leq z, \text{ if and only if } \frac{\sqrt{n}}{f(x_p)}(p - F_n(x_p)) \leq z + \sqrt{n}\frac{A_n(z)}{f(x_p)}. \tag{7.27}$$

PROOF. Define

$$A_n(z) = F_n\left(x_p + \frac{z}{\sqrt{n}}\right) - F_n(x_p) - \frac{z}{\sqrt{n}}f(x_p) = \frac{B_n}{n} - \frac{z}{\sqrt{n}}f(x_p) + U_n, \tag{7.28}$$

where B_n is the number of observations in the interval $(x_p, x_p + z/\sqrt{n}]$, and U_n satisfies $\Pr(|U_n| \leq 2/n) = 1$. In particular, $U_n = \mathcal{O}_P(1/n)$. The conditional distribution of B_n given $\mathbf{P} = P$ is $\text{Bin}(n, \theta_n)$, where

$$\theta_n = F\left(x_p + \frac{z}{\sqrt{n}}\right) - F(x_p) = \frac{z}{\sqrt{n}}f(x_p) + o\left(\frac{1}{\sqrt{n}}\right) = \mathcal{O}\left(\frac{1}{\sqrt{n}}\right).$$

The characteristic function of $\sqrt{n}(A_n(z) - U_n)$ is

$$\begin{aligned}
\mathrm{E}\exp\left\{it\sqrt{n}\left(A_n(z) - U_n\right)\right\} \\
= \exp\left\{-itzf(x_p)\right\}\mathrm{E}\exp\left(it\frac{B_n}{\sqrt{n}}\right) \\
= \left(1 - \theta_n + \theta_n \exp\left\{\frac{it}{\sqrt{n}}\right\}\right)^n \exp\{-itzf(x_p)\}.
\end{aligned}$$

We can write

$$\exp\left(\frac{it}{\sqrt{n}}\right) = 1 + \frac{it}{\sqrt{n}} - \frac{t^2}{2n} + o\left(\frac{1}{n}\right).$$

It follows that

$$\begin{aligned}
\left(1 - \theta_n + \theta_n \exp\left\{\frac{it}{\sqrt{n}}\right\}\right)^n &= \left(1 + it\frac{\theta_n}{\sqrt{n}} + o\left(\frac{1}{n}\right)\right)^n \\
&= \left(1 + i\frac{zt}{n}f(x_p) + o\left(\frac{1}{n}\right)\right)^n \\
&\to \exp\{iztf(x_p)\},
\end{aligned}$$

as $n \to \infty$. So
$$\lim_{n \to \infty} E\exp\{i\sqrt{n}t\,(A_n(z) - U_n)\} = 1,$$
for all t. So, $\sqrt{n}[A_n(z) - U_n] \xrightarrow{D} 0$ by the continuity theorem B.93, and $\sqrt{n}[A_n(z) - U_n] \xrightarrow{P} 0$ by Theorem B.90. $A_n(z) = U_n + o_P(1/\sqrt{n})$. Since $U_n = \mathcal{O}_P(1/n) = o_P(1/\sqrt{n})$, it follows that $A_n(z) = o_P(1/\sqrt{n})$.

Finally, we prove (7.27). The following inequalities are all equivalent:
$$\sqrt{n}(Y_p^{(n)} - x_p) \leq z,$$
$$Y_p^{(n)} \leq x_p + \frac{z}{\sqrt{n}},$$
$$F_n(Y_p^{(n)}) \leq F_n\left(x_p + \frac{z}{\sqrt{n}}\right),$$
$$p \leq F_n\left(x_p + \frac{z}{\sqrt{n}}\right),$$
$$p \leq A_n(z) + F_n(x_p) + \frac{z}{\sqrt{n}}f(x_p),$$
$$\frac{\sqrt{n}}{f(x_p)}[p - F_n(x_p)] \leq z + \sqrt{n}\frac{A_n(z)}{f(x_p)}.$$

The equivalence of the first and last of these is (7.27). □

Now, we are ready to prove Theorem 7.25.

PROOF OF THEOREM 7.25. From Lemma 7.26, we know that
$$\Pr(\sqrt{n}(Y_p^{(n)} - x_p) \leq z) = \Pr\left(\frac{\sqrt{n}}{f(x_p)}(p - F_n(x_p)) - \sqrt{n}\frac{A_n(z)}{f(x_p)} \leq z\right).$$

We will prove that the right-hand side of this equation converges to the necessary normal probability. We have that $F_n(x_p) = C_n/n + D_n$, where C_n is the number of observations less than or equal to x_p and $D_n = \mathcal{O}_P(1/n)$. Also, $A_n(z) = o_P(1/\sqrt{n})$, so
$$\frac{\sqrt{n}}{f(x_p)}(p - F_n(x_p)) - \sqrt{n}\frac{A_n(z)}{f(x_p)} = -\frac{\sqrt{n}}{f(x_p)}\left(\frac{C_n}{n} - p\right) + o_P(1). \quad (7.29)$$

The central limit theorem B.97 tells us that $\sqrt{n}(C_n/n - p) \xrightarrow{D} N(0, p(1-p))$. This, together with Lemma 7.19 applied to (7.29), completes the proof. □

Example 7.30. Suppose that F has derivative
$$f(x) = \frac{1}{\sigma\pi}\left(1 + \left(\frac{x-\mu}{\sigma}\right)^2\right)^{-1},$$

where $\sigma > 0$ and μ are some numbers. If $p = 1/2$, $x_p = \mu$ and $f(x_p) = (\sigma\pi)^{-1}$. It follows that the sample median $Y_{1/2}^{(n)}$ has asymptotic distribution (given $\mathbf{P} = P$)

$$\sqrt{n}(Y_{\frac{1}{2}}^{(n)} - \mu) \xrightarrow{D} N\left(0, \frac{\sigma^2 \pi^2}{4}\right).$$

The asymptotic variance of $Y_{1/2}^{(n)}$ is $2.467\sigma^2/n$.

Example 7.31. Suppose that F has derivative

$$f(x) = \frac{1}{\sigma\sqrt{2\pi}} \exp\left\{-\frac{1}{2\sigma^2}(x - \mu)^2\right\},$$

where $\sigma > 0$ and μ are some numbers. If $p = 1/2$, $x_p = \mu$ and $f(x_p) = (\sigma\sqrt{2\pi})^{-1}$. It follows that the sample median $Y_{1/2}^{(n)}$ has asymptotic distribution (given $\mathbf{P} = P$)

$$\sqrt{n}(Y_{\frac{1}{2}}^{(n)} - \mu) \xrightarrow{D} N\left(0, \frac{\sigma^2 \pi}{2}\right).$$

The asymptotic variance of $Y_{1/2}^{(n)}$ is $1.571\sigma^2/n$.

For distributions that are bounded above or below, a different sort of result holds for the $p = 1$ or $p = 0$ quantile. The following theorems are examples.

Theorem 7.32. *Suppose that $t \in \mathbb{R}$, $\alpha > 0$, and*

$$\lim_{x \uparrow t}(t - x)^{-\alpha}[1 - F(x)] = c > 0.$$

Let $\{X_n\}_{n=1}^{\infty}$ be IID with CDF F and let $X_{(n)} = \max\{X_1, \ldots, X_n\}$. Then $n^{1/\alpha}(t - X_{(n)})$ converges in distribution to a distribution with CDF $G(x) = 1 - \exp(-cx^\alpha)$, for $x > 0$.

PROOF. Write

$$\Pr(n^{\frac{1}{\alpha}}[t - X_{(n)}] \geq x) = \Pr\left(X_{(n)} \leq t - \frac{x}{n^{\frac{1}{\alpha}}}\right) = F\left(t - \frac{x}{n^{\frac{1}{\alpha}}}\right)^n$$

$$= \left[1 - \frac{x^\alpha}{n}\left(\frac{n^{\frac{1}{\alpha}}}{x}\right)^\alpha \left\{1 - F\left(t - \frac{x}{n^{\frac{1}{\alpha}}}\right)\right\}\right]^n.$$

Since

$$\lim_{n \to \infty} x^\alpha \left(\frac{n^{\frac{1}{\alpha}}}{x}\right)^\alpha \left\{1 - F\left(t - \frac{x}{n^{\frac{1}{\alpha}}}\right)\right\} = cx^\alpha,$$

it follows that

$$\lim_{n \to \infty} \Pr(n^{\frac{1}{\alpha}}[t - X_{(n)}] \geq x) = \exp(-cx^\alpha). \qquad \square$$

Example 7.33. Suppose that $\{X_n\}_{n=1}^{\infty}$ are conditionally IID $U(0,\theta)$ given $\Theta = \theta$. The CDF of X_i (given $\Theta = \theta$) is x/θ for $0 < x < \theta$ and 1 for $x \geq \theta$. With $t = \theta$ we get $\lim_{x \uparrow t}(t-x)^{-1}[1 - F(x)] = 1/\theta$. So Theorem 7.32 says that $n(\theta - X_{(n)}) \xrightarrow{D} Exp(1/\theta)$.

A similar theorem can be proven for distributions bounded below.

Proposition 7.34. *Suppose that $t \in \mathbb{R}$, $\alpha > 0$, and $\lim_{x \downarrow t}(x-t)^{-\alpha} F(x) = c > 0$. Let $\{X_n\}_{n=1}^{\infty}$ be IID with CDF F and let $X_{(1)} = \min\{X_1, \ldots, X_n\}$. Then $n^{1/\alpha}(X_{(1)} - t)$ converges in distribution to a distribution with CDF $G(x) = 1 - \exp(-cx^\alpha)$, for $x > 0$.*

Krem (1963) proves that extreme order statistics (like the min and max) are asymptotically independent of the central order statistics (like the quantiles).

7.2.2 Several Quantiles

We can prove a theorem similar to Theorem 7.25 for several sample quantiles simultaneously.

Theorem 7.35. *Let $0 < p_1 < \cdots < p_k < 1$. Suppose that $\{X_n\}_{n=1}^{\infty}$ are conditionally IID with distribution P given $\mathbf{P} = P$ and suppose that P has CDF F with derivative f in a neighborhood of each x_{p_i} ($i = 1, \ldots, k$), where $F(x_{p_i}) = p_i$, $0 < f(x_{p_i}) < \infty$, and $0 < p < 1$. Define $Y_{p_i}^{(n)} = F_n^{-1}(p_i)$, where F_n is the empirical CDF of $(X_1, \ldots, X_n)$. Then*

$$\sqrt{n}(Y_{p_1}^{(n)} - x_{p_1}, \ldots, Y_{p_k}^{(n)} - x_{p_k})^\top \xrightarrow{D} N_k(\mathbf{0}, \mathbf{\Psi}),$$

where $\mathbf{\Psi} = ((\psi_{ij}))$ and $\psi_{ij} = p_{\min\{i,j\}} - p_i p_j / [f(x_{p_i}) f(x_{p_j})]$.

PROOF. Define

$$Z_{i,n} = \sqrt{n}(Y_{p_i}^{(n)} - x_{p_i}), \quad W_{i,n} = \sqrt{n} \frac{p_i - F_n(x_{p_i})}{f(x_{p_i})}.$$

Let $z_1, \ldots, z_k$ be real numbers and let $A_{i,n}(z_i)$ equal (7.28) with $p = p_i$ for $i = 1, \ldots, k$. Then,

$$Z_{i,n} \leq z_i \text{ if and only if } W_{i,n} - \sqrt{n}\frac{A_{i,n}(z_i)}{f(x_{p_i})} \leq z_i.$$

Since $(A_{1,n}(z_1), \ldots, A_{k,n}(z_k)) = o_P(1/\sqrt{n})$, it follows from Lemmas 7.19 and 7.26 that the two vectors Z_n and W_n converge in distribution to the same thing if either one of them converges. It is easier to find what W_n converges to, so that is what we will do.

We can write

$$F_n(x_{p_i}) = \frac{1}{n} \sum_{j=1}^{i} M_j + o_P\left(\frac{1}{\sqrt{n}}\right),$$

where M_j is the number of observed values in the interval $(x_{p_{j-1}}, x_{p_j}]$. For convenience, set $p_0 = 0$, $p_{k+1} = 1$, $x_{p_0} = -\infty$, and $x_{p_{k+1}} = \infty$. It is clear that the conditional distribution of $(M_1, \ldots, M_{k+1})$ given $\mathbf{P} = P$ is multinomial, $Mult(n; q_1, \ldots, q_{k+1})$, where $q_i = p_i - p_{i-1}$ for $i = 1, \ldots, k+1$. Set $G_n = (M_1, \ldots, M_{k+1})^\top$ and $q = (q_1, \ldots, q_{k+1})$. The multivariate central limit theorem B.99 implies that, conditional on $\mathbf{P} = P$,

$$\sqrt{n}(\frac{1}{n}G_n - q) \xrightarrow{D} N_k(\mathbf{0}, \Sigma)$$

(a multivariate normal distribution), where

$$\Sigma = \begin{pmatrix} q_1 & 0 & 0 \\ 0 & \ddots & 0 \\ 0 & 0 & q_{k+1} \end{pmatrix} - \begin{pmatrix} q_1 \\ \vdots \\ q_{k+1} \end{pmatrix} (q_1, \ldots, q_{k+1}).$$

Next, note that

$$\begin{pmatrix} F_n(x_{p_1}) \\ \vdots \\ F_n(x_{p_k}) \end{pmatrix} = \frac{1}{n} A G_n + o_P\left(\frac{1}{\sqrt{n}}\right), \tag{7.36}$$

where A is the $k \times (k+1)$ matrix with 1 on and below the diagonal and 0 above the diagonal:

$$A = \begin{pmatrix} 1 & 0 & 0 & \cdots & 0 \\ 1 & 1 & 0 & \cdots & 0 \\ \vdots & \vdots & \ddots & \cdots & \vdots \\ 1 & 1 & 1 & \cdots & 0 \end{pmatrix}.$$

Call the vector on the left of (7.36) R. Then the conditional mean of R given $\mathbf{P} = P$ is $p = Aq = (p_1, \ldots, p_k)^\top$, and $\sqrt{n}(p - R) \xrightarrow{D} N_k(\mathbf{0}, A\Sigma A^\top)$. All that remains is to compute $A\Sigma A^\top$. This can be seen to equal

$$A\Sigma A^\top = \begin{pmatrix} p_1 & p_1 & \cdots & p_1 \\ p_1 & p_2 & \cdots & p_2 \\ \vdots & \vdots & \ddots & \vdots \\ p_1 & p_2 & \cdots & p_k \end{pmatrix} - \begin{pmatrix} p_1 \\ p_2 \\ \vdots \\ p_k \end{pmatrix} (p_1, \cdots, p_k). \qquad \square$$

Since the definition of F_n is arbitrary between observed values, the asymptotic distribution in Theorem 7.35 applies to every vector of random variables whose ith coordinate is between $X_{(j-1)}$ and $X_{(j)}$ when $(j-1)/n < p_i < j/n$.

An analogue to Theorems 7.32 and 7.34 can be proven for the joint distribution of the smallest and largest order statistics.

Proposition 7.37. *Suppose that $t_1, t_2 \in \mathbb{R}$, $\alpha_1, \alpha_2 > 0$, and*

$$\lim_{x \downarrow t_1}(x - t_1)^{-\alpha_1} F(x) = c_1 > 0,$$

$$\lim_{x \uparrow t_2}(t_2 - x)^{-\alpha_2}[1 - F(x)] = c_2 > 0.$$

Let $\{X_n\}_{n=1}^{\infty}$ be IID with CDF F, and let $X_{(1)} = \min\{X_1, \ldots, X_n\}$ and $X_{(n)} = \max\{X_1, \ldots, X_n\}$. Then the asymptotic joint CDF of $n^{1/\alpha_1}(X_{(1)} - t_1)$ and $n^{1/\alpha_2}(t_2 - X_{(n)})$ is $(1 - \exp(-c_1 x_1^{\alpha_1}))(1 - \exp(-c_2 x_2^{\alpha_2}))$.

7.2.3 Linear Combinations of Quantiles*

A linear combination of sample quantiles is called an *L-estimator*. Suppose that f is the derivative of the conditional CDF of the X_i given $\mathbf{P} = P$ and that f is symmetric about $g(\theta)$. Here, we suppose that $F = \Theta^{-1}(\theta)$ and $f(x) = h(x - g(\theta))$. Let $Z_i = X_i - g(\theta)$. Then the conditional density of the Z_i is $h(\cdot)$. If we choose to sample quantiles symmetric about the median, say p and $1-p$, then $x_p = 2g(\theta) - x_{1-p}$ by symmetry. Let $z_p = x_p - g(\theta)$ for all p so that $z_p = -z_{1-p}$. Let $W_p^{(n)} = Y_p^{(n)} - g(\theta)$, so that $W_p^{(n)} - z_p = Y_p^{(n)} - x_p$.

$$\sqrt{n}\begin{pmatrix} W_p^{(n)} - z_p \\ W_{\frac{1}{2}}^{(n)} \\ W_{1-p}^{(n)} + z_p \end{pmatrix} \xrightarrow{\mathcal{D}} N_3\left(\mathbf{0}, \begin{pmatrix} \frac{p(1-p)}{h^2(z_p)} & \frac{p}{2h(z_p)h(0)} & \frac{p^2}{h^2(z_p)} \\ \frac{p}{2h(z_p)h(0)} & \frac{1}{4h^2(0)} & \frac{p}{2h(z_p)h(0)} \\ \frac{p^2}{h^2(z_p)} & \frac{p}{2h(z_p)h(0)} & \frac{(1-p)p}{h^2(z_p)} \end{pmatrix}\right).$$

If the goal is to estimate $g(\Theta)$, it might be good if the asymptotic mean were $g(\theta)$ given $\Theta = \theta$. The asymptotic conditional mean of $a_1 Y_p^{(n)} + a_2 Y_{1/2}^{(n)} + a_3 Y_{1-p}^{(n)}$ is $(a_1 + a_2 + a_3)g(\theta) + (a_3 - a_1)EW_{1-p}^{(n)}$, since h is symmetric around 0. For $p < 1/2$, this will equal $g(\theta)$ for all θ if and only if $a_3 = a_1$ and $a_1 + a_2 + a_3 = 1$. Hence our estimator must be

$$\phi(X) = aY_p^{(n)} + (1 - 2a)Y_{\frac{1}{2}}^{(n)} + aY_{1-p}^{(n)}, \tag{7.38}$$

for some a.

Example 7.39 (Continuation of Example 7.30; see page 406). As an example, consider the case of Cauchy distributions with a location parameter Θ. Then $Z_i = X_i - \theta$, $h(x) = (\pi[1 + x^2])^{-1}$, and $z_{1-p} = \tan[\pi(1/2 - p)]$. So, for example, if $p = 1/3$, then $z_p = -1/\sqrt{3}$, $z_{1-p} = 1/\sqrt{3}$, and $h(z_p) = 3/(4\pi) = h(z_{1-p})$. The asymptotic covariance matrix of the three sample quantiles is then

$$\Sigma = \frac{\pi^2}{n}\begin{pmatrix} \frac{32}{81} & \frac{2}{9} & \frac{16}{81} \\ \frac{2}{9} & \frac{1}{4} & \frac{2}{9} \\ \frac{16}{81} & \frac{2}{9} & \frac{32}{81} \end{pmatrix}.$$

---

*This section may be skipped without interrupting the flow of ideas.

The asymptotic variance of the estimator in (7.38) is

$$(a, 1-2a, a)\Sigma \begin{pmatrix} a \\ 1-2a \\ a \end{pmatrix} = \frac{\pi^2}{n}\left(\frac{1}{4} - \frac{a}{9} + a^2\frac{11}{27}\right).$$

This variance is minimized at $a = 3/22$. The minimum asymptotic variance is $8\pi^2/(33n) = 2.39/n$, which is not much better than we got with the median alone $(2.467/n)$.

Perhaps improvement can be made in Example 7.30 by altering p. The general method for doing this is illustrated by continuing the example.

Example 7.40 (Continuation of Example 7.39; see page 410). For general $p < 1/2$,

$$h(z_p) = \frac{1}{\pi}\cos^2\pi\left(p - \frac{1}{2}\right) = \frac{1}{\pi}c(p),$$

say, where $c(0) = 1$. The asymptotic covariance matrix of the three quantiles is

$$\Sigma = \frac{\pi^2}{n}\begin{pmatrix} \frac{p(1-p)}{c(p)^2} & \frac{p}{2c(p)} & \frac{p^2}{c(p)^2} \\ \frac{p}{2c(p)} & \frac{1}{4} & \frac{p}{2c(p)} \\ \frac{p^2}{c(p)^2} & \frac{p}{2c(p)} & \frac{p(1-p)}{c(p)^2} \end{pmatrix}.$$

The asymptotic variance of the estimator in (7.38) is

$$(a, 1-2a, a)\Sigma\begin{pmatrix} a \\ 1-2a \\ a \end{pmatrix}$$
$$= \frac{\pi^2}{n}\left[\frac{1}{4} - 2a\left(\frac{1}{2} - \frac{p}{c(p)}\right) + a^2\left(1 + \frac{2p}{c(p)^2} - \frac{4p}{c(p)}\right)\right].$$

The variance is minimized at

$$a = a^*(p) = \frac{c(p)^2 - 2pc(p)}{2[c(p)^2 - 4pc(p) + 2p]},$$

and the minimum variance is

$$\frac{\pi^2}{4n}\left[1 - \frac{(c(p) - 2p)^2}{c(p)^2 - 4pc(p) + 2p}\right] = s(p).$$

We can numerically minimize $s(p)$ and find the minima occur at $p = 0.42085$ and at $p = 0.07915$. The minimum $s(p)$ is 2.302, which is only slightly better than using $p = 1/3$.

Example 7.41. Suppose that the distributions are double exponential (also known as Laplace distribution). That is, $h(x) = \exp(-|x|)/2$. Then

$$z_p = \begin{cases} \log 2p & \text{if } p \leq \frac{1}{2}, \\ -\log 2(1-p) & \text{if } p > \frac{1}{2}, \end{cases} \quad h(z_p) = \begin{cases} p & \text{if } p \leq \frac{1}{2}, \\ 1-p & \text{if } p > \frac{1}{2}. \end{cases}$$

The asymptotic covariance matrix of three symmetric sample quantiles is

$$\Sigma = \frac{1}{n}\begin{pmatrix} \frac{1}{p}-1 & 1 & 1 \\ 1 & 1 & 1 \\ 1 & 1 & \frac{1}{p}-1 \end{pmatrix}.$$

The asymptotic variance of the estimator in (7.38) is

$$(a, 1-2a, a)\Sigma\begin{pmatrix} a \\ 1-2a \\ a \end{pmatrix} = \frac{a^2}{n}\left(\frac{2}{p}-4\right)+1,$$

which has a minimum at $a = 0$ and the minimum value is 1. This means that, no matter what p is, it is better to use just the median.

7.3 Large Sample Estimation

7.3.1 Some Principles of Large Sample Estimation

One would hope that if a large sample were available, then better knowledge of $\mathbf{P}$ would be available, and we would be close to the situation of having independent observations. Since predictive inference in usually not the goal for classical statistics, the issue becomes how well we have estimated Θ. There is the belief that an estimator ought to get Θ correct eventually. That is, the estimator should be *consistent* (see Definition 7.9). If more than one estimator is consistent, then one might ask, "Which is better?" Without a loss function or some indication of how we plan to use the estimator, this question is not interesting. There are, nonetheless, answers to the question.

Let Θ be k-dimensional, and suppose that the FI regularity conditions (see Definition 2.78) hold. Then the Fisher information matrix (based on a single observation) $\mathcal{I}_{X_1}(\theta)$ can be calculated. Suppose also that an estimator $\hat{\Theta}_n$ of Θ converges in distribution, say $\sqrt{n}(\hat{\Theta}_n - \theta) \xrightarrow{D} N_k(\mathbf{0}, V_\theta)$ given $\Theta = \theta$. If we wish to estimate $g(\Theta)$, with g continuous, then the delta method tells us that $\sqrt{n}[g(\hat{\Theta}_n) - g(\Theta)] \xrightarrow{D} N(0, c_\theta^\top V_\theta c_\theta)$, where

$$c_\theta^\top = \left(\frac{\partial}{\partial \theta_1}g(\theta),\ldots,\frac{\partial}{\partial \theta_k}g(\theta)\right). \tag{7.42}$$

Corollary 5.23 says that the smallest possible variance for an unbiased estimator of $g(\Theta)$ is $c_\theta^\top \mathcal{I}_{X_1}(\theta)^{-1} c_\theta$. Since $g(\hat{\Theta}_n)$ is asymptotically unbiased, the ratio of these two variances might be used as a measure of how good a consistent estimator is.

Definition 7.43. If G_n is an estimator of $g(\Theta)$ for each n and

$$\sqrt{n}(G_n - g(\theta)) \xrightarrow{D} N(0, v_\theta), \text{ for all } \theta,$$

then the ratio $c_\theta^\top \mathcal{I}_{X_1}(\theta)^{-1} c_\theta / v_\theta$ is called the *asymptotic efficiency of* G_n at θ. If the ratio is 1, the sequence $\{G_n\}_{n=1}^\infty$ is called *asymptotically efficient*.

Suppose that $\{G_n\}_{n=1}^\infty$ and $\{G'_n\}_{n=1}^\infty$ are sequences of estimators of $g(\Theta)$, and we have a specific criterion that we require of our estimator, such as variance equal to ϵ. Suppose that G_{n_0} and $G'_{n'_0}$ satisfy this criterion. Then the *relative efficiency of* $\{G_n\}_{n=1}^\infty$ *to* $\{G'_n\}_{n=1}^\infty$ for the specific criterion is n'_0/n_0. Suppose that the criterion is allowed to change in such a way that the sample sizes required to satisfy it go to ∞, for example, variance equal to ϵ with ϵ going to 0. If the ratio n'_0/n_0 converges to a value r, then r is called the *asymptotic relative efficiency (ARE)*[11] *of* $\{G_n\}_{n=1}^\infty$ *to* $\{G'_n\}_{n=1}^\infty$.

Example 7.44. Let $\{X_n\}_{n=1}^\infty$ be conditionally IID with $N(\mu, \sigma^2)$ distribution given $\Theta = (\mu, \sigma)$. Let $g(\theta) = \mu$. Let $G_n = \overline{X}_n$, the sample average, and let G'_n be the sample median. Let our specific criterion be that the asymptotic variance of the estimator must equal ϵ. Since the central limit theorem B.97 says that $\sqrt{n}(G_n - \mu) \xrightarrow{\mathcal{D}} N(0, \sigma^2)$, and Example 7.31 on page 407 shows that $\sqrt{n}(G'_n - \mu) \xrightarrow{\mathcal{D}} N(0, \sigma^2\pi/2)$, we have the relative efficiency equal to $\sqrt{2/\pi} = 0.798$ for all ϵ. If we let $\epsilon \to 0$, the ARE of the sample median to the sample mean is 0.798 as well.

The idea of ARE is to compare the sizes of samples needed to make comparable inferences from the two sequences.

Example 7.45. Suppose that $\{X_n\}_{n=1}^\infty$ are conditionally IID $U(0, \theta)$ given $\Theta = \theta$. The MLE is $\hat{\Theta}_n = \max X_i$. Another estimator is twice the sample average, $2\overline{X}_n$. Suppose that our criterion is that the actual variance of the estimator must equal $\theta^2 \epsilon$. Since $\hat{\Theta}_n/\theta$ has $Beta(n,1)$ distribution, the variance of $\hat{\Theta}_n$ is $\theta^2 n/[(n+1)^2(n+2)]$. The variance of $2\overline{X}_n$ is $\theta^2/(3n)$. Let n_0 be the sample size at which $\hat{\Theta}_n$ has variance $\theta^2 \epsilon$, and let n'_0 be the sample size at which $2\overline{X}_n$ has variance $\theta^2 \epsilon$. It is easy to see that we must have $n'_0 = (n_0+1)^2(n_0+2)/(3n_0)$. So, $n'_0/n_0 = (n+1)^2(n+2)/(3n^2)$ for all ϵ. As $\epsilon \to 0$, $n \to \infty$ and the ratio n'_0/n_0 goes to ∞. That is, the ARE of $\hat{\Theta}_n$ to $2\overline{X}_n$ is ∞.

Example 7.46. Let H and H' be nondegenerate distributions that have some common scale feature (like the same finite standard deviation or the same interquartile range). Suppose that $a_n(G_n - g(\theta)) \xrightarrow{\mathcal{D}} H$ and $b_n(G'_n - g(\theta)) \xrightarrow{\mathcal{D}} H'$. Suppose also that $\lim_{n\to\infty} a_n/b_n = r$. Then r is the *relative rate of convergence* of $\{G_n\}_{n=1}^\infty$ to $\{G'_n\}_{n=1}^\infty$.[12] Note that when H and H' are both normal and a_n and b_n are both $\mathcal{O}(\sqrt{n})$, the relative rate of convergence is the square root of the ARE for asymptotic variance.

In Section 7.3.2, we show that the class of maximum likelihood estimators (see Section 5.1.3) are efficient under quite general conditions. At first, it

---

[11] This definition of ARE is taken from Serfling (1980, pp. 50–52). Serfling's definition actually applies to more types of inference than estimators, but we will not pursue that generality here.

[12] Solve Problem 22 on page 470 to show that the relative rate of convergence is uniquely defined. Relative rate of convergence is not an example of a criterion for ARE, but it has a similar nature.

might seem that achieving asymptotic efficiency of 1 would be the best possible, but sometimes efficiency greater than 1 is possible.[13]

Example 7.47.[14] Suppose that $\{X_n\}_{n=1}^\infty$ are conditionally IID $N(\theta, 1)$ given $\Theta = \theta$. We already know that $\mathcal{I}_{X_1}(\theta) = 1$ and $\sqrt{n}(\overline{X}_n - \theta) \sim N(0,1)$, so $\overline{X}_n$ is asymptotically efficient. (Actually, it is efficient in finite samples.) Let θ_0 be arbitrary, and define a new estimator of Θ:

$$\delta_n = \begin{cases} \overline{X}_n & \text{if } |\overline{X}_n - \theta_0| \geq n^{-\frac{1}{4}}, \\ \theta_0 + a(\overline{X}_n - \theta_0) & \text{if } |\overline{X}_n - \theta_0| < n^{-\frac{1}{4}}, \end{cases}$$

where $0 < a < 1$. This is like using $\overline{X}_n$ when $\overline{X}_n$ is not close to θ_0, but using the posterior mean of Θ from a prior centered at θ_0 when $\overline{X}$ is close to θ_0.

We will now calculate the efficiency of δ_n. Suppose that $\theta \neq \theta_0$. Then

$$\sqrt{n}|\overline{X}_n - \delta_n| = \sqrt{n}(1-a)|\overline{X}_n - \theta_0|I_{[0, n^{-\frac{1}{4}})}(|\overline{X}_n - \theta_0|).$$

Hence, for $\epsilon > 0$, $P'_\theta(\sqrt{n}|\overline{X}_n - \delta_n| > \epsilon)$ is at most

$$P'_\theta\left(|\overline{X}_n - \theta_0| \leq n^{-\frac{1}{4}}\right) = P'_\theta\left(\theta_0 - n^{-\frac{1}{4}} \leq \overline{X}_n \leq \theta_0 + n^{-\frac{1}{4}}\right)$$
$$= P'_\theta\left((\theta_0 - \theta)\sqrt{n} - n^{\frac{1}{4}} \leq Z \leq (\theta_0 - \theta)\sqrt{n} + n^{\frac{1}{4}}\right),$$

where $Z = \sqrt{n}(\overline{X}_n - \theta)$ has $N(0,1)$ distribution given $\Theta = \theta$. This last probability goes to 0 as n goes to infinity because both of the endpoints either go to $+\infty$ or $-\infty$. Hence, if $\theta \neq \theta_0$, $\delta_n = \overline{X}_n + o_P(1/\sqrt{n})$.

Now, suppose that $\theta = \theta_0$. Then

$$\sqrt{n}|a(\overline{X}_n - \theta_0) + \theta_0 - \delta_n| = (1-a)\sqrt{n}|\overline{X} - \theta_0|I_{[n^{-\frac{1}{4}}, \infty)}(|\overline{X}_n - \theta_0|).$$

Hence, for $\epsilon > 0$, $P'_\theta(\sqrt{n}|a(\overline{X}_n - \theta_0) + \theta_0 - \delta_n| > \epsilon)$ is at most

$$P'_\theta\left(|\overline{X}_n - \theta_0| > n^{-\frac{1}{4}}\right) = P'_\theta\left(\sqrt{n}|\overline{X}_n - \theta_0| > n^{\frac{1}{4}}\right) \to 0,$$

as $n \to \infty$. So, if $\theta = \theta_0$, $\delta_n = \theta_0 + a(\overline{X} - \theta_0) + o_P(1/\sqrt{n})$. It follows that $\sqrt{n}(\delta_n - \theta) \xrightarrow{D} N(0, v_\theta)$, where $v_{\theta_0} = a^2$ and $v_\theta = 1$ for all other θ. Efficiency is $1/a^2 > 1$ at $\theta = \theta_0$.

The phenomenon of Example 7.47 is called *superefficiency*. It is easy to see how one could arrange for an estimator to be superefficient at several different possible θ. LeCam (1953) proved that, under conditions a little stronger than the FI regularity conditions, superefficiency can only occur at a set of zero Lebesgue measure.

---

[13] When an estimator is efficient, or when two estimators have ARE equal to 1, more detailed comparisons are often made in a study of *second-order efficiency*. We will not study second-order efficiency in this text.

[14] This example is due to Hodges; see LeCam (1953).

7.3.2 Maximum Likelihood Estimators

In Section 5.1.3, we defined maximum likelihood estimators (MLE) to be estimators that maximize the likelihood function $L(\theta) = f_{X|\Theta}(x|\theta)$. That is, an MLE of Θ after observing $X = x$ is any θ at which $L(\theta)$ achieves its maximum, if there are any such θ. In this section, we prove some large sample properties of these estimators.

Theorem 7.48.[15] *Assume that $\{X_n\}_{n=1}^{\infty}$ are conditionally IID given $\Theta = \theta$ each with density $f_{X_1|\Theta}(x|\theta)$. Then, for each θ_0 and each $\theta \neq \theta_0$,*

$$\lim_{n \to \infty} P'_{\theta_0}\left[\prod_{i=1}^{n} f_{X_1|\Theta}(X_i|\theta_0) > \prod_{i=1}^{n} f_{X_1|\Theta}(X_i|\theta)\right] = 1.$$

PROOF. With P_{θ_0} measure 1, $\prod_{i=1}^{n} f_{X_1|\Theta}(x_i|\theta_0) > \prod_{i=1}^{n} f_{X_1|\Theta}(x_i|\theta)$ if and only if

$$R(x) = \frac{1}{n}\sum_{i=1}^{n} \log \frac{f_{X_1|\Theta}(x_i|\theta)}{f_{X_1|\Theta}(x_i|\theta_0)} < 0.$$

By the weak law of large numbers B.95, under P_{θ_0},

$$R(X) \xrightarrow{P} E_{\theta_0} \log \frac{f_{X_1|\Theta}(X|\theta)}{f_{X_1|\Theta}(X|\theta_0)} = -\mathcal{I}_{X_1}(\theta_0;\theta),$$

where $\mathcal{I}_{X_1}(\theta_0;\theta)$ is the Kullback–Leibler information from Definition 2.89. By Proposition 2.92, we know that $-\mathcal{I}_{X_1}(\theta_0;\theta) < 0$ if $\theta \neq \theta_0$. It follows that $\lim_{n \to \infty} P'_{\theta_0}(R(X) < 0) = 1$. □

Theorem 7.48 suggests that the MLE should be consistent, since the P_θ probability goes to 1 that the likelihood function is higher at θ than at some other parameter value. Some further conditions are required to prove consistency. Wald (1949) proved almost sure convergence of the MLE under the assumption that the likelihood function was continuous. Theorems 7.49 and 7.54 are very much like Wald's result.

Theorem 7.49. *Let $\{X_1\}_{n=1}^{\infty}$ be conditionally IID given $\Theta = \theta$ with density $f_{X_1|\Theta}(x|\theta)$ with respect to a measure ν on a space $(\mathcal{X}^1, \mathcal{B}^1)$. Fix $\theta_0 \in \Omega$, and define, for each $M \subseteq \Omega$ and $x \in \mathcal{X}^1$,*

$$Z(M,x) = \inf_{\psi \in M} \log \frac{f_{X_1|\Theta}(x|\theta_0)}{f_{X_1|\Theta}(x|\psi)}.$$

Assume that for each $\theta \neq \theta_0$ there is an open set N_θ such that $\theta \in N_\theta$ and $E_{\theta_0} Z(N_\theta, X_i) > 0$. If Ω is not compact, assume further that there is a compact $C \subseteq \Omega$ such that $\theta_0 \in C$ and $E_{\theta_0} Z(\Omega \setminus C, X_i) > 0$. Then, $\lim \hat{\Theta}_n = \theta_0$, a.s. $[P_{\theta_0}]$.

---

[15] This theorem can be strengthened to an almost sure result. See Problem 28 on page 471.

PROOF. If Ω is compact, let $C = \Omega$. It suffices to prove that for every $\epsilon > 0$,

$$P'_{\theta_0}(\limsup_{n \to \infty} \|\hat{\Theta}_n - \theta_0\| \geq \epsilon) = 0. \tag{7.50}$$

Let $\epsilon > 0$ and let N_0 be the open ball of radius ϵ around θ_0. Since $C \setminus N_0$ is a compact set, and $\{N_\theta : \theta \in C \setminus N_0\}$ is an open cover, we may extract a finite subcover, $N_{\theta_1}, \ldots, N_{\theta_\ell}$. Rename these sets and C^C to $\Omega_1, \ldots, \Omega_m$, so that $\Omega = N_0 \cup \left(\cup_{j=1}^m \Omega_j\right)$, and $E_{\theta_0} Z(\Omega_j, X_i) > 0$.

Let $\mathcal{X}^\infty$ be the infinite product space of copies of $\mathcal{X}^1$. Let $x \in \mathcal{X}$ denote a generic sequence of possible data values. Let $E_{\theta_0} Z(\Omega_j, X_i) = c_j$. By the strong law of large numbers 1.63, $\sum_{i=1}^n Z(\Omega_j, X_i)/n \to c_j$, a.s. $[P_{\theta_0}]$. Let $B_j \subseteq \mathcal{X}^\infty$ be the set of data sequences such that convergence holds, and let $B = \cap_{j=1}^m B_j$. Then $P_{\theta_0}(B) = 1$ and $\sum_{i=1}^n Z(\Omega_j, x_i)/n \to c_j > 0$ for each $x = (x_1, x_2, \ldots) \in B$. Now, notice that

$$\{x : \limsup_{n \to \infty} \|\hat{\Theta}_n(x_1, \ldots, x_n) - \theta_0\| \geq \epsilon\}$$

$$\subseteq \bigcup_{j=1}^m \{x : \hat{\Theta}_n(x_1, \ldots, x_n) \in \Omega_j, \text{ infinitely often}\}$$

$$\subseteq \bigcup_{j=1}^m \left\{x : \inf_{\psi \in \Omega_j} \frac{1}{n} \sum_{i=1}^n \log \frac{f_{X_1|\Theta}(x_i|\theta_0)}{f_{X_1|\Theta}(x_i|\psi)} \leq 0, \text{ infinitely often}\right\}$$

$$\subseteq \bigcup_{j=1}^m \left\{x : \frac{1}{n} \sum_{i=1}^n Z(\Omega_j, x_i) \leq 0, \text{ infinitely often}\right\} \subseteq \bigcup_{j=1}^m B_j^C.$$

Since this last set is B^C and $P_{\theta_0}(B^C) = 0$, (7.50) follows. □

The hard part of using this theorem is verifying the conditions.

Example 7.51. Suppose that $\{X_n\}_{n=1}^\infty$ given $\Theta = \theta$ are IID with $U(0, \theta)$ distribution. Then $f_{X_1|\Theta}(x|\theta) = 1/\theta$ for $0 \leq x \leq \theta$. We need $E_{\theta_0} \inf_{\psi \in N_\theta} g(X_i) > 0$ where $g(x)$ is the function

$$\log \frac{f_{X_1|\Theta}(x|\theta_0)}{f_{X_1|\Theta}(x|\psi)} = \begin{cases} \log \frac{\psi}{\theta_0} & \text{if } x \leq \min\{\theta_0, \psi\}, \\ \infty & \text{if } \psi < x \leq \theta_0, \\ -\infty & \text{if } \theta_0 < x \leq \psi, \\ \text{undefined} & \text{otherwise.} \end{cases}$$

Since the last two cases have 0 probability under P_{θ_0}, we can choose $N_\theta = ([\theta + \theta_0]/2, \infty)$ when $\theta > \theta_0$. In this case, $Z(N_\theta, x) = \log([\theta + \theta_0]/[2\theta_0]) > 0$, a.s. $[P_{\theta_0}]$. If $\theta < \theta_0$, choose $N_\theta = (\theta/2, [\theta + \theta_0]/2)$. In this case $Z(N_\theta, x) = \infty$ if $x > [\theta + \theta_0]/2$. Hence, $E_{\theta_0} Z(N_\theta, X_i) > 0$ in either case.

We also need a compact set C such that $E_{\theta_0} Z(\Omega \setminus C, X_i) > 0$. Let $C = [\theta_0/a, a\theta_0]$, for some $a > 1$. Then

$$\inf_{\theta \in \Omega \setminus C} \log \frac{f_{X_1|\Theta}(X_i|\theta_0)}{f_{X_1|\Theta}(X_i|\theta)} = \begin{cases} \log \frac{X_i}{\theta_0} & \text{if } X_i < \frac{1}{a}\theta_0, \\ \log a & \text{if } X_i \geq \frac{1}{a}\theta_0. \end{cases}$$

The conditional mean of this given $\Theta = \theta_0$ is

$$\frac{1}{\theta_0}\left[\int_0^{\frac{1}{a}\theta_0} \log\frac{x}{\theta_0}dx + \int_{\frac{1}{a}\theta_0}^{\theta_0} \log a\,dx\right].$$

The first integral goes to 0 and the second goes to ∞ as $a \to \infty$. This means that there is some $a > 1$ such that the mean is positive. It follows from Theorem 7.49 that the MLE is consistent.

In this example, it would have been easier to find the distribution of $\hat{\Theta}_n$ and prove directly that it was consistent, but we will need the above calculation in Example 7.82 on page 432.

Example 7.52. Suppose that $\{X_n\}_{n=1}^\infty$ given $\Theta = \theta$ are IID with $N(\theta, 1)$ distribution. It is easy to calculate

$$\log\frac{f_{X_1|\Theta}(x|\theta_0)}{f_{X_1|\Theta}(x|\theta)} = x(\theta_0 - \theta) + \frac{1}{2}(\theta^2 - \theta_0^2) \equiv g(x,\theta). \tag{7.53}$$

The minimum of this over any set occurs at θ equal to the value in the set closest to x. So, if $N_\theta = (\theta - \epsilon, \theta + \epsilon)$, then $E_{\theta_0} Z(N_\theta, x) = \mathcal{I}_{X_1}(\theta_0; \theta) + E_{\theta_0}(R)$, where

$$R = \begin{cases} \epsilon(x - \theta) + \frac{\epsilon^2}{2} & \text{if } x < \theta - \epsilon, \\ x(\theta - x) + \frac{x^2 - \theta^2}{2} & \text{if } \theta - \epsilon \leq x \leq \theta + \epsilon, \\ \epsilon(\theta - x) + \frac{\epsilon^2}{2} & \text{if } x > \theta + \epsilon. \end{cases}$$

Clearly, $E_{\theta_0}(R)$ can be made arbitrarily small by choosing ϵ small. Similarly, if $C = [\theta_0 - u, \theta_0 + u]$, for large u, then

$$Z(C^C, x) = \begin{cases} x(\theta_0 - x) + \frac{x^2 - \theta_0^2}{2} & \text{if } x < \theta_0 - u, \\ u(x - \theta_0) + \frac{u^2}{2} & \text{if } \theta_0 - u \leq x \leq \theta_0, \\ u(\theta_0 - x) + \frac{u^2}{2} & \text{if } \theta_0 < x \leq \theta_0 + u, \\ x(\theta_0 - x) + \frac{x^2 - \theta_0^2}{2} & \text{if } x > \theta_0 + u. \end{cases}$$

We can make the integrals over the first and last portions of this arbitrarily small by choosing u large enough. The integral over the two middle portions equals $u^2/2 - \exp(-u^2/2 - 1)\sqrt{2/\pi}$, which is positive for large u.

Unfortunately, if the parameter is $\Theta = (M, \Sigma)$ and $X_i \sim N(\mu, \sigma^2)$ given $\Theta = (\mu, \sigma)$, it is not possible to find a compact set C such that $E_{\theta_0} Z(C^C, X_i) > 0$. Berk (1966) replaces this condition with a weaker condition that first appeared in Kiefer and Wolfowitz (1956). The proof that this weaker condition suffices for convergence of the MLE involves martingales and is deferred to Lemma 7.83. (Also, see Problem 45 on page 474 and Example 7.85 on page 434.)

One of the conditions of Theorem 7.49 can be weakened if $f_{X_1|\Theta}(x|\cdot)$ is continuous.[16]

---

[16] A slightly more general result can be proved by assuming that $f_{X_1|\Theta}$ is upper semicontinuous (USC). A function $f : \Omega \to \mathbb{R}$ is *upper semicontinuous* if $\limsup_{n\to\infty} f(\theta_n) \leq f(\theta)$ whenever $\theta_n \to \theta$. Usc functions possess two properties that are needed in the proof of Theorem 7.49. The sum of two USC functions is USC, and the maximum of a USC function is attained on a compact set.

Lemma 7.54. *Assume the same conditions as in Theorem 7.49, except that we now only require that $E_{\theta_0} Z(N_\theta, X_i) > -\infty$. Assume further that $f_{X|\Theta}(x|\cdot)$ is continuous at θ for every θ, a.s. $[P_{\theta_0}]$. Then, $\lim \hat{\Theta}_n = \theta_0$, a.s. $[P_{\theta_0}]$.*

PROOF. If Ω is compact, let $C = \Omega$. For each $\theta \neq \theta_0$ in C, let $N_\theta^{(k)}$ be a closed ball centered at θ with radius at most $1/k$ such that, for each k, $N_\theta^{(k+1)} \subseteq N_\theta^{(k)} \subseteq N_\theta$. This ensures that $\cap_{k=1}^\infty N_\theta^{(k)} = \{\theta\}$. So, for each x, $Z(N_\theta^{(k)}, x)$ increases with k. For each x such that $f_{X_1|\Theta}$ is continuous, $\log[f_{X_1|\Theta}(x|\theta_0)/f_{X_1|\Theta}(x|\psi)]$ is continuous in ψ. So, for each k, there exists $\theta_k \in N_\theta^{(k)}$ ($\{\theta_k\}_{k=1}^\infty$ might depend on x) such that[17] $Z(N_\theta^{(k)}, x) = \log[f_{X_1|\Theta}(x|\theta_0)/f_{X_1|\Theta}(x|\theta_k)]$. Since $\theta_k \to \theta$,

$$\lim_{k \to \infty} Z(N_\theta^{(k)}, x) = \log \frac{f_{X_1|\Theta}(x|\theta_0)}{f_{X_1|\Theta}(x|\theta)}. \quad (7.55)$$

Since $N_\theta^{(k)} \subseteq N_\theta$, it follows that $Z(N_\theta^{(k)}, x) \geq Z(N_\theta, x)$. If $E_{\theta_0} Z(N_\theta, X_i) = \infty$, then we have $E_{\theta_0} Z(N_\theta^{(k)}, X_i) = \infty$, for all k. If $E_{\theta_0} Z(N_\theta, X_i)$ is finite, then apply Fatou's lemma A.50 to $\{Z(N_\theta^{(k)}, x) - Z(N_\theta, x)\}_{k=1}^\infty$ and use (7.55) to get

$$\liminf_{k \to \infty} E_{\theta_0} Z(N_\theta^{(k)}, X_i) \geq E_{\theta_0} \lim_{k \to \infty} Z(N_\theta^{(k)}, X_i) = \mathcal{I}_{X_1}(\theta_0; \theta) > 0, \quad (7.56)$$

where $\mathcal{I}_{X_1}$ is the Kullback–Leibler information. Either way, we can now choose $k^*(\theta)$ so that $E_{\theta_0} Z(N_\theta^{(k^*)}, X_i) > 0$, and apply Theorem 7.49. □

7.3.3 MLEs in Exponential Families

In exponential families, MLEs exist (with probability tending to 1 as $n \to \infty$) and are asymptotically normally distributed, and differentiable functions of the MLE are asymptotically efficient estimators

Theorem 7.57. *Suppose that $\{X_n\}_{n=1}^\infty$ are conditionally IID given $\Theta = \theta$ with nondegenerate exponential family distribution whose density with respect to a measure ν is*

$$f_{X_1|\Theta}(x|\theta) = c(\theta) \exp(\theta^\top x).$$

Suppose that the natural parameter space is Ω an open subset of $\mathbb{R}^k$. Let $\hat{\Theta}_n$ be the MLE of Θ based on $X_1, \ldots, X_n$ if it exists. Then

- $\lim_{n \to \infty} P_\theta(\hat{\Theta}_n \text{ exists}) = 1$,

---

[17]If $f_{X_1|\Theta}$ is USC, θ_k still exists. One must change the lim to lim inf wherever it appears in front of Z, and one must change the equality to $\geq$ in (7.55) and (7.56) to make the rest of the proof work.

- under P_θ, $\sqrt{n}(\hat{\Theta}_n - \theta) \xrightarrow{D} N_k(\mathbf{0}, \mathcal{I}_{X_1}(\theta)^{-1})$, where $\mathcal{I}_{X_1}(\theta)$ is the Fisher information matrix.

PROOF. If the MLE exists, it will satisfy the equation that says that the partial derivatives of the log-likelihood function with respect to each coordinate of θ are 0, since the parameter space is open. Since

$$\frac{\partial}{\partial \theta_i} \log f_{X|\Theta}(x|\theta) = n\bar{x}_i + n\frac{\partial}{\partial \theta_i} \log c(\theta),$$

the resulting equation is $\bar{x}_n = v(\theta)$, where the ith coordinate of $v(\theta)$ is $\partial \log c(\theta)/\partial \theta_i$. It follows from Proposition 2.70 that

$$\mathrm{Cov}_\theta(X_i, X_j) = -\frac{\partial^2}{\partial \theta_i \partial \theta_j} \log c(\theta) = \sigma_{ij}.$$

Since a covariance matrix is positive semidefinite, it follows that $-\log c(\theta)$ is a convex function. Since each X_i has nondegenerate exponential family distribution, their coordinates are not linearly dependent, hence the matrix $\Sigma = ((\sigma_{ij}))$ will be positive definite. It follows that $v(\cdot)$ has a differentiable inverse $h(\cdot)$ in the sense that for each θ there is a neighborhood N of θ such that $h(v(\psi)) = \psi$ for $\psi \in N$, and the derivatives of h are continuous. (See the inverse function theorem C.2.) If $\bar{x}_n$ is in the image of v, then, for at least one such function h, the MLE equals $h(\bar{x}_n)$. By the weak law of large numbers B.95 $\overline{X}_n \xrightarrow{P} E_\theta X$ under P_θ and $E_\theta X = v(\theta)$ by Proposition 2.70. It follows that $\overline{X}_n$ will be in the range of v with probability tending to 1 as $n \to \infty$. It follows that the MLE exists with probability tending to 1. The multivariate central limit theorem B.99 says that under P_θ, $\sqrt{n}(\overline{X}_n - v(\theta)) \xrightarrow{D} N(\mathbf{0}, \Sigma)$. Using the delta method, we get that under P_θ, $\sqrt{n}(\hat{\Theta}_n - \theta) \xrightarrow{D} N_k(\mathbf{0}, A\Sigma A^\top)$, where $A = ((a_{ij}))$ with $a_{ij} = \partial h_i(t)/\partial t_j$ evaluated at $t = v(\theta)$, which is the (i,j) element of Σ^{-1}. So, $A = \Sigma^{-1}$. It is also easy to see that Σ is the Fisher information matrix $\mathcal{I}_{X_1}(\theta)$, so $\sqrt{n}(\hat{\Theta}_n - \theta) \xrightarrow{D} N(\mathbf{0}, \mathcal{I}_{X_1}(\theta)^{-1})$. □

The following corollary is trivial.

Corollary 7.58. *Under the conditions of Theorem 7.57, the MLE of Θ is consistent.*

Another corollary says that differentiable functions of MLEs are asymptotically efficient estimators.

Corollary 7.59. *Assume the conditions of Theorem 7.57. Suppose that $g : \Omega \to \mathbb{R}$ has continuous partial derivatives. Then $g(\hat{\Theta}_n)$ is an asymptotically efficient estimator of $g(\Theta)$.*

PROOF. Let c_θ be defined in (7.42). Using the delta method, we get

$$\sqrt{n}(g(\hat{\Theta}_n) - g(\theta)) \xrightarrow{D} N(0, c_\theta^\top \mathcal{I}_{X_1}(\theta)^{-1} c_\theta).$$

It follows that $g(\hat{\Theta}_n)$ is an asymptotically efficient estimator of $g(\Theta)$. □

7.3.4 Examples of Inconsistent MLEs

There are some curious examples of MLEs that are inconsistent. Each of these examples fails one or more of the conditions of the theorems on consistency that are proved in this chapter.

Example 7.60. This example was introduced by Neyman and Scott (1948) and discussed by Barnard (1970).[18] Suppose that $(X_i, Y_i) \sim N_2(\mu_i \mathbf{1}, \sigma^2 I)$, given $\Theta = (\sigma, \mu_1, \mu_2, \ldots)$, and the individual vectors are conditionally independent. These observations are not conditionally IID, so that none of our theorems applies as stated. Nonetheless, we can write a likelihood function

$$L(\theta) = \frac{1}{(2\pi)^n} \sigma^{-2n} \exp\left\{-\frac{1}{2\sigma^2} \sum_{i=1}^{n}[(x_i - \mu_i)^2 + (y_i - \mu_i)^2]\right\}.$$

The logarithm of this is (aside from an additive constant)

$$-2n \log \sigma - \frac{1}{2\sigma^2}\left\{2\sum_{i=1}^{n}\left(\frac{x_i + y_i}{2} - \mu_i\right)^2 + \frac{1}{2}\sum_{i=1}^{n}(x_i - y_i)^2\right\}.$$

The MLEs are easily calculated as $\hat{M}_{i,n} = (X_i + Y_i)/2$ and $\hat{\Sigma}_n^2 = \sum_{i=1}^{n}(X_i - Y_i)^2/[4n]$. Since the conditional distribution of $X_i - Y_i$ given $\Theta = \theta$ is $N(0, 2\sigma^2)$, it follows that $\hat{\Sigma}_n^2 \xrightarrow{P} \sigma^2/2$ under P_θ. The MLE of Σ^2 is inconsistent.

Barnard (1970) suggests an empirical Bayes approach, which is to choose a distribution for the M_i with a fixed finite number of parameters, call them Ψ. Then treat (Ψ, Σ) as the parameters and integrate the M_i out of the problem. See Section 8.4 for a discussion of empirical Bayes methods. Kiefer and Wolfowitz (1956) let the distribution of the M_i be more general, but they assume that the distribution lies in a compact set of distributions so that methods like those of Theorem 7.49 and Lemma 7.54 can be used.

Example 7.61. Let $\Omega = (1/2, 1]$ and suppose that for $x = 0, 1, 2, \ldots$,

$$f_{X_1|\Theta}(x|\theta) = \begin{cases} \theta(1-\theta)^x & \text{if } \theta \neq 1, \\ 2^{-(x+1)} & \text{if } \theta = 1. \end{cases}$$

This is a family of geometric distributions with $\theta = 1/2$ renamed to $\theta = 1$. The density is neither continuous nor USC at $\theta = 1$. So, Lemma 7.54 will not apply. We can write

$$\log \frac{f_{X_1|\Theta}(x|1)}{f_{X_1|\Theta}(x|\theta)} = -\log(2\theta) - x \log[2(1-\theta)], \tag{7.62}$$

and we note that for every compact subset C of Ω, the infimum of (7.62) over $\theta \in C^C$ is 0 for $x > 0$ and is negative for $x = 0$. So the conditions of Theorem 7.49

---

[18] Barnard claims that Neyman presented him with the example in a taxicab in Paris in 1946. Barnard had just met Neyman for the first time and "was arguing for the broad general validity of the method of maximum likelihood" when Neyman asked him what he would do in this example.

fail as well. Let $\overline{X}_n$ be the average of the first n observations. The MLE based on the first n observations is

$$\hat{\Theta}_n = \begin{cases} (1+\overline{X}_n)^{-1} & \text{if } \overline{X}_n < 1, \\ 1 & \text{if } \overline{X}_n \geq 1. \end{cases}$$

Under P_1, $\sqrt{n}(\overline{X}_n - 1) \xrightarrow{D} N(0, 2)$, since the mean of each X_i is 1 and the variance is 2. It follows that $\lim_{n\to\infty} P_1'(\overline{X}_n \geq 1) = 1/2$ and $1/(1+\overline{X}_n) \xrightarrow{P} 1/2$ under P_1. (This is not surprising, since $\Theta = 1$ should have been $\Theta = 1/2$.) So,

$$\lim_{n\to\infty} P_1'\left(|\hat{\Theta}_n - 1| > \frac{1}{4}\right) = \frac{1}{2},$$

and $\hat{\Theta}_n$ is inconsistent because it does not converge appropriately for $\Theta = 1$.

7.3.5 Asymptotic Normality of MLEs

Outside of exponential families, the proof that MLEs are asymptotically normal is a bit more complicated than the proof of Theorem 7.57. The following theorem gives conditions under which the MLE is asymptotically normal in general parametric families.

Theorem 7.63. *Let Ω be a subset of $\mathbb{R}^p$, and let $\{X_n\}_{n=1}^{\infty}$ be conditionally IID given $\Theta = \theta$ each with density $f_{X_1|\Theta}(\cdot|\theta)$. Let $\hat{\Theta}_n$ be an MLE. Assume that $\hat{\Theta}_n \xrightarrow{P} \theta$ under P_θ for all θ. Assume that $f_{X_1|\Theta}(x|\theta)$ has continuous second partial derivatives with respect to θ and that differentiation can be passed under the integral sign. Assume that there exists $H_r(x, \theta)$ such that, for each $\theta_0 \in \text{int}(\Omega)$ and each k, j,*

$$\sup_{\|\theta-\theta_0\|\leq r}\left|\frac{\partial^2}{\partial\theta_k\partial\theta_j}\log f_{X_1|\Theta}(x|\theta_0) - \frac{\partial^2}{\partial\theta_k\partial\theta_j}\log f_{X_1|\Theta}(x|\theta)\right| \leq H_r(x, \theta_0), \tag{7.64}$$

with $\lim_{r\to 0} E_{\theta_0} H_r(X, \theta_0) = 0$. Assume that the Fisher information matrix $\mathcal{I}_{X_1}(\theta)$ is finite and nonsingular. Then, under P_{θ_0},

$$\sqrt{n}(\hat{\Theta}_n - \theta_0) \xrightarrow{D} N(0, \mathcal{I}_{X_1}^{-1}(\theta_0)).$$

Before we prove Theorem 7.63, here is an example in which condition (7.64) is met.

Example 7.65. Suppose that $f_{X_1|\Theta}(x|\theta) = (\pi[1+(x-\theta)^2])^{-1}$. Then

$$\frac{\partial^2}{\partial\theta^2}\log f_{X_1|\Theta}(x|\theta) = -2\frac{1-(x-\theta)^2}{[1+(x-\theta)^2]^2}.$$

This is differentiable and the derivative has finite mean. Hence H_r exists as in Theorem 7.63.

The idea of the proof of Theorem 7.63 is the following. We work with the vector $\ell'_\theta(X)$ of partial derivatives of the logarithm of the likelihood function divided by n. We evaluate a Taylor expansion of $\ell'_\theta(X)$ around θ_0 at the point $\hat{\Theta}_n$. Since $\ell'_{\hat{\Theta}_n}(X)$ should be 0, we get that $\ell'_{\theta_0}(X)$ is essentially the matrix B_n of second partial derivatives of the logarithm of the likelihood times $\hat{\Theta}_n - \theta_0$ divided by n. Since $\ell'_{\theta_0}(X)$ is the average of IID random vectors with mean 0 and covariance matrix $\mathcal{I}_{X_1}(\theta_0)$, $\sqrt{n}\ell'_{\theta_0}(X)$ is asymptotically normal with covariance $\mathcal{I}_{X_1}(\theta_0)$. Similarly, B_n is nearly the average of IID random matrices, so $B_n \xrightarrow{P} -\mathcal{I}_{X_1}(\theta_0)$. Setting the two sides of the Taylor expansion equal, we get that $\sqrt{n}\mathcal{I}_{X_1}(\theta_0)(\hat{\Theta}_n - \theta_0)$ is asymptotically normal with covariance matrix $\mathcal{I}_{X_1}(\theta_0)$. Multiplying by $\mathcal{I}_{X_1}(\theta_0)^{-1}$ gives the desired result.

PROOF OF THEOREM 7.63. Let

$$\ell_\theta(x) = \frac{1}{n}\sum_{i=1}^{n} \log f_{X_1|\Theta}(x_i|\theta).$$

The ith coordinate of the gradient $\ell'_\theta(x)$ is $\left(\sum_{i=1}^{n} \partial \log f_{X_1|\Theta}(x_i|\theta)/\partial\theta_j\right)/n$. Since $\theta_0 \in \text{int}(\Omega)$, there is an open neighborhood of θ_0 in the interior of Ω. Since $\hat{\Theta}_n \xrightarrow{P} \theta_0$ under P_{θ_0}, it follows that $Z_n I_{\text{int}(\Omega)^C}(\hat{\Theta}_n) = o_P(1/\sqrt{n})$ as $n \to \infty$ for every sequence $\{Z_n\}_{n=1}^{\infty}$ of random variables.[19] Note that $\ell'_{\hat{\Theta}_n}(X) = \mathbf{0}$ for $\hat{\Theta}_n \in \text{int}(\Omega)$. It follows that

$$\ell'_{\hat{\Theta}_n}(X) = \ell'_{\hat{\Theta}_n}(X)I_{\text{int}(\Omega)^C}(\hat{\Theta}_n) = o_P\left(\frac{1}{\sqrt{n}}\right).$$

Using a one-term Taylor expansion (see Theorem C.1) of each coordinate of $\ell'_{\hat{\Theta}_n}(X)$ around θ_0, we get

$$\ell'_{\theta_0}(X) + \left(\left(\frac{\partial^2}{\partial\theta_j\partial\theta_k}\ell_\theta(X)\bigg|_{\theta=\theta^*_{n,j}}\right)\right)(\hat{\Theta}_n - \theta_0) = o_P\left(\frac{1}{\sqrt{n}}\right), \quad (7.66)$$

where $\theta^*_{n,j}$ is between $\theta_{0,j}$ and $\hat{\Theta}_{n,j}$ for each j. Since $\hat{\Theta}_n \xrightarrow{P} \theta_0$ under P_{θ_0}, $\theta^*_{n,j} \xrightarrow{P} \theta_{0,j}$ for each j. Set B_n equal to the matrix in (7.66). Then

$$\ell'_{\theta_0}(X) + B_n(\hat{\Theta}_n - \theta_0) = o_P\left(\frac{1}{\sqrt{n}}\right). \quad (7.67)$$

By passing derivatives under the integral sign in the equation

$$0 = \frac{\partial}{\partial\theta_j}\int f_{X_1|\Theta}(x|\theta)d\nu(x),$$

---

[19]In fact, $Z_n I_{\text{int}(\Omega)^C}(\hat{\Theta}_n) = o_P(r_n)$ for every r_n. The reason is that it equals 0 with probability tending to 1 and it doesn't matter what it equals when it isn't 0.

we see that $E_{\theta_0}\ell'_{\theta_0}(X) = 0$. Similarly, we get that the conditional covariance matrix given $\Theta = \theta_0$ of $\ell'_{\theta_0}(X)$ is $\mathcal{I}_{X_1}(\theta_0)$. The multivariate central limit theorem B.99 gives us that $\sqrt{n}\ell'_{\theta_0}(X) \xrightarrow{D} N(0, \mathcal{I}_{X_1}(\theta_0))$. So $\sqrt{n}\ell'_{\theta_0}(X) = \mathcal{O}_P(1)$. It follows from (7.67) that

$$\sqrt{n}B_n(\hat{\Theta}_n - \theta_0) = \mathcal{O}_P(1). \tag{7.68}$$

Next, note that $B_{n(j,k)} = \left(\sum_{i=1}^n \partial^2 \log f_{X_1|\Theta}(X_i|\theta_0)/\partial\theta_j\partial\theta_k\right)/n + \Delta_n$, and (7.64) ensures that $|\Delta_n| \le \sum_{i=1}^n H_r(X_i, \theta_0)/n$, when $\|\theta_0 - \theta_n^*\| \le r$. The weak law of large numbers B.95 says that

$$\frac{1}{n}\sum_{i=1}^n H_r(X_i, \theta_0) \xrightarrow{P} E_{\theta_0} H_r(X_i, \theta_0).$$

Let $\epsilon > 0$ and let r be small enough so that $E_{\theta_0} H_r(X_i, \theta_0) < \epsilon/2$. Then

$$\begin{aligned} P'_{\theta_0}(|\Delta_n| > \epsilon) &\le P'_{\theta_0}\left(\frac{1}{n}\sum_{i=1}^n H_r(X_i, \theta_0) > \epsilon\right) + P'_{\theta_0}(\|\theta_0 - \theta_n^*\| \ge r) \\ &\le P'_{\theta_0}\left(\left|\frac{1}{n}\sum_{i=1}^n H_r(X_i, \theta_0) - E_{\theta_0} H_r(X_i, \theta_0)\right| > \frac{\epsilon}{2}\right) \\ &\quad + P'_{\theta_0}(\|\theta_0 - \theta_n^*\| \ge r). \end{aligned}$$

The last two probabilities go to 0 as $n \to \infty$, hence it follows that $\Delta_n = o_P(1)$. Hence $B_n \xrightarrow{P} -\mathcal{I}_{X_1}(\theta_0)$, and $B_n = \mathcal{O}_P(1)$ but $B_n \ne o_P(1)$. It follows from (7.68) that $\sqrt{n}(\hat{\Theta}_n - \theta_0) = \mathcal{O}_P(1)$. Now, write $B_n = -\mathcal{I}_{X_1}(\theta_0) + C_n$ where $C_n = o_P(1)$. Then $C_n(\hat{\Theta}_n - \theta_0) = o_P(1/\sqrt{n})$, and we can rewrite (7.67) as

$$\sqrt{n}\ell'_{\theta_0}(X) - \mathcal{I}_{X_1}(\theta_0)\sqrt{n}(\hat{\Theta}_n - \theta_0) = o_P(1).$$

By Lemma 7.19, we get that $-\mathcal{I}_{X_1}(\theta_0)\sqrt{n}(\hat{\Theta}_n - \theta_0) \xrightarrow{D} N(0, \mathcal{I}_{X_1}(\theta_0))$. Since multiplication by a matrix is a continuous function, the result is proven. □

When applying Theorems 7.57 and 7.63 with observed data, it is common to replace $\mathcal{I}_{X_1}(\theta_0)$ by a matrix that does not depend on the unknown parameter. One possibility is $\mathcal{I}_{X_1}(\hat{\Theta}_n)$, which is often called the *expected Fisher information*. In the proof of Theorem 7.63, we saw that $\mathcal{I}_{X_1}(\hat{\Theta}_n) \xrightarrow{P} \mathcal{I}_{X_1}(\theta_0)$ given $\Theta = \theta_0$. We also saw, however, that $\mathcal{I}_{X_1}(\theta_0)$ arose in the theorem as an approximation to $-1/n$ times the matrix of second partial derivatives of the log-likelihood function at a point near $\hat{\Theta}_n$ (and near θ_0). It has been suggested [see, for example, Efron and Hinkley (1978)] that one use $1/n$ times

$$-\left(\left(\frac{\partial^2}{\partial\theta_i\partial\theta_j}\log f_{X|\Theta}(x|\theta)\big|_{\theta=\hat{\Theta}_n}\right)\right) \tag{7.69}$$

in place of $\mathcal{I}_{X_1}(\hat{\Theta}_n)$ when $X = x$ is observed and one wishes to use the MLE to make inference about Θ. The quantity in (7.69) is called the *observed*

Fisher information. We will see later (in Section 7.4.2) that the observed Fisher information is indeed the appropriate matrix to use when the goal is to approximate the posterior distribution of a parameter by a normal distribution. Efron and Hinkley (1978) say that the reason for preferring observed over expected information is that the inverse of observed information is closer to the conditional variance of the MLE given an ancillary.

Example 7.70.[20] Assume that $(X_1, Z_1), \ldots, (X_n, Z_n)$ are conditionally IID with Z_i having $Ber(1/2)$ distribution and $X_i | Z_i = z$ having $N(\theta, 1/[z+1])$ distribution given $\Theta = \theta$. That is, we flip a fair coin before observing each X_i and if the coin comes up tails, we get an $N(\theta, 1)$ observation. If the coin comes up heads, we get an $N(\theta, 1/2)$ observation. The log-likelihood function is a constant plus

$$\frac{1}{2} \sum_{i=1}^{n} \log(Z_i + 1) - \frac{1}{2} \sum_{i=1}^{n} (Z_i + 1)(X_i - \theta)^2.$$

The MLE is the weighted average $\hat{\Theta}_n = \sum_{i=1}^{n}(Z_i + 1)X_i / \sum_{i=1}^{N}(Z_i + 1)$. The Fisher information is $\mathcal{I}_{X_1}(\theta) = 3n/2$, which is also the expected Fisher information. The approximation to the distribution of $\hat{\Theta}_n$ given $\Theta = \theta$ using the expected Fisher information is $N(\theta, 2/[3n])$. On the other hand, the observed Fisher information is $J = \sum_{i=1}^{n}(Z_i + 1)$. A natural ancillary upon which to condition is $Z = (Z_1, \ldots, Z_n)$. The conditional distribution of $\hat{\Theta}_n$ given Z and $\Theta = \theta$ is $N(\theta, 1/J)$, which is the same as the approximation based on the observed Fisher information.

LeCam (1970) proves asymptotic normality of MLEs under ostensibly weaker conditions than Theorem 7.63. The conditions do not require the existence of continuous second derivatives. They do, however, require the existence of functions that behave very much like second derivatives. Also, a condition very much like (7.64) is required, where the second derivatives are replaced by these other functions that behave like second derivatives.[21]

7.3.6 Asymptotic Properties of *M*-Estimators*

In Section 5.1.5, we introduced a class of estimators called *M*-estimators. These estimators can be thought of as being chosen to maximize the log of some alternative likelihood or some (nearly) arbitrary function rather than the likelihood function. For example, suppose that we choose some function $\rho(a, b)$ and maximize $\sum_{i=1}^{n} \rho(X_i, \theta)$ as a function of θ. If $\rho(a, b) = \log f_{X_1|\Theta}(a|b)$, then we get maximum likelihood as a special case.

Not every function ρ will be appropriate, however. The following conditions will be assumed throughout this section:

1. For each θ_0 and for all $\theta \neq \theta_0$, $\mathrm{E}_{\theta_0}[\rho(X_i, \theta_0) - \rho(X_i, \theta)] > 0$.

---

[20]This is a modification of an example of Cox (1958).
[21]The paper is worth locating if only to read the author's footnote.
*This section may be skipped without interrupting the flow of ideas.

2. For each θ_0 and each $\theta \neq \theta_0$, there exists an open set N_θ containing θ such that $E_{\theta_0} \inf_{\theta' \in N_\theta}[\rho(X_i, \theta_0) - \rho(X_i, \theta')] > -\infty$.

3. For each θ_0, there exists a compact set C containing θ_0 such that $E_{\theta_0} \inf_{\theta \notin C}[\rho(X_i, \theta_0) - \rho(X_i, \theta)] > 0$.

Condition 1 says that ρ allows one to distinguish possible values of Θ from each other and that $\rho(X_i, \theta_0)$ tends to be larger when $\Theta = \theta_0$ than when Θ equals some other value. Condition 2 says that it cannot be the case that even when $\Theta = \theta_0$, there are some other possible values θ' of Θ that lead to $\rho(X_i, \theta')$ being much larger than $\rho(X_i, \theta_0)$. Condition 3 says that there is a region around θ_0 such that all values of $\rho(X_i, \theta)$, for θ not in that region, tend to be less than $\rho(X_i, \theta_0)$ simultaneously. If $\rho(a, b) = \log f_{X_1|\Theta}(a|b)$, then conditions 2 and 3 are two of the conditions of Lemma 7.54. In fact, the method of proof for that lemma can be applied to prove the following proposition.

Proposition 7.71. *Suppose that $\rho(X, \cdot)$ is continuous, a.s. $[P_{\theta_0}]$. Also, assume conditions 1–3. If $\hat{\Theta}_n$ is the value of θ that maximizes $\sum_{i=1}^{n} \rho(X_i, \theta)$, then $\hat{\Theta}_n \to \theta_0$, a.s. $[P_{\theta_0}]$.*

Next, suppose that ρ is differentiable with respect to the second coordinate. Then, set $\psi(X_i, \theta) = \partial \rho(X_i, \theta)/\partial \theta$, and assume that $E_\theta \psi(X_i, \theta) = 0$ for all θ. (This is the same as condition (5.39) on page 312.) Also, suppose that ψ is continuous. Now, we can try to solve $\sum_{i=1}^{n} \psi(X_i, \theta) = 0$. If there is more than one solution, we can choose the one closest to some reasonable (with any luck, consistent) estimator of Θ. The next theorem says that as n increases, the probability that there is a solution to this equation near θ goes to 1, given $\Theta = \theta$.

Theorem 7.72. *Assume that $\Omega \subseteq \mathbb{R}$. Let $\psi : \mathcal{X} \times \Omega \to \mathbb{R}$ be such that*

- $E_\theta \psi(X_i, \theta) = 0$,

- $\psi(x, \theta)$ *is continuous in θ,*

- *for each θ_0, there exists $\delta > 0$ such that $E_{\theta_0} \psi(X_i, \theta)$ is strictly decreasing as a function of θ for $|\theta - \theta_0| < \delta$.*

Then, for each $\epsilon > 0$,

$$\lim_{n \to \infty} P'_{\theta_0}\left(\exists \text{ a solution of } \sum_{i=1}^{n} \psi(X_i, \theta) = 0 \text{ in } (\theta_0 - \epsilon, \theta_0 + \epsilon)\right) = 1.$$

PROOF. If $\theta_1 \in (\theta_0 - \delta, \theta_0) \cap (\theta_0 - \epsilon, \theta_0)$ and $\theta_2 \in (\theta_0, \theta_0 + \delta) \cap (\theta_0, \theta_0 + \epsilon)$, then $E_{\theta_0} \psi(X_i, \theta_1) > 0$ and $E_{\theta_0} \psi(X_i, \theta_2) < 0$. By the weak law of large numbers B.95, under P_{θ_0}, $\sum_{i=1}^{n} \psi(X_i, \theta_1)/n \xrightarrow{P} E_{\theta_0} \psi(X_i, \theta_1) > 0$ and

$\sum_{i=1}^{n} \psi(X_i, \theta_2)/n \xrightarrow{P} \mathrm{E}_{\theta_0} \psi(X_i, \theta_2) < 0$. We now note that the probability that there is a solution equals

$$P'_{\theta_0}\left(\frac{1}{n}\sum_{i=1}^{n}\psi(X_i,\theta) = 0, \text{ for some } \theta\right)$$
$$\geq P'_{\theta_0}\left(\frac{1}{n}\sum_{i=1}^{n}\psi(X_i,\theta_1) > 0 \text{ and } \frac{1}{n}\sum_{i=1}^{n}\psi(X_i,\theta_2) < 0\right),$$

and the last of these goes to 1 as $n \to \infty$. $\square$

A similar result can be proven if $\mathrm{E}_{\theta_0}\psi(X_i,\theta)$ is nondecreasing.

Corollary 7.73. *If G_n is the closest solution to a consistent estimator, then G_n is consistent.*

Example 7.74. Suppose that $f_{X_1|\Theta}(x|\theta) = \{\pi[1 + (x - \theta)^2]\}^{-1}$. A consistent estimator is $Y_{1/2}$, the median. The likelihood equation is

$$\sum_{i=1}^{n}\frac{X_i - \theta}{1 + (X_i - \theta)^2} = 0,$$

which has several solutions in general. To check the conditions of Theorem 7.72, we note that $\psi(x,\theta) = (x - \theta)/[1 + (x - \theta)^2]$. Clearly, $\mathrm{E}_\theta \psi(X,\theta) = 0$ for all θ. The derivative of $\mathrm{E}_{\theta_0}\psi(X,\theta)$ evaluated at $\theta = \theta_0$ is

$$\int \frac{x^2 - 1}{(x^2 + 1)^3}dx = \int \frac{dx}{(x^2+1)^2} - 2\int \frac{dx}{(x^2+1)^3} = -\frac{1}{2}\int \frac{dx}{(x^2+1)} < 0.$$

It follows that $\mathrm{E}_{\theta_0}(X,\theta)$ is strictly decreasing in θ for θ near θ_0. If we choose the solution to the likelihood equation closest to the median, we have another consistent estimator.

Freedman and Diaconis (1982) give examples in which M-estimators are inconsistent. Basically, the distribution from that the data arise has a density that is designed to be particularly incompatible with the function ρ (or ψ) that one uses for the M-estimation.

The next theorem says that we can get efficient estimators without actually finding MLEs by starting with any $\sqrt{n}$-consistent estimator and then using one step of Newton's method to try to solve the likelihood equation. The theorem is stated in terms of general M-estimators.

Theorem 7.75. *Let $\Omega \subseteq \mathrm{I\!R}^k$ be open. Let $\tilde{\Theta}_n - \theta_0 = \mathcal{O}_P(1/\sqrt{n})$ under P_{θ_0}. Let $\psi : \mathcal{X} \times \Omega \to \mathrm{I\!R}^k$ be such that $\mathrm{E}_{\theta_0}\psi(X,\theta_0) = \mathbf{0}$, and $\partial^2 \psi(x,\theta)/\partial \theta^2$ is continuous in θ. Define two matrices $J_\ell = ((J_{\ell;j,t}))$ for $\ell = 1, 2$ where*

$$J_{1;j,t} = \mathrm{E}_{\theta_0}\frac{\partial}{\partial \theta_t}\psi_j(X_i,\theta_0),$$
$$J_{2;j,t} = \mathrm{Cov}_{\theta_0}(\psi_j(X_i,\theta_0), \psi_t(X_i,\theta_0)).$$

7.3. Large Sample Estimation

Assume that J_1 is nonsingular. Also, assume that there exists $H_r(x)$ such that

$$\sup_{\|\theta-\theta_0\|\leq r}\left|\frac{\partial}{\partial\theta_t}\psi_j(x,\theta)-\frac{\partial}{\partial\theta_t}\psi_j(x,\theta_0)\right|\leq H_r(x),$$

where $\lim_{r\to 0} E_{\theta_0} H_r(X_i) = 0$. Let $M(\theta) = ((m_{j,t}(\theta)))$, where

$$m_{j,t}(\theta) = \sum_{i=1}^{n}\frac{\partial}{\partial\theta_t}\psi_j(X_i,\theta).$$

Let $\hat{\Theta}_n^* = \tilde{\Theta}_n - M^{-1}(\tilde{\Theta}_n)\sum_{i=1}^{n}\psi(X_i,\tilde{\Theta}_n)$. Then, under P_{θ_0},

$$\sqrt{n}(\hat{\Theta}_n^* - \theta_0) \xrightarrow{D} N_k(0, J_1^{-1}J_2 J_1^{-1\top}).$$

PROOF. Let $p(\theta) = \sum_{i=1}^{n}\psi(X_i,\theta)/n$, so that $\hat{\Theta}_n^* = \tilde{\Theta}_n - M^{-1}(\tilde{\Theta}_n)p(\tilde{\Theta}_n)$. We can write

$$p(\tilde{\Theta}_n) = p(\theta_0) + M(\theta^*)(\tilde{\Theta}_n - \theta_0),$$

where θ^* is between θ_0 and $\tilde{\Theta}_n$. It follows that under P_{θ_0}, $\theta^* \xrightarrow{P} \theta_0$. By hypothesis,

$$|m_{j,t}(\theta^*) - m_{j,t}(\theta_0)| = \frac{1}{n}\left|\sum_{i=1}^{n}\frac{\partial}{\partial\theta_t}\psi_j(X_i,\theta^*) - \frac{\partial}{\partial\theta_t}\psi_j(X_i,\theta_0)\right|$$

$$\leq \frac{1}{n}\sum_{i=1}^{n}|H_{\|\theta^*-\theta_0\|}(X_i)|.$$

Since $\|\theta^* - \theta_0\| = \mathcal{O}_P(1)$, $|m_{j,t}(\theta^*) - m_{j,t}(\theta_0)| = \mathcal{O}_P(1)$. By hypothesis $(\tilde{\Theta}_n - \theta_0)^2 = \mathcal{O}_P(1/n)$, so $p(\tilde{\Theta}_n) = p(\theta_0) + M(\theta_0)(\tilde{\Theta}_n - \theta_0) + \mathcal{O}_P(1/n)$. Also,

$$M(\tilde{\Theta}_n) = M(\theta_0) + \mathcal{O}_P\left(\frac{1}{\sqrt{n}}\right) = M(\theta_0)\left[I + \mathcal{O}_P\left(\frac{1}{\sqrt{n}}\right)\right].$$

It follows that

$$\hat{\Theta}_n^* = \tilde{\Theta}_n - p(\theta_0) + M(\theta_0)^{-1}(M(\theta_0)\tilde{\Theta}_n - \theta_0) + \mathcal{O}_P\left(\frac{1}{n}\right)$$

$$= \tilde{\Theta}_n - M(\theta_0)^{-1}p(\theta_0) - (\tilde{\Theta}_n - \theta_0) + o_P\left(\frac{1}{\sqrt{n}}\right).$$

The last equality holds since $E_{\theta_0}\psi(X_i,\theta_0) = \mathbf{0}$, implying that $p(\theta_0) = o_P(1)$ and because $\tilde{\Theta}_n - \theta_0 = o_P(1)$. So,

$$\sqrt{n}(\hat{\Theta}_n - \theta_0) = -\sqrt{n}M(\theta_0)^{-1}p(\theta_0) + o_P(1). \qquad (7.76)$$

The weak law of large numbers B.95 says that under P_{θ_0},

$$M(\theta_0) \xrightarrow{P} \left(\left(E_{\theta_0}\frac{\partial}{\partial \theta_t}\psi_j(X,\theta_0)\right)\right) = J_1.$$

Also, the multivariate central limit theorem B.99 says that under P_{θ_0},

$$\sqrt{n}p(\theta_0) \xrightarrow{D} N_k(0, J_2).$$

The result now follows easily from (7.76) and Theorem 7.22. □

Example 7.77. Consider the Cauchy distribution with a location parameter. The likelihood equation is very difficult to solve, but there is a simple $\sqrt{n}$-consistent estimator, namely, the sample median $\tilde{\Theta}_n$. Set

$$\psi(x,\theta) = \frac{\partial}{\partial \theta}\log f_{X_1|\Theta}(x|\theta) = 2\frac{x-\theta}{1+(x-\theta)^2}.$$

We now calculate

$$\begin{aligned}
E_{\theta_0}\frac{\partial}{\partial \theta}\psi(X,\theta_0) &= -2E_{\theta_0}\frac{1-(X-\theta_0)^2}{[1+(X-\theta_0)^2]^2} = -2E_0\frac{1-X^2}{(1+X^2)^2} \\
&= -\frac{2}{\pi}\int\frac{1-x^2}{(1+x^2)^3}dx = .5 \neq 0.
\end{aligned}$$

The other conditions of Theorem 7.75 can also be verified. The following estimator is asymptotically efficient:

$$\hat{\Theta}_n^* = \tilde{\Theta}_n + \frac{\sum_{i=1}^n \frac{X_i-\tilde{\Theta}_n}{1+(X_i-\tilde{\Theta}_n)^2}}{\sum_{i=1}^n \frac{1-(X_i-\tilde{\Theta}_n)^2}{[1+(X_i-\tilde{\Theta}_n)^2]^2}}.$$

A theorem like Theorem 7.63 can be proven for consistent M-estimators. The asymptotic distribution is the same as the asymptotic distribution of $\hat{\Theta}_n^*$ in Theorem 7.75. The proof proceeds either by showing that the M-estimator differs from $\hat{\Theta}_n^*$ by $o_p(1/\sqrt{n})$ or by rewriting the proof of Theorem 7.63 using ψ in place of ℓ'_θ. The details are left to the interested reader. Huber (1967) also proves theorems of this sort.

7.4 Large Sample Properties of Posterior Distributions

There are three kinds of large sample properties of posterior distributions which we explore in this section. One kind is classical properties, such as consistency and asymptotic normality conditional on a parameter. Examples include Theorem 7.80 and the asymptotic normality of posterior distributions as presented in Section 7.4.2. Another kind is prior properties, where the probability statements concern the prior joint distributions

7.4. Large Sample Properties of Posterior Distributions

of all random quantities of interest. Some examples include Theorems 7.78 and 7.120. A third kind is pointwise properties. These concern limits along certain data sequences, and an example is in Section 7.4.3.

One thing that must be kept in mind about the prior properties is that, without further effort and/or conditions, these properties do not usually imply corresponding non-Bayesian properties. For example, in Section 7.4.1 we will prove that, under certain conditions, the posterior distribution concentrates around the actual value of the parameter as n increases, with probability 1 under the prior joint distribution of the data and the parameter. But if the prior distribution of the parameter is concentrated on a small portion of the parameter space, one cannot expect the posterior distribution to concentrate near θ given $\Theta = \theta$ for values of θ not in that small portion. Some examples are given in Section 7.4.1.

7.4.1 Consistency of Posterior Distributions[+]

Doob (1949) proved a theorem that says that if there exists a consistent estimator of a parameter Θ, then the posterior distribution of Θ is consistent in the following sense. Let $\mu_{\Theta|X_1,\ldots,X_n}(\cdot|x_1,\ldots,x_n)$ denote the posterior probability measure over (Ω, τ) given $(X_1,\ldots,X_n) = (x_1,\ldots,x_n)$. Let $A \in \tau$ and let I_A be the indicator function of A. The theorem says that $\mu_{\Theta|X_1,\ldots,X_n}(A|x_1,\ldots,x_n)$ converges almost surely, as $n \to \infty$, to $I_A(\Theta)$. The proof given below is adapted from Schwartz (1965, Theorem 3.2) and is similar to the proof of Theorem 2 of Schervish and Seidenfeld (1990).

Theorem 7.78.[22] *Let $(S, \mathcal{A}, \mu)$ be a probability space. Let $(\mathcal{X}^1, \mathcal{B}^1)$ be a Borel space, and let (Ω, τ) be a finite-dimensional parameter space with Borel σ-field. Let $\Theta : S \to \Omega$ and $X_n : S \to \mathcal{X}^1$, for $n = 1, 2, \ldots$ be measurable functions. Suppose that there exists a sequence of functions $h_n : \mathcal{X}^n \to \Omega$ such that $h_n(X_1,\ldots,X_n)$ converges in probability to Θ. Let $\mu_{\Theta|X_1,\ldots,X_n}(\cdot|x_1,\ldots,x_n)$ denote the posterior probability measure on (Ω, τ) given $(X_1,\ldots,X_n) = (x_1,\ldots,x_n)$. For each $A \in \tau$,*

$$\lim_{n \to \infty} \mu_{\Theta|X_1,\ldots,X_n}(A|X_1,\ldots,X_n) = I_A(\Theta), \text{ a.s. } [\mu].$$

PROOF. According to Theorem B.90, there is a subsequence $\{n_k\}_{k=1}^{\infty}$ such that $Z_k = h_{n_k}(X_1,\ldots,X_{n_k})$ converges to Θ, a.s. Let $A \in \tau$, and let $Z = \lim_{k \to \infty} Z_k$ when the limit exists and $Z = \theta_0$ otherwise, where $\theta_0 \in \Omega$. Then $I_A(\Theta) = I_A(Z)$ a.s. Since Z is measurable with respect to the σ-field generated by $\{X_n\}_{n=1}^{\infty}$, part I of Lévy's theorem B.118 says that

---

[+]This section contains results that rely on the theory of martingales. It may be skipped without interrupting the flow of ideas.

[22]The proof relies on martingale theory. The proof in Doob (1949) does not require that a consistent estimator exists, but it has a slightly stronger assumption which implies that a consistent estimator exists.

$E(I_A(Z)|X_1,\ldots,X_n)$ converges almost surely to $I_A(Z)$. Since $I_A(Z) = I_A(\Theta)$, a.s., $E(I_A(Z)|X_1,\ldots,X_n) = \mu_{\Theta|X_1,\ldots,X_n}(A|X_1,\ldots,X_n)$, a.s., and the result is proven. □

The intuitive meaning of this theorem is that when a consistent estimator of Θ exists, the posterior distribution of Θ will tend to concentrate near the true value of Θ, with probability 1 under the joint distribution of the data and the parameter. A consistent estimator of Θ fails to exist if the parameter is not identifiable through the sequence of data values. The parameter M in Problem 16 on page 75 is an example of such a nonidentifiable parameter. Note that there was no explicit mention of the prior distribution of the parameter in Theorem 7.78. Since the theorem makes claims only about probabilities calculated under the joint distribution of the data and the parameter, it holds for all prior distributions. If the prior "misses the true value," then the conclusion to the theorem is not very interesting.

Example 7.79. Suppose that person A has a prior for Θ that is concentrated on a set $C \subseteq \Omega$ and person B believes that Θ is concentrated on a set $D \subseteq \Omega$. Suppose that E is an open set such that $C \subseteq E$ and that the closures of D and E are disjoint. Person B believes that $I_E(\Theta) = 0$ with probability 1, and person A believes that $I_E(\Theta) = 1$ with probability 1. Oddly enough, they both believe that the limit of the posterior probability of E is $I_E(\Theta)$ with probability 1; they just don't agree on which of the two possible values $I_E(\Theta)$ will equal.

Example 7.79 raises an interesting question. If two people (A and B) have different models for data, what does person A believe about the asymptotic behavior of the posterior distribution calculated by person B? Berk (1966) proved a theorem like Theorem 7.80 for the finite-dimensional parameter case. Strasser (1981) used a result of Perlman (1972) to prove a theorem for more general parameters. Theorem 7.80 says that under conditions similar to those that guarantee consistency of MLEs, if person B uses a parametric family with parameter Θ and person A believes that the data are IID with the distribution corresponding to $\Theta = \theta_0$ in the parametric family, then person A believes that person B's posterior will asymptotically be concentrated on the set of θ values such that $\mathcal{I}_{X_1}(\theta_0;\theta)$ is small, where $\mathcal{I}_{X_1}(\theta_0;\theta)$ is the Kullback–Leibler information.

Theorem 7.80. *Assume the conditions of Theorem 7.49 or of Lemma 7.54. For $\epsilon > 0$, define $C_\epsilon = \{\theta : \mathcal{I}_{X_1}(\theta_0;\theta) < \epsilon\}$. Let μ_Θ be a prior distribution such that $\mu_\Theta(C_\epsilon) > 0$, for every $\epsilon > 0$. Then, for every $\epsilon > 0$ and open set N_0 containing C_ϵ, the posterior satisfies $\lim_{n\to\infty} \mu_{\Theta|X^n}(N_0|x^n) = 1$, a.s. $[P_{\theta_0}]$, where $X^n = (X_1,\ldots,X_n) = (x_1,\ldots,x_n) = x^n$ are the data.*

PROOF. For each $x \in \mathcal{X}^\infty$, the infinite product space of copies of $\mathcal{X}^1$, define

$$D_n(\theta, x) = \frac{1}{n}\sum_{i=1}^n \log \frac{f_{X_1|\Theta}(x_i|\theta_0)}{f_{X_1|\Theta}(x_i|\theta)}.$$

7.4. Large Sample Properties of Posterior Distributions

Write the posterior odds of N_0 as

$$\frac{\int_{N_0} d\mu_{\Theta|X^n}(\theta|x^n)}{\int_{N_0^C} d\mu_{\Theta|X^n}(\theta|x^n)} = \frac{\int_{N_0} \prod_{i=1}^n f_{X_1|\Theta}(x_i|\theta)d\mu_{\Theta}(\theta)}{\int_{N_0^C} \prod_{i=1}^n f_{X_1|\Theta}(x_i|\theta)d\mu_{\Theta}(\theta)}$$

$$= \frac{\int_{N_0} \exp(-nD_n(\theta,x))d\mu_{\Theta}(\theta)}{\int_{N_0^C} \exp(-nD_n(\theta,x))d\mu_{\Theta}(\theta)}. \quad (7.81)$$

The idea behind the remainder of the proof is the following. For each x in a set with probability 1, we find a lower bound on the numerator of the last expression in (7.81) and an upper bound on the denominator such that the ratio of these bounds goes to ∞.

First, look at the denominator of the last expression in (7.81). Just as in the proof of Theorem 7.49, construct the sets $\Omega_1,\ldots,\Omega_m$ so that $\Omega = N_0 \cup (\cup_{j=1}^m \Omega_j)$, and $E_{\theta_0} Z(\Omega_j, X_i) = c_j > 0$. It is easy to see that for $M \subseteq \Omega$,

$$\inf_{\theta \in M} D_n(\theta, x) \geq \frac{1}{n}\sum_{i=1}^n Z(M, x_i).$$

So the denominator of the last expression in (7.81) is at most

$$\sum_{j=1}^m \int_{\Omega_j} \exp(-nD_n(\theta,x))d\mu_{\Theta}(\theta) \leq \sum_{j=1}^m \sup_{\theta \in \Omega_j} \exp(-nD_n(\theta,x))\mu_{\Theta}(\Omega_j)$$

$$\leq \sum_{j=1}^m \exp\left(-\sum_{i=1}^n Z(\Omega_j, x_i)\right)\mu_{\Theta}(\Omega_j).$$

For each j, the strong law of large numbers 1.63 says there exists $B_j \subseteq \mathcal{X}^\infty$ such that $P_{\theta_0}(B_j) = 1$ and such that, for every $x \in B_j$, there exists an integer $K_j(x)$ such that $n \geq K_j(x)$ implies $\sum_{i=1}^n Z(\Omega_j, x_i)/n > c_j/2 > 0$. Let $c = \min\{c_1,\ldots,c_m\}$, $B = \cap_{j=1}^m B_j$, and $N(x) = \max\{K_1,\ldots,K_m\}$. For each $x \in B$ and $n \geq N(x)$, the denominator of the last expression in (7.81) is at most $\exp(-nc/2)$.

For the numerator, let $0 < \delta < \min\{\epsilon, c/2\}/4$. For each $x \in \mathcal{X}^\infty$ or $\theta \in \Omega$, let

$$W_n(x) = \{\theta : D_\ell(\theta, x) \leq \mathcal{I}_{X_1}(\theta_0; \theta) + \delta, \text{ for all } \ell \geq n\},$$
$$V_n(\theta) = \{x : D_\ell(\theta, x) \leq \mathcal{I}_{X_1}(\theta_0; \theta) + \delta, \text{ for all } \ell \geq n\}.$$

For each θ, the strong law of large numbers 1.63 says that $D_n(\theta, x) \to \mathcal{I}_{X_1}(\theta_0; \theta)$, a.s. $[P_{\theta_0}]$, so $P_{\theta_0}(\cup_{n=1}^\infty V_n(\theta)) = 1$. Now use this fact together with the fact that the sets $V_n(\theta)$ are increasing and the fact that $x \in V_n(\theta)$ if and only if $\theta \in W_n(x)$ to write

$$\mu_{\Theta}(C_\delta) = \int_{C_\delta} P_{\theta_0}\left(\bigcup_{n=1}^\infty V_n(\theta)\right)d\mu_{\Theta}(\theta)$$

$$= \lim_{n\to\infty} \int_{C_\delta} P_{\theta_0}(V_n(\theta))d\mu_\Theta(\theta)$$

$$= \lim_{n\to\infty} \int_{C_\delta} \int_{\mathcal{X}^\infty} I_{V_n(\theta)}(x)dP_{\theta_0}(x)d\mu_\Theta(\theta)$$

$$= \lim_{n\to\infty} \int_{\mathcal{X}^\infty} \int_{C_\delta} I_{V_n(\theta)}(x)d\mu_\Theta(\theta)dP_{\theta_0}(x)$$

$$= \lim_{n\to\infty} \int_{\mathcal{X}^\infty} \int_{C_\delta} I_{W_n(x)}(\theta)d\mu_\Theta(\theta)dP_{\theta_0}(x)$$

$$= \lim_{n\to\infty} \int_{\mathcal{X}^\infty} \mu_\Theta(C_\delta \cap W_n(x))dP_{\theta_0}(x)$$

$$= \int_{\mathcal{X}^\infty} \lim_{n\to\infty} \mu_\Theta(C_\delta \cap W_n(x))dP_{\theta_0}(x).$$

Since $\mu_\Theta(C_\delta \cap W_n(x)) \leq \mu_\Theta(C_\delta)$ for all x and n, we have $\lim_{n\to\infty} \mu_\Theta(C_\delta \cap W_n(x)) = \mu_\Theta(C_\delta)$, a.s. $[P_{\theta_0}]$, because strict inequality with positive probability would contradict the above string of equalities. So, there is a set $B' \subseteq \mathcal{X}^\infty$ with $P_{\theta_0}(B') = 1$ and for every $x \in B'$, there exists $N'(x)$ such that $n \geq N'(x)$ implies $\mu_\Theta(C_\delta \cap W_n(x)) > \mu_\Theta(C_\delta)/2$. So, if $x \in B'$ and $n \geq N'(x)$, the numerator of the last expression in (7.81) is at least

$$\int_{C_\delta \cap W_n(x)} \exp(-n[\mathcal{I}_{X_1}(\theta_0;\theta)+\delta])d\mu_\Theta(\theta) \geq \frac{1}{2}\exp(-2n\delta)\mu_\Theta(C_\delta)$$

$$\geq \frac{1}{2}\exp\left(-\frac{nc}{4}\right)\mu_\Theta(C_\delta),$$

since $\mathcal{I}_{X_1}(\theta_0;\theta) \leq \delta$ for $\theta \in C_\delta$. It follows that if $x \in B \cap B'$ and $n \geq \max\{N(x), N'(x)\}$, then the ratio in (7.81) is at least $\mu_\Theta(C_\delta)\exp(nc/4)/2$, which goes to ∞ with n. □

Example 7.82 (Continuation of Example 7.51; see page 416). Suppose that $\{X_n\}_{n=1}^\infty$ given $\Theta = \theta$ are IID with $U(0,\theta)$ distribution. We saw earlier that the conditions of Theorem 7.49 are satisfied. The Kullback–Leibler information is

$$\mathcal{I}_{X_1}(\theta_0;\theta) = \begin{cases} \log\frac{\theta}{\theta_0} & \text{if } \theta \geq \theta_0, \\ \infty & \text{if } \theta < \theta_0. \end{cases}$$

The set C_ϵ is the interval $[\theta_0, \exp(\epsilon)\theta_0]$. An open set N_0 containing this interval will need to contain an open interval $(\theta_0 - \delta, \theta_0)$ for $\delta > 0$. So long as the prior distribution assigns postive mass to every open interval, then, for every θ_0, every open interval around θ_0 will have posterior probability going to 1, a.s. $[P_{\theta_0}]$. This is a much stronger claim than one could infer from Theorem 7.78.[23]

As we noted earlier, the conditions of Theorem 7.49 fail in some multiparameter problems. Berk (1966) proves the following lemma, giving a

---

[23] See Problem 48 on page 474 to see why the posterior probability of C_ϵ does not go to 1 almost surely.

7.4. Large Sample Properties of Posterior Distributions

slightly weaker condition that holds in more cases.[24]

Lemma 7.83. *In the notation of Theorems 7.49 and 7.80, instead of assuming that there is a compact set C such that $E_{\theta_0} Z(\Omega \backslash C, X_i) > 0$, assume that there exist an integer p and a compact set C such that*

$$E_{\theta_0} \inf_{\psi \in \Omega \backslash C} \sum_{i=1}^{p} \log \frac{f_{X|\Theta}(X_i|\theta_0)}{f_{X|\Theta}(X_i|\psi)} > 0. \tag{7.84}$$

Then $D_n(\theta, x)$ is bounded below, uniformly in θ, by a random variable that converges almost surely to a positive value.

PROOF. Let $\aleph_{p,n}$ be the collection of all subsets of size p of distinct elements of the set $\{1, \ldots, n\}$. Denote such subsets $\alpha = \{\alpha_1, \ldots, \alpha_p\}$. For $y \in \mathcal{X}^1$, let $g(\theta, y)$ stand for $\log f_{X|\Theta}(y|\theta_0)/f_{X|\Theta}(y|\theta)$. It is clear that for every $\alpha \in \aleph_{p,n}$ and each $\theta \in C^C$,

$$\frac{1}{p} \sum_{i=1}^{p} g(\theta, x_{\alpha_i}) \geq \inf_{\psi \in C^C} \frac{1}{p} \sum_{i=1}^{p} g(\psi, x_{\alpha_i}).$$

If we add both sides of this inequality for all $\alpha \in \aleph$ and divide by $\binom{n}{p}$, we get

$$D_n(\theta, x) \geq \binom{n}{p}^{-1} \sum_{\alpha \in \aleph} \inf_{\psi \in C^C} \frac{1}{p} \sum_{i=1}^{p} g(\psi, x_{\alpha_i}).$$

Call the right-hand side of this expression $G_{n,p}(x)$. (Note that $G_{p,p}(X)$ is the random variable in (7.84) and that $G_{n,p}(x)$ does not depend on θ, so that it is a uniform lower bound.) Due to the symmetry with respect to permutations of coordinates, it is clear that $X_1, \ldots, X_n$ are conditionally exchangeable given $\mathcal{F}_n$, the σ-field generated by $\{G_{n+i,p}\}_{i=0}^{\infty}$. It follows that for every $\alpha \in \aleph_{p,n}$,

$$\begin{aligned} E_{\theta_0}(G_{p,p}(X)|\mathcal{F}_n) &= E_{\theta_0}\left(\inf_{\psi \in C^C} \frac{1}{p} \sum_{i=1}^{p} g(\psi, X_{\alpha_i}) \middle| \mathcal{F}_n\right) \\ &= E_{\theta_0}(G_{n,p}(X)|\mathcal{F}_n) = G_{n,p}(X). \end{aligned}$$

Now, apply part II of Lévy's theorem B.124 to conclude that $G_{n,p}(X)$ converges almost surely to $E_{\theta_0}(G_{p,p}(X)|\mathcal{F}_\infty)$, where $\mathcal{F}_\infty = \cap_{n=p}^{\infty} \mathcal{F}_n$. Since $\mathcal{F}_\infty$ is a sub-σ-field of the tail σ-field of the sequence $\{X_n\}_{n=1}^{\infty}$, the Kolmogorov zero–one law B.68 says that $E_{\theta_0}(G_{p,p}(X)|\mathcal{F}_\infty)$ is constant a.s. $[P_{\theta_0}]$. The constant must be $E_{\theta_0}(G_{p,p}(X)) > 0$. So, $\lim_{n \to \infty} G_{n,p}(X) = E_{\theta_0}(G_{p,p}(X)) > 0$, a.s. $[P_{\theta_0}]$. □

---

[24] The proof of Lemma 7.83 involves martingale theory. Lemma 7.83 also provides a weaker condition under which the MLE converges a.s. $[P_{\theta_0}]$.

Example 7.85 (Continuation of Example 7.52; see page 417). Suppose that $\{X_n\}_{n=1}^\infty$ given $\Theta = (\mu, \sigma)$ are IID with $N(\mu, \sigma^2)$ distribution. If $\theta_0 = (\mu_0, \sigma_0)$, it is easy to calculate

$$\log \frac{f_{X_1, X_2 | \Theta}(x_1, x_2 | \theta_0)}{f_{X_1, X_2 | \Theta}(x_1, x_2 | \theta)} = 2\log \frac{\sigma}{\sigma_0} + s^2 \left(\frac{1}{\sigma^2} - \frac{1}{\sigma_0^2}\right) - \frac{(\overline{x} - \mu_0)^2}{\sigma_0^2} + \frac{(\overline{x} - \mu)^2}{\sigma^2}, \quad (7.86)$$

where $\overline{x} = (x_1 + x_2)/2$ and $s^2 = (x_1 - x_2)^2/4$. (Without loss of generality, assume $\mu_0 = 0$ and $\sigma_0 = 1$ for the rest of this example.) Let C be the rectangle where $\overline{x} \in [-u, u]$ and $s^2 \in [1/v, v]$. The integral of (7.86) for $(\overline{x}, s^2) \notin C$ can be made negligible by choosing v and u large. For $(\overline{x}, s^2) \in C$, the minimum of (7.86) will occur at one of the points (i) $(\overline{x}, v)$, (ii) $(\overline{x}, 1/v)$, (iii) $(u, s^2 + (\overline{x} - u)^2)$, or (iv) $(-u, s^2 + (\overline{x} + u)^2)$. By choosing v large enough, one can check that case (ii) can be ignored and that (7.86) is very large in case (i). In case (iii), (7.86) equals $1 + \log(s^2 + (\overline{x} - u)^2) - s^2 - \overline{x}^2$. The integral of the last two terms (over the entire sample space) is -1.5. By choosing u sufficiently large, the integral of the first two terms over the region where they add to less than 1.5 can be made negligibly small. This is similar for case (iv). So we can ensure that the minimum of (7.86) over C^C has positive integral.

A famous example, in which the conditions of Theorem 7.80 fail, was given by Diaconis and Freedman (1986a, 1986b). It concerns an infinite-dimensional parameter space and a prior distribution constructed from the Dirichlet process described in Section 1.6.1. It is shown that, conditional on the data arising from a cleverly chosen continuous distribution, the posterior mean of a particular function Y of the parameter is not consistent. Initially, what is surprising about this example is the following. Since continuous distribution functions can be approximated arbitrarily closely by discrete distribution functions, one might think that continuous distributions are "close" in some sense to those distributions on which the Dirichlet process concentrates. The problem is that the sense of closeness is not Kullback–Leibler information. Rather, it is based on convergence in distribution.[25] Although Theorem 7.78 can be used to show that the posterior mean of Y converges in probability to Y given distributions in a set C with prior probability 1, the convergence does not extend to parameter values that are "close" to C in the sense of convergence in distribution. (See

---

[25]There is a simple way to understand why inconsistency arises in the Diaconis and Freedman (1986a, 1986b) example. As Barron (1986) pointed out, when the data come from a continuous distribution, the posterior for Y is the same (with probability 1) as what one would get if one assumed that the data were conditionally IID given Y with the distribution given by the normalized base measure of the Dirichlet process. (See Lemma 1.104.) Since the normalized base measure that Diaconis and Freedman (1986a, 1986b) use looks absolutely nothing like the distribution that actually generates the data, it is not surprising that the posterior mean of Y is not consistent. In fact the distribution that generates the data is chosen to be particularly incompatible with the base measure of the Dirichlet process in much the same way that the examples of inconsistent M-estimators were constructed by Freedman and Diaconis (1982).

Problem 47 on page 474.) Theorem 7.80 suggests that the type of closeness that implies consistency in Bayesian problems is much stronger.[26]

7.4.2 Asymptotic Normality of Posterior Distributions

Walker (1969) first proved that, under some conditions, the posterior distribution of a one-dimensional parameter would look more and more like a normal distribution as more conditionally IID data were collected. Dawid (1970) proved a similar result under weaker conditions. Heyde and Johnstone (1979) later extracted the essence of Walker's proof to show that it could extend to sequences of data that were not necessarily conditionally IID.[27] Johnstone (1978) contains a multiparameter version of the theorem of Heyde and Johnstone (1979). Still others—Brenner, Fraser, and McDunnough (1982), Fraser and McDunnough (1984), and Chen (1985)—prove that, under certain conditions, the likelihood function (or the posterior density) converges (in probability or almost surely) to a normal density. The type of asymptotic normality proven by Walker (1969) and by Heyde and Johnstone (1979) follows from the convergence of the posterior density.

In this section, we present a hybrid of the various theorems mentioned above. First, we prove that the posterior density of a suitable transformation of the parameter vector converges to a normal density in probability. We will then use this to conclude that posterior probabilities converge in probability to multivariate normal probabilities. The general situation involves a sequence of random quantities $X_n : S \to \mathcal{X}_n$, for $n = 1, 2, \ldots$ and a parameter $\Theta : S \to \mathbb{R}^k$ such that the conditional distribution of X_n given $\Theta = \theta$ has a density $f_{X_n|\Theta}(\cdot|\theta)$ with respect to a σ-finite measure ν_n on $\mathcal{X}_n$. We use the notation

$$\ell_n(\theta) = \log f_{X_n|\Theta}(X_n|\theta), \quad \ell_n''(t) = \left(\left(\frac{\partial^2}{\partial \theta_i \partial \theta_j} \ell_n(\theta)\bigg|_{\theta=t}\right)\right). \quad (7.87)$$

Let $\hat{\Theta}_n$ stand for the MLE of Θ if it exists, and let[28]

$$\Sigma_n = \begin{cases} -\ell_n''^{-1}(\hat{\Theta}_n) & \text{if the inverse and } \hat{\Theta}_n \text{ exist}, \\ I_k & \text{if not.} \end{cases} \quad (7.88)$$

The following regularity conditions are used in the general theorems.

---

[26] Barron (1988) gives necessary and sufficient conditions for the posterior distribution to concentrate on sets "close" to the distribution that generates the data. His results even apply in nonparametric settings.

[27] What Heyde and Johnstone (1979) did (whether intentionally or not) was to take the conclusions Walker (1969) derived from the assumption that the data were conditionally IID, and use them as assumptions. To use Heyde and Johnstone's result for the conditionally IID case, one need only repeat the portion of Walker's proof in which the assumptions of Heyde and Johnstone's theorem are proven directly. Alternatively, one could prove the assumptions independently.

[28] Notice that Σ_n^{-1} is the observed Fisher information matrix.

General Regularity Conditions:

1. The parameter space is $\Omega \subseteq \mathbb{R}^k$ for some finite k.

2. θ_0 is a point interior to Ω.

3. The prior distribution of Θ has a density with respect to Lebesgue measure that is positive and continuous at θ_0.

4. There exists a neighborhood $N_0 \subseteq \Omega$ of θ_0 on which $\ell_n(\theta)$ is twice continuously differentiable with respect to all coordinates of θ, a.s. $[P_{\theta_0}]$.

5. The largest eigenvalue of Σ_n goes to 0 in probability.

6. For $\delta > 0$, define $N_0(\delta)$ to be the open ball of radius δ around θ_0. Let λ_n be the smallest eigenvalue of Σ_n. If $N_0(\delta) \subseteq \Omega$, then there exists $K(\delta) > 0$ such that

$$\lim_{n \to \infty} P'_{\theta_0}\left(\sup_{\theta \in \Omega \setminus N_0(\delta)} \lambda_n[\ell_n(\theta) - \ell_n(\theta_0)] < -K(\delta)\right) = 1.$$

7. For each $\epsilon > 0$, there exists $\delta(\epsilon) > 0$ such that

$$\lim_{n \to \infty} P'_{\theta_0}\left(\sup_{\theta \in N_0(\delta(\epsilon)), \|\gamma\|=1} \left|1 + \gamma^\top \Sigma_n^{\frac{1}{2}} \ell''_n(\theta) \Sigma_n^{\frac{1}{2}} \gamma\right| < \epsilon\right) = 1.$$

A few words of explanation of these conditions is in order. The first speaks for itself. The second avoids having likelihood functions that are largest near the boundary of Ω and hence cannot look like normal densities. The third ensures that the prior density doesn't destroy the asymptotic normality of the likelihood function. The fourth is one of two smoothness conditions which also rules out distributions for which the support of the distribution depends on θ. Condition 5 ensures that the amount of information in the data about all aspects of Θ increases without bound. Condition 6 ensures that the MLE is consistent and that the likelihood function can be ignored for values not near θ_0. Condition 7 is a smoothness condition on the amount of information in the data about Θ.

To be specific about what we mean by saying that the posterior distribution will look more like a normal distribution as more data are collected, consider the posterior probability that $\Sigma_n^{-1/2}(\Theta - \hat{\Theta}_n) \in B$ as a statistic T_n (function of the data) prior to observing the data. Then T_n converges in probability (under P_{θ_0}) to the multivariate normal probability of B. We can also make corresponding claims about the posterior density of Θ.

7.4.2.1 Posterior Densities

First, we prove that the posterior densities of a sequence of transformations of the parameter converge in probability uniformly on compact sets to a multivariate normal density. Since we expect that the posterior density of Θ will become more and more concentrated around θ_0 given $\Theta = \theta_0$, the posterior density of Θ itself should not have an interesting asymptotic behavior. Rather, if we rescale Θ so that its variance is approximately constant (as a function of sample size), then perhaps the transformed random variable will have an interesting posterior density asymptotically.[29]

Theorem 7.89. *Assume the general regularity conditions, and let $\hat{\Theta}_n$ be an MLE of Θ. Define ℓ_n'' as in (7.87), and let Σ_n be defined by (7.88). Let $\Psi_n = \Sigma_n^{-1/2}(\Theta - \hat{\Theta}_n)$. Then the posterior density of Ψ_n given X_n converges in probability uniformly on compact sets to the $N_k(\mathbf{0}, I_k)$ density ϕ given $\Theta = \theta_0$. That is, for each compact subset B of $\mathbb{R}^k$, and each $\epsilon > 0$,*

$$\lim_{n \to \infty} P'_{\theta_0}\left(\sup_{\psi \in B} |f_{\Psi_n|X_n}(\psi|X_n) - \phi(\psi)| > \epsilon\right) = 0.$$

PROOF. First, note that general regularity condition 6 guarantees that $\hat{\Theta}_n$ is consistent, since, for each δ the probability goes to 1 that $\hat{\Theta}_n$ is inside of $N_0(\delta)$. Use Taylor's theorem C.1 to write

$$\begin{aligned}
f_{X_n|\Theta}(X_n|\theta) &= f_{X_n|\Theta}(X_n|\hat{\Theta}_n)\exp\{\ell_n(\theta) - \ell_n(\hat{\Theta}_n)\} \\
&= f_{X_n|\Theta}(X_n|\hat{\Theta}_n)\exp\Big\{-\frac{1}{2}(\theta - \hat{\Theta}_n)^\top \Sigma_n^{-\frac{1}{2}} \quad (7.90)\\
&\quad (I_k - R_n(\theta, X_n))\Sigma_n^{-\frac{1}{2}}(\theta - \hat{\Theta}_n) + \Delta_n\Big\},
\end{aligned}$$

where

$$\Delta_n = (\theta - \hat{\Theta}_n)^\top \ell_n'(\hat{\Theta}_n) I_{\text{int}(\Omega)}(\hat{\Theta}_n),$$

$$R_n(\theta, X) = I_k + \Sigma_n^{\frac{1}{2}} \ell_n''(\theta_n^*)\Sigma_n^{\frac{1}{2}},$$

with θ_n^* between θ and $\hat{\Theta}_n$. Since $\theta_0 \in \text{int}(\Omega)$ and $\hat{\Theta}_n$ is consistent, it follows that $\lim_{n \to \infty} P'_{\theta_0}(\Delta_n = 0, \text{ for all } \theta) = 1$. Now we can write the posterior density of Θ as

$$f_{\Theta|X_n}(\theta|X_n) = f_\Theta(\theta)\frac{f_{X_n|\Theta}(X_n|\theta)}{f_{X_n}(X_n)},$$

---

[29] It is interesting to note that LeCam (1970) proves asymptotic normality of MLEs by first showing that the logarithm of the likelihood function (as a function of $t = \sqrt{n}(\theta - \theta_0)$) is asymptotically quadratic with the same distribution as $-t^\top \mathcal{I}_{X_1}(\theta_0)t/2 + t^\top Y$, where $Y \sim N_k(\mathbf{0}, \mathcal{I}_{X_1}(\theta_0))$. The maximum of this function is $t = \mathcal{I}_{X_1}(\theta_0)^{-1}Y$, which has $N_k(\mathbf{0}, \mathcal{I}_{X_1}(\theta_0)^{-1})$ distribution.

where
$$f_{X_n}(\cdot) = \int_\Omega f_\Theta(\theta) f_{X_n|\Theta}(\cdot|\theta) d\theta.$$

The posterior density of Ψ_n, $f_{\Psi_n|X_n}(\psi|X_n)$, can be written as

$$\frac{|\Sigma_n|^{\frac{1}{2}} f_{X_n|\Theta}(X_n|\hat{\Theta}_n) f_\Theta(\Sigma_n^{\frac{1}{2}}\psi + \hat{\Theta}_n)}{f_{X_n}(X_n)} \frac{f_{X_n|\Theta}(X_n|\Sigma_n^{\frac{1}{2}}\psi + \hat{\Theta}_n)}{f_{X_n|\Theta}(X_n|\hat{\Theta}_n)}. \tag{7.91}$$

Our first step is to see how the first factor in (7.91) behaves as $n \to \infty$. Choose $0 < \epsilon < 1$ and let η be such that

$$1 - \epsilon \leq \frac{1-\eta}{(1+\eta)^{\frac{k}{2}}}, \quad 1 + \epsilon \geq \frac{1+\eta}{(1-\eta)^{\frac{k}{2}}}.$$

Since the prior is continuous at θ_0, there exists $\delta_1 > 0$ such that $\|\theta - \theta_0\| < \delta_1$ implies $|f_\Theta(\theta) - f_\Theta(\theta_0)| < \eta f_\Theta(\theta_0)$. By general regularity condition 7, there exists $\delta_2 > 0$ such that

$$\lim_{n \to \infty} P'_{\theta_0}\left(\sup_{\theta \in N_0(\delta_2), \|\gamma\|=1} \left|1 + \gamma^\top \Sigma_n^{\frac{1}{2}} \ell''_n(\theta) \Sigma_n^{\frac{1}{2}} \gamma \right| < \eta \right) = 1. \tag{7.92}$$

Let $\delta = \min\{\delta_1, \delta_2\}$. Write $f_{X_n}(X_n) = J_1 + J_2$, where

$$J_1 = \int_{N_0(\delta)} f_\Theta(\theta) f_{X_n|\Theta}(X_n|\theta) d\theta, \quad J_2 = \int_{\Omega \setminus N_0(\delta)} f_\Theta(\theta) f_{X_n|\Theta}(X_n|\theta) d\theta.$$

Use (7.90) to write

$$\begin{aligned} J_1 &= f_{X_n|\Theta}(X_n|\hat{\Theta}_n) \int_{N_0(\delta)} f_\Theta(\theta) \exp\left\{-\frac{1}{2}(\theta - \hat{\Theta}_n)^\top \Sigma_n^{-\frac{1}{2}} \right. \\ & \left. (I_k - R_n(\theta, X_n)) \Sigma_n^{-\frac{1}{2}}(\theta - \hat{\Theta}_n) + \Delta_n \right\} d\theta. \end{aligned}$$

Because $\delta \leq \delta_1$, it follows that

$$(1-\eta)J_3 < \frac{J_1}{f_\Theta(\theta_0) f_{X_n|\Theta}(X_n|\hat{\Theta}_n)} < (1+\eta)J_3, \tag{7.93}$$

where

$$J_3 = \int_{N_0(\delta)} \exp\left\{-\frac{1}{2}(\theta - \hat{\Theta}_n)^\top \Sigma_n^{-\frac{1}{2}}(I_k - R_n(\theta, X_n))\Sigma_n^{-\frac{1}{2}}(\theta - \hat{\Theta}_n) \right. \\ \left. + \Delta_n \right\} d\theta.$$

7.4. Large Sample Properties of Posterior Distributions

It follows from (7.92) and the consistency of $\hat{\Theta}_n$ that the limit as $n \to \infty$ of P'_{θ_0} of the intersection of $\{\Delta_n = 0\}$ with the following event is 1:

$$\left\{ \int_{N_0(\delta)} \exp\left[-\frac{1+\eta}{2}(\theta - \hat{\Theta}_n)^\top \Sigma_n^{-1}(\theta - \hat{\Theta}_n)\right] d\theta \leq J_3 \right.$$

$$\left. \leq \int_{N_0(\delta)} \exp\left[-\frac{1-\eta}{2}(\theta - \hat{\Theta}_n)^\top \Sigma_n^{-1}(\theta - \hat{\Theta}_n)\right] d\theta \right\}.$$

We can write the two integrals that bound J_3 above as

$$\int_{N_0(\delta)} \exp\left\{-\frac{1 \pm \eta}{2}(\theta - \hat{\Theta}_n)^\top \Sigma_n^{-1}(\theta - \hat{\Theta}_n)\right\} d\theta$$
$$= (2\pi)^{\frac{k}{2}}(1 \pm \eta)^{-\frac{k}{2}}|\Sigma_n|^{\frac{1}{2}}\Phi(C_n),$$

where $\Phi(C_n)$ is the probability that an $N_k(\mathbf{0}, I_k)$ vector is in C_n, and

$$C_n = \{t : \hat{\Theta}_n + (1 \pm \eta)^{-\frac{1}{2}}\Sigma_n^{\frac{1}{2}} t \in N_0(\delta)\}.$$

By general regularity condition 5, $\Sigma_n^{1/2} t = o_P(1)$ for all t, so $\Phi(C_n) \xrightarrow{P} 1$. Hence,

$$\lim_{n \to \infty} P'_{\theta_0}\left[(2\pi)^{\frac{k}{2}} \frac{|\Sigma_n|^{\frac{1}{2}}}{(1+\eta)^{\frac{k}{2}}} < J_3 < (2\pi)^{\frac{k}{2}} \frac{|\Sigma_n|^{\frac{1}{2}}}{(1-\eta)^{\frac{k}{2}}}\right] = 1. \quad (7.94)$$

By the way we chose η related to ϵ, we get from (7.93) and (7.94) that

$$\lim_{n \to \infty} P'_{\theta_0}\left[(2\pi)^{\frac{k}{2}}|\Sigma_n|^{\frac{1}{2}}(1-\epsilon) < \frac{J_1}{f_\Theta(\theta_0) f_{X_n|\Theta}(X_n|\hat{\Theta}_n)}\right.$$
$$\left. < (2\pi)^{\frac{k}{2}}|\Sigma_n|^{\frac{1}{2}}(1+\epsilon)\right] = 1.$$

In other words,

$$\frac{J_1}{|\Sigma_n|^{\frac{1}{2}} f_{X_n|\Theta}(X_n|\hat{\Theta}_n)} \xrightarrow{P} (2\pi)^{\frac{k}{2}} f_\Theta(\theta_0). \quad (7.95)$$

Next, we show that

$$\frac{J_2}{|\Sigma_n|^{\frac{1}{2}} f_{X_n|\Theta}(X_n|\hat{\Theta}_n)} \xrightarrow{P} 0. \quad (7.96)$$

Using (7.90), we can write

$$J_2 = f_{X_n|\Theta}(X_n|\hat{\Theta}_n) \exp[\ell_n(\theta_0) - \ell_n(\hat{\Theta}_n)]$$
$$\times \int_{\Omega \setminus N_0(\delta)} f_\Theta(\theta) \exp[\ell_n(\theta) - \ell_n(\theta_0)] d\theta. \quad (7.97)$$

Now, refer to general regularity condition 6. Since $\lambda_n \leq |\Sigma_n|^{1/k}$, if $\theta \notin N_0(\delta)$, then $\ell_n(\theta) - \ell_n(\theta_0) < -|\Sigma_n|^{-1/k} K(\delta)$ with probability tending to 1. Hence, the integral on the right-hand side of (7.97) is less than

$$\exp[-|\Sigma_n|^{-\frac{1}{k}} K(\delta)] \int_{\Omega \setminus N_0(\delta)} f_\Theta(\theta) d\theta \leq \exp\left(-|\Sigma_n|^{-\frac{1}{k}} K(\delta)\right),$$

with probability tending to 1. Since $\hat{\Theta}_n$ is an MLE, $\exp[\ell_n(\theta_0) - \ell_n(\hat{\Theta}_n)] \leq 1$, and general regularity condition 5 says

$$\frac{\exp[-|\Sigma_n|^{-\frac{1}{k}} K(\delta)]}{|\Sigma_n|^{\frac{1}{2}}} \xrightarrow{P} 0.$$

So (7.96) holds. Combining (7.95) with (7.96), we get

$$\frac{f_{X_n}(X_n)}{|\Sigma_n|^{\frac{1}{2}} f_{X_n|\Theta}(X_n|\hat{\Theta}_n)} \xrightarrow{P} (2\pi)^{\frac{k}{2}} f_\Theta(\theta_0). \tag{7.98}$$

Since $\hat{\Theta}_n$ is consistent, and the prior is continuous at θ_0, we have that $f_\Theta(\Sigma_n^{1/2}\psi + \hat{\Theta}_n) \xrightarrow{P} f_\Theta(\theta_0)$ uniformly for ψ in a compact set. It follows that

$$\frac{|\Sigma_n|^{\frac{1}{2}} f_{X_n|\Theta}(X_n|\hat{\Theta}_n) f_\Theta(\Sigma_n^{\frac{1}{2}}\psi + \hat{\Theta}_n)}{f_{X_n}(X_n)} \xrightarrow{P} (2\pi)^{-\frac{k}{2}},$$

uniformly on compact sets.

To complete the proof, we need to show that the second fraction in (7.91) converges in probability to $\exp(-\|\psi\|^2/2)$ uniformly on compact sets. From (7.90), we get

$$\frac{f_{X_n|\Theta}(X_n|\Sigma_n^{\frac{1}{2}}\psi + \hat{\Theta}_n)}{f_{X_n|\Theta}(X_n|\hat{\Theta}_n)}$$
$$= \exp\left\{-\frac{1}{2}\psi^\top (I_k - R_n(\Sigma_n^{\frac{1}{2}}\psi + \hat{\Theta}_n, X_n))\psi + \Delta_n\right\}.$$

Let $\eta, \epsilon > 0$, and let B be a compact subset of $\mathbb{R}^k$. Let b be a bound on $\|\psi\|^2$ for $\psi \in B$. By general regularity condition 7, there exists δ and M such that $n \geq M$ implies

$$P'_{\theta_0}\left(\sup_{\theta \in N_0(\delta), \|\gamma\|=1} \left|1 + \gamma^\top \Sigma_n^{\frac{1}{2}} \ell''_n(\theta) \Sigma_n^{\frac{1}{2}} \gamma\right| < \frac{\eta}{b}\right) > 1 - \frac{\epsilon}{2}.$$

Let $N \geq M$ be large enough so that $n \geq N$ implies

$$P'_{\theta_0}\left(\Sigma_n^{\frac{1}{2}}\psi + \hat{\Theta}_n \in N_0(\delta)\right) > 1 - \frac{\epsilon}{2}.$$

7.4. Large Sample Properties of Posterior Distributions

Then, if $n \geq N$,

$$P'_{\theta_0}\left(|\psi^T(I_k - R_n(\Sigma_n^{\frac{1}{2}}\psi + \hat{\Theta}_n, X_n))\psi - \|\psi\|^2| < \eta \text{ for all } \psi \in B\right) > 1 - \epsilon.$$

Since $P'_{\theta_0}(\Delta_n = 0,$ for all $\psi) \to 1$, it follows that the second fraction in (7.91) is between $\exp(-\eta)\exp(-\|\psi\|^2/2)$ and $\exp(\eta)\exp(-\|\psi\|^2/2)$ with probability tending to 1, uniformly on compact sets. Since η is arbitrary, the desired result follows. □

We now give two examples in which X_n does not consist of conditionally IID coordinates, but the general regularity conditions still hold.

Example 7.99. Let Ω be the interval $(-1, 1)$. Let $\{Z_n\}_{n=1}^\infty$ be IID $N(0,1)$ and let $Y_0 = 0$. Define $Y_n = \theta Y_{n-1} + Z_n$ for $n = 1, 2, \ldots$. The sequence $\{Y_n\}_{n=1}^\infty$ is called a *first-order autoregressive process*. The Y_i are clearly not conditionally IID given $\Theta = \theta$ except for the case $\theta = 0$. Let $X_n = (Y_1, \ldots, Y_n)$. Then $\ell_n(\theta)$ is a constant plus $-\left(Y_n^2 + (1+\theta^2)\sum_{i=1}^{n-1} Y_i^2 - 2\theta\sum_{i=1}^n Y_i Y_{i-1}\right)/2$. The MLE is easily calculated as $\hat{\Theta}_n = \sum_{i=1}^n Y_i Y_{i-1} / \sum_{i=1}^n Y_{i-1}^2$. The first four general regularity conditions are trivially satisfied if the prior has a continuous density. Also, $\ell_n''(\theta) = -\sum_{i=1}^n Y_{i-1}^2$. Since ℓ_n'' does not depend on θ, general regularity condition 7 is satisfied. Since

$$\text{Cov}_\theta(Y_i^2, Y_{i-k}^2) = \theta^{2k}\text{Var}_\theta(Y_{i-k}^2) \leq \frac{2\theta^{2k}}{(1-\theta^2)^2},$$

it is easy to show that $\lim_{n\to\infty} \text{Var}_\theta(\sum_{i=1}^n Y_{i-1}^2/n) = 0$. Hence $\Sigma_n \xrightarrow{P} 0$ and $n\Sigma_n$ converges in probability. Thus, general regularity condition 5 holds. Finally, note that, given $\Theta = \theta_0$, $\sum_{i=1}^n Y_i Y_{i-1}/n \xrightarrow{P} \theta_0$. Since

$$\ell_n(\theta) - \ell_n(\theta_0) = -\frac{\theta - \theta_0}{2}\left((\theta + \theta_0)\sum_{i=1}^n Y_i^2 - 2\sum_{i=1}^n Y_i Y_{i-1}\right),$$

it follows that general regularity condition 6 holds.

Example 7.100. Let $Y_1, Y_2, \ldots$ be conditionally independent given $\Theta = \theta$ with $Y_i \sim N(\theta, i)$. Let $X_n = (Y_1, \ldots, Y_n)$. The logarithm of the likelihood is a constant plus $\ell_n(\theta) = -\left(\sum_{i=1}^n \log(i) + (y_i - \theta)^2/i\right)/2$. The MLE is easily seen to be $\hat{\Theta}_n = \left(\sum_{i=1}^n Y_i/i\right) / \left(\sum_{i=1}^n 1/i\right)$. The second derivative of $\ell_n(\theta)$ is $-\sum_{i=1}^n 1/i$, which does not depend on θ, so general regularity condition 7 holds. Since $\Sigma_n = 1/[\sum_{i=1}^n 1/i] = \mathcal{O}(1/\log(n))$, general regularity condition 5 holds. Since

$$\lambda_n[\ell_n(\theta) - \ell_n(\theta_0)] = \frac{\theta - \theta_0}{\sum_{i=1}^n \frac{1}{i}}\left[\sum_{i=1}^n \frac{1}{i}\left(Y_i - \frac{\theta + \theta_0}{2}\right)\right],$$

and

$$\frac{\sum_{i=1}^n \frac{Y_i}{i}}{\sum_{i=1}^n \frac{1}{i}} \sim N\left(\theta_0, \left[\sum_{i=1}^n \frac{1}{i}\right]^{-1}\right),$$

it follows that $\lambda_n[\ell_n(\theta) - \ell_n(\theta_0)] \xrightarrow{P} -(\theta - \theta_0)^2/2$, so general regularity condition 6 holds. The first four general regularity conditions hold if the prior is continuous. Note that, in this example, the MLE is not $\sqrt{n}$-consistent, but the posterior distribution is still asymptotically normal.

7.4.2.2 Posterior Probabilities

We have proven that the sequence of posterior densities of Ψ_n converges in probability uniformly on compact sets to the $N(\mathbf{0}, I_k)$ density. This makes it easy to conclude that posterior probabilities converge in probability as well.

Theorem 7.101. *Assume the general regularity conditions, and let $\hat{\Theta}_n$ be an MLE of Θ. Let $B \subset \mathbb{R}^k$ be a Borel set. Define ℓ_n'' as in (7.87), let Σ_n be defined by (7.88), and let $\Psi_n = \Sigma_n^{-1/2}(\Theta - \hat{\Theta}_n)$. Then $\Pr(\Psi_n \in B|X_n) \xrightarrow{P} \Phi(B)$, under P_{θ_0}, where $\Phi(B)$ stands for the probability that an $N_k(\mathbf{0}, I_k)$ vector lies in B.*

PROOF. First, suppose that B is a subset of a compact set, and let c be the Lebesgue measure of B. Then

$$\begin{aligned}|\Pr(\Psi_n \in B|X_n) - \Phi(B)| &\leq \int_B |f_{\Psi_n|X_n}(\psi|X_n) - \phi(\psi)|\, d\psi \\ &\leq c \sup_{\psi \in B} |f_{\Psi_n|X_n}(\psi|X_n) - \phi(\psi)|.\end{aligned}$$

This goes to 0 in probability by Theorem 7.89.

Next, let B be an arbitrary Borel set, and let $\epsilon > 0$ be given. Let B_ϵ be a compact set such that $\Phi(B_\epsilon) > 1 - \epsilon/3$. Let N be large enough so that $n \geq N$ implies both of the following:

$$P_{\theta_0}\left(|\Pr(\Psi_n \in B_\epsilon|X_n) - \Phi(B_\epsilon)| > \frac{\epsilon}{3}\right) < \frac{\epsilon}{2},$$
$$P_{\theta_0}\left(|\Pr(\Psi_n \in B \cap B_\epsilon|X_n) - \Phi(B \cap B_\epsilon)| > \frac{\epsilon}{3}\right) < \frac{\epsilon}{2}.$$

We know that

$$\begin{aligned}|\Pr(\Psi_n \in B|X_n) - \Phi(B)| &\leq |\Pr(\Psi_n \in B \cap B_\epsilon|X_n) - \Phi(B \cap B_\epsilon)| \\ &\quad + \Pr(\Psi_n \in B_\epsilon^C|X_n) + \Phi(B_\epsilon^C).\end{aligned}$$

So, if $n \geq N$,

$$P_{\theta_0}(|\Pr(\Psi_n \in B|X_n) - \Phi(B)| > \epsilon) < \epsilon. \qquad \square$$

7.4.2.3 Conditionally IID Random Quantities

Walker (1969) proves a result like Theorem 7.101 in the case of conditionally IID random quantities. He gives a long list of regularity conditions and then proves that they imply the general regularity conditions. These conditions are very similar to those of the Cramér–Rao inequality and those used to prove asymptotic normality of MLEs. Rather than repeat those conditions here, we will use the conditions already stated elsewhere in this book.

7.4. Large Sample Properties of Posterior Distributions

Theorem 7.102. *Let $\{Y_n\}_{n=1}^{\infty}$ be conditionally IID given Θ, and let $X_n = (Y_1, \ldots, Y_n)$. Suppose that (7.64) and the conditions of either Theorem 7.49 or Lemma 7.54[30] hold for $\{Y_n\}_{n=1}^{\infty}$, and that the first four general regularity conditions hold. Also, suppose that the Fisher information $\mathcal{I}_{X_1}(\theta_0)$ is positive definite. Then the remaining general regularity conditions hold.*

PROOF. That general regularity condition 5 holds follows from the fact that $n\Sigma_n \xrightarrow{P} \mathcal{I}_{X_1}(\theta_0)^{-1}$. For general regularity condition 6, let $\delta > 0$ and let $Z(\cdot, \cdot)$ be as in Theorem 7.49. Then

$$\sup_{\theta \in \Omega \setminus N_0(\delta)} [\ell_n(\theta) - \ell_n(\theta_0)] = - \inf_{\theta \in \Omega \setminus N_0(\delta)} [\ell_n(\theta_0) - \ell_n(\theta)] \quad (7.103)$$

$$\leq - \min\left\{ \inf_{\theta \in N_1^*}[\ell_n(\theta_0) - \ell_n(\theta)], \ldots, \inf_{\theta \in N_m^*}[\ell_n(\theta_0) - \ell_n(\theta)], \inf_{\theta \notin C}[\ell_n(\theta_0) - \ell_n(\theta)] \right\}$$

$$\leq - \min\left\{ \sum_{i=1}^{n} Z(\Omega_1, X_i), \ldots, \sum_{i=1}^{n} Z(\Omega_m, X_i), \right\},$$

where $\Omega_1, \ldots, \Omega_m$ are as in the proof of Theorem 7.49. Since

$$\frac{1}{n} \sum_{i=1}^{n} Z(\Omega_j, X_i) \xrightarrow{P} E_{\theta_0} Z(\Omega_j, X_i)$$

and these means are all positive for $j = 1, \ldots, m$, it follows that if

$$K(\delta) \leq \frac{1}{2\lambda} \min\left\{ E_{\theta_0} Z(\Omega_1, X_i), \ldots, E_{\theta_0} Z(\Omega_m, X_i) \right\},$$

where λ is the smallest eigenvalue of $\mathcal{I}_{X_1}(\theta_0)$, then general regularity condition 6 holds. For general regularity condition 7, let $\epsilon > 0$ and let δ be small enough so that $E_{\theta_0} H_\delta(Y_i, \theta_0) < \epsilon/(\mu + \epsilon)$, where H_δ comes from (7.64) and μ is the largest eigenvalue of $\mathcal{I}_{X_1}(\theta_0)$. Let μ_n stand for the largest eigenvalue of Σ_n. For $\theta \in N_0(\delta)$, we have

$$\sup_{\|\gamma\|=1} \left| 1 + \gamma^\top \Sigma_n^{\frac{1}{2}} \ell_n''(\theta) \Sigma_n^{\frac{1}{2}} \gamma \right| = \sup_{\|\gamma\|=1} \left| \gamma^\top \Sigma_n^{\frac{1}{2}} [\Sigma_n^{-1} + \ell_n''(\theta)] \Sigma_n^{\frac{1}{2}} \gamma \right|$$

$$\leq \mu_n \sup_{\|\gamma\|=1} \left| \gamma^\top [\Sigma_n^{-1} + \ell_n''(\theta)] \gamma \right|$$

$$\leq \mu_n \left(\sup_{\|\gamma\|=1} \left| \gamma^\top [\Sigma_n^{-1} + \ell_n''(\theta_0)] \gamma \right| + \sup_{\|\gamma\|=1} \left| \gamma^\top [\ell_n''(\theta_0) - \ell_n''(\theta)] \gamma \right| \right).$$

---

[30] Note that the condition of Lemma 7.83 could be used instead.

If $\hat{\Theta}_n \in N_0(\delta)$ and $|\mu - n\mu_n| < \epsilon$, it follows from (7.64) that the last expression above is no greater than

$$(\mu + \epsilon)\frac{2}{n}\sum_{i=1}^{n} H_\delta(Y_i, \theta_0).$$

By the weak law of large numbers B.95 and our choice of δ, this last expression converges in probability to something no greater than ϵ. This implies general regularity condition 7. □

To make the theorems of this section apply to prior probabilities *not* conditional on the parameter, suppose that the prior distribution satisfies $\Pr(\Theta \in \text{int}(\Omega)) = 1$. We can now apply the result from Problem 6 on page 468 to conclude that the prior probability that the posterior after n observations will be within ϵ of the normal approximation goes to 1 as n goes to infinity.

Example 7.104. Suppose that $X_1, \ldots, X_{10}$ are conditionally IID given $\Theta = \theta$ with $Cau(\theta, 1)$ distribution. Suppose that the observations are

-5, -3, 0, 2, 4, 5, 7, 9, 11, 14.

Then the MLE of Θ is $\hat{\Theta}_{10} = 4.531$, and $\ell''_{10}(4.531) = -1.23116$. So, Theorems 7.101 and 7.102 suggest that $N(4.531, 0.813)$ is approximately the distribution of Θ. To see how good this approximation is, look at Figure 7.105. The solid line is an approximation to the posterior by numerically integrating the likelihood times the prior (trapezoidal rule from $\theta = -1$ to $\theta = 11$). The dotted line is the normal approximation. The functions are not particularly similar. For example, the normal approximation to $\Pr(\Theta \geq 5|X = x)$ is 0.3015, while the numerical integral under the posterior curve is 0.3560.

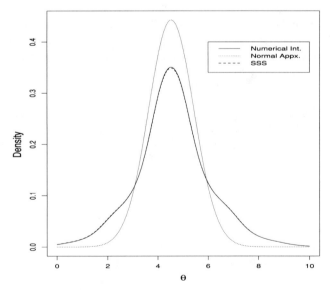

FIGURE 7.105. Posterior Density for Θ in Cauchy Example

7.4.2.4 Loss of Information*

For the case of conditionally IID random quantities, we can try to answer the question "How much information do we lose by not knowing Θ?" The Kullback–Leibler information is $E_{\theta_0} \log[f_{X_n|\Theta}(X_n|\theta_0)/f_{X_n}(X_n)]$ for comparing the distribution of X_n given $\Theta = \theta_0$ to the prior predictive distribution. Alternatively, we can examine $\log[f_{X_n|\Theta}(X_n|\theta_0)/f_{X_n}(X_n)]$ and see how it behaves for large n.

Theorem 7.106. *Assume the conditions of Theorem 7.102. Then, given $\Theta = \theta_0$,*

$$-2\log \frac{f_{X_n|\Theta}(X_n|\theta_0)}{f_{X_n}(X_n)} + k\log\left(\frac{n}{2\pi}\right) - 2\log f_\Theta(\theta_0) + \log |\mathcal{I}_{X_1}(\theta_0)| \xrightarrow{D} \chi_k^2, \tag{7.107}$$

where $\mathcal{I}_{X_1}(\theta_0)$ is the Fisher information matrix for one observation.

PROOF. We have assumed enough conditions to conclude that the general regularity conditions hold and that $\Sigma_n^{-1/2}(\hat{\Theta}_n - \theta_0) \xrightarrow{D} N_k(0, I_k)$, given $\Theta = \theta_0$. Hence, we can use whatever steps we wish from the proof of Theorem 7.89. As in (7.90), we can write

$$\log f_{X_n|\Theta}(X_n|\theta_0) = \log f_{X_n|\Theta}(X_n|\hat{\Theta}_n)$$

$$- \frac{1}{2}\left[(\theta_0 - \hat{\Theta}_n)^\top \Sigma_n^{-\frac{1}{2}}(I_k - R_n(\theta, X_n))\Sigma_n^{-\frac{1}{2}}(\theta - \hat{\Theta}_n) + \Delta_n\right].$$

Using (7.92) and the consistency of the MLE, we conclude that

$$\frac{\log f_{X_n|\Theta}(X_n|\theta_0) - \log f_{X_n|\Theta}(X_n|\hat{\Theta}_n)}{-\frac{1}{2}(\theta_0 - \hat{\Theta}_n)^\top \Sigma_n^{-1}(\theta_0 - \hat{\Theta}_n)} = 1 + o_P(1)$$

given $\Theta = \theta_0$. Since the denominator of this expression converges in distribution to $-.5$ times χ_k^2, it follows that the numerator does also. It follows from (7.98) that

$$\log f_{X_n}(X_n) - \log f_{X_n|\Theta}(X_n|\hat{\Theta}_n) - \frac{1}{2}\log|\Sigma_n| \xrightarrow{P} \frac{k}{2}\log(2\pi) + \log f_\Theta(\theta_0).$$

Since the determinant is a continuous function of a matrix, and $n\Sigma_n \xrightarrow{P} \mathcal{I}_{X_1}(\theta_0)^{-1}$, we get

$$\log f_{X_n}(X_n) - \log f_{X_n|\Theta}(X_n|\hat{\Theta}_n) + \frac{k\log(n)}{2} \xrightarrow{P}$$
$$\frac{k}{2}\log(2\pi) + \log f_\Theta(\theta_0) - \frac{1}{2}\log|\mathcal{I}_{X_1}(\theta_0)|.$$

---

*This section may be skipped without interrupting the flow of ideas.

The conclusion now follows. □

Theorem 7.106 says that the amount of Kullback–Leibler information lost for not knowing Θ, when observing n conditionally IID random variables, tends to be about $k \log(n)/2$, for all continuous priors. The effect of the prior distribution is of a lower order of magnitude. Notice that choosing $f_\Theta(\theta) = |\mathcal{I}_{X_1}(\theta)|^{1/2}$ makes the last two terms on the left-hand side of (7.107) cancel. Suppose that $|\mathcal{I}_{X_1}(\theta)|^{1/2}$ is integrable with respect to Lebesgue measure. In that case, $c|\mathcal{I}_{X_1}(\theta)|^{1/2}$ would then be Jeffreys' prior (for some c) as described in Section 2.3.4. Note also that

$$\int \log \frac{f_\Theta(\theta)}{c|\mathcal{I}_{X_1}(\theta)|^{\frac{1}{2}}} f_\Theta(\theta) d\theta \geq 0,$$

for every prior f_Θ, with equality only when f_Θ is Jeffreys' prior. It follows that if Jeffreys' prior describes someone's beliefs, then that person believes that his or her predictive density for the data $f_{X_n}(X_n)$ will (asymptotically) be smaller, relative to $f_{X_n|\Theta}(X_n|\theta_0)$, than that of someone who believes any other prior. An alternative way to say this is that a person believing Jeffreys' prior thinks that he or she has more to learn about Θ from the data than does someone believing a different prior. This informal description can be made more rigorous, as Clarke and Barron (1994) do. They consider a decision problem in which the action space is the set of continuous prior densities, and the loss is

$$L(\theta, f_\Theta) = \mathrm{E}_\theta \log \frac{f_{X_n|\Theta}(X_n|\theta)}{\int_\Omega f_{X_n|\Theta}(X_n|\theta) f_\Theta(\theta) d\theta}.$$

Note that this loss is precisely the Kullback–Leibler information for comparing the distribution of X_n given $\Theta = \theta$ to the prior predictive distribution. They show that Jeffreys' prior is asymptotically least favorable in this decision problem.

7.4.3 Laplace Approximations to Posterior Distributions*

In Section 7.4.2, we calculated the asymptotic distribution, given $\Theta = \theta$, of the integral of the posterior density over some set. Sometimes, one is only interested in the value of such an integral. The method of Laplace gives us a way to calculate approximations to such integrals together with an order of magnitude of the error. This discussion is a hybrid of the papers by Tierney and Kadane (1986) and Kass, Tierney, and Kadane (1990).

Suppose that we are interested in the posterior mean of a positive function g of Θ. For example, if $g(\theta) = f_{X_1|\Theta}(y|\theta)$ for some fixed value y, then $\mathrm{E}(g(\Theta)|X = x)$ is the predictive density of a future observation. In general,

---

*This section may be skipped without interrupting the flow of ideas.

7.4. Large Sample Properties of Posterior Distributions

we can write

$$E(g(\Theta)|X=x) = \frac{\int g(\theta) f_\Theta(\theta) f_{X|\Theta}(x|\theta) d\theta}{\int f_\Theta(\theta) f_{X|\Theta}(x|\theta) d\theta}.$$

The method of Laplace provides approximations to each of these integrals for specific values of x. Some conditions and notation are needed to state the approximations precisely.

Theorem 7.108. *For each n, let $(\mathcal{X}_n, \mathcal{B}_n)$ be a Borel space, and let X_n be a random quantity taking values in $\mathcal{X}_n$. Let $X^n = (X_1, \ldots, X_n)$ and let $(\mathcal{X}^n, \mathcal{B}^n)$ be the product space of $\mathcal{X}_1, \ldots, \mathcal{X}_n$. Let $\{P_\theta : \theta \in \Omega\}$ be a parametric family of distributions for $\{X_n\}_{n=1}^\infty$ with $\Omega \subseteq \mathbb{R}$. Suppose that the distribution of X^n given $\Theta = \theta$ is absolutely continuous with respect to a measure ν_n on $(X^n, \mathcal{B}^n)$ for all n with density $f_{X^n|\Theta}(\cdot|\theta)$. Let $g : \Omega \to \mathbb{R}^+$ be a function. Let $f_\Theta(\theta)$ be the prior density of Θ with respect to Lebesgue measure. Assume that $f_{X^n|\Theta}(x^n|\theta)$ for all n and $x^n \in \mathcal{X}^n$, $g(\theta)$, and $f_\Theta(\theta)$ are all continuously differentiable with respect to θ six times. Assume that $\int g(\theta) f_\Theta(\theta) d\theta < \infty$. Define*

$$\ell_n(\theta; x^n) = \log f_{X^n|\Theta}(x^n|\theta),$$
$$H_n(\theta; x^n) = \frac{1}{n}[\ell_n(\theta; x^n) + \log f_\Theta(\theta)],$$
$$H_n^*(\theta; x^n) = H_n(\theta; x^n) + \frac{1}{n} \log g(\theta).$$

Now let $\mathcal{Y} = \prod_{n=1}^\infty \mathcal{X}^n$ and define the set $A \subseteq \mathcal{Y}$ as the set of all $x = (x^1, x^2, x^3, \ldots) \in \mathcal{Y}$ with the following properties:

- *The integrals $\int g(\theta) f_\Theta(\theta) f_{X^n|\Theta}(x^n|\theta) d\theta$ and $\int f_\Theta(\theta) f_{X^n|\Theta}(x^n|\theta) d\theta$ are finite for all n.*

- *ℓ_n achieves its maximum at a point $\theta'_n(x^n)$ for each n.*

- *For each n, H_n and H_n^* achieve their maxima at points $\hat{\theta}_n(x^n)$ and $\theta_n^*(x^n)$, respectively, where the first derivatives are zero.*

- *$\hat{\theta}_n(x^n)$ and $\theta_n^*(x^n)$ converge as $n \to \infty$.*

- *The second derivatives of H_n and H_n^* at their maxima converge to negative numbers.*

- *There exists $\epsilon > 0$ such that for every $0 < \delta \leq \epsilon$,*

$$\limsup_{n \to \infty} \frac{1}{n} \sup_{|\theta - \hat{\theta}_n(x^n)| > \delta} \ell_n(\theta; x^n) - \ell_n(\theta'_n(x^n); x^n) < 0. \quad (7.109)$$

For each $(x_1, x_2, x_3, \ldots) \in A$, define
$$\sigma_n^2(x^n) = -\frac{1}{H_n''(\hat{\theta}_n(x^n); x^n)}, \quad \sigma_n^{*2}(x^n) = -\frac{1}{H_n^{*''}(\theta_n^*(x^n); x^n)}.$$
For each $(x_1, x_2, x_3, \ldots) \in A$, $\mathrm{E}(g(\Theta)|X^n = x^n)$ equals
$$\frac{\sigma_n^*(x^n)}{\sigma_n(x^n)} \exp(n[H_n^*(\theta^*(x^n); x^n) - H_n(\hat{\theta}_n(x^n); x^n)]) \left[1 + \mathcal{O}(n^{-2})\right].$$

PROOF. Since virtually everything depends on n and x^n in the statement of the theorem, we will simplify the notation by not explicitly expressing that dependence. For example, $H(\hat{\theta})$ will stand for $H_n(\hat{\theta}_n(x^n); x^n)$. Now, let $(x_1, x_2, \ldots) \in A$ and write

$$\mathrm{E}(g(\Theta)|X^n = x^n) = \frac{\int_\Omega \exp(nH^*(\theta))d\theta}{\int_\Omega \exp(nH(\theta))d\theta}. \tag{7.110}$$

We assumed that $\hat{\theta}$ and θ^* both converge. We now show that they converge to the same thing and that so does θ'. Suppose that $\hat{\theta}$ converges to θ_0 and θ^* converges to θ_1. If there exists $\delta > 0$ such that $|\theta' - \hat{\theta}| > \delta$ for infinitely many n, then (7.109) says that there is $\eta > 0$ such that $\ell(\hat{\theta}) < \ell(\theta') - \eta$ infinitely often. Since $H(\hat{\theta}) = \ell(\hat{\theta}) + \mathcal{O}(n^{-1})$, and similarly for θ', it follows that $H(\hat{\theta}) < H(\theta') - \eta$, infinitely often. This contradicts the definition of $\hat{\theta}$ being the location of the maximum of H for all n. It follows that, for each $\delta > 0$, there exists N such that $n \geq N$ implies $|\theta' - \hat{\theta}| < \delta$. Hence θ' converges to θ_0 also. A similar argument shows that θ' converges to θ_1, hence $\theta_0 = \theta_1$. Let $\epsilon > 0$ and $\delta < \epsilon$. Let $\Omega' = \Omega \cap (\theta_0 - \delta, \theta_0 + \delta)$. Since $\exp(nH(\theta)) = \exp(\ell(\theta))f_\Theta(\theta)$, condition (7.109) implies that $\int_{\Omega \setminus \Omega'} \exp(nH(\theta))d\theta$ and $\int_{\Omega \setminus \Omega'} \exp(nH^*(\theta))d\theta$ are exponentially small. For this reason, we can replace Ω by Ω' in (7.110) without incurring an error larger than $\mathcal{O}(n^{-2})$.

If we expand $H(\theta)$ in a Taylor series (see Theorem C.1) around $\theta = \hat{\theta}$, we get

$$H(\theta) = H(\hat{\theta}) + (\theta - \hat{\theta})H'(\hat{\theta}) + \frac{1}{2}(\theta - \hat{\theta})^2 H''(\hat{\theta}) + \frac{1}{6}(\theta - \hat{\theta})^3 H'''(\hat{\theta})$$
$$+ \frac{1}{24}(\theta - \hat{\theta})^4 H^{(iv)}(\hat{\theta}) + \frac{1}{120}(\theta - \hat{\theta})^5 H^{(v)}(\hat{\theta}) + \frac{1}{720}(\theta - \hat{\theta})^6 H^{(vi)}(\tilde{\theta}),$$

where $\tilde{\theta}$ is between θ and $\hat{\theta}$, and $H^{(iv)}$, $H^{(v)}$, and $H^{(vi)}$ respectively stand for the fourth, fifth, and sixth derivatives of H. Use the Taylor series of $\exp(x)$ around $x = 0$ and the fact that $H'(\hat{\theta}) = 0$ to write

$$\exp(nH(\theta)) = \exp(nH(\hat{\theta})) \exp\left(\frac{n}{2}(\theta - \hat{\theta})^2 H''(\hat{\theta})\right)$$
$$\times \left[1 + \frac{n}{6}(\theta - \hat{\theta})^3 H'''(\hat{\theta}) + \frac{n}{24}(\theta - \hat{\theta})^4 H^{(iv)}(\hat{\theta})\right.$$
$$\left. + \frac{n}{120}(\theta - \hat{\theta})^5 H^{(v)}(\hat{\theta}) + \frac{n^2}{72}(\theta - \hat{\theta})^6 H'''(\hat{\theta})^2 + R_n(\theta)\right],$$

7.4. Large Sample Properties of Posterior Distributions   449

where
$$\int_{\Omega'} R_n(\theta) \exp\left(\frac{n}{2}(\theta - \hat{\theta})^2 H''(\hat{\theta})\right) d\theta = \mathcal{O}(n^{-2})$$

as $n \to \infty$, because $R_n(\theta)$ is bounded on the bounded set Ω'. We can also show that
$$\int_{\Omega'} (\theta - \hat{\theta})^k \exp\left(-\frac{n}{2\sigma^2}(\theta - \hat{\theta})^2\right) d\theta = \mathcal{O}(n^{-2}),$$

for all odd k. (In fact, these last integrals are exponentially small.) This implies that

$$\int_\Omega \exp(nH(\theta)) d\theta = \exp(nH(\hat{\theta})) \Bigg(\int_\Omega \exp\left(-\frac{n}{2\sigma^2}(\theta - \hat{\theta})^2\right)$$
$$\times \left[1 + \frac{n}{24}(\theta-\hat{\theta})^4 H^{(iv)}(\hat{\theta}) + \frac{n^2}{72}(\theta-\hat{\theta})^6 H'''(\hat{\theta})^2\right] d\theta + \mathcal{O}(n^{-2})\Bigg)$$
$$= \sqrt{2\pi}\frac{\sigma}{\sqrt{n}} \exp(nH(\hat{\theta})) \left[1 + \frac{\sigma^4}{8n} H^{(iv)}(\hat{\theta}) + \frac{5\sigma^6}{24n} H'''(\hat{\theta})^2 + \mathcal{O}(n^{-2})\right].$$
(7.111)

A similar argument shows that

$$\int_\Omega \exp(nH^*(\theta)) d\theta = \sqrt{2\pi}\frac{\sigma^*}{\sqrt{n}} \exp(nH(\theta^*)) \qquad (7.112)$$
$$\times \left[1 + \frac{\sigma^{*4}}{8n} H^{*(iv)}(\theta^*) + \frac{5\sigma^{*6}}{24n} H^{*'''}(\theta^*)^2 + \mathcal{O}(n^{-2})\right].$$

Next, we prove that $\hat{\theta}$ and θ^* differ by $\mathcal{O}(n^{-1})$. Since H^* has 0 derivative at θ^*, we can write

$$\begin{aligned}
0 &= H^{*'}(\theta^*) = H'(\theta^*) + \mathcal{O}(n^{-1}) \\
&= H'(\hat{\theta}) + (\theta^* - \hat{\theta})H''(\hat{\theta}) + \mathcal{O}(n^{-1}) + o(\hat{\theta} - \theta^*) \\
&= -(\theta^* - \hat{\theta})\frac{1}{\sigma^2} + \mathcal{O}(n^{-1}) + o(\hat{\theta} - \theta^*).
\end{aligned}$$

It follows that $\theta^* - \hat{\theta} = \mathcal{O}(n^{-1}) + o(\hat{\theta} - \theta^*)$. Since $\hat{\theta} - \theta^* = o(1)$, it follows that $\theta^* - \hat{\theta} = \mathcal{O}(n^{-1})$. It also follows that the kth derivative of H^* at θ^* differs from the kth derivative of H at $\hat{\theta}$ by $\mathcal{O}(n^{-1})$. In particular, $\sigma^{*2} = \sigma^2 + \mathcal{O}(n^{-1})$. Now, take the ratio (7.112) divided by (7.111) to get that $E(g(\Theta)|X = x)$ equals

$$\frac{\sigma^*}{\sigma} \exp(n[H^*(\theta^*) - H(\hat{\theta})]) \frac{1 + \frac{\sigma^{*4}}{8n} H^{*(iv)}(\theta^*) + \frac{5\sigma^{*6}}{24n} H^{*'''}(\theta^*)^2 + \mathcal{O}(n^{-2})}{1 + \frac{\sigma^4}{8n} H^{(iv)}(\hat{\theta}) + \frac{5\sigma^6}{24n} H'''(\hat{\theta})^2 + \mathcal{O}(n^{-2})}$$
$$= \frac{\sigma^*}{\sigma} \exp(n[H^*(\theta^*) - H(\hat{\theta})]) \left[1 + \mathcal{O}(n^{-2})\right]. \qquad \square$$

Theorem 7.108 makes claims about the conditional distribution of Θ given $X^n = x^n$ for a sequence of x^n values having certain properties (namely being in the set A). Of course, we will only ever get to observe the beginning of such a sequence. If our model says that the set A occurs in probability, that is, $\mathcal{P}(A)$, then we might feel comfortable believing that the unobserved tail of the sequence will continue to produce a point in A. All that is required (in addition to the conditions of Theorem 7.108) is that the Fisher information from one observation be positive and finite for all θ and that the MLE is interior to Ω with probability tending to 1.

Example 7.113. Let $\{X_n\}_{n=1}^{\infty}$ be a sequence of IID $N(\mu, \sigma^2)$ random variables conditional on $(M, \Sigma, \Lambda) = (\mu, \sigma, \lambda)$. Let the prior for M given $\Sigma = \sigma, \Lambda = \lambda$ be $N(\mu_0, \sigma^2/\lambda)$. Let Σ and Λ be independent with $\Sigma^2 \sim \Gamma^{-1}(a_0/2, b_0/2)$ and $\Lambda \sim Exp(c_0)$. Conditional on $\Lambda = \lambda_0$, this is precisely the same as the natural conjugate prior distribution, which was given in Example 1.24 on page 14. Hence, the conditional density of $X = (X_1, \ldots, X_n)$ given $\Lambda = \lambda$ is the same as the marginal density of the data in that example. This is easily calculated as

$$\sqrt{\frac{n\lambda}{n+\lambda}} \left(b_0 + w + \frac{n\lambda}{n+\lambda}(\bar{x} - \mu_0)^2\right)^{-\frac{a_0+n}{2}}.$$

Suppose that we want to calculate the predictive density of a future observation Y. Conditional on $\Lambda = \lambda$,

$$Y \sim t_{a_0+n}\left(\frac{\lambda\mu_0 + n\bar{x}}{\lambda+n}, \frac{b_0 + w + \frac{n\lambda}{n+\lambda}(\bar{x}-\mu_0)^2}{a_0+n}\left[1+\frac{1}{\lambda}\right]\right).$$

So, for each value of y, we can let $g(\lambda)$ be the density of Y at y and apply Laplace's method for many values of y.

As an example, suppose that we observe $n = 10$ observations with $\bar{x} = 14.7$ and $w = 52.2$. Suppose that the prior had $a_0 = 1 = b_0 = c_0$ and $\mu_0 = 10$. Then $\hat{\theta} = 0.1598$ provides the maximum of the function H. For each value of y between 0 and 30, say, we can let $g(\lambda)$ be the t_{11} density as described above, and we get the plot in Figure 7.114. For comparison, Figure 7.114 includes the predictive density that would have been obtained had $\Pr(\Lambda = 1) = 1$ been assumed. (The prior mean of Λ is 1.)

A naïve alternative to the Laplace approximation is to use the MLE $g(\hat{\theta})$ to approximate the posterior mean. First, note that $\exp(nH^*(\theta^*)) = g(\theta^*)\exp(nH(\theta^*))$. Since $\hat{\theta} - \theta^* = \mathcal{O}(n^{-1})$, we get that $g(\theta^*) = g(\hat{\theta}) + \mathcal{O}(n^{-1})$ and (as in the proof) $\sigma^{*2} = \sigma^2 + \mathcal{O}(n^{-1})$. Combining these facts together with the fact that $H(\theta^*) = H(\hat{\theta}) + \mathcal{O}([\theta^* - \hat{\theta}]^2)$ we get that the difference between $g(\hat{\theta})$ and the Laplace approximation is $\mathcal{O}(n^{-1})$. Since the Laplace approximation differs from the $E(g(\Theta)|X^n = x^n)$ by $\mathcal{O}(n^{-2})$, we get that $g(\hat{\theta})$ differs from $E(g(\Theta)|X^n = x^n)$ by $\mathcal{O}(n^{-1})$. So, the Laplace approximation can be thought of as a higher-order correction to the use of the MLE as an approximation to $E(g(\Theta)|X^n = x^n)$.

7.4. Large Sample Properties of Posterior Distributions

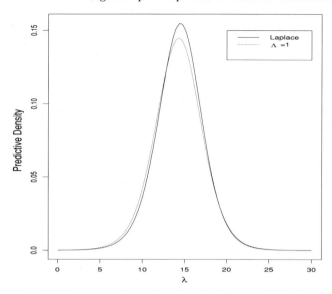

FIGURE 7.114. Predictive Density for Y in Example 7.113

The Laplace method is most useful in hierarchical models, which will be discussed in more detail in Chapter 8. An example in which the Laplace method does not work so well is Example 7.104.

Example 7.115 (Continuation of Example 7.104; see page 444). We have observed ten random variables with $Cau(\theta, 1)$ distribution given $\Theta = \theta$. Suppose that our prior distribution for Θ was very flat, say $N(0, 1000)$. We are now in position to approximate the mean of any positive function of Θ. For example, for each t, we can approximate the mean of $\exp(t\Theta)$. This would be the moment generating function of Θ. For each x we could approximate the mean of $[\pi(1 + (x - \Theta)^2)]^{-1}$. This would be the predictive density of a future observation at x. A serious problem arises with this data set. For some values of t or x, the associated function H^* is not unimodal. This makes the use of the Laplace approximation unsatisfactory.

Nevertheless, we are able to approximate the moment generating function for small values of t at least. This is done by using the function $g(\Theta) = \exp(t\Theta)$ for several different small values of t. Kass, Tierney, and Kadane (1988) suggest using numerical derivatives of the moment generating function for approximating moments of the parameter. For example, if we use $t = k \times 10^{-5}$ for $k = 0, 1, 2, 3, 4$, we can approximate two derivatives of the moment generating function at 0. Laplace's method uses $g(\theta) = \exp(t\theta)$ for the values of t listed above and gives

t	$E(\exp(t\Theta)) - 1.0$
0.0	0.0
1.0×10^{-5}	4.49078×10^{-5}
2.0×10^{-5}	8.98176×10^{-5}
3.0×10^{-5}	13.47294×10^{-5}
4.0×10^{-5}	17.96432×10^{-5}

We can fit a quartic polynomial to these values, and its first two derivatives at

0 will approximate the first two moments. The fitted quartic is $-31677440x^4 + 2972.6x^3 + 9.85075x^2 + 4.49068x + 1.0$. The first derivative is 4.49068 and the second is 19.7015. Unfortunately, the estimated variance would be negative, so these moment estimates are not very good. The problem here is that, for some values of t, the function H^* is not far from being multimodal.

Using numerical integration (trapezoidal rule for $-20 < \theta < 40$ with 6000 intervals), we approximate the mean to be 4.58994 and the variance to be 2.20456. The moment generating function can also be approximated by numerical integration. It is

t	$E(\exp(t\Theta)) - 1.0$
0.0	0.0
1.0×10^{-5}	4.59005×10^{-5}
2.0×10^{-5}	9.18034×10^{-5}
3.0×10^{-5}	13.77086×10^{-5}
4.0×10^{-5}	18.36161×10^{-5}

If we fit a quartic to this, we get

$$-226671x^4 + 38.17317x^3 + 11.63566x^2 + 4.58994x + 1.0,$$

from which we approximate the mean as 4.58994 and the variance as 2.20380.

For multiparameter problems, the situation is much the same except that second and higher derivatives are more complicated objects. For a function f of k variables with p derivatives and for points x and y in $\mathbb{R}^k$, we define

$$D^{(p)}(f;x,y) = \sum_{j_1=1}^{k} \cdots \sum_{j_p=1}^{k} \prod_{s=1}^{p} y_{j_s} \left. \frac{\partial^p}{\partial z_{j_1} \cdots \partial z_{j_p}} f(z) \right|_{z=x}.$$

This is the analogue to the pth derivative evaluated at x times y to the power p. In particular, $D^{(2)}(f;x,y) = y^\top M y$, where M is the matrix of second partial derivatives of f evaluated at x. All of the above reasoning applies as well in the case of k-dimensional θ. For example,

$$\begin{aligned}\exp(nH(\theta)) &= \exp(nH(\hat{\theta})) \exp\left(-\frac{n}{2}(\theta-\hat{\theta})^\top \Sigma^{-1}(\theta-\hat{\theta})\right) \Big[R_n(\theta) \\ &\quad + 1 + \frac{n}{6}D^{(3)}(H;\hat{\theta},\theta-\hat{\theta}) + \frac{n}{24}D^{(4)}(H;\hat{\theta},\theta-\hat{\theta}) \\ &\quad + \frac{n}{120}D^{(5)}(H;\hat{\theta},\theta-\hat{\theta}) + \frac{n^2}{72}[D^{(3)}(H;\hat{\theta},\theta-\hat{\theta})]^2\Big],\end{aligned}$$

where Σ is minus the inverse of the matrix of second partials of H evaluated at $\hat{\theta}$ and

$$\int_\Omega R_n(\theta) \exp\left(-\frac{n}{2}(\theta-\hat{\theta})^\top \Sigma^{-1}(\theta-\hat{\theta})\right) d\theta = \mathcal{O}(n^{-2}),$$

as $n \to \infty$. The net effect of these modifications is that

$$E(g(\Theta)|X=x) = \frac{|\Sigma^*|^{\frac{1}{2}}}{|\Sigma|^{\frac{1}{2}}} \exp(n[H^*(\theta^*) - H(\hat{\theta})]) \left[1 + \mathcal{O}(n^{-2})\right],$$

7.4. Large Sample Properties of Posterior Distributions

where Σ^* is minus the inverse of the matrix of second partials of H^* evaluated at θ^*.

One additional thing that we can do in the multiparameter case is approximate marginal densities of subsets of the parameter vector. Such densities are ratios of integrals of different dimensions, and we will not obtain $\mathcal{O}(n^{-2})$ approximations in this case.

Theorem 7.116. *For each n, let $(\mathcal{X}_n, \mathcal{B}_n)$ be a Borel space, and let X_n be a random quantity taking values in $\mathcal{X}_n$. Let $X^n = (X_1, \ldots, X_n)$ and let $(\mathcal{X}^n, \mathcal{B}^n)$ be the product space of $\mathcal{X}_1, \ldots, \mathcal{X}_n$. Let $\{P_\theta : \theta \in \Omega\}$ be a parametric family of distributions for $\{X_n\}_{n=1}^\infty$ with $\Omega \subseteq \mathbb{R}^k$ with $k > 1$. Suppose that the distribution of X^n given $\Theta = \theta$ is absolutely continuous with respect to a measure ν_n on $(X^n, \mathcal{B}^n)$ for all n with density $f_{X^n|\Theta}(\cdot|\theta)$. Let $f_\Theta(\theta)$ be the prior density of Θ with respect to Lebesgue measure. Assume that $f_{X^n|\Theta}(x^n|\theta)$ for all n and $x^n \in \mathcal{X}^n$, and $f_\Theta(\theta)$ are all continuously differentiable with respect to θ four times. Write a typical point $\theta \in \Omega$ as $\theta = (\gamma, \psi)$, where $\gamma \in \mathbb{R}^p$ and $\psi \in \mathbb{R}^{k-p}$ with $1 \leq p < k$. For each γ, let $\Omega(\gamma) = \{\psi : (\gamma, \psi) \in \Omega\}$. Define*

$$\ell_n(\theta; x^n) = \log f_{X^n|\Theta}(x^n|\theta),$$
$$H_n(\theta; x^n) = \frac{1}{n}[\ell_n(\theta; x^n) + \log f_\Theta(\theta)],$$
$$H_n^*(\psi; x^n, \gamma) = H_n((\gamma, \psi); x^n).$$

Now let $\mathcal{Y} = \prod_{n=1}^\infty \mathcal{X}^n$, and define the set $A \subseteq \mathcal{Y}$ as the set of all $x = (x^1, x^2, x^3, \ldots) \in \mathcal{Y}$ with the following properties:

- *The integrals $\int f_\Theta(\theta) f_{X^n|\Theta}(x^n|\theta) d\theta$ and $\int f_\Theta(\theta, \psi) f_{X^n|\Theta}(x^n|\gamma, \psi) d\psi$ are finite for all n and γ.*

- *ℓ_n achieves its maximum at a point $\theta_n'(x^n)$ for each n.*

- *For each n and γ, $\ell_n(\gamma, \psi)$ achieves its maximum at a point that we will call $\psi_n'(x^n; \gamma)$.*

- *For each n, H_n achieves its maximum at a point $\hat{\theta}_n(x^n)$ where the first derivative is zero.*

- *For each n and γ, $H_n^*(\cdot; x^n, \gamma)$ achieves its maximum at $\psi_n^*(x^n; \gamma)$, a point where the first derivative is zero.*

- *$\hat{\theta}_n(x^n)$ and $\psi_n^*(x^n; \gamma)$ converge as $n \to \infty$.*

- *The second derivatives of H_n and H_n^* at their maxima converge to negative definite matrices.*

- *There exists $\epsilon > 0$ such that, for every $0 < \delta \leq \epsilon$,*

$$\limsup_{n \to \infty} \frac{1}{n} \sup_{\|\theta - \hat{\theta}_n(x^n)\| > \delta} \ell_n(\theta; x^n) - \ell_n(\theta_n'(x^n); x^n) < 0. \quad (7.117)$$

For each $(x_1, x_2, x_3, \ldots) \in A$, define

$$\sigma_n^2(x^n) = -\left(\left(\frac{\partial^2}{\partial \theta_i \partial \theta_j} H_n(\theta; x^n)\bigg|_{\theta=\hat{\theta}_n(x^n)}\right)\right)^{-1},$$

$$\sigma_n^{*2}(x^n; \gamma) = -\left(\left(\frac{\partial^2}{\partial \psi_i \partial \psi_j} H_n^*(\psi; x^n, \gamma)\bigg|_{\psi=\psi_n^*(x^n;\gamma)}\right)\right)^{-1}.$$

For each $(x_1, x_2, x_3, \ldots) \in A$, the marginal posterior density of Γ (the first p coordinates of Θ) given $X^n = x^n$ is $f_{\Gamma|X^n}(\gamma|x^n)$ equal to

$$\frac{n^{\frac{p}{2}}|\sigma_n^*(x^n;\gamma)|^{\frac{1}{2}}}{(2\pi)^{\frac{p}{2}}|\sigma_n(x^n)|^{\frac{1}{2}}}$$
$$\times \ \exp(n[H_n^*(\psi^*(x^n); x^n, \gamma) - H_n(\hat{\theta}_n(x^n); x^n)]) \times \left[1 + \mathcal{O}(n^{-1})\right].$$

PROOF. As in the proof of Theorem 7.108, we will suppress the dependence on n and x^n. Let $(x_1, x_2, \ldots) \in A$, and write

$$f_{\Gamma|X^n}(\gamma|x^n) = \frac{\int_{\Omega(\gamma)} \exp(nH^*(\psi;\gamma))d\psi}{\int_{\Omega} \exp(nH(\theta))d\theta}. \tag{7.118}$$

As in the proof of Theorem 7.108, θ' converges to the same thing to which $\hat{\theta}$ converges, and $\psi'(\gamma)$ converges to the same thing to which $\psi^*(\gamma)$ converges. Let $\epsilon > 0$ and $\delta < \epsilon$. Let Ω' be that part of Ω inside the ball of radius δ around θ_0. Since $\exp(nH(\theta)) = \exp(\ell_n(\theta))f_\Theta(\theta)$, condition (7.117) implies that $\int_{\Omega \setminus \Omega'} \exp(nH(\theta))d\theta$ is exponentially small. For this reason, we can replace Ω by Ω', the part of Ω inside a ball of radius δ around the limit θ_0 of $\hat{\theta}$. We can also replace $\Omega(\gamma)$ by $\Omega'(\gamma)$, the part of $\Omega(\gamma)$ inside a ball of radius δ around the limit of $\psi^*(\gamma)$. The error in (7.118) for doing this is no larger than $\mathcal{O}(n^{-1})$.

By expanding $H(\theta)$ in a Taylor series (see Theorem C.1) around $\theta = \hat{\theta}$, as in the proof of Theorem 7.108, we can obtain

$$\int_\Omega \exp(nH(\theta))d\theta = (2\pi)^{\frac{k}{2}} \frac{|\sigma|^{\frac{1}{2}}}{n^{\frac{k}{2}}} \exp(nH(\hat{\theta})) \times \left[1 + \mathcal{O}(n^{-1})\right].$$

Similarly,

$$\int_{\Omega(\gamma)} \exp(nH^*(\psi;\gamma))d\psi$$
$$= (2\pi)^{\frac{k-p}{2}} \frac{|\sigma^*|^{\frac{1}{2}}}{n^{\frac{k-p}{2}}} \exp(nH^*(\psi^*(\gamma))) \times \left[1 + \mathcal{O}(n^{-1})\right].$$

Taking the ratio of these two gives the result. □

The proof of Theorem 7.116 can be adapted to show that the approximate Bayes factor in (4.27) on page 227 is an $\mathcal{O}(n^{-1})$ approximation to the true Bayes factor when the hypothesis is $H : \Gamma = \gamma_0$. In this case, the parameter under the hypothesis is Ψ, the last $k-p$ coordinates of Θ. One must replace θ' by $\hat{\theta}$ and $\psi'(\gamma)$ by $\psi^*(\gamma)$ in the approximation of Theorem 7.116, but this does not alter the order of the approximation. One can also show that the $\mathcal{O}(n^{-1})$ term in Theorem 7.116 is uniform for γ in compact sets.

7.4.4 Asymptotic Agreement of Predictive Distributions[+]

The next theorem is due to Blackwell and Dubins (1962). It concerns the difference between posterior predictive distributions calculated under two different models.[31] If the two models are somewhat similar (in the sense that they at least assign positive probabilities to the same events), then the posterior probabilities (calculated from the two models) of every event will become uniformly closer as the amount of data increases.

To be precise, we will need to set up some notation. Let $(\mathcal{X}_n, \mathcal{B}_n)$ be measurable spaces, let $\mathcal{X} = \prod_{i=1}^{\infty} \mathcal{X}_i$, and let $\mathcal{B}$ be the product σ-field, $\mathcal{B} = \mathcal{B}_1 \otimes \mathcal{B}_2 \otimes \cdots$. Suppose that P and Q are probability measures on $(\mathcal{X}, \mathcal{B})$. Let P_n and Q_n be the respective marginal distributions on $(\mathcal{Y}_n, \mathcal{C}_n)$, where $\mathcal{Y}_n = \mathcal{X}_1 \times \cdots \mathcal{X}_n$ and $\mathcal{C}_n = \mathcal{B}_1 \otimes \cdots \otimes \mathcal{B}_n$. Let P^n and Q^n stand for versions of the conditional distributions on $(\mathcal{Y}^n, \mathcal{C}^n)$ given the first n coordinates, where $\mathcal{Y}^n = \mathcal{X}_{n+1} \times \mathcal{X}_{n+2} \times \cdots$ and $\mathcal{C}^n = \mathcal{B}_{n+1} \otimes \mathcal{B}_{n+2} \otimes \cdots$. That is, for each $B \in \mathcal{C}^n$, there is a $\mathcal{C}_n$ measurable function $P^n(B|x_1, \ldots, x_n)$ such that, for each $(x_1, \ldots, x_n) \in \mathcal{Y}_n$, $P^n(\cdot|x_1, \ldots, x_n)$ is a probability measure over $\mathcal{C}^n$ and for each bounded measurable $\phi : \mathcal{X} \to \mathbb{R}$,

$$\int_{\mathcal{X}} \phi(x) dP(x) = \int_{\mathcal{Y}_n} \int_{\mathcal{Y}^n} \phi(x) dP^n(x_{n+1}, \ldots | x_1, \ldots, x_n) dP_n(x_1, \ldots, x_n)$$

(and similarly for Q^n). For arbitrary probability measures S and T on the same σ-field $\mathcal{C}$, let

$$\rho(S, T) = \sup_{A \in \mathcal{C}} |S(A) - T(A)|.$$

Before we state and prove the main theorem, we will give conditions under which the hypotheses of the theorem hold.

Lemma 7.119. *Let $\pi_2 \ll \pi_1$ be probability measures on a parameter space (Ω, τ) with parametric family $\{P_\theta : \theta \in \Omega\}$. Suppose that for every $B \in \mathcal{B}$,*

$$P(B) = \int_\Omega P_\theta(B) d\pi_1(\theta), \quad Q(B) = \int_\Omega P_\theta(B) d\pi_2(\theta).$$

[+]This section contains results that rely on the theory of martingales. It may be skipped without interrupting the flow of ideas.
[31]The proof relies on martingale theory.

Then $Q \ll P$.

PROOF. If $Q(B) > 0$, then there exists a set $C \subseteq \Omega$ such that $P_\theta(B) > 0$ for all $\theta \in C$ and $\pi_2(C) > 0$. It follows that $\pi_1(C) > 0$ and then that $P(B) > 0$. □

The importance of Lemma 7.119 is that it applies to the popular model in which data are conditionally IID given some parameter Θ with a distribution in a parametric family. Lemma 7.119 says that if two Bayesians agree on the parametric family but disagree on the prior distribution, then Theorem 7.120 will apply to them, so long as one of the prior distributions is absolutely continuous with respect to the other.

Theorem 7.120. *If $Q \ll P$, then for each P^n there exists a version of Q^n such that*

$$Q\left[(x_1, x_2, \ldots): \lim_{n \to \infty} \rho(P^n(\cdot|x_1, \ldots, x_n), Q^n(\cdot|x_1, \ldots, x_n)) = 0\right] = 1.$$

The proof of this theorem requires some lemmas and corollaries.

Lemma 7.121. *Let Q be a probability measure, and let E denote expectation with respect to Q. Let $\{Y_n\}_{n=1}^\infty$ be a sequence of random variables such that $\lim_{n \to \infty} Y_n = Y$ a.s. $[Q]$ and $|Y_n| \leq m$ for all n and some nonnegative m. Let $\{\mathcal{U}_j\}_{j=1}^\infty$ be an increasing sequence of σ-fields. Let $\mathcal{U}$ be the smallest σ-field containing all of the $\mathcal{U}_j$. Then*

$$\lim_{j \to \infty, n \to \infty} E(Y_n|\mathcal{U}_j) = E(Y|\mathcal{U}).$$

PROOF. Let $G_k = \sup_{n \geq k} Y_n$. For fixed k and $n \geq k$, $Y_n \leq G_k$ and $E(Y_n|\mathcal{U}_i) \leq E(G_k|\mathcal{U}_i)$ a.s. for each i. Define

$$Z = \lim_{j \to \infty} \sup_{i \geq j, n \geq j} E(Y_n|\mathcal{U}_i).$$

Then

$$Z \leq \lim_{j \to \infty} \sup_{i \geq j} E(G_k|\mathcal{U}_i) = \lim_{i \to \infty} E(G_k|\mathcal{U}_i) = E(G_k|\mathcal{U}) \to E(Y|\mathcal{U}),$$

as $k \to \infty$. The first equality holds because the supremum decreases as j increases. The limit follows from the martingale convergence theorem B.117. Similarly, we can show that

$$\lim_{j \to \infty} \inf_{i \geq j, n \geq j} E(Y_n|\mathcal{U}_i) \geq E(Y|\mathcal{U}).$$

Together these imply the lemma. □

Corollary 7.122. *If the probability is 1 that only finitely many of $\{E_n\}_{n=1}^\infty$ occur, then*

$$\lim_{n \to \infty, j \to \infty} Q(\cup_{k=n}^\infty E_k | \mathcal{U}_j) = 0, \quad a.s.$$

7.4. Large Sample Properties of Posterior Distributions   457

PROOF. The condition in the statement of the theorem is $Q(\cup_{n=1}^{\infty} \cap_{k=n}^{\infty} E_k^C) = 1$. This is equivalent to $Q(\cap_{n=1}^{\infty} \cup_{k=n}^{\infty} E_k) = 0$. Let Y_n be the indicator of the event $\cup_{k=n}^{\infty} E_k$. Then $Y = \lim_{n \to \infty} Y_n$ is the indicator of $\cap_{n=1}^{\infty} \cup_{k=n}^{\infty} E_k$, and $E(Y|\mathcal{U}) = 0$. Now apply Lemma 7.121. □

Corollary 7.123. If $\lim_{n \to \infty} T_n = 0$ a.s., then for each $\epsilon > 0$,
$$\lim_{n \to \infty, j \to \infty} Q(\sup_{k \geq n} |T_k| > \epsilon | \mathcal{U}_j) = 0, \quad a.s.$$

PROOF. Let $E_k = \{|T_k| > \epsilon\}$, so that $\{\sup_{k \geq n} |T_k| > \epsilon = \cup_{k=n}^{\infty} E_k\}$. Now apply Corollary 7.122. □

Lemma 7.124. Let $Q \ll P$ and let $q = dQ/dP$. Define
$$q_n(x_1, \ldots, x_n) = \int q(x_1, x_2, \ldots) dP^n(x_{n+1}, \ldots | x_1, \ldots, x_n),$$
$$d_n(x_1, x_2, \ldots) = \begin{cases} \frac{q(x_1, x_2, \ldots)}{q_n(x_1, \ldots, x_n)} & \text{if } q_n > 0, \\ 1 & \text{if } q_n = 0. \end{cases}$$

Then, $q_n = dQ_n/dP_n$, and for each $\epsilon > 0$,
$$\lim_{n \to \infty} Q^n(|d_n - 1| > \epsilon | x_1, \ldots, x_n) = 0, \quad a.s. \ [Q].$$

PROOF. Since, for every $B \in \mathcal{B}_n$,
$$\int_B q_n(x_1, \ldots, x_n) dP_n(x_1, \ldots, x_n)$$
$$= \int_B \int q(x_1, x_2, \ldots) dP^n(x_{n+1}, \ldots | x_1, \ldots, x_n) dP_n(x_1, \ldots, x_n)$$
$$= \int_{B \times \mathcal{Y}^n} q(x_1, x_2, \ldots) dP(x_1, x_2, \ldots) = \int_{B \times \mathcal{Y}^n} dQ(x_1, x_2, \ldots),$$

which equals $Q_n(B)$. Hence, $q_n = dQ_n/dP_n$.

Under probability P, $E[q(X_1, X_2, \ldots) | x_1, \ldots, x_n] = q_n(x_1, \ldots, x_n)$, so $\{q_n(X_1, \ldots, X_n)\}_{n=1}^{\infty}$ is a martingale. By Part I of Lévy's theorem B.118, we conclude $q_n \to q$ a.s. $[P]$. Since $q_n > 0$ a.s. $[Q_n]$, this implies that $d_n \to 1$ a.s. $[Q]$. Now apply Corollary 7.123 with $\mathcal{U}_j = \mathcal{C}_j$. □

PROOF OF THEOREM 7.120. For convenience, let u denote $(x_1, \ldots, x_n)$ and let v denote $(x_{n+1}, \ldots)$. For each P^n, let
$$Q^n(C|u) = \int_C d_n(u, v) dP^n(v|u),$$

which is a version of the conditional distribution of Q given $\mathcal{C}_n$ since, for each $A \in \mathcal{C}_n$ and $C \in \mathcal{C}^n$,
$$\int_A Q^n(C|u) dQ_n(u) = \int_A \int_C \frac{q(u, v)}{q_n(u)} dP^n(v|u) q_n(u) dP_n(u)$$
$$= \int_A \int_C q(u, v) dP(u, v) = Q(A \cap C).$$

For each u and $\epsilon > 0$, let

$$A(u) = \{v : d_n(u,v) > 1\}, \quad A(u,\epsilon) = \{v : d_n(u,v) > 1 + \epsilon\}.$$

Then

$$\begin{aligned}
\rho(P^n(\cdot|u), Q^n(\cdot|u)) &= \int_{A(u)} [d_n(u,v) - 1] dP^n(v|u) \\
&\leq \epsilon + \int_{A(u,\epsilon)} [d_n(u,v) - 1] dP^n(v|u) \\
&\leq \epsilon + \int_{A(u,\epsilon)} d_n(u,v) dP^n(v|u) \\
&= \epsilon + Q^n(A(u,\epsilon)|u).
\end{aligned}$$

Now, write $\{\lim_{n\to\infty} \rho(P^n(\cdot|u_n), Q^n(\cdot|u_n)) = 0\}$ as

$$\begin{aligned}
&\bigcap_{\epsilon > 0} \bigcup_{N=1}^{\infty} \bigcap_{n=N}^{\infty} \{\rho(P^n(\cdot|u_n), Q^n(\cdot|u_n)) \leq 2\epsilon\} \\
&\supseteq \bigcap_{\epsilon > 0} \bigcup_{N=1}^{\infty} \bigcap_{n=N}^{\infty} \{Q^n(A(u_n, \epsilon)|u_n) \leq \epsilon\} \\
&\supseteq \bigcap_{\epsilon > 0} \bigcup_{N=1}^{\infty} \bigcap_{n=N}^{\infty} \{Q^n(|d_n - 1| > \epsilon|u_n) \leq \epsilon\} \\
&= \left\{\lim_{n\to\infty} Q^n(|d_n - 1| > \epsilon|u_n) = 0\right\}.
\end{aligned}$$

The first containment is what we just proved, and the second is trivial. Lemma 7.124 says that Q of the last of these sets is 1, hence Q of the first of these sets is 1. □

7.5 Large Sample Tests

7.5.1 Likelihood Ratio Tests

Asymptotic theory can provide approximate tests in complicated situations. Let $\Omega \subseteq \mathbb{R}^p$, and assume that $\Omega_H = \{\theta : g(\theta) = c\}$. Reparameterize, if necessary, so that the first k coordinates of θ are $g(\theta)$, for $k \leq p$. Let $\hat{\Theta}_{n,H}$ be the MLE of Θ assuming Ω_H is the parameter space, and let $\hat{\Theta}_n$ be the unrestricted MLE. Then the likelihood ratio (LR) criterion (as introduced in Section 4.5.5) is

$$L_n = \frac{\sup_{\theta \in \Omega_H} f_{X|\Theta}(X|\theta)}{\sup_{\theta \in \Omega} f_{X|\Theta}(X|\theta)} = \frac{f_{X|\Theta}(X|\hat{\Theta}_{n,H})}{f_{X|\Theta}(X|\hat{\Theta}_n)}.$$

7.5. Large Sample Tests

We will first consider the special case in which $p = k = 1$ and $g(\theta) = \theta$. Then

$$\begin{aligned}-2\log L_n &= -2\log f_{X|\Theta}(X|c) + 2\log f_{X|\Theta}(X|\hat{\Theta}_n) \\ &= -2\ell_n(c) + 2\ell_n(\hat{\Theta}_n).\end{aligned}$$

Suppose that ℓ_n has two continuous derivatives. Then

$$\ell_n(c) = \ell_n(\hat{\Theta}_n) + (c - \hat{\Theta}_n)\ell_n'(\hat{\Theta}_n) + \frac{1}{2}(c - \hat{\Theta}_n)^2 \ell_n''(\theta_n^*),$$

where θ_n^* is between c and $\hat{\Theta}_n$. Also, $\ell_n'(\hat{\Theta}_n) = 0$. So

$$-2\log L_n = -\ell_n''(\theta_n^*)(c - \hat{\Theta}_n)^2.$$

Now, suppose that $\sqrt{n}(c - \hat{\Theta}_n) \xrightarrow{D} N(0, 1/\mathcal{I}_{X_1}(c))$ under P_c. Then

$$\frac{1}{n}\ell_n''(\theta_n^*) \xrightarrow{P} -\mathcal{I}_{X_1}(c), \quad -2\log L_n \xrightarrow{D} \chi_1^2,$$

under P_c, that is, under H.

For the more general (higher-dimensional) cases, we have the following theorem.

Theorem 7.125. *Assume the conditions of Theorem 7.63. Let L_n be the LR criterion for testing $H : \Theta_i = c_i$ for $i = 1, \ldots, k$. Then $-2\log L_n \xrightarrow{D} \chi_k^2$ under H.*

PROOF. Let $c^\top = (c_1, \ldots, c_k)$, and let $\theta_0 \in \Omega$ be of the form $\theta_0^\top = (c^\top, \psi_0^\top)$, where ψ_0 has dimension $p - k$. Under H, the parameter is $\Psi^\top = (\Theta_{k+1}, \ldots, \Theta_p)$, and the conditions of Theorem 7.63 hold in this smaller problem.

We will find the asymptotic distribution of $-2\log L_n$ under P_{θ_0} and see that it does not depend on ψ_0. Let $\hat{\Theta}_{n,H}$ be the MLE assuming that Ω_H is the parameter space. Then

$$\hat{\Theta}_{n,H} = \begin{pmatrix} c \\ \hat{\Psi}_{n,H} \end{pmatrix}.$$

We will also write the overall MLE in partitioned form as

$$\hat{\Theta}_n = \begin{pmatrix} \hat{c} \\ \hat{\Psi}_n \end{pmatrix}.$$

Then

$$\sup_{\theta \in \Omega_H} f_{X|\Theta}(X|\theta) = f_{X|\Theta}\left(X \middle| \begin{bmatrix} c \\ \hat{\Psi}_{n,H} \end{bmatrix}\right).$$

Use Taylor's theorem C.1 to write

$$\ell_n\begin{pmatrix} c \\ \hat{\Psi}_{n,H} \end{pmatrix} = \ell_n(\hat{\Theta}_n) + \left[\begin{pmatrix} c \\ \hat{\Psi}_{n,H} \end{pmatrix} - \hat{\Theta}_n \right]^\top \begin{pmatrix} \vdots \\ \frac{\partial}{\partial \theta_i}\ell_n(\hat{\Theta}_n) \\ \vdots \end{pmatrix}$$

$$+ \frac{1}{2}\left[\begin{pmatrix} c \\ \hat{\Psi}_{n,H} \end{pmatrix} - \hat{\Theta}_n \right]^\top \left(\left(\frac{\partial^2}{\partial \theta_i \partial \theta_j}\ell_n(\theta_n^*)\right)\right)\left[\begin{pmatrix} c \\ \hat{\Psi}_{n,H} \end{pmatrix} - \hat{\Theta}_n \right], \quad (7.126)$$

with θ_n^* coordinatewise between $\hat{\Theta}_n$ and $\hat{\Theta}_{n,H}$. Next, we use Taylor's theorem C.1 to expand the gradient vector of ℓ_n at both $\hat{\Theta}_n$ and $\hat{\Theta}_{n,H}$ around θ_0. The vector of partial derivatives around $\hat{\Theta}_n$ is the p-dimensional vector

$$\mathbf{0}_p = \begin{pmatrix} \vdots \\ \frac{\partial}{\partial \theta_i}\ell_n(\theta_0) \\ \vdots \end{pmatrix} + \left(\left(\frac{\partial^2}{\partial \theta_i \partial \theta_j}\ell_n(\theta_n^\dagger)\right)\right)(\hat{\Theta}_n - \theta_0), \quad (7.127)$$

where $\theta_n^\dagger$ is coordinatewise between θ_0 and $\hat{\Theta}_n$. The vector of partial derivatives around $\hat{\Theta}_{n,H}$ is the $(p-k)$-dimensional vector

$$\mathbf{0}_{p-k} = \begin{pmatrix} \vdots \\ \frac{\partial}{\partial \psi_i}\ell_n(\theta_0) \\ \vdots \end{pmatrix} + \left(\left(\frac{\partial^2}{\partial \psi_i \partial \psi_j}\ell_n(\theta_n^{\dagger\dagger})\right)\right)(\hat{\Psi}_{n,H} - \psi_0), \quad (7.128)$$

where $\theta_n^{\dagger\dagger}$ is coordinatewise between θ_0 and $\hat{\Theta}_{n,H}$. It follows from (7.64) that

$$\begin{pmatrix} \hat{A}_n & \hat{B}_n \\ \hat{B}_n^\top & \hat{D}_n \end{pmatrix} = \frac{1}{n}\left(\left(\frac{\partial^2}{\partial \theta_i \partial \theta_j}\ell_n(\theta_n^\dagger)\right)\right) \xrightarrow{P} -\mathcal{I}_{X_1}(\theta_0) = \begin{pmatrix} A_0 & B_0 \\ B_0^\top & D_0 \end{pmatrix},$$

$$\hat{D}_{n,H} = \frac{1}{n}\left(\left(\frac{\partial^2}{\partial \psi_i \partial \psi_j}\ell_n(\theta_n^{\dagger\dagger})\right)\right) \xrightarrow{P} D_0.$$

Equating the last $p - k$ coordinates of the two $\mathbf{0}$ vectors in (7.127) and (7.128), we get

$$\hat{D}_{n,H}(\hat{\Psi}_{n,H} - \psi_0) = \hat{B}_n^\top(\hat{c} - c) + \hat{D}_n(\hat{\Psi}_n - \psi_0).$$

We know from Theorem 7.63 that $\hat{\Psi}_{n,H} - \psi_0 = \mathcal{O}_P(1/\sqrt{n})$, $\hat{\Psi}_n - \psi_0 = \mathcal{O}_P(1/\sqrt{n})$, and $\hat{c} - c = \mathcal{O}_P(1/\sqrt{n})$. Also, we just proved that $\hat{D}_n - D_0 = o_P(1)$, $\hat{B}_n^\top - B_0 = o_P(1)$, and $\hat{D}_{n,H} - D_0 = o_P(1)$. It follows that

$$D_0(\hat{\Psi}_{n,H} - \psi_0) = B_0^\top(\hat{c} - c) + D_0(\hat{\Psi}_n - \psi_0) + o_P\left(\frac{1}{\sqrt{n}}\right).$$

Hence,
$$\hat{\Psi}_{n,H} = \hat{\Psi}_n + D_0^{-1} B_0^\top (\hat{c} - c) + o_P\left(\frac{1}{\sqrt{n}}\right). \qquad (7.129)$$

Now, combine (7.126) and (7.129) to conclude that

$$\ell_n\begin{pmatrix} c \\ \hat{\Psi}_{n,H} \end{pmatrix} - \ell_n(\hat{\Theta}_n)$$

$$= -\frac{n}{2}\begin{bmatrix} c - \hat{c} \\ D_0^{-1} B_0^\top (\hat{c} - c) \end{bmatrix}^\top \mathcal{I}_{X_1}(\theta_0) \begin{bmatrix} c - \hat{c} \\ D_0^{-1} B_0^\top (\hat{c} - c) \end{bmatrix} + o_P(1)$$

$$= \frac{n}{2}(\hat{c} - c)^\top [A_0 - B_0 D_0^{-1} B_0^\top](\hat{c} - c) + o_P(1).$$

The matrix $A_0 - B_0 D_0^{-1} B_0^\top$ is the negative of the inverse of the upper-left $k \times k$ corner of $\mathcal{I}_{X_1}(\theta_0)^{-1}$, which, in turn, is minus the inverse of the asymptotic covariance matrix of $\hat{c}$. Since

$$-2 \log L_n = \ell_n(\hat{\Theta}_n) - \ell_n\begin{pmatrix} c \\ \hat{\Psi}_{n,H} \end{pmatrix},$$

it follows that $-2 \log L_n \xrightarrow{D} \chi_k^2$. Note that the choice of ψ_0 is irrelevant. $\square$

When appealing to the asymptotic distribution of the LR criterion, the tradition is to choose α and reject H if $-2 \log L_n$ is greater than the $1 - \alpha$ quantile of the χ_k^2 distribution.

Example 7.130 (Continuation of Example 7.104; see page 444). Using the same data as in the previous Cauchy example, suppose that we wish to test $H : \Theta = 5$. The two values of the likelihood function are

$$\ell_{10}(4.531) = -10 \log(\pi) - 27.36 \text{ and } \ell_{10}(5) = -10 \log(\pi) - 27.50.$$

So $-2 \log L_n = 0.28$, which is too small to reject H at any popular level.

7.5.2 Chi-Squared Goodness of Fit Tests

Another large sample test is the chi-squared (χ^2) goodness of fit test motivated as asymptotically UMP invariant. If Ω is the set of all distributions and θ_0 is one element of Ω, then we can test $H : \Theta = \theta_0$ asymptotically as follows. Choose a fixed dimension p, and divide $\mathcal{X}$ into p disjoint regions $R_1, \ldots, R_p$. Let $Q_i = P_\theta(R_i)$ and $q_{i,0} = P_{\theta_0}(R_i)$. We replace $H : \Theta = \theta_0$ by $H^* : Q = q_0$. H implies H^*, but Ω_{H^*} is bigger than Ω_H. The general result is the following.

Theorem 7.131. *Suppose that $\{X_n\}_{n=1}^\infty$ are IID with distribution P. Let $(R_1, \ldots, R_p)$ be a partition of $\mathcal{X}$. Define Y_j for $j = 1, \ldots, p$ to be the number of the first n X_i that are in R_j, and define $q_i = P(R_i)$. If*

$$C_n = \sum_{i=1}^p \frac{(Y_i - nq_i)^2}{nq_i},$$

then $C_n \xrightarrow{D} \chi^2_{p-1}$ as $n \to \infty$.

PROOF. The distribution of $Y = (Y_1, \ldots, Y_p)^\top$ is $Mult(n; q_1, \ldots, q_p)$ and we know that

$$\sqrt{n} \begin{pmatrix} \frac{Y_1}{n} - q_1 \\ \vdots \\ \frac{Y_p}{n} - q_p \end{pmatrix} \xrightarrow{D} N_p(\mathbf{0}, \Sigma),$$

where $\Sigma = ((\sigma_{i,j}))$, with

$$\sigma_{i,j} = \begin{cases} -q_i q_j & \text{if } i \neq j, \\ q_i(1-q_i) & \text{if } i = j. \end{cases}$$

Let Σ_* be the upper-right $p-1 \times p-1$ corner of Σ. Define

$$D_n = \frac{1}{n}(Y_1 - nq_{1,0}, \ldots, Y_{p-1} - nq_{p-1,0})\Sigma_*^{-1} \begin{pmatrix} Y_1 - nq_{1,0} \\ \vdots \\ Y_{p-1} - nq_{p-1,0} \end{pmatrix},$$

and note that $D_n \xrightarrow{D} \chi^2_{p-1}$. We can rewrite D_n by using

$$\Sigma_*^{-1} = \left(\begin{bmatrix} q_1 & 0 & 0 \\ 0 & \ddots & 0 \\ 0 & 0 & q_{p-1} \end{bmatrix} - \begin{bmatrix} q_1 \\ \vdots \\ q_{p-1} \end{bmatrix} [q_1, \ldots, q_{p-1}] \right)^{-1}.$$

The inverse of a matrix of the form $A - bb^\top$ is

$$A^{-1} + A^{-1}bb^\top A^{-1} \frac{1}{1 - b^\top a^{-1} b}.$$

This means that D_n can be written as

$$\frac{1}{n} \left[\sum_{i=1}^{p-1} \frac{(Y_i - nq_i)^2}{q_i} + \frac{1}{q_{p,0}} \left(\sum_{i=1}^{p-1} \{Y_i - nq_i\} \right)^2 \right] = \sum_{i=1}^{p} \frac{(Y_i - nq_i)^2}{nq_i} = C_n.$$

□

The traditional χ^2 goodness of fit test is to reject the hypothesis that the distribution of the data is P if C_n is greater than the $1 - \alpha$ quantile of the χ^2_{p-1} distribution.

Example 7.132. Bortkiewicz (1898) reports data on the number of men killed by horsekick in the Prussian army.[32] The data were collected from 14 army units for 20 years.

---

[32] See Bishop, Fienberg, and Holland (1975) for a more complete analysis of this data.

Number killed	0	1	2	3	4	≥ 5
Count	144	91	32	11	2	0

These data are clearly not uniformly distributed over the six categories, but we illustrate the χ^2 test with each $q_i = 1/6$. The value of C_{280} is 366.4, which far exceeds the 0.9999 quantile of the χ_5^2 distribution.

A possible Bayesian approach to this problem is to try to measure how close the distribution **P** is to P. Of course there are many measures of closeness. We could let $Q = (Q_1, \ldots, Q_p)$, where $Q_i = \mathbf{P}(R_i)$, and find a large sample approximation to the posterior distribution of Q based on just the data $(Y_1, \ldots, Y_p)$. Theorems 7.102 and 7.101 give one such approximation as

$$N_p\left(\begin{bmatrix} \frac{y_1}{n} \\ \vdots \\ \frac{y_p}{n} \end{bmatrix}, S\right),$$

where $S = ((s_{i,j}))$, with

$$s_{i,j} = \begin{cases} -\frac{y_i y_j}{n^3} & \text{if } i \neq j, \\ \frac{y_i(n-y_i)}{n^3} & \text{if } i = j. \end{cases}$$

We could then examine the distribution of $Q - q$, where $q = (q_1, \ldots, q_p)$, or specifically of $\|Q - q\|$, or whatever. For example, let S_* be the upper-left $p-1 \times p-1$ corner of S and consider

$$(Q_1 - q_1, \ldots, Q_{p-1} - q_{p-1})^\top S_*^{-1} \begin{pmatrix} Q_1 - q_1 \\ \vdots \\ Q_{p-1} - q_{p-1} \end{pmatrix}.$$

This quantity would have approximately an $NC\chi^2_{p-1}(\sum_{i=1}^p (y_i - nq_i)^2/y_i)$ distribution.

A different type of hypothesis might be $H : \Theta \in \mathcal{P}_0$, where $\mathcal{P}_0$ is a parametric family with k-dimensional parameter space Γ with $k < p-1$. This case was considered by Fisher (1924).

Theorem 7.133. *Let Γ be a k-dimensional parameter space with parameter Ψ and $k < p-1$. Let $R_1, \ldots, R_p$ be a partition of $\mathcal{X}$. Let Y_i be the number of observations in R_i for $i = 1, \ldots, p$. Call $Y = (Y_1, \ldots, Y_p)$ the reduced data. Let S_ψ for $\psi \in \Gamma$ stand for the conditional distribution of the reduced data given $\Psi = \psi$. Define $q_i(\psi) = S_\psi(R_i)$ and $q(\psi) = (q_1(\psi), \ldots, q_p(\psi))$. Assume that q has at least two derivatives and is one-to-one. Let $\hat{\Psi}_n$ be the MLE based on the reduced data, and let $\mathcal{I}_{X_1}(\psi)$ be the Fisher information matrix. Assume that $\hat{\Psi}_n$ is asymptotically normal $N_k(\psi, \mathcal{I}_{X_1}(\psi)^{-1}/n)$. Define $\hat{q}_{i,n} = q_i(\hat{\Psi}_n)$ and*

$$C_n = \sum_{i=1}^p \frac{(Y_i - n\hat{q}_{i,n})^2}{n\hat{q}_{i,n}}.$$

464   Chapter 7. Large Sample Theory

Then $C_n \xrightarrow{D} \chi^2_{p-k-1}$ as $n \to \infty$.

PROOF. The likelihood function for the reduced data is $\ell(\psi) = \prod_{i=1}^{p} q_i^{y_i}(\psi)$. Setting the partial derivatives of the log of the likelihood equal to 0 gives the equations

$$0 = \sum_{i=1}^{p} \frac{y_i}{q_i(\psi)} \frac{\partial}{\partial \psi_j} q_i(\psi),$$

for $j = 1, \ldots, k$. Since $\hat{\Psi}_n$ is $\sqrt{n}$-consistent and q is continuous, it follows that $\hat{q}_{i,n}$ is a $\sqrt{n}$-consistent estimator of $q_i(\Psi)$. Since $Y_i/[n\hat{q}_{i,n}] \xrightarrow{P} 1$ for each i, it follows from the likelihood equations (and Problem 7 on page 468) that

$$\sum_{i=1}^{p} \frac{Y_i^2}{n^2 q_j^2(\hat{\Psi}_n)} \frac{\partial}{\partial \psi_j} q_i(\hat{\Psi}_n) = o_P(1). \tag{7.134}$$

The argument we just finished for the case in which the hypothesis is simple shows that for every $\psi \in \Gamma$,

$$C(\psi) = \sum_{i=1}^{p} \frac{(Y_i - nq_i(\psi))^2}{nq_i(\psi)} \xrightarrow{D} \chi^2_{p-1},$$

under S_ψ. Then

$$C(\psi) - C_n = \sum_{i=1}^{p} \frac{Y_i^2}{n} \left(\frac{1}{q_i(\psi)} - \frac{1}{\hat{q}_{i,n}} \right).$$

Use the delta method to write

$$q_i(\psi) = \hat{q}_{i,n} + \sum_{j=1}^{k} (\psi_j - \hat{\Psi}_{n,j}) \frac{\partial}{\partial \psi_j} q_i(\hat{\Psi}_n)$$
$$+ \frac{1}{2} \sum_{j=1}^{k} \sum_{t=1}^{k} (\psi_j - \hat{\Psi}_{n,j})(\psi_t - \hat{\Psi}_{n,t}) \frac{\partial^2}{\partial \psi_j \partial \psi_t} q_i(\hat{\Psi}_n) + o_p\left(\frac{1}{n}\right).$$

It follows that

$$\frac{1}{q_i(\psi)} - \frac{1}{\hat{q}_{i,n}} = -\frac{1}{q_i(\psi)\hat{q}_{i,n}} \sum_{j=1}^{k} (\psi_j - \hat{\Psi}_{n,j}) \frac{\partial}{\partial \psi_j} q_i(\hat{\Psi}_n)$$
$$- \frac{1}{2q_i(\psi)\hat{q}_{i,n}} \sum_{j=1}^{k} \sum_{t=1}^{k} (\psi_j - \hat{\Psi}_{n,j})(\psi_t - \hat{\Psi}_{n,t}) \frac{\partial^2}{\partial \psi_j \partial \psi_t} q_i(\hat{\Psi}_n) + o_p\left(\frac{1}{n}\right)$$
$$= \frac{1}{\hat{q}_{i,n}^3} \left(\sum_{j=1}^{k} (\psi_j - \hat{\Psi}_{n,j}) \frac{\partial}{\partial \psi_j} q_i(\hat{\Psi}_n) \right)^2 - \frac{1}{\hat{q}_{i,n}^2} \sum_{j=1}^{k} (\psi_j - \hat{\Psi}_{n,j}) \frac{\partial}{\partial \psi_j} q_i(\hat{\Psi}_n)$$

7.5. Large Sample Tests

$$-\frac{1}{2\hat{q}_{i,n}^2}\sum_{j=1}^{k}\sum_{t=1}^{k}(\psi_j-\hat{\Psi}_{n,j})(\psi_t-\hat{\Psi}_{n,t})\frac{\partial^2}{\partial\psi_j\partial\psi_t}q_i(\hat{\Psi}_n)+o_p\left(\frac{1}{n}\right).$$

So, we can write

$$C(\psi)-C_n = -\sum_{i=1}^{p}\frac{Y_i^2}{n}\left\{\frac{1}{\hat{q}_{i,n}^2}\sum_{j=1}^{k}(\psi_j-\hat{\Psi}_{n,j})\frac{\partial}{\partial\psi_j}q_i(\hat{\Psi}_n)\right.$$

$$+\frac{1}{\hat{q}_{i,n}^3}\left(\sum_{j=1}^{k}(\psi_j-\hat{\Psi}_{n,j})\frac{\partial}{\partial\psi_j}q_i(\hat{\Psi}_n)\right)^2$$

$$\left.-\frac{1}{2\hat{q}_{i,n}^2}\sum_{j=1}^{k}\sum_{t=1}^{k}(\psi_j-\hat{\Psi}_{n,j})(\psi_t-\hat{\Psi}_{n,t})\frac{\partial^2}{\partial\psi_j\partial\psi_t}q_i(\hat{\Psi}_n)\right\}+o_p(1).$$

We can rearrange the sum of the first set of terms inside the large brace and use (7.134) to remove these terms from the sum. Then

$$C(\psi)-C_n = -\sum_{i=1}^{p}\frac{Y_i^2}{n}\left\{\frac{1}{\hat{q}_{i,n}^3}\left(\sum_{j=1}^{k}(\psi_j-\hat{\Psi}_{n,j})\frac{\partial}{\partial\psi_j}q_i(\hat{\Psi}_n)\right)^2\right.$$

$$\left.\frac{1}{2\hat{q}_{i,n}^2}\sum_{j=1}^{k}\sum_{t=1}^{k}(\psi_j-\hat{\Psi}_{n,j})(\psi_t-\hat{\Psi}_{n,t})\frac{\partial^2}{\partial\psi_j\partial\psi_t}q_i(\hat{\Psi}_n)\right\}+o_p(1).$$

Since $Y_i = \mathcal{O}_P(n)$ and the inner summations are both $\mathcal{O}_P(1/n)$, we can use the fact that $Y_i/[n\hat{q}_{i,n}] \xrightarrow{P} 1$ for every i (and Problem 7 on page 468) to rewrite $C(\psi) - C_n$ as

$$-\sum_{j=1}^{k}\sum_{t=1}^{k}(\psi_j-\hat{\Psi}_{n,j})(\psi_t-\hat{\Psi}_{n,t})\sum_{i=1}^{p}Y_i\left\{\frac{1}{\hat{q}_{i,n}^2}\frac{\partial}{\partial\psi_j}q_i(\hat{\Psi}_n)\frac{\partial}{\partial\psi_t}q_i(\hat{\Psi}_n)\right.$$

$$\left.-\frac{1}{2\hat{q}_{i,n}}\frac{\partial^2}{\partial\psi_j\partial\psi_t}q_i(\hat{\Psi}_n)\right\}+o_p(1).$$

Next, notice that

$$\frac{1}{n}\frac{\partial^2}{\partial\psi_j\partial\psi_t}\log\ell(\hat{\Psi}_n) = \sum_{i=1}^{p}\frac{Y_i}{n}\left\{\frac{1}{\hat{q}_{i,n}}\frac{\partial^2}{\partial\psi_j\partial\psi_t}q_i(\hat{\Psi}_n)\right.$$

$$\left.-\frac{1}{q_i^2(\hat{\Psi}_n)}\frac{\partial}{\partial\psi_j}q_i(\hat{\Psi}_n)\frac{\partial}{\partial\psi_t}q_i(\hat{\Psi}_n)\right\}\xrightarrow{P}-\mathcal{I}_{X_1}(\psi)_{j,t},$$

the (j,t) element $\mathcal{I}_{X_1}(\psi)$. Combining this with the previous equation, we get

$$C(\psi)-C_n = n(\psi-\hat{\Psi}_n)^{\top}\mathcal{I}_{X_1}(\psi)(\psi-\hat{\Psi}_n)+o_P(1).$$

Since $\mathcal{I}_{X_1}(\psi)^{-1}/n$ is the asymptotic covariance matrix of $\hat{\Psi}_n$, we have that $C(\psi) - C_n \xrightarrow{D} \chi_k^2$. Next we prove that $C(\psi) - C_n$ is asymptotically independent of C_n. This will make the asymptotic characteristic function of C_n equal to the ratio of the asymptotic characteristic functions of $C(\psi)$ and $C(\psi) - C_n$. Since the former is that of χ_{p-1}^2 and the latter is that of χ_k^2, the ratio is that of χ_{p-k-1}^2.

Define $\hat{q}_n = (\hat{q}_{1,n}, \ldots \hat{q}_{p,n})^\top$. Use the delta method to write

$$\sqrt{n}(\hat{q}_n - q(\psi)) = V[\hat{\Psi}_n - \psi] + o_P(1),$$

where $V = ((v_{i,j}))$ is the $p \times k$ matrix, with $v_{i,j} = \partial q_i(\psi)/\partial \psi_j$. It follows that $\sqrt{n}(\hat{q}_n - q(\psi))$ is asymptotically $N_p(\mathbf{0}, V\mathcal{I}_{X_1}(\psi)^{-1}V^\top)$. Since q is one-to-one, V has rank k and $V^- = (V^\top V)^{-1}V^\top$ exists. It is easy to see that $V^- V$ is a $k \times k$ identity matrix. Hence,

$$\begin{aligned} C(\psi) - C_n &= n[V(\psi - \hat{\Psi}_n)]^\top V^{-\top}\mathcal{I}_{X_1}(\psi)V^-V(\psi - \hat{\Psi}_n) + o_P(1) \\ &= n(\hat{q}_n - q)^\top V^{-\top}\mathcal{I}_{X_1}(\psi)V^-(\hat{q}_n - q) + o_P(1). \end{aligned}$$

Also, since $(Y_i - n\hat{q}_{i,n})^2/n = \mathcal{O}_P(1)$ for each i and $1/\hat{q}_{i,n} \xrightarrow{P} 1/q_i(\psi)$, we can use Problem 7 on page 468 to conclude

$$C_n = \sum_{i=1}^p \frac{(Y_i - n\hat{q}_{i,n})^2}{nq_i(\psi)} + o_P(1).$$

The proof will be complete if we can show that $Y - \hat{q}_n$ and $\hat{q}_n - q(\psi)$ are asymptotically independent. Since they are jointly asymptotically multivariate normal, it suffices to show that they are asymptotically uncorrelated. Since $Y - q(\psi) = Y - \hat{q}_n + (\hat{q}_n - q(\psi))$, we need only show that the asymptotic covariance matrix of $\hat{q}_n$ (namely, $V\mathcal{I}_{X_1}(\psi)V^\top$) is the same as the asymptotic covariance between Y and $\hat{q}_n$. We find the latter as follows. First, note that (following some tedious algebra)

$$E\left[(Y_t - nq_t(\psi))\sum_{i=1}^p Y_i \frac{\frac{\partial}{\partial \psi_j}q_i(\psi)}{q_i(\psi)}\right] = \frac{\partial}{\partial \psi_j}q_t(\psi).$$

Hence the asymptotic covariance between Y and the vector $D(\psi)$ of partial derivatives of $\log \ell(\psi)$ is V. Next, use the delta method to write

$$\begin{aligned} 0 &= \frac{1}{\sqrt{n}}\sum_{i=1}^p Y_i \frac{\frac{\partial}{\partial \psi_j}q_i(\hat{\Psi}_n)}{q_i(\hat{\Psi}_n)} \\ &= \frac{1}{\sqrt{n}}\sum_{i=1}^p Y_i \frac{\frac{\partial}{\partial \psi_j}q_i(\psi)}{q_i(\psi)} + \sum_{t=1}^k m_{n,t,j}\sqrt{n}(\hat{\Psi}_n - \psi) + o_P(1), \end{aligned}$$

where

$$m_{n,s,t} = \frac{1}{n}\frac{\partial^2}{\partial \psi_s \partial \psi_t}\log \ell(\psi).$$

Set $M_n = ((m_{n,s,t}))$, and note that $M_n \xrightarrow{P} \mathcal{I}_{X_1}(\psi)$. It follows that

$$\frac{1}{n}D(\psi) = \sqrt{n}\mathcal{I}_{X_1}(\psi)(\hat{\Psi}_n - \psi) + o_P(1).$$

Hence, $\sqrt{n}(\hat{\Psi}_n - \psi) = \mathcal{I}_{X_1}(\psi)^{-1}D(\psi)/n + o_P(1)$. So, the asymptotic covariance between Y and $\hat{\Psi}_n$ is $V\mathcal{I}_{X_1}(\psi)^{-1}$. Since $\hat{q}_n - V\hat{\Psi}_n = o_P(1)$, it follows that the asymptotic covariance between Y and $\hat{q}_n$ is $V\mathcal{I}_{X_1}(\psi)^{-1}V^\top$, and the proof is complete. □

In applying Theorem 7.133, one must be careful to calculate the MLE of Ψ based on the reduced data Y, not on the original data X.

Example 7.135 (Continuation of Example 7.132; see page 462). A more reasonable hypothesis to test in the horsekick data example is that the distribution of horsekicks is a member of the Poisson family. Because there are no data in the "≥ 5" category, the likelihood function for the reduced data is the same as for the original data, and the MLE is the sample average $\hat{\Psi}_{280} = 0.7$. The six values of $\hat{q}_{i,280}$ corresponding to $i = 0, 1, 2, 3, 4, 5$ are $(139.0, 97.3, 34.1, 7.9, 1.4, 0.2)$, respectively, and $C_{280} = 2.346$, which is the 0.3276 quantile of χ^2_4 distribution.

Example 7.136. Let $\{X_n\}_{n=1}^\infty$ be exchangeable. Suppose that we want to test the hypothesis that they have normal distribution. Let $R_1 = (-\infty, r_1]$, $R_i = (r_{i-1}, r_i]$ for $i = 2, \ldots, p-1$, and $R_p = (r_{p-1}, \infty)$. (For convenience, define $r_0 = -\infty$ and $r_p = \infty$.) Then $q_i(\psi) = \Phi([r_i - \mu]/\sigma)$ if $\psi = (\sigma, \mu)$. The likelihood function to maximize is $L(\psi) = \prod_{i=1}^p q_i(\psi)^{Y_i}$, where $Y_i = \sum_{j=1}^n I_{R_i}(X_j)$. The MLE will not equal the sample average and sample standard deviation in general.

Example 7.137. The usual χ^2 test of independence in a two-way ($r \times c$) contingency table is an example of Theorem 7.133. In this case, the data are not reduced. That is, each R_i contains only one element of $\mathcal{X}$. In fact, the R_i are the cells themselves, which would be better denoted $R_{i,j}$ for the cell in row i and column j ($i = 1, \ldots, r$, $j = 1, \ldots, c$). The parameter Ψ consists of two marginal probability vectors, one for the rows Ψ^R and one for the columns Ψ^C. Then $q_{i,j}(\psi) = \psi_i^R \psi_j^C$. The MLE $\hat{\Psi}_n$ is easily seen to consist of $\hat{\Psi}_i^R$ equal to the row i total divided by n and $\hat{\Psi}_j^C$ equal to the column j total divided by n. One easily verifies that C_n is the usual χ^2 statistic, and the appropriate degrees of freedom are $(r-1)(c-1)$.

7.6 Problems

Section 7.1:

1. Prove Proposition 7.4 on page 396.

2. Prove Proposition 7.14 on page 398.

3. Let X and $\{X_n\}_{n=1}^\infty$ be random variables, and suppose that $X_n \xrightarrow{D} X$. Prove that $X_n = \mathcal{O}_P(1)$.

4. Let the conditional distribution of $\{X_n\}_{n=1}^\infty$ be that of IID $N(\mu, \sigma^2)$ random variables given $\Theta = (\mu, \sigma)$. Define

$$\overline{X}_n = \frac{1}{n}\sum_{i=1}^n X_i, \quad S_n = \sqrt{\frac{1}{n-1}\sum_{i=1}^n (X_i - \overline{X}_n)^2}.$$

Find the asymptotic distribution of $\overline{X}_n + 1.96 S_n$. That is, find a_n and b_n so that $a_n[\overline{X}_n + 1.96 S_n - b_n]$ converges in distribution to a nondegenerate distribution.

5. Suppose that for each $\theta \in \Omega$, $(\mathcal{X}_n, \mathcal{B}_n, P_{\theta,n})$ is a sequence of probability spaces. Let $Y_n : \mathcal{X}_n \to \mathbb{R}^k$ be random vectors for each n. Suppose that $Y_n = o_P(1)$ for each θ. Let τ be a sigma field of subsets of Ω such that for every n and every $A \in \mathcal{B}_n$, $P_{\theta,n}(A)$ is τ measurable as a function of θ. Let Q be a probability measure over (Ω, τ). Define $Q_n(\cdot) = \int_\Omega P_{\theta,n}(\cdot) dQ(\theta)$ for each n so that $(\mathcal{X}_n, \mathcal{B}_n, Q_n)$ is a probability space. Show that $Y_n = o_P(1)$ with respect to the sequence $\{Q_n\}_{n=1}^\infty$.

6. Consider the setup in Problem 5 above. Let $\mathcal{X}_n = \mathcal{X}$ for all n. If $P_{\theta,n} \xrightarrow{W} P_\theta$ as $n \to \infty$ for each θ, let $Q_0(\cdot) = \int_\Omega P_\theta(\cdot) dQ(\theta)$. Prove that $Q_n \xrightarrow{W} Q_0$.

7. Suppose that $Z_{i,n} \xrightarrow{P} c_i$ and $X_{i,n} = \mathcal{O}_P(1)$ for $i = 1, \ldots, p$ as $n \to \infty$. Show that $\sum_{i=1}^p Z_{i,n} X_{i,n} - \sum_{i=1}^p c_i X_{i,n} = o_P(1)$.

8. Suppose that $\rho(t) \geq 0$ is an even function of t with $\rho(t) > 0$ for $t \neq 0$ and $\rho(t)$ a strictly increasing function of $|t|$. If

$$\lim_{n\to\infty} E_\theta \rho(X_n - g(\theta)) = 0$$

for all θ, then show that X_n is consistent for $g(\Theta)$.

9.*Suppose that $\sqrt{n}(Y_n - \mu) \xrightarrow{D} N_{k+1}(\mathbf{0}, \Sigma)$. Let U_n^* be the smallest root of the polynomial $p_n(u) = \sum_{i=0}^k Y_{n,i} u^i$, where $Y_n^\top = (Y_{n,0}, \ldots, Y_{n,k})$. Let u_0 be the smallest root of $p(u) = \sum_{i=0}^k \mu_i u^i$. Assume that u_0 has multiplicity exactly 3 (i.e., $p(u_0) = p'(u_0) = p''(u_0) = 0$, but $p'''(u_0) \neq 0$.) Find a_n so that $a_n(U_n^* - u_0)$ converges in distribution to a nondegenerate distribution, and find the asymptotic distribution.

10. Let $\{X_n\}_{n=1}^\infty$ be conditionally IID with $N(\theta, 1)$ given $\Theta = \theta$. Let $Y_n = \Phi((c - \overline{X}_n)/\sqrt{1 - 1/n})$, where Φ is the standard normal distribution function and c is a constant. Find a_n and b_n such that $a_n(Y_n - b_n)$ has a nondegenerate limiting distribution, and find the distribution.

11. Let $\{X_n\}_{n=1}^\infty$ be conditionally IID $Ber(\theta)$ given $\Theta = \theta$. Let $Y_n = n^{-1}\sum X_i$, and let $g(y) = 2\sin^{-1}\sqrt{y}$ (i.e., $g^{-1}(z) = \sin^2(z/2)$.) Suppose the prior for Θ has density given by

$$f_\Theta(\theta) = c\theta^{-\frac{1}{2}}(1-\theta)^{-\frac{1}{2}}\exp\left(-\frac{1}{2\sigma}(g(\theta) - \mu)^2\right),$$

where $0 < \theta < 1$, μ and σ are constants, and

$$c = \left[\Phi\left(\frac{\pi - \mu}{\sigma}\right) - \Phi\left(-\frac{\mu}{\sigma}\right)\right]^{-1}.$$

Let $Z_n = g(Y_n)$. Find a_n and b_n such that $a_n Z_n + b_n$ converges in distribution to a nondegenerate distribution with respect to the marginal distribution of the data. (*Hint:* Recall that $d\sin^{-1}(u)/du = \{1-u^2\}^{-1/2}$.)

Section 7.2:

12. Suppose that $F(t) = 1$ and $F(t - \epsilon) < 1$ for all $\epsilon > 0$ and that F is differentiable at all values less than t with derivative f such that $\lim_{x \uparrow t} f(x) = c$ with $0 < c < \infty$, and F is continuous at t. Let $\{X_n\}_{n=1}^{\infty}$ be IID with CDF F and let $X_{(n)} = \max\{X_1, \ldots, X_n\}$. Prove that $n(t - X_{(n)}) \xrightarrow{D} Exp(c)$.

13. Prove Proposition 7.34 on page 408.

14. Prove Proposition 7.37 on page 410.

15. Suppose that $\{X_n\}_{n=1}^{\infty}$ are conditionally IID with Cauchy distribution having median θ given $\Theta = \theta$. Calculate the asymptotic efficiency of the sample median and of the best linear combination of three symmetrically placed sample quantiles.

16. Let $F(x) = [1 + \exp(-x)]^{-1}$. Assume that $\{X_n\}_{n=1}^{\infty}$ are conditionally IID given $\Theta = \theta$ with CDF $F(x - \theta)$.

    (a) Prove that the density is symmetric about θ.

    (b) If we wish to use the L-estimator based on $Y_p^{(n)}$, $Y_{1/2}^{(n)}$, and $Y_{1-p}^{(n)}$, with $p < 1/2$, find the best p and the best coefficients.

17. *Let $\{X_n\}_{n=1}^{\infty}$ be conditionally IID given $\Theta = \theta$ with density equal to

$$f_{X|\Theta}(x|\theta) = \begin{cases} 4\left(x - \theta + \frac{1}{2}\right) & \text{if } \theta - \frac{1}{2} < x < \theta, \\ 4\left(\frac{1}{2} - x + \theta\right) & \text{if } \theta \le x < \theta + \frac{1}{2}. \end{cases}$$

    (a) Find the asymptotic joint distribution of the p, $1/2$, and $1-p$ sample quantiles of a sample of size n as $n \to \infty$.

    (b) Find the best linear combination of the three sample quantiles p, $1/2$, and $1 - p$ for estimating Θ.

    (c) Try to find the best p if one wishes to estimate Θ using a linear combination of the three sample quantiles p, $1/2$, $1 - p$, and show that the usual analysis fails.

18. In Problem 17 above, find the asymptotic joint distribution of the largest and smallest order statistics from a sample of size n.

19. Let the conditional distribution of $\{X_n\}_{n=1}^{\infty}$ given $\Theta = \theta$ be IID $U(0, \theta)$. Let $X_{(k)}^{(n)}$ denote the kth order statistic based on $X_1, \ldots, X_n$. Find a_n and b_n such that $a_n(X_{(k)}^{(n)} - b_n)$ converges in distribution to a nondegenerate distribution as $n \to \infty$ for fixed k.

Section 7.3:

20. Return to Problem 16 above.

    (a) Find the asymptotic variance of the estimator found in part (b).

(b) Compute the Fisher information $\mathcal{I}_{X_1}(\theta)$ and the efficiency of the estimator found in part (b) of Problem 16.

(c) Compute the efficiency of $\overline{X}_n = \sum_{i=1}^n X_i/n$ as an estimator of Θ.

21. Let $\{X_n\}_{n=1}^\infty$ be conditionally IID given $\Theta = \theta$ with distribution $U(\theta^2, \theta)$ where the parameter space is the interval $(0, 1)$.

    (a) Find the MLE of Θ.

    (b) Find a nondegenerate asymptotic distribution for the MLE.

22. Prove that the relative rate of convergence is unique by first showing the following. Let $a_n, a'_n > 0$ and let H, H' be CDFs. If $a_n(G_n - g(\theta)) \xrightarrow{D} H$ and $a'_n(G_n - g(\theta)) \xrightarrow{D} H'$, then $\lim_{n\to\infty} a'_n/a_n = c \in (0, \infty)$ and $H'(x) = H(x/c)$.

23. Prove the claim at the end of Example 7.46 on page 413 about the relative rate of convergence being the square root of the ARE for asymptotic variance.

24. Let $\{X_n\}_{n=1}^\infty$ be conditionally IID with $N(\theta, 1)$ distribution given $\Theta = \theta$. Let $\overline{X}_n = n^{-1}\sum_{i=1}^n X_i$ and $S_n = \sum_{i=1}^n (X_i - \overline{X})^2$. Let a_n be such that $Pr(S_n > a_n) = 1/n$. Let k_n be the largest integer less than or equal to $\sqrt{n}$. Consider the following two estimators:

$$T_n = \begin{cases} \overline{X}_n & \text{if } S_n \leq a_n \\ n & \text{if } S_n > a_n \end{cases}, \quad U_n = \frac{1}{k_n}\sum_{i=1}^{k_n} X_i.$$

(a) Show that the ARE of U_n to T_n is 0 using the criterion of rate of convergence from Example 7.46 on page 413.

(b) Show that for any fixed $\epsilon > 0$,

$$P'_\theta(|T_n - \theta| > \epsilon) = \frac{1}{n} + o\left(\frac{1}{n}\right),$$
$$P'_\theta(|U_n - \theta| > \epsilon) = o\left(\frac{1}{n}\right).$$

Comment on this in light of part (a). (Hint: If $X \sim N(0, 1)$, then $Pr(|X| \geq c) \leq 2\phi(c)/c$.[33])

(c) What happens if we replace ϵ by $a/\sqrt{n}$ in part (b)?

25. Let $\Theta > 0$ be a parameter, and $\{X_n\}_{n=1}^\infty$ be a conditionally IID sample (given Θ) with exponential distribution $Exp(\theta)$, and let $\widetilde{X}_n$ be the sample median. Let $\overline{X}_n$ be the sample average.

    (a) Find $a(\theta)$ such that conditional on $\Theta = \theta$, $\sqrt{n}(\log(2)/\widetilde{X}_n - \theta) \xrightarrow{D} N(0, a(\theta))$.

    (b) Using the same criteria as in Example 7.44 on page 413, find the ARE of $\log(2)/\widetilde{X}_n$ to $1/\overline{X}_n$ as estimators of Θ.

---

[33]This inequality is equivalent to *Mill's ratio*.

26. Let $\{X_n\}_{n=1}^{\infty}$ be conditionally IID with $N(\theta, 1)$ distribution given $\Theta = \theta$. We observe $Y_1, \ldots, Y_n$ where
$$Y_i = \begin{cases} 0 & \text{if } X_i \leq 0, \\ X_i & \text{if } 0 < X_i < 1, \\ 1 & \text{if } X_i \geq 1. \end{cases}$$

   (a) Find a minimal sufficient statistic.

   (b) Construct two different (i.e., different by more than $o_P(1/\sqrt{n})$) consistent, asymptotically normal estimates of θ, and compute their ARE using the same criterion as in Example 7.44 on page 413.

27. Let the parameter space be two-dimensional. Suppose that $\{T_n\}_{n=1}^{\infty}$ are conditionally IID given $\Theta = (\theta_1, \theta_2)$ with density $f_{T|\Theta}(t|\theta_1, \theta_2)$. Suppose that the conditions of Theorem 7.63 hold. Let $\hat{\Theta}_1$ be the MLE of the first coordinate of Θ. Let $\tilde{\Theta}_1(\theta_2)$ be the MLE of the first coordinate of Θ if it is assumed that the second coordinate is known to equal θ_2. Use the same criteria as in Example 7.44 on page 413 to find the ARE of these two estimators when it is assumed that the second coordinate of Θ is known to equal θ_2. Express your answer in terms of the Fisher information matrix.

28. Under the conditions of Theorem 7.48, prove that
$$P'_{\theta_0}\left[\prod_{i=1}^{n} f_{X_1|\Theta}(X_i|\theta_0) \leq \prod_{i=1}^{n} f_{X_1|\Theta}(X_i|\theta), \text{ infinitely often}\right] = 0.$$

29. Verify the conditions of Theorem 7.49 in the case in which the observations are IID given $\Theta = \theta$ with density $f_{X|\Theta}(x|\theta) = \theta e^{-\theta x}$, for $x > 0$.

30. Assume that $\Omega = \{\theta_1, \ldots, \theta_m\}$ is finite, and let the prior distribution be $(\pi_1, \ldots, \pi_m)$, with $\pi_i = \Pr(\Theta = \theta_i)$. Let μ be a measure such that $P_\theta \ll \mu$ for each $\theta \in \Omega$, and let
$$f_i(x) = \frac{dP_{\theta_i}}{d\mu}(x).$$
Assume that $\{X_n\}_{n=1}^{\infty}$ are conditionally IID given $\Theta = \theta_i$ with density $f_i(x)$ for each i. Let $\hat{\Theta}_n$ be the MLE after n observations. Prove
$$\lim_{n\to\infty} \Pr(\hat{\Theta}_n = \theta_i) = \pi_i.$$

31. Let $\{X_n\}_{n=1}^{\infty}$ be conditionally IID given $\Theta = \theta$ with $N(\theta, 1)$ distribution. Let Ω be the set of all integers.

    (a) Find the MLE $\hat{\Theta}_n$ of Θ and prove that it is unbiased.

    (b) Show that there exist positive constants a and b such that, for all sufficiently large n, $\text{Var}_\theta(\hat{\Theta}_n) \leq a \exp(-bn)$ for every integer θ. (Hint: Use Mill's ratio from Problem 24(b) on page 470.)

32. Return to the situation in Problem 16 on page 140. Find the MLE of Θ and its nondegenerate asymptotic distribution.

33. Consider Example 7.60 (page 420) once again. This time, assume that we observe $k \geq 2$ observations with conditional mean μ_i for every i. Find the MLE of Σ^2 and what it converges to in probability.

34. Prove Proposition 7.71 on page 425.

35. Suppose that $\{X_n\}_{n=1}^\infty$ are conditionally IID given $\Theta = \theta$ with the following discrete logistic conditional density with respect to counting measure on the integers:

$$f_{X_1|\Theta}(x|\theta) = \frac{\exp(x\theta)}{1 + \exp(\theta) + \exp(2\theta)}, \quad \text{for } x = 0, 1, 2.$$

   (a) Find a $\sqrt{n}$-consistent estimator of Θ.

   (b) Find an explicit form for an asymptotically efficient, asymptotically normal estimator of Θ based on $X_1, \ldots, X_n$.

36. Let

$$\psi(x, \theta) = \begin{cases} -1 & \text{if } x \leq \theta - 1, \\ \sin\left[\frac{\pi}{2}(x - \theta)\right] & \text{if } \theta - 1 < x < \theta + 1, \\ 1 & \text{if } x \geq \theta + 1. \end{cases}$$

   (a) Prove that there is always a solution to $\sum_{i=1}^n \psi(X_i, \theta) = 0$.

   (b) Assume that the X_i are conditionally IID $N(\theta, 1)$ given $\Theta = \theta$. Now prove that for each $\epsilon > 0$,

$$\lim_{n \to \infty} P_{\theta_0}(\exists \text{ a solution to } \sum_{i=1}^n \psi(X_i, \theta) = 0 \text{ in } [\theta_0 - \epsilon, \theta_0 + \epsilon]) = 1.$$

37. Suppose that $\{X_n\}_{n=1}^\infty$ are conditionally IID given $\Theta = \theta$ with $U(-\theta, \theta)$ and $\Omega = (0, \infty)$. Find the MLE $\hat{\Theta}_n$ of Θ based on n observations, and find a_n so that $a_n(\hat{\Theta}_n - \theta)$ has nondegenerate asymptotic distribution given $\Theta = \theta$. Also find that distribution.

38. Suppose that n arrows are fired at a circular target of radius a whose center is at the point $(0, 0, 0) \in \mathbb{R}^3$. The target lies in the plane where the third coordinate is 0. Suppose that arrow i passes through the point $(X_i, Y_i, 0)$. Let $R_i = \sqrt{X_i^2 + Y_i^2}$. Suppose that (X_i, Y_i) are conditionally IID $N_2(0, \theta I_2)$ given $\Theta = \theta$. The data we observe are all (X_i, Y_i) pairs for those arrows that hit the target. We also know n.

   (a) Find the distribution of R_i^2/Θ for an arbitrary arrow (whether or not it will hit the target).

   (b) Find the conditional probability $P_\theta(R_i \leq a)$ that arrow i hits the target.

   (c) Find the MLE $\hat{\Theta}_n$ of Θ.

   (d) Find the asymptotic distribution of $\hat{\Theta}_n$ as $n \to \infty$.

39. *Suppose that Y_1, Y_2 are conditionally independent with $Y_i \sim Bin(n_i, p_i)$ given $P_1 = p_1, P_2 = p_2$, where n_1 and n_2 are known sample sizes. The parameter space is $\Omega = \{(p_1, p_2) : p_2 \geq p_1\}$.

   (a) Find the MLE $(\hat{P}_1, \hat{P}_2)$ of (P_1, P_2).

   (b) Find the asymptotic distributions of $\hat{P}_1$ and of $\hat{P}_2$ as n_1 and n_2 go to infinity. (*Hint:* Consider the case $p_1 = p_2$ separate from the case $p_1 < p_2$.)

40. Let $\{X_n\}_{n=1}^\infty$ be conditionally IID with $N(\theta,1)$ distribution given $\Theta = \theta$. Let $\hat{\Theta}_n$ be any estimator whatsoever of Θ. Let ψ be the derivative (with respect to θ) of the log of the conditional density of each X_i given $\Theta = \theta$. Find the asymptotically efficient estimator of Theorem 7.75.

41. *Suppose that $\{X_n\}_{n=1}^\infty$ are conditionally IID $U(0,\theta)$ random variables given $\Theta = \theta$. The parameter space is the interval $[1,2]$. Suppose that we only get to observe $Y_i = I_{[0,1]}(X_i)$ for each i. That is, we only see whether or not each observation is between 0 and 1.

    (a) Find the MLE of Θ based on $Y_1, \ldots, Y_n$.

    (b) Find the asymptotic (as $n \to \infty$) distribution of the MLE found above.

    (c) In terms of asymptotic efficiency, how does the MLE found above compare to the MLE based on observing the actual X_i values?

42. Suppose that $\{X_n\}_{n=1}^\infty$ are conditionally IID given $\Theta = \theta$ with conditional density
$$f_{X_1|\Theta}(x|\theta) = \frac{\theta^\alpha \alpha}{x^{\alpha+1}} I_{[\theta,\infty)}(x),$$
where α is known and the parameter space is $\Omega = (0,\infty)$.

    (a) Find the MLE $\hat{\Theta}_n$ of Θ based on $X_1, \ldots, X_n$.

    (b) Prove that $\hat{\Theta}_n$ is inadmissible if the loss is squared error.

    (c) Find a_n and b_n such that $a_n\hat{\Theta}_n + b_n$ has nondegenerate asymptotic distribution, and find that distribution.

43. Let $\{X_n\}_{n=1}^\infty$ be conditionally IID given Θ, a one-dimensional parameter. Assume the conditions of Theorem 7.63 and that there are no superefficient estimators. Let $\hat{\Theta}_n$ be the MLE of Θ, and let $\{T_n\}_{n=1}^\infty$ be another sequence of estimators with $\sqrt{n}(T_n - \theta) \xrightarrow{D} N(0, v(\theta))$ given $\Theta = \theta$ for all θ and T_n a function of $(X_1, \ldots, X_n)$. Consider the joint asymptotic distribution of $\sqrt{n}([\hat{\Theta}_n, T_n]^T - \theta\mathbf{1})$ given $\Theta = \theta$. Prove that the asymptotic covariance is $1/\mathcal{I}_{X_1}(\theta)$. (*Hint:* Look at the proof of Theorem 5.9 on page 298.)

44. *A psychologist is studying paired subjects. Each person in each pair is asked a yes-no question. Let $X_{i,j} = 1$ if person i in pair j answers yes ($X_{i,j} = 0$ otherwise) for $i = 1, 2$ and $j = 1, 2, \ldots$. The psychologist wants to assume that there are parameters Θ such that all of the $X_{i,j}$ are conditionally independent given Θ. Suppose that the psychologist believes that there is a number α such that
$$\frac{\Pr(X_{2,j} = 1|\Theta = \theta)}{\Pr(X_{2,j} = 0|\Theta = \theta)} = \alpha \frac{\Pr(X_{1,j} = 1|\Theta = \theta)}{\Pr(X_{1,j} = 0|\Theta = \theta)}.$$

    (a) Prove that there exist numbers $\beta_1, \beta_2, \ldots$ such that
    $$\Pr(X_{1,j} = x, X_{2,j} = y|\Theta = \theta) = \frac{(\alpha\beta_j)^y}{1 + \alpha\beta_j} \frac{\beta_j^x}{1 + \beta_j}. \tag{7.138}$$

(b) The psychologist decides to let $\Theta = (A, B_1, B_2, \ldots)$ so that (7.138) gives the conditional probability of observing (x, y) for pair j given $\Theta = (\alpha, \beta_1, \beta_2, \ldots)$. Observations are then made for $j = 1, \ldots, m$. Let $Z_j = X_{1,j} + X_{2,j}$, $T = \#\{j : Z_j = 1\}$, and $S = \sum_{j=1}^m X_{2,j} I_{\{1\}}(Z_j) = \sum_{j:Z_j=1} X_{2,j}$. Write the likelihood function.

(c) Find the MLEs of A and $B_1, \ldots, B_m$. (*Hint:* First, find the MLEs of the B_i for fixed value of A, and then find the MLE of A.)

(d) Since the MLE of A depends only on the pairs with $Z_j = 1$, find the conditional density of the data given $(Z_1, \ldots, Z_m)$. Also find the conditional MLE of A based on this distribution.

(e) Show that the conditional MLE found in part (d) is consistent as $m \to \infty$.

45. Assume that $\{X_n\}_{n=1}^\infty$ are conditionally IID with $N(\mu, \sigma^2)$ distribution given $\Theta = (\mu, \sigma)$. (See the end of Example 7.52 on page 417.) Prove that for every θ_0 and every compact set $C \subseteq \Omega$, $E_{\theta_0} Z(C^C, X_i) \leq 0$ in the notation of Theorem 7.49.

Section 7.4:

46. Use the problem description in Problem 16 on page 75. Show that the posterior distribution of M given $(X_1, \ldots, X_n)$ is not consistent in the sense of Theorem 7.78.

47. Assume the conditions of Theorem 7.78. Prove that there exists a subset $A \subseteq \Omega$ with $\mu_\Theta(A) = 1$ such that for every $\theta \in A$,

$$P'_\theta \left(\lim_{n \to \infty} \mu_{\Theta|X_1,\ldots,X_n}(A|X_1, \ldots, X_n) = I_A(\theta) \right) = 1.$$

48. Return to the situation in Example 7.82 on page 432. Prove that the posterior probability of C_ϵ does not almost surely converge to 1 given $\Theta = \theta_0$. (*Hint:* Rewrite the posterior probability of $(0, \theta_0)$ in terms of the random variable $n(\theta_0 - X_{(n)})$, where $X_{(n)}$ is the largest observation. Then use the result in Example 7.33 on page 408.)

49. Suppose that $X_1, \ldots, X_n$ are conditionally IID with exponential distribution $Exp(\theta)$ given $\Theta = \theta$. Let Θ have a $\Gamma(a, b)$ prior distribution. Use Laplace's method to construct a formula for the approximation to the posterior mean of Θ. How does this compare to the exact posterior mean?

50. Suppose that $X_1, \ldots, X_n$ are conditionally IID with Laplace distribution $Lap(\theta, 1)$ given $\Theta = \theta$. Let the prior distribution of Θ be $Lap(0, 1)$. We wish to approximate the predictive density of a future observation, namely $\int [\exp(-|x - \theta|)/2] f_{\Theta|X}(\theta|x) d\theta$ for various values of x.

(a) Use Laplace's method to construct a formula for the approximation.

(b) Describe how to use importance sampling to do the approximation.

51. Let $\Theta = (\Gamma, \Psi)$. Suppose that one wishes to test the hypothesis $H : \Gamma = \gamma_0$. Let the prior probability of H be positive, and suppose that the prior for Ψ given that H is true is the conditional prior of Ψ given $\Gamma = \gamma_0$ calculated from the prior on Θ given that H is false. Prove that the approximate Bayes factor in (4.27) is the same as the Laplace approximation of Theorem 7.116 divided by $f_\Gamma(\gamma_0)$ when the hypothesis is $H : \Gamma = \gamma_0$.

52. Let $\{X_n\}_{n=1}^\infty$ be conditionally IID given $(P_1, P_2) = (p_1, p_2)$ with $Ber(p_1 + p_2)$ distribution. Let the prior distribution be $(P_1, P_2) \sim Dir_3(\alpha_1, \alpha_2, \alpha_3)$.

    (a) Find the posterior distribution of P_1 given $(X_1, \ldots, X_n)$.

    (b) Conditional on $(P_1, P_2) = (p_1, p_2)$, say what happens to the posterior distribution found in part (a) as $n \to \infty$.

Section 7.5:

53. Let the parameter be $\Theta = (M_1, M_2, \Sigma)$, and suppose that conditional on $\Theta = (\mu_1, \mu_2, \sigma)$ $X_{1,1}, \ldots, X_{1,n_1}, X_{2,1}, \ldots, X_{2,n_2}$ are independent with $X_{i,j}$ having $N(\mu_i, \sigma^2)$ distribution for $j = 1, \ldots, n_i$ and $i = 1, 2$. Prove that the size α likelihood ratio test of $H : M_1 = M_2$ versus $A : M_1 \neq M_2$ is also the UMPU level α.

54.*Let α be a known number strictly between 0 and $1/2$. The parameter space is $\Omega = \{(\theta_1, \theta_2) : 0 \leq \theta_1 \leq \alpha, 0 \leq \theta_2 \leq 1\}$. Let D be a discrete random variable with conditional density given $(\Theta_1, \Theta_2) = (\theta_1, \theta_2)$

$$f_{D|\Theta_1,\Theta_2}(d|\theta_1, \theta_2) = \begin{cases} \theta_1(1-\theta_2) & \text{if } d = -2, \\ \left(\frac{1}{2} - \alpha\right)\frac{1-\theta_1}{1-\alpha} & \text{if } d = -1, \\ \alpha\frac{1-\theta_1}{1-\alpha} & \text{if } d = 0, \\ \left(\frac{1}{2} - \alpha\right)\frac{1-\theta_1}{1-\alpha} & \text{if } d = 1, \\ \theta_1\theta_2 & \text{if } d = 2. \end{cases}$$

The hypothesis of interest is $H : \Theta_1 = \alpha, \Theta_2 = 1/2$ versus $A : \Theta_1 < \alpha, \Theta_2 \neq 1/2$. We will observe only one D value, and α will be the level for every test below.

    (a) Find the likelihood ratio test and its power function.

    (b) Consider the group consisting of two transformations $g_+ D = D$ and $g_- D = -D$. Show that the testing problem is invariant under the action of this group.

    (c) Show that the likelihood ratio test is invariant.

    (d) Find a uniformly most powerful invariant test and compare its power function to that of the likelihood ratio test.

    (e) Find the least powerful invariant test.

55. Suppose that $X \sim N(\theta, 1)$ given $\Theta = \theta$. Let Ω_H be the set of rational numbers, and let Ω_A be the set of irrational numbers. Prove that the likelihood ratio test with level α of $H : \Theta \in \Omega_H$ versus $A : \Theta \in \Omega_A$ is the trivial test $\phi(x) \equiv \alpha$.

CHAPTER 8
Hierarchical Models

When a model has many parameters, it may be the case that we can consider them as a sample from some distribution. In this way we model the parameters with another set of parameters and build a model with different levels of hierarchy. In this chapter we will discuss situations in which it is natural to model in this way.

8.1 Introduction

8.1.1 General Hierarchical Models

We turn our attention now to a situation in which the observations are not exchangeable. Suppose, for example, that several treatments are being administered in a clinical trial. From each treatment group, we will make some observations. It may be plausible to model the observations within each treatment group as exchangeable, but it would seem strange to model all observations as exchangeable. For each treatment group, we might develop a parametric model as we have done elsewhere in this text. A *hierarchical model* for this example involves treating the set of parameters corresponding to the different treatment groups as a sample from another population. Prior to seeing any observations, we can model the parameters as exchangeable.[1] This would mean that we could introduce another set of parameters to model their joint distribution. These second-level parameters

---

[1] It is not essential that we model the parameters as exchangeable a priori, but it is mathematically convenient. Such a model corresponds to treating the different goups symmetrically prior to observing any data.

are called *hyperparameters*. We would then need to specify a distribution for these hyperparameters. Here are some examples.

Example 8.1. Suppose that there are k treatment groups. Let $X_{i,j}$ stand for the observed response of subject j in treatment group i. We might invent parameters $M_1, \ldots, M_k$ and model the $X_{i,j}$ as conditionally independent given $(M_1, \ldots, M_k) = (\mu_1, \ldots, \mu_k)$ with $X_{i,j} \sim N(\mu_i, 1)$. We might then model $M_1, \ldots, M_k$ as a priori exchangeable with distribution $N(\Theta, 1)$ distribution given Θ. Here, Θ is a hyperparameter. We should also specify a distribution for Θ.[2] Note that we have only one Θ regardless of what k is.

Example 8.2. A survey is conducted in three different cities. Each person surveyed is asked a yes–no question. Treat "yes" as $X = 1$ and "no" as $X = 0$. Then, to each person i in city j, there corresponds a Bernoulli random variable $X_{i,j}$. It might seem plausible to treat the $X_{i,j}$ observations from a single city i as exchangeable. Suppose that we invent three parameters P_1, P_2, P_3. Then we can model the $X_{i,j}$ for fixed i as conditionally IID $Ber(p)$ given $P_i = p$. We would then need to construct a joint distribution for (P_1, P_2, P_3). For instance, we could model the P_i as exchangeable with $Beta(\alpha, \beta)$ distribution conditional on $A = \alpha$ and $B = \beta$. Here, A and B are the hyperparameters. We would then need a joint distribution for (A, B). Note that we only use a single pair (A, B) no matter how many P_i we have in this simple model.

The intuitive concept of how hierarchical models work is the following. Suppose that the data comprise several groups, each of which we consider to be a collection of exchangeable random variables. From the data in each group, we obtain direct information about the corresponding parameters. Thinking of the hyperparameters as known for the time being, we then update the distributions of the parameters using the data, to get posterior distributions for the parameters via Bayes' theorem 1.31. Future data (in each group) are still exchangeable with the same conditional distributions given the parameters, but the distributions of the parameters have changed. In fact, the distribution of each parameter (given the hyperparameters) has now been updated using *only* the data from its corresponding group. Hence the parameters are no longer exchangeable. Now, we can also update the distribution of the hyperparameters. To do this, we first find the conditional distribution of the data given the hyperparameters. Then, we can use Bayes' theorem 1.31 again to find the posterior distribution of the hyperparameters given the data. The marginal posterior of the parameters given the data is found by integrating the hyperparameters out of the joint posterior of the parameters and hyperparameters. This is how the data from all groups combine to provide information about all of the parameters, not just the ones corresponding to their own group. It is the common dependence of all parameters on the hyperparameters that allows us to make use of common information in updating the distributions of all parameters. A diagram of the directions of influence is given in Figure 8.3. A random variable at the

---

[2] We set all variances equal to 1 for simplicity in this first example. In any real application, the variances would be unknown parameters as well.

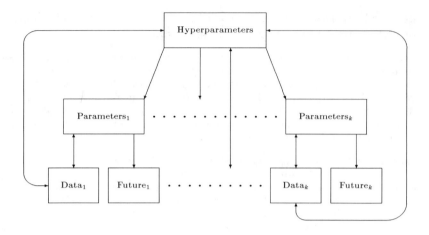

FIGURE 8.3. Schematic Diagram of Hierarchical Model

pointed end of an arrow has its conditional distribution calculated given the random variable at the other end of the arrow. Double-headed arrows indicate places where Bayes' theorem 1.31 is used. Future data are included in the diagram to indicate how observed data in all groups affect predictive distributions of all future data through the hyperparameters.

In theory, the updating can be performed as follows. We can denote the data to be observed as X, the parameters as Θ, the hyperparameters as Ψ, and future data as Y. Assume that X and Y are conditionally independent given Θ and Ψ. Then the conditional posterior density of the parameters given the hyperparameters is

$$f_{\Theta|X,\Psi}(\theta|x,\psi) = \frac{f_{X|\Theta,\Psi}(x|\theta,\psi)f_{\Theta|\Psi}(\theta|\psi)}{f_{X|\Psi}(x|\psi)},$$

where the density of the data given the hyperparameters alone is

$$f_{X|\Psi}(x|\psi) = \int f_{X|\Theta,\Psi}(x|\theta,\psi)f_{\Theta|\Psi}(\theta|\psi)d\theta.$$

The marginal posterior distributions of the parameters can be found from

$$f_{\Theta|X}(\theta|x) = \int f_{\Theta|X,\Psi}(\theta|x,\psi)f_{\Psi|X}(\psi|x)d\psi,$$

where the posterior density of Ψ given $X = x$ is

$$f_{\Psi|X}(\psi|x) = \frac{f_{X|\Psi}(x|\psi)f_{\Psi}(\psi)}{f_X(x)},$$

and the marginal density of the X is

$$f_X(x) = \int f_{X|\Psi}(x|\psi) f_\Psi(\psi).$$

Finally, the predictive distribution of future data Y is found from the posterior density of the parameters:

$$f_{Y|X}(y|x) = \int \int f_{Y|\Theta,\Psi}(y|\theta,\psi) f_{\Theta|X,\Psi}(\theta|x,\psi) f_{\Psi|X}(\psi|x) d\theta d\psi.$$

Hierarchical models were first popularized by Lindley and Smith (1972) and Smith (1973) for the special cases of multivariate normal observations. Hierarchical models are special cases of *partial exchangeability*, which we consider in more detail in the remainder of this seciton.

8.1.2 Partial Exchangeability*

A natural generalization of both hierarchical models and exchangeability is the concept of *partial exchangeability*. There are several types of partial exchangeability. Diaconis and Freedman (1980b) present a good overview with some examples.

There are two ways to think about what exchangeability means, and each of them leads to a way of extending the concept to partial exchangeability. Let $X_1, X_2, \ldots$ be a (possibly finite) sequence of random quantities.

1. They are exchangeable if, for each n, every permutation of n of them has the same joint distribution as every other permutation of n of them.

2. They are exchangeable if, for each n, each sequence $\{z_i\}_{i=1}^n$ of n possible outcomes has the same probability of being the observed value of $\{X_i\}_{i=1}^n$ as every permutation of $\{z_i\}_{i=1}^n$.

The second description has the drawback that it only makes sense, as stated, for discrete random variables. There are ways to make it precise for more general cases, but they lose some intuitive appeal in the translation. Oddly enough, it is this second description that has the greatest potential for generalizing the concept of exchangeability.

Based on the first description, we have the following restrictive extension of exchangeability.

Definition 8.4. A sequence $X_1, X_2, \ldots$ is *marginally partially exchangeable* if it can be partitioned into subsequences $X_1^{(k)}, X_2^{(k)}, \ldots$, for $k = 1, 2, \ldots$ such that the random quantities in each subsequence are exchangeable. To which subsequence each X_i belongs must be known in advance, and the subsequences (as well as the original sequence) may be finite or infinite.

---

*This section may be skipped without interrupting the flow of ideas.

If the subsequences are infinite, DeFinetti's representation theorem 1.49 can be applied to each of the subsequences to conclude that there must exist probability measures corresponding to each subsequence (with some unspecified joint distribution) such that the random variables in each subsequence are conditionally independent given their corresponding probability measures. Similarly, we can introduce finite-dimensional parametric families for each subsequence and hence reduce the problem of finding joint distributions for the probability measures to finite-dimensional joint distributions. This is basically what hierarchical models do. (See Examples 8.1 and 8.2 on page 477.)

As an example of an attempt to extend the second description of exchangeability, consider a Markov chain $\{X_n\}_{n=1}^{\infty}$ of Bernoulli random variables. Clearly, not every permutation of a sequence of possible outcomes has the same probability. But there are some permutations that do have the same probability. In particular, two sequences with the same first value and the same numbers of the four types of transitions (0 to 1, 0 to 0, 1 to 1, and 1 to 0) will have the same probability. For example:

$$(1,1,0,1,0,0,0,1,0,0,1,1) \text{ and } (1,1,1,0,0,0,0,1,0,1,0,1).$$

This is certainly not a case of marginal partial exchangeability. It does, however, have the intuitive appearance of being a generalization of the second description of exchangeability. Diaconis and Freedman (1980c) give an in-depth treatment of the type of partial exchangeability that characterizes Markov chains.

The most general type of partial exchangeability is described by Diaconis and Freedman (1984). In fact, it is so general that it is satisfied by arbitrary joint distributions. (See Example 2.118 on page 129.) Nevertheless, each specific instance of partial exchangeability leads to a representation theorem of the type of Theorem 2.111. In Example 2.116 on page 127 we saw that Theorem 2.111 contains a reformulation of DeFinetti's representation theorem 1.49 as a special case. In Section 8.1.3, we give examples of Theorem 2.111 for partially exchangeable random quantities that are not exchangeable.

8.1.3 Examples of the Representation Theorem*

In this section, we give examples of the representation theorem 2.111 to cases of partially exchangeable random quantities.

Example 8.5. This example is the one-way analysis of variance with only two groups and equal variances. To handle more groups is a simple matter, but the notation gets in the way of an initial understanding. Let $\mathcal{X}_n = \mathbb{R}^n$ and $\mathcal{T}_n = \mathbb{R}^2 \times \mathbb{R}^{+0}$. Suppose that there is a deterministic sequence $\{j_n\}_{n=1}^{\infty}$ with $j_n \in \{0, 1\}$ for

---

*This section may be skipped without interrupting the flow of ideas.

all n. The j_n sequence tells us from which group the nth observation comes. Let

$$T_n(x_1,\ldots,x_n) = \left(\sum_{i=1}^n j_i x_i, \sum_{i=1}^n (1-j_i)x_i, \sum_{i=1}^n x_i^2\right).$$

Let $k_n = \sum_{i=1}^n j_i$. Let $r_n(\cdot, t)$ be the uniform distribution on the surface of the sphere of radius $\sqrt{t_3 - t_1^2/k_n - t_2^2/(n-k_n)}$ around the point whose ith coordinate is $j_i t_1/k_n + (1-j_i)t_2/(n-k_n)$. One can check that the conditions of Theorem 2.111 are met.

We would like to proceed as in Example 2.117 on page 128, but we cannot assume that the coordinates are IID in the limit distributions. So, we will construct the joint distribution of s_0 observations from group 0 and s_1 observations from group 1 (for fixed s_0 and s_1) given $T_n = t$, and see what happens as $n \to \infty$. Call these observations $Z = (Z_1, \ldots, Z_{s_0})$ and $W = (W_1, \ldots, W_{s_1})$. Let

$$\sigma_n^2 = \frac{1}{n}\left(t_3 - \frac{1}{k_n}t_1^2 - \frac{1}{n-k_n}t_2^2\right), \quad \mu_n = \frac{1}{k_n}t_1, \quad \eta_n = \frac{1}{n-k_n}t_2.$$

Then

$$f_{Z,W|T_n}(z,w|t_1,t_2,t_3) = \frac{\Gamma\left(\frac{n}{2}\right)}{\Gamma\left(\frac{n-s_0-s_1}{2}\right)(n\pi)^{\frac{s_0+s_1}{2}}\sigma_n^{s_0+s_1}}$$

$$\times \left(1 - \frac{\sum_{i=1}^{s_0}(z_i-\mu_n)^2}{n\sigma_n^2} - \frac{\sum_{i=1}^{s_1}(w_i-\eta_n)^2}{n\sigma_n^2}\right)^{\frac{n-s_0-s_1}{2}}.$$

Since

$$\lim_{n\to\infty} \frac{\Gamma\left(\frac{n}{2}\right)}{\Gamma\left(\frac{n-s_0-s_1}{2}\right) n^{\frac{s_0+s_1}{2}}} = 2^{-\frac{s_0+s_1}{2}},$$

we have that $f_{Z,W|T_n}$ is asymptotically equivalent to

$$(2\pi)^{-\frac{s_0+s_1}{2}} \sigma_n^{-(s_0+s_1)} \left(1 - \frac{\sum_{i=1}^{s_0}(z_i-\mu_n)^2}{n\sigma_n^2} - \frac{\sum_{i=1}^{s_1}(w_i-\eta_n)^2}{n\sigma_n^2}\right)^{\frac{n-s_0-s_1}{2}}.$$

If σ_n converges to $\sigma \in (0, \infty)$ and μ_n and ν_n converge to μ and ν, respectively, this function converges uniformly on compact sets to the density of s_0 $N(\mu, \sigma^2)$ and s_1 $N(\nu, \sigma^2)$ random variables all independent. If σ^2 goes to ∞, there is no limit distribution. If σ^2 goes to 0, and $\mu_n \to \mu$ and $\nu_n \to \nu$, the function converges to 0 uniformly outside of every open neighborhood of the point with ith coordinate $j_i\mu + (1-j_i)\nu$. In this case the limit distribution is point masses at μ and ν depending on whether $j_i = 1$ or 0. Finally, if σ_n goes to a finite value and either μ_n or ν_n diverges to $\pm\infty$, there is no limit distribution. The extreme distributions either have all coordinates degenerate or have the coordinates being independent normal random variables with common variance. In either case, there are two different means depending on the values of the sequence $\{j_n\}_{n=1}^\infty$.

Lauritzen (1984, 1988) shows how to characterize much more general normal linear models using the representation of Theorem 2.111. See Problem 1 on page 532 for an example of the characterization of a conditionally partially exchangeable sequence by means of Theorem 2.111.

482   Chapter 8. Hierarchical Models

Aldous (1981) introduces a special kind of partial exchangeability that arises in the two-way analysis of variance.

Definition 8.6. Let $X = ((X_{i,j}))_{i=1,j=1}^{\infty,\infty}$ be an array of random variables. Let $R_i = (X_{i,1}, X_{i,2}, \ldots)$ and $C_j = (X_{1,j}, X_{2,j}, \ldots)$. We say that X is *row and column exchangeable* if both $\{R_n\}_{n=1}^{\infty}$ and $\{C_n\}_{n=1}^{\infty}$ are exchangeable sequences.

Row and column exchangeability is a special case of the conditions of Theorem 2.111.

Example 8.7. Let X be a row and column exchangeable array. Let $\{(r_n, c_n)\}_{n=1}^{\infty}$ be a sequence of pairs of integers such that $r_{n+1} \geq r_n$ and $c_{n+1} \geq c_n$ with at least one inequality strict for each n. Let $\mathcal{X}_n$ be $\mathbb{R}^{r_n c_n}$, and let X_n be the first r_n rows and c_n columns of X. That is, we add at least one row and/or at least one column each time we increase n.[3] Let $T_n = (T_n^r, T_n^c)$, where T_n^r and T_n^c are defined as follows. For each row of X_n, construct the order statistic (smallest to largest) of the numbers in that row and then arrange these order statistics according to the smallest value in the row. Call the result T_n^r. Define T_n^c by doing the same thing to the columns. For example, suppose that

$$X_n = \begin{pmatrix} -1 & 3 & 2 & 0 \\ 4 & -2 & 1 & 3 \\ 1 & 0 & -1 & 2 \end{pmatrix}.$$

Then

$$T_n^r = \begin{cases} (-2, 1, 3, 4) \\ (-1, 0, 2, 3) \\ (-1, 0, 1, 2) \end{cases}, \quad T_n^c = \begin{cases} (-2, 0, 3) \\ (-1, 1, 4) \\ (-1, 1, 2) \\ (0, 2, 3) \end{cases}.$$

Basically, you throw away the information about in which row and in which column each number was, but you keep the information about which other numbers were in the same row and column with each number. Each of the $r_n!c_n!$ matrices that can be obtained from X_n by permuting the rows and then the columns will have the same value of T_n. Similarly, all of those $r_n!c_n!$ arrays can be constructed from T_n by a somewhat more tedious algorithm. Clearly, $r_n(\cdot, t)$ must be uniform over those $r_n!c_n!$ arrays to preserve row and column exchangeability.

Finding all of the $Q(\cdot, x)$ distributions is no small task. Aldous (1981) proves the following result. An array X is row and column exchangeable if and only if there exists a measurable function $f : [0,1]^4 \to \mathbb{R}$ such that X has the same distribution as $Y = ((Y_{i,j}))$, with $Y_{i,j} = f(M, A_i, B_j, G_{i,j})$, where $M, A_1, A_2, \ldots, B_1, B_2, \ldots, G_{1,1}, \ldots$ are all IID $U(0,1)$ random variables.

---

[3]Technically, there is a way to write Theorem 2.111 so that it applies to partially ordered sets like the set of all (r, c) pairs, but the author thought that the proof of Theorem 2.111 was complicated enough without introducing this added level of mathematical detail.

8.2 Normal Linear Models

A particularly simple case with which to work is that of linear models in which the observables are modeled as having normal distributions given parameters and the parameters are also modeled as jointly normal (except for the scale parameters).

8.2.1 One-Way ANOVA

Suppose that we will observe data from k different treatment groups. Let the jth observation in the ith group be $X_{i,j}$. Suppose that we model the $X_{i,j}$ as conditionally independent $N(\mu_i, \sigma^2)$ random variables given M $=$ $(\mu_1, \ldots, \mu_k)$ and $\Sigma = \sigma$ for $j = 1, \ldots, n_i$ and $i = 1, \ldots, k$. Next, suppose that we model M $= (M_1, \ldots, M_k)$ as a vector of IID $N(\psi, \tau^2)$ random variables given $\Psi = \psi$ and T $= \tau$. To be precise, we should have said that the $X_{i,j}$ have $N(\mu_i, \sigma^2)$ distribution given M $= (\mu_1, \ldots, \mu_k)$, $\Sigma = \sigma$, $\Psi = \psi$, and T $= \tau$. Next, we model Ψ as $N(\psi_0, \tau^2/\zeta_0)$ given T $= \tau$ and $\Sigma = \sigma$. Finally, we need a joint distribution for (Σ, T), which will remain unspecified for now.

The joint distribution of all quantities can be summarized as in Table 8.9. Future observations have a distribution like the first stage. The posterior distribution has only the last three stages. Let X stand for the entire data vector, and let x be the observed value.

Conditional on $\Sigma = \sigma$, T $= \tau$, and $\Psi = \psi$, the posterior of the M_i is found from simple normal distribution updating. The M_i, given $\Sigma = \sigma$, T $= \tau$, and $\Psi = \psi$, are independent with M_i having distribution

$$N\left(\mu_i(\psi, \sigma, \tau), \frac{\tau^2 \sigma^2}{n_i \tau^2 + \sigma^2}\right), \tag{8.8}$$

where

$$\mu_i(\psi, \sigma, \tau) = \frac{n_i \bar{x}_i \tau^2 + \psi \sigma^2}{n_i \tau^2 + \sigma^2}.$$

The conditional joint distribution of the data given $\Psi = \psi$, $\Sigma = \sigma$, and

TABLE 8.9. Hierarchical Model for One-Way ANOVA

Stage	Density
Data	$(2\pi\sigma^2)^{-\frac{n}{2}} \exp\left\{-\frac{1}{2\sigma^2} \sum_{i=1}^{k}[n_i(\bar{x}_i - \mu_i)^2 + (n_1 - 1)s_i^2]\right\}$
Parameter	$(2\pi\tau^2)^{-\frac{k}{2}} \exp\left\{-\frac{1}{2\tau^2} \sum_{i=1}^{k}(\mu_i - \psi)^2\right\}$
Hyperparameter	$\sqrt{\zeta_0}(2\pi\tau^2)^{-\frac{1}{2}} \exp\left\{-\frac{\zeta_0}{2\tau^2}(\psi - \psi_0)^2\right\}$
Variance	$f_{\Sigma,T}(\sigma, \tau)$

Chapter 8. Hierarchical Models

$T = \tau$, is that the $\overline{X}_i$ and S_i are independent with[4]

$$\overline{X}_i \sim N\left(\psi, \frac{\sigma^2}{n_i} + \tau^2\right), \quad S_i^2 \sim \frac{\sigma^2}{n_i - 1}\chi_{n_i-1}^2. \tag{8.10}$$

It follows that the posterior of Ψ conditional on $\Sigma = \sigma$ and $T = \tau$ is

$$\Psi | X = x, \Sigma = \sigma, T = \tau \sim N\left(\psi_1(\sigma, \tau), \left[\frac{\zeta_0}{\tau^2} + \sum_{i=1}^{k} \frac{n_i}{\sigma^2 + \tau^2 n_i}\right]^{-1}\right), \tag{8.11}$$

where

$$\psi_1(\sigma, \tau) = \frac{\frac{\zeta_0 \psi_0}{\tau^2} + \sum_{i=1}^{k} \frac{n_i \overline{x}_i}{\sigma^2 + \tau^2 n_i}}{\frac{\zeta_0}{\tau^2} + \sum_{i=1}^{k} \frac{n_i}{\sigma^2 + \tau^2 n_i}}.$$

To find the posterior distribution of (Σ, T), let $\overline{X}$ stand for the vector with coordinates $\overline{X}_1, \ldots, \overline{X}_k$. Then, the conditional distribution of the data given $\Sigma = \sigma$ and $T = \tau$ is

$$\overline{X} \sim N_k(\psi_0 \mathbf{1}, W(\sigma, \tau)), \quad S_i^2 \sim \frac{\sigma^2}{n_i - 1}\chi_{n_i-1}^2, \tag{8.12}$$

where

$$W(\sigma, \tau) = \begin{bmatrix} \frac{\sigma^2}{n_1} + \tau^2\left[1 + \frac{1}{\zeta_0}\right] & \frac{\tau^2}{\zeta_0} & \frac{\tau^2}{\zeta_0} \\ \frac{\tau^2}{\zeta_0} & \ddots & \frac{\tau^2}{\zeta_0} \\ \frac{\tau^2}{\zeta_0} & \frac{\tau^2}{\zeta_0} & \frac{\sigma^2}{n_k} + \tau^2\left[1 + \frac{1}{\zeta_0}\right] \end{bmatrix}.$$

It follows that $f_{\Sigma,T|\overline{X},S_1^2,\ldots,S_k^2}(\sigma, \tau | \overline{x}, s_1^2, \ldots, s_k^2)$ is proportional to

$$f_{\Sigma,T}(\sigma, \tau)\sigma^{-[n_1 + \cdots + n_k - k]}|W(\sigma, \tau)|^{-\frac{1}{2}}$$

$$\times \exp\left(-\sum_{i=1}^{k}\left[\frac{s_i^2(n_i - 1)}{2\sigma^2}\right] - \frac{1}{2}(\overline{x} - \psi_0 \mathbf{1})^\top W^{-1}(\sigma, \tau)(\overline{x} - \psi_0 \mathbf{1})\right).$$

There is one special case in which the above formulas simplify tremendously. For fixed λ, suppose that

$$\Pr\left(T = \frac{\Sigma}{\sqrt{\lambda}}\right) = 1.$$

---

[4]The reduced model in (8.10) is sometimes called a *variance components model*. In this model, the vector M is not of interest, but rather only the variance T of its coordinates. The two terms involving τ and σ are components of the variance of the observations. Hill (1965) gives a Bayesian analysis of such models.

That is, T is just a known scalar multiple of Σ. In this case, it is convenient to define $\gamma_0 = \lambda \zeta_0$ so that

$$\mu_i(\psi, \sigma, \tau) = \frac{n_i \bar{x}_i + \psi \lambda}{\lambda_i} \equiv \mu_i(\psi),$$

$$\frac{\tau^2 \sigma^2}{n_i \tau^2 + \sigma^2} = \frac{\sigma^2}{\lambda_i},$$

$$\psi_1(\sigma, \tau) = \frac{\sum_{i=1}^{k} \gamma_i \bar{x}_i + \gamma_0 \psi_0}{\gamma_0 + \sum_{i=1}^{k} \gamma_i} \equiv \psi_1,$$

$$\frac{\zeta_0}{\tau^2} + \sum_{i=1}^{k} \frac{n_i}{\sigma^2 + \tau^2 n_i} = \gamma_0 + \sum_{i=1}^{k} \gamma_i,$$

$$W(\sigma, \tau) = \sigma^2 \begin{bmatrix} \frac{1}{n_1} + \frac{1}{\lambda} + \frac{1}{\gamma_0} & \frac{1}{\gamma_0} & & \frac{1}{\gamma_0} \\ \frac{1}{\gamma_0} & \ddots & & \frac{1}{\gamma_0} \\ \frac{1}{\gamma_0} & & \frac{1}{\gamma_0} & \frac{1}{n_k} + \frac{1}{\lambda} + \frac{1}{\gamma_0} \end{bmatrix},$$

where

$$\lambda_i = \lambda + n_i, \quad \gamma_i = \frac{n_i \lambda}{\lambda_i}.$$

Note how μ_i and ψ_1 no longer depend on σ and τ. In fact, we can use Proposition 8.13 below to show that

$$|W(\sigma, \tau)| = \sigma^{2k} \prod_{i=1}^{k} \frac{1}{\gamma_i} \left(1 + \frac{1}{\gamma_0} \sum_{i=1}^{k} \gamma_i \right).$$

Proposition 8.13.[5] *Let B be a positive definite $k \times k$ matrix, and let x be a vector of dimension k. Define $A = B + cxx^\top$, then*

$$A^{-1} = B^{-1} - \frac{c}{1 + cx^\top B^{-1} x} B^{-1} xx^\top B^{-1}$$

and $|A| = |B|(1 + cx^\top B^{-1} x)$.

In this case, there is a simple conjugate prior for Σ which makes the posterior of Σ of the same form. That would be that Σ^2 has inverse gamma distribution $\Gamma^{-1}(a_0/2, b_0/2)$. The posterior of Σ^2 would be $\Gamma^{-1}(a_1/2, b_1/2)$, where $a_1 = a_0 + \sum_{i=1}^{k} n_i$, $\gamma_* = \sum_{i=1}^{k} \gamma_i$, $u = \sum_{i=1}^{k} \gamma_i \bar{x}_i / \gamma_*$, and

$$b_1 = b_0 + \sum_{i=1}^{k} (n_i - 1) s_i^2 + \sigma^2 (\bar{x} - \psi_0 \mathbf{1})^\top W^{-1}(\sigma, \tau)(\bar{x} - \psi_0 \mathbf{1})$$

$$= b_0 + \sum_{i=1}^{k} \{(n_i - 1) s_i^2 + \gamma_i (\bar{x}_i - u)^2\} + \frac{\gamma_* \gamma_0}{\gamma_0 + \gamma_*} (u - \psi_0)^2.$$

---

[5] This proposition is also used in the analysis of the two-way ANOVA in Section 8.2.2.

Posterior distributions for linear functions of location parameters are now t distributions as are predictive distributions of future observations. For example, if Y is the average of m future observations from population i, then some of the various posterior and predictive distributions are

$$\Psi \sim t_{a_1}\left(\psi_1, \frac{b_1}{a_1}\frac{1}{\gamma_0 + \gamma_*}\right),$$

$$M_i \sim t_{a_1}\left(\mu_i(\psi_1), \frac{b_1}{a_1}\left[\frac{1}{\lambda_i} + \left(\frac{\lambda}{\lambda_i}\right)^2 \frac{1}{\gamma_0 + \gamma_*}\right]\right),$$

$$Y \sim t_{a_1}\left(\mu_i(\psi_1), \frac{b_1}{a_1}\left[\frac{1}{m} + \frac{1}{\lambda_i} + \left(\frac{\lambda}{\lambda_i}\right)^2 \frac{1}{\gamma_0 + \gamma_*}\right]\right).$$

When T is not a known scalar multiple of Σ, there is still a way to simplify the formulas slightly. That is, introduce $\Lambda = \Sigma^2/T^2$ as a replacement for T in the hyperparameter. Then, the simplified formulas are still correct as long as they are understood to represent conditional distributions given $\Lambda = \lambda$. In this case, it is also possible to let the values a_0, b_0, ζ_0, and ψ_0 depend on λ, if one wishes. The posterior for Λ is not particularly simple, but it is the only part of the posterior that is not simple. It is proportional to $f_\Lambda(\lambda)$ times

$$\left(\frac{b_0}{2}\right)^{\frac{a_0}{2}} \frac{\sqrt{\zeta_0}\Gamma\left(\frac{a_1}{2}\right)}{\Gamma\left(\frac{a_0}{2}\right)} \left(\prod_{i=1}^k \gamma_i\right)^{\frac{1}{2}} \left[\left(1 + \frac{1}{\lambda\zeta_0}\sum_{i=1}^k \gamma_i\right)\left(\frac{b_1}{2}\right)^{a_1}\right]^{-\frac{1}{2}}.$$

Numerical integration is required to make any marginal (not conditional on Λ) or predictive inferences.

The model just described can be used to find a solution to the problem that gave rise to the James–Stein estimator in Section 3.2.3. In that problem $\Pr(\Sigma = 1) = 1$, but otherwise it is the same. Assuming Σ to be unknown, the above model gives the posterior mean of M_j to be

$$E(M_j|X=x) = \int_0^\infty \frac{n_j\bar{x}_j + \frac{\sum_{i=1}^k \gamma_i\bar{x}_i + \zeta_0\lambda\psi_0}{\zeta_0\lambda + \sum_{i=1}^k \gamma_i}\lambda}{\lambda_j} f_{\Lambda|X}(\lambda|x)d\lambda.$$

Since the integration on the right-hand side is over λ, we should see explicitly where λ is. So, we rewrite the formula as

$$\begin{aligned}E(M_j|X=x) &= \int_0^\infty \frac{n_j\bar{x}_j + \frac{\sum_{i=1}^k \frac{n_i}{\lambda+n_i}\bar{x}_i + \zeta_0\psi_0}{\zeta_0 + \sum_{i=1}^k \frac{n_i}{\lambda+n_i}}\lambda}{\lambda + n_j} f_{\Lambda|X}(\lambda|x)d\lambda \\ &= E[\alpha_j(\Lambda)|X=x]\bar{x}_j + E[\{1 - \alpha_j(\Lambda)\}v(\Lambda)|X=x],\end{aligned}$$

where $\alpha_j(\lambda) = n_j/(n_j + \lambda)$ and $v(\lambda) = \psi_1(\sigma, \tau)$ is itself a weighted average of ψ_0 and a weighted average of all of the sample averages. This is similar to the empirical Bayes modification to the James–Stein estimator, namely (3.55) on page 165. The way it behaves can be understood as follows. Λ is a measure of how much more spread there is within each group relative to the spread between groups. Now, suppose that all $n_i = 1$. Then $v(\lambda)$ is $k/(1+\lambda)$ times $\bar{x}$ plus ζ_0 times ψ_0 all divided by $k/(1+\lambda) + \zeta_0$. If the posterior distribution of Λ is concentrated near 0 (that is, there is far less variation within groups than between), then $v(\Lambda)$ will get very little weight, since $\alpha(\Lambda)$ will be close to 1. This makes sense because the large spread between the means suggests that the information from $\bar{x}_j$ is much more valuable than the other $\bar{x}_i$s. If, however, Λ has lots of mass for large values, then there will be a great deal of shrinkage, and $v(\Lambda)$ will be near ψ_0. For distributions of Λ concentrated on intermediate values, $\bar{x}_j$, $\bar{x}$, and ψ_0 all receive moderate weight.

Example 8.14. Consider the following data gathered from three groups:

i	1	2	3
n_i	10	12	15
$\bar{x}_i$	27.9268	18.1622	19.5475
s_i^2	23.8227	57.6736	32.3858

Suppose that we want to have Σ^2 and T^2 be independent in the prior distribution. Suppose that the prior for Σ^2 is $\Gamma^{-1}(a_0'/2, b_0'/2)$ and the prior for T^2 is $\Gamma^{-1}(c_0/2, d_0/2)$. Then Λ has the distribution of $b_0 c_0/(a_0 d_0)$ times an F_{c_0, a_0} random variable. The conditional distribution of Σ^2 given $\Lambda = \lambda$ can be shown (see Problem 11 on page 534) to be $\Gamma^{-1}(a_0, b_0(\lambda))$, where $a_0 = [a_0' + c + 0]/2$ and $b_0(\lambda) = [b_0' + \lambda d_0]/2$. Suppose that the rest of the prior distribution is specified by $\psi_0 = 10$, $\zeta_0 = 0.1$, $a_0' = 1$, $b_0' = 10$, $c_0 = 1$, and $d_0 = 1$. The posterior distribution of Λ can be found approximately, its mode is around 1.07, and it has probability of about 0.94 of $\Lambda \leq 10$. Hence $\alpha_j(\Lambda) = n_j/[\Lambda + n_j]$ is close to 1 with high probability for all j, and there will be little shrinkage toward the overall mean.

We can numerically calculate the posterior distributions of the three M_i using either Laplace's method (Section 7.4.3) or importance sampling (Section B.7). For Laplace's method, the "Θ" is Λ, and the function $g(\lambda)$ is one of the posterior densities of the M_i given $\Lambda = \lambda$ evaluated at various values of μ. These densities are t_{a_1} with location and scale

$$\frac{n_i \bar{x}_i + \psi_1(\lambda)\lambda}{\lambda + n_i}, \text{ and } \sqrt{\frac{b_1(\lambda)}{a_1}}\sqrt{\frac{1}{\lambda + n_i} + \frac{\lambda^2}{(\lambda + n_i)^2} \frac{1}{\zeta_0 + \sum_{i=1}^{3} \gamma_i}},$$

where $\psi_1(\lambda) = (\zeta_0 \lambda \psi_0 + \sum_{i=1}^{3} \gamma_i \bar{x}_i)/(\zeta_0 + \sum_{i=1}^{3} \gamma_i)$, and $b_1(\lambda)$ is the same as b_1 with b_0 replaced by $b_0(\lambda)$.

For importance sampling, we sampled 1000 values from the prior distribution of Λ and used these to approximate the integrals that equal the posterior densities at various μ values. We also used the delta method to calculate standard deviations for the density values and found these to be at most 0.09 times the density values in all cases (less than 0.05 times the density in 80% of the cases).

The three posterior means were calculated from the posterior densities and were found to equal 26.08, 18.86, and 19.86. We see that some shrinkage has occurred. The numerically evaluated densities are shown in Figure 8.15 together with the results of an *empirical Bayes* analysis to be described in Section 8.4 and a *successive substitution sampling* analysis to be described in Section 8.5.

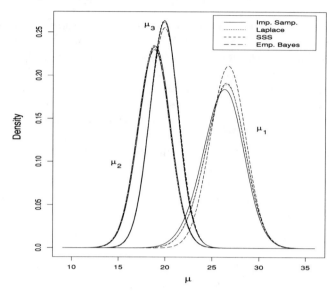

FIGURE 8.15. Numerical Approximations to Posterior Densities

One could generalize this model to the case in which the variance of $X_{i,j}$ conditional on the parameters is Σ_i^2. In this case, one can only obtain closed-form posteriors conditional on all of the variance parameters, $\Sigma_1^2, \ldots, \Sigma_k^2, T^2$. Numerical integration over all $k+1$ variance parameters would then be needed. We postpone illustration of this until Section 8.5, at which time we introduce an alternative method of solution that is better suited to this type of problem.

8.2.2 Two-Way Mixed Model ANOVA*

In this section, we examine a two-way analysis of variance with one random effect and one fixed effect and equal numbers of observations per cell. The recommended analysis of this model will be described in Section 8.5. The analysis given here is mainly motivational as well as illustrative.

Suppose that

$$Y_{i,j,k} = M + A_i + B_j + (AB)_{i,j} + \epsilon_{i,j,k}, \qquad (8.16)$$

where A stands for the random effect and B stands for the fixed effect and

$$\sum_{j=1}^{b} B_j = 0 = \sum_{j=1}^{b} (AB)_{i,j}, \text{ for all } i, \qquad (8.17)$$

for $i = 1, \ldots, a$, $j = 1, \ldots, b$, and $k = 1, \ldots, m$. We suppose that the $\epsilon_{i,j,k}$

---

*This section may be skipped without interrupting the flow of ideas.

are conditionally IID $N(0, \sigma_e^2)$ given $\Sigma_e^2 = \sigma_e^2$ (and all other parameters). It is also traditional to assume that, conditional on other parameters, the A_i are independent of each other and of the B_j and the $(AB)_{i,j}$, that the B_j are independent of the $(AB)_{i,j}$, and that the $(AB)_{i,j}$ for different i are independent of each other. We can let

$$M_{i,j} = M + A_i + B_j + (AB)_{i,j},$$

and put these into vectors $M_i = (M_{i,1}, \ldots, M_{i,b})^\top$. We can then express the model described above by saying that the M_i are conditionally independent $N_b(\theta, \Sigma)$ vectors given $\Theta = \theta$ and $\Sigma = \sigma$. Here Σ is a $b \times b$ matrix and $\Theta = (M + B_1, \ldots M + B_b)^\top$. In order to ensure that (8.17) is reflected in the conditional distribution of M_i given M, we assume that Σ has the form

$$\Sigma = \Sigma_A^2 \mathbf{1}\mathbf{1}^\top + \Sigma_{AB}^2 \left(I - \frac{1}{b}\mathbf{1}\mathbf{1}^\top\right),$$

where $\mathbf{1}$ is b-dimensional. At the next stage of the hierarchy, assume that the coordinates of Θ, namely $\Theta_1, \ldots, \Theta_b$, are conditionally IID $N(\mu, \sigma_B^2)$ given $M = \mu$ and $\Sigma_B = \sigma_B$ (and other parameters), since $\Theta_j = M + B_j$ in the notation of (8.16). At the next stage, model $M \sim N(\mu_0, \sigma_B^2/\tau)$ given $\Sigma_B = \sigma_B$ (and other parameters). Finally, $\Sigma_e, \Sigma_A, \Sigma_{AB}$, and Σ_B have some joint distribution. In summary, we have Table 8.18.

One way to proceed, after collecting data, would be to march through the levels of the hierarchy, finding all of the posterior distributions. This is done in much the same way as in the simpler model of Section 8.2.1, but with an extra level in the hierarchy. Alternatively, we could take an approach that is typically done in the classical analysis of this model. That approach is to pretend that some of the parameters are not of interest and integrate them out of the model. In particular, the M_i and Θ_j are usually integrated out of the classical analysis. This is easy to do in the model of

TABLE 8.18. Hierarchical Model for Two-Way ANOVA

Stage	Random Variables	Distribution	Conditional on
Data	$Y_{i,j,k}$, for all i, j, k	independent $N(\mu_{i,j}, \sigma_e^2)$	All $M_{i,j} = \mu_{i,j}$, $\Sigma_e = \sigma_e$, $M = \mu$, $\Theta = \theta$, $\Sigma_A = \sigma_A$, $\Sigma_B = \sigma_B$, $\Sigma_{AB} = \sigma_{AB}$
Parameter	M_i for all i	independent $N_b(\theta, \sigma_A^2 \mathbf{1}\mathbf{1}^\top + \sigma_{AB}^2[I - \frac{1}{b}\mathbf{1}\mathbf{1}^\top])$	$M = \mu$, $\Sigma_e = \sigma_e$, $\Sigma_A = \sigma_A$, $\Theta = \theta$, $\Sigma_B = \sigma_B$, $\Sigma_{AB} = \sigma_{AB}$
Hyper-parameter	Θ_j, $j = 1, \ldots, b$	independent $N(\mu, \sigma_B^2)$	$\Sigma_A = \sigma_A$, $M = \mu$, $\Sigma_{AB} = \sigma_{AB}$, $\Sigma_B = \sigma_B$, $\Sigma_e = \sigma_e$
Hyperhyper-parameter	M	$N\left(\mu_0, \sigma_B^2 \tau^{-1}\right)$	$\Sigma_A = \sigma_A$, $\Sigma_e = \sigma_e$, $\Sigma_B = \sigma_B$, $\Sigma_{AB} = \sigma_{AB}$
Variance	$\Sigma_e, \Sigma_B, \Sigma_A, \Sigma_{AB}$	Whatever	

Table 8.18. To do this, we note first note that sufficient statistics are

$$RSS = \sum_{i=1}^{a}\sum_{j=1}^{b}\sum_{k=1}^{m}(y_{i,j,k} - \bar{y}_{i,j})^2, \quad \overline{Y}_i = \frac{1}{m}\sum_{k=1}^{m}\begin{pmatrix} Y_{i,1,k} \\ \vdots \\ Y_{i,b,k} \end{pmatrix}.$$

The distribution of the $\overline{Y}_i$ is that of independent b-variate normals, with distribution $N_b(\mu_i, \frac{\sigma_e^2}{m}I)$. To integrate the M_i out of the model conditional on everything else, we note that the distribution of RSS depends only on Σ_e, so only the distribution of the $\overline{Y}_i$ changes to

$$N_b\left(\theta, \frac{\sigma_e^2}{m}I + \sigma_A^2 \mathbf{1}\mathbf{1}^\top + \sigma_{AB}^2\left[I - \frac{1}{b}\mathbf{1}\mathbf{1}^\top\right]\right), \tag{8.19}$$

and they are still conditionally independent. This means that we can reduce the sufficient statistic even further. We will still need RSS, and of course we will need $\overline{Y}_{.} = \sum_{i=1}^{a}\overline{Y}_i/a$. Because of the special form of the covariance matrix of the $\overline{Y}_i$, we do not need the whole matrix $\sum_{i=1}^{a}(\overline{Y}_i - \overline{Y}_.)(\overline{Y}_i - \overline{Y}_.)^\top$, which would be required if the covariance matrix were unconstrained. Instead, one can use the fact that

$$\left(\frac{\sigma_e^2}{m}I + \sigma_A^2 \mathbf{1}\mathbf{1}^\top + \sigma_{AB}^2\left[I - \frac{1}{b}\mathbf{1}\mathbf{1}^\top\right]\right)^{-1}$$

$$= \frac{1}{\frac{\sigma_e^2}{m} + \sigma_{AB}^2}\left[I - \frac{\sigma_A^2 - \frac{\sigma_{AB}^2}{b}}{\frac{\sigma_e^2}{m} + \sigma_{AB}^2}\mathbf{1}\mathbf{1}^\top\right] \tag{8.20}$$

to write the conditional density of the $\overline{Y}_i$ in terms of $\overline{Y}_{.}$ and

$$SSA = \sum_{i=1}^{a} bm(\overline{Y}_{i,.} - \overline{Y}_{.,.})^2,$$

$$SSI = \sum_{i=1}^{a}\sum_{j=1}^{b} m(\overline{Y}_{i,j} - \overline{Y}_{i,.} - \overline{Y}_{.,j} + \overline{Y}_{.,.})^2$$

$$= \sum_{i=1}^{a} m(\overline{Y}_i - \overline{Y}_.)^\top(\overline{Y}_i - \overline{Y}_.) - SSA,$$

where $\overline{Y}_{i,.}$ is the average of the coordinates of $\overline{Y}_i$ and $\overline{Y}_{.,.}$ is the average of the coordinates of $\overline{Y}_.$. These two sums of squares are the usual sums of squares for the random effect and for interaction, respectively. The conditional distribution of $\overline{Y}_.$ given the parameters is the same as (8.19) except that the covariance matrix must be divided by a because we averaged a independent vectors with the same distribution.

To integrate Θ out of the distribution, we note that SSA and SSI depend only on Σ_e, Σ_A, and Σ_{AB}. Using (8.20) once again, we can write the conditional density of $\overline{Y}_{..}$ in terms of $\overline{Y}_{..}$ and $SSB = \sum_{j=1}^{b} ma(\overline{y}_{.j} - \overline{y}_{..})^2$, where $\overline{Y}_{.j}$ is coordinate j of $\overline{Y}_{.}$, and SSB is the usual sum of squares for the fixed effect. In summary, the sufficient statistics for the model that involves only the parameters Σ_e, Σ_A, Σ_{AB}, Σ_B, and M are RSS, SSA, SSI, SSB, and $\overline{Y}_{...}$. The conditional distributions of these quantities given the parameters are easily calculated using the fact that they are functions of orthogonal transformations of the original data. Hence, they are all conditionally independent given the parameters, and their distributions are

$$\overline{Y}_{...} \sim N\left(\mu, \frac{\sigma_e^2}{bam} + \frac{\sigma_A^2}{a}\right), \qquad RSS \sim \sigma_e^2 \chi^2_{ba[m-1]},$$
$$SSB \sim (\sigma_e^2 + m\sigma_{AB}^2 + am\sigma_B^2)\chi^2_{b-1}, \qquad SSA \sim (\sigma_e^2 + m\sigma_A^2)\chi^2_{a-1},$$
$$SSI \sim (\sigma_e^2 + m\sigma_{AB}^2)\chi^2_{[a-1][b-1]}.$$

At this point, the classical analysis differs from any further Bayesian analysis. The classical analysis usually ignores $\overline{Y}_{...}$ and makes inference based on the sums of squares. Since the distribution of $\overline{Y}_{...}$ depends on some of the variance parameters, a Bayesian would still make use of it, even if interest was solely in the variance parameters. In particular, we could integrate M out of the problem and see that the conditional distribution of $\overline{Y}_{...}$ given the variance parameters alone is

$$N\left(\mu_0, \frac{\sigma_e^2}{bam} + \frac{\sigma_A^2}{a} + \frac{\sigma_B^2}{\tau}\right).$$

8.2.3 Hypothesis Testing

On page 384, we found a UMPUI[6] test for the hypothesis of equal means in a one-way analysis of variance. This test was the usual F-test. In Section 4.5.6, we illustrated how the usual F-test was a Bayes rule in a decision problem. This decision problem had the property that the prior probability was positive that all the means were equal. It may be that we do not feel that exact equality between the means has positive probability, but we are still interested in how far apart they are. In fixed-effects models, there is a straightforward way to measure the differences between the means which resembles the F-test but uses a prior in which the probability is zero that any two means are equal.

Suppose that X has $N_n(g\beta, \sigma^2 v)$ distribution given $(B, \Sigma) = (\beta, \sigma)$, where g and v are known matrices with g being $n \times p$ and v being $n \times n$, respectively. Let the prior distribution be that $B \sim N_p(\beta_0, \sigma^2 w_0^{-1})$ given $\Sigma = \sigma$ and $\Sigma^2 \sim \Gamma^{-1}(a_0/2, b_0/2)$, where w_0 is a known, nonsingular $p \times p$

---

[6]Uniformly most powerful unbiased invariant.

matrix. The sufficient statistics from a sample $X = x$ are
$$\hat{\beta} = (g^\top v^{-1} g)^{-1} g^\top v^{-1} x,$$
$$RSS = (x - g\hat{\beta})^\top v^{-1} (x - g\hat{\beta}).$$
The posterior distribution given $X = x$ has the form $B \sim N_p(\beta_1, \sigma^2 w_1^{-1})$ given $\Sigma = \sigma$ and $\Sigma^2 \sim \Gamma^{-1}(a_1/2, b_1/2)$, where
$$a_1 = a_0 + n, \qquad w_1 = w_0 + g^\top v^{-1} g,$$
$$\beta_1 = w_1^{-1}(w_0 \beta_0 + (g^\top v^{-1} g)\hat{\beta}),$$
$$b_1 = b_0 + RSS + (\beta_0 - \hat{\beta})^\top w_0 w_1^{-1} (g^\top v^{-1} g)(\beta_0 - \hat{\beta}).$$

Let $B_0 = \{\beta : a\beta = \psi_0\}$ where a is a $q \times p$ matrix of rank $q \leq p$, and ψ_0 is some q-dimensional vector in the column space of a. Suppose that we are interested in how far B is from B_0. Let $\beta_0 \in B_0$, and define $H = (B - \beta_0)^\top h (B - \beta_0)$, where $h = a^\top (a w_1^{-1} a^\top)^{-1} a$. Note that
$$H = (aB - \psi_0)^\top (a w_1^{-1} a^\top)^{-1} (aB - \psi_0)$$
is the same no matter which $\beta_0 \in B_0$ is used in its definition. There are two natural ways to measure the distance between B and B_0. One is by $\rho(B, B_0) = H/\text{trace}(h)$, and the other is by $\rho(B, B_0)/\Sigma^2$. The reason for the $trace(h)$ in the denominator is two-fold. First, there is a sense in which h measures the precision of that part of B that lies in B_0. Because of this, there are two factors that contribute to $(B - \beta_0)^\top h (B - \beta_0)$ being large. One is how far B is from β_0 and the other is how precisely we know B. Only the former should contribute to the distance between B and B_0. The latter can be used to judge how well we know the distance between B and B_0, but not to increase the actual distance. This means that we must adjust H somehow to remove the effect of the precision. The trace of h is a natural way to do that. The second reason for the trace of h is that it is invariant under alternative representations for the set B_0. That is, if $B_0 = \{\beta : c\beta = \psi_0\}$ also, then $\text{trace}(c^\top (c w_1^{-1} c^\top)^{-1} c) = \text{trace}(h)$.[7]

To express our uncertainty about the distance between B and B_0, we need the distribution of H and/or the distribution of H/Σ^2.

Theorem 8.21. *Suppose that, conditional on $Y = y$, the distribution of Z is $NC\chi_q^2(y)$. Suppose also that Y has $\Gamma(a_1/2, b_1/[2c])$ distribution, then the marginal distribution of Z is $ANC\chi^2(q, a_1, \gamma)$, where $\gamma = c/(c + b_1)$. (See page 668.) The mean of Z is $E(Z) = q + ca_1/b_1 = q + a_1\gamma/(1 - \gamma)$.*

PROOF. The conditional density of Z given $Y = y$ is
$$f_{Z|Y}(z|y) = \sum_{i=0}^{\infty} \exp\left(-\frac{y}{2}\right) \frac{\left(\frac{y}{2}\right)^i}{i!} \frac{2^{-\left(\frac{q}{2}+i\right)}}{\Gamma\left(\frac{q}{2}+i\right)} z^{\frac{q}{2}+i-1} \exp\left(-\frac{z}{2}\right).$$

---

[7] For those with a background in multivariate analysis, it is possible to show that $H/\text{trace}(h)$ is the weighted average of the squares of the principal components of the projection of $w^{1/2}(B - \beta_0)$ into the space $w^{1/2} B_0$.

The marginal density of Y is

$$f_Y(y) = \frac{\left(\frac{b_1}{2c}\right)^{\frac{a_1}{2}}}{\Gamma\left(\frac{a_1}{2}\right)} y^{\frac{a_1}{2}-1} \exp\left(-\frac{b_1 y}{2c}\right).$$

The joint density is

$$f_{Z,Y}(z,y) = \sum_{i=0}^{\infty} \frac{2^{-\left(\frac{q}{2}+2i\right)} \left(\frac{b_1}{2c}\right)^{\frac{a_1}{2}}}{i!\,\Gamma\left(\frac{q}{2}+i\right)\Gamma\left(\frac{a_1}{2}\right)} y^{i+\frac{a_1}{2}-1}$$

$$\times \exp\left(-\frac{y}{2}\left[\frac{b_1}{c}+1\right]\right) z^{\frac{q}{2}+i-1} \exp\left(-\frac{z}{2}\right).$$

Integrating out y gives the marginal density of Z

$$f_Z(z) = \sum_{i=0}^{\infty} \frac{\left(\frac{b_1}{c}\right)^{\frac{a_1}{2}} \Gamma\left(\frac{a_1}{2}+i\right)}{i!\,\Gamma\left(\frac{a_1}{2}\right)\left(1+\frac{b_1}{c}\right)^{\frac{a_1}{2}+i}} \frac{2^{-\left(\frac{q}{2}+i\right)}}{\Gamma\left(\frac{q}{2}+i\right)} z^{\frac{q}{2}+i-1} \exp\left(-\frac{z}{2}\right).$$

Use the formula for γ to complete the proof that the distribution is $ANC\chi^2$. The mean of Z given $Y = y$ is $q + y$, and the mean of Y is ca_1/b_1. □

Theorem 8.22. *Suppose that the conditional distribution of ZY given $Y = y$ is noncentral χ^2, $NC\chi_q^2(cy)$. Also, suppose that the distribution of Y is $\Gamma(a_1/2, b_1/2)$, then the marginal distribution of $a_1 Z/[q(b_1 + c)]$ is $ANCF(q, a_1, \gamma)$, where $\gamma = c/(c+b_1)$. (See page 669.) If $a_1 > 2$, the mean of Z is $E(Z) = c + qb_1/(a_1 - 2)$.*

PROOF. The conditional density of Z given $Y = y$ is

$$f_{Z|Y}(z|y) = \sum_{i=0}^{\infty} \exp\left(-\frac{cy}{2}\right) \frac{\left(\frac{cy}{2}\right)^i}{i!} \frac{y^{\frac{q}{2}+i}}{2^{\frac{q}{2}+i}\Gamma\left(\frac{q}{2}+i\right)} z^{\frac{q}{2}+i-1} \exp\left(-\frac{zy}{2}\right).$$

The marginal density of Y is

$$f_Y(y) = \frac{\left(\frac{b_1}{2}\right)^{\frac{a_1}{2}}}{\Gamma\left(\frac{a_1}{2}\right)} y^{\frac{a_1}{2}-1} \exp\left(-\frac{b_1 y}{2}\right).$$

The joint density is $f_{Z,Y}(z,y)$ equal to $\exp(-y[b_1 + c + z]/2)$ times

$$\sum_{i=0}^{\infty} \frac{\left(\frac{c}{2}\right)^i \left(\frac{b_1}{2}\right)^{\frac{a_1}{2}}}{i!} \frac{y^{\frac{a_1}{2}+\frac{q}{2}+2i-1} z^{\frac{q}{2}+i-1}}{2^{2q+i}\Gamma\left(\frac{q}{2}+i\right)\Gamma\left(\frac{a_1}{2}\right)}.$$

Integrating y out of this gives the density of Z:

$$f_Z(z) = \sum_{i=0}^{\infty} \frac{c^i b_1^{\frac{a_1}{2}}}{i!(b_1+c+z)^{\frac{a_1}{2}+\frac{q}{2}+2i}} \frac{\Gamma\left(\frac{a_1}{2}+\frac{q}{2}+2i\right)}{\Gamma\left(\frac{q}{2}+i\right)\Gamma\left(\frac{a_1}{2}\right)} z^{\frac{q}{2}+i-1}.$$

Now, make the change of variables from z to $u = z/(z+b_1+c)$. The inverse is $z = (b_1+c)u/(1-u)$. The derivative is $(b_1+c)/(1-u)^2$. The density of $U = Z/(Z+b_1+c)$ is

$$f_U(u) = \sum_{i=0}^{\infty} \frac{b_1^{\frac{a_1}{2}} c^i}{(b_1+c)^{\frac{a_1}{2}+i} i!} \frac{\Gamma(\frac{a_1}{2}+\frac{q}{2}+2i)}{\Gamma(\frac{q}{2}+i)\Gamma(\frac{a_1}{2})} u^{\frac{q}{2}+i-1}(1-u)^{\frac{a_1}{2}+i-1}.$$

Setting $\gamma = c/(c+b_1)$ and rearranging Γ function values produces the $ANCB(q, a_1, \gamma)$ density. We know that

$$\frac{a_1 U}{q(1-U)} = \frac{a_1 Z}{q(b_1+c)},$$

which must have $ANCF(q, a_1, \gamma)$ distribution. The mean is obtained by noting that $E(Z|Y=y) = c + q/y$ and $E(1/Y) = b_1/(a_1-2)$ if $a_1 > 2$. □

In Theorem 8.21, let $Z = H/\Sigma^2$. Since $aB - \psi_0$ has multivariate normal distribution $N_q(a\beta_1 - \psi_0, \sigma^2 a w_1^{-1} a^\top)$ given $\Sigma = \sigma$, it follows that Z has noncentral χ^2 distribution with q degrees of freedom and noncentrality parameter $y = (a\beta_1 - \psi_0)^\top (a w_1^{-1} a^\top)^{-1} (a\beta_1 - \psi_0)/\sigma^2$ given $\Sigma = \sigma$. Since

$$c = (a\beta_1 - \psi_0)^\top (a w_1^{-1} a^\top)^{-1} (a\beta_1 - \psi_0) \qquad (8.23)$$

is a constant in the posterior distribution, we can let $Y = c/\Sigma^2$, which has $\Gamma(a_1/2, b_1/[2c])$ distribution. It follows that the distribution of H/Σ^2 is $ANC\chi^2(q, a_1, \gamma)$.

In Theorem 8.22, let $Z = H$ and $Y = 1/\Sigma^2$. Now, ZY has noncentral χ^2 distribution with q degrees of freedom and noncentrality parameter cy given $Y = y$. Also, $Y \sim \Gamma(a_1/2, b_1/c)$. It follows that the distribution of $a_1 H/[q(b_1+c)]$ is $ANCF(q, a_1, \gamma)$.

Example 8.24. We will use the same data as in Example 8.14 on page 487, but we will use a conjugate prior for the parameters Σ and $B = (M_1, M_2, M_3)^\top$. The design matrix g is particularly simple and v is the identity matrix. We get $g^\top v^{-1} g$ to be the 3×3 diagonal matrix with 10, 12, and 15 on the diagonal. Suppose that the prior has hyperparameters

$$a_0 = 1, \qquad b_0 = 10,$$

$$\beta_0 = \begin{pmatrix} 10 \\ 10 \\ 10 \end{pmatrix}, \qquad w_0 = \begin{pmatrix} 6.7742 & -3.2258 & -3.2258 \\ -3.2258 & 6.7742 & -3.2258 \\ -3.2258 & -3.2258 & 6.7742 \end{pmatrix}.$$

The posterior distribution has hyperparameters

$$a_1 = 38, \qquad b_1 = 1729.7,$$

$$\beta_1 = \begin{pmatrix} 24.44734 \\ 19.43751 \\ 20.11565 \end{pmatrix}, \qquad w_1 = \begin{pmatrix} 16.7742 & -3.2258 & -3.2258 \\ -3.2258 & 18.7742 & -3.2258 \\ -3.2258 & -3.2258 & 21.7742 \end{pmatrix}.$$

Now, suppose that $B_0 = \{\beta : \beta_1 = \beta_2 = \beta_3\}$. This can be represented by the matrix and vector

$$a = \begin{pmatrix} 1 & 0 & -1 \\ 0 & 1 & -1 \end{pmatrix}, \qquad \psi_0 = \begin{pmatrix} 0 \\ 0 \end{pmatrix}.$$

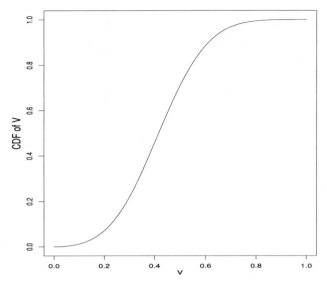

FIGURE 8.25. CDF of $V = \sqrt{H/[44.3331\Sigma^2]}$

The noncentrality parameter of the alternate noncentral distributions can be calculated to equal $\gamma = 307.5124/(1729.7 + 307.5124) = 0.1509$. The trace of h is 44.3331.

If we want to describe our uncertainty about how far apart the M_i are, we could look at the CDF of some function of H or of H/Σ^2. For example, Figure 8.25 gives the graph of the CDF of $V = \sqrt{H/[44.3331\Sigma^2]}$. We see that it is almost certain that the average distance between the M_i is less than Σ and there is a 95% chance that the average distance is at least 0.18Σ.

8.3 Nonnormal Models*

Hierarchical models are useful for problems in which data have any sort of distribution. We will give two examples in this section.

8.3.1 Poisson Process Data

Suppose that several stochastic processes are being compared. For example, each process may be registering the occurrence of defects produced by one of several machines. Or, each process may be registering the times at which a criminal is arrested. Suppose that we model the processes as Poisson processes conditional on parameters $\Theta_1, \ldots, \Theta_k$, so that process i has rate θ_i given $\Theta_i = \theta_i$. We could then model the Θ_i as a priori exchangeable

---

*This section may be skipped without interrupting the flow of ideas.

496   Chapter 8. Hierarchical Models

random variables with $\Gamma(\alpha, \beta)$ distribution given $A = \alpha$ and $B = \beta$. We would then need a distribution for (A, B). Suppose that the data for process i consist of T_i units of time and N_i occurrences. The posterior distributions of the Θ_is given $A = \alpha$, $B = \beta$, $T_i = t_i$, and $N_i = n_i$ are of independent random variables with Θ_i having $\Gamma(\alpha+n_i, \beta+t_i)$ distribution. The posterior density of (A, B) is proportional to

$$f_{A,B}(\alpha, \beta) \frac{\beta^{k\alpha}}{\Gamma(\alpha)^k} \prod_{i=1}^{k} \frac{\Gamma(\alpha + n_i)}{(\beta + t_i)^{\alpha+n_i}}. \tag{8.26}$$

This would require numerical integration or approximations in order to make use of it.

Example 8.27. Suppose that N_i is the number of times an individual is arrested in T_i units of time (months). We will assume that T_i is independent of the parameters, and that conditional on $T_i = t_i$ and $\Theta_1 = \theta_1, \ldots, \Theta_n = \theta_k, A = \alpha, B = \beta$, the N_i are independent $Poi(t_i \theta_i)$. The Θ_i are modeled as IID $\Gamma(\alpha, \beta)$ given $A = \alpha, B = \beta$. We will use the following prior distribution for (A, B):

$$B|A = \alpha \sim \Gamma\left(b^{(0)}, \frac{c^{(0)}}{\alpha}\right), \quad A \sim \Gamma(a^{(0)}, d^{(0)}).$$

In this prior, B/A is independent of A. Suppose that we use the prior hyperparameters $a^{(0)} = 1/2$, $b^{(0)} = 1$, $c^{(0)} = 13$, and $d^{(0)} = 1$. The data consist of $k = 6$ individuals with the following observations:

Subject (i)	1	2	3	4	5	6
Time (t_i)	36	27	14	6	20	30
Number (n_i)	2	3	1	1	2	2

We will illustrate two numerical techniques for drawing inferences from this data and model. Suppose that we want the predictive distributions of the numbers of arrests in a future 24-month period for two different individuals. One of them is the second observed individual in the data set, and the other is an individual not in the data set but deemed to be a priori exchangeable with them. Denote these individuals by $i = 2$ and $i = 7$, respectively, and denote the numbers of arrests by M_2 and M_7 to distinguish them from the observed data. What we seek is $f_{M_i|X}(n|x)$ for $i = 2, 7$ and $n = 0, 1, \ldots$, where $X = x$ is the observed data. We can write

$$f_{M_i|X}(n|x) = \int\int f_{M_i|X,A,B}(n|x, \alpha, \beta) f_{A,B|X}(\alpha, \beta|x) d\alpha d\beta,$$

$$f_{M_i|X,A,B}(n|x, \alpha, \beta) = \int f_{M_i|\Theta_i}(n|\theta) f_{\Theta_i|X,A,B}(\theta|x, \alpha, \beta) d\theta,$$

$$f_{M_i|\Theta_i}(n|\theta) = \exp(-24\theta)\frac{(24\theta)^n}{n!},$$

$$f_{\Theta_2|X,A,B}(\theta|x, \alpha, \beta) = \frac{(\beta + 27)^{\alpha+3}}{\Gamma(\alpha+3)} \theta^{\alpha+2} \exp[-\theta(\beta + 27)],$$

$$f_{\Theta_7|X,A,B}(\theta|x, \alpha, \beta) = \frac{\beta^\alpha}{\Gamma(\alpha)} \theta^{\alpha-1} \exp(-\theta\beta),$$

$$f_{M_2|X,A,B}(n|x,\alpha,\beta) = \frac{24^n(\beta+27)^{\alpha+3}\Gamma(\alpha+3+n)}{n!\Gamma(\alpha+3)(\beta+51)^{\alpha+3+n}},$$

$$f_{M_7|X,A,B}(n|x,\alpha,\beta) = \frac{24^n\beta^\alpha\Gamma(\alpha+n)}{n!\Gamma(\alpha)(\beta+24)^{\alpha+n}}.$$

Therefore, we need to be able to integrate these last two expressions times the expression in (8.26) renormalized to be a density. The normalization constant is $f_X(x)$, the integral of (8.26) over α and β.

First, we used Laplace's method from Section 7.4.3, since all the functions being integrated are positive. The "Θ" in this example is (A, B), and $g(\theta)$ is one of the several functions obtained by fixing n in either $f_{M_2|X,A,B}(n|x,\alpha,\beta)$ or $f_{M_7|X,A,B}(n|x,\alpha,\beta)$ from above. Due to the form of the prior, it seemed sensible to transform to $(A, B/A)$ before applying Laplace's method.

Second, we used importance sampling (see Section B.7) to integrate numerically. We used a single set of 100,000 pseudorandom pairs drawn from the prior distribution of (A, B) to perform all of the integrals. We also calculated variances using the delta method for each ordinate. The results for M_2 are shown in Figure 8.28, and those for M_7 are shown in Figure 8.29. The standard deviations of the importance sample ordinates were all at least two orders of magnitude smaller than the ordinates themselves. As we can see in the figures, the two methods produce nearly the same results.

8.3.2 Bernoulli Process Data

Suppose that we can collect counts from several different sources. For example, we might be administering several treatments and we count how many recoveries occur in each treatment group. The data from group i will consist of n_i, the number of subjects in the group, and X_i, the number of

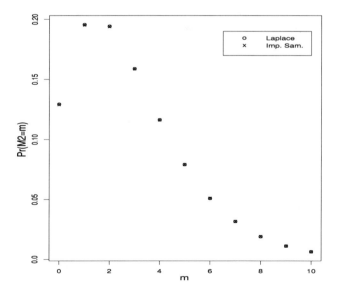

FIGURE 8.28. Numerical Approximations to Density of M_2

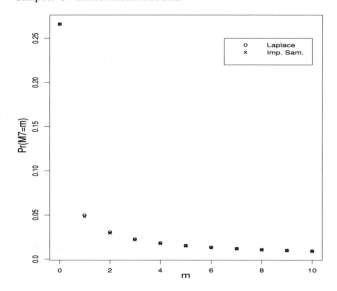

FIGURE 8.29. Numerical Approximations to Density of M_7.

successes, for $i = 1, \ldots, k$. We model the successes as Bernoulli processes conditional on parameters P_i, with the probability of success in group i being p_i given $P_i = p_i$. We can model the P_i as exchangeable random variables with $Beta(\theta r, [1-\theta]r)$ conditional on $\Theta = \theta$ and $R = r$. Here, Θ is like the average probability and R is like a measure of similarity. The larger R is, the more similar the P_i are. The posterior distribution of the P_i given $\Theta = \theta$, $R = r$, and $X_i = x_i$ is that of independent random variables with P_i having $Beta(\theta r + x_i, [1-\theta]r + n_i - x_i)$ distribution. The posterior density of (Θ, R) would be proportional to

$$f_{\Theta,R}(\theta, r) \frac{\Gamma(r)^k}{\Gamma(\theta r)^k \Gamma([1-\theta]r)^k} \prod_{i=1}^{k} \frac{\Gamma(\theta r + x_i)\Gamma([1-\theta]r + n_i - x_i)}{\Gamma(r + n_i)}.$$

This would require numerical integration or approximations in order to make use of it.

One possible approximation that is available puts this problem into the normal model framework. If the n_i will be large, we can model $Y_i = 2\arcsin\sqrt{X_i/n_i}$ as approximately $N(2\arcsin\sqrt{p_i}, 1/n_i)$ random variables given $P_i = p_i$. We could then use the same transformation on the P_i to model the $M_i = 2\arcsin\sqrt{P_i}$ as approximately $N(\mu, 1/\tau)$ given $M = \mu$ and $T = \tau$. Then M can be modeled as $N(\mu^{(0)}, 1/(\lambda\tau))$ given $T = \tau$, and T can be given some distribution. Here, M plays the role of $2\arcsin\sqrt{\Theta}$ and T plays the role of R from the earlier model. The posterior distribution of the M_i given $M = \mu$ and $T = \tau$ is that of independent random variables with M_i having $N(\psi_i(\mu), 1/(\tau + n_i))$ distribution, where

$$\psi_i(\mu) = \frac{\mu\tau + n_i y_i}{\tau + n_i}.$$

The posterior of M given $T = \tau$ is $N(\mu^{(1)}(\tau), 1/[\tau\gamma(\tau)])$, where

$$\gamma(\tau) = \lambda + \sum_{i=1}^{k} \frac{n_i}{n_i + \tau}, \quad \mu^{(1)}(\tau) = \frac{\lambda\mu^{(0)} + \sum_{i=1}^{k} \frac{n_i}{n_i+\tau} y_i}{\gamma(\tau)}.$$

The posterior for T cannot be given in closed form, but the density is proportional to $f_T(\tau)$ times

$$\tau^{\frac{k}{2}} \prod_{i=1}^{k}(\tau + n_i)^{-\frac{1}{2}} \gamma(\tau)^{-\frac{1}{2}} \exp\left(-\frac{\tau}{2}\left\{w(\tau) + \frac{\hat{\lambda}(\tau)\lambda}{\gamma(\tau)}(\mu^{(0)} - \hat{y}(\tau))^2\right\}\right),$$

where

$$\hat{\lambda}(\tau) = \sum_{i=1}^{k} \frac{n_i}{n_i + \tau}, \quad \hat{y}(\tau) = \frac{\sum_{i=1}^{k} \frac{n_i y_i}{n_i + \tau}}{\hat{\lambda}(\tau)}, \quad w(\tau) = \sum_{i=1}^{k} \frac{n_i}{n_i + \tau}(y_i - \hat{y}(\tau))^2.$$

Once again, if n_i is large for each i, then $n_i/(n_i + \tau) \approx 1$ and $n_i + \tau \approx n_i$ for each i. It follows that $\hat{\lambda}(\tau)$ is approximately k and that $\hat{y}(\tau)$ is approximately the average of the y_i, say $\bar{y}$. Hence, the posterior density of T is approximately proportional to

$$f_T(\tau)\tau^{\frac{k}{2}} \exp\left(-\frac{\tau}{2}\left[w + \frac{k\lambda}{k+\lambda}(\mu^{(0)} - \bar{y})^2\right]\right),$$

where $w = \sum_{i=1}^{k}(y_i - \bar{y})^2$. Also, the conditional posterior for M given $T = \tau$ is approximately

$$N\left(\frac{\lambda\mu^{(0)} + \sum_{i=1}^{k} y_i}{\lambda + k}, \frac{1}{\lambda + k}\right).$$

If T has a $\Gamma(a^{(0)}/2, b^{(0)}/2)$ prior, then the approximate posterior of T is $\Gamma(a^{(1)}/2, b^{(1)}/2)$, where

$$\begin{aligned} a^{(1)} &= a^{(0)} + k, \\ b^{(1)} &= b^{(0)} + w + \frac{k\lambda}{k+\lambda}(\mu^{(0)} - \bar{y})^2. \end{aligned}$$

Of course, using these same approximations, the conditional distribution of M_i given M and T would be $N(y_i, 1/k)$, which is independent of M and T anyway.

8.4 Empirical Bayes Analysis*

Classical statisticians try to make use of hierarchical models either by leaving the hyperparameters at various stages of the hierarchy unspecified or by not specifying a distribution for the hyperparameters at certain stages. This allows them to treat these values as "unknown parameters" in much the same way that they treat parameters in other models. For example, the hierarchical model in Table 8.9 could be altered by letting Ψ, T, and Σ be unknown parameters to be estimated without specifying distributions. In Table 8.18, we could let M, Σ_e^2, Σ_B^2, Σ_A^2, and Σ_{AB}^2 be the parameters by integrating the other parameters out the way we did in Section 8.2.2. A good introduction to empirical Bayes analysis was given by Morris (1983). Robbins (1951, 1955, 1964) first introduced the term "empirical Bayes" and the general methodology.

8.4.1 Naïve Empirical Bayes

The naïve approach to empirical Bayes analysis is to estimate the hyperparameters at some level of the hierarchical model and then pretend as if these were known a priori and use the resulting posterior distributions for parameters at lower levels in the hierarchy. For example, in the one-way ANOVA (see Table 8.9), we could use (8.12) to specify the joint density of the data given the parameters Ψ, T, and Σ. Then we could let $\Lambda = \Sigma^2/T^2$ so that the likelihood of Ψ, Σ^2, and Λ is

$$\prod_{i=1}^{k}\left(\frac{1}{n_i}+\frac{1}{\lambda}\right)^{-\frac{1}{2}} (\sigma^2)^{-\frac{n}{2}} \exp\left(-\frac{1}{2\sigma^2}\sum_{i=1}^{k}\left[\frac{(\overline{x}_i-\psi)^2}{\frac{1}{n_i}+\frac{1}{\lambda}}+(n_i-1)s_i^2\right]\right),$$

where $n = \sum_{i=1}^{k} n_i$. For fixed σ^2 and λ, this is maximized over ψ by choosing

$$\hat{\Psi}(\lambda) = \frac{1}{\sum_{i=1}^{k}\left(\frac{1}{n_i}+\frac{1}{\lambda}\right)^{-1}} \sum_{i=1}^{k} \frac{\overline{x}_i}{\frac{1}{n_i}+\frac{1}{\lambda}}.$$

If we plug this value for ψ into the likelihood and maximize over σ^2 for fixed λ, we get

$$\hat{\Sigma^2}(\lambda) = \frac{1}{n}\sum_{i=1}^{k}\left(\frac{[\overline{x}_i - \hat{\Psi}(\lambda)]^2}{\frac{1}{n_i}+\frac{1}{\lambda}}+(n_i-1)s_i^2\right).$$

---

*This section may be skipped without interrupting the flow of ideas.

If we plug this value for σ^2 into the likelihood, we get the following function of λ to maximize:
$$\prod_{i=1}^{k}\left(\frac{1}{n_i}+\frac{1}{\lambda}\right)^{-\frac{1}{2}}[\hat{\Sigma}^2(\lambda)]^{-\frac{n}{2}}. \tag{8.30}$$

This would produce the MLE of Λ, call it $\hat{\Lambda}$. Then set $\hat{\Sigma}^2 = \hat{\Sigma}^2(\hat{\Lambda})$ and $\hat{\Psi} = \hat{\Psi}(\hat{\Lambda})$ to get the overall MLEs. Then, we can make inference about the M_is by using the conditional distribution in (8.8).

In the special case in which all $n_i = m$, $\hat{\Psi}(\lambda) = \sum_{i=1}^{k} \overline{x}_i/k$ and is not a function of λ. Also $\hat{\Sigma}^2(\lambda)$ simplifies and the derivative of (8.30) can actually be set equal to zero to solve for the maximum. If we let $g = 1/m + 1/\lambda$, then
$$\hat{\Sigma}^2(\lambda) = (ng)^{-1}\sum_{i=1}^{k}(\overline{x}_i - \hat{\Psi})^2 + \frac{m-1}{n}\sum_{i=1}^{k}s_i^2,$$
and the derivative of the log of (8.30) becomes
$$-\frac{k}{2g} + \frac{n}{2\hat{\Sigma}^2(\lambda)}\frac{\sum_{i=1}^{k}(\overline{x}_i - \hat{\Psi})^2}{ng^2}. \tag{8.31}$$

Setting this equal to 0 gives $g = \sum_{i=1}^{k}(\overline{x}_i - \hat{\Psi})^2/[k\hat{\Sigma}^2(\lambda)]$. Solving for g yields g to equal a multiple of the usual F statistic for testing the hypothesis of no difference between groups:
$$g = \frac{(n-k)\sum_{i=1}^{k}(\overline{x}_i - \hat{\Psi})^2}{k(m-1)\sum_{i=1}^{k}s_i^2} = \frac{k-1}{km}F.$$

Of course, $g \geq 1/m$ is required. If $F < k/(k-1)$, the derivative in (8.31) is negative at $g = 1/m$, so the maximum occurs at $g = 1/m$. Hence, the MLE of Λ is
$$\hat{\Lambda} = \begin{cases} \frac{mk}{(k-1)F-k} & \text{if } F > \frac{k}{k-1}, \\ \infty & \text{otherwise.} \end{cases}$$

This means that $\hat{T}^2 = 0$ if $F \leq k/(k-1)$.

Example 8.32 (Continuation of Example 8.14; see page 487). Using the data in this example, we can calculate the likelihood function for Λ and maximize it. The maximum occurs at $\hat{\Lambda} = 2.614$. The other MLEs are $\hat{\Psi} = 21.78441$ and $\hat{\Sigma}^2 = 38.37032$. This makes $\hat{T}^2 = 14.67878$. Now, we could use (8.8) to say that the approximate distribution of M_i is
$$N\left(\frac{n_i\overline{x}_i + \hat{\Psi}\hat{\Lambda}}{\hat{\Lambda}+n_i}, \frac{\hat{\Sigma}^2}{n_i + \hat{\Lambda}}\right).$$

For the three groups, these distributions are respectively
$$N(26.6539, 3.04188), \quad N(18.8101, 2.62559), \quad N(19.8795, 2.17840).$$

Example 8.33 (Continuation of Example 7.60; see page 420). In this example, each observation is a pair (X_i, Y_i) that are conditionally IID $N(\mu_i, \sigma^2)$ and the pairs are conditionally independent given $(\Sigma, M_1, M_2, \ldots)$. Suppose that we model $(M_1, M_2, \ldots)$ as conditionally IID $N(\mu, \sigma^2/\lambda)$ given $(M, \Lambda) = (\mu, \lambda)$. The empirical Bayes approach might treat (Σ, M, Λ) as the parameter to be estimated by maximum likelihood. The likelihood function for these parameters is

$$f_{X|\Sigma,M,\Lambda}(x|\sigma,\mu,\lambda) = (2\pi)^{-n} \left(\frac{\lambda}{\lambda+2}\right)^{\frac{n}{2}} \sigma^{-2n} \times \exp\left(-\frac{1}{2\sigma^2}\left[\sum_{i=1}^n (x_i - y_i)^2\right.\right.$$

$$\left.\left. + \frac{2\lambda}{\lambda+2}\sum_{i=1}^n \left(\frac{x_i+y_i}{2} - \frac{\bar{x}+\bar{y}}{2}\right)^2 + \frac{2\lambda n}{\lambda+2}\left(\mu - \frac{\bar{x}+\bar{y}}{2}\right)^2\right]\right).$$

The MLE for M is $\hat{M} = (\overline{X} + \overline{Y})/2$. The MLE for Σ^2 as a function of λ is

$$\hat{\Sigma}^2(\lambda) = \frac{1}{2n}\left[\sum_{i=1}^n (X_i - Y_i)^2 + \frac{2\lambda}{\lambda+2}\sum_{i=1}^n \left(\frac{X_i+Y_i}{2} - \frac{\overline{X}+\overline{Y}}{2}\right)^2\right].$$

The MLE of Λ can be found from

$$\frac{\hat{\Lambda}}{\hat{\Lambda}+2} = \min\left\{1, \frac{\sum_{i=1}^n (X_i-Y_i)^2}{\sum_{i=1}^n \left(\frac{X_i+Y_i}{2} - \frac{\overline{X}+\overline{Y}}{2}\right)^2}\right\}.$$

Since the "observations" $(X_i+Y_i)/2$ are conditionally IID $N(\mu, \sigma^2[(1/2)+(1/\lambda)])$ given the parameters, it follows that

$$\frac{1}{n}\sum_{i=1}^n \left(\frac{X_i+Y_i}{2} - \frac{\overline{X}+\overline{Y}}{2}\right)^2 \xrightarrow{P} \frac{(\lambda+2)\sigma^2}{2\lambda}.$$

This implies that $\hat{\Lambda}$ is consistent and, in turn, that $\hat{\Sigma}^2(\hat{\Lambda})$ is consistent. The extra terms added due to the empirical Bayes analysis make $\hat{\Sigma}^2(\hat{\Lambda})$ consistent (relative to the empirical Bayes model).

It is not required that one use maximum likelihood estimates in a naïve empirical Bayes analysis. For example, in the one-way ANOVA example, we could use

$$\tilde{\Psi} = \frac{1}{n}\sum_{i=1}^k n_i \bar{x}_i, \qquad \tilde{\Sigma}^2 = \frac{1}{n-k}\sum_{i=1}^k (n_i-1)s_i^2,$$

$$\tilde{T}^2 = \max\left\{0, \frac{\sum_{i=1}^k n_i(\bar{x}_i - \tilde{\Psi})^2 - (k-1)\tilde{\Sigma}^2}{n - \frac{1}{n}\sum_{i=1}^k n_i^2}\right\},$$

which are based on unbiased estimators.

8.4.2 Adjusted Empirical Bayes

It is generally recognized that naïve empirical Bayes analyses underestimate the variances of parameters because they do not take into account the fact that estimated hyperparameters were not really known a priori. For example, in the empirical Bayes version of Example 8.14, we treat $\hat{\Psi}$ as if it were Ψ and were known a priori. To reflect the fact that we really do not know Ψ a priori, the posterior variance of M_i should be increased by

$$\frac{\Sigma^4}{(n_i T^2 + \Sigma^2)^2} \text{Var}(\Psi) = \frac{\Lambda^2}{(n_i + \Lambda)^2} \text{Var}(\Psi).$$

We would already have an estimate of Λ from the naïve analysis. We could use (8.11) with $\zeta_0 = 0$ and estimate $\text{Var}(\Psi)$ by

$$\left[\sum_{i=1}^{k} \frac{n_i}{\hat{\Sigma}^2 + n_i \hat{T}^2} \right]^{-1}.$$

The value of this estimate would depend on how we estimated T and Σ, of course. We should also increase the variance of M_i to reflect the fact that Σ and T were estimated. An easy way to do this is to replace the normal distribution in the posterior by a t distribution with appropriate degrees of freedom. Morris (1983) chooses, instead, to replace the naïve variance expression $\hat{\Sigma}^2/(n_i + \hat{\Lambda})$ by

$$\frac{\hat{\Sigma}^2}{n_i} \left(1 - \frac{k-1}{k} \frac{\hat{\Lambda}}{\hat{\Lambda} + n_i} \right). \tag{8.34}$$

This amounts to estimating the shrinkage factor $\Lambda/(\Lambda + n_i)$ by a smaller value.

Example 8.35 (Continuation of Example 8.14; see page 487). We can estimate $\text{Var}(\Psi)$ by

$$\left(\frac{10}{38.37032 + 10 \times 14.67878} + \frac{12}{38.37032 + 12 \times 14.67878} \right.$$
$$\left. + \frac{15}{38.37032 + 15 \times 14.67878} \right)^{-1} = 5.95368.$$

The additional variance terms for the three groups are 0.25568, 0.19048, and 0.13112, respectively. The adjustments specified by (8.34) are 3.30693, 2.81623, and 2.30494, respectively. Adding these together gives the adjusted variances to be 3.56261, 3.00671, and 2.43606, respectively, all somewhat larger than the naïve variances.

We might now ask how the adjusted empirical Bayes posteriors compare to the posteriors calculated from a hierarchical model with prior distributions for all parameters. Such a model exists in the original description on page 487. Plots of the posterior densities from these models were drawn in Figure 8.15 together with the adjusted empirical Bayes distributions. Two of the M_i have empirical Bayes distributions that are very close to the posteriors, but M_1 has a noticeably smaller variance in the empirical Bayes analysis than in the Bayesian analysis.

TABLE 8.36. Hierarchical Model for One-Way ANOVA with Unequal Variances

Stage	Density
Data	$(2\pi\sigma_i^2)^{-\frac{n}{2}} \exp\left\{-\frac{1}{2\sigma_i^2}\sum_{i=1}^{k}[n_i(\overline{x}_i - \mu_i)^2 + (n_i-1)s_i^2]\right\}$
Parameter	$(2\pi\tau^2)^{-\frac{k}{2}} \exp\left\{-\frac{1}{2\tau^2}\sum_{i=1}^{k}(\mu_i - \psi)^2\right\}$
Hyperparameter	$\sqrt{\zeta_0}(2\pi\tau^2)^{-\frac{1}{2}} \exp\left\{-\frac{\zeta_0}{2\tau^2}(\psi - \psi_0)^2\right\}$
Variance	$f_{\Sigma_1,\ldots,\Sigma_k,T}(\sigma_1,\ldots,\sigma_k,\tau)$

8.4.3 Unequal Variance Case

The case of a one-way ANOVA with unequal variances can also be handled by empirical Bayes analysis. Suppose that we begin with the model in Table 8.36, which is a generalization of the model of Section 8.2.1. The posterior mean of M_i for fixed values of the variance parameters and Ψ is

$$\hat{M}_i(\psi, \sigma_i, \tau) = \frac{\frac{n_i \overline{x}_i}{\sigma_i^2} + \frac{\psi}{\tau}}{\frac{n_i}{\sigma_i^2} + \frac{1}{\tau}}. \quad (8.37)$$

The posterior mean of Ψ for fixed values of the variance parameters is

$$\begin{aligned}\hat{\Psi}(\sigma_1,\ldots,\sigma_k,\tau) &= \left(\sum_{i=1}^{k} \frac{\overline{x}_i}{\frac{\sigma_i^2}{n_i} + \tau}\right)\left(\sum_{i=1}^{k} \frac{1}{\frac{\sigma_i^2}{n_i} + \tau}\right)^{-1} \\ &= \left(\sum_{i=1}^{k} \frac{n_i \overline{x}_i}{\sigma_i^2 + n_i\tau}\right)\left(\sum_{i=1}^{k} \frac{n_i}{\sigma_i^2 + n_i\tau}\right)^{-1}. \quad (8.38)\end{aligned}$$

The resulting likelihood function for $T, \Sigma_1, \ldots, \Sigma_k$ is τ^{-k} times

$$\prod_{i=1}^{k}\sigma_i^{-n_i} \exp\left(-\frac{1}{2}\sum_{i=1}^{k}\left\{\frac{(n_i-1)s_i^2}{\sigma_i^2} + \frac{n_i}{\sigma_i^2 + n_i\tau}(\overline{x}_i - \hat{\Psi}(\sigma_1,\ldots,\sigma_k,\tau))^2\right\}\right).$$

The MLEs of the variance parameters must be either found numerically or approximated. Morris (1983) suggests using approximately unbiased estimates instead. For example, $\hat{\Sigma}_i^2 = (n_i - 1)s_i^2/n_i$, and

$$\hat{T} = \frac{\sum_{i=1}^{k} \frac{n_i}{\hat{\Sigma}_i^2 + n_i \hat{T}}\left\{\frac{k}{k-1}(\overline{x}_i - \hat{\Psi}(\hat{\Sigma}_1,\ldots,\hat{\Sigma}_k,\hat{T}))^2 - \frac{\hat{\Sigma}_i^2}{n_i}\right\}}{\sum_{i=1}^{k} \frac{n_i}{\hat{\Sigma}_i^2 + n_i \hat{T}}}, \quad (8.39)$$

where (8.38) and (8.39) must be solved iteratively. One can choose a starting $\hat{T}$ value and plug it into (8.38) (together with $\hat{\Sigma}_1^2, \ldots, \hat{\Sigma}_k^2$) to produce a $\hat{\Psi}$ to plug into (8.39) to produce a new $\hat{T}$, and so on, until the estimates converge.[8] Morris (1983) also suggests replacing $\hat{M}_i$ by $(1-B_i)\bar{x}_i + B_i\hat{\Psi}(\hat{\Sigma}_1,\ldots,\hat{\Sigma}_k,\hat{T})$, where

$$B_i = \frac{k-3}{k-2}\frac{\hat{\Sigma}_i^2}{\hat{\Sigma}_i^2 + n_i\hat{T}^2}$$

causes there to be less shrinkage toward a common mean.[9] The recommended variance for M_i is given as

$$\frac{1}{n_i}\hat{\Sigma}_i^2\left[1-\left(1-\frac{\frac{n_i}{\hat{\Sigma}_i^2+n_i\hat{T}}}{\sum_{j=1}^{k}\frac{n_j}{\hat{\Sigma}_j^2+n_j\hat{T}}}\right)B_i\right]$$

$$+(\bar{x}_i-\hat{\Psi}(\hat{\Sigma}_1,\ldots,\hat{\Sigma}_k,\hat{T}))^2\frac{2}{k-3}B_i^2\frac{kn_i}{(\hat{\Sigma}_i^2+n_i\hat{T}^2)\sum_{j=1}^{k}\frac{n_j}{\hat{\Sigma}_j^2+n_j\hat{T}^2}}.$$

Kass and Steffey (1989) present an alternative treatment of this case from a Bayesian viewpoint. They find a normal approximation to the posterior distribution of the parameters $V = (\Sigma_1,\ldots,\Sigma_k,\Psi,T)$ (in a manner similar to the method of Laplace) and then use the delta method to approximate the mean and variance of (8.37), thought of as a function of V. The posterior variance of M_i is $E(\Sigma_i^2)/n_i$ plus the variance of (8.37).

8.5 Successive Substitution Sampling

The model analyzed in Section 8.2.2 is an example of one that got out of hand very quickly, even though it started out in a fairly straightforward manner. Another method for finding posterior distributions can be used for such models without getting bogged down in such messy calculation. The method is a simulation version of the method of *successive substitution* used to solve fixed-point problems.

8.5.1 The General Algorithm

In general, if $g : A \to A$, and we are interested in finding an x such that $g(x) = x$, we could proceed as follows. Pick $x_0 \in A$. For $n = 1, 2, \ldots$, define $x_n = g(x_{n-1})$. If $\{x_n\}_{n=1}^{\infty}$ converges and g is continuous, then the limit x is a *fixed point* of g, that is, $g(x) = x$.

---

[8] The algorithm described here is an example of *successive substitution*, which will be described in Section 8.5.

[9] In the case $k = 3$ there is no shrinkage, and when $k = 2$, I don't know what is recommended, although it seems clear from the formula for the adjusted variance that $k > 3$ is required for this analysis.

The type of fixed-point problem we will study is the following. Suppose that $Y_1, \ldots, Y_k$ are random quantities and that we know the conditional distribution of Y_i given the others, for each i. Suppose that the conditional distribution of Y_i given the others has density $f_{Y_i|\{Y_j : j \neq i\}}$ with respect to a measure λ_i. (It will prove convenient to use the notation $Y_{\setminus i}$ to stand for $\{Y_j : j \neq i\}$, so that this last density can be written $f_{Y_i|Y_{\setminus i}}$.) We wish to find the joint distribution of $(Y_1, \ldots, Y_k)$. Suppose that the joint distribution has density f_Y with respect to the product measure $\lambda = \lambda_1 \times \cdots \times \lambda_k$. Let $X' = (X'_1, \ldots, X'_k)$ have a distribution with density $f_{X'}$ with respect to λ. Define the distribution of a new random quantity $X = (X_1, \ldots, X_k)$ as follows. Suppose that $X' = (x'_1, \ldots, x'_k)$ is observed. The density of X_1 with respect to λ_1 is $f_{Y_1|Y_{\setminus 1}}(\cdot | x'_2, \ldots, x'_k)$. The conditional density of X_2 given $X_1 = x_1$ is $f_{Y_2|Y_{\setminus 2}}(\cdot | x_1, x'_3, \ldots, x'_k)$. Continue until we get the conditional density of X_k given $X_1 = x_1, \ldots, X_{k-1} = x_{k-1}$ to be $f_{Y_k|Y_{\setminus k}}(\cdot | x_1, \ldots, x_{k-1})$. In words, when we derive the conditional distribution of X_j given $X_1, \ldots, X_{j-1}$, we use the observed values of $X'_{j+1}, \ldots, X'_k$ in the conditional distributions. When we get to $j = k$, we are using only the X_i values. If we define $z^1 = (x'_2, \ldots, x'_k)$, $z^i = (x_1, \ldots, x_{i-1}, x'_{i+1}, \ldots, x'_k)$ for $i = 2, \ldots, k-1$, and $z^k = (x_1, \ldots, x_{k-1})$, then the following equation is satisfied:

$$f_X(x) = \int \left[\prod_{i=1}^k f_{Y_i|Y_{\setminus i}}(x_i | z^i)\right] f_{X'}(x') d\lambda(x').$$

We can define the operator T from the set of densities with respect to λ to itself by

$$T(f)(x) = \int \left[\prod_{i=1}^k f_{Y_i|Y_{\setminus i}}(x_i | z^i)\right] f(x') d\lambda(x').$$

It is easy to see that $T(f_Y) = f_Y$, so the joint density of Y is a fixed point of T.

The method of successive substitution applied to the fixed-point problem just described would be to pick an initial density f_0, say, and then let $f_n = T(f_{n-1})$ for $n = 1, 2, \ldots$. This would require the calculation of a great many integrals that may not have closed-form expressions. An alternative is to draw samples from the various conditional distributions instead of calculating the integrals. In the notation just used, suppose that $X' = (x'_1, \ldots, x'_k)$ is generated from the distribution with density $f_{X'}$. Then suppose that $X = (X_1, \ldots, X_k)$ is generated as follows. Generate X_1 from the distribution with density $f_{Y_1|Y_{\setminus 1}}(\cdot | x'_2, \ldots, x'_k)$. Let x_1 be the generated value. Generate X_2 from the distribution with density $f_{Y_2|Y_{\setminus 2}}(\cdot | x_1, x'_3, \ldots, x'_k)$. Continue until we generate X_k from the distribution with density $f_{Y_k|Y_{\setminus k}}(\cdot | x_1, \ldots, x_{k-1})$. The joint density of X is $T(f_{X'})$. So, we can take a starting density f_0 and generate X^0 from the distribution with this density. Then, using the method just described for $n = 1, 2, \ldots$, generate X^n from the distribution with density $T(f_{n-1})$. This method has been called *successive substitution sampling* (abbreviated SSS) because it is just a sampling version of successive substitution.[10]

One must, of course, stop the iteration at some point using the sample with density $T(f_n)$ in lieu of a sample with density f_Y. There are several ways to prove that SSS converges as n goes to infinity. The following theorem is proven by Schervish and Carlin (1992). Its proof, which is given for completeness, relies heavily on operator theory in Hilbert space.[11] Readers unfamiliar with this theory can safely skip over the proof. The necessary theorems from operator theory are stated in Appendix C.[12]

Theorem 8.40. *In the notation of this section, let*

$$K(x', x) = \prod_{i=1}^{k} f_{Y_i | Y_{\setminus i}}(x_i | z^i).$$

Assume that

$$\int \int |K(x', x)|^2 \frac{f_Y(x')}{f_Y(x)} d\lambda(x') d\lambda(x) < \infty \tag{8.41}$$

and that $K > 0$ almost everywhere with respect to $\lambda \times \lambda$. Let $\mathcal{H}$ be the set of functions f such that[13] $\|f\|^2 = \int |f(x)|^2 / f_Y(x) d\lambda(x) < \infty$. *There exists a number $c \in [0, 1)$ such that for every density $f_0 \in \mathcal{H}$, the sequence of functions $f_n = T(f_{n-1}) = T^n(f_0)$ for $n = 1, 2, \ldots$ satisfies $\|f_n - f_Y\| \leq \|f_0\| c^n$ for all n.*

---

[10] Many authors call this method *Gibbs sampling*. This is actually a misnomer. Geman and Geman (1984) described this method as a way to generate a sample from a Gibbs distribution, and they called their particular implementation the Gibbs sampler. Gelfand and Smith (1990) generalized the method to arbitrary distributions but continued to call it Gibbs sampling, even though they were no longer sampling Gibbs distributions. The SSS algorithm is a special case of the broad class of *Markov chain Monte Carlo* methods. Note that the sequence $\{X^n\}_{n=0}^{\infty}$ is a Markov chain (see Definition B.125). A good survey of general Markov Chain Monte Carlo methods is given by Tierney (1994).

[11] An alternative is to notice that the sequence $X^1, X^2, \ldots$ is a Markov chain (see Definition B.125 on page 650). One then applies a theorem like the one given by Doob (1953, Section V.5). The conditions of such theorems are often difficult, if not impossible, to verify in specific applications.

[12] Some good treatments of operator theory can be found in Berberian (1961) and Dunford and Schwartz (1963).

[13] We use the symbol $\|f\|$ for the norm of an element of a Hilbert space. The norm $\|T\|$ of an operator T is the supremum of $\|T(f)\|/\|f\|$. Dunford and Schwartz (1963) use the symbols $|f|$ and $|T|$ for these norms. They use the symbol $\|T\|$ for the Hilbert–Schmidt norm or *double norm* of a Hilbert–Schmidt-type operator. We only mention this here in case the reader decides to refer to Dunford and Schwartz (1963) for some of the proofs of auxiliary results.

PROOF. We will use Hilbert space notation and define the inner product

$$\langle g, h \rangle = \int g(x)h(x)d\mu(x),$$

where $\mu(A) = \int_A [1/f_Y(x)]d\lambda(x)$. It follows that $\mathcal{H}$ is the Hilbert space $L^2(\mu)$. The norm in this space is $\|g\| = \sqrt{\langle g, g \rangle}$. If we let $K_0(x', x) = K(x', x)f_Y(x')$, then $T(f)(x) = \int K_0(x', x)d\mu(x')$ is the operator that takes a density for observations at one iteration of SSS to the density of observations as the next iteration, and (8.41) becomes

$$\int \int |K_0(x', x)| d\mu(x') d\mu(x) < \infty.$$

In fact, it is clear that $K_0(x', x)$ is a joint density of two successive iterations of SSS, x' and x, if the first iteration has the solution density f_Y. Furthermore, by writing each of the conditional density factors in K_0 as the ratio of the joint density f_Y to a joint density for all but one of the observations, and then rearranging the factors, one can show that $T^*(f)(x) = \int K_0(x, x')d\mu(x')$ is the operator that takes a density for observations at one iteration of SSS to the density of observations at the next iteration if the order of updating coordinates is reversed. For this reason, it is easy to see that for each g and h in $\mathcal{H}$ that are integrable with respect to λ,

$$\int g(x)d\lambda(x) = \int T(g)(x)d\lambda(x),$$
$$\int g(x)d\lambda(x) = \int T^*(g)(x)d\lambda(x),$$
$$\int T(g)(x)h(x)d\mu(x) = \int g(x)T^*(h)(x)d\mu(x).$$

The last equation is the definition of what it means to say that T^* is the *adjoint* of the operator T. It also follows from this equation that the adjoint of the composition $U = T(T^*)$ is itself, U. That is to say, U is *self-adjoint*. Since U is two applications of successive substitution, it follows that

$$\int f(x)d\lambda(x) = \int U(f)(x)d\lambda(x). \tag{8.42}$$

According to Theorem C.10 the operator T is of Hilbert–Schmidt type because (8.41) holds. Theorem C.11 says that such an operator is completely continuous.[14] It follows then that the adjoint operator T^* is also

---

[14] An operator T is *completely continuous* if every bounded set $B \subseteq \mathcal{H}$ is mapped by T to a set whose closure is sequentially compact. (That is, every sequence in $T(B)$ has a convergent subsequence.)

completely continuous as is U. Since U is self-adjoint and completely continuous, $\mathcal{H}$ has an orthonormal basis of eigenfunctions of U. Also, Theorem C.12 says that a self-adjoint completely continuous operator has an eigenvalue whose absolute value is equal to the norm of the operator.

Let V be the operator defined by $V(f) = U(f) - f_Y \langle f_Y, f \rangle$. In particular, $V(f_Y) = 0$ because $T(f_Y) = T^*(f_Y) = f_Y$ and $\langle f_Y, f_Y \rangle = 1$. It is easy to see that $V = W^*W$, where $W(f) = T(f) - f_Y \langle f_Y, f \rangle$, and W^* is the adjoint of W. It follows from Theorem C.13 that $\|V\| = \|W^*W\| = \|W\|^2$. The remainder of the proof will be to show that $\|V\|$ and hence $\|W\|$ are strictly less than 1, and then to show that this implies the conclusion to the theorem.

Since V is self-adjoint and completely continuous, we can show that $\|V\| < 1$ by showing that the absolute value of its largest eigenvalue is strictly less than 1. Let r be the largest eigenvalue of V, which is real since V is self-adjoint. Let $V(g) = rg$. If $r = 0$, the result holds, so suppose that $r \neq 0$. Then

$$\langle g, f_Y \rangle = \frac{1}{r} \langle V(g), f_Y \rangle = \frac{1}{r} \langle g, V(f_Y) \rangle = 0,$$

since $V(f_Y) = 0$. Since g is not identically 0, we can write $g = g^+ - g^-$ where g^+ and g^- are respectively the positive and negative parts of g. Let B be the set of x such that $g(x) > 0$, and let C be the set of x such that $g(x) < 0$. Then $\lambda(B) > 0$ and $\lambda(C) > 0$ since $\langle f_Y, g \rangle = 0$ but g is not identically 0. We will show that $|r| < 1$ by means of contradiction. If $|r| = 1$, then

$$V(g) = U(g^+ - g^-) = U(g^+) - U(g^-) = \begin{cases} g^+ - g^- & \text{if } r = 1, \\ g^- - g^+ & \text{if } r = -1. \end{cases}$$

Since $K_0 > 0$, it follows that $U(g^+)(x) > 0$ and $U(g^-)(x) > 0$ for all x. Hence,

$$g^+(x) < \begin{cases} U(g^+)(x) & \text{if } r = 1, \\ U(g^-)(x) & \text{if } r = -1, \end{cases} \text{ for } x \in B,$$

$$g^-(x) < \begin{cases} U(g^-)(x) & \text{if } r = 1, \\ U(g^+)(x) & \text{if } r = -1, \end{cases} \text{ for } x \in C.$$

It follows that $U(g^+ + g^-) > g^+ + g^-$ for all x. In other words, $U(|g|) > |g|$, which would imply $\int U(|g|) d\lambda > \int |g| d\lambda$, which contradicts (8.42). Hence, $|r| < 1$ and the largest eigenvalue of V has absolute value $|r| < 1$. It follows that $\|V\| = |r| < 1$.

Now, we know that $\|W\| = |r|^{1/2} = c < 1$. If f is a density, then $\langle f_Y, f \rangle = 1$ and

$$W(f) = T(f) - f_Y = T(f - f_Y).$$

Similarly, if $\langle f_Y, g \rangle = 0$, then $W(g) = T(g)$ and $\langle f_Y, W(g) \rangle = 0$, from which it follows that $W^n(g) = T^n(g)$ for all n. Since $\langle f_Y, f - f_Y \rangle = 0$ for every density f, it follows that, for all n,

$$W^n(f) = W^n(f - f_Y) = T^n(f - f_Y) = T^n(f) - f_Y.$$

So, for all n,

$$\|T^n(f_0) - f_Y\| = \|W^n(f_0)\| \leq \|W\|^n \|f_0\| = c^n \|f_0\|. \qquad \square$$

Although it appears that one needs to know the solution f_Y in order to check the conditions of this theorem, one often knows the function f_Y up to a multiplicative constant. Hence, one could, at least in principle, check the finiteness of the various integrals in Theorem 8.40.

Example 8.43. Suppose that the posterior density of (Y_1, Y_2, Y_3) is proportional to

$$f(y) = y_3^{-4} \exp\left(-\frac{1}{2y_3}\left\{\frac{(y_2 - 0.9y_1)^2}{0.19} + y_1^2 + 4\right\}\right).$$

It is not difficult to see that the three conditional distributions are

$$\begin{aligned}
Y_1 | Y_2 = y_2, Y_3 = y_3 &\sim N(0.9y_2, 0.19y_3), \\
Y_2 | Y_1 = y_1, Y_3 = y_3 &\sim N(0.9y_1, 0.19y_3), \\
Y_3 | Y_1 = y_1, Y_2 = y_2 &\sim \Gamma^{-1}\left(3, \frac{4 + y_1^2 + \frac{(y_2 - 0.9y_1)^2}{0.19}}{2}\right).
\end{aligned}$$

The integrand in (8.41) is a constant times $x_3'^{6} x_3^{-5}$ times e to the power

$$-\frac{1}{2x_3'}\left\{4 + \frac{(x_1 - 0.9x_2')^2}{0.19} + \frac{(x_2 - 0.9x_1)^2}{0.19} + x_1'^2 + \frac{(x_2' - 0.9x_1')^2}{0.19}\right\}$$

$$-\frac{1}{2x_3}\left\{4 + x_1^2 + \frac{(x_2 - 0.9x_1)^2}{0.19}\right\}.$$

By collecting terms here, it is not difficult to show that this function is integrable over the six variables $x_1, x_2, x_3, x_1', x_2', x_3'$.

After one stops the iteration, one has a vector Y from approximately the correct distribution. One can repeat the process and produce $Y^1, \ldots, Y^m$ for some value m. If one wants the marginal density of Y_i, one can let $Y_{\setminus i}^s$ stand for the $(k-1)$-dimensional vector formed from Y^s by removing Y_i^s, and then calculate

$$\hat{f}_{Y_i}(y) = \frac{1}{m} \sum_{s=1}^{m} f_{Y_i | Y_{\setminus i}}(y | Y_{\setminus i}^s). \qquad (8.44)$$

This estimator is based on the simple fact that, for each s,

$$f_Y(y) = \mathrm{E}\left(f_{Y_i|Y_{\setminus i}}\left(y|Y_{\setminus i}^s\right)\right).$$

If one wants the mean of Y_i, one can calculate

$$\hat{Y}_i = \frac{1}{m}\sum_{s=1}^{m} \mathrm{E}(Y_i|Y_{\setminus i} = Y_{\setminus i}^s), \qquad (8.45)$$

assuming that the conditional mean of Y_i given the others is easily available. Equation (8.45) should be better than $\sum_{s=1}^{m} Y_i^s/m$, since the variance of the simple average is the variance of (8.45) plus $1/m$ times the mean of the conditional variance of Y_i given $Y_{\setminus i}$. Similarly, the variance of Y_i can be approximated by

$$\frac{1}{m}\sum_{s=1}^{m} \mathrm{Var}(Y_i|Y_{\setminus i} = Y_{\setminus i}^s) + \frac{1}{m}\sum_{s=1}^{m}(\mathrm{E}(Y_i|Y_{\setminus i} = Y_{\setminus i}^s) - \hat{Y}_i)^2, \qquad (8.46)$$

which should be a better estimate than the sample variance of the Y_i^s.

The SSS algorithm, as described, assumes that the random quantities $Y_1, \ldots, Y_k$ are in a fixed order for every iteration. This is not actually required for convergence of the algorithm. The proof of convergence is simplified by making this assumption however. Note also that each Y_i need not be a single random variable. Some of them might themselves be vectors. The question of how to arrange the coordinates is important for the rate of convergence. The more dependence that exists between successive iterations, the slower the convergence will be. One can understand why this is true intuitively by realizing that convergence "occurs" when an iteration is "independent" of the starting iteration. The more dependence lingers from one iteration to the next, the longer it takes to get an iteration that is essentially independent of the start. Example 8.43 can be used to illustrate how the choice of coordinate arrangement affects the dependence between iterates.

Example 8.47 (Continuation of Example 8.43; see page 510). It is not difficult to see that every order of the three coordinates is essentially equivalent in this example. Instead, let us compare the natural order Y_1, Y_2, Y_3 to the alternative arrangement $X_1 = (Y_1, Y_2)^\top$, $X_2 = Y_3$. That is, let the first random quantity be a two-dimensional vector consisting of both Y_1 and Y_2. To illustrate the effect of this change on the amount of dependence between iterations, we will calculate the conditional distribution of Y_1 at the next iteration given the variables at the current iteration for both arrangements.

In the natural order, we generate Y_1 with $N(0.9y_2', 0.19y_3')$ distribution given $Y_1' = y_1'$, $Y_2' = y_2'$, and $Y_3' = y_3'$. In the vector arrangement, we generate the whole vector (Y_1, Y_2) at once with distribution $N_2(\mathbf{0}, y_3' A)$ where

$$A = \begin{pmatrix} 1 & 0.9 \\ 0.9 & 1 \end{pmatrix}.$$

The conditional distribution of Y_1 given $Y_1' = y_1', Y_2' = y_2'$, and $Y_3' = y_3'$ is $N(0, y_3')$ in this case. The dependence on the previous iteration is greatly reduced in the vector arrangement. In the natural order, Y_1 is much more constrained by the values y_2' and y_3' than in the vector arrangement. A similar calculation shows that, in the natural order, the conditional distribution of Y_2 at the next iteration given $Y_1' = y_1', Y_2' = y_2'$, and $Y_3' = y_3'$ is $N(0.81 y_2', 0.3439 y_3')$, while in the vector arrangement it is $N(0, y_3')$. Although Y_2 is less dependent on the previous iteration than is Y_1, it is still more dependent in the natural order than in the vector arrangement.

As a rule of thumb, if one knows that several random variables are highly dependent, it will be better, if possible, to treat them as a single random quantity in the SSS algorithm rather than to treat each one as a separate coordinate.

8.5.2 Normal Hierarchical Models

Take the model in Section 8.2.1 as an example. The vector Y in the discussion of this section will be the collection of all parameters of the model, namely $M_1, \ldots, M_k, \Psi, \Sigma^2, T^2$. We will use the prior in which Σ^2 and T^2 are independent with inverse gamma distributions $\Gamma^{-1}(a_0/2, b_0/2)$ and $\Gamma^{-1}(c_0/2, d_0/2)$, respectively. All distributions will be conditional on the data. It is easy to calculate the various conditional distributions we need. In the following list, each distribution is to be understood as conditional on both the data and on all of the other parameters.

$$M_i \sim N\left(\frac{\frac{n_i \bar{x}_i}{\sigma^2} + \frac{\psi}{\tau^2}}{\frac{n_i}{\sigma^2} + \frac{1}{\tau^2}}, \frac{1}{\frac{n_i}{\sigma^2} + \frac{1}{\tau^2}}\right)$$

$$\Psi \sim N\left(\frac{\sum_{i=1}^k \mu_i + \zeta_0 \psi_0}{k + \zeta_0}, \frac{\tau^2}{k + \zeta_0}\right)$$

$$T^2 \sim \Gamma^{-1}\left(\frac{c_0 + k + 1}{2}, \frac{d_0 + \sum_{i=1}^k (\mu_i - \psi)^2 + \zeta_0(\psi - \psi_0)^2}{2}\right)$$

$$\Sigma^2 \sim \Gamma^{-1}\left(\frac{a_0 + n_1 + \cdots + n_k}{2}, \frac{b_0 + \sum_{i=1}^k \{(n_i - 1)s_i^2 + n_i(\bar{x}_i - \mu_i)^2\}}{2}\right)$$

It is easy to generate pseudorandom numbers from each of the above distributions, so that SSS could be implemented without much trouble.

Example 8.48 (Continuation of Example 8.14; see page 487). We used naïve empirical Bayes estimates of the parameters as starting values and then ran twenty thousand iterations, taking every twentieth iteration as a sampled value. We also ran forty thousand iterations where we took every fortieth iteration as a sampled value. The differences were negligible. Then we calculated (8.44) for each of the three population means. These densities are plotted in Figure 8.15. The respective posterior means were calculated to be 26.30, 18.78, and 19.82 using (8.45). The posterior variances were calculated using (8.46) to be 4.525, 2.998, and 2.387, respectively.

The SSS algorithm is also well suited to handle the case in which each population has its own variance, Σ_i^2. Suppose that the Σ_i^2 are independent with $\Gamma^{-1}(a_0/2, b_0/2)$ distribution in the prior. In this case, we replace two of the above distributions with

$$M_i \sim N\left(\frac{\frac{n_i \bar{x}_i}{\sigma_i^2} + \frac{\psi}{\tau^2}}{\frac{n_i}{\sigma_i^2} + \frac{1}{\tau^2}}, \frac{1}{\frac{n_i}{\sigma_i^2} + \frac{1}{\tau^2}}\right),$$

$$\Sigma_i^2 \sim \Gamma^{-1}\left(\frac{a_0 + n_i}{2}, \frac{b_0 + (n_i - 1)s_i^2 + n_i(\bar{x}_i - \mu_i)^2}{2}\right).$$

This model has $k - 1$ more parameters than the equal variance model.

An intermediate case between equal variance and independent variances is to have a hierarchical model for the variances. Suppose that the Σ_i^2 are conditionally independent with $\Gamma^{-1}(a_0/2, a_0\sigma^2/2)$ distribution given $\Sigma^2 = \sigma^2$. Then suppose that the prior for Σ^2 is $\Gamma(f_0/2, g_0/2)$. This model has k more parameters than the equal variance model, and the conditional distributions required for SSS must change to include the M_i distributions just given for the other unequal variance model together with

$$\Sigma_i^2 \sim \Gamma^{-1}\left(\frac{a_0 + n_i}{2}, \frac{a_0\sigma^2 + (n_i - 1)s_i^2 + n_i(\bar{x}_i - \mu_i)^2}{2}\right),$$

$$\Sigma^2 \sim \Gamma\left(\frac{f_0 + ka_0}{2}, \frac{g_0 + a_0 \sum_{i=1}^{k} \sigma_i^{-2}}{2}\right).$$

Example 8.49. Suppose that we use the same data as in Example 8.14 on page 487, but we use the hierarchical model for the population variances. We continue to use $a_0 = 1$, $c_0 = 1$, $d_0 = 1$, $\psi_0 = 10$, and $\zeta_0 = 0.1$, but we include $f_0 = 1$ and $g_0 = 0.1$. The resulting posterior densities are plotted in Figure 8.50. The posterior means and variances from (8.45) and (8.46) are

i	1	2	3
mean	27.1392	18.7672	19.7399
variance	3.1634	4.6745	2.2667

We could also do a naïve empirical Bayes analysis. (Recall that the adjusted empirical Bayes analysis requires $k > 3$.) We will use $\hat{\Sigma}_i^2 = s_i^2$. We will also adjust the variance of M_i by adding on $\hat{T}^2$ times the square of the coefficient of ψ in (8.37). This results in M_i having variance

$$\hat{\Sigma}_i^2 \left(\frac{1}{n_i} + \frac{\hat{T}^2 \hat{\Sigma}_i^2}{(\hat{\Sigma}_i^2 + n_i \hat{T}^2)^2}\right).$$

We need to iterate between (8.38) and (8.39). Starting with $\hat{T} = 1$, it took four iterations to get no difference between the iterations. The results were $\hat{\Psi} = 21.9809$

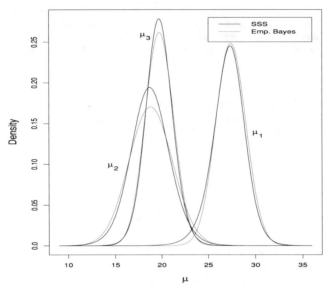

FIGURE 8.50. Numerical Approximations to Posterior Densities

and $\hat{T} = 23.4547$. This leads to the following naïve empirical Bayes posteriors: $N(27.3786, 2.5817)$, $N(18.8116, 5.4845)$, and $N(19.7526, 2.3257)$. These three densities are also plotted in Figure 8.50. Notice that the hierarchical model for the variances brings the estimated variance for the second population (the largest of the three) down quite a bit from the empirical Bayes value while it brings the other two variances up. Although the hierarchical model variance for M_3 is slightly larger than the empirical Bayes variance, the density (in Figure 8.50) is more peaked. The additional variance comes from heavier tails.

The more complicated two-way ANOVA described in Section 8.2.2 can be handled using SSS in a much simpler fashion than the analytical discussion in Section 8.2.2. Let the parameters be M, A_i, B_j, $(AB)_{i,j}$ (for $i = 1, \ldots, a$ and $j = 1, \ldots, b$), Σ_e^2, Σ_A^2, Σ_B^2, and Σ_{AB}^2. We also have the constraints (8.17). Because of the constraints, we cannot apply SSS in the most naïve manner. For example, the random variables $B_1, \ldots, B_b$ have the property that the conditional distribution of each one given the others is concentrated on a single value (namely, minus the sum of the others) with probability one. This is an extreme case of dependence. No matter what starting values one generates for $B_1, \ldots, B_b$, one will never change them, no matter how many iterations one performs! Clearly, convergence cannot occur in this case.[15] There are two ways to circumvent this problem. One is to drop one of the parameters from the algorithm and just calculate it when needed. That is, just calculate $B_b = -\sum_{j=1}^{b-1} B_j$ when it appears in

---

[15]In fact, the condition that the function K be strictly positive in Theorem 8.40 is violated.

some conditional distribution, but treat $B_1, \ldots, B_{b-1}$ as the parameters. Another approach is to treat the entire vector $B = (B_1, \ldots, B_b)$ as one parameter (one of the Y_i) as in the vector arrangement in Example 8.47 on page 511. Also, $(AB)_i = ((AB)_{i,1}, \ldots, (AB)_{i,b})$ could be treated as one parameter for each i. We choose the vector approach here because the constraints introduce dependence among the coordinates, which will slow down convergence.

Suppose that our model says that M, $A_1, \ldots, A_a$, B, $(AB)_1, \ldots, (AB)_a$ are all conditionally independent given $\Sigma_e^2, \Sigma_A^2, \Sigma_B^2$, and Σ_{AB}^2 with

$$M \sim N\left(\theta_0, \tfrac{\sigma_B^2}{\tau}\right), \qquad B \sim N_b\left(0, \sigma_B^2 \left[I - \tfrac{1}{b}\mathbf{1}\mathbf{1}^\top\right]\right),$$
$$A_i \sim N(0, \sigma_A^2), \qquad (AB)_i \sim N_b\left(0, \sigma_{AB}^2 \left[I - \tfrac{1}{b}\mathbf{1}\mathbf{1}^\top\right]\right).$$

This produces the same model as in Section 8.2.2. Next, assume that the variance parameters are independent with inverse gamma distributions

$$\Sigma_e^2 \sim \Gamma^{-1}\left(\tfrac{a_e}{2}, \tfrac{b_e}{2}\right), \qquad \Sigma_A^2 \sim \Gamma^{-1}\left(\tfrac{a_A}{2}, \tfrac{b_A}{2}\right),$$
$$\Sigma_B^2 \sim \Gamma^{-1}\left(\tfrac{a_B}{2}, \tfrac{b_B}{2}\right), \qquad \Sigma_{AB}^2 \sim \Gamma^{-1}\left(\tfrac{a_{AB}}{2}, \tfrac{b_{AB}}{2}\right).$$

Note that the prior distributions of the constrained parameters are the conditional distributions of independent random variables given that the constraint holds. That is, for example, if $B_1, \ldots, B_b$ were IID $N(0, \sigma_B^2)$ and we found the conditional distribution of B given that $\sum_{j=1}^{b} B_j = 0$, that conditional distribution would be the distribution given above for B. (See Problem 13 on page 535.) Also, if $\tau = b$, then this model is the same as saying that the B_j are IID $N(0, \sigma_B^2)$ and $M = \sum_{j=1}^{b} B_j / b$.

The posterior distributions of the parameters conditional on the other parameters are now easily calculated after we introduce some notation. Suppose that there are $n_{i,j}$ observations with the A factor at level i and the B factor at level j. We do not need to assume that the cells all have the same sample size as we did in Section 8.2.2. In fact, there can even be empty cells in this analysis. Define

$$n_{.,j} = \sum_{i=1}^{a} n_{i,j}, \qquad \overline{y}_{.,j,.} = \tfrac{1}{n_{.,j}} \sum_{i=1}^{a} \sum_{k=1}^{n_{i,j}} y_{i,j,k},$$
$$n_{i,.} = \sum_{j=1}^{b} n_{i,j}, \qquad \overline{y}_{i,.,.} = \tfrac{1}{n_{i,.}} \sum_{j=1}^{b} \sum_{k=1}^{n_{i,j}} y_{i,j,k},$$
$$n_{.,.} = \sum_{i=1}^{a} \sum_{j=1}^{b} n_{i,j}, \qquad \overline{y}_{.,.,.} = \tfrac{1}{n_{.,.}} \sum_{j=1}^{b} \sum_{i=1}^{a} \sum_{k=1}^{n_{i,j}} y_{i,j,k},$$
$$\overline{\alpha}_{.} = \tfrac{1}{n_{.,.}} \sum_{i=1}^{a} n_{i,.}\alpha_i, \qquad \overline{y}_{i,j,.} = \tfrac{1}{n_{i,j}} \sum_{k=1}^{n_{i,j}} y_{i,j,k},$$
$$\overline{\alpha}_j = \tfrac{1}{n_{.,j}} \sum_{i=1}^{a} n_{i,j}\alpha_i, \qquad \overline{(\alpha\beta)}_{.,.} = \tfrac{1}{n_{.,.}} \sum_{i=1}^{a} \sum_{j=1}^{b} n_{i,j}(\alpha\beta)_{i,j},$$
$$\overline{\beta}_{.} = \tfrac{1}{n_{.,.}} \sum_{j=1}^{b} n_{.,j}\beta_j, \qquad \overline{(\alpha\beta)}_{.,j} = \tfrac{1}{n_{.,j}} \sum_{i=1}^{a} n_{i,j}(\alpha\beta)_{i,j},$$
$$\overline{\beta}_i = \tfrac{1}{n_{i,.}} \sum_{j=1}^{b} n_{i,j}\beta_j, \qquad \overline{(\alpha\beta)}_{i,.} = \tfrac{1}{n_{i,.}} \sum_{j=1}^{b} n_{i,j}(\alpha\beta)_{i,j},$$
$$w = \sum_{i=1}^{a} \sum_{j=1}^{b} \sum_{k=1}^{n_{i,j}} (y_{i,j,k} - \overline{y}_{i,j,.})^2.$$

By our seeking the conditional posterior distribution of one parameter given the others, the data can be viewed as having a simple structure. For example, if we want the conditional posterior distribution of M given the other parameters, we construct $Y^*_{i,j,k} = Y_{i,j,k} - A_i - B_j - (AB)_{i,j}$. Then the $Y^*_{i,j,k}$ are IID $N(\mu, \sigma_e^2)$ given $M = \mu$. The conditional posterior distribution of M can easily be shown to be

$$M \sim N\left(\frac{\frac{n_{\cdot,\cdot,\cdot}(\overline{y}_{\cdot,\cdot,\cdot} - \overline{\alpha}_{\cdot} - \overline{\beta}_{\cdot} - \overline{(\alpha\beta)}_{\cdot,\cdot})}{\sigma_e^2} + \frac{\tau\theta_0}{\sigma_B^2}}{\frac{n_{\cdot,\cdot,\cdot}}{\sigma_e^2} + \frac{\tau}{\sigma_B^2}}, \frac{\sigma_B^2 \sigma_e^2}{n_{\cdot,\cdot,\cdot}\sigma_B^2 + \tau\sigma_e^2}\right).$$

A similar analysis works for each A_i. The result is

$$A_i \sim N\left(\frac{\frac{n_{i,\cdot,\cdot}(\overline{y}_{i,\cdot,\cdot} - \mu - \overline{\beta}_{\cdot} - \overline{(\alpha\beta)}_{i,\cdot})}{\sigma_e^2}}{\frac{n_{i,\cdot,\cdot}}{\sigma_e^2} + \frac{1}{\sigma_A^2}}, \frac{\sigma_A^2 \sigma_e^2}{n_{i,\cdot,\cdot}\sigma_A^2 + \sigma_e^2}\right).$$

Similarly, if we want the conditional posterior of the vector $(AB)_i$ given the other parameters, we construct $Y^*_{i,j,k} = Y_{i,j,k} - M - A_i - B_j$. Then the $Y^*_{i,j,k}$ are independent with $Y^*_{i,j,k}$ having $N((\alpha\beta)_{i,j}, \sigma_e^2)$ distribution given $(AB)_i = ((\alpha\beta)_{i,1}, \ldots, (\alpha\beta)_{i,b})^\top$. The posterior can be found by first calculating the posterior as if the parameters were unconstrained and then conditioning on the constraint as in Problem 13 on page 535. Define the b-dimensional vectors

$$z^i = \frac{1}{\sigma_e^2}\begin{pmatrix} n_{i,1}(\overline{y}_{i,1,\cdot} - \mu - \alpha_i - \beta_1) \\ \vdots \\ n_{i,b}(\overline{y}_{i,b,\cdot} - \mu - \alpha_i - \beta_b) \end{pmatrix}, \quad v^i = \begin{pmatrix} \frac{\sigma_e^2 \sigma_{AB}^2}{\sigma_e^2 + n_{i,1}\sigma_{AB}^2} \\ \vdots \\ \frac{\sigma_e^2 \sigma_{AB}^2}{\sigma_e^2 + n_{i,b}\sigma_{AB}^2} \end{pmatrix}.$$

Let $\mathrm{diag}(v^i)$ stand for the diagonal matrix with diagonal entries equal to the coordinates of v^i. Then the conditional posterior of $(AB)_i$ is $N_b(C^i z^i, C^i)$, where $C^i = \mathrm{diag}(v^i) - v^i v^{i\top}/(\mathbf{1}^\top v^i)$. A similar analysis works for the B vector. In this case, define the vectors

$$z = \frac{1}{\sigma_e^2}\begin{pmatrix} n_{\cdot,1}(\overline{y}_{\cdot,1,\cdot} - \mu - \overline{\alpha}_1 - \overline{(\alpha\beta)}_{\cdot,1}) \\ \vdots \\ n_{\cdot,b}(\overline{y}_{\cdot,b,\cdot} - \mu - \overline{\alpha}_b - \overline{(\alpha\beta)}_{\cdot,b}) \end{pmatrix}, \quad v = \begin{pmatrix} \frac{\sigma_e^2 \sigma_B^2}{\sigma_e^2 + n_{\cdot,1}\sigma_B^2} \\ \vdots \\ \frac{\sigma_e^2 \sigma_B^2}{\sigma_e^2 + n_{\cdot,b}\sigma_B^2} \end{pmatrix}.$$

The posterior of B is $N_b(Cz, C)$, where $C = \mathrm{diag}(v) - vv^\top/(\mathbf{1}^\top v)$.

Since the variance parameters (except for Σ_e^2) are conditionally independent of the data given the other parameters, their conditional posteriors will not depend on the data. We can calculate the conditional posteriors making use of Proposition 8.13:

$$\Sigma_A^2 \sim \Gamma^{-1}\left(\frac{a_A + a}{2}, \frac{b_A + \sum_{i=1}^a \alpha_i^2}{2}\right),$$

$$\Sigma_B^2 \sim \Gamma^{-1}\left(\frac{a_B + b}{2}, \frac{b_B + \tau(\mu - \theta_0)^2 + \sum_{j=1}^b B_j^2}{2}\right),$$

$$\Sigma_{AB}^2 \sim \Gamma^{-1}\left(\frac{a_{AB} + ab - a}{2}, \frac{b_{AB} + \sum_{i=1}^a \sum_{j=1}^b (AB)_{i,j}^2}{2}\right),$$

and Σ_e^2 has distribution

$$\Gamma^{-1}\left(\frac{a_e + n_{\cdot,\cdot}}{2}, \frac{b_e + w + \sum_{i=1}^a \sum_{j=1}^b n_{i,j}(\overline{y}_{i,j,\cdot} - \mu - \alpha_i - \beta_j - (\alpha\beta)_{i,j})^2}{2}\right).$$

The only remaining problem for implementing SSS in this example is how to simulate from a multivariate normal distribution $N_b(Cz, C)$ with a singular covariance matrix $C = \text{diag}(v) - vv^\top/(\mathbf{1}^\top v)$. The most straightforward way is to find a $b \times (b-1)$ matrix D such that $DD^\top = C$, then generate an $N_{b-1}(\mathbf{0}, I)$ vector V, and use $D(V + D^\top z)$. Let v_* be the first $b-1$ coordinates of v, let $\sqrt{v_*}$ be the vector with jth coordinate $\sqrt{v_j}$, let $\text{diag}(\sqrt{v_*})$ be the diagonal matrix with (j,j) element equal to $\sqrt{v_j}$, and let

$$h = \frac{1 - \sqrt{\frac{v_k}{\mathbf{1}^\top v}}}{\mathbf{1}^\top v}.$$

Then the following matrix satisfies $DD^\top = C$:

$$D = \begin{pmatrix} \text{diag}(\sqrt{v_*}) - hv_*\sqrt{v_*}^\top \\ -\sqrt{\frac{v_b}{\mathbf{1}^\top v}}\sqrt{v_*}^\top \end{pmatrix}.$$

One should also note that, just as in the one-way ANOVA, we could have had unequal variances in the cells. That is, we could have had the conditional variance of $Y_{i,j,k}$ be $\Sigma_{i,j}^2$ instead of Σ_e^2 for all i and j. This would have introduced $ab - 1$ additional variance parameters, but the conditional distributions would have been only slightly more complicated. The serious reader should work this case out in detail. In addition, a hierarchical model for the $\Sigma_{i,j}^2$ could be introduced. Intermediately, one could model the $Y_{i,j,k}$ as having variance Σ_i^2 or Σ_j^2 so that some cells have the same variance and others do not. All such models can be handled in nearly the same fashion as above.

8.5.3 Nonnormal Models

There is a large class of problems to which the SSS methodology can apply. We will not attempt to catalogue this class. We give only a few more examples to show how, with a little imagination, the methodology can apply even

518   Chapter 8. Hierarchical Models

where one would not normally think. In Example 7.104 on page 444, the observables X_i were modeled as having $Cau(\theta, 1)$ distribution given $\Theta = \theta$. Suppose now that we introduce an extra parameter Y_i for each observation and say that X_i given $Y_i = y$ and $\Theta = \theta$ has $N(\theta, y)$ distribution and the Y_i are independent of Θ and of each other with $\Gamma^{-1}(1/2, 1/2)$ distribution.[16] It follows that $X_i \sim Cau(\theta, 1)$ given $\Theta = \theta$; hence this model is equivalent to the original model. However, this new model is easily handled via SSS. In Example 7.104, we supposed that Θ had $N(0, 1000)$ as a prior. The conditional posterior of Θ given the Y_i is

$$\Theta \sim N\left(\frac{\sum_{i=1}^{n} \frac{x_i}{y_i}}{0.001 + \sum_{i=1}^{n} \frac{1}{y_i}}, \frac{1}{0.001 + \sum_{i=1}^{n} \frac{1}{y_i}}\right).$$

The conditional posterior of Y_i given Θ is $\Gamma^{-1}(1, [1 + (x_i - \theta)^2]/2)$.

Example 8.51 (Continuation of Example 7.104; see page 444). After 40 iterations, we constructed 10,000 vectors of the 11 parameters $Y_1, \ldots, Y_{10}, \Theta$. The estimated mean and variance of Θ were 4.585 and 2.233, respectively. The posterior density of Θ is plotted in Figure 7.105 on page 444 together with the normal approximation of Theorem 7.101 and an approximation by numerical integration.

The same principle as described here can be used if one wishes to use a Cauchy or t distribution for the prior distribution of the location parameter for normally distributed data. In fact, a t distribution for the prior combined with data having t distribution can be handled using SSS and simple normal/inverse-gamma posteriors.

The next example is the model described in Section 8.3.2 in which there are k groups of subjects with n_i subjects in group i. The data are $X_i = x_i$, the number of subjects with a positive response to some query. The X_i are modeled as conditionally independent $Bin(n_i, p_i)$ given $P_1 = p_1, \ldots, P_k = p_k$. The P_i are modeled as conditionally independent $Beta(\theta r, [1 - \theta]r)$ given $\Theta = \theta$, $R = r$. Finally, we will suppose that Θ and R are independent with discrete prior distributions having densities f_Θ and f_R with respect to counting measures on sets $\{\theta_1, \ldots, \theta_a\}$ and $\{r_1, \ldots, r_b\}$, respectively. The conditional posteriors of the P_i given the other parameters were already seen to be $Beta(\theta r + x_i, [1 - \theta]r + n_i - x_i)$. The posterior of Θ given the others has probability of $\Theta = \theta_j$ proportional to

$$f_\Theta(\theta_j)\Gamma(\theta_j r)^{-k}\Gamma([1 - \theta_j]r)^{-k}\prod_{i=1}^{k} p_i^{r\theta_j - 1}(1 - p_i)^{r[1-\theta_j]-1}.$$

---

[16] This distribution is also known as χ_1^2.

The conditional posterior probability of $R = r_j$ given the other parameters is proportional to

$$f_R(r_j)\Gamma(r_j)^k \Gamma(\theta r_j)^{-k} \Gamma([1-\theta]r_j)^{-k} \prod_{i=1}^{k} p_i^{r_j \theta - 1}(1-p_i)^{r_j[1-\theta]-1}.$$

Simulating random variables with a discrete distribution can be done by the following tedious but straightforward method. Let X have density f_X with respect to counting measure on the set $\{x_1, x_2, \ldots\}$. Generate a $U(0,1)$ variable U. Set $X = x_j$ where j is the first n such that $\sum_{i=1}^{n} f_x(x_i) \geq U$. This is the discrete version of the *probability integral transform*.

As an alternative to sampling from discrete distributions, we could introduce some latent variables that make the problem look like a normal hierarchical model.[17] Let $X_i = \sum_{j=1}^{n_i} X_{i,j}$, where the $X_{i,j}$ are modeled as IID $Ber(p_i)$ given $P_i = p_i$ for each i. Let $Z_{i,j}$ be IID with $N(\mu_i, 1)$ distribution given $M_i = \mu_i$ where $P_i = \Phi(M_i)$, and assume that $X_{i,j} = I_{[0,\infty)}(Z_{i,j})$. We can treat the $Z_{i,j}$ as parameters or missing data. Let the prior for the M_i be that they are conditionally IID with $N(\mu, \tau^2)$ distribution given $M = \mu$ and $T = \tau$. Let T^2 have an inverse gamma distribution. Either let M be independent of T with $N(\mu_0, \sigma_0^2)$ distribution or let M given $T = \tau$ have $N(\mu_0, \tau^2/\lambda_0)$ distribution. The conditional distribution of the $Z_{i,j}$ given the M_i, M, T, and the $X_{i,j}$ is that of independent, truncated $N(M_i, 1)$ random variables. (Those $Z_{i,j}$ corresponding to $X_{i,j} = 1$ are truncated to the interval $[0, \infty)$ and the others are truncated to the interval $(-\infty, 0)$.) The conditional distribution of the M_i given the $Z_{i,j}$, the $X_{i,j}$, M, and T as well as the conditional distributions of M and T given the others are all obtained as in the appropriate normal hierarchical model.

8.6 Mixtures of Models

8.6.1 General Mixture Models

A different type of hierarchical model is one in which one contemplates several different models for the same data but does not wish to condition on just one of them. For example, consider a case in which one observes pairs (X_i, Y_i) and one wishes to predict the Y coordinate from the X coordinate (often called *regression*). One typical model is that there are parameters $\Theta = (B_0, B_1, \Sigma)$ such that conditional on $\Theta = (\beta_0, \beta_1, \sigma)$ and $X = x$, $Y \sim N(\beta_0 + \beta_1 x, \sigma^2)$ and X is independent of Θ. Another model says that there are parameters $\Theta = (B_0, B_1, \Sigma)$ such that conditional on $\Theta = (\beta_0, \beta_1, \sigma)$ and $X = x$, $\log(Y) \sim N(\beta_0 + \beta_1 x, \sigma^2)$ and X is independent of Θ. The two

---

[17] This model and some generalizations of it are discussed by Albert and Chibb (1993).

Θs are not the same random quantities. In fact, it is common to believe that at most one of them actually exists. Let Ψ be a random quantity such that conditional on $\Psi = 0$, there is a Θ_0 such that conditional on $\Theta_0 = (\beta_0, \beta_1, \sigma)$ and $X = x$, $\log(Y) \sim N(\beta_0 + \beta_1 x, \sigma^2)$ and conditional on $\Psi = 1$, there is a Θ_1 such that conditional on $\Theta_1 = (\beta_0, \beta_1, \sigma)$ and $X = x$, $Y \sim N(\beta_0 + \beta_1 x, \sigma^2)$. In a sense, the parameters are now $(\Psi, \Theta_0, \Theta_1)$, but the joint distribution of Θ_0 and Θ_1 is of no interest since there are no data that depend on both of them. In fact, we don't even need to believe that they coexist. One would need to specify a prior distribution for Ψ, a conditional prior for Θ_0 given $\Psi = 0$, and a conditional prior for Θ_1 given $\Psi = 1$. After observing data, one could calculate conditional posteriors for Θ_0 and Θ_1 given $\Psi = 0$ and $\Psi = 1$, respectively. One could also construct prior predictive distributions for the data given Ψ alone and use these to get the posterior for Ψ. In symbols, we need f_Ψ, $f_{\Theta_0|\Psi}(\theta_0|0)$, and $f_{\Theta_1|\Psi}(\theta_1|1)$. The original models give $f_{Y,X|\Theta_0}$ and $f_{Y,X|\Theta_1}$, where we assume that

$$f_{Y,X|\Theta_0}(y,x|\theta_0) = f_{Y,X|\Theta_0,\Psi}(y,x|\theta_0,0) = f_{Y,X|\Theta_0,\Theta_1,\Psi}(y,x|\theta_0,\theta_1,0),$$

$$f_{Y,X|\Theta_1}(y,x|\theta_1) = f_{Y,X|\Theta_1,\Psi}(y,x|\theta_1,1) = f_{Y,X|\Theta_0,\Theta_1,\Psi}(y,x|\theta_0,\theta_1,1),$$

so that (Y, X) is conditionally independent of Θ_{1-i} given Θ_i and $\Psi = i$ for $i = 0, 1$. The predictive density of (Y, X) given Ψ is

$$f_{Y,X|\Psi}(y,x|\psi) = \int_{\Omega_\psi} f_{Y,X|\Theta_\psi}(y,x|\theta_\psi) f_{\Theta_\psi|\Psi}(\theta_\psi|\psi) d\theta_\psi,$$

where Ω_ψ is the parameter space given $\Psi = \psi$, for $\psi = 0, 1$. The conditional posteriors are

$$f_{\Theta_\psi|X,Y,\Psi}(\theta_\psi|x,y,\psi) = \frac{f_{Y,X|\Theta_\psi}(y,x|\theta_\psi) f_{\Theta_\psi|\Psi}(\theta_\psi|\psi)}{f_{Y,X|\Psi}(y,x|\psi)},$$

for $\psi = 0, 1$. The posterior of Ψ is

$$f_{\Psi|X,Y}(\psi|x,y) = \frac{f_{X,Y|\Psi}(x,y|\psi) f_\Psi(\psi)}{f_{X,Y|\Psi}(x,y|0) f_\Psi(0) + f_{X,Y|\Psi}(x,y|1) f_\Psi(1)}.$$

If there are future data (Y', X') that are conditionally independent of (Y, X) given the parameters, then predictive inference is available.

$$f_{Y',X'|Y,X}(y',x'|y,x) = \sum_{\psi=0}^{1} f_{Y',X'|\Psi}(y',x'|\psi) f_{\Psi|X,y}(\psi|x,y)$$

$$= \sum_{\psi=0}^{1} f_{\Psi|X,y}(\psi|x,y) \int_{\Omega_\psi} f_{Y',X'|\Theta_\psi}(y',x'|\theta_\psi) f_{\Theta_\psi|X,Y}(\theta_\psi|x,y) d\theta_\psi.$$

Notice that the predictive density $f_{Y',X'|Y,X}$ is a weighted average of the two predictive densities one would have used if one had believed each of the two models. The weights are the posterior probabilities of the two models. If, for example, model 0 looks orders of magnitude better than model 1 based on the (Y, X) data (that is, $f_{\Psi|X,Y}(0|x,y)$ is much much larger than $f_{\Psi|X,Y}(1|x,y)$), then the predictive distribution of the future data will be almost the same as if only model 0 had been used from the start.[18] The real advantage to this approach arises when neither model appears much better than the other based on the data. In this case, we can hedge our predictions to allow for the possibility that one or the other model will later turn out to appear better.

Of course, the above description can be extended to apply to arbitrary data X and an arbitrary number of models. For example, the two models in the regression example considered above can be embedded in a family of models in which, conditional on $\Psi = \psi$, $\Theta_\psi = (\beta_0, \beta_1, \sigma)$, and $X = x$,

$$\frac{Y^\psi - 1}{\psi} \sim N(\beta_0 + \beta_1 x, \sigma^2),$$

where $\psi = 0$ is defined by continuity (taking a limit). This is the familiar Box–Cox family of transformations introduced by Box and Cox (1964). If uncountably many values of ψ are being considered, the sums over ψ must be replaced by integrals that, presumably, must be evaluated numerically.

8.6.2 Outliers

One popular use of mixtures of models is to allow for the possibility of *outliers* in a data set. An outlier is an observation whose distribution (before seeing the data) is not like that of the other observations. This "definition" of outlier is intentionally vague. Consider an example to help to clarify the concept.

Example 8.52. Suppose that $X_1, \ldots, X_n$ are potential observations, but we believe that some of them may not have the same distributions as the others. Box and Tiao (1968) describe a model similar to the following. Let $\Theta = (M, \Sigma)$ and suppose that the conditional distribution of each X_i given $\Theta = (\mu, \sigma)$ is $N(\mu, \sigma^2)$ with probability $1 - \alpha$ and is $N(\mu, c\sigma^2)$ with probability α, where $c > 1$ and α are constants chosen a priori. Suppose that the conditional distribution of M given $\Sigma = \sigma$ is $N(\mu_0, \sigma^2/\lambda_0)$ and Σ^2 has $\Gamma^{-1}(a_0/2, b_0/2)$ distribution. There is missing data here, namely the indicators of whether each observation has variance $c\sigma^2$ or not. Let Ψ stand for the subset of $\{1, \ldots, n\}$ such that the observations with subscripts in Ψ have larger variance. For each possible value ψ of Ψ, let n_ψ indicate

---

[18] The predictive density has been used by many authors as a means for selecting and comparing models. See Geisser and Eddy (1979) and Dawid (1984) for different perspectives.

522   Chapter 8. Hierarchical Models

the number of elements of ψ. Let

$$Z_i = \begin{cases} 1 & \text{if } i \notin \Psi, \\ \frac{1}{c} & \text{if } i \in \Psi. \end{cases}$$

Then $f_{X|\Psi}(x|\psi)$ equals a constant times

$$c^{-\frac{n_\psi}{2}} \lambda_1^{-\frac{1}{2}} \left(b_0 + \sum_{i=1}^n z_i (x_i - \hat{x}_\psi)^2 + \frac{\lambda_0 \sum_{i=1}^n z_i}{\lambda_1} (\hat{x}_\psi - \mu_0)^2 \right)^{-\frac{a_0+n}{2}}, \qquad (8.53)$$

where

$$\lambda_1 = \lambda_0 + \sum_{i=1}^n z_i, \quad \hat{x}_\psi = \frac{\sum_{i=1}^n z_i x_i}{\sum_{i=1}^n z_i}.$$

If we model the Z_i as independent, then the prior probability of $\Psi = \psi$ is $\alpha^{n_\psi}(1-\alpha)^{n-n_\psi}$, so we can calculate the posterior probability of each possible subset of outliers.

For example, suppose that $n = 15$ and our prior has $a_0 = 1$, $b_0 = 100$, $\mu_0 = 0$, $\lambda_0 = 1$, $\alpha = 0.02$, and $c = 25$. (For computational simplicity, we truncate the distribution of Ψ to a maximum of six elements.) The data are the infamous Darwin data [see Fisher (1966), p. 37] shown below

$$-67, -48, 6, 8, 14, 16, 23, 24, 28, 29, 41, 49, 56, 60, 75$$

The possible values of ψ with the highest posterior probability are given in Table 8.54 under the column "Model 1." The set of size three with the highest posterior probability of being outliers is $\{1, 2, 15\}$ with probability 0.0030. We can also add the probabilities of all subsets that contain a specific observation to get the marginal probabilities that each observation is an outlier. See Table 8.55. The observations not listed in Table 8.55 each have probability less than 0.005 of being an outlier.

Of course, one need not choose a single value of c or a single value of α. One could treat these as further mixing parameters like Ψ and compute a posterior

TABLE 8.54. Posterior Probabilities of Outlier Sets

ψ	Model 1	Model 2
$\emptyset$	0.7192	0.7767
$\{1\}$	0.1295	0.0885
$\{1,2\}$	0.0483	0.0448
$\{2\}$	0.0241	0.0166
$\{15\}$	0.0114	0.0080
$\{14\}$	0.0060	0.0042
$\{13\}$	0.0052	0.0037
$\{12\}$	0.0043	0.0031
$\{11\}$	0.0036	0.0026
$\{3\}$	0.0033	0.0023
$\{1,15\}$	0.0032	0.0031
$\{4\}$	0.0032	0.0023

8.6. Mixtures of Models   523

TABLE 8.55. Posterior Probabilities of Outlier Observations

i	1	2	11	12	13	14	15
Model 1	0.1970	0.0816	0.0050	0.0060	0.0075	0.0088	0.0195
Model 2	0.1627	0.0803	0.0047	0.0057	0.0074	0.0089	0.0214

distribution for them. For example, (8.53) is now $f_{X|\Psi,C,A}(x|\psi,c,\alpha)$, which does not depend on α. This is because X is conditionally independent of A given Ψ and C. Suppose that we let A have a Beta distribution. It is difficult to have a Beta distribution with mean 0.02 which is neither extremely concentrated near its mean nor extremely concentrated near 0. Suppose that we choose $Beta(1,49)$, which has $\Pr(A \leq 0.02) = 0.6358$. Also, let C have probability 0.05 of being one of the numbers $5, 10, \ldots, 100$. The posterior distribution of C is almost the same as the prior, meaning that the different values of C do not lead to much difference in the predictive density of the data, although $C = 10$ has the highest posterior probability. The posterior distribution of A has mean 0.0204. The posterior probabilities of the various Ψ sets is given in Table 8.54 under the column "Model 2." The probability of the set $\{1,2,15\}$ is now 0.0058. The probabilities that each of the observations is an outlier is in Table 8.55. Although the probability that $\Psi = \{1\}$ is smaller in Model 2, the probability that observation 1 is an outlier is still quite large, because there are many other sets ψ containing 1 which now have higher probability of equaling Ψ. For example, the probability of three outliers is twice as high in Model 2 as in Model 1, and the probability of four outliers is six times as high.

Before we leave this example, we offer another variation. Suppose that we give C a continuous prior distribution, say $\Gamma^{-1}(c_0/2, d_0/2)$ truncated below at $c = 1$ and independent of (Σ, M, A). We could find the posterior distributions of whatever we wanted by using successive substitution sampling (see Section 8.5). The following conditional posterior distributions are easy to find and are easy to simulate:

$$C \sim \Gamma^{-1}\left(\frac{c_0 + n_\psi}{2}, \frac{d_0 + \sum_{i \in \psi}(x_i - \mu)^2}{2\sigma^2}\right),$$

$$\Sigma^2 \sim \Gamma^{-1}\left(\frac{a_0 + n + 1}{2}, \frac{b_0 + \frac{1}{c}\sum_{i \in \psi}(x_i - \mu)^2 + \sum_{i \notin \psi}(x_i - \mu)^2 + \lambda_0(\mu - \mu_0)^2}{2}\right),$$

$$M \sim N\left(\frac{\lambda_0 \mu_0 + \frac{1}{c}\sum_{i \in \psi} x_i + \sum_{i \notin \psi} x_i}{\lambda_0 + n - n_\psi + \frac{1}{c}n_\psi}, \frac{\sigma^2}{\lambda_0 + n - n_\psi + \frac{1}{c}n_\psi}\right),$$

$$A \sim Beta(\alpha_0 + n_\psi, \beta_0 + n - n_\psi),$$

$$Z_i \sim Ber\left(\frac{\alpha}{\alpha + \sqrt{c}(1-\alpha)\exp\left(-\frac{c-1}{2c\sigma^2}(x_i - \mu)^2\right)}\right),$$

where, as before, $\Psi = \{i : Z_i = 1\}$, and the distribution of C is still truncated below at $c = 1$. Since the inverse of the Γ CDF is available in many subroutine libraries, the truncated distribution can be simulated using the probability integral transform. Notice that the Z_i are independent of each other given the other parameters and the data. An analysis like the one just described is developed by Verdinelli and Wasserman (1991).

Notice that in the last variation, the model no longer resembles a mixture of models. In fact, it is just a more highly parameterized model with parameter (Σ, M, Ψ, C, A). Alternatively, the parameter could be taken as (Σ, M, C, A) with Z being considered as missing data.

This example is not meant to be a prescription for how to handle outliers, but merely an example of how mixtures of models can be used for such a problem. Freeman (1980) describes several other methods for detecting outliers. West (1984) describes hierarchical models for accommodating outliers in linear regression.

In fact, any situation in which there is uncertainty is amenable to analysis using a mixture of models. Even the simplest univariate one-sample problem can admit several prior distributions and/or parametric families. The different combinations of parametric family and prior distribution can be mixed using the general theory outlined above. See Problem 14 on page 75 for a simple example.

8.6.3 Bayesian Robustness

In Section 5.1.5, we introduced M-estimators as robust estimators that might be less sensitive to anomalies in the data. Because Bayesian solutions depend on prior distributions for parameters in addition to the conditional distributions of data given parameters, one might be interested in prior distributions that provide a measure of robustness. Also, one might be interested in the degree of robustness that a particular choice of prior exhibits when compared with several others.

A straightforward way of comparing a particular prior distribution μ_Θ to several others is to compute whatever one would normally compute using μ_Θ as the prior and then recompute the same quantities using all of the other priors. This activity often goes by the name of *sensitivity analysis*. If the number of alternative priors is too large, one might be able to compute bounds on the various quantities of interest as the prior ranges over the alternatives. One popular way of specifying a set of alternative priors is by means of ϵ-*contamination*. For a given μ_Θ, $\epsilon > 0$, and set $\mathcal{C}$ of probability measures on (Ω, τ), one forms the collection

$$\mathcal{C}_\epsilon = \{(1-\epsilon)\mu_\Theta + \epsilon\eta : \eta \in \mathcal{C}\} \tag{8.56}$$

of alternative prior distributions. The set $\mathcal{C}_\epsilon$ is called an ϵ-contamination class. If $\mu_\Theta \in \mathcal{C}$, then $\mu_\Theta \in \mathcal{C}_\epsilon$ also. Note that each element of $\mathcal{C}_\epsilon$ is a mixture of two possible prior distributions. The largest set $\mathcal{C}$ one could use is the set of all probability distributions on (Ω, τ). Suppose that one is interested in posterior probabilities of sets $C \in \tau$. It is possible to calculate bounds on the posterior probabilities of such sets as the prior ranges over $\mathcal{C}_\epsilon$.

Theorem 8.57.[19] *Suppose that X has conditional density $f_{X|\Theta}(x|\theta)$ with respect to ν given $\Theta = \theta$. Let $\mathcal{C}$ be the set of all distributions on (Ω, τ), and let $\mathcal{C}_\epsilon$ be as in (8.56). For each $\pi \in \mathcal{C}_\epsilon$, let $\pi(\cdot|x)$ denote the posterior distribution of Θ given $X = x$ calculated as if π were the prior distribution. Similarly, let $\mu_{\Theta|X}(\cdot|x)$ denote the posterior calculated as if μ_Θ were the prior. For each $C \in \tau$,*

$$\inf_{\pi \in \mathcal{C}_\epsilon} \pi(C|x) = \frac{(1-\epsilon)f_X(x)\mu_{\Theta|X}(C|x)}{(1-\epsilon)f_X(x) + \epsilon \sup_{\theta \in C^c} f_{X|\Theta}(x|\theta)},$$

$$\sup_{\pi \in \mathcal{C}_\epsilon} \pi(C|x) = 1 - \frac{(1-\epsilon)f_X(x)[1 - \mu_{\Theta|X}(C|x)]}{(1-\epsilon)f_X(x) + \epsilon \sup_{\theta \in C} f_{X|\Theta}(x|\theta)},$$

where f_X denotes the marginal density of X under the assumption that μ_Θ is the prior.

PROOF. For $\pi \in \mathcal{C}_\epsilon$ with $\pi = (1-\epsilon)\mu_\Theta + \epsilon\eta$, it is easy to see that

$$\pi(C|x) = \frac{(1-\epsilon)f_X(x)\mu_{\Theta|X}(C|x) + \epsilon \int_C f_{X|\Theta}(x|\theta)d\eta(\theta)}{(1-\epsilon)f_X(x) + \epsilon g(x)}, \quad (8.58)$$

where $g(x) = \int f_{X|\Theta}(x|\theta)d\eta(\theta)$ is the marginal density of X under the assumption that η is the prior. The expression in (8.58) will get smaller if η is replaced by any η^* such that $\eta^*(C) = 0$ and $\eta^*(D) \geq \eta(D)$ for $D \subseteq C^C$. (For example, rescale $\eta(\cdot \cap C^C)$ to be a probability.) It follows that the smallest values of (8.58) occur when $\eta(C) = 0$. When $\eta(C) = 0$, (8.58) becomes

$$\pi(C|x) = \frac{(1-\epsilon)f_X(x)\mu_{\Theta|X}(C|x)}{(1-\epsilon)f_X(x) + \epsilon g(x)}. \quad (8.59)$$

This can be minimized by making $g(x)$ as large as possible. But since $g(x)$ is an average of values of $f_{X|\Theta}(x|\theta)$, its supremum is clearly equal to $\sup_{\theta \in C^c} f_{X|\Theta}(x|\theta)$. So the infimum of $\pi(C|x)$ equals (8.59) with $g(x)$ replaced by $\sup_{\theta \in C^c} f_{X|\Theta}(x|\theta)$. The supremum is obtained by applying the same argument to C^C. □

Example 8.60. Let $X \sim Exp(\theta)$ given $\Theta = \theta$, and let μ_Θ be the $\Gamma(a,b)$ distribution. Let C be the interval $(0, c(x)]$ where $c(x)$ is the γ quantile of the posterior distribution of Θ. In this case, the posterior is $\Gamma(a+1, b+x)$ and the marginal density of the data is $f_X(x) = ab^a/(b+x)^{a+1}$. The likelihood function is $\theta \exp(-x\theta)$, which increases for $\theta < 1/x$ and decreases thereafter. So, we have

$$\sup_{\theta \in C} f_{X|\Theta}(x|\theta) = \begin{cases} \frac{\exp(-1)}{x} & \text{if } \frac{1}{x} \leq c(x), \\ c(x)\exp(-c(x)x) & \text{if } \frac{1}{x} > c(x), \end{cases}$$

---

[19]This theorem appears in Berger (1985).

526   Chapter 8. Hierarchical Models

$$\sup_{\theta \in C^C} f_{X|\Theta}(x|\theta) = \begin{cases} \frac{\exp(-1)}{x} & \text{if } \frac{1}{x} \geq c(x), \\ c(x)\exp(-c(x)x) & \text{if } \frac{1}{x} < c(x). \end{cases}$$

The value of $\mu_{\Theta|X}(C|x) = \gamma$ by design. The bounds given by Theorem 8.57 for the ϵ-contamination class using $\mathcal{C}$ equal to all distributions are, for $1/x \leq c(x)$,

$$\frac{(1-\epsilon)\frac{\gamma a b^a}{(b+x)^{a+1}}}{(1-\epsilon)\frac{a b^a}{(b+x)^{a+1}} + \epsilon c(x)\exp(-c(x)x)} \leq \pi(C|x) \leq 1 - \frac{(1-\epsilon)(1-\gamma)\frac{a b^a}{(b+x)^{a+1}}}{(1-\epsilon)\frac{a b^a}{(b+x)^{a+1}} + \epsilon\frac{\exp(-1)}{x}}.$$

For example, with $\gamma = 0.5$, $a = b = 1$, and $\epsilon = 0.1$, we have $c(x) = 1.678/(1+x)$, which is greater than or equal to $1/x$ for $x \geq 1.474$. Figure 8.61 shows a plot of the lower and upper bounds on the posterior probabilities of the interval $[0, 1.678/(1+x)]$ as a function of x. Notice how the degree of robustness depends on the observed data. When x is very small, the likelihood function is quite large for large values of θ outside of the interval $(0, c(x)]$, since $c(x)$ never gets bigger than 1.678. A prior that assigned probability 1 to such a large θ value would be consistent with a very small observed x and would give low probability to every subinterval of $[0, 1.678]$. If such priors seem unreasonable, then perhaps the class $\mathcal{C}_\epsilon$ is too large.

Additionally, we may wish to find bounds for the posterior mean of a measurable function g of Θ as the prior distribution varies over a class such as an ϵ-contamination class. The following theorem, which is helpful in this regard, is due to Lavine, Wasserman, and Wolpert (1991, 1993).

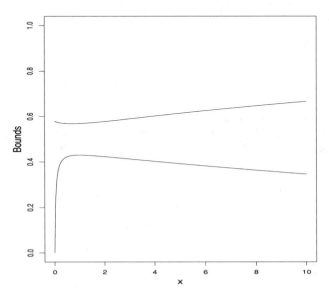

FIGURE 8.61. Lower and Upper Bounds on Posterior Probabilities

8.6. Mixtures of Models

Theorem 8.62. *Let Γ be a class of prior distributions on (Ω, τ), and let $g : \Omega \mapsto \mathbb{R}$ be a measurable function. Suppose that $\inf_{\pi \in \Gamma} \int f_{X|\Theta}(x|\theta) d\pi(\theta) > 0$. For each $\pi \in \Gamma$, define $s_\pi(\lambda) = \int f_{X|\Theta}(x|\theta)[g(\theta) - \lambda]d\pi(\theta)$, and let*

$$\bar{s}(\lambda) = \sup_{\pi \in \Gamma} s_\pi(\lambda).$$

Then for finite λ, the least upper bound on the posterior means of $g(\Theta)$ is λ if and only if $\bar{s}(\lambda) = 0$.

PROOF. Let

$$\lambda_0 = \sup_{\pi \in \Gamma} \frac{\int f_{X|\Theta}(x|\theta) g(\theta) d\pi(\theta)}{\int f_{X|\Theta}(x|\theta) d\pi(\theta)},$$

and assume that λ_0 is finite. For the "if" direction, suppose that $\bar{s}(\lambda) = 0$. We need to prove that $\lambda = \lambda_0$. Since $\bar{s}(\lambda) = 0$, we know that $s_\pi(\lambda) \leq 0$ for all $\pi \in \Gamma$ and that there exists a sequence $\{\pi_n\}_{n=1}^\infty$ of elements of Γ such that, for each n, $s_{\pi_n}(\lambda) > -1/n$. This last claim can be written as $\int f_{X|\Theta}(x|\theta) g(\theta) d\pi_n(\theta) > \lambda \int f_{X|\Theta}(x|\theta) d\pi_n(\theta) - 1/n$, which implies

$$\frac{\int f_{X|\Theta}(x|\theta) g(\theta) d\pi_n(\theta)}{\int f_{X|\Theta}(x|\theta) d\pi_n(\theta)} > \lambda - \frac{1}{n \int f_{X|\Theta}(x|\theta) d\pi_n(\theta)},$$

for all n. We know that

$$\lambda_0 \geq \sup_n \frac{\int f_{X|\Theta}(x|\theta) g(\theta) d\pi_n(\theta)}{\int f_{X|\Theta}(x|\theta) d\pi_n(\theta)} \geq \lambda - \frac{1}{\sup_n \int f_{X|\Theta}(x|\theta) d\pi_n(\theta)}. \quad (8.63)$$

Because $\inf_{\pi \in \Gamma} \int f_{X|\Theta}(x|\theta) d\pi(\theta) > 0$, the far right-hand side of (8.63) equals λ, so $\lambda_0 \geq \lambda$. We can rewrite $s_\pi(\lambda) \leq 0$ as $\int f_{X|\Theta}(x|\theta) g(\theta) d\pi(\theta) \leq \lambda \int f_{X|\Theta}(x|\theta) d\pi(\theta)$, which implies

$$\frac{\int f_{X|\Theta}(x|\theta) g(\theta) d\pi(\theta)}{\int f_{X|\Theta}(x|\theta) d\pi(\theta)} \leq \lambda.$$

Since this is true for all $\pi \in \Gamma$, it follows that $\lambda_0 \leq \lambda$, and we conclude $\lambda_0 = \lambda$.

For the "only if" part, we must show that $\bar{s}(\lambda_0) = 0$. From the fact that

$$\frac{\int f_{X|\Theta}(x|\theta) g(\theta) d\pi(\theta)}{\int f_{X|\Theta}(x|\theta) d\pi(\theta)} \leq \lambda_0,$$

for all $\pi \in \Gamma$, it easily follows that $\bar{s}(\lambda_0) \leq 0$. Suppose that $\bar{s}(\lambda_0) = -\epsilon$ for some $\epsilon > 0$. We will derive a contradiction. We know that there exists a sequence $\{\pi_n\}_{n=1}^\infty$ of elements of Γ such that, for each n,

$$\frac{\int f_{X|\Theta}(x|\theta) g(\theta) d\pi_n(\theta)}{\int f_{X|\Theta}(x|\theta) d\pi_n(\theta)} > \lambda_0 - \frac{1}{n}.$$

Since $\bar{s}(\lambda_0) = -\epsilon$, it follows that, for every n,

$$\frac{\int f_{X|\Theta}(x|\theta)g(\theta)d\pi_n(\theta)}{\int f_{X|\Theta}(x|\theta)d\pi_n(\theta)} \leq \lambda_0 - \frac{\epsilon}{\int f_{X|\Theta}(x|\theta)d\pi_n(\theta)}.$$

These two inequalities imply that $\int f_{X|\Theta}(x|\theta)d\pi_n(\theta) > n\epsilon$ for every n. This contradicts $\sup_{\pi \in \Gamma} \int f_{X|\Theta}(x|\theta)d\pi(\theta) < \infty$. □

To use Theorem 8.62 to find bounds on posterior means, we first note that lower bounds can be obtained by replacing g by $-g$ and finding another upper bound. For fixed λ, $s_\pi(\lambda)$ is a linear function of π. In the case of ϵ-contamination classes ($\Gamma = \mathcal{C}_\epsilon$), it follows that $\bar{s}(\lambda)$ is the supremum over the set of contaminations of the form $\pi(B) = (1-\epsilon)\mu_\Theta + \epsilon I_B(\theta_0)$ for $\theta_0 \in \Omega$. For such a π, the posterior mean of $g(\Theta)$ is

$$\frac{(1-\epsilon)\int g(\theta)f_{X|\Theta}(x|\theta)d\mu_\Theta(\theta) + f_{X|\Theta}(x|\theta_0)\epsilon g(\theta_0)}{(1-\epsilon)\int f_{X|\Theta}(x|\theta)d\mu_\Theta(\theta) + \epsilon f_{X|\Theta}(x|\theta_0)}.$$

One can usually find the supremum and infimum of this expression as a function of θ_0 using standard numerical methods. The two integrals, $\int g(\theta)f_{X|\Theta}(x|\theta)d\mu_\Theta(\theta)$ and $\int f_{X|\Theta}(x|\theta)d\mu_\Theta(\theta)$, are constants in these numerical problems.

Example 8.64 (Continuation of Example 8.60; see page 525).We have $X \sim Exp(\theta)$ given $\Theta = \theta$, and μ_Θ is the $\Gamma(a,b)$ distribution. Let $g(\theta) = \theta$. So

$$\int g(\theta)f_{X|\Theta}(x|\theta)d\mu_\Theta(\theta) = \frac{a(a+1)b^a}{(b+x)^{a+2}},$$

$$\int f_{X|\Theta}(x|\theta)d\mu_\Theta(\theta) = \frac{ab^a}{(b+x)^{a+1}}.$$

The function for which we need to find extremes is then

$$h(\theta) = \frac{(1-\epsilon)\frac{a(a+1)b^a}{(b+x)^{a+2}} + \epsilon\theta^2 \exp(-x\theta)}{(1-\epsilon)\frac{ab^a}{(b+x)^{a+1}} + \epsilon\theta \exp(-x\theta)}.$$

If we let $a = b = 1$ and $\epsilon = 0.1$ as before, we can find the extremes of $h(\theta)$ for every possible x. Figure 8.65 shows the lower and upper bounds on $E(\Theta|X = x)$ for x between 0.1 and 10. As $x \to 0$, the upper bound goes to ∞ because x close to 0 is most consistent with very large values for Θ. The bounds get very close together and small as $x \to \infty$ because large x values are most consistent with very small values of Θ.

In addition to sensitivity analysis, one can try to find prior distributions such that resulting inferences exhibit some degree of robustness to changes in the prior. Consider the case in which $X_1, \ldots, X_n \sim N(\mu, \sigma^2)$ given $\Theta = (\mu, \sigma)$. The natural conjugate prior is one of the form $M \sim N(\mu_0, \sigma^2/\lambda_0)$ given $\Sigma = \sigma$ and $\Sigma^2 \sim \Gamma^{-1}(a_0/2, b_0/2)$. Such priors have the property that the posterior mean of M is $\mu_1 = (n\bar{x} + \lambda_0\mu_0)/(n+\lambda_0)$, which has a component $\lambda_0\mu_0/(n+\lambda_0)$ that remains the same no matter what the data values

8.6. Mixtures of Models

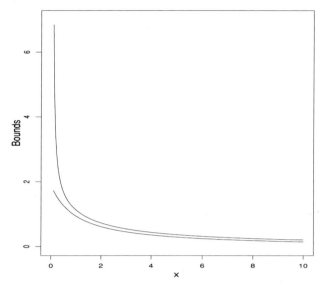

FIGURE 8.65. Lower and Upper Bounds on Posterior Mean of Θ

are. It is sometimes desirable to have the influence of the prior become less pronounced as the data move away from what would be predicted by the prior. Alternatives to natural conjugate priors, which are less influential, are ones in which M is independent of Σ^2 with $t_{c_0}(\mu_0, \tau^2)$ distribution. We will assume that $\Sigma^2 \sim \Gamma^{-1}(a_0/2, b_0/2)$ in this prior also. At first, it may seem difficult to work with such a prior because the posterior cannot be written in closed form. However, we can use the following trick to make the problem more tractable: Invent a random variable Y with $\Gamma^{-1}(c_0/2, c_0\tau_0^2/2)$ distribution independent of Σ, and pretend as if $M \sim N(\mu_0, Y)$ given Y. The marginal distribution of M is then $t_{c_0}(\mu_0, \tau_0^2)$ as earlier prescribed. But now, if we treat (M, Σ, Y) as the parameter, we can use successive substitution sampling (SSS) because the following conditional distributions are obtained:

$$M|\Sigma = \sigma, Y = y \sim N(\mu_1(y, \sigma), \tau_1(y, \sigma)),$$
$$\Sigma^2|M = \mu, Y = y \sim \Gamma^{-1}\left(\frac{a_0 + n}{2}, \frac{b_1(\mu, y)}{2}\right),$$
$$Y|M = \mu, \Sigma = \sigma \sim \Gamma^{-1}\left(\frac{c_0 + 1}{2}, \frac{d_1(\mu)}{2}\right),$$

where

$$\mu_1(y, \sigma) = \frac{\frac{\mu_0}{y} + \frac{n\bar{x}}{\sigma^2}}{\frac{1}{y} + \frac{n}{\sigma^2}}, \qquad \tau_1(y, \sigma) = \left(\frac{1}{y} + \frac{n}{\sigma^2}\right)^{-1}$$

530   Chapter 8. Hierarchical Models

$$b_1(\mu, y) = b_0 + \sum_{i=1}^{n}(x_i - \bar{x})^2 + n(\bar{x} - \mu)^2, \qquad d_1(\mu) = c_0\tau_0^2 + (\mu - \mu_0)^2.$$

In fact, since the prior density for M will tend to be very flat relative to the likelihood function with even a small amount of data, this prior is a lot like using an improper prior such as Lebesgue measure.

Of course, Bayesians can be interested in the same aspects of robustness in which classical statisticians are interested, namely robustness with respect to unexpected observations. In the classical framework, we introduced M-estimators (see Section 5.1.5) to be less sensitive to extreme observations. In the Bayesian framework, this would correspond to using alternative conditional distributions for the data given Θ to reflect the opinion that occasional extreme observations might arise. Consider the case in which $X_1, \ldots, X_n \sim N(\mu, \sigma)$, given $\Theta = (\mu, \sigma)$, most of the time but in which an observation with higher variance is occasionally observed. Alternatively, suppose that each observation X_i comes with its own variance Σ_i^2 and that the Σ_i^2 are exchangeable. If the conditional distribution of Σ_i^2 were $\Gamma^{-1}(a_0/2, a_0T/2)$ given T, this would be equivalent to saying that the X_i had $t_{a_0}(\mu, \tau^2)$ distribution given $M = \mu$ and $T = \tau$. That is, we would have changed the likelihood from normal to t_{a_0}. Once again, it may seem difficult to work with such a likelihood because the posterior cannot be written in closed form. However, we can again use SSS to make the problem more tractable. Suppose that we model M as $N(\mu_0, Y)$ given $Y, \Sigma_1, \ldots, \Sigma_n, T$, and Y as independent of the Σ_i and T with $\Gamma^{-1}(c_0/2, c_0\tau_0^2/2)$ distribution, and $T \sim \Gamma(d_0/2, f_0/2)$. The conditional posterior distributions needed are

$$M \sim N(\mu_1(\sigma_1, \ldots, \sigma_n, y), \tau_1(\sigma_1, \ldots, \sigma_n, y)),$$

$$\Sigma_i^2 \sim \Gamma^{-1}\left(\frac{a_0+1}{2}, \frac{(x_i-\mu)^2 + a_0\tau}{2}\right),$$

$$Y \sim \Gamma^{-1}\left(\frac{c_0+1}{2}, \frac{c_0\tau_0^2 + (\mu-\mu_0)^2}{2}\right),$$

$$T \sim \Gamma\left(\frac{d_0+n}{2}, \frac{f_0 + \sum_{i=1}^{n}\frac{a_0}{\sigma_i^2}}{2}\right),$$

where

$$\tau_1(\sigma_1, \ldots, \sigma_n, y) = \left(\frac{1}{y} + \sum_{i=1}^{n}\frac{1}{\sigma_i^2}\right)^{-1},$$

$$\mu_1(\sigma_1, \ldots, \sigma_n, y) = \left(\frac{\mu_0}{y} + \sum_{i=1}^{n}\frac{x_i}{\sigma_i^2}\right)\tau_1(\sigma_1, \ldots, \sigma_n, y).$$

Alternatively, one could integrate numerically.

Example 8.66. Suppose that we model the data as above with $\mu_0 = 0$, $a_0 = 5$, $c_0 = 1$, $\tau_0 = 2$, $d_0 = 1$, and $f_0 = 1/2$. This prior has the property that the density of the data given the parameters has thinner tails than the prior density of M. This is because the degrees of freedom is 5 for the t distribution of the data given the parameters, but the degrees of freedom is only 1 for the t distribution of M. This will allow the posterior to resemble the likelihood to a large extent. Consider the following 10 observations:

$$1.66,\ 1.07,\ 0.640,\ 0.310,\ 0.295,\ -0.070,\ -0.107,\ -1.67,\ -1.90,\ -1.97.$$

The posterior mean of M is -0.077, and the posterior standard deviation of M is 0.4202. (The sample average is -0.173, and the sample standard deviation over $\sqrt{10}$ is 0.4017.) We could now consider what changes if one of the observations moves off to ∞. For example, suppose that we take the smallest observation, -1.97, and let it move to 0 and then to $+\infty$. Figure 8.67 shows a plot of the posterior mean of M as a function of the moving observation. Notice that the mean of M increases almost linearly with the moving observation for some time and then begins to decrease again. The decrease is due to the moving observation's having reached a level beyond which it is more likely to be coming from the tail of the distribution than from a large value of M.

The other curves in Figure 8.67 correspond to models with different degrees of freedom. All four of the cases with $a_0, c_0 \in \{1, 5\}$ are illustrated. The value of c_0 does not have nearly as much influence as the value of a_0. When a_0 changes to 1 (with $c_0 = 1$ and the original data), the posterior mean and standard deviation of M become 0.1258 and 0.1388, respectively. (The MLE of M would be 0.2064, but the likelihood is not very peaked.) In this case, as one observation changes, the posterior mean of M is affected most by the average of those observations in the middle of the data set. When the moving observation enters the middle of the data set, the posterior mean of M varies linearly with the observation. But when it moves out of the middle, the posterior mean moves back down again.

Of course, one straightforward way to develop robust models is to form mixtures of all sensible models for the data. Those models with high prior

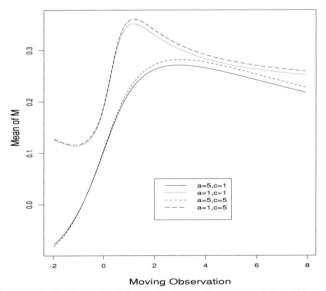

FIGURE 8.67. Posterior Mean of M as a Function of One Observation

TABLE 8.68. Relative Values of Prior Predictive Density

a_0	X_{10}			
	−1.970	0.531	3.030	8.030
1	0.252	0.918	0.367	1.501
2	0.553	1.048	0.663	1.412
5	1.000	1.000	1.000	1.000
10	1.187	1.001	1.091	0.945
20	1.283	1.003	1.110	0.623
60	1.348	1.003	1.105	0.221
∞	1.381	1.003	1.097	0.096
average	1.001	0.997	0.911	0.828

predictive density will surface as the ones that contribute most to the posterior predictive distribution of future data.

Example 8.69 (Continuation of Example 8.66; see page 530). Suppose that we are not sure which degrees of freedom to use for the conditional distribution of X_i given (M, T). We might try a mixture of models with model i having $a_0 = i$ for i in some set where $a_0 = \infty$ means that the conditional distribution is $N(M, T^2)$ rather than $t_{a_0}(M, T^2)$. Table 8.68 lists the relative values of the prior predictive densities of the data for a few values of a_0 with $a_0 = 5$ taken as 1.0. Four different data sets are used; they differ only in the value of the last observation, which is listed in the column heading. The t_5 distribution is relatively robust as one observation increases, and the equal mixture of the seven models (the row labeled "average") has predictive density remarkably close to that of the t_5 model. One might argue that the equal mixture of the seven other models is not itself sensible. Putting 3/7 of the probability on $\{20, 60, \infty\}$ degrees of freedom is saying that one is somewhat confident that the data will be approximately normal. On the other hand, putting 3/7 of the probability on $\{1, 2, 5\}$ degrees of freedom is saying that one is equally confident that the data will likely have an occasional "outlier."

8.7 Problems

Section 8.1:

1. Let $\mathcal{X}_n$ be the space of sequences of 0s and 1s of length $n+1$ which start with 0. Let $T_n(0, x_1, \ldots, x_n)$ be the four counts of transitions (from 0 to 1, from 0 to 0, from 1 to 1, from 1 to 0) in $(0, x_1, \ldots, x_n)$. Call these counts $(T_{n,0,1}, T_{n,0,0}, T_{n,1,1}, T_{n,1,0})$. Let $r_n(A, t)$ be uniform over all sequences with the appropriate numbers of transitions (that is, $t_{0,1}$ transitions from 0 to 1, etc.).

    (a) Show that the conditions of Theorem 2.111 hold.

    (b) Find the extreme points of the set $\mathcal{M}$.

8.7. Problems   533

(c) Write the representation of Theorem 2.111 as an integral over a finite-dimensional space.

2. Suppose that $\{Y_{ij} : j = 1, \ldots, n_i; i = 1, \ldots, k\}$ are conditionally independent with $Y_{ij} \sim N(\theta_i, 1)$ given $\Theta = (\theta_1, \ldots, \theta_k)$ and $M = \mu$. Suppose that $\Theta_1, \ldots, \Theta_k$ are IID with $N(\mu, 1)$ distribution given $M = \mu$ and that $M \sim N(\mu_0, 1)$.

   (a) Find the marginal distribution of each Y_{ij}.
   (b) Show that the Y_{ij} are not exchangeable.
   (c) Find the posterior distribution of Θ and M.

3. Suppose that $\{Y_{ij} : j = 1, \ldots, n_i; i = 1, \ldots, k\}$ are conditionally independent with $Y_{ij} \sim N(\theta_i, 1)$ given $\Theta = (\theta_1, \ldots, \theta_k)$, $M = \mu$, and $T = \tau$. Suppose that $\Theta_1, \ldots, \Theta_k$ are IID with $N(\mu, \tau^2)$ distribution given $M = \mu$ and $T = \tau$. Show that the improper prior with density τ^{-1} (with respect to Lebesgue measure) leads to an improper posterior for T.

Section 8.2:

4. Prove Proposition 8.13 on page 485.

5.*Consider the data in the following table:

i	n_i	$X_{i,j}, j = 1, \ldots, n_i$			
1	3	9.549	10.274	7.142	
2	2	11.430	11.890		
3	3	6.898	4.329	6.905	
4	4	12.620	13.050	12.530	11.890

   Model the $X_{i,j}$ using the one-way ANOVA model of Section 8.2.1 with prior hyperparameters $\psi_0 = 15$, $\zeta_0 = 0.25$, $b_0 = 1$, $a_0 = 1$, and $\Lambda \sim Exp(0.2)$.

   (a) Find the product of the prior density of Λ and the "marginal likelihood" function of Λ.
   (b) Use numerical integration to find the normalizing constant for the posterior density.
   (c) Find the posterior predictive density of a future observation from the $i = 4$ group using a numerical integration method, importance sampling, or the method of Laplace. (Approximate the density at the points from 9 to 15 in steps of 0.05.)

6. Prove that (8.20) on page 490 is true.

7. Prove the following matrix theorem: If A and B are nonsingular matrices, then

$$(A + B)^{-1} = A^{-1} - A^{-1}(A^{-1} + B^{-1})^{-1}A^{-1},$$
$$A(A + B)^{-1}B = (A^{-1} + B^{-1})^{-1}.$$

534   Chapter 8. Hierarchical Models

8. Prove the following matrix theorem: Let $\Sigma_* = \Sigma_1 + \Sigma_2$, where Σ_1 and Σ_2 are nonsingular symmetric matrices, and let a and b be vectors. Then

$$a^\top \Sigma_1 a + b^\top \Sigma_2 b - (\Sigma_1 a + \Sigma_2 b)^\top \Sigma_*^{-1}(\Sigma_1 a + \Sigma_2 b)$$
$$= (a-b)^\top (\Sigma_1^{-1} + \Sigma_2^{-1})^{-1}(a-b). \qquad (8.70)$$

(*Hint:* Use Problem 7 above.)

Section 8.4:

9. Each scientific paper published by a particular author receives a random number of citations from other authors in the years following its publication. For $i = 1, \ldots, k$, and $j = 1, \ldots, n$, let $X_{i,j}$ denote the number of citations paper j received i years after publication. Let $r_1, \ldots, r_k$ be known positive numbers. Model the $X_{i,j}$ as conditionally independent given $M_1, \ldots, M_n, \Theta$ with $X_{i,j}$ having $Poi(M_j r_i)$ distribution. We model the M_j as IID with $Exp(\theta)$ distribution conditional on $\Theta = \theta$.

   (a) Find minimal sufficient statistics for the parameters $M_1, \ldots, M_n, \Theta$.

   (b) Supposing that Θ is the only parameter, find a one-dimensional sufficient statistic.

   (c) Find the MLE $\hat\Theta$ of Θ.

   (d) Use the naïve empirical Bayes approach (assuming $\Theta = \hat\Theta$) to find the posterior distribution of the parameters $M_1, \ldots, M_n$.

10. Suppose that $X_i \sim Bin(n_i, \theta_i)$, $i = 1, \ldots, k$ are conditionally independent given $\Theta = (\theta_1, \ldots, \theta_k)$. Suppose that we model the Θ_i as conditionally IID with $Beta(\alpha, \beta)$ distribution given $(A, B) = (\alpha, \beta)$.

    (a) Although the formulas cannot be written out completely, describe how one would implement the naïve empirical Bayes approach using MLEs for A and B.

    (b) Use the following data, and compute the naïve empirical Bayes posterior distributions and posterior means for the parameters. We have $k = 2$, $n_1 = 5$, $n_2 = 10$, $X_1 = 3$, and $X_2 = 3$.

Section 8.5:

11. Suppose that $X \sim \Gamma^{-1}(a,b)$ and $Y \sim \Gamma^{-1}(c,d)$ are independent. Let $Z = X/Y$. Prove that the conditional distribution of X given $Z = z$ is $\Gamma^{-1}(a + c, b + dz)$.

12. Using the notation of Problem 9 above, suppose that Θ has a prior distribution, which is $\Gamma(a,b)$, with a and b known constants.

    (a) Find the posterior density of Θ except for the normalizing constant.

    (b) Set up a successive substitution sampling scheme to generate a sample of Θ and $M_1, \ldots, M_n$ from the joint posterior distribution.

    (c) Write a formula for an approximation to the posterior density of Θ and of $M_1, \ldots, M_n$ based on the successive substitution sample.

(d) Describe the similarities and differences between the above approximations to the joint density for the M_j and the approximation found via the empirical Bayes approach.

13. Let v and z be k-dimensional vectors with $v_i > 0$ for $i = 1, \ldots, k$. Suppose that $X \sim N_k(\text{diag}(v)z, \text{diag}(v))$, where $\text{diag}(v)$ is a diagonal matrix with (i,i) element equal to v_i. Prove that the conditional distribution of X given $\mathbf{1}^\top X = c$ is

$$N_k\left([\text{diag}(v) - \frac{1}{\mathbf{1}^\top v}vv^\top]z, \text{diag}(v) - \frac{1}{\mathbf{1}^\top v}vv^\top\right).$$

14.*Prove that condition (8.41) holds if $Y \sim N_k(\mu, \Sigma)$ and SSS is applied to the coordinates in their natural order. (*Hint:* First, prove that the conditional distribution of the next iterate X given the current iterate X' is multivariate normal with constant covariance matrix and mean that is a linear function of X'. You can now integrate over x' analytically using facts from the theory of the multivariate normal distribution. The integral over x becomes the integral of the ratio of two normal densities. You can use problems 7 and 8 in this chapter to show that this integral is a constant times the integral of a normal density.)

Section 8.6:

15. The SSS algorithm allows one to approximate posterior distributions without calculating the marginal density of the data $f_X(x)$. When fitting mixture models, it is important to compute $f_{X|\Psi}(x|\psi)$, where Ψ is the parameter indexing models. Consider a single model with parameter $\Theta = (\Theta_1, \ldots, \Theta_p)$. Each iteration of SSS starts with a simulated vector $\Theta^{(i)} = (\Theta_1^{(i)}, \ldots, \Theta_p^{(i)})$ and then simulates $\Theta^{(i+1)}$ one coordinate at a time using the conditional posterior distribution of each coordinate given the others:

$$f_{\Theta_j|\Theta_{\setminus j}, X}(\Theta_j | \Theta_1^{(i+1)}, \ldots, \Theta_{j-1}^{(i+1)}, \Theta_{j+1}^{(i)}, \ldots, \Theta_p^{(i)}, x). \tag{8.71}$$

Call the expression in (8.71) $V_j^{(i+1)}$. (Note that $V_1^{(i+1)}$ and $V_p^{(i+1)}$ have slightly different formulas due to the effect of being at the ends of the vector.) Prove that

$$E\left(\frac{f_{X|\Theta}(x|\Theta^{(i+1)})f_\Theta(\Theta^{(i+1)})}{\prod_{j=1}^p V_j^{(i+1)}}\right) = f_X(x),$$

where the $E(\cdot)$ refers to the joint distribution of $(\Theta^{(i)}, \Theta^{(i+1)})$ in the simulation.

16. Suppose that one wishes to fit a mixture of k models, but that one is required to use SSS to fit each model. Let $\Psi \in \{1, \ldots, k\}$ index the different models, and let Θ_i be the parameters of model i, for $i = 1, \ldots, k$. Explain how one could use Problem 15 above to help estimate $f_{\Psi|X}(\psi|x)$ for all values of ψ.

CHAPTER 9
Sequential Analysis

Most of the results of earlier chapters concern situations in which a particular data set is to be observed and the decisions, if any, to be made concern the values of future observations. It sometimes happens that as we observe, we get to decide what, if any, data to collect next. In this chapter, we will describe some theory and methods for dealing with such situations.[1]

9.1 Sequential Decision Problems

As a simple example of a situation in which we need to decide whether or not to collect more data, consider the following.

Example 9.1. We are considering purchasing a shipment of parts. Prior to observing any data, we believe that the proportion P of defective parts has a $U(0,1)$ distribution. We believe that the individual parts ($X_i = 1$ if part i is defective) are conditionally independent $Ber(p)$ random variables given $P = p$. We decide that we can sample at most 10 parts, and we will reject the shipment if the posterior mean of P is greater than 0.6. This will occur if there are 7 or more defectives out of a sample of 10. Suppose that the first seven parts are defective. Clearly, there is no need to sample any more parts. Similarly, if the first six parts were defective, it might seem highly unlikely that the shipment would be acceptable. Whether or not to continue sampling would depend on the relative costs of sampling and of rejecting a good shipment.

The general sequential decision problem can be defined as follows.

---

[1] The discussion in Section 9.1 is largely adapted from Chapter 12 of DeGroot (1970).

9.1. Sequential Decision Problems

Definition 9.2. Let $(S, \mathcal{A}, \mu)$ be a probability space, let $(\mathcal{V}, \mathcal{D})$ be a measurable space, and let $V : S \to \mathcal{V}$ be a random quantity. For $i = 1, 2, \ldots$, let $(\mathcal{X}_i, \mathcal{B}_i)$ be measurable spaces and let $(\mathcal{X}_0, \mathcal{B}_0) = (\{0\}, \{\emptyset, \mathcal{X}_0\})$ be a trivial space. For $i = 0, 1, \ldots$, let $X_i : S \to \mathcal{X}_i$ be random quantities. Let $\mathcal{X} = \prod_{i=0}^{\infty} \mathcal{X}_i$ with product σ-field $\mathcal{B}^{\infty}$. Let $\mathcal{B}^n$ be the sub-σ-field generated by the first $n+1$ coordinates (including 0). That is, $B \in \mathcal{B}^n$ if and only if $B = C \times \prod_{i=n+1}^{\infty} \mathcal{X}_i$, where $C \in \mathcal{B}_0 \otimes \cdots \otimes \mathcal{B}_n$. Let $X = (X_0, X_1, X_2, \ldots)$. A *stopping time* is a nonnegative, extended integer-valued function N (i.e., $N \in \mathcal{N} = \{0, 1, \ldots\} \cup \{\infty\}$) defined on $\mathcal{X}$ such that, for every finite n, $\{x : N(x) = n\}$ is measurable with respect to $\mathcal{B}^n$. Let the action space be $\aleph = \aleph' \times \mathcal{N}$, and let the loss be $L : \mathcal{V} \times \aleph \to \mathbb{R}$ such that $L(v, (a, n)) = \sum_{i=0}^{n} c_i + L'(v, a)$, where $c_i \geq 0$ for all i ($c_0 = 0$). Let α be a σ-field of subsets of $\aleph'$, and let $\mathcal{P}_A$ be the collection of probability measures on $(\aleph', \alpha)$. A *randomized sequential decision rule* is a pair $\delta = (\delta^*, N)$ where $\delta^* : \mathcal{X} \to \mathcal{P}_A$ and N is a stopping time. The function δ^* is called the *terminal decision rule*. A *nonrandomized sequential decision rule* is a randomized sequential decision rule such that, for each $x \in \mathcal{X}$, there is $\delta'(x) \in \aleph'$ such that for each $A \in \alpha$ $\delta^*(x)(A) = 1$ if $\delta'(x) \in A$ and $\delta^*(x)(A) = 0$ if $\delta'(x) \notin A$. If δ is nonrandomized, then δ' is called the *terminal decision rule*.

A convenient notation will be to let $X^n = (X_0, \ldots, X_n)$ for finite n and $X^{\infty} = X$ if necessary. Also, $x^n = (x_0, x_1, \ldots, x_n)$ and $x^{\infty} = x$ for $x \in \mathcal{X}$. Note that we have assumed that a stopping time might be infinite. If there exists x such that $N(x) = \infty$, then $\delta^*(x)$ must still be defined, even though it is hardly a "terminal" decision. For convenience, we will often write summations like $\sum_{n=1}^{\infty+}$ to indicate that one extra term for $n = \infty$ is to be included in the usual sum $\sum_{n=1}^{\infty}$. One way to prevent decision rules from taking infinite samples with positive probability is to require that $\sum_{i=1}^{\infty} c_i = \infty$ so that any rule that takes infinite samples with positive probability must have infinite risk.

If we can restrict attention to decision rules (δ', N) such that $N \leq n$, then there is an intuitively simple method of finding the optimal sequential decision rule. The idea is to decide what would be the optimal decision and its risk after observing X^n, then compare this to what the risk would be if we only observed X^{n-1}. Whether $N(x) = n$ or $N(x) = n-1$ is decided is based on which is smaller. We now know what the optimal procedure is after observing $n-1$ observations and we know its risk. Compare this to what would be optimal if we stopped after X^{n-2}, and so on. This procedure is called *backward induction*. Consider an illustration.

Example 9.3. Suppose that $\{X_n\}_{n=1}^{\infty}$ are conditionally IID $Ber(\theta)$ given $\Theta = \theta$ and Θ has $U(0,1)$ distribution. Suppose that we can take at most four observations. The action space has $\aleph' = \{0, 1\}$, and the loss function is $L(\theta, (a, n)) = $

$0.01n + L'(\theta, a)$ with

$$L'(\theta, a) = \begin{cases} 1 & \text{if } \theta > 0.4 \text{ and } a = 0, \\ 1 & \text{if } \theta \leq 0.4 \text{ and } a = 1, \\ 0 & \text{otherwise.} \end{cases}$$

It follows that the optimal action, after N is determined, is to choose $a = 1$ if the posterior probability of $\Theta \leq 0.4$ is less than 0.5. If we observe $X^4 = x^4$, there are five possible posteriors depending on the value of $y_4 = \sum_{i=1}^{4} x_i$. The risks are the probabilities of wrong decision plus 0.04 for the four observations.

y_4	Posterior	$\Pr(\Theta \leq 0.4 \mid X = x)$	a	Risk
0	$Beta(1, 5)$	0.9222	0	0.1178
1	$Beta(2, 4)$	0.6630	0	0.3770
2	$Beta(3, 3)$	0.3174	1	0.3574
3	$Beta(4, 2)$	0.0870	1	0.1270
4	$Beta(5, 1)$	0.0102	1	0.0502

Next, suppose that we only observe $X^3 = x^3$. Let $y_3 = x_1 + x_2 + x_3$. The posterior will be $Beta(y_3 + 1, 4 - y_3)$, and the predictive distribution for X_4 is $Ber([y_3 + 1]/5)$. The risk for stopping is just 0.03 (for the three observations) plus the probability of wrong decision based on three observations. The risk for continuing is the weighted average of the two possible risks that could occur depending on the value of X_4. For example, if $y_3 = 2$, the predictive distribution for X_4 is $Ber(0.6)$, and the risk for continuing is $0.6 \times 0.1270 + 0.4 \times 0.3574 = 0.2192$. For the other values we calculate

	y_3			
	0	1	2	3
Posterior	$Beta(1, 4)$	$Beta(2, 3)$	$Beta(3, 2)$	$Beta(4, 1)$
$\Pr(\Theta \leq 0.4 \mid X = x)$	0.8704	0.5248	0.1792	0.0256
a	0	0	1	1
Risk(stop)	0.1596	0.5052	0.2092	0.0556
$\Pr(X_4 = 1)$	0.2	0.4	0.6	0.8
Risk(continue)	0.1696	0.3692	0.2192	0.0656
Stop	yes	no	yes	yes
Risk	0.1596	0.3692	0.2092	0.0556

So, only if $y_3 = 1$ would we continue to observe X_4. Next, suppose that we only observe $X^2 = x^2$, and let $y_2 = x_1 + x_2$.

	y_2		
	0	1	2
Posterior	$Beta(1, 3)$	$Beta(2, 2)$	$Beta(3, 1)$
$\Pr(\Theta \leq 0.4 \mid X = x)$	0.7840	0.3520	0.0640
a	0	1	1
Risk(stop)	0.2360	0.3720	0.0840
$\Pr(X_3 = 1)$	0.25	0.5	0.75
Risk(continue)	0.2120	0.2892	0.0940
Stop	no	no	yes
Risk	0.2120	0.2892	0.0840

We would continue if $y_2 \in \{0, 1\}$. Next, suppose that we only observe $X_1 = x_1$.

	x_1	
	0	1
Posterior	$Beta(1,2)$	$Beta(2,1)$
$\Pr(\Theta \leq 0.4\|X=x)$	0.6400	0.1600
a	0	1
Risk(stop)	0.3700	0.1700
$\Pr(X_2=1)$	1/3	2/3
Risk(continue)	0.2377	0.1524
Stop	no	no
Risk	0.2377	0.1524

If we take one observation, we will take two. Finally, before we take any observations, $\Pr(\Theta \leq 0.4) = 0.4$, so the terminal decision would be $a=1$ and the risk would be 0.4. On the other hand, $\Pr(X_1=1)=0.5$, so the risk of continuing is $0.5 \times 0.1524 + 0.5 \times 0.2377 = 0.1951$. Hence, we should take the first observation.

To summarize, the optimal procedure is

Data	(1,1,.,.)	(0,0,0,.)	(1,0,1,.)	(0,1,1,.)	(1,0,0,0)
N	2	3	3	3	4
a	1	0	1	1	0

Data	(0,1,0,0)	(0,0,1,0)	(1,0,0,1)	(0,1,0,1)	(0,0,1,1)
N	4	4	4	4	4
a	0	0	1	1	1

where the dots stand for observations that do not need to be taken.

To compare with other procedures, there is the fixed sample size procedure with $n=4$, which has risk 0.2239. This risk is the average of the five possible risks after four observations because each of the five possibilities has probability 1/5. This procedure rejects H if $y_4 \in \{2,3,4\}$. The optimal procedure which takes at most three observations has risk 0.2232 (see Problem 1 on page 567).

After reviewing Example 9.3, it is clear that if δ is a sequential decision rule such that $\Pr(N=0) > 0$, then $\Pr(N=0) = 1$, since we have not allowed any randomization in the decision of whether or not to take observations. The decision as to whether to take any observations is based on the prior distribution and the various costs. No randomness is involved; hence $\{N=0\}$ is either $\emptyset$ or all of S.

For a general problem, let Q be a probability on $\mathcal{V}$ (usually the parameter space Ω).[2] Define

$$\rho_0(Q) = \min_{a \in \aleph'} \int_{\mathcal{V}} L'(u,a) dQ(u),$$

the minimum risk possible without taking any observations if the prior is Q. If Q denotes a prior distribution, then for each n, let $Q_n(\cdot|x)$ denote the conditional distribution obtained from Q by conditioning on $X_0 = x_0, \ldots, X_n = x_n$. (For $n = \infty$, $Q_\infty(\cdot|x)$ denotes conditional probability given $X = x$.) In particular, $Q_0(\cdot|x) = Q$. If N is a stopping time, then

---

[2] It may be that Q is already the conditional distribution obtained from some other probability P after conditioning on some observations.

$Q_N(\cdot|x)$ will denote $\sum_{n=1}^{\infty+} Q_n(\cdot|x) I_{\{n\}}(N(x))$. (See Problem 3 on page 567 for an alternative understanding of Q_n and Q_N.) Suppose that we observe X^n and make the best possible decision. Then $\rho_0(Q_n(\cdot|x)) + \sum_{i=0}^{n} c_i$ is the risk including the cost of observations.

Definition 9.4. Let Q be a prior distribution on $\mathcal{V}$. Suppose that $\delta = (\delta^*, N)$ is a sequential decision rule such that for every n (finite or infinite) and every $x \in \{x : N(x) = n\}$,

$$\int_{\mathcal{V}} L'(u, \delta^*(x)) dQ_n(u|x) = \rho_0(Q_n(\cdot|x)).$$

Then δ is said to *decide optimally after stopping*.

Another way to describe what it means for $\delta = (\delta^*, N)$ to decide optimally after stopping is to say that if $N(x) = n$, then $\delta^*(x)$ is the same as the formal Bayes rule for a sample of size n. A decision rule that decides optimally after stopping may not have an optimal stopping time, but once the decision to stop is made, the optimal terminal decision is made. Clearly, the formal Bayes rule in a sequential decision problem will decide optimally after stopping (see Problem 2 on page 567). For a decision rule that decides optimally after stopping, the Bayes risk is

$$\rho(Q, \delta) = \mathrm{E}\left\{\rho_0(Q_N(\cdot|X)) + \sum_{i=0}^{N} c_i\right\}.$$

Definition 9.5. Suppose that δ decides optimally after stopping and Q is a prior on $\mathcal{V}$. We say that δ is *regular* if $\rho(Q, \delta) \leq \rho_0(Q)$ and if, for every finite $n > 0$ and every $x \in \{x : N(x) > n\}$,

$$\mathrm{E}\left\{\rho_0(Q_N(\cdot|X)) + \sum_{i=0}^{N} c_i \middle| X^n = x^n\right\} < \rho_0(Q_n(\cdot|x)) + \sum_{i=0}^{n} c_i. \quad (9.6)$$

In words, a decision rule is regular if, whenever the stopping time has not yet occurred, the risk of stopping is larger than the risk of continuing. The rule in Example 9.3 on page 537 is regular, as is every backward induction rule. (See Problem 4 on page 568.)

Theorem 9.7. *If δ decides optimally after stopping, then there is a regular δ_1 such that $\rho(Q, \delta_1) \leq \rho(Q, \delta)$.*

PROOF. Define δ_1 as follows. The terminal decision rule for δ_1 is to decide optimally after stopping (just like δ). The stopping time N_1 for δ_1 is the smaller of N (the stopping time for δ) and the first time at which (9.6) fails. Clearly, this is finite and is a stopping time since both sides of (9.6) are $\mathcal{B}^n$ measurable. If δ is regular, then (9.6) never fails and $\delta_1 = \delta$. Next, note that both sides of (9.6) are equal for each x such that $N(x) = n$. So, we can compute $\rho(Q, \delta_1)$ as

$$\sum_{n=0}^{\infty+} \int_{\{x: N_1(x)=n\}} \left[\rho_0(Q_n(\cdot|x)) + \sum_{i=1}^{n} c_i \right] dF_{X^n}(x^n)$$

$$\leq \sum_{n=0}^{\infty+} \int_{\{x: N_1(x)=n\}} \mathrm{E}\left\{ \rho_0(Q_N(\cdot|X)) + \sum_{i=0}^{N} c_i \,\bigg|\, X^n = x^n \right\} dF_{X^n}(x^n)$$

$$= \sum_{n=0}^{\infty+} \mathrm{E}\left[\rho(Q, \delta) | N_1 = n\right] \Pr(N_1 = n) = \rho(Q, \delta). \qquad \square$$

Regular decision rules do not sample too many observations, but they may not sample enough. That is, whenever a regular decision rule continues sampling, the risk for continuing is smaller than the risk for stopping. However, when a regular decision rule stops, the risk for continuing may still be smaller than the risk for stopping. For example, the optimal rule from a class of rules whose stopping times are all bounded by the same n is regular.

Proposition 9.8. *The optimal rule from the class of sequential decision rules that sample no more than n observations is regular.*

Definition 9.9. *If $\delta_i = (\delta_i^*, N_i)$ is a regular decision rule for $i = 1, \ldots, k$, the maximum of $\delta_1, \ldots, \delta_k$, denoted $\max\{\delta_1, \ldots, \delta_k\}$, is the decision rule with stopping time $N = \max\{N_1, \ldots, N_k\}$ and terminal decision rule to decide optimally after stopping.*

Theorem 9.10. *Let Q be a prior on $\mathcal{V}$. If $\delta_1, \ldots, \delta_k$ are regular with finite risk, then $\delta_0 = \max\{\delta_1, \ldots, \delta_k\}$ is regular and $\rho(Q, \delta_0) \leq \rho(Q, \delta_i)$, for $i = 1, \ldots, k$.*

PROOF. We need only prove this for $k = 2$ because the general case follows easily by induction. It is clear that

$$\mathcal{X} = \{x : N_1(x) = N_0(x)\} \cup \{x : N_1(x) < N_2(x)\}.$$

First, suppose that $N_1(x) < N_2(x)$. Then $N_0(x) = N_2(x)$. Let $n = N_1(x)$. Then

$$\mathrm{E}\left\{ \rho_0(Q_{N_0}(\cdot|X)) + \sum_{i=0}^{N_0} c_i \,\bigg|\, X^n = x^n \right\}$$

$$= \mathrm{E}\left\{ \rho_0(Q_{N_2}(\cdot|X)) + \sum_{i=0}^{N_2} c_i \,\bigg|\, X^n = x^n \right\}$$

$$< \rho_0(Q_n(\cdot|x)) + \sum_{i=0}^{n} c_i = \mathrm{E}\left\{\rho_0(Q_{N_1}(\cdot|X)) + \sum_{i=0}^{N_1} c_i \,\bigg|\, X^n = x^n\right\}. \quad (9.11)$$

The first equality is true because δ_0 and δ_2 agree for all x such that $N_2(x) = N_0(x)$. The inequality follows since δ_2 is regular and $N_2(x) > n$. The last equality follows since $N_1(x) = n$. Next, suppose that $N_1(x) = N_0(x)$ and $n = N_2(x) \le N_0(x)$:

$$\begin{aligned}
\mathrm{E}&\left\{\rho_0(Q_{N_0}(\cdot|X)) + \sum_{i=0}^{N_0} c_i \,\bigg|\, X^n = x^n\right\} \\
&= \mathrm{E}\left\{\rho_0(Q_{N_1}(\cdot|X)) + \sum_{i=0}^{N_1} c_i \,\bigg|\, X^n = x^n\right\} \\
&\le \rho_0(Q_n(\cdot|x)) + \sum_{i=0}^{n} c_i \quad (9.12) \\
&= \mathrm{E}\left\{\rho_0(Q_{N_1}(\cdot|X)) + \sum_{i=0}^{N_1} c_i \,\bigg|\, X^n = x^n\right\}.
\end{aligned}$$

The reasons for each line are the same as before except that the inequality is only strict if $N_1(x) > n$. (Note that (9.12) holds even if $n = \infty$.) Together (9.11) and (9.12) show that δ_0 satisfies (9.6). In both of (9.11) and (9.12), $n = \min\{N_1(x), N_2(x)\}$. Write

$$C_n = \{x : \min\{N_1(x), N_2(x)\} = n\}, \quad \mathcal{X} = \bigcup_{n=0}^{\infty} C_n, \quad (9.13)$$

$$\rho(Q, \delta_j) = \sum_{n=0}^{\infty} \int_{C_n} \mathrm{E}\left\{\rho_0(Q_{N_j}(\cdot|X)) + \sum_{i=0}^{N_j} c_i \,\bigg|\, X^n = x^n\right\} dF_{X^n}(x^n)$$

for $j = 0, 1, 2$. Together (9.11) and (9.12) say that the integrand in the second line of (9.13) for $j = 0$ is no greater than for either $j = 1$ or $j = 2$. The inequalities in the conclusion to the theorem follow. □

If $r = \inf_\delta \rho(Q, \delta)$, then there is a sequence $\{\delta_i\}_{i=1}^{\infty}$ such that $r = \lim_{i \to \infty} \rho(Q, \delta_i)$. Finding such a sequence is not as difficult as it may seem.

Definition 9.14. Let $\delta = (\delta^*, N)$ be a regular sequential decision rule. Let N' be a stopping time. The *truncation of δ at N'* is the decision rule with stopping time $\min\{N, N'\}$ and terminal decision optimal after stopping.

Lemma 9.15.[3] *Let δ_0 be the optimal rule in a sequential decision problem, and suppose that δ_0 has finite risk. For each $n = 1, 2, \ldots$, let δ_n be the*

---

[3]This lemma is used to help prove Corollary 9.17.

truncation of δ_0 to at most n observations. Define

$$p_n = \int_{\{x:N_0(x)>n\}} \rho_0(Q_n(\cdot|x))dF_{X^n}(x^n).$$

If $\lim_{n\to\infty} p_n = 0$, then $\lim_{n\to\infty} \rho(Q,\delta_n) = \rho(Q,\delta_0)$.

PROOF. If $N = 0$, the result is trivial, so suppose that $N \geq 1$. For a general decision rule $\delta = (\delta^*, N)$, define

$$T_n(\delta) = \int_{\{x:N(x)=n\}} \rho_0(Q_n(\cdot|x))dF_{X^n}(x^n) + \Pr(N=n)\sum_{i=0}^{n} c_i.$$

We know that for $k = 1, \ldots, n-1$,

$$\{x : N_0(x) = k\} = \{x : N_n(x) = k\}$$

and

$$\{x : N_n(x) = n\} = \{x : N_0(x) = n\} \cup \{x : N_0(x) > n\}.$$

So $T_k(\delta_n) = T_k(\delta_0)$ for $k = 1, \ldots, n-1$ and

$$T_n(\delta_n) = T_n(\delta_0) + p_n + \Pr(N_0 > n)\sum_{i=1}^{n} c_i.$$

So, we can write

$$\rho(Q,\delta_0) = \sum_{n=1}^{\infty} T_n(\delta_0) = \lim_{n\to\infty} \sum_{i=1}^{n} T_i(\delta_0)$$

$$\rho(Q,\delta_n) = \sum_{i=1}^{n-1} T_i(\delta_0) + T_n(\delta_n) = \sum_{i=1}^{n} T_i(\delta_0) + p_n + \Pr(N_0 > n)\sum_{i=1}^{n} c_i.$$

Since $\lim_{n\to\infty} \Pr(N_0 > n) = 0$ and $\lim_{n\to\infty} p_n = 0$, the result follows. □

Lemma 9.16.[4] *Suppose that $L' \geq 0$ and $\lim_{n\to\infty} E\rho_0(Q_n(\cdot|X)) = 0$. Then $\lim_{n\to\infty} p_n = 0$, where p_n is defined in Lemma 9.15.*

PROOF. Since p_n is the integral of $\rho_0(Q_n(\cdot|x))$ over a subset of $\mathcal{X}_0 \times \cdots \times \mathcal{X}_n$ and the integrand is nonnegative, p_n is less than $E\rho_0(Q_n(\cdot|X))$. □

These last two results combine into a corollary that provides a sequence of decision rules with risk converging to the optimal risk.

Corollary 9.17. *Suppose that $L' \geq 0$ and $\lim_{n\to\infty} E\rho_0(Q_n(\cdot|X)) = 0$. Let $\delta_{n,0}$ be the optimal rule among those that take at most n observations. Then $\lim_{n\to\infty} \rho(Q,\delta_{n,0}) = \rho(Q,\delta_0)$.*

---

[4]This lemma is used to help prove Corollary 9.17.

Example 9.18 (Continuation of Example 9.3; see page 537). If $X^n = x^n$ is observed, let $y_n = \sum_{i=1}^n x_i$. Then $\rho_0(Q_n(\cdot|x))$ is the smaller of the two probabilities that a $Beta(y_n + 1, n - y_n + 1)$ random variable is at most 0.4 or is at least 0.4. If y_n/n converges to anything other than 0.4, one of the two probabilities will go to 0 and the other to 1. Since y_n/n will converge to something other than 0.4 with probability 1 and ρ_0 is bounded, the dominated convergence theorem A.57 says that $\lim_{n\to\infty} E\rho_0(Q(\cdot|X_0,\ldots,X_n)) = 0$. Hence, we could find rules with approximately optimal risk by taking a sequence of optimal rules among the classes of those that take at most n observations for $n = 1, 2, \ldots$. The method used for $n = 4$ is easily generalized to arbitrary n. Here are the computed risks for the optimal rules δ_n for several values of n:

n	5	10	20	50	100	200
Risk	0.1921	0.1720	0.1643	0.1631	0.1631	0.1631

After $n = 75$, the risk did not change in the first eight significant digits. After $n = 125$, sixteen significant digits remained constant.

Example 9.19. Suppose that $\{X_n\}_{n=1}^\infty$ are conditionally IID $N(\mu, \sigma^2)$ given $(M, \Sigma) = (\mu, \sigma)$ and $M \sim N(\mu_0, \sigma^2/\lambda_0)$ given $\Sigma = \sigma$ and $\Sigma^2 \sim \Gamma^{-1}(a_0/2, b_0/2)$, with $a_0 > 2$. Let $\aleph' = \mathbb{R}$, and let the loss be $L((\mu, \sigma), (a, n)) = cn + (\mu - a)^2$. The posterior distribution of M given $X^n = x^n$ is $t_{a_n}(\mu_n, b_n/[\lambda_n a_n])$, where

$$\lambda_n = \lambda_0 + n, \qquad \mu_n = \frac{\lambda_0 \mu_0 + n\bar{x}_n}{\lambda_n},$$

$$a_n = a_0 + n, \qquad b_n = b_0 + \sum_{i=1}^n (x_i - \bar{x}_n)^2 + \frac{n\lambda_0}{\lambda_1}(\bar{x}_n - \mu_0)^2.$$

Hence, the optimal decision after stopping at $N = n$ is $a = \mu_n$ and

$$\rho_0(Q_n(\cdot|x)) = \frac{b_n}{(a_n - 2)\lambda_n}.$$

The prior mean of this is $b_0/[(a_0 - 2)(\lambda_0 + n)]$, which goes to 0. It follows that the risk of the optimal procedure that takes at most n observations converges to the optimal risk. If $a_0 \leq 2$, then $\rho_0(Q) = \infty$, and it pays to take one or two observations until $a_n > 2$. At this point, pretend that the problem starts over and use the above reasoning.

If we modify the problem to have loss $L((\mu, \sigma), (a, n)) = cn + (\mu - a)^2/\sigma^2$, then $\rho_0(Q_n(\cdot|x)) = 1/\lambda_n$, which depends on the data only through n. Hence, it is easy to see that the optimal rule has $N = n$ with probability 1, where n provides a minimum to $cn + 1/(\lambda_0 + n)$.

Proposition 9.20. *If there exists finite n such that $\rho_0(Q_n(\cdot|x)) < c_{n+1}$ for all x, then the optimal procedure takes no more than n observations. The optimal procedure is a fixed sample size procedure if $\rho_0(Q_n(\cdot|x))$ depends on the data only through n.*

In general, it is quite difficult to specify the optimal sequential decision procedure. The first part of Example 9.19 is one such case. To find or approximate the optimal rule in general, we will suppose that the cost of each observation is the same and that the available observations are exchangeable. That is, assume that $c_n = c$ for all n and $\{X_n\}_{n=1}^\infty$ are exchangeable. If we let $\rho^*(Q) = \inf_\delta \rho(Q,\delta)$ denote the risk of the optimal rule, then it is not difficult to see that

$$\rho^*(Q) = \min\{\rho_0(Q), E(\rho^*(Q_1(\cdot|X))) + c\}, \tag{9.21}$$

since the second term is just the mean of the optimal risk of continuing after the first observation given the first observation. If this is smaller than the optimal risk for no data, then it is the optimal risk. Otherwise, the optimal decision is to take no data and $\rho_0(Q)$ is the optimal risk. Clearly, the optimal sequential decision rule is to stop sampling at $N(x) = n$, where n is the first time that $\rho_0(Q_n(\cdot|X)) = \rho^*(Q_n(\cdot|X))$. This prescription is only useful if we know ρ^*. As we will demonstrate in the next theorem, we can approximate ρ^* by using successive substitution (see Section 8.5).

Theorem 9.22. *Let Q be a probability measure and suppose that $L' \geq 0$ and $\lim_{n\to\infty} E\rho_0(Q_n(\cdot|X)) = 0$. Define*

$$\rho_{n+1}(Q) = \min\{\rho_0(Q), E(\rho_n(Q_1(\cdot|X))) + c\},$$

for $n = 0, 1, \ldots$. Then $\lim_{n\to\infty} \rho_n(Q) = \rho^*(Q)$ and $\rho_n(Q)$ is the risk of the optimal rule among those that take at most n observations.

PROOF. Clearly, in light of Corollary 9.17, we need only prove that ρ_n is the optimal risk for rules that take at most n observations. We will use induction. We know that ρ_0 is the optimal risk among rules that take no observations. Suppose that ρ_k is the optimal risk among rules that take at most k observations for some $k \geq 0$. Then

$$E(\rho_k(Q_1(\cdot|X))) + c \tag{9.23}$$

is the risk for taking at least one observation and then using the optimal rule that takes at most k more observations. The optimal rule that takes at most $k+1$ observations must either take at least one observation or take no observations. Hence, the risk of the optimal rule that takes at most $k+1$ observations is the smaller of (9.23) and $\rho_0(Q)$. That is,

$$\min\{\rho_0(Q), E(\rho_k(Q_1(\cdot|X))) + c\} = \rho_{k+1}(Q). \qquad \square$$

Theorem 9.22 can be applied to $Q_k(\cdot|X)$ to produce the following corollary.

Corollary 9.24. *Let Q be a probability measure, and suppose that $L' \geq 0$ and $\lim_{n\to\infty} E\rho_0(Q_n(\cdot|X)) = 0$. For each n and k, the conditional mean of the risk of the optimal rule among those that take at most $n+k$ observations given the first k observations and given that the optimal rule takes at least k observations is $\rho_n(Q_k(\cdot|X)) + ck$.*

Corollary 9.24 can be used to define an alternative decision rule.

546   Chapter 9. Sequential Analysis

Definition 9.25. The decision rule that continues to sample until the first n such that $\rho_0(Q_n(\cdot|x)) = \rho_k(Q_n(\cdot|x))$ is called the *k-step look-ahead rule*.

Example 9.26 (Continuation of Example 9.19; see page 544). It is incredibly difficult to calculate ρ_n for $n > 2$. We illustrate here how to calculate ρ_n for $n = 1, 2$. The posterior distribution is determined by four hyperparameters (a, b, μ, λ) and
$$\rho_0(a, b, \mu, \lambda) = \frac{b}{\lambda(a-2)}.$$

After observing $X_1 = x$, let the posterior hyperparameters be
$$\begin{aligned}(a, b, \mu, \lambda)(x) &= \left(a+1, b + \frac{\lambda}{\lambda+1}(x-\mu)^2, \frac{\lambda\mu+x}{\lambda+1}, \lambda+1\right) \\ &= (a+1, b(x), \mu(x), \lambda+1).\end{aligned}$$

We can write $b(X_1) = (1+Y^2)b$, where
$$Y = \frac{X_1 - \mu}{\sqrt{b}}\sqrt{\frac{\lambda}{\lambda+1}} \sim t_a\left(0, \frac{1}{a}\right). \tag{9.27}$$

In particular, $E(Y^2) = 1/(a-2)$. It follows that
$$E(\rho_0((a, b, \mu, \lambda)(X_1))) = bE\frac{1+Y^2}{(\lambda+1)(a-1)} = \frac{b}{(\lambda+1)(a-2)}.$$

So,
$$\begin{aligned}\rho_1(a, b, \mu, \lambda) &= \min\left\{\frac{b}{\lambda(a-2)}, c + \frac{b}{(\lambda+1)(a-2)}\right\} \\ &= \frac{b}{(\lambda+1)(a-2)} + \min\left\{\frac{b}{\lambda(\lambda+1)(a-2)}, c\right\},\end{aligned}$$

$$\begin{aligned}\rho_1((a, b, \mu, \lambda)(x)) &= \frac{b(x)}{(\lambda+2)(a-1)} \\ &\quad + \begin{cases} c & \text{if } c \leq \frac{b(x)}{(\lambda+1)(\lambda+2)(a-1)}, \\ \frac{b(x)}{(\lambda+1)(\lambda+2)(a-1)} & \text{if not} \end{cases} \\ &= b\frac{1+y^2}{(\lambda+2)(a-1)} + \begin{cases} c & \text{if } |y| \geq r, \\ b\frac{1+y^2}{(\lambda+1)(\lambda+2)(a-1)} & \text{if not,} \end{cases}\end{aligned}$$

where y is as in (9.27) once again, and
$$r = \begin{cases} \sqrt{\frac{c(\lambda+1)(\lambda+2)(a-1)}{b} - 1} & \text{if } c(\lambda+1)(\lambda+2)(a-1) \geq b, \\ 0 & \text{if not.} \end{cases}$$

It follows that $E(\rho_1((a, b, \mu, \lambda)(X_1)))$ equals
$$\frac{b}{(\lambda+2)(a-2)} + c(1-p) + \frac{b}{(\lambda+1)(\lambda+2)(a-1)}\int_{-r}^{r}(1+y^2)f_Y(y)dy$$

$$= \frac{b}{(\lambda+2)(a-2)} + c(1-p)$$
$$+ \frac{b}{(\lambda+1)(\lambda+2)(a-1)} \int_{-r}^{r} \frac{\Gamma\left(\frac{a+1}{2}\right)}{\Gamma\left(\frac{a}{2}\right)\sqrt{\pi}} (1+y^2)^{-\frac{a-1}{2}} dy$$
$$= \frac{b}{(\lambda+2)(a-2)} + c(1-p) + \frac{b}{(\lambda+1)(\lambda+2)(a-2)} q,$$

where $p = \Pr(|Y| \le r)$ and $q = \Pr(|Z| \le r)$, where $Z \sim t_{a-2}(0, 1/[a-2])$.

We could now calculate $\rho_2(a, b, \mu, \lambda)$ after each observation. If $\rho_0(a, b, \mu, \lambda)$ is greater than ρ_2, we should continue to sample. If $\rho_0(a, b, \mu, \lambda)$ equals ρ_2, the two-step look-ahead rule would stop. We could, however, try to achieve a better approximation to ρ^*. One way to do this might be to numerically integrate $\rho_2((a, b, \mu, \lambda)(x))$ times the predictive density of the next observation in order to approximate $\rho_3(a, b, \mu, \lambda)$.

Consider the results in Table 9.30. We used a prior with $a_0 = 3$, $b_0 = 8$, $\mu_0 = 0$, and $\lambda_0 = 1$. The cost per observation was $c = 0.1$. After the fourth observation, we do not know whether or not $\rho_0 = \rho^*$. If we numerically integrate ρ_2, we get $\rho_3 = \rho_2$. This means that we would have to consider at least four more observations before there was any chance that the optimal rule would continue sampling. But four more observations would cost 0.4 more without taking into account the loss from squared error. Since the mean of ρ_0 with four more observations is just 5/9 times the current ρ_0, which equals 0.337076, it seems unlikely that four more observations would bring the risk down enough to justify continuing. In fact, the lowest possible posterior risk we could obtain from sampling four more observations would occur if all four of them were equal to the current posterior mean, and then the risk would be 0.587264, which is barely less than ρ_0.

Another way to approximate ρ^* is from below. It is possible (see Problem 6 on page 568) to show that if $0 \le \gamma_0 \le \rho^*$ (for example, $\gamma_0 = 0$) and
$$\gamma_n(Q) = \min\{\rho_0(Q), \mathrm{E}(\gamma_{n-1}(Q_1(\cdot|X))) + c\} \qquad (9.28)$$
for $n = 1, 2, \ldots$, then $\gamma_n \le \rho^*$ for all n and $\lim_{n\to\infty} \gamma_n(Q) = \rho^*(Q)$.

Example 9.29 (Continuation of Example 9.3; see page 537). Suppose that we observe $X_1 = X_2 = 1$ and we are concerned with whether or not the optimal rule stops at this point. We already saw that the optimal rule that takes at most four observations stops at this point, but the optimal rule might continue. The terminal risk is 0.064 (not counting cost of observations). The posterior is $Beta(3, 1)$. Treating $Beta(3, 1)$ as the prior, we can compute ρ_n and γ_n for as many n as we desire. We get $\rho_n = 0.064$ for all n and $\gamma_n = 0.064$ for $n \ge 33$.

TABLE 9.30. Two-Step Look-Ahead Rule for Example 9.26

i	X_i	ρ_0	ρ_2
0		8.000000	2.866667
1	-0.129354	2.002092	1.201046
2	-2.158607	1.214599	0.928760
3	1.558454	0.935753	0.915571
4	-0.677818	0.606737	0.606737

This means that the optimal risk for continuing from this point is 0.064 and we should stop now.

Suppose that we observe $X_1 = X_4 = 1$ and $X_2 = X_3 = 0$. The optimal rule that takes at most four observations has to stop at this point, and the terminal risk is 0.3174. The posterior is $Beta(3,3)$. Treating this as the prior, we could calculate ρ_n and γ_n for many n. At $n = 100$, they are both 0.2274. This means that the optimal rule would continue sampling and that the optimal risk for continuing (not counting cost of current observations) is 0.2274.

9.2 The Sequential Probability Ratio Test

Just as hypothesis tests can be introduced as special cases of decision rules, sequential hypothesis tests are special cases of sequential decision rules. Just as sequential decision rules require a more general setup than fixed sample size rules, sequential hypothesis tests require a slightly more general setup than fixed sample size tests.

Definition 9.31. Suppose that $X_i \in \mathcal{X}_i$ are random quantities for $i = 1, 2, \ldots$. Let $X = (X_1, X_2, \ldots)$.[5] Let $\mathcal{P}_0$ be a parametric family of distributions for X with parameter space Ω. Let $\Omega_H \cap \Omega_A = \emptyset$ and $\Omega_H \cup \Omega_A = \Omega$. A *sequential test* of a hypothesis $H : \Theta \in \Omega_H$ versus $A : \Theta \in \Omega_A$ is a pair of functions (ϕ, N) where N is a stopping time and $\phi : \mathcal{X} \to [0,1]$ gives the conditional probability of rejecting H given $X = x$.

Example 9.32. Let $\{X_n\}_{n=1}^\infty$ be conditionally IID with $N(\theta, 1)$ distribution given $\Theta = \theta$. Let $\Omega_H = (-\infty, \theta_0]$ and $\Omega_A = (\theta_0, \infty)$. Let $\{v_n\}_{n=1}^\infty$ and $\{w_n\}_{n=1}^\infty$ be sequences of positive real numbers. The following is a sequential test of H versus A:

$$N = \min\{n : \overline{x}_n - \theta_0 \notin (-w_n, v_n)\},$$
$$\phi(x) = \begin{cases} 1 & \text{if } N < \infty \text{ and } \overline{x}_N \geq v_n, \\ 0 & \text{if } N < \infty \text{ and } \overline{x}_N \leq -w_n, \text{ or if } N = \infty, \end{cases}$$

where $\overline{x}_n$ is the average of the first n coordinates of x.

The Neyman–Pearson lemma 4.37 was the starting point from which the theory of hypothesis testing originated. In sequential testing problems, there is a similar starting point. We need to begin with a parameter space consisting of only two points $\Omega = \{0, 1\}$. Suppose that P_i has a density f_i with respect to some measure ν (such as $P_0 + P_1$). That is, $\{X_n\}_{n=1}^\infty$ are conditionally IID with density f_i given $\Theta = i$. When we have observed $X_1 = x_1, \ldots, X_n = x_n$, we will calculate the likelihood ratio

$$L_n(x) = \frac{\prod_{i=1}^n f_1(x_i)}{\prod_{i=1}^n f_0(x_i)},$$

---

[5] Classical decision rules cannot stop at $N = 0$, because prior information is not used. Hence, we have dispensed with the X_0 term in this setting.

which tells us how much more likely the data are under P_1 than under P_0. The *sequential probability ratio test* [see Wald (1947)] SPRT(B, A) is, for each n, to reject $H : \Theta = 0$ if $L_n(x) \geq A$, accept H if $L_n(x) \leq B$, and to continue sampling if $B < L_n(x) < A$, where $0 < B < 1 < A$. Another way to write this is to let

$$N(x) = \inf\{n : L_n(x) \notin (B, A)\},$$

and reject H if $L_{N(x)}(x) \geq A$, accept H if $L_{N(x)}(x) \leq B$.

It is clear that $\{x : N(x) = n\}$ is measurable with respect to the correct σ-field. We would like to show that N is finite, a.s.

Theorem 9.33. *Let $\{Z_n\}_{n=1}^\infty$ be IID with $\mathrm{Var}(Z_i) > 0$. Let $S_n = \sum_{i=1}^n Z_i$ and $N = \inf\{n : S_n \notin (b, a)\}$, where $b < a$. Then $\Pr(N < \infty) = 1$.*

PROOF. Let $c = |a| + |b|$. Choose r large enough so that $r\mathrm{Var}(Z_i) > c^2$. For each multiple of r, that is $n = rk$, write

$$\Xi_i = \sum_{j=(i-1)r+1}^{ir} Z_j, \quad S_n = \Xi_1 + \cdots + \Xi_k.$$

If $|\Xi_m| \geq c$ for some m, then $N \leq rm$ because S_i would have to move across one of the boundaries between $i = r(m-1)$ and $i = rm$ if it has not done so already, since c is the distance between the boundaries. It follows that

$$\{N = \infty\} \subseteq \{|\Xi_j| < c, \text{ for all } j\}.$$

We know that $\mathrm{E}\Xi_j^2 \geq r\mathrm{Var}(Z_j) > c^2$. From this it follows that $p = \Pr(|\Xi_j| \geq c) > 0$. Since the Ξ_j are IID,

$$\Pr(N = \infty) \leq \Pr(|\Xi_j| < c, i = 1, 2, \ldots) = \prod_{i=1}^\infty (1-p) = 0. \quad \square$$

When we apply Theorem 9.33, we will let $Z_i = \log[f_1(X_i)/f_0(X_i)]$, $a = \log A$, and $b = \log B$.

Theorem 9.34. *If $\alpha = P_0(L_N \geq A)$ and $\beta = P_1(L_N \leq B)$, then $\alpha \leq (1-\beta)/A$ and $\beta \leq (1-\alpha)B$.*

PROOF. Since $\{N = n\}$ is in the σ-field generated by $X_1, \ldots, X_n$, it follows that

$$\alpha = \sum_{n=1}^\infty P_0(N = n, L_n \geq A)$$

$$= \sum_{n=1}^\infty \int_{\{N=n, L_n \geq A\}} \prod_{i=1}^n f_0(x_i) d\nu(x_1) \cdots d\nu(x_n)$$

$$\begin{aligned}
&= \sum_{n=1}^{\infty} \int_{\{N=n, L_n \geq A\}} \prod_{i=1}^{n} \frac{f_0(x_i)}{f_1(x_i)} \prod_{i=1}^{n} f_1(x_i) d\nu(x_1) \cdots d\nu(x_n) \\
&= \sum_{n=1}^{\infty} \int_{\{N=n, L_n \geq A\}} \frac{1}{L_n} \prod_{i=1}^{n} f_1(x_i) d\nu(x_1) \cdots d\nu(x_n) \\
&\leq \frac{1}{A} \sum_{n=1}^{\infty} \int_{\{N=n, L_n \geq A\}} \prod_{i=1}^{n} f_1(x_i) d\nu(x_1) \cdots d\nu(x_n) \\
&= \frac{1}{A} P_1(L_N \geq A) = \frac{1}{A}(1-\beta).
\end{aligned}$$

Similarly, $\beta = P_1(L_N \leq B) \leq B P_0(L_N \leq B) = B(1-\alpha)$. $\square$

If we ignore the overshoot of the boundaries, we can replace the inequalities by equalities and solve the equations for

$$\alpha \approx \frac{1-B}{A-B}, \qquad \beta \approx B\frac{A-1}{A-B},$$
$$A \approx \frac{1-\beta}{\alpha}, \qquad B \approx \frac{\beta}{1-\alpha}.$$

Theorem 9.35. *Let α^* and β^* be strictly between 0 and 1. The SPRT with $A = (1-\beta^*)/\alpha^*$ and $B = \beta^*/(1-\alpha^*)$ has operating characteristics $\alpha = P_0(L_N \geq A)$ and $\beta = P_1(L_N \leq B)$, which satisfy $\alpha + \beta \leq \alpha^* + \beta^*$.*

PROOF. If $\alpha \leq \alpha^*$ and $\beta \leq \beta^*$, the result is clearly true. So, suppose that either $\beta > \beta^*$ or $\alpha > \alpha^*$. (We will see shortly that both inequalities cannot occur simultaneously.) If $\beta > \beta^*$, then $1-\beta < 1-\beta^*$ and

$$\alpha \leq \frac{1}{A}(1-\beta) = \frac{\alpha^*}{1-\beta^*}(1-\beta) < \alpha^*.$$

It now follows that

$$\beta^* < \beta \leq B(1-\alpha) = \beta^* \frac{1-\alpha}{1-\alpha^*}.$$

Hence,

$$0 < \beta - \beta^* \leq \beta^* \left(\frac{1-\alpha}{1-\alpha^*} - 1\right) = \beta^* \frac{\alpha^* - \alpha}{1-\alpha^*}.$$

It follows that

$$\begin{aligned}
\alpha^* + \beta^* - \alpha - \beta &= (\alpha^* - \alpha) + (\beta^* - \beta) \\
&\geq \alpha^* - \alpha - \beta^* \frac{\alpha^* - \alpha}{1-\alpha^*} \\
&= (\alpha^* - \alpha)(1 - B) > 0.
\end{aligned}$$

Similarly, if $\alpha > \alpha^*$, we can show that $\beta^* > \beta$ and

$$\alpha^* + \beta^* - \alpha - \beta \geq (\beta^* - \beta)\left(1 - \frac{1}{A}\right) > 0.$$ □

Example 9.36. Suppose that $X_i \sim Ber(\theta)$ given $\Theta = \theta$. Suppose that $\Theta \in \{0.25, 0.75\}$ and $H : \Theta = 0.25$. Then

$$Z_i = \log \frac{f_1(X_i)}{f_0(X_i)} = \begin{cases} -\log 3 & \text{if } X_i = 0, \\ \log 3 & \text{if } X_i = 1. \end{cases}$$

There will be no overshoot of the boundaries if $a = k_1 \log 3$ and $b = -k_2 \log 3$ for k_1 and k_2 integers. Here are some examples:

k_1	k_2	α	β	A	B
1	1	0.25	0.25	3	0.3333
2	2	0.1	0.1	9	0.1111
2	1	0.077	0.308	9	0.3333
1	2	0.308	0.077	3	0.1111
3	3	0.036	0.036	27	0.0370

Suppose that we choose the level 0.1 test with $k_1 = k_2 = 2$. We could calculate the mean of N, the expected number of observations needed. It is clear that N is even and that $N = 2k$ if and only if the first $2k - 2$ observations come in pairs 0,1 or 1,0 and the last two are 1,1 or 0,0. So,

$$P_0(N = 2k) = 0.625 \times 0.375^{k-1}, \quad k = 1, 2, \ldots.$$

It follows that $E_0(N) = \sum_{k=1}^{\infty} 2k P_0(N = 2k) = 3.2$. It is easy to see that $E_1(N) = 3.2$ also.

To compare this to a fixed sample size procedure, it takes $n = 6$ to have $\alpha = \beta = 0.1035$ (not quite as good as the sequential procedure). The test has test function

$$\phi_6(x) = \begin{cases} 1 & \text{if } \sum_{i=1}^{6} x_i \in \{4,5,6\}, \\ 0.5 & \text{if } \sum_{i=1}^{6} x_i = 3, \\ 0 & \text{if } \sum_{i=1}^{6} x_i \in \{0,1,2\}. \end{cases}$$

This test takes nearly twice as many observations and has higher error probabilities. One can calculate that $P_0(N \leq 6) = 0.947$, so there is some chance that the sequential procedure will need more observations. But this will only occur for data sets in which ϕ_6 randomizes. In fact, the two tests make almost all the same decisions based on six observations. The only disagreements come when two 1s are followed by four 0s or when two 0s are followed by four 1s, although ϕ_6 does randomize sometimes when the sequential procedure makes a terminal decision. For example, if the first six observations are 0,1,1,1,0,0, then the sequential procedure would reject H after four observations, but ϕ_6 would randomize.

Suppose that a Bayesian believed that $\Pr(\Theta = 0.25) = 0.5$ before seeing any data. Then, after n observations with x successes,

$$\Pr\left(\Theta = 0.25 \left| \sum_{i=1}^{n} X_i = x\right.\right) = \frac{0.5 \times 0.25^x 0.75^{n-x}}{0.5 \times 0.25^x 0.75^{n-x} + 0.5 \times 0.75^x 0.25^{n-x}}$$

$$= \left(1 + 3^{2x-n}\right)^{-1}.$$

For even n, $x = 1 + n/2$ leads to a posterior probability of $\Theta = 0.25$ equal to 0.1. Similarly, $x = n/2 - 1$ leads to a posterior probability equal to 0.9. The SPRT with $\alpha = \beta = 0.1$ turns out to be to reject H as soon as the posterior probability that H is true falls to 0.1 and accept to H as soon as it rises to 0.9, if we have equal prior probabilities to start.

An interesting calculation can be done in Example 9.36. We found that $\mathrm{E}_0(N) = 3.2$. Notice also that

$$\mathrm{E}_0(S_N) = 0.1 \times 2\log 3 + 0.9 \times (-2\log 3) = -1.6 \log 3.$$

Since $\mathrm{E}_0(Z_i) = 0.25 \log 3 - 0.75 \log 3 = -0.5 \log 3$, we see that $\mathrm{E}_0(S_N) = \mathrm{E}_0(Z_i)\mathrm{E}_0(N)$. It is as if N were fixed in advance!

Theorem 9.37 (Wald's Lemma). *Let $\{Z_n\}_{n=1}^\infty$ be IID such that $\mathrm{E}(Z_i)$ exists. Let N be a stopping time such that $\mathrm{E}(N) < \infty$. If $S_N = \sum_{i=1}^N Z_i$, then $\mathrm{E}(S_N) = \mathrm{E}(Z_i)\mathrm{E}(N)$.*

PROOF. We can write $S_N = \sum_{n=1}^\infty Z_n I_{\{n,n+1,\ldots\}}(N)$. Now write

$$\begin{aligned}
\mathrm{E}(S_N) &= \mathrm{E}\left(\sum_{n=1}^\infty Z_n I_{\{n,n+1,\ldots\}}(N)\right) \\
&= \mathrm{E}\left(\sum_{n=1}^\infty Z_n^+ I_{\{n,n+1,\ldots\}}(N)\right) - \mathrm{E}\left(\sum_{n=1}^\infty Z_n^- I_{\{n,n+1,\ldots\}}(N)\right) \\
&= \sum_{n=1}^\infty \mathrm{E}(Z_n I_{\{n,n+1,\ldots\}}(N)).
\end{aligned}$$

Since $I_{\{n,n+1,\ldots\}}(N) = 1 - I_{\{0,1,\ldots,n-1\}}$ is a function of $Z_1,\ldots,Z_{n-1}$, it is independent of Z_n. Hence

$$\mathrm{E}(Z_n I_{\{n,n+1,\ldots\}}(N)) = \mathrm{E}(Z_n)\Pr(N \geq n) = \mathrm{E}(Z_1)\Pr(N \geq n).$$

It follows that $\mathrm{E}(S_N) = \sum_{n=1}^\infty \mathrm{E}(Z_1)\Pr(N \geq n) = \mathrm{E}(Z_1)\mathrm{E}(N)$. $\square$

Wald's lemma can be used to help approximate the expected value of N under distributions other than the hypothesis and alternative. If we approximate by assuming that there is no overshoot, then

$$S_N = \begin{cases} a & \text{if reject } H, \\ b & \text{if accept } H. \end{cases}$$

All we need to complete the approximation is $\Pr(\text{reject } H)$.

Lemma 9.38. *If $\{X_n\}_{n=1}^\infty$ are IID with distribution P, and there exists $h \neq 0$ such that*

$$\int \left(\frac{f_1(x)}{f_0(x)}\right)^h dP(x) = 1,$$

9.2. The Sequential Probability Ratio Test

then, to the approximation of no overshoot, for the SPRT(B, A),

$$P(\text{reject } H) = \frac{1 - B^h}{A^h - B^h}.$$

PROOF. If $h > 0$, consider the $\text{SPRT}(B^h, A^h)$ as a test of the hypothesis that the density of each observation (with respect to P) is 1 versus the alternative that the density is $(f_1/f_0)^h$. The likelihood ratio is $L_n^* = L_n^h$ and $B^h < L_n^* < A^h$ if and only if the original likelihood ratio satisfies $B < L_n < A$. So

$$P(\text{reject } H) = P(L_N \geq A) = P(L_N^* \geq A^*) \approx \frac{1 - B^h}{A^h - B^h}.$$

If $h < 0$, consider the $\text{SPRT}(A^{-h}, B^{-h})$ as a test of H^* that the density is $(f_1/f_0)^h$ versus the alternative that the density is 1. Then the likelihood ratio is $L_n^* = L_n^{-h}$ and

$$\begin{aligned} P(\text{reject } H) &= P(L_N \geq A) = P(L_N^* \leq B^*) \\ &\approx A^{-h} \left(\frac{B^{-h} - 1}{B^{-h} - A^{-h}} \right) = \frac{1 - B^h}{A^h - B^h}. \end{aligned}$$

$\square$

Using the no-overshoot approximation,

$$\begin{aligned} \mathrm{E}(S_N) &= aP(\text{reject } H) + bP(\text{accept } H) \\ &= b + (a - b)P(\text{reject } H) \\ &\approx \log(B) + \log\left(\frac{A}{B}\right)\left(\frac{1 - B^h}{A^h - B^h}\right). \end{aligned}$$

So, if $\mathrm{E}(Z_1) \neq 0$, we get

$$\mathrm{E}(N) \approx \frac{\log(B) + \log\left(\frac{A}{B}\right)\left(\frac{1-B^h}{A^h-B^h}\right)}{\mathrm{E}(Z_1)}.$$

Example 9.39 (Continuation of Example 9.36; see page 551). Suppose that $\{X_n\}_{n=1}^\infty$ are IID $Ber(0.6)$, but we are testing $H : \Theta = 0.25$ versus $A : \Theta = 0.75$. We have

$$\frac{f_1(x)}{f_0(x)} = \begin{cases} \frac{1}{3} & \text{if } x = 0, \\ 3 & \text{if } x = 1. \end{cases}$$

If $h = -\log_3(1.5) = -0.36907$, then $0.4(1/3)^h + 0.6 \times 3^h = 1$. We can now calculate

$$\mathrm{E}_{0.6}(S_N) = \log\left(\frac{1}{9}\right) + \log(81)\left(\frac{1 - \left(\frac{1}{9}\right)^{-0.36907}}{9^{-0.36907} - \left(\frac{1}{9}\right)^{-0.36907}}\right) = 0.8451,$$

$$\mathrm{E}_{0.6}(Z_1) = 0.6 \times \log(3) - 0.4 \times \log(3) = 0.2197,$$

$$\mathrm{E}_{0.6}(N) = \frac{0.8451}{0.2197} = 3.846.$$

Notice that the mean stopping time is longer when Θ is between the hypothesis and the alternative.

If $\{X_n\}_{n=1}^\infty$ are IID $Ber(0.5)$, then $h = 0$ is the only value that works in the equation in Lemma 9.38. Hence, Lemma 9.38 has nothing to say about this case.

The following result has a proof similar to that of Wald's lemma, but applies to the case not yet handled in Example 9.36.

Proposition 9.40. *Suppose that $\{Z_n\}_{n=1}^\infty$ are IID, with $\mathrm{E}(Z_i) = 0$ and $\mathrm{E}(Z_i^2) = \sigma^2$. Suppose that N is a stopping time such that $\mathrm{E}(N) < \infty$. Then $\mathrm{E}(S_N^2) = \sigma^2 \mathrm{E}(N)$.*

Example 9.41 (Continuation of Example 9.36; see page 551). If $\{X_n\}_{n=1}^\infty$ are IID $Ber(0.5)$, then $\mathrm{E}(Z_i) = 0$ and $\mathrm{E}(Z_i^2) = (\log(3))^2 = 1.2069$. Also, $\mathrm{E}(S_N^2) = 4(\log(3))^2 = 4.8278$. It follows that $\mathrm{E}(N) = 4$. Of course, this example is simple enough that we could calculate $\mathrm{E}_\theta(N)$ for all θ without any of these theorems. See Problem 9 on page 568.

The SPRT has an optimal property in terms of expected sample size which follows from its being a Bayes rule in a sequential decision problem. This is very much like the Neyman–Pearson fundamental lemma 3.87 in which a minimal complete class of Bayes rules was found in the fixed sample size problem for a simple hypothesis and a simple alternative.

Lemma 9.42. *Suppose that $0 < \gamma_1 < \gamma_2 < 1$ and that f_0 and f_1 are two different densities with respect to a measure ν. There exist $0 < w < 1$ and $c > 0$, such that for every $\gamma \in [\gamma_1, \gamma_2]$, the SPRT$(B, A)$ with*

$$B = \frac{\gamma}{1-\gamma}\frac{1-\gamma_2}{\gamma_2}, \quad A = \frac{\gamma}{1-\gamma}\frac{1-\gamma_1}{\gamma_1}$$

is a Bayes rule in the sequential decision problem with action space $\aleph = \{0,1\} \times \{1, 2, \ldots\}$, parameter space $\{0, 1\}$, prior distribution $\Pr(\Theta = 0) = \gamma$, and loss function

$$L(f_i, (j, n)) = cn + \begin{cases} w_i & \text{if } i \neq j, \\ 0 & \text{otherwise,} \end{cases}$$

where $w_0 = 1 - w$ and $w_1 = w$.

PROOF. First we will find the general solution of the sequential decision problem, and then we will show that there is one whose solution is SPRT(B, A). To put the problem in testing form, let $\Omega_H = \{f_0\}$. Suppose that a sequential test is $\delta = (\phi, N)$. Define

$$\alpha_0(\delta) = \mathrm{E}_0(\phi(X)), \quad \alpha_1(\delta) = 1 - \mathrm{E}_1(\phi(X)).$$

Then the Bayes risk of δ with respect to prior probability $\gamma = \Pr(\Theta = 0)$ is

$$\rho(\gamma, \delta) = \gamma(w_0 \alpha_0(\delta) + c\mathrm{E}_0[N(X)]) + (1 - \gamma)(w_1 \alpha_1(\delta) + c\mathrm{E}_1[N(X)]).$$

Define, for each $0 \leq \gamma \leq 1$,

$$U(\gamma) = \inf_{\delta} \rho(\gamma, \delta).$$

Since $N(x) \geq 1$ for all x, it follows that $U(\gamma) > 0$ for all γ. Since $\rho(\gamma, \delta)$ is a positive linear function of γ for each δ, it follows that U is the infimum of a collection of positive linear functions; hence, it is concave and continuous on $(0, 1)$ and positive at the two endpoints. Define

$$f_{i,n}(x) = \prod_{j=1}^{n} f_i(x_j),$$

for $i = 0, 1$ and $n = 1, 2, \ldots$, where $x = (x_1, x_2, \ldots)$. The posterior probability of $\{\Theta = 0\}$ given $X_1 = x_1, \ldots, X_n = x_n$ is

$$\gamma_n(x) = \frac{\gamma f_{0,n}(x)}{\gamma f_{0,n}(x) + (1-\gamma) f_{1,n}(x)} = \frac{1}{1 + \frac{1-\gamma}{\gamma} L_n(x)},$$

where $L_n(x)$ is the likelihood ratio after n observations (used in every SPRT). After observing $X_1 = x_1, \ldots, X_n = x_n$, the posterior mean of the loss to be incurred if $N = n$ is $h(\gamma_n(x)) + cn$, where

$$h(\gamma) = \min\{w_0 \gamma, w_1(1-\gamma)\}.$$

The posterior mean of the loss to be incurred if $N > n$ is at least $U(\gamma_n(x)) + cn$. Hence, the Bayes rule is to continue sampling so long as $h(\gamma_n(x)) > U(\gamma_n(x))$ and to stop at $N(x)$ equal to the first n such that $h(\gamma_n(x)) \leq U(\gamma_n(x))$. Note that $h(\gamma)$ is continuous, has a graph shaped like a triangle, and satisfies $h(0) = h(1) = 0$. Since U is concave, it follows that $h(\gamma) > U(\gamma)$ for γ in some interval (g_1, g_2). Figure 9.43 shows the U and h functions for a typical example.[6] (If $h(\gamma) \leq U(\gamma)$ for all γ, define $g_1 = g_2$ to be the value of γ at which h is maximized.) Hence, the Bayes rule continues sampling so long as

$$B_* = \frac{\gamma}{1-\gamma} \frac{1-g_2}{g_2} < L_n(x) < \frac{\gamma}{1-\gamma} \frac{1-g_1}{g_1} = A_*;$$

it rejects H if $L_N(x) \geq A_*$; and it accepts H if $L_N(x) \leq B_*$. Therefore, the Bayes rule is SPRT(B_*, A_*).

The g_1 and g_2 found above depend on the particular decision problem only through w and c, where we assume that $w_1 = w$ and $w_0 = 1 - w$. This is true because the functions h and U depend on the decision problem only

---

[6]The example in Figure 9.43 has f_0 being the $Ber(0.6)$ distribution and f_1 being the $Ber(0.8)$ distribution. Also, $w_0 = 0.4$ and $c = 0.02$. In this example, $g_1 = 0.31$ and $g_2 = 0.48$.

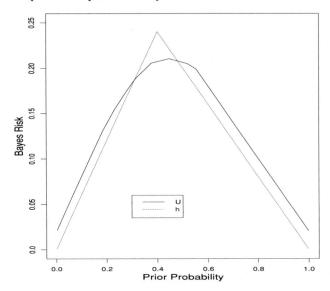

FIGURE 9.43. Typical U and h Functions for SPRT

through these values. So we call the two values $g_1(w,c)$ and $g_2(w,c)$. To finish the proof, we need only find c and w so that $\gamma_i = g_i(w,c)$ for both $i = 1, 2$. Define

$$\beta(w,c) = \frac{g_2(w,c)}{1 - g_2(w,c)},$$

$$\lambda(w,c) = \frac{g_1(w,c)}{\beta(w,c)(1 - g_1(w,c))}.$$

It is easy to see that g_1 and g_2 are also functions of β and λ. Set

$$\lambda_0 = \frac{\gamma_1}{1 - \gamma_1} \frac{1 - \gamma_2}{\gamma_2}, \quad \beta_0 = \frac{\gamma_2}{1 - \gamma_2}.$$

Then, we need to find w and c so that $\lambda(w,c) = \lambda_0$ and $\beta(w,c) = \beta_0$. As $c \downarrow 0$, $U(0)$ and $U(1)$ both approach 0, hence $g_1(w,c)$ tends to 0 and $g_2(w,c)$ tends to 1 as $c \downarrow 0$ for fixed w. Hence, for fixed w, $\lim_{c \downarrow 0} \lambda(w,c) = 0$. Since $\rho(\gamma, \delta)$ increases as c increases for every γ and δ, it follows that $U(\gamma)$ increases as c increases for every γ. Hence $g_1(w,c)$ is a decreasing function of c and $g_2(w,c)$ is increasing in c for fixed w. It follows that λ is strictly increasing in c for fixed w. As $c \to \infty$, eventually, $h(\gamma) \leq U(\gamma)$ for all γ. Let $c_0(w) = \inf\{c : g_1(w,c) = g_2(w,c)\}$. Then $\lim_{c \uparrow c_0(w)} \lambda(w,c) = 1$. Since $0 < \lambda_0 < 1$, there exists a unique $c = c(w)$ such that $\lambda(w,c) = \lambda_0$. As w approaches 0 for fixed c, U approaches the constant function c and h approaches the constant function 0 while the peak of the triangle in the graph of h moves toward $\gamma = 0$. Hence $g_1(w,c)$ and $g_2(w,c)$ approach 0.

Also $c_0(w)$ and $c(w)$ approach 0 as w approaches 0. Hence

$$\lim_{w\downarrow 0} g_2(w,c(w)) = 0, \quad \lim_{w\downarrow 0} \beta(w,c(w)) = 0.$$

As w approaches 1 for fixed c, U approaches the constant function c and h approaches the constant function 0 while the peak of the triangle in the graph of h moves toward $\gamma = 1$. Hence $g_1(w,c)$ and $g_2(w,c)$ approach 1. Also $c_0(w)$ and $c(w)$ approach 0 as w approaches 0. Hence

$$\lim_{w\downarrow 0} g_2(w,c(w)) = 1, \quad \lim_{w\downarrow 0} \beta(w,c(w)) = \infty.$$

Since $\beta(w,c(w))$ is continuous in w, there exists w such that $\beta(w,c(w)) = \beta_0$. □

We are now in position to prove a theorem of Wald and Wolfowitz (1948) that says that the SPRT has the smallest expected sample size of all tests with the same error probabilities. Since fixed sample size tests can be considered as sequential tests in which N is not a function of x, we state the theorem in terms of sequential tests only.

Theorem 9.44. *Let $\Omega_H = \{f_0\}$ and $\Omega_A = \{f_1\}$. Let $B_0 < 1 < A_0$. Let $\delta_0 = (\phi_0, N_0)$ be $SPRT(B_0, A_0)$, and suppose that $E_0(\phi_0(X)) = \alpha$ and $1 - E_1(\phi_0(X)) = \beta$. Among all sequential tests $\delta = (\phi, N)$ for which $N \geq 1$, $E_0(\phi(X)) \leq \alpha$, $1 - E_1(\phi(X)) \leq \beta$, and $E_i(N) < \infty$ for $i = 0, 1$, δ_0 minimizes $E_i(N)$ for $i = 0, 1$.*

PROOF. Let $\delta = (\phi, N)$ be a sequential test of H versus A with $E_0(\phi(X)) = \alpha_1 \leq \alpha$, $1 - E_1(\phi(X)) = \beta_1 \leq \beta$, and $E_i(N) < \infty$ for $i = 0, 1$. Pick $0 < \gamma < 1$ and define

$$\gamma_1 = \frac{\gamma}{A_0(1-\gamma)+\gamma}, \quad \gamma_2 = \frac{\gamma}{B_0(1-\gamma)+\gamma}.$$

Then $0 < \gamma_1 < \gamma < \gamma_2 < 1$ and

$$B_0 = \frac{\gamma}{1-\gamma}\frac{1-\gamma_2}{\gamma_2}, \quad A_0 = \frac{\gamma}{1-\gamma}\frac{1-\gamma_1}{\gamma_1}.$$

Lemma 9.42 says that there exist $0 < w < 1$ and $c > 0$ such that δ_0 is a Bayes rule in the sequential decision problem with action space $\aleph = \{0,1\} \times \{1,2,\ldots\}$, parameter space $\{0,1\}$, prior distribution $\Pr(\Theta = 0) = \gamma$, and loss function

$$L(f_i,(j,n)) = cn + \begin{cases} w_i & \text{if } i \neq j, \\ 0 & \text{otherwise,} \end{cases}$$

where $w_0 = 1 - w$ and $w_1 = w$. It follows that

$$\gamma(w_0\alpha_0 + cE_0 N_0) + (1-\gamma)(w_1\beta_0 + cE_1 N_0)$$
$$\leq \gamma(w_0\alpha_1 + cE_0 N) + (1-\gamma)(w_1\beta_1 + cE_1 N).$$

Since $\alpha_0 \leq \alpha_1$, $\beta_0 \leq \beta_1$, $c > 0$, $w_0 > 0$, and $w_1 > 0$, it follows that

$$\gamma(E_0 N_0 - E_0 N) + (1 - \gamma)(E_1 N_0 - E_1 N) \leq 0.$$

Since this is true for every $\gamma \in (0,1)$, the limit as γ goes to 0 or to 1 of the left-hand side is also less than or equal to 0. These two limits are respectively $E_0 N_0 - E_0 N$ and $E_1 N_0 - E_1 N$. □

The reason for the extra conditions that $E_i(N) < \infty$ for $i = 0, 1$ is that there may be another test with $E_0(N) = \infty$ and very small value of $E_1(N)$.

9.3 Interval Estimation*

In Section 5.2.5, we gave examples of loss functions associated with interval estimation. For example, let the (terminal) action space be the set of all pairs of real numbers (a, b) with $a \leq b$. We could set

$$L'(\theta, (a,b)) = k(b - a) + 1 - I_{[a,b]}(g(\theta)),$$

or

$$L'(\theta, (a,b)) = k(b - a)^2 + 1 - I_{[a,b]}(g(\theta)),$$

or

$$L'(\theta, (a,b)) = k(b - a) + \begin{cases} a - g(\theta) & \text{if } g(\theta) < a, \\ g(\theta) - b & \text{if } g(\theta) > b, \\ 0 & \text{otherwise,} \end{cases}$$

where $k > 0$. The first two loss functions above lead to intervals with equal posterior density at a and b. The third one leads to intervals with equal posterior probability below a and above b. Another alternative is to let $\aleph' = \mathbb{R}$ and set

$$L'(\theta, a) = 1 - I_{[a-d, a+d]}(g(\theta)),$$

where $d > 0$ is some fixed half-width for the interval. In this case, the interval has a fixed width and the coverage probability is determined (along with the center of the interval) by the data.

Example 9.45. Suppose that $\{X_n\}_{n=1}^\infty$ are IID $N(\mu, \sigma^2)$ given $\Theta = (\mu, \sigma)$. Let $g(\theta) = \mu$. Suppose that we want an interval of half-width d and the cost of each observation is c. Suppose that the prior is conjugate. The posterior distribution of M after n observations will be $t_{a_n}(\mu_n, b_n/[a_n \lambda_n])$, where the posterior hyperparameters a_n, b_n, μ_n, and λ_n are as in Example 9.19 on page 544. The optimal decision upon stopping is μ_n. The terminal risk (not counting cost of observation) is

$$\rho_0(a_n, b_n, \mu_n, \lambda_n) = 2\left[1 - T_{a_n}\left(d\sqrt{\frac{a_n \lambda_n}{b_n}}\right)\right],$$

---

*This section may be skipped without interrupting the flow of ideas.

where T_{a_n} stands for the CDF of the $t_{a_n}(0,1)$ distribution. To implement the one-step look-ahead rule, we need to calculate

$$\rho_1(a_n, b_n, \mu_n, \lambda_n) = \min\left\{2\left[1 - T_{a_n}\left(d\sqrt{\frac{a_n\lambda_n}{b_n}}\right)\right],\right.$$
$$\left. c + 2\left[1 - ET_{a_n+1}\left(d\sqrt{\frac{(a_n+1)(\lambda_n+1)}{b_n(1+Y^2)}}\right)\right]\right\},$$

where

$$Y = (X_{n+1} - \mu_n)\frac{\lambda_n}{b_n(\lambda_n+1)} \sim t_{a_n}\left(0, \frac{1}{a_n}\right).$$

Even this formula requires numerical integration to compute. For example, suppose that we use a prior with $a_0 = 2$, $b_0 = 7$, $\mu_0 = 1$, and $\lambda_0 = 1$. The cost of each observation is 0.005 and the half-width of the interval is $d = 1.96$. Table 9.47 contains some data along with ρ_0 and ρ_1. The terminal decision is $\mu_9 = -1.011531$, and so the interval is $[-2.9715, , 0.9485]$. The posterior probability in the interval is 0.9810. The probability is so high because the cost of each observation is so low. If the cost had been 0.01 instead, the one-step look-ahead rule would have stopped after seven observations with the interval $[-3.0214, 0.8986]$ and posterior probability 0.9640.

The classical approach to fixed-width interval estimation has traditionally not been through a loss function. Rather, one requires that sampling continue until one has an interval of a fixed width with the desired confidence coefficient or greater. No cost of observation is taken into account, except possibly for comparing different procedures.

The most naïve procedure would be to compute a fixed sample size coefficient γ confidence interval for each n, and stop at the first n such that the interval has half-width at most d.

Definition 9.46. Let $\{X_n\}_{n=1}^\infty$ be conditionally IID $N(\mu, \sigma^2)$ given $\Theta = (\mu, \sigma)$. Let $\overline{X}_n = \sum_{i=1}^n X_i/n$ and $S_n^2 = \sum_{i=1}^n (X_i - \overline{X}_n)^2/(n-1)$ for each n. Let N_1 be the smallest $n \geq 2$ such that $S_n T_{n-1}^{-1}(1-\alpha/2) \leq \sqrt{n}d$,

TABLE 9.47. One-Step Look-Ahead Rule for Example 9.45

i	X_i	ρ_0	ρ_1
0		0.4047363	0.2816774
1	−1.745003	0.2396304	0.1774288
2	0.793758	0.1178335	0.0881298
3	−4.385832	0.1475386	0.1190706
4	−4.708225	0.1268197	0.1053043
5	−0.363233	0.0797290	0.0675238
6	1.162189	0.0599002	0.0521162
7	−0.244931	0.0360019	0.0330767
8	1.455418	0.0265989	0.0258814
9	−3.079450	0.0189839	0.0189839

where T_{n-1} is the CDF of the $t_{n-1}(0,1)$ distribution. Then the interval $[\overline{X}_{N_1} - d, \overline{X}_{N_1} + d]$ is called the *naïve coefficient $1-\alpha$ sequential confidence interval for M*.

It is not surprising that the naïve confidence interval does not have coverage probability $1 - \alpha$. That is, $P_\theta(\overline{X}_{N_1} - d \leq \mu \leq \overline{X}_{N_1} + d) \neq 1 - \alpha$. We can write an expression for the coverage probability, however:

$$P_\theta(\overline{X}_{N_1} - d \leq \mu \leq \overline{X}_{N_1} + d) \qquad (9.48)$$
$$= \sum_{n=2}^{\infty} P_\theta(\overline{X}_n - d \leq \mu \leq \overline{X}_n + d | N_1 = n) P_\theta(N_1 = n).$$

First, recall that N_1 is the first $n \geq 2$ such that $S_n^2 \leq k_n$ for some sequence $\{k_n\}_{n=1}^{\infty}$. We will prove that the event $\{N_1 = n\}$ is independent of $\overline{X}_n$. This will allow some simplification in (9.48). It is sufficient to show that $S_2^2, \ldots, S_n^2$ are all independent of $\overline{X}_n$. For each $k = 1, 2, \ldots$, consider the $k \times k$ matrix Γ_k whose rows are all unit vectors and whose ith row (for $i < k$) is proportional to the vector with 1 in the first i places and $-i$ in the $(i+1)$st place. The kth row is proportional to the vector of all 1s. It is easy to see that these rows are orthogonal to each other, so Γ_k is orthogonal for each k. For $i < n$, define W_i to be the inner product of the ith row of Γ_n with $(X_1, \ldots, X_n)$.[7] Note that the inner product of the last row of Γ_k and the vector $X^k = (X_1, \ldots, X_k)$ is $\sqrt{k}\overline{X}_k$. So, $\overline{X}_n$ is independent of $W_1, \ldots, W_{n-1}$. Also, since $\|\Gamma_k X^k\| = \sum_{i=1}^{k} X_i^2$, it follows that $S_k^2 = \sum_{i=1}^{k-1} W_i^2$ for $k = 2, 3, \ldots$. Hence $\overline{X}_n$ is independent of $S_2, \ldots, S_n$. This means that we can write

$$P_\theta(\overline{X}_n - d \leq \mu \leq \overline{X}_n + d | N_1 = n) = P_\theta(\overline{X}_n - d \leq \mu \leq \overline{X}_n + d)$$
$$= 2\left[1 - \Phi\left(\sqrt{n}\frac{d}{\sigma}\right)\right].$$

Hence,

$$P_\theta(\overline{X}_{N_1} - d \leq \mu \leq \overline{X}_{N_1} + d) = \sum_{n=2}^{\infty} 2\left[1 - \Phi\left(\sqrt{n}\frac{d}{\sigma}\right)\right] P_\theta(N_1 = n).$$

Some numerical method would be required to calculate $P_\theta(N_1 = n)$, but the argument above can be used to show that it depends on θ only through d/σ.

It should be noted that a Bayesian does not have the same problem with naïve posterior probability intervals. So long as the decision of whether or not to take more data is a measurable function solely of the available data,

---

[7]Note that, for fixed i, W_i is the same no matter which $n > i$ is chosen for its definition.

the posterior distribution of a parameter given the data does not depend on whether or not any more data will be taken. Hence, the Bayesian can declare, after every observation, what is the posterior probability that the parameter lies in any set he or she wishes. If, after n observations, there is an interval of half-width d such that the posterior probability is $1-\alpha$ that M is in that interval, the Bayesian can declare that and stop sampling.

There are some classical procedures that do actually have the desired coverage probability. We will present one such procedure here. It is a two-stage sampling procedure. One chooses an initial sample size n_0 and estimates Σ^2 by $S_{n_0}^2$. Then one collects a second sample whose size depends on $S_{n_0}^2$. Define

$$c = \frac{d}{T_{n_0-1}^{-1}\left(1-\frac{\alpha}{2}\right)}, \quad N_2 = \max\left\{n_0, \left\lfloor \frac{S_{n_0}^2}{c^2}\right\rfloor + 1\right\},$$

where $\lfloor z \rfloor$ denotes the largest integer less than or equal to z. Use the interval centered at $\overline{X}_{N_2}$ with half-width d.

Lemma 9.49. *With the above notation, the conditional distribution of*

$$\sqrt{N_2}\frac{\overline{X}_{N_2} - \mu}{S_{n_0}} \tag{9.50}$$

given $\Theta = (\mu, \sigma)$ *is* $t_{n_0-1}(0,1)$.

PROOF. We can write

$$\overline{X}_{N_2} = \frac{1}{N_2}\left(\sum_{i=1}^{n_0} X_i + \sum_{i=n_0+1}^{N_2} X_i\right).$$

We know that $\sum_{i=1}^{n_0} X_i$ and $\{X_{n_0+i}\}_{i=1}^\infty$ are independent of $S_{n_0}^2$. Since N_2 is constant conditional on $S_{n_0}^2$, we have that the conditional distribution of $\overline{X}_{N_2}$ given $S_{n_0}^2$ is $N(\mu, \sigma^2/N_2)$. So, the conditional distribution of $\sqrt{N_2}(\overline{X}_{N_2} - \mu)/\sigma$ is $N(0,1)$, which is independent of $S_{n_0}^2$. It follows that (9.50) has $t_{n_0-1}(0,1)$ distribution. □

Technically, we could use the interval with half-width

$$\frac{1}{\sqrt{N_2}} S_{n_0} T_{n_0-1}^{-1}\left(1-\frac{\alpha}{2}\right),$$

but this should be pretty close to d.

The problems with this procedure derive from the first stage of sampling. First, there is the question of how to choose n_0, which turns out to be crucial to the performance of the method as we will see in Example 9.52. Second, there is the fact that the estimate of Σ^2 is based only on the first n_0 observations. In order to get a good estimate, n_0 must be large, but then we might sample too much if Σ^2 is small. If we choose n_0 too small, then c is small and N_2 is large.

TABLE 9.51. Classical Fixed-Width Confidence Interval Sample Sizes for Example 9.52

n_0	$S_{n_0}^2$	c	N_2
2	3.222653	0.0238	136
3	6.707906	0.2075	33
4	6.616990	0.3793	18
5	5.885602	0.4983	12
6	6.462292	0.5814	12
7	5.625235	0.6416	9
8	5.809566	0.6870	9
9	5.561759	0.7224	9

Example 9.52 (Continuation of Example 9.45; see page 558). Suppose that we desire a classical sequential confidence interval with half-width 1.96 and coefficient 0.95. We will use the same sequence of data as we had earlier. To correspond more closely with a classical analysis, we should change to an improper prior. In this case, the interval based on the first nine observations is the first one to have posterior probability greater than 0.95 and so this would be the naïve sequential interval. For various values of n_0, we can implement the classical procedure; the results are summarized in Table 9.51. If $n_0 = 7, 8$, or 9, we will get essentially the same result as the naïve interval.

9.4 The Relevance of Stopping Rules

In Section 9.1, we introduced sequential decision problems and examined rules that decide optimally after stopping. In Problem 2 on page 567, the reader is asked to prove that the formal Bayes rule in a sequential decision problem decides optimally after stopping. What exactly is the meaning of deciding optimally after stopping? In words, it means that after the stopping rule says to take no more observations, we then behave exactly as we would if we had observed whatever data we now have under a fixed sample size scheme. This is in stark contrast to classical sequential procedures like the level α SPRT. When the SPRT stops at $N = n$, the terminal decision is most definitely not the same as that of a level α test based on a fixed-sized sample of size n. On the other hand, when the SPRT is viewed as a formal Bayes rule, it is true that the terminal decision is exactly the same as what the formal Bayes rule would be after observing a fixed-sized sample of size n.

In light of Problem 2 on page 567, it makes perfect sense for the Bayesian statistician to make the same decision after stopping as he or she would have made had the data arrived in a nonsequential fashion. Why, then, does the classical statistician behave differently in the two situations? The easiest way to answer this question is to see what would happen if the classical statistician tried to use a fixed sample size terminal decision after stopping.

9.4. The Relevance of Stopping Rules

For a simple example, suppose that $\{X_n\}_{n=1}^{\infty}$ are IID $N(\theta, 1)$ given $\Theta = \theta$. We will consider the problem of testing the hypothesis $H : \Theta = \theta_0$ versus $A : \Theta \neq \theta_0$. Let $\overline{X}_n$ be the average of the first n of the X_i. Given $\Theta = \theta_0$, $\sqrt{n}(\overline{X}_n - \theta_0)$ has $N(0, 1)$ distribution for every n. Hence, for every c,

$$P'_{\theta_0}(\sqrt{n}(\overline{X}_n - \theta_0) > c) = 1 - \Phi(c) > 0.$$

It follows that

$$P'_{\theta_0}\left(\limsup_{n \to \infty} \sqrt{n}(\overline{X}_n - \theta_0) > c\right) > 0.$$

However, the event $\{\limsup_{n \to \infty} \sqrt{n}(\overline{X}_n - \theta_0) > c\}$ is in the tail σ-field $\mathcal{C}$, so the Kolmogorov zero–one Law B.68 says

$$P'_{\theta_0}(\limsup_{n \to \infty} \sqrt{n}(\overline{X}_n - \theta_0) > c) = 1.$$

Similarly, for every c,

$$P'_{\theta_0}(\liminf_{n \to \infty} \sqrt{n}(\overline{X}_n - \theta_0) < c) = 1.$$

Hence, given $\Theta = \theta_0$, for every c, the probability is 1 that there will exist n such that $|\sqrt{n}(\overline{X}_n - \theta_0)| > c$. Let

$$N = \inf\{n : |\sqrt{n}(\overline{X}_n - \theta_0)| > c\},$$

where c is the $1 - \alpha/2$ quantile of the standard normal distribution. It follows that N is a stopping time. Suppose that a classical statistician were to use this stopping time. Suppose that after observing $N = n$, he or she were to use a terminal decision that was the usual level α test of $H : \Theta = \theta_0$ versus $A : \Theta \neq \theta_0$ based on a sample of size n. This test would be to reject H if $|\sqrt{n}(\overline{X}_n - \theta_0)| > c$. This person would reject the hypothesis with probability 1 given $\Theta = \theta_0$. This would not be a level α sequential test. Clearly, from a classical viewpoint, the terminal decision rule has to depend on which stopping time was used and not just the observed data.

The phenomenon just illustrated is often called "sampling to a foregone conclusion." That is, the stopping time is designed so that, if a fixed sample size terminal decision is used upon stopping, the conclusion to be drawn is determined in advance. A good deal of discussion of this phenomenon exists in the literature. It is particularly pertinent to clinical trials in which researchers would like to stop before the original study plan is finished if the results seem overwhelming. [See, for example, Cornfield (1966).] The concern is raised that this might allow unscrupulous researchers to sample to a foregone conclusion. The methods of sequential analysis are designed to prevent that, at least in the classical setting, by making the terminal decision rule depend on which stopping time is used. From a Bayesian point of view, so long as the stopping time is a function of the observed data, and one conditions on the observed data, the terminal decision rule should be whatever would be optimal if that data were observed from a fixed sample size rule. If this is so, can a Bayesian be tricked into sampling to a foregone conclusion? The answer is no, if the Bayesian uses a proper prior.[8]

To see why a Bayesian cannot sample to a foregone conclusion,[9] suppose that N is a strictly positive,[10] integer-valued random variable that might equal ∞. Suppose also that N is a function of observable data $\{X_n\}_{n=1}^{\infty}$ in such a way that, for every finite n, $I_{\{n\}}(N)$ is a function of $X_1, \ldots, X_n$. Let

$$B_n = \{(x_1, \ldots, x_n) : N = n\}.$$

Let Z be a random variable of interest (perhaps the indicator of some subset of the parameter space or anything else) with finite mean c. Suppose that, for every n and every $(x_1, \ldots, x_n) \in B_n$, $E(Z|X_1 = x_1, \ldots, X_n = x_n) > d \geq c$. (A similar argument works for $d \leq c$.) It follows from the law of total probability B.70 that

$$E(Z) = \sum_{n=1}^{\infty} E(E(Z|X_1, \ldots, X_n)|N = n) \Pr(N = n) \quad (9.53)$$
$$+ E(E(Z|X_1, X_2, \ldots)|N = \infty) \Pr(N = \infty).$$

If we suppose that $\Pr(N = \infty) = 0$, then the right-hand side of (9.53) is greater than d and the left-hand side equals $c \leq d$, which is a contradiction. Hence, $\Pr(N = \infty) > 0$. This means that, if a Bayesian does not believe a priori the conclusion (in this case that the mean of Z is greater than c), then it cannot be guaranteed that he or she will believe it after sequential sampling.

A more useful result is available if $Z = I_B(Y)$ for some random variable Y. In this case, we can calculate a bound on the conditional probability of stopping given $Y \notin B$. First, note that $E(Z) = \Pr(Y \in B) = c$, and rewrite (9.53) as

$$\Pr(Y \in B) \geq \sum_{n=1}^{\infty} E(\Pr(Y \in B|X_1, \ldots, X_n)|N = n) \Pr(N = n).$$

The right-hand side of this is greater than $d \Pr(N < \infty)$. It follows that

---

[8]The reason that a proper prior is needed is subtle. The argument given below depends on the law of total probability B.70. Kadane, Schervish, and Seidenfeld (1993) show that when improper priors are viewed as finitely additive probabilities, sampling to a foregone conclusion is possible because finitely additive probabilities do not satisfy the law of total probability.

[9]This argument is like one given by Kerridge (1963).

[10]We saw in Example 9.3 on page 537 that if $\Pr(N = 0) > 0$, then $\Pr(N = 0) = 1$.

$\Pr(N < \infty) < c/d$. Now, write

$$\Pr(N < \infty | Y \notin B) = \frac{\Pr(Y \notin B | N < \infty) \Pr(N < \infty)}{\Pr(Y \notin B)}.$$

By design, $\Pr(Y \notin B | N < \infty) \leq 1 - d$ and $\Pr(Y \notin B) = 1 - c$. Combining these results gives

$$\Pr(N < \infty | Y \notin B) < \frac{c(1-d)}{d(1-c)}. \tag{9.54}$$

The claim of (9.54) was made without proof by Savage (1962) with reference to a specific example. Cornfield (1966) proves the result in another specific example. A similar claim, without an explicit bound, was made by Good (1956).

It should be noted that the proof that a Bayesian cannot sample to a foregone conclusion involves probabilities calculated under the joint distribution of all random quantities. For example, if Θ has a continuous distribution and $Z = I_B(\Theta)$ for some set B, it might be the case that for some $\theta \in B$, $P'_\theta(N < \infty) = 1$ (see Problem 12 on page 569, for example), but the set of all such θ must have small prior probability.

This discussion is not meant to say that Bayesians can ignore all stopping rules. All it says is that so long as the stopping rule is a function of the observed data[11] and the Bayesian conditions on the observed data, no further account need be taken of the stopping rule. An example is given by Berger and Berry (1988) of a situation in which one or the other of the two criteria is not met. We give a modified version here. Other examples are given by Roberts (1967).

Example 9.55. Suppose that $\{X_n\}_{n=1}^\infty$ are IID $Ber(\theta)$ conditional on $\Theta = \theta$ and $\Theta \in \{0.49, 0.51\}$. The observations cannot be taken at will; rather they arrive according to a Poisson process with rate $\lambda(\theta)$ conditional on $\Theta = \theta$, but the X_i are independent of the arrival times conditional on Θ. Suppose that $\lambda(0.49)$ is one observation per second and $\lambda(0.51)$ is one observation per hour. The stopping rule will be to sample for one minute and then stop when the next observation arrives.

Suppose that we observe $N = 60$. If we accidentally consider $X_1, \ldots, X_{60}$ to be the observed data and condition on it alone, we will be ignoring valuable information. Specifically, the time it took to observe the 60 values contains valuable information about Θ. Furthermore, the very fact that 60 observations were observed contains information about Θ, even if we don't know how long it took to

---

[11] We mean that, for each n, the event $\{N = n\}$ must be measurable with respect to the σ-field generated by $X_1, \ldots, X_n$. In a trivial sense, the stopping time is always a function of the observed data if the observed data are defined to include all and only those observations with subscripts up to and including N. If you know that the observed data are $X_1, \ldots, X_n$, then you know that $N = n$. What is required is that, for each n, if you are merely told the values of $X_1, \ldots, X_n$ (but not N), you would be able to figure out whether or not $N > n$.

get them. Also, the criterion that the stopping rule be a function of the observed data would not be met in this case, since one cannot tell from looking at the first 60 X_i values that the experiment would stop at $N = 60$. One also needs to look at the clock (and possibly the calendar).

This is not to say that one could not make inference based solely on $X = (X_1, \ldots, X_N)$ in this case. To obtain the density of X given Θ, we first introduce $\{Y_n\}_{n=1}^\infty$, the interarrival times of the Poisson process. Now, $\{N = n\}$ is a function of $Y_1, \ldots, Y_n$, and we can write (assuming that $\lambda(\theta)$ has units of "observations per second") the conditional joint density of X and $Y = (Y_1, \ldots, Y_N)$ given $\Theta = \theta$ as

$$f_{X,Y|\Theta}(x, y|\theta) = \theta^k(1-\theta)^{n-k}\lambda(\theta)^n \exp(-\lambda(\theta)t),$$

where $t = \sum_{i=1}^n y_i$, n is the observed value of N, and $k = \sum_{i=1}^n x_i$. We can integrate Y out of this to obtain the conditional density of X given Θ.[12] To integrate out Y for fixed n, transform from $y_1, \ldots, y_n$ to $t, y_{n-1}, \ldots, y_1$. The Jacobian is 1, and the ranges of integration are (with t innermost and y_1 outermost)

$$t > 60,$$
$$0 < y_i < 60 - \sum_{j=1}^{i-1} y_j, \text{ for } i = n-1, \ldots, 2,$$
$$0 < y_1 < 60.$$

The result of integrating out these variables is

$$\frac{(60\lambda(\theta))^{n-1} \exp(-60\lambda(\theta))}{(n-1)!}.$$

So the likelihood is

$$f_{X|\Theta}(x|\theta) = \theta^k(1-\theta)^{n-k}\frac{(60\lambda(\theta))^{n-1} \exp(-60\lambda(\theta))}{(n-1)!}. \qquad (9.56)$$

For example, suppose that the prior for Θ puts probability q on $\Theta = 0.51$. Then the posterior probabilities of the two values of Θ are in the ratio:

$$\frac{f_{\Theta|X}(0.51|x)}{f_{\Theta|X}(0.49|x)} = \frac{q}{1-q} 2.778 \times 10^{-4} \exp(59.983)(1.0408)^{2k}(2.6689 \times 10^{-4})^n.$$

If, for example, $n = 60$ and $k = 30$ are the only observed data and q is not essentially 0 or 1, then the posterior probability of $\Theta = 0.49$ will be essentially

---

[12] An alternative method for calculating the conditional density of X given Θ is the following. The conditional joint density of the X_i given $\Theta = \theta$ and $N = n$ is still that of n IID $Ber(\theta)$ random variables since N is independent of the values of the X_i given Θ. The conditional density of N given $\Theta = \theta$ is

$$f_{N|\Theta}(n|\theta) = \frac{(60\lambda(\theta))^{n-1}\exp(-60\lambda(\theta))}{(n-1)!}, \text{ for } n = 1, 2, \ldots.$$

That is, N is just one plus a $Poi(60\lambda(\theta))$ random variable. So the likelihood function for observing X alone would be as given in (9.56).

1 because $N = 60$ is orders of magnitude more likely when the Poisson process has rate 1 than when it has rate $2.778 \times 10^{-4} = \lambda(0.51)$. On the other hand, if $t = 216,000$ (2.5 days) is also observed, then the posterior probability of $\Theta = 0.51$ is essentially 1.

A classical statistician could also make inference based solely on X without much trouble. First notice that

$$\log \frac{f_{X|\Theta}(x|0.51)}{f_{X|\Theta}(x|0.49)} = 0.08k - 8.2287n$$

plus a constant. If a level $\alpha = 0.05$ test of $H : \Theta = 0.49$ versus $A : \Theta = 0.51$ is desired, we need to add up the probabilities (given $\Theta = 0.49$) of all k and n values with small n and large k until the sum exceeds 0.05. This happens at $n = 48$ and $k = 29$. So the MP level 0.05 test is

$$\phi(n, k) = \begin{cases} 1 & \text{if } n < 48 \text{ or if } n = 48 \text{ and } k = 29, \\ 0.3077 & \text{if } n = 48 \text{ and } k = 29, \\ 0 & \text{otherwise.} \end{cases}$$

If $n = 48$ and $k = 29$ are observed and neither prior probability is close to 0, then the posterior probability of $\Theta = 0.51$ is essentially 0. In fact the Bayes factor (likelihood ratio) drops below 1 when $N > 6$. So, the evidence is more in favor of the hypothesis than the alternative whenever $N > 6$ is observed, yet the MP level 0.05 test continues to reject H even when 47 observations have been observed. The size of the test that rejects H if and only if $N \leq 6$ is 6.3×10^{-19}. The type II error probability of this test is about the same.

What happened at the end of Example 9.55 is illustrative of the faulty reasoning that leads to the choice of a test by its level. The reasoning is that we wish to protect against the more costly error (type I error) so we make the probability of type I error small and then choose the test with the smallest type II error. What happened in Example 9.55 is that the data are so well able to distinguish the two hypotheses that making the type I error probability as large as 0.05 makes the type II error probability drop to 0. This is just the opposite effect as what was desired.

9.5 Problems

Section 9.1:

1. Take the situation described in Example 9.3 on page 537 and find the optimal procedure that takes at most three observations. (It is not the truncation of the rule in Example 9.3 to three observations.) Comment on why it differs from the optimal rule in Example 9.3 even before the third observation.

2. Prove that the formal Bayes rule in a sequential decision problem decides optimally after stopping almost surely.

3. Refer to the setup in Definition 9.2. Let $\mathcal{C}$ be the collection of all $A \in \mathcal{B}^\infty$ such that, for every n, $A \cap \{N = n\} \in \mathcal{B}^n$. Let Q be the probability on $(\mathcal{V}, \mathcal{D})$ induced by V from μ.

(a) Prove that $\mathcal{C}$ is a σ-field.

(b) For each n and each $D \in \mathcal{D}$, prove that $Q_n(D|x) = \Pr(V^{-1}(D) | X^{-1}(\mathcal{B}^n))$, a.s.

(c) Prove that, for each $D \in \mathcal{D}$, $Q_N(D|x) = \Pr(V^{-1}(D) | X^{-1}(\mathcal{C}))$, a.s.

4. Prove that a rule computed via backward induction is regular.

5. Prove Proposition 9.8 on page 541.

6. *Suppose that $L' \geq 0$ and $\lim_{n \to \infty} \mathrm{E}\rho_0(Q(\cdot | X_0, \ldots, X_n)) = 0$. Let γ_n be defined by (9.28) where $0 \leq \gamma_0 \leq \rho^*$.

   (a) Prove that $\gamma_n \leq \rho^*$ for all n.

   (b) Prove that
   $$|\gamma_{n+m}(Q) - \gamma_n(Q)| \\ \leq \mathrm{E}\left(|\gamma_m(Q(\cdot | X_1, \ldots, X_n)) - \gamma_0(Q(\cdot | X_1, \ldots, X_n))|\right).$$

   (c) Prove that $\lim_{n \to \infty} \gamma_n(Q)$ converges to some quantity $\gamma^*(Q)$ that satisfies
   $$\gamma^*(Q) = \min\{\rho_0(Q), \mathrm{E}(\gamma^*(Q(\cdot|X_1))) + c\}. \tag{9.57}$$

   (d) Suppose that γ_1^* and γ_2^* both satisfy (9.57) for all probabilities Q. Show that
   $$|\gamma_1^*(Q) - \gamma_2^*(Q)| \leq \mathrm{E}\left(|\gamma_1^*(Q(\cdot|X_1, \ldots, X_n)) - \gamma_2^*(Q(\cdot|X_1, \ldots, X_n))|\right).$$

   (e) Prove that $\gamma_1^* = \gamma_2^*$.

Section 9.2:

7. Stein (1946) proves that if $\Pr(f_1(X_i) \neq f_0(X_i)) > 0$, then there exist c and $\rho < 1$ such that $\Pr(N > n) \leq c\rho^n$ for the SPRT. Define $Z_i = \log[f_1(X_i)/f_0(X_i)]$.

   (a) Prove that $\mathrm{Var}(Z_i) > 0$ implies $\Pr(f_1(X_i) \neq f_0(X_i)) > 0$.

   (b) Find an example in which $\Pr(f_1(X_i) \neq f_0(X_i)) > 0$ but $\mathrm{Var}(Z_i) = 0$.

   (c) Under the conditions of Theorem 9.33, prove that there is a subsequence $\{n_k\}_{k=1}^\infty$ and a c and $\rho < 1$ such that $\Pr(N > n_k) < c\rho^{n_k}$.

8. Prove Proposition 9.40 on page 554.

9. Suppose that $\{X_n\}_{n=1}^\infty$ are IID $Ber(\theta)$ given $\Theta = \theta$. Let N be the first n such that $\left|\sum_{i=1}^n X_i - n/2\right| \geq 2$. Prove that $\mathrm{E}_\theta(N) = 2/[\theta^2 + (1-\theta)^2]$.

Section 9.3:

10. Let $\{(\mathcal{X}_n, \mathcal{B}_n)\}_{n=1}^\infty$ be a sequence of sample spaces, and let f_n, g_n be two densities on $(\mathcal{X}_n, \mathcal{B}_n)$ for every n. Let $X_n : S \to \mathcal{X}_n$ be a random quantity for every n. Define $Z_n = g_n(X_n)/f_n(X_n)$. Let P be the probability that says that X_n has density f_n for every n. Let $k > 0$ and $N = \inf\{n : Z_n \geq k\}$. Prove that $P(N < \infty) \leq 1/k$.

11.*An alternative to fixed-width confidence intervals is to form a *confidence sequence*. A coefficient γ confidence sequence for Θ is a sequence of sets $\{R_n\}_{n=1}^{\infty}$ such that $P_\theta(\theta \in R_n,$ for all $n) \geq \gamma$ for all θ. Let $\{X_n\}_{n=1}^{\infty}$ be conditionally IID with $N(\theta, 1)$ distribution given $\Theta = \theta$. Use the result from Problem 10 above to find a coefficient γ confidence sequence for Θ. (*Hint:* Let X_n in Problem 10 be $(X_1, \ldots, X_n)$ in this problem. Let P be P_θ and let g_n be the prior predictive density of $(X_1, \ldots, X_n)$ under a suitable prior for Θ.)

Section 9.4:

12. Suppose that $\{X_n\}_{n=1}^{\infty}$ are conditionally IID $N(\theta, 1)$ given $\Theta = \theta$ and that $\Theta \sim N(0, 1)$. Show that, given $\Theta = 0$, for every α the probability is 1 that there will exist n such that $\Pr(\Theta \leq 0 | X_1, \ldots, X_n) > \alpha$.

13.*Suppose that $\{X_n\}_{n=1}^{\infty}$ are conditionally IID $N(\theta, 1)$ given $\Theta = \theta$ and that $\Theta \sim N(\theta_0, 1/\lambda)$. In this problem, we will prove that, given $\Theta = \theta > \theta_0$, the probability is less than 1 that there will exist n such that $\Pr(\Theta \leq \theta_0 | X_1, \ldots, X_n) > \alpha > 1/2$, so long as $\Pr(\Theta \leq \theta_0) < \alpha$. In what follows, let $\theta > \theta_0$ and $\alpha > 1/2$.

   (a) Define $Z_i = X_i - \theta_0$ and $\psi(h) = E_\theta \exp(hZ_i)$. Show that there exists $h < 0$ such that $\psi(h) = 1$.

   (b) Let $p = \Pr(\Theta \leq \theta_0) < \alpha$, and let $S_n = \sum_{i=1}^{n} Z_i$. Prove that $\Pr(\Theta \leq \theta_0 | X_1, \ldots, X_n) > \alpha$ is equivalent to $S_n \leq c_n$, where
   $$c_n = -\sqrt{n+\lambda}\Phi^{-1}(\alpha) + \sqrt{\lambda}\Phi^{-1}(p).$$

   (c) Prove that $c_n < c_1 < 0$ for all n.

   (d) Let f_1 be the $N(\theta - \theta_0, 1)$ density (the conditional density of Z_i given $\Theta = \theta$), and let $f_0 = f_1$ times $\exp(hx)$, where h was found in part (a). Let $b > 0$, and consider the $\text{SPRT}(B, A)$ for testing the hypothesis $\{f_0\}$ against the alternative $\{f_1\}$ with $A = \exp(-hb)$ and $B = \exp(-hc_1)$. Let M_b be the stopping time of this test. Use Theorem 9.35 to show that
   $$P_\theta(\text{SPRT}(B, A) \text{ accepts hypothesis}) \leq \frac{1 - \exp(hb)}{\exp(hc_1) - \exp(hb)}.$$

   (e) Define
   $$N = \inf\{n : \Pr(\Theta \leq 0 | X_1, \ldots, X_n) > \alpha\},$$
   $$M = \inf\{n : S_n < c_1\}.$$
   Show that $P_\theta(N < \infty) \leq P_\theta(M < \infty)$.

   (f) Prove that
   $$P_\theta(M < \infty) = \lim_{b \to \infty} P_\theta(\text{SPRT}(B, A) \text{ accepts hypothesis}) < 1.$$

Appendix A

Measure and Integration Theory

This appendix contains an introduction to the theory of measure and integration. The first section is an overview. It could serve either as a refresher for those who have previously studied the material or as an informal introduction for those who have never studied it.

A.1 Overview

A.1.1 Definitions

In many introductory statistics and probability courses, one encounters discrete and continuous random variables and vectors. These are all special cases of a more general type of random quantity that we will study in this text. Before we can introduce the more general type of random quantity, we need to generalize the sums and integrals that figure so prominently in the distributions of discrete and continuous random variables and vectors. The generalization is through the concept of a *measure* (to be defined shortly), which is a way of assigning numerical values to the "sizes" of sets.

Example A.1. Let S be a nonempty set, and let $A \subseteq S$. Define $\mu(A)$ to be the number of elements of A. Then $\mu(S) > 0$, $\mu(\emptyset) = 0$, and if $A_1 \cap A_2 = \emptyset$, $\mu(A_1 \cup A_2) = \mu(A_1) + \mu(A_2)$. Note that $\mu(A) = \infty$ is possible if S has infinitely many elements. The measure μ described here is called *counting measure on S*.

Example A.2. Let A be an interval of real numbers. If A is bounded, let $\mu(A)$ be the length of A. If A is unbounded, let $\mu(A) = \infty$. It is easy to see that $\mu(\mathbb{R}) = \infty$,[1] $\mu(\emptyset) = 0$, and if $A_1 \cap A_2 = \emptyset$ and $A_1 \cup A_2$ is an interval, then

---

[1] By $\mathbb{R}$, we mean the set of real numbers.

$\mu(A_1 \cup A_2) = \mu(A_1) + \mu(A_2)$. The measure μ described here is called *Lebesgue measure*.

Example A.3. Let $f : \mathbb{R} \to \mathbb{R}^+$ be a continuous function.[2] Define, for each interval A, $\mu(A) = \int_A f(x)dx$. Then $\mu(\mathbb{R}) > 0$, $\mu(\emptyset) = 0$, and if $A_1 \cap A_2 = \emptyset$ and $A_1 \cup A_2$ is an interval, then $\mu(A_1 \cup A_2) = \mu(A_1) + \mu(A_2)$.

Since measure will be used to give sizes to sets, the domain of a measure will be a collection of sets. In general, we cannot assign sizes to all sets, but we need enough sets so that we can take unions and complements. A collection of sets that is closed under taking complements and finite unions is called a *field*. A field that is closed under taking countable unions is called a *σ-field*.

Example A.4. Let S be any set. Let $\mathcal{A} = \{S, \emptyset\}$. This σ-field is called the *trivial σ-field*. As a second example, let $A \subset S$. Let $\mathcal{A} = \{S, A, A^C, \emptyset\}$. Let B be another subset of S, and let $\mathcal{A} = \{S, A, B, A^C, B^C, A \cap B, A \cap B^C, \ldots\}$. Such examples grow rapidly. The largest σ-field is the collection of all subsets of S, called the *power set of S* and denoted 2^S.

Example A.5. One field of subsets of $\mathbb{R}$ is the collection of all unions of finitely many disjoint intervals (unbounded intervals are allowed). This collection is not a σ-field, however.

It is easy to prove that the intersection of an arbitrary collection of σ-fields is itself a σ-field. Since 2^S is a σ-field, it is easy to see that, for every collection of subsets $\mathcal{C}$ of S, there is a smallest σ-field $\mathcal{A}$ that contains $\mathcal{C}$, namely the intersection of all σ-fields that contain $\mathcal{C}$. This smallest σ-field is called the σ-field *generated by $\mathcal{C}$*.

The most commonly used σ-field in this book will be the one generated by the collection $\mathcal{C}$ of open subsets of a topological space.[3] This σ-field is called the *Borel σ-field*. It is easy to see that the Borel σ-field $\mathcal{B}^1$ for $\mathbb{R}$ is the σ-field generated by the intervals of the form $[b, \infty)$. It is also the σ-field generated by the intervals of the form $(-\infty, a]$ and the σ-field generated by the intervals of the form (a, b). Since multidimensional Euclidean spaces are topological spaces, they also have Borel σ-fields.

An alternative way to generate the Borel σ-fields of $\mathbb{R}^k$ spaces is by means of product spaces. The σ-field generated by all product sets (one factor from each σ-field) in a product space is called the *product σ-field*. In $\mathbb{R}^k$, the product σ-field of one-dimensional Borel sets $\mathcal{B}^1$ is the same as the Borel σ-field $\mathcal{B}^k$ in the k-dimensional space (Proposition A.38).

Sometimes, we need to extend $\mathbb{R}$ to include points at infinity. *The extended real numbers* are the points in $\mathbb{R} \cup \{\infty, -\infty\}$. The Borel σ-field $\mathcal{B}^+$ of the extended real numbers consists of $\mathcal{B}^1$ together with all sets of the form $B \cup \{\infty\}$, $B \cup \{-\infty\}$, and $B \cup \{\infty, -\infty\}$ for $B \in \mathcal{B}^1$. It is easy to check that $\mathcal{B}^+$ is a σ-field. (See Problem 4 on page 603.)

---

[2] By $\mathbb{R}^+$, we mean the open interval $(0, \infty)$.

[3] A space $\mathcal{X}$ is a *topological space* if it has a collection $\mathcal{D}$ of subsets called a *topology* which satisfies the following conditions: $\emptyset \in \mathcal{D}$, $\mathcal{X} \in \mathcal{D}$, the intersection of finitely many elements of $\mathcal{D}$ is in $\mathcal{D}$, and the union of arbitrarily many elements of $\mathcal{D}$ is in $\mathcal{D}$. The sets in $\mathcal{D}$ are called *open sets*.

If $\mathcal{A}$ is a σ-field of subsets of a set S, then a *measure* μ on S is a function from $\mathcal{A}$ to the extended real numbers that satisfies

- $\mu(\emptyset) = 0$,
- $\{A_n\}_{n=1}^{\infty}$ mutually disjoint implies $\mu\left(\cup_{i=1}^{\infty} A_i\right) = \sum_{i=1}^{\infty} \mu(A_i)$.

If μ is a measure, the triple $(S, \mathcal{A}, \mu)$ is called a *measure space*. If $(S, \mathcal{A}, \mu)$ is a measure space and $\mu(S) = 1$, then μ is called a *probability* and $(S, \mathcal{A}, \mu)$ is called a *probability space*.

Some examples of measures were given earlier. The Caratheodory extension theorem A.22 shows how to construct measures by first defining countably additive set functions on fields and then extending them to the generated σ-field. Lebesgue measure is defined in this manner by starting with length for unions of disjoint intervals.

Sets with measure zero are ubiquitous in measure theory, so there is a special term that allows us to refer to them more easily. If E is some statement concerning the points in S, and μ is a measure on S, we say that E is true *almost everywhere with respect to* μ, written a.e. $[\mu]$, if the set of s such that E is not true is contained in a set A with $\mu(A) = 0$. If μ is a probability, then almost everywhere is often expressed as *almost surely* and denoted a.s. $[\mu]$.

Example A.6. It is well known that a nondecreasing function can have at most a countable number of discontinuities. Since countable sets have Lebesgue measure (length) 0, it follows that nondecreasing functions are continuous almost everywhere with respect to Lebesgue measure.

Infinite measures are difficult to deal with unless they behave like finite measures in certain important ways. If there exists a countable partition of the set S such that each element of the partition has finite μ measure, then we say that μ is σ-*finite*.

A.1.2 Measurable Functions

There are certain types of functions with which we will be primarily concerned. Suppose that S is a set with a σ-field $\mathcal{A}$ of subsets, and let T be another set with a σ-field $\mathcal{C}$ of subsets. Suppose that $f : S \to T$ is a function. We say f is *measurable* if for every $B \in \mathcal{C}$, $f^{-1}(B) \in \mathcal{A}$. When there are several possible σ-fields of subsets of either S or T, we will need to say explicitly with respect to which σ-field f is measurable. If f is measurable, one-to-one, and onto and f^{-1} is measurable, we say that f is *bimeasurable*. If the two sets S and T are topological spaces with Borel σ-fields, a measurable function is *Borel measurable*.

As examples, all continuous functions are Borel measurable. But many discontinuous functions are also measurable. For example, step functions are measurable. All monotone functions are measurable. In fact, it is very difficult to describe a nonmeasurable function without using some heavy mathematics.

If S and T are sets, $\mathcal{C}$ is a σ-field of subsets of T, and $f : S \to T$ is a function, then it is easy to show that $f^{-1}(\mathcal{C})$ is a σ-field of subsets of S. In fact, it is the smallest σ-field of subsets of S such that f is measurable, and it is called the σ-*field generated by* f.

Some useful properties of measurable functions are in Theorem A.38. To summarize, multivariate functions with measurable coordinates are measurable; compositions of measurable functions are measurable; sums, products, and ratios of measurable functions are measurable; limits, suprema, and infima of sequences of measurable functions are measurable.

As an application of the preceding results, we have Theorem A.42, which says that one function g is a function of another f if and only if g is measurable with respect to the σ-field generated by f.

Many theorems about measurable functions are proven first for a special class of measurable functions called *simple functions* and then extended to all measurable functions using some limit theorems. A measurable function f is called *simple* if it assumes only finitely many distinct values. The most fundamental limit theorem is Theorem A.41, which says that every nonnegative measurable function can be approached from below (pointwise) by a sequence of nonnegative simple functions.

A.1.3 Integration

The integral of a function with respect to a measure is a way to generalize the Reimann integral. The interested readers should be able to convince themselves that the integral as defined here is an extension of the Riemann integral. That is, if the Riemann integral of a function exists, then so does the integral as defined here, and the two are equal. We define the integral in stages. We start with nonnegative simple functions. If f is a nonnegative simple function represented as $f(s) = \sum_{i=1}^{k} a_i I_{A_i}(s)$, with the a_i distinct and the A_i mutually disjoint, then the *integral of f with respect to μ* is $\int f(s)d\mu(s) = \sum_{i=1}^{k} a_i \mu(A_i)$. If 0 times ∞ occurs in such a sum, the result is 0 by convention. The integral of a nonnegative simple function is allowed to be ∞.

For general nonnegative measurable functions, we define the integral of f with respect to μ as $\int f(s)d\mu(s) = \sup_{g \leq f, g \text{ simple}} \int g(s)d\mu(s)$. For general functions f, let $f^+(s) = \max\{f(s), 0\}$ and $f^-(s) = -\min\{f(s), 0\}$ (the positive and negative parts of f, respectively). Then $f(s) = f^+(s) - f^-(s)$. The integral of f with respect to μ is

$$\int f(s)d\mu(s) = \int f^+(s)d\mu(s) - \int f^-(s)d\mu(s),$$

if at least one of the two integrals on the right is finite. If both are infinite, the integral is *undefined*. We say that f is *integrable* if the integral of f is defined and is finite. The integral is defined above in terms of its values at all points in S. Sometimes we wish to consider only a subset of $A \subseteq S$. The integral of f over A with respect to μ is

$$\int_A f(s)d\mu(s) = \int I_A(s)f(s)d\mu(s).$$

Several important properties of integrals will be needed in this text. Proposition A.49 and Theorem A.53 state a few of the simpler ones, namely that functions that are almost everywhere equal have the same integral, that the integral of a linear combination of functions is the linear combination of the integrals, that smaller functions have smaller integrals, and that two integrable functions that have the same integral over every set are equal almost everywhere. Another useful property, given in Theorem A.54, is that a nonnegative integrable function leads to a new measure ν by means of the equation $\nu(A) = \int_A f(s) d\mu(s)$.

The most important theorems concern the interchange of limits with integration. Let $\{f_n\}_{n=1}^{\infty}$ be a sequence of measurable functions such that $f_n(x) \to f(x)$ a.e. $[\mu]$. The monotone convergence theorem A.52 says that if the f_n are nonnegative and $f_n(x) \leq f(x)$ a.e. $[\mu]$, then

$$\lim_{n \to \infty} \int f_n(x) d\mu(x) = \int f(x) d\mu(x). \tag{A.7}$$

The dominated convergence theorem A.57 says that if there exists an integrable function g such that $|f_n(x)| \leq g(x)$, a.e. $[\mu]$, then (A.7) holds.

Part 1 of Theorem A.38 says that measurable functions into each of two measurable spaces combine into a jointly measurable function. Measures and integration can also be extended from several spaces into the product space. For example, suppose that μ_i is a measure on the space $(S_i, \mathcal{A}_i)$ for $i = 1, 2$. To define a measure on $(S_1 \times S_2, \mathcal{A}_1 \otimes \mathcal{A}_2)$, we can proceed as follows. For each product set $A = A_1 \times A_2$, define $\mu_1 \times \mu_2(A) = \mu_1(A_1)\mu_2(A_2)$. The Caratheodory extension theorem A.22 allows us to extend this definition to all of the product space. Lebesgue measure on $\mathbb{R}^2$, denoted $dxdy$, is such a product measure. Not every measure on a product space is a product measure. Product probability measures will correspond to independent random variables.

Extending integration to product spaces proceeds through two famous theorems. Tonelli's theorem A.69 says that a nonnegative function f satisfies

$$\int f(x,y) d\mu_1 \times \mu_2(x,y) = \int \left[\int f(x,y) d\mu_1(x) \right] d\mu_2(y)$$
$$= \int \left[\int f(x,y) d\mu_2(y) \right] d\mu_1(x).$$

Fubini's theorem A.70 says that the same equations hold if f is integrable with respect to $\mu_1 \times \mu_2$. These results also extend to finite product spaces $S_1 \times \cdots \times S_n$.

A.1.4 Absolute Continuity

A special type of relationship between two measures on the same space is called *absolute continuity*. If μ_1 and μ_2 are two measures on the same space, we say that μ_2 is absolutely continuous with respect to μ_1, denoted $\mu_2 \ll \mu_1$, if $\mu_1(A) = 0$ implies $\mu_2(A) = 0$. When $\mu_2 \ll \mu_1$, we say that μ_1 is a dominating measure for μ_2. Here are some examples:

Example A.8.

- Let f be any nonnegative measurable function and let μ_1 be a measure. Define $\mu_2(A) = \int_A f(s) d\mu_1(s)$. (See Theorem A.54.) Then, $\mu_2 \ll \mu_1$.
- Let S be the natural numbers and let $a_1, a_2, \ldots$ be any sequence of nonnegative numbers. Define μ_1 to be counting measure on S, and let $\mu_2(A) = \sum_{a_i \in A} a_i$. Then $\mu_2 \ll \mu_1$.

- Let $\mu_1, \mu_2, \ldots$ be a collection of measures on the same space $(S, \mathcal{A})$. Let $a_1, a_2, \ldots$ be a collection of positive numbers. Then $\mu = \sum_{\text{All } i} a_i \mu_i$ is a measure and $\mu_i \ll \mu$ for all i.

The last example above is important because it tells us that for every countable collection of measures, there is a single measure such that all measures in the collection are absolutely continuous with respect to it.

The Radon–Nikodym theorem A.74 says that the first part of Example A.8 is the most general form of absolute continuity with respect to σ-finite measures. That is, if μ_1 is σ-finite and $\mu_2 \ll \mu_1$, then there exists an extended real-valued measurable function f such that $\mu_2(A) = \int_A f(x) d\mu_1(x)$. In addition, if g is μ_2 integrable, then $\int g(x) d\mu_2(x) = \int g(x) f(x) d\mu_1(x)$. The function f is called the *Radon–Nikodym derivative* of μ_2 with respect to μ_1 and is usually denoted $(d\mu_2/d\mu_1)(s)$.

A similar theorem, A.81, relates integrals with respect to measures on two different spaces. It says that a function $f : S_1 \to S_2$ induces a measure on the range S_2. If μ_1 is a measure on S_1, then define $\mu_2(A) = \mu_1\left(f^{-1}(A)\right)$. Integrals with respect to μ_2 can be written as integrals with respect to μ_1 in the following way: $\int g(y) d\mu_2(y) = \int g(f(x)) d\mu_1(x)$. The measure μ_2 is called the measure *induced* on S_2 by f from μ_1.

A.2 Measures

A *measure* is a way of assigning numerical values to the "sizes" of sets. The collection of sets whose sizes are given by a measure is a σ-field. (See Examples A.4 and A.5 on page 571.)

Definition A.9. A nonempty collection of subsets $\mathcal{A}$ of a set S is called a *field* if

- $A \in \mathcal{A}$ implies[4] $A^C \in \mathcal{A}$,
- $A_1, A_2 \in \mathcal{A}$ implies $A_1 \cup A_2 \in \mathcal{A}$.

A field $\mathcal{A}$ is called a σ-field if $\{A_n\}_{n=1}^\infty \in \mathcal{A}$ implies $\cup_{i=1}^\infty A_i \in \mathcal{A}$.

Proposition A.10. *Let $\aleph$ be an arbitrary set of indices, and let $\mathcal{Y} = \{\mathcal{A}_\alpha : \alpha \in \aleph\}$ be an arbitrary collection of σ-fields of subsets of a set S. Then $\cap_{\alpha \in \aleph} \mathcal{A}_\alpha$ is also a σ-field of subsets of S.*

Because of Proposition A.10 and the fact that 2^S is a σ-field, it is easy to see that, for every collection of subsets $\mathcal{C}$ of S, there is a smallest σ-field $\mathcal{A}$ that contains $\mathcal{C}$, namely the intersection of all σ-fields that contain $\mathcal{C}$.

Definition A.11. Let $\mathcal{C}$ be the collection of intervals in $\mathbb{R}$. The smallest σ-field containing $\mathcal{C}$ is called the *Borel σ-field*. In general, if S is a topological space, and $\mathcal{B}$ is the smallest σ-field that contains all of the open sets, then $\mathcal{B}$ is called the *Borel σ-field*.

---

[4] The symbol A^C stands for the complement of the set A.

In addition to the Borel σ-field, the *product σ-field* is also generated by a simple collection of sets.

Definition A.12.

- Let $\aleph$ be an index set, and let $\{S_\alpha\}_{\alpha \in \aleph}$ be a collection of sets. Define $S = \prod_{\alpha \in \aleph} S_\alpha$. We call S a *product space*.
- For each $\alpha \in \aleph$, let $\mathcal{A}_\alpha$ be a σ-field of subsets of S_α. Define the *product σ-field* as follows. $\otimes_{\alpha \in \aleph} \mathcal{A}_\alpha$ is the smallest σ-field that contains all sets of the form $\prod_{\alpha \in \aleph} A_\alpha$, where $A_\alpha \in \mathcal{A}_\alpha$ for all α and all but finitely many A_α are equal to S_α.

In the special case in which $\aleph = \{1, 2\}$, we use the notation $S = S_1 \times S_2$ and the product σ-field is denoted $\mathcal{A}_1 \otimes \mathcal{A}_2$.

Proposition A.13.[5] *The Borel σ-field $\mathcal{B}^k$ of $\mathbb{R}^k$ is the same as the product σ-field of k copies of $(\mathbb{R}, \mathcal{B}^1)$.*

There are other types of collections of sets that are related to σ-fields. Sometimes it is easier to prove results about these other collections and then use the theorems that follow to infer similar results about σ-fields.

Definition A.14. Let S be a set. A collection Π of subsets of S is called a *π-system* if $A, B \in \Pi$ implies $A \cap B \in \Pi$. A collection Λ is called a *λ-system* if $S \in \Lambda$, $A \in \Lambda$ implies $A^C \in \Lambda$, and $\{A_n\}_{n=1}^\infty \in \Lambda$ with $A_i \cap A_j = \emptyset$ for $i \neq j$ implies $\cup_{i=1}^\infty A_i \in \Lambda$.

As in Proposition A.10, the intersection of arbitrarily many π-systems is a π-system, and so too with λ-systems. The following propositions are also easy to prove.

Proposition A.15. *If S is a set and $\mathcal{C}$ is a collection of subsets of S such that $\mathcal{C}$ is a π-system and a λ-system, then $\mathcal{C}$ is a σ-field.*

Proposition A.16. *If S is a set and Λ is a λ-system of subsets, then $A, A \cap B \in \Lambda$ implies $A \cap B^C \in \Lambda$.*

The following lemma is the key to a useful uniqueness theorem.

Lemma A.17 (π–λ theorem).[6] *Suppose that Π is a π-system, that Λ is a λ-system, and that $\Pi \subseteq \Lambda$. Then the smallest σ-field containing Π is contained in Λ.*

PROOF. Define $\lambda(\Pi)$ to be the smallest λ-system containing Π, and define $\sigma(\Pi)$ to be the smallest σ-field containing Π. For each $A \subseteq S$, define $\mathcal{G}_A$ to be the collection of all sets $B \subseteq S$ such that $A \cap B \in \lambda(\Pi)$.

First, we show that $\mathcal{G}_A$ is a λ-system for each $A \in \lambda(\Pi)$. To see this, note that $A \cap S \in \lambda(\Pi)$, so $S \in \mathcal{G}_A$. If $B \in \mathcal{G}_A$, then $A \cap B \in \lambda(\Pi)$, and Proposition A.16 says that $A \cap B^C \in \lambda(\Pi)$, so $B^C \in \mathcal{G}_A$. Finally, $\{B_n\}_{n=1}^\infty \in \mathcal{G}_A$ with the B_n

---

[5] This proposition is used in the proof of Theorem A.38.
[6] This lemma is used in the proofs of Theorems A.26 and B.46 and Lemma A.61.

disjoint implies that $A \cap B_n \in \lambda(\Pi)$ with $A \cap B_n$ disjoint, so their union is in $\lambda(\Pi)$. But their union is $A \cap (\cup_{n=1}^{\infty} B_n)$. So $\cup_{n=1}^{\infty} B_n \in \mathcal{G}_A$.

Next, we show that $\lambda(\Pi) \subseteq \mathcal{G}_C$ for every $C \in \lambda(\Pi)$. Let $A, B \in \Pi$, and notice that $A \cap B \in \Pi$, so $B \in \mathcal{G}_A$. Since $\mathcal{G}_A$ is a λ-system containing Π, it must contain $\lambda(\Pi)$. It follows that $A \cap C \in \lambda(\Pi)$ for all $C \in \lambda(\Pi)$. If $C \in \lambda(\Pi)$, it then follows that $A \in \mathcal{G}_C$. So, $\Pi \subseteq \mathcal{G}_C$ for all $C \in \lambda(\Pi)$. Since $\mathcal{G}_C$ is a λ-system containing Π, it must contain $\lambda(\Pi)$.

Finally, if $A, B \in \lambda(\Pi)$, we just proved that $B \in \mathcal{G}_A$, so $A \cap B \in \lambda(\Pi)$ and hence $\lambda(\Pi)$ is also a π-system. By Proposition A.15, $\lambda(\Pi)$ is a σ-field containing Π and hence must contain $\sigma(\Pi)$. Since $\lambda(\Pi) \subseteq \Lambda$, the proof is complete. □

We are now in a position to give a precise definition of measure.

Definition A.18.

- A pair $(S, \mathcal{A})$, where S is a set and $\mathcal{A}$ is a σ-field, is called a *measurable space*.

- A function $\mu : \mathcal{A} \to [0, \infty]$ is called a *measure* if
  - $\mu(\emptyset) = 0$,
  - $\{A_n\}_{n=1}^{\infty}$ mutually disjoint implies $\mu(\cup_{i=1}^{\infty} A_i) = \sum_{i=1}^{\infty} \mu(A_i)$.

- A function $\mu : \mathcal{A} \to [-\infty, \infty]$ that satisfies the above two conditions and does not assume both of the values ∞ and $-\infty$ is called a *signed measure*.[7]

- If μ is a measure, the triple $(S, \mathcal{A}, \mu)$ is called a *measure space*.

- If $(S, \mathcal{A}, \mu)$ is a measure space and $\mu(S) = 1$, then μ is called a *probability* and $(S, \mathcal{A}, \mu)$ is called a *probability space*.

Some examples of measures were given in Section A.1.

Theorem A.19.[8] *If $(S, \mathcal{A}, \mu)$ is a measure space and $\{A_n\}_{n=1}^{\infty}$ is a monotone sequence,[9] then $\mu(\lim_{i \to \infty} A_i) = \lim_{i \to \infty} \mu(A_i)$ if either of the following holds:*

- *the sequence is increasing,*
- *the sequence is decreasing and $\mu(A_1) < \infty$.*

PROOF. If the sequence is increasing, then let $B_1 = A_1$ and $B_k = A_k \setminus A_{k-1}$ for $k > 1$.[10] Then $\{B_n\}_{n=1}^{\infty}$ are disjoint and the following are true:

$$\bigcup_{i=1}^{k} B_i = A_k, \quad \bigcup_{i=1}^{\infty} B_i = \lim_{k \to \infty} A_k, \quad \mu(A_k) = \sum_{i=1}^{k} \mu(B_i)$$

---

[7] Signed measures will only be used in Section A.6.

[8] This theorem is used in the proofs of Theorems A.50 and B.90 and Lemma A.72.

[9] A sequence of sets $\{A_n\}_{n=1}^{\infty}$ is *monotone* if either $A_1 \subseteq A_2 \subseteq \ldots$ or $A_1 \supseteq A_2 \supseteq \ldots$. In the first case, we say that the sequence is *increasing* and $\lim_{n \to \infty} A_n = \cup_{i=1}^{\infty} A_i$. In the second case, we say that the sequence is *decreasing* and $\lim_{n \to \infty} A_n = \cap_{i=1}^{\infty} A_i$.

[10] The symbol $A \setminus B$ is another way of saying $A \cap B^C$.

$$\lim_{k\to\infty} \mu(A_k) = \sum_{i=1}^{\infty} \mu(B_i) = \mu\left(\bigcup_{i=1}^{\infty} B_i\right) = \mu\left(\lim_{k\to\infty} A_k\right).$$

If the sequence is decreasing, then let $B_i = A_i \setminus A_{i+1}$, for $i = 1, 2, \ldots$. It follows that

$$A_1 = \lim_{k\to\infty} A_k \bigcup \left(\bigcup_{i=1}^{\infty} B_i\right),$$

and all of the sets on the right-hand side are disjoint. It follows that

$$A_k = A_1 \setminus \bigcup_{i=1}^{k-1} B_i,$$

$$\mu(A_1) = \mu\left(\lim_{k\to\infty} A_k\right) + \sum_{i=1}^{\infty} \mu(B_i),$$

$$\mu(A_k) = \mu(A_1) - \sum_{i=1}^{k-1} \mu(B_i),$$

$$\lim_{k\to\infty} \mu(A_k) = \mu(A_1) - \sum_{i=1}^{\infty} \mu(B_i) = \mu\left(\lim_{k\to\infty} A_k\right). \quad \square$$

Another useful theorem concerning sequences of sets is the following.

Theorem A.20 (First Borel–Cantelli lemma).[11] *If $\sum_{n=1}^{\infty} \mu(A_n) < \infty$, then $\mu\left(\bigcap_{i=1}^{\infty} \bigcup_{n=i}^{\infty} A_n\right) = 0$.*

PROOF. Let $B_i = \bigcup_{n=i}^{\infty} A_n$ and $B = \bigcap_{i=1}^{\infty} B_i$. Since $B \subseteq B_i$ for each i, it follows that $\mu(B) \leq \mu(B_i)$ for all i. Since $\mu(B_i) \leq \sum_{n=i}^{\infty} \mu(A_n)$, it follows that $\lim_{i\to\infty} \mu(B_i) = 0$. Hence $\mu(B) = 0$. $\square$

Theorem A.22 below is used in several places for extending measures defined on a field to the smallest σ-field containing the field. A definition is required first.

Definition A.21. *Let S be a set, $\mathcal{A}$ a collection of subsets of S, and $\mu : \mathcal{A} \to \mathbb{R} \cup \{\pm\infty\}$ a set function. Suppose that $S = \bigcup_{i=1}^{\infty} A_i$ with $\mu(A_i) < \infty$ for each i. Then we say μ is σ-finite. If μ is a σ-finite measure on $(S, \mathcal{A})$, then $(S, \mathcal{A}, \mu)$ is called a σ-finite measure space.*

The proof of Theorem A.22 is adapted from Royden (1968).

Theorem A.22 (Cratheodory extension theorem).[12] *Let μ be a set function defined on a field $\mathcal{C}$ of subsets of a set S that is σ-finite, nonnegative, extended*

---

[11] This theorem is used in the proofs of Lemma A.72 and Theorems B.90 and 1.61. There is a second Borel–Cantelli lemma, which involves probability measures, but we will not use it in this text. See Problem 20 on page 663. The set whose measure is the subject of this theorem is sometimes called A_n *infinitely often* because it is the set of points that are in infinitely many of the A_n.

[12] This theorem is used to prove the existence of many common measures (including product measure) and in the proofs of Lemma A.24 and of Theorems B.118, B.131, and B.133.

real-valued, and countably additive and satisfies $\mu(\emptyset) = 0$. Then there is a unique extension of μ to a measure on a measure space[13] $(S, \mathcal{A}, \mu^*)$. (That is, $\mathcal{C} \subseteq \mathcal{A}$ and $\mu(A) = \mu^*(A)$ for all $A \in \mathcal{C}$.)

PROOF. The proof will proceed as follows. First, we will define μ^* and $\mathcal{A}$. Then we will show that μ^* is monotone and subadditive, that $\mathcal{C} \subseteq \mathcal{A}$, that $\mathcal{A}$ is a σ-field, that μ^* is countably additive on $\mathcal{A}$, that μ^* extends μ, and finally that μ^* is the unique extension.

For each $B \in 2^S$, define

$$\mu^*(B) = \inf \sum_{i=1}^{\infty} \mu(A_i), \qquad (A.23)$$

where the inf is taken over all $\{A_i\}_{i=1}^{\infty}$ such that $B \subseteq \cup_{i=1}^{\infty} A_i$ and $A_i \in \mathcal{C}$ for all i. Let

$$\mathcal{A} = \{B \in 2^S : \mu^*(C) = \mu^*(C \cap B) + \mu^*(C \cap B^C), \text{ for all } C \in 2^S\}.$$

First, we show that μ^* is monotone and subadditive. Clearly, $\mu^*(A) \leq \mu(A)$ for all $A \in \mathcal{C}$ and $B_1 \subseteq B_2$ implies $\mu^*(B_1) \leq \mu^*(B_2)$. It is also easy to see that $\mu^*(B_1 \cup B_2) \leq \mu^*(B_1) + \mu^*(B_2)$ for all $B_1, B_2 \in 2^S$. In fact, if $\{B_n\}_{n=1}^{\infty} \in 2^S$, then $\mu^*(\cup_{i=1}^{\infty} B_i) \leq \sum_{i=1}^{\infty} \mu^*(B_i)$. The proof is to notice that the collection of numbers whose inf is μ^* of the union includes all of the sums of the numbers whose infima are the μ^* values being added together.

Next, we show that $\mathcal{C} \subseteq \mathcal{A}$. Let $A \in \mathcal{C}$ and $C \in 2^S$. Since μ^* is subadditive, we only need to show that $\mu^*(C) \geq \mu^*(C \cap A) + \mu^*(C \cap A^C)$. If $\mu^*(C) = \infty$, this is clearly true. So let $\mu^*(C) < \infty$. From the definition of μ^*, for every $\epsilon > 0$, there exists a collection $\{A_i\}_{i=1}^{\infty}$ of elements of $\mathcal{C}$ such that $\sum_{i=1}^{\infty} \mu(A_i) < \mu^*(C) + \epsilon$. Since $\mu(A_i) = \mu(A_i \cap A) + \mu(A_i \cap A^C)$ for every i, we have

$$\mu^*(C) + \epsilon > \sum_{i=1}^{\infty} \mu(A_i \cap A) + \sum_{i=1}^{\infty} \mu(A_i \cap A^C)$$
$$\geq \mu^*(C \cap A) + \mu^*(C \cap A^C).$$

Since this is true for every $\epsilon > 0$, it must be that $\mu^*(C) \geq \mu^*(C \cap A) + \mu^*(C \cap A^C)$, hence $A \in \mathcal{A}$.

Next, we show that $\mathcal{A}$ is a σ-field. It is clear that $\emptyset \in \mathcal{A}$ and $A \in \mathcal{A}$ implies $A^C \in \mathcal{A}$ by the symmetry in the definition of $\mathcal{A}$. Let $A_1, A_2 \in \mathcal{A}$ and $C \in 2^S$. We can write

$$\mu^*(C) = \mu^*(C \cap A_1) + \mu^*(C \cap A_1^C)$$
$$= \mu^*(C \cap A_1) + \mu^*(C \cap A_1^C \cap A_2) + \mu^*(C \cap A_1^C \cap A_2^C)$$
$$\geq \mu^*(C \cap [A_1 \cup A_2]) + \mu^*(C \cap [A_1 \cup A_2]^C),$$

where the first two equalities follow from $A_1, A_2 \in \mathcal{A}$, and the last follows from the subadditivity of μ^*. So, $A_1 \cup A_2 \in \mathcal{A}$. Let $\{A_n\}_{n=1}^{\infty} \in \mathcal{A}$; then we can write

---

[13] The usual statement of this theorem includes the additional claim that the measure space $(S, \mathcal{A}, \mu^*)$ is complete. A measure space is *complete* if every subset of every set with measure 0 is in the σ-field.

$A = \cup_{i=1}^{\infty} A_i = \cup_{i=1}^{\infty} B_i$, where each $B_i \in \mathcal{A}$ and the B_i are disjoint. (This just makes use of complements and finite unions of elements of $\mathcal{A}$ being in $\mathcal{A}$.) Let $D_n = \cup_{i=1}^{n} B_i$ and $C \in 2^S$. Since $A^C \subseteq D_n^C$ and $D_n \in \mathcal{A}$ for each n, we have

$$\begin{aligned}
\mu^*(C) &= \mu^*(C \cap D_n) + \mu^*(C \cap D_n^C) \\
&\geq \mu^*(C \cap D_n) + \mu^*(C \cap A^C) \\
&= \sum_{i=1}^{n} \mu^*(C \cap B_i) + \mu^*(C \cap A^C).
\end{aligned}$$

Since this is true for every n,

$$\begin{aligned}
\mu^*(C) &\geq \sum_{i=1}^{\infty} \mu^*(C \cap B_i) + \mu^*(C \cap A^C) \\
&\geq \mu^*(C \cap A) + \mu^*(C \cap A^C),
\end{aligned}$$

where the last inequality follows from subadditivity. So, $\mathcal{A}$ is a σ-field.

Next, we show that μ^* is countably additive when restricted to $\mathcal{A}$. If A_1, A_2 are disjoint elements of $\mathcal{A}$, then $A_1 = (A_1 \cup A_2) \cap A_1$ and $A_2 = (A_1 \cup A_2) \cap A_1^C$. It follows that

$$\mu^*(A_1 \cup A_2) = \mu^*(A_1) + \mu^*(A_2).$$

By induction, μ^* is finitely additive on $\mathcal{A}$. Let $A = \cup_{i=1}^{\infty} A_i$, where each $A_i \in \mathcal{A}$ and the A_i are disjoint. Since $\cup_{i=1}^{n} A_i \subseteq A$, we have, for every n, $\mu^*(A) \geq \sum_{i=1}^{n} \mu^*(A_i)$, which implies $\mu^*(A) \geq \sum_{i=1}^{\infty} \mu^*(A_i)$. By subadditivity, we get the reverse inequality, hence μ^* is countably additive on $\mathcal{A}$.

Next, we prove that μ^* extends μ. Since μ^* is countably additive on $\mathcal{A}$, we can let $B \in \mathcal{C}$ and $\{A_n\}_{n=1}^{\infty} \in \mathcal{C}$ be disjoint, such that $B \subseteq \cup_{n=1}^{\infty} A_n$. Then $B \subseteq \cup_{n=1}^{\infty} (A_n \cap B) = B$, and $\mu^*(B) \leq \sum_{n=1}^{\infty} \mu(A_n \cap B) = \mu(B)$, since μ is countably additive on $\mathcal{C}$.

To prove uniqueness, suppose that μ' also extends μ to $\mathcal{A}$. Then $\mu'(B) \leq \sum_{n=1}^{\infty} \mu(A_n)$ if $B \subseteq \cup_{n=1}^{\infty} A_n$. Hence, $\mu'(B) \leq \mu^*(B)$ for all $B \in \mathcal{A}$. If there exists B such that $\mu'(B) < \mu^*(B)$, let $\{A_n\}_{n=1}^{\infty} \in \mathcal{C}$ be disjoint and such that $\mu(A_n) < \infty$ and $\cup_{n=1}^{\infty} A_n = S$. Then, there exists n such that $\mu'(B \cap A_n) < \mu^*(B \cap A_n)$. Since $\mu'(A_n) = \mu^*(A_n)$, it must be that $\mu'(B^C \cap A_n) > \mu^*(B^C \cap A_n)$, but this is a contradiction. □

Here are some examples:

- Let $S = \mathbb{R}$ and let $\mathcal{B}$ be the Borel σ-field. Define $\mu((a,b]) = b - a$ for intervals, and extend μ to finite unions of disjoint intervals by addition. Theorem A.22 will extend μ to the σ-field $\mathcal{B}$. This measure is called *Lebesgue measure on the real line*.

- Let F be any monotone increasing function on $\mathbb{R}$ which is continuous from the right. Let $S = \mathbb{R}$ and let $\mathcal{B}$ be the Borel σ-field. Define $\mu((a,b]) = F(b) - F(a)$. This can be extended to all of $\mathcal{B}$. In particular, if F is a CDF, then μ is a probability.

In the examples above, the claim was made that μ could be extended to the Borel σ-field. To do this by way of the Caratheodory extension theorem A.22, we need μ to be defined on a field, countably additive, and σ-finite. For the cases described above, this can be arranged as follows. Suppose that μ is defined on intervals of the form $(a, b]$ with $a = -\infty$ and/or $b = \infty$ possible.[14] The collection $\mathcal{C}$ of all unions of finitely many disjoint intervals of this form is easily seen to be a field. If $(a_1, b_1], \ldots, (a_n, b_n]$ are mutually disjoint, set

$$\mu\left(\bigcup_{i=1}^n (a_i, b_i]\right) = \sum_{i=1}^n \mu((a_i, b_i]).$$

It is not hard to see that this extension of μ to $\mathcal{C}$ is well defined. This means that if $\cup_{i=1}^n (a_i, b_i] = \cup_{i=1}^m (c_i, d_i]$, where $(c_1, d_1], \ldots, (c_m, d_m]$ are also mutually disjoint, then $\sum_{i=1}^n \mu((a_i, b_i]) = \sum_{i=1}^m \mu((c_i, d_i])$. If μ is finite for every interval, then it is σ-finite. To see that μ is countably additive on $\mathcal{C}$, suppose that $\mu((a, b]) = F(b) - F(a)$, where F is nondecreasing and continuous from the right. If $\{(a_n, b_n]\}_{n=1}^\infty$ is a sequence of disjoint intervals and $(a, b]$ is an interval such that $\cup_{n=1}^\infty (a_n, b_n] \subseteq (a, b]$, then it is not difficult to see that $\sum_{n=1}^\infty \mu((a_n, b_n]) \leq \mu((a, b])$. If $(a, b] \subseteq \cup_{n=1}^\infty (a_n, b_n]$, we can also prove that $\sum_{n=1}^\infty \mu((a_n, b_n]) \geq \mu((a, b])$ (see Problem 7 on page 603). Together these facts will imply that μ is countably additive on $\mathcal{C}$.

The proof of Theorem A.22 leads us to the following useful result. Its proof is adapted from Halmos (1950).

Lemma A.24.[15] *Let $(S, \mathcal{A}, \mu)$ be a σ-finite measure space. Suppose that $\mathcal{C}$ is a field such that $\mathcal{A}$ is the smallest σ-field containing $\mathcal{C}$. Then, for every $A \in \mathcal{A}$ and $\epsilon > 0$, there is $C \in \mathcal{C}$ such that $\mu(C \Delta A) < \epsilon$.*[16]

PROOF. Clearly, μ and $\mathcal{C}$ satisfy the conditions of Theorem A.22, so that μ is equal to the μ^* in the proof of that theorem. Let $A \in \mathcal{A}$ and $\epsilon > 0$ be given. It follows from (A.23) that there exists a sequence $\{A_i\}_{i=1}^\infty$ in $\mathcal{C}$ such that $A \subseteq \cup_{i=1}^\infty A_i$ and

$$\mu(A) > \sum_{i=1}^\infty \mu(A_i) - \frac{\epsilon}{2}.$$

Since μ is countably additive,

$$\lim_{n \to \infty} \mu\left(\bigcup_{i=1}^n A_i\right) = \mu\left(\bigcup_{i=1}^\infty A_i\right),$$

so that there exists n such that

$$\mu\left(\bigcup_{i=1}^\infty A_i\right) < \mu\left(\bigcup_{i=1}^n A_i\right) + \frac{\epsilon}{2}.$$

Let $C = \cup_{i=1}^n A_i$, which is clearly in $\mathcal{C}$. Now

$$\mu\left(A \bigcap C^C\right) \leq \mu\left(\bigcup_{i=1}^\infty A_i \bigcap C^C\right) = \mu\left(\bigcup_{i=1}^\infty A_i\right) - \mu(C) < \frac{\epsilon}{2}.$$

---

[14] If $b = \infty$, we mean (a, ∞) by $(a, b]$. That is, we do not intend ∞ to be a point in the space S.

[15] This lemma is used in the proof of the Kolmogorov zero-one law B.68.

[16] The symbol Δ here refers to the symmetric difference operator on pairs of sets. We define $C \Delta A$ to be $(C \cap A^C) \cup (C^C \cap A)$.

Similarly,

$$\mu\left(C\bigcap A^C\right) \le \mu\left(\bigcup_{i=1}^{\infty} A_i \bigcap A^C\right) = \mu\left(\bigcup_{i=1}^{\infty} A_i\right) - \mu(A) < \frac{\epsilon}{2}.$$

It now follows that $\mu(A \Delta C) < \epsilon$. □

Sets with measure zero are ubiquitous in measure theory, so there is a special definition that allows us to refer to them more easily.

Definition A.25. Let E be some statement concerning the points in S such that for each point $s \in S$ E is either true or false but not both. Suppose that there exists a set $A \in \mathcal{A}$ such that $\mu(A) = 0$ and that for all $s \in A^C$, E is true. Then we say that E is true *almost everywhere with respect to* μ, written a.e. $[\mu]$. If μ is a probability, then almost everywhere is often expressed as *almost surely* and denoted a.s. $[\mu]$.

The following theorem implies uniqueness of measures with certain properties.

Theorem A.26.[17] *Suppose that μ_1 and μ_2 are measures on $(S, \mathcal{A})$ and $\mathcal{A}$ is the smallest σ-field containing the π-system Π. If μ_1 and μ_2 are both σ-finite on Π and they agree on Π, then they agree on $\mathcal{A}$.*

PROOF. First, let $C \in \Pi$ be such that $\mu_1(C) = \mu_2(C) < \infty$, and define $\mathcal{G}_C$ to be the collection of all $B \in \mathcal{A}$ such that $\mu_1(B \cap C) = \mu_2(B \cap C)$. Using simple properties of measures, we see that $\mathcal{G}_C$ is a λ-system that contains Π, hence it equals $\mathcal{A}$ by Lemma A.17. (For example, if $B \in \mathcal{G}_C$,

$$\mu_1(B^C \cap C) = \mu_1(C) - \mu_1(B \cap C) = \mu_2(C) - \mu_2(B \cap C) = \mu_2(B^C \cap C),$$

so $B^C \in \mathcal{G}_C$.)

Next, if μ_1 and μ_2 are not finite, there exists a sequence $\{C_n\}_{n=1}^{\infty} \in \Pi$ such that $\mu_1(C_n) = \mu_2(C_n) < \infty$, and $S = \cup_{n=1}^{\infty} C_n$. (Since Π is only a π-system, we cannot assume that the C_n are disjoint.) For each $A \in \mathcal{A}$,

$$\mu_j(A) = \lim_{n \to \infty} \mu_j\left(\bigcup_{i=1}^{n} [C_i \cap A]\right) \text{ for } j = 1, 2.$$

Since $\mu_j\left(\cup_{i=1}^{n}[C_i \cap A]\right)$ can be written as a linear combination of values of μ_j at sets of the form $A \cap C$, where $C \in \Pi$ is the intersection of finitely many of $C_1, \ldots, C_n$, it follows from $A \in \mathcal{G}_C$ that $\mu_1\left(\cup_{i=1}^{n}[C_i \cap A]\right) = \mu_2\left(\cup_{i=1}^{n}[C_i \cap A]\right)$ for all n, hence $\mu_1(A) = \mu_2(A)$. □

A.3 Measurable Functions

There are certain types of functions with which we will be primarily concerned.

---

[17]This theorem is used in the proofs of Theorems B.32, B.46, B.118, B.131, and 1.115, Lemma A.64, and Corollary B.44.

A.3. Measurable Functions

Definition A.27. Suppose that S is a set with a σ-field $\mathcal{A}$ of subsets, and let T be another set with a σ-field $\mathcal{C}$ of subsets. Suppose that $f : S \to T$ is a function. We say f is *measurable* if for every $B \in \mathcal{C}$, $f^{-1}(B) \in \mathcal{A}$. If f is measurable, one-to-one, and onto and f^{-1} is measurable, we say that f is *bimeasurable*. If $T = \mathbb{R}$, the real numbers, and $\mathcal{C} = \mathcal{B}$, the Borel σ-field, then if f is measurable, we say that f is *Borel measurable*.

Proposition A.28. *Suppose that $(S, \mathcal{A})$ and $(T, \mathcal{C})$ are measurable spaces. Suppose that $f : S \to T$ is a function.*

- *If $\mathcal{A} = 2^S$, then f is measurable.*
- *If $\mathcal{C} = \{T, \emptyset\}$, then f is measurable.*
- *If $\mathcal{A} = \{S, \emptyset\}$, $\{y\} \in \mathcal{C}$ for every $y \in T$, and f is measurable, then f is constant.*

As examples, if $S = T = \mathbb{R}$ and $\mathcal{A} = \mathcal{B}$ is the Borel σ-field, then all continuous functions are measurable. But many discontinuous functions are also measurable. For example, step functions are measurable. All monotone functions are measurable. In fact, it is very difficult to describe a nonmeasurable function without using some heavy mathematics.

The following theorems make it easier to show that a function is measurable.

Theorem A.29.[18] *Let $\aleph$, S, and T be arbitrary sets. Let $\{A_\alpha : \alpha \in \aleph\}$ be a collection of subsets of T, and let A be an arbitrary subset of T. Let $f : S \to T$ be a function. Then*

$$f^{-1}\left(\bigcup_{\alpha \in \aleph} A_\alpha\right) = \bigcup_{\alpha \in \aleph} f^{-1}(A_\alpha),$$

$$f^{-1}\left(\bigcap_{\alpha \in \aleph} A_\alpha\right) = \bigcap_{\alpha \in \aleph} f^{-1}(A_\alpha),$$

$$f^{-1}(A^C) = f^{-1}(A)^C.$$

PROOF. For the union, if $s \in f^{-1}(\cup_{\alpha \in \aleph} A_\alpha)$, then $f(s) \in \cup_{\alpha \in \aleph} A_\alpha$, hence there exists α such that $f(s) \in A_\alpha$, so $s \in f^{-1}(A_\alpha)$ and $s \in \cup_{\alpha \in \aleph} f^{-1}(A_\alpha)$. If $s \in \cup_{\alpha \in \aleph} f^{-1}(A_\alpha)$, then there exists α such that $s \in f^{-1}(A_\alpha)$, hence $f(s) \in A_\alpha$, hence $f(s) \in \cup_{\alpha \in \aleph} A_\alpha$, hence $s \in f^{-1}(\cup_{\alpha \in \aleph} A_\alpha)$. This proves the first equality. The second is almost identical in that "there exists α" is merely replaced by "for all α" in the above proof. For the complement, if $s \in f^{-1}(A^C)$, then $f(s) \in A^C$ and $f(s) \notin A$. Hence, $s \notin f^{-1}(A)$ and $s \in f^{-1}(A)^C$. If $s \in f^{-1}(A)^C$, then $s \notin f^{-1}(A)$ and $f(s) \notin A$. So, $f(s) \in A^C$ and $s \in f^{-1}(A^C)$. □

Corollary A.30.[19] *If S and T are sets and $\mathcal{C}$ is a σ-field of subsets of T and $f : S \to T$ is a function, then $f^{-1}(\mathcal{C})$ is a σ-field of subsets of S. In fact, it is the smallest σ-field of subsets of S such that f is measurable.*

---

[18] This theorem is used in the proof of Theorem A.34.

[19] This corollary is used in the proof of Theorem A.42, and it is used to define the σ-field generated by a function.

Definition A.31. The σ-field $f^{-1}(\mathcal{C})$ in Corollary A.30 is called the *σ-field generated by f*.

A measurable function also generates a σ-field of subsets of its image.

Proposition A.32. *Let $(T, \mathcal{C})$ be a measurable space. Let $U \subseteq T$ be arbitrary (possibly not even in $\mathcal{C}$). Define $\mathcal{C}_* = \{U \cap B : B \in \mathcal{C}\}$. Then $\mathcal{C}_*$ is a σ-field of subsets of U.*

Definition A.33. The σ-field $\mathcal{C}_*$ in Proposition A.32 is called the *restriction of the σ-field $\mathcal{C}$ to U*. If $f : S \to T$ and $U = f(S)$, then $\mathcal{C}_*$ is called the *image σ-field of f*.

Theorem A.34.[20] *Let $(S, \mathcal{A})$ be a measurable space and let $f : S \to T$ be a function. Let $\mathcal{C}^*$ be a nonempty collection of subsets of T, and let $\mathcal{C}$ be the smallest σ-field that contains $\mathcal{C}^*$. If $f^{-1}(\mathcal{C}^*) \subseteq \mathcal{A}$, then $f^{-1}(\mathcal{C}) \subseteq \mathcal{A}$.*

PROOF. Let $\mathcal{C}_2$ be the collection of all subsets B of T such that $f^{-1}(B) \in \mathcal{A}$. By assumption, $\mathcal{C}^* \subseteq \mathcal{C}_2$. We will now prove that $\mathcal{C}_2$ is a σ-field; hence it must contain $\mathcal{C}$, which implies the conclusion of the theorem. Clearly, $\mathcal{C}_2$ is nonempty, since $\mathcal{C}^*$ is nonempty. Let $A \in \mathcal{C}_2$. Theorem A.29 implies $f^{-1}(A^C) = f^{-1}(A)^C \in \mathcal{A}$, since $\mathcal{A}$ is a σ-field. This means that $A^C \in \mathcal{C}_2$. Let $A_1, A_2, \ldots \in \mathcal{C}_2$. Then Theorem A.29 implies

$$f^{-1}\left(\bigcup_{i=1}^{\infty} A_i\right) = \bigcup_{i=1}^{\infty} f^{-1}(A_i) \in \mathcal{A},$$

since $\mathcal{A}$ is a σ-field. So $\mathcal{C}_2$ is a σ-field. □

To use this theorem to show that a function $f : S \to T$ is measurable when T has a σ-field of subsets $\mathcal{C}$, we can find a smaller collection of subsets $\mathcal{C}^*$ such that $\mathcal{C}$ is the smallest σ-field containing $\mathcal{C}^*$ and prove that $f^{-1}(\mathcal{C}^*) \subseteq \mathcal{A}$. Theorem A.34 would then imply $f^{-1}(\mathcal{C}) \subseteq \mathcal{A}$ and f is measurable. As an example, consider the next lemma.

Lemma A.35.[21] *Let $(S, \mathcal{A})$ be a measurable space, and let $f : S \to \mathbb{R}$ be a function. Then f is measurable if and only if $f^{-1}((b, \infty)) \in \mathcal{A}$ for all $b \in \mathbb{R}$.*

PROOF. The "only if" part is trivial. For the "if" part, let $\mathcal{C}^*$ be the collection of all subsets of $\mathbb{R}$ of the form (b, ∞). The smallest σ-field containing these is the Borel σ-field $\mathcal{B}$, so $f^{-1}(\mathcal{B}) \subseteq \mathcal{A}$ by Theorem A.34. □

There are versions of Lemma A.35 that apply to intervals of the form $(-\infty, a]$ and those of the form (a, b), and so on. Similarly, there is a version for general topological spaces.

Proposition A.36.[22] *Let $(S, \mathcal{A})$ be a measurable space, and let $(T, \mathcal{C})$ be a topological space with Borel σ-field. Then $f : S \to \mathbb{R}$ is measurable if and only if $f^{-1}(C) \in \mathcal{A}$ for all open C (or for all closed C).*

---

[20] This theorem is used in the proofs of Lemma A.35, Proposition A.36, Corollary A.37, Theorems A.38, B.75, and B.133, and to prove that stochastic processes are measurable.

[21] This lemma is used in the proofs of Theorems A.38 and A.74.

[22] This proposition is used in the proof of Theorem A.38.

A.3. Measurable Functions   585

Another example of the use of Theorem A.34 is the proof that all continuous functions are measurable. The result follows because the Borel σ-field is the smallest σ-field containing open sets.

Corollary A.37. *Let $(S, \mathcal{A})$ and $(T, \mathcal{B})$ be topological spaces with their Borel σ-fields. If $f : S \to T$ is continuous, then f is measurable.*

Here are some properties of measurable functions that will prove useful.

Theorem A.38. *Let $(S, \mathcal{A})$ be a measurable space.*

1. *Let $\aleph$ be an index set, and let $\{(T_\alpha, \mathcal{C}_\alpha)\}_{\alpha \in \aleph}$ be a collection of measurable spaces. For each $\alpha \in \aleph$, let $f_\alpha : S \to T_\alpha$ be a function. Define $f : S \to \prod_{\alpha \in \aleph} T_\alpha$ by $f(s) = \{f_\alpha(s)\}_{\alpha \in \aleph}$. Then f is measurable (with respect to the product σ-field) if and only if each f_α is measurable.*

2. *If $(V, \mathcal{C}_1)$ and $(U, \mathcal{C}_2)$ are measurable spaces and $f : S \to V$ and $g : V \to U$ are measurable, then $g(f) : S \to U$ is measurable.*

3. *Let f and g be measurable functions from S to $\mathbb{R}^n$, and let a be a constant scalar and let $b \in \mathbb{R}^n$ be constant. Then the following functions are also measurable: $f + g$ and $af + b$. If $n = 1$, then $f \cdot g$ and f/g are also measurable, where f/g can be set equal to an arbitrary constant when $g = 0$.*

4. *If, for each n, f_n is a measurable, extended real-valued function, then $\sup_n f_n$, $\inf_n f_n$, $\limsup_n f_n$, and $\liminf_n f_n$ are all measurable.*

5. *Let $(T, \mathcal{C})$ be a metric space with Borel σ-field. If $f_k : S \to T$ is a measurable function for each $k = 1, 2, \ldots$ and $\lim_{k \to \infty} f_k(s) = f(s)$ for all s, then f is measurable.*

6. *Let $(T, \mathcal{C})$ be a metric space with Borel σ-field, and let μ be a measure on $(S, \mathcal{A})$. If $f_k : S \to T$ is a measurable function for each $k = 1, 2, \ldots$ and $\lim_{k \to \infty} f_k(s)$ exists a.e. $[\mu]$, then there is a measurable $f : S \to T$ such that $\lim_{k \to \infty} f_k(s) = f(s)$, a.e. $[\mu]$.*

PROOF. (1) Suppose that f is measurable. To show that f_α is measurable, let $B_\alpha \in \mathcal{C}_\alpha$ and let $B_\beta = T_\beta$ for $\beta \neq \alpha$. Set $C = \prod_{\beta \in \aleph}^n B_\beta$, which is in the product σ-field, because all but finitely many B_β equal the entire space T_β. Then $f_\alpha^{-1}(B_\alpha) = f^{-1}(C)$. Since f is measurable, $f^{-1}(C) \in \mathcal{A}$. Now, suppose that each f_α is measurable, and let $B = \prod_{\alpha \in \aleph} B_\alpha$, with $B_\alpha \in \mathcal{C}_\alpha$ for all α and all but finitely many B_α (say $B_{\alpha_1}, \ldots, B_{\alpha_n}$) equal to T_α. Then $f^{-1}(B) = \cap_{i=1}^n f_{\alpha_i}^{-1}(B_{\alpha_i}) \in \mathcal{A}$. Since the sets of the form B generate the product σ-field, $f^{-1}(B) \in \mathcal{A}$ for all B in the product σ-field according to Theorem A.34.

(2) Let $A \in \mathcal{C}_2$. We need to prove that $g(f)^{-1}(A) \in \mathcal{A}$. First, note that $g(f)^{-1} = f^{-1}(g^{-1})$. Since g is measurable, $g^{-1}(A) \in \mathcal{C}_1$. Since f is measurable, $f^{-1}(g^{-1}(A)) \in \mathcal{A}$. So $g(f)^{-1}(A) \in \mathcal{A}$.

(3) The arithmetic parts of the theorem are all similar. They all follow from parts 2 and 1. For example, $h(x, y) = x + y$ is a measurable function from $\mathbb{R}^2$ to $\mathbb{R}$, so $h(f, g) = f + g$ is measurable. For the quotient, a little more care is needed. Let $h(x, y) = x/y$ when $y \neq 0$ and let it be an arbitrary constant when $y = 0$. Then h is measurable since $\{(x, y) : y = 0\}$ is in $\mathcal{B}^2$. It follows that $h(f, g)$ is measurable.

(4) Let $f = \sup_n f_n$. Then, for each finite b, $\{s : f(s) \leq b\} = \cap_{n=1}^\infty \{s : f_n(s) \leq b\} \in \mathcal{A}$. Also $\{s : f(s) = -\infty\} = \cap_{n=1}^\infty \{s : f_n(s) = -\infty\} \in \mathcal{A}$, and

$\{s : f(s) = \infty\} = \cap_{i=1}^{\infty} \cup_{n=1}^{\infty} \{s : f_n(s) > i\} \in \mathcal{A}$. Similar arguments work for inf. Since $\limsup_n f_n = \inf_k \sup_{n \geq k} f_n$ and $\liminf_n f_n = \sup_k \inf_{n \geq k} f_n$, these are also measurable.

(5) Let d be the metric in T. For each closed set $C \in \mathcal{C}$, and each m, let $C_m = \{t : d(t, C) < 1/m\}$. For each closed C, define

$$A_*(C) = \bigcap_{m=1}^{\infty} \bigcup_{n=1}^{\infty} \bigcap_{k=n}^{\infty} f_k^{-1}(C_m). \tag{A.39}$$

It is easy to see that $A_*(C) \in \mathcal{A}$ is the set of all s such that $\lim_{n \to \infty} f_n(s) \in C$. Obviously, $f^{-1}(C)$ consists of those s such that $\lim_{n \to \infty} f_n(s) \in C$. Hence, $f^{-1}(C) = A_*(C) \in \mathcal{A}$, and Proposition A.36 says that f is measurable.

(6) Let $G = \{s : \lim_{k \to \infty} f_k(s) \text{ does not exist}\}$, and let $G \subseteq C$ with $\mu(C) = 0$. Let $t \in T$, and define $f(s) = t$ for $s \in C$ and $f(s) = \lim_{k \to \infty} f_k(s)$ for $s \in C^C$. Apply part 5 to the restrictions of the functions $\{f_k\}_{k=1}^{\infty}$ to C^C to conclude that f restricted to C^C (call the restriction g) is measurable. If $A \in \mathcal{C}$, $f^{-1}(A) = g^{-1}(A) \in \mathcal{A}$ if $t \notin A$ and $f^{-1}(A) = g^{-1}(A) \cup C \in \mathcal{A}$ if $t \in A$. So f is measurable. □

Part 6 is particularly useful in that it allows us to treat the limit of a sequence of measurable functions as a measurable function even if the limit only exists almost everywhere. This is only useful, however, if we can show that functions that are equal almost everywhere have similar properties.

Many theorems about measurable functions are proven first for a special class of measurable functions called *simple functions* and then extended to all measurable functions using some limit theorems.

Definition A.40. A measurable function f is called *simple* if it assumes only finitely many distinct values.

A simple function is often expressed in terms of its values. Let f be a simple function taking values in $\mathbb{R}^n$ for some n. Suppose that $\{a_1, \ldots, a_k\}$ are the distinct values assumed by f, and let $A_i = f^{-1}(\{a_i\})$. Then $f(s) = \sum_{i=1}^{k} a_i I_{A_i}(s)$. The most fundamental limit theorem is the following.

Theorem A.41. *If f is a nonnegative measurable function, then there exists a sequence of simple functions $\{f_i\}_{i=1}^{\infty}$ such that for all $s \in S$, $f_i(s) \uparrow f(s)$.*

PROOF. For $k = 1, \ldots, i2^i$, let $A_{k,i} = \{s : (k-1)/2^i \leq f(s) < k/2^i\}$. Define $A_{0,i} = \{s : f(s) \geq i\}$. Then $A_{0,i}, A_{1,i}, \ldots, A_{i2^i,i}$ are disjoint and their union is S. Define

$$f_i(s) = \begin{cases} \frac{k-1}{2^i} & \text{if } s \in A_{k,i} \text{ for } k > 0, \\ i & \text{if } s \in A_{0,i}. \end{cases}$$

It is clear that $f_i(s) \leq f(s)$ for all i and s, and each f_i is a simple function. Since, for $k > 0$, $A_{k,i} = A_{2k-1,i+1} \cup A_{2k,i+1}$, and $A_{0,i} = A_{0,i+1} \cup A_{i2^{i+1}+1,i+1} \cup \cdots \cup A_{(i+1)2^{i+1},i+1}$, it is easy to see that $f_i(s) \leq f_{i+1}(s)$ for all i and all s. It is also easy to see that, for each s, there exists n such that for $i \geq n$, $|f(s) - f_i(s)| \leq 2^{-i}$. Hence $f_i(s) \uparrow f(s)$. □

The following theorem will be very useful throughout the study of statistics. It says that one function g is a function of another f if and only if g is measurable with respect to the σ-field generated by f.

Theorem A.42. *Let $(S_1, \mathcal{A}_1)$, $(S_2, \mathcal{A}_2)$, and $(S_3, \mathcal{A}_3)$ be measurable spaces such that $\mathcal{A}_3$ contains all singletons. Suppose that $f : S_1 \to S_2$ is measurable. Let $\mathcal{A}_{1f}$ be the σ-field generated by f. Let T be the image of f and let $\mathcal{A}_*$ be the image σ-field of f. Let $g : S_1 \to S_3$ be a measurable function. Then g is $\mathcal{A}_{1f}$ measurable if and only if there is a measurable function $h : T \to S_3$ such that for each $s \in S_1$, $g(s) = h(f(s))$.*

PROOF. For the "if" part, assume that there is a measurable $h : T \to S_3$ such that $g(s) = h(f(s))$ for all $s \in S_1$. Let $B \in \mathcal{A}_3$. We need to show that $g^{-1}(B) \in \mathcal{A}_{1f}$. Since h is measurable, $h^{-1}(B) \in \mathcal{A}_*$, so $h^{-1}(B) = T \cap A$ for some $A \in \mathcal{A}_2$. Since $f^{-1}(A) = f^{-1}(T \cap A)$ and $g^{-1}(B) = f^{-1}(h^{-1}(B))$, it follows that $g^{-1}(B) = f^{-1}(A) \in \mathcal{A}_{1f}$.

For the "only if" part, assume that g is $\mathcal{A}_{1f}$ measurable. For each $t \in S_3$, let $C_t = g^{-1}(\{t\})$. Since g is measurable with respect to $\mathcal{A}_{1f}$, let $A_t \in \mathcal{A}_{1f}$ be such that $C_t = f^{-1}(A_t)$. (Such A_t exists because of Corollary A.30.) Define $h(s) = t$ for all $s \in A_t \cap T$. (Note that if $t_1 \neq t_2$, then $A_{t_1} \cap A_{t_2} \cap T = \emptyset$, so h is well defined.) To see that $g(s) = h(f(s))$, let $g(s) = t$, so that $s \in C_t = f^{-1}(A_t)$. This means that $f(s) \in A_t \cap T$, which in turn implies $h(f(s)) = t = g(s)$.

To see that h is measurable, let $A \in \mathcal{A}_3$. We must show that $h^{-1}(A) \in \mathcal{A}_*$. Since g is $\mathcal{A}_{1f}$ measurable, $g^{-1}(A) \in \mathcal{A}_{1f}$, so there is some $B \in \mathcal{A}_2$ such that $g^{-1}(A) = f^{-1}(B)$. We will show that $h^{-1}(A) = B \cap T \in \mathcal{A}_*$ to complete the proof. If $s \in h^{-1}(A)$, then $t = h(s) \in A$ and $s = f(x)$ for some $x \in C_t \subseteq g^{-1}(A) = f^{-1}(B)$, so $f(x) \in B$. Hence, $s \in B \cap T$. This implies that $h^{-1}(A) \subseteq B \cap T$. Lastly, if $s \in B \cap T$, $s = f(x)$ for some $x \in f^{-1}(B) = g^{-1}(A)$ and $h(s) = h(f(x)) = g(x) \in A$. So, $h(s) \in A$ and $s \in h^{-1}(A)$. This implies $B \cap T \subseteq h^{-1}(A)$. □

The condition that $\mathcal{A}_3$ contain singletons is needed to avoid the situation in the following example.

Example A.43. Let $S_1 = S_2 = S_3 = \mathbb{R}$ and let $\mathcal{A}_1 = \mathcal{A}_2$ be the Borel σ-field, while $\mathcal{A}_3$ is the trivial σ-field. Then every function $g : S_1 \to S_3$ is $\mathcal{A}_{1f}$ measurable no matter what f is, for example, $g(s) = s$. If $f(s) = s^2$, then g is not a function of f.

A.4 Integration

The integral of a function with respect to a measure is a way to generalize the notion of weighted average. We define the integral in stages. We start with nonnegative simple functions.

Definition A.44. *Let f be a nonnegative simple function represented as $f(s) = \sum_{i=1}^{k} a_i I_{A_i}(s)$, with the a_i distinct and the A_i mutually disjoint. Then, the integral of f with respect to μ is $\int f(s) d\mu(s) = \sum_{i=1}^{k} a_i \mu(A_i)$. If 0 times ∞ occurs in such a sum, the result is 0 by convention.*

The integral of a nonnegative simple function is allowed to be ∞. It turns out that the formula for the integral of a nonnegative simple function is more general than in Definition A.44.

Proposition A.45.[23] *If $(S, \mathcal{A}, \mu)$ is a measure space, $A_i \in \mathcal{A}$ and $a_i \geq 0$ for $i = 1, \ldots, n$, and $f(s) = \sum_{i=1}^{n} a_i I_{A_i}(s)$, then $\int f(s) d\mu(s) = \sum_{i=1}^{n} a_i \mu(A_i)$.*

Next, we consider general nonnegative measurable functions. If f is a nonnegative simple function, then for every nonnegative simple function $g \leq f$, it follows easily from Definition A.44 that $\int g(s) d\mu(s) \leq \int f(s) d\mu(s)$. Hence, the following definition contains no contradiction with Definition A.44.

Definition A.46. *If f is a nonnegative measurable function, then the* integral *of f with respect to μ is $\int f(s) d\mu(s) = \sup_{g \leq f, g \text{ simple}} \int g(s) d\mu(s)$.*

For general functions f, define the *positive part* as $f^+(s) = \max\{f(s), 0\}$ and define the *negative part* as $f^-(s) = -\min\{f(s), 0\}$. Then $f(s) = f^+(s) - f^-(s)$. If $f \geq 0$, then $f^- \equiv 0$ and $\int f^-(s) d\mu(s) = 0$; hence the following definition contains no contradiction with the previous definitions.

Definition A.47. *If f is a measurable function, then the* integral *of f with respect to μ is*

$$\int f(s) d\mu(s) = \int f^+(s) d\mu(s) - \int f^-(s) d\mu(s),$$

if at least one of the two integrals on the right is finite. If both are infinite, the integral is undefined. *We say that f is* integrable *if the integral of f is defined and is finite.*

The integral is defined above in terms of its values at all points in S. Sometimes we wish to consider only a subset of S.

Definition A.48. *If $A \subseteq S$ and f is measurable, the* integral *of f over A with respect to μ is*

$$\int_A f(s) d\mu(s) = \int I_A(s) f(s) d\mu(s).$$

Here are a few simple facts about integrals.

Proposition A.49. *Let $(S, \mathcal{A}, \mu)$ be a probability space, and let $f, g : S \to \mathbb{R}$ be measurable.*

1. *If $f = g$ a.e. $[\mu]$, then $\int f(s) d\mu(s) = \int g(s) d\mu(s)$ if either integral is defined.*

2. *If $\int f(s) d\mu(s)$ is defined and a is a constant, then*

$$\int a f(s) d\mu(s) = a \int f(s) d\mu(s).$$

3. *If f and g are integrable with respect to μ, and $f \leq g$, a.e. $[\mu]$, then*

$$\int f(s) d\mu(s) \leq \int g(s) d\mu(s).$$

4. *If f and g are integrable and $\int_A f(s) d\mu(s) = \int_A g(s) d\mu(s)$ for all $A \in \mathcal{A}$, then $f = g$, a.e. $[\mu]$.*

---

[23] This proposition is used in the proof of Theorem A.53.

The proofs of the next few theorems are essentially borrowed from Royden (1968).

Theorem A.50 (Fatou's lemma).[24] *Let $\{f_n\}_{n=1}^\infty$ be a sequence of nonnegative measurable functions. Then*

$$\int \liminf_{n\to\infty} f(s)d\mu(s) \leq \liminf_{n\to\infty} \int f_n(s)d\mu(s).$$

PROOF. Let $f(s) = \liminf_{n\to\infty} f_n(s)$. Since

$$\int f(s)d\mu(s) = \sup_{\text{simple } \phi \leq f} \int \phi(s)d\mu(s),$$

we need only prove that, for every simple $\phi \leq f$,

$$\int \phi(s)d\mu(s) \leq \liminf_{n\to\infty} \int f_n(s)d\mu(s).$$

Since this is clearly true if $\phi(s) = 0$, a.s. $[\mu]$, we will assume that $\mu(A) > 0$, where $A = \{s : \phi(s) > 0\}$. Let $\phi \leq f$ be simple, let $\epsilon > 0$, and let δ and M be the smallest and largest positive values that ϕ assumes. For each n, define

$$A_n = \{s \in A : f_k(s) > (1-\epsilon)\phi(s), \text{ for all } k \geq n\}.$$

Since $(1-\epsilon)\phi(s) < f(s)$ for all $s \in A$, $\cup_{n=1}^\infty A_n = A$ and $A_n \subseteq A_{n+1}$ for all n. Let $B_n = A \cap A_n^C$.

$$\int f_n(s)d\mu(s) \geq \int_{A_n} f_n(s)d\mu(s) \geq (1-\epsilon)\int_{A_n} \phi(s)d\mu(s). \qquad (A.51)$$

If $\mu(B_n) = \infty$ for $n = n_0$, then $\mu(A) = \infty$ and $\int \phi(s)d\mu(s) = \infty$, since ϕ takes on only finitely many different values. The rightmost integral in (A.51) is at least $\delta\mu(A_n)$, which goes to ∞ as n increases, hence $\liminf_{n\to\infty} \int f_n(s)d\mu(s) = \infty$ and the result is true. So, assume $\mu(B_n) < \infty$ for all n. Since $\cap_{n=1}^\infty B_n = \emptyset$, it follows from Theorem A.19 that $\lim_{n\to\infty} \mu(B_n) = 0$. So, there exists N such that $n \geq N$ implies $\mu(B_n) < \epsilon$. Since

$$\int \phi(s)d\mu(s) = \int_A \phi(s)d\mu(s) = \int_{A_n} \phi(s)d\mu(s) + \int_{B_n} \phi(s)d\mu(s)$$

$$\leq \int_{A_n} \phi(s)d\mu(s) + M\epsilon,$$

(A.51) implies that, for $n \geq N$,

$$\int f_n(s)d\mu(s) \geq (1-\epsilon)\int \phi(s)d\mu(s) - \epsilon(1-\epsilon)M.$$

---

[24]This theorem is used in the proofs of Theorems A.52, A.57, A.60, B.117, and 7.80.

If $\int \phi(s)d\mu(s) = \infty$, the result is true again. If $\int \phi(s)d\mu(s) = K < \infty$, then for every $n \geq N$,

$$\int f_n(s)d\mu(s) \geq \int \phi(s)d\mu(s) - \epsilon[(1-\epsilon)M + K],$$

hence

$$\liminf_{n\to\infty} \int f_n(s)d\mu(s) \geq \int \phi(s)d\mu(s) - \epsilon[(1-\epsilon)M + K].$$

Since this is true for every $\epsilon > 0$,

$$\liminf_{n\to\infty} \int f_n(s)d\mu(s) \geq \int \phi(s)d\mu(s).$$

□

Theorem A.52 (Monotone convergence theorem). Let $\{f_n\}_{n=1}^{\infty}$ be a sequence of measurable nonnegative functions, and let f be a measurable function such that $f_n(x) \leq f(x)$ a.e. $[\mu]$ and $f_n(x) \to f(x)$ a.e. $[\mu]$. Then,

$$\lim_{n\to\infty} \int f_n(x)d\mu(x) = \int f(x)d\mu(x).$$

PROOF. Since $f_n \leq f$ for all n, $\int f_n(x)d\mu(x) \leq \int f(x)d\mu(x)$ for all n. Hence

$$\liminf_{n\to\infty} \int f_n(x)d\mu(x) \leq \limsup_{n\to\infty} \int f_n(x)d\mu(x) \leq \int f(x)d\mu(x).$$

By Fatou's lemma A.50, $\int f(x)d\mu(x) \leq \liminf_{n\to\infty} \int f_n(x)d\mu(x)$. □

Theorem A.53. If $\int f(s)d\mu(s)$ and $\int g(s)d\mu(s)$ are defined and they are not both infinite and of opposite signs, then $\int [f(s) + g(s)]d\mu(s) = \int f(s)d\mu(s) + \int g(s)d\mu(s)$.

PROOF. If $f, g \geq 0$, then by Theorem A.41, there exist sequences of nonnegative simple functions $\{f_n\}_{n=1}^{\infty}$ and $\{g_n\}_{n=1}^{\infty}$ such that $f_n \uparrow f$ and $g_n \uparrow g$. Then $(f_n + g_n) \uparrow (f + g)$ and $\int [f_n(s) + g_n(s)]d\mu(s) = \int f_n(s)d\mu(s) + \int g_n(s)d\mu(s)$ by Proposition A.45. The result now follows from the monotone convergence theorem A.52. For integrable f and g, note that $(f+g)^+ + f^- + g^- = (f+g)^- + f^+ + g^+$. What we just proved for nonnegative functions implies that

$$\int (f+g)^+(s)d\mu(s) + \int f^-(s)d\mu(s) + \int g^-(s)d\mu(s)$$

$$= \int [(f+g)^+(s) + f^-(s) + g^-(s)]d\mu(s)$$

$$= \int [(f+g)^-(s) + f^+(s) + g^+(s)]d\mu(s)$$

$$= \int (f+g)^-(s)d\mu(s) + \int f^+(s)d\mu(s) + \int g^+(s)d\mu(s).$$

Rearranging the terms in the first and last expressions gives the desired result. If both f and g have infinite integral of the same sign, then it follows easily using

Proposition A.49, that $f + g$ has infinite integral of the same sign. Finally, if only one of f and g has infinite integral, it also follows easily from Proposition A.49 that $f + g$ has infinite integral of the same sign. □

A nonnegative function can be used to create a new measure.

Theorem A.54. *Let $(S, \mathcal{A}, \mu)$ be a measure space, and let $f : S \to \mathbb{R}$ be nonnegative and measurable. Then $\nu(A) = \int_A f(s)d\mu(s)$ is a measure of $(S, \mathcal{A})$.*

PROOF. Clearly, ν is nonnegative and $\nu(\emptyset) = 0$, since $f(s)I_\emptyset(s) = 0$, a.e. $[\mu]$. Let $\{A_n\}_{n=1}^\infty$ be disjoint. For each n, define $g_n(s) = f(s)I_{A_n}(s)$ and $f_n(s) = \sum_{i=1}^n g_i(s)$. Define $A = \cup_{n=1}^\infty A_n$. Then $0 \leq f_n \leq fI_A$, a.e. $[\mu]$ and f_n converges to fI_A, a.e. $[\mu]$. So, the monotone convergence theorem A.52 says that

$$\lim_{n \to \infty} \int f_n(s)d\mu(s) = \nu(A). \tag{A.55}$$

Also, $\nu(A_i) = \int g_i(s)d\mu(s)$, for each i. It follows from Theorem A.53 that

$$\nu\left(\bigcup_{i=1}^n A_i\right) = \int f_n(s)d\mu(s) = \sum_{i=1}^n \int g_i(s)d\mu(s) = \sum_{i=1}^n \nu(A_i). \tag{A.56}$$

Take the limit as $n \to \infty$ of the second and last terms in (A.56) and compare to (A.55) to see that ν is countably additive. □

Theorem A.57 (Dominated convergence theorem). *Let $\{f_n\}_{n=1}^\infty$ be a sequence of measurable functions, and let f and g be measurable functions such that $f_n(x) \to f(x)$ a.e. $[\mu]$, $|f_n(x)| \leq g(x)$ a.e. $[\mu]$, and $\int g(x)d\mu(x) < \infty$. Then,*

$$\lim_{n \to \infty} \int f_n(x)d\mu(x) = \int f(x)d\mu(x).$$

PROOF. We have $-g(x) \leq f_n(x) \leq g(x)$ a.e. $[\mu]$, hence

$$\begin{aligned} g(x) + f_n(x) &\geq 0, \quad \text{a.e. } [\mu], \\ g(x) - f_n(x) &\geq 0, \quad \text{a.e. } [\mu], \\ \lim_{n \to \infty}[g(x) + f_n(x)] &= g(x) + f(x) \quad \text{a.e. } [\mu], \\ \lim_{n \to \infty}[g(x) - f_n(x)] &= g(x) - f(x) \quad \text{a.e. } [\mu]. \end{aligned}$$

It follows from Fatou's lemma A.50 and Theorem A.53 that

$$\begin{aligned} \int [g(x) + f(x)]d\mu(x) &\leq \liminf_{n \to \infty} \int [g(x) + f_n(x)]d\mu(x) \\ &= \int g(x)d\mu(x) + \liminf_{n \to \infty} \int f_n(x)d\mu(x), \\ \int f(x)d\mu(x) &\leq \liminf_{n \to \infty} \int f_n(x)d\mu(x). \end{aligned}$$

Similarly, it follows that

$$\int [g(x) - f(x)]d\mu(x) \leq \liminf_{n\to\infty} \int [g(x) - f_n(x)]d\mu(x)$$
$$= \int g(x)d\mu(x) - \limsup_{n\to\infty} \int f_n(x)d\mu(x),$$
$$\int f(x)d\mu(x) \geq \limsup_{n\to\infty} \int f_n(x)d\mu(x).$$

Together, these imply the conclusion of the theorem. □

An alternate version of the dominated convergence theorem is the following.

Proposition A.58.[25] *Let $\{f_n\}_{n=1}^\infty$, $\{g_n\}_{n=1}^\infty$ be sequences of measurable functions such that $|f_n(x)| \leq g_n(x)$, a.e. $[\mu]$. Let f and g be measurable functions such that $\lim_{n\to\infty} f_n(x) = f(x)$ and $\lim_{n\to\infty} g_n(x) = g(x)$, a.e. $[\mu]$. Suppose that $\lim_{n\to\infty} \int g_n(x)d\mu(x) = \int g(x)d\mu(x) < \infty$. Then, $\lim_{n\to\infty} \int f_n(x)d\mu(x) = \int f(x)d\mu(x)$.*

The proof is the same as the proof of Theorem A.57, except that g_n replaces g in the first three lines and wherever g appears with f_n and a limit is being taken.

For σ-finite measure spaces, the minimal condition that guarantees convergence of integrals is *uniform integrability*.

Definition A.59. A sequence of integrable functions $\{f_n\}_{n=1}^\infty$ is *uniformly integrable* (with respect to μ) if $\lim_{c\to\infty} \sup_n \int_{\{x:|f_n(x)|>c\}} |f_n(x)|d\mu(x) = 0$.

Theorem A.60.[26] *Let μ be a σ-finite measure. Let $\{f_n\}_{n=1}^\infty$ be a sequence of integrable functions such that $\lim_{n\to\infty} f_n = f$ a.e. $[\mu]$. Then $\lim_{n\to\infty} \int f_n(x)d\mu(x) = \int f(x)d\mu(x)$ if $\{f_n\}_{n=1}^\infty$ is uniformly integrable.*[27]

PROOF. Let f_n^+, f_n^-, f^+, and f^- be the positive and negative parts of f_n and f. We will prove that the result holds for nonnegative functions and take the difference to get the general result. Let $\epsilon > 0$ and let c be large enough so that $\sup_n \int_{\{x:f_n(x)>c\}} f_n(x)d\mu(x) < \epsilon$. The functions

$$g_n(x) = \begin{cases} f_n(x) & \text{if } f_n(x) \leq c, \\ c & \text{if } f_n(x) > c \end{cases}$$

converge a.e. $[\mu]$ to

$$g(x) = \begin{cases} f(x) & \text{if } f(x) \leq c, \\ c & \text{if } f(x) > c. \end{cases}$$

We now have

$$\int f(x)d\mu(x) \geq \int g(x)d\mu(x) = \lim_{n\to\infty} \int g_n(x)d\mu(x)$$
$$\geq \limsup_{n\to\infty} \int f_n(x)d\mu(x) - \epsilon,$$

---

[25] This proposition is used in the proof of Scheffé's theorem B.79.
[26] This theorem is used in the proofs of Theorems 1.121 and B.118.
[27] One could replace "if" by "if and only if," but we will never need the "only if" part of the theorem in this book.

where the second line follows from the dominated convergence theorem A.57 and the third from our choice of c. Since this is true for every ϵ, we have $\int f(x)d\mu(x) \geq \limsup \int f_n(x)d\mu(x)$. Combining this with Fatou's lemma A.50 gives

$$\int f(x)d\mu(x) = \lim_{n\to\infty} \int f_n(x)d\mu(x).$$

□

A.5 Product Spaces

In Definition A.12, we introduced product spaces and product σ-fields. We would like to be able to define measures on $(S_1 \times S_2, \mathcal{A}_1 \otimes \mathcal{A}_2)$ in terms of measures on $(S_1, \mathcal{A}_1)$ and $(S_2, \mathcal{A}_2)$. The derivation of product measure given here resembles the derivation in Billingsley (1986, Section 18).

Lemma A.61.[28] *Let $(S_1, \mathcal{A}_1, \mu_1)$ and $(S_2, \mathcal{A}_2, \mu_2)$ be σ-finite measure spaces, and let $\mathcal{A}_1 \otimes \mathcal{A}_2$ be the product σ-field.*

- *For every $B \in \mathcal{A}_1 \otimes \mathcal{A}_2$ and every $x \in S_1$, $B_x = \{y : (x,y) \in B\} \in \mathcal{A}_2$ and $\mu_2(B_x)$ is a measurable function from $(S_1, \mathcal{A}_1)$ to $\mathbb{R} \cup \{\infty\}$.*
- *For every $B \in \mathcal{A}_1 \otimes \mathcal{A}_2$ and every $y \in S_2$, $B^y = \{x : (x,y) \in B\} \in \mathcal{A}_1$ and $\mu_1(B^y)$ is a measurable function from $(S_2, \mathcal{A}_2)$ to $\mathbb{R} \cup \{\infty\}$.*

PROOF. Clearly, we need only prove one of the two sets of assertions. First, let $B = A_1 \times A_2$ with $A_i \in \mathcal{A}_i$ for $i = 1, 2$ and $x \in S_1$. Then

$$B_x = \begin{cases} A_2 & \text{if } x \in A_1, \\ \emptyset & \text{otherwise.} \end{cases}$$

So, $B_x \in \mathcal{A}_2$. Let $\mathcal{C}$ be the collection of all sets $B \subseteq S_1 \times S_2$ such that $B_x \in \mathcal{A}_2$. If $B \in \mathcal{C}$, then $(B^C)_x = \{y : (x,y) \notin B\} = (B_x)^C$, so $B^C \in \mathcal{C}$. Let $\{B_n\}_{n=1}^\infty \in \mathcal{C}$. Then it is easy to see that

$$\left(\bigcup_{n=1}^\infty B_n\right)_x = \left\{y : (x,y) \in \bigcup_{n=1}^\infty B_n\right\} = \bigcup_{n=1}^\infty \{y : (x,y) \in B_n\} = \bigcup_{n=1}^\infty (B_n)_x \in \mathcal{C}. \tag{A.62}$$

Clearly, $S_1 \times S_2 \in \mathcal{C}$, so $\mathcal{C}$ is a σ-field containing all product sets; hence it contains $\mathcal{A}_1 \otimes \mathcal{A}_2$. Next, let $f_B(x) = \mu_2(B_x)$ for $B \in \mathcal{A}_1 \otimes \mathcal{A}_2$. Write $S_1 \times S_2 = \cup_{n=1}^\infty E_n$ with $E_n = A_{1n} \times A_{2n}$ and $\mu_i(A_i n) < \infty$ for all n and $i = 1, 2$ and with the E_n disjoint. Then let $f_{B,n} = \mu_2((B \cap E_n)_x)$. It follows that $f_B = \sum_{n=1}^\infty f_{B,n}$. If we can show that $f_{B,n}$ is measurable for each n, then so is f_B, since they are nonnegative, and the sum is well defined. If $B = B_1 \times B_2$, then $f_{B,n}(x) = I_{A_{1n} \cap B_1}(x) \mu_2(A_{2n} \cap B_2)$, which is a measurable function. Let $\mathcal{D}$ be the collection of all sets $D \subseteq S_1 \times S_2$

---

[28] This lemma is used in the proofs of Lemmas A.64 and A.67 and Theorems A.69 and B.46.

such that $f_{D,n}$ is measurable. If $D \in \mathcal{D}$, then $f_{D^C,n} = \mu_2(A_{2n}) - f_{D,n}$, which is measurable, so $D^C \in \mathcal{D}$. If $\{D_m\}_{m=1}^\infty \in \mathcal{D}$ with the D_m disjoint, then

$$f_{\cup_{m=1}^\infty D_m, n}(x) = \mu_2\left(\bigcup_{m=1}^\infty (D_m \cap E_n)_x\right) = \sum_{m=1}^\infty \mu_2(D_m \cap E_n)_x$$

$$= \sum_{m=1}^\infty f_{D_m,n}(x),$$

which is a measurable function, so $\cup_{m=1}^\infty D_m \in \mathcal{D}$. Clearly, $S_1 \times S_2 \in \mathcal{D}$, so $\mathcal{D}$ is a λ-system (see Definition A.14) that contains the π-system of product sets. By the π–λ theorem A.17, $\mathcal{D}$ contains $\mathcal{A}_1 \otimes \mathcal{A}_2$. □

The following corollary to Lemma A.64 is a sort of dual to part 1 of Theorem A.38.

Corollary A.63. *Let $(S_1, \mathcal{A}_1)$, $(S_2, \mathcal{A}_2)$, and $(\mathcal{X}, \mathcal{B})$ be measurable spaces. If $f : S_1 \times S_2 \to \mathcal{X}$ is measurable, then for every $s_1 \in S_1$, $f_{s_1}(s_2) = f(s_1, s_2)$ is a measurable function from S_2 to $\mathcal{X}$.*

Lemma A.64.[29] *Suppose that $(S_1, \mathcal{A}_1, \mu_1)$ and $(S_2, \mathcal{A}_2, \mu_2)$ are σ-finite measure spaces. For each $x \in S_1$, $y \in S_2$, and $B \in \mathcal{A}_1 \otimes \mathcal{A}_2$, define B_x and B^y as in Lemma A.61. Then $\nu_1(B) = \int_{S_1} \mu_2(B_x) d\mu_1(x)$ and $\nu_2(B) = \int_{S_2} \mu_1(B^y) d\mu_2(y)$ both define the same measure on $(S_1 \times S_2, \mathcal{A}_1 \otimes \mathcal{A}_2)$. If $A_i \in \mathcal{A}_i$ for $i = 1, 2$, then $\nu_1(A_1 \times A_2) = \mu_1(A_1)\mu_2(A_2)$.*

PROOF. First, prove that ν_1 is a measure. The proof that ν_2 is a measure is identical. Clearly, $\nu_1(B) \geq 0$ for all B and $\nu_1(\emptyset) = 0$. If $\{B_n\}_{n=1}^\infty$ are disjoint, then

$$\nu_1\left(\bigcup_{n=1}^\infty B_n\right) = \int_{S_1} \sum_{n=1}^\infty \mu_2((B_n)_x) d\mu_1(x) = \sum_{n=1}^\infty \int_{S_1} \mu_2((B_n)_x) d\mu_1(x)$$

$$= \sum_{n=1}^\infty \nu_1((B_n)_x),$$

where the first equality follows from the definition of ν_1, the fact that μ_2 is countably additive, and (A.62); the second equality follows from the monotone convergence theorem A.52 and the fact that $\sum_{n=1}^m \mu_2((B_n)_x) \leq \sum_{n=1}^\infty \mu_2((B_n)_x)$ for all m; and last equality follows from the definition of ν_1. This proves that ν_1 (and so too ν_2) is a measure. Note that if $B = A_1 \times A_2$, then

$$\nu_1(B) = \int_{S_1} I_{A_1}(x)\mu_2(A_2) d\mu_1(x) = \mu_1(A_1)\mu_2(A_2)$$

$$= \int_{S_2} I_{A_2}(y)\mu_1(A_1) d\mu_2(y) = \nu_2(B).$$

So, $\nu_1 = \nu_2$ on the π-system consisting of product sets. Since each of μ_1 and μ_2 is σ-finite, there exists a countable collection of product sets whose union is $S_1 \times S_2$

---

[29]This lemma is used in the proof of Lemma A.67.

and such that each one has finite $\nu_1 = \nu_2$ measure. By Theorem A.26, ν_1 agrees with ν_2 on all of $\mathcal{A}_1 \otimes \mathcal{A}_2$. □

Definition A.65. Let $(S_i, \mathcal{A}_i, \mu_i)$ for $i = 1, 2$ be σ-finite measure spaces. Define the *product measure* $\mu_1 \times \mu_2$ on $(S_1 \times S_2, \mathcal{A}_1 \otimes \mathcal{A}_2)$ as the common value of the two measures ν_1 and ν_2 in Lemma A.64.

Lebesgue measure on $\mathbb{R}^2$, denoted $dxdy$, is a product measure. Not every measure on a product space is a product measure. Product probability measures will correspond to independent random variables. (See Theorem B.66.)

Proposition A.66. *Let μ be a measure on a product space $(S_1 \times S_2, \mathcal{A}_1 \otimes \mathcal{A}_2)$. Then μ is a product measure if and only if there exist set functions $\mu_i : \mathcal{A}_i \to \mathbb{R}$ for $i = 1, 2$ such that, for every $A_1 \in \mathcal{A}_1$ and $A_2 \in \mathcal{A}_2$, $\mu(A_1 \times A_2) = \mu_1(A_1)\mu_2(A_2)$.*

Lemma A.67.[30] *Let f be a measurable function from $S_1 \times S_2$ to $\mathbb{R}$ such that either $\{x \in S_1 : \int |f(x,y)|d\mu_2(y) = \infty\} \subseteq A \in \mathcal{A}_1$, where $\mu_1(A) = 0$, or $f \geq 0$. Then, there is a measurable (possibly extended real-valued) function $g : S_1 \to \mathbb{R} \cup \{\pm\infty\}$ such that $g(x) = \int f(x,y)d\mu_2(y)$, a.e. $[\mu_1]$. If f is the indicator of a measurable set B, then*

$$\int g(x)d\mu_1(x) = \mu_1 \times \mu_2(B). \quad (A.68)$$

PROOF. For each $B \in \mathcal{A}_1 \otimes \mathcal{A}_2$, note that $\int I_B(x,y)d\mu_2(y) = \mu_2(B_x)$, where B_x is defined in Lemma A.61. It was shown there that $\mu_2(B_x)$ is a measurable function of x. It follows from Lemma A.64 that (A.68) holds. It now follows from the linearity of integrals that if f is a nonnegative simple function, then $g(x) = \int f(x,y)d\mu_2(y)$ is a measurable function of x. If f is a nonnegative measurable function, let $\{f_n\}_{n=1}^\infty$ be a sequence of nonnegative simple functions such that $f_n \leq f$ for all n and $\lim_{n\to\infty} f_n(x,y) = f(x,y)$ for all (x,y). Then, the monotone convergence theorem A.52 says that $\lim_{n\to\infty} \int f_n(x,y)d\mu_2(y) = \int f(x,y)d\mu_2(y) = g(x)$ for all x. By part 5 of Theorem A.38, g is measurable. If $\mu_1\{x \in S_1 : \int |f(x,y)|d\mu_2(y) = \infty\} = 0$, then the argument just given applies to both f^+ and f^- and the difference $\int f^+(x,y)d\mu_2(y) - \int f^-(x,y)d\mu_2(y)$ is defined a.e. $[\mu_1]$ and equals $\int f(x,y)d\mu_2(y)$, a.e. $[\mu_1]$. If we let $g(x) = \int f^+(x,y)d\mu_2(y) - \int f^-(x,y)d\mu_2(y)$ for all $x \notin A$, and let $g(x)$ be constant on A, then $g(x) = \int f(x,y)d\mu_2(x)$, a.e. $[\mu_1]$, and g is measurable. □

The following two theorems will be used many times in the study of product spaces.

Theorem A.69 (Tonelli's theorem). *Let $(S_1, \mathcal{A}_1, \mu_1)$ and $(S_2, \mathcal{A}_2, \mu_2)$ be σ-finite measure spaces. Let $f : S_1 \times S_2 \to \mathbb{R}$ be a nonnegative measurable function. Then*

$$\int f(x,y)d\mu_1 \times \mu_2(x,y) = \int \left[\int f(x,y)d\mu_1(x)\right]d\mu_2(y)$$

---

[30]This lemma is used in the proofs of Theorem A.70 and of Lemmas 6.48 and B.46.

$$= \int \left[\int f(x,y)d\mu_2(y)\right] d\mu_1(x).$$

PROOF. As in the proof of Lemma A.67, let $\{f_n\}_{n=1}^\infty$ be a sequence of nonnegative simple functions such that $f_n \leq f$ for all n and $\lim_{n\to\infty} f_n(x,y) = f(x,y)$ for all (x,y). If $f_n(x,y) = \sum_{i=1}^{k_n} a_{i,n} I_{B_{i,n}}(x,y)$, then $\int f_n(x,y)d\mu_2(y) = \sum_{i=1}^{k_n} a_{i,n}\mu_2(B_{i,n,x})$ by Lemma A.61 and

$$\int \left[\int f_n(x,y)d\mu_2(y)\right] d\mu_1(x) = \int f(x,y)d\mu_1 \times \mu_2(x,y)$$

by (A.68). Since $0 \leq \int f_n(x,y)d\mu_2(y) \leq \int f(x,y)d\mu_2(y)$ for all x and n, and $\lim_{n\to\infty} \int f_n(x,y)d\mu_2(y) = \int f(x,y)d\mu_2(y)$ as in the proof of Lemma A.67, it follows from the monotone convergence theorem A.52 that

$$\int f(x,y)d\mu_1 \times \mu_2(x,y) = \lim_{n\to\infty} \int f_n(x,y)d\mu_1 \times \mu_2(x,y)$$
$$= \lim_{n\to\infty} \int \left[\int f_n(x,y)d\mu_2(y)\right]d\mu_1(x)$$
$$= \int \left[\lim_{n\to\infty} \int f_n(x,y)d\mu_2(y)\right]d\mu_1(x)$$
$$= \int \left[\int f(x,y)d\mu_2(x,y)\right]d\mu_1(x).$$

The proof that the iterated integrals can be calculated in the other order is similar. □

Theorem A.70 (Fubini's theorem). *Let $(S_1, \mathcal{A}_1, \mu_1)$ and $(S_2, \mathcal{A}_2, \mu_2)$ be σ-finite measure spaces. If $f : S_1 \times S_2 \to \mathbb{R}$ is integrable with respect to $\mu_1 \times \mu_2$, then*

$$\int f(x,y)d\mu(x,y) = \int \left[\int f(x,y)d\mu_1(x)\right]d\mu_2(y) = \int \left[\int f(x,y)d\mu_2(y)\right]d\mu_1(x).$$

PROOF. Let $g(x) = \int |f(x,y)|d\mu_2(y)$, a.e. $[\mu_1]$ be measurable. Then

$$\int g(x)d\mu_1(x) = \int \left[\int |f(x,y)|d\mu_2(y)\right]d\mu_1(x) = \int |f(x,y)|d\mu_1 \times \mu_2(x,y) < \infty$$

follows from Tonelli's theorem A.69 applied to $|f|$. It follows that

$$\left\{x : \int |f(x,y)|d\mu_2(y) = \infty\right\} \subset A \in \mathcal{A}_1$$

implies $\mu_1(A) = 0$. Apply Tonelli's theorem A.69 to f^+ and f^- and note that the set of all x such that $\int f^+(x,y)d\mu_2(y) - \int f^-(x,y)d\mu_2(y)$ is undefined is a subset of $\{x : \int |f(x,y)|d\mu_2(y) = \infty\}$. It follows that this difference of integrals is defined a.e. $[\mu_1]$ and the integral (with respect to μ_1) of the difference (which equals $\int[\int f(x,y)d\mu_2(y)]d\mu_1(x)$) is the difference of the integrals (which equals $\int f(x,y)d\mu_1 \times \mu_2(x,y)$). □

All of the results of this section can be extended to finite product spaces $S_1 \times \cdots \times S_n$ by simple inductive arguments.

A.6 Absolute Continuity

It is also common to consider two different measures on the same space.

Definition A.71. Let μ_1 and μ_2 be two measures on the same space $(S, \mathcal{A})$. Suppose that, for all $A \in \mathcal{A}$, $\mu_1(A) = 0$ implies $\mu_2(A) = 0$. Then, we say that μ_2 is *absolutely continuous with respect to* μ_1, denoted $\mu_2 \ll \mu_1$. When $\mu_2 \ll \mu_1$, we say that μ_1 is a *dominating measure* for μ_2.

Consider next a function f and a measure μ such that $\int f(x)d\mu(x)$ is defined. Then $\nu(A) = \int_A f(x)d\mu(x)$ is defined for all measurable A. If f takes on negative values with positive measure, then ν is not a measure because it assigns negative values to some sets, such as $A = \{x : f(x) < 0\}$. However, ν is still a signed measure.

If one of a pair of two measures is finite, there is a necessary and sufficient condition for absolute continuity which resembles the definition of continuity of functions.

Lemma A.72.[31] *Let μ_1 and μ_2 be measures on a space $(S, \mathcal{A})$. Consider the following condition:*

For every $\epsilon > 0$, there is δ_ϵ such that $\mu_1(A) < \delta_\epsilon$ implies $\mu_2(A) < \epsilon$. (A.73)

- *If condition (A.73) holds, then $\mu_2 \ll \mu_1$.*
- *If $\mu_2 \ll \mu_1$ and μ_2 is finite, then condition (A.73) holds.*

PROOF. For the first part, let $\epsilon > 0$ and suppose that $\mu_1(A) = 0$. Then $\mu_1(A) < \delta_\epsilon$ and $\mu_2(A) < \epsilon$. Since this is true for all $\epsilon > 0$, $\mu_2(A) = 0$. For the second part, suppose that $\mu_2 \ll \mu_1$, that μ_2 is finite, and that (A.73) fails. Then there exists $\epsilon > 0$ such that, for every integer n, there is A_n with $\mu_1(A_n) < 1/n^2$ but $\mu_2(A_n) \geq \epsilon$. Let $A = \cap_{k=1}^\infty \cup_{n=k}^\infty A_n$. By the first Borel–Cantelli lemma A.20, $\mu_1(A) = 0$ so $\mu_2(A) = 0$. Since μ_2 is finite, Theorem A.19 implies that

$$\mu_2(A) = \lim_{k \to \infty} \mu_2 \left(\bigcup_{n=k}^\infty A_n \right) \geq \epsilon.$$

This is a contradiction. □

The following theorem says that the first part of Example A.8 on page 574 is the most general form of absolute continuity with respect to σ-finite measures. The proof is mostly borrowed from Royden (1968).

Theorem A.74 (Radon–Nikodym theorem). *Let μ_1 and μ_2 be measures on $(S, \mathcal{A})$ such that $\mu_2 \ll \mu_1$ and μ_1 is σ-finite. Then there exists an extended real-valued measurable function $f : S \to [0, \infty]$ such that for every $A \in \mathcal{A}$,*

$$\mu_2(A) = \int_A f(x)d\mu_1(x). \qquad (A.75)$$

---

[31] This lemma is used in the proof of Lemma B.119.

Also, if $g : S \to \mathbb{R}$ is μ_2 integrable, then

$$\int g(x) d\mu_2(x) = \int g(x) f(x) d\mu_1(x). \tag{A.76}$$

The function f is called the Radon–Nikodym derivative of μ_2 with respect to μ_1 and it is unique a.e. $[\mu_1]$. The Radon–Nikodym derivative is sometimes denoted $(d\mu_2/d\mu_1)(s)$. If μ_2 is σ-finite, then f is finite a.e. $[\mu_1]$.

PROOF. First, we prove uniqueness a.e. $[\mu_1]$. Suppose that such an f exists. Let g be another function such that f and g are not a.e. $[\mu_1]$ equal. Let $A_n = \{x : f(x) > g(x) + 1/n\}$ and $B_n = \{x : f(x) < g(x) - 1/n\}$. Since f and g are not equal a.e. $[\mu_1]$, then there exists n such that either $\mu_1(A_n) > 0$ or $\mu_1(B_n) > 0$. Let A be a subset of either A_n or B_n with finite positive measure. Then $\int_A f(x) d\mu_1(x) \neq \int_A g(x) d\mu_1(x)$. Hence $g \neq d\mu_2/d\mu_1$.

The proof of existence proceeds as follows. First, we show that we can reduce to the case in which μ_1 is finite. Then, we create a collection of signed measures ν_α indexed by a real number α. For each α we find a set A^α such that every subset of A^α has positive ν_α measure and every subset of the complement B^α has negative ν_α measure. We then show that $B^\beta \subseteq B^\alpha$ for $\beta \geq \alpha$, which allows us to define $f(x) = \sup\{\alpha : x \in B^\alpha\}$. Finally, we show that f satisfies (A.75) and (A.76).

Now, we prove that we need only consider finite μ_1. Since μ_1 is σ-finite, let $\{A_n\}_{n=1}^\infty$ be disjoint elements of $\mathcal{A}$ such that $\mu_1(A_i) < \infty$ and $S = \cup_{i=1}^\infty A_i$. Let $\mu_{j,i}$ be μ_j restricted to A_i for $j = 1, 2$ and each i. Then $\mu_{2,i} \ll \mu_{1,i}$ for each i and each $\mu_{1,i}$ is finite. Suppose that for each i we can find f_i as in the theorem with μ_j replaced by $\mu_{j,i}$ for $j = 1, 2$. Then $f(x) = \sum_{i=1}^\infty I_{A_i}(x) f_i(x)$ is the function required by the theorem as stated. Hence, we prove the theorem only for the case in which μ_1 is finite.

Suppose that μ_1 is finite, and define the signed measure $\nu_\alpha = \alpha \mu_1 - \mu_2$ for each nonnegative rational number α. (Note that $\nu_\alpha(A)$ never equals ∞, although it may equal $-\infty$.) For each α, define

$$P_\alpha = \{A \in \mathcal{A} : \nu_\alpha(B) \geq 0, \text{ for every } B \subseteq A\},$$
$$\lambda_\alpha = \sup_{A \in P_\alpha} \nu_\alpha(A).$$

That is, λ_α is the supremum of the signed measures of sets all of whose subsets have nonnegative signed measure.[32] Since $\emptyset \in P_\alpha$, $\lambda_\alpha \geq 0$. Let $\{A_n\}_{n=1}^\infty$ be such that $\lambda_\alpha = \lim_{i \to \infty} \nu_\alpha(A_i)$, and let $A^\alpha = \cup_{i=1}^\infty A_i$. Since every subset of A^α can be written as a union of subsets of the A_i, it follows that $A^\alpha \in P_\alpha$, hence $\lambda_\alpha \geq \nu_\alpha(A^\alpha)$. Since $A^\alpha \setminus A_i \subseteq A^\alpha$, it follows that $\nu_\alpha(A^\alpha \setminus A_i) \geq 0$ for all i and $\nu_\alpha(A^\alpha) = \nu_\alpha(A^\alpha \setminus A_i) + \nu_\alpha(A_i) \geq \nu_\alpha(A_i)$ for all i. It follows that $\lambda_\alpha \leq \nu_\alpha(A^\alpha)$. Hence $\lambda_\alpha = \nu_\alpha(A^\alpha) < \infty$. Define $B^\alpha = (A^\alpha)^C$.

Next, we prove that every subset of B^α has nonpositive measure.[33] If not, let $B \subseteq B^\alpha$ such that $\nu_\alpha(B) > 0$. If B has no subsets with negative signed measure,

---

[32] The sets in P_α are often called the *positive sets* relative to the signed measure ν_α.

[33] Such sets are called *negative sets* relative to the signed measure ν_α.

then $B\cup A^\alpha \in P_\alpha$ and $\nu_\alpha(A^\alpha \cup B) > \lambda_\alpha$, a contradiction. So, let n_1 be the smallest positive integer such that there is a subset $B_1 \subseteq B$ with $\nu_\alpha(B_1) < -1/n_1$. For each $k > 1$, let n_k be the smallest positive integer such that there exists a subset $B_k \subseteq B \setminus \cup_{i=1}^{k-1} B_i$ with $\nu_\alpha(B_k) < -1/n_k$. Now, let $C = B \setminus \cup_{k=1}^{\infty} B_k$. Clearly $\nu_\alpha(C) > 0$. If we prove that C has no subsets with negative signed measure, then $C \in P_\alpha$ and we have another contradiction. So, suppose that $D \subseteq C$ has $\nu_\alpha(D) = -\epsilon < 0$. Since $\nu_\alpha(B) > 0$, it must be that $\sum_{k=1}^{\infty} \nu_\alpha(B_k) > -\infty$. Hence $\lim_{k\to\infty} n_k = \infty$. So, there is k such that $1/(n_{k+1} - 1) < \epsilon$. Notice that $D \subseteq C \subseteq B \setminus \cup_{i=1}^{k} B_k$. Since $\nu_\alpha(D) < -1/(n_{k+1}-1)$, this contradicts the definition of n_{k+1}.

If $\beta > \alpha$, we have

$$\nu_\alpha(A^\alpha \cap B^\beta) \geq 0, \quad \nu_\beta(A^\alpha \cap B^\beta) \leq 0.$$

Subtract the first inequality from the second to get $(\beta - \alpha)\mu_1(A^\alpha \cap B^\beta) \leq 0$, from which it follows that $\mu_1(A^\alpha \cap B^\beta) = 0$. Since $\nu_\beta(A) \geq \nu_\alpha(A)$ for $\beta \geq \alpha$, we can assume that $A^\alpha \subseteq A^\beta$ if $\beta \geq \alpha$. It follows that $B^\beta \subseteq B^\alpha$ for $\beta \geq \alpha$, and we can define $f(x) = \sup\{\alpha : x \in B^\alpha\}$. Since $B^0 = S$, $f(x) \geq 0$ for all x. It is easy to see that $f(x) \geq \alpha$ if $x \in B^\alpha$ and $f(x) \leq \alpha$ if $x \in A^\alpha$. It is also easy to see that $\{x : f(x) \geq b\} = \cup_{\alpha \geq b} B^\alpha$. Since this is a countable union of measurable sets, it is measurable. By Lemma A.35, f is measurable.

Next, we prove that (A.75) holds for every $A \in \mathcal{A}$. Let $A \in \mathcal{A}$ be arbitrary and let $\epsilon > 0$ be given. Let $N > \mu_1(A)/\epsilon$ be a positive integer. Define $E_k = A \cap B^{k/N} \cap A^{(k+1)/N}$ and $E_\infty = A \setminus \cup_{k=1}^{\infty} A^{k/N}$. Then $A = \cup_{k=1}^{\infty} E_k \cup E_\infty$ and the E_j are all disjoint. So $\mu_2(A) = \mu_2(E_\infty) + \sum_{k=0}^{\infty} \mu_2(E_k)$. By construction $f(x) \in [k/N, (k+1)/N]$ for all $x \in E_k$ and $f(x) = \infty$ for all $x \in E_\infty$. Since $\nu_{k/N}(E_k) \leq 0$ and $\nu_{(k+1)/N}(E_k) \geq 0$, we have, for finite k,

$$\left| \mu_2(E_k) - \int_{E_k} f(x) d\mu_1(x) \right| \leq \frac{1}{N} \mu_1(E_k). \tag{A.77}$$

If $\mu_1(E_\infty) > 0$, then $\mu_2(E_\infty) = \infty$ since $\nu_\alpha(E_\infty) < 0$ for all α. If $\mu_1(E_\infty) = 0$, then $\mu_2(E_\infty) = 0$ by absolute continuity. Either way, $\mu_2(E_\infty) = \int_{E_\infty} f(x) d\mu_1(x)$. Adding this into the sum of (A.77) over all finite k gives

$$\left| \mu_2(E) - \int_E f(x) d\mu_1(x) \right| \leq \frac{1}{N} \mu_1(E) < \epsilon.$$

Since this is true for every $\epsilon > 0$, (A.75) is established.

To prove (A.76), we note that it is true if g is an indicator function, hence it is true for all simple functions. By the monotone convergence theorem A.52, it is true for all nonnegative functions and by subtraction it is true for all integrable functions.

Finally, if $f(x) = \infty$ for all $x \in A$ with $\mu_1(A) > 0$, then $\mu_2(B) = \infty$ for every $B \subseteq A$ such that $\mu_1(B) > 0$. It is now impossible for μ_2 to be σ-finite. □

In statistical applications, we will often have a class of measures, each of which is absolutely continuous with respect to a single σ-finite measure. It would be nice if the single dominating measure were in the original class or could be constructed from the class. The following theorem addresses this problem. The proof is borrowed from Lehmann (1986).

Theorem A.78.[34] *Let μ be a σ-finite measure on $(S, \mathcal{A})$. Suppose that $\aleph$ is a collection of measures on $(S, \mathcal{A})$ such that for every $\nu \in \aleph$, $\nu \ll \mu$. Then there exists a sequence of nonnegative numbers $\{c_i\}_{i=1}^{\infty}$ and a sequence of elements of $\aleph$, $\{\nu_i\}_{i=1}^{\infty}$ such that $\sum_{i=1}^{\infty} c_i = 1$ and $\nu \ll \sum_{i=1}^{\infty} c_i \nu_i$ for every $\nu \in \aleph$.*

PROOF. If $\aleph$ is a countable collection, the result is trivially true. If μ is finite, let $\lambda = \mu$. If μ is not finite, then there exists a countable partition of S into $\{S_i\}_{i=1}^{\infty}$ such that $0 < \mu(S_i) = d_i < \infty$. For each $B \in \mathcal{A}$, let $\lambda(B) = \sum_{i=1}^{\infty} \mu(B \cap S_i)/(2^i d_i)$. In either case λ is finite and $\nu \ll \lambda$ for every $\nu \in \aleph$. Define $\mathcal{Q}$ to be the collection of all measures of the form $\sum_{i=1}^{\infty} a_i \nu_i$ where $\sum_{i=1}^{\infty} a_i = 1$ and each $\nu_i \in \aleph$. Clearly $\beta \in \mathcal{Q}$ implies $\beta \ll \lambda$.

Next, let $\mathcal{D}$ be the collection of sets C in $\mathcal{A}$ such that there exists $Q \in \mathcal{Q}$ satisfying $\lambda(\{x \in C : dQ/d\lambda(x) = 0\}) = 0$ and $Q(C) > 0$. To see that $\mathcal{D}$ is nonempty, let ν be a measure in $\aleph$ that is not identically 0 and let $C = \{x : d\nu/d\lambda(x) > 0\}$. Then with $Q = \nu$, we have $\{x \in C : dQ/d\lambda(x) = 0\} = \emptyset$ and $Q(C) = \nu(C) = \nu(S) > 0$, so $C \in \mathcal{D}$. Since λ is finite, $\sup_{C \in \mathcal{D}} \lambda(C) = c < \infty$, so there exist $\{C_n\}_{n=1}^{\infty}$ such that $\lim_{n \to \infty} \lambda(C_n) = c$ and $C_n \in \mathcal{D}$ for all n. Let $C_0 = \cup_{n=1}^{\infty} C_n$ and let $Q_n \in \mathcal{Q}$ be such that $Q_n(C_n) > 0$ and $\lambda(\{x \in C_n : dQ_n/d\lambda(x) = 0\}) = 0$. Let $Q_0 = \sum_{n=1}^{\infty} 2^{-n} Q_n \in \mathcal{Q}$, so that $dQ_0/d\lambda = \sum_{n=1}^{\infty} 2^{-n} dQ_n/d\lambda$ and

$$\left\{x \in C_0 : \frac{dQ_0}{d\lambda}(x) = 0\right\} \subseteq \bigcup_{n=1}^{\infty} \left\{x \in C_n : \frac{dQ_n}{d\lambda}(x) = 0\right\},$$

which implies that $C_0 \in \mathcal{D}$ and $\lambda(C_0) = c$.

Since $Q_0 \in \mathcal{Q}$, we now need only prove that $\nu \ll Q_0$ for all $\nu \in \aleph$ to finish the proof. Suppose that $Q_0(A) = 0$ and $\nu \in \aleph$. We must prove $\nu(A) = 0$. Since $Q_0(A \cap C_0) = 0$ and $dQ_0/d\lambda(x) > 0$ for all $x \in C_0$, it follows that $\lambda(A \cap C_0) = 0$ and hence $\nu(A \cap C_0) = 0$. Let $C = \{x : d\nu/d\lambda(x) > 0\}$. Then, $\nu(A \cap C_0^C \cap C^C) = 0$ since $d\nu/d\lambda(x) = 0$ for $x \in C^C$. Let $D = A \cap C_0^C \cap C$, which is disjoint from C_0. If $\lambda(D) > 0$, then $\lambda(C_0 \cup D) > \lambda(C_0)$ and $D \in \mathcal{D}$. It follows easily that $C_0 \cup D \in \mathcal{D}$ and $\lambda(C_0 \cup D) > \lambda(C_0)$ contradicts $\lambda(C_0) = c$. Hence $\lambda(D) = 0$ and $\nu(D) = 0$, which implies $\nu(A) = \nu(A \cap C_0) + \nu(A \cap C_0^C \cap C^C) + \nu(D) = 0$. □

There is a chain rule for Radon–Nikodym derivatives.

Theorem A.79 (Chain rule).[35] *Let ν and η be σ-finite measures and suppose that $\mu \ll \nu \ll \eta$. Then*

$$\frac{d\mu}{d\eta}(s) = \frac{d\mu}{d\nu}(s) \frac{d\nu}{d\eta}(s), \quad a.e. \ [\eta]. \tag{A.80}$$

PROOF. It is easy to see that $\mu \ll \eta$ so that $d\mu/d\eta$ exists. For every set A, it follows from (A.76) that

$$\mu(A) = \int_A \frac{d\mu}{d\nu}(s) d\nu(s) = \int_A \frac{d\mu}{d\nu}(s) \frac{d\nu}{d\eta}(a) d\eta(s).$$

---

[34] This theorem is used in the proofs of Lemmas 2.15 and 2.24. It appears as Theorem 2 in Appendix 3 of Lehmann (1986) and is attributed to Halmos and Savage (1949).

[35] This theorem is used in the proof of Lemma 2.15.

By the uniqueness of Radon–Nikodym derivatives, (A.80) holds. □

The Radon–Nikodym theorem A.74 relates integrals with respect to two different measures on the same space. There are also theorems that relate integrals with respect to two different measures on two different spaces.

Theorem A.81. *A measurable function f from one measure space $(S_1, \mathcal{A}_1, \mu_1)$ to a measurable space $(S_2, \mathcal{A}_2)$, $f : S_1 \to S_2$, induces a measure on the range S_2. For each $A \in \mathcal{A}_2$, define $\mu_2(A) = \mu_1\left(f^{-1}(A)\right)$. Integrals with respect to μ_2 can be written as integrals with respect to μ_1 in the following way: If $g : S_2 \to \mathbb{R}$ is integrable, then*

$$\int g(y) d\mu_2(y) = \int g(f(x)) d\mu_1(x). \tag{A.82}$$

PROOF. What needs to be proven is that μ_2 is indeed a measure and that (A.82) holds. To see that μ_2 is a measure, note that if $A, B \in \mathcal{A}_2$ are disjoint, then so too are $f^{-1}(A)$ and $f^{-1}(B)$. The fact that μ_2 is nonnegative and countably additive now follows directly from the same fact about μ_1.

If $g : S_2 \to \mathbb{R}$ is the indicator function of a set A, then

$$\begin{aligned}
\int g(y) d\mu_2(y) &= \mu_2(A) = \mu_1(f^{-1}(A)) \\
&= \int I_{f^{-1}(A)}(x) d\mu_1(x) = \int g(f(x)) d\mu_1(x).
\end{aligned}$$

That (A.82) is true for all nonnegative simple functions follows by adding the far ends of this equation (multiplied by positive constants). The monotone convergence theorem A.52 allows us to extend the equality to all nonnegative integrable functions. By subtraction, we can extend to all integrable functions. □

Definition A.83. The measure μ_2 in Theorem A.81 is called *the measure induced on $(S_2, \mathcal{A}_2)$ by f from μ_1*.

If the measure μ_1 in Theorem A.81 is not finite, and the function f is not one-to-one, the measure μ_2 may not be very interesting.

Example A.84. Let $S_1 = \mathbb{R}^2$, $S_2 = \mathbb{R}$, μ_1 equal Lebesgue measure on $\mathbb{R}^2$, and $f(x, y) = x$. Let the two σ-fields be Borel σ-fields. The measure μ_2 that f induces on $(S_2, \mathcal{A}_2)$ from μ_1 is the following. If $A \in \mathcal{A}_2$ and the Lebesgue measure of A is 0, then $\mu_2(A) = 0$. Otherwise, $\mu_2(A) = \infty$. Although μ_2 is absolutely continuous with respect to Lebesgue measure, it is not σ-finite. The only functions g that are integrable with respect to μ_2 are those that are almost everywhere 0.

If μ_1 is σ-finite, there is a way to avoid the problem in Example A.84 by making use of the following result.

Theorem A.85.[36] *A measure μ on a space $(S, \mathcal{A})$ is σ-finite if and only if there exists an integrable function $f : S \to \mathbb{R}$ such that $f > 0$, a.e. $[\mu]$.*

---

[36]This theorem is used in the proof of Theorem B.46.

PROOF. For the "if" part, let f be as in the statement of the theorem. Let $0 < \int f(s)d\mu(s) = c < \infty$. Let $A_n = \{s : 1/n \leq f(s) < 1/(n-1)\}$, for $n = 1, 2, \ldots$. We see that $A_1 = \{s : f(s) \geq 1\}$ and $S = \cup_{n=1}^{\infty} A_n$. We can write

$$c = \int f(s)d\mu(s) = \sum_{n=1}^{\infty} \int_{A_n} f(s)d\mu(s) \geq \sum_{n=1}^{\infty} \int_{A_n} \frac{1}{n}d\mu(s) = \sum_{n=1}^{\infty} \mu(A_n)\frac{1}{n}.$$

It follows that $\mu(A_n) \leq nc$ for all n. Hence μ is σ-finite.

For the "only if" part, assume that μ is σ-finite, and let $\{A_n\}_{n=1}^{\infty}$ be mutually disjoint sets such that $S = \cup_{n=1}^{\infty} A_n$ and $\mu(A_n) < \infty$ for all n. Define $f(s)$ to equal $2^{-n}/\mu(A_n)$ for all $s \in A_n$ and for all n such that $\mu(A_n) > 0$. For n such that $\mu(A_n) = 0$, set $f(s) = 0$ if $s \in A_n$. Then

$$\int f(s)d\mu(s) = \sum_{n=1}^{\infty} \frac{2^{-n}}{\mu(A_n)}\mu(A_n) \leq 1. \qquad \square$$

Example A.86 (Continuation of Example A.84; see page 601). Let $h(x, y) = \exp(-[x^2 + y^2]/2)$. It is known that h is integrable with respect to μ_1 and h is everywhere strictly positive. Let $\mu'_1(C) = \int_C h(x, y)d\mu_1(x, y)$. Then $\mu'_1 \ll \mu_1$ and $\mu_1 \ll \mu'_1$. The measure μ'_2 induced on $(S_2, \mathcal{A}_2)$ from μ'_1 by $f(x, y) = x$ is $\mu'_2(B) = \sqrt{2\pi}\int_B \exp(-x^2/2)dx$. A function $g : S_2 \to \mathbb{R}$ is integrable with respect to μ'_2 if and only if $\exp(-x^2/2)g(x)$ is integrable with respect to Lebesgue measure.

As a sort of reverse version of Theorem A.81, functions from a measurable space to a measure space induce measures on the domain space.

Proposition A.87. *Let f be a measurable function from a measurable space $(S_1, \mathcal{A}_1)$ to a measure space $(S_2, \mathcal{A}_2, \mu_2)$, $f : S_1 \to S_2$. Let $\mathcal{A}_{1f} \subseteq \mathcal{A}_1$ be the σ-field generated by f, and let T be the image of f. Suppose that $T \in \mathcal{A}_2$. Then f induces a measure μ_1 on $(S_1, \mathcal{A}_{1f})$ defined by $\mu_1(A) = \mu_2(T \cap B)$ if $A = f^{-1}(B)$. Furthermore, if $g : (S_1, \mathcal{A}_{1f}) \to \mathbb{R}$ is integrable with respect to μ_1, then*

$$\int g(x)d\mu_1(x) = \int_T h(y)d\mu_2(y), \qquad (A.88)$$

where h satisfying $h(f(x)) = g(x)$ is guaranteed to exist by Theorem A.42.

A.7 Problems

Section A.2:

1. Let S be a set and let $\mathcal{A}$ be the collection of all subsets of S that either are countable or have countable complement. Prove that $\mathcal{A}$ is a σ-field.

2. Prove Proposition A.10 on page 575.

3. Prove Proposition A.13 on page 576. (*Hint:* First, show that every open ball in $\mathbb{R}^k$ is the union of countably many open rectangles. Then prove that the smallest σ-field containing open balls must be the same as the smallest σ-field containing open rectangles.)

4. Prove that $\mathcal{B}^+$ defined on page 571 is a σ-field of subsets of the extended real numbers.

5. Prove Proposition A.15 on page 576.

6. Prove Proposition A.16 on page 576.

7. *Let $F : \mathbb{R} \to \mathbb{R}$ be a nondecreasing function that is continuous from the right. For each interval $(a, b]$, define $\mu((a, b]) = F(b) - F(a)$.

   (a) Suppose that $\{(a_n, b_n]\}_{n=1}^\infty$ is a sequence of disjoint intervals such that $\cup_{n=1}^\infty (a_n, b_n] \subseteq (a, b]$. Prove that $\sum_{n=1}^\infty \mu((a_n, b_n]) \leq \mu((a, b])$. (*Hint:* Prove it for finite collections and take a limit.)

   (b) Suppose that $\{(a_n, b_n]\}_{n=1}^\infty$ is a sequence of disjoint intervals such that $(a, b] \subseteq \cup_{n=1}^\infty (a_n, b_n]$. Prove that $\sum_{n=1}^\infty \mu((a_n, b_n]) \geq \mu((a, b])$. (*Hint:* First, prove it for finite collections by induction. For infinite collections, let $\mu((a, b]) > \epsilon > 0$. Cover a compact interval $[a + \delta, b]$ with finitely many open intervals $(a_n, b_n + \delta_n)$ such that $|\mu((a, b]) - \mu((a + \delta, b])| < \epsilon/2$ and $|\sum_{n=1}^\infty \mu((a_n, b_n]) - \sum_{n=1}^\infty \mu((a_n, b_n + \delta_n])| < \epsilon/2$. This can be done by using continuity from the right.)

   (c) Prove that μ is countably additive on the smallest field containing intervals of the form $(a, b]$. (*Hint:* Deal separately with finite and semi-infinite intervals)

8. A measure space $(S, \mathcal{A}, \mu)$ is complete if $A \subseteq B \in \mathcal{A}$ and $\mu(B) = 0$ implies $A \in \mathcal{A}$. Let $(S, \mathcal{C}, \mu)$ be a measure space, and let $\mathcal{A} = \mathcal{C} \cup \mathcal{D}$, where $\mathcal{D} = \{D : \exists A, C \in \mathcal{C} \text{ with } D \Delta A \subseteq C \text{ and } \mu(C) = 0\}$. For each $D \in \mathcal{D}$, define $\mu^*(D) = \mu(A)$ where $D \Delta A \subseteq C$ and $\mu(C) = 0$. For $C \in \mathcal{C}$, define $\mu^*(C) = \mu(C)$. Show that μ^* is well defined and that $(S, \mathcal{D}, \mu^*)$ is a complete measure space.

Section A.3:

9. Prove Proposition A.28 on page 583.

10. Prove Proposition A.32 on page 584.

11. Prove Proposition A.36 on page 584.

12. Let $(S, \mathcal{A}, \mu)$ be a measure space, and let $\{f_n\}_{n=1}^\infty$ be a sequence of measurable functions from S to $\mathbb{R}$. Suppose that for every $\epsilon > 0$, $\sum_{n=1}^\infty \mu(\{s : f_n(s) > \epsilon\}) < \infty$. Prove that $\lim_{n \to \infty} f_n(s) = 0$, a.e. $[\mu]$. (*Hint:* Use the first Borel–Cantelli lemma A.20.)

13. Let $(S_j, \mathcal{A}_j)$ for $j = 0, 1, 2, 3$ be measurable spaces. Let $f_j : S_0 \to S_j$ be measurable for $j = 1, 2, 3$. Let $\mathcal{A}_{0,j}$ be the σ-field generated by f_j for $j = 1, 2$. Prove that f_3 is measurable with respect to $\mathcal{A}_{0,1} \cap \mathcal{A}_{0,2}$ if and only if there exist measurable $g_j : S_j \to S_3$ for $j = 1, 2$ such that $f_3 = g_1(f_1) = g_2(f_2)$.

Section A.4:

14. If $f \geq 0$ is measurable and $\int f(s)d\mu(s) = 0$, then show that $f(s) = 0$, a.e. $[\mu]$.

15. If $f(s) > 0$ for all $s \in A$ and $\mu(A) > 0$, prove that $\int_A f(s)d\mu(s) > 0$.

16. Prove Proposition A.45 on page 588. (*Hint:* Use induction on n.)

17. Prove Proposition A.49 on page 588. (*Hint:* For part 4, use Problem 14 on page 604.)

18. Let $S = \mathbb{R}$ and let $\mathcal{A}$ be the σ-field of sets that are either countable or have countable complement. (See Problem 1 on page 602.) Let μ be Lebesgue measure. Suppose that $f : S \to \mathbb{R}$ is integrable. Prove that $f = 0$, a.e. $[\mu]$.

19. Let $(S, \mathcal{A})$ be a measurable space, and let f be a bounded measurable function. (That is, there exist a and b such that $a \leq f(x) \leq b$ for all $x \in S$.)

    (a) Let μ be a measure on $(S, \mathcal{A})$ such that $\mu(S) = 1$. Prove that
    $$a \leq \int f(x)d\mu(x) \leq b.$$

    (b) Let $\epsilon > 0$. Prove that there exists a simple function g such that for all measures μ satisfying $\mu(S) = 1$, $|\int f(x)d\mu(x) - \int g(x)d\mu(x)| < \epsilon$.

20. Prove the following alternative type of monotone convergence theorem: Let $\{f_n\}_{n=1}^{\infty}$ be a sequence of integrable functions such that $f_n(x)$ converges monotonically to $f(x)$ a.e. $[\mu]$. Then $\int f(x)d\mu(x)$ is defined and $\int f(x)d\mu(x) = \lim_{n \to \infty} \int f_n(x)d\mu(x)$. (*Hint:* Use the dominated convergence theorem A.57 on the positive parts of f_n and the monotone convergence theorem A.52 on the negative parts, or vice versa, depending on whether the convergence is from above or below.)

21. Let $(S, \mathcal{A}, \mu)$ be a measure space, let $\{g_n\}_{n=1}^{\infty}$ be a sequence of integrable functions that converges a.e. $[\mu]$, and let g be another integrable function. Suppose that for all $C \in \mathcal{A}$,
    $$\lim_{n \to \infty} \int_C g_n(s)d\mu(s) = \int_C g(s)d\mu(s).$$
    Prove that $\lim_{n \to \infty} g_n = g$, a.e. $[\mu]$.

Section A.5:

22. Prove Proposition A.66 on page 595.

23. Let $(S_1, \mathcal{A}_1)$ and $(S_2, \mathcal{A}_2)$ be measurable spaces, and define the product space $(S_1 \times S_2, \mathcal{A}_1 \otimes \mathcal{A}_2)$. Prove that $A \times B \in \mathcal{A}_1 \otimes \mathcal{A}_2$ with $A \subseteq S_1$ and that $B \subseteq S_2$ implies $A \in \mathcal{A}_1$ and $B \in \mathcal{A}_2$. (*Hint:* For each $C \in \mathcal{A}_1 \otimes \mathcal{A}_2$, define $C_y = \{x : (x, y) \in C\}$. Then let $\mathcal{C} = \{C : C_y \in \mathcal{A}_1, \text{ for all } y \in S_2\}$. Prove that $\mathcal{C}$ is a σ-field containing all product sets.)

Section A.6:

24. Suppose that $\mu_1 \ll \mu_2$ and $\mu_2 \ll \mu_1$.

    (a) Show that a.e. $[\mu_1]$ means the same thing as a.e. $[\mu_2]$.
    (b) Show that
    $$\frac{d\mu_1}{d\mu_2}(s) = \left(\frac{d\mu_2}{d\mu_1}(s)\right)^{-1}, \quad \text{a.e. } [\mu_1] \text{ and a.e. } [\mu_2].$$

25. If μ_1 is a measure and f is a nonnegative measurable function, then define the measure μ_2 by $\mu_2(A) = \int_A f(s)d\mu_1(s)$. Prove that $\mu_2 \ll \mu_1$.

26. Let λ be Lebesgue measure on $\mathbb{R}$ and define
    $$\mu(A) = \lambda(A) + cI_A(x_0),$$
    for some fixed $c > 0$ and $x_0 \in \mathbb{R}$.

    (a) Prove that μ is a measure.
    (b) Show that $\lambda \ll \mu$, but that $\mu \not\ll \lambda$.
    (c) Show that $\int f(x)d\mu(x) = \int f(x)d\lambda(x) + cf(x_0)$.

27. *In the proof of Theorem A.74, we proved the *Hahn decomposition theorem* for signed measures, namely that if ν is a signed measure on $(S, \mathcal{A})$, then there exists $A \in \mathcal{A}$ such that A is a positive set and A^C is a negative set relative to ν.

    (a) Let ν be a signed measure on $(S, \mathcal{A})$. Suppose that there are two different Hahn decompositions. That is, A_1 and A_2 are both positive sets and A_1^C and A_2^C are both negative sets. Prove that every measurable subset B of $A_1 \cap A_2^C$ has $\nu(B) = 0$.
    (b) If ν is a signed measure on $(S, \mathcal{A})$, use the Hahn decomposition theorem to create definitions for the following:
        i. The integral with respect to ν of a measurable function.
        ii. When a function is integrable with respect to ν.
    (c) If there are two different Hahn decompositions for a signed measure ν, prove that the definition of integral with respect to ν produces the same value for both decompositions.

28. In the statement of Proposition A.87 on page 602, prove that the measure μ_1 is well defined. (That is, suppose that $A = f^{-1}(B_1) = f^{-1}(B_2)$, and prove that $\mu_2(B_1 \cap T) = \mu_2(B_2 \cap T)$.) Also prove that μ_1 is a measure.

29. In the statement of Proposition A.87 on page 602, assuming that μ_1 is a well-defined measure, prove that (A.88) holds.

Appendix B
Probability Theory

This appendix builds on Appendix A but is otherwise self-contained. It contains an introduction to the theory of probability. The first section is an overview. It could serve either as a refresher for those who have previously studied the material or as an informal introduction for those who have never studied it.

B.1 Overview

B.1.1 Mathematical Probability

The measure theoretic definition of probability is that a measure space $(S, \mathcal{A}, \mu)$ is called a *probability space* and μ is called a *probability* if $\mu(S) = 1$. Each element of $\mathcal{A}$ is called an *event*. A measurable function X from S to some other space $(\mathcal{X}, \mathcal{B})$ is called a *random quantity*. The most popular type of random quantity is a *random variable*, which occurs when $\mathcal{X}$ is $\mathbb{R}$ with the Borel σ-field. The probability measure μ_X induced on $(\mathcal{X}, \mathcal{B})$ by X from μ is called the *distribution of X*.

Example B.1. Let $S = \mathcal{X} = \mathbb{R}$ with Borel σ-field. Let f be a nonnegative function such that $\int f(x)dx = 1$. Define $\mu(A) = \int_A f(x)dx$ and $X(s) = s$. Then X is a continuous random variable with density f, and $\mu_X = \mu$. If we let ν denote Lebesgue measure, then $\mu_X \ll \nu$ with $d\mu_X/d\nu = f$.

Example B.2. Let $S = \mathbb{R}$ with Borel σ-field. Let $\mathcal{X} = \{x_1, x_2, \ldots\}$, a countable set. Let f be a nonnegative function defined on $\mathcal{X}$ such that $\sum_{i=1}^{\infty} f(x_i) = 1$. Define $\mu(A) = \sum_{\{i : x_i \in A\}} f(x_i)$. Then X is a discrete random variable with probability mass function f, and $\mu_X = \mu$. If we let ν denote counting measure on $\mathcal{X}$, then $\mu \ll \nu$ with $d\mu/d\nu = f$.

In both of these examples, we will say that f is the *density* of X with respect to ν.

When there is one probability space $(S, \mathcal{A}, \mu)$ from which all other probabilities are induced by way of random quantities, then the probability in that one space will be denoted Pr. So, for example, if μ_X is the distribution of a random quantity X and if $B \in \mathcal{B}$, then $\Pr(X \in B) = \mu(X^{-1}(B)) = \mu_X(B)$.

The *expected value* or *mean* or *expectation* of a random variable X is defined (and denoted) as $\mathrm{E}(X) = \int x d\mu_X(x)$, if the integral exists, where μ_X is the distribution of X. If X is a vector of random variables (called a *random vector*), then $\mathrm{E}(X)$ will stand for the vector with coordinates equal to the means of the coordinates of X.

The (in)famous law of the unconscious statistician, B.12, is very useful for calculating means of functions of random quantities. It says that $\mathrm{E}[f(X)] = \int f(x) d\mu_X(x)$. For example, the *variance* of a random variable X with mean c is $\mathrm{Var}(X) = \mathrm{E}([X-c]^2)$, which can be calculated as $\int (x-c)^2 d\mu_X(x)$. The *covariance* between two random variables X and Y with means c_X and c_Y, respectively, is $\mathrm{Cov}(X,Y) = \mathrm{E}([X-c_X][Y-c_Y])$.

B.1.2 Conditioning

We begin with a heuristic derivation of the important concepts using the special case of discrete random quantities. Afterwards, we define the important terms in a more rigorous way.

Consider the case of two random quantities X and Y, each of which assumes at most countably many distinct values, $X \in \mathcal{X} = \{x_1, \ldots\}$ and $Y \in \mathcal{Y} = \{y_1, \ldots\}$. Let $p_{ij} = \Pr(X = x_i, Y = y_j)$. Then

$$\Pr(X = x_i) = \sum_{j=1}^{\infty} p_{ij} = p_{i\cdot}, \quad \text{and}$$

$$\Pr(Y = y_j) = \sum_{i=1}^{\infty} p_{ij} = p_{\cdot j}.$$

These equations give the *marginal distributions of X and Y*, respectively. We can define the *conditional probability that $X = x_i$ given $Y = y_j$* by

$$\Pr(X = x_i | Y = y_j) = \frac{p_{ij}}{p_{\cdot j}} = p_{i|j}.$$

Note that for each j, $\sum_{i=1}^{\infty} p_{i|j} = 1$ so that the numbers $\{p_{i|j}\}_{i=1}^{\infty}$ define a probability distribution on $\mathcal{X}$ known as the *conditional distribution of X given $Y = y_j$*. We can calculate the *conditional mean (expectation)* of a function f of X given $Y = y_j$ by

$$\mathrm{E}(f(X) | Y = y_j) = \sum_{i=1}^{\infty} f(x_i) p_{i|j}.$$

From the conditional distribution, we could define a measure on $(\mathcal{X}, 2^{\mathcal{X}})$ by

$$\mu_{X|Y}(A | y_j) = \sum_{x_i \in A} p_{i|j}.$$

It follows that, for each j, $\mathrm{E}(f(X)|Y = y_j) = \int f(x)d\mu_{X|Y}(x|y_j)$. We can think of this conditional mean as a function of y:

$$g(y) = \mathrm{E}(f(X)|Y = y).$$

The marginal distribution of Y is a measure on $(\mathcal{Y}, 2^\mathcal{Y})$ defined by

$$\mu_Y(B) = \sum_{y_j \in B} p_{.j}, \quad \text{for all } B \in 2^\mathcal{Y}.$$

Similarly, the joint distribution of (X, Y) induces a measure on $(\mathcal{X} \times \mathcal{Y}, 2^\mathcal{X} \otimes 2^\mathcal{Y})$ by $\mu_{X,Y}(C) = \sum_{(x_i, y_j) \in C} p_{ij}$, for all $C \in 2^\mathcal{X} \otimes 2^\mathcal{Y}$. The point of all of these measures and distributions is the following. We can write the integral of g over any set $B \in 2^\mathcal{Y}$ as

$$\begin{aligned}\int_B g(y)d\mu_Y(y) &= \sum_{y_j \in B} g(y_j)p_{.j} = \sum_{y_j \in B} \sum_{i=1}^{\infty} f(x_i)p_{i|j}p_{.j} \\ &= \int f(x)I_B(y)d\mu_{X,Y}(x,y) = \mathrm{E}\left(f(X)I_B(Y)\right).\end{aligned}$$

The overall equation

$$\int_B g(y)d\mu_Y(y) = \mathrm{E}\left(f(X)I_B(Y)\right)$$

will be used as the property that defines conditional expectation in general. Through the definition of conditional expectation, we will define conditional probability and conditional distributions in general.

Theorem B.21 says that, in general, if a random variable X has finite mean and if $\mathcal{C}$ is a sub-σ-field of $\mathcal{A}$, then a function $g : S \to \mathbb{R}$ exists which is measurable with respect to the σ-field $\mathcal{C}$ and such that

$$\mathrm{E}(XI_B) = \int_B g(s)d\mu(s), \quad \text{for all } B \in \mathcal{C}. \tag{B.3}$$

This is the general version of what we worked out above for discrete random variables in which $\mathcal{C}$ was the σ-field generated by Y. We will use the symbol $\mathrm{E}(X|\mathcal{C})$ to stand for the function g. The two important features that $\mathrm{E}(X|\mathcal{C})$ possesses are that it is measurable with respect to the σ-field $\mathcal{C}$ and that it satisfies (B.3). Any function that equals $\mathrm{E}(X|\mathcal{C})$ a.s. $[\mu]$ will also satisfy (B.3), so there may be many functions that satisfy the definition of conditional expectation. All such functions are called *versions* of the conditional expectation. When we say that a random variable equals $\mathrm{E}(X|\mathcal{C})$, we will mean that it is a version of $\mathrm{E}(X|\mathcal{C})$.

Notice that we can set $B = S$ in (B.3) and the equation becomes $\mathrm{E}(X) = \mathrm{E}[\mathrm{E}(X|\mathcal{C})]$. This result is called the law of total probability. A useful generalization is given in Theorem B.70.

If $\mathcal{C}$ is the σ-field generated by another random quantity Y, then the symbol $\mathrm{E}(X|Y)$ is usually used instead of $\mathrm{E}(X|\mathcal{C})$. For the case in which $\mathcal{C}$ is the σ-field generated by Y, some special notation is introduced. We saw in Theorem A.42 that a function is measurable with respect to the σ-field generated by Y if and only if it is a function of Y. Hence, there is a function h defined on the space $\mathcal{Y}$ where Y takes its values such that $\mathrm{E}(X|Y) = h(Y)$. We use the notation $\mathrm{E}(X|Y = t)$ to stand for $h(t)$. (See Corollary B.22.) In this notation, we have, for all $B \in \mathcal{C}$, $\mathrm{E}(XI_B) = \int_C \mathrm{E}(X|Y = y) d\mu_Y(t)$, where μ_Y is the distribution of Y.

Example B.4. Let $S = \mathbb{R}^2$ and let $\mathcal{A}$ be the two-dimensional Borel sets. Let
$$\mu(A) = \int_A \frac{1}{\sqrt{3}\pi} \exp\left\{-\frac{2}{3}(s_1^2 + s_2^2 - s_1 s_2)\right\} ds_1 ds_2.$$

Suppose that $X(s) = s_1$ and $Y(s) = s_2$ when $s = (s_1, s_2)$. Now $\mathrm{E}(|X|) = \sqrt{2/\pi} < \infty$. We claim that $g(s) = s_2/2$ and $h(t) = t/2$ satisfy the conditions required to be $\mathrm{E}(X|Y)(s)$ and $\mathrm{E}(X|Y=t)$, respectively. First, note that the σ-field generated by Y is $\mathcal{A}_Y = \{\mathbb{R} \times C : C$ is Borel measurable$\}$, and μ_Y is the measure with density $\exp(-t^2/2)/\sqrt{2\pi}$. It is clear that any measurable function of s_2 alone is $\mathcal{A}_Y$ measurable. Let $B = \mathbb{R} \times C$, so that $\mathrm{E}(XI_B)$ equals

$$\int_{-\infty}^{\infty} \int_C s_1 \frac{1}{\sqrt{3}\pi} \exp\left\{-\frac{2}{3}(s_1^2 + s_2^2 - s_1 s_2)\right\} ds_2 ds_1$$
$$= \int_C \int_{-\infty}^{\infty} s_1 \frac{\sqrt{2}}{\sqrt{3}\pi} \exp\left\{-\frac{2}{3}\left(s_1 - \frac{1}{2}s_2\right)^2\right\} \frac{1}{\sqrt{2\pi}} \exp\left\{-\frac{1}{2}s_2^2\right\} ds_1 ds_2$$
$$= \int_C \frac{1}{2}s_2 \frac{1}{\sqrt{2\pi}} \exp\left\{-\frac{1}{2}s_2^2\right\} ds_2$$
$$= \int_C \int_{-\infty}^{\infty} \frac{1}{2}s_2 \frac{\sqrt{2}}{\sqrt{3}\pi} \exp\left\{-\frac{2}{3}\left(s_1 - \frac{1}{2}s_2\right)^2\right\} \frac{1}{\sqrt{2\pi}} \exp\left\{-\frac{1}{2}s_2^2\right\} ds_1 ds_2$$
$$= \int_B \frac{1}{2}s_2 \frac{1}{\sqrt{3}\pi} \exp\left\{-\frac{2}{3}(s_1^2 + s_2^2 - s_1 s_2)\right\} ds_1 ds_2 = \int_B g(s) d\mu(s).$$

Note also that the third line in the above string equals $\int_C h(s_2) d\mu_Y(s_2)$.

It is easy to see that if X is already measurable with respect to $\mathcal{C}$, then $\mathrm{E}(X|\mathcal{C}) = X$.

Conditional probability turns out to be the special case of conditional expectation in which $X = I_A$. That is, we define $\Pr(A|\mathcal{C}) = \mathrm{E}(I_A|\mathcal{C})$. A conditional probability is *regular* if $\Pr(\cdot|\mathcal{C})(s)$ is a probability measure for all s. It turns out that, under very general conditions (see Theorem B.32), we can choose the functions $\Pr(A|\mathcal{C})(\cdot)$ in such a way that they are regular conditional probabilities. In particular, the space $(\mathcal{X}, \mathcal{B})$ needs to be sufficiently like the real numbers with the Borel σ-field. Such spaces are called *Borel spaces* as defined in Definition B.31. All of the most common spaces are Borel spaces. In particular, $\mathbb{R}^k$ for all finite k and $\mathbb{R}^\infty$. For those readers with more mathematical background, complete separable metric spaces are also Borel spaces. Also, finite and countable products of Borel spaces are Borel spaces.

In the future, we will assume that all versions of conditional probabilities are regular when they are on Borel spaces. If $\mathcal{C}$ is the σ-field generated by Y, then $\Pr(A|Y = y)$ will be used to stand for $\mathrm{E}(I_A|Y = y)$.

If $X : S \to \mathcal{X}$ is a random quantity, its *conditional distribution* is the collection of conditional probabilities on $\mathcal{X}$ induced from the restriction of conditional probabilities on S to the σ-field generated by X. If the $P(\cdot|\mathcal{C})$ are regular conditional probabilities, then we say that the version of the conditional distribution of X given $\mathcal{C}$ is a *regular conditional distribution*. When we refer to a conditional distribution without the word "version," we will mean a version of the conditional distribution. Occasionally, we will need to choose a version that satisfies some other condition. In those cases, we will try to be explicit about versions.

Because conditional distributions are probability measures, many of the theorems from Appendix A which apply to such measures apply to conditional distributions. For example, the monotone convergence theorem A.52 and the dominated convergence theorem A.57 apply to conditional means because limits of measurable functions are still measurable. Also, most of the properties of probability measures from this appendix apply as well.

We now turn our attention to the existence and calculation of densities for conditional distributions. If the joint distribution of two random quantities has a density with respect to a product measure, then the conditional distributions have densities that can be calculated in the usual way, as the joint density divided by the marginal density of the conditioning variable. Theorem B.46 allows us to extend this result to joint distributions that are not absolutely continuous with respect to product measures, such as when one of the quantities is a function of the other. Here, we merely give an example of how such conditional densities are calculated.

Example B.5. Let $X = (X_1, X_2)$ have bivariate normal distribution with density

$$f_X(x_1, x_2) = \frac{1}{2\pi\sigma_1\sigma_2\sqrt{1-\rho^2}} \exp\left(-\frac{1}{2(1-\rho^2)}\left[\frac{(x_1-\mu_1)^2}{\sigma_1^2} - 2\frac{(x_1-\mu_1)(x_2-\mu_2)}{\sigma_1\sigma_2} + \frac{(x_2-\mu_2)^2}{\sigma_2^2}\right]\right)$$

with respect to Lebesgue measure on $\mathbb{R}^2$. The marginal density of $Y = X_1 + X_2$ with respect to Lebesgue measure is

$$f_Y(y) = \frac{1}{\sqrt{2\pi}\sigma} \exp\left(-\frac{1}{2\sigma^2}[y-(\mu_1+\mu_2)]^2\right),$$

where $\sigma^2 = \sigma_1^2 + \sigma_2^2 + 2\rho\sigma_1\sigma_2$. The pair (X, Y) does not have a joint density with respect to Lebesgue measure on $\mathbb{R}^3$, but it does have a joint density with respect to the measure ν on $\mathbb{R}^3$ defined as follows. For each $A \subseteq \mathbb{R}^3$, let $A^* = \{(x_1, x_2) : (x_1, x_2, x_1 + x_2) \in A\}$. Let $\nu(A) = \lambda_2(A^*)$, where λ_k is Lebesgue measure on $\mathbb{R}^k$ for $k = 1, 2$. Then $f_{X,Y}(x, y) = f_X(x)$ is the joint density of (X, Y) with respect to ν, and

$$\frac{f_X(x)}{f_Y(y)} = \frac{1}{\sqrt{2\pi}\sigma^*} \exp\left(-\frac{1}{2\sigma^{*2}}\left(x_1 - \mu_1 - \frac{[\sigma_1^2 + \rho\sigma_1\sigma_2](y-\mu_1-\mu_2)}{\sigma^2}\right)^2\right),$$

if $y = x_1 + x_2$, is the conditional density of X given $Y = y$ with respect to the measure $\nu_{X|Y}(A|y) = \lambda_1(A_y^*)$, where $A_y^* = \{x_1 : (x_1, y - x_1) \in A\}$.

The concept of *conditional independence* will turn out to be central to the development of statistical models. A collection $\{X_n\}_{n=1}^{\infty}$ of random quantities is conditionally independent given another quantity Y if the conditional distribution (given Y) of every finite subset is a product measure. If, in addition, Y is constant almost surely, we say that $\{X_n\}_{n=1}^{\infty}$ are *independent*. We will call random quantities (conditionally) IID if they are (conditionally) independent and they all have the same conditional distribution.

B.1.3 Limit Theorems

There are three types of convergence which we consider for sequences of random quantities: almost sure convergence, convergence in probability, and convergence in distribution. The weakest of these is the last. (See Theorem B.90.) A sequence $\{X_n\}_{n=1}^{\infty}$ *converges in distribution* to X if $\lim_{n\to\infty} \mathrm{E}\left(f\left(X_n\right)\right) = \mathrm{E}\left(f\left(X\right)\right)$ for every bounded continuous function f. We denote this type of convergence $X_n \xrightarrow{\mathcal{D}} X$. If $\mathcal{X} = \mathbb{R}$, a more common way to express $X_n \xrightarrow{\mathcal{D}} X$ is that $\lim_{n\to\infty} F_n(x) = F(x)$ for all x at which F is continuous, where F_n is the CDF of X_n and F is the CDF of X.[1]

If $\mathcal{X}$ is a metric space with metric d, we say that a sequence $\{X_n\}_{n=1}^{\infty}$ *converges in probability* to X if, for every $\epsilon > 0$, $\lim_{n\to\infty} \Pr(d(X_n, X) > \epsilon) = 0$. We write this as $X_n \xrightarrow{P} X$. Almost sure convergence is the same as almost everywhere convergence of functions, and it is the strongest of the three.

A popular method for proving convergence in distribution involves the use of characteristic functions. The *characteristic function* of a random vector X is the complex-valued function

$$\phi_X(t) = \mathrm{E}\left(\exp[it^\top X]\right).$$

It is easy to see that the characteristic function exists for every random vector and has complex absolute value at most 1 for all t. Other facts that follow directly from the definition are the following. If $Y = aX + b$, then $\phi_Y(t) = \phi_X(at) \exp(itb)$. If X and Y are independent, $\phi_{X+Y} = \phi_X \phi_Y$.

The importance of characteristic functions is that they characterize distributions (see the uniqueness theorem B.106) and they are "continuous" as a function of the distribution in the sense of convergence in distribution (see the continuity theorem B.93).

Two of the more useful limit theorems are the weak law of large numbers B.95 and the central limit theorem B.97. If $\{X_n\}_{n=1}^{\infty}$ are IID random variables with finite mean μ, then the weak law of large numbers says that the sample average $\overline{X}_n = \sum_{i=1}^{n} X_i/n$ converges in probability to μ. If, in addition, they have finite variance σ^2, the central limit theorem B.97 says that $\sqrt{n}(\overline{X}_n - \mu) \xrightarrow{\mathcal{D}} N(0, \sigma^2)$, the normal distribution with mean 0 and variance σ^2.

---

[1] See Problem 25 on page 664. If $\mathcal{X} = \mathbb{R}^k$, the same idea can be used. That is, $X_n \xrightarrow{\mathcal{D}} X$ if and only if the joint CDFs F_n of X_n converge to the joint CDF F of X at all points at which F is continuous. Since we will not need to use this characterization, we will not prove it.

B.2 Mathematical Probability

In this chapter, we will present the basic framework of the measure theoretic probability calculus. Most of the concepts like random quantities, distributions, an so forth. will be special cases of measure theoretic concepts introduced in Appendix A.

B.2.1 Random Quantities and Distributions

We begin by introducing the basic building blocks of probability theory.

Definition B.6. A *probability space* is a measure space $(S, \mathcal{A}, \mu)$ with $\mu(S) = 1$. Each element of $\mathcal{A}$ is called an *event*. If $(S, \mathcal{A}, \mu)$ is a probability space, $(\mathcal{X}, \mathcal{B})$ is a measurable space, and $X : S \to \mathcal{X}$ is measurable, then X is called a *random quantity*. If $\mathcal{X} = \mathbb{R}$ and $\mathcal{B}$ is the Borel or Lebesgue σ-field, then X is called a *random variable*. Let μ_X be the probability measure induced on $(\mathcal{X}, \mathcal{B})$ by X from μ (see Definition A.83). This probability measure is called *the distribution of X*. The distribution of X is said to be *discrete* if there exists a countable set $A \subseteq \mathcal{X}$ such that $\mu_X(A) = 1$. The distribution of X is *continuous* if $\mu_X(\{x\}) = 0$ for all $x \in \mathcal{X}$.

The distribution of X is easily seen to be equivalent to the restriction of μ to the σ-field generated by X, $\mathcal{A}_X$.

When there is one probability space from which all other probabilities are induced by way of random quantities, then the probability in that one space will be denoted Pr. So, for example, in the above definition of the distribution of a random quantity X, if $B \in \mathcal{B}$, then $\Pr(X \in B) = \mu(X^{-1}(B)) = \mu_X(B)$.

The distribution of a random variable can be described by its *cumulative distribution function*.

Definition B.7. A function F is a *(cumulative) distribution function (CDF)* if it has the following properties:

- F is nondecreasing;
- $\lim_{x \to -\infty} F(x) = 0$;
- $\lim_{x \to \infty} F(x) = 1$;
- F is continuous from the right.

Proposition B.8. *If X is a random variable, then the function $F_X(x) = \Pr(X \leq x)$ is a CDF. In this case, F_X is called the CDF of X.*

A distribution function F can be used to create a measure on $(\mathbb{R}, \mathcal{B})$ as follows. Set $\mu((a, b]) = F(b) - F(a)$, and extend this to the whole σ-field using the Caratheodory extension theorem A.22.[2]

We can also construct a distribution function from a probability measure on the real numbers. If μ is a probability measure on $(\mathbb{R}, \mathcal{B}^1)$, the CDF associated with it is $F(x) = \mu((-\infty, x])$. If f is a Borel measurable function from $\mathbb{R}$ to $\mathbb{R}$, we will write $\int f(x) dF(x)$ and $\int f(x) d\mu(x)$ interchangeably.

---

[2] See the discussion on page 581 and Problem 7 on page 603.

If μ is a probability measure on $(\mathbb{R}^n, \mathcal{B}^n)$, a joint CDF can be defined as

$$F(x_1, \ldots, x_n) = \mu\left((-\infty, x_1] \times \cdots \times (-\infty, x_n]\right),$$

the measure of an orthant. For every joint CDF, there is a random vector X with that CDF and we call the CDF F_X.

Definition B.9. Let $(S, \mathcal{A}, \mu)$ be a probability space, and let $(\mathcal{X}, \mathcal{B}, \nu)$ be a measure space. Suppose that $X : S \to \mathcal{X}$ is measurable. Let μ_X be the measure induced on $(\mathcal{X}, \mathcal{B})$ by X from μ. Suppose that $\mu_X \ll \nu$. Then we call the Radon–Nikodym derivative $f_X = d\mu_X/d\nu$ the *density of X with respect to ν*.

Proposition B.10. *If $h : X :\to \mathbb{R}$ is measurable and f_X is the density of X with respect to ν, then $\int h(x) dF_X(x) = \int h(x) f_X(x) d\nu(x)$.*

Definition B.11. If X is a random variable with CDF $F_X(\cdot)$, then the *expected value (or mean, or expectation) of X* is $E(X) = \int x dF_X(x)$. If X is a random vector, then $E(X)$ will stand for the vector with coordinates equal to the means of the coordinates of X.

The following theorem is often called the law of the unconscious statistician, because some people forget that it is not really the definition of expected value.

Theorem B.12.[3] *If $X : S \to \mathcal{X}$ is a random quantity and $f : \mathcal{X} \to \mathbb{R}$ is a measurable function, then $E[f(X)] = \int f(x) d\mu_X(x)$, where μ_X is the distribution of X.*

PROOF. If we let $Y = f(X)$, then Y induces a measure (with CDF F_Y) on $(\mathbb{R}, \mathcal{B}^1)$ according to Theorem A.81. The definition of $E(Y)$ is $\int y dF_Y(y)$, and Theorem A.81 says that $\int y dF_Y(y) = \int f(x) dF_X(x)$. □

Definition B.13. If X is a random variable with finite mean c, then the *variance of X* is the mean of $(X - c)^2$ and is denoted $\text{Var}(X)$. If X is a random vector with finite mean vector c, then the *covariance matrix* of X is the mean of $(X - c)(X - c)^\top$ and is also denoted $\text{Var}(X)$. The *covariance* of two random variables X and Y with finite means c_X and c_Y is $E([X - c_X][Y - c_Y])$ and is denoted $\text{Cov}(X, Y)$.

It is possible for a random variable to have finite mean and infinite variance.

Proposition B.14. *If X has finite mean μ, then $\text{Var}(X) = E(X^2) - \mu^2$.*

B.2.2 Some Useful Inequalities

Although there are theoretical formulas for calculating means of functions of random variables, often they are not analytically tractable. We may, on the other hand, only need to know that a mean is less than some value. For this reason, we present some well-known inequalities concerning means of random variables.

---

[3]This theorem is used in making sense of the notation E_θ when introducing parametric models.

614   Appendix B. Probability Theory

Theorem B.15 (Markov inequality). [4] *Suppose that X is a nonnegative random variable with finite mean μ. Then, for all $c > 0$, $\Pr(X \geq c) \leq \mu/c$.*

PROOF. Let F be the CDF of X. Then, we can write

$$\mu = \int x dF(x) \geq \int_{[c,\infty)} x dF(x) \geq c \int_{[c,\infty)} dF(x) = c \Pr(X \geq c).$$

Divide the extreme parts by c to get the result. □

The following well-known inequality follows trivially from the Markov inequality B.15.

Corollary B.16 (Tchebychev's inequality).[5] *Suppose that X is a random variable with finite variance σ^2 and finite mean μ. Then, for all $c > 0$,*

$$\Pr(|X - \mu| \geq c) \leq \frac{\sigma^2}{c^2}.$$

Another well-known inequality involves *convex functions*.[6] The proof of this theorem resembles the proofs in Ferguson (1967) and Berger (1985).

Theorem B.17 (Jensen's inequality).[7] *Let g be a convex function defined on a convex subset $\mathcal{X}$ of $\mathbb{R}^k$ and suppose that $\Pr(X \in \mathcal{X}) = 1$. If $\mathrm{E}(X)$ is finite, then $\mathrm{E}(X) \in \mathcal{X}$ and $g(\mathrm{E}(X)) \leq \mathrm{E}(g(X))$.*

PROOF. First, we prove that $\mathrm{E}(X) \in \mathcal{X}$ by induction on the dimension of $\mathcal{X}$. Without loss of generality, we can assume that $\mathrm{E}(X) = 0$, since we can subtract $\mathrm{E}(X)$ from X and from every element of $\mathcal{X}$, and $\mathrm{E}(X) \in \mathcal{X}$ if and only if $0 \in \mathcal{X} - \mathrm{E}(X)$. If $k = 0$, then $\mathcal{X} = \{0\}$ and $\mathrm{E}(X) = 0$. Suppose that $0 \in \mathcal{X}$ for all $\mathcal{X}$ with dimension strictly less than $m \leq k$. Now suppose that X and $\mathcal{X}$ have dimension m and $0 \notin \mathcal{X}$. Since $\mathcal{X}$ and $\{0\}$ are disjoint convex sets, the separating hyperplane theorem C.5 says that there is a nonzero vector v and a constant c such that, for every $x \in \mathcal{X}$, $v^\top x \leq c$ and $0 \geq c$.[8] If we let $Y = v^\top X$, then we have $\Pr(Y \leq c) = 1$ and $\mathrm{E}(Y) = 0 \geq c$. It follows that $\Pr(Y = c) = 1$ and $c = 0$. Hence, X lies in the $(m-1)$-dimensional convex set $\mathcal{Z} = \mathcal{X} \cap \{x : v^\top x = 0\}$. It follows that $0 \in \mathcal{Z} \subseteq \mathcal{X}$.

Next, we prove the inequality by induction on k. For $k = 0$, $\mathrm{E}(g(X)) = g(\mathrm{E}(X))$, since X is degenerate. Suppose that the inequality holds for all dimensions up to $m - 1 < k$. Let X have dimension m. Define the subset of $\mathbb{R}^{m+1}$,

$$\mathcal{X}' = \{(x, z) : x \in \mathcal{X}, z \in \mathbb{R}, \text{ and } g(x) \leq z\}.$$

Let (x_1, z_1) and (x_2, z_2) be in $\mathcal{X}'$ and define

$$(y, w) = (\alpha x_1 + (1 - \alpha) x_2, \alpha z_1 + (1 - \alpha) z_2).$$

---

[4] This theorem is used in the proofs of Corollary B.16 and Lemma 1.61.
[5] This corollary is used in the proof of Theorem 1.59.
[6] Let $\mathcal{X}$ be a linear space. A function $f : \mathcal{X} \to \mathbb{R}$ is *convex* if $f(\lambda x + (1 - \lambda) y) \leq \lambda f(x) + (1 - \lambda) f(y)$ for all $x, y \in \mathcal{X}$ and all $\lambda \in [0, 1]$.
[7] This theorem is used in the proofs of Lemma B.114 and Theorems B.118 and 3.20.
[8] The symbol $v^\top$ stands for the transpose of the vector v.

Since $\alpha g(x_1) + (1-\alpha)g(x_2) \geq g(y)$ and $w \geq \alpha g(x_1) + (1-\alpha)g(x_2)$, it follows that $(y, w) \in \mathcal{X}'$, so $\mathcal{X}'$ is convex. It is also clear that $(\mathrm{E}(X), g(\mathrm{E}(X)))$ is a boundary point of $\mathcal{X}'$. The supporting hyperplane theorem C.4 says that there is a vector $v = (v_x, v_z)$ such that, for all $(x, z) \in \mathcal{X}'$, $v_x^\top x + v_z z \geq v_x^\top \mathrm{E}(X) + v_z g(\mathrm{E}(X))$. Since $(x, z_1) \in \mathcal{X}'$ implies $(x, z_2) \in \mathcal{X}'$ for all $z_2 > z_1$, it cannot be that $v_z < 0$, since then $\lim_{z \to \infty} v_x^\top x + v_z z = -\infty$, a contradiction. Since $(x, g(x)) \in \mathcal{X}'$ for all $x \in \mathcal{X}$, it follows that $v_x^\top X + v_z g(X) \geq v_x^\top \mathrm{E}(X) + v_z g(\mathrm{E}(X))$, from which we conclude

$$v_z g(\mathrm{E}(X)) \leq v_x^\top [X - \mathrm{E}(X)] + v_z g(X). \tag{B.18}$$

Taking expectations of both sides of this gives $v_z g(\mathrm{E}(X)) \leq v_z g(X)$. If $v_z > 0$, the proof is complete. If $v_z = 0$, then (B.18) becomes $0 \leq v^\top[X - \mathrm{E}(X)]$ which implies $v^\top[X - \mathrm{E}(X)] = 0$ with probability 1. Hence X lies in an $(m-1)$-dimensional space, and the induction hypothesis finishes the proof. □

The famous Cauchy–Schwarz inequality for vectors[9] has a probabilistic version.

Theorem B.19 (Cauchy–Schwarz inequality).[10] *Let X_1 and X_2 be two random vectors of the same dimension such that $\mathrm{E}(\|X_i\|^2)$ is finite for $i = 1, 2$. Then*

$$\mathrm{E}(|X_1^\top X_2|) \leq \sqrt{\mathrm{E}\|X_1\|^2 \mathrm{E}\|X_2\|^2}. \tag{B.20}$$

PROOF. Let $Z = 1$ if $X_1^\top X_2 \geq 0$ and $Z = -1$ if $X_1^\top X_2 < 0$. Let $Y = \|X_1 + cZX_2\|^2$, where $c = -\sqrt{\mathrm{E}\|X_1\|^2/\mathrm{E}\|X_2\|^2}$. Then $Y \geq 0$ and $Z^2 = 1$. So

$$\begin{aligned}
0 &\leq \mathrm{E}(Y) = \mathrm{E}\|X_1\|^2 + c^2 \mathrm{E}\|X_2\|^2 + 2c\mathrm{E}(|X_1^\top X_2|) \\
&= 2\mathrm{E}\|X_1\|^2 - 2\frac{\mathrm{E}(|X_1^\top X_2|)\sqrt{\mathrm{E}\|X_1\|^2}}{\sqrt{\mathrm{E}\|X_2\|^2}}.
\end{aligned}$$

The desired result follows immediately from this inequality. □

B.3 Conditioning

B.3.1 Conditional Expectations

Section B.1.2 contains a heuristic derivation of the important concepts in conditioning using the special case of discrete random quantities. We now turn to a more general presentation.

Theorem B.21.[11] *Let $(S, \mathcal{A}, \mu)$ be a probability space, and suppose that $X : S \to \mathbb{R}$ is a measurable function with $\mathrm{E}(|X|) < \infty$. Let $\mathcal{C}$ be a sub-σ-field of $\mathcal{A}$. Then there exists a $\mathcal{C}$ measurable function $g : S \to \mathbb{R}$ which satisfies*

$$\mathrm{E}(XI_B) = \int_B g(s) d\mu(s), \quad \text{for all } B \in \mathcal{C}.$$

---

[9] That is, if x_1 and x_2 are vectors, then $|x_1^\top x_2| \leq \|x_1\| \|x_2\|$.
[10] This theorem is used in the proofs of Theorems 3.44, 5.13, and 5.18.
[11] This theorem is used to help define the general concept of conditional expectation.

PROOF. Use Theorem A.54 to construct two measures μ_+ and μ_- on $(S, \mathcal{C})$:

$$\mu_+(B) = \int_B X^+(s)d\mu(s), \quad \mu_-(B) = \int_B X^-(s)d\mu(s).$$

It is clear that $\mu_+ \ll \mu$ and $\mu_- \ll \mu$. The Radon–Nikodym theorem A.74 tells us that there are $\mathcal{C}$ measurable functions g_+ and g_- such that

$$\mu_+(B) = \int_B g_+(s)d\mu(s), \quad \mu_-(B) = \int_B g_-(s)d\mu(s).$$

Since $\mathrm{E}(XI_B) = \mu_+(B) - \mu_-(B)$, the result follows with $g = g_+ - g_-$. □

We will use the symbol $\mathrm{E}(X|\mathcal{C})$ to stand for the function g. If $\mathcal{C}$ is the σ-field generated by another random quantity Y, then the symbol $\mathrm{E}(X|Y)$ is usually used instead of $\mathrm{E}(X|\mathcal{C})$. For the case in which $\mathcal{C}$ is the σ-field generated by Y, the next corollary follows from Theorem B.21 with the help of Theorem A.42.

Corollary B.22. *Let $(S, \mathcal{A}, \mu)$ be a probability space, and let $(\mathcal{Y}, \mathcal{C})$ be a measurable space such that $\mathcal{C}$ contains all singletons. Suppose that $X : S \to \mathbb{R}$ and $Y : S \to \mathcal{Y}$ are measurable functions and $\mathrm{E}(|X|) < \infty$. Let μ_Y be the measure induced on $(\mathcal{Y}, \mathcal{C})$ by Y from μ (see Theorem A.81). Let $\mathcal{A}_Y$ be the sub-σ-field of $\mathcal{A}$ generated by Y. Then there exists a function $h : \mathcal{Y} \to \mathbb{R}$ that satisfies the following: If $B \in \mathcal{A}_Y$ equals $Y^{-1}(C)$ for $C \in \mathcal{C}$, then $\mathrm{E}(XI_B) = \int_C h(t)d\mu_Y(t)$.*

We will use the symbol $\mathrm{E}(X|Y = t)$ to stand for the $h(t)$ in Corollary B.22. At this point the reader might wish to review Example B.4 on page 609.

To summarize the above results, we state the following.

Definition B.23. *Let $(S, \mathcal{A}, \mu)$ be a probability space, and suppose that $X : S \to \mathbb{R}$ is measurable and $\mathrm{E}(|X|) < \infty$. Let $\mathcal{C}$ be a sub-σ-field of $\mathcal{A}$. We define the* conditional mean (conditional expectation) *of X given $\mathcal{C}$ denoted $\mathrm{E}(X|\mathcal{C})$ to be any $\mathcal{C}$ measurable function $g : S \to \mathbb{R}$ that satisfies*

$$\mathrm{E}(XI_B) = \int_B g(s)d\mu(s), \quad \text{for all } B \in \mathcal{C}.$$

Each such function is called a version *of the conditional mean. If $Y : S \to \mathcal{Y}$ and $\mathcal{C}$ is the sub-σ-field generated by Y, then $\mathrm{E}(X|\mathcal{C})$ is also called the* conditional mean of X given Y, *denoted $\mathrm{E}(X|Y)$. If, in addition, the σ-field of subsets of $\mathcal{Y}$ contains singletons, let $h : \mathcal{Y} \to \mathbb{R}$ be the function such that $g = h(Y)$. Then $h(t)$ is denoted by $\mathrm{E}(X|Y = t)$.*

When we say that a random variable equals $\mathrm{E}(X|Y)$, we will mean that it is a version of $\mathrm{E}(X|Y)$. The following propositions are immediate consequences of the above definitions.

Proposition B.24. *Let $(S, \mathcal{A}, \mu)$ be a probability space, and let $(\mathcal{Y}, \mathcal{C})$ be a measurable space such that $\mathcal{C}$ contains singletons. Let $X : S \to \mathbb{R}$ and $Y : S \to \mathcal{Y}$ be measurable. Let μ_Y be the measure on $\mathcal{Y}$ induced from μ by Y. A function $g : \mathcal{Y} \to \mathbb{R}$ is a version of $\mathrm{E}(X|Y = t)$ if and only if for all $B \in \mathcal{C}$, $\int_B g(t)d\mu_Y(t) = \mathrm{E}(XI_B(Y))$.*

Proposition B.25.

- If Z and W are both versions of $\mathrm{E}(X|\mathcal{C})$, then $Z = W$, a.s.
- If X is $\mathcal{C}$ measurable, then $\mathrm{E}(X|\mathcal{C}) = X$, a.s.

Proposition B.26. If $\mathcal{C} = \{S, \emptyset\}$, the trivial σ-field, then $\mathrm{E}(X|\mathcal{C}) = \mathrm{E}(X)$.

Proposition B.27.[12] Let $(S, \mathcal{A}, \mu)$ be a probability space, and let $(\mathcal{Y}, \mathcal{C})$ be a measurable space. Let $X : S \to \mathbb{R}$ and $Y : S \to \mathcal{Y}$ be measurable, and let $g : \mathcal{Y} \to \mathbb{R}$ be such that $g(Y)X$ is integrable. Let μ_Y be the measure on $\mathcal{Y}$ induced from μ by Y. Then $\mathrm{E}(g(Y)X) = \int g(t)\mathrm{E}(X|Y = t)d\mu_Y(t)$.

Proposition B.28.[13] Let $(S, \mathcal{A}, \mu)$ be a probability space and let $X : S \to \mathbb{R}$, $Y : S \to (\mathcal{Y}, \mathcal{B}_1)$, and $Z : S \to (\mathcal{Z}, \mathcal{B}_2)$ be measurable functions. Let μ_Y and μ_Z be the measures induced on $\mathcal{Y}$ and $\mathcal{Z}$ by Y and Z, respectively, from μ. Suppose that $\mathrm{E}(|X|) < \infty$ and that Z is a one-to-one function of Y, that is, there exists a bimeasurable $h : \mathcal{Y} \to \mathcal{Z}$ such that $Z = h(Y)$. Then $\mathrm{E}(X|Y = y) = \mathrm{E}(X|Z = h(y))$, a.s. $[\mu_Y]$.

Conditional probability is the special case of conditional expectation in which $X = I_A$.

Definition B.29. Let $(S, \mathcal{A}, \mu)$ be a probability space. For each $A \in \mathcal{A}$, the *conditional probability of A given $\mathcal{C}$ (or given Y if $\mathcal{C}$ is the σ-field generated by Y) is* $\Pr(A|\mathcal{C}) = \mathrm{E}(I_A|\mathcal{C})$. If $\Pr(\cdot|\mathcal{C})(s)$ is a probability on $(S, \mathcal{A})$ for all $s \in S$, then the conditional probabilities given $\mathcal{C}$ are called *regular conditional probabilities*.

It turns out that under very general conditions (see Theorem B.32), we can choose the functions $\Pr(A|\mathcal{C})$ in such a way that they are regular conditional probabilities. In the future, we will assume that this is done in all such cases. If $\mathcal{C}$ is the σ-field generated by Y, then $\Pr(A|Y = y)$ will be used to stand for $\mathrm{E}(I_A|Y = y)$ as in the discussion following Corollary B.22.

If $X : S \to \mathcal{X}$ is a random quantity, its conditional distribution is the collection of conditional probabilities on $\mathcal{X}$ induced from the restriction of conditional probabilities on S to the σ-field generated by X.

Definition B.30. Let $(S, \mathcal{A}, \mu)$ be a probability space and let $(\mathcal{X}, \mathcal{B})$ be a measurable space. Suppose that $X : S \to \mathcal{X}$ is a measurable function. Let P be the probability on $(\mathcal{X}, \mathcal{B})$ induced by X from μ. Let $\mathcal{C}$ be a sub-σ-field of $\mathcal{A}$. For each $B \in \mathcal{B}$, let $P(B|\mathcal{C}) = \Pr(A|\mathcal{C})$, where $A = X^{-1}(B)$. We say that any set of functions from S to $[0, 1]$ of the form

$$\{P(B|\mathcal{C})(\cdot), \text{ for all } B \in \mathcal{B}\}$$

is a *version of the conditional distribution of X given $\mathcal{C}$*. If $\mathcal{C}$ is the σ-field generated by another random quantity $Y : S \to \mathcal{Y}$, a version of the *conditional*

---

[12] This proposition is used in the proof of Theorem B.64.

[13] This proposition is used to facilitate the transition from spaces of probability measures to subsets of Euclidean space when parametric models are introduced. It is also used in the proof of Theorem 2.114.

distribution of X given Y is specified by any collection of probability functions of the form

$$\{\Pr(\cdot|Y = t), \quad \text{for all } t \in \mathcal{Y}\}.$$

If the $P(\cdot|\mathcal{C})$ are regular conditional probabilities, then we say that the version of the conditional distribution of X given $\mathcal{C}$ is a *regular conditional distribution*.

When we refer to a conditional distribution without the word "version," we will mean a version of the conditional distribution. Occasionally, we will need to choose a version that satisfies some other condition. In those cases, we will try to be explicit about versions.

If $\mathcal{X}$ is sufficiently like the real numbers, there will be versions of conditional distributions that are regular. We make that precise with the following definition.

Definition B.31. Let $(\mathcal{X}, \mathcal{B})$ be a measurable space. If there exists a bimeasurable function $\phi : \mathcal{X} \to R$, where R is a Borel subset of $\mathbb{R}$, then $(\mathcal{X}, \mathcal{B})$ is called a *Borel space*.

In particular, we can show that all Euclidean spaces with the Borel σ-fields are Borel spaces. (See Lemma B.36.) First, we prove that regular conditional distributions exist on Borel spaces. The proof is borrowed from Breiman (1968, Section 4.3).

Theorem B.32. *Let $(S, \mathcal{A}, \mu)$ be a probability space and let $\mathcal{C}$ be a sub-σ-field of $\mathcal{A}$. Let $(\mathcal{X}, \mathcal{B})$ be a Borel space. Let $X : S \to \mathcal{X}$ be a random quantity. Then there exists a regular conditional distribution of X given $\mathcal{C}$.*

PROOF. Let $\phi : \mathcal{X} \to R$ be the function guaranteed by Definition B.31. Define the random variable $Z = \phi(X) : S \to R \subseteq \mathbb{R}$. First we prove that the σ-field generated by X, $\mathcal{A}_X$, is contained in the σ-field generated by Z, $\mathcal{A}_Z$. Let $B \in \mathcal{A}_X$; then there is $C \in \mathcal{B}$ such that $B = X^{-1}(C)$. Since ϕ is one-to-one, $\phi^{-1}(\phi(C)) = C$. Since ϕ^{-1} is measurable, $\phi(C)$ is a Borel subset of R. Now, $Z^{-1}(\phi(C)) = X^{-1}(C) = B$, hence $B \in \mathcal{A}_Z$. It is also easy to see that $\mathcal{A}_Z$ is contained in $\mathcal{A}_X$, so they are equal. If Z has a regular conditional distribution, then so does X. The remainder of the proof is to show that Z has a regular conditional distribution.

For each rational number q, choose a version of $\Pr(Z \leq q|\mathcal{C})$ and let

$$M_{q,r} = \{s : \Pr(Z \leq q|\mathcal{C})(s) < \Pr(Z \leq r|\mathcal{C})(s)\}, \quad M = \bigcup_{q > r} M_{q,r}.$$

According to Problem 3 on page 662 and countable additivity, $\mu(M) = 0$. Next, define

$$N_q = \{s : \lim_{r \downarrow q,\ r \text{ rational}} \Pr(Z \leq r|\mathcal{C})(s) \neq \Pr(Z \leq q|\mathcal{C})\}, \quad N = \bigcup_{\text{All } q} N_q.$$

We can use Problem 3 on page 662 once again to prove that $\mu(N_q) = 0$ for all q, hence $\mu(N) = 0$. Similarly, we can show that $\mu(L) = 0$, where L is the set

$$\left\{ s : \lim_{\substack{r \to -\infty \\ r \text{ rational}}} \Pr(Z \leq r|\mathcal{C})(s) \neq 0 \right\} \cup \left\{ s : \lim_{\substack{r \to \infty \\ r \text{ rational}}} \Pr(Z \leq r|\mathcal{C})(s) \neq 1 \right\}.$$

If G is an arbitrary CDF, we can define

$$F(z|\mathcal{C})(s) = \begin{cases} G(z) & \text{if } s \in M \cup N \cup L, \\ \lim_{r \downarrow z,\, r \text{ rational}} \Pr(Z \le r|\mathcal{C}) & \text{otherwise.} \end{cases}$$

$F(\cdot|\mathcal{C})(s)$ is a CDF for every s (see Problem 2 on page 661), and it is easy to check that $F(z|\mathcal{C})$ is a version of $\Pr(Z \le z|\mathcal{C})$ for every z. If we extend $F(\cdot|\mathcal{C})(s)$ to a probability measure $\eta(\cdot; s)$ on the Borel σ-field for every s, we only need to check that, for every Borel set B, $\eta(B; \cdot)$ is a version of $\Pr(Z \in B|\mathcal{C})$. That is, for every $C \in \mathcal{C}$, we need

$$\int_C \eta(B; s) d\mu(s) = \Pr(\{Z \in B\} \cap C). \tag{B.33}$$

By construction, (B.33) is true if B is an interval of the form $(-\infty, z]$. Such intervals form a π-system Π such that $\mathcal{B}$ is the smallest σ-field containing Π. If we define

$$Q_1(B) = \frac{\int_C \eta(B; s) d\mu(s)}{\Pr(C)}, \quad Q_2(B) = \frac{\Pr(\{Z \in B\} \cap C)}{\Pr(C)},$$

we see that Q_1 and Q_2 agree on Π. Tonelli's theorem A.69 can be used to see that Q_1 is countably additive, while Q_2 is clearly a probability. It follows from Theorem A.26 that Q_1 and Q_2 agree on $\mathcal{B}$. □

Note that the only condition required for regular conditional distributions to exist is a condition on the space of the random quantity for which we desire a regular conditional distribution. The σ-field $\mathcal{C}$, or the random quantity on which we condition, can be quite general. In the future, if we assume that $(\mathcal{X}, \mathcal{B})$ is a Borel space, we can construct regular conditional distributions given anything we wish. Also, since the function in the definition of Borel space is one-to-one and the Borel σ-field of $\mathbb{R}$ contains singletons, it follows that the σ-field of a Borel space contains singletons (cf. Theorem A.42).

B.3.2 Borel Spaces*

In this section we prove that there are lots of Borel spaces. First, we prove that every space satisfying some general conditions is a Borel space, and then we will show that Euclidean spaces satisfy those conditions. Then, we show that finite and countable products of Borel spaces are Borel spaces. The most general type of Borel space in which we shall be interested is a complete separable metric space (sometimes called a Polish space).

Definition B.34. Let $\mathcal{X}$ be a topological space. A subset D of $\mathcal{X}$ is *dense* if, for every $x \in \mathcal{X}$ and every open set U containing x, there is an element of D in U. If there exists a countable dense subset of $\mathcal{X}$, then $\mathcal{X}$ is *separable*. Suppose that $\mathcal{X}$ is a metric space with metric d. A sequence $\{x_n\}_{n=1}^\infty$ is *Cauchy* if, for every ϵ, there exists N such that $m, n \ge N$ implies $d(x_n, x_m) < \epsilon$. A metric space $\mathcal{X}$ is *complete* if every Cauchy sequence converges. A complete and separable metric space is called a *Polish space*.

---

*This section may be skipped without interrupting the flow of ideas.

We would like to prove that all Polish spaces are Borel spaces. First, we prove that $\mathbb{R}^\infty$ is a Borel space (Lemma B.36). Then we prove that there exist bimeasurable maps between Polish spaces and measurable subsets of $\mathbb{R}^\infty$ (Lemma B.40.) The following simple proposition pieces these results together.

Proposition B.35. *If $\mathcal{X}$ is a Borel space and there exists a bimeasurable function $f : \mathcal{Y} \to \mathcal{X}$, then $\mathcal{Y}$ is a Borel space.*

Lemma B.36. *The infinite product space $\mathbb{R}^\infty$ is a Borel space.*

PROOF. The idea of the proof[14] is the following. We start by transforming each coordinate to the interval $(0, 1)$ using a continuous function with continuous inverse. For each number in $(0, 1)$ we find a base 2 expansion, which is a sequence of 0s and 1s. We then take these sequences (one for each coordinate) and merge them into a single sequence, which we then interpret as the base 2 expansion of a number in $(0, 1)$. If this sequence of transformations is bimeasurable, we have our function ϕ.

Let $\psi : \mathbb{R}^\infty \to (0, 1)^\infty$ be defined by

$$\psi(x_1, x_2, \ldots) = \left(\frac{1}{2} + \frac{\tan^{-1}(x_1)}{\pi}, \frac{1}{2} + \frac{\tan^{-1}(x_2)}{\pi}, \ldots \right),$$

which is bimeasurable. For each $x \in [0, 1)$, set $y_0(x) = x$ and for $j = 1, 2, \ldots$, define

$$z_j(x) = \begin{cases} 1 & \text{if } 2y_{j-1}(x) \geq 1, \\ 0 & \text{if not,} \end{cases}$$

$$y_j(x) = 2y_{j-1}(x) - z_j(x).$$

For each j, z_j is a measurable function. It is easy to see that $z_j(x)$ is the jth digit in a base 2 expansion of x with infinitely many 0s. Note also that $y_j(x) \in [0, 1)$ for all j and x.

Create the following triangular array of integers:

```
 1
 2  3
 4  5  6
 7  8  9 10
11 12 13 14 15
 .  .  .  .  .
 .  .  .  .  .
```

Let the jth integer from the top of the ith column be $\ell(i, j)$. Then

$$\ell(i, j) = \frac{i(i+1)}{2} + i(j-1) + \frac{(j-1)(j-2)}{2}.$$

---

[14] This proof is adapted from Breiman (1968, Theorem A.47).

Clearly, each integer t appears once and only once as $\ell(i,j)$ for some i and j.[15] Define
$$h(x_1, x_2, \ldots) = \sum_{i=1}^{\infty} \sum_{j=1}^{\infty} \frac{z_j(x_i)}{2^{\ell(i,j)}}. \tag{B.37}$$
Then h is clearly a measurable function from $(0,1)^{\infty}$ to a subset R of $(0,1)$. There is a countable subset of $(0,1)$ which is not in the image of h. These are the numbers with only finitely many 0s in one or more of the subsequences $\{\ell(i,j)\}_{j=1}^{\infty}$ of their base 2 expansion for $i = 1, 2, \ldots$. For example, the number $c = \sum_{i=0}^{\infty} 2^{-i(i+1)/2-1}$ is not in R.[16] Since the complement of a countable set is measurable, the set R is measurable.

We define $\phi = h(\psi)$. If we can show that h has a measurable inverse, the proof is complete. For each $x \in R$, define
$$\phi_i(x) = \sum_{j=1}^{\infty} \frac{z_{\ell(i,j)}(x)}{2^j}. \tag{B.38}$$
Clearly, each ϕ_i is measurable. Note that, for each i and j,
$$z_j(\phi_i(x)) = z_{\ell(i,j)}(x). \tag{B.39}$$
Combining (B.37), (B.38), (B.39), and the fact that every integer appears once and only once as $\ell(i,j)$ for some i and j, we see that $h(\phi_1(x), \phi_2(x), \ldots) = x$, so that $(\phi_1, \phi_2, \ldots)$ is the inverse of h and it is measurable. □

Lemma B.40. *If $(\mathcal{X}, \mathcal{B})$ is a Polish space with the Borel σ-field and metric d, then it is a Borel space.*[17]

PROOF. All we need to prove is that there exists a bimeasurable $f : \mathcal{X} \to G$, where G is a measurable subset of $\mathbb{R}^{\infty}$. We then use Lemma B.36 and Proposition B.35.

Let $\{x_n\}_{n=1}^{\infty}$ be a countable dense subset of $\mathcal{X}$, and let d be the metric on $\mathcal{X}$. Define the function $f : \mathcal{X} \to \mathbb{R}^{\infty}$ by
$$f(x) = (d(x, x_1), d(x, x_2), \ldots).$$
We will first show that f is continuous, which will make it measurable. Suppose that $\{y_n\}_{n=1}^{\infty}$ is a sequence in $\mathcal{X}$ that converges to $y \in \mathcal{X}$. The kth coordinate of $f(y_n)$ is $d(y_n, x_k)$, which converges to $d(y, x_k)$ because the metric is continuous. Hence, each coordinate of f is continuous, and f is continuous. Next, we prove that f is one-to-one. Suppose that $f(x) = f(y)$. Then $d(x, x_n) = d(y, x_n)$ for

---

[15] It is easy to check the following. For each integer t, let $k = \inf\{n : t \leq n(n+1)/2\}$. Then $r(t) = 1 + k(k+1)/2 - t$ and $s(t) = k + 1 - r(t)$ have the property that $\ell(r(t), s(t)) = t$, $r(\ell(i,j)) = i$, and $s(\ell(i,j)) = j$.

[16] This number corresponds to having 1s in the first column of the triangular array but nowhere else. Clearly, $0 < c < 1$, but it is impossible to have 1s in the entire first column, since this would require $x_1 = 1$. Even if $x_1 = 1$ had been allowed, its base 2 expansion would have ended in infinitely many 0s rather than infinitely many 1s.

[17] This proof is adapted from p. 219 of Billingsley (1968) and Theorem 15.8 of Royden (1968).

all n. Since $\{x_n\}_{n=1}^{\infty}$ is dense, there exists a subsequence $\{x_{n_j}\}_{j=1}^{\infty}$ such that $\lim_{j\to\infty} x_{n_j} = x$. It follows that $0 = \lim_{j\to\infty} d(x, x_{n_j}) = \lim_{j\to\infty} d(y, x_{n_j})$; hence $\lim_{j\to\infty} x_{n_j} = y$, and $y = x$.

Next, we prove that $f^{-1} : f(\mathcal{X}) \to \mathcal{X}$ is continuous. Suppose that a sequence of points $\{f(y_n)\}_{n=1}^{\infty}$ converges to $f(y)$. Let $\lim_{j\to\infty} x_{n_j} = y$. Then $\lim_{j\to\infty} d(y, x_{n_j}) = 0$. But $d(y, x_{n_j})$ is the n_j coordinate of $f(y)$, which in turn is the limit (as $n \to \infty$) of the n_j coordinate of $f(y_n)$. For each j, $d(y_n, y) \leq d(y_n, x_{n_j}) + d(y, x_{n_j})$. Let $\epsilon > 0$ and let j be large enough so that $d(y, x_{n_j}) < \epsilon/2$. Now, let N be large enough so that $n \geq N$ implies $d(y_n, x_{n_j}) < d(y, x_{n_j}) + \epsilon/2$. It follows that, if $n \geq N$, $d(y_n, y) < \epsilon$. Hence $\lim_{n\to\infty} y_n = y$ and f^{-1} is continuous, hence measurable.

Finally, we will prove that the image G of f is a measurable subset of $\mathbb{R}^{\infty}$. We will do this by proving that G is the intersection of countably many open subsets of $\overline{G}$.[18] Let G_n be the following set:

$$\{x \in \mathbb{R}^{\infty} : \exists\, O_x \text{ a neighborhood of } x \text{ with } d(a, b) \leq 1/n \text{ for all } a, b \in f^{-1}(O_x)\}.$$

Since $O_x \subseteq G_n$ for each $x \in G_n$, G_n is open. Also, since f and f^{-1} are continuous, it is easy to see that $G \subseteq G_n$ for all n. Let $G' = \overline{G} \cap \bigcap_{n=1}^{\infty} G_n$. For each $x \in G'$, let $O_{x,n} \subseteq G_n$ be such that $O_{x,1} \supseteq O_{x,2} \supseteq \cdots$ and that $d(a, b) \leq 1/n$ for all $a, b \in f^{-1}(O_{x,n})$. Note that $f^{-1}(O_{x,n}) \supseteq f^{-1}(O_{x,n+1})$ for all n. If $y_n \in f^{-1}(O_{x,n})$ for every n, then $\{y_n\}_{n=1}^{\infty}$ is a Cauchy sequence, since $n, m \geq N$ implies $d(y_n, y_m) \leq 1/N$. Hence, there is a limit y to the sequence. It is easy to see that if there were two such sequences with limits y and y', then $d(y, y') < \epsilon$ for all $\epsilon > 0$, hence $y = y'$. So we can define a function $h : G' \to \mathcal{X}$ by $h(x) = y$. If $x \in G$, then clearly $h(x) = f^{-1}(x)$. If $x' \in O_{x,n}$, then $d(h(x), h(x')) \leq 1/n$, so h is continuous. We now prove that $G' \subseteq G$, which implies that $G = G'$ and the proof will be complete. Let $x \in G'$, and let $x_n \in G$ be such that $x_n \to x$. (This is possible since $G' \subseteq \overline{G}$.) Since h is continuous, $f^{-1}(x_n) \to h(x)$. If $y_n = f^{-1}(x_n)$ and $y = h(x)$, then $y_n \to y$ and $f(y_n) \to f(y) \in G$, since f is continuous. But $f(y_n) = x_n$, so $f(y) = x$, and the proof is complete. □

Next, we show that products of Borel spaces are Borel spaces.

Lemma B.41. *Let $(\mathcal{X}_n, \mathcal{B}_n)$ be a Borel space for each n. The product spaces $\prod_{i=1}^{n} \mathcal{X}_i$ for all finite n and $\prod_{n=1}^{\infty} \mathcal{X}_n$ with product σ-fields are Borel spaces.*

PROOF. We will prove the result for the infinite product. The proofs for finite products are similar. If $\mathcal{X}_n = \mathbb{R}$ for all n, the result is true by Lemma B.36. For general $\mathcal{X}_n$, let $\phi_n : \mathcal{X}_n \to R_n$ and $\phi_* : \mathbb{R}^{\infty} \to R_*$ be bimeasurable, where R_n and R_* are measurable subsets of $\mathbb{R}$. Then, it is easy to see that

$$\phi : \prod_{n=1}^{\infty} \mathcal{X}_n \to \phi_*\left(\prod_{n=1}^{\infty} R_n\right)$$

is bimeasurable, where $\phi(x_1, x_2, \ldots) = \phi_*(\phi_1(x_1), \phi_2(x_2), \ldots)$. □

Next, we show that the set of bounded continuous functions from $[0, 1]$ to the real numbers is also a Polish space.

---

[18] We use symbol $\overline{G}$ to stand for the *closure* of the set G. The closure of a subset G of a topological space is the smallest closed set containing G. A set is *closed* if and only if its complement is open.

Lemma B.42.[19] *Let $C[0,1]$ be the set of all bounded continuous functions from $[0,1]$ to $\mathbb{R}$. Let $\rho(f,g) = \sup_{x \in [0,1]} |f(x) - g(x)|$. Then, ρ is a metric on $C[0,1]$ and $C[0,1]$ is a Polish space.*

PROOF. That ρ is a metric is easy to see. To see that C is separable, let D_k be the set of functions that take on rational values at the points $0, 1/k, \ldots, (k-1)/k, 1$ and are linear between these values. Let $D = \cup_{k=1}^{\infty} D_k$. The set D is countable. Every continuous function on a compact set is uniformly continuous, so let $f \in C[0,1]$ and $\epsilon > 0$. Let δ be small enough so that $|x-y| < \delta$ implies $|f(x) - f(y)| < \epsilon/4$. Let k be larger than $4/\epsilon$. There exists $g \in D_k$ such that $|g(i/k) - f(i/k)| < \epsilon/4$ for each $i = 0, \ldots, k$. For $i/k < x < (i+1)/k$, $|f(x) - f(i/k)| < \epsilon/4$, and $|g(x) - g(i/k)| < \epsilon/2$, so $|f(x) - g(x)| < \epsilon$. To see that $C[0,1]$ is complete, let $\{f_n\}_{n=1}^{\infty}$ be a Cauchy sequence. Then, for all x, $\{f_n(x)\}_{n=1}^{\infty}$ is a Cauchy sequence of real numbers that converges to some number $f(x)$. We need to show that the convergence of f_n to f is uniform. To the contrary, assume that there exists ϵ such that, for each n there is x_n such that $|f_n(x_n) - f(x_n)| > \epsilon$. We know that there exists n such that $m > n$ implies $|f_n(x) - f_m(x)| < \epsilon/2$ for all x. In particular, $|f_n(x_n) - f_m(x_n)| < \epsilon/2$ for all $m > n$. Since $\lim_{m \to \infty} f_m(x_n) = f(x_n)$, it follows that there exists m such that $|f_m(x_n) - f(x_n)| < \epsilon/2$, a contradiction. □

Because Borel spaces have σ-fields that look just like the Borel σ-field of the real numbers, their σ-fields are generated by countably many sets. The countable field that generates the Borel σ-field of $\mathbb{R}$ is the collection of all sets that are unions of finitely many disjoint intervals (including degenerate ones and infinite ones) with rational endpoints.

Proposition B.43.[20] *Let $(\mathcal{X}, \mathcal{B})$ be a Borel space. Then there exists a countable field C such that $\mathcal{B}$ is the smallest σ-field containing C.*

Because a field is a π-system, Theorem A.26 and Proposition B.43 imply the following.

Corollary B.44. *Let $(\mathcal{X}, \mathcal{B})$ be a Borel space, and let C be a countable field that generates $\mathcal{B}$. If μ_1 and μ_2 are σ-finite measures on $\mathcal{B}$ that agree on C, then they agree on $\mathcal{B}$.*

B.3.3 Conditional Densities

Because conditional distributions are probability measures, many of the theorems from Appendix A which apply to such measures apply to conditional distributions. For example, the monotone convergence theorem A.52 and the dominated convergence theorem A.57 apply to conditional means because limits of measurable functions are still measurable. Also, most of the properties of probability measures from this appendix apply as well. In this section, we focus on the existence and calculation of densities for conditional distributions.

If the joint distribution of two random quantities has a density with respect to a product measure, then the conditional distributions have densities that can

---

[19] This lemma is used in the proof of Lemma 2.121.
[20] This proposition is used in the proofs of Lemmas 2.124 and 2.126 and Theorem 3.110.

be calculated in the usual way.

Proposition B.45. *Let $(S, \mathcal{A}, \mu)$ be a probability space and let $(\mathcal{X}, \mathcal{B}_1, \nu_\mathcal{X})$ and $(\mathcal{Y}, \mathcal{B}_2, \nu_\mathcal{Y})$ be σ-finite measure spaces. Let $X : S \to \mathcal{X}$ and $Y : S \to \mathcal{Y}$ be measurable functions. Let $\mu_{X,Y}$ be the probability induced on $(\mathcal{X} \times \mathcal{Y}, \mathcal{B}_1 \otimes \mathcal{B}_2)$ by (X, Y) from μ. Suppose that $\mu_{X,Y} \ll \nu_\mathcal{X} \times \nu_\mathcal{Y}$. Let the density be $f_{X,Y}(x, y)$. Let the probability induced on $(\mathcal{Y}, \mathcal{B}_2)$ by Y from μ be denoted μ_Y. Then μ_Y is absolutely continuous with respect to $\nu_\mathcal{Y}$ with density*

$$f_Y(y) = \int_\mathcal{X} f_{X,Y}(x, y) d\nu_\mathcal{X}(x),$$

and the conditional distribution of X given Y has densities

$$f_{X|Y}(x|y) = \frac{f_{X,Y}(x, y)}{f_Y(y)}$$

with respect to $\nu_\mathcal{X}$.

This proposition can be proven directly using Tonelli's theorem A.69 or as a special case of Theorem B.46 (see Problem 15 on page 663).

Theorem B.46. *Let $(\mathcal{X}, \mathcal{B}_1)$ be a Borel space, let $(\mathcal{Y}, \mathcal{B}_2)$ be a measurable space, and let $(\mathcal{X} \times \mathcal{Y}, \mathcal{B}_1 \otimes \mathcal{B}_2, \nu)$ be a σ-finite measure space. Then, there exists a measure $\nu_\mathcal{Y}$ on $(\mathcal{Y}, \mathcal{B}_2)$ and for each $y \in \mathcal{Y}$, there exists a measure $\nu_{\mathcal{X}|\mathcal{Y}}(\cdot|y)$ on $(\mathcal{X}, \mathcal{B}_1)$ such that for each integrable or nonnegative $h : \mathcal{X} \times \mathcal{Y} \to \mathbb{R}$, $\int h(x, y) d\nu_{\mathcal{X}|\mathcal{Y}}(x|y)$ is $\mathcal{B}_2$ measurable and*

$$\int h(x, y) d\nu(x, y) = \int \left[\int h(x, y) d\nu_{\mathcal{X}|\mathcal{Y}}(x|y) \right] d\nu_\mathcal{Y}(y). \qquad (B.47)$$

PROOF. Let f be the strictly positive integrable function guaranteed by Theorem A.85. Without loss of generality, assume that $\int f(x, y) d\nu(x, y) = 1$. The measure $\mu(A) = \int_A f(x, y) d\nu(x, y)$ is a probability, $\nu \ll \mu$, and $(d\nu/d\mu)(x, y) = 1/f(x, y)$. Let $\mu_{\mathcal{X}|\mathcal{Y}}$ be a regular conditional distribution on $(\mathcal{X}, \mathcal{B}_1)$ constructed from μ, and let $\nu_\mathcal{Y}$ be the marginal distribution on $(\mathcal{Y}, \mathcal{B}_2)$. Define

$$\nu_{\mathcal{X}|\mathcal{Y}}(A|y) = \int_A \frac{1}{f(x, y)} d\mu_{\mathcal{X}|\mathcal{Y}}(x|y).$$

Note that

$$\int I_{A \times B}(x, y) d\mu_{\mathcal{X}|\mathcal{Y}}(x|y) = I_B(y) \mu_{\mathcal{X}|\mathcal{Y}}(A|y), \qquad (B.48)$$

which is a measurable function of y because $\mu_{\mathcal{X}|\mathcal{Y}}$ is a regular conditional distribution. Just as in the proof of Lemma A.61, we can use the π–λ theorem A.17 to show that $\int g d\mu_{\mathcal{X}|\mathcal{Y}}$ is measurable if g is the indicator of an element of the product σ-field. It follows that $\int g d\mu_{\mathcal{X}|\mathcal{Y}}$ is measurable for every nonnegative simple function g. By the monotone convergence theorem A.52, letting $\{g_n\}_{n=1}^\infty$ be nonnegative simple functions increasing to g everywhere, it follows that $\int g(x, y) d\mu_{\mathcal{X}|\mathcal{Y}}(x|y)$ is measurable for all nonnegative measurable functions, and hence $\int h d\nu_{\mathcal{X}|\mathcal{Y}} = \int h/f d\mu_{\mathcal{X}|\mathcal{Y}}$ is measurable if h is nonnegative.

Next, define a probability η on $(\mathcal{X} \times \mathcal{Y}, \mathcal{B}_1 \otimes \mathcal{B}_2)$ by

$$\eta(C) = \int \left[\int I_C(x,y) d\mu_{\mathcal{X}|\mathcal{Y}}(x|y) \right] d\nu_{\mathcal{Y}}(y).$$

It follows from (B.48) that η and μ agree on the collection of all product sets (a π-system that generates $\mathcal{B}_1 \otimes \mathcal{B}_2$). Theorem A.26 implies that they agree on $\mathcal{B}_1 \otimes \mathcal{B}_2$. By linearity of integrals and the monotone convergence theorem A.52, if g is nonnegative, then

$$\int g(x,y) d\eta(x,y) = \int \left[\int g(x,y) d\mu_{\mathcal{X}|\mathcal{Y}}(x|y) \right] d\nu_{\mathcal{Y}}(y)$$
$$= \int \left[\int g(x,y) f(x,y) d\nu_{\mathcal{X}|\mathcal{Y}}(x|y) \right] d\nu_{\mathcal{Y}}(y). \quad \text{(B.49)}$$

For every nonnegative h,

$$\int h(x,y) d\nu(x,y) = \int \frac{h(x,y)}{f(x,y)} f(x,y) d\nu(x,y) = \int \frac{h(x,y)}{f(x,y)} d\mu(x,y) \quad \text{(B.50)}$$
$$= \int \frac{h(x,y)}{f(x,y)} d\eta(x,y) = \int \left[\int h(x,y) d\nu_{\mathcal{X}|\mathcal{Y}}(x|y) \right] d\nu_{\mathcal{Y}}(y),$$

where the second equality follows from the fact that $d\mu/d\nu = f$, the third follows from the fact that μ and η are the same measure, and the fourth follows from (B.49). If h is integrable with respect to ν, then (B.50) applies to h^+, h^-, and $|h|$, and all three results are finite. Also, $\int |h(x,y)| d\nu_{\mathcal{X}|\mathcal{Y}}(x|y)$ is measurable and $\nu_{\mathcal{Y}}(\{y : \int |h(x,y)| d\nu_{\mathcal{X}|\mathcal{Y}}(x|y) = \infty\}) = 0$. So $\int h^+(x,y) d\nu_{\mathcal{X}|\mathcal{Y}}(x|)$ and $\int h^-(x,y) d\nu_{\mathcal{X}|\mathcal{Y}}(x|y)$ are both finite almost surely, and their difference is $\int h(x,y) d\nu_{\mathcal{X}|\mathcal{Y}}(x|y)$, a measurable function. It now follows that (B.47) holds. □

The measures $\nu_{\mathcal{Y}}$ and $\nu_{\mathcal{X}|\mathcal{Y}}$ in Theorem B.46 are not unique. In the proof, we could easily have defined $\nu_{\mathcal{Y}}$ several ways, such as $\nu_{\mathcal{Y}}(A) = \int_A g(y) \mu_{\mathcal{Y}}(y)$ for any strictly positive function g with finite $\mu_{\mathcal{Y}}$ integral. A corresponding adjustment would have to be made to the definition of $\nu_{\mathcal{X}|\mathcal{Y}}$:

$$\nu_{\mathcal{X}|\mathcal{Y}}(A|y) = \int_A \frac{1}{f(x,y)g(y)} d\mu_{\mathcal{X}|\mathcal{Y}}(x|y).$$

In the special case in which ν is a product measure $\nu_1 \times \nu_2$, it is easy to show that ν_1 can play the role of $\nu_{\mathcal{X}|\mathcal{Y}}(\cdot|y)$ for all y and that ν_2 can play the role of $\nu_{\mathcal{Y}}$ in Theorem B.46. (See Problem 15 on page 663.)

There is a familiar application of Theorem B.46 to cases in which $\mathcal{X}$ and $\mathcal{Y}$ are Euclidean spaces but ν is concentrated on a lower-dimensional manifold defined by a function $y = g(x)$.

Proposition B.51. *Suppose that $\mathcal{X} = \mathbb{R}^n$ and $\mathcal{Y} = \mathbb{R}^k$, with $k < n$. Let $g : \mathcal{X} \to \mathcal{Y}$ be such that there exists $h : \mathcal{X} \to \mathbb{R}^{n-k}$ such that $v(x) = (g(x), h(x))$ is one-to-one, is differentiable, and has a differentiable inverse. For $y \in \mathbb{R}^k$ and $w \in \mathbb{R}^{n-k}$, define $J(y,w)$ to be the Jacobian, that is, the determinant of the matrix of partial derivatives of the coordinates of $v^{-1}(y,w)$ with respect to the coordinates of y and of w. Let λ_i be Lebesgue measure on $\mathbb{R}^i$, for each i. Define* a measure ν on $\mathcal{X} \times \mathcal{Y}$ by $\nu(C) = \lambda_n(\{x : (x, g(x)) \in C\})$. Then, $\nu_\mathcal{Y}$ equal to Lebesgue measure on $\mathbb{R}^k$ and $\nu_{\mathcal{X}|\mathcal{Y}}(A|y) = \int_{A_y^*} J(y,w) d\lambda_{n-k}(w)$ satisfy (B.47), where $A_y^* = \{w : v^{-1}(y,w) \in A\}$.

We are now in position to derive a formula for conditional densities in general.[21]

Theorem B.52. *Let $(S, \mathcal{A}, \mu)$ be a probability space, let $(\mathcal{X}, \mathcal{B}_1)$ be a Borel space, let $(\mathcal{Y}, \mathcal{B}_2)$ be a measurable space, and let $(\mathcal{X} \times \mathcal{Y}, \mathcal{B}_1 \otimes \mathcal{B}_2, \nu)$ be a σ-finite measure space. Let $\nu_\mathcal{Y}$ and $\nu_{\mathcal{X}|\mathcal{Y}}$ be as guaranteed by Theorem B.46. Let $X : S \to \mathcal{X}$ and $Y : S \to \mathcal{Y}$ be measurable functions. Let $\mu_{X,Y}$ be the probability induced on $(\mathcal{X} \times \mathcal{Y}, \mathcal{B}_1 \otimes \mathcal{B}_2)$ by (X, Y) from μ. Suppose that $\mu_{X,Y} \ll \nu$. Let the density be $f_{X,Y}(x, y)$. Let the probability induced on $(\mathcal{Y}, \mathcal{B}_2)$ by Y from μ be denoted μ_Y. Then, $\mu_Y \ll \nu_\mathcal{Y}$; for each $y \in \mathcal{Y}$,*

$$\frac{d\mu_Y}{d\nu_\mathcal{Y}}(y) = f_Y(y) = \int_\mathcal{X} f_{X,Y}(x,y) d\nu_{\mathcal{X}|\mathcal{Y}}(x|y); \tag{B.53}$$

and the conditional distribution of X given $Y = y$ has density

$$f_{X|Y}(x|y) = \frac{f_{X,Y}(x,y)}{f_Y(y)} \tag{B.54}$$

with respect to $\nu_{\mathcal{X}|\mathcal{Y}}(\cdot|y)$.

PROOF. It follows from Theorem B.46 that for all $B \in \mathcal{B}_2$,

$$\begin{aligned}
\mu_Y(B) &= \int I_B(y) f_{X,Y}(x,y) d\nu(x,y) \\
&= \int I_B(y) \left[\int f_{X,Y}(x,y) d\nu_{\mathcal{X}|\mathcal{Y}}(x|y) \right] d\nu_\mathcal{Y}(y).
\end{aligned}$$

The fact that $\mu_Y \ll \nu_\mathcal{Y}$ and (B.53) both follow from this equation. Let $\mu_{X|Y}(\cdot|y)$ denote a regular conditional distribution of X given $Y = y$. For each $A \in \mathcal{B}_1$ and $B \in \mathcal{B}_2$, apply Theorem B.46 with $h(x,y) = I_A(x) I_B(y) f_{X|Y}(x|y) f_Y(y)$ to conclude

$$\mu_{X,Y}(A \times B) = \int_B \left[\int_A f_{X|Y}(x|y) d\nu_{\mathcal{X}|\mathcal{Y}}(x|y) \right] d\mu_Y(y).$$

Since this is true for all $B \in \mathcal{B}_2$, we conclude that

$$\mu_{X|Y}(A|y) = \int_A f_{X|Y}(x|y) d\nu_{\mathcal{X}|\mathcal{Y}}(x|y).$$

Hence (B.54) gives the density of $\mu_{X|Y}(\cdot|y)$ with respect to $\nu_{\mathcal{X}|\mathcal{Y}}(\cdot|y)$. □

The point of Theorem B.52 is that we can calculate conditional densities for random quantities even if the measure that dominates the joint distribution is not a product measure. When the joint distribution is dominated by a product

---

[21] The condition that the joint distribution have a density with respect to a measure ν in Theorem B.52 is always met since ν can be taken equal to the joint distribution. The theorem applies even if ν is not the joint distribution, however.

measure, the conditional distributions are all dominated by the same measure. (See Problem 15 on page 663.) In general, however, the conditional distribution of X given $Y = y$ is dominated by a measure that depends on y. For example, if $Y = g(X)$, the joint distribution of (X, Y) is not dominated by a product measure even if the distribution of X is dominated. (See also Problem 7 on page 662.) Nevertheless, we have the following result.

Corollary B.55.[22] *Let $(S, \mathcal{A}, \mu)$ be a probability space, let $(\mathcal{Y}, \mathcal{B}_2)$ be a measurable space such that $\mathcal{B}_2$ contains all singletons, and let $(\mathcal{X}, \mathcal{B})$ be a Borel space with $\nu_\mathcal{X}$ a σ-finite measure on $(\mathcal{X}, \mathcal{B})$. Let $X : S \to \mathcal{X}$ and $g : \mathcal{X} \to \mathcal{Y}$ be measurable functions. Let $Y = g(X)$. Suppose that the distribution of X has density f_X with respect to $\nu_\mathcal{X}$. Define ν on $(\mathcal{X} \times \mathcal{Y}, \mathcal{B}_1 \otimes \mathcal{B}_2)$ by $\nu(C) = \nu_\mathcal{X}(\{x : (x, g(x)) \in C\})$. Let $\mu_{X,Y}$ be the probability induced on $(\mathcal{X} \times \mathcal{Y}, \mathcal{B}_1 \otimes \mathcal{B}_2)$ by (X, Y) from μ. Let the probability induced on $(\mathcal{Y}, \mathcal{B}_2)$ by Y from μ be denoted μ_Y. Then $\mu_{X,Y} \ll \nu$ with Radon–Nikodym derivative $f_{X,Y}(x, y) = f_X(x) I_{\{g(x)\}}(y)$. Also, the conditions of Theorem B.46 hold, and we can write*

$$\frac{d\mu_Y}{d\nu_\mathcal{Y}}(y) = f_Y(y) = \int_\mathcal{X} I_{\{g(x)\}}(y) f_X(x) d\nu_{X|Y}(x|y),$$

$$f_{X|Y}(x|y) = \begin{cases} \frac{f_X(x)}{f_Y(y)} & \text{if } y = g(x), \\ 0 & \text{otherwise.} \end{cases}$$

Also, the conditional distribution of Y given X is given by $\mu_{Y|X}(C|x) = I_C(g(x))$.

PROOF. Since $\nu_\mathcal{X}$ is σ-finite, ν is also. Since Y is a function of X, Theorem A.81 implies that for all integrable h, $\int h(x, y) d\nu(x, y) = \int h(x, g(x)) d\nu_\mathcal{X}(x)$. The facts that $f_{X,Y}$ has the specified form and that $\mu_{Y|X}$ is the conditional distribution of Y given X follow easily from this equation. □

The point of Corollary B.55 is that if $Y = g(X)$, then we can assume that the conditional distribution of X given $Y = y$ is concentrated on $g^{-1}(\{y\})$.

Example B.56.[23] Let f be a spherically symmetric density with respect to λ_n, Lebesgue measure on $\mathbb{R}^n$. That is, $f(x) = h(x^\top x)$ for some function $h : \mathbb{R} \to \mathbb{R}^{+0}$ (the interval $[0, \infty)$) and $\int h(x^\top x) d\lambda_n(x) = 1$. Let X have density f and let $V = X^\top X$. Let $R = V^{-1/2}$, and transform to spherical coordinates:

$$\begin{aligned} x_1 &= r\cos(\theta_1), \\ x_2 &= r\sin(\theta_1)\cos(\theta_2), \\ &\vdots \\ x_{n-1} &= r\sin(\theta_1)\cdots\cos(\theta_{n-1}), \\ x_n &= r\sin(\theta_1)\cdots\sin(\theta_{n-1}). \end{aligned}$$

The Jacobian is $r^{n-1}j(\theta)$, where j is some function of θ alone. The Jacobian for the transformation to v and θ is $v^{(n/2)-1}j(\theta)/2$. The integral of $j(\theta)$ over all θ

---

[22] This corollary is used in the proof of Theorem 2.86 and in Example 3.106.
[23] The calculation in this example is used again in Example 4.121.

values is $\pi^{n/2}/\Gamma(n/2)$. So, the marginal density of V is

$$f_V(v) = \pi^{\frac{n}{2}} v^{\frac{n}{2}-1} \frac{h(v)}{2\Gamma\left(\frac{n}{2}\right)}.$$

The conditional density of X given $V = v$ is then

$$f_{X|V}(x|v) = \frac{2\Gamma\left(\frac{n}{2}\right) v^{1-\frac{n}{2}}}{\pi^{\frac{n}{2}}} I_{\{v\}}(x^\top x)$$

with respect to the measure $\nu_{X|V}(C|v) = \int_{C^*} v^{(n/2)-1} j(\theta) d\lambda_{n-1}(\theta)/2$, where

$$C^* = \{\theta : v^{\frac{1}{2}}(\cos(\theta_1), \ldots, \sin(\theta_1) \cdots \sin(\theta_{n-1})) \in C\}.$$

It follows that the conditional distribution of X given $V = v$ is given by

$$\mu_{X|V}(C|v) = \frac{\Gamma\left(\frac{n}{2}\right)}{\pi^{\frac{n}{2}}} \int_{C^*} j(\theta) d\lambda_{n-1}(\theta).$$

It is easy to see that $\mu_{X|V}(\cdot|v)$ is the uniform distribution over the sphere of radius v in n dimensions.

Another example was given in Example B.5 on page 610.

B.3.4 Conditional Independence

The concept of *conditional independence* will turn out to be central to the development of statistical models.

Definition B.57. Let $\aleph$ be an index set, let Y and $\{X_i\}_{i \in \aleph}$ be random quantities, and let $\mathcal{A}_i$ be the σ-field generated by X_i. We say that $\{X_i\}_{i \in \aleph}$ *are conditionally independent given Y* if, for every n and every set of distinct indices $i_1, \ldots, i_n$ and every collection of sets $A_1 \in \mathcal{A}_{i_1}, \ldots, A_n \in \mathcal{A}_{i_n}$, we have

$$\Pr\left(\bigcap_{j=1}^n A_j \bigg| Y\right) = \prod_{j=1}^n \Pr(A_j|Y), \text{ a.s.} \tag{B.58}$$

If, in addition, Y is constant almost surely, we say $\{X_i\}_{i \in \aleph}$ are independent.

Under the same conditions as above, if all of the conditional distributions of the X_i given Y are the same, then we say $\{X_i\}_{i \in \aleph}$ *are conditionally IID given Y*. If, in addition, Y is constant almost surely, we say $\{X_i\}_{i \in \aleph}$ are IID.

Example B.59. Let F be a joint CDF of n random variables $X_1, \ldots, X_n$, and let μ be the corresponding measure on $\mathbb{R}^n$. Then μ is a product measure if and only if $X_1, \ldots, X_n$ are independent (see Proposition B.66).

Example B.60 (Continuation of Example B.56; see page 627).[24] Transform to (Y, V), where $Y = X/V^{1/2}$. Then, the conditional distribution of Y given V is given by

$$\mu_{Y|V}(D|v) = \Gamma\left(\frac{n}{2}\right) \pi^{-\frac{n}{2}} \int_{D'} j(\theta) d\lambda_{n-1}(\theta),$$

---

[24]This calculation is used again in Example 4.121.

where $D' = \{\theta : (\cos(\theta_1), \ldots, \sin(\theta_1) \cdots \sin(\theta_{n-1})) \in D\}$. We note that this formula does not depend on v; hence Y is independent of V. In addition, it is easy to see that $\mu_{Y|V}(y|v)$ is just the uniform distribution over the sphere of radius 1 in n dimensions.

The use of conditional independence in predictive inference is based on the following theorem.

Theorem B.61.[25] *Let $\aleph$ be an index set, let Y and $\{X_i\}_{i \in \aleph}$ be a collection of random quantities, and let $\mathcal{A}_i$ be the σ-field generated by X_i. Then $\{X_i\}_{i \in \aleph}$ are conditionally independent given Y if and only if for every n and m and every set of distinct indices $i_1, \ldots, i_n, j_1, \ldots, j_m$ and every collection of sets $A_1 \in \mathcal{A}_{i_1}, \ldots, A_n \in \mathcal{A}_{i_n}$, we have*

$$\Pr\left(\bigcap_{i=1}^n A_i \middle| Y, X_{j_1}, \ldots, X_{j_m}\right) = \Pr\left(\bigcap_{i=1}^n A_i \middle| Y\right), \quad a.s. \tag{B.62}$$

PROOF. For the "if" part, we will assume (B.62) and prove (B.58) by induction on n. For $n = 1$, there is nothing to prove. Assuming (B.58) is true for all $n \leq k$, we now prove it for $n = k + 1$. Let $A_j \in \mathcal{A}_{i_j}$ for $j = 1, \ldots, k + 1$. According to (B.62) and (B.58) for $n = k$, we have

$$\Pr\left(\bigcap_{i=1}^k A_i \middle| Y, X_{i_{k+1}}\right) = \Pr\left(\bigcap_{i=1}^n A_i \middle| Y\right) = \prod_{i=1}^k \Pr(A_i|Y).$$

It follows that for all $B \in \mathcal{A}_Y$, the σ-field generated by Y,

$$\Pr\left(B \bigcap_{i=1}^{k+1} A_i\right) = \Pr\left(B \cap A_{k+1} \bigcap_{i=1}^k A_i\right)$$

$$= \int_{B \cap A_{k+1}} \Pr\left(\bigcap_{i=1}^k A_i \middle| Y, X_{k+1}\right)(s) d\mu(s)$$

$$= \int_{B \cap A_{k+1}} \prod_{i=1}^k \Pr(A_i|Y)(s) d\mu(s) = \int_B I_{A_{k+1}}(s) \prod_{i=1}^k \Pr(A_i|Y)(s) d\mu(s)$$

$$= \int_B \Pr(A_{k+1}|Y)(s) \prod_{i=1}^k \Pr(A_i|Y)(s) d\mu(s) = \int_B \prod_{i=1}^{k+1} \Pr(A_i|Y)(s) d\mu(s).$$

The equality of the first and last terms above for all $B \in \mathcal{A}_Y$ means that $\prod_{i=1}^{k+1} \Pr(A_i|Y) = \Pr(\cap_{i=1}^{k+1} A_i|Y)$, a.s., which is what we need to complete the induction.

For the "only if" part, we will assume (B.58) and prove (B.62). For a function g to be the left-hand side of (B.62), it must be measurable with respect to the σ-field $\mathcal{A}_{Y,m}$ generated by $Y, X_{j_1}, \ldots, X_{j_m}$, and satisfy

$$\int_C g(s) d\mu(s) = \Pr\left(C \bigcap_{i=1}^n A_i\right), \tag{B.63}$$

---

[25]This theorem is used in the proofs of Theorems 2.14 and 2.20.

for all $C \in \mathcal{A}_{Y,m}$. Clearly, the right-hand side of (B.62) is measurable with respect to $\mathcal{A}_{Y,m}$. If $C = C_Y \cap C_X$, where $C_Y \in \mathcal{A}_Y$ and C_X is in the σ-field generated by $X_{j_1}, \ldots, X_{j_m}$, then

$$\Pr\left(X \bigcap_{i=1}^n A_i\right) = \int_{C_Y} \Pr\left(C_X \bigcap_{i=1}^n A_i \middle| Y\right)(s) d\mu(s)$$

$$= \int_{C_Y} I_{C_X}(s) \Pr\left(\bigcap_{i=1}^n A_i \middle| Y\right)(s) d\mu(s)$$

$$= \int_C \Pr\left(\bigcap_{i=1}^n A_i \middle| Y\right)(s) d\mu(s).$$

This means that (B.63) holds with $g = \Pr(\bigcap_{i=1}^n A_i | Y)$ so long as C is of the specified form. To show that it holds for all $C \in \mathcal{A}_{Y,m}$, we first note that $\mathcal{A}_{Y,m}$ is the smallest σ-field containing all sets of the specified form. Clearly, (B.63) holds for all sets that are unions of finitely many disjoint sets of the specified form by linearity of integrals. These sets form a field $\mathcal{C}$. According to Lemma A.24, for each $\epsilon > 0$, there is $C_\epsilon \in \mathcal{C}$ such that $\Pr(C_\epsilon \Delta C) < \epsilon/2$. The following facts follow trivially:

$$\int_{C_\epsilon} g(s) d\mu(s) = \Pr\left(C_\epsilon \bigcap_{i=1}^n A_i\right),$$

$$\left|\Pr\left(C \bigcap_{i=1}^n A_i\right) - \Pr\left(C_\epsilon \bigcap_{i=1}^n A_i\right)\right| < \frac{\epsilon}{2},$$

$$\left|\int_C g(s) d\mu(s) - \int_{C_\epsilon} g(s) d\mu(s)\right| < \frac{\epsilon}{2}.$$

Combining these gives that $\left|\int_C g(s) d\mu(s) - \Pr(C \cap_{i=1}^n A_i)\right| < \epsilon$. Since ϵ is arbitrary, (B.63) holds for all $C \in \mathcal{A}_{Y,m}$. □

A particular case of interest involves three random quantities. Theorem B.64 says that when there are only two Xs in Theorem B.61, we can check conditional independence by checking only one of the equations of the form (B.62).

Theorem B.64.[26] *Let X, Y, and Z be three random quantities, and let $\mathcal{A}_X$, $\mathcal{A}_Y$, and $\mathcal{A}_Z$ be the σ-fields generated by each of them. Suppose that for all $A \in \mathcal{A}_X$, $\Pr(A|Y, Z) = \Pr(A|Y)$. Then X and Z are conditionally independent given Y.*

PROOF. We need to check that for every $A \in \mathcal{A}_X$ and $B \in \mathcal{A}_Z$, $\Pr(A \cap B|Y) = \Pr(A|Y)\Pr(B|Y)$. Equivalently, for all such A and B, and all $C \in \mathcal{A}_Y$, we must show

$$\Pr(A \cap B \cap C) = \int I_C(s) \Pr(A|Y)(s) \Pr(B|Y)(s) d\mu(s). \tag{B.65}$$

---

[26]This theorem is used in the proofs of Theorems 2.14 and 2.20.

Since we have assumed that $\Pr(A|Y,Z) = \Pr(A|Y)$, we have that, for all $B \in \mathcal{A}_Z$ and $C \in \mathcal{A}_Y$,

$$\Pr(A \cap B \cap C) = \int I_C(s) I_B(s) \Pr(A|Y)(s) d\mu(s).$$

We can use Proposition B.27 with $g(Y) = I_C \Pr(A|Y)$ and $X = I_B$ to see that

$$\int I_C(s) I_B(s) \Pr(A|Y)(s) d\mu(s) = \int I_C(s) \Pr(A|Y)(s) \Pr(B|Y)(s) d\mu(s).$$

Together, these last two equations prove (B.65). □

The following result relates product measure on a product space to independent random variables.

Proposition B.66. *Let $(S, \mathcal{A}, \mu)$ be a probability space and let $(T_i, \mathcal{B}_i)$ ($i = 1, \ldots, n$) be measurable space. Let $X_i : S \to T_i$ be measurable for $i = 1, \ldots, n$. Let μ_i be the measure that X_i induces on T_i for each i, and let $T^n = T_1 \times \cdots \times T_n$, $\mathcal{B}^n = \mathcal{B}_1 \otimes \cdots \otimes \mathcal{B}_n$. Let μ^* be the measure that $(X_1, \ldots, X_n)$ induces on $(T^n, \mathcal{B}^n)$ from μ. Then μ^* is the product measure $\mu^n = \mu_1 \times \cdots \times \mu_n$, if and only if the X_i are independent.*

The same result holds for conditional independence.

Corollary B.67. *Random quantities $X_1, \ldots, X_n$ are conditionally independent given Y if and only if the product measure of the conditional distributions of $X_1, \ldots, X_n$ given Y is a version of the conditional distribution of $(X_1, \ldots, X_n)$ given Y.*

There is an interesting theorem that applies to sequences of independent random variables, even if they are not identically distributed.

Theorem B.68 (Kolmogorov zero-one law).[27] *Suppose that $(S, \mathcal{A}, \mu)$ is a probability space. Let $\{X_n\}_{n=1}^{\infty}$ be a sequence of independent random quantities. For each n, let $\mathcal{C}_n$ be the σ-field generated by $(X_n, X_{n+1}, \ldots)$ and let $\mathcal{C} = \cap_{n=1}^{\infty} \mathcal{C}_n$. Then every set in $\mathcal{C}$ has probability 0 or probability 1.*

PROOF. Let $\mathcal{A}_n$ be the σ-field generated by $(X_1, \ldots, X_n)$. Then $\mathcal{C}_* = \cup_{n=1}^{\infty} \mathcal{A}_n$ is a field. It is easy to see that $\mathcal{C}$ is contained in the smallest σ-field containing $\mathcal{C}_*$. Let $A \in \mathcal{C}$. By Lemma A.24, for every $k > 0$, there exists n and $C_k \in \mathcal{A}_n$ such that $\mu(A \triangle C_k) < 1/k$. It follows that

$$\lim_{k \to \infty} \mu(C_k) = \mu(A),$$
$$\lim_{k \to \infty} \mu(C_k \cap A) = \mu(A). \tag{B.69}$$

Since $A \in \mathcal{C}$, it follows that $A \in \mathcal{C}_{n+1}$; hence A and C_k are independent for every k. It follows that $\mu(C_k \cap A) = \mu(A)\mu(C_k)$. It follows from (B.69) that $\mu(A) = \mu(A)^2$, and hence either $\mu(A) = 0$ or $\mu(A) = 1$. □

---

[27]This theorem is used in the proofs of Corollary 1.63 and Lemma 7.83, and in the discussion of "sampling to a foregone conclusion" in Section 9.4.

The σ-field $\mathcal{C}$ in Theorem B.68 is often called the *tail σ-field* of the sequence $\{X_n\}_{n=1}^{\infty}$. An interesting feature of the tail σ-field is that limits are measurable with respect to it.[28] (See Problem 21 on page 663.)

B.3.5 The Law of Total Probability

Next, we introduce some theorems that are very simple to state for discrete random variables but appear to be rather unwieldy in the general case. We will, however, need them often.

Theorem B.70 (Law of total probability). *Let $(S, \mathcal{A}, \mu)$ be a probability space, and let Z be a random variable with $\mathrm{E}(|Z|) < \infty$. Let $\mathcal{C} \subseteq \mathcal{B}$ be sub-σ-fields of $\mathcal{A}$. Then $\mathrm{E}(Z|\mathcal{C}) = \mathrm{E}(\mathrm{E}(Z|\mathcal{B})|\mathcal{C})$, a.e. $[\mu]$.*

PROOF. Define $T = \mathrm{E}(Z|\mathcal{B}) : S \to \mathbb{R}$, which is any $\mathcal{B}$ measurable function satisfying $\mathrm{E}(ZI_B) = \int_B T(s) d\mu(s)$, for all $B \in \mathcal{B}$. We need to show that $\mathrm{E}(Z|\mathcal{C}) = \mathrm{E}(T|\mathcal{C})$ a.s. $[\mu]$. The function $\mathrm{E}(T|\mathcal{C})$ is any $\mathcal{C}$ measurable function satisfying $\int_C \mathrm{E}(T|\mathcal{C})(s) d\mu(s) = \mathrm{E}(TI_C)$, for all $C \in \mathcal{C}$. But, since $\mathcal{C} \subseteq \mathcal{B}$, $C \in \mathcal{C}$ implies $C \in \mathcal{B}$. So, for $C \in \mathcal{C}$,

$$\int_C \mathrm{E}(T|\mathcal{C})(s) d\mu(s) = \mathrm{E}(TI_C) = \int I_C(s)T(s) d\mu(s) = \int_C T(s) d\mu(s) = \mathrm{E}(ZI_C),$$

where the last equality follows since $T = \mathrm{E}(Z|\mathcal{B})$ and $C \in \mathcal{B}$. Since $\mathrm{E}(T|\mathcal{C})$ is $\mathcal{C}$ measurable, equating the first and last entries of the above string of equations means that $\mathrm{E}(T|\mathcal{C})$ satisfies the condition required for it to equal $\mathrm{E}(Z|\mathcal{C})$. □

When $\mathcal{B}$ and $\mathcal{C}$ are the σ-fields generated by two random quantities X and Y, respectively, $\mathcal{C} \subseteq \mathcal{B}$ means Y is a function of X. So, Theorem B.70 can be rewritten in this case.

Corollary B.71. *Let $X : S \to U_1$, $Y : S \to U_2$, and $Z : S \to \mathbb{R}$ be measurable functions such that $\mathrm{E}(|Z|) < \infty$. Suppose that Y is a function of X. Then,*

$$\mathrm{E}(Z|Y) = \mathrm{E}\{\mathrm{E}(Z|X)|Y\}, \text{ a.s. } [\mu].$$

The most popular special case of this corollary occurs when Y is constant.

Corollary B.72.[29] *Let $(S, \mathcal{A}, \mu)$ be a probability space. Let $X : S \to U_1$ and $Z : S \to \mathbb{R}$ be measurable functions such that $\mathrm{E}(|Z|) < \infty$. Then, $\mathrm{E}(Z) = \mathrm{E}\{\mathrm{E}(Z|X)\}$.*

This is the special case of Theorem B.70 when $\mathcal{C}$ is the trivial σ-field.

The following theorem implies that if a conditional mean given X depends on X only through $h(X)$, then it is also the conditional mean given $h(X)$.

Theorem B.73.[30] *Let $(S, \mathcal{A}, \mu)$ be a probability space and let $\mathcal{B}$ and $\mathcal{C}$ be sub-σ-fields of $\mathcal{A}$ with $\mathcal{C} \subseteq \mathcal{B}$. Let $Z : S \to \mathbb{R}$ be measurable such that $\mathrm{E}(|Z|) <$*

---

[28] The tail σ-field will play a role in the proofs of Corollary 1.63 and Theorem 1.49.

[29] This corollary is used in the proof of Theorem B.75.

[30] This theorem is used in the proofs of Theorems 1.49 and 2.6.

B.3. Conditioning   633

∞. *Then there exists a version of* $\mathrm{E}(Z|\mathcal{B})$ *that is* $\mathcal{C}$ *measurable if and only if* $\mathrm{E}(Z|\mathcal{B}) = \mathrm{E}(Z|\mathcal{C})$, *a.s.* $[\mu]$.

PROOF. For the "if" direction, if $\mathrm{E}(Z|\mathcal{B}) = \mathrm{E}(Z|\mathcal{C})$, a.s. $[\mu]$, then $\mathrm{E}(Z|\mathcal{C})$ is measurable with respect to both $\mathcal{C}$ and $\mathcal{B}$, and hence it is a $\mathcal{C}$ measurable version of $\mathrm{E}(Z|\mathcal{B})$. For the "only if" direction, if W is a $\mathcal{C}$ measurable version of $\mathrm{E}(Z|\mathcal{B})$, then $W = \mathrm{E}(W|\mathcal{C})$, a.s. $[\mu]$ by the second part of Proposition B.25. By the law of total probability B.70, $\mathrm{E}(W|\mathcal{C}) = \mathrm{E}(Z|\mathcal{C})$, a.s. $[\mu]$. □

A useful corollary is the following.

Corollary B.74.[31] *Let* $(S, \mathcal{A}, \mu)$ *be a probability space. Let* $(S_1, \mathcal{A}_1)$ *and* $(S_2, \mathcal{A}_2)$ *be measurable spaces, and let* $X : S \to S_1$ *and* $h : S_1 \to S_2$ *be measurable functions. Let* $Z : S \to \mathbb{R}$ *be measurable such that* $\mathrm{E}(|Z|) < \infty$. *Define* $Y = h(X)$. *Then* $\mathrm{E}(Z|X = x, Y = y) = \mathrm{E}(Z|X = x)$ *a.s. with respect to the measure on* $(S_1 \times S_2, \mathcal{A}_1 \otimes \mathcal{A}_2)$ *induced by* $(X, Y) : S \to S_1 \times S_2$ *from* μ.

The following theorem deals with conditioning on two random quantities at the same time. In words it says that the conditional mean of a random variable Z given two random quantities X_1 and X_2 can be calculated two ways. One is to condition on both X_1 and X_2 at once, and the other is to condition on one of them, say X_2, and then find the conditional mean of Z given X_1, but starting from the conditional distribution of (Z, X_1) given X_2.

Theorem B.75.[32] *Let* $(S, \mathcal{A}, \mu)$ *be a probability space and let* $(\mathcal{X}_i, \mathcal{B}_i)$ *for* $i = 1, 2$ *be measurable spaces. Let* $X_i : S \to \mathcal{X}_i$ *for* $i = 1, 2$ *and* $Z : S \to \mathbb{R}$ *be random quantities such that* $\mathrm{E}(|Z|) < \infty$. *Let* $\mu_{1,2,Z}$ *denote the measure on* $(\mathcal{X}_1 \times \mathcal{X}_2 \times \mathbb{R}, \mathcal{B}_1 \otimes \mathcal{B}_2 \otimes \mathcal{B})$ *induced by* (X_1, X_2, Z) *from* μ. *(Here,* $\mathcal{B}$ *denotes the Borel* σ-*field.) For each* $(x, y) \in \mathcal{X}_1 \times \mathcal{X}_2$, *let* $g(x, y)$ *denote* $\mathrm{E}(Z|(X_1, X_2) = (x, y))$. *For each* $A \in \mathcal{A}$ *and* $y \in \mathcal{X}_2$, *let* $\mu^{(2)}(A|y)$ *denote* $\mathrm{Pr}(A|X_2 = y)$. *For each* $y \in \mathcal{X}_2$, *let* $h(x, y)$ *denote the conditional mean of* Z *given* $X_1 = x$ *calculated in the probability space* $(S, \mathcal{A}, \mu^{(2)}(\cdot|y))$. *Then* $h = g$ *a.s.* $[\mu_{1,2,Z}]$.

PROOF. Saying that $h = g$ a.s. $[\mu_{1,2,Z}]$ is equivalent to saying that

$$h(X_1(s), X_2(s)) = g(X_1(s), X_2(s)), \quad \text{a.s. } [\mu].$$

To prove this we first note that $f(s) = h(X_1(s), X_2(s))$ is measurable with respect to the σ-field generated by (X_1, X_2), $\mathcal{A}_{X_1, X_2}$. All that remains is to show that it satisfies the integral condition required to be $\mathrm{E}(Z|X_1, X_2)$. That is, for all $C \in \mathcal{A}_{X_1, X_2}$,

$$\mathrm{E}(ZI_C) = \int_C f(s) d\mu(s). \tag{B.76}$$

Let μ_2 be the measure on $(\mathcal{X}_2, \mathcal{B}_2)$ induced by X_2 from μ. First, suppose that $C = A \cap B$, where $A \in \mathcal{A}_{X_1}$ and $B \in \mathcal{A}_{X_2}$. The last hypothesis of the theorem says that for all $A \in \mathcal{A}_{X_1}$, $\mathrm{E}(ZI_A|X_2 = y) = \int_A h(X_1(s), y) d\mu^{(2)}(s|y)$. If $\mu_{1|2}(\cdot|y)$ is the probability on $(\mathcal{X}_1, \mathcal{B}_1)$ induced by X_1 from $\mu^{(2)}(\cdot|y)$, then $\mu_{1|2}(\cdot|y)$ is also the conditional distribution of X_1 given $X_2 = y$ as in Theorem B.46. Suppose

---

[31]This corollary is used in the proof of Theorem 2.14.
[32]This theorem is used in the proof of Lemma 2.120, and it is used in making sense of the notation E_θ when introducing parametric models.

that $A = X_1^{-1}(D)$ and $B = X_2^{-1}(F)$. Then $A \cap B = (X_1, X_2)^{-1}(D \times F)$ and $\mathrm{E}(ZI_A | X_2 = y) = \int_D h(x,y) d\mu_1(x)$. By Corollary B.72 and Theorem B.46, we can write

$$\mathrm{E}(ZI_A I_B) = \int_F \int_D h(x,y) d\mu_{1|2}(x|y) d\mu_2(y)$$
$$= \int_{D \times F \times \mathbb{R}} h(x,y) d\mu_{1,2,Z}(x,y,z) = \int_{A \cap B} f(s) d\mu(s).$$

This proves (B.76) for $C = A \cap B$. Let $\mathcal{C}$ be the collection of all sets C in $\mathcal{A}$ such that (B.76) holds. Clearly $S \in \mathcal{C}$. If $C \in \mathcal{C}$, then $C^C \in \mathcal{C}$ since $\int_S f(s) d\mu(s) = \mathrm{E}(Z)$. By additivity of integrals, if $\{C_i\}_{i=1}^{\infty} \in \mathcal{C}$, then $\cup_{i=1}^{\infty} C_i \in \mathcal{C}$, hence $\mathcal{C}$ contains the smallest σ-field containing all sets of the form $A \cap B$ for $A \in \mathcal{A}_{X_1}$ and $B \in \mathcal{A}_{X_2}$. Theorem A.34 can be used to show that this σ-field is $\mathcal{A}_{X_1, X_2}$. □

If a random variable has finite second moment, then there is a concept of conditional variance.

Definition B.77. Let $X : S \to \mathbb{R}^k$ have finite second moment, and let $\mathcal{C}$ be a sub-σ-field of $\mathcal{A}$. Then the *conditional covariance matrix of X given $\mathcal{C}$* is defined as $\mathrm{Var}(X|\mathcal{C}) = \mathrm{E}[(X - \mathrm{E}(X|\mathcal{C}))(X - \mathrm{E}(X|\mathcal{C}))^\top | \mathcal{C}]$.

The following result is easy to prove.

Proposition B.78.[33] Let $X : S \to \mathbb{R}^k$ have finite second moment, and let $\mathcal{C}$ be a sub-σ-field of $\mathcal{A}$. Then $\mathrm{Var}(X) = \mathrm{EVar}(X|\mathcal{C}) + \mathrm{Var}[\mathrm{E}(X|\mathcal{C})]$.

B.4 Limit Theorems

There are several types of convergence that will be of interest to us. They involve sequences of random quantities or sequences of distributions.

B.4.1 Convergence in Distribution and in Probability

The simplest type of convergence occurs when the distributions have densities with respect to a common measure. The following theorem is due to Scheffé (1947).

Theorem B.79 (Scheffé's theorem).[34] *Let $\{p_n\}_{n=1}^{\infty}$ and p be nonnegative functions from a measure space $(\mathcal{X}, \mathcal{B}, \nu)$ to $\mathbb{R}$ such that the integral of each function is 1 and $\lim_{n \to \infty} p_n(x) = p(x)$, a.e. $[\nu]$. Then*

$$\lim_{n \to \infty} \int_B p_n(x) d\nu(x) = \int_B p(x) d\nu(x), \text{ for all } B \in \mathcal{B}.$$

PROOF. Let $\delta_n(x) = p_n(x) - p(x)$, and let δ_n^+ and δ_n^- be its positive and negative parts. Clearly, both $\lim_{n \to \infty} \delta_n^+ = 0$ and $\lim_{n \to \infty} \delta_n^- = 0$, a.e. $[\nu]$. Since

---

[33] This proposition is used in the proofs of Theorems 2.36 and 2.86.
[34] This theorem is used in the proofs of Lemma 1.113 and Theorem 1.121.

$0 \leq \delta_n^- \leq p$ is true, it follows from the dominated convergence theorem A.57 that $\lim_{n\to\infty} \int_B \delta_n^-(x) d\nu(x) = 0$ for all B. Since both p_n and p are densities, $\int_{\mathcal{X}} \delta_n(x) d\nu(x) = 0$ for all n. It follows that $\lim_{n\to\infty} \int_{\mathcal{X}} \delta_n^+(x) d\nu(x) = 0$. Since $I_B(x)\delta_n^+(x) \leq \delta_n^+(x)$ for all x, it follows from Proposition A.58 that

$$\lim_{n\to\infty} \int_B \delta_n^+(x) d\nu(x) = 0.$$

So, $\lim_{n\to\infty} \int_B [p_n(x) - p(x)] d\nu(x) = 0$ for all B. □

Since defining convergence requires a topology, the following definitions require that the random quantities lie in various types of topological spaces.

Definition B.80. Let $\{X_n\}_{n=1}^\infty$ be a sequence of random quantities and let X be another random quantity, all taking values in the same topological space $\mathcal{X}$. Suppose that $\lim_{n\to\infty} E(f(X_n)) = E(f(X))$ for every bounded continuous function $f : \mathcal{X} \to \mathbb{R}$, then we say that X_n *converges in distribution* to X, which is written $X_n \xrightarrow{D} X$.

Convergence in distribution is sometimes defined in terms of probability measures. The reason is that if $X_n \xrightarrow{D} X$, the actual values of X_n and of X do not play any role in the convergence. All that matters is the distributions of X_n and of X.

Definition B.81. Let $\{P_n\}_{n=1}^\infty$ be a sequence of probability measures on a topological space $(\mathcal{X}, \mathcal{B})$ where $\mathcal{B}$ contains all open sets. Let P be another probability on $(\mathcal{X}, \mathcal{B})$. We say that P_n *converges weakly*[35] to P (denoted $P_n \xrightarrow{W} P$) if, for each bounded continuous function $g : \mathcal{X} \to \mathbb{R}$, $\lim_{n\to\infty} \int g(x) dP_n(x) = \int g(x) dP(x)$.

---

[35]This is not exactly the same as the concept of weak convergence in normed linear spaces [see, for example, Dunford and Schwartz (1957), p. 419]. The collection of all probability measures on a space $(\mathcal{X}, \mathcal{B})$ can be considered a subset of a normed linear space $\mathcal{L}$ consisting of all finite signed measures ν (see Definition A.18) with the norm being $\sup_{B\in\mathcal{B}} |\nu(B)|$. Weak convergence of a sequence $\{\nu_n\}_{n=1}^\infty$ in this space would require the convergence of $L(\nu_n)$ for every bounded linear functional L on $\mathcal{L}$. Every bounded measurable function g on $(\mathcal{X}, \mathcal{B})$ determines a bounded linear functional L_g on $\mathcal{L}$ by $L_g(\nu) = \int g(x) d\nu(x)$, where the integral with respect to a signed measure can be defined as in Problem 27 on page 605. Hence, weak convergence of a sequence of probability measures would require convergence of the means of all bounded measurable functions. In particular, $\lim_{n\to\infty} P_n(B) = P(B)$ for all measurable sets B, not just those for which P assigns 0 probability to the boundary (see the portmanteau theorem B.83 on page 636). Alternatively, we can consider the set of bounded continuous functions $f : \mathcal{X} \to \mathbb{R}$ as a normed linear space $\mathcal{N}$ with $\|f\| = \sup_x |f(x)|$. Then the set of finite signed measures $\mathcal{L}$ is a set of bounded linear functionals on $\mathcal{N}$ using the definition $\nu(f) = \int f(x) d\nu(x)$. *Weak* convergence* of a sequence $\{\nu_n\}_{n=1}^\infty$ in $\mathcal{L}$ to ν is defined as the convergence of $\nu_n(f)$ to $\nu(f)$ for all $f \in \mathcal{N}$. This is precisely convergence in distribution. Hence, it would make more sense to call convergence in distribution weak* convergence rather than weak convergence. Since the tradition in probability theory is to call it weak convergence, we will continue to do so.

It is easy to see that these two types of convergence are the same.

Proposition B.82. *Let P_n be the distribution of X_n, and let P be the distribution of X. Then, $X_n \xrightarrow{D} X$ if and only if $P_n \xrightarrow{W} P$.*

Since we will usually be dealing with $\mathcal{X}$ spaces that are metric spaces, there are some equivalent ways to define convergence in distribution or weak convergence. The proofs of Theorems B.83 and B.88 are adapted from Billingsley (1968).

Theorem B.83 (Portmanteau theorem).[36] *The following are all equivalent in a metric space:*

1. $P_n \xrightarrow{W} P$;
2. $\limsup_{n\to\infty} P_n(B) \leq P(B)$ for each closed B;
3. $\liminf_{n\to\infty} P_n(A) \geq P(A)$, for each open A;
4. $\lim_{n\to\infty} P_n(C) = P(C)$, for each C with $P(\partial C) = 0$.[37]

PROOF. Let d be the metric in the metric space. First, assume (1) and let B be a closed set. Let $\delta > 0$ be given. For each $\epsilon > 0$, define $C_\epsilon = \{x : d(x,B) \leq \epsilon\}$, where $d(x,B) = \inf_{y \in B} d(x,y)$. Since $|d(x,B) - d(y,B)| \leq d(x,y)$, we see that $d(x,B)$ is continuous in x. Each C_ϵ is closed and $\cap_{\epsilon > 0} C_\epsilon = B$. Let ϵ be small enough so that $P(C_\epsilon) \leq P(B) + \delta$. Let $f : \mathbb{R} \to \mathbb{R}$ be

$$f(t) = \begin{cases} 1 & \text{if } t \leq 0, \\ 1-t & \text{if } 0 < t < 1, \\ 0 & \text{if } t \geq 1, \end{cases}$$

and define $g_\epsilon(x) = f(d(x,B)/\epsilon)$. Then g_ϵ is bounded and continuous. So,

$$\lim_{n\to\infty} \int g_\epsilon(x) dP_n(x) = \int g_\epsilon(x) dP(x).$$

It is easy to see that $0 \leq g_\epsilon(x) \leq 1$, $g_\epsilon(x) = 1$ for all $x \in B$, and $g_\epsilon(x) = 0$ for all $x \notin C_\epsilon$. Hence, for every $\delta > 0$,

$$P_n(B) = \int I_B(x) dP_n(x) \leq \int g_\epsilon(x) dP_n(x) \to \int g_\epsilon(x) dP(x)$$
$$\leq \int I_{C_\epsilon}(x) dP(x) = P(C_\epsilon) \leq P(B) + \delta.$$

It follows that $\limsup_{n\to\infty} P_n(B) \leq P(B)$, which is (2).

That (2) and (3) are equivalent follows easily from the facts that if A is open, then $B = A^C$ is closed and $P_n(A) = 1 - P_n(B)$. It is also easy to see that (2) and (3) together imply (4). Next assume (4), let B be a closed set, and define C_ϵ as above. The boundary of C_ϵ is a subset of $\{x : d(x,B) = \epsilon\}$. There can be at most countably many ϵ such that these sets have positive probability. Hence, there

---

[36]This theorem is used in the proofs of Theorem B.88 and Lemma 7.19.

[37]We use the symbol ∂ in front of the name of a subset of a topological space to refer to the *boundary* of the set. The boundary of a set C in a topological space is the intersection of the closure of the set with the closure of the complement.

exists a sequence $\{\epsilon_k\}_{k=1}^\infty$ converging to 0 such that $P(d(X,B) = \epsilon_k) = 0$ for all k. It follows that $\lim_{n\to\infty} P_n(C_{\epsilon_k}) = P(C_{\epsilon_k})$ for all k. Since $P_n(B) \leq P_n(C_{\epsilon_k})$ for every n and k, we have, for every k,

$$\limsup_{n\to\infty} P_n(B) \leq \lim_{n\to\infty} P_n(C_{\epsilon_k}) = P(C_{\epsilon_k}).$$

Since $P(B) = \lim_{k\to\infty} P(C_{\epsilon_k})$, we have (2). So, (2), (3), and (4) are equivalent and (1) implies (2).

All that remains is to prove that (3) implies (1). Assume (3), and let f be a bounded continuous function. Let $m < f(x) < M$ for all x. For each k, let $F_{i,k} = \{x : f(x) < m + (M-m)i/k\}$ for $i = 1, \ldots, k$. Let $F_{0,k} = \emptyset$. Each $F_{i,k}$ is open, since f is continuous. Let $G_{i,k} = F_{i,k} \setminus F_{i-1,k}$ for $i = 1, \ldots, k$. It is easy to see that for every probability Q,

$$m + (M-m) \sum_{i=1}^k \frac{i-1}{k} Q(G_{i,k}) \leq \int f(x) dQ(x) < m + (M-m) \sum_{i=1}^k \frac{i}{k} Q(G_{i,k}).$$

Since $Q(G_{i,k}) = Q(F_{i,k}) - Q(F_{i-1,k})$ for every i and k, we get

$$m + \frac{M-m}{k} \sum_{i=1}^k Q(F_{i,k}) \leq \int f(x) dQ(x) < m + \frac{M-m}{k} + \frac{M-m}{k} \sum_{i=1}^k Q(F_{i,k}). \tag{B.84}$$

For each i,

$$\liminf_{n\to\infty} P_n(F_{i,k}) \geq P(F_{i,k}). \tag{B.85}$$

It follows that, for every k,

$$\int f(x) dP(x) \leq m + \frac{M-m}{k} + \frac{M-m}{k} \sum_{i=1}^k P(F_{i,k})$$

$$\leq m + \frac{M-m}{k} + \frac{M-m}{k} \sum_{i=1}^k \liminf_{n\to\infty} P_n(F_{i,k})$$

$$\leq \frac{M-m}{k} + \liminf_{n\to\infty} \int f(x) dP_n(x),$$

where the first inequality follows from the second inequality in (B.84) with $Q = P$, the second inequality follows from (B.85), and the third inequality follows from the first inequality in (B.84) with $Q = P_n$. Letting k be arbitrarily large, we get

$$\int f(x) dP(x) \leq \liminf_{n\to\infty} \int f(x) dP_n(x). \tag{B.86}$$

Now, apply the same reasoning to $-f$ to get

$$-\int f(x) dP(x) \leq \liminf_{n\to\infty} \int -f(x) dP_n(x) = -\limsup_{n\to\infty} \int f(x) dP_n(x),$$

$$\int f(x) dP(x) \geq \limsup_{n\to\infty} \int f(x) dP_n(x). \tag{B.87}$$

Together, (B.86) and (B.87) imply (1). □

Theorem B.88 (Continuous mapping theorem).[38] *Let $\{X_n\}_{n=1}^{\infty}$ be a sequence of random quantities, and let X be another random quantity all taking values in the same metric space $\mathcal{X}$. Suppose that $X_n \xrightarrow{D} X$. Let $\mathcal{Y}$ be a metric space and let $g : \mathcal{X} \to \mathcal{Y}$. Define*

$$C_g = \{x : g \text{ is continuous at } x\}.$$

Suppose that $\Pr(X \in C_g) = 1$. Then $g(X_n) \xrightarrow{D} g(X)$.

PROOF. Let P_n be the distribution of $g(X_n)$ and let P be the distribution of $g(X)$. Let B be a closed subset of $\mathcal{Y}$. If $x \in \overline{g^{-1}(B)}$ but $x \notin g^{-1}(B)$, then g is not continuous at x. It follows that $\overline{g^{-1}(B)} \subseteq g^{-1}(B) \cup C_g^C$. Now write

$$\begin{aligned}\limsup_{n \to \infty} P_n(B) &= \limsup_{n \to \infty} \Pr(X_n \in g^{-1}(B)) \le \limsup_{n \to \infty} \Pr(X_n \in \overline{g^{-1}(B)}) \\ &\le \Pr(X \in \overline{g^{-1}(B)}) \le \Pr(X \in g^{-1}(B)) + \Pr(X \in C_g^C) \\ &= \Pr(X \in g^{-1}(B)) = P(B),\end{aligned}$$

and the result now follows from the portmanteau theorem B.83. □

Another type of convergence is convergence in probability.

Definition B.89. *If $\{X_n\}_{n=1}^{\infty}$ and X are random quantities in a metric space with metric d, and if, for every $\epsilon > 0$, $\lim_{n \to \infty} \Pr(d(X_n, X) > \epsilon) = 0$, then we say that X_n converges in probability to X, which is written $X_n \xrightarrow{P} X$.*

The following theorem is useful in that it relates convergence in distribution, convergence in probability, and the simpler concept of convergence almost surely.

Theorem B.90.[39] *Let $\{X_n\}_{n=1}^{\infty}$ be a sequence of random vectors and let X be a random vector.*

1. *If $\lim_{n \to \infty} X_n = X$ a.s., then $X_n \xrightarrow{P} X$.*
2. *If $X_n \xrightarrow{P} X$, then $X_n \xrightarrow{D} X$.*
3. *If X is degenerate and $X_n \xrightarrow{D} X$, then $X_n \xrightarrow{P} X$.*
4. *If $X_n \xrightarrow{P} X$, then there is a subsequence $\{n_k\}_{k=1}^{\infty}$ such that $\lim_{k \to \infty} X_{n_k} = X$, a.s.*

PROOF. First, assume that X_n converges a.s. to X. For each n and ϵ, let $A_{n,\epsilon} = \{s : d(X_n(s), X(s)) \le \epsilon\}$. Then $X_n(s)$ converges to $X(s)$ if and only if

$$s \in \bigcap_{\text{All } \epsilon} \left(\bigcup_{N=1}^{\infty} \left[\bigcap_{n=N}^{\infty} A_{n,\epsilon} \right] \right).$$

Since this set must have probability 1, then so too must $\bigcup_{N=1}^{\infty} \left(\bigcap_{n=N}^{\infty} A_{n,\epsilon} \right)$ for all ϵ. By Theorem A.19, it follows that for every ϵ, $\lim_{N \to \infty} \Pr\left(\bigcap_{n=N}^{\infty} A_{n,\epsilon} \right) = 1$.

---

[38] This theorem is used to provide a short proof of DeFinetti's representation theorem for Bernoulli random variables in Example 1.82 on page 46.

[39] This theorem is used in the proofs of Theorems B.95, 1.49, 7.26, and 7.78.

Hence, for each $\epsilon > 0$, $\lim_{n\to\infty} \Pr(A_{n,\epsilon}^C) = 0$, which is precisely what it means to say that $X_n \xrightarrow{P} X$.

Next, assume that $X_n \xrightarrow{P} X$. Let $g : \mathcal{X} \to \mathbb{R}$ be bounded and continuous with $|g(x)| \leq K$ for all x. Let $\epsilon > 0$, and let A be a compact set with $\Pr(X \in A) > 1 - \epsilon/[6K]$. A continuous function (like g) on a compact set is uniformly continuous. So let $\delta > 0$ be such that $x \in A$ and $d(x,y) < \delta$ implies $|g(x) - g(y)| < \epsilon/3$. Since $X_n \xrightarrow{P} X$, there exists N such that $n \geq N$ implies $\Pr(d(X_n, X) < \delta) > 1 - \epsilon/[6K]$. Let $B = \{X \in A, d(X_n, X) < \delta\}$. It follows that $|g(X)I_B - g(X_n)I_B| < \epsilon/3$ and, for all $n \geq N$, $\Pr(B) > 1 - \epsilon/[3K]$. Also, note that $n \geq N$ implies

$$|\mathrm{E}g(X) - \mathrm{E}[g(X)I_B]| < \frac{\epsilon}{3}, \quad |\mathrm{E}g(X_n) - \mathrm{E}[g(X_n)I_B]| < \frac{\epsilon}{3}.$$

So, $n \geq N$ implies

$$\begin{aligned}
|\mathrm{E}g(X) - \mathrm{E}g(X_n)| &\leq |\mathrm{E}g(X) - \mathrm{E}[g(X)I_B]| + |\mathrm{E}[g(X)I_B] - \mathrm{E}[g(X_n)I_B]| \\
&\quad + |\mathrm{E}[g(X_n)I_B] - \mathrm{E}g(X_n)| \\
&\leq \frac{\epsilon}{3} + \frac{\epsilon}{3} + \frac{\epsilon}{3} = \epsilon.
\end{aligned}$$

Thus, $\lim_{n\to\infty} \mathrm{E}g(X_n) = \mathrm{E}g(X)$, and we have proven $X_n \xrightarrow{D} X$.

Next, suppose that X is degenerate at x_0 and $X_n \xrightarrow{D} X$. Let $\epsilon > 0$, and define

$$g(x) = \begin{cases} 1 & \text{if } d(x, x_0) \leq \frac{\epsilon}{2}, \\ 0 & \text{if } d(x, x_0) \geq \epsilon, \\ 2\frac{d(x,x_0) - \frac{\epsilon}{2}}{\epsilon} & \text{otherwise.} \end{cases}$$

Since g is bounded and continuous, $\mathrm{E}g(X_n)$ converges to $\mathrm{E}g(X)$. But $\mathrm{E}g(X) = 1$ since $\Pr(g(X) = 1) = 1$, and $\mathrm{E}g(X_n) \leq \Pr(d(X_n, X) < \epsilon)$, since $0 \leq g(x) \leq 1$ for all x. So $\lim_{n\to\infty} \Pr(d(X_n, x_0) < \epsilon) = 1$, and $X_n \xrightarrow{P} X$.

Finally, assume that $X_n \xrightarrow{P} X$. Let n_k be such that $n \geq n_k$ implies

$$\Pr\left(d(X_n, X) \geq \frac{1}{k}\right) < 2^{-k}.$$

Define $A_k = \{d(X_{n_k}, X) \geq 1/k\}$. By the first Borel–Cantelli lemma A.20, we have $\Pr(B) = 0$, where $B = \bigcap_{i=1}^{\infty} \bigcup_{k=i}^{\infty} A_k$. It is easy to check that B is the event that $d(X_{n_k}, X)$ is at least $1/k$ for infinitely many different k. Hence $B^C \subseteq \{\lim_{k\to\infty} X_{n_k} = X\}$, and $\lim_{k\to\infty} X_{n_k} = X$, a.s. □

B.4.2 Characteristic Functions

There is a very important method for proving convergence in distribution which involves the use of *characteristic functions*.

Definition B.91. Let X be a random vector. The complex-valued function

$$\phi_X(t) = \mathrm{E}\left(\exp[it^\top X]\right)$$

is called the *characteristic function of X*. If F is a k-dimensional distribution function, the function $\phi_F(t) = \int \exp[it^\top x] dF(x)$ is called the *characteristic function of F*.

Example B.92. Let X have standard normal distribution. Then

$$\begin{aligned}\phi_X(t) &= \int \exp(itx)\frac{1}{\sqrt{2\pi}}\exp\left(-\frac{x^2}{2}\right)dx = \frac{1}{\sqrt{2\pi}}\int \exp\left(-\frac{[x-it]^2+t^2}{2}\right)dx \\ &= \exp\left(-\frac{t^2}{2}\right).\end{aligned}$$

Similarly, for other normal distributions, $N(\mu, \sigma^2)$, the characteristic functions are $\phi_X(t) = \exp(-\sigma^2 t^2/2 + it\mu)$.

By Theorem B.12, if X has CDF F, then $\phi_X = \phi_F$. It is easy to see that the characteristic function exists for every random vector and it has complex absolute value at most 1 for all t. Other facts that follow directly from the definition are the following. If $Y = aX + b$, then $\phi_Y(t) = \phi_X(at)\exp(itb)$. If X and Y are independent, $\phi_{X+Y} = \phi_X \phi_Y$.

The reason that characteristic functions are so useful for proving convergence in distribution is two-fold. First, for each characteristic function ϕ, there is only one CDF F such that $\phi_F = \phi$. (See the uniqueness theorem B.106.) Second, characteristic functions are "continuous" as a function of the distribution in the sense of convergence in distribution. That is, $X_n \xrightarrow{D} X$ if and only if $\lim_{n\to\infty} \phi_{X_n}(t) = \phi_X(t)$ for all t.[40] (See the continuity theorem B.93.)

Theorem B.93 (Continuity theorem).[41] *For finite-dimensional random vectors, convergence in distribution is equivalent to convergence of characteristic functions. That is, $X_n \xrightarrow{D} X$ if and only if $\lim_{n\to\infty} \phi_{X_n}(t) = \phi_X(t)$ for all t.*

PROOF. The "only if" part follows from Definition B.80 and the fact that one can write $\exp(it^\top x)$ as two bounded, continuous, real-valued functions of x for every t.

For the "if" part, suppose that X is k-dimensional and that $\lim_{n\to\infty} \phi_{X_n}(t) = \phi_X(t)$ for all t. To prove that for each bounded continuous g, $\lim_{n\to\infty} Eg(X_n) = Eg(X)$, we will truncate g to a bounded rectangle and then approximate the truncated function by a function g' whose mean is a linear combination of values of the characteristic function. The mean of $g'(X_n)$ will then converge to the mean of $g'(X)$. We then need to show that the means of $g'(X)$ and $g'(X_n)$ approximate the means of $g(X)$ and $g(X_n)$, respectively.

First, we need to find a bounded rectangle on which to do the truncation. For each coordinate X^ℓ of X, we will show that if a and b are continuity points of the CDF F_{X^ℓ} of X^ℓ, and $F_{X^\ell}(b) - F_{X^\ell}(a) > q$, then there is $b' > b$ and $a' < a$ such that $\lim_{n\to\infty} F_{X_n^\ell}(b') - F_{X_n^\ell}(a') \geq q$. For each a, b, δ, define

$$f_{a,b,\delta}(x) = \begin{cases} 1 & \text{if } a < x < b, \\ 1 - \frac{a-x}{\delta} & \text{if } a - \delta < x \leq a, \\ 1 - \frac{x-b}{\delta} & \text{if } b \leq x < b + \delta, \\ 0 & \text{otherwise.} \end{cases} \quad (B.94)$$

---

[40]This presentation is a hybrid of the presentations given by Breiman (1968), Chapter 8) and Hoel, Port, and Stone (1971, Chapter 8).
[41]This theorem is used in the proofs of Theorems B.95, B.97, and 7.20.

Note that this function has equal values at $a - \delta$ and $b + \delta$. Consider the interval $[a - \delta, b + \delta]$ as a circle identifying the two endpoints. Now, use the Stone–Weierstrass theorem C.3 to approximate uniformly $f_{a,b,\delta}$ to within ϵ on the circle by $f'_{a,b,\delta,\epsilon}(x) = \sum_{j=-\ell}^{\ell} b_j \exp(2\pi i j x/c)$, where $c = b - a + 2\delta$. If Y is a random variable, then $\mathrm{E} f'_{a,b,\delta,\epsilon}(Y)$ is a linear combination of values of the characteristic function of Y. So, we have $\lim_{n\to\infty} \mathrm{E} f'_{a,b,\delta,\epsilon}(X_n) = \mathrm{E} f'_{a,b,\delta,\epsilon}(X)$. Let $q > 0$, and let a and b be continuity points of F_{X^ℓ} such that $F_{X^\ell}(b) - F_{X^\ell}(a) = v > q$. Let $w = v - q$. Let $\delta > 0$ be arbitrary, and define $a' = a - \delta$ and $b' = b + \delta$. Let N be large enough so that $n \geq N$ implies $|\mathrm{E} f'_{a,b,\delta,w/3}(X_n^\ell) - \mathrm{E} f'_{a,b,\delta,w/3}(X^\ell)| < w/3$. If $n \geq N$, then

$$\begin{array}{rclcl}
F_{X_n^\ell}(b') - F_{X_n^\ell}(a') & \geq & \mathrm{E} f_{a,b,\delta}(X_n^\ell) & \geq & \mathrm{E} f'_{a,b,\delta,\frac{w}{3}}(X_n^\ell) - \frac{w}{3} \\
& \geq & \mathrm{E} f'_{a,b,\delta,\frac{w}{3}}(X^\ell) - \frac{2w}{3} & \geq & \mathrm{E} f_{a,b,\delta}(X^\ell) - w \\
& \geq & F_{X^\ell}(b) - F_{X^\ell}(a) - w & = & q.
\end{array}$$

Now, let g be a bounded continuous function, and suppose that $|g(x)| < K$ for all x. Let $\epsilon > 0$. For each coordinate X^ℓ of X, let a_ℓ and b_ℓ be continuity points of F_{X^ℓ} such that $F_{X^\ell}(b_\ell) - F_{X^\ell}(a_\ell) > 1 - \epsilon/(7[K + \epsilon/7]k)$. Let $\delta > 0$ be arbitrary, and define $a'_\ell = a_\ell - \delta$, $b'_\ell = b_\ell - \delta$, and $g^*(x) = g(x) \prod_{j=1}^{k} f_{a'_\ell, b'_\ell, \delta}(x_\ell)$. Use the Stone-Weierstrass theorem C.3 to uniformly approximate g^* to within $\epsilon/7$ on the rectangle $\{x : a'_\ell - \delta \leq x_\ell \leq b'_\ell + \delta\}$ by

$$g'(x) = \sum_{j_1=-m_1}^{m_1} \cdots \sum_{j_k=-m_k}^{m_k} a_{j_1,\ldots,j_k} \exp(2\pi i j^\top x),$$

where j is the vector with ℓth coordinate $j_\ell/[b'_\ell - a'_\ell + 2\delta]$. Then,

$$\lim_{n\to\infty} \mathrm{E} g'(X_n) = \mathrm{E} g'(X).$$

Let N_1 be large enough so that $n \geq N_1$ implies $F_{X_n^\ell}(b'_\ell) - F_{X_n^\ell}(a'_\ell) \geq 1 - \epsilon/(7[K + \epsilon/7]k)$ for all j. Let N_2 be large enough so that $n \geq N_2$ implies $|\mathrm{E} g'(X_n) - \mathrm{E} g'(X)| < \epsilon/7$. Let R be the rectangle $R = \{x : a'_\ell < x_\ell \leq b'_\ell\}$. Since g' is periodic in every coordinate, it is bounded by $K + \epsilon/7$ on all of $\mathbb{R}^k$. If $n \geq \max\{N_1, N_2\}$, then $|\mathrm{E} g(X_n) - \mathrm{E} g(X)|$ is no greater than

$$\mathrm{E}|g(X_n)I_{R^C}(X_n)| + \mathrm{E}|g(X)I_{R^C}(X)| + \mathrm{E}|g'(X_n)I_{R^C}(X_n)|$$
$$+ |\mathrm{E} g(X_n)I_R(X_n) - \mathrm{E} g'(X_n)I_R(X_n)| + |\mathrm{E} g'(X_n) - \mathrm{E} g'(X)|$$
$$+ \mathrm{E}|g'(X)I_{R^C}(X)| + |\mathrm{E} g'(X)I_R(X) - \mathrm{E} g(X)I_R(X)| \leq \epsilon. \qquad \square$$

We will prove two more limit theorems that make use of the continuity theorem B.93. Suppose that X has finite mean. Since $|\exp(itx) - 1| \leq \min\{|tx|, 2\}$ for all t, x,[42] and

$$\lim_{t\to 0} \frac{\exp(itx) - 1}{t} = ix$$

---

[42] See Problem 26 on page 664.

for all x, it follows from the dominated convergence theorem that

$$\frac{d}{dt}\phi_X(t)\bigg|_{t=0} = i\mathrm{E}(X).$$

Similarly, if X has finite variance, it can be shown that

$$\frac{d^2}{dt^2}\phi_X(t)\bigg|_{t=0} = -\mathrm{E}(X^2).$$

Using these two facts, we can prove the *weak law of large numbers* and the *central limit theorem*.

Theorem B.95 (Weak law of large numbers). *Suppose that $\{X_n\}_{n=1}^\infty$ are IID random variables with finite mean μ. Then, $\overline{X}_n = \sum_{i=1}^n X_i/n$ converges in probability to μ.*

PROOF. First, we will prove that the characteristic function of $\overline{X}_n - \mu$ converges to 1 for all t. Let $Y_i = X_i - \mu$. Since $\phi_{Y_i}(0) = 1$, $\log \phi_{Y_i}(t)$ exists and is differentiable near $t = 0$, and we know that

$$\frac{d}{dt}\log\phi_{Y_i}(0) = 0 = \lim_{t\to 0}\frac{\log\phi_{X_i}(t)}{t}. \tag{B.96}$$

The characteristic function of $\overline{X}_n - \mu$ is $\phi_*(t) = \phi_{Y_i}(t/n)^n$. For fixed t, let n be large enough so that t/n is close enough to 0 for $\log \phi_{Y_i}(t/n)$ to be well defined. We know that

$$\log \phi_*(t) = n\log\phi_{Y_i}\left(\frac{t}{n}\right) = t\frac{\log\phi_{Y_i}\left(\frac{t}{n}\right)}{\frac{t}{n}}.$$

The limit of this quantity, as $n \to \infty$, is 0 by (B.96). It follows that for all t, $\lim_{n\to\infty}\phi_*(t) = 1$. By the continuity theorem B.93, $\overline{X}_n - \mu \xrightarrow{D} 0$. By Theorem B.90, $\overline{X}_n - \mu \xrightarrow{P} 0$. □

In Chapter 1, we prove a strong law of large numbers 1.62, which has a stronger conclusion and a weaker hypothesis. There is also a weak law of large numbers for the case of infinite means. (See Problem 27 on page 664.)

The following theorem is very useful for approximating distributions.

Theorem B.97 (Central limit theorem). *Suppose that $\{X_i\}_{i=1}^\infty$ is a sequence that is IID with finite mean μ and finite variance σ^2. Let $\overline{X}_n$ be the average of the first n X_is. Then $\sqrt{n}(\overline{X}_n - \mu) \xrightarrow{D} N(0, \sigma^2)$, the normal distribution with mean 0 and variance σ^2.*

PROOF. Set $Y_n = \sqrt{n}(\overline{X}_n - \mu)$. We might as well assume that $\mu = 0$, since we have just subtracted it from each X_i. Since the second derivative of the characteristic function at $t = 0$ of each X_i is $-\sigma^2$, we can apply l'Hôpital's rule twice to conclude

$$\lim_{t\to 0}\frac{\log\phi_{X_i}(t)}{t^2} = -\frac{\sigma^2}{2}. \tag{B.98}$$

The characteristic function of Y_n is $\phi_{Y_n}(t) = \phi_{X_i}(t/\sqrt{n})^n$. We will prove that this converges to $\exp(-t^2\sigma^2/2)$ for each t. Since $\log\phi_{Y_n}(t) = n\log\phi_{X_i}(t/\sqrt{n})$, we use (B.98) to note that

$$\lim_{n\to\infty} n\frac{\log \phi_{X_i}\left(\frac{t}{\sqrt{n}}\right)}{t^2} = -\frac{\sigma^2}{2}.$$

It follows that $\lim_{n\to\infty} \phi_{Y_n}(t) = \exp(-t^2\sigma^2/2)$, and the continuity theorem B.93 finishes the proof. □

There is also a multivariate version of the central limit theorem.

Theorem B.99 (Multivariate central limit theorem).[43] *Let $\{X_n\}_{n=1}^{\infty}$ be a sequence of IID random vectors in $\mathbb{R}^p$ with mean μ and covariance matrix Σ. Then $\sqrt{n}(\overline{X}_n - \mu) \xrightarrow{D} N_p(\mathbf{0}, \Sigma)$, a multivariate normal distribution.*

PROOF. Let $Y_n = \sqrt{n}(\overline{X}_n - \mu)$ and let $Y \sim N_p(\mathbf{0}, \Sigma)$. Then $Y_n \xrightarrow{D} Y$ if and only if the characteristic function of Y_n converges to that of Y. That is, if and only if, for each $\lambda \in \mathbb{R}^p$, $E \exp\{i\lambda^\top Y_n\} \to E \exp\{i\lambda^\top Y\}$. This occurs if and only if, for each λ, $\lambda^\top Y_n \xrightarrow{D} \lambda^\top Y$. The distribution of $\lambda^\top Y$ is $N(0, \lambda^\top \Sigma \lambda)$, and $\lambda^\top Y_n$ is $\sqrt{n}$ times the average of the $\lambda^\top (X_n - \mu)$. By the univariate central limit theorem B.97, $\lambda^\top Y_n \xrightarrow{D} \lambda^\top Y$. □

There are inversion formulas for characteristic functions which allow us to obtain or approximate the original distributions from the characteristic functions.

Example B.100 (Continuation of Example B.92; see page 640). Let X have distribution $N(0, \sigma^2)$. Then $\int |\phi_X(t)| dt < \infty$. In fact,

$$\frac{1}{2\pi} \int \exp(-ixt)\phi_X(t)dt = \frac{1}{2\pi} \int \exp\left(-\frac{\sigma^2}{2}\left[t + \frac{i}{\sigma^2}x\right]^2 - \frac{1}{2\sigma^2}x^2\right) dt$$

$$= \frac{1}{\sqrt{2\pi}\sigma} \exp\left(-\frac{1}{2\sigma^2}x^2\right) = f_X(x).$$

Example B.100 says that the following inversion formula applies to normal distributions with 0 mean. It is equally easy to see that it applies to $N_k(\mathbf{0}, I_k)$ distributions.[44]

Lemma B.101 (Continuous inversion formula).[45] *Let $X \in \mathbb{R}^k$ have integrable characteristic function. Then the distribution of X has a bounded density f_X with respect to Lebesgue measure given by*

$$f_X(x) = \frac{1}{(2\pi)^k} \int \exp(-it^\top x)\phi_X(t)dt. \tag{B.102}$$

PROOF. Clearly, the function in (B.102) is bounded since ϕ_X is integrable. Let Y_σ have $N_k(\mathbf{0}, \sigma^2 I_k)$ distribution. The characteristic function of $X + Y_\sigma$ is $\phi_X \phi_{Y_\sigma}$.

$$\frac{1}{(2\pi)^k} \int \exp(-it^\top x)\phi_X(t)\phi_{Y_\sigma}(t)dt$$

---

[43] This theorem is used in the proofs of Theorems 7.35 and 7.57.
[44] We use the symbol I_k to stand for the $k \times k$ identity matrix.
[45] This lemma is used in the proofs of Lemma B.105 and Corollary B.106.

$$= \frac{1}{(2\pi)^k} \int \int \exp(-it^\top x) \exp(it^\top z) \phi_{Y_\sigma}(t) dF_X(z) dt \quad \text{(B.103)}$$

$$= \int f_{Y_\sigma}(x-z) dF_X(z) = f_{X+Y_\sigma}(x),$$

where the second equality follows from the fact that (B.102) applies to normal distributions. Now suppose that we let σ go to zero. Since ϕ_X is integrable and $\phi_{Y_\sigma}(t)$ goes to 1 for all t, it follows that the left-hand side of (B.103) converges to the right-hand side of (B.102). It also follows that f_{X+Y_σ} is bounded uniformly in σ and x. Let B be a hypercube such that the probability is 0 that X is in the boundary of B. Then

$$\int_B \lim_{\sigma \to 0} f_{X+Y_\sigma}(x) dx = \lim_{\sigma \to 0} \int_B f_{X+Y_\sigma}(x) dx = \int_B f_X(x) dx, \quad \text{(B.104)}$$

where the first equality follows from the boundedness of f_{X+Y_σ}, and the second is proven as follows. The difference between $\int_B f_{X+Y_\sigma}(x) dx$ and $\int_B f_X(x) dx$ is the sum over the 2^k corners of the hypercube B of terms like

$$\sum_{i=1}^k \Pr(b_i - Y_{\sigma,i} < X_i \leq b_i, Y_{\sigma,i} > 0) + \Pr(b_i < X_i \leq b_i - Y_{\sigma,i}, Y_{\sigma,i} < 0),$$

where b_i is the ith coordinate of the corner. We can write

$$\Pr(b_i - Y_{\sigma,i} < X_i \leq b_i, Y_{\sigma,i} > 0) = \int_0^\infty \Pr(b_i - y < X_i \leq b_i, y > 0) dF_{Y_{\sigma,i}}(y).$$

This last expression goes to 0 as $\sigma \to 0$ since b_i is a continuity point for F_{X_i}. A similar argument applies to the other probability. The equality of the first and last expressions in (B.104) is what it means to say that $\lim_{\sigma \to 0} f_{X+Y_\sigma}(x)$ is the density of X with respect to Lebesgue measure. This, in turn, equals the right-hand side of (B.102). □

Lemma B.105.[46] *Let Y be a random variable such that ϕ_Y is integrable. Let X be an arbitrary random variable independent of Y. For all finite $a < b$ and c,*

$$\Pr(a < X + cY \leq b) = \frac{1}{2\pi} \int \left(\frac{\exp(-ibt) - \exp(-iat)}{-it} \right) \phi_X(t) \phi_Y(ct) dt.$$

PROOF. Since ϕ_Y is integrable and $\phi_{X+cY}(t) = \phi_X(t) \phi_Y(ct)$, it follows that $X + cY$ has integrable characteristic function. Lemma B.101 says that (B.102) applies to $X + cY$, hence

$$f_{X+cY}(x) = \frac{1}{2\pi} \int \phi_X(t) \phi_Y(ct) \exp(-itx) dt$$

$$\Pr(a < X + cY \leq b) = \int_a^b f_{X+cY}(x) dx$$

---

[46] This lemma is used in the proof of Corollary B.106.

$$= \frac{1}{2\pi} \int_a^b \int \phi_X(t)\phi_Y(ct)\exp(-itx)dtdx$$

$$= \frac{1}{2\pi} \int \phi_Y(ct)\phi_X(t) \int_a^b \exp(-itx)dxdt$$

$$= \frac{1}{2\pi} \int \phi_Y(ct)\phi_X(t) \left(\frac{\exp(-itb) - \exp(-ita)}{-it} \right) dt. \quad \square$$

Corollary B.106 (Uniqueness theorem).[47] *Let F and G be two univariate CDFs such that $\phi_F = \phi_G$. Then $F = G$.*

PROOF. In the proof of Lemma B.101, we proved that if $Y \sim N(0,1)$, and if a and b are continuity points of F, and X has CDF F, then $\lim_{c \to 0} \Pr(a < X + cY \leq b) = \Pr(a < X \leq b)$. The same is true of G. Hence, $F = G$ by Lemma B.105. $\square$

An obvious consequence of the uniqueness theorem is the following.

Corollary B.107.[48] *Suppose that F and G are k-dimensional CDFs such that for every bounded continuous f, $\int f(x)dF(x) = \int f(x)dG(x)$. Then $F = G$.*

B.5 Stochastic Processes

B.5.1 Introduction

Sometimes we wish to specify a joint distribution for an infinite sequence of random variables. Let $(S, \mathcal{A}, \mu)$ be a probability space. If $X_n : S \to \mathbb{R}$ for every n and each X_n is measurable with respect to the Borel σ-field $\mathcal{B}$, we can define a σ-field of subsets of $\mathbb{R}^\infty$ such that the infinite sequence $X = (X_1, X_2, \ldots)$ is measurable. Let $\mathcal{B}^\infty$ be the smallest σ-field that contains all finite-dimensional orthants, that is, every set B of the form

$$\{x : x_{i_1} \leq c_1, \ldots, x_{i_n} \leq c_n, \text{ for some } n \text{ and some integers } i_1, \ldots, i_n$$

and some numbers $c_1, \ldots, c_n\}.$

It is clear that $X^{-1}(B) \in \mathcal{A}$ since it is the intersection of finitely many sets in $\mathcal{A}$. By Theorem A.34, it follows that $X^{-1}(\mathcal{B}^\infty) \subseteq \mathcal{A}$, so X is measurable with respect to this σ-field.

B.5.2 Martingales[+]

A particular type of stochastic process that is sometimes of interest is a *martingale*. [For more discussion of martingales, see Doob (1953), Chapter VII.]

---

[47]This corollary is used in the proof of Theorem 2.74.
[48]This corollary is used in the proof of DeFinetti's representation theorem 1.49.
[+]This section contains results that rely on the theory of martingales. It may be skipped without interrupting the flow of ideas.

Definition B.108. Let $(S, \mathcal{A}, \mu)$ be a probability space. Let $\mathcal{N}$ be a set of consecutive integers. For each $n \in \mathcal{N}$, let $\mathcal{F}_n$ be a sub-σ-field of $\mathcal{A}$ such that $\mathcal{F}_n \subseteq \mathcal{F}_{n+1}$ for all n such that n and $n+1$ are in $\mathcal{N}$. Let $\{X_n\}_{n \in \mathcal{N}}$ be a sequence of random variables such that X_n is measurable with respect to $\mathcal{F}_n$ for all n. The sequence of pairs $\{(X_n, \mathcal{F}_n)\}_{n \in \mathcal{N}}$ is called a *martingale* if, for all n such that n and $n+1$ are in $\mathcal{N}$, $\mathrm{E}(X_{n+1}|\mathcal{F}_n) = X_n$. It is called a *submartingale* if, for every n, $\mathrm{E}(X_{n+1}|\mathcal{F}_n) \geq X_n$.

Note that a martingale is also a submartingale.

Example B.109. A simple example of a martingale is the following. Let $\mathcal{N} = \{1, 2, \ldots\}$ and let $\{Y_n\}_{n=1}^{\infty}$ be independent random variables with mean 0. Let $X_n = \sum_{i=1}^{n} Y_i$. Let $\mathcal{F}_n$ be the σ-field generated by $Y_1, \ldots, Y_n$. Then,

$$\mathrm{E}(Y_1 + \cdots + Y_{n+1}|\mathcal{F}_n) = Y_1 + \cdots + Y_n = X_n,$$

since $\mathrm{E}(Y_{n+1}|\mathcal{F}_n) = 0$ by independence. If each Y_i has nonnegative finite mean, then $\mathrm{E}(X_{n+1}|\mathcal{F}_n) \geq X_n$, and we have a submartingale.

Example B.110. Another example of a martingale is the following. Let $\mathcal{N}$ be a collection of consecutive integers, and let $\{\mathcal{F}_n\}_{n \in \mathcal{N}}$ be an increasing sequence of σ-fields. Let X be a random variable with $\mathrm{E}(|X|) < \infty$. Set $X_n = \mathrm{E}(X|\mathcal{F}_n)$. By the law of total probability B.70,

$$\mathrm{E}(X_{n+1}|\mathcal{F}_n) = \mathrm{E}[\mathrm{E}(X|\mathcal{F}_{n+1})|\mathcal{F}_n] = \mathrm{E}(X|\mathcal{F}_n) = X_n,$$

so $\{(X_n, \mathcal{F}_n)\}_{n \in \mathcal{N}}$ is a martingale.

Example B.111. If $\{(X_n, \mathcal{F}_n)\}_{n \in \mathcal{N}}$ is a martingale, then

$$|X_n| = |\mathrm{E}(X_{n+1}|\mathcal{F}_n)| \leq \mathrm{E}(|X_{n+1}||\mathcal{F}_n), \tag{B.112}$$

hence $\{(|X_n|, \mathcal{F}_n)\}_{n \in \mathcal{N}}$ is a submartingale.

The following result is proven using the same argument as in Example B.111.

Proposition B.113.[49] *If $\{(X_n, \mathcal{F}_n)\}_{n \in \mathcal{N}}$ is a martingale, then $\mathrm{E}|X_n|$ is nondecreasing in n.*

The reader should note that if $\{(X_n, \mathcal{F}_n)\}_{n \in \mathcal{N}}$ is a submartingale and if $\mathcal{M} \subseteq \mathcal{N}$ is a string of consecutive integers, then $\{(X_n, \mathcal{F}_n)\}_{n \in \mathcal{M}}$ is also a submartingale. Similarly, if k is an integer (positive or negative) and $\mathcal{M} = \{n : n+k \in \mathcal{N}\}$, then $\{(X'_n, \mathcal{F}'_n)\}_{n \in \mathcal{M}}$ is a submartingale, where $X'_n = X_{n+k}$ and $\mathcal{F}'_n = \mathcal{F}_{n+k}$. This latter is just a shifting of the index set.

There are important convergence theorems that apply to many martingales and submartingales. They say that if the set $\mathcal{N}$ is infinite, then limit random variables exist. A lemma is needed to prove these theorems.[50] It puts a bound on how often a submartingale can cross an interval between two numbers. It is used to show that such crossings cannot occur infinitely often with high probability. (Infinitely many crossings of a nondegenerate interval would imply divergence of the submartingale.)

---

[49] This proposition is used in the proof of Theorem B.122.
[50] This lemma is proven by Doob (1953, Theorem VII, 3.3).

Lemma B.114 (Upcrossing lemma).[51] *Let $\mathcal{N} = \{1, \ldots, N\}$, and suppose that $\{X_n, \mathcal{F}_n\}_{n=1}^N$ is a submartingale. Let $r < q$, and define V to be the number of times that the sequence $X_1, \ldots, X_N$ crosses from below r to above q. Then*

$$E(V) \leq \frac{1}{q-r}\left(E|X_N| + |r|\right). \tag{B.115}$$

PROOF. Let $Y_n = \max\{0, X_n - r\}$ for every n. Since $g(x) = \max\{0, x\}$ is a nondecreasing convex function of x, it is easy to see (using Jensen's inequality B.17) that $\{Y_n, \mathcal{F}_n\}_{n=1}^N$ is a submartingale. Note that a consecutive set of $X_i(s)$ cross from below r to above q if and only if the corresponding consecutive set of $Y_i(s)$ cross from 0 to above $q - r$. Let $T_0(s) = 0$ and define T_m for $m = 1, 2, \ldots$ as

$$\begin{aligned} T_m(s) &= \inf\{k \leq N : k > T_{m-1}(s), Y_k(s) = 0\}, \text{ if } m \text{ is odd}, \\ T_m(s) &= \inf\{k \leq N : k > T_{m-1}(s), Y_k(s) \geq q - r\}, \text{ if } m \text{ is even}, \\ T_m(s) &= N + 1, \text{ if the corresponding set above is empty}. \end{aligned}$$

Now $V(s)$ is one-half of the largest even m such that $T_m(s) \leq N$. Define, for $i = 1, \ldots, N$,

$$R_i(s) = \begin{cases} 1 & \text{if } T_m(s) < i \leq T_{m+1}(s) \text{ for } m \text{ odd}, \\ 0 & \text{otherwise}. \end{cases}$$

Then $(q - r)V(s) \leq \sum_{i=1}^N R_i(s)(Y_i(s) - Y_{i-1}(s)) = \hat{X}$, where $Y_0 \equiv 0$ for convenience. First, note that for all m and i, $\{T_m(s) \leq i\} \in \mathcal{F}_i$. Next, note that for every i,

$$\{s : R_i(s) = 1\} = \bigcup_{m \text{ odd}} \left(\{T_m \leq i - 1\} \cap \{T_{m+1} \leq i - 1\}^C\right) \in \mathcal{F}_{i-1}. \tag{B.116}$$

$$\begin{aligned} E(\hat{X}) &= \sum_{i=1}^N \int_{\{s:R_i(s)=1\}} (Y_i(s) - Y_{i-1}(s))d\mu(s) \\ &= \sum_{i=1}^N \int_{\{s:R_i(s)=1\}} (E(Y_i|\mathcal{F}_{i-1})(s) - Y_{i-1}(s))d\mu(s) \\ &\leq \sum_{i=1}^N \int (E(Y_i|\mathcal{F}_{i-1})(s) - Y_{i-1}(s))d\mu(s) \\ &= \sum_{i=1}^N (E(Y_i) - E(Y_{i-1})) = E(Y_N), \end{aligned}$$

where the second equality follows from (B.116) and the inequality follows from the fact that $\{Y_n, \mathcal{F}_n\}_{n=1}^N$ is a submartingale. It follows that $(q-r)E(V) \leq E(Y_N)$. Since $E(Y_N) \leq |r| + E(|X_N|)$, it follows that (B.115) holds. □

The proof of the following convergence theorem is adapted from Chow, Robbins, and Siegmund (1971).

---

[51] This lemma is used in the proofs of Theorems B.117 and B.122.

Theorem B.117 (Martingale convergence theorem: part I).[52] *Suppose that $\{(X_n, \mathcal{F}_n)\}_{n=1}^{\infty}$ is a submartingale such that $\sup_n \mathrm{E}|X_n| < \infty$. Then $X = \lim_{n \to \infty} X_n$ exists a.s. and $\mathrm{E}|X| < \infty$.*

PROOF. Let $X^* = \limsup_{n \to \infty} X_n$ and $X_* = \liminf_{n \to \infty} X_n$. Let $B = \{s : X_*(s) < X^*(s)\}$. We will prove that $\mu(B) = 0$. We can write

$$B = \bigcup_{r < q,\ r, q \text{ rational}} \{s : X^*(s) \geq q > r \geq X_*(s)\}.$$

Now, $X^*(s) > q > r \geq X_*(s)$ if and only if the values of $X_n(s)$ cross from being below r to being above q infinitely often. For fixed r and q, we now prove that this has probability 0; hence $\mu(B) = 0$. Let V_n equal the number of times that $X_1, \ldots, X_n$ cross from below r to above q. According to Lemma B.114,

$$\sup_n \mathrm{E}(V_n) \leq \frac{1}{q-r} \left(\sup_n \mathrm{E}(|X_n|) + |r| \right) < \infty.$$

The number of times the values of $\{X_n(s)\}_{n=1}^{\infty}$ cross from below r to above q equals $\lim_{n \to \infty} V_n(s)$. By the monotone convergence theorem A.52,

$$\infty > \sup_n \mathrm{E}(V_n) = \mathrm{E}(\lim_{n \to \infty} V_n).$$

It follows that $\mu(\{s : \lim_{n \to \infty} V_n(s) = \infty\}) = 0$.

Since $\mu(B) = 0$, we have that $X = \lim_{n \to \infty} X_n$ exists a.s. Fatou's lemma A.50 says $\mathrm{E}(|X|) \leq \liminf_{n \to \infty} \mathrm{E}(|X_n|) \leq \sup_n \mathrm{E}(|X_n|) < \infty$. □

For the particular martingale in which $X_n = \mathrm{E}(X|\mathcal{F}_n)$ for a single X, we have an expression for the limit.

Theorem B.118 (Lévy's theorem: part I).[53] *Let $\{\mathcal{F}_n\}_{n=1}^{\infty}$ be an increasing sequence of σ-fields. Let $\mathcal{F}_{\infty}$ be the smallest σ-field containing all of the $\mathcal{F}_n$. Let $\mathrm{E}(|X|) < \infty$. Define $X_n = \mathrm{E}(X|\mathcal{F}_n)$ and $X_{\infty} = \mathrm{E}(X|\mathcal{F}_{\infty})$. Then $\lim_{n \to \infty} X_n = X_{\infty}$, a.s.*

The proof of this theorem requires a lemma that will also be needed later.

Lemma B.119.[54] *Let $\{\mathcal{F}_n\}_{n=1}^{\infty}$ be a sequence of σ-fields. Let $\mathrm{E}(|X|) < \infty$. Define $X_n = \mathrm{E}(X|\mathcal{F}_n)$. Then $\{X_n\}_{n=1}^{\infty}$ is a uniformly integrable sequence.*

PROOF. Since $\mathrm{E}(X|\mathcal{F}_n) = \mathrm{E}(X^+|\mathcal{F}_n) - \mathrm{E}(X^-|\mathcal{F}_n)$, and the sum of uniformly integrable sequences is uniformly integrable, we will prove the result for nonnegative X. Let $A_{c,n} = \{X_n \geq c\} \in \mathcal{F}_n$. So $\int_{A_{c,n}} X_n(s) d\mu(s) = \int_{A_{c,n}} X(s) d\mu(s)$. If we can find, for every $\epsilon > 0$, a C such that $\int_{A_{c,n}} X(s) d\mu(s) < \epsilon$ for all n and all $c \geq C$, we are done. Define $\eta(A) = \int_A X(s) d\mu(s)$. We have $\eta \ll \mu$ and η is finite.

---

[52] This theorem is used in the proof of Theorems B.118 and 1.121.

[53] This theorem is used in the proofs of Theorem 7.78 and Lemma 7.124.

[54] This lemma is used in the proofs of Theorems B.118, B.122, and B.124. It is borrowed from Billingsley (1986, Lemma 35.2).

By Lemma A.72, we have that for every $\epsilon > 0$ there exists δ such that $\mu(A) < \delta$ implies $\eta(A) < \epsilon$. By the Markov inequality B.15,

$$\mu(A_{c,n}) \leq \frac{1}{c}\mathrm{E}(X_n) = \frac{1}{c}\mathrm{E}(X),$$

for all n. Let $C = \mathrm{E}(X)/[2\delta]$. Then $c \geq C$ implies $\mu(A_{c,n}) < \delta$ for all n, so $\eta(A_{c,n}) < \epsilon$ for all n. □

PROOF OF THEOREM B.118. By Lemma B.119, $\{X_n\}_{n=1}^{\infty}$ is a uniformly integrable sequence. Let Y be the limit of the martingale guaranteed by Theorem B.117. Since Y is a limit of functions of the X_n, it is measurable with respect to $\mathcal{F}_\infty$. It follows from Theorem A.60 that for every event A, $\lim_{n\to\infty} \mathrm{E}(X_n I_A) = \mathrm{E}(Y I_A)$. Next, note that, for every $A \in \mathcal{F}_n$,

$$\int_A Y(s) d\mu(s) = \lim_{n\to\infty} \int_A \mathrm{E}(X|\mathcal{F}_n)(s) d\mu(s) = \int_A X(s) d\mu(s),$$

where the last equality follows from the definition of conditional expectation. Since this is true for every n and every $A \in \mathcal{F}_n$, it is true for all A in the field $\mathcal{F} = \cup_{n=1}^{\infty} \mathcal{F}_n$. Since $|X|$ is integrable, we can apply Theorem A.26 to conclude that the equality holds for all $A \in \mathcal{F}_\infty$, the smallest σ-field containing $\mathcal{F}$. The equality $\mathrm{E}(X I_A) = \mathrm{E}(Y I_A)$ for all $A \in \mathcal{F}_\infty$ together with the fact that Y is $\mathcal{F}_\infty$ measurable is precisely what it means to say that $Y = \mathrm{E}(X|\mathcal{F}_\infty) = X_\infty$. □

For negatively indexed martingales, there is also a convergence theorem. Some authors refer to negatively indexed martingales in a different fashion, which is often more convenient.

Definition B.120. Let $(S, \mathcal{A}, \mu)$ be a probability space. For each $n = 1, 2, \ldots$, let $\mathcal{F}_n$ be a sub-σ-field of $\mathcal{A}$ such that $\mathcal{F}_{n+1} \subseteq \mathcal{F}_n$ for all n. Let $\{X_n\}_{n=1}^{\infty}$ be a sequence of random variables such that X_n is measurable with respect to $\mathcal{F}_n$ for all n. The sequence of pairs $\{(X_n, \mathcal{F}_n)\}_{n=1}^{\infty}$ is called a *reversed martingale* if for all n $\mathrm{E}(X_n|\mathcal{F}_{n+1}) = X_{n+1}$.

Example B.121. As in Example B.110, we can let $\{\mathcal{F}_n\}_{n=1}^{\infty}$ be a decreasing sequence of σ-fields, and let $\mathrm{E}(|X|) < \infty$. Define $X_n = \mathrm{E}(X|\mathcal{F}_n)$. It follows from the law of total probability B.70 that $\{(X_n, \mathcal{F}_n)\}_{n=1}^{\infty}$ is a reversed martingale.

The following theorem is proven by Doob (1953, Theorem VII 4.2).

Theorem B.122 (Martingale convergence theorem: part II).[55] *Suppose that $\{(X_n, \mathcal{F}_n)\}_{n<0}$ is a martingale with $\mathrm{E}|X_{-1}| < \infty$. Then $X = \lim_{n\to-\infty} X_n$ exists a.s. and is finite with probability 1.*

PROOF. Just as in the proof of Theorem B.117, we let V_n be the number of times that the finite sequence $X_n, X_{n+1}, \ldots, X_{-1}$ crosses from below a rational r to above another rational q (for $n < 0$). The upcrossing lemma B.114 says that

$$\mathrm{E}(V_n) \leq \frac{1}{q-r}\left(\mathrm{E}(|X_{-1}|) + |r|\right) < \infty.$$

---

[55] This theorem is used in the proof of Theorem B.124.

As in the proof of Theorem B.117, it follows that $X = \lim_{n \to -\infty} X_n$ exists with probability 1. From (B.112) and Lemma B.119, it follows that

$$\lim_{n \to -\infty} E(|X_n|) = E(|X|).$$

By Proposition B.113, it follows that $E(|X|) < \infty$, and so X is finite with probability 1. □

It is usually more convenient to express Theorem B.122 in terms of reversed martingales.

Corollary B.123.[56] *If $\{(X_n, \mathcal{F}_n)\}_{n=1}^{\infty}$ is a reversed martingale with $E|X_1| < \infty$, then $X = \lim_{n \to \infty} X_n$ exists a.s. and is finite with probability 1.*

There is also a version of Lévy's theorem B.118 for reversed martingales.

Theorem B.124 (Lévy's theorem: part II).[57] *Let $\{\mathcal{F}_n\}_{n=1}^{\infty}$ be a decreasing sequence of σ-fields. Let $\mathcal{F}_\infty$ be the intersection $\cap_{n=1}^{\infty} \mathcal{F}_n$. Let $E(|X|) < \infty$. Define $X_n = E(X|\mathcal{F}_n)$ and $X_\infty = E(X|\mathcal{F}_\infty)$. Then $\lim_{n \to \infty} X_n = X_\infty$ a.s. and X_∞ is finite a.s.*

PROOF. It is easy to see that $\{(X_n, \mathcal{F}_n)\}_{n=1}^{\infty}$ is a reversed martingale and that $E(|X_1|) < \infty$. By Theorem B.122, it follows that $\lim_{n \to -\infty} X_n = Y$ exists and is finite a.s. To prove that $Y = X_\infty$ a.s., note that $X_\infty = E(X_1|\mathcal{F}_\infty)$ since $\mathcal{F}_\infty \subseteq \mathcal{F}_1$. So, we must show that $Y = E(X_1|\mathcal{F}_\infty)$. Let $A \in \mathcal{F}_\infty$. Then

$$\int_A X_n(s) d\mu(s) = \int_A X_1(s) d\mu(s),$$

since $A \in \mathcal{F}_n$ and $X_n = E(X_1|\mathcal{F}_n)$. Once again, using (B.112) and Lemma B.119, it follows that $\int_A Y(s) d\mu(s) = \int_A X_1(s) d\mu(s)$; hence $Y = E(X_1|\mathcal{F}_\infty)$. □

B.5.3 Markov Chains*

Another type of stochastic process we will occasionally meet is a *Markov chain*.[58]

Definition B.125. Let $\{X_n\}_{n=1}^{\infty}$ be a sequence of random variables taking values in a space $\mathcal{X}$ with σ-field $\mathcal{B}$. The sequence is called a *Markov chain (with stationary transition distributions)*[59] if there exists a function $p : \mathcal{B} \times \mathcal{X} \to [0,1]$ such that

- for all $x \in \mathcal{X}$, $p(\cdot, x)$ is a probability measure on $\mathcal{B}$;
- for all $B \in \mathcal{B}$, $p(B, \cdot)$ is $\mathcal{B}$ measurable;

---

[56] This corollary is used in the proof of Theorem B.124.

[57] This theorem is used in the proofs of Theorem 1.62, Corollary 1.63, Lemma 2.121, and Lemma 7.83.

*This section may be skipped without interrupting the flow of ideas.

[58] In this text, we only use Markov chains as occasional examples of sequences of random variables that are not exchangeable.

[59] There are more general definitions of Markov chains and Markov processes in which the transition distribution from X_n to X_{n+1} is allowed to depend on n. We will not need these more general processes in this book.

- for each n and each $B \in \mathcal{B}$,

$$p(B,x) = \Pr(X_{n+1} \in B | X_1 = x_1, X_2 = x_2, \ldots, X_{n-1} = x_{n-1}, X_n = x),$$

almost surely with respect to the joint distribution of $(X_1, \ldots, X_n)$.

The last condition in the definition of a Markov chain says that the conditional distribution of X_{n+1} given the past depends only on the most recent past X_n. In other words, X_{n+1} is conditionally independent of $X_1, \ldots, X_{n-1}$ given X_n.

Example B.126. A sequence $\{X_n\}_{n=1}^\infty$ of IID random variables is a Markov chain with $p(B,x) = \Pr(X_i \in B)$ for all x.

Example B.127. Let $\{X_n\}_{n=1}^\infty$ be Bernoulli random variables such that

$$\Pr(X_{n+1} = 1 | X_1 = x_1, \ldots, X_n = x_n) = p_{x_n, 1},$$

for $x_n \in \{0, 1\}$. The entire joint distribution of the sequence is determined by the numbers $p_{0,1}$, $p_{1,1}$, and $\Pr(X_1 = 1)$.

B.5.4 General Stochastic Processes

Occasionally, we will have to deal with more complicated stochastic processes. What makes them more complicated is that they consist of more than countably many random quantities.

Example B.128. Let $\mathcal{F}$ be a set of real-valued functions of a real vector. That is, there exists k such that $F \in \mathcal{F}$ means $F : \mathbb{R}^k \to \mathbb{R}$. Suppose that $X : S \to \mathcal{F}$ is a random quantity whose values are functions themselves. We would like to be able to discuss the distribution of X. We will need a σ-field of subsets of $\mathcal{F}$ in order to discuss measurability. A natural σ-field is the smallest σ-field that contains all sets of the form $A_{t,x} = \{F \in \mathcal{F} : F(t) \leq x\}$, for all $t \in \mathbb{R}^k$ and all $x \in \mathbb{R}$. It can be shown (see below) that X is measurable with respect to this σ-field if, for every $t \in \mathbb{R}^k$, the real-valued function $G_t : S \to \mathbb{R}$ is Borel measurable, where $G_t(s) = F(t)$ when $X(s) = F$.

A general stochastic process can be defined, and it resembles the above example in all important aspects.

Definition B.129. Let $(S, \mathcal{A}, \mu)$ be a probability space, and let R be some set. For each $r \in R$, let $(\mathcal{X}_r, \mathcal{B}_r)$ be a Borel space, and let $X_r : S \to \mathcal{X}_r$ be measurable. The collection of random variables $X = \{X_r : r \in R\}$ is called a *stochastic process*.

Example B.130. If every $(\mathcal{X}_r, \mathcal{B}_r)$ is the same space $(\mathcal{X}, \mathcal{B})$, then X can be thought of as a "random function" from R to $\mathcal{X}$ as follows. For each $s \in S$, define the function $F_s : R \to S$ by $F_s(r) = X_r(s)$. In order to make this a true random function, we need a σ-field on the set of functions from R to $\mathcal{X}$. Since this set of functions is the product set $\mathcal{X}^R$, a natural σ-field is the product σ-field $\mathcal{B}^R$. The product σ-field is easily seen to be the smallest σ-field containing all sets of the form $A_{r,B} = \{F : F(r) \in B\}$, for $r \in R$ and $B \in \mathcal{B}$. Now, let $\mathbf{F} : S \to \mathcal{X}^R$ be defined by $\mathbf{F}(s) = F_s$. Then $\mathbf{F}$ is measurable because

$$\mathbf{F}^{-1}(A_{r,B}) = \{s : F_s(r) \in B\} = \{s : X_r(s) \in B\} \in \mathcal{A},$$

because X_r is measurable.

The important theorem about stochastic processes is that their distribution is determined by the joint distributions of all finite collections of the X_r.

Theorem B.131.[60] *Let R be a set and, for each $r \in R$, let $(\mathcal{X}_r, \mathcal{B}_r)$ be a Borel space. Let $X = \{X_r : r \in R\}$ and $X' = \{X'_r : r \in R\}$ be two stochastic processes. Suppose that for every k and every k-tuple $(r_1, \ldots, r_k) \in R^k$, the joint distribution of $(X_{r_1}, \ldots, X_{r_k})$ is the same as that of $(X'_{r_1}, \ldots, X'_{r_k})$. Then the distribution of X is the same as that of X'.*

PROOF. Define $\mathcal{X} = \prod_{r \in R} \mathcal{X}_r$ and let $\mathcal{B}$ be the product σ-field. Say that a set $C \in \mathcal{B}$ is a *finite-dimensional cylinder set* if there exists k and $r_1, \ldots, r_k \in R$ and a measurable $D \subseteq \prod_{i=1}^{k} \mathcal{X}_{r_i}$ such that

$$C = \{x \in \mathcal{X} : (x_{r_1}, \ldots, x_{r_k}) \in D\}.$$

It is easy to see that if $\{r_1, \ldots, r_k\} \subseteq \{t_1, \ldots, t_m\}$ for $m \geq k$, then there exists a measurable subset D' of $\prod_{j=1}^{m} \mathcal{X}_{s_j}$ such that

$$C = \{x \in \mathcal{X} : (x_{s_1}, \ldots, x_{s_m}) \in D'\},$$

by taking the Cartesian product of D times the product of those $\mathcal{X}_r$ for $r \in \{s_1, \ldots, s_m\} \setminus \{r_1, \ldots, r_k\}$ and then possibly rearranging the coordinates of all points in this set to match the order of $r_1, \ldots, r_k$ among $s_1, \ldots, s_m$. So, if C and G are both finite-dimensional cylinder sets with

$$G = \{x \in \mathcal{X} : (x_{h_1}, \ldots, x_{h_\ell}) \in E\},$$

then we can let $\{t_1, \ldots, t_m\} = \{r_1, \ldots, r_k\} \cup \{h_1, \ldots, h_\ell\}$ and write

$$\begin{aligned} C &= \{x \in \mathcal{X} : (x_{t_1}, \ldots, x_{t_m}) \in D'\}, \\ G &= \{x \in \mathcal{X} : (x_{t_1}, \ldots, x_{t_m}) \in G'\}. \end{aligned}$$

It follows that

$$C \cap G = \{x \in \mathcal{X} : (x_{t_1}, \ldots, x_{t_m}) \in D' \cap G'\}.$$

So the finite-dimensional cylinder sets form a π-system. By assumption, the distributions of X and X' agree on this π-system. Since $\mathcal{X} = \{x \in \mathcal{X} : x_r \in \mathcal{X}_r\}$ for arbitrary $r \in R$ and since the distributions of X and X' are finite measures, we can apply Theorem A.26 to conclude that the distributions are the same. □

Another important fact about general stochastic processes is that it is possible to specify a joint distribution for the entire process by merely specifying all of the finite-dimensional joint distributions, so long as they obey a consistency condition.

Definition B.132. Let $\mathcal{X} = \prod_{r \in R} \mathcal{X}_r$ with the product σ-field, where $(\mathcal{X}_r, \mathcal{B}_r)$ is a Borel space for every r. For each finite k and each k-tuple $(i_1, \ldots, i_k)$ of distinct elements of R, let $P_{i_1, \ldots, i_k}$ be a probability measure on $\prod_{j=1}^{k} \mathcal{X}_{i_j}$. We say that these probabilities are *consistent* if the following conditions hold for each k and distinct $i_1, \ldots, i_k \in R$ and each A in the product σ-field of $\prod_{j=1}^{k} \mathcal{X}_{i_j}$:

---

[60]This theorem is used in the proofs of Theorem B.133 and DeFinetti's representation theorem 1.49.

- For each permutation π of k items, $P_{i_1,\ldots,i_k}(A) = P_{i_{\pi(1)},\ldots,i_{\pi(k)}}(B)$, where
$$B = \{(x_{\pi(1)},\ldots,x_{\pi(k)}) : (x_1,\ldots,x_k) \in A\}.$$

- For each $\ell \in R \setminus \{i_1,\ldots,i_k\}$, $P_{i_1,\ldots,i_k}(A) = P_{i_1,\ldots,i_k,\ell}(B)$, where
$$B = \{(x_1,\ldots,x_k,x_{k+1}) : (x_1,\ldots,x_k) \in A, x_{k+1} \in \mathcal{X}_\ell\}.$$

Since the set R may not be ordered, the first condition ensures that it does not matter in what order one writes a finite set of indices. The second condition is the substantive one, and it ensures that the marginal distributions of subsets of coordinates are the probability measures associated with those subsets.

To avoid excessive notation, it will be convenient to refer to P_J as the probability measure associated with a finite subset $J \subseteq R$ without specifying the order of the elements of J. When the consistency conditions in Definition B.132 hold, this should not cause any confusion.

The proof of the following theorem is adapted from Loève (1977, pp. 94–5). The theorem says that consistent finite-dimensional distributions determine a unique joint distribution on the product space.

Theorem B.133.[61] *Let $\mathcal{X} = \prod_{r \in R} \mathcal{X}_r$ with the product σ-field, where $\mathcal{X}_r$ is a Borel space for every r. For each finite subset $J \subseteq R$, let P_J be a probability measure on $\prod_{r \in J} \mathcal{X}_r$. Suppose that the P_J are consistent as defined in Definition B.132. Then there exists a unique distribution on $\mathcal{X}$ with finite-dimensional marginals given by the P_J.*

PROOF. The uniqueness follows from Theorem B.131, if we can prove existence. We begin by proving the existence in the special case $\mathcal{X}_r = \mathbb{R}$ for all r. Let $\mathcal{C}$ be the class of all unions of finitely many finite-dimensional cylinder sets of the form $C = \prod_{r \in R} C_r$, where all but finitely many of the C_r equal $\mathbb{R}$ and the others are unions of finitely many intervals. The class $\mathcal{C}$ is a field. For $C \in \mathcal{C}$, define $P(C) = P_J(\prod_{r \in J} C_r)$. We need to show that P is countably additive. Equivalently, we need to show that if $\{A_n\}_{n=1}^\infty$ is a decreasing sequence of elements of $\mathcal{C}$ such that $P(A_n) > \epsilon$ for all n, then $A = \cap_{n=1}^\infty A_n$ is nonempty. Suppose that $P(A_n) > \epsilon$ for all n. Let J_n be the set of all subscripts involved in $A_1,\ldots,A_n$ and J be the union of these sets. Let $A_n = B_n \times \prod_{r \notin J_n} \mathcal{X}_r$. Then $P(A_n) = P_{J_n}(B_n)$, and B_n is the union of finitely many products of intervals. For each product of intervals H that constitutes B_n, we can find a product of bounded closed intervals contained in H such that the P_{J_n} probability of the union of these H is as close as we wish to $P_{J_n}(B_n)$. Let C_n be a finite union of products of closed bounded intervals contained in B_n such that $P_{J_n}(B_n \setminus C_n) < \epsilon/2^{n+1}$. Let D_n be the cylinder set corresponding to C_n. Then
$$P_{J_n}(A_n \setminus D_n) = P_{J_n}(B_n \setminus C_n) < \frac{\epsilon}{2^{n+1}}.$$

Now, let $E_n = A_n \cap_{i=1}^n D_i$, so that $P(A_n \setminus E_n) < \epsilon/2$. It follows that $P(E_n) > \epsilon/2$, so each E_n is nonempty. Let $x^n = (x_1^n, x_2^n, \ldots) \in E_n$. Since $E_1 \supseteq E_2 \supseteq \cdots$, it follows that for every $k \geq 0$, $x^{n+k} \in E_n \subseteq D_n$. Hence $(x_i^{n+k}; i \in J_n) \in C_n$. Since

---

[61] This theorem is used in the proof of Lemma 2.123.

each C_n is bounded, there is a subsequence of $\{(x_i^n; i \in J_1)\}_{n=1}^{\infty}$ that converges to a point $(x_i; i \in J_1) \in C_1$. Let the subsequence be $\{(x_i^{n'_k}; i \in J_1)\}_{k=1}^{\infty}$. Then there is a subsequence of $\{(x_i^{n''_k}; i \in J_2)\}_{k=1}^{\infty}$ that converges to a point $(x_i; i \in J_2) \in C_2$. Continue extracting subsequences to get a limit point $x_J = (x_i; i \in J) \in D_n$ for all n. Hence, every point that extends x_J to an element of $\mathcal{X}$ is in A_n for all n, and A is nonempty. Now apply the Caratheodory extension theorem A.22 to extend P to the entire product σ-field.

For general Borel spaces, let $\phi_r : \mathcal{X}_r \to F_r$ be a bimeasurable mapping to a Borel subset of $\mathbb{R}$ for each r. It follows easily by using Theorem A.34 that the function $\phi : \mathcal{X} \to \prod_{r \in R} F_r$ is bimeasurable, where $\phi(x) = (\phi_r(x_r); r \in R)$. For each finite subset J, ϕ induces a probability on $\prod_{i \in J} \mathbb{R}$ from P_J, and these are clearly consistent. By what we have already proven there is a probability P on $\prod_{r \in R} \mathbb{R}$ with the desired marginals. Then ϕ^{-1} induces a probability on $\mathcal{X}$ from P with the desired marginals. □

B.6 Subjective Probability

It is not obvious for what purpose a mathematical probability, as described in this chapter and defined in Definition A.18, would ever be useful. In this section, we try to show how the mathematical definition of probability is just what one would want to use to describe one's uncertainty about unknown quantities if one were forced to gamble on the outcomes of those unknown quantities.[62]

DeFinetti (1974) suggests that probability be *defined* in terms of those gambles an agent is willing to accept. Others, like DeGroot (1970), would only require that probabilities be subjective degrees of belief. Either way, we might ask, "Why should degrees of belief or gambling behavior satisfy the measure theoretic definition of probability?" In this section, we will try to motivate the measure theoretic definition of probability by considering gambling behavior. We begin by adopting the viewpoint of DeFinetti (1974).[63]

For the purposes of this discussion, let a *random variable* be any number about which we are uncertain. For each bounded random variable X, assume that there is some fair price p such that an agent is indifferent between all gambles that pay $c(X - p)$, where c is in some sufficiently small symmetric interval around 0 such that the maximum loss is still within the means of the agent to pay. For example, suppose that $X = x$ is observed. If $c(x - p) > 0$, then the agent would receive this amount. If $c(x - p) < 0$, then the agent would lose $-c(x - p)$. It must be that $-c(x - p)$ is small enough for the agent to be able to pay. Surely, for x in a

---

[62] In Section 3.3, we give a much more elaborate motivation for the entire apparatus of Bayesian decision theory, which includes mathematical probability as one of its components. An alternative derivation of mathematical probability from operational considerations is given in Chapter 6 of DeGroot (1970).

[63] There are a few major differences between the approach in this section and DeFinetti's approach, which DeFinetti, were he alive, would be quick to point out. Out of respect for his memory and his followers, we will also try to point out these differences as we encounter them.

bounded set, c can be made small enough for this to hold, so long as the agent has some funds available.

Definition B.134. The fair price p of a random quantity is called its *prevision* and is denoted $P(X)$. It is assumed, for a bounded random quantity X, that the agent is indifferent between all gambles whose net gain (loss if negative) to the agent is $c(X - P(X))$ for all c in some symmetric interval around 0.

The symmetric interval around 0 mentioned in the definition of prevision may be different for different random variables. For example, it might stand to reason that the interval corresponding to the random variable $2X$ would be half as wide as the interval corresponding to X.

Another assumption we make is that if an agent is willing to accept each of a countable collection of gambles, then the agent is willing to accept all of them at once, so long as the maximum possible loss is small enough for the agent to pay.[64] An example of countably many gambles, each of which is acceptable but cannot be accepted together, is the famous St. Petersburg paradox.

Example B.135. Suppose that a fair coin is tossed until the first head appears. Let N be the number of tosses until the first head appears. For $n = 1, 2, \ldots$, define
$$X_n = \begin{cases} 2^n & \text{if } N = n, \\ 0 & \text{otherwise}. \end{cases}$$
Suppose that our agent says that $P(X_n) = 1$ for all n. For each n, there is $c_n < 0$ such that the agent is willing to accept $c_n(X_n - 1)$. If $-\sum_{n=1}^{\infty} c_n 2^n$ is too big, however, the agent cannot accept all of the gambles at once. Similarly, there are $c_n > 0$ such that the agent is willing to accept $c_n(X_n - 1)$. If $\sum_{n=1}^{\infty} c_n$ is too big, the agent cannot accept all of these gambles. The St. Petersburg paradox corresponds to the case in which $c_n = 1$ for all n. In this case, the agent pays ∞ and only receives 2^N in return. We have ruled out this possibility by requiring that the agent be able to afford the worst possible loss.

The following example illustrates how it is possible to accept infinitely many gambles at once.

Example B.136. Suppose that a random quantity X could possibly be any one of the positive integers. For each positive integer x, let
$$I_x = \begin{cases} 1 & \text{if } X = x, \\ 0 & \text{if not}. \end{cases}$$
Suppose that our agent is indifferent between all gambles of the form $c(I_x - 2^{-x})$ for all $-1 \leq c \leq 1$ and all integers x. Then, we assume that the agent is also indifferent between all gambles of the form $\sum_{x=1}^{\infty} c_x(I_x - 2^{-x})$, so long as $-1 \leq c_x \leq 1$ for all x. (Note that the largest possible loss is no more than 1.) Let $Y = \sum_{x=1}^{\infty} c_x I_x$ with $-1 \leq c_x \leq 1$ for all x. Note that Y is a bounded random

---

[64]DeFinetti would not require an agent to accept countably many gambles at once, but rather only finitely many. We introduce this stronger requirement to avoid mathematical problems that arise when the weaker assumption holds but the stronger one does not. Schervish, Seidenfeld, and Kadane (1984) describe one such problem in detail.

quantity, and that the agent has implicitly agreed to accept all gambles of the form $c(Y - \mu)$ for $-1 \leq c \leq 1$, where $\mu = \sum_{x=1}^{\infty} c_x 2^{-x}$. If the agent were foolish enough to be indifferent between all gambles of the form $d(Y - p)$ for $-a \leq d \leq a$ where $p \neq \mu$, then a clever opponent could make money with no risk. For example, if $p > \mu$, let $f = \min\{1, a\}$. The opponent would ask the agent to accept the gamble $f(Y-p)$ as well as the gambles $-fc_x(I_x - 2^{-x})$ for $x = 1, 2, \ldots$. The net effect to the agent of these gambles is $-f(p - \mu) < 0$, no matter what value X takes! A similar situation arises if $p < \mu$. Only $p = \mu$ protects the agent from this sort of problem, which is known as *Dutch book*.

To avoid Dutch book, we introduce the following definition.

Definition B.137. Let $\{X_\alpha : \alpha \in A\}$ be a collection of random variables. Suppose that, for each α, an agent gives a prevision $P(X_\alpha)$ and is indifferent between all gambles of the form $c(X_\alpha - P(X_\alpha))$ for $-d_\alpha \leq c \leq d_\alpha$. These previsions are *coherent* if there exists no countable subset $B \subseteq A$ and $\{c_b : -d_b \leq c_b \leq d_b,$ for all $b \in B\}$ such that $\sum_{b \in B} c_b(X_b - P(X_b)) < 0$ under all circumstances.[65] If a collection of previsions is not coherent, we say that it is *incoherent*.

Coherence of a sufficiently rich collection of previsions is equivalent to a probability assignment.

Theorem B.138.[66] *Let $(S, \mathcal{A})$ be a measurable space. Suppose that, for each $C \in \mathcal{A}$, the agent assigns a prevision $P(I_C)$, where I_C is the indicator of C. Define $\mu : \mathcal{C} \to \mathbb{R}$ by $\mu(C) = P(I_C)$. Then the previsions are coherent if and only if μ is a probability on $(S, \mathcal{A})$.*

PROOF. Without loss of generality, suppose that the agent is indifferent between all gambles of the form $c(I_C - P(I_C))$, for all $-1 \leq c \leq 1$. For the "if" part, assume that μ is a probability. Let $\{C_n\}_{n=1}^{\infty} \in \mathcal{A}$ and $c_i \in [-1, 1]$ be such that with

$$X = \sum_{n=1}^{\infty} c_n(I_{C_n} - \mu(C_n)),$$

the maximum losses from X and from $-X$ are small enough for the agent to afford. Since this makes X bounded, it follows from Fubini's theorem A.70 that $E(X) = 0$; hence it is impossible that $X < 0$ under all circumstances, and the previsions are coherent.

For the "only if" part, assume that the previsions are coherent. Clearly, $\mu(\emptyset) = 0$, since $I_\emptyset = 0$ and $-c\mu(\emptyset) \geq 0$ for both positive and negative c. It is also easy to see that $\mu(A) \geq 0$ for all A. If $\mu(A) < 0$, then for all negative c, $c(I_A - \mu(A)) < 0$ and we have incoherence. Countable additivity follows in a similar fashion. Let $\{A_n\}_{n=1}^{\infty}$ be mutually disjoint, and let $A = \cup_{n=1}^{\infty} A_n$. If $\mu(A) < \sum_{n=1}^{\infty} \mu(A_n)$,

---

[65] When only finitely many gambles are required to be combined at once, as by DeFinetti (1974), incoherence requires that the sum be strictly less than some negative number under all circumstances. That is, DeFinetti would allow a strictly negative gamble to be called coherent, so long as the least upper bound was 0.

[66] This theorem is used in the proof of Theorem B.139.

then the following gamble is always negative:

$$\sum_{n=1}^{\infty}(I_{A_n} - \mu(A_n)) - (I_A - \mu(A)).$$

If $\mu(A) > \sum_{n=1}^{\infty} \mu(A_n)$, then the negative of the above gamble is always negative. Either way there is incoherence. □

Theorem B.138 says that if an agent insists on dealing with a σ-field of subsets of some set S, then expressing coherent previsions for gambles on events is equivalent to choosing probabilities.[67] Similar claims can be made about bounded random variables.

Theorem B.139. *Let $\mathcal{C}$ be the collection of all bounded measurable functions from a measurable space $(S, \mathcal{A})$ to $\mathbb{R}$. Suppose that, for each $X \in \mathcal{C}$, an agent assigns a prevision $P(X)$. The previsions are coherent if and only if there exists a probability μ on $(S, \mathcal{A})$ such that $P(X) = \mathrm{E}(X)$ for all $X \in \mathcal{C}$.*

PROOF. Suppose that the agent is indifferent between all gambles of the form $c(X - P(X))$ for $-d_X \leq c \leq d_X$. For the "if" direction, the proof is virtually identical to the corresponding part of the proof of Theorem B.138. For the "only if" part, note that $I_A \in \mathcal{C}$ for every $A \in \mathcal{A}$. It follows from Theorem B.138 that a probability μ exists such that $\mu(A) = P(I_A)$ for all $A \in \mathcal{A}$. Hence $P(X) = \mathrm{E}(X)$ for all simple functions X. Let $X > 0$ and let $X_1 \leq X_2 \leq \cdots$ be simple functions less than or equal to X such that $\lim_{n \to \infty} X_n = X$. Then $X = \sum_{n=1}^{\infty}(X_{n+1} - X_n)$, so

$$P(X) = \sum_{n=1}^{\infty} P(X_{n+1} - X_n) = \lim_{n \to \infty} \mathrm{E}(X_{n+1}) = \mathrm{E}(X),$$

from coherence and the monotone convergence theorem A.52. For general X, let X^+ and X^- be, respectively, the positive and negative parts of X. Since $P(X) = P(X^+) - P(X^-)$ follows easily from coherence, the proof is complete. □

We conclude this "motivation" of probability theory from gambling considerations by trying to motivate conditional probability. Suppose that, in addition to assigning previsions to gambles involving arbitrary bounded random variables, the agent is also required to assign *conditional previsions* in the following way. Let $\mathcal{C}$ be a sub-σ-field of $\mathcal{A}$, and suppose that gambles of the form $cI_A(X - p)$, for all nonempty $A \in \mathcal{C}$, are being considered.[68] The fair price would be that value of p, denoted $P(X|A)$, such that the agent was indifferent between all gambles of the form $cI_A(X - P(X|A))$ for all c in some symmetric interval around 0. Rather than choose a different $P(X|A)$ for each A, the agent has the option of choosing a single function $Q : S \to \mathbb{R}$ such that Q is measurable with respect to the σ-field $\mathcal{C}$. The conditional gambles would then be $cI_A(X - Q)$.

Example B.140. For the simple case in which $\mathcal{C} = \{\emptyset, A, A^C, S\}$, Q is measurable if and only if it takes on only two values, one on A and the other on A^C. In

---

[67]In the theory of DeFinetti (1974), one obtains *finitely additive probabilities* without assuming that probabilities have been assigned to all elements of a σ-field.

[68]DeFinetti (1974) would only require that such conditional gambles be considered one at a time rather than a σ-field at a time.

this case, there are only two sets of conditional gambles (other than the "unconditional" gambles $c[X-P(X)]$), namely $cI_A(X-P(X|A))$ and $cI_{A^C}(X-P(X|A^C))$. Here, $Q = P(X|A)I_A + P(X|A^C)I_A^C$. Note that the previsions $P(XI_A)$ and $P(I_A) = \mu(A)$ are already expressed. It is easy to see that

$$cI_A(X - P(X|A))$$
$$= c(XI_A - \mathrm{E}(XI_A)) - cP(X|A)(I_A - \mu(A)) + c[P(X|A)\mu(A) - \mathrm{E}(XI_A)].$$

Clearly, the only coherent choices of $P(X|A)$ satisfy $P(X|A)\mu(A) = \mathrm{E}(XI_A)$. If $\mu(A) > 0$, then $P(X|A) = \mathrm{E}(XI_A)/\mu(A)$, the usual conditional mean of X given A. Similarly, $P(X|A^C)\mu(A^C) = \mathrm{E}(XI_A^C)$ must hold.

The general situation is not much different from Example B.140.

Theorem B.141. *Suppose that an agent must choose a function Q that is measurable with respect to a sub-σ-field $\mathcal{C}$ so that for each nonempty $A \in \mathcal{C}$, he or she is indifferent between all gambles of the form $cI_A(X - Q)$. The choice of Q is coherent if and only if $\mathrm{E}(QI_A) = \mathrm{E}(XI_A)$, for all $A \in \mathcal{C}$.*

PROOF. As in Example B.140, note that

$$cI_A(X - Q) = c(XI_A - \mathrm{E}(XI_A)) - c(QI_A - \mathrm{E}(QI_A)) + c[\mathrm{E}(QI_A) - \mathrm{E}(XI_A)].$$

The choice of Q can be coherent if and only if $\mathrm{E}(QI_A) = \mathrm{E}(XI_A)$. □

The reader should note the similarity between the conditions in Theorem B.141 and Definition B.23. The function Q must be a version of the conditional mean of X given $\mathcal{C}$.

Example B.142. Let (X, Y) be random variables with a traditional joint density with respect to Lebesgue measure $f_{X,Y}$. That is, for all $C \in \mathbb{R}^2$,

$$\Pr((X,Y) \in C) = \int_C f_{X,Y}(x,y)dxdy,$$

and for all bounded measurable functions $g : \mathbb{R}^2 \to \mathbb{R}$,

$$\mathrm{E}(g(X,Y)) = \int g(x,y)f_{X,Y}(x,y)dxdy. \tag{B.143}$$

Let $\mathcal{C}$ be the σ-field generated by Y. That is, $\mathcal{C} = \{Y^{-1}(A) : A \in \mathcal{B}\}$, where $\mathcal{B}$ is the Borel σ-field of subsets of $\mathbb{R}$. It is straightforward to check that for all $A \in \mathcal{C}$, $\mathrm{E}(XI_A) = \mathrm{E}(QI_A)$, where $Q(s) = h(Y(s))$, and

$$h(y) = \int x \frac{f_{X,Y}(x,y)}{f_Y(y)} dx,$$

and $f_Y(y) = \int f_{X,Y}(x,y)dx$ is the usual marginal density of Y. (Just apply (B.143) with $g(x,y) = xh(y)$ and with $g(x,y) = xI_C(y)$, where $A = Y^{-1}(C)$.)

What we have done in this section is give a motivation for the use of the mathematical probability calculus to express uncertainty for the purposes of gambling. We assume that an agent chooses which gambles to accept in such a way that he or she is not subject to Dutch book, which is a combination of acceptable gambles that produces a loss no matter what happens. We were also able to use this approach to motivate the mathematical definition of conditional expectation by introducing conditional gambles and requiring that the same coherence condition apply to conditional and unconditional gambles alike.

B.7 Simulation*

Several times in this text, we will want to generate observations that have a desired distribution. Such observations will be called *pseudorandom numbers* because samples appear to have the properties of random variables, but they are actually generated by a complicated deterministic process. We will not go into detail on how pseudorandom numbers with uniform $U(0,1)$ distribution are generated. In this section, we wish to prove a couple of useful theorems about how to generate pseudorandom numbers with other distributions under the assumption that pseudorandom numbers with $U(0,1)$ distribution can be generated.

Theorem B.144. *Let F be a CDF and define the inverse of F by*

$$F^{-1}(q) = \begin{cases} \inf\{x : F(x) \geq q\} & \text{if } q > 0, \\ \sup\{x : F(x) > 0\} & \text{if } q = 0. \end{cases}$$

If U has $U(0,1)$ distribution, then $X = F^{-1}(U)$ has CDF F.

PROOF. We will calculate $\Pr(X \leq t)$ for all t. First, let t be a continuity point of F. Then

$$\Pr(X \leq t) = \Pr(F^{-1}(U) \leq t) = \Pr(U \leq F(t)) = F(t),$$

where the second equality follows from the fact that, at a continuity point t, $X \leq t$ if and only if $U \leq F(t)$, and the third equality follows from the fact that U has $U(0,1)$ distribution. Finally, let t be a jump point of F and let $F(t) - \lim_{x \uparrow t} F(x) = c$. Then $X = t$ if and only if $t - c < U \leq t$, so

$$\Pr(X = t) = \Pr(t - c < U \leq t) = c.$$

So, X has CDF F at continuity points of F and its distribution has the same sized jumps as F at the same points. So the CDF of X is F. □

This theorem allows us to generate pseudorandom variables with arbitrary CDF F, if we can find F^{-1}. The method described in this theorem is called the *probability integral transform*. Note that the probability integral transform has a surprising theoretical implication.

Proposition B.145. *Let U have $U(0,1)$ distribution, and let X be a random quantity taking values in a Borel space $\mathcal{X}$. Then there exists a measurable function $f : [0,1] \to \mathcal{X}$ such that $f(U)$ has the same distribution as X.*

The next theorem allows us to find pseudorandom variables with arbitrary density f if we can generate pseudorandom variables with another density g such that $f(x) \leq kg(x)$ for some number k and all x.

Theorem B.146 (Acceptance–rejection). *Let f be a nonnegative integrable function, and let g be a density function. Let $k > 0$ and suppose that $f(x) \leq kg(x)$ for all x. Suppose that $\{Y_i\}_{i=1}^{\infty}$ and $\{U_i\}_{i=1}^{\infty}$ are all independent and that the Y_i have density g and the U_i are $U(0,1)$. Define $Z = Y_N$, where*

$$N = \min\left\{i : U_i \leq \frac{f(Y_i)}{kg(Y_i)}\right\}.$$

---

*This section may be skipped without interrupting the flow of ideas.

Then Z has density proportional to f.

PROOF. We can write the CDF of Z as

$$\Pr(Z \le t) = \Pr\left(Y_i \le t \bigg| U_i \le \frac{f(Y_i)}{kg(Y_i)}\right) = \frac{\Pr\left(Y_i \le t, U_i \le \frac{f(Y_i)}{kg(Y_i)}\right)}{\Pr\left(U_i \le \frac{f(Y_i)}{kg(Y_i)}\right)}$$

$$= \frac{\mathrm{E}\left[\Pr\left(Y_i \le t, U_i \le \frac{f(Y_i)}{kg(Y_i)} \bigg| Y_i\right)\right]}{\mathrm{E}\left[\Pr\left(U_i \le \frac{f(Y_i)}{kg(Y_i)} \bigg| Y_i\right)\right]},$$

where we have used the law of total probability B.70 in the last equation. The conditional probability in the numerator is

$$\Pr\left(Y_i \le t, U_i \le \frac{f(Y_i)}{kg(Y_i)} \bigg| Y_i\right) = \begin{cases} 0 & \text{if } Y_i > t, \\ \frac{f(Y_i)}{kg(Y_i)} & \text{if } Y_i \le t. \end{cases}$$

The mean of this is

$$\int_{-\infty}^{t} \frac{f(y)}{kg(y)} g(y) dy = \frac{1}{k} \int_{-\infty}^{t} f(y) dy,$$

since Y_i has PDF $g(\cdot)$. Similarly, the denominator conditional probability can be written as

$$\Pr\left(U_i \le \frac{f(Y_i)}{kg(Y_i)} \bigg| Y_i\right) = \frac{f(Y_i)}{kg(Y_i)}.$$

The mean of this is likewise seen to be $\int f(y) dy / k$. The ratio of these is

$$\Pr(Z \le t) = \frac{\int_{-\infty}^{t} f(y) dy}{\int f(y) dy},$$

hence Z has density proportional to f. □

Next, we prove a theorem that allows us to simulate from distributions with bounded densities and sufficiently thin tails even when we only know the density up to a normalizing constant. The theorem is due to Kinderman and Monahan (1977).

Theorem B.147 (Ratio of uniforms method). *Let $f : \mathbb{R} \to [0, \infty)$ be an integrable function. Define*

$$A = \left\{(u, v) \in \mathbb{R}^2 : 0 \le u \le \sqrt{f\left(\frac{v}{u}\right)}\right\}.$$

If (U, V) has uniform distribution over the set A, then V/U has density proportional to f.

PROOF. Let (U, V) be uniformly distributed on the set A. Then $f_{U,V}(u, v) = I_A(u, v)/c$, where c is the area of A. Define $X = U$ and $Y = V/U$. The Jacobian for the transformation is x and the joint density of (X, Y) is

$$f_{X,Y}(x, y) = \frac{x}{c} I_A(x, xy) = \frac{x}{c} I_{[0, \sqrt{f(y)}]}(x).$$

It follows that $f_Y(y) = \int_0^{\sqrt{f(y)}} \frac{x}{c} dx = \frac{1}{2c} f(y)$. □

If both $f(x) \leq b$ and $a \leq x\sqrt{f(x)} \leq b$ for all x, then A is contained in the rectangle with opposite corners $(0, a)$ and (b, c). We can then generate $U \sim U(0, b)$ and $V \sim U(a, c)$. We set $X = V/U$, and if $U^2 \leq f(X)$, take X as our desired random variable. If $U^2 < f(X)$, try again.

An important application of simulation is to the numerical integration technique called *importance sampling*. Suppose that we wish to know the value of the ratio of two integrals

$$\frac{\int v(\theta) h(\theta) d\theta}{\int h(\theta) d\theta}, \tag{B.148}$$

where θ can be a vector. Suppose that f is a density function such that h/f is nearly constant and it is easy to generate pseudorandom numbers with density f. Let $\{X_i\}_{i=1}^\infty$ be an IID sequence of pseudorandom numbers with density f. Then

$$\int h(\theta) d\theta = E\left(\frac{h(X_i)}{f(X_i)}\right),$$

$$\int v(\theta) h(\theta) d\theta = E\left(v(X_i) \frac{h(X_i)}{f(X_i)}\right),$$

where the expectations are with respect to the pseudo-distribution of X_i. If we let $W_i = h(X_i)/f(X_i)$ and $Z_i = v(X_i) W_i$, then the weak law of large numbers B.95 says that $\overline{Z}_n/\overline{W}_n$ converges in probability to (B.148).[69] The reason that we want h/f to be nearly constant is so that the variance of W_i is small. In Section 7.1.3, we will show how to approximate the variance of $\overline{Z}_n/\overline{W}_n$ as an estimate of (B.148).

B.8 Problems

Section B.2:

1. Suppose that an urn contains $m \geq 3$ white balls and $n \geq 3$ black balls. Suppose that the urn is well mixed so that at any time, the probability that any one of the remaining balls in the urn is as likely to be drawn as any other. We will draw three balls without replacement and set $X_i = 1$ if the ith ball drawn is black, $X_i = 0$ if the ith ball is white. Show that

$$\Pr(X_1 = 1, X_2 = 0, X_3 = 1) = \Pr(X_1 = 0, X_2 = 1, X_3 = 1)$$
$$= \Pr(X_1 = 1, X_2 = 1, X_3 = 0).$$

2. Suppose that H is a nondecreasing function and

$$F(x) = \inf_{\substack{t > x \\ t \text{ rational}}} H(t).$$

---

[69] The strong law of large numbers 1.63 says that $\overline{Z}_n/\overline{W}_n$ converges a.s. to (B.148).

(a) Prove that F is continuous from the right.

(b) Prove that $\inf_{\text{all } x} H(x) = \inf_{\text{all } x} F(x)$.

(c) Prove that $\sup_{\text{all } x} H(x) = \sup_{\text{all } x} F(x)$.

Section B.3:

3. Using the definition of conditional probability, show that $A \cap B = \emptyset$ implies
$$\Pr(A|\mathcal{C}) + \Pr(B|\mathcal{C}) = \Pr(A \cup B|\mathcal{C}), \text{ a.s.}$$
Use this to help prove that $\{A_n\}_{n=1}^{\infty}$ disjoint implies
$$\Pr\left(\bigcup_{n=1}^{\infty} A_n \Big| \mathcal{C}\right) = \sum_{n=1}^{\infty} \Pr(A_n|\mathcal{C}), \text{ a.s.}$$

4.*Let X_1 and X_2 be IID random variables with $U(0, 1)$ distribution. Let
$$T = \max\{X_1, X_2\}.$$
Using the definition of conditional distribution, show that the conditional distribution of X_1 given $T = t$ is a mixture of a point mass at t and a $U(0, t)$ distribution. Also, find the mixture.

5. Let $(S, \mathcal{A}, \mu)$ be a probability space. Let $\mathcal{C}$ be a sub-σ-field such that $\mu(C) \in \{0, 1\}$ for all $C \in \mathcal{C}$. Let $E|X| < \infty$. Prove that $E(X|\mathcal{C}) = E(X)$, a.s. $[\mu]$.

6. Let $(S, \mathcal{A}, \mu)$ be a probability space. Let $\{A_n\}_{n=1}^{\infty}$ be a partition of S, and let $\mathcal{C}$ be the smallest σ-field containing $\{A_n\}_{n=1}^{\infty}$. Let X be a random variable. Show that $E(X|\mathcal{C}) = \sum_{1}^{\infty} I_{A_n} w_n$, where
$$w_n = \begin{cases} \frac{\int_{A_n} X d\mu}{\mu(A_n)} & \text{if } \mu(A_n) > 0, \\ 0 & \text{otherwise.} \end{cases}$$

7. Let Φ denote the standard normal CDF, and let the joint CDF of random variables (X, Y) be
$$F_{X,Y}(x, y) = \begin{cases} \frac{\Phi(y)}{2} & \text{if } y - 1 \leq x < y + 1, \\ \Phi(y) & \text{if } x \geq y + 1, \\ 0 & \text{otherwise.} \end{cases}$$

(a) Find the conditional distribution of X given Y.

(b) Find the conditional distribution of Y given X.

8. Prove Proposition B.25 on page 617. (*Hint:* Use part 4 of Proposition A.49.)

9. Prove Proposition B.26 on page 617.

10. Prove Proposition B.27 on page 617. (*Hint:* Prove it for g an indicator function, then for simple functions, then for nonnegative measurable functions, then for all integrable functions.)

11. Prove Proposition B.28 on page 617.

12. Suppose that $X_1, \ldots, X_n$ are independent, each with distribution $N(c, 1)$. Find the conditional distribution of $X_1, \ldots, X_n$ given $\overline{X}_n = x$, where $\overline{X}_n = \sum_{i=1}^n X_i/n$.

13. Let $\mathcal{B}_1 \subseteq \mathcal{B}_2 \subseteq \cdots$ be a sequence of σ-fields, and let $X \geq 0$. Suppose that $E(X|\mathcal{B}_n) = Y$ for all n. Let $\mathcal{B}$ be the smallest σ-field containing all of the $\mathcal{B}_n$. Show that $E(X|\mathcal{B}) = Y$, a.s. (*Hint:* Show that the union of the $\mathcal{B}_n$ is a π-system, and use Theorem A.26.)

14. Prove Proposition B.43 on page 623.

15. Assume the conditions of Theorem B.46. Also, suppose that $(\mathcal{X}, \mathcal{B}_1, \nu_1)$ and $(\mathcal{Y}, \mathcal{B}_2, \nu_2)$ are σ-finite measure spaces and $\nu = \nu_1 \times \nu_2$. Prove that ν_1 can play the role of $\nu_{\mathcal{X}|\mathcal{Y}}(\cdot|y)$ for all y and that ν_2 can play the role of $\nu_{\mathcal{Y}}$ in the statement of Theorem B.46.

16. Prove Proposition B.51 on page 625. (*Hint:* Notice that $I_A(v^{-1}(y, w)) = I_{A_y^*}(w)$.)

17. Prove Proposition B.66 on page 631. (*Hint:* Prove the result for product sets first, and then use Theorem A.26.)

18. Prove Corollary B.67 on page 631.

19. Prove Corollary B.74 on page 633.

20. Prove the second Borel–Cantelli lemma: If $\{A_n\}_{n=1}^\infty$ are mutually independent and $\sum_{n=1}^\infty \Pr(A_n) = \infty$, then $\Pr(\cap_{i=1}^\infty \cup_{n=i}^\infty A_n) = 1$. (This set is sometimes called A_n infinitely often.)(*Hint:* Find the probability of the complement by using the fact that $1 - x \leq \exp(-x)$ for $0 \leq x \leq 1$.)

21. *Suppose that $(S, \mathcal{A}, \mu)$ is a measure space. Let $\{f_n\}_{n=1}^\infty$ be a sequence of measurable functions $f_n : S \to T$, where $(T, \mathcal{B})$ is a metric space with Borel σ-field. Let $\mathcal{C}$ be the tail σ-field of $\{f_n\}_{n=1}^\infty$. If $\lim_{n\to\infty} f_n(s) = f(s)$, for all s, then prove that f is measurable with respect to $\mathcal{C}$. (*Hint:* Refer to the proof of part 5 of Theorem A.38. Show that the set $A_* \in \mathcal{C}$ by showing that the union in (A.39) does not need to start at 1.)

22. Let $(S, \mathcal{A}, \mu)$ be a probability space, and let $\mathcal{C}$ be the tail σ-field of a sequence of random quantities $\{X_n\}_{n=1}^\infty$, where $X_n : S \to \mathcal{X}$ for all n. Let $\mathcal{D}$ be the σ-field generated by $\{X_n\}_{n=1}^\infty$. Let $X = (X_1, X_2, \ldots) \in \mathcal{X}^\infty$. If π is a permutation of a finite set of integers $\{1, \ldots, n\}$, let $\pi X = (X_{\pi(1)}, \ldots, X_{\pi(n)}, X_{n+1}, \ldots)$. We say that $A \in \mathcal{D}$ is *symmetric* if $A = X^{-1}(B)$ and for every permutation π of finitely many coordinates, $A = (\pi X)^{-1}(B)$ as well.

    (a) Prove that every $C \in \mathcal{C}$ is symmetric.

    (b) Show that there can be symmetric events that are not in $\mathcal{C}$.

23. Prove Proposition B.78 on page 634.

Section B.4:

24. Find a sequence of random variables that converges in probability to 0 but does not converge a.s. to 0. (*Hint:* Consider the countable collection of all subsets of $[0,1]$ of the form $[k/2^n, (k+1)/2^n]$ with k and n integers. Arrange them in an appropriate sequence.)

25. Let $\{X_n\}_{n=1}^{\infty}$ be a sequence of random variables, and let X be another random variable. Let F_n be the CDF of X_n and let F be the CDF of X. Prove that $X_n \xrightarrow{D} X$ if and only if $\lim_{n\to\infty} F_n(x) = F(x)$ for every x such that F is continuous at x.

26. Prove that $|\exp(iy) - 1| \leq \min\{|y|, 2\}$ for all y. (*Hint:* Show that $\exp(iy) = 1 + i \int_0^y \exp(is) ds$ for $y \geq 0$ and a similar formula for $y < 0$.)

27. Prove the weak law of large numbers for infinite means: Suppose that $\{X_i\}_{i=1}^{\infty}$ are IID with mean ∞. Then, for all real x, $\lim_{n\to\infty} \Pr(\overline{X}_n > x) = 1$, where $\overline{X}_n = \sum_{i=1}^n X_i/n$. (*Hint:* Define $Y_{i,t} = \min\{X_i, t\}$. Prove that $E(Y_{i,t}) < \infty$ for all t, but $\lim_{t\to\infty} E(Y_{i,t}) = \infty$.)

28. *Suppose that X is a random vector having bounded density with respect to Lebesgue measure. Prove that the characteristic function of X is integrable. (*Hint:* Run the proof of Lemma B.101 in reverse.)

Section B.5:

29. Let $\{i_n\}_{n=1}^{\infty}$ be a sequence of numbers in $\{0,1\}$. Suppose that $\{X_n\}_{n=1}^{\infty}$ is a sequence of Bernoulli random variables such that

$$\Pr(X_1 = i_1, \ldots, X_n = i_n) = \frac{12}{x+2} \frac{1}{\binom{n+4}{x+2}},$$

where $x = \sum_{j=1}^n i_j$. Show that this specifies a consistent set of joint distributions for $n = 1, 2, \ldots$.

30. Let μ be a finite measure on $(\mathbb{R}, \mathcal{B})$, where $\mathcal{B}$ is the Borel σ-field. Suppose that $\{X(t) : -\infty < t < \infty\}$ is a stochastic process such that $X(t)$ has $Beta(\mu(-\infty, t], \mu(t, \infty))$ distribution for each t, $X(t) > X(s)$ if $t > s$, and $X(\cdot)$ is continuous from the right.

    (a) Prove that $\Pr(\lim_{t\to\infty} X(t) = 1) = 1$.

    (b) Let $U = \inf\{t : X(t) \geq 1/2\}$. Prove that the median of U is $\inf\{t : \mu(-\infty, t] \geq \mu(t, \infty)\}$. (*Hint:* Write $\{U \leq s\}$ in terms of $X(\cdot)$.)

31. Let R be a set, and let $(\mathcal{X}_r, \mathcal{B}_r)$ be a Borel space for every $r \in R$. Let $\mathcal{X} = \prod_{r \in R} \mathcal{X}_r$ and let $\mathcal{B}$ be the product σ-field. For each $r \in R$, let $X_r : \mathcal{X} \to \mathcal{X}_r$ be the projection function $X_r(x) = x_r$. Prove that $\mathcal{B}$ is the union of all of the σ-fields generated by all of the countable collections of X_r functions. That is, let Q be the set of all countable subsets of R, and for each $q \in Q$ let $X^q = \{X_r\}_{r \in q}$ and let $\mathcal{B}^q$ be the σ-field generated by X^q. Then show that $\mathcal{B} = \cup_{q \in Q} \mathcal{B}^q$.

Section B.7:

32. Prove Proposition B.145 on page 659.

Appendix C
Mathematical Theorems Not Proven Here

There are several theorems of a purely mathematical nature which we use on occasion in this text, but which we do not wish to prove here because their proofs involve a great deal of mathematical background of which we will not make use anywhere else.

C.1 Real Analysis

Theorem C.1 (Taylor's theorem).[1] *Suppose that $f : \mathbb{R}^m \to \mathbb{R}$ has continuous partial derivatives of all orders up to and including $k+1$ with respect to all coordinates in a convex neighborhood D of a point x_0. For $x \in D$ and $i = 1, \ldots, k+1$, define*

$$D^{(i)}(f, x, y) = \sum_{j_1=1}^{m} \cdots \sum_{j_i=1}^{m} \left(\left. \frac{\partial^i}{\partial z_{j_1} \cdots \partial z_{j_i}} f(z) \right|_{z=x} \prod_{s=1}^{i} y_{j_s} \right),$$

where we allow notation like $\partial^3 / \partial z_1 \partial z_1 \partial z_4$ to stand for $\partial^3 / \partial z_1^2 \partial z_4$. Then, for $x \in D$,

$$f(x) = f(x_0) + \sum_{i=1}^{k} \frac{1}{i!} D^{(i)}(f; x_0, x - x_0) + \frac{1}{(k+1)!} D^{(k+1)}(f; x^*, x - x_0),$$

where x^* is on the line segment joining x and x_0.

---

[1] This theorem is used in the proofs of Theorems 7.63, 7.89, 7.108, and 7.125. For a proof (with $m = 2$), see Buck (1965), Theorem 16 on page 260.

Theorem C.2 (Inverse function theorem).[2] *Let f be a continuously differentiable function from an open set in $\mathbb{R}^n$ into $\mathbb{R}^n$ such that $((\partial f_i/\partial x_j))$ is a nonsingular matrix at a point x. If $y = f(x)$, then there exist open sets U and V such that $x \in U$, $y \in V$, f is one-to-one on U, and $f(U) = V$. Also, if $g : V \to U$ is the inverse of f on U, then g is continuously differentiable on V.*

Theorem C.3 (Stone–Weierstrass theorem).[3] *Let $\mathcal{A}$ be a collection of continuous complex functions defined on a compact set C and satisfying these conditions:*

- *If $f \in \mathcal{A}$, then the complex conjugate of f is in $\mathcal{A}$.*
- *If $x_1 \neq x_2 \in C$, then there exists $f \in \mathcal{A}$ such that $f(x_1) \neq f(x_2)$.*
- *If $f, g \in \mathcal{A}$, then $f + g \in \mathcal{A}$ and $fg \in \mathcal{A}$.*
- *If $f \in \mathcal{A}$ and c is a constant, then $cf \in \mathcal{A}$.*
- *For each $x \in C$, there exists $f \in \mathcal{A}$ such that $f(x) \neq 0$.*

Then, for every continuous complex function f on C, there exists a sequence $\{f_n\}_{n=1}^{\infty}$ in $\mathcal{A}$ such that f_n converges uniformly to f on C.

Theorem C.4 (Supporting hyperplane theorem).[4] *If S is a convex subset of a finite-dimensional Euclidean space, and x_0 is a boundary point of S, then there is a nonzero vector v such that for every $x \in S$, $v^\top x \geq v^\top x_0$.*

Theorem C.5 (Separating hyperplane theorem).[5] *If S_1 and S_2 are disjoint convex subsets of a finite-dimensional Euclidean space, then there is a nonzero vector v and a constant c such that for every $x \in S_1$, $v^\top x \leq c$ and for every $y \in S_2$, $v^\top y \geq c$.*

Theorem C.6 (Bolzano–Weierstrass theorem).[6] *Suppose that B is a closed and bounded subset of a finite-dimensional Euclidean space. Then every infinite subset of B has a cluster point in B.*

C.2 Complex Analysis

Theorem C.7.[7] *Let f be an analytic function in a neighborhood of a point z. Then the derivatives of f of every order exist and are analytic in a neighborhood*

---

[2] This theorem is used in the proof of Theorem 7.57. For a proof, see Rudin (1964), Theorem 9.17.

[3] This theorem is used in the proofs of DeFinetti's representation theorem 1.49 and 1.47 and Theorem B.93. For a proof, see Rudin (1964), Theorem 7.31.

[4] This theorem is used in the proof of Theorem B.17. For a proof, see Berger (1985), Theorem 12 on page 341, or Ferguson (1967), Theorem 1 on page 73.

[5] This theorem is used in the proof of Theorems B.17, 3.77 and 3.95. For a proof, see Berger (1985), Theorem 13 on page 342, or Ferguson (1967), Theorem 2 on page 73.

[6] This theorem is used in the proof of Theorem 3.77. For a proof, see Dugundji (1966), Theorems 3.2 and 4.3 of Chapter XI.

[7] This theorem is used to show that certain estimators are UMVUE, and in the proof of Theorem 2.74. For a proof, see Churchill (1960, Sections 52 and 56).

of z. If $f^{(k)}$ denotes the kth derivative of f, then

$$f(x) = \sum_{k=0}^{\infty} (x-z)^k \frac{f^{(k)}(z)}{k!}$$

for all x in some circle around z.

Theorem C.8 (Maximum modulus theorem).[8] *Let f be an analytic function in an open set D which is continuous on the closure of D. Let the maximum value of $|f(z)|$ for z in the closure of D be c. Then $|f(z)| < c$ for all $z \in D$ unless f is constant on D.*

Theorem C.9 (Cauchy's equation).[9] *Let G be a Borel subset of $\mathbb{R}^k$ with positive Lebesgue measure. Let $f : G \to \mathbb{R}$ be measurable. Let $H_1 = G$ and $H_n = H_{n-1} + G$ for each n. For each n, let $g_n : H_n \to \mathbb{R}$ be measurable such that $g_n \left(\sum_{i=1}^n x_i \right) = \sum_{i=1}^n f(x_i)$, for almost all $(x_1, \ldots, x_n) \in G^n$. Then there is a real number a and a vector $b \in \mathbb{R}^k$ such that $f(x) = a + b^\top x$ a.e. in G.*

C.3 Functional Analysis

Theorem C.10.[10] *If T is an operator with finite norm on the Hilbert space $L^2(\mu)$ given by $T(f)(x) = \int K(x', x) d\mu(x')$, then T is of Hilbert–Schmidt type if and only if*

$$\int \int |K(x', x)|^2 d\mu(x') d\mu(x) < \infty.$$

Theorem C.11.[11] *Every operator of Hilbert–Schmidt type is completely continuous.*

Theorem C.12.[12] *If T is a completely continuous self-adjoint operator, then T has an eigenvalue λ with $|\lambda| = \|T\|$.*

Theorem C.13.[13] *If T is a linear operator with finite norm and T^* is its adjoint operator, then $\|T^*T\| = \|T\|^2$.*

---

[8]This theorem is used in the proof of Theorem 2.64. For a proof, see Churchill (1960), Section 54, or Ahlfors (1966), Theorem 12' on page 134.

[9]This theorem is used in the proof of Theorem 2.114. For a proof, see Diaconis and Freedman (1990), Theorem 2.1.

[10]This theorem is used in the proof of Theorem 8.40. For a proof, see Section XI.6 of Dunford and Schwartz (1963). By $L^2(\mu)$ we mean $\{f : \int f^2(x)d\mu(x) < \infty\}$.

[11]This theorem is used in the proof of Theorem 8.40. For a proof, see Theorem 6 of Section XI.6 of Dunford and Schwartz (1963). The reader should note that Dunford and Schwartz (1963) use the term *compact* instead of *completely continuous*.

[12]This theorem is used in the proof of Theorem 8.40. For a proof, see Lemma 1 in Section VIII.3 of Berberian (1961).

[13]This theorem is used in the proof of Theorem 8.40. For a proof, see part (5) of Theorem 2 on p. 132 of Berberian (1961).

Appendix D
Summary of Distributions

The distributions used in this book are listed here. We give the name and symbol used to describe each distribution. Each distribution is absolutely continuous with respect to some measure or other. In most cases the mean and variance are given. In some cases, the symbol for the CDF is given.

D.1 Univariate Continuous Distributions

Alternate noncentral beta
Symbol: $ANCB(q, a, \gamma)$
Density: $f_X(x) = \sum_{k=0}^{\infty} \frac{\Gamma(\frac{a}{2}+k)}{k!\,\Gamma(\frac{a}{2})} (1-\gamma)^{\frac{a}{2}} \gamma^k \frac{\Gamma(\frac{a}{2}+\frac{q}{2}+2k)}{\Gamma(\frac{q}{2}+k)\Gamma(\frac{a}{2}+k)} x^{\frac{q}{2}+k-1}(1-x)^{\frac{a}{2}+k-1}$
Dominating measure: Lebesgue measure on $[0, 1]$

Alternate noncentral chi-squared
Symbol: $ANC\chi^2(q, a, \gamma)$[1]
Density: $f_X(x) = \sum_{k=0}^{\infty} \frac{\Gamma(k+\frac{a}{2})}{k!\,\Gamma(\frac{a}{2})} \gamma^k (1-\gamma)^{\frac{a}{2}} \frac{x^{\frac{q}{2}+k-1}}{2^{\frac{q}{2}+k}\Gamma(\frac{q}{2}+k)} \exp\left(-\frac{x}{2}\right)$
Dominating measure: Lebesgue measure on $[0, \infty)$
Mean: $q + a\frac{\gamma}{1-\gamma}$
Variance: $2\left[q + a\frac{\gamma(2-\gamma)}{1-\gamma}\right]$

---

[1] This distribution was derived without a name by Geisser (1967). It was named L^2 by Lecoutre and Rouanet (1981).

Alternate noncentral F

Symbol: $ANCF(q, a, \gamma)^2$

Density: $f_X(x) = \sum_{k=0}^{\infty} \frac{\Gamma(k+\frac{a}{2})}{k!\Gamma(\frac{a}{2})} \gamma^k (1-\gamma)^{\frac{a}{2}} \frac{\Gamma(2k+\frac{q+a}{2})}{\Gamma(k+\frac{q}{2})\Gamma(k+\frac{a}{2})} \frac{q^{k+\frac{q}{2}} a^{k+\frac{a}{2}} x^{\frac{q}{2}+k-1}}{(a+qx)^{\frac{q+a}{2}+2k}}$

Dominating measure: Lebesgue measure on $[0, \infty)$

Mean: $(1-\gamma)\frac{a}{a-2} + \gamma\frac{a}{q}$, if $a > 2$

Variance: $\frac{2a^2(a-2+q)(1-\gamma)^2}{(a-2)^2(a-4)q} + \frac{4a^2\gamma(1-\gamma)}{(a-2)q^2}$, if $a > 4$

Beta

Symbol: $Beta(\alpha, \beta)$

Density: $f_X(x) = \frac{\Gamma(\alpha+\beta)}{\Gamma(\alpha)\Gamma(\beta)} x^{\alpha-1}(1-x)^{\beta-1}$

Dominating measure: Lebesgue measure on $[0, 1]$

Mean: $\frac{\alpha}{\alpha+\beta}$

Variance: $\frac{\alpha\beta}{(\alpha+\beta)^2(\alpha+\beta+1)}$

Cauchy

Symbol: $Cau(\mu, \sigma^2)$

Density: $f_X(x) = \pi^{-1}\left(1 + \frac{(x-\mu)^2}{\sigma^2}\right)^{-1}$

Dominating measure: Lebesgue measure on $(-\infty, \infty)$

Mean: Does not exist

Variance: Does not exist

Chi-squared

Symbol: χ_a^2

Density: $f_X(x) = \frac{x^{\frac{a}{2}-1}}{2^{\frac{a}{2}}\Gamma(\frac{a}{2})} \exp\left(-\frac{x}{2}\right)$

Dominating measure: Lebesgue measure on $[0, \infty)$

Mean: a

Variance: $2a$

---

[2] The alternate noncentral F distribution, with a different scaling factor, was called the ψ^2 distribution by Rouanet and Lecoutre (1983). See also Lecoutre (1985). The distribution was derived without a name by Ferrándiz (1985). Schervish (1992) gives additional details concerning the $ANC\chi^2$, $ANCB$, and $ANCF$ distributions.

Exponential
Symbol: $Exp(\theta)$
Density: $f_X(x) = \theta \exp(-x\theta)$
Dominating measure: Lebesgue measure on $[0, \infty)$
Mean: $\frac{1}{\theta}$
Variance: $\frac{1}{\theta^2}$

F
Symbol: $F_{q,a}$
Density: $f_X(x) = \frac{\Gamma(\frac{q+a}{2}) q^{\frac{q}{2}} a^{\frac{a}{2}}}{\Gamma(\frac{q}{2})\Gamma(\frac{a}{2})} x^{\frac{q}{2}-1}(a+qx)^{-\frac{q+a}{2}}$
Dominating measure: Lebesgue measure on $[0, \infty)$
Mean: $\frac{a}{a-2}$, if $a > 2$
Variance: $2a^2 \frac{q+a-2}{q(a-4)(a-2)^2}$, if $a > 4$

Gamma
Symbol: $\Gamma(\alpha, \beta)$
Density: $f_X(x) = \frac{\beta^\alpha}{\Gamma(\alpha)} x^{\alpha-1} \exp(-\beta x)$
Dominating measure: Lebesgue measure on $[0, \infty)$
Mean: $\frac{\alpha}{\beta}$
Variance: $\frac{\alpha}{\beta^2}$

Inverse gamma
Symbol: $\Gamma^{-1}(\alpha, \beta)$
Density: $f_X(x) = \frac{\beta^\alpha}{\Gamma(\alpha)} x^{-\alpha-1} \exp\left(-\frac{\beta}{x}\right)$
Dominating measure: Lebesgue measure on $[0, \infty)$
Mean: $\frac{\beta}{\alpha-1}$, if $\alpha > 1$
Variance: $\frac{\beta^2}{(\alpha-1)^2(\alpha-2)}$, if $\alpha > 2$

Laplace
Symbol: $Lap(\mu, \sigma)$
Density: $f_X(x) = \frac{1}{2\sigma} \exp\left(-\frac{|x-\mu|}{\sigma}\right)$.
Dominating measure: Lebesgue measure on $\mathbb{R}$
Mean: μ
Variance: $2\sigma^2$

Noncentral beta
Symbol: $NCB(\alpha, \beta, \psi)$
Density: $f_X(x) = \sum_{k=0}^{\infty} \left(\frac{\psi}{2}\right)^k \exp\left(-\frac{\psi}{2}\right) \frac{\Gamma(\alpha+\beta)}{k!\,\Gamma(\alpha+k)\Gamma(\beta)} x^{\alpha+k-1}(1-x)^{\beta-1}$
Dominating measure: Lebesgue measure on $[0, 1]$

Noncentral chi-squared
Symbol: $NC\chi_q^2(\psi)$
Density: $f_X(x) = \sum_{k=0}^{\infty} \left(\frac{\psi}{2}\right)^k \exp\left(-\frac{\psi}{2}\right) \frac{x^{\frac{q}{2}+k-1}}{k!\,2^{\frac{q}{2}+k}\Gamma\left(\frac{q}{2}+k\right)} \exp\left(-\frac{x}{2}\right)$
Dominating measure: Lebesgue measure on $[0, \infty)$
Mean: $q + \psi$
Variance: $2q + 4\psi$

Noncentral F
Symbol: $NCF(q, a, \psi)$
Density: $f_X(x) = \sum_{k=0}^{\infty} \left(\frac{\psi}{2}\right)^k \exp\left(-\frac{\psi}{2}\right) \frac{\Gamma\left(k+\frac{q+a}{2}\right)}{k!\,\Gamma\left(k+\frac{q}{2}\right)\Gamma\left(\frac{a}{2}\right)} \frac{q^{k+\frac{q}{2}} a^{\frac{a}{2}} x^{\frac{q}{2}+k-1}}{(a+qx)^{\frac{q+a}{2}+k}}$
Dominating measure: Lebesgue measure on $[0, \infty)$
Mean: $\left(1 + \frac{\psi}{q}\right) \frac{a}{a-2}$, if $a > 2$
Variance: $2\left(\frac{a}{q}\right)^2 \frac{(q+\psi)^2+(q+2\psi)(a-2)}{(a-2)^2(a-4)}$ if $a > 4$

Noncentral t
Symbol: $NCt_a(\delta)$
Density: $f_X(x) = \sum_{k=0}^{\infty} \frac{(\delta x)^k}{k!} \exp\left(-\frac{\delta^2}{2}\right) \frac{\Gamma\left(\frac{a+k+1}{2}\right)}{\sqrt{a\pi}\,\Gamma\left(\frac{a}{2}\right)\left(\frac{a}{2}\right)^{\frac{k}{2}}} \left(1 + \frac{x^2}{a}\right)^{-\frac{a+k+1}{2}}$
Dominating measure: Lebesgue measure on $\mathbb{R}$
Mean: $\delta \frac{\Gamma\left(\frac{a-1}{2}\right)}{\Gamma\left(\frac{a}{2}\right)} \sqrt{\frac{a}{2}}$, if $a > 1$
Variance: $\frac{a(\delta^2+1)}{a-2} - \frac{a\delta^2}{2}\left[\frac{\Gamma\left(\frac{a-1}{2}\right)}{\Gamma\left(\frac{a}{2}\right)}\right]^2$, if $a > 2$
CDF: $NCT_a(\cdot\,;\delta)$

Normal
Symbol: $N(\mu, \sigma^2)$
Density: $f_X(x) = (\sqrt{2\pi}\sigma)^{-1} \exp\left(-\frac{(x-\mu)^2}{2\sigma^2}\right)$
Dominating measure: Lebesgue measure on $(-\infty, \infty)$

Mean: μ
Variance: σ^2
CDF: $\Phi(\cdot)$ (For $N(0,1)$ distribution)

Pareto

Symbol: $Par(\alpha, c)$
Density: $f_X(x) = \frac{c^\alpha \alpha}{x^{\alpha+1}}$
Dominating measure: Lebesgue measure on $[c, \infty)$
Mean: $\frac{c\alpha}{\alpha-1}$, if $\alpha > 1$
Variance: $\frac{c^2 \alpha}{(\alpha-2)(\alpha-1)^2}$, if $\alpha > 2$

t

Symbol: $t_a(\mu, \sigma^2)$
Density: $f_X(x) = \frac{\Gamma\left(\frac{a+1}{2}\right)}{\sqrt{a\pi}\,\Gamma\left(\frac{a}{2}\right)\sigma} \left(1 + \frac{(x-\mu)^2}{a\sigma^2}\right)^{-\frac{a+1}{2}}$
Dominating measure: Lebesgue measure on $(-\infty, \infty)$
Mean: μ, if $a > 1$
Variance: $\sigma^2 \frac{a}{a-2}$, if $a > 2$
CDF: $T_a(\cdot)$ (For $t_a(0,1)$ distribution)

Uniform

Symbol: $U(a, b)$
Density: $f_X(x) = (b-a)^{-1}$
Dominating measure: Lebesgue measure on $[a, b]$
Mean: $\frac{a+b}{2}$
Variance: $\frac{(b-a)^2}{12}$

D.2 Univariate Discrete Distributions

Bernoulli

Symbol: $Ber(p)$
Density: $f_X(x) = p^x(1-p)^{1-x}$
Dominating measure: Counting measure on $\{0, 1\}$
Mean: p
Variance: $p(1-p)$

Binomial

Symbol: $Bin(n,p)$
Density: $f_X(x) = \binom{n}{x} p^x (1-p)^{1-x}$
Dominating measure: Counting measure on $\{0, \ldots, n\}$
Mean: np
Variance: $np(1-p)$

Geometric

Symbol: $Geo(p)$
Density: $f_X(x) = p(1-p)^x$
Dominating measure: Counting measure on $\{0, 1, 2, \ldots\}$
Mean: $\frac{1-p}{p}$
Variance: $\frac{1-p}{p^2}$

Hypergeometric

Symbol: $Hyp(N, n, k)$
Density: $f_X(x) = \frac{\binom{n}{x}\binom{N-n}{k-x}}{\binom{N}{k}}$
Dominating measure: Counting measure on
$\{\max\{0, n - N + k\}, \ldots, \min\{n, k\}\}$
Mean: $\frac{nk}{N}$
Variance: $n\left(\frac{k}{N}\right)\left(\frac{N-k}{N}\right)\left(\frac{N-n}{N-1}\right)$

Negative binomial

Symbol: $Negbin(a, p)$
Density: $f_X(x) = \frac{\Gamma(a+x)}{\Gamma(a)x!} p^a (1-p)^x$
Dominating measure: Counting measure on $\{0, 1, 2, \ldots\}$
Mean: $a\frac{1-p}{p}$
Variance: $a\frac{1-p}{p^2}$

Poisson

Symbol: $Poi(\lambda)$
Density: $f_X(x) = \exp(-\lambda)\frac{\lambda^x}{x!}$
Dominating measure: Counting measure on $\{0, 1, 2, \ldots\}$
Mean: λ
Variance: λ

D.3 Multivariate Distributions

Dirichlet
Symbol: $Dir_k(\alpha_1, \ldots, \alpha_k)$
Density: $f_{X_1, \ldots, X_{k-1}}(x_1, \ldots, x_{k-1}) = \frac{\Gamma(\alpha_0)}{\Gamma(\alpha_1) \cdots \Gamma(\alpha_k)} x_1^{\alpha_1 - 1} \cdots x_{k-1}^{\alpha_{k-1} - 1} (1 - x_1 - \cdots - x_{k-1})^{\alpha_k - 1}$, where $\alpha_0 = \sum_{i=1}^{k} \alpha_i$
Dominating measure: Lebesgue measure on
$\{(x_1, \ldots, x_{k-1}) : \text{all } x_i \geq 0 \text{ and } x_1 + \cdots x_{k-1} \leq 1\}$
Mean: $E(X_i) = \frac{\alpha_i}{\alpha_0}$
Variance: $\text{Var}(X_i) = \frac{\alpha_i(\alpha_0 - \alpha_i)}{\alpha_0^2(\alpha_0 + 1)}$
Covariance: $\text{Cov}(X_i, X_j) = -\frac{\alpha_i \alpha_j}{\alpha_0^2(\alpha_0 + 1)}$

Multinomial
Symbol: $Mult_k(n, p_1, \ldots, p_k)$
Density: $f_{X_1, \ldots, X_k}(x_1, \ldots, x_k) = \binom{n}{x_1, \ldots, x_k} p_1^{x_1} \cdots p_k^{x_k}$
Dominating measure: Counting measure on
$\{(x_1, \ldots, x_k) : \text{all } x_i \in \{0, \ldots, n\} \text{ and } x_1 + \cdots + x_k = n\}$
Mean: $E(X_i) = np_i$
Variance: $\text{Var}(X_i) = np_i(1 - p_i)$
Covariance: $\text{Cov}(X_i, X_j) = -np_i p_j$

Multivariate Normal
Symbol: $N_p(\mu, \sigma)$
Density: $f_X(x) = (2\pi)^{-\frac{p}{2}} |\sigma|^{-\frac{1}{2}} \exp\left(-\frac{1}{2}(x - \mu)^\top \sigma^{-1}(x - \mu)\right)$
Dominating measure: Lebesgue measure on $\mathbb{R}^p$
Mean: $E(X_i) = \mu_i$
Variance: $\text{Var}(X_i) = \sigma_{i,i}$
Covariance: $\text{Cov}(X_i, X_j) = \sigma_{i,j}$

References

AHLFORS, L. (1966). *Complex Analysis* (2nd ed.). New York: McGraw-Hill.

AITCHISON, J. and DUNSMORE, I. R. (1975). *Statistical Prediction Analysis*. Cambridge: Cambridge University Press.

ALBERT, J. H. and CHIBB, S. (1993). Bayesian analysis of binary and polychotomous response data. *Journal of the American Statistical Association*, **88**, 669–679.

ALDOUS, D. J. (1981). Representations for partially exchangeable random variables. *Journal of Multivariate Analysis*, **11**, 581–598.

ALDOUS, D. J. (1985). Exchangeability and related topics. In P. L. HENNEQUIN (Ed.), *École d'Été de Probabilités de Saint-Flour XIII–1983* (pp. 1–198). Berlin: Springer-Verlag.

ANDERSON, T. W. (1984). *An Introduction to Multivariate Statistical Analysis* (2nd ed.). New York: Wiley.

ANSCOMBE, F. J. and AUMANN, R. J. (1963). A definition of subjective probability. *Annals of Mathematical Statistics*, **34**, 199–205.

ANTONIAK, C. E. (1974). Mixtures of Dirichlet processes with applications to Bayesian nonparametric problems. *Annals of Statistics*, **2**, 1152–1174.

BAHADUR, R. R. (1957). On unbiased estimates of uniformly minimum variance. *Sankhyā*, **18**, 211–224.

BARNARD, G. A. (1970). Discussion on paper by Dr. Kalbfleisch and Dr. Sprott. *Journal of the Royal Statistical Society (Series B)*, **32**, 194–195.

BARNARD, G. A. (1976). Conditional inference is not inefficient. *Scandinavian Journal of Statistics*, **3**, 132–134.

BARNDORFF-NIELSEN, O. E. (1988). *Parametric Statistical Models and Likelihood*. Berlin: Springer-Verlag.

BARNETT, V. (1982). *Comparative Statistical Inference* (2nd ed.). New York: Wiley.

BARRON, A. R. (1986). Discussion of "On the consistency of Bayes estimates" by Diaconis and Freedman. *Annals of Statistics*, **14**, 26–30.

BARRON, A. R. (1988). The exponential convergence of posterior probabilities with implications for Bayes estimators of density functions. Technical Report 7, Department of Statistics, University of Illinois, Champaign, IL.

BASU, D. (1955). On statistics independent of a complete sufficient statistic. *Sankhyā*, **15**, 377–380.

BASU, D. (1958). On statistics independent of sufficient statistics. *Sankhyā*, **20**, 223–226.

BAYES, T. (1764). An essay toward solving a problem in the doctrine of chances. *Philosophical Transactions of the Royal Society of London*, **53**, 370–418.

BECKER, R. A., CHAMBERS, J. M., and WILKS, A. R. (1988). *The New S Language: A Programming Environment for Data Analysis and Graphics*. Pacific Grove, CA: Wadsworth and Brooks/Cole.

BERBERIAN, S. K. (1961). *Introduction to Hilbert Space*. New York: Oxford University Press.

BERGER, J. O. (1985). *Statistical Decision Theory and Bayesian Analysis* (2nd ed.). New York: Springer-Verlag.

BERGER, J. O. (1994). An overview of robust Bayesian analysis (with discussion). *Test*, **3**, 5–124.

BERGER, J. O. and BERRY, D. A. (1988). The relevance of stopping rules in statistical inference (with discussion). In S. S. GUPTA and J. O. BERGER (Eds.), *Statistical Decision Theory and Related Topics IV* (pp. 29–72). New York: Springer-Verlag.

BERGER, J. O. and SELLKE, T. (1987). Testing a point null hypothesis: The irreconcilability of P values and evidence (with discussion). *Journal of the American Statistical Association*, **82**, 112–122.

BERK, R. H. (1966). Limiting behavior of posterior distributions when the model is incorrect. *Annals of Mathematical Statistics*, **37**, 51–58.

BERKSON, J. (1942). Tests of significance considered as evidence. *Journal of the American Statistical Association*, **37**, 325–335.

BERTI, P., REGAZZINI, E., and RIGO, P. (1991). Coherent statistical inference and Bayes theorem. *Annals of Statistics*, **19**, 366–381.

BICKEL, P. J. and FREEDMAN, D. A. (1981). Some asymptotic theory for the bootstrap. *Annals of Statistics*, **9**, 1196–1217.

BILLINGSLEY, P. (1968). *Convergence of Probability Measures*. New York: Wiley.

BILLINGSLEY, P. (1986). *Probability and Measure* (2nd ed.). New York: Wiley.

BISHOP, Y. M. M., FIENBERG, S. E., and HOLLAND, P. W. (1975). *Discrete Multivariate Analysis: Theory and Practice*. Cambridge, MA: MIT Press.

BLACKWELL, D. (1947). Conditional expectation and unbiased sequential estimation. *Annals of Mathematical Statistics*, **18**, 105–110.

BLACKWELL, D. (1973). Discreteness of Ferguson selections. *Annals of Statistics*, **1**, 356–358.

BLACKWELL, D. and DUBINS, L. (1962). Merging of opinions with increasing information. *Annals of Mathematical Statistics*, **33**, 882–886.

BLACKWELL, D. and RAMAMOORTHI, R. V. (1982). A Bayes but not classically sufficient statistic. *Annals of Statistics*, **10**, 1025–1026.

BLYTH, C. R. (1951). On minimax statistical decision procedures and their admissibility. *Annals of Mathematical Statistics*, **22**, 22–42.

BONDAR, J. V. (1988). Discussion of "Conditionally acceptable frequentist solutions" by George Casella. In S. S. GUPTA and J. O. BERGER (Eds.), *Statistical Decision Theory and Related Topics IV* (pp. 91–93). New York: Springer-Verlag.

BORTKIEWICZ, L. V. (1898). *Das Gesetz der Kleinen Zahlen.* Leipzig: Teubner.

BOX, G. E. P. and COX, D. R. (1964). An analysis of transformations (with discussion). *Journal of the Royal Statistical Society (Series B)*, **26**, 211–246.

BOX, G. E. P. and TIAO, G. C. (1968). A Bayesian approach to some outlier problems. *Biometrika*, **55**, 119–129.

BREIMAN, L. (1968). *Probability.* Reading, MA: Addison-Wesley.

BRENNER, D., FRASER, D. A. S., and MCDUNNOUGH, P. (1982). On asymptotic normality of likelihood and conditional analysis. *Canadian Journal of Statistics*, **10**, 163–172.

BROWN, L. D. (1967). The conditional level of Student's t test. *Annals of Mathematical Statistics*, **38**, 1068–1071.

BROWN, L. D. (1971). Admissible estimators, recurrent diffusions, and insoluble boundary value problems. *Annals of Mathematical Statistics*, **42**, 855–903. (See also correction, *Annals of Statistics*, **1**, 594–596.)

BROWN, L. D. and HWANG, J. T. (1982). A unified admissibility proof. In S. S. GUPTA and J. O. BERGER (Eds.), *Statistical Decision Theory and Related Topics III* (pp. 205–230). New York: Academic Press.

BUCK, C. (1965). *Real Analysis* (2nd ed.). New York: McGraw-Hill.

BUEHLER, R. J. (1959). Some validity criteria for statistical inferences. *Annals of Mathematical Statistics*, **30**, 845–863.

BUEHLER, R. J. and FEDDERSON, A. P. (1963). Note on a conditional property of Student's t. *Annals of Mathematical Statistics*, **34**, 1098–1100.

CASELLA, G. and BERGER, R. L. (1987). Reconciling Bayesian and frequentist evidence in the one-sided testing problem (with discussion). *Journal of the American Statistical Association*, **82**, 106–111.

CHALONER, K., CHURCH, T., LOUIS, T. A., and MATTS, J. P. (1993). Graphical elicitation of a prior distribution for a clinical trial. *The Statistician*, **42**, 341–353.

CHANG, T. and VILLEGAS, C. (1986). On a theorem of Stein relating Bayesian and classical inferences in group models. *Canadian Journal of Statistics*, **14**, 289–296.

CHAPMAN, D. and ROBBINS, H. (1951). Minimum variance estimation without regularity assumptions. *Annals of Mathematical Statistics*, **22**, 581–586.

CHEN, C.-F. (1985). On asymptotic normality of limiting density functions with Bayesian implications. *Journal of the Royal Statistical Society (Series B)*, **47**, 540–546.

CHOW, Y. S., ROBBINS, H., and SIEGMUND, D. (1971). *Great Expectations: The Theory of Optimal Stopping.* New York: Houghton Mifflin.

CHURCHILL, R. V. (1960). *Complex Variables and Applications* (2nd ed.). New York: McGraw Hill.

CLARKE, B. S. and BARRON, A. R. (1994). Jeffreys' prior is asymptotically least favorable under entropy risk. *Journal of Statistical Planning and Inference*, **41**, 37–60.

CORNFIELD, J. (1966). A Bayesian test of some classical hypotheses—with applications to sequential clinical trials. *Journal of the American Statistical Association*, **61**, 577–594.

COX, D. R. (1958). Some problems connected with statistical inference. *Annals of Mathematical Statistics*, **29**, 357–372.

COX, D. R. (1977). The role of significance tests. *Scandinavian Journal of Statistics*, **4**, 49–70.

COX, D. R. and HINKLEY, D. V. (1974). *Theoretical Statistics*. London: Chapman and Hall.

CRAMÉR, H. (1945). *Mathematical Methods of Statistics*. Princeton: Princeton University Press.

CRAMÉR, H. (1946). Contributions to the theory of statistical estimation. *Skandinavisk Aktuarietidsk*, **29**, 85–94.

DAVID, H. A. (1970). *Order Statistics*. New York: Wiley.

DAWID, A. P. (1970). On the limiting normality of posterior distributions. *Proceedings of the Cambridge Philosophical Society*, **67**, 625–633.

DAWID, A. P. (1982). Intersubjective statistical models. In G. KOCH and F. SPIZZICHINO (Eds.), *Exchangeability in Probability and Statistics* (pp. 217–232). Amsterdam: North-Holland.

DAWID, A. P. (1984). Statistical theory: The prequential approach. *Journal of the Royal Statistical Society (Series A)*, **147**, 278–292.

DAWID, A. P., STONE, M., and ZIDEK, J. V. (1973). Marginalization paradoxes in Bayesian and structural inference. *Journal of the Royal Statistical Society (Series B)*, **35**, 189–233.

DEFINETTI, B. (1937). Foresight: Its logical laws, its subjective sources. In H. E. KYBURG and H. E. SMOKLER (Eds.), *Studies in Subjective Probability* (pp. 53–118). New York: Wiley.

DEFINETTI, B. (1974). *Theory of Probability, Vols. I and II*. New York: Wiley.

DEGROOT, M. H. (1970). *Optimal Statistical Decisions*. New York: Wiley.

DEMOIVRE, A. (1756). *The Doctrine of Chance* (3rd ed.). London: A. Millar.

DIACONIS, P. and FREEDMAN, D. A. (1980a). Finite exchangeable sequences. *Annals of Probability*, **8**, 745–764.

DIACONIS, P. and FREEDMAN, D. A. (1980b). DeFinetti's generalizations of exchangeability. In R. C. JEFFREY (Ed.), *Studies in Inductive Logic and Probability, II* (pp. 233–249). Berkeley: University of California.

DIACONIS, P. and FREEDMAN, D. A. (1980c). DeFinetti's theorem for Markov chains. *Annals of Probability*, **8**, 115–130.

DIACONIS, P. and FREEDMAN, D. A. (1984). Partial exchangeability and sufficiency. In J. K. GHOSH and J. ROY (Eds.), *Statistics: Applications and New Directions* (pp. 205–236). Calcutta: Indian Statistical Institute.

DIACONIS, P. and FREEDMAN, D. A. (1986a). On the consistency of Bayes estimates (with discussion). *Annals of Statistics*, **14**, 1–26.

DIACONIS, P. and FREEDMAN, D. A. (1986b). On inconsistent Bayes estimates of location. *Annals of Statistics*, **14**, 68–87.

DIACONIS, P. and FREEDMAN, D. A. (1990). Cauchy's equation and DeFinetti's theorem. *Scandinavian Journal of Statistics*, **17**, 235–250.

DIACONIS, P. and YLVISAKER, D. (1979). Conjugate priors for exponential families. *Annals of Statistics*, **7**, 269–281.

DICKEY, J. M. (1980). Beliefs about beliefs, a theory of stochastic assessments of subjective probabilities. In J. M. BERNARDO, M. H. DEGROOT, D. V. LINDLEY, and A. F. M. SMITH (Eds.), *Bayesian Statistics* (pp. 471–487). Valencia, Spain: University Press.

DOOB, J. L. (1949). Application of the theory of martingales. In *Le Calcul des Probabilités et ses Applications* (pp. 23–27). Paris: Colloques Internationaux du Centre National de la Recherche Scientifique.

DOOB, J. L. (1953). *Stochastic Processes*. New York: Wiley.

DUBINS, L. E. and FREEDMAN, D. A. (1963). Random distribution functions. *Bulletin of the American Mathematical Society*, **69**, 548–551.

DUGUNDJI, J. (1966). *Topology*. Boston: Allyn and Bacon.

DUNFORD, N. and SCHWARTZ, J. T. (1957). *Linear Operators, Part I: General Theory*. New York: Interscience.

DUNFORD, N. and SCHWARTZ, J. T. (1963). *Linear Operators, Part II: Spectral Theory*. New York: Interscience.

EBERHARDT, K. R., MEE, R. W., and REEVE, C. P. (1989). Computing factors for exact two-sided tolerance limits for a normal distribution. *Communications in Statistics—Simulation and Computation*, **18**, 397–413.

EDWARDS, W., LINDMAN, H., and SAVAGE, L. J. (1963). Bayesian statistical inference for psychological research. *Psychological Review*, **70**, 193–242.

EFRON, B. (1979). Bootstrap methods: Another look at the jackknife. *Annals of Statistics*, **7**, 1–26.

EFRON, B. (1982). *The Jackknife, the Bootstrap and Other Resampling Plans*. Philadelphia: Society for Industrial and Applied Mathematics.

EFRON, B. and HINKLEY, D. V. (1978). Assessing the accuracy of the maximum likelihood estimator: Observed versus expected Fisher information. *Biometrika*, **65**, 457–487.

EFRON, B. and MORRIS, C. N. (1975). Data analysis using Stein's estimator and its generalizations. *Journal of the American Statistical Association*, **70**, 311–319.

EFRON, B. and TIBSHIRANI, R. J. (1993). *An Introduction to the Bootstrap*. London: Chapman and Hall.

ESCOBAR, M. D. (1988). *Estimating the Means of Several Normal Populations by Nonparametric Estimation of the Distribution of the Means*. Ph.D. thesis, Yale University.

FABIUS, J. (1964). Asymptotic behavior of Bayes' estimates. *Annals of Mathematical Statistics*, **35**, 846–856.

FERGUSON, T. S. (1967). *Mathematical Statistics: A Decision Theoretic Approach*. New York: Academic Press.

FERGUSON, T. S. (1973). A Bayesian analysis of some nonparametric problems. *Annals of Statistics*, **1**, 209–230.

FERGUSON, T. S. (1974). Prior distributions on spaces of probability measures. *Annals of Statistics*, **2**, 615–629.

FERRÁNDIZ, J. R. (1985). Bayesian inference on Mahalanobis distance: An alternative to Bayesian model testing. In J. M. BERNARDO, M. H. DEGROOT, D. V. LINDLEY, and A. F. M. SMITH (Eds.), *Bayesian Statistics 2: Proceedings of the Second Valencia International Meeting* (pp. 645–653). Amsterdam: North Holland.

FIELLER, E. C. (1954). Some problems in interval estimation. *Journal of the Royal Statistical Society (Series B)*, **16**, 175–185.

FISHBURN, P. C. (1970). *Utility Theory for Decision Making*. New York: Wiley.

FISHER, R. A. (1922). On the mathematical foundations of theoretical statistics. *Philosophical Transactions of the Royal Society of London, Series A*, **222A**, 309–368.

FISHER, R. A. (1924). The conditions under which χ^2 measures the discrepancy between observation and hypothesis. *Journal of the Royal Statistical Society*, **87**, 442–450.

FISHER, R. A. (1925). Theory of statistical estimation. *Proceedings of the Cambridge Philosophical Society*, **22**, 700–725.

FISHER, R. A. (1934). Two new properties of mathematical likelihood. *Proceedings of the Royal Society of London, A*, **144**, 285–307.

FISHER, R. A. (1935). The fiducial argument in statistical inference. *Annals of Eugenics*, **6**, 391–398.

FISHER, R. A. (1936). Has Mendel's work been rediscovered? *Annals of Science*, **1**, 115–137.

FISHER, R. A. (1943). Note on Dr. Berkson's criticism of tests of significance. *Journal of the American Statistical Association*, **38**, 103–104.

FISHER, R. A. (1966). *The Design of Experiments* (8th ed.). New York: Hafner.

FRASER, D. A. S. and MCDUNNOUGH, P. (1984). Further remarks on asymptotic normality of likelihood and conditional analyses. *Canadian Journal of Statistics*, **12**, 183–190.

FREEDMAN, D. A. (1963). On the asymptotic behavior of Bayes' estimates in the discrete case. *Annals of Mathematical Statistics*, **34**, 1386–1403.

FREEDMAN, D. A. (1977). A remark on the difference between sampling with and without replacement. *Journal of the American Statistical Association*, **72**, 681.

FREEDMAN, D. A. and DIACONIS, P. (1982). On inconsistent M-estimators. *Annals of Statistics*, **10**, 454–461.

FREEDMAN, L. S. and SPIEGELHALTER, D. J. (1983). The assessment of subjective opinion and its use in relation to stopping rules of clinical trials. *The Statistician*, **32**, 153–160.

FREEMAN, P. R. (1980). On the number of outliers in data from a linear model. In J. M. BERNARDO, M. H. DEGROOT, D. V. LINDLEY, and A. F. M. SMITH (Eds.), *Bayesian Statistics* (pp. 349–365). Valencia, Spain: University Press.

GABRIEL, K. R. (1969). Simultaneous test procedures—some theory of multiple comparisons. *Annals of Mathematical Statistics*, **40**, 224–250.

GARTHWAITE, P. and DICKEY, J. (1988). Quantifying expert opinion in linear regression problems. *Journal of the Royal Statistical Society (Series B)*, **50**, 462–474.

GARTHWAITE, P. H. and DICKEY, J. M. (1992). Elicitation of prior distributions for variable-selection problems in regression. *Annals of Statistics*, **20**, 1697–1719.

GAVASAKAR, U. K. (1984). *A Study of Elicitation Procedures by Modelling the Errors in Responses*. Ph.D. thesis, Carnegie Mellon University.

GEISSER, S. (1967). Estimation associated with linear discriminants. *Annals of Mathematical Statistics*, **38**, 807–817.

GEISSER, S. and EDDY, W. F. (1979). A predictive approach to model selection. *Journal of the American Statistical Association*, **74**, 153–160.

GELFAND, A. E. and SMITH, A. F. M. (1990). Sampling-based approaches to calculating marginal densities. *Journal of the American Statistical Association*, **85**, 398–409.

GEMAN, S. and GEMAN, D. (1984). Stochastic relaxation, Gibbs distributions and the Bayesian restoration of images. *IEEE Trans. on Pattern Analysis and Machine Intelligence*, **6**, 721–741.

GNANADESIKAN, R. (1977). *Methods for Statistical Data Analysis of Multivariate Observations*. New York: Wiley.

GOOD, I. J. (1956). Discussion of "Chance and control: Some implications of randomization" by G. Spencer Brown. In C. CHERRY (Ed.), *Information Theory: Third London Symposium* (pp. 13–14). London: Butterworths.

HALL, P. (1992). *The Bootstrap and Edgeworth Expansion*. New York: Springer-Verlag.

HALL, W. J., WIJSMAN, R. A., and GHOSH, J. K. (1965). The relationship between sufficiency and invariance with applications in sequential analysis. *Annals of Mathematical Statistics*, **36**, 575–614.

HALMOS, P. R. (1950). *Measure Theory*. New York: Van Nostrand.

HALMOS, P. R. and SAVAGE, L. J. (1949). Application of the Radon–Nikodym theorem to the theory of sufficient statistics. *Annals of Mathematical Statistics*, **20**, 225–241.

HAMPEL, F. R., RONCHETTI, E. M., ROUSSEEUW, P. J., and STAHEL, W. A. (1986). *Robust Statistics: The Approach Based on Influence Functions*. New York: Wiley.

HARTIGAN, J. (1983). *Bayes Theory*. New York: Springer-Verlag.

HEATH, D. and SUDDERTH, W. D. (1976). DeFinetti's theorem on exchangeable variables. *American Statistician*, **30**, 188–189.

HEATH, D. and SUDDERTH, W. D. (1989). Coherent inference from improper priors and from finitely additive priors. *Annals of Statistics*, **17**, 907–919.

HEWITT, E. and SAVAGE, L. J. (1955). Symmetric measures on cartesian products. *Transactions of the American Mathematical Society*, **80**, 470–501.

HEYDE, C. C. and JOHNSTONE, I. M. (1979). On asymptotic posterior normality for stochastic processes. *Journal of the Royal Statistical Society (Series B)*, **41**, 184–189.

HILL, B. M. (1965). Inference about variance components in the one-way model. *Journal of the American Statistical Association*, **60**, 806–825.

HILL, B. M., LANE, D., and SUDDERTH, W. D. (1987). Exchangeable urn processes. *Annals of Probability*, **15**, 1586–1592.

HOEL, P. G., PORT, S. C., and STONE, C. J. (1971). *Introduction to Probability Theory*. Boston: Houghton Mifflin.

HOGARTH, R. M. (1975). Cognitive processes and the assessment of subjective probability distributions (with discussion). *Journal of the American Statistical Association*, **70**, 271–294.

HUBER, P. J. (1964). Robust estimation of a location parameter. *Annals of Mathematical Statistics*, **35**, 73–101.

HUBER, P. J. (1967). The behaviour of maximum likelihood estimates under nonstandard conditions. In L. M. LECAM and J. NEYMAN (Eds.), *Proceedings of the Fifth Berkeley Symposium on Mathematical Statistics and Probability*, volume 1 (pp. 221–233). Berkeley: University of California.

HUBER, P. J. (1977). *Robust Statistical Procedures*. Philadelphia: Society for Industrial and Applied Mathematics.

HUBER, P. J. (1981). *Robust Statistics*. New York: Wiley.

JAMES, W. and STEIN, C. M. (1960). Estimation with quadratic loss. In J. NEYMAN (Ed.), *Proceedings of the Fourth Berkeley Symposium on Mathematical Statistics and Probability*, volume 1 (pp. 361–379). Berkeley: University of California.

JAYNES, E. T. (1976). Confidence intervals vs. Bayesian intervals (with discussion). In W. L. HARPER and C. A. HOOKER (Eds.), *Foundations of Probability Theory, Statistical Inference, and Statistical Theories of Science* (pp. 175–257). Dordrecht: D. Reidel.

JEFFREYS, H. (1961). *Theory of Probability* (3rd ed.). Oxford: Oxford University Press.

JOHNSTONE, I. M. (1978). Problems in limit theory for martingales and posterior distributions from stochastic processes. Master's thesis, Australian National University.

KADANE, J. B., DICKEY, J. M., WINKLER, R. L., SMITH, W., and PETERS, S. C. (1980). Interactive elicitation of opinion for a normal linear model. *Journal of the American Statistical Association*, **75**, 845–854.

KADANE, J. B., SCHERVISH, M. J., and SEIDENFELD, T. (1985). Statistical implications of finitely additive probability. In P. GOEL and A. ZELLNER

(Eds.), *Bayesian Inference and Decision Techniques with Applications: Essays in Honor of Bruno DeFinetti* (pp. 59–76). Amsterdam: Elsevier Science Publishers.

KADANE, J. B., SCHERVISH, M. J., and SEIDENFELD, T. (1993). Reasoning to a foregone conclusion. Technical Report 580, Department of Statistics, Carnegie Mellon University, Pittsburgh, PA.

KAGAN, A. M., LINNIK, Y. V., and RAO, C. R. (1965). On a characterization of the normal law based on a property of the sample average. *Sankhyā, Series A*, **32**, 37–40.

KAHNEMAN, D., SLOVIC, P., and TVERSKY, A. (Eds.) (1982). *Judgment Under Uncertainty: Heuristics and Biases*. Cambridge: Cambridge University Press.

KASS, R. E. and RAFTERY, A. E. (1995). Bayes factors. *Journal of the American Statistical Association*, **90**, 773–795.

KASS, R. E. and STEFFEY, D. (1989). Approximate Bayesian inference in conditionally independent hierarchical models (parametric empirical Bayes models). *Journal of the American Statistical Association*, **84**, 717–726.

KASS, R. E., TIERNEY, L., and KADANE, J. B. (1988). Asymptotics in Bayesian computation. In J. M. BERNARDO, M. H. DEGROOT, D. V. LINDLEY, and A. F. M. SMITH (Eds.), *Bayesian Statistics 3* (pp. 261–278). Oxford: Clarendon Press.

KASS, R. E., TIERNEY, L., and KADANE, J. B. (1990). The validity of posterior expansions based on Laplace's method. In S. GEISSER, J. S. HODGES, S. J. PRESS, and A. ZELLNER (Eds.), *Bayesian and Likelihood Methods in Statistics and Econometrics* (pp. 473–488). Amsterdam: Elsevier (North Holland).

KEIFER, J. and WOLFOWITZ, J. (1956). Consistency of the maximum likelihood estimator in the presence of infinitely many incidental parameters. *Annals of Mathematical Statistics*, **27**, 887–906.

KERRIDGE, D. (1963). Bounds for the frequency of misleading Bayes inferences. *Annals of Mathematical Statistics*, **34**, 1109–1110.

KINDERMAN, A. J. and MONAHAN, J. F. (1977). Computer generation of random variables using the ratio of uniform deviates. *ACM Transactions on Mathematical Software*, **3**, 257–260.

KINGMAN, J. F. C. (1978). Uses of exchangeability. *Annals of Probability*, **6**, 183–197.

KNUTH, D. E. (1984). *The TEXbook*. Reading, MA: Addison-Wesley.

KRAFT, C. H. (1964). A class of distribution function processes which have derivatives. *Journal of Applied Probability*, **1**, 385–388.

KRASKER, W. and PRATT, J. W. (1986). Discussion of "On the consistency of Bayes estimates" by Diaconis and Freedman. *Annals of Statistics*, **14**, 55–58.

KREM, A. (1963). On the independence in the limit of extreme and central order statistics. *Publications of the Mathematical Institute of the Hungarian Academy of Science*, **8**, 469–474.

KSHIRSAGAR, A. M. (1972). *Multivariate Analysis*. New York: Marcel Dekker.

KULLBACK, S. (1959). *Information Theory and Statistics*. New York: Wiley.

LAMPORT, L. (1986). *LaTeX: A Document Preparation System*. Reading, MA: Addison-Wesley.

LAURITZEN, S. L. (1984). Extreme point models in statistics (with discussion). *Scandinavian Journal of Statistics*, **11**, 65–91.

LAURITZEN, S. L. (1988). *Extremal Families and Systems of Sufficient Statistics*. Berlin: Springer-Verlag.

LAVINE, M. (1992). Some aspects of Polya tree distributions for statistical modelling. *Annals of Statistics*, **20**, 1222–1235.

LAVINE, M., WASSERMAN, L., and WOLPERT, R. L. (1991). Bayesian inference with specified prior marginals. *Journal of the American Statistical Association*, **86**, 964–971.

LAVINE, M., WASSERMAN, L., and WOLPERT, R. L. (1993). Linearization of Bayesian robustness problems. *Journal of Statistical Planning and Inference*, **37**, 307–316.

LECAM, L. M. (1953). On some asymptotic properties of maximum likelihood estimates and related Bayes estimates. *University of California Publications in Statistics*, **1**, 277–330.

LECAM, L. M. (1970). On the assumptions used to prove asymptotic normality of maximum likelihood estimates. *Annals of Mathematical Statistics*, **41**, 802–828.

LECOUTRE, B. (1985). Reconsideration of the F-test of the analysis of variance: The semi-Bayesian significance test. *Communications in Statistics—Theory and Methods*, **14**, 2437–2446.

LECOUTRE, B. and ROUANET, H. (1981). Deux structures statistiques fondamentales en analyse de la variance univariée et mulitvariée. *Mathématiques et Sciences Humaines*, **75**, 71–82.

LEHMANN, E. L. (1958). Significance level and power. *Annals of Mathematical Statistics*, **29**, 1167–1176.

LEHMANN, E. L. (1983). *Theory of Point Estimation*. New York: Wiley.

LEHMANN, E. L. (1986). *Testing Statistical Hypotheses* (2nd ed.). New York: Wiley.

LEHMANN, E. L. and SCHEFFÉ, H. (1955). Completeness, similar regions and unbiased estimates. *Sankhyā*, **10**, 305–340. (Also **15**, 219–236, and correction **17**, 250.)

LINDLEY, D. V. (1957). A statistical paradox. *Biometrika*, **44**, 187–192.

LINDLEY, D. V. and NOVICK, M. R. (1981). The role of exchangeability in inference. *Annals of Statistics*, **9**, 45–58.

LINDLEY, D. V. and PHILLIPS, L. D. (1976). Inference for a Bernoulli process (a Bayesian view). *American Statistician*, **30**, 112–119.

LINDLEY, D. V. and SMITH, A. F. M. (1972). Bayes estimates for the linear model. *Journal of the Royal Statistical Society (Series B)*, **34**, 1–41.

LOÈVE, M. (1977). *Probability Theory I* (4th ed.). New York: Springer-Verlag.

MAULDIN, R. D., SUDDERTH, W. D., and WILLIAMS, S. C. (1992). Polya trees and random distributions. *Annals of Statistics*, **20**, 1203–1221.

MAULDIN, R. D. and WILLIAMS, S. C. (1990). Reinforced random walks and random distributions. *Proceedings of the American Mathematical Society*, **110**, 251–258.

MENDEL, G. (1866). Versuche über pflanzenhybriden. *Verhandlungen Naturforschender Vereines in Brünn*, **10**, 1.

MÉTIVIER, M. (1971). Sur la construction de mesures aléatoires presque sûrement absolument continues par rapport à une mesure donnée. *Zeitschrift fur Wahrscheinlichkeitstheorie*, **20**, 332–344.

MORRIS, C. N. (1983). Parametric empirical Bayes inference: Theory and applications (with discussion). *Journal of the American Statistical Association*, **78**, 47–65.

NACHBIN, L. (1965). *The Haar Integral*. Princeton: Van Nostrand.

NEYMAN, J. (1935). Su un teorema concernente le cosiddette statistiche sufficienti. *Giornale Dell'Istituto Italiano degli Attuari*, **6**, 320–334.

NEYMAN, J. and PEARSON, E. S. (1933). On the problem of the most efficient test of statistical hypotheses. *Philosophical Transactions of the Royal Society of London, Series A*, **231**, 289–337.

NEYMAN, J. and SCOTT, E. L. (1948). Consistent estimates based on partially consistent observations. *Econometrica*, **16**, 1–32.

PEARSON, K. (1900). On the criterion that a given system of deviations from the probable in the case of a correlated system of variables is such that it can be reasonably supposed to have arisen from random sampling. *Philosophical Magazine (5thSeries)*, **50**, 339–357. (See also correction, *Philosophical Magazine (6thSeries)*, **1**, 670–671.)

PERLMAN, M. (1972). On the strong consistency of approximate maximum likelihood estimators. In L. M. LECAM, J. NEYMAN, and E. L. SCOTT (Eds.), *Proceedings of the Sixth Berkeley Symposium on Mathematical Statistics and Probability*, volume 1 (pp. 263–281). Berkeley: University of California.

PIERCE, D. A. (1973). On some difficulties in a frequency theory of inference. *Annals of Statistics*, **1**, 241–250.

PITMAN, E. (1939). The estimation of location and scale parameters of a continuous population of any given form. *Biometrika*, **30**, 391–421.

PRATT, J. W. (1961). Review of "Testing Statistical Hypotheses" by E. L. Lehmann. *Journal of the American Statistical Association*, **56**, 163–167.

PRATT, J. W. (1962). Discussion of "On the foundations of statistical inference" by Allan Birnbaum. *Journal of the American Statistical Association*, **57**, 314–316.

RAO, C. R. (1945). Information and the accuracy attainable in the estimation of statistical parameters. *Bulletin of the Calcutta Mathematical Society*, **37**, 81–91.

RAO, C. R. (1973). *Linear Statistical Inference and Its Applications* (2nd ed.). New York: Wiley.

ROBBINS, H. (1951). Asymptotically subminimax solutions of compound statistical decision problems. In J. NEYMAN (Ed.), *Proceedings of the Second Berkeley Symposium on Mathematical Statistics and Probability* (pp. 131–148). Berkeley: University of California.

ROBBINS, H. (1955). An empirical Bayes approach to statistics. In J. NEYMAN (Ed.), *Proceedings of the Third Berkeley Symposium on Mathematical Statistics and Probability*, volume 1 (pp. 157–164). Berkeley: University of California.

ROBBINS, H. (1964). The empirical Bayes approach to statistical decision problems. *Annals of Mathematical Statistics*, **35**, 1–20.

ROBERT, C. P. (1993). A note on Jeffreys–Lindley paradox. *Statistica Sinica*, **3**, 601–608.

ROBERTS, H. V. (1967). Informative stopping rules and inferences about population size. *Journal of the American Statistical Association*, **62**, 763–775.

ROUANET, H. and LECOUTRE, B. (1983). Specific inference in ANOVA: From significance tests to Bayesian procedures. *British Journal of Mathematical and Statistical Psychology*, **36**, 252–268.

ROYDEN, H. L. (1968). *Real Analysis*. London: Macmillan.

RUBIN, D. B. (1981). The Bayesian bootstrap. *Annals of Statistics*, **9**, 130–134.

RUDIN, W. (1964). *Principles of Mathematical Analysis* (2nd ed.). New York: McGraw-Hill.

SAVAGE, L. J. (1954). *The Foundations of Statistics*. New York: Wiley.

SAVAGE, L. J. (1962). *The Foundations of Statistical Inference*. London: Methuen.

SCHEFFÉ, H. (1947). A useful convergence theorem for probability distributions. *Annals of Mathematical Statistics*, **18**, 434–438.

SCHERVISH, M. J. (1983). User-oriented inference. *Journal of the American Statistical Association*, **78**, 611–615.

SCHERVISH, M. J. (1992). Bayesian analysis of linear models (with discussion). In J. M. BERNARDO, J. O. BERGER, A. P. DAWID, and A. F. M. SMITH (Eds.), *Bayesian Statistics 4: Proceedings of the Second Valencia International Meeting* (pp. 419–434). Oxford: Clarendon Press.

SCHERVISH, M. J. (1994a). Discussion of "Bootstrap: More than a stab in the dark?" by G. A. Young. *Statistical Science*, **9**, 408–410.

SCHERVISH, M. J. (1994b). A significance paradox. Technical Report 598, Department of Statistics, Carnegie Mellon University, Pittsburgh, PA.

SCHERVISH, M. J. and CARLIN, B. P. (1992). On the convergence of successive substitution sampling. *Journal of Computational and Graphical Statistics*, **1**, 111–127.

SCHERVISH, M. J. and SEIDENFELD, T. (1990). An approach to consensus and certainty with increasing evidence. *Journal of Statistical Planning and Inference*, **25**, 401–414.

SCHERVISH, M. J., SEIDENFELD, T., and KADANE, J. B. (1984). The extent of non-conglomerability of finitely additive probabilities. *Zeitschrift fur Wahrscheinlichkeitstheorie*, **66**, 205–226.

SCHERVISH, M. J., SEIDENFELD, T., and KADANE, J. B. (1990). State dependent utilities. *Journal of the American Statistical Association*, **85**, 840–847.

SCHWARTZ, L. (1965). On Bayes procedures. *Zeitschrift fur Wahrscheinlichkeitstheorie*, **4**, 10–26.

SEIDENFELD, T. and SCHERVISH, M. J. (1983). A conflict between finite additivity and avoiding Dutch Book. *Philosophy of Science*, **50**, 398–412.

SEIDENFELD, T., SCHERVISH, M. J., and KADANE, J. B. (1992). A representation of partially ordered preferences. Technical Report 453, Department of Statistics, Carnegie Mellon University, Pittsburgh, PA.

SERFLING, R. J. (1980). *Approximation Theorems of Mathematical Statistics*. New York: Wiley.

SETHURAMAN, J. (1994). A constructive definition of Dirichlet priors. *Statistica Sinica*, **4**, 639–650.

SINGH, K. (1981). On the asymptotic accuracy of Efron's bootstrap. *Annals of Statistics*, **9**, 1187–1195.

SMITH, A. F. M. (1973). A general Bayesian linear model. *Journal of the Royal Statistical Society, Ser. B*, **35**, 67–75.

SPJØTVOLL, E. (1983). Preference functions. In P. J. BICKEL, K. DOKSUM, and J. L. HODGES, JR. (Eds.), *A Festschrift for Erich L. Lehmann* (pp. 409–432). Belmont, CA: Wadsworth.

STATSCI (1992). *S-PLUS, Version 3.1 (software package)*. Seattle: StatSci Division, MathSoft, Inc.

STEIN, C. M. (1946). A note on cumulative sums. *Annals of Mathematical Statistics*, **17**, 498–499.

STEIN, C. M. (1956). Inadmissibility of the usual estimator for the mean of a multivariate normal distribution. In J. NEYMAN (Ed.), *Proceedings of the Third Berkeley Symposium on Mathematical Statistics and Probability*, volume 1 (pp. 197–206). Berkeley: University of California.

STEIN, C. M. (1965). Approximation of improper prior measures by prior probability measures. In J. NEYMAN and L. M. LECAM (Eds.), *Bernoulli, Bayes, Laplace: Anniversary Volume* (pp. 217–240). New York: Springer-Verlag.

STEIN, C. M. (1981). Estimation of the mean of a multivariate normal distribution. *Annals of Statistics*, **9**, 1135–1151.

STIGLER, S. M. (1986). *The History of Statistics: The Measurement of Uncertainty before 1900*. Cambridge, MA: Belknap.

STONE, M. (1976). Strong inconsistency from uniform priors. *Journal of the American Statistical Association*, **71**, 114–125.

STONE, M. and DAWID, A. P. (1972). Un-Bayesian implications of improper Bayes inference in routine statistical problems. *Biometrika*, **59**, 369–375.

STRASSER, H. (1981). Consistency of maximum likelihood and Bayes estimates. *Annals of Statistics*, **9**, 1107–1113.

STRAWDERMAN, W. E. (1971). Proper Bayes minimax estimators of the multivariate normal mean. *Annals of Mathematical Statistics*, **42**, 385–388.

TAYLOR, R. L., DAFFER, P. Z., and PATTERSON, R. F. (1985). *Limit Theorems for Sums of Exchangeable Random Variables*. Totowa, NJ: Rowman and Allanheld.

TIERNEY, L. (1994). Markov chains for exploring posterior distributions (with discussion). *Annals of Statistics*, **22**, 1701–1762.

TIERNEY, L. and KADANE, J. B. (1986). Accurate approximations for posterior moments and marginal densities. *Journal of the American Statistical Association*, **81**, 82–86.

VENN, J. (1876). *The Logic of Chance* (2nd ed.). London: Macmillan.

VERDINELLI, I. and WASSERMAN, L. (1991). Bayesian analysis of outlier problems using the Gibbs sampler. *Statistics and Computing*, **1**, 105–117.

VON MISES, R. (1957). *Probability, Statistics and Truth*. London: Allen and Unwin.

VON NEUMANN, J. and MORGENSTERN, O. (1947). *Theory of Games and Economic Behavior* (2nd ed.). Princeton: Princeton University Press.

WALD, A. (1947). *Sequential Analysis*. New York: Wiley.

WALD, A. (1949). Note on the consistency of the maximum likelihood estimate. *Annals of Mathematical Statistics*, **20**, 595–601.

WALD, A. and WOLFOWITZ, J. (1948). Optimum character of the sequential probability ratio test. *Annals of Mathematical Statistics*, **19**, 326–339.

WALKER, A. M. (1969). On the asymptotic behaviour of posterior distributions. *Journal of the Royal Statistical Society (Series B)*, **31**, 80–88.

WALLACE, D. L. (1959). Conditional confidence level properties. *Annals of Mathematical Statistics*, **30**, 864–876.

WELCH, B. L. (1939). On confidence limits and sufficiency, with particular reference to parameters of location. *Annals of Mathematical Statistics*, **10**, 58–69.

WEST, M. (1984). Outlier models and prior distributions in Bayesian linear regression. *Journal of the Royal Statistical Society (Series B)*, **46**, 431–439.

WILKS, S. S. (1941). Determination of sample sizes for setting tolerance limits. *Annals of Mathematical Statistics*, **12**, 91–96.

YOUNG, G. A. (1994). Bootstrap: More than a stab in the dark? (with discussion). *Statistical Science*, **9**, 382–415.

ZELLNER, A. (1971). *An Introduction to Bayesian Inference in Econometrics*. New York: Wiley.

Notation and Abbreviation Index

0 (vector of 0s), 385

1 (vector of 1s), 345

2^S (power set), 571

$\ll$ (absolutely continuous), 574, 597
a.e. (almost everywhere), 582
$ANCB(\cdot,\cdot,\cdot)$ (distribution), 668
$ANC\chi^2(\cdot,\cdot,\cdot)$ (distribution), 668
$ANCF(\cdot,\cdot,\cdot)$ (distribution), 669
ANOVA (analysis of variance), 384
ARE (asymptotic relative efficiency), 413
a.s. (almost surely), 582
$\mathcal{A}_X$ (σ-field generated by X), 51, 82
$\aleph$ (action space), 144
α (action space σ-field), 144

$\setminus$ (remove one set from another), 577
$\overline{B}$ (closure of set), 622
$Ber(\cdot)$ (distribution), 672
$Beta(\cdot,\cdot)$ (distribution), 669
$Bin(\cdot,\cdot)$ (distribution), 673
$\mathcal{B}^k$ (Borel σ-field), 576
$\mathcal{B}$ (Borel σ-field), 575

$Cau(\cdot,\cdot)$ (distribution), 669
CDF (cumulative distribution function), 612
c'_g (constant related to RHM), 367
c_g (constant related to LHM), 367
χ_a^2 (distribution), 669
$\xrightarrow{\mathcal{D}}$ (converges in distribution), 635
$\xrightarrow{P}$ (converges in probability), 638
$\xrightarrow{W}$ (converges weakly), 635
$Cov_\theta(\cdot,\cdot)$ (conditional covariance given $\Theta = \theta$), 19
Cov (covariance), 613
$\mathcal{C}_\mathcal{P}$ (σ-field on set of probability measures), 27
A^C (complement of set), 575

$Dir(\cdot)$ (Dirichlet process), 54

$Dir_k(\ldots)$ (distribution), 674
$d\mu_2/d\mu_1$ (Radon–Nikodym derivative), 575, 598
Δ (symmetric difference), 581

$E_\theta(\cdot)$ (conditional mean given $\Theta = \theta$), 19
$Exp(\cdot)$ (distribution), 670
$E(\cdot)$ (expected value), 607, 613
$E(\cdot|\cdot)$ (conditional mean), 616

f^+ (positive part), 588
f^- (negative part), 588
$f_{X|\Theta}$ (conditional density of X given Θ), 13
$f_{X|Y}$ (conditional density), 13
$F_{q,a}$ (distribution), 670

$\Gamma^{-1}(\cdot,\cdot)$ (distribution), 670
$\Gamma(\cdot,\cdot)$ (distribution), 670
$Geo(\cdot)$ (distribution), 673

HPD (highest posterior density), 327
$Hyp(\cdot,\cdot,\cdot)$ (distribution), 673

IID (independent and identically distributed), 611
I_k (identity matrix), 643
$\mathcal{I}_{X|T}(\cdot;\cdot|\cdot)$ (conditional Kullback–Leibler information), 115
$\mathcal{I}_{X|T}(\cdot|\cdot)$ (conditional Fisher information), 111
$\mathcal{I}_X(\cdot)$ (Fisher information), 111
$\mathcal{I}_X(\cdot;\cdot)$ (Kullback–Leibler information), 115
$I_A(\cdot)$ (indicator function), 9

$Lap(\cdot,\cdot)$ (distribution), 670
λ_g (measure constructed from LHM), 367
LHM (left Haar measure), 363
LMP (locally most powerful), 245
LMVUE (locally minimum variance unbiased estimator), 300

Notation and Abbreviation Index

LR (likelihood ratio), 274

MC (most cautious), 230
MLE (maximum likelihood estimator), 307
MLR (monotone likelihood ratio), 239
MP (most powerful), 230
MRE (minimum risk equivariant), 347
$Mult_k(\ldots)$ (distribution), 674
$\mu_{\Theta|X}(\cdot|\cdot)$ (posterior distribution), 16

$NCB(\cdot,\cdot,\cdot)$ (distribution), 671
$NC\chi_q^2(\cdot)$ (distribution), 671
$NCF(\cdot,\cdot,\cdot)$ (distribution), 671
$NCt_a(\cdot)$ (distribution), 671
$NCT_a(\cdot;\cdot)$ (CDF of NCt distribution), 671
$Negbin(\cdot,\cdot)$ (distribution), 673
ν (dominating measure), 13
$N(\cdot,\cdot)$ (distribution), 671
$N_p(\cdot,\cdot)$ (distribution), 674
$\mathcal{N}$ (integers plus ∞), 537

Ω (parameter space), 13, 82
o_P (stochastic small order), 396
$\mathcal{O}_P$ (stochastic large order), 396
o (small order), 394
$\mathcal{O}$ (large order), 394

$\mathcal{P}_0$ (parametric family), 50, 82
$Par(\cdot,\cdot)$ (distribution), 672
∂B (boundary of set), 636
$\Phi(\cdot)$ (CDF of normal distribution), 672
$\mathbf{P}_n$ (empirical probability measure), 12
$Poi(\cdot)$ (distribution), 673
$\Pr(\cdot)$ (probability), 612
$\Pr(\cdot|\cdot)$ (conditional probability), 617
$P_{\theta,T}(\cdot)$ (conditional distribution of T given $\Theta = \theta$), 84
$P'_\theta(\cdot)$ (conditional probability given $\Theta = \theta$), 51, 83
$P_\theta(\cdot)$ (conditional distribution given $\Theta = \theta$), 51, 83
$\mathbf{P}$ (random probability measure), 25
$\mathcal{P}$ (set of all probability measures), 27

$Q_n(\cdot|x)$ (conditional distribution given n observations), 539

$\mathbb{R}$ (real numbers), 570
$\mathbb{R}^+$ (positive reals), 571
$\mathbb{R}^{+0}$ (nonnegative reals), 627
ρ_g (measure constructed from RHM), 367
RHM (right Haar measure), 363
$r(\eta,\delta)$ (Bayes risk), 149
$R(\theta,\delta)$ (risk function), 149

$(S, \mathcal{A}, \mu)$ (measure space), 577
SPRT (sequential probability ratio test), 549
SSS (successive substitution sampling), 507

τ (parameter space σ-field), 13
Θ' (parametric index), 50
Θ (parameter), 51
T (statistic), 84
$T_a(\cdot)$ (CDF of t distribution), 672
$t_a(\cdot,\cdot)$ (distribution), 672
$v^\top$ (transpose of vector), 614

UMA (uniformly most accurate), 317
UMAU (uniformly most accurate unbiased), 321
UMC (uniformly most cautious), 230
UMCU (uniformly most cautious unbiased), 254
UMP (uniformly most powerful), 230
UMPU (uniformly most powerful unbiased), 254
UMPUAI (uniformly most powerful unbiased almost invariant), 384
UMVUE (uniformly minimum variance unbiased estimator), 297
USC (upper semicontinuous), 417
$U(\cdot,\cdot)$ (distribution), 672

$\text{Var}_\theta(\cdot,\cdot)$ (conditional variance given $\Theta = \theta$), 19
Var (variance), 613

x (element of sample space), 82
$\mathcal{X}$ (sample space), 13, 82

Name Index

Ahlfors, L., 667, 675
Aitchison, J., 325, 675
Albert, J., 519, 675
Aldous, D., 46, 79, 482, 675
Anderson, T., 386, 675
Andrews, C., ix
Anscombe, F., 181, 675
Antoniak, C., 59, 675
Aumann, R., 181, 675

Bahadur, R., 94, 675
Barnard, G., 320, 420, 675
Barndorff-Nielsen, O., 307, 675
Barnett, V., vii, 675
Barron, A., 434–435, 446, 675, 677
Basu, D., 99–100, 675
Bayes, T., 16, 29, 676
Becker, R., x, 676
Berberian, S., 507, 667, 676
Berger, J., 22, 173, 284, 525, 565, 614, 666, 676
Berger, R., 283, 677
Berk, R., 417, 430, 432, 676
Berkson, J., 218, 281, 676
Berry, D., 565, 676
Berti, P., 21, 676
Bhattacharyya, A., 305
Bickel, P., 330–331, 676
Billingsley, P., 46, 621, 636, 648, 676
Bishop, Y., 462, 676
Blackwell, D., 56, 86, 152, 455, 676
Blyth, C., 158, 676
Bohrer, R., x
Bondar, J., 236, 676
Bortkiewicz, L., 462, 677
Box, G., 21, 521, 677
Breiman, L., 618, 640, 677
Brenner, D., 435, 677
Brown, L., 99, 160, 167, 677
Buck, C., 665, 677
Buehler, R., 99, 677

Carlin, B., 507, 686
Casella, G., 283, 677

Chaloner, K., 24, 677
Chambers, J., x, 676
Chang, T., 379, 677
Chapman, D., 303, 677
Chen, C., 435, 677
Chibb, S., 519, 675
Chow, Y., 647, 677
Church, T., 24, 677
Churchill, R., 666–667, 677
Clarke, B., 446, 677
Cornfield, J., 563, 565, 678
Cox, D., 21, 218, 424, 521, 677–678
Cramér, H., 301, 678

Daffer, P., 33, 688
David, H., 404, 678
Dawid, A., 21, 125, 435, 521, 678, 687
DeFinetti, B., ix, 6, 21, 25, 28, 654, 656–657, 678
DeGroot, M., ix, 91, 98, 181, 362, 536, 654, 678
DeMoivre, A., 8, 678
Diaconis, P., ix, 15, 28, 41, 46, 108, 123, 126, 426, 434, 479–480, 667, 678–680
Dickey, J., 24, 679, 681–682
Doob, J., 36, 429, 507, 645–646, 679
Doytchinov, B., ix
Dubins, L., 70, 455, 676, 679
Dugundji, J., 666, 679
Dunford, N., 507, 635, 667, 679
Dunsmore, I., 325, 675

Eberhardt, K., 326, 679
Eddy, W., 521, 681
Edwards, W., 222, 284, 679
Efron, B., 166, 330–331, 335–336, 423, 679
Escobar, M., 60, 679

Fabius, J., 61, 679
Fedderson, A., 99, 677
Ferguson, T., 52, 56, 61, 173, 179, 181, 248, 258, 614, 666, 680

Ferrándiz, J., 669, 680
Fieller, E., 321, 680
Fienberg, S., 462, 676
Fishburn, P., 181, 680
Fisher, R., 89, 96, 217–218, 307, 370, 373, 522, 680
Fraser, D., 435, 677, 680
Freedman, D., 15, 28, 40–41, 46, 61, 70, 123, 126, 330–331, 426, 434, 479–480, 667, 676, 678–680
Freedman, L., 24, 680
Freeman, P., 524, 681

Gabriel, K., 252, 681
Garthwaite, P., 24, 681
Gavasakar, U., 24, 681
Geisser, S., 521, 668, 681
Gelfand, A., 507, 681
Geman, D., 507, 681
Geman, S., 507, 681
Ghosh, J., 381–382, 681
Gnanadesikan, R., 22, 681
Good, I., 565, 681

Hadjicostas, P., ix
Hall, P., 337–338, 681
Hall, W., 381–382, 681
Halmos, P., 364, 600, 681
Hampel, F., 315, 681
Hartigan, J., 20–21, 33, 681
Heath, D., 21, 46, 681–682
Hewitt, E., 46, 682
Heyde, C., 435, 682
Hill, B., 9, 484, 682
Hinkley, D., 218, 423, 678–679
Hodges, J., 414
Hoel, P., 640, 682
Hogarth, R., 24, 682
Holland, P., 462, 676
Huber, P., 310, 315, 428, 682
Hwang, J., 160, 677

James, W., 163, 682
Jaynes, E., 379, 682
Jeffreys, H., 122, 229, 284, 682
Jiang, T., ix
Johnstone, I., 435, 682

Kadane, J., 21, 24, 183–184, 446, 564, 655, 682–683, 687–688
Kagan, A., 349, 683
Kahneman, D., 23, 683
Kass, R., ix, 226, 446, 505, 683
Kerridge, D., 564, 683
Kiefer, J., 417, 420, 683
Kinderman, A., 660, 683
Kingman, J., 36, 683
Knuth, D., x, 683
Kraft, C., 66, 683
Krasker, W., 56, 683
Krem, A., 408, 683
Kshirsagar, A., 386, 684
Kullback, S., 116, 684

Lamport, L., x, 684
Lane, D., 9, 682
Lauritzen, S., 28, 123, 481, 684
Lavine, M., 69, 526, 684
LeCam, L., 414, 437, 684
Lecoutre, B., 668–669, 684, 686
Lehmann, E., 231, 280, 285, 298, 350, 684
Lévy, P., 648, 650
Lindley, D., 6, 229, 284, 479, 684
Lindman, H., 222, 284, 679
Linnik, Y., 349, 683
Loève, M., 34, 653, 685
Louis, T., 24, 677

Matts, J., 24, 677
Mauldin, R., 66, 69, 685
McDunnough, P., 435, 677, 680
Mee, R., 326, 679
Mendel, G., 217, 685
Métivier, M., 66, 685
Monahan, J., 660, 683
Morgenstern, O., 181–182, 688
Morris, C., 166, 500, 679, 685

Nachbin, L., 364, 685
Neyman, J., 89, 175, 231, 247, 420, 685
Nobile, A., ix
Novick, M., ix, 6, 684

Oue, S., ix

Patterson, R., 33, 688
Pearson, E., 175, 231, 247, 685
Pearson, K., 216, 685
Perlman, M., 430, 685
Peters, S., 24, 682
Phillips, L., 6, 684
Pierce, D., 99, 685
Pitman, E., 347, 685
Port, S., 640, 682
Portnoy, S., x
Pratt, J., 56, 98, 683, 685

Raftery, A., 226, 683
Ramamoorthi, R., 86, 676
Rao, C., 152, 301, 349, 683, 685–686
Reeve, C., 326, 679
Regazzini, E., 21, 676
Rigo, P., 21, 676
Robbins, H., 303, 647, 677, 686
Robert, C., 225, 686
Roberts, H., 565, 686
Ronchetti, E., 315, 681
Rouanet, H., 668–669, 684, 686
Rousseeuw, P, 315, 681
Royden, H., 578, 589, 597, 621, 686
Rubin, D., 332, 686
Rudin, W., 666, 686

Savage, L., 46, 181, 222, 284, 565, 600, 679, 681–682, 686
Scheffé, H., 298, 634, 684, 686
Schervish, M., v
Schwartz, J., 507, 635, 667, 679
Schwartz, L., 429, 687
Scott, E., 420, 685
Seidenfeld, T., ix, 21, 183–184, 187, 429, 564, 655, 682–683, 686–687
Sellke, T., 284, 676
Serfling, R., 413, 687
Sethuraman, J., 56, 687
Short, T., ix
Shurlow, N., v
Siegmund, D., 647, 677
Singh, K., 331, 687
Slovic, P., 23, 683

Smith, A., 479, 507, 681, 684, 687
Smith, W., 24, 682
Spiegelhalter, D., 24, 680
Spjøtvoll, E., 283, 687
Stahel, W., 315, 681
Steffey, D., 505, 683
Stein, C., 163, 379, 382, 568, 682, 687
Stigler, S., 8, 687
Stone, C., 640, 682
Stone, M., 21, 678, 687
Strasser, H., 430, 688
Strawderman, W., ix, 166, 688
Sudderth, W., 9, 21, 46, 66, 69, 681–682, 685

Taylor, R., 33, 688
Tiao, G., 521, 677
Tibshirani, R., 336, 679
Tierney, L., 225, 446, 507, 683, 688
Tversky, A., 23, 683

Venn, J., 8, 688
Verdinelli, I., 524, 688
Villegas, C., 379, 677
Von Mises, R., 10, 688
Von Neumann, J., 181–182, 688

Wald, A., 415, 549, 552, 557, 688
Walker, A., 435, 442, 688
Wallace, D., 99, 688
Wasserman, L., ix, 524, 526, 684, 688
Welch, B., 320, 688
West, M., 524, 688
Wijsman, R., x, 381–382, 681
Wilks, A., x, 676
Wilks, S., 325, 688
Williams, S., 66, 69, 685
Winkler, R., 24, 682
Wolfowitz, J., 417, 420, 557, 683, 688
Wolpert, R., 526, 684

Ylvisaker, D., 108, 679
Young, G., 329, 688

Zellner, A., 16, 688
Zidek, J., 21, 678

Subject Index*

Abelian group, *353*
Absolutely continuous, 574, *597*, 668
Absolutely continuous function, *211*
Accept hypothesis, *214*
Acceptance–rejection, *659*
Action space, *144*
Admissible, *154*–157, 162, 167, 174
  λ, *154*–156, 162
Almost everywhere, 572, *582*
Almost invariant function, *383*
Almost surely, 572, *582*
Alternate noncentral beta
  distribution, *668*
Alternate noncentral χ^2 distribution, *668*
Alternate noncentral F distribution, *669*
Alternative, 2, *214*
  composite, *215*
  simple, *215*, 233
Analysis of variance, 384, 491
Analytic function, *105*
Ancillary statistic, *95*, 99, 119
  maximal, *97*
ANOVA, 384, 491
Archemedian condition, 192
ARE, *413*
Asymptotic distribution, *399*
Asymptotic efficiency, *413*
Asymptotic relative efficiency, *413*
Asymptotic variance, *402*
Autoregression, 141
Autoregressive process, 441
Axioms of decision theory, 183–184, 296

Backward induction, *537*
Bahadur's theorem, 94
Base measure, *54*
Base of test, *215*–216

Basu's theorem, 99
Bayes factor, *221*, 238, 262–263, 274
Bayes risk, *149*
Bayes rule, *150*, 154–155, 167–168, 178
  extended, *169*
  formal, 146, *150*, 157, 348, 351, 369
  generalized, *156*–157
  partial, 147, *150*
Bayes' theorem, 4, 16
Bayesian bootstrap, 332
Bernoulli distribution, *672*
Beta distribution, 54, *669*
Bhattacharyya lower bounds, 305
Bias, *296*
Bimeasurable function, 572, *583*, 618
Binomial distribution, *673*
Bolzano–Weierstrass theorem, 666
Bootstrap, 329
  Bayesian, 332
  nonparametric, 329
  parametric, 330
Borel σ-field, 571, *575*
Borel space, 609, *618*
Borel–Cantelli lemma:
  first, 578
  second, 663
Boundary, *636*
Boundedly complete statistic, *94*, 99
Box–Cox transformations, 521

Called-off preference, *184*
Caratheodory extension theorem, 578
Cauchy distribution, *669*
Cauchy sequence, *619*
Cauchy's equation, 667
Cauchy–Schwarz inequality, 615
CDF, *612*
  empirical, *404*–405, 408
Central limit theorem, 642
  multivariate, 643
Chain rule, 600
Chapman–Robbins lower bound, 304

---

*Italicized page numbers indicate where a term is defined.

Subject Index

Characteristic function, 611, *639*
Chi-squared distribution, *669*
Chi-squared test of independence, 467
Closed set, *622*
Closure, *622*
Coherent tests, *252*
Complete class, *174*
    essentially, *174*, 244, 251, 256
        minimal, *174*
    minimal, *174–175*
Complete class theorem, 179
Complete measure space, *579*, 603
Complete metric space, *619*
Complete statistic, *94*, 298
    boundedly, *94*, 99
Composite alternative, *215*
Composite hypothesis, *215*
Conditional distribution, 13, 16, 607, 609, *617*
    regular, 610, *618*
    version, *617*
Conditional expectation, 19, 607, *616*
    version, 608, *616*
Conditional Fisher information, *111*, 119
Conditional independence, 9, 610, *628*
Conditional Kullback–Leibler information, *115*, 119
Conditional mean, 607, *616*
    version, *616*
Conditional preference, *185*
    consistent, *186*
Conditional probability, 607, 609, *617*
    regular, 609, *617*
Conditional score function, *111*
Conditionally sufficient statistic, *95*
Confidence coefficient, *315*, 325
Confidence interval, 3
    fixed-width, *559*
    sequential, 559
Confidence sequence, *569*
Confidence set, 279, *315*, 379
    conservative, *315*
    exact, *315*
    randomized, *316*
    UMA, *317*
    UMAU, *321*
Conjugate prior, *92*
Conservative confidence set, *315*

Conservative prediction set, *324*
Conservative tolerance set, *325*
Consistent, *397*, 412
Consistent conditional preference, *186*
Consistent distributions, *652*
Contingency table, 467
Continuity axiom, 184
Continuity theorem, 640
Continuous distribution, *612*
Continuous mapping theorem, 638
Convergence:
    pointwise, *184*
    weak, 399, *635*
Convergence in distribution, 399, 611, *635*
Convergence in probability, 396, 611, *638*
Convex function, *614*
Counting measure, *570*
Covariance, 607, *613*
Cramér–Rao lower bound, 301
    multiparameter, 306
Credible set, *327*
Cumulative distribution function (see CDF), *612*
Cylinder set, *652*

Data, 82
Decide optimally after stopping, *540*
Decision rule, *145*
    maximum, *541*
    nonrandomized, *145*, 151, 153
    nonrandomized sequential, *537*
    randomized, *145*, 151
    randomized sequential, *537*
    regular, *540*–541
    sequential, *537*
        nonrandomized, *537*
        randomized, *537*
    terminal, *537*
    truncated, *542*
Decision theory, 144, 181
    axioms, 183–184
Decreasing sequence of sets, *577*
DeFinetti's representation theorem, 28
Degenerate exponential family, *104*
Degenerate weak order, *183*
Delta method, 401, 464, 466

696 Subject Index

Dense, *619*
Density, 607, *613*
Dirichlet distribution, 52, 54, *674*
Dirichlet process, 52, *54*, 332, 434
Discrete distribution, *612*
Distribution:
    alternate noncentral beta, *668*
    alternate noncentral χ^2, *668*
    alternate noncentral F, *669*
    asymptotic, 399
    Bernoulli, *672*
    beta, 54, *669*
    binomial, *673*
    Cauchy, *669*
    chi-squared, *669*
    conditional, 13, 16
    consistent, *652*
    continuous, 612
    Dirichlet, 52, 54, *674*
    discrete, 612
    empirical, *12*, 38
    exponential, *670*
    F, *670*
    fiducial, 370, 373
    gamma, *670*
    geometric, *673*
    half-normal, 389
    hypergeometric, *673*
    inverse gamma, *670*
    Laplace, *670*
    least favorable, 168
    marginal, *14*
    multinomial, *674*
    multivariate normal, 643, *674*
    negative binomial, *673*
    noncentral beta, 289, *671*
    noncentral χ^2, *671*
    noncentral F, 289, *671*
    noncentral t, 289, 325, *671*
    normal, 21, 349, 611, 640, 642, *671*
        multivariate, 643, *674*
    Pareto, *672*
    Poisson, *673*
    posterior, *16*
    predictive, *14*
        posterior, *18*
        prior, *14*
    prior, *13*
    improper, *20*
    t, *672*
    uniform, 659, *672*
Distribution function (see CDF), *612*
Dominance axiom, 185
Dominated convergence theorem, 591
Dominates, *154*
Dominating measure, 574, *597*
Dutch book, 656

Efficiency:
    asymptotic, *413*
    asymptotic relative, *413*
    second-order, 414
Elicitation of probabilities, 22–23
Empirical Bayes, 166, 420, 500
Empirical CDF, *404*–405, 408
Empirical distribution, *12*, 38
Empirical probability measure, *12*
ϵ-contamination class, *524*, 526, 528
Equal-tailed test, 263
Equivalence class, *140*
Equivalence relation, *140*
Equivariant rule, *357*
    location, *346*–347, 351
    minimum risk (see MRE), *347*
    scale, *350*
Essentially complete class, *174*, 244, 251, 256
    minimal, *174*
Estimator, 3, *296*
    maximum likelihood, 3, *307*
    MRE, 347, 351, 363
    Pitman, *347*, 363
    point, 3, 296
    unbiased, 3, *296*
Event, 606, *612*
Exact confidence set, *315*
Exact prediction set, *324*
Exchangeable, 7, 27–28
    partially, 125, *479*
    row and column, *482*
Expectation, 607, *613*
    conditional, 19, *616*
Expected Fisher information, *423*
Expected loss principle, 146, 181
Expected value (see Expectation), *613*
Exponential distribution, *670*

Exponential family, *102*–103, 105, 109, 155, 239, 249
   degenerate, *104*
   nondegenerate, *104*
Extended Bayes rule, *169*
Extended real numbers, *571*
Extremal family, 123, *125*

F distribution, *670*
Fatou's lemma, 589
FI regularity conditions, 111
Fiducial distribution, 370, 373
Field, 571, *575*
Finite population sampling, 74
Finitely additive probability, 21, 281, 564, 657
Fisher information, *111*, 113, 301, 412, 463
   conditional, *111*, 119
   expected, *423*
   observed, 226, *424*, 435
Fisher–Neyman factorization theorem, 89
Fixed point, *505*
Fixed-point problem, 505
Fixed-width confidence interval, *559*
Floor of test, *215*–216
Formal Bayes rule, 146, *150*, 157, 348, 351, 369
Fubini's theorem, 596
Function:
   absolutely continuous, *211*
   bimeasurable, *583*
   measurable, 572, *583*
   simple, *586*

Gamma distribution, *670*
General linear group, *354*
Generalized Bayes rule, *156*–157, 159
Generalized Neyman–Pearson lemma, 247
Generated σ-field, 571–572, *584*
Geometric distribution, *673*
Gibbs sampling, *507*
Goodness of fit test, 218, 461
Gross error sensitivity, *312*
Group, *353*, 355–356
   abelian, *353*
   general linear, *354*
   location, *354*
   location-scale, *354*, 357, 368
   permutation, 355
   scale, *354*

Haar measure:
   left, *363*
      related, *366*
   right, *363*
      related, *366*
Hahn decomposition theorem, 605
Half-normal distribution, *389*
Hierarchical model, 166, 476
Highest posterior density region (see HPD), *327*
Hilbert space, 507
Hilbert–Schmidt-type operator, 507, *667*
Horse lottery, *182*
Hotelling's T^2, 388
HPD region, *327*, 329, 343
Hypergeometric distribution, *673*
Hyperparameters, *477*
Hypothesis, 2, *214*
   composite, *215*
   one-sided, *241*
   simple, *215*, 233
Hypothesis test, 2
   predictive, 219, 325
   randomized, 3
Hypothesis-testing loss, *214*

Identity element of group, *353*
Ignorable statistic, *142*
IID, 2, 8, 611, *628*
   conditionally, 9–10, 83, 611, *628*
Image sigma field, *584*
Importance sampling, 403, *661*
Improper prior, *20*, 122, 223, 263
Inadmissible, *154*
Increasing sequence of sets, *577*
Independence, 610, *628*
   conditional, 9, 610, 628
Indifferent, *183*
Induced measure, 575, *601*
Infinitely often, 578, *663*
Influence function, *311*
Information:
   Fisher, *111*, 113, 463

Kullback–Leibler, *115*–116
Integrable, *588*
  uniformly, *592*
Integral, 573, *587–588*
  over a set, 588
Invariance of distributions, *355*
Invariant function, *357*
  almost, *383*
  location, 346
  maximal, *358*
  scale, *350*
Invariant loss, *356*
  location, *346*
  scale, *350*–351
Invariant measure, *363*
Inverse function theorem, 666
Inverse gamma distribution, *670*
Inverse of group element, *353*

Jacobian, 625
James–Stein estimator, 163, 486
Jeffreys' prior, *122*, 446
Jensen's inequality, 614

Kolmogorov zero–one law, 631
Kullback–Leibler divergence, *116*
Kullback–Leibler information, 115–116
  conditional, *115*, 119

Lévy's theorem, 648, 650
λ-admissible, *154*–156, 162
Laplace approximation, 226, *446*
Laplace distribution, *670*
Large order, *394*
  stochastic, *396*
Law of large numbers:
  strong, 34–36
  weak, 642
Law of the unconscious statistician, 607, 613
Law of total probability, 632
Least favorable distribution, *168*
Lebesgue measure, 571, *580*
Left Haar measure, *363*
  related, *366*
Lehmann–Scheffé theorem, 298
L-estimator, *410*
Level of test, *215*–216

LHM, *363*
Likelihood function, 2, *13*, 307
Likelihood ratio test (see LR test), *274*
Linear regression, 276, 321
LMP test, *245*, 265, 289
LMPU test, 265, 292
LMVUE, *300*
Locally minimum variance unbiased estimator, *300*
Locally most powerful test (see LMP), *245*
Location equivariant rule, *346*
Location estimation, *346*
Location group, *354*
Location invariant function, *346*
Location invariant loss, *346*
Location parameter, *344*
Location-scale group, *354*
Location-scale parameter, *345*
Look-ahead decision rule, *546*
Loss function, *144*, 162, 189, 296
  convex, 349
  hypothesis-testing, *214*
  squared-error, 146, 297
  0–1, *215*
  0–1–c, *215*, 218
Lower boundary, *170*, 179, 233–235, 287
LR test, 223, 273–*274*, 458–459

Marginal distribution, 14, *607*
Marginalization paradox, 21
Markov chain, 15, 507, *650*
Markov chain Monte Carlo, *507*
Markov inequality, 614
Martingale, 645–*646*
  reversed, 33, *649*
Martingale convergence theorem, 648–649
Maximal ancillary, *97*
Maximal invariant, *358*
Maximin strategy, *168*
Maximin value, *168*
Maximum likelihood estimator, 3, *307*, 415, 418–421
Maximum modulus theorem, 667
Maximum of decision rules, *541*
MC test, *230*

Mean, 607, *613*
  conditional, *616*
  trimmed, *314*
Measurable function, 572, *583*
Measure, 570, 572, 575, *577*
  induced, *601*
  Lebesgue, 571, *580*
  product, *595*
  σ-finite, 572, *578*, 601
  signed, *577*, 597
Measure space, 572, *577*
M-estimator, *313*–315, 424–428, 434
Method of Laplace, 226, *446*
Method of moments, *340*
Mill's ratio, 470
Minimal complete class, *174*–175
Minimal essentially complete class, *174*
Minimal sufficient statistic, *92*
Minimax principle, 167, 189
Minimax rule, *167*–169
Minimax theorem, 172
Minimax value, *168*
Minimum risk equivariant (see MRE), *347*
MLE, 3, *307*, 415, 418–421
MLR, *239*–244
Monotone convergence theorem, 590
Monotone likelihood ratio, *239*–244
Monotone sequence of sets, *577*
Most cautious test, *230*
Most powerful test, *230*
MP test, *230*
MRE, *347*, 349, 351, 363
Multinomial distribution, *674*
Multiparameter Cramér-Rao lower bound, 306
Multivariate central limit theorem, 643
Multivariate normal distribution, 643, *674*

Natural parameter, *103*, 105
Natural parameter space, *103*, 105
Natural sufficient statistic, *103*
Negative binomial distribution, *673*
Negative part, 573, *588*
Negative set, *598*
Neyman structure, *266*

Neyman–Pearson fundamental lemma, 175, 231
NM-lottery, *182*
Noncentral beta distribution, 289, *671*
Noncentral χ^2 distribution, *671*
Noncentral F distribution, 289, *671*
Noncentral t distribution, 289, 325, *671*
Nondegenerate exponential family, *104*
Nondegenerate weak order, *183*
Nonnull states, *184*
Nonparametric, 52
Nonparametric bootstrap, 329
Nonrandomized decision rule, *145*, 151, 153
Nonrandomized sequential decision rule, *537*
Normal distribution, 21, 349, 611, 640, 642, *671*
  multivariate, 643, *674*
Null states, *184*

Observed Fisher information, 226, *424*, 435
One-sided hypothesis, *241*
One-sided test, 239, *243*
Open set, *571*
Operating characteristic, *215*
Orbit, *358*
Order statistics, *86*
Outliers, 521

Parameter, 1, 6, 50–*51*, 82
  location, *344*
  location-scale, *345*
  natural, *103*, 105
  scale, *345*
Parameter space, 1, *50*, 82
  natural, *103*, 105
Parametric bootstrap, 330
Parametric family, 1, *50*, 102
Parametric index, 33, *50*
Parametric models, 12
Parametric Models, 49
Pareto distribution, *672*
Partial Bayes rule, 147, *150*
Partially exchangeable, 125, *479*

Percentile-t bootstrap confidence interval, 336
Permutations, 355
π–λ theorem, 576
Pitman's estimator, *347*, 363
Pivotal, *316*, 370, 373
Point estimation, 296
Point estimator, *296*
Pointwise convergence, *184*
Poisson distribution, *673*
Polish space, *619*
Polya tree distribution, *69*
Polya urn scheme, 9
Portmanteau theorem, 636
Positive part, 573, *588*
Positive set, *598*
Posterior distribution, 4, *16*
    asymptotic normality, 435, 437, 442–443
    consistency, 429–430
Posterior predictive distribution, *18*
Posterior risk, 146, *150*
Power function, 2, *215*, 240
Power set, *571*
Prediction set, *324*–325
    conservative, *324*
    exact, *324*
Predictive distribution, *14*, 455
    posterior, *18*
    prior, *14*
Predictive hypothesis test, 219, 325
Preference, 182
    conditional, *185*
        consistent, *186*
Prevision, 655
Prior distribution, 4, *13*
    improper, *20*, 223, 263
    natural conjugate family, 92
Prize, *181*
Probability, 572, *577*
    empirical, *12*
    random, 27
Probability integral transform, 519, *659*
Probability space, 572, *577*, 606, 612
Product measure, *595*
Product σ-field, *576*
Product space, *576*
Pseudorandom numbers, 659

Pure significance test, *217*
P-value, *279*, 375, 380

Quantile:
    sample, 404–405, 408

Radon–Nikodym derivative, 575, *598*
Radon–Nikodym theorem, 597
Random probability measure, *27*
Random quantity, 82, 606, *612*
Random variables, 606, *612*
    exchangeable, 27
    IID, 8
Randomized confidence set, *316*
Randomized decision rule, *145*, 151
Randomized sequential decision rule, *537*
Randomized test, 3
Rao–Blackwell theorem, 152
Ratio of uniforms, *660*
Regression, 276, 321, 519
Regular conditional distribution, 610, *618*
Regular conditional probabilities, 609, *617*
Regular decision rule, *540*
Reject hypothesis, *214*
Rejection region, 2
Related LHM, *366*
Related RHM, *366*
Relative rate of convergence, *413*, 470
Restriction of σ-field, *584*
Reversed martingale, *649*
RHM, *363*
Right Haar measure, *363*
    related, *366*
Risk function, 149–*150*, 153, 155, 167, 216, 233, 297–298
Risk set, *170*–172, 179, 233, 235, 287
Robustness, 310
    Bayesian, 524
Row and column exchangeable, *482*

Sample quantile, *404*–405, 408
Sample space, 2, *82*
Scale equivariant rule, *350*
Scale estimation, 350
Scale group, *354*
Scale invariant function, *350*

Scale invariant loss, *350*–351
Scale parameter, *345*
Scheffé's theorem, 634
Score function, *111*, *122*, 302, 305
    conditional, *111*
Second-order efficiency, 414
Sensitivity analysis, 524
Separable space, *619*
Separating hyperplane theorem, 666
Sequential decision rule, *537*
Sequential probability ratio test, *549*
Sequential test, *548*
Set estimation, 296
Shrinkage estimator, *163*
σ-field, *575*
    Borel, 571, *575*
    generated, 571–572, *584*
    image, *584*
    restriction, *584*
    tail, *632*
σ-finite measure, 572, *578*, 601
Signed measure, *577*, 597, 605, 635
Significance probability, *217*, 228, 280
Significance test, *217*
Simple alternative, *215*
Simple function, *586*
Simple hypothesis, *215*
Size of test, 2, *215*–216
Small order, *394*
    stochastic, *396*
SPRT, *549*
Squared-error loss, 146, 297
$\sqrt{n}$-consistent, *401*
SSS, *507*
St. Petersburg paradox, 655
State independence, 184, 205
State-dependent utility, *205*–206
States of Nature, *181*, 189, 205
Statistic, *83*
    ancillary, *95*, 99, 119
    boundedly complete, *94*
    complete, *94*, 298
    sufficient, *84*–*85*–86, 99, 103, 150–151, 298
Stein estimator (see James–Stein estimator), 163
Stochastic large order, *396*
Stochastic small order, *396*
Stone–Weierstrass theorem, 666

Stopping time, *537*, 548, 552, 554
Strict preference, *183*
Strong law of large numbers, 34–36
Strongly unimodal, *329*
Submartingale, *646*
Successive substitution, 505–*506*, 545
Successive substitution sampling, *507*
Sufficient statistic, *84*–*85*–86, 99, 103, 109, 150–151, 298
    conditionally, *95*
    minimal, *92*
    natural, *103*
Superefficiency, 414
Supporting hyperplane theorem, 666
Sure-thing principle, 184

t distribution, *672*
Tail σ-field, *632*
Tailfree process, *60*
Taylor's theorem, 665
Tchebychev's inequality, 614
Terminal decision rule, *537*
Test:
    goodness of fit, *218*, 461
    one-sided, 239, *243*
    two-sided, *256*, 273
Test function, 175, *215*
Theorem:
    Bahadur, 94
    Basu, 99
    Bayes, 4, 16
    Bhattacharyya lower bounds, 305
    Bolzano–Weierstrass, 666
    Caratheodory extension, 578
    Cauchy's equation, 667
    central limit, 642
        multivariate, 643
    chain rule, 600
    Chapman–Robbins bound, 304
    complete class, 179
    continuity, 640
    continuous mapping, 638
    Cramér–Rao lower bound, 301
    DeFinetti, 27–28
    dominated convergence, 591
    Fatou's lemma, 589
    Fisher–Neyman, 89
    Fubini, 596

Hahn decomposition, 605
inverse function, 666
Kolmogorov zero–one law, 631
Lévy, 648, 650
law of total probability, 632
Lehmann–Scheffé, 298
martingale convergence, 648–649
maximum modulus, 667
minimax, 172
monotone convergence, 590
multivariate central limit, 643
Neyman–Pearson, 175, 231
  generalized, 247
π–λ, 576
portmanteau, 636
Radon–Nikodym, 597
Rao–Blackwell, 152
Scheffé, 634
separating hyperplane, 666
Stone–Weierstrass, 666
strong law of large numbers, 36
supporting hyperplane, 666
Taylor, 665
Tonelli, 595
uniqueness, 645
upcrossing, 647
weak law of large numbers, 642
Tolerance coefficient, *325*
Tolerance set, 219, *325*
  conservative, *325*
Tonelli's theorem, 595
Topological space, *571*, 575
Topology, *571*
Transformation, *354*
Transition kernel, *124*
Trimmed mean, *314*
Trivial σ-field, *571*
Truncated decision rule, *542*
Two-sided alternative, *246*
Two-sided hypothesis, *246*
Two-sided test, *256*, 273
Type I error, *214*
Type II error, *214*

UMA confidence set, *317*
UMAU confidence set, *321*
UMC test, *230*–231, 239, 244, 255, 257
UMCU test, *254*–256

UMP test, *230*, 240, 243–244, 255, 257
UMPU test, *254*–256
UMPUAI test, *384*
UMVUE, *297*–299
Unbiased estimator, 3, *296*–302
Unbiased test, *254*
Uniform distribution, 659, *672*
Uniformly integrable, *592*
Uniformly minimum variance
  unbiased estimator (see
  UMVUE), *297*
Uniformly most accurate confidence
  set (see UMA), *317*
Uniformly most accurate unbiased
  confidence set (See UMAU),
  321
Uniformly most cautious test (see
  UMC), *230*
Uniformly most cautious unbiased
  test (see UMCU), *254*
Uniformly most powerful test (see
  UMP), *230*
Uniformly most powerful unbiased
  test (see UMPU), *254*
Uniqueness theorem, 645
Upcrossing lemma, 647
Upper semicontinuous, *417*
USC, *417*
Utility function, 181, *188*
  state-dependent, *205*–206

Variance, 607, *613*
Variance components, 484
Variance stabilizing transformation,
  402
Version of conditional distribution,
  617
Version of conditional expectation,
  608, *616*
Version of conditional mean, *616*

Wald's lemma, 552
Weak convergence, 399, *635*
Weak* convergence, *635*
Weak law of large numbers, 642, 664
Weak order, *183*, 216–217, 280
  degenerate, *183*
  nondegenerate, *183*
Weak preference, *182*

Springer Series in Statistics

(continued from p. ii)

Read/Cressie: Goodness-of-Fit Statistics for Discrete Multivariate Data.
Reinsel: Elements of Multivariate Time Series Analysis.
Reiss: A Course on Point Processes.
Reiss: Approximate Distributions of Order Statistics: With Applications to Nonparametric Statistics.
Rieder: Robust Asymptotic Statistics.
Ross: Nonlinear Estimation.
Sachs: Applied Statistics: A Handbook of Techniques, 2nd edition.
Salsburg: The Use of Restricted Significance Tests in Clinical Trials.
Särndal/Swensson/Wretman: Model Assisted Survey Sampling.
Schervish: Theory of Statistics.
Seneta: Non-Negative Matrices and Markov Chains, 2nd edition.
Shedler: Regeneration and Networks of Queues.
Siegmund: Sequential Analysis: Tests and Confidence Intervals.
Tanner: Tools for Statistical Inference: Methods for the Exploration of Posterior Distributions and Likelihood Functions, 2nd edition.
Todorovic: An Introduction to Stochastic Processes and Their Applications.
Tong: The Multivariate Normal Distribution.
Vapnik: Estimation of Dependences Based on Empirical Data.
Weerahandi: Exact Statistical Methods for Data Analysis.
West/Harrison: Bayesian Forecasting and Dynamic Models.
Wolter: Introduction to Variance Estimation.
Yaglom: Correlation Theory of Stationary and Related Random Functions I: Basic Results.
Yaglom: Correlation Theory of Stationary and Related Random Functions II: Supplementary Notes and References.